Avian Physiology

Fourth Edition

Avian Physiology

Fourth Edition

Edited by
P.D. Sturkie

With Contributions by C.A. Benzo, K.I. Brown, G.E. Duke
M.R. Fedde, B. Glick, P. Griminger, A.L. Harvey
S. Harvey, R.L. Hazelwood, A.L. Johnson, M.R. Kare
A.D. Kenny, J.R. Mason, I.G. Marshall, D.B. Meyer
D.C. Meyer, R.K. Ringer, C.G. Scanes, P.D. Sturkie
B.C. Wentworth, G.C. Whittow

With 199 Illustrations

Springer-Verlag New York Berlin Heidelberg Tokyo

P.D. Sturkie, Ph.D.
Emeritus Professor of Physiology, Department of Animal Sciences, Cook College, Rutgers University, New Brunswick, New Jersey 08903, U.S.A.

Library of Congress Cataloging-in-Publication Data
Main entry under title:
Avian physiology.
Bibliography: p.
Includes index.
1. Birds—Physiology. I. Sturkie, Paul D.
QL698.A85 1986 598.2′1 85-26049

Typeset by Kingsport Press, Kingsport, Tennessee.
Printed and bound by Halliday Lithograph, West Hanover, Massachusetts.
Printed in the United States of America.

9 8 7 6 5 4 3 2 1

ISBN 0-387-96195-X Springer-Verlag New York Berlin Heidelberg Tokyo
ISBN 3-540-96195-X Springer-Verlag Berlin Heidelberg New York Tokyo

To Dr. H.H. Dukes
Emeritus Professor of Veterinary Physiology,
Cornell University
Teacher, adviser and friend
Who introduced me to Physiology and encouraged me
To publish the first edition of *Avian Physiology*

Preface to the Fourth Edition

Since the publication of earlier editions, there has been a considerable increase in research activity in a number of areas, with each succeeding edition including new chapters and an expansion of knowledge in older chapters.

The fourth edition contains two new chapters, on muscle and immunophysiology, the latter an area where research on Aves has contributed significantly to our general knowledge of the subject.

The new edition has a number of new contributors, who have written on the nervous system, sense organs, muscle, endocrines, reproduction, digestion and immunophysiology. Contributors from previous editions have expanded their offerings considerably.

The authors are indebted to various investigators, journals and books for the many illustrations used. Individual acknowledgement is made in the legends and references.

Preface to the Third Edition

Since the publication of the first and second editions, there has been a considerable increase of research activity in avian physiology in a number of areas, including endocrinology and reproduction, heart and circulation, respiration, temperature regulation, and to a lesser extent in some other areas.

There appeared in 1972–1974 a four volume treatise entitled "Avian Biology," including material on physiology, and earlier a three volume presentation on the biology of the domestic fowl that also involves a great deal on the physiology and biochemistry of certain organ systems.

However, "Avian Physiology" remains a one-volume contribution that gives a balanced account of most of the principal organs and systems of Aves in a classical manner. The type size in this edition is smaller and the page size larger, so actually more material is covered in considerably fewer pages than in previous editions. The aim of this and previous editions has been to serve as a textbook and a source of reference for the experimental physiologist and to provide pertinent material for courses in comparative physiology, zoology, ecology, and ornithology.

The third edition contains two new chapters by P. Griminger on lipid and protein metabolism, which emphasize the differences in the metabolic products and pathways of birds and mammals. New contributors include M.R. Fedde and T.B. Bolton, who have completely revised and expanded the chapters on respiration and the nervous system, respectively, and J.G. Rogers, Jr., W.J. Mueller, H. Opel, and D.C. Meyer, who have made contributions to Chapters 2, 16, 17, and 19, respectively.

R.L. Hazelwood has revised his chapter on carbohydrate metabolism and has written a separate one on the pancreas. R.K. Ringer has revised the material on the thyroid and has contributed the chapter on adrenals. New material on temperature regulation and energy metabolism, particularly on wild species, has been added by C.G. Whittow.

The sections on the blood, heart, and circulation have been revised by the Editor under three chapters. The chapter on the chemical constituents of blood in the earlier editions has been omitted and such data are presented elsewhere. The editor has also contributed the chapters on digestion and absorption (Chapters 9 and 10), kidneys and urine (Chapter 14), hypophysis (Chapter 15), and the greater part of chapters on reproduction in the female (Chapter 16) and the male (Chapter 17).

July 1975 P.D.S.

Preface to the Second Edition

Since the publication of the first edition in 1954 there has been a considerable increase in research activity in avian endocrinology and reproduction and a modest increase in research in other areas. Much work, however, remains to be done on such systems as respiration, muscle, nerve, and digestion.

New features of the second edition include a chapter on the nervous system by Dr. Jasper ten Cate, Professor of Comparative Physiology, University of Amsterdam, Holland, and contributions from other authors active in various fields. An expanded chapter on chemical constituents has been written, mainly by Dr. D.J. Bell, Head of Biochemistry Section, Poultry Research Center, Edinburgh, Scotland. The section on coagulation of blood was written by Dr. Paul Griminger, Associate Professor of Nutrition, Rutgers University.

Expanded chapters on temperature regulation and on energy metabolism have been contributed by Dr. G.C. Whittow, formerly physiologist at the Hannah Dairy Research Institute, and now Associate Professor of Physiology, Rutgers University.

The chapter on carbohydrate metabolism was completely revised by Dr. R.L. Hazelwood, Associate Professor of Physiology, University of Houston.

The chapter on sense organs was revised by Dr. M.R. Kare, Professor of Physiology, North Carolina State College. Chapter 19, "Thyroids," has been considerably expanded by Dr. Robert K. Ringer, Professor of Avian Physiology, Michigan State University.

Chapter 15, "Reproduction in the Female," has also been expanded. That part of it relating to calcium metabolism and egg laying was contributed by Dr. T.G. Taylor, Reader in Physiological Chemistry, University of Reading, England, and Dr. D.A. Stringer, Unilever Research Laboratory, Bedford, England.

Most of the chapters that I revised have also been enlarged. The revision has resulted in a substantial increase in the size of the book.

The authors are indebted to various investigators, journals, and books for many of the illustrations used. Individual acknowledgment is made in the legends.

May 1965 P.D.S.

Preface to the First Edition

Physiology may be divided into three main categories: cellular, comparative, and special—i.e., the physiology of special groups of organisms. The physiology of special groups has received the most attention. In the animal field, interest has centered largely on mammalian physiology, with particular emphasis on human physiology and its relationship to medicine. By comparison, the physiology of birds has been neglected. Knowledge in certain areas of avian physiology is limited, fragmentary, and often confused, and little or no new research is being conducted. Much of the physiological research on the bird has been conducted from the comparative viewpoint, which is concerned more with broad functional relationships between groups of animals than with details of a special group. In some areas, however, these fundamental functions have not been definitely established. Even in certain fields, such as endocrinology, where there is considerably more research activity on the bird, there are wide gaps in our knowledge.

This book is the first one in any language devoted to the specialized physiology of birds. It deals mainly with the chicken, the duck, and the pigeon, because most of the research has been conducted on these species and they represent species of economic importance to man.

Inasmuch as physiology provides a rational basis for much of animal husbandry and veterinary medicine, this book should be of especial interest to teachers, students, and research workers in poultry science and husbandry and in veterinary medicine. More knowledge and research in avian physiology, particularly on the domestic species, should have important applications to the poultry industry, which is rapidly expanding in this country. Although few poultry departments at present offer course work on the physiology of birds, it is hoped that this book may be instrumental in increasing the number of institutions offering such work and in stimulating more research. It may serve, also, as a source of reference for the experimental physiologist and should provide pertinent physiological material for courses in comparative physiology, ecology, and ornithology.

The bibliography is extensive but not exhaustive. An attempt was made to select the most important and more recent references, with minor considerations given to priority. The references are cited at the end of each chapter.

The writer is indebted to investigators, journals, and books for many of the illustrations used. Separate acknowledgment is made in the legends to the authors and books or journals from which illustrations came. The original drawings and modifications of illustrations of others were prepared by my wife, to whom I am grateful.

Special thanks are extended to colleagues who read one or more chapters and made helpful suggestions. These are Drs. H.H. Dukes, J.A. Dye, F.B. Hutt, R.M. Fraps, C.S. Shaffner, A.V. Nalbandov, T.C. Byerly, J.H. Leathem, J.B. Allison, W.C. Russell, and H.J. Metzger.

July 1953 P.D.S.

Contents

Contributors

Camillo A. Benzo, Ph.D.
Associate Professor, Department of Anatomy and Cell Biology, State University of New York, Upstate Medical Center, Syracuse, New York, U.S.A.

Keith I. Brown, Ph.D.
Professor, Department of Poultry Science, Ohio State University and Ohio Agricultural Research and Development Center, Wooster, Ohio, U.S.A.

Gary E. Duke, Ph.D.
Professor, Department of Veterinary Biology, College of Veterinary Medicine, University of Minnesota, St. Paul, Minnesota, U.S.A.

M. Roger Fedde, Ph.D.
Professor of Physiology, Department of Anatomy and Physiology, College of Veterinary Medicine, Kansas State University, Manhattan, Kansas, U.S.A.

Bruce Glick, Ph.D.
Professor, Poultry Immunology and Physiology, Poultry Science Department, College of Agriculture, Mississippi State University, Mississippi State, Mississippi, U.S.A.

P. Griminger, Ph.D.
Professor of Nutrition, Department of Nutrition, Cook College, Rutgers University, New Brunswick, New Jersey, U.S.A.

A.L. Harvey, Ph.D.
Department of Physiology and Pharmacology, University of Strathclyde, Glasgow, Scotland, United Kingdom

Stephen Harvey, Ph.D.
Wolfson Institute, University of Hull, Hull Hub, England

R.L. Hazelwood, Ph.D.
Professor of Physiology, Department of Biology, University of Houston, Houston, Texas, U.S.A.

A.L. Johnson, Ph.D.
Associate Professor of Physiology, Department of Animal Sciences, Cook College, Rutgers University, New Brunswick, New Jersey, U.S.A.

Morley R. Kare, Ph.D.
Monell Chemical Senses Center, Philadelphia, Pennsylvania, U.S.A.

Alexander D. Kenny, Ph.D., D.Sc.
Professor and Chairman, Department of Pharmacology, Texas Tech University Health Sciences Center, Lubbock, Texas, U.S.A.

J. Russell Mason, Ph.D.
Monell Chemical Senses Center and Department of Biology, University of Pennsylvania, Philadelphia, Pennsylvania, U.S.A.

I.G. Marshall, Ph.D.
Department of Physiology and Pharmacology, University of Strathclyde, Glasgow, Scotland, United Kingdom

David B. Meyer, Ph.D.
Professor of Anatomy, Department of Anatomy, Wayne State University—School of Medicine, Detroit, Michigan, U.S.A.

Donald C. Meyer, Ph.D.
Associate Professor of Physiology, Eastern Virginia Medical School, Norfolk, Virginia, U.S.A.

R.K. Ringer, Ph.D.
Professor of Physiology, Department of Animal Sciences, Michigan State University, East Lansing, Michigan, U.S.A.

Colin G. Scanes, Ph.D.
Professor of Physiology and Chairman, Department of Animal Sciences, Cook College, Rutgers University, New Brunswick, New Jersey, U.S.A.

P.D. Sturkie, Ph.D.
Emeritus Professor of Physiology, Department of Animal Sciences, Cook College, Rutgers University, New Brunswick, New Jersey, U.S.A.

B.C. Wentworth, Ph.D.
Chairperson, Department of Poultry Science, University of Wisconsin, Madison, Wisconsin, U.S.A.

G. Causey Whittow, Ph.D.
Professor and Chairman, Department of Physiology, John A. Burns School of Medicine, University of Hawaii, Honolulu, Hawaii, U.S.A.

1
Nervous System

C.A. Benzo

Introduction and Neurohistology

Function of the Nervous System

The vertebrate nervous system is responsible both for maintaining contact between the animal and its external and internal environments and for the proper adjustments of the animal to the changes in these environments. The animal maintains contact with the external environment through sensory receptors at the surface of the body. The internal environment is monitored by receptors located in muscles, joints, ligaments, and visceral organs. Basically, adjustments to changes in either environment are brought about by reflex arcs consisting of afferent (sensory) neurons, centers within the spinal cord or brain, and efferent (motor) neurons. Afferent neurons carry sensory information to the central nervous system, and efferent neurons convey motor impulses from the central nervous system to various effector mechanisms, such as muscles and glands. The nervous system works in harmony with the endocrine system to coordinate the many complex activities involved in normal body functions. The nervous system is the rapid coordinator in response to a given stimulus, whereas the endocrine system is more deliberate in its action and is brought into play for conditions that require a more intense or prolonged response.

Divisions of the Nervous System

As in mammals, the nervous system in birds can be subdivided anatomically into the central nervous system and the peripheral nervous system and functionally into the *somatic* and the visceral or *autonomic* nervous systems.

The central nervous system (CNS) includes the brain and spinal cord. The peripheral nervous system (PNS) includes the cranial and spinal nerves that emanate from the brain and spinal cord, respectively, and the many ganglia and plexi that are associated with visceral innervation.

The somatic nervous system includes those neural elements of the PNS and CNS that are involved with transmitting both conscious and unconscious sensory information from the head, extremities, and body wall to appropriate processing centers within the CNS, and with the motor control of voluntary (striated) musculature.

The visceral (autonomic) nervous system consists of those neural structures of the PNS and CNS that are involved with conveying sensory input, usually unconscious, from visceral systems to the CNS, and with motor activities controlling involuntary (smooth) and cardiac muscle and glands. Anatomically, the autonomic nervous system is considered by many authors to be a visceral motor system, exclusively. Functionally, however, it should be remembered that all visceral motor responses are triggered by appropriate sensory stimuli.

Structure of Nervous Tissue

The nervous system is composed of the nerve cells proper, or neurons, and supportive cells called *neuroglia* or simply glial cells. The *neuron* is the morphologic and functional unit of the nervous system. Each neuron is in contact, or *synapse,* with other neurons such that it is an interconnecting segment in the vast neural network constituting the nervous system. The neuron is specifically designed to function as a conveyor of information and is incapable of reproducing itself. It requires other cells, the neuroglia, to provide it with essential nutrients and other supportive maintenance. Neuroglia also serve to cluster neurons into functionally similar groups within the brain and spinal cord.

In general, each neuron consists of a cell body, the *perikaryon* or soma, from which extends a single process called the *axon* and one or more branching processes called *dendrites* (Figure 1–1). The perikaryon contains the nucleus of the nerve cell. The axon transmits impulses away from the perikaryon, and the dendrites convey impulses to the perikaryon and axon.

Perikaryon. Perikarya of neurons reside in the brain, spinal cord, dorsal root ganglia, the ganglia of the cranial nerves, and in ganglia associated with the visceral motor system.

The perikaryon consists of the nucleus and cytoplasm. The nucleus contains a prominent nucleolus that is composed of RNA and is associated with protein synthesis. The cytoplasm contains mitochondria, the Golgi apparatus, Nissl bodies, and other inclusions (Figure 1–1). Nissl bodies are aggregations of the granular, or rough, endoplasmic reticulum with attached ribosomes, and they are the protein-synthesizing machinery of the neuron. Within the cytoplasm also are found neurofibrils, which are bundles of neurofilaments that extend into the axon and dendrites, and microtubules. Microtubules are believed to be involved in the transport of enzyme-containing vesicles from the perikaryon down the axon. The enzymes that are responsible for the synthesis of transmitter substance are produced by the ribosomes and packaged into vesicles within the Golgi apparatus. These vesicles then move from the Golgi region of the perikaryon down the axon to the nerve terminal. Transmitter substance is synthesized at the nerve terminal, and to some extent, during axonal transport of the vesicles.

Axon. The axon arises from the axon hillock of the perikaryon, at a site called the initial segment. After the initial segment, the axon is of uniform diameter and extends for a variable distance before branching into telodendria. The length of axons varies from a fraction of a millimeter to a meter or more. The axon

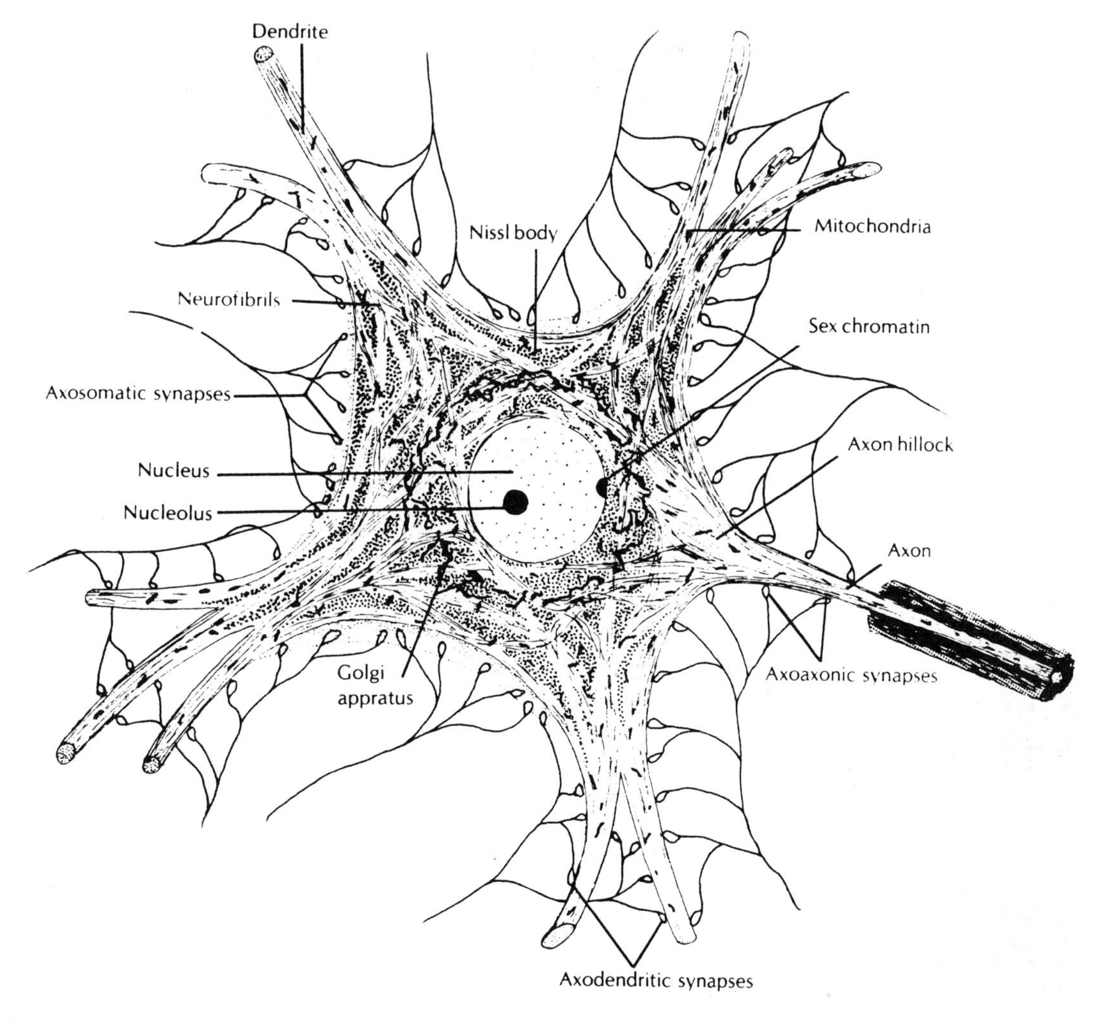

FIGURE 1–1. Diagrammatic representation of the constituents of a nerve cell. (From Barr and Kiernan, 1983.)

hillock, initial segment, and the axon itself lack Nissl bodies and ribosomes, but contain mitochondria, neurofilaments, microtubules, and vesicles. Each telodendron terminates as a bouton that is apposed to the cell membrane of another nerve cell, a muscle cell, or a gland cell to form a synapse. The axon functions in the transmission of neural messages from the dendritic–perikaryal region of its cell to synaptic junctions.

In birds, as in mammals, the axons of myelinated peripheral nerves are surrounded by a sheath of myelin that is formed by the spiral growth of the Schwann cell about the axon (Figure 1–2). In unmyelinated nerves, several axons are embedded within a Schwann cell that does not form a myelin sheath. The myelin sheath of axons within the CNS are produced and maintained by a glial cell type, the oligodendrocyte. Depending also on their size, myelinated axons generally conduct the nerve impulse faster than do unmyelinated axons. The nerve impulse in myelinated fibers is propogated by discontinuous spread, or saltatory conduction, a "skipping" between the junctions of myelin along the axonal membrane (Figure 1–3). In unmyelinated nerves, the impulse is propagated without decrement as a continuous, not saltatory, spread.

Myelinated and unmyelinated axons are major constituents of peripheral nerves. Within the spinal cord and brain, myelinated axons and their associated glial cells are the predominant components of the white matter, whereas unmyelinated axons, dendrites, resident perikarya, and their supportive cells comprise the gray matter.

Dendrites. Dendrites are true extensions of the perikaryon, and they contain the same cytoplasmic organelles that are present in the perikaryon. The dendrite–perikaryal region of the neuron functions as the receptive and integrative unit in impulse conduction. Within the CNS, dendrites extend from the perikaryon and undergo extensive branching. These dendrites are unmyelinated, and are synaptically apposed to other nerve

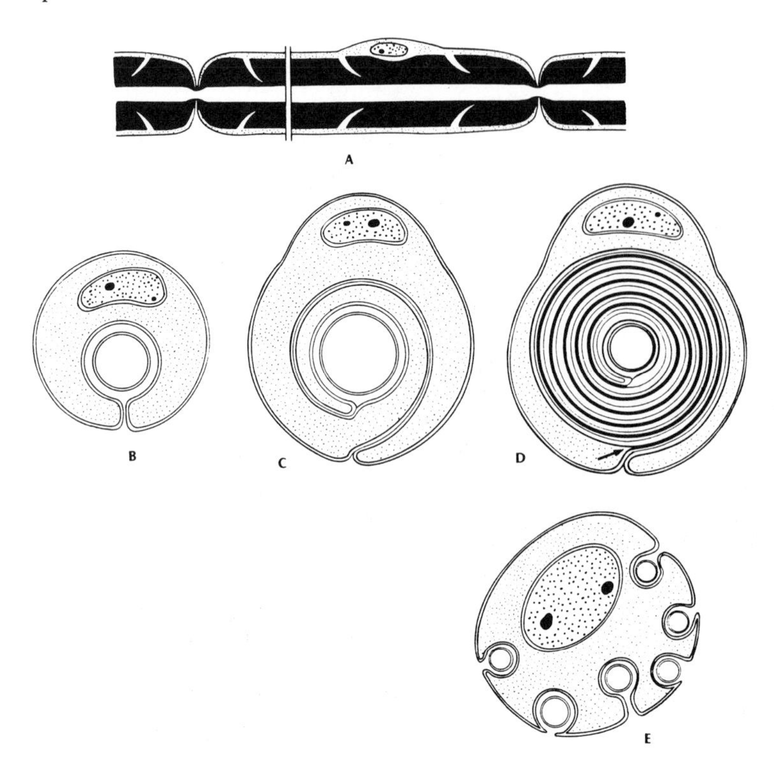

FIGURE 1–2. (**A**) The myelin sheath of an axon and its Schwann cell. (**B,C,D**) Successive stages in the development of the myelin sheath from the plasma membrane of a Schwann cell. (**E**) Relationship of a Schwann cell to several unmyelinated axons. (From Barr and Kiernan, 1983.)

cell membranes. The three-dimensional region through which the dendrites of a single neuron arborize is known as the dendritic field of that nerve cell. Outside the CNS, the peripheral dendrites of afferent neurons arise in either intero- or exteroreceptors in various parts of the body. When stimulated, these dendrites convey impulses to the afferent axons, which carry sensory information from peripheral structures to the CNS.

Axonal Transport. Because most metabolic processes necessary to maintain the viability of the nerve cell occur within the perikaryon, the cytoplasm of the neuron is in continuous movement, allowing an uninterrupted flow of substances between the perikaryon and its axon and dendrites. The transport of materials down the axon, away from the perikaryon, is termed anterograde or orthograde flow, whereas the movement of axoplasm toward the perikaryon is called retrograde flow (see Lubinska, 1975). The synthesis and transport of perikaryal proteins and glycoproteins and their incorporation in axonal and synaptic structures have been examined in the visual system of the pigeon (Marko and Cuénod, 1973) and in the ciliary ganglion of the chicken (Koenig et al., 1973). Transport of material down the axon is effected at two rates, fast (100–400 mm/day) transport and slow (1–10 mm/day) transport. The rapid transport mechanism carries certain essential proteins, lipids, and the enzyme-containing vesicles involved in neurotransmission, whereas slow transport is involved with the movement of materials that are necessary for sustaining and rebuilding axonal structures (Banks and Mayor, 1972; Couraud and Di-Giamberardino, 1980; Couraud et al., 1982). Less is

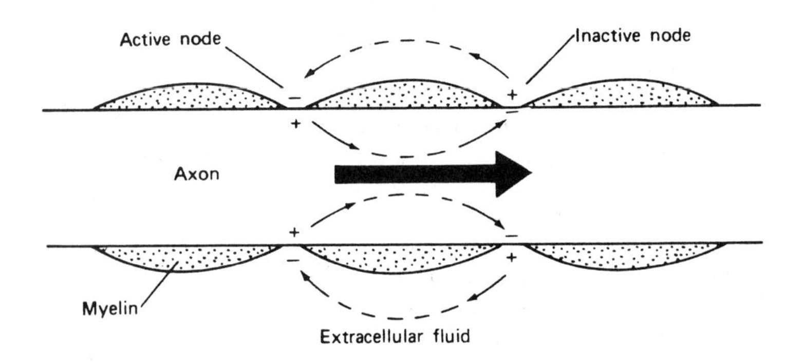

FIGURE 1–3. Saltatory conduction in a myelinated fiber. (From Chaplin and Demers, 1978. Reprinted by permission of John Wiley & Sons, Inc.)

known about the mechanisms that may be involved in retrograde flow. It appears, however, that information about the functional state of the axon is fed back continuously to the perikaryon, and that certain proteins that are picked up by the axon terminals are transported to the perikaryon for catabolism (see Filliatreau and DiGiamberardino, 1982).

Neurophysiology and Neurochemistry

The Synapse

The synapse is the site of contact of one neuron with another. For transfer of information to occur within the nervous system, the nerve impulse must pass from one neuron to another across a gap, the synaptic cleft, between adjacent nerve cell membranes (Figure 1–4).

Because neurons, when excited, theoretically can conduct the nerve impulse in either direction with equal facility, synapses determine the direction taken by excitation through the nervous system. Synapses with other nerve cells may be between the bouton of one neuron and the perikaryon of another neuron (axosomatic), between the bouton of one neuron and the dendrites of another neuron (axodendritic), between the bouton of one neuron and another axon (axoaxonal), or between the dendrites of one neuron and those of another neuron (dendrodendritic) (Figure 1–5). Synapses between neurons and their supportive cells (axoglial) have been found in both the avian (DeGennaro and Benzo, 1976) and the mammalian spinal cord (Hendrikson and Vaughn, 1974; Morales and Duncan, 1975). Although the functional significance of axoglial synapses is not yet clear, their presence suggests a morphologic basis for intercommunication between adult derivitives of the embryonic neural tube.

The axon of one neuron may terminate in only a few or in many thousands of synapses, and the dendritic–perikaryal region of one neuron may receive up to several thousand synaptic contacts from many different neurons. Axon terminations in muscle cells (neuromuscular junctions) and in gland cells (neuroglandular junctions) basically are similar to the synapses between two nerve cells. The synapse of each terminal bouton of a voluntary motor neuron is called a motor endplate, a specialization that is absent or less obvious in smooth muscle, cardiac muscle, and gland cells.

The cell membrane of the axon terminal at the synapse is known as the presynaptic membrane, and the cell membrane of the dendritic–perikaryal region of the other nerve cell, or that of the muscle or gland cell, is called the postsynaptic membrane. Within the cytoplasm of each terminal bouton are found mitochondria and presynaptic vesicles. These vesicles contain the active neurotransmitter substances. Communication between neurons and other cells is effected through these chemical agents that are released at the bouton into the synaptic cleft and influence the excitability of the postsynaptic membrane. The postsynaptic membrane is chemosensitive, and it responds to the released neurotransmitter in a characteristic manner. Depending on the chemical nature of the neurotrans-

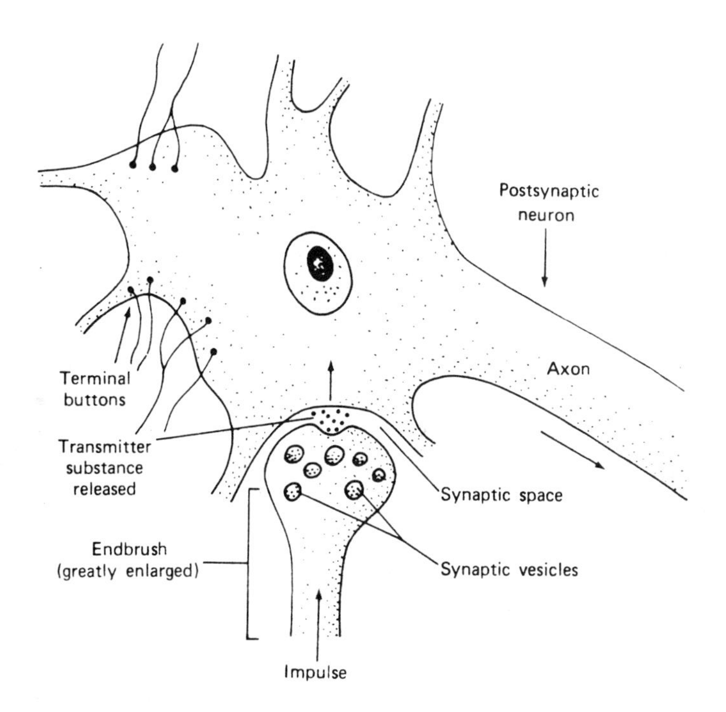

FIGURE 1–4. The structure of the synapse. The terminal bouton (button or endbrush) is greatly enlarged to show some details. (From Chaplin and Demers, 1978. Reprinted by permission of John Wiley & Sons, Inc.)

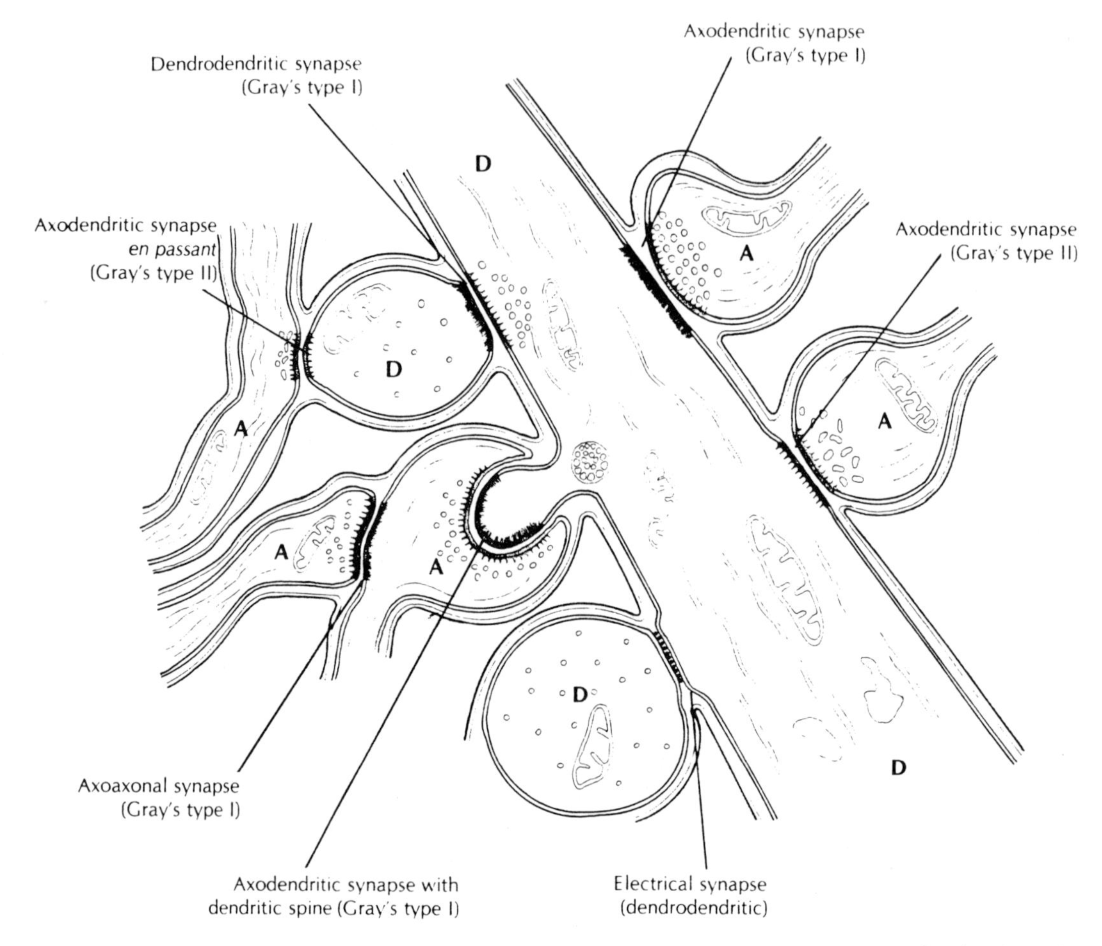

FIGURE 1–5. The structure of various types of synapses (*A*, axon; *D*, dendrite) (see text for details). (From Barr and Kiernan, 1983.)

mitter and the receptor properties of the postsynaptic membrane, the ensuing permeability changes in the postsynaptic membrane result in either excitation or inhibition of the affected cell. Because a given cell may receive very many synaptic contacts, synapses are regions where the activities of several neurons can interact in the production of a final response. For this reason, synapses are considered to be the sites of the integrative function of the nervous system.

Not all synapses involve the ability of the neuron to release a chemical transmitter and that of another cell to respond. Some synapses display electrical coupling in which current from the axon terminal depolarizes the postsynaptic membrane without the need of a chemical agent (Figure 1–5). Electrical synapses occur mainly in invertebrates, but they have been found in the ciliary ganglion of the bird (Martin and Pilar 1963a, b, 1964; Hess et al., 1969; Cantino and Mugnaini, 1975).

The Nerve Impulse

Conduction of the nerve impulse is accomplished through the progressive depolarization of the semipermeable and selectively permeable membrane that surrounds each neuron. A resting nerve cell is a charged neuron that is not conducting an impulse. Polarization of the nerve cell membrane occurs when the neuron is at rest. Neurons, like muscle cells, concentrate potassium within themselves and exclude sodium. Because the nerve cell membrane is permeable to potassium, a diffusion potential arises from the unequal concentrations of potassium on either side (Figure 1–6). This diffusion potential primarily is responsible for the increased negativity in the intracellular neuroplasm compared to the interstitial fluid of the extracellular space. This negative potential across the resting nerve cell membrane is called the membrane potential. A variety of stimuli can alter the permeability of the

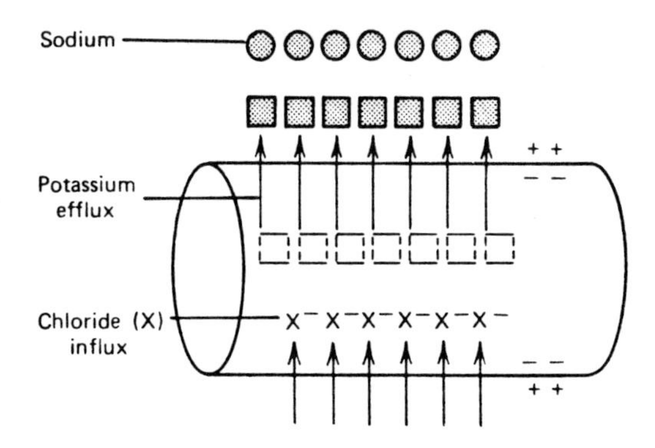

FIGURE 1–6. Membrane polarization (see text for details). (From Chaplin and Demer, 1978. Reprinted by permission of John Wiley & Sons, Inc.)

cell membrane to certain ions, resulting in changes in the membrane potential. During excitation of the axon, there is a sudden reversal of the membrane potential and the membrane at that site is depolarized, producing an action potential. Repetition of this process by adjacent sites on the membrane produces a wave of depolarization that moves along the nerve fiber and constitutes conduction of the nerve impulse (Figure 1–3). Once generated in an axon, the nerve impulse is conducted along its entire length. The conduction rate of the nerve impulse is directly proportional to the cross-sectional diameter of the fiber and to the thickness of its myelin sheath. Generally, impulse conduction in myelinated nerve fibers is exceedingly rapid compared with that in unmyelinated fibers of similar size.

Carpenter and Bergland (1957) measured the impulse velocity in the sciatic nerves of chick embryos, young chicks, and adult chickens. They found an increase in the velocity of conduction of the fastest component during development from 0.5 to 50 m/sec, concomitant with the deposition of myelin. Brown et al. (1972) studied fiber size and conduction velocity in the vagus nerve of the chicken. They found 10,000 myelinated fibers; the largest was 6–7 μm and 87% less than 3 μm in diameter. There were also about 5,000 unmyelinated fibers. The compound action potential in the cervical vagus showed two main peaks, with conduction velocities of 0.8–1.2 and 2.2–32.4 m/msec, which very likely correspond to impulse conduction in unmyelinated and myelinated fibers, respectively.

The Reflex Arc

All life forms are challenged continually by the demands of their environments. Survival largely depends on the ability of the animal to respond to those demands in an adaptive way. Such behavior, in simple animals, is characterized by sterotyped responses to given stimuli. In higher animals, however, responses to stimuli vary depending on a number of interacting environmental factors and on the animal's memory of past events that may be related to the stimuli.

Any behavior of the animal in response to some change in the environment is called a *reflex,* and a *reflex arc* describes the network of neurons that mediate the reflex (Figures 1–7 and 1–8). The reflex arc, in abstract, involves three essential components: (1) The *afferent limb* of the reflex arc consists of the peripheral sensory receptor organs and the afferent axons that supply them; (2) the *efferent limb* of the reflex arc consists of a motor neuron that arises from the spinal cord or brainstem and a peripheral effector mechanism, such as a muscle or gland cell; and (3) some number of

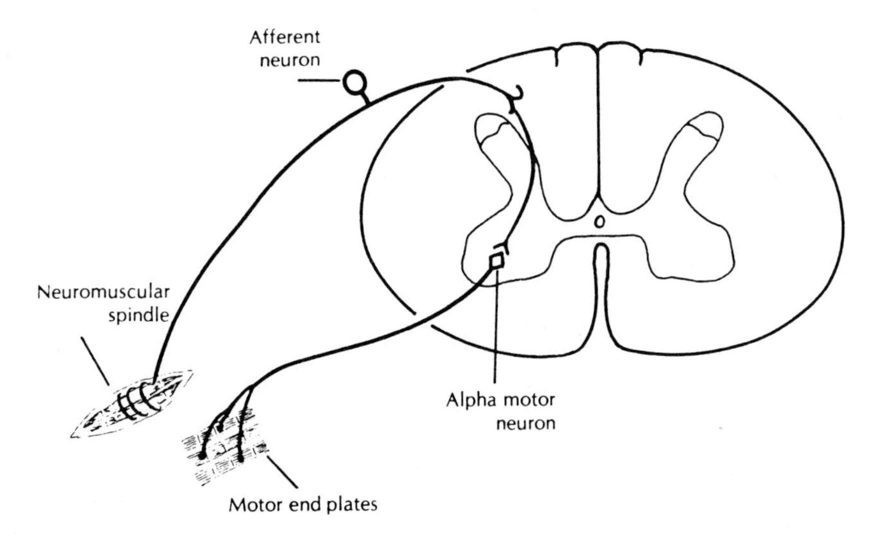

FIGURE 1–7. The stretch reflex arc (monosynaptic). (From Barr and Kiernan, 1983.)

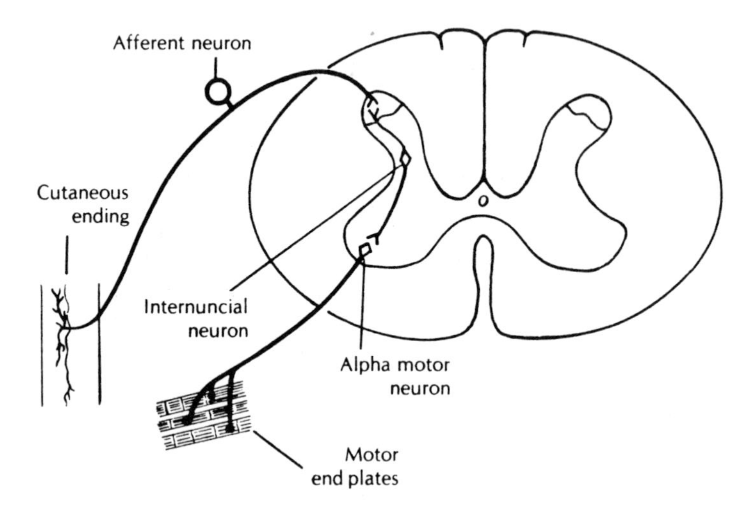

FIGURE 1–8. The flexor reflex arc (polysynaptic). (From Barr and Kiernan, 1983.)

CNS neurons in between the afferent and efferent limbs constitute the *center* of the reflex arc. These neurons calculate the appropriate transformations between the afferent (input) and efferent (output) limbs. At its simplest, and for some very basic reflexes, only one synapse need lie between the sensory event and the motor response. In such cases, the reflex arc is termed monosynaptic (Figure 1–7). These, of course, will be simple, unchangeable, and fairly gross responses of the animal to the environment. Other more complex reflexes require a staggering number of calculations involving perhaps millions of intervening neurons (polysynaptic reflex arc) (Figure 1–8).

Reflexes usually are classified as exteroceptive or interoceptive. The exteroceptive reflexes are generated by stimulation of the cutaneous receptors, which are the sense organs of the skin for perception of pain, touch, cold, and warmth (Figure 1–8); the chemoreceptors, which are involved in taste and smell, and the distance receptors, which are used in vision and hearing. The interoceptive reflexes arise with stimulation of the visceroceptors, the sensory receptors in the visceral organs, and the proprioceptors, the sense organs that are stimulated by actions of the body itself (Figure 1–7). The latter are found in the muscle spindles and tendons and also in the semicircular canals of the inner ear.

Thus, the response of the animal as a whole to some environmental event usually involves a series of various types of reflexes with modifying influences originating in the higher centers of the CNS. This is accomplished through a feedback control system. For example, when an animal performs some movement, one or more receptor organs responds to the effect of the movement. These receptors generate impulses in afferent axons that inform the CNS of the degree of movement. The CNS then can modify the movement so that the animal will be able to perform some function more efficiently. Such modification involves the comparison of the pattern of receptor impulses (input), whether from cutaneous receptors in the skin or auditory stimulation from the inner ear, with a preferred or desired pattern (output).

Neurotransmitters

Of prime significance in the integrative activities of the nervous system is the ability of one neuron to exert its influence on another. Such communication between neurons occurs at the synapse, and is mediated by neurotransmitters. During the nerve impulse, the depolarization of the terminal bouton by the action potential results in the release of the neurotransmitter by exocytosis from the vesicle into the synaptic cleft, where the chemical can influence the excitability of the postsynaptic cell.

Acetylcholine and *noradrenaline* are two chemical substances that have been positively identified as neurotransmitters released from peripheral nerve terminals in mammals, and there is strong evidence that these chemicals, among others, act as transmitters at synapses within the avian nervous system as well (see Chapter 7). Acetylcholine is known to be released from preganglionic terminals in both sympathetic and parasympathetic ganglia. It is also the transmitter released by postganglionic parasympathetic fibers and by somatic (skeletal) motor neurons at the motor endplate. Harvey and van Helden (1981) identified acetylcholine receptors in singly- and multiply-innervated skeletal muscle fibers in the chick embryo. Noradrenaline and adrenaline are released at postganglionic sympathetic nerve terminals (DeSantis et al., 1975), and Chapter 7).

In addition to acetylcholine, noradrenaline, and adrenaline, a number of other substances, such as γ-aminobutyric acid (GABA), serotonin (5-hydroxytryptamine), and dopamine, have been suggested as putative neurotransmitters within the avian CNS. Some evidence for their possible roles as transmitters in the bird is presented below. In attempts to establish that a given chemical has a transmitter role in the avian CNS, some investigators have detected the presence of the substance itself within CNS neurons; others have taken as evidence the presence within the neuron of the enzymes responsible for the synthesis and degradation of that substance.

Acetylcholine. Aprison and Takahashi (1965) determined the concentration of acetylcholine in the various parts of the pigeon brain. The highest levels were found in the optic lobes and the lowest levels in the cerebellum. The enzyme synthesizing acetylcholine, choline acetyltransferase, is present in the avian brain, and its activity was found to be high in the optic lobes, with lower levels in the cerebellum and olfactory lobes (Hebb, 1955; Hebb and Ratković, 1964). The activity of cholinesterase, the enzyme responsible for degradation of acetylcholine, was found to be high in the avian brain, including the cerebellum (Aprison et al., 1964).

The Catecholamines. Three important catecholamines, noradrenaline, adrenaline, and dopamine, have been found in the avian brain. Although it is probable that dopamine and noradrenaline represent two different nerve types, adrenaline and noradrenaline may be interchangeable (Juorio and Vogt, 1970) in that each may be released in varying proportions by the same nerve terminals and act on the same receptors (DeSantis et al., 1975).

Catecholamines are formed from the amino acid phenylalanine by an hydroxylase and dopa decarboxylase. (See also Chapers 14 and 22.) Burack and Badger (1964) reported that these enzymes are present in the primitive streak of the embryonic chick. Enemar et al. (1965) also found dopa decarboxylase activity in the sympathetic chain. Dopamine β-hydroxylase converts dopamine to noradrenaline, and this enzyme has been found in chick embryo hearts (Ignarro and Shideman, 1968). Phenylethanolamine-*N*-methyltransferase, the enzyme that is responsible for the synthesis of adrenaline, has been detected in the hen brain (Pohorecky et al., 1968). Monoamine oxidase, which is responsible in part for metabolizing catecholamines, is found in avian brain (Aprison et al., 1964). The enzymes responsible for synthesizing and degrading catecholamines are at least present in various avian tissues, even if all of them have not yet been positively identified in the avian brain. However, the presence of degrading enzymes notwithstanding, reuptake of catecholamines by the presynaptic bouton is thought to be the predominant method of transmitter inactivation.

Juorio and Vogt (1967) measured the concentration of catecholamines in different areas of the pigeon brain. The highest levels of dopamine were detected in the nucleus basalis, and very low levels were found in the optic lobes, cerebellum, and spinal cord. These regions also were deficient in adrenaline and noradrenaline. The hypothalamus, however, was the region that was richest in these two amines. The amount of adrenaline relative to noradrenaline was found to be higher than that in mammalian brains. Gunne (1962) obtained a value of 17% for the ratio of adrenaline to adrenaline plus noradrenaline. Juorio and Vogt (1970) confirmed this, but showed that the ratio was quite variable between species and even between breeds of fowl. Callingham and Sharman (1970) found that the amounts of dopamine, adrenaline, and noradrenaline increased with age in the fowl, as did the relative amount of adrenaline (see Dube and Parent, 1981).

Other Substances. Serotonin (5-hydroxytryptamine) is synthesized from tryptophan by an hydroxylase and decarboxylase, and it is destroyed by an amine oxidase. (See also Chapter 6). These enzymes have been detected in the pigeon brain (Aprison et al., 1964; Gal and Marshal, 1964). The concentration of serotonin is highest in the nucleus basalis and lowest in the cerebellum (Juorio and Vogt, 1967).

Studies on the effects of serotonin on the young chick have produced conflicting data. Several investigators have observed that serotonin produced sleep (Hehman et al., 1961; Spooner and Winters, 1967), while others have observed that it initially produces alertness and then sedation (Dewhurst and Marley, 1965).

Although serotonin is suspected by some investigators to be an anorexigenic agent in mammals, the role of serotonin in the control of food intake in birds is yet unknown. Denbow et al. (1982) observed that injections of serotonin into the lateral brain ventricles of young chicks decreased food intake in fully fed chicks, but had no effect on food consumption in 24-hr-fasted birds. Similarly, serotonin had no effect on water intake in fasted chicks, and it produced equivocal results regarding the drinking responses in fed chicks.

Gamma-aminobutyric acid (GABA) is present in chicken brain at a much higher concentration than that found in the mammalian brain (Tsukada et al., 1962). GABA is synthesized by the enzyme glutamic acid decarboxylase. This enzyme, together with the degradative enzyme, GABA transaminase, have been found in the chick cerebellum (Kuriyama et al., 1968), but as with catecholamines, uptake rather than degradation is probably the more important method of transmitter inactivation.

GABA, or its precursor, glutamic acid, when injected into chicks produced a state resembling natural sleep (Kramer and Seifter, 1966) and altered tectal electrical activity (Scholes and Roberts, 1964). GABA and glutamate have also been implicated as important neurotransmitters in the avian visual pathway (Bondy and Purdy, 1977; Cuénod et al., 1981; Toggenburger et al., 1982).

A number of amino acids, some of which have been suggested as being transmitters in the mammalian CNS, have been detected in the hen brain (Frontali, 1964), and have putative transmitter roles in the avian CNS (Nistico, 1980; Cuénod et al., 1981; Toggenburger et al., 1982; DePlazas, 1982).

A number of peptides that were originally isolated from the gut are present in specific neuronal populations within the nervous system. Somatostatin (Shiosaka et al., 1981), substance P (Erichsen et al., 1982; Gamlin et al., 1982; Saffrey et al., 1982), vasoactive intestinal polypeptide, and neurotensin (Yamada and Mikami, 1982; Saffrey et al., 1982) have been detected in the avian CNS and PNS. It is thought that these substances may act as neurotransmitters, or more probably as "neuromodulators," since they have been found to coexist in certain neurons with other putative transmitter substances.

Prostaglandins have been extracted from chick nervous tissue, and they have been found to have pharmacologic effects if injected into young chicks (Horton, 1971).

Enkephalin-like immunoreactivity has been detected in various regions of the avian brain (Ryan et al., 1981; Erichsen et al., 1982; Reiner et al., 1982) and PNS (Saffrey et al., 1982).

Further details pertaining to putative transmitters in the avian CNS, particularly on the effects of drugs in modifying their concentrations, can be found in Bolton (1971b) and in Pearson (1972).

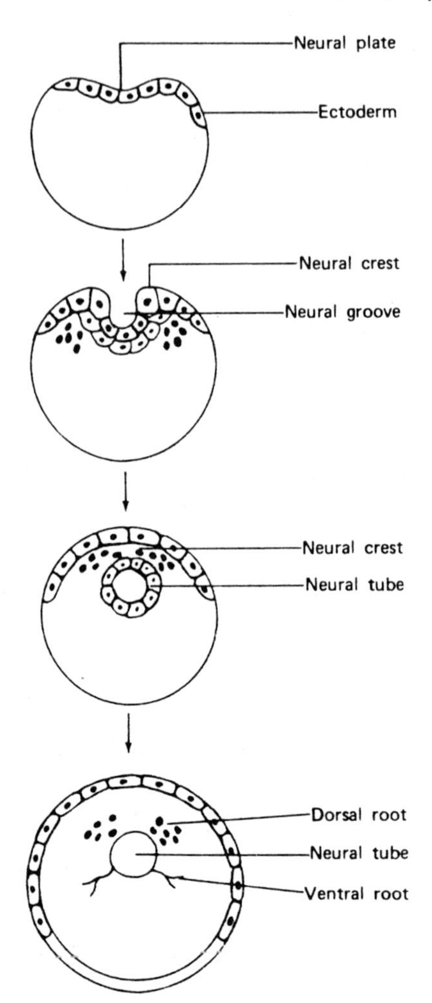

FIGURE 1–9. The formation of the neural tube during the early development of the spinal cord. (From Chaplin and Demers, 1978. Reprinted by permission of John Wiley & Sons, Inc.)

Development of the Nervous System

The Spinal Cord and Peripheral Nerves

The earliest indication of the formation of the central nervous system in the bird appears in chick embryos between 16 and 18 hr of egg incubation. This occurs as a local thickening of the surface ectoderm that forms the neural plate (Figure 1–9). The neural plate then becomes folded longitudinally to form the neural groove. The neural groove is bounded laterally by two elevations of the lateral portions of the neural plate known as the neural folds. With subsequent fusion of the margins of the neural folds, first in the cephalic region of the embryo and later caudally, the neural groove is closed to form the neural tube. The cephalic portion of the neural tube becomes dilated and eventually will form the brain; the remainder of the neural tube will develop into the spinal cord.

In the closure of the neural tube, the superficial ectoderm also becomes fused in the midline and separates from the developing neural tube to form a continuous ectodermal covering. There are cells lying at the edges of the neural folds that are not involved in the fusion of either the superficial ectoderm or the neural folds (Figure 1–9). These cells form a pair of longitudinal aggregations extending on either side of the midline in the angles between the superficial ectoderm and the neural tube. These cellular aggregations are known as the neural crest, and they only temporarily maintain a position dorsal to the neural tube. As development proceeds, the cells of the neural crest migrate ventrolaterally on either side of the spinal cord and become segmentally clustered. The segmentally arranged groups derived from the neural crest ultimately give rise to the dorsal root ganglia of the spinal nerves, the sensory ganglia of the cranial nerves, the autonomic (visceral motor) ganglia, Schwann cells, the cells of the suprarenal medulla, and melanocytes (see Ziller and Smith, 1982).

By 55 hr of egg incubation, closure of the neural tube is completed throughout its entire length. The last regions to close are at the cephalic and caudal ends of the neural groove. In younger stages, these regions remain open, and are known as the anterior neuropore and the sinus rhomboidalis, respectively. When first established, the spinal cord region of the neural tube has a lumen that is elliptical in cross section. As development progresses, the lateral wall of the cord becomes greatly thickened compared with the dorsal and ventral walls, which remain thin. In this process, the lumen (central canal) becomes compressed laterally so that it appears as a vertical slit in cross section. The thin dorsal wall of the tube is called the roof plate, and the thin ventral wall, the floor plate. The thickened side walls are known as the lateral plates.

The nerve roots of the spinal nerves begin to form by the fourth embryonic day. Lateral to the spinal cord, the dorsal and ventral roots unite to form the spinal nerve (Figure 1–10). The dorsal root ganglion is located on the dorsal root between the spinal cord and the point where dorsal and ventral roots unite. Distal to the union of the dorsal and ventral roots is a branch (the ramus communicans) that connects the spinal nerve to a paravertebral ganglion of the sympathetic nerve trunk (Figure 1–11). The dorsal root is established by the growth of nerve fibers from sensory neurons of the dorsal root ganglion medially into the dorsal portion of the lateral plate of the spinal cord. At the same time, nerve fibers grow distally from these neurons to form the peripheral part of the nerve (Figure 1–10). The fibers that arise from the dorsal root ganglion convey sensory impulses toward the spinal cord. Concurrent with the development of the dorsal root, the ventral root is formed by fibers that grow out from neurons located in the ventral portion of the lateral plate of the spinal cord (Figure 1–10). These fibers pass out through the ventral root and conduct motor impulses from the CNS to appropriate peripheral muscles.

The sympathetic ganglia arise from cells of the neural crest. By the end of the fourth embryonic day, the primordia of the sympathetic ganglia have become interconnected longitudinally by slender strands. The developing sympathetic ganglia appear as local enlargements on paired, cordlike structures. These are the paravertebral sympathetic trunks. Each sympathetic ganglion is connected with the corresponding spinal nerve by a thin cord that is the primordium of the ramus communicans. Later, both sensory and motor fibers appear in the rami communicantes, establishing communication between the sympathetic ganglia and the spinal nerve. From certain sympathetic ganglia, cells migrate still farther ventrally and establish the primordia of the prevertebral sympathetic ganglia on either side of the midline at the level of the dorsal aorta. Preganglionic sympathetic fibers arise from mo-

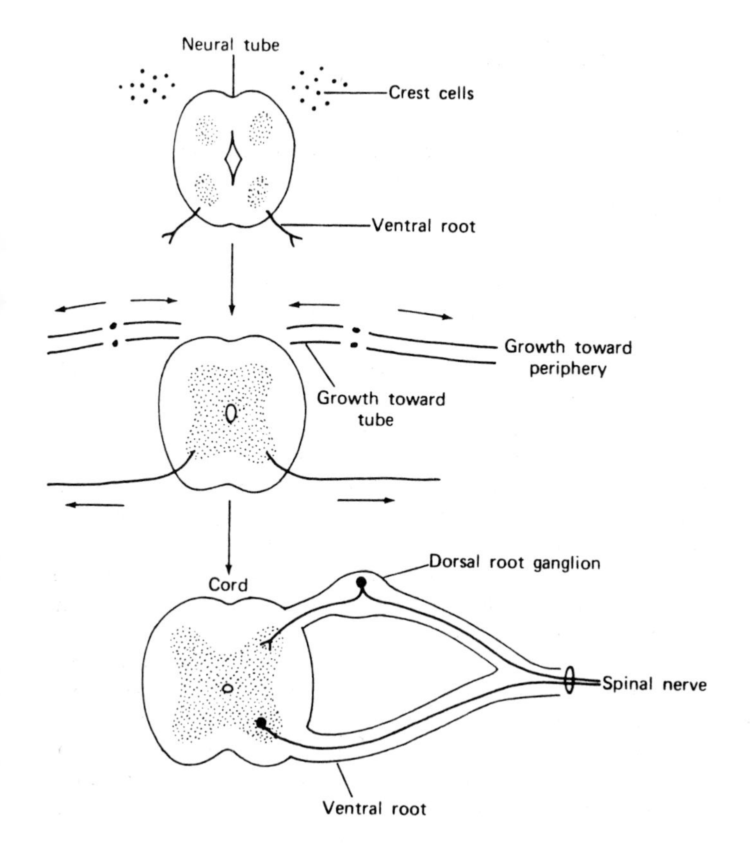

FIGURE 1–10. The development of the dorsal and ventral roots of the spinal nerve. (From Chaplin and Demers, 1978. Reprinted by permission of John Wiley & Sons, Inc.)

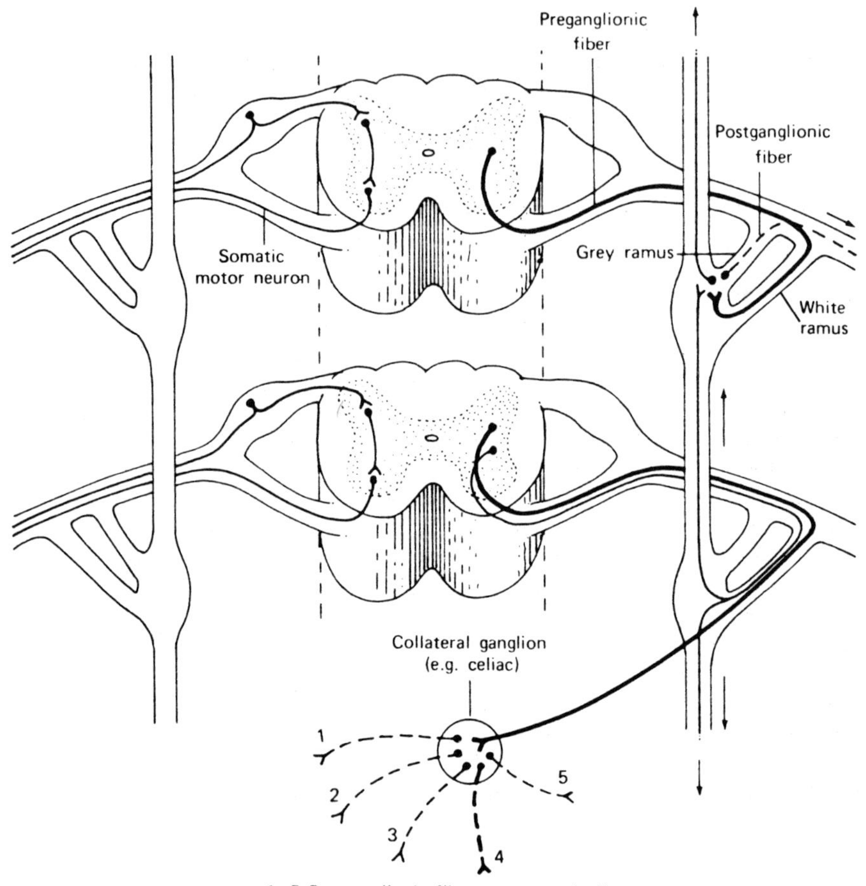

FIGURE 1–11. The connections of the sympathetic division of the autonomic nervous system with the spinal cord. The solid lines represent preganglionic fibers and the broken lines represent postganglionic fibers. (From Chaplin and Demers, 1978. Reprinted by permission of John Wiley & Sons, Inc.)

tor neurons residing within the spinal cord, whereas postganglionic sympathetic fibers stem from nerve cells within the paravertebral or prevertebral sympathetic ganglia (Figure 1–11).

The Brain

By 27 hr of egg incubation, the cephalic part of the neural tube in the chick embryo is markedly enlarged compared with the more caudal parts. Its thickened walls and dilated lumen mark the region that will develop into the brain (Figure 1–12). Three divisions, known as the primary brain vesicles, can be distinguished in the enlarged cephalic region of the neural tube. Most rostrally is a conspicuous dilation known as the forebrain or *prosencephalon.* Posterior to the prosencephalon is the midbrain or *mesencephalon.* Posterior to the mesencephalon is the hindbrain or *rhombencephalon.* The rhombencephalon is continuous posteriorly with the spinal cord region, or *myelon,* of the neural tube without any definite point of transition (Figure 1–12).

The organs of special sense arise early in avian development. By about 33 hr of egg incubation, the optic vesicles, which later become the sensory parts of the eyes, are established as paired, lateral outgrowths of the prosencephalon. The first indication of the formation of the sensory part of the ear becomes evident at about 35 hr of egg incubation. At this stage, a pair of thickenings, termed the auditory placodes, arise in the superficial ectoderm on the dorsolateral surface of the rhombencephalon.

In chick embryos of about 38 hr of egg incubation, indications of the impending division of the three primary vesicles to form the five regions characteristic of the adult brain are beginning to appear. In the establishment of the five-vesicle condition of the brain, the prosencephalon is subdivided to form the telen-

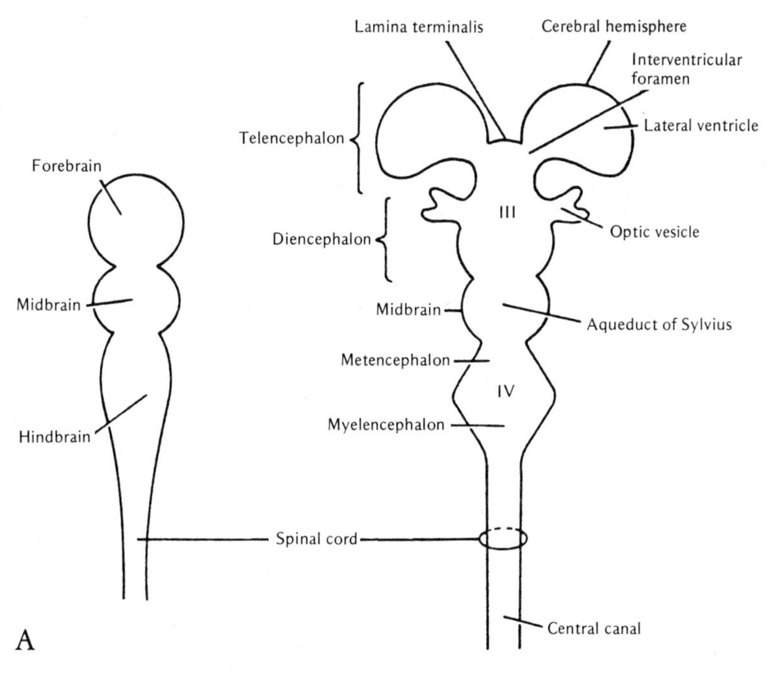

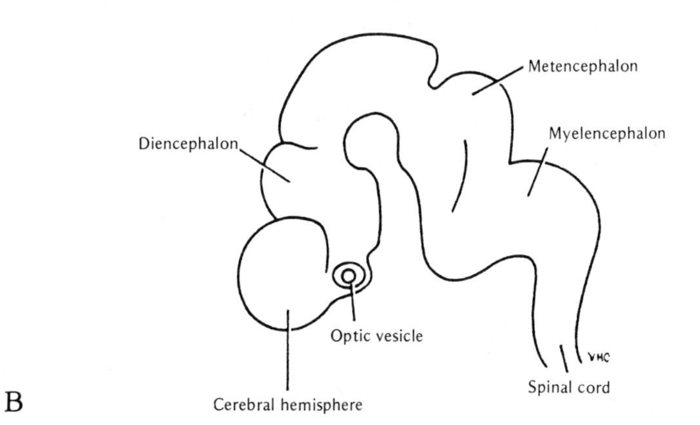

FIGURE 1–12. The divisions of the primary brain vesicles. **A** Dorsal view. **B** lateral view. (From Snell, 1975. Reprinted by permission of Little, Brown and Company, © 1975.)

cephalon and diencephalon, the mesencephalon remains undivided, and the rhombencephalon divides to form the metencephalon and myelencephalon (Figure 1–12).

By the fourth embryonic day, the beginnings of differentiation of the various brain regions into their adult derivatives are evident. The telencephalic vesicles develop into the cerebral hemispheres, and their cavities become the paired, lateral ventricles of the adult brain. The hemispheres undergo substantial enlargement later in development, and they extend dorsally, posteriorly, and rostrally, eventually covering the entire diencephalon and mesencephalon under their posterior lobes. They contain motor and sensory integration areas and the brain centers for memory and for all behaviors that are conditioned by past experience.

The diencephalon gives rise to the various components of the epithalamus, thalamus, and hypothalamus in the adult brain, and it encloses the major portion of the third brain ventricle. The anterior part of the roof of the diencephalon becomes richly vascular, and later invaginates into the third brain ventricle to form the anterior choroid plexus.

The dorsal walls of the mesencephalon increase in thickness and give rise to four symmetrically placed elevations, the corpora quadrigemina of the adult brain. The anterior pair, the superior colliculi, constitute the brain center for visual reflexes; the posterior pair, the inferior colliculi, are the center for auditory reflexes. The floor of the mesencephalon also becomes thickened and develops into the crura cerebri of the adult. It serves as the main pathway for the fiber tracts that connect the cerebral hemispheres with the posterior part of the brain and the spinal cord. The now thickened walls of the mesencephalon enclose a narrow canal, the cerebral aqueduct or aqueduct of Sylvius, that communicates with the fourth brain ventricle (Figure 1–12).

The roof of the metencephalon undergoes enlargement and becomes the cerebellum of the adult brain, the coordinating center for complex muscular movements. Extensive invaginations of fiber tracts ventrally and laterally give rise to the pons and to the cerebellar peduncles.

The ventral and lateral walls of the myelencephalon become the floor and side walls, respectively, of the medulla oblongata. Functionally, the medulla serves both as a conduction path between the spinal cord and brain and as a reflex center for involuntary activities such as respiration. The thin roof of the medulla receives a rich supply of blood vessels that grow into the fourth ventricle to form the posterior choroid plexus. The fourth ventricle extends through both the metencephalon and myelencephalon, and it is continuous with the central canal of the spinal cord (Figure 1–12). The ventricular system of the brain contains cerebrospinal fluid, which is produced by the choroid plexi.

In the region of the developing brain, cells derived from the cephalic portion of the neural crest aggregate and give rise to the sensory ganglia that are associated with the cranial nerves. Recent evidence suggests that cephalic neural crest cells also give rise to mesenchymal cells in the head and neck (see Noden, 1983a, b).

The Peripheral Nervous System

The peripheral nervous system (PNS) consists of those parts of the nervous system other than the spinal cord and brain. The PNS includes the spinal and cranial nerves and the ganglia and plexi involved with visceral innervation. The general arrangement of the PNS in birds does not differ in its basic organization from that in mammals.

Spinal Nerves

In the bird, each spinal nerve arises by a ventral (motor) root and a dorsal (sensory) root from the spinal cord (Figure 1–13). Attached to each dorsal root is a sensory ganglion in which the perikarya of afferent neurons reside. After fusion of the dorsal and ventral roots to form the spinal nerve, the nerve leaves the spinal canal. Each spinal nerve then divides into a dorsal branch (ramus) that supplies the dorsal muscles and skin of the back, and a ventral branch that supplies the ventral muscles and the skin of the anterior body wall. The spinal nerve is a mixed nerve with both sensory and motor fibers that carry information to and from the CNS, respectively. The functional components of a typical avian spinal nerve include two types of sensory fibers and two types of motor fibers. General somatic afferent fibers conduct impulses to the CNS from exteroceptors (on the surface of the body) or proprioceptors (in muscles, joints, and tendons), and general visceral afferent fibers bring interoceptive in-

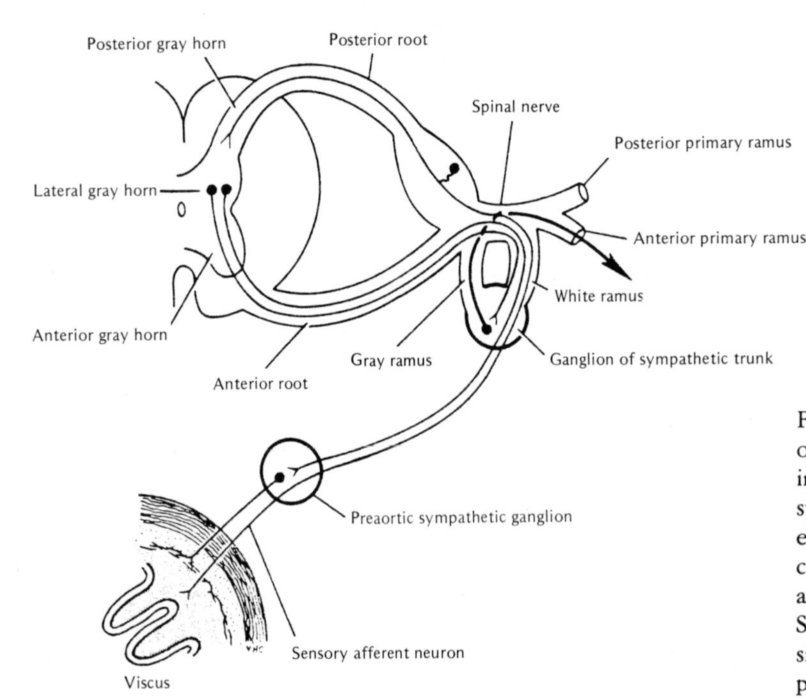

FIGURE 1–13. A transverse section of the thoracic spinal cord, showing two examples of preganglionic sympathetic fibers leaving the lateral gray matter. An afferent (visceral sensory) neuron passes from a viscus into the spinal cord. (From Snell, 1975. Reprinted by permission of Little, Brown and Company, © 1975.)

formation to the CNS from visceroceptors that are located in the visceral organs. Motor impulses from CNS centers to skeletal muscle are conducted by general somatic efferent fibers, whereas general visceral efferent (autonomic) fibers convey impulses to smooth and cardiac muscle and to glands.

The number of pairs of spinal nerves arising from the spinal cord varies among different birds. In the chicken, there are 15 pairs of cervical nerves, 7 pairs of thoracic nerves, and 14 pairs of lumbosacral nerves. In the cervical, thoracic, and lumbosacral regions, the ventral branch of the spinal nerve communicates with the corresponding segmental visceral motor (autonomic) ganglion (Figures 1–11 and 1–13). These ganglia are part of the sympathetic portion of the autonomic nervous system.

In the regions of the wings and hind limbs, the ventral rami of certain spinal nerves divide and then reunite to form the brachial and lumbosacral plexi, respectively. Different segmental nerves may be involved in the formation of these plexi in different birds (see Bennett, 1974). In the chicken, the brachial plexus is formed by the ventral rami of the last two to four cervical nerves and by the first, and sometimes the second, thoracic nerve. This plexus gives rise to nerves that supply the shoulder region and the muscles and skin of the wing. The innervation of the forelimb of the chicken has been documented by several investigators (see Yasuda, 1960; Bennett, 1974; Ohmori et al., 1982; Straznicky and Tay, 1983).

The lumbosacral plexus arises from the first to the eighth lumbosacral nerves in the chicken. This plexus gives rise to nerves that supply the skin and muscles of the hip region and the lower limb. Yasuda (1961) has described the innervation of the muscles of the chicken hind limb.

Cranial Nerves

The cranial nerves in the bird are less regular in their arrangement than are the spinal nerves, and their functional components vary as well. Some are mixed nerves; some have only sensory fibers; and at least one, the hypoglossal (12th), is thought to be purely motor in function (however, see Bottjer and Arnold, 1982). The sensory ganglia that are associated with some of the cranial nerves are structural and functional analogs of the dorsal root ganglia of the spinal nerves. Similarly, sensory fibers terminate in, and motor fibers arise from, areas of the gray matter in the brain stem that are functionally analogous to the gray areas of the spinal cord.

Like mammals, birds have 12 pairs of cranial nerves arising from the brain stem (Figure 1–14). The gross anatomy of avian cranial nerves is very similar to that of mammals. Watanabe and colleagues have detailed the anatomic distribution of the cranial nerves in the chicken (see Watanabe, 1960, 1964; Watanabe and Yasuda, 1968, 1970; Watanabe et al., 1967).

As in mammals, the olfactory (first) and optic (second) nerves are not true "nerves," but rather are extensions (tracts) of special sensory areas of the brain. The oculomotor (third), trochlear (fourth), and abducent (sixth) nerves supply motor and sensory (proprioceptive) fibers to the extrinsic muscles of the eye. The trigeminal (fifth) nerve is the major general sensory nerve for the head region and is the motor nerve for the muscles of mastication (see Noden, 1980). The extensive fiber distribution of the trigeminal nerve has been examined in several avian species (see Dubbeldam and Karten, 1978; Berkhout et al., 1982; Arends and Dubbeldam, 1982). The facial (seventh) and glossopharyngeal (ninth) nerves are mixed nerves that together are responsible for the sensory and autonomic innervation of the mucous membranes and glands of the tongue, mouth, pharynx, larynx, and esophagus.

The visceral motor fibers of the facial nerve supply the lacrimal gland and, together with those of the ninth nerve, innervate the salivary glands. Each nerve has a small skeletal motor component that is responsible for supplying several small muscles in the head and neck region, including the cutaneous muscles of the face and neck (facial nerve) and portions of the upper esophagus (glossopharyngeal nerve).

The acoustic (eighth) nerve is a special sensory nerve that conveys auditory (cochlear branch) and equilibrium sense (vestibular branch) impulses into the avian brain.

Skeletal motor fibers in the vagus (10th) nerve, and those of the cranial root of the accessory (11th) nerve, supply the esophageal and laryngeal muscles. The spinal or cervical root of the accessory nerve innervates several muscles in the neck. The visceral motor fibers of the vagus nerve distribute partially with those of the glossopharyngeal in supplying much of the digestive tract musculature and associated glands. The sensory fibers of the vagus convey afferent impulses mainly from the thoracic and abdominal viscera. The hypoglossal (12th) nerve in the bird supplies the lingual, hyoid, laryngeal, and syringeal muscles.

Sensory Receptor Organs

In addition to those receptors concerned with the special senses, birds possess a variety of general sensory receptors that are similar to those found in mammals. These include proprioceptors, such as muscle spindles and Golgi tendon organs, various cutaneous receptors (exteroceptors), and visceroceptors (interoceptors). These receptor organs are scattered throughout the tissues of the body and are relatively simple in structure. They convey information to the CNS via afferent fibers that travel in both spinal and cranial nerves. In contrast to these general sensory endings, receptors

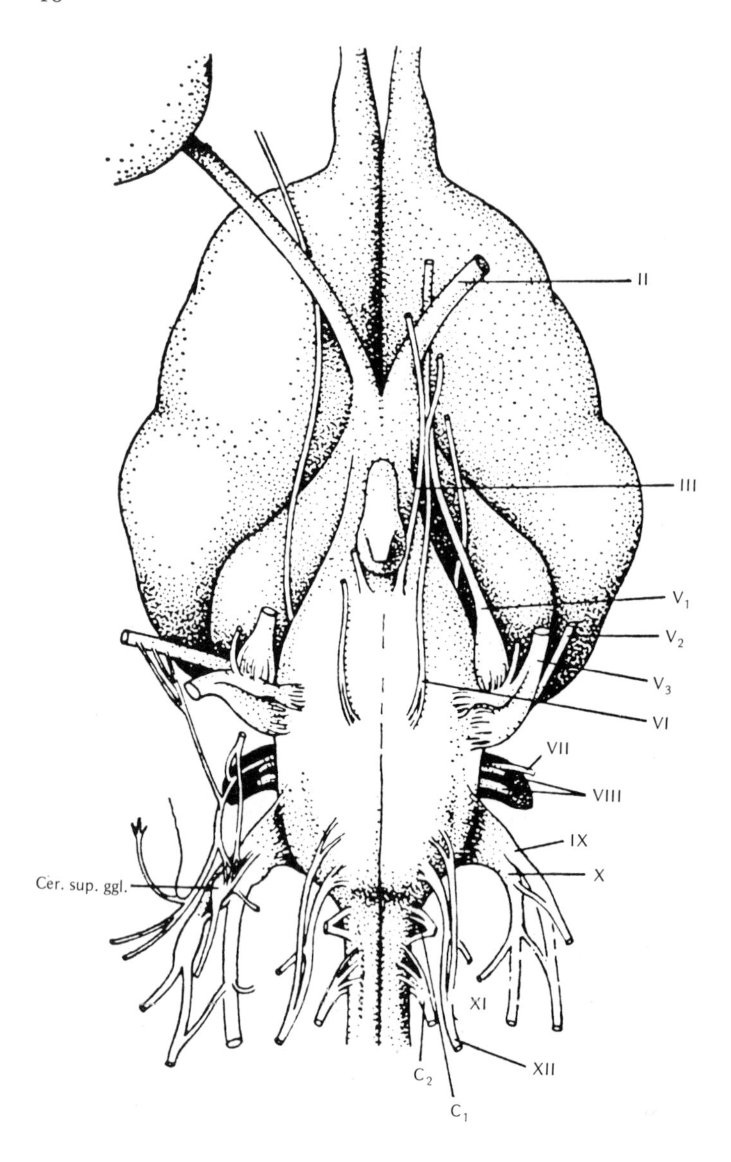

FIGURE 1–14. A ventral view of the brain of the goose showing the roots of the cranial nerves. V_1, V_2, and V_3 are branches of the trigeminal nerve; C_1 and C_2 are cervical nerves; *cer. sup. ggl.* is the superior cervical ganglion. (From Pearson, 1972.)

for the special senses are localized and highly specialized. Their afferent axons are carried by various cranial nerves into the CNS for processing of olfactory, visual, auditory, vestibular, and gustatory information. These special afferent pathways are described separately below.

Muscle spindles are general sensory receptors that are found exclusively in vertebrate skeletal muscle (Chapter 3). They lie between and in parallel with skeletal muscle fibers (extrafusal fibers) and receive a separate and rich motor and sensory innervation. Muscle spindles consist of highly specialized muscle fibers (intrafusal fibers) enclosed within a prominent capsule. De Anda and Rebollo (1967) described three types of intrafusal fibers based on fiber size in avian muscle spindles; small-diameter fibers with nuclear chains, large-diameter fibers with nuclear bag, and fibers of intermediate type. Large-diameter intrafusal fibers are innervated by larger diameter nerve fibers, small intrafusal fibers by smaller nerves (see also Chapter 3). Both have a basket-like arrangement of nerve fibers near the nuclei. Two types of motor (de Anda and Rebollo, 1967) and sensory (Adal and Chew-Cheng, 1980) endings generally are observed. Other investigators have classified these intrafusal fibers into three types using ultrastructural criteria (Adal, 1973; Peinone and Daneo, 1978). Some investigators contend, however, that most avian muscle spindles are monofibril spindles with only a single intrafusal fiber type (Maier and Eldred, 1971; James and Meek, 1973).

Golgi tendon organs are encapsulated mechanoreceptors that are located along the myotendinous junction of skeletal muscles in birds and mammals. They respond to both active contraction and passive stretch of the muscle. Haiden (1981) has documented the ultrastructural characteristics of the avian Golgi tendon organ.

Cutaneous receptors are either free nerve endings or encapsulated organs. Merckel's disks occur in the skin and buccal cavity (see Pac, 1982). Structures resembling Krause's end bulbs have also been described, as have Herbst's lamellar corpuscles, which appear to be analogous to the Vater–Pacinian corpuscles of mammals (see Bennett, 1974; Malinovsky and Pac, 1980). They occur in the skin of the bill, tongue, and palate, and have been found also in the articular capsule of the pigeon's shoulder joint (Halata and Munger, 1980).

Thermoreceptors, and *vibration-sensitive* and *vibration-insensitive mechanoreceptors,* have been examined in the bill and tongue of the duck (Gregory, 1973; Leitner and Roumy, 1974a, b) and in the beak skin of the goose (Gottschaldt et al., 1982). Vibration-sensitive Pacinian corpuscles also have been identified in the interosseous tissue of the hind limb of the chicken (Skoglund, 1960) and of the duck (Dorward and McIntyre, 1971).

Visceral sensory receptors are located primarily in the heart, lungs, and bronchi, but also are found in other tissues of the thorax and abdomen. Their function is associated with the control of circulation (Jones and Johansen, 1972) and respiration (Jukes, 1971; King and Molony, 1971; Lasjewski, 1972).

Somatic Motor Innervation

Avian somatic (skeletal) muscles are innervated by axons of somatic efferent neurons that are carried by spinal or cranial nerves. These axons terminate in motor endplates on the voluntary muscle fibers. Acetylcholine is the neurotransmitter that is released at the effector site (see Harvey and Van Helden, 1981).

Two types of skeletal muscle fibers have been identified in birds (see Chapter 3, on muscle). One type has an orderly structure of fibrils and straight Z disks running across the fiber (*Fibrillenstruktur* fibers, which are adapted for quick, intermittent contractions). They usually are innervated by one, or only a few, axon branches (*en plaque* terminals). The other muscle fiber type shows no organized structural pattern (*Feldenstruktur* fibers, which are innervated by many axon terminals) (*en grappe* terminals).

The iris in the bird is composed also of striated *Fibrillenstruktur* fibers, but these fibers receive *en grappe* endings (see Sivak, 1983).

In addition, the nictitating membrane in the bird, unlike that in the mammal, is operated by two striated muscles (of the *Fibrillenstrucktur* type) that are supplied by the abducent nerve (see also Chapter 2).

The Autonomic Nervous System

The autononomic nervous system (ANS) includes those efferent nerves that innervate smooth and cardiac muscle and glands. While visceral afferent fibers are not, by anatomic definition, "autonomic nerves," they travel with visceral efferent fibers and are integral functional components in most autonomic responses (Cabot and Cohen, 1977; Abdalla and Kind, 1979; Koley et al., 1979). In spite of its name, the ANS actually does not function in complete autonomy, apart from the rest of the nervous system; higher CNS centers often modify or initiate many autonomic mechanisms. The ANS functions in maintaining homeostasis in the various organ systems of the body. It is divided somewhat artificially into sympathetic and parasympathetic portions. The two divisions often will have contrasting effects on some target organ, such as increasing or decreasing blood flow to it, thereby tuning the animal's visceral response to its changing environment. "Normal" activity (e.g., heart rate, respiration) represents a balance between parasympathetic and sympathetic input into a given organ system (see Chapters 6 and 7).

A third component of the ANS, distinct from the sympathetic and parasympathetic divisions, is the enteric nervous system. This portion of the ANS consists of interconnected populations of neurons (plexi) contained within the walls of the alimentary tract. The myenteric plexus (of Auerbach) is located between the longitudinal and circular muscle layers of the gut wall, and the submucous plexus (of Meissner) lies in the connective tissue between the circular muscle layer and the muscularis mucosae (see also Chapter 11). These intrinsic neurons of the alimentary tract may receive both sympathetic and parasympatheitc input, but they are capable of regulating peristaltic and related spontaneous activity of the gut in the absence of any extrinsic innervation (see Ali and McLelland, 1978). See Chapter 11 for more details on birds.)

Thus, the efferent innervation of smooth and cardiac muscle and of the glands of the body differs from that of skeletal muscle in that the connection between the CNS and the target tissue involves at least two motor neurons rather than a single motor neuron. In autonomic innervation, the perikaryon of the first motor neuron (preganglionic neuron) is found in a visceral motor nucleus located in the brain stem or spinal cord. The axon of the preganglionic neuron usually will synapse with a second motor neuron (postganglionic neuron) located in an autonomic ganglion. The axon of the postganglionic neuron, in turn, will terminate on the effector cell (muscle or gland), or, if appropriate, on a third neuron that resides within one of the enteric plexi of the gut (see Figure 1–13).

Anatomy and Distribution. The sympathetic division of the ANS in avians is composed of those visceral efferent fibers leaving the spinal cord within the ventral roots of the spinal nerves in the thoracic and lumbar regions (preganglionic fibers) and entering the paravertebral or the prevertebral ganglia, in which they synapse with postganglionic neurons (Figure 1–11). Some sympathetic fibers may synapse in the paravertebral ganglion of the vertebral segment where they leave the spinal cord. The postganglionic fiber then may join with somatic efferent fibers and be distributed via the peripheral nerves to blood vessels and to feather muscles (Jeikouski and Drenckhahn, 1981). Other fibers do not synapse in the paravertebral ganglion, but pass beyond to synapse in another paravertebral or often in an unpaired prevertebral ganglion. Sympathetic preganglionic fibers that synapse within prevertebral ganglia travel in the splanchnic nerves. The anatomy of sympathetic nerve distribution distinguishes such fibers from parasympathetic fibers, which leave in certain cranial nerves and in sacral nerves and synapse in scattered, small ganglia that are usually in or on the target organ.

In the bird, two sympathetic nerve trunks bearing segmental (paravertebral) ganglia extend along the vertebral column from the base of the skull to the level of the sixth coccygeal vertebra, where the trunks fuse. Unlike in mammals, there is no ganglion impar at this point in birds. In most birds there are *37 ganglion pairs;* 14 cervical, 7 thoracic, 13 lumbosacral, and 3 coccygeal. The most anterior pair are the anterior cervical ganglia. They send sympathetic postganglionic fibers to all the cranial nerves except the optic and the acoustic. These ganglia also give sympathetic contributions to the medulla oblongata, and send fibers that form nerve plexi on the major arteries of the head and neck. This is an alternate pathway for sympathetic nerve distribution to visceral structures in this region.

The splanchnic nerves arising from the thoracic and lumbosacral regions of the spinal cord contain preganglionic sympathetic fibers and visceral afferent axons that are involved with the innervation of abdominal and pelvic structures. In the chicken, the greater splanchnic nerves are formed from the second to fifth thoracic ganglia and travel to the abdomen where they form the coeliac plexus. The lesser splanchnic nerves are derived from the fifth to seventh thoracic and first and second lumbosacral ganglia. They are the sympathetic contributions to the aortic plexus. Sympathetic fibers from the sixth to the 12th lumbosacral ganglia contribute to the hypogastric plexus, the posterior mesenteric plexus, the pelvic plexus, and the cloacal plexus. From sacral levels of the spinal cord, parasympathetic fibers (pelvic splanchnic nerves) join with sympathetic fibers to form the hypogastric plexus. This plexus contributes to the posterior mesenteric and pelvic plexi, which supply the lower alimentary tract. The nerve of Remak is a ganglionated trunk in avians that is comparable to the hypogastric or pelvic plexus in mammals. This nerve trunk connects with the anterior mesenteric, pancreaticoduodenal, aortic, posterior mesenteric, and hypogastric plexi. Other plexi, designated by appropriate names, are found in the region of the heart, the stomachs, the coeliac artery, the liver, the spleen, the adrenals, and the kidney (see Bennett, 1974; Ali and McLelland, 1978; Tindall, 1979).

There are four pairs of ganglia in the head that are associated with the distribution of autonomic nerves to that region; these ganglia, however, are involved with distributing parasympathetic rather than sympathetic fibers. There are ciliary, ethmoidal, sphenopalatine, and submandibular ganglia; in the bird, there are no otic ganglia as found in the mammal. The details of the communications between these ganglia and the cranial nerves are given by Bennett (1974). The preganglionic parasympathetic fibers arise from visceral motor nuclei in various regions of the brain stem and distribute with branches of the oculomotor, facial, glossopharyngeal, and vagus nerves.

Visceral Motor Innervation. Physiologically, sympathetic and parasympathetic neurons differ from one another with respect to the neurotransmitter that is released at the terminal effector site. Both pre- and postganglionic parasympathetic fibers (cholinergic fibers) release acetylcholine. Acetylcholine also is the transmitter released by preganglionic sympathetics fibers, but, with some exceptions, noradrenaline is the neurotransmitter liberated by postganglionic sympathetic nerve endings (adrenergic fibers) (see also Chapter 7). The picture is less clear regarding the synaptic physiology of neurons within the enteric plexi. These neurons have been found to be structurally and chemically quite different than those residing in sympathetic or parasympathetic ganglia. Many enteric neurons are "nonadrenergic" and "noncholinergic," containing instead several different peptides or nucleotides that have pharmacologically demonstrable actions on the gut. It is likely that some of these substances may serve as enteric neurotransmitters.

In the alimentary tract of the bird, the tongue, pharynx, and upper esophagus are innervated by sensory and parasympathetic fibers via the glossopharyngeal nerve (see also Chapter 11). Its fibers communicate with those of the vagus nerve. The vagus supplies sensory and visceral efferent fibers to the crop, proventriculus, gizzard, and small intestine (Abdalla and Kind, 1979). The large intestine and rectum are supplied by parasympathetic fibers from the sacral region via Remak's nerve. Several anatomical studies have suggested that sympathetic nerves via the various abdominal and pelvic plexi supply all parts of the alimentary tract (see Bolton, 1971a, b; Bennett, 1974). Sympathetic nerves are excitatory to some parts of the alimentary tract such as the gizzard, lower esophagus, and proventriculus, whereas inhibitory effects of vagal nerve stimulation have been reported (Sato et al., 1970).

Studies of the autonomic innervation of the heart and cardiovascular system have revealed that both the atria and the ventricles of the bird heart receive adrenergic and cholinergic nerve fibers (Bolton and Bowman, 1969; Bennett et al., 1970; Cabot and Cohen, 1977). Akester et al. (1969) and Bennett and Malmfors (1970) demonstrated an especially dense innervation of the sinoatrial and atrioventricular nodes. (See also Chapters 6 and 7.)

Section of the vagus nerves produces an increase in heart rate in most birds, and bradycardia follows the stimulation of the peripheral end of either vagus in the chicken and duck (Chapters 6 and 7). The chronic pulmonary alterations that follow vagotomy in the chicken were reported by Burger and Fedde (1964).

Autonomic innervation to these and other visceral systems in the bird was reviewed extensively by Bennett (1974) and is detailed elsewhere in this volume. A review of the actions of various drugs on the autonomic control of the cardiovascular and digestive systems in birds can be found in Bolton (1971b, 1976).

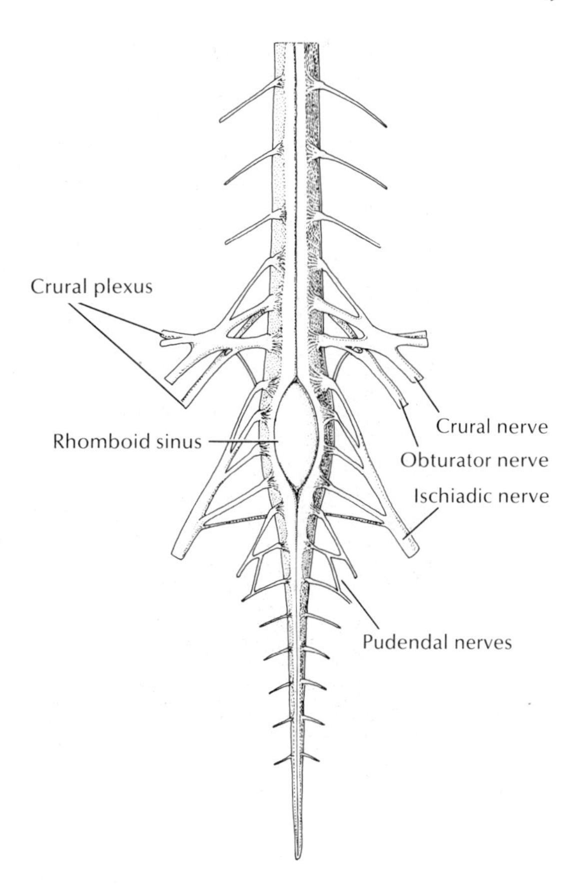

FIGURE 1–15. Lumbosacral part of the spinal cord of the flamingo showing three nerve plexi and the rhomboid sinus which contains the glycogen body. (From Bolton, 1976.)

The Spinal Cord

Gross Anatomy of the Spinal Cord

The avian spinal cord is contained within the spinal canal of the vertebral column. The spinal cord of birds, unlike that of mammals, extends throughout the whole length of the canal and has neither a cauda equina nor a pronounced filum terminale. The cervical, thoracic, and lumbosacral regions correspond to these areas of the vertebral column, and compared with the mammalian cord, the cervical and lumbosacral regions are longer and the thoracic region shorter in the avian spinal cord. The spinal cord is enlarged in the cervical and lumbosacral regions because of the many neurons residing in the gray matter at these cord levels. These neurons give rise to axons that form the brachial and lumbosacral plexi that supply the limbs (see Martin, 1979). In addition, the lumbosacral region of the avian spinal cord presents some unique and interesting features. It exhibits peculiar formations of tissues that store glycogen, specifically the glycogen body and the accessory lobes of Lachi (Figures 1–15 and 1–16).

The Glycogen Body and Accessory Lobes. The glycogen body is a prominent, gelatinous structure that occupies the lumbosacral or rhomboid sinus in the dorsomedial portion of the cord at the level of the lumbosacral plexus. The accessory lobes of Lachi, much less conspicuous than the glycogen body, consist of segmentally arranged protrusions that extend bilaterally from the periphery of the lumbosacral cord (see Ariëns-Kappers et al., 1936; DeGennaro and Benzo, 1976, 1978). First described by Lachi in 1889 (cited by Ariëns-Kappers et al., 1936), the accessory lobes are also known as Hofmann's nuclei, Hofmann–Kolliker nuclei, and the major marginal nuclei. Structures resembling the accessory lobes, but not the glycogen body, have also been identified in the reptilian spinal cord (see Ariëns-Kappers et al., 1936). Except for the presence of numerous large nerve cells in the accessory lobes, the histologic appearance of this tissue is similar to that of the glycogen body; both tissues are filled with glycogen-rich cells (DeGennaro and Benzo, 1978). (see also Chapter 13). The nerve cells found within the accessory lobes have been termed "paragriseal" neurons because their perikarya are found outside the gray matter of the spinal cord. Extensive deposits of glycogen are found within the accessory lobe cells, which resemble neuroglial astrocytes with extensive processes that envelop the perikarya and axons of the paragriseal cells and blood vessels throughout this tissue. The presence of axoglial synapses and other junctional complexes

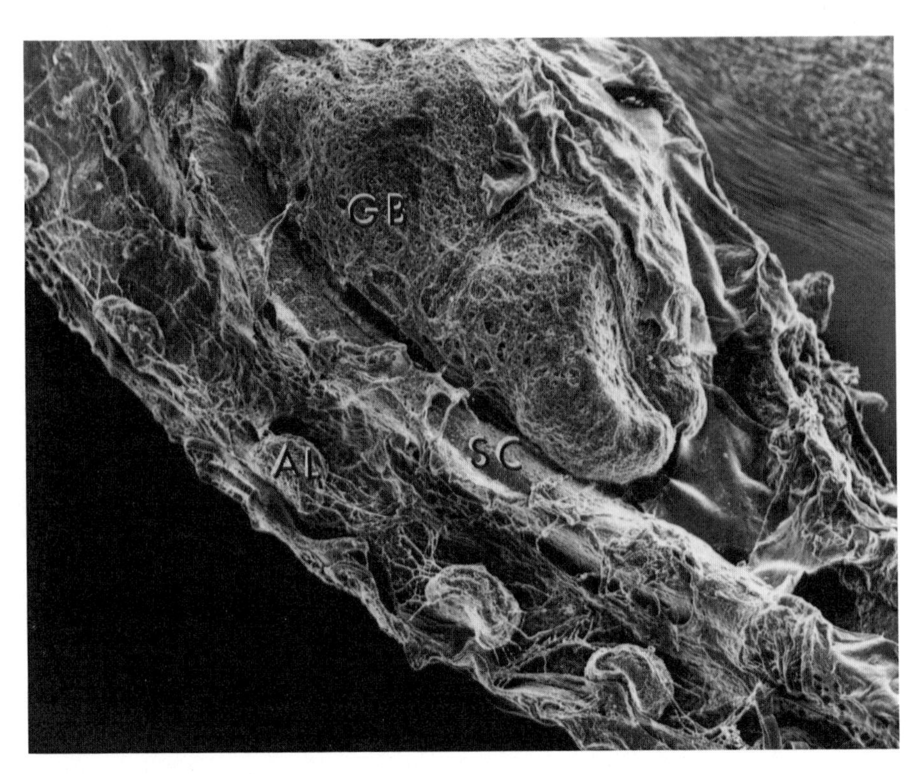

FIGURE 1–16. A scanning electron micrograph showing a lateral view of the lumbosacral region of the spinal cord from a newly hatched chick. The accessory lobes of Lachi (*AL*) are ventrolateral to the glycogen body (*GB*) that resides in the rhomboid sinus of the spinal cord (*SC*). ×45. (From Benzo and DeGennaro, 1981.)

between neurons and the glycogen-rich cells in the accessory lobes (DeGennaro and Benzo, 1976, 1978) suggests that some possible functional relationship exists between the glycogen cells and the CNS. Similar glycogen-filled cells, which surround the ependyma of the central canal and extend the entire length of the avian spinal cord, have been described (Sansone and Lebeda, 1976; Sansone, 1977, 1980; Uehara et al., 1982).

Although the functional implications of such extensive glycogen deposits in the accessory lobes and the glycogen body remain obscure, several studies have indicated that these tissues are metabolically very active, not only in the synthesis and degradation of glycogen (Benzo and DeGennaro, 1974, 1981) but also in glucose utilization via glycolyis and the pentose phosphate pathway (Benzo et al., 1975; Fink et al., 1975). Interestingly, both tissues lack the enzyme glucose-6-phosphatase, which is required if these tissues are to have a nutritive function similar to that of the liver in avian metabolism. It has been suggested, however, that the abundant glycogen stores in these tissues may be metabolically linked to the support of lipid synthesis and myelin formation in the CNS of the avian embryo in addition to serving as a source of organic acids, such as pyruvate and lactate, which might provide alternate substrates to the CNS under conditions of metabolic stress (for review, see DeGennaro, 1982; Benzo and DeGennaro, 1983).

Functional Anatomy of the Spinal Cord

The spinal cord, like the brain, functions as a processing, relay, and integrative center of the nervous system. Sensory signals originating in the various sensory receptors are transmitted through the nervous system by ascending sensory pathways (or tracts). The neural messages that regulate motor activity are processed in and conveyed through the nervous system by descending motor pathways. Both the ascending sensory and the descending motor pathways are hierarchically organized within the CNS; processing centers for each pathway, such as ganglia, nuclei, laminae, etc., are located at different anatomic levels of the spinal cord and brain.

In cross section, the avian spinal cord shows an arrangement that is very similar to that in the mammal (Figure 1–17). An X-shaped area of gray matter, containing perikarya and predominantly unmyelinated axons of nerve cells, is surrounded by white matter. The white matter consists of numerous tracts, or funiculi,

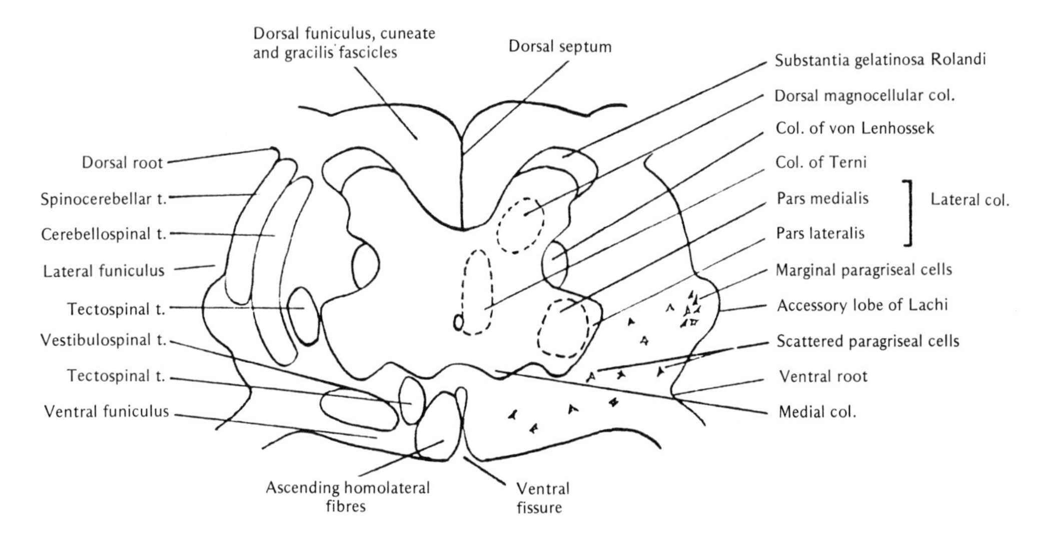

FIGURE 1–17. The fiber tracts and histological features of the avian spinal cord (cross section) (see text for details). *t.,* tract; *col.,* column. (With permission from Bolton, 1971a.)

of myelinated fibers. The relative sizes of the dorsal and ventral horns of the gray matter vary considerably, depending on the level of the cord sectioned. The dorsal horns are primarily sensory regions of the gray matter, whereas the ventral horns are predominantly motor areas. Sensory fibers enter via the dorsal root and pass into the dorsal horn, where some synapse. Cell bodies in this region form a dorsal magnocullular column. Motor fibers arise in the ventral horn from large multipolar neurons. This horn is divided into medial and lateral portions. Levi-Montalcini (1950) showed that preganglionic visceral efferent neurons originate from perikarya in a medial column (of Terni) close to the central spinal canal, in contrast to their origin in mammals. Some motor fibers (fibers of von Lenhossék) in the cervical cord are thought by some authors to leave via the dorsal root, and these may represent the avian forerunners of the mammalian spinal accessory nerve. The ventral root is believed to be composed exclusively of motor fibers (see Ariëns-Kappers et al., 1936). Scattered paragriseal cells also are found in the white matter, lateral and ventral to the ventral horn. These cells are thought by some investigators to be displaced ventral horn or anterior commissural neurons (Figure 1–17).

Ascending Sensory Pathways. Little information is available regarding the arrangement of ascending tracts in the avian spinal cord. Incoming dorsal root sensory impulses pass via monosynaptic and polysynaptic pathways to the dorsal funiculus of the same side (ipsilateral tracts). Here they are carried in tracts that correspond to the cuneate and gracilis columns of mammals to small cuneate and gracilis nuclei in the medulla (Karten, 1963). Fibers entering progressively rostrad are placed laterally to those entering more caudally. Friedlander (1898, cited by Ariëns-Kappers et al., 1936) showed that following hemisection of the lumbar cord, contralateral degeneration occurred at the lumbar level in the spinomesencephalic tract situated in the ventral funiculus. Ipsilateral degeneration occurred in the more posterior cord in the ventral funiculus close to the ventral fissure and in the peripheral dorsolateral funiculus. The peripheral dorsolateral funicular degeneration corresponds to the position of the spinocerebellar tract (Whitlock, 1952). The degeneration in the ventral funiculus corresponds to tracts identified by Ariëns-Kappers et al. (1936) as forerunners of the mammalian spinothalamic tracts for tactile, pain, and thermal sensibilities. Karten (1963) observed that the ventral and lateral funiculi contain ascending fibers that pass to the medulla, midbrain, tectum, and thalamus of the same side (see also Karten and Revzin, 1966).

Oscarsson et al. (1963) conducted an electrophysiologic study in which electrical recordings from fiber tracts of the spinal cord were obtained while stimulating muscle, skin, and mixed leg and wing nerves in the duck. These investigators found that dorsal funicular tracts were activated only ipsilaterally by mono- and polysynaptic pathways. Ventral tracts were monosynaptically activated only by contralateral pathways, but were polysynaptically activated both ipsi- and contralaterally. Their results confirmed that dorsal tracts are uncrossed, whereas ventral and lateral tracts were crossed. Easily stimulated (low-threshold) muscle afferents from the leg activated contralateral tracts, the discharge of which was inhibited by skin and muscle afferent stimulations that required exaggerated (high-threshold) stimulation. Low-threshold muscle afferents from the wings were uncrossed and not so inhibited.

Descending Motor Pathways. A number of descending tracts have been identified in the avian spinal cord (reviewed by Ariens-Kappers et al., 1936; Bolton, 1971a; and Pearson, 1972). A cerebellospinal tract occurs medial to the spinocerebellar. Tectospinal tracts occur in the lateral funiculus, accompanied by the vestibulospinal tract in the ventral funiculus (Figure 1–17). Rubrospinal tracts also are found in the lateral funiculus (Zecha, 1961). Although a number of investigators contend that a system comparable to the pyramidal tracts of mammals is absent in birds, there is some evidence that at least the upper cervical region of the avian cord receives direct corticospinal efferents. Zecha (1962) traced such fibers originating in the telencephalic region of the forebrain into the posterolateral funiculi of the cord in pigeons and parakeets.

The Brain

Gross Anatomy of the Brain

The avian brain contains five major subdivisions: the *telencephalon, diencephalon, mesencephalon, metencephalon,* and *myelencephalon.* In the fully developed brain, the *cerebrum* consists of the telencephalon and diencephalon; the two cerebral hemispheres are derived from the telencephalon. The metencephalon includes the *cerebellum* and the *pons,* and the myelencephalon is the *medulla oblongata.* The brainstem consists of the mesencephalon (midbrain), pons, and medulla, with the pons and medulla comprising the lower brain stem or bulbar region.

Although some minor structural differences are found among the various avian species, a common variation from the mammalian brain is the marked development of the optic tectum in birds. In addition, during its phylogenetic history, the bird has enlarged the internal portions of its cerebral hemispheres rather than the surface as did the mammal. Thus, the bird's forebrain is distinguished by the striking development of the striatum, a collection of nuclei and fiber tracts (white matter), whereas the cortex (gray matter) is relatively thin when compared with the mammal.

Internal Anatomy of the Brain

The Forebrain Region. The main subdivisions of the avian forebrain are the cerebral hemispheres, the thalamus, and the hypothalamus (Figures 1–18 and 1–19).

The cerebral hemispheres consist of cortex, hyperstriatum, neostriatum, ectostriatum, archistriatum, and paleostriatum. The hyperstriatum, ectostriatum, and paleostriatum are believed to be important forebrain areas associated with visual integration, pattern discrimination, and visually controlled defensive reflexes. The neostriatum appears to be involved in hearing.

The upper laminae of the hyperstriatum together with the overlying cortex constitute the "Wulst" region, or eminentia sagittalis, of the avian brain. The Wulst is a gross thickening of the hyperstriatal complex along the anterior and dorsal surface of the cerebral hemispheres. The relative size of the Wulst and its rostral–caudal extent vary among different species.

The thalamus in birds contains a number of nuclei that are the chief relay stations between the cortex of the forebrain and the sensory systems of the whole body. Particularly prominent are the nucleus rotundus (a visual relay center), the nucleus ovoidalis (an auditory relay center), and the dorsal thalamic nuclei, which receive the medial optic tract fibers. Other thalamic nuclei have important connections with different cortical and subcortical regions of the brainsteam (see Kitt and Brauth, 1982; Miceli and Reperant, 1982; Reiner

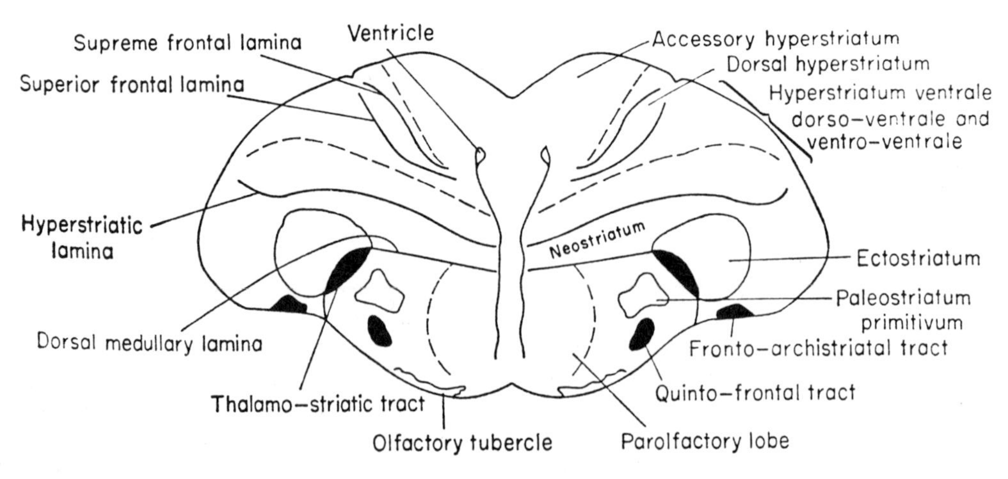

FIGURE 1–18. A diagrammatic transverse section through the rostral portion of the avian forebrain. (From Pearson, 1972.)

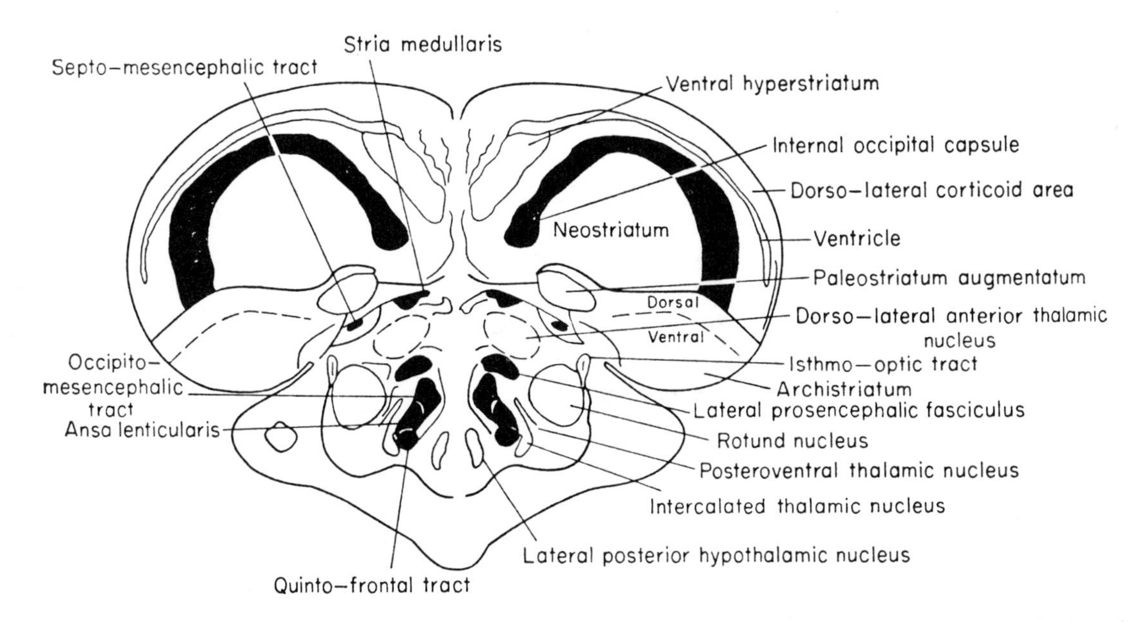

FIGURE 1–19. A diagrammatic transverse section through the avian forebrain showing diencephalonic and overlying structures (see text for details). (From Pearson, 1972.)

et al., 1982). the detailed neuroanatomy of the avian thalamus can be found in the stereotaxic atlases of van Tienhoven and Juhász (1962) and Karten and Hodos (1967).

The hypothalamus is situated just ventral to the thalamus; like the thalamus, it consists of several well-defined nuclei. The hypothalamus is the main forebrain center for the integrative control of the autonomic nervous system and of several endocrine glands. The hypothalamic nuclei receive afferent input from various regions of the body and in turn have efferent connections with the thalamus, adjacent forebrain regions, and the autonomic nuclei of the brain stem and spinal cord. Important connections of the hypothalamus with the hypophysis that exist in the bird are considered in detail in other chapters in this volume (see also Berk and Butler, 1981; Kuenzel and van Tienhoven, 1982).

The Midbrain Region. The avian midbrain is a complex region involved in orienting the bird's eyes, head, and body towards sights and sounds. The midbrain receives the optic nerves and contains the nuclei of origin of the oculomotor and trochlear nerves. The predominant midbrain structure is the optic tectum (Figure 1–20). Medial to the tectum reside several important nuclei, particularly the nucleus mesencephalicus lateralis, pars dorsalis (an auditory relay center). In addition, two important fiber tracts connecting both halves of the brain occur in the midbrain, the posterior commissure and the supraoptic decussation, which connect the tectal regions and other adjacent nuclei.

The Cerebellum. The avian cerebellum is well developed and consists of a large central area, the vermis, and two lateral lobes (Figure 1–21). The central vermis is subdivided into anterior, middle, and posterior lobes by two deep fissures; the middle lobe is separated from the anterior lobe by the fissure primura and from the posterior lobe by the fissure secunda. Although the cerebellum has an abundant sensory input, it is essentially a major motor center of the avian brain that functions in maintaining equilibrium and in coordinating both stereotyped and nonstereotyped muscle movements.

The histology of the bird's cerebellum is similar to that of the mammal. There are four central nuclei, which give rise to efferent axons to the spinal cord; those to the medulla end in relation to the vestibular nuclei (Wold, 1981). Other cerebellar tracts terminate in association with the inferior olive of the medulla and the nuclei of the oculomotor, trochlear, trigeminal, abducent, and facial nerves.

Several important afferent tracts terminate in the cerebellum. These fibers bring sensory information from the spinal cord, vestibular nuclei, pontine nuclei, trigeminal nuclei, the optic tectum, and different areas of the forebrain. Whitlock (1952) has detailed the somatotopic representation of the parts of the bird's body on the cerebellum.

The Medulla. The medulla is the part of the lower brain stem that connects the spinal cord with the midbrain. Through the medulla pass numerous prominent fiber tracts that link the spinal cord with areas of the

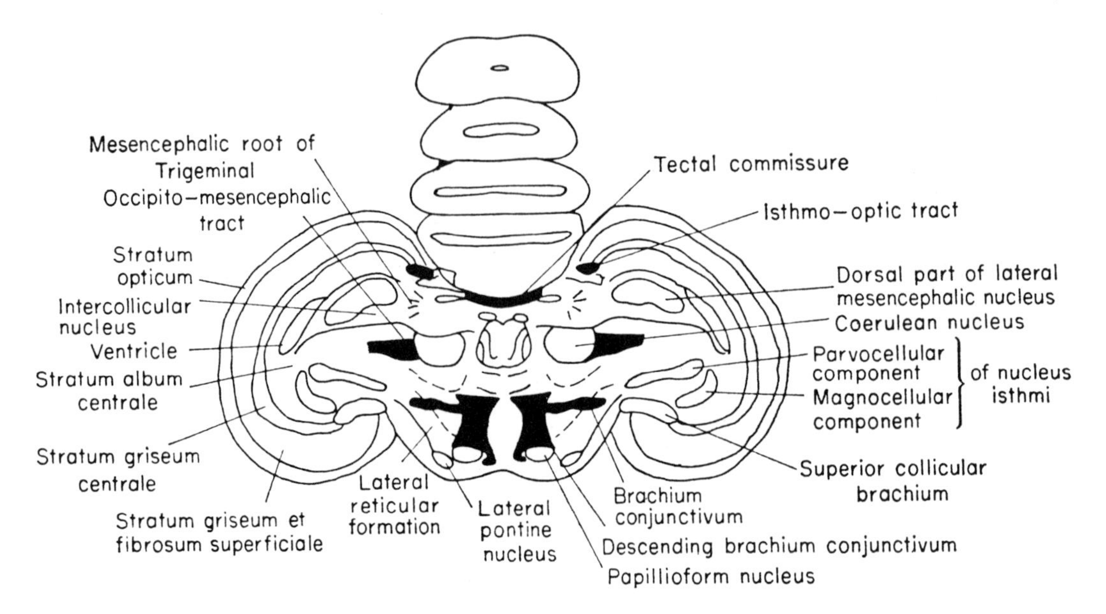

FIGURE 1–20. A diagrammatic transverse section through the avian midbrain region. (From Pearson, 1972.)

cerebellum, midbrain, and forebrain (Figures 1–22 and 1–23). The medulla is situated beneath the floor of the fourth brain ventricle and the cerebellum. It is connected to the cerebellum by the cerebellar peduncles.

The medulla contains a number of important nuclei, including those of cranial nerves five through twelve. The inferior olive is a medullary center that connects with the cerebellum and is believed to be involved with proprioception. A substantial portion of the physiologically important reticular formation is found within the medulla. In the bird, the medullary reticular forma-

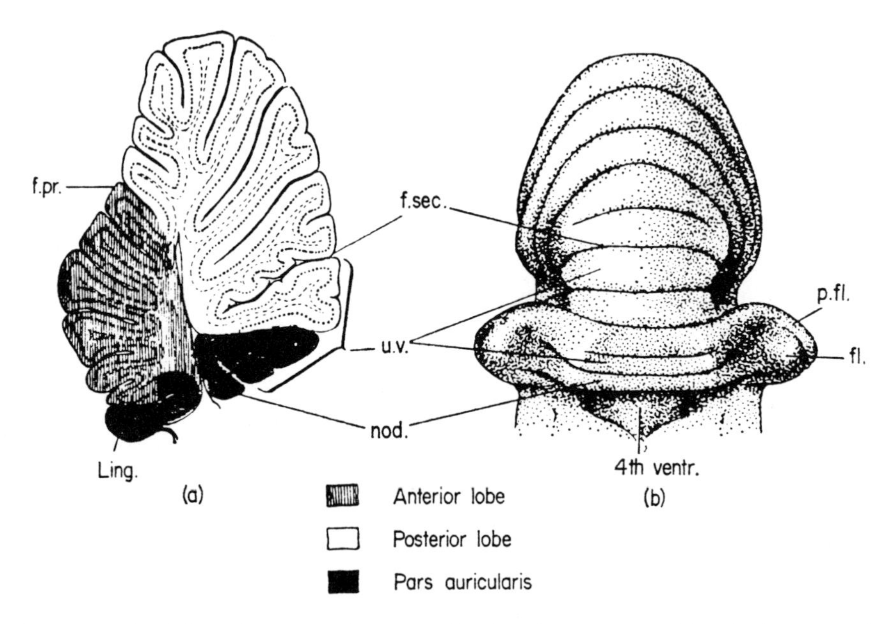

FIGURE 1–21. The avian cerebellum. (**A**) Section through the midsagittal plane. (**B**) A posterior view. Abbreviations: *f.pr.,* primary fissure (of Larsell); *f.sec.,* secondary fissure; *fl.,* flocculus; *p.fl.,* paraflocculus; *Ling.,* Lingula; *nod,* nodulus; *u.v.,* uvula; *4th ventr.,* 4th ventricle. (From Pearson, 1972.)

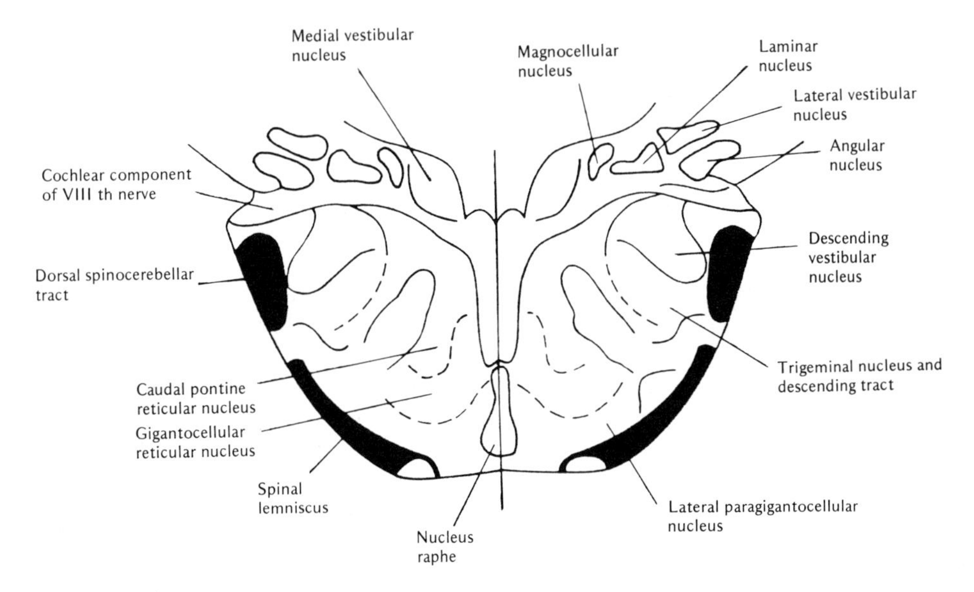

FIGURE 1–22. A diagrammatic transverse section through the rostral portion of the avian medulla (see text). (From Pearson, 1972.)

tion is composed of the inferior, middle, and superior reticular nuclei. Axons from these nuclei enter a prominent fiber tract, the medial longitudinal fasciculus. The reticular formation receives sensory input from most of the sensory organs of the body, and connects, directly or indirectly, with all levels of the central nervous system. The reticular formation has a significant influence over the activity of several systems, including the sleep–arousal mechanism, the motor systems of the spinal cord and brain, and the control of visceral organ function. Within this area of the brainstem also reside the centers that control respiration and cardiovascular activity (see Chapters 6, 7, and 8).

Cranial Sensory Systems

In addition to the vast amount of sensory information that reaches the brain via the spinal nerves and spinal cord pathways, brain centers also receive an abundance of sensory impulses that are carried by certain cranial nerves from the organs of special sense, as well as

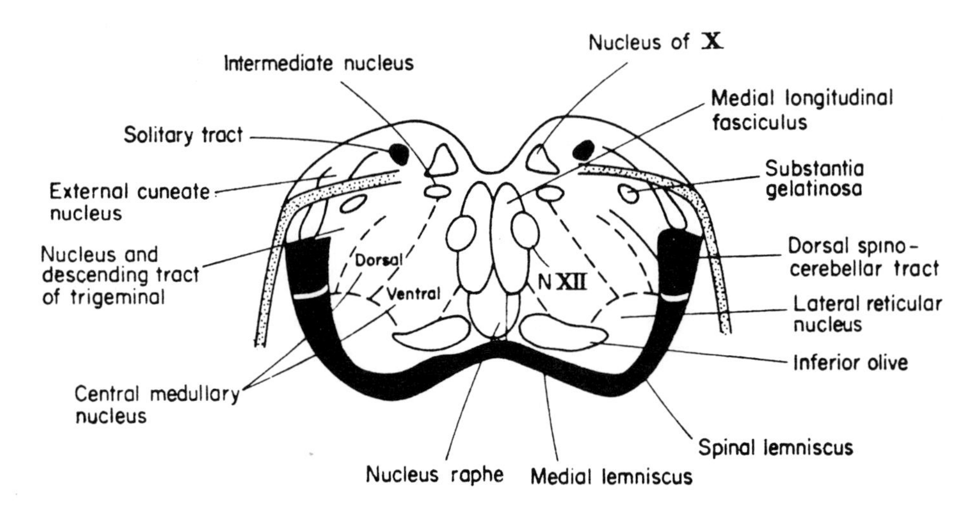

FIGURE 1–23. A diagrammatic transverse section through the posterior region of the avian medulla (see text for details). (From Pearson, 1972.)

from the proprioceptors and cutaneous receptors of the head region.

Visual Pathways. The unusually large and prominent avian eye and the highly developed optic tectum underscore the fact that, with few exceptions, the bird is a vision-oriented animal (see Chapter 2).

Visual impulses from the retina are carried in the optic nerve. In birds, the optic nerve axons decussate (cross over) virtually completely. There is some evidence, however, that a few axons do not cross (Knowlton, 1964; Takatsuji et al., 1983). The optic nerve forms two main roots, medial and lateral, and a minor basal root. The basal root, or accessory optic tract, terminates in the main accessory optic nucleus (nucleus ectomammillaris). This nucleus has efferent projections to the oculomotor nuclei, the vestibular nuclei, and cerebellum. It is believed to be an important relay center in the control of oculomotor function (Brecha and Karten, 1981; Hodos et al., 1982).

The medial root forms the isthmooptic tract (Figure 1–19). This tract carries retinal afferent fibers to the nucleus lateralis anterior and the nucleus dorsolateralis anterior of the thalamus. This dorsal thalamic nuclear complex is also known as the "nucleus opticus principalis thalami" (Karten et al., 1973). Besides carrying afferent optic axons, the medial root also brings back important efferent fibers from the nucleus isthmoopticus to the retina. The lateral root or marginal afferent tract terminates in the optic tectum, the lateral geniculate nucleus, and several other thalamic nuclei (Cowan et al., 1961). Those dorsal thalamic nuclei that receive axons from the retina have connections with the hyperstriatum of the opposite side via a prominent fiber tract, the lateral forebrain bundle. This tract connects not only with the contralateral hyperstriatum, but with the thalamic nuclei from which it received fibers, and with the neostriatum, periectostriatum, tectum, and lateral geniculate nucleus of both sides as well (Karten et al., 1973).

The visual field and retina are represented topographically on the contralateral optic tectum (McGill et al., 1966; Duff et al., 1981). The tectum, in turn, has many connections within the brain. Important tectal efferents project to the nucleus isthmoopticus. Several investigators have demonstrated that fibers from the tectum and lateral optic tract create an organized representation of the retina on the nucleus isthmoopticus (McGill et al., 1966; Holden and Powell, 1972). The isthmoopticus nucleus, in turn, has efferent projections to those areas of the contralateral retina from which it received afferent fibers from the tectum. Hayes and Webster (1981) described a population of "ectopic" neurons that reside outside of the nucleus isthmoopticus and also project to the contralateral retina. Their exact retinal targets and their functional role, if any, in modifying retinal input are presently unknown. The tectum also provides the nucleus rotundus of the thalamus with important afferent fibers (Karten, 1969). This nucleus is a visual relay center that connects with the ectostriatum and paleostriatum via the lateral forebrain bundle (Powell and Cowan, 1961; Revzin and Karten, 1967).

In summary, there are two major paths by which visual stimuli reach forebrain areas. One is via the medial optic tract and the dorsal thalamic nuclear complex to the hyperstriatum, and the other is via the lateral or marginal optic tract, the tectum, and the nucleus rotundus to the ectostriatum and paleostriatum. Information from the optic tectum can modify the receptive field properties of the retinal ganglion cells via efferent fibers carried in the medial optic tract from neurons of the isthmooptic nucleus. All the important regions of the thalamus and midbrain that are associated with vision in the bird receive projections from the hyperstriatum (see Ziegler, 1963; Cohen, 1967a, b; Hodos and Karten, 1966; Hodos et al., 1973; Showers, 1982; Watanabe et al., 1983).

Vestibular Pathways. Axons within the vestibular branch of the eighth cranial nerve (acoustic) carry information associated with equilibrium sense into the avian brain. The cell bodies of these axons reside in the vestibular ganglion. The peripheral axons of these neurons innervate the anterior and exterior (lateral) ampullae of the semicircular canals and the utriculus in the avian labyrinth. Their central axons are distributed to the vestibular nuclei (Figure 1–22). The vestibular centers are located in the medulla and contain six nuclei; the superior, dorsolateral, ventrolateral (also known as Deiter's nucleus), dorsomedial, tangential, and descending (see Wold, 1979, 1981). These nuclei, particularly the superior and dorsolateral, have important projections to the cerebellum. The ventrolateral nucleus connects with the oculomotor and trochlear nuclei, and, via the vestibulospinal tract, with the motor areas of the spinal cord. As a group, the vestibular nuclei are of particular importance in the bird, because they are involved in the moment-to-moment coordination of the body musculature during swift bipedal locomotion and flight.

There is some evidence that efferent fibers are present also in the vestibular branch (Desmedt and Delwaide, 1963; Boord, 1969; Strutz and Schmidt, 1982). These fibers originate from motor nuclei in the medullary reticular formation, enter the vestibular root of the acoustic nerve, and terminate in the vestibular labyrinth and cochlea (see Desmedt and Delwaide, 1963; Whitehead and Morest, 1981; Eden and Correia, 1982; Firbas and Muller, 1983). Although it has been suggested by some investigators that they act to dampen auditory input, the possible role of these efferent fibers in modifying hearing or equilibrium sense in the bird is unknown.

Auditory Pathways. Auditory impulses are carried by fibers within the cochlear branch of the eighth cranial nerve. The cell bodies of these fibers are located in the cochlear ganglion. The peripheral axons in the cochlear nerve innervate the posterior ampulla of that semicircular canal, the sacculus, and the papillae of the cochlea and lagena. The central axons of these neurons enter the medulla, where the vestibular components are separated from those that arise from the cochlea itself, and pass to vestibular nuclei. The cochlear axons are distributed to three cochlear nuclei; the magnocellular, the angular, and the laminar (Parks, 1981; Jhaveri and Morest, 1982; Rylander and Snow, 1982; Parks et al., 1983) (Figure 1–22). Fibers from these nuclei pass to the ipsilateral superior olive, decussate, and form the lateral lemniscus. The lateral lemniscus terminates in the major auditory projection center, the nucleus mesencephalicus lateralis, pars dorsalis. This nucleus has its major projection to the ipsilateral thalamic nucleus ovoidalis and a minor connection via the dorsal supraoptic decussation to the contralateral ovoid nucleus (Karten, 1966, 1967a, b, 1968). From these thalamic relay centers, auditory impulses are conveyed principally to the ipsilateral neostriatum where they terminate in the mediocaudal region (Field L of Rose) (see also Zaretsky, 1978; Kirsch et al., 1980).

Olfactory and Gustatory Pathways. In achieving flight, birds have become more dependent on their highly developed senses of vision and balance rather than on the ability to smell or to taste. Although less developed in the bird, the perceptual capacity for olfactory and gustatory stimuli most likely exists in all avian species.

There is considerable variation both in the structure and in the degree of development of the olfactory bulb among birds. Relatively large olfactory bulbs tend to be found in aquatic birds, or in those subsisting on diets of flesh or fruit. The diameter of the olfactory bulbs relative to the cerebral hemispheres is variable as well, and this ratio has been used as one criterion for classifying birds into the microsmatic–macrosmatic series (see Pearson, 1972). Hummel (1979) has examined the ultrastructure of the olfactory bulb in the chicken.

The olfactory nerve axons convey impulses from the specialized olfactory mucosa of the posterior-superior nasal chamber to the mitral neurons of the olfactory bulb. Fibers arising from the mitral neurons form the medial and lateral olfactory tracts. Axons in the medial tract terminate in the anterior olfactory nucleus (prepyriform area), and those fibers of the lateral olfactory tract end in the prepyriform area and in the ventral portion of the nucleus basalis in the paleostriatum (Jones and Levi-Montalcini, 1958). The lateral olfactory tract is very small in microsmatic birds, such as the sparrow.

The extent to which various birds use the sense of smell to find food or for other purposes varies greatly (see Chapter 2). Like olfaction, the ability to taste is relatively unimportant in most avian species (Chapter 2).

Cranial Motor Systems

In addition to the efferent outflow from motor centers in the brain that descend in spinal cord tracts, numerous motor fibers are carried by the cranial nerves from brainstem nuclei to various somatic and visceral structures. In the head region, the extrinsic muscles of the eye, the muscles of mastication, and the various secretory glands are supplied by cranial nerves. Also, fibers within the glossopharyngeal and vagus nerves leave the head region to bring visceral motor innervation to the heart, lungs, and the smooth muscle and glands of much of the digestive system (see Chapters 6, 7, 8 and 11).

The extrinsic muscles of the eye are innervated by three cranial nerves, the oculomotor, trochlear, and abducent (Chapter 2). The oculomotor and trochlear axons arise from motor nuclei in the midbrain, and the abducent fibers orginate from two nuclei that reside in the medulla. Each of these nuclei receives fibers from the tectum, the vestibular nuclei, and the cerebellum. This input is responsible for controlling fixation of the bird's visual field during movements of the body. These three cranial nerves also carry afferent impulses to the brain from proprioceptors within the muscle spindles of the extrinsic eye muscles.

Residing near the chief nucleus (somatic motor) of the oculomotor nerve is an accessory (Edinger–Westphal) nucleus (visceral motor), which contains preganglionic parasympathetic neurons. Axons arising from these neurons are carried by the oculomotor nerve to the ciliary ganglion (Narayanan and Narayanan, 1979). From this ganglion, postganglionic fibers leave to innervate the iris and choroid of the eye. The accessory nucleus receives fibers from the tectum which are involved in the pupillary response to light (Reiner et al., 1983).

The jaw muscles are innervated by both motor and sensory fibers. Proprioceptive afferents from muscle spindles within the jaw muscles convey impulses back to the motor neurons that supply those muscles. The trigeminal nerve also has a large cutaneous sensory distribution. Most cutaneous sensation is conveyed to the chief sensory nucleus of the trigeminal nerve. This nucleus receives impulses from various specialized sensory corpuscles, such as the corpuscles of Grandy that are distributed around the beak or bill. Because of the connections of the chief sensory nucleus with the forebrain (see Pearson, 1972), together with the production of feeding deficits by lesions of this nucleus, Zeigler and Karten (1973) suggested that this nucleus may play an important role in avian feeding behavior.

The facial nerve supplies motor fibers (skeletal) to several muscles of the lateral and ventral areas of the head and neck region. This nerve also participates in the innervation (visceral motor) of the salivary glands.

The glossopharyngeal, vagus, accessory and hypoglossal nerves originate from a collection of motor nuclei in the medulla. Besides supplying both visceral and skeletal muscles, the glossopharyngeal and the vagus nerves supply the glands associated with the alimentary tract. In birds, neurons of the facial and glossopharyngeal nerves function in controlling the salivary glands of the head. Further details on the control of salivation, cardiovascular and respiratory activity, and gut motility are found in appropriate chapters of this volume.

The Blood–Brain Barrier

The concept of a "blood–brain barrier" has developed from observations that certain substances, when injected intravenously, are virtually excluded from the cerebrospinal fluid (CSF) and tissues of the CNS, although they pass quite freely from the blood into the usual interstitial fluids and other body tissues. Such barriers between the blood and the CSF and the nervous tissue of the CNS exist in the choroid plexi and in essentially all areas of the brain except the hypothalamus, where blood-borne substances diffuse with ease into the extracellular spaces. The absence of a blood–brain barrier in the hypothalamus is very important, because this region of the brain responds to many different changes in blood chemistry, such as glucose concentration and electrolyte levels, and these responses are essential signals for feedback control of various homeostatic mechanisms (for review, see Benzo, 1983).

In general, the blood–brain barrier is highly permeable to water, oxygen, and carbon dioxide, moderately permeable to electrolytes such as chloride, potassium, and sodium, and virtually impermeable to large molecules and substances, such as gold, sulfur, and arsenic. Although the latter substances are not physiologically important, they are often used in conjunction with certain drugs, and the presence of the blood–brain barrier makes it almost impossible to achieve effective concentrations of such drugs in the tissues of the CNS.

The blood–brain barrier is not a simple or well-defined anatomic structure, but rather is a "structural complex" including the capillary endothelial cells, their basement membranes, and glial (astrocytic) processes that intervene between the nerve cells and those capillaries that supply them. The structural features of the avian blood–brain barrier have been examined by Stewart and Wiley (1981). As in mammals, the endothelial cells in the capillaries of the avian brain are connected by tight junctions and lack fenestrae. Furthermore, these cells contain only a few pinocytotic vesicles. These features give the endothelial cells primary control over the passage of large, polar molecules across the capillary wall, and, together with the other structural components of the blood–brain barrier, account for the low permeability of brain vessels.

The newly hatched chick and the immature domestic fowl have been popular subjects for the study of drug action on the CNS because of the belief that although the chick is considered to be "behaviorally mature," the ease with which various substances diffuse into the nervous tissue of the brain has not yet declined to the level found in the adult bird. Several studies have demonstrated that substances injected into the newly hatched chick enter the brain more rapidly and attain higher concentrations than in the adult bird. These substances include electrolytes such as chloride (Waelsch, 1955; Lajtha, 1957) and various biologically active compounds such as noradrenaline (Spooner et al., 1966; Spooner and Winters, 1966b), serotonin (Hanig and Seifter, 1968; Bulat and Supek, 1968), and GABA (Scholes and Roberts, 1964). After approximately 1 month of age, however, the amounts of these substances entering the young chick brain have declined to levels similar to those found in the adult.

The Avian Brain and Behavior

Functional Localization in the Nervous System

Studies of the functions of the spinal cord and other parts of the CNS have been made by transecting the cord or brain at different levels or by removing some part of it. Separation of a lower part of the CNS from a higher part causes the loss of certain functions because of the interruption of the pathways between the higher and lower centers. After transection of the thoracic or lumbar cord, birds and other animals may survive for months or even years, and the local reactions in the spinal cord can be studied without interfering influences from higher centers of the CNS.

An animal with a transected spinal cord is called a *spinal animal.* The animal immediately shows a bilateral loss of mobility and sensation in the areas of the body supplied by the spinal nerves below the transection. This is called *spinal shock,* during which time all reflexes are absent. After a given period, typical spinal reflexes reappear; later other reflexes of the limbs and trunk are restored. After complete transection of the thoracic cord, pigeons were found to lie passively with their legs drawn up toward the body. When attempting to fly, they beat their wings violently (ten Cate, 1960).

Spinal birds have been shown to exhibit great reflex capacity. After transection of the thoracic cord, it is possible, by passively bending and stretching one foot of the bird, to produce antagonistic stretching and bending reflexes of the opposite foot. When reflex activity has increased, lifting the pigeon causes rapid, rhythmical motions of both feet. Various reflex movements of the tail may be provoked in a spinal bird. If such a bird, with head and body extended vertically, is suddenly raised upward to the horizontal position, there is a fanlike spreading of the main tail feathers. If the bird is pushed back to its original position, there is only a brief upward flick of the tail. These reflexes normally serve to maintain the bird's balance.

A spinal bird usually is unable to walk because it can no longer maintain the upright posture in the absence of the control of higher CNS centers. When such a bird was fixed in a small four-wheeled carriage in a normal position, in such a way that its extended legs could reach the ground, a true walking movement in the hind limbs was evoked by applying painful stimuli to the hind part of the body and the bird could move over relatively great distances. This demonstrated that the lumbosacral cord of birds shows a high degree of autonomy in the coordination of walking movements. Higher CNS centers, however, initiate and regulate these movements (ten Cate, 1960).

The reflex movements of the wings in a spinal bird generally are identical and simultaneous. After unilateral denervation of a wing, the loss of afferent stimuli from one side is compensated for by those of the other side. Such birds can fly and maintain their balance. If the bird is forced into a position in which normally only one wing is used, then abnormalities arise if the wing needed for correction is the denervated one. The rhythmic beating of the wings was shown to be dependent on the reflex stimulus exerted on each wingbeat by the preceding one.

If the dorsal roots of the nerves supplying a leg are sectioned unilaterally, sensory stimuli from the normal side cannot compensate for the contralateral loss, as is the case with the wings. This indicates a difference in the patterns of innervation between the upper and lower extremities in the bird, alternating in the wings and simultaneous in the legs. If only one side of the spinal cord is sectioned, the ipsilateral leg is paralyzed, and the tail is twisted toward the normal side. Sensation is only slightly diminished. All of these phenomena gradually subside. Within a week or two, the bird can again stand and walk (ten Cate, 1960).

Removal of the entire lumbosacral region of the pigeon's spinal cord resulted in the loss of sensation and movement in the caudal part of the body (Sammartino, 1933). Partial recovery of function was noted to occur in the cloacal sphincter muscle (skeletal muscle), but smooth muscle of the viscera did not regain its tone.

Studies to determine the nature and extent of nerve fiber regeneration following transection of the spinal cord in birds have shown that complete, structural restitution of the cord occurred only in chick embryos of 2–5 days of egg incubation (Clearwaters, 1954). Ten Cate (1960) showed that regeneration of the whole cord did not occur in adult spinal pigeons, but growth of nerve fibers in the proximal as well as in the distal part of the cord was found. These abortive attempts at regeneration appeared as fine fibers growing in various directions. Regeneration and restoration of functional activity following transection of the spinal cord in the higher vertebrates has never been observed. It is generally believed that the basis of this failure to recover from neural injury is a poor environment for axon regeneration. Most CNS axons simply cannot sustain regeneration over substantial distances within the spinal cord.

Studies in which parts of the brain were ablated (brain injury) or electrically stimulated have been performed in attempts at understanding the precise, functional roles of various centers of the avian brain. Meaningful interpretation of many of the observed responses in terms of specific nuclei or tracts often has been difficult. For example, both eye and limb movements can be produced by electrical stimulation of the cerebral hemispheres, tectum, or cerebellum. Each of these regions is involved in the production or control of muscle movement, but stimulation experiments have revealed little about the precise role of any region.

The cerebellum apparently is involved in the regulation of muscle tone and the coordination of movement. After the removal of the avian cerebellum, the neck and leg muscles, but not the wing muscles, are subject to spasm. Swaying and jerking movements of the head occurred in such birds when feeding. Lesions of the vestibular nuclei on one side lead to a deviation of the eyes, rotation and lateral flexion of the head, and spiral rotation of the neck and trunk toward the side of the lesion. A loss in muscle tone of the limbs on the affected side and an increase in tone on the opposite side are produced as well.

Threatening or defensive reactions were produced in pigeons after stimulation of the diencephalon or anterior hypothalamic region (Åkerman, 1966). Putkonen (1967) induced similar reactions in the chicken, and also produced fear and attacking reactions by stimulating certain areas of the forebrain. For further details pertaining to these studies, see Bolton (1971b) and Pearson (1972).

Song production or *vocalization* resulting from stimulating discrete brain areas in the bird has been the subject of a number of investigations. The most effective point in producing vocalization, located in the midbrain, is called the torus semicircularis (DeLanerolle and Andrew, 1974). Most investigators agree that the lowest threshold points from which song can be evoked by electrical stimulation lie in a nucleus medial and adjacent to the nucleus mesencephalicus lateralis pars dorsalis, known to be an ascending auditory relay nucleus. This vocalization region, called the torus externus or nucleus intercollicularis, has been shown to produce song in the red-wing blackbird (Brown, 1969, 1971;

Newman, 1972), the Japanese quail (Potash, 1970), and the Java sparrow (Seller, 1980). Other brain regions related to the production of song have been found in the hypothalamus, septum, and archistriatum (McCasland and Konishi, 1981; Nottenbohm et al., 1982). Considerable progress in attempting to understand the neurological basis of song production has been made by Nottenbohm and co-workers, who examined the neural correlates of vocalization in canaries and chaffinches (see Nottenbohm, 1980a, for review). Two particularly important contributions from this group have been the discoveries of the neural lateralization of vocalization and sexual dimorphism of song-related brain areas in these species. Apparently, the production of song in canaries is dominated by left-sided brain structures, as is language in humans. Neural areas concerned with the control of vocalization are much less developed in the female bird (see DeVoogd and Nottenbohm, 1981; Gurney, 1982; Goldman and Nottenbohm, 1983). If, however, the female is given testosterone, the relevant brain structures develop more fully, and the female bird acquires song production (Nottenbohm, 1980b).

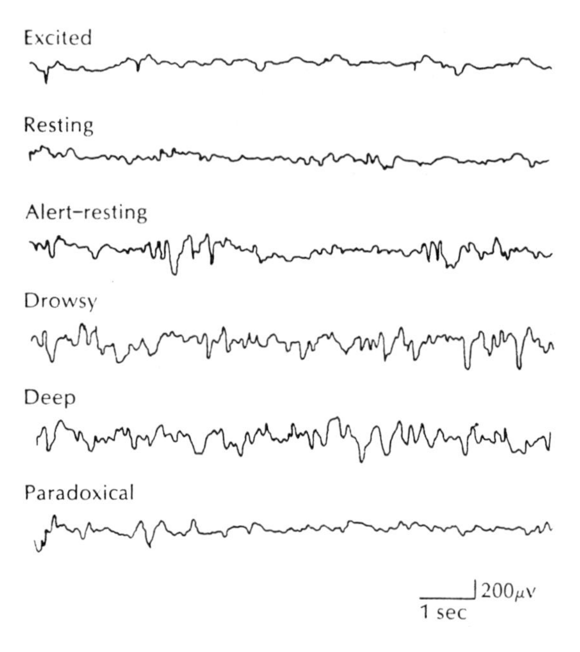

FIGURE 1–24. EEG patterns during wakefulness and sleep in an adult hen. The EEGs were recorded from the Wulst region. (Data from Ookawa, 1972b. Figure reproduced from Bolton, 1976.)

Electroencephalographic Studies

Behavioral studies have utilized the electroencephalogram (EEG) to monitor the effects of drugs or of metabolic or environmental stress on avian brain activity. This method generally involves the recording of the electrical activity of the cerebral hemispheres by means of electrodes inserted through the skull bones so that their tips lie just outside the cerebral dura mater. Although other brain areas can be monitored, the electrodes usually are placed to record from the Wulst region, which is the projection of the dorsal surface of the cerebral hemisphere formed by the underlying hyperstriatal regions.

Most avian EEG studies have been done on the domestic fowl. Several states of brain activity have been recognized, which were found to correlate well with certain behavioral changes. Alert and sleep states have been distinguished in the bird, and each state has been subdivided into characteristic stages (Figure 1–24).

The alert state has been divided into excited and unexcited stages. The alert excited stage is characterized by 20- to 50-μV, 30- to 60-Hz waves. The alert unexcited stage shows 50- to 150μV, 17- to 24-Hz waves. The sleep state has three stages; stage 1 (resting) and stages 2 and 3 (true sleep). Stage 1 sleep is an alert resting stage and is associated with desultory activities, such as preening. The EEG tract consists primarily of 50- to 150-μV, 17- to 24-Hz waves in which are interspersed bursts of 200- to 300-μV, 3- to 12-Hz waves. Stage 2 sleep is the beginning of true sleep. The bird crouches and the head droops. The EEG consists mainly of 200- to 400-μV, 6- to 12-Hz waves with brief periods of slower, 3- to 4-Hz waves. Stage 3 sleep represents activity in which the head is generally placed under the wing and both eyes are closed. Most of the EEG activity is 200- to 400-μV, 2- to 4-Hz waves, but bursts of 50-μV, 30- to 60-Hz activity do occur, probably representing paradoxical sleep. At such times, twitching has been noticed. Episodes of paradoxical or fast sleep usually last about 6 sec.

The frequency of EEG activity is reduced and the wave size increased with a reduction in the visual and auditory stimuli in the bird's environment. When the bird is asleep, stimuli insufficient for arousal produce a reduction in wave size and an increase in the frequency of the EEG. Strong stimuli awaken the bird, and EEG activity then becomes fast and small (Ookawa et al., 1962; Ookawa and Gotoh, 1965; Ookawa, 1972b).

Prior to 2 months of age, young chicks show a slightly atypical EEG. At 1 day old, the EEG of the alert excited stage exhibits 25-μV and 30- to 60-Hz waves, and the alert unexcited EEG shows 25-μV and 17- to 25-Hz waves. Stage 3 sleep is characterized by 100-μV, 1.5- to 5-Hz waves with fast-sleep, 25-μV, 30- to 60-Hz waves, bursts. Such activity persists up to 1 month of age, when stage 1 and stage 2 sleep can be distinguished (Ookawa and Takagi, 1968). The patterns of EEG activity in the chick embryo (Corner et al., 1967), in the young adult chicken (Corner et al., 1972), and in the 1-day-old Japanese quail (Kovacs et al., 1981) have been documented.

Sugihara and Gotoh (1973) recorded and compared the EEG activities from different forebrain areas and the optic tectum in the fowl during states of arousal, slow-wave sleep, and paradoxical sleep. They found that no regular slow waves, such as those which are seen in the hippocampus of the mammal during arousal and paradoxical sleep, could be observed in the hippocampus of chickens, nor did high-voltage fast waves appear in the archistriatum during arousal. Regular 9- to 13-Hz waves did appear in the paleostriatum primitivum during slow-wave sleep. They recorded monophasic spikes in the optic tectum during the paradoxical sleep stage, and these were found to be associated with rapid movements.

A considerable literature exists on the effects of abnormal physiologic states, such as nutritional deficiencies (Tokaji and Gerard, 1939; Swank and Jasper, 1942; Sheff and Tureen, 1962; Pomeroy and Welch, 1967), hypothermia and hyperthermia (Peters et al., 1961, 1964, 1968, 1969; Peters and Hilgerford, 1971), catalepsy (Gilman et al., 1950; Ookawa, 1972a), and epilepsy (Ookawa, 1977), on the EEG in the bird. Similarly, investigators have studied the effects of drugs on the EEG in avians (Garcia-Aust, 1954; Tuge et al., 1960; Katori, 1962; Spooner and Winters, 1966a; Ookawa, 1973a, c, d). Much of the literature on the EEG in abnormal states in birds has been reviewed by Bolton (1971b), Pearson (1972), and Ookawa (1977).

References

Abdalla, A.B., and A.S. Kind. (1979). The afferent and efferent myelinated fibers of the avian cervical vagus. J. Anat., 128, 135.

Adal, M.N. (1973). The fine structure of the intrafusal muscle fibres of muscle spindles in the domestic fowl. J. Anat., 115, 407.

Adal, M.N., and S.B. Chew-Cheng. (1980). The sensory ending of duck muscle spindles. J. Anat., 131, 657.

Åkerman, B. (1966). Behavioral effects of electrical stimulation in the forebrain of the pigeon. II: Protective behavior. Behaviour, 26, 339.

Akester, A.R., B. Akester, and S.P. Mann. (1969). Catecholamines in the avian heart. J. Anat., 104, 591.

Ali, H.A., and J. McLelland. (1978). Avian enteric nerve plexuses. A histochemical study. Cell Tissue Res., 189, 537.

Aprison, M.H., and R. Takahaski. (1965). Biochemistry of the avian central nervous system: II. 5-Hydroxytryptamine, acetylcholine, 3,4, dihydroxyphenylethylamine and norepinephrine in several discrete areas of the pigeon brain. J. Neurochem., 12, 221.

Aprison, M.H., R. Takahaski, and T.L. Folkerth. (1964). Biochemistry of the avian central nervous system: I. 5-Hydroxytryptophan decarboxylase, monoamine oxidase and choline acetylase–acetylcholinesterase systems in several discrete areas of the pigeon brain. J. Neurochem., 11, 341.

Arends, J.J., and J.L. Dubbeldam. (1982). Exteroceptive and proprioceptive afferents of the trigeminal and facial motor nuclei in the mallard (*Anas platyrhynchos*). J. Comp. Neurol., 209, 313.

Ariëns-Kappers, C.U., G.C. Huber, and E.C. Crosby. (1936). "The Comparative Anatomy of the Nervous System of Vertebrates Including Man," Vols. 1 and 2. New York: Macmillan.

Banks, P., and D. Mayor. (1972). Intra-axonal transport in noradrenergic neurons in the sympathetic nervous system. In "Neurotransmitters and Metabolic Regulation," Vol. 36 (R.M.S. Smellie, Ed.). Biochemical Society Symposium, p. 133.

Barr, M.L., and J.A. Kiernan. (1983). "The Human Nervous System," 4th ed. Philadelphia: Harper and Row.

Bennett, T. (1974). The peripheral and autonomic nervous systems. In "Avian Biology," Vol. 4 (D.S. Farner and J.R. King, Eds.). New York: Academic Press, p. 1.

Bennett, T., and T. Malmfors. (1970). The adrenergic nervous system of the domestic fowl (*Gallus domesticus* L.). Z. Zellforsch. Mikrosk. Anat., 106, 22.

Bennett, T., G. Burnstock, J.L.S. Cobb, and T. Malmfors. (1970). An ultrastructural and histochemical study of the short-term effects of 6-hydroxydopamine on adrenergic nerves in the domestic fowl. Br. J. Pharmacol., 38, 802.

Benzo, C.A. (1983). The hypothalamus and blood glucose regulation. Life Sci., 32, 2509.

Benzo, C.A., and L.D. DeGennaro. (1974). Glycogen synthase and phosphorylase in the developing chick glycogen body. J. Exp. Zool., 188, 375.

Benzo, C.A., and L.D. DeGennaro. (1981). Glycogen metabolism in the developing accessory lobes of Lachi in the nerve cord of the chick: Metabolic correlations with the avian glycogen body. J. Exp. Zool., 215, 47.

Benzo, C.A., and L.D. DeGennaro. (1983). An hypothesis of function for the avian glycogen body: A novel role for glycogen in the central nervous system. Med. Hypotheses, 10, 69.

Benzo, C.A., L.D. DeGennaro, and S.B. Stearns. (1975). Glycogen metabolism in the developing chick glycogen body: Functional significance of the direct oxidative pathway. J. Exp. Zool., 193, 161.

Berk, M.L., and A.B. Butler. (1981). Efferent projections of the medial preoptic nucleus and medial hypothalamus in the pigeon. J. Comp. Neurol., 203, 379.

Berkhoudt, H., B.G. Klein, and H.P. Zeigler. (1982). Afferents to the trigeminal and facial motor nuclei in the pigeon (*Columba livia*): Central connections of jaw motoneurons. J. Comp. Neurol., 209, 301.

Bolton, T.B. (1971a). The structure of the nervous system. In "Physiology and Biochemistry of the Domestic Fowl" (D.J. Bell and B.M. Freeman, Eds.). New York: Academic Press, p. 641.

Bolton, T.B. (1971b). The physiology of the nervous system. In "Physiology and Biochemistry of the Domestic Fowl" (D.J. Bell and B.M. Freeman, Eds.). New York: Academic Press, p. 675.

Bolton, T.B. (1976). The Nervous System. In "Avian Physiology," 3rd ed. (P.D. Sturkie, Ed.). New York: Springer-Verlag, p. 4.

Bolton, T.B., and W.C. Bowman. (1969). Adrenoreceptors in the cardiovascular system of the domestic fowl. Eur. J. Pharmacol., 5, 121.

Bondy, S.C., and J.L. Purdy. (1977). Putative neurotransmitters of the avian visual pathway. Brain Res., 119, 417.

Boord, R.L. (1969). The anatomy of the avian auditory system. Ann. N.Y. Acad. Sci., 167, 186.

Bottjer, S.W., and A.P. Arnold. (1982). Afferent neurons in the hypoglossal nerve of the zebra finch (*Poephila guttata*): Localization with horseradish peroxidase. J. Comp. Neurol., 210, 190.

Brecha, N.C., and H.J. Karten. (1981). Organization of the avian accessory optic system. Ann. N.Y. Acad. Sci., 374, 215.

Brown, J.L. (1969). The control of avian vocalization by the central nervous system. In "Bird Vocalizations" (R.A. Hinde, Ed.). Cambridge: Cambridge University Press, p. 79.

Brown, J.L. (1971). An exploration study of vocalization areas in the brain of the redwinged blackbird (*Angelaius phoeniceus*). Behavior, 39, 91.

Brown, C.M., V. Molony, A.S. King, and R.D. Cook. (1972). Fibre size and conduction velocity in the vagus of the domestic fowl (*Gallus domesticus*). Acta Anat., 83, 451.

Bulat, M., and Z. Supek. (1968). Passage of 5-hydroxytryptamine through the blood–brain barrier, its metabolism in the brain and elimination of 5-hydroxyindolacetic acid from the brain tissue. J. Neurochem., 15, 383.

Burack, W.R., and A. Badger. (1964). Sequential appearance of dopa decarboxylase, dopamine β-oxidase and norepinephrine *N*-methyltransferase activities in the embryonic chick. Fed. Proc. Fed. Am. Soc. Exp. Biol., 23, 561.

Burger, R.E., and M.R. Fedde. (1964). Physiological and pharmacological factors which influence the incidence of acute pulmonary alterations following vagotomy in the domestic cock. Poult. Sci., 43, 384.

Cabot, J.B., and D.H. Cohen. (1977). Avian sympathetic cardiac fibers and their cells or origin: Anatomical and electrophysical characteristics. Brain Res., 131, 73.

Callingham, B.A., and D.F. Sharman (1970). The concentration of catecholamines in the brain of the domestic fowl (*Gallus domesticus*). Br. J. Pharmacol., 40, 1.

Cantino, D., and E. Mugnaini. (1975). The structural basis for electrotonic coupling in the avian ciliary ganglion. A study with thin sectioning and freeze-fracturing. J. Neurocytol., 4, 505.

Carpenter, F.G., and R.M. Bergland. (1957). Excitation and conduction in immature nerve fibres of the developing chick. Am. J. Physiol., 190, 371.

Chaplin, J.P., and A. Demers. (1978). "Primer of Neurology and Neurophysiology." New York: John Wiley and Sons.

Clearwaters, K. (1954). Regeneration of the spinal cord of the chick. J. Comp. Neurol., 101, 317.

Cohen, D.H. (1967a). Visual intensity discrimination in pigeons following unilateral and bilateral tectal lesions. J. Comp. Physiol. Psychol., 63, 172.

Cohen, D.H. (1967b). The hyperstriatal region of the avian forebrain. A lesion study of possible functions including its role in cardiac and respiratory conditioning. J. Comp. Neurol., 131, 559.

Corner, M.A., J.P. Schadé, J. Sedláček, R. Stoeckart, and A.P.C. Bot. (1967). Developmental patterns in the central nervous system of birds. I. Electrical activity in the cerebral hemisphere, optic lobe & cerebellum. Prog. Brain Res., 26, 145.

Corner, M.A., W.L. Bakhuis, and C. van Wingerden. (1973). Sleep and wakefulness during early life in the domestic chicken and their relationship to hatching and embryonic mortality. In "Prenatal Ontogeny of the Central Nervous System and Behaviour" (Gottlieb G., Ed.). Chicago: University of Chicago Press.

Couraud, J.Y., and L. DiGiamberardino. (1980). Axonal transport of the molecular forms of acetylcholinesterase in chick sciatic nerve. J. Neurochem., 35, 1053.

Couraud, J.Y., L. DiGiamberardino, and R. Hassig. (1982). Slow transport of the molecular forms of butyrlcholinesterase in a peripheral nerve. Neuroscience (Oxford), 7:1015.

Cowan, W.M., L. Adamson, and T.P.S. Powell. (1961). An experimental study of the avian visual system. J. Anat., 95, 545.

Cuénod, R.J., A. Beaudet, V. Canzek, P. Streit, and J.C. Reubi. (1981). Glutamatergic pathways in the pigeon and the rat brain. Adv. Biochem. Psychopharmacol., 27, 57.

de Anda, G., and M.A. Rebollo. (1967). The neuromuscular spindles in the adult chicken. 1. Morphology. Acta Anat., 67, 437.

DeGennaro, L.D. (1982). The Glycogen Body. In "Avian Biology," Vol. 6 (D.S. Farner and J.R. King, Eds.). New York: Academic Press, p. 341.

DeGennaro, L.D., and C.A. Benzo. (1976). Ultrastructural characterization of the accessory lobes of Lachi (Hofmann's Nuclei) in the nerve cord of the chick. I. Axoglial synapses. J. Exp. Zool., 198, 97.

DeGennaro, L.D., and C.A. Benzo. (1978). Ultrastructural characterization of the accessory lobes of Lachi (Hofmann's nuclei) in the nerve cord of the chick. II. Scanning and transmission electron microscopy with observations on the glycogen body. J. Exp. Zool., 206, 229.

DeLanerolle, N., and R.J. Andrew. (1974). Midbrain structures controlling vocalization in the domestic chick. Brain Behav. Evol., 10, 354.

Denbow, D.M., H.P. Van Krey, and J.A. Cherry. (1983). Feeding and drinking response of young chicks to injections of serotonin into the lateral entricle of the brain. Poultr. Sci., 61, 150.

DePlazas, S.F. (1982). Ontogenesis of gamma amino butyric acid receptor sites in chick embryo cerebellum. Dev. Brain Res., 3, 263.

DeSantis, V.P., W. Längsfeld, R. Lindmar, and K. Loffelholz. (1975). Evidence for noradrenaline and adrenaline as sympathetic transmitters in the chicken. Br. J. Pharmacol., 55, 345.

Desmedt, J.E., and P.J. Delwaide. (1963). Neuronal inhibition in a bird. Effect of strychnine and picrotoxin. Nature (London), 200, 585.

DeVoogd, T.J., and F. Nottenbohm. (1981). Sex differences in dendritic morphology of a song control nucleus in the canary: A quantitative Golgi study. J. Comp. Neurol., 196, 309.

Dewhurst, W.G., and E. Marley. (1965). The effects of α-methyl derivatives of noradrenaline, phenylethylamine and tryptamine on the central nervous system of the chicken. Br. J. Pharmacol., 25, 682.

Dorward, P.K., and A.K. McIntyre. (1971). Responses of vibration-sensitive receptors in the interosseous region of the duck's hind limb. J. Physiol. (London), 219, 77.

Dubbeldam, J.L., and H.J. Karten. (1978). The trigeminal system in the pigeon (*Columba livia*). I. Projections of the Gasserian ganglion. J. Comp. Neurol., 180, 661.

Dube, L., and A. Parent. (1981). The monoamine-containing neurons in avian brain 1. A study of the brain stem of the chicken (*Gallus gallus domesticus*) by means of fluorescence and acetylcholinesterase (E.C.3.1.1.7) histochemistry. J. Comp. Neurol., 196, 695.

Duff, T.A., G. Scott, and R. Mai. (1981). Regional differences in pigeon optic tract, chiasm, and retino-receptive layers of the optic tectum. J. Comp. Neurol., 198, 231.

Eden, A.R., and M.J. Correia. (1982). Identification of multiple groups of efferent vestibular neurons in the adult pigeon using horseradish peroxidase and DAPI. Brain Res., 248, 201.

Enemar, A., B. Falck, and R. Håkanson. (1965). Observations on the appearance of norepinephrine in the sympathetic nervous system of the chick embryo. Dev. Biol., 11, 268.

Erichsen, J.T., A. Reiner, and H.J. Karten. (1982). Co-occurrence of substance P-like and leucine enkephalin-like immunoreactivities in neurons and fibers of the avian nervous system. Nature (London), 295, 407.

Filliatreau, G., and L. DiGiamberardino. (1982). Quantitative analysis of axonal transport of cytoskeletal proteins in chicken oculomotor nerve. J. Neurochem., 39, 1033.

Fink, A.S., P.M. Hefferan, and R.R. Howell. (1975). Enzymatic and biochemical characterization of the avian glycogen body. Comp. Biochem. Physiol. A, 508, 525.

Firbas, W., and G. Muller. (1983). The efferent innervation of the avian cochlea. Hear. Res., 10, 109.

Frontali, N. (1964). Brain glutamic acid decarboxylase and synthesis of γ-aminobutyric acid in vertebrate and invertebrate species. In "Comparative Neurochemistry" (D. Richter, Ed.). Proc. Int. Neurochem. Symp. 5th, 1962. Oxford: Pergamon Press, p. 185.

Gal, E.M., and F.D. Marshal. (1964). The hydroxylation of tryptophan by pigeon brain *in vitro*. Prog. Brain Res., 8, 56.

Gamlin, P.D., A. Reiner, and H.J. Karten. (1982). Substance P-containing neurons of the avian suprachiasmic nucleus project directly to the nucleus of Edinger-Westphal. Proc. Natl. Acad. Sci. U.S.A., 79, 3891.

Garcia-Austt, E. (1954). Development of electrical activity in cerebral hemispheres of the chick embryo. Proc. Soc. Exp. Biol. Med., 86, 348.

Gilman, T.T., F.L. Marcuse, and A.U. Moore. (1950). Animal hypnosis: A study in the induction of tonic immobility in chickens. J. Physiol. Psychol., 43, 99.

Goldman, S.A., and F. Nottenbohm. (1983). Neuronal production, migration, and differentiation in a vocal control nucleus of the adult female canary brain. Proc. Natl. Acad. Sci. U.S.A., 80, 2390.

Gottschaldt, K.M., H. Fruhstorfer, M. Schmidt, and I. Kraft. (1982). Thermosensitivity and its possible fine-structural basis in mechanoreceptors in the beak skin of geese. J. Comp. Neurol., 105, 219.

Gregory, J.E. (1973). An electrophysiological investigation of the receptor apparatus of the duck's bill. J. Physiol. (London), 229, 151.

Gunne, L.M. (1963). Relative adrenaline content in brain tissue. Acta Physiol. Scand., 56, 324.

Gurney, M.E. (1983). Behavioral correlates of sexual differentiation in the zebra finch song system. Brain Res., 231, 153.

Haiden, G.J. (1981). The ultrastructure of the avian Golgi tendon organ. Anat. Rec., 200, 153.

Halata, Z., and B.L. Munger. (1980). The ultrastructure of Ruffini and Herbst corpuscles in the articular capsule of the domestic pigeon. Anat. Rec., 198, 681.

Hanig, J.P., and J. Seifter. (1968). Amines in the brain of neonate chicks after parenteral injection of biogenic and other amines. Fed. Proc. Fed. Am. Soc. Exp. Biol., 27, 651.

Harvey, A.L., and D. Van Helden. (1981). Acetylcholine receptors in singly and multiply innervated skeletal muscle fibres of the chicken during development. J. Physiol. (London), 317, 397.

Hayes, B.P., and K.E. Webster. (1981). Neurons situated outside the isthmo-optic nucleus and projecting to the eye in adult birds. Neurosci. Lett., 26, 107.

Hebb, C.O. (1955). Choline acetylase in mammalian and avian sensory systems. Q. J. Exp. Physiol., 40, 176.

Hebb, C.O., and D. Ratković. (1964). Choline acetylase in the evolution of the brain in vertebrates. In "Comparative Neurochemistry" (D. Richter, Ed.), Proc. Int. Neurochem. Symp., 5th. Oxford: Pergamon Press, p. 347.

Hehman, K.N., A.R. Vonderahe, and J.J. Peters. (1961). Effect of serotonin on the behavior, electrical activity in the brain, seizure threshold in the newly hatched chick. Neurology, 11, 1011.

Hendrikson, C.K., and J.E. Vaughn. (1974). Fine structural relationships between neurites and radial glial processes in developing mouse spinal cord. J. Neurocytol., 3, 659.

Hess, A., G. Pilar, and J.N. Weakly. (1969). Correlation between transmission and structure in avian ciliary ganglion synapses. J. Physiol. (London), 202, 339.

Hodos, W., and H.J. Karten. (1966). Brightness and pattern discrimination deficits in the pigeon after lesions of nucleus rotundus. Exp. Brain Res., 2, 151.

Hodos, W., H.J. Karten, and J.C. Bonbright. (1973). Visual intensity and pattern discrimination after lesions of the thalamo fugal visual pathway in pigeons. J. Comp. Neurol., 148, 447.

Hodos, W., K.A. Macko, and D.I. Sommers. (1982). Interactions between components of the avian visual system. Behav. Brain Res., 5, 157.

Holden, A.L., and T.P.S. Powell. (1972). The functional organization of the isthmo-optic nucleus in the pigeon. J. Physiol. (London), 223, 419.

Horton, E.W. (1971). Prostaglandins. In "Physiology and Biochemistry of Domestic Fowl," Vol. 1 (D.J. Bell and B.M. Freeman, Eds.). New York: Academic Press, p. 589.

Hummel, G. (1979). The fine structure of the bulbus olfactorius of the hen. Acta Histol. Embryol., 8, 289.

Ignarro, L.J., and F.E. Shideman. (1968). Appearance and concentrations of catecholamines and their biosynthesis in the embryonic and developing chick. J. Pharmacol. Exp. Ther., 159, 38.

James, N.T., and G.A. Meek. (1973). An electron microscopic study of avian muscle spindles. J. Ultrastruct. Res., 43, 193.

Jeikouski, H., and D. Drenckhahn. (1981). Evidence for exclusive adrenergic innervation of feather muscles (mm. pennati) in the chicken. Histochemical studies and experiments with 5-hydroxydopamine. Cell Tissue Res., 221, 157.

Jhaveri, S., and D.K. Morest. (1982). Neuronal architecture in nucleus magnocellularis of the chicken auditory system with observations on nucleus laminaris: A light and electron microscope study. Neuroscience (Oxford), 7, 809.

Jones, A.W., and R. Levi-Montalcini. (1958). Patterns of differentiation of the nerve centers and fiber tracts of the avian cerebral hemispheres. Arch. Ital. Biol., 96, 231.

Jones, D.R., and K. Johansen. (1972). The blood vascular system of birds. In "Avian Biology," Vol. 2 (D.S. Farner and J.R. King, Eds.). New York: Academic Press, p. 158.

Jukes, M.G.M. (1971). In "Biochemistry and Physiology of the Domestic Fowl," Vol. 1 (D.J. Bell and B.M. Freeman, Eds.). New York: Academic Press, p. 171.

Juorio, A.V., and M. Vogt. (1967). Monoamines and their metabolites in the avian brain. J. Physiol. (London), 189, 489.

Juorio, A.V., and M. Vogt. (1970). Adrenaline in bird brain. J. Physiol. (London), 209, 757.

Karten, H.J. (1963). Ascending pathways from the spinal cord in the pigeon (*Columba livia*). Proc. Int. Cong. Zool., 16th, 2, 23.

Karten, H.J. (1966). Efferent projections of the nucleus mesencephalicus lateralis, pars dorsalis (MLD) in the pigeon (*Columba livia*). Anat. Rec., 154, 365.

Karten, H.J. (1967a). The organization of the ascending auditory pathway in the pigeon (*Columba livia*) 1. Diencephalic projections of the inferior colliculus (nucleus mesencephalicus lateralis pars dorsalis). Brain Res., 6, 409.

Karten, H.J. (1967b). Telencephalic projections of the nucleus ovoidalis in the pigeon (*Columba livia*). Anat. Rec., 157, 268.

Karten, H.J. (1968). The ascending auditory pathway in the pigeon. II. Telencephalic projections of the nucleus ovoidalis thalami. Brain Res., 11, 134.

Karten, H.J. (1969). The organization of the avian telencephalon and some speculations on the phylogeny of amniote telencephalon. Ann. N.Y. Acad. Sci., 167, 164.

Karten, H.J., and A.M. Revzin. (1966). The afferent connections of the nucleus rotundus in the pigeon. Brain Res., 2, 368.

Karten, H.J., and W. Hodos. (1967). "A Stereotaxic Atlas of the Brain of the Pigeon (*Columba livia*)." Baltimore: Johns Hopkins Press.

Karten, H.J., W. Hodos, W.J.H. Nauta, and A.M. Revzin. (1973). Neural connections of the "Visual Wulst" of the avian telencephalon. Experimental studies in the pigeon (*Columba livia*) and owl (*Speotyto cunicularia*). J. Comp. Neurol., 150, 253.

Katori, M. (1962). The development of the spontaneous electrical activity in the brain of a chick embryo and the effects of several drugs on it. Jpn. J. Pharmacol., 12, 9.

King, A.S., and V. Molony. (1971). The anatomy of respiration. In "Physiology and Biochemistry of the Domestic Fowl" (D.J. Bell and B.M. Freeman, Eds.). New York: Academic Press, p. 93.

Kirsch, M., R.B. Coles, and J.H. Leppelsack. (1980). Unit recordings from a new auditory area in the frontal neostriatum of the awake starling (*Sturnus vulgaris*). Exp. Brain Res., 38, 375.

Kitt, C.A., and S.E. Brauth. (1982). A paleostriatal-thalamic-telencephalic path in pigeons. Neuroscience (Oxford), 7, 2735.

Knowlton, V.Y. (1964). Abnormal differentiation of embryonic avian brain centres associated with unilateral anophthalmia. Acta Anat., 58, 222.

Koenig, H.L., L. DiGiamberadino, and G. Bennett. (1973). Renewal of proteins and glycoproteins of synaptic constituents by means of axonal transport. Brain Res., 62, 413.

Koley, J., J. SenGupta, S.P. Sarkar, and B.N. Koley. (1979). Sympathetic afferents from the avian abdomen. Int. Res. Comm. Symp. Med. Sci. Lib. Compend., 7, 245.

Kovacs, S.A., G.C. Wilson, and J.K. Kovach. (1981). Normal EEG of the restrained twenty-four-hour-old Japanese quail (*Coturnix coturnix japonica*). Poult. Sci., 60, 243.

Kramer, S.Z., and J. Seifter. (1966). The effects of GABA and biogenic amines on behavior and brain electrical activity in chicks. Life Sci., 5, 527.

Kuenzel, W.J., and A. van Tienhoven. (1982). Nomenclature and location of avian hypothalamic nuclei and associated circumventricular organs. J. Comp. Neurol., 206, 293.

Kuriyama, K., B. Sisken, J. Ito, D.G. Simonsen, B. Haber, and E. Roberts. (1968). The γ-aminobutyric acid system in the developing chick embryo cerebellum. Brain Res., 11, 412.

Lajtha, A. (1957). The development of the blood–brain barrier. J. Neurochem., 1, 216.

Lasjewski, R.C. (1972). Respiratory function in birds. In "Avian Biology," Vol. 2 (D.S. Farner and J.R. King, Eds.). New York: Academic Press, p. 288.

Leitner, L.-M., and M. Roumy. (1974a). Mechanosensitive units in the upper bill and in the tongue of the domestic duck. Pfleugers Arch., 346, 141.

Leitner, L.-M., and M. Roumy. (1974b). Thermosensitive units in tongue and in the skin of the duck's bill. Pfleugers Arch., 346, 151.

Levi-Montalcini, R. (1950). The origin and development of the visceral system in the spinal cord of the chick. J. Morphol., 86, 253.

Lubinska, L. (1975). On axoplasmic flow. Int. Rev. Neurobiol., 17, 241.

Maier, A. and E. Eldred. (1971). Comparisons in the structure of avian muscle spindles. J. Comp. Neurol., 143, 25.

Malinovski, L., and L. Pac. (1980). Ultrastructure of the Herbst corpuscle from beak skin of the pigeon. Z. Mikrosk. Anat. Forsch., 94, 292.

Manni, E., G.M. Azzena, and R. Bortolani. (1965). Jaw muscle proprioception and mesencephalic trigeminal cells in birds. Exp. Neurol., 12, 320.

Marko, P., and M. Cuénod. (1973). Contribution of the nerve cell body to renewal of axonal and synaptic glycoproteins in the pigeon visual system. Brain Res., 62, 419.

Martin. A.H. (1979). A cytoarchitectonic scheme for the spinal cord of the domestic fowl (*Gallus gallus domesticus*): Lumbar region. Acta Morphol. Neerl.-Scand., 17, 105.

Martin, A.R., and G. Pilar. (1963a). Dual mode of synaptic transmission in the avian ciliary ganglion. J. Physiol. (London), 168, 443.

Martin, A.R., and G. Pilar. (1963b). Transmission through the ciliary ganglion of the chick. J. Physiol. (London), 168, 464.

Martin, A.R., and G. Pilar. (1964). An analysis of electrical coupling at synapses in the avian ciliary ganglion. J. Physiol. (London), 171, 454.

McCasland, J.S., and M. Konishi. (1981). Interaction between auditory and motor activities in an avian song control nucleus. Proc. Natl. Acad. Sci. U.S.A., 78, 7815.

McGill, J.J., T.P.S. Powell, and W.M. Cowan. (1966). The retinal representation upon the optic tectum and isthmo-optic nucleus in the pigeon. J. Anat., 100, 5.

Miceli, D., and J. Reperant. (1982). Thalamo-hyperstriatal projections in the pigeon (*Columba livia*) as demonstrated by retrograde double labeling with fluorescent tracers. Brain Res., 245, 365.

Morales, R., and D. Duncan. (1975). Specialized contacts of astrocytes with astrocytes and with other cell types in the spinal cord of the cat. Anat. Rec., 182, 255.

Narayanan, C.H., and Y. Narayanan. (1976). An experimental inquiry into the central source of preganglionic fibers to the chick ciliary ganglion. J. Comp. Neurol., 166, 101.

Newman, J.D. (1972). Midbrain control of vocalization in redwinged blackbirds (*Agelaius phoeniceus*). Brain Res., 48, 227.

Nistico, T. (1980). Relations between dopaminergic and gamma amino butyric acid-ergic mechanisms in avian brain. Pharmacol. Res. Commun., 12, 507.

Noden, D.M. (1980). Somatotopic and functional organization of the avian trigeminal ganglion. Horseradish peroxidase analysis in the hatchling chick. J. Comp. Neurol., 190, 405.

Noden, D.M. (1983a). The role of the neural crest in patterning of avian cranial skeletal, connective and muscle tissues. Dev. Biol., 96, 144.

Noden, D.M. (1983b). The embryonic origins of avian cephalic and cervical muscles and associated connective tissues. Am. J. Anat., 168, 257.

Nottenbohm, F. (1980a). Brain pathways for vocal learning in birds: A review of the first 10 years. In "Progress in Psychobiology and Physiological Psychology," Vol. 9 (J.M.S. Sprague and A.N.E. Epstein, Eds.). New York: Academic Press, p. 84.

Nottenbohm, F. (1980b). Testosterone triggers growth of brain vocal control nuclei in adult female canaries. Brain Res., 189, 429.

Nottenbohm, F., D.B. Kelley, and J.A. Paton. (1982). Connections of vocal control nuclei in the canary telencephalon. J. Comp. Neurol., 207, 344.

Ohmori, Y., T. Watanabe, and T. Fujioka. (1982). Localization of the motoneurons innervating the forelimb muscles in the spinal cord of the domestic fowl. Anat. Histol. Embryol., 11, 124.

Ookawa, T. (1972a). Polygraphic recording during adult hen hypnosis. Poult. Sci., 51, 853.

Ookawa, T. (1972b). Avian wakefulness and sleep on the basis of recent electroencephalographic observations. Poult. Sci., 51, 1565.

Ookawa, T. (1973a). Notes of abnormal electroencephalograms in the telencephalon of the chicken and pigeon. Poult. Sci., 52, 182.

Ookawa, T. (1973b). Effect of strychnine on the electroencephalogram recorded from the Wulst of curarized adult chickens. Poult. Sci., 52, 1090.

Ookawa, T. (1973c). Effect of intravenously administered strychnine on the EEG recorded from the deep structure of the adult chicken telencephalon. Poults. Sci., 52, 806.

Ookawa, T. (1973d). Effect of some convulsant drugs on the electroencephalogram recorded from the Wulst of the adult chicken and pigeon under curarized conditions. Poult. Sci., 52, 1704.

Ookawa, T. (1977). Behavioral and electroencephalographic manifestations of avian epilepsy: A review of the literature. Poult. Sci., 56, 773.

Ookawa, T., and J. Gotoh. (1965). Electroencephalogram of the chicken recorded from the skull under various conditions. J. Comp. Neurol., 124, 1.

Ookawa, T., and K. Takagi. (1968). Electroencephalograms of free behavioral chicks at various developmental ages. Jpn. J. Physiol., 18, 87.

Ookawa, T., J. Gotoh, T. Kumazawa, and K. Takagi. (1962). Electroencephalogram of chickens. Proc. Meeting Jpn. Soc. Vet. Sci., 53rd, Jpn. J. Vet. Sci., 24 (Suppl.), 438.

Oscarsson, O., I. Rosen, and N. Uddenberg. (1963). Organization of the ascending tracts in the spinal cord of the duck. Acta Physiol. Scand., 59, 143.

Pac, L. (1982). Contribution to the study of Merkel corpuscles in the domestic fowl. Folia Morphol. (Warsaw), 30, 340.

Parks, T.N. (1981). Morphology of axosomatic endings in an avian cochlear nucleus: Nucleus magnocellularis of the chicken. J. Comp. Neurol., 203, 425.

Parks, T.N., P. Collins, and J.W. Conlee. (1983). Morphology and origin of axonal endings in the nucleus laminaris of the chicken. J. Comp. Neurol., 214, 32.

Pearson, R. (1972). "The Avian Brain." New York: Academic Press.

Peinone, S.M. and L.S. Daneo. (1978). Ultrastructural observations on avian muscle spindles: Evidence of three intrafusal fibre types. Riv. Biol., 71, 3.

Peters, J.J., and E.J. Hilgeford. (1971). EEG episodes of rhythmic waves and seizure patterns following hypothermic hypoxia in the chick embryo. Electroencephalogr. Clin. Neurophysiol., 31, 631.

Peters, J.J., C.J. Cusick, and A.R. Vonderahe. (1961). Electrical studies of hypothermic effects on the eye, cerebrum and skeletal muscles of the developing chick. J. Exp. Zool., 148, 31.

Peters, J.J., A.R. Vonderahe, and J.J. McDonough. (1964). Electrical changes in brain and eye of the developing chick during hyperthermia. Am. J. Physiol., 207, 260.

Peters, J.J., T.P. Bright, and A.R. Vonderahe. (1968). Electroencephalographic studies of survival following hypothermic hypoxia in developing chicks. J. Exp. Zool., 167, 179.

Peters, J.J., A.R. Vonderahe, and E.J. Hilgeford. (1969). Electroencephalographic episodes of 1 to 7 per second rhythmic waves following hypothermic hypoxia in developing chicks. J. Exp. Zool., 170, 427.

Pohorecky, L.A., M.J. Zigmond, H.J. Karten, and R.J. Wurtman. (1968). Phenylethanolamine-*N*-methyltransferase activity (PNMT) in mammalian, avian and reptilian brain. Fed. Proc. Fed. Am. Soc. Exp. Biol., 27, 239.

Pomeroy, L.R., and A.J. Welch. (1967). Computer-assisted electroencephalograph analysis of chick pyridoxine deficiency states. Technical Report 36, The University of Texas, Austin, p. 1.

Potash, L.M. (1970). Neuroanatomical regions relevant to production and analysis of vocalization within the avian torus semicircularis. Experientia, 26, 1104.

Powell, T.P.S., and W.M. Cowan. (1961). The thalamic projection upon the telencephalon in the pigeon (*Columba livia*). J. Anat., 95, 78.

Putkonen, P.T.S. (1967). Electrical stimulation of the avian brain. Ann. Acad. Sci. Fenn., Ser. A5, 130, 1.

Reiner, A., N.C. Brecha, and H.J. Karten. (1982). Basal ganglia pathways to the tectum. The afferent and efferent connections of the lateral spiriform nucleus of the pigeon. J. Comp. Neurol., 208, 16.

Reiner, A., H.J. Karten, and N.C. Brecha. (1982). Enkephalin-mediated basal ganglia influences over the optic tectum: Immunohistochemistry of the tectum and lateral spiriform nucleus in the pigeon. J. Comp. Neurol., 208, 37.

Reiner, A., H.J. Karten, P.D.R. Gamlin, and J.T. Erichsen. (1983). Parasympathetic ocular control. Functional subdivision and circuitry of the avian nucleus of Edinger–Westphal. Trends Neurosci., 6, 140.

Revzin, A.M., and H. Karten. (1967). Rostral projections of the optic tectum and the nucleus rotundus in the pigeon. Brain Res., 3, 264.

Ryan, S.M., A.P. Arnold, and R.P. Elde. (1981). Enkephalin-like immunoreactivity in vocal control regions of the zebra finch brain. Brain Res., 229, 236.

Rylander, M.K., and J. Snow. (1982). Cytoarchitectonic of some nuclei in the avian auditory and visual systems. Anat. Anz., 151, 421.

Saffrey, M.J., J.M. Polak, and G. Burnstock. (1982). Distribution of vasoactive intestinal polypeptide-, substance P-, enkephalin- and neurotensin-like immunoreactive nerves in the chicken gut during development. Neuroscience, 7, 279.

Sammartino, U. (1933). Sugli animali a midolla spinale accrociato. Arch. Farmacol. Sper., 55, 219.

Sansone, F.M. (1977). The craniocaudal extent of the glycogen body in the domestic chicken. J. Morphol., 153, 87.

Sansone, F.M. (1980). An ultrastructural study of the craniocaudal continuation of the glycogen body. J. Morphol., 163, 45.

Sanson, F.M. and F.J. Lebeda. (1976). A brachial glycogen body in the spinal cord of the domestic chicken. J. Morphol. 148, 23.

Sato, H., A. Ohga, and Y. Nakazato. (1970). The excitatory and inhibitory innervation of the stomachs of the domestic fowl. Jpn. J. Pharmacol., 20, 382.

Scholes, N.W., and E. Roberts. (1964). Pharmacological studies of the optic system of the chick: Effect of γ-aminobutyric acid and pentobarbital. Biochem. Pharmacol., 13, 1319.

Seller, T.J. (1980). Midbrain regions involved in call production in Java sparrows. Behav. Brain Res., 1, 257.

Sheff, A.G., and L.L. Tureen. (1962). EEG studies of normal and encephalomalacia chicks. Proc. Soc. Exp. Biol. Med., 111, 407.

Shiosaka, S., K. Takatsuki, S. Inagaki, M. Sakanaka, H. Takaji, E. Senba, T. Matsugaki, and M. Tohyama. (1981). Topographic atlas of somatostatin-containing neuron system. I. Telencephalon and diencephalon. J. Comp. Neurol., 202, 103.

Showers, M.C. (1982). Telencephalon of birds. In "Comparative Correlative Neuroanatomy of the Vertebrate Telencephalon" (E.C. Crosby and H.N. Schnitzlein, Eds.). New York: Macmillan, p. 218.

Sivak, J.G. (1983). Ultrastructure of the avian iris dilator muscle. Rev. Can. Biol. Exp., 42, 57.

Skoglund, C.R. (1960). Properties of pacinian corpuscles of ulnar and tibial location in cat and fowl. Acta Physiol. Scand., 50, 385.

Snell, R.S. (1975). "Clinical Embryology for Medical Students," 2nd ed. Boston: Little, Brown.

Spooner, C.E., and W.D. Winters. (1966a). Neuropharmacological profile of the young chick. Int. J. Neuropharmacol., 5, 217.

Spooner, C.E., and W.D. Winters. (1966b). Distribution of monoamines and regional uptake of DL-norepinephrine-7-H^3 and dopamine-1-H^3 in the avian brain. Pharmacologist, 8, 189.

Spooner, C.E., and W.D. Winters. (1967). Evoked responses during spontaneous and monoamine-induced states of wakefulness. Brain Res., 4, 189.

Spooner, C.E., and W.D. Winters, and A.J. Mandell. (1966). DL-Norepinephrine-7-H^3 uptake, water content and thiocyanate space in the brain during maturation. Fed. Proc. Fed. Am. Soc. Exp. Biol., 25, 451.

Stewart, P.A., and M.J. Wiley. (1981). Structural and histochemical features of the avian blood–brain barrier. J. Comp. Neurol., 202, 157.

Straznicky, C., and D. Tay. (1983). The localization of motoneuron pools innervating wing muscles in the chick. Acta Embryol., 166, 209.

Strutz, J., and C.L. Schmidt. (1982). Acoustic and vestibular efferent neurons in the chicken (*Gallus domesticus*). A horseradish peroxidase study. Acta Otolaryngol. (Stockh.), 94, 45.

Sugihara, K., and J. Gotoh. (1973). Depth electroencephalograms of chickens in wakefulness and sleep. Jpn. J. Physiol., 23, 371.

Swank, R.L., and H.H. Jasper. (1942). Electroencephalograms of thiamine-deficient pigeons. Arch. Neurol. Psychiatry, 47, 821.

Takatsuji, K., H. Ito, and H. Masai. (1983). Ipsilateral retinal projections in the Japanese quail, *Coturnix coturnix japonica.* Brain Res. Bull., 10, 53.

ten Cate, J. (1960). Locomotor movements in the spinal pigeon. J. Exp. Biol., 37, 609.

Tindall, A.R. (1979). The innervation of the hind gut of the domestic fowl. Br. Poult. Sci., 20, 473.

Toggenburger, G., D. Felix, M. Cuenod, and H. Henke. (1982). In vitro release of endogenous beta-alanine, GABA, and glutamate, and electrophysiological effect of beat-alanine in pigeon optic tectum. J. Neurochem., 39, 176.

Tokaji, E., and R.W. Gerard. (1939). Avitaminosis B, and pigeon brain potentials. Proc. Soc. Exp. Biol. Med., 41, 653.

Tsukada, Y., K. Uemura, S. Hirano, and Y. Nagata. (1962). In "Comparative Neurochemistry" (D. Richter, Ed.). Proc. 5th Int. Neurochem. Symp., 5th. Oxford: Pergamon Press, p. 179.

Tuge, H., Y. Kanayama, and C.H. Yueh. (1960). Comparative studies on the development of the EEG. Jpn. J. Physiol., 10, 211.

Uehara, M., T. Veshima, and N. Kudo. (1982). The fine structure of glycogen-containing cells in the chicken spinal cord. Jpn. J. Vet. Res., 30, 1.

van Tienhoven, A., and L.P. Juhász. (1962). The chicken telencephalon, diencephalon and mesencephalon in sterotaxic coordinates. J. Comp. Neurol., 118, 185.

Waelsch, H. (1955). In "Biochemistry of the Developing Nervous System" (H. Waelsch, Ed.). New York: Academic Press, p. 187.

Watanabe, T. (1960). On the peripheral course of the vagus nerve in the fowl. Jpn. J. Vet. Sci., 22, 145.

Watanabe, T. (1964). Peripheral courses of the hypoglossal, accessory and glossopharayngeal nerves. Jpn. J. Vet. Sci., 26, 249.

Watanabe, T., and M. Yasuda. (1968). Peripheral course of the olfactory nerve in the fowl. J. Vet. Sci., 30, 275.

Watanabe, T., and M. Yasuda. (1970). Peripheral course of the trigeminal nerve in the fowl. Jpn. J. Vet. Sci., 32, 43.

Watanabe, T., G. Isomura, and M. Yasuda. (1967). Distribution of nerves in the oculomotor and ciliary muscles. Jpn. J. Vet. Sci., 29, 151.

Watanabe, M., H. Ito, and H. Masai. (1983). Cytoarchitecture and visual receptive neurons in the Wulst of the Japanese quail (*Coturnix coturnix japonica*). J. Comp. Neurol., 213, 188.

Whitehead, M.C., and D.K. Morest. (1981). Dual populations of efferent and afferent cochlear axons in the chicken (*Gallus gallus*). Neuroscience, 6, 2351.

Whitlock, D.G. (1952). A neurohistological and neurophysiological study of afferent fiber tracts and receptive areas of the avian cerebellum. J. Comp. Neurol., 97, 567.

Wold, J.E. (1979). The vestibular nuclei in the domestic hen (*Gallus domesticus*). VII. Afferents from the spinal cord. Arch. Ital. Biol., 117, 30.

Wold, J.E. (1981). The vestibular nuclei in the domestic hen (*Gallus domesticus*): VI. Afferents from the cerebellum. J. Comp. Neurol., 201, 319.

Yamada, S., and S.I. Mikami. (1982). Immunohistochemical localization of vasoactive intestinal polypeptide-containing neurons in the hypothalamus of the Japanese quail, *Coturnix coturnix.* Cell Tissue Res., 226, 13.

Yasuda, M. (1960). On the nervous supply of the thoracic limb in the fowl. Jpn. J. Vet. Sci., 22, 89.

Yasuda, M. (1961). On the nervous supply of the hind limb of the fowl. Jpn. J. Vet. Sci., 23, 145.

Zaretsky, M.D. (1978). A new auditory area of the song bird forebrain: A connection between auditory and song control centers. Exp. Brain Res., 32, 267.

Zecha, A. (1961). Bezit een vogel een fasciculus rubro-bulbospinalis? Ned. Tijdschr. Geneeskd., 105, 2373.

Zecha, A. (1962). The "pyramidal tract" and other telencephalic efferents in birds. Acta Morph. Neerl.-Scand., 5, 194.

Zeigler, H.P. (1963). Effects of endbrain lesions upon visual discrimination learning in pigeons. J. Comp. Neurol., 120, 161.

Zeigler, H.P., and H.J. Karten. (1973). Brain mechanisms and feeding behaviour in the pigeon (*Columba livia*). I. Quintofrontal structures. J. Comp. Neurol., 152, 59.

Ziller, C., and J. Smith. (1982). Migration and differentiation of neural crest cells and their derivitives: In vivo and in vitro studies on the early development of the avian peripheral nervous system. Reprod. Nutr. Dev., 22, 153.

2
Sense Organs

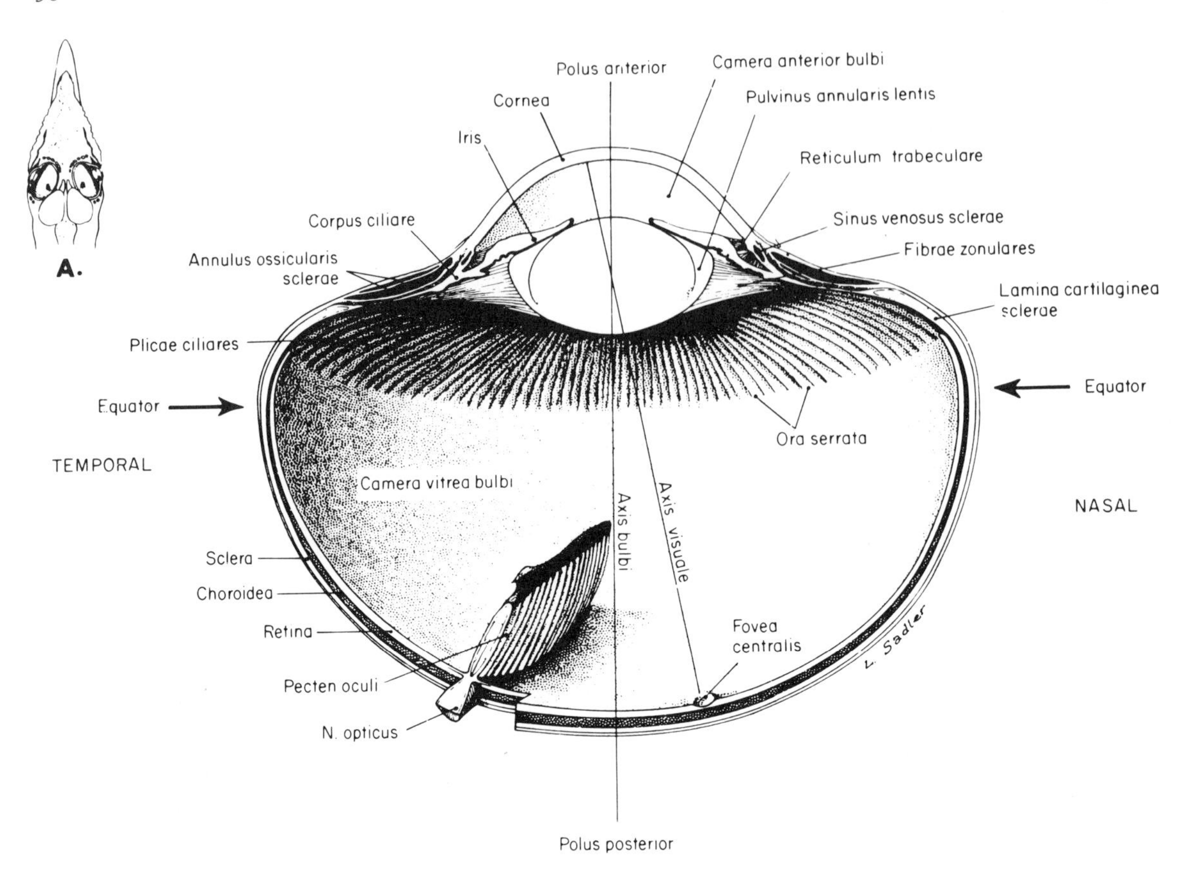

FIGURE 2–1. Drawing of transverse section (horizontal plane) of chicken eye. Insert (A) indicates position of eyes within head. (From Evans, 1979.)

The Avian Eye

D.B. MEYER

Introduction

The avian eye (Figure 2–1) displays the basic pattern of organization found in all vertebrate eyes (Walls, 1942; Rochon-Duvigneaud, 1943; Duke-Elder, 1958), and incorporates many adaptations that have greatly improved its visual abilities to the point where it is now considered to be the finest ocular organ in the animal kingdom. Its wall, like that of other vertebrate eyes, is constructed of three concentric layers or tunics: an internal *sensory tunic,* or *retina;* a middle *vascular tunic* (*uvea*) including the *iris, ciliary body,* and *choroid,* and an external *fibrous tunic* composed of a transparent *cornea* and cartilaginous *sclera.*

In all vertebrate eyes, the visual process converts light energy into nerve impulses that must be delivered to the brain for further integration into visual images. The retina, well protected at the back of the interior of the eye, represents the visual receptor; its optic nerve is its communicating link with the higher centers of the brain. (See also Chapter 1.) The other structures lying in the path of the incoming light (visual axis; Figure 2–1) represent the transparent dioptric media; owing to their position, these participate either directly or indirectly in the refraction of the light rays. The cornea, aqueous humor, lens, and vitreous body make up these media, the amount of their refractive ability depending on the differences in their optical density. Although the greatest refraction occurs at the external surface of the cornea, the lens represents the most important refractive organ because of its ability to change shape (alter its curvature) and thus actively focus the light rays on the retinal receptors (rods and cones) for both near and distant objects. The light image received by the rods and cones is inverted, and produces electrical changes that can be recorded both from the retina as a whole and from individual optic nerve fibers (electroretinography). The production of an electrical potential involves the selective absorption of the light rays by specific visual pigments housed in the rods and cones and the bleaching of these pigments. Exact details of the discharge of impulses in the retina are obscure.

The lens, biconvex and encapsulated, is suspended from the ciliary body by delicate annular ligaments (zonular fibers) (Figure 2–1), which, together with the lens, separate the large, gelatinous vitreous body from the fluid-filled aqueous chambers (anterior and posterior). The anterior chamber lies directly behind the cornea, between it and the iris (and pupil); the posterior chamber lies behind the iris, between it and the lens with its zonular fibers.

Aqueous humor fills both anterior and posterior chambers and is responsible for regulating the intraocular pressure, which maintains the proper shape and rigidity of the eyeball. This intraocular fluid is slowly but continually secreted into the posterior chamber by the ciliary body; it circulates through the pupil and gains access to the anterior chamber where it is discharged, also continually, into the annular sinus venosus sclerae (canal of Schlemm) at the periphery of the anterior chamber (Figure 2–1).

More detailed information on the histological organization and function of these accessory ocular tissues is available in Walls (1942), Duke-Elder and Wybar (1961), and Hogan et al. (1971). In this chapter, emphasis will be placed on the several uniquely avian features, both structural and functional, which have been perfected in the evolvement of the exceptional avian visual abilities. According to Pumphrey (1948a), the attainment of these features represents the culmination of phylogenetic development toward diurnal vision. Chief among them are the large size of the eye, which permits the formation of a large retinal image, the development of a superb accommodative apparatus for increasing the depth of focus, and the presence of elaborate and diverse retinal areas (and foveas) responsibile for providing a relatively high degree of acute monocular and binocular vision. A bizarre, heavily vascularized organ, the pecten (Figure 2–1), also characterizes the avian eye and has been implicated in many diverse functions.

The extremely large avian eye leaves little room in the head for anything else. For example, the ostrich eye is 50 mm in diameter, the largest of any land vertebrate (Walls, 1942), and the eyes of the tawny owl comprise one-third of the entire weight of the head (Leuckart, 1876). In general, the largest eyes are possessed by nocturnal predators, and the smallest by waterfowl and swamp birds. The avian eye has been classified into three groups (Walls, 1942); flat, globose, and tubular (Figure 2–2). The flat shape (Figure 2–2a) is characteristic for most birds, whereas the globose shape (Figure 2–2b) is found in the diurnal species, which require high resolution for great distances (e.g., crow). The greatest visual acuity, however, is attained by the tubular shape (most owls and some eagles) (Figure 2–2c), which permits an increased "throw" of the image and is characterized by a broad concave ciliary region containing a ring of thin overlapping bones, the *scleral ossicles,* which encircle the cornea (Figure 2–1). Numbering between 10 and 18 (14 or 15 being the most common), these ossicles reinforce the ciliary body, and together with the support provided by the cartilages of the sclera, permit a more powerful accommodative function.

Accommodation

The ability to adjust, or focus, the eye so that objects of various distances from the retina can be seen sharply is called accommodation. The mechanism to carry out this important visual function is extremely well developed in birds, and permits a greater range than any other vertebrate group. Most birds, for example, have a range of 20 diopters, which is twice the maximum accommodation of a 20-year-old man (Pumphrey, 1948a). Aquatic birds, especially the cormorant (Hess, 1909–1910) and the dipper (Goodge, 1960), exhibit the greatest range (almost 50 diopters), whereas nocturnal birds, which rely more heavily on sonic acuity, accommodate the least (2–4 diopters). Avian accommodation is always for near vision, and utilizes a variety of mechanisms that involve both anatomic and physio-

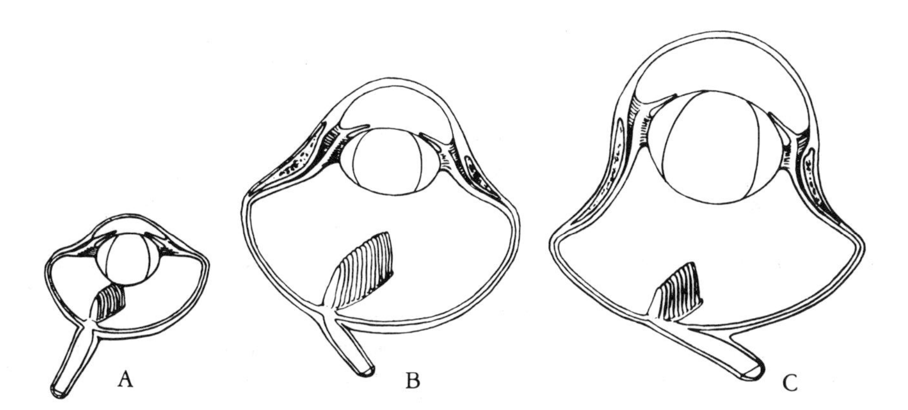

FIGURE 2–2. Shapes of avian eyes as revealed by the ventral half of left eyeball: (**A**) flat; (**B**) globose; (**C**) tubular. (After Soemmerring, 1818.)

logic adaptations. Basically, two phenomena are involved: alteration in the curvature of the cornea (*corneal accommodation*) and deformation of the lens (*lenticular accommodation*). Retinal movement involves another possible, but highly controversial, mechanism in some birds (Walls, 1942) whereby choroidal musculature can draw the retina backward during sharp focusing like the fine adjustment of a microscope.

Both corneal and lenticular accommodation depend on the action of the ciliary musculature, which is skeletal in type (thus voluntary), most often meridional in disposition, and quite variable in its composition. One to three separate muscle masses is the general appearance; four independent muscles have been observed in the red-tailed hawk (*Buteo borealis*) (Lord, 1956) (see also Chapter 1).

Corneal Accommodation. In the case of corneal accommodation, contraction of the ciliary muscles draws the corneal margin inward, thus changing its radius of curvature. On the basis of ophthalmometric measurements of the pigeon cornea during drug-induced accommodation, Gundlach et al. (1945) found that the cornea represents the chief accommodating organ in this species. It provides the pigeon with such a great range of accommodation (17 diopters) that these authors find little need for any other mechanism. Levy (1979), on the other hand, criticizes the validity of these findings and reports a much smaller range of accommodation (5–6 diopters) in the pigeon and mallard duck. He observes no changes in corneal curvature in these species when examined either by TV videotape photokeratoscopy or by the photographic study of fine Purkinje images before and after drug-induced accommodation.

Lenticular Accommodation. The principal means of accommodation in birds, reptiles, and mammals is by deformation of the lens. In birds, this is brought about by direct squeezing of the lens by the ciliary body and iris. The avian lens is soft and malleable, easily deformed, and capable of considerable alterations in its curvature. It has also developed an extensive equatorial thickening, the annular pad (Figure 2–1) for contact with the ciliary body. The annular pad is best developed in birds with a high degree of accommodation, less in nocturnal species, still more poorly formed in diving and flightless birds, and smallest or vestigial in running birds. Contraction of the ciliary muscles is believed to be responsible for thrusting the ciliary body inward against the annular pad, which then acts as an architectural column transmitting the stress directly to the softer lenticular center (Walls, 1942). This action occurs evenly along the entire extent of the equatorial lens and pushes its anterior surface forward against the iris and into the pupil. Forward movement of the lens itself is prevented by the zonular fibers and pectinate ligament, whereas posterior displacement is prevented by a relatively unyielding vitreous body. The iridial sphincter also has an accommodative function and it attains its best development (of all vertebrates) in amphibious birds, where it causes the most extreme anterior deformation of the lens and undoubtedly helps to account for the great range of accommodation in these species.

Vision

Retina. The avian retina is typically vertebrate in its organization and stratification. Arising as a direct continuation of the brain, it consists of two portions: an external, nonsensory single layer of cuboidal epithelium containing pigment, the *pigment epithelium,* and an internal, transparent, much thicker neuroepithelium (*sensory retina*) containing several types of neurons and glial cells. The neuron cell bodies and their synaptic connections occupy specific and separate localities within this sensory retina, producing a very characteristic stratification. The basic organization involves a three-neuron chain that spans the entire thickness of the sensory retina and forms a direct and simple pathway for the light-induced nerve impulses to reach the optic nerve (Figure 2–3). Visual receptors (*rods and cones,* R) represent the first neurons in this chain. They lie in the deepest (most external) portion of the sensory retina and consist of light-sensitive dendritic endings (outer segments) and axons that synapse at another locale with dendrites from the second neuron in the chain, *bipolar cells* (B). Axons from the bipolar cells synapse with dendrites from the third link in the chain, the *ganglion cells* (G). Axons from these cells then take a tangential course parallel to the retinal surface and pass out of the eye as the optic nerve. Integrated with these three neurons are two associative neurons, the *horizontal* (H) and *amacrine* (A) cells, which participate in important regulatory synaptic relationships along this chain. The horizontal cells lie most external, and communicate with both the receptors (rods and cones) and the bipolar cell dendrites. The amacrine cells synapse with bipolar cell axons and ganglion cell dendrites. (See Chapter 1 for further details on visual neural pathways.)

The avian retina is relatively thick compared to that observed in other vertebrates, contains a fascinating array of photoreceptors, and exhibits several possible combinations of areas (and foveas) specialized for more acute (and often stereoscopic) vision. Completely devoid of blood vessels that may possibly interfere with the incoming light rays, the avian retina receives its nutrients from two outside sources: capillaries within the choroid (choriocapillaris; C, Figure 2–3) lying immediately deep (external) to the pigment epithelium and the well-vascularized pecten situated within the vitreous body.

FIGURE 2–3. Diagrammatic representation of the vertebrate retina showing the basic organization of its neuroepithelium: a three-neuron chain including receptor (visual) cells (*R*), bipolar neurons (*B*), and ganglion cells (*G*) under the influence of two associative neurons, horizontal (*H*) and amacrine (*A*) cells. Externally, the pigment epithelial cells (*P*) lie in close relationship to its vascular supply, the choriocapillaris (*C*), and the outer segments of the receptor cells. Light must traverse the entire thickness of the transparent, sensory retina (left arrow) before striking the photosensitive outer segments. Nerve impulses generated at this site travel in the opposite direction (right arrow) to leave the eye via the optic nerve (Modified from Dowling, 1968.)

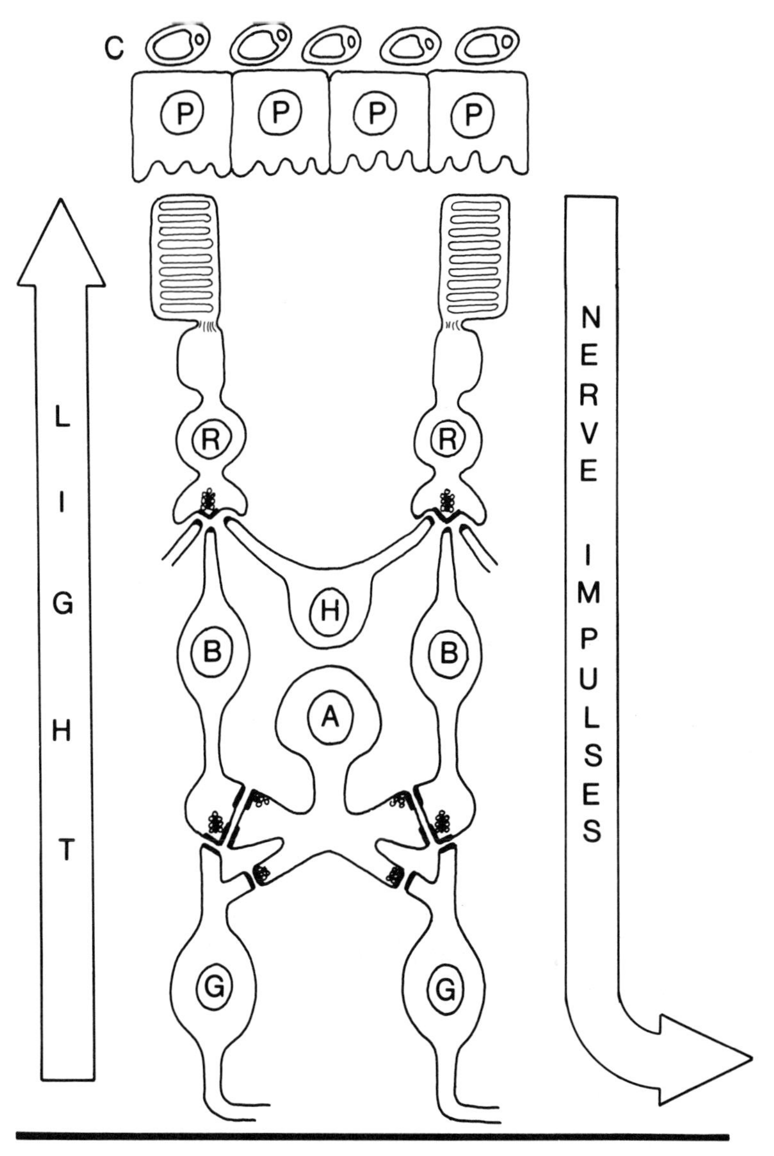

Visual Acuity. Visual acuity data of different birds tabulated by Donner (1951) and Fischer (1969) reveal a wide range of resolving power culminating in the superb acuity of diurnal birds of prey. According to Rochon-Duvigneaud (1919), the buzzard possesses four times the visual acuity of man; Fischer (1969) reported twice the resolving power in Old World vultures. Oehme (1964), however, reported that the maximal acuity of the common buzzard is close to that of the human; that of the kestrel is close to the rhesus monkey.

Color Discrimination. The avian retina mediates color vision. Initially demonstrated in the chicken (Watson, 1915; Lashley, 1916) and pigeon (Watson, 1915; Hamilton and Coleman, 1933), color discriminatory abilities have been reported to exist in other diurnal birds, e.g., emus (Neumann, 1962), hummingbirds (Poley, 1968), honeycreepers (Winkel, 1969), gulls (Thompson, 1971), weaverbirds (Mayr, 1972a), daws (Wessels, 1974), and even some nocturnal species, e.g., the little owl (*Athene noctua*) (Meyknecht, 1941), and tawny owl (*Strix aluco*) (Ferens, 1947; Martin, 1974). The spectral sensitivities of these birds are not the same, however, and several visual systems have generally been described: trichromatic in the pigeon (Hamilton and Coleman, 1933; Riggs et al., 1972), chicken (Bowmaker and Knowles, 1977), and tawny owl (Bowmaker and Martin, 1978), and tetrachromatic in the daw (Wessels, 1974). More recent data also suggest a tetrachromatic system in pigeons, as well as a possible pentachromatic system (Emmerton, 1983) involving ultraviolet light. Behavioral studies have shown that hummingbirds (Goldsmith, 1980) as well as pigeons (Kreithen and Eisner, 1978; Emmerton and Remy, 1983) are responsive to ultraviolet light, suggesting not only that these birds see quite differently from man, but that such a sensitivity may be important in their migratory activities.

Movement Detection. The avian retina is also capable of a great range of movement detection. Pigeons, for example, can discriminate the rate of movement of an object as slow as 15°/hr (Meyer, 1964) and yet have the highest flicker fusion frequency recorded in vertebrates (150/sec) (Dodt and Wirth, 1953). Such exceptional discriminatory ability is undoubtedly needed for controlling rapid flight movements and developing navigational skills for migration, e.g., detection of movements of the sun, changes in altitude, etc. The pecten and fovea, for example, have been implicated as possible mechanisms in movement detection, and more recent studies have provided evidence that the retinal ganglion cells are highly specialized in birds for the detection of different types of movements (Maturana, 1962; Maturana and Frenk, 1963).

Visual Cells. The avian retina features three distinct types of visual elements (rods, single cones, double cones (Figure 2–4), which are characterized by their structural and chemical compartmentation. Their outer segments resemble those of other vertebrate visual cells, and consist of stacks of membranous discs housing the photosensitive visual pigments. Within the inner segments, the mitochondria are concentrated distally to form the so-called *ellipsoid*, whereas most of the other organelles, e.g., Golgi apparatus and endoplasmic reticulum, reside in the proximal portion or *myoid*. The

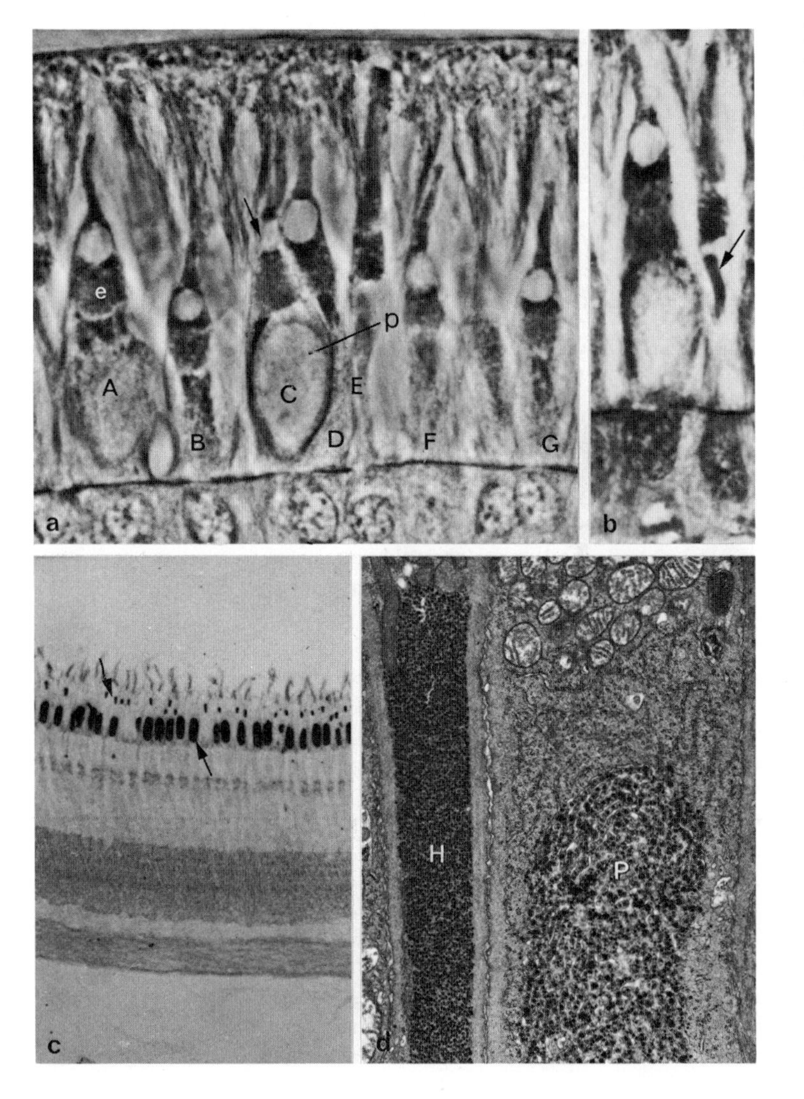

FIGURE 2–4. Avian photoreceptors. **(a)** Phase-contrast photomicrograph of chicken visual cells. (From Meyer and Cooper, 1966). *A*, Double cone seen from side. Large oil droplet vacuole resides in chief cone immediately external to ellipsoid (*e*). *B*, Single cone with oil droplet vacuole; *C*, accessory; *D*, chief components of double cone, each with oil droplet vacuole: oval in accessory (*arrow*), round in chief cone. *E*, rod; *F* and *G*, single cones. Note large, oval, unstained paraboloid (*p*) occupying most of inner segment of accessory cone. Rod hyperboloid is not in focus. Unstained, Kolmer's fixative, bleached. ×2,000. **(b)** Phase-contrast photomicrograph of double cone and rod in chicken. Arrow indicates rod hyperboloid. Same preparation as **a.** **(c)** Periodic acid-Schiff preparation of chicken retina showing positive reactions (glycogen) in rod hyperboloid (upper arrow) and accessory cone paraboloid (lower arrow). ×82. **(d)** Electron micrograph of hyperboloid (*H*) and paraboloid (*P*) in rod and accessory cone, respectively, of Japanese quail retina. Note smooth endoplasmic reticulum in paraboloid and absence of membrane system in hyperboloid. ×12,000.

myoid represents the principal site of protein synthesis in visual cells; most of the ribonucleic acid is localized here. Further compartmentation may involve the presence of brightly colored or colorless oil droplet inclusions at the distal extremities of the ellipsoids (Figures 2–4a and 2–4b) and characteristically shaped glycogen deposits within the myoids (O'Rahilly and Meyer, 1963) (Figure 2–4c).

Rods. Rods are concerned with dim-light vision (scotopic) and are very sensitive to light, functioning primarily therefore at low levels of illumination. In birds the rods are characterized by relatively large, cylindrical outer segments (Figures 2–4a and 2–4b), the visual pigment-containing discs of which undergo continual formation and destruction (Meyer et al., 1973).

Avian rod inner segments lack oil droplets, but can be easily recognized in light microscopic preparations by applying the periodic acid-Schiff (PAS) reaction (Figure 2–4c), which selectively stains the glycogen-containing *hyperboloids* residing in the myoids (Meyer and May, 1973). As the name implies, the myoid has the ability to shorten (contract) or elongate (relax) in response to illumination and in coordination with the movement of pigment into and out of the processes of the pigment epithelium surrounding the visual cells' outer segments. In the light, the rod myoids elongate and the cone myoids (except the accessory member of the double cone) contract concomitantly with the migration of pigment toward them. Opposite reactions occur in the dark. These photomechanical changes, which represent adaptations for both diurnal and nocturnal vision, are quite rapid and extensive in birds.

Cones. Cones serve bright-light vision and function in both color discrimination and high visual acuity. In birds, the cones are distinguished by the presence of oil droplets (most often brightly colored, but also colorless), and relatively short, conical outer segments (Figure 2–4a).

Avian single cones (Figure 2–4a) contain a relatively large, circular oil droplet of variable coloration. It is red or yellow in weaverbirds (Mayr, 1972b) and red or yellow-green in Bonnelli's eagle and the common kite (Gallego et al., 1975), whereas in the chicken it has been reported to be red or yellow (Schultze, 1867; Morris and Shorey, 1967), red or yellow-green (Gallego et al., 1975), predominantly red (Meyer and Cooper, 1966), or a multitude of colors.

Double cones, present in all vertebrate groups except placental mammals, represent one of the most obscure elements of the vertebrate retina (Walls, 1942). Each consists of two intimately associated but independent cones that are dissimilar in size, shape, and structure: a tall, thin *chief cone* and a broad short *accessory cone* (Figure 2–4a). They are best visualized when both components lie side by side in the plane of section (Figure 2–4a; C, D) rather than perpendicular to it (Figure 2–4a; A). The significance of double cones is not known, although many theories have been advanced including the obvious economical advantage for both metabolism and space. They do respond differently to light; the chief cones contract, whereas the accessory cone is unresponsive. In the chicken, they are twice as numerous as single cones throughout the retina, and equal in number to rods in all regions except a circumscribed area posterior to the pecten (Meyer and May, 1973). According to Matsusaka (1963a), these double cones occupy four times the area of a single cone and 15 times that of a rod. At the level of their inner segments, he calculates that 82% of the retinal area consists of double cones. In the great tit (*Parus major*), the distribution of single and double cones assumes a definite pattern, each single cone being surrounded by four double cones (Engström, 1958). Between each group are rods and other single cones.

The chief cone is the tallest and thinnest of the components of the double cone. It always possesses a large circular oil droplet which in the chicken has been described as yellow (Schultze, 1867; Hoffmann, 1877; Meyer and Cooper, 1966; Gallego et al., 1975) or green (Morris and Shorey, 1967). In weaverbirds it is green (Mayr, 1972b); in two nocturnal species (Carina *noctua* and *Asio flammeus*), it appears yellow orange.

The accessory component of the double cone is readily identified in PAS preparations owing to the conspicuous barrel-shaped deposit of glycogen (*paraboloid*) within its myoid (Figure 2–4c). The presence of an oil droplet, however, has been very controversial. It has not been observed in most birds, including the pigeon (Cohen, 1963) and weaverbirds (Mayr, 1972b), but in the chicken, it has been pictorially documented from light microscopy as a small, oval, yellowish-green droplet (Meyer and Cooper, 1966). In the Japanese quail, from one to five small droplets constitute the accessory cone oil droplet (Hazlett et al., 1974); in Bonnelli's eagle and the common kite, it is small and colorless (Gallego et al., 1975).

Oil Droplets. Oil droplets are not confined to birds but have also been described in amphibians, reptiles, and nonplacental mammals. Each colored droplet consists of one or more stable carotenoid pigments dissolved in a mixture of lipids, the majority of which are neutral (Hudson et al., 1971). Astaxanthin has been repeatedly substantiated as the carotenoid in the red droplets, whereas considerable controversy still exists with respect to the identification of the pigments found in the other droplets (Meyer et al., 1965).

Almost every color of the visible spectrum has been reported in the cone oil droplets of birds, but it is now generally accepted that many of these colors, particularly those in the blue-violet range, are merely the result of optical phenomena elicited by crude lenses used in the past (see Walls and Judd, 1933). In diurnal birds they are generally colored red, orange, yellow, or green, whereas in nocturnal species they appear colorless or pale yellow. In the chicken, Meyer and Cooper (1966) observed and documented pictorially that the three different colored droplets characteristic for this species are associated with specific cone types: red in the single, yellow in the chief component of the double cone, and yellowish-green in its accessory member. Mariani and Leure-duPree (1978) recognized a similar, but slightly modified relationship in the pigeon: red, orange, or yellowish green in single cones, yellow in chief, and colorless in accessory cones.

These colorful oil droplet mosaics have aroused considerable interest regarding the possible role they play in the visual process of birds. They have been implicated as protective screening devices against ultraviolet light (Goldsmith, 1980) and as color discriminatory devices, or more simply as band-pass filters for absorbing visible light rays. In combination with different visual pigments, they have recently been shown to increase color contrast by cutting off the short wave branch of the pigment sensitivity curve and displacing the sensitivity maximum to longer wavelengths (Barlow, 1982; Govardovskii, 1983), the net result being an improvement in color discrimination. Oil droplet pigments are not essential for color vision, however. Behavioral studies utilizing Japanese quail with experimentally produced carotenoid-free (colorless) oil droplets (Meyer et al., 1971) have reported normal color discrimination (Kovach et al., 1976; Duecker and Schulze, 1977). Since animals are unable to synthesize these pigments, such carotenoid-free retinas are easily produced and maintained in newly hatched quail by the exclusion of carotenoid pigments from the diet of their parents (Meyer, 1971).

Glycogen Bodies. The rod hyperboloids and accessory cone paraboloids (Figure 4c) have been regarded as possible adaptations for the improvement of visual acuity. Considered as refractile bodies by early investigators, it is not well established that they are composed primarily of glycogen particles, although the ultrastructural organization (Figure 2–4d) and the chemical reactivity of these bodies are not the same (Cohen, 1963; Eichner and Themann, 1963; Matsusaka, 1963a, b).

Morphological changes have been observed in these bodies during both light and dark adaptation (van Genderen Stort, 1887). In the light, accessory cone inner segments display an increase in paraboloid membrane density (Matsusaka, 1963a) and contract, causing a swelling of the paraboloids and permitting a greater amount of light to be concentrated upon, and absorbed by, the cone outer segments. It must be emphasized, however, that most authorities regard these glycogen concentrations as merely energy sources for visual cell metabolism. Both glycogen synthetase and phosphorylase activities have been demonstrated within the paraboloid (Amemiya, 1975), and their glycogen content has been shown to decrease in the light (Eichner and Themann, 1963).

Visual Pigments. The outer segments of the photoreceptors contain the visual pigments, which are responsible for the absorption of light incident upon the retina. These pigments are conjugated proteins, consisting of an opsin prosthetic group joined by means of a Schiff-base linkage to a Vitamin A chromophore. The chromophore, which may be based either on the Vitamin A_1 aldehyde (retinal) or on the vitamin A_2 aldehyde (dehydroretinal), is always present in the native pigment in the 11-cis configuration. The action of light is to isomerize the chromophore from the 11-cis to the all-trans form, thereby initiating a complex sequence of conformational changes that eventually results in excitation of the photoreceptor.

The wavelengths of light most efficiently absorbed by visual pigments of different birds, i.e., their peak absorbance (λ_{max}), have been tabulated by Sillman (1973) and reflect the type of visual pigment utilized by these birds. For example, those with maxima around 500 nm are considered most likely to be rod pigments, whereas pigments in the general range of 560–575 nm belong to cones. The first cone pigment, *iodopsin,* was isolated from the chicken retina by Wald (1937) and characterized as having an absorbance peak of 562 nm by Bliss (1946) and Wald et al. (1955). Subsequent attempts to extract the cone pigments have been frustrating, although Crescitelli et al. (1964) succeeded in identifying iodopsin in the turkey. Utilizing a refined technique of microspectrophotometry, however, three different cone pigments have been detected in the pigeon (Bowmaker, 1977) and tawny owl (Bowmaker and Martin, 1978), and two in the chicken (Bowmaker and Knowles, 1977). Iodopsin-like pigments have also been found in the emu and tinamou (Sillman et al., 1981).

The presence of one or more cone pigments within the avian retina is important in understanding the mechanism of color vision in birds. Extensive behavioral and electrophysiologic data have already established this function, but have not explained the mechanism involved. The integration of one cone pigment with three different oil droplet colorations has been considered a plausible mechanism to provide three unique spectral absorbances (Schultze, 1866; Wald and Zussman, 1938) and account for color discrimination in some birds. However, combinations of several cone pigments and oil droplet colorations are now known to exist in several birds, which no doubt implies a different color discriminatory mechanism.

In contrast to cone pigments, rod pigments are readily extracted, and therefore substantially more information is available with respect to their distribution and properties. All of the prevalent avian rod visual pigments examined either in solution or microspectrophotometrically absorb maximally at 500–506 nm. Thus, they are relatively constant in their spectral location and do not seem to vary with the behavior or environment of the species.

Areas and Foveas. Improved visual acuity in birds may be correlated with the presence, within the cone-rich avian retina, of well-developed areas and foveas (often one or two). These so-called "areas of acute vision" represent circumscribed thickenings of the sensory retina involving thinner and longer visual cells that improve the resolving power, combined with a concomitant increase in the number of bipolar and ganglion cells. Areas such as these are present in some representatives of all vertebrate classes, but not all, and attain their greatest perfection in diurnal birds, where almost always one, and sometimes two or even three, distinct areas may be developed. The most common area is located in the central region of the fundus slightly above and at the nasal side of the optic nerve (and thus pecten) (Figure 2–1), and is termed the *central* (nasal, principal, posterior) *area;* a second area occupies a more lateral, or eccentric position, temporal to and slightly above the optic nerve and is designated as the *lateral* (temporal, eccentric, anterior) *area.* The shape of these two areas may be oval or circular, whereas a third area (*linear area*) has the form of a long band or ribbon extending horizontally for a variable distance across the central part of the retina.

A depression or fovea is generally observed within either a central or lateral area or within both. An area may exist without a fovea, but a fovea is only found within an area. Foveas are caused by a radial displacement of the more internal layers of the retinal area, resulting in a shallow saucer-shaped or deeper funnel-shaped (convexiclivate) cavities. Visual cell density is greater in the fovea than elsewhere in an area. Cell density counts of photoreceptor and ganglion cells in the avian central fovea have revealed the lowest convergence ratio (receptors per ganglion cell) clearly at this site (Fite, 1974); a one-to-one relationship between cones and ganglion cells apparently exists here.

The central fovea in swift-flying diurnal birds of prey has a deep, well-excavated pit (Figure 2–5) containing extremely thin cones, which in some hawks and eagles produce a resolving power at least eight times greater than that of man (Walls, 1942).

Several functions have been attributed to the fovea. Walls (1937, 1940, 1942) and Snyder and Miller (1978) contend that the margins of the fovea, particularly the convexiclivate type, act as a convex lens and magnify the retinal image (Figure 2–5), thus providing improved resolution. Therefore, the deeper the fovea, the greater its magnifying power. Pumphrey (1948b), however, questions its magnifying effect. He believes that such a fovea is capable of the maintenance of accurate fixation and the sensitive appreciation of angular movements of a fixed object. More recently, Harkness and Bennett-Clark (1978) characterized this fovea as a sensitive directional focus indicator.

The shallow, saucer-like fovea involves minimal refraction and functions in binocular vision only (Pumphrey, 1948b); it cannot tolerate aberration and loss of definition (Pumphrey, 1948a). Kreithen and Keeton (1974) presented evidence that it functions in the pigeon for the detection of polarized light and is therefore indirectly involved in nagivation.

The complete absence of an area is rare, although Walls (1942) stated that areas are often absent in do-

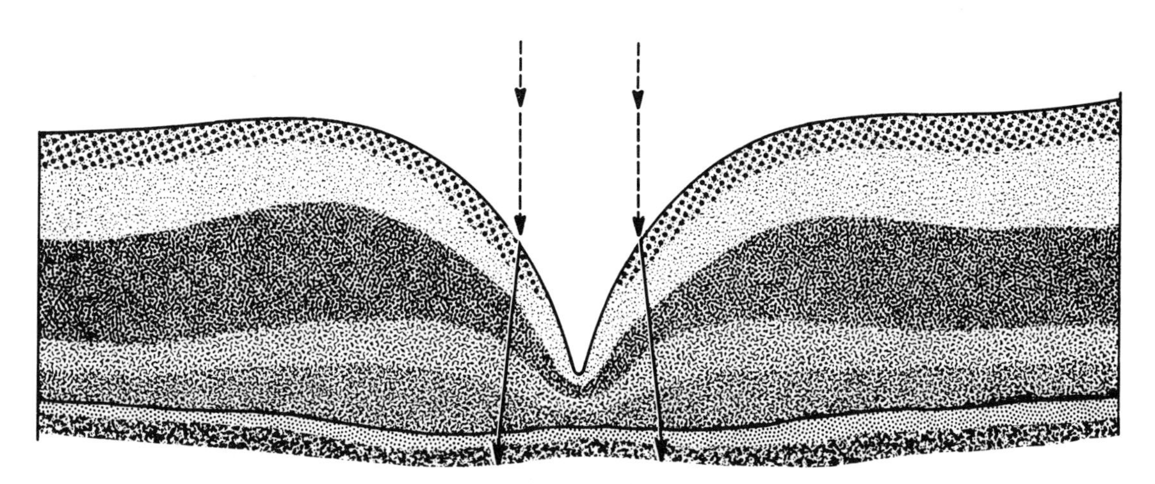

FIGURE 2–5. Magnifying effect of convexiclivate fovea as depicted by Walls (1942). Arrows represent incident light, which, when falling upon the fovea, becomes refracted laterally resulting in enlargement of image at visual cell level. Foveal depression and retinal stratification as theorized by Rochon-Duvigneaud (1943) in *Astur palumbarius.*

mestic birds. Nevertheless, Morris (1982) and others reported a central area in the chicken, and Meyer and May (1973) detected the greatest concentration of chicken cones in a topographical region corresponding to the site of a lateral area.

A central area is characteristic of most birds and most often contains a fovea, which is generally conceded to be involved in monocular vision. It functions independently of the central fovea of the other eye and is apparently of use to flying birds only for seeing and avoiding obstacles. Double cones predominate over single cones in the fovea of several members of the weaverbird family (Mayr, 1972b), whereas blackbirds and starlings lack double cones (Oehme, 1962). In diurnal birds, foveal rods have been reported by some investigators (Fritsch, 1911; Oehme, 1961; Fite, 1973) and denied by others (Slonaker, 1918).

When both central and lateral areas are present each contains a fovea; this is characteristic of many birds that obtain their food "on the wing" (e.g., passerines, hummingbirds, kingfishers, and bitterns) and thus require accurate perception of distance and speed. The location, depths, and relative positions of the two foveas exhibit considerable variation depending on the species. The central area and its foveal depression are usually better developed than its lateral counterpart. In birds of prey, the lateral fovea provides direct forward vision for its rapid descent toward its prey (Delage, 1920). In the kingfisher, which relies on both aerial and aquatic vision, the lateral fovea is positioned very eccentrically and seems to function in binocular vision under water (Kolmer, 1924).

A ribbon-like or linear area is characteristic of many avian species. Several specialized functions have been attributed to it: the minimization of eye movement (Walls, 1942), movement detection (Pumphrey, 1961; Munk, 1970), spatial orientation (by determining the normal position of the eye with respect to the horizontal plane) (Duijm, 1958), and the localization of the clear visual horizon (Pennycuick, 1960).

Linear areas can occur alone (shearwaters, flamingos), but are most often associated with a central area (and fovea) and sometimes with both central and lateral foveas. The presence of three distinct retinal areas (central, lateral, and linear), two of which (central and lateral) possess a fovea, is a unique avian adaptation that permits the formation of three separate and distinct visual fields (*visual tridents*) (Figure 2–6), two lateral monocular fields (one for each eye), and a central binocular field, the linear area no doubt enhancing the ability to follow moving objects. Such an arrangement is characteristic of diurnal birds of prey, which require a very high degree of visual acuity and stereoscopic vision, e.g., eagles, hawks, falcons, vultures, and buzzards. Rochon-Duvigneaud (1920) theorized that in these bifoveate conditions both foveas function independently, constituting a sort of compass by giving separate points of distinct vision: This is a marvelous apparatus for the estimation of distances.

Pecten

The complete avascularity of the avian retina appears to be associated with the presence of a bizarre supplemental nutritional device—the pecten oculi. Arising from the site of exit of the optic nerve and projecting a variable distance into the vitreous body in the ventral and temporal quadrant of the fundus (Figure 2–1),

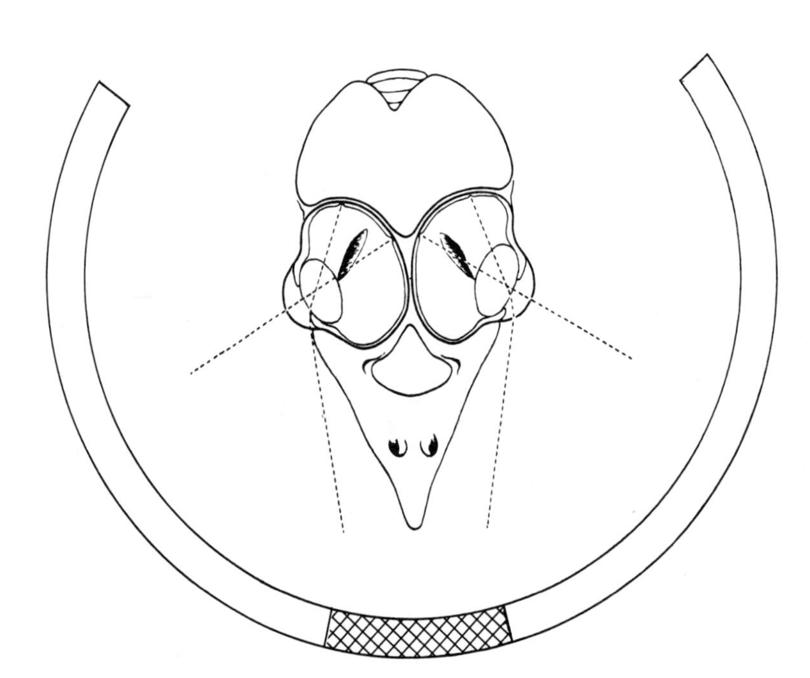

FIGURE 2–6. Projection of visual fields of bifoveate avian eye indicating visual trident. (After Walls, 1942.) Head and eyes are those of barn swallow showing central and lateral foveas with their visual axes (modified from Polyak, 1957) and visual fields: binocular (cross-hatched) for two lateral foveas, monocular for each central fovea.

the pecten consists almost exclusively of capillaries and extravascular pigmented stromal cells. Lacking both muscular and nervous tissues, it is unique to birds and has been implicated in many functions, the most plausible of which is the nutrition of the inner layers of the avascular avian retina.

The size, location, and morphology of the pecten, as well as the relative location of the fovea, has been examined in a variety of avian species by Wood (1917), who provided beautifully reproduced ophthalmoscopic pictures of the avian fundus. Three morphological types of pectens are discerned: conical, vaned, and pleated.

Conical Pecten. The conical pecten has been described only in the kiwi (Lindsay-Johnson, 1901; Wood, 1907). Resembling the conus papillaris of reptiles, it arises trumpet shaped from the center of a circular optic disc and extends into the vitreous as a heavily pigmented (brown-black) cone, without pleats or vanes, nearly as far as the lens. Walls (1942) believed it is degenerative.

Vaned Pecten. The vaned pecten occurs in the rhea (Vrabec, 1958), ostrich (Walls, 1942), and tinamou (personal observation). It consists of a central vertical lamina, which takes origin from a linear optic disc, and a series of lateral vanes (25–30) that arise from it (Meyer, 1977).

Pleated Pecten. The pleated pecten is by far the most common, found in most avian families including those of the cassowaries (Kajikawa, 1923) and emus (Vrabec, 1958). It arises from a linear optic disc as a single accordion-pleated lamina held in place at its free (apical) end (except in a few species, e.g., owls) by a more heavily pigmented crest or bridge of tissue. The mass, shape, number, and arrangement of pleats, their extent of pigmentation, and relationship to the ventral ciliary body varies considerably in different birds. Within a particular species, however, these features remain relatively constant, so much so that individual birds can often be identified merely by the morphology of their pecten and its association with other ocular components.

The size of the pecten and the number of its pleats do not necessarily coincide with the size of the eye, but appear to be directly related to the behavior of the bird toward illumination and its general level of activity. Active, diurnal birds with high visual acuity and monocular vision tend to possess a larger and more pleated pecten, and nocturnal species with poor vision a smaller one of simpler morphology. When the pecten is unfolded (by cutting it from the optic disc and bridge), its length varies between one-half and one-third of the greatest diameter of the eyeball (Giebel, 1857). The highest number of pleats (30) has been recorded in *Corvus.* The singing birds have the most pleats (20–30) and nocturnal predators (e.g., *Caprimulgus, Podargus*) the least (3–4).

Functions of Pecten. Published accounts of the pecten first appeared about 300 years ago. Regarded as "one of the greatest puzzles in comparative ophthalmology" (Walls, 1942), many functions have been attributed to it, most of which lack critical experimental substantiation. In the most interesting survey, Wingstrand and Munk (1965) recognized 30 different functional theories, which they organized into a table of nine major groups provided with the most pertinent references.

Nutrition. According to the most popular and longstanding theory, the pecten acts as a nutritive organ, providing oxygen and nutrients for the inner portion of the retina. Its richness in capillaries, its pleated arrangement to increase its surface area, and its advantageous intraocular location in close proximity to a retina devoid of blood vessels were ample grounds to persuade early investigators of this important function. Convincing evidence that fluids (e.g., fluorescent markers) can readily pass into the vitreous from the pecten (Bellhorn and Bellhorn, 1975) supports this supposition, although experimental evidence that nutrients derived from the pecten are important for the maintenance of the retina has been controversial. Wingstrand and Munk (1965) showed that operative elimination of the arterial supply to the pecten not only results in a complete degeneration of this organ, but also produces well-developed retinal degenerative changes beginning with, and sometimes confined to, its innermost layers. They also detected an oxygen diffusion gradient within the vitreous sufficient to be metabolically significant for retinal maintenance. Experimental destruction of the pecten by photocoagulation (Francois and Neetens, 1974) and intraocular electrocautery (Brach, 1975), on the other hand, has had no visible effect on the histology of the retina. Brach (1975), who claimed to have produced total ablation of the chick pecten, contended that the retinal atrophy observed by Wingstrand and Munk (1965) is a result of operational damage to the optic nerve axons owing to accidental destruction of the blood supply to the optic nerve head. He believed, therefore, that the pecten is not concerned primarily with the nutritional supply of the retina.

Secretion. Secretory function has been attributed to the pecten, primarily involving stromal cells (Lindsay-Johnson, 1901; Fischlschweiger and O'Rahilly, 1968). Glycosaminoglycans have been detected histochemically as one of the secretory products (Agarwal et al., 1966).

Regulation of Intraocular Pressure. The fact that fluid may be given off into the vitreous body by the pecten has led some investigators, e.g., Kajikawa (1923) and Mann (1924a, b), to believe that the pecten is involved in the regulation of intraocular pressure, a function that is normally attributed to the ciliary body because of its continual production of aqueous humor.

Light Absorption. The dense packaging of melanin pigment granules within the stromal cells has also implicated the pecten in a light-absorbing function, which would reduce internal reflections that may possibly interfere with the production of a clear image (Hanzely et al., 1975).

Vision. Many believe that the pecten is concerned with the perception of movement. Agreeing with Mann (1924a, b) that the pecten casts a shadow upon the retina, Menner (1938) maintained that these shadows are never completely symmetrical, but are of significant dimensions to influence retinal responses and that the pecten and its shadow increases the retina's sensitivity for the perception of movement. In support of this theory, Crozier and Wolf (1944a, b) and Thomson (1928, 1929) suggested that the pecten could serve to orient birds in space by acting as a sextant, casting a shadow upon the retina and permitting the latter to estimate the angular position and movement of the sun.

The pecten has been implicated in many other functions for which there is no apparent anatomical basis, e.g., an intraocular shade against the glare of the sun (Barlow and Ostwald, 1972) and a magnetic sensor during orientation (navigation) (Danilov et al., 1970).

The Avian Ear and Hearing

D.B. MEYER

Introduction

Birds have long been known for their keen sense of hearing and high degree of equilibration. Their excellent voice production and remarkable ability to imitate sounds has inferred an exceptional degree of sound analysis (pitch discrimination) within a wide range of auditory frequencies. Their aerial mode of life, in particular, has demanded this sonic acuity, in addition to a well-coordinated balance and position sense. The avian ear carries out these essential functions in a manner closely resembling the mammalian ear, although it is simpler in structure and reptilian in design. To understand how the ear, or any organ for that matter, can carry out its functions demands first of all a thorough and intimate knowledge of its structural components, as well as physical interrelationships that can theoretically occur. Fortunately, considerable information is available on the anatomy of the avian and mammalian ear, and this has provided the impetus for the interesting speculations on the possible functional mechanisms involved in the mechanoreception of both sound and position. To appreciate these, the structure and function of the mammalian ear is conveniently employed as an instructive yardstick for comparison, even though such comparisons stress differences rather than similarities (Pumphrey, 1948). These differences in structure necessitate a meaningful descriptive terminology geared toward avian anatomy. An International Committee on Avian Anatomical Nomenclature has standardized such a terminology (Baumel et al., 1979), and whenever possible, the terms recommended by it have been adopted in this chapter.

The ear, be it avian or mammalian, includes three separate but contiguous anatomic segments (external, middle, and inner) (Figure 2–7), which develop from completely independent and different embryonic primordia and yet combine intimately to form a synchronized functional unit. The *external ear* serves to collect the sound waves from the outside air and conduct them to the middle ear. For this purpose, it consists of a collecting device (the *auricle* or pinna in man) and a simple conducting tube, the *external acoustic meatus,* which develops as a simple epidermal invagination representing the first branchial cleft. It terminates medially as a partition, the *tympanic membrane* (eardrum), which completely separates the external ear from the middle ear. The middle ear is an air-filled, ossicle-containing space (*tympanic cavity*) that receives the sound waves as mechanical vibrations of the tympanic membrane in its lateral wall and transfers them in an amplified form to the inner ear at its medial wall.

In contrast to the mammalian middle ear, which utilizes three ossicles (malleus, incus, and stapes) to affect such sound transfers, birds accomplish this function with a single skeletal element, the *columella,* which extends medially across the tympanic cavity to form a direct connection between tympanic membrane and the fluid within the inner ear. The tympanic membrane and columella thus function as a mechanical transformer that matches the impedance of air and inner-ear fluid, facilitating the transfer of sound energy (Wever et al., 1948). The columella develops completely separately from the other ear anlagen as a mesenchymal condensation (probably with neural crest contributions) within the second branchial arch. The tympanic cavity, which eventually comes to house it, represents the persisting first (and possibly second) pharyngeal pouch.

The *inner ear* is responsible for the initial analysis and characterization of the sound vibrations, as well as for maintaining equilibrium. It consists of two very complex, fluid-filled components or labyrinths, one membranous, the other bony. The *membranous labyrinth* constitutes the diversified sense organ. It is a completely enclosed, epithelial-lined series of spaces filled with *endolymph* and containing several localized sensory areas of thickened epithelium in the form of spots (*maculae*), ridges (*crests*), or elongated nipples (*papil-*

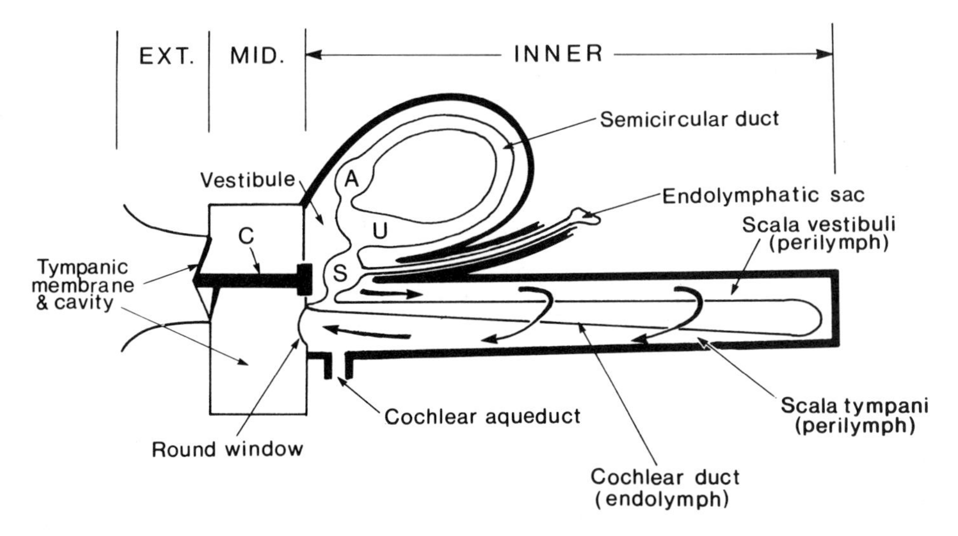

FIGURE 2–7. Schematic representation of the organization of the avian ear. *A,* Ampulla; *C,* columella; *S,* saccule; *U,* utricle. Arrows indicate course of perilymph displacement brought about by vibrations of columellar footplate. (Modified from von Békésy, 1960.)

lae). The complete epithelial lining including these sensory areas is derived from the embryonic otic vesicle, which forms the definitive components of the membranous labyrinth: the *three semicircular ducts, utricle, saccule, endolymphatic duct and sac, utriculosaccular duct, sacculocochlear duct,* and *cochlear duct* (Figure 2–8). This completely self-contained epithelial complex is surrounded by fluid, the *perilymph,* which in turn is enclosed in a shell of spongy bone, the *osseous labyrinth* (Figure 2–9). The osseous labyrinth closely follows the contours of the membranous semicircular and cochlear ducts, but forms a common chamber, the vestibule (Figure 2–7) for the utricle and saccule. To avoid confusion, the osseous semicircular channels are termed *canals,* and the membranous channels, *ducts.* The part of the osseous labyrinth enclosing the membranous cochlear duct is called the cochlea.

The perilymph-filled vestibule is a very important component of the auditory apparatus. Its lower, lateral wall houses the oval window with its columellar footplate, and its lower, anterior part is continuous with a perilymph channel (the scala vestibuli), which traverses the cochlea in close proximity to the cochlear duct (Figure 2–7). In this way, mechanical vibrations received by the tympanic membrane and transported to the oval window by the columella are converted into fluid displacements, which are eventually transmitted to the sensory cells of the cochlear duct via the perilymph within the vestibule and its scala.

In birds the voluminous development of the eye has influenced the shape and position of the inner ear (Werner, 1963), which is displaced far caudally and reaches almost into the posterior cranial wall. The semi-

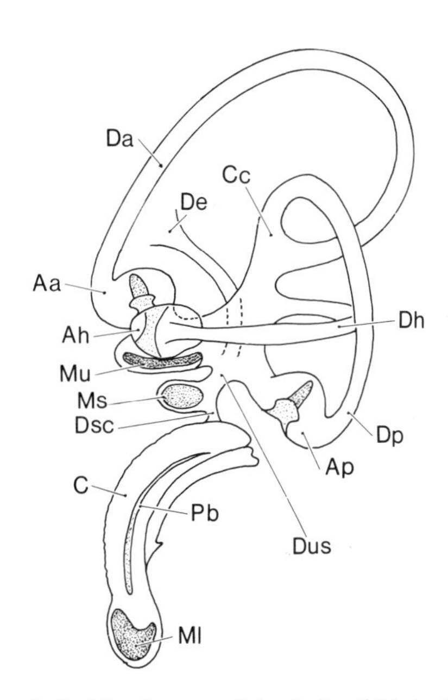

FIGURE 2–8. Membranous labyrinth of birds (lateral view, left side) showing relative positions of sensory areas (ampullary crests, vestibular maculae, and basilar papilla). *Aa, Ah, Ap,* Ampullae: anterior, horizontal, posterior, respectively; *C,* cochlear duct; *Cc,* common crus; *Da, Dh, Dp,* Semicircular ducts: anterior, horizontal, posterior, respectively; *De,* endolymphatic duct; *Dsc,* sacculocochlear duct; *Dus,* utriculosaccular duct; *Ml, Ms, Mu,* maculae: lagenar, saccular, utricular, respectively; *Pb,* basilar papilla. (After Schwartzkopff, 1973; combined from Werner, 1939, and Lüdtke and Schölzel, 1966).

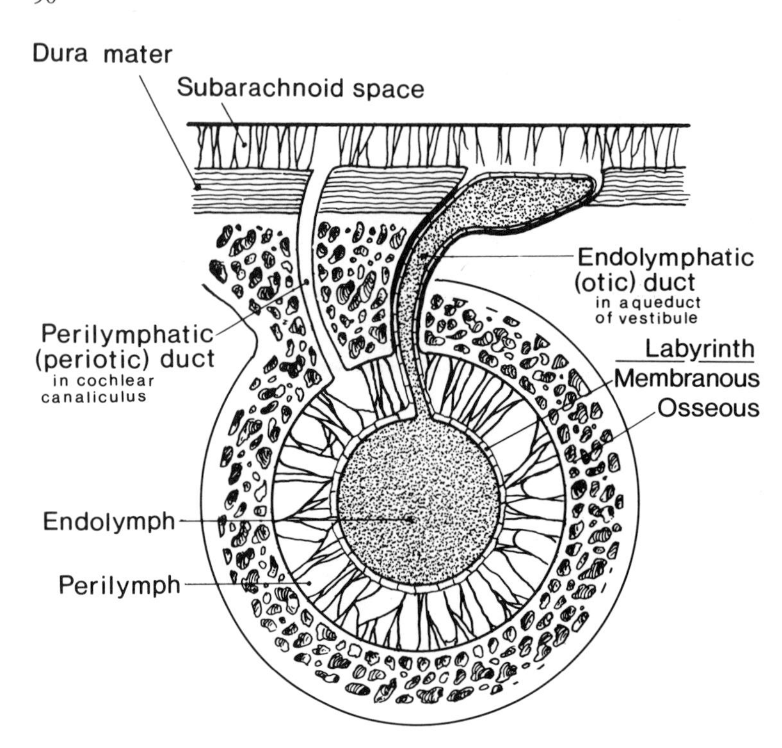

FIGURE 2–9. Basic scheme of inner ear. Note epithelial lining (from otocyst) of membranous labyrinth separating endolymph from perilymph. (Modified from Gardner et al., 1975.)

circular ducts become extended more laterally and dorsally, and the cochleae are elongated ventrally so as to approach one another closely.

Structure of the Ear

External Ear. The external ear of birds is inconspicuous and lacks an auricle. Instead, it usually possesses specialized feathers, ear (skin) flaps, and opercula. Auricular feathers are absent in only a few avian groups (Struthioniformes, some gallinaceous species, vultures) (Freye, 1952–1953; Schwartzkopff, 1973). In addition to minimizing turbulence during flight and protecting the ear opening itself, the feathers are modified to reflect sound and are able to enhance sound collection by forming an ear funnel directed toward the meatal opening (Freye, 1953). Ear flaps may likewise form sound-gathering funnels (Stellbogen, 1930), as well as protecting the ear by blocking the opening, e.g., during crowing (Stresemann, 1934). An especially well-developed anterior ear flap, or *operculum,* provided with skeletal muscle is found in some owls. According to Payne (1971), movements of these flaps alter the overall directional sensitivity pattern by redirecting the regions of maximum sensitivity.

According to Schwartzkopff (1968), several Russian investigators have found interesting correlations between auricular feather structure and ecological performance. The width of the feathers, for example, is related to sound localization and their fine structure varies in accordance with auditory efficiency. In some birds, the right and left ears may differ in height and direction of their external openings. Such ear asymmetry is a remarkable adaptation, which may enhance the localization of sound and thus assist nocturnal birds in their quest for prey (Stresemann, 1934; Pumphrey, 1948).

Schwartzkopff (1962) believed that movements of the ear flaps associated with such ear asymmetries play a very important role. Time and loudness differences in the sound reaching both asymmetric ears (of owls) provide the basis for sound location (Schwartzkopff, 1955b).

External Acoustic Meatus. The external acoustic meatus of birds generally is a short, curved tube with an almost circular or oval external opening. Its total form, length, and diameter vary with the species. In the chicken, duck, and goose, for example, it measures approximately 12–14 mm in length (Freye, 1952–1953); in diving birds it is especially narrow (Schwartzkopff, 1955b) and capable of being closed to afford direct protection against violence and external pressure (Pohlman, 1921).

Tympanic Membrane. The tympanic membrane (eardrum) represents an important constituent of both the external and middle ear since its development involves a close union of the terminal portions of their tubular anlagen. The external surface of the membrane is covered by epidermis directly continuous with the lining of the external acoustic meatus, whereas the internal surface is lined by an epithelium continuous with that lining the tympanic cavity and pharyngotympanic tube. In birds, the tympanic membrane is well developed, but variable in size, thickness, and shape (Freye-Zumpfe, 1952–1953) depending on the species, and somewhat inclined from the median plane. It projects outward into the meatus like a flat cone, giving it a convex appearance caused by the central protrusion of the columella (Figure 2–7). Sound waves from the meatus impinging upon this convex membrane cause it to vibrate at a frequency determined by the frequency of the sound waves and at an amplitude determined by the intensity of the sound.

Middle Ear. The middle ear (Figure 2–7) an air-filled tympanic cavity lined by an epithelium that also covers its contents: a simple columellar apparatus, a single muscle, and ligaments, as well as the tympanic membrane and cochlear (round) window (accessory tympanic membrane).

Tympanic Cavity. The middle ear cavity is directly continuous with the pharynx via the pharyngotympanic tube, and communicates with a large group of accessory air cavities occupying the surrounding skull bones and even extending into the mandible (Portmann, 1950) and beak (Tiedemann, 1910) in some species. Right and left tympanic cavities often communicate with each other via interconnecting air sinuses (Pohlman, 1921; Wada, 1924; de Burlet, 1934), a feature that is strongly developed in owls and has been implicated in the transference of pressure fluctuations (Stresemann, 1934) emanating at the round window (Pumphrey, 1961).

Columella. The internal surface of the tympanic membrane is in intimate contact with the columella, which extends across the tympanic cavity at variable angles, depending on the species, and ends at the oval window. The columella consists of two segments, a lateral cartilaginous *extracolumella* and a medial, bony *columella proper* (Figure 2–10). The extracolumella consists of three cartilagenous processes that are at right angles to each other and resemble a tripod.

The columella proper consists of a rodlike shaft or body (*stapes*) that is continuous laterally with the cartilaginous extracolumella and terminates medially in a flat, oval base or footplate occupying the vestibular (oval) window (Figure 2–10). The osseocartilaginous junction of the extracolumella and columella proper is abrupt.

Since the area of the tympanic membrane is always much greater than that of the footplate, the pressure generated at the oval window is substantially amplified to produce the relatively large pressure fluctuations required by the inner ear. The amount of such amplification (ratio of tympanic membrane to footplate) in a variety of birds is shown in Table 2–1. Additional data have been published by Schwartzkopff (1957) that indicate that the transformation quotients (in adult birds) vary between 15 and 40, values that correspond somewhat with data obtained from several mammals. Avian species with particularly sensitive hearing have higher ratios.

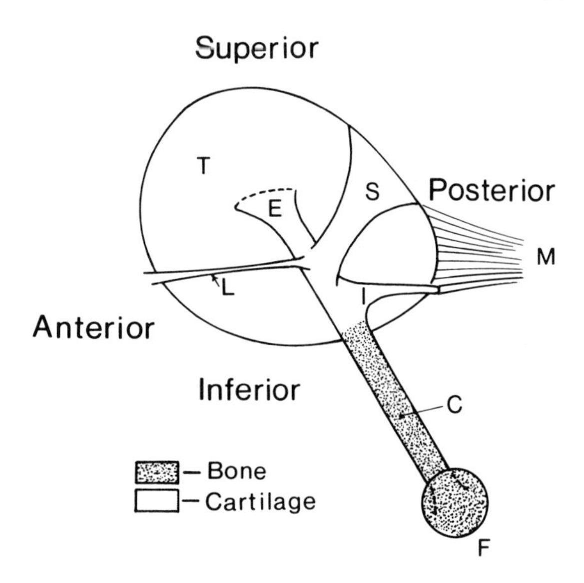

FIGURE 2–10. Diagrammatic representation of the avian columella and its relationship to the tympanic membrane. *C,* Columella proper; *E,* extracolumellar process; *F,* footplate of columella; *I,* infracolumellar process; *L,* columellosquamosal ligament; *M,* columellar muscle; *S,* supracolumellar process; *T,* tympanic membrane. (Modified from Hodges, 1974, and Pohlman, 1921.)

The general configurations of the avian columella are quite consistent, but their shape and size vary greatly with the species. The footplate is primarily oval, and its external surface (facing the inner ear) is species specific in its appearance.

Columellar Muscle. The single muscle of the middle ear arises from the occipital bone completely outside the middle ear cavity and enters the cavity by means of a foramen (Smith, 1904). It is inserted at the apex of the infracolumellar process and the posterior margin of the tympanic membrane, between this process and the supracolumellar process (Fig. 2–10). It is considered to be homologous to the stapedius on the basis of its branchial origin (second pharyngeal arch) and its innervation by the facial nerve.

Paratympanic Organ. Within the thin mucosal lining of the dorsomedial wall of the tympanic cavity of most birds (but not owls or parrots) is found a small, epithelial vesicle derived from the first branchial cleft (Vitali, 1911). Termed the *paratympanic organ,* it contains tall sensory cells in contact with nerve endings from the geniculate ganglion (Stresemann, 1934). It has been

TABLE 2–1. Body surface, tympanic membrane (movable part), and footplate areas and their ratios in some birds[a]

Species	Body area (= weight ⅔) (cm²)	Relative size of tympanic membrane/ body surface	Tympanic membrane area (cm²)	Columella base area (cm²)	Area ratio: tympanic membrane/footplate
Chiffchaff (*Phylloscopus collybita*)	4.0	0.020	0.078	0.0036	22
Willow warbler (*Phylloscopus trochilus*)	4.5	0.021	0.004	0.0034	28
Common tit (*Parus communis*)	5.0	0.018	0.089	0.0039	23
Blue tit (*Parus caeruleus*)	5.1	0.016	0.084	0.0032	26
Icterine warbler (*Hippolais icterina*)	5.7	0.015	0.086	0.0030	29
Black-cap warbler (*Sylvia atricapilla*)	6.6	0.019	0.126	0.0044	29
Barn swallow (*Hirundo rustica*)	7.4	0.010	0.171	0.0038	19
Chaffinch (*Fringilla coelebs*)	7.9	0.015	0.114	0.0041	28
Great tit (*Parus major*)	7.9	0.013	0.104	0.0042	25
Common bullfinch (*Pyrrhula pyrrhula*)	9.0	0.013	0.117	0.0048	24
House sparrow (*Passer domesticus*)	9.6	0.0094	0.091	0.0042	22
Blackbird (*Turdus merula*)	20.9	0.0077	0.160	0.0073	22
Black-billed magpie (*Pica pica*)	35.5	0.0075	0.265	0.0116	23
Carion crow (*Corvus corone*)	65.5	0.0053	0.347	0.0151	23
Long-eared owl (*Asio otus*)	44.9	0.0107	0.480	0.0120	40
Tawny owl (*Strix aluco*)	66.4	0.0089	0.593	0.0198	30
Pigeon (*Columba livia*)	47.8	0.0043	0.204	0.0116	14
Common gallinule (*Gallinula chloropus*)	41.7	0.0032	0.132	0.0078	16
Common coot (*Fulica atra*)	84.0	0.0025	0.209	0.0106	19
Mallard (*Anas platyrhynchos*)	82.5	0.0034	0.285	0.0109	26
Juvenile (10 days)	15.2	0.0054	0.082	0.0055	15
Common buzzard (*Buteo buteo*)	86.1	0.0039	0.330	0.0180	18
Great crested grebe (*Podiceps cristatus*)	86.0	0.0016	0.140	0.0095	16
Ring-necked pheasant (*Phasianus colchicus*)	113.0	0.0033	0.368	0.0133	28
Chicken (*Gallus domesticus*)	153.0	0.0019	0.291	0.0133	22
Juvenile (40 days)	27.6	0.0052	0.144	0.0083	17
Juvenile (1 day)	10.5	0.0066	0.069	0.0060	11
Common crane (*Grus grus*)	245.0	0.0017	0.418	0.0169	25

[a] After Schwartzkopff (1955b).

implicated as a sense organ, although its exact function has yet to be determined experimentally.

Inner Ear. The sensory portion of the inner ear is the membranous labyrinth, which includes several nonauditory receptive areas composing the *vestibular labyrinth* and a single organ of hearing, the *cochlear labyrinth*. The vestibular labyrinth of birds possesses mechanoreceptors capable of providing information to the brain regarding position and movement of the head. They also initiate reflexes involving muscles of eyes (vestibuloocular), neck, trunk, and limbs.

Vestibular Labyrinth. SEMICIRCULAR DUCTS. As in the human, three semicircular ducts (anterior, posterior, and horizontal) are present in birds (Figure 2–8) and serve as detectors of angular acceleration in their three spatial planes.

Each semicircular duct arises from and terminates into the utricle, and enlarges at a point close to the utricle to form a dilatation or *ampulla* (Figure 2–8). The ducts are oriented at right angles to each other; the *anterior* (rostral, superior) and *posterior* (caudal) lie in the vertical plane, and the *horizontal* (lateral) in the horizontal plane. Considerable variation exists in the overall configuration and position of the ducts in different bird species, (Turkewitsch, 1934; Werner, 1960); the axis of the beak is parallel to the plane of the horizontal semicircular duct in many species, whereas in others this axis lies at right angles to the same plane (Duijm, 1951).

The ducts are filled with a fluid, *endolymph,* which (in the pigeon, for example) has a specific gravity slightly greater than water (1.0017) and a twofold increase in viscosity (1.19 cp) (Money et al., 1971). Each duct occupies a corresponding bony *semicircular canal* (anterior, posterior, horizontal) belonging to the *osseous labyrinth* (Figure 2–9). The ducts occupy about one-fourth to one-third of the potential cross-sectional space of the osseous canal, the remainder being filled with endosteum and a supporting fibrous reticulum containing a circulating fluid, the *perilymph.* Pigeon perilymph has a specific gravity closely approximating water (1.0006) and a viscosity considerably less than that of endolymph (0.76 centipoises) (Money et al., 1971).

Each membranous duct is lined by a nonsensory, simple squamous epithelium throughout most of its extent, except within the ampulla. There it thickens to form a narrow, crescent-shaped elevation, the *planum semilunatum,* along its lateral wall, and a neuroepithelial covering for a prominent, elongated ridge, the *ampullary crest,* which occupies its floor.

The *plana semilunata* consist of nonsensory, tall columnar cells that have secretory capabilities and are involved somehow in the excitation of sensory hairs (Dohlman, 1959). The *ampullary crests* are perpendicular to the long axis of the ampulla. They represent the sensory areas of the semicircular duct system, which is specialized for the perception of fluid movements within the ducts. This is accomplished by sensory hair cells that are capable of evoking an impulse when adequately stimulated by a moving force. The *neuroepithelium* of the ampullary crest consists of sensory cells and supporting cells organized into stratified columnar epithelium.

Two morphological types of *sensory cells* are observed in the ampullary crests of birds: flask-shaped *Type I* and cylindrical-shaped *Type II* (Wersäll, 1956) (Figure 2–11). The apical margin (that facing the endolymph) of both Type I and Type II cells possesses a dense terminal web (cuticular plate) except on one side where two centrioles are present, one of which forms the basal body for an elongated true cilium, the *kinocilium.* A bundle of nonmotile sensory hairs (stereocilia) [70–80 in the chicken (Friedmann, 1962); 35–50 in the pigeon (Landolt et al., 1975)] arise from the terminal web, and together with the much longer kinocilium confer a common pattern to the sensory cells (Lowenstein and Wersäll, 1959). The hairs occupy parallel rows within a bundle, the row nearest the kinocilium containing the longest hairs. The length of the hairs within adjacent rows then decrease in steplike fashion with increasing distance from kinocilium (Figure 2–11).

The sensory hairs and kinocilia of both Type I and Type II cells are embedded in the cupula, a wedge-shaped, rather rigid gelatinous mass that covers the crest as it extends perpendicularly across the ampulla and reaches above it as far as the roof. Displacement of the endolymph caused by head rotation results in the movement of the cupula, and, at the same time, a bending or shearing of the attached sensory hairs and kinocilia. Three types of cupular motion have been described: swinging door, sliding, and drum membrane; they are discussed in detail by Landolt et al.,

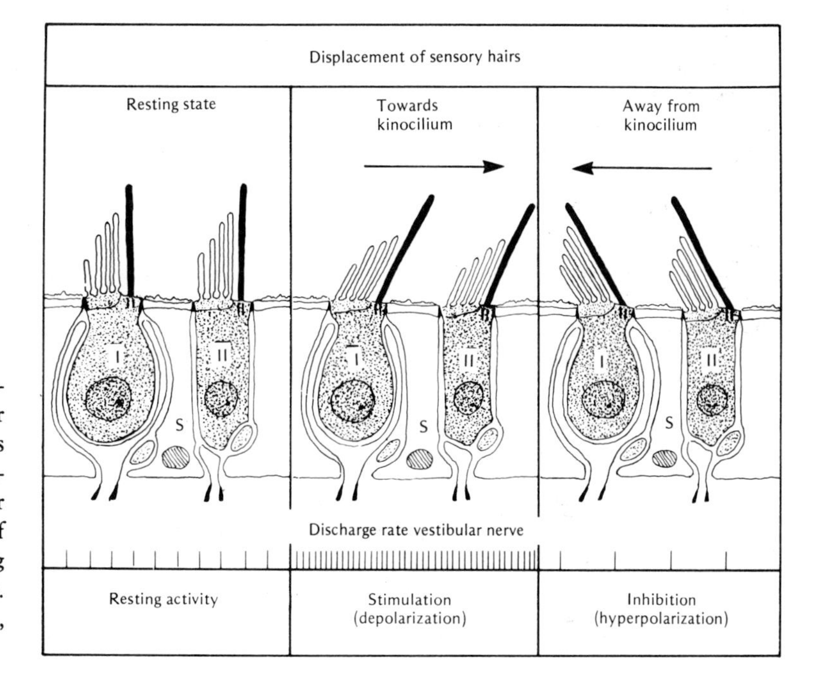

FIGURE 2–11. Diagrammatic representation of the vestibular hair cells (I and II), the nerve endings associated with them and the relationship that exists between hair cell orientation and the pattern of stimulation of the innervating nerve fibers. *S,* Supporting cells. (Modified from Wersäll et al., 1966, and Flock, 1971.)

(1975). The type of response will depend on the morphologic polarization of the hairs and kinocilia (Figure 2–11) as well as the angular velocity of head rotation in the plane of the semicircular duct; the latter, in turn, depends on the inertial movement of the endolymph. Since excitation of the sensory hair cells (*depolarizing response*) is believed to correspond to the displacement of the hair toward the kinocilium (Figure 2–11) (Wersäll et al., 1966), it follows that the hair cells within the *lateral* semicircular duct will be stimulated by displacement of the cupula toward the utricle (the kinocilia face in that direction), whereas displacement of the cupula away from the utricle will be inhibitory there (*hyperpolarizing response*) but excitatory in the vertically oriented ampullae. Therefore, head rotation that excites the right horizontal duct will, at the same time, inhibit the left, and excitation of the right *anterior* duct will be counteracted by inhibition of the left *posterior* duct. *Transduction,* i.e., the conversion of such mechanical information into nervous activity within the sensory hair cells, involves several components, e.g., the kinocilium, particularly its base (Hillman, 1969), the sensory hairs (Hudspeth and Jacobs, 1979; Hudspeth, 1982), the terminal web, and microtubules.

In the Type I cell, the afferent nerve forms a large cuplike ending, or *chalice,* which encloses most of its cell body (Figure 2–11) and usually many sensory cells (Rosenhall, 1970), whereas the Type II cell is contacted by numerous small nerve endings (both afferent and efferent) along their basal surface. Impulses received by the afferent terminals are carried along their peripheral dendritic axons toward their origin from the bipolar vestibular ganglion cells. Owing to the surrounding nerve chalice, efferent endings directed toward the Type I cells cannot contact their surfaces directly but form synaptic connections (varicose enlargements, boutons) with the nerve chalices themselves, or with their afferent dendrites. The efferent nerves originate in the medulla (Boord, 1960). Cellular processes initiated by these nerves may control the transduction mechanisms of the sensory hair cells (Ashmore and Russell, 1983) or be involved in postsynaptic inhibition (Flock, 1971).

The ampullary crest within the horizontally placed ampulla is in the form of a simple ridge, similar to that observed in man.

DARK AND LIGHT CELLS. In the ampullary crest and the floor of the ampulla, at least two cell types are present; dark (osmiophilic) and light (osmiophobic). The morphological features of these cells coupled with experimental data indicated to Dohlman (1964) and Dohlman and Boord (1964) that the dark cells absorb fluid and certain ions (Na^+ and K^+) and that the light cells secrete fluid and possibly other ions and substances (Dohlman, 1965) into the endolympyh. Thus, these cells undoubtedly serve to maintain the ionic composition and volume of the endolymph (Ishiyama et al., 1970).

UTRICLE. The utricle receives the openings of the three semicircular ducts and is in communication with the sacculus via the utriculosaccular duct, which opens at its base (Figure 2–8). The epithelium forming the wall of the utricle is very thin except at the expanded floor, which thickens to form the neuroepithelia for two sensory areas: the macula utriculi (containing types I and II cells) and the crista neglecta. Each macula (utricular, saccular, or lagenar) is characterized by the spatial distribution of sensory cell types, the direction of its morphological polarization and the location, as well as the cellular composition, of its striola (Table 2-2).

Statoconia (otoconia, otoliths) are perfect single crystals of calcite (Carlström et al., 1953), a form of calcium carbonate, which vary considerably in size depending on the species (Jørgensen, 1970). In the chicken they are formed in the endolymphatic sac and transported to the maculae via the endolymph (Vasquez, 1955; de Vincentiis and Marmo, 1966). Displaced statoconia are often seen in a degenerative state lying on the surface of dark cells that surround the macula (Lim, 1974; Hunter-Duvar, 1983). These dark cells, therefore, have been implicated as sites of decalcification of statoconia, a process which is perhaps involved in the homeostasis of calcium ions and the maintenance of the proper environment of the endolymph.

The *crista neglecta* has the appearance of a small flat crest covered with a thickened epithelium. It occurs in all vertebrate classes, including man, and is particularly well developed in owls (Money and Correia, 1972). It is relatively large in good flyers and small in poor ones (chickens, parrots). Its epithelium consists of randomly distributed sensory hair cells (Types I and

TABLE 2–2. Sensory cell distribution and polarization (arrows) in the pigeon macula[a]

Macula	Pars externa	Striola	Pars interna
Utricular	—Type II→	Two bands of Type I	←Type II—
Saccular	←Type I— Type II peripherally	Type I	—Type I →
Lagenar	←Type II—	Two bands of Type I	—Type II→

[a] After Rosenhall (1970).

II) with no apparent striola and supporting cells covered by a gelatinous membrane (Correia et al., 1974).

The function of the crista neglecta is not well understood. Correia et al. (1974) surmised that it is affected by vibratory motion of endolymph and is able to transduce this energy into an electrical impulse. The association of its nerve supply with the vestibular ganglion implicates a vestibular function, and an auditory function has been attributed to it in amphibians (Okajima, 1913). It is believed to be homologous to the ampullary crest (Kuwamoto and Altmann, 1967).

SACCULE. The avian saccule is considerably smaller than the utricle and quite variable in shape. Its upper posteromedial part gives off the endolymphatic duct, and its posterior part, the sacculocochlear duct (ductus reuniens). The epithelium forming its floor is thickened to form a typical macula, which occupies an inclined position (Werner, 1939).

Both Type I and II cells are found in the saccular macula. Several functions have been attributed to this macula. In addition to being considered a gravity receptor because of its statoconial membrane, it may play a role in eye movements (Benjamins and Huizinga, 1927; Money and Correia, 1972) and in regulating the production and pressure of endolymph (Werner, 1939).

LAGENA. The lagena represents the blind termination of the cochlear duct that is distal to, and completely independent of, the auditory sensory cells (basilar papilla). It is found in most vertebrates but is absent in mammals, except monotremes. Since the avian cochlear duct bends hooklike toward the middle of the cranial base, the two lagenae lie quite close to one another. The *macular region* of the lagena occupies the terminal area so that its long axis is oriented vertically, thus perpendicular to the plane of the utricular macula. Both Type I and II cells occupy the macula, but their distribution and concentration vary and differ, according to Jørgensen (1970) and Rosenhall (1970).

Considerable controversy exists regarding the function of the lagenar macula. Proponents of both an auditory and equilibrium function refer to the association of its nerve (lagenar) with the cochlear nerve and its axonal terminations within the cochlear and vestibular nerve centers of the brain (Boord and Rasmussen, 1963; Boord, 1969). Pumphrey (1948) believes it is responsive to low frequencies, and Schwartzkopff (1949) noted that its experimental removal causes disturbances of spatial and temporal movement coordinators.

ENDOLYMPHATIC DUCT AND SAC. The tubular endolymphatic duct terminates in the vicinity of the subdural space as an enlarged, flattened endolymphatic sac. In pigeons and sparrows, this membranous sac is lined by an epithelium consisting basically of two cell types, *light* and *dark*, which have been implicated in secretion (Vigh et al., 1963) and absorption, as well as in the cleaning up (phagocytosis) of the endolymph (Dohlman, 1966). Endolymph flows into the endolymphatic sac from the direction of the cochlear duct.

Cochlear Labyrinth. COCHLEA. The avian cochlea is a weakly curved osseous tube directed downward, forward, and inward, which serves to support the membranous cochlear duct (with its lagena) within its perilymph-filled cavity. It is considerably shorter (but wider) than the mammalian cochlea. Singing birds possess the shortest, owls the longest cochlea. Unlike the spiral mammalian cochlear duct, the avian cochlea suspends the relatively straight duct (*scala media*) within its center by two cartilaginous shelves that divide the perilymphatic space into a narrow upper channel, the *scala vestibuli,* and a much larger lower one, the *scala tympani* (Figures 2–12 and 2–13).

SCALA VESTIBULI AND TYMPANI. Both perilymphatic scalae are lined by a flat epithelium (mesodermal in origin), which becomes intimately applied to the cochlear duct (Figure 2–13). The scala vestibuli communicates with the scala tympani by two ducts (Figure 2–12): a *basal interscalar canal* (Schwartzkopff, 1973) and an *apical interscalar canal* (helicotrema) within the apical cartilage. The scala tympani terminates basally in the vicinity of the round window (*fenestra cochlearis*) (Figures 2–7 and 2–12). In most birds the round window has twice the surface area as the oval window (fenestra vestibularis); in Strigidae, it is five times greater (Bremond, 1963). These perilymphatic channels form an uninterrupted system for the circulation of perilymph from the vestibule to the round window in the scala tympani where the volume displacements can be equalized by compensatory movements of the secondary tympanic membrane.

COCHLEAR DUCT. The cochlear duct is directly continuous with the saccule via a sacculocochlear duct (ductus reuniens) (Figures 2–8 and 2–12). The epithelium composing the roof of the cochlear duct is highly modified to form the *tegmentum vasculosum,* whereas that on the duct floor has differentiated to form a neuroepithelial auditory sense organ, the *basilar papilla* (Figure 2–13). The distal termination of the duct is occupied by the lagena, which has been described as part of the vestibular labyrinth.

The *tegmentum vasculosum* consists of a heavily vascularized epithelium composed of two different and highly specialized cell types, light and dark, which are believed to be specialized for active secretion (light cells) and absorption (dark cells).

The *neuroepithelium* consists of two basic types of sensory hair cells (Takasaka and Smith, 1969: Rosenhall, 1971), which vary in height, location, ultrastructure, and innervation: *tall* and *short,* with intermediate or transitional types (Takasaka and Smith, 1969; Firbas and Müller, 1983). Other cells include columnar supporting cells within the neuroepithelium, as well as *hyaline cells.*

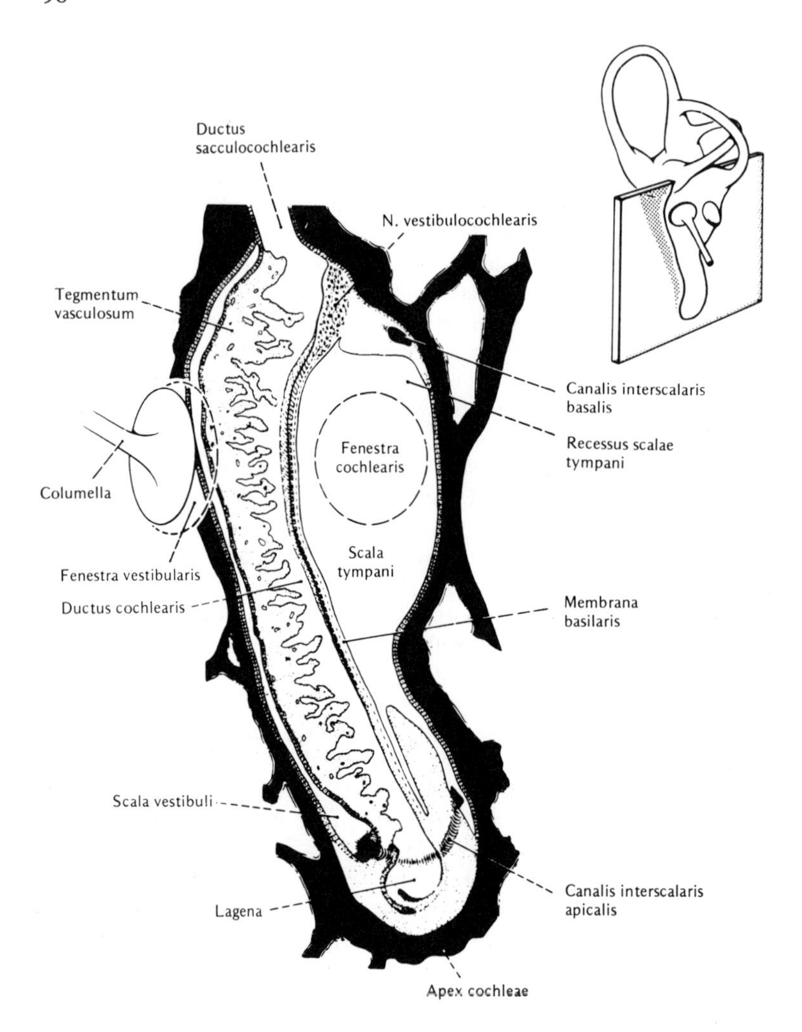

FIGURE 2–12. Schematic representation of the avian cochlea. The plane of section is indicated in the upper right drawing. Note the small size of the scala vestibuli in comparison with the scala tympani. (From Evans, 1979.)

The *sensory hair cells* occupy most of the surface area of the basilar papilla, the supporting elements being relegated at this site to well-marked microvillar borders (MV) surrounding all sensory cells. The total number of sensory cells per papilla has been reported to range between 1,200 and 1,300 (Katsuki, 1965), although approximately 5,000 have been counted in the 7- to 10-day-old chick (Tilney and Saunders, 1983), and an average of 10,400 has been recorded in the pigeon (Goodley and Boord, 1966).

Only one structural hair cell type has been observed in the pigeon (Vinnikov et al., 1965), in the sparrow, finch, and thrush (Yashiki, 1968), and in the cock (Jahnke et al., 1969), but others recognize two types in the chicken (Tanaka and Smith, 1978) and pigeon (Rosenhall, 1971), which differ in shape, size, and location. The sensory hairs in the chicken vary between 50 and 170 per bundle per cell (Tanaka and Smith, 1978). A third type of hair cell has also been described in the pigeon (Takasaka and Smith, 1971) and budgerigar (Firbas and Müller, 1983).

Further evidence of the diversity of sensory cell types has been provided by Firbas and Müller (1983). Utilizing a histochemical method for acetylcholinesterase, these authors localized three different types of efferent axosomatic synapses, each associated with a specific hair cell type.

Hearing

Mechanisms. The essential basis for hearing in birds, as well as in mammals, involves the discrimination of frequencies of sound (pitch), an analysis that has conjured up many theories regarding its location and mechanism. Understandably, the complexity of the phenomenon of hearing has made the establishment of a truly general theory extremely difficult, and no hypothesis as yet has been proven free of justified criticism. A very readable assessment of the more significant theories, including their major limitations, is presented by Gulick (1971). For the most part, they involve primarily the localization of the sound analysis

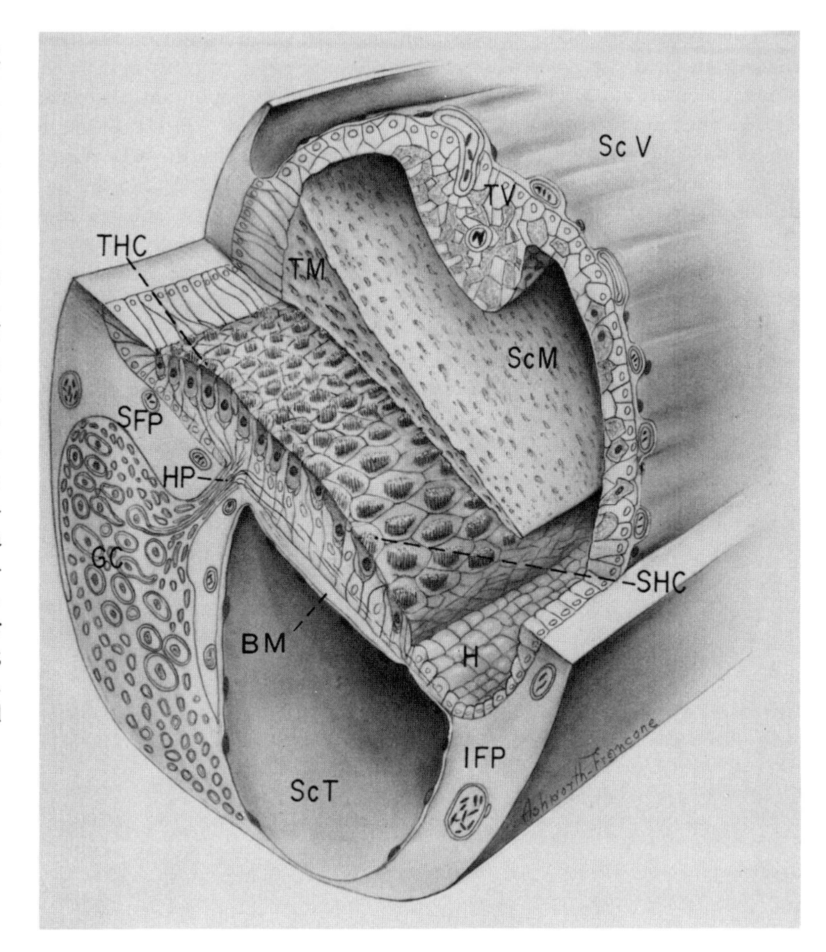

FIGURE 2–13. Schematic, three-dimensional drawing of pigeon cochlear duct. The lateral wall and roof of the scala vestibuli have been removed. A portion of the tectorial membrane has also been removed to show the surface appearance of the basilar papilla with its stereocilia. *BM,* Basilar membrane; *GC,* ganglion cells of cochlear nerve; *H,* hyaline cells; *HP,* habenula perforata (site of penetration of nerve fibers through superior (anterior) cartilage shelf (*SFP*) into basilar papilla; *IFP,* inferior cartilage shelf; *ScM,* scala media (cochlear duct); *ScT,* scala tympani; *ScV,* scala vestibuli; *SHC,* short hair cells; THC, tall hair cells; *TM,* tectorial membrane; *TV,* tegmentum vasculosum. (Courtesy of Takasaka and Smith, 1971.)

(peripherally and/or centrally), the vibratory activity of the basilar membrane, a ciliary restraint imposed by a fibrous tectorial membrane (Wever, 1971), and the synchrony of nervous discharges (volleying) from the sensory hair cells. According to the most prevalent theory, the vibrations produced within the perilymph by the movement of the columellar footplate somehow cause a traveling wave of displacement along the basilar membrane. Since the mechanical properties (mass and stiffness) of this membrane vary from base to apex, its displacement will likewise vary with the frequency of the sound wave. Low frequencies, e.g., 25 Hz, will cause a maximum displacement apically where the membrane is widest and most flexible and stimulate the nerve endings there, whereas high frequencies will bring about a similar effect basally, albeit involving a stiffer and thinner basilar membrane and a different sensory cell population.

Von Békésy (1960) cites ter Kuile in 1900 as providing the clearest explanation for the mechanism of nerve cell stimulation by the traveling wave. The frequency (pitch) can be determined, he says, ". . . by the length of the membrane up to the point at which the wave continues to be propagated with a certain amplitude." (von Békésy, 1960). Unfortunately, as this mechanical selectivity by the relatively short and heavily damped basilar membrane is not able to account for the psychologically measured resolving capacity of the avian auditory system, additional refinements are deemed necessary. Among the conceivable possibilities are the gradation in sterociliary number and dimensions along the basilar membrane (Tilney and Saunders, 1983) and the numerous synaptic connections constituting the sensory pathways from the basilar papilla to the cerebral cortex. According to Sachs et al. (1978), the information received by the avian brain from the peripheral auditory system is comparable to that transmitted to the mammalian brain.

Nerve excitation within the cochlear duct is likewise not completely understood. The displacement of the basilar membrane undoubtedly results in an up-and-down movement of the adjoining hair cells (mechanoreceptors) and a synchronous vibration of the tectorial membrane, depending on the frequency (Mygind, 1965). This results in an anterior-posterior (superior-inferior) deflection or shearing of the sterocilia owing to the swinging action of the tectorial membrane, which is firmly attached along one margin. An excitatory response (depolarization) of the hair cells is subsequently elicited when their sterocilia are displaced in the direction of their morphologic polarization (posterior) and an inhibitory reaction (hyperpolarization) during anterior displacement (Figure 2–11). Depolarization produces a receptor potential (cochlear microphonic), which in turn elicits an action potential within the afferent nerve endings. The efferent system functions to balance this output through direct presynaptic inhibition of the dendrites of the cochlear nerve cells (Wersäll, 1968), the net result being a decrease in the acoustic output to the higher auditory centers. (See Chapter 1 for further details on nervous pathways.)

Range of Hearing. The hearing range of a variety of birds is collated in Table 2–3. It reveals that the optimum performance is achieved between 1,000 and 4,000 Hz, except for two owls that reach 6,000, and that the maximum sensitivity extends to almost 30,000 Hz. Within the ultrasonic range, birds do not exhibit significant hearing abilities, and virtually no hearing has been demonstrated above the upper frequency limit of the human ear (Schwartzkopff, 1968). The frequency range within a single avian species, however, is narrower than in mammals, and the overall picture of avian hearing ability suggests that it is similar to the human in general range and ability to discriminate pitch (Thorpe, 1961). There are certain auditory characteristics, however, which exceed those of man, and

TABLE 2–3. Hearing range of various birds and man[a]

Species	Lower limit (Hz)[b]	Highest sensitivity (Hz)	Upper limit (Hz)	Method[c]	Reference
Man (*Homo sapiens*)	16	1000–2800	10000–24000		
Mallard (*Anas platyrhynchos*)	300	2000–3000	8000	C	Trainer (1946)
Long-eared owl (*Asio otus*)	100	6000	18000	C	Schwartzkopff (1955a)
Canvasback (*Aythya valisineria*)	190		5200	C	Edwards (1943)
Great horned owl (*Bubo virginianus*)	60		7000	C	Edwards (1943)
Eagle owl (*Bubo bubo*)	60	1000	8000	C	Trainer (1946)
Greenfinch (*Chloris chloris*)			20000	C	Granit (1941)
Pigeion (*Columbia livia*)			12000	C	Wassiljew (1933)
	50	1800–2400	11500	E	Wever and Bray (1936)
	200		7500	C	Brand and Kellogg (1939a)
	300	1000–4000		E	Heise (1953)
	300	1000–2000		C	Trainer (1946)
Crow (*Corvus brachyrhynchos*)	300	1000–2000	8000	C	Trainer (1946)
Robin (*Erithacus rubecula*)			21000	C	Granit (1941)
Sparrow hawk (*Falco sparverius*)	300	2000	10000	C	Trainer (1946)
Chaffinch (*Fringilla coelebs*)	200	3200	29000	E	Schwartzkopff (1955a)
Ring-billed gull (*Larus delawarensis*)	100	500–800	3000	—	Schwartzkopff (1973)
Red crossbill (*Loxia curvirostra*)			20000	C	Knecht (1940)
Budgerigar (*Mellopsittacus undulatus*)	40	2000	14000	C	Knecht (1940)
Prairie horned lark (*Otocoris alpestris*)	350		7600	C	Edwards (1943)
Sparrow (*Passer domesticus*)			18000	C	Granit (1941)
	675		11500	C	Brand and Kellogg (1939a)
Ring-necked pheasant (*Phasianus colchicus*)	250		10500	C	Stewart (1955)
Black-billed magpie (*Pica pica*)	100	800–1600	21000	E	Schwartzkopff (1955a)
Snow bunting (*Plectrophenax nivalis*)	400		7200	C	Edwards (1943)
Bullfinch (*Pyrrhula pyrrhula*)	100	3200		C	Schwartzkopff (1949).
	200	3200	20000–25000	E	Schwartzkopff (1952)
			21000	C	Granit (1941)
Canary (*Serinus canaria*)	1100		10000	C	Brand and Kellogg (1939b)
	250	2800	10000	C	Dooling *et al.* (1971)
Cape penguin (*Spheniscus demersus*)	100	600–4000	15000	E	Wever *et al.* (1969)
Tawny owl (*Strix aluco*)	100	3000–6000	21000	E	Schwartzkopff (1955a)
Starling (*Sturnus vulgaris*)	700		15000	C	Brand and Kellogg (1939a)
		2000		C	Trainer (1946)

[a] Collated from Schwartzkopff (1955a), Frings and Slocum (1958), and Schwartzkopff (1973).
[b] Hz (cps).
[c] C, conditioning; E, electrophysiology.

these are exemplified by the extremely rapid frequency changes within a time frame that is too rapid for human discrimination. Pumphrey (1961) notes that a small bird accomplishes this 10 times better than humans. Furthermore, Greenewalt (1968) points out that the syrinx of song birds enables them to produce two notes or phrases simultaneously and that "these sounds can be modulated, in either frequency, or in amplitude, or more usually in both, with extraordinary rapidity; so rapidly in fact that human ears cannot preceive the modulations as such, receiving instead an impression of notes of varying quality or timbre."

Quite fascinating is the recent disclosure that low-frequency sounds, termed *infrasounds,* can be detected as low as 0.05 Hz by homing pigeons in a sound-isolated chamber (Kreithen and Quine, 1979). Such sounds have been recorded in thunderstorms, earthquakes, auroras, ocean waves, and even mountain ranges, and are believed to be an important source of navigational and meteorological information during migration (Kreithen, 1979). At this level, pigeons are at least 50 dB more sensitive than humans (Kreithen and Quine, 1979). In addition, a few birds, e.g., oilbirds *(Steatornis caripensis)* (Griffin, 1953) and swiftlets *(Collocallia brevirostra unicolor)* (Novick, 1959), have successfully developed the ability to use their own sound (short clicks between 4 and 7 Hz) for *echolocation* as bats do, for acoustic orientation and navigation in the dark when pursuing prey and avoiding obstacles.

The Chemical Senses in Birds

M.R. Kare and J.R. Mason

The chemical senses are commonly thought to fall into three classes: (1) olfaction (smell), (2) gustation (taste), and (3) the common chemical sense. In birds, as in most other vertebrates, olfaction is usually thought to be a telereceptor, capable of receiving airborne chemical stimuli in extreme dilution over relatively great distances. Gustation, on the other hand, usually requires more intimate contact of higher concentrations of the chemical stimuli with the taste receptors. Gustatory receptors are most often located in the taste buds of the oral cavity, although functional taste buds are found outside the oral cavity and on the body surface of some fish (i.e., the channel catfish, *Ictalurus punctatus*) (Pfaffmann, 1978). The common chemical sense is usually reserved for nonspecific stimuli, which are often irritating.

Common Chemical Sense

"Parker (1922) suggests that the common chemical sense is relatively primitive and that taste and olfaction are later differentiations. The prevalence of the common chemical sense among vertebrates, and the diverse, relatively unspecialized nature of the receptors (i.e., free nerve endings), support this interpretation. In higher vertebrates, a major component of the common chemical sense is the trigeminal system. An extensive review of this system with special emphasis on trigeminal chemoreception in the nasal and oral cavities has been provided by Silver and Maruniak (1980).

Irritants such as ammonia and acids stimulate the free nerve endings of numerous surfaces, including those in the nasal chambers, mouth, and eyelids of vertebrates. The organization of the trigeminal system in birds does not appear to be essentially different from that found in mammals (Dubbledam and Karten, 1978), although its extent is exaggerated in some aquatic forms, such as ducks and flamingos (Welty, 1975). The well-developed nature of the trigeminal system in aquatic species may serve in the initiation of diving reflexes, or in the detection of tactile properties of foods. That latter possibility is consistent with evidence that pigeons use oral trigeminal cues for this purpose (Ziegler, 1977).

The pigeon and gray partridge are indifferent to strong ammonia solutions that stimulate trigeminal receptors in mammals (Soudek, 1929). Likewise, parrots consume *Capsicum* peppers that are rejected by mammals (Mason and Reidinger, 1983a), and red-winged blackbirds are relatively insensitive to capsaicin, the pungent principle in *Capsicum* peppers (Mason and Maruniak, 1983). From such results, one might conclude that the avian trigeminal system is relatively insensitive to chemical stimuli, although capsaicin and ammonia are probably not characteristic of irritants that birds are likely to encounter. Perhaps the use of ecologically and evolutionarily more relevant irritants (e.g., saponins in plants) would lead to different conclusions. Also, the avian trigeminal system may serve purposes different from those of the mammalian system. Pigeons may home using trigeminal cues, when other sensory inputs are blocked (Wallraff, 1980), and the European starling readily avoids nonirritating concentrations of phenethyl alcohol on the basis of trigeminally mediated information (Mason and Silver, 1983). These data suggest that some birds may be able to make qualitative discriminations between odors using only the trigeminal system, but this possibility remains controversial. Walker et al. (1979), using conditioned suppression, found that after bilateral section of the olfactory nerves pigeons could detect but no longer discriminate between the chemically similar compounds butyl and pentyl acetate. More systematic work is needed to elucidate the function(s) of the common chemical sense in birds.

Smell

The question of whether or not birds possess olfactory capabilities was a controversial one for many years.

Nineteenth-century naturalists (e.g., Audubon, 1826) carried out experiments designed to test the olfactory ability of vultures. Both positive and negative results were obtained, and the reports of these early investigators were followed by others whose conclusions regarding the sense of smell in birds were as contradictory as they were numerous. Anatomic investigations carried out over the last two decades have indicated that birds possess olfactory systems whose complexity and development vary widely among species. Neural events, presumably the result of stimulus–receptor interaction, have been studied electrophysiologically in birds, and some species have been reported to regulate their behavior on the basis of olfactory information (Archer and Glen, 1969; Wenzel, 1973, 1980).

The Olfactory Organ. Bang (1971) has summarized the functional anatomy of the olfactory system of birds representing 23 orders. Birds possess several nasal conchae and lack a vomeronasal (Jacobsen's) organ, although the latter has been identified in the very early embryonic life of some birds (Matthes, 1934). Typically, the avian olfactory system consists of external nares (nostrils), nasal chambers (conchae), internal nares (choane), and olfactory nerves, the peripheral terminals of which lie in the olfactory epithelium, and the olfactory bulbs of the brain. There are three nasal chambers, but reportedly only the turbinates of the third (posteriosuperior) chamber possess olfactory epithelium (Bang, 1971). The first two chambers serve to moisten, warm, and cleanse inspired air. Other possible functions of these chambers are suggested by studies of "dynamic gliders," such as albatroses, petrels, and fulmars (Welty, 1975). Dissections of the nasal chambers of these birds reveal a pair of small forward-opening pockets of the middle chamber, which may act as organs for detecting variable pressures produced by differing external airstream velocities (see Figure 2–14).

In pelicans and their allies, the external nares are small or closed and there is a reduction in size of other parts of the olfactory system, but the choane between the third chamber and the mouth are relatively large. These openings may be adapted to provide retronasal access for volatile materials held in the mouth (Welty, 1975). The comparative anatomy of the nose and nasal airstreams is discussed by Bang and Bang (1959) (see Figures 2–15 and 2–16). Much general information on olfactory receptor cells, nerves, and central projections is found in Biedler (1971). (See Chapter 1 for further details on olfactory neural pathways.)

Numerous negative reports on olfaction in birds have probably discouraged the use of this animal class in olfactory research, and this may explain why so little work on the mechanism of olfaction deals with avian species. Yet olfaction is important for some birds (Wenzel, 1980), and intriguing evolutionary questions remain to be addressed. For example, there is as yet no explanation for the observation that diving petrels and auks, well-known examples of convergent evolution, differ sharply in respect to olfactory system. Very likely, olfactory development (or the lack of it) among avian species reflects the importance of this sense in locating food or in homing.

Methods of Detecting Olfaction. Two general laboratory methods have been used to detect olfactory perception in birds. The neurophysiological methods have involved recording from the olfactory nerve (Tucker, 1965) or directly from the olfactory bulb (Wenzel and Sieck, 1972) during odorant stimulus presentation in an olfactometer. Behavioral study of olfaction in birds involves two techniques. In the first, birds are required to discriminate between air and an odor that previous training has made relevant to the test (Shumake et al., 1970; Walker et al., 1979). The second technique involves continuous monitoring of heart rate and/or respiration during intermittent presentation of odorous stimuli (Wenzel, 1968). Which of these latter techniques is the more accurate depends in part on the species under investigation (Shallenberger, 1973, 1975).

Olfactory Development in Various Species. The olfactory system is well developed in kiwi, vulture, albatross, and petrels; moderately developed in the fowl, pigeon, and most birds of prey; and poorly developed

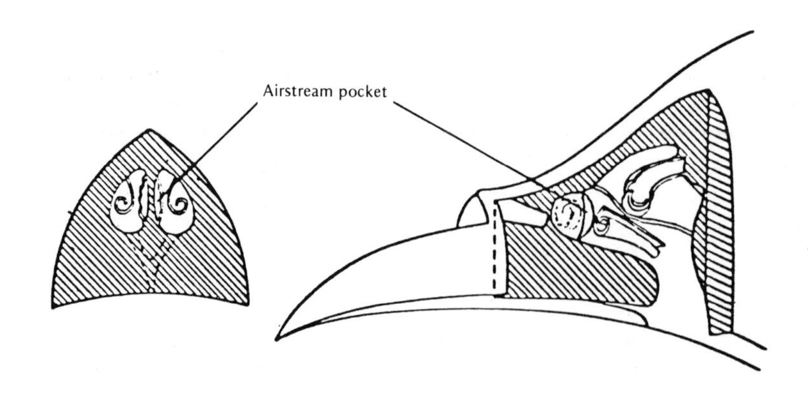

FIGURE 2–14. Cross and longitudinal sections of the nasal chambers of a fulmar, showing the location of the valvelike pockets that may serve sea birds as airvelocity sense organs to aid them in exploiting winds of varying speeds during dynamic gliding. (After Mangold, 1946.)

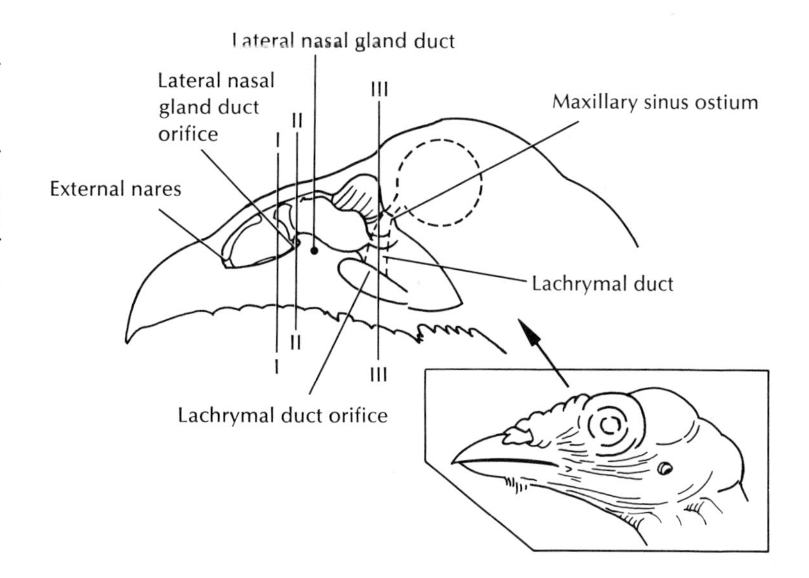

FIGURE 2–15. Diagramatic sagittal section of right medial surface of the nasal chambers of a chicken. Olfactory sensory area indicated by diagonal shading; position of eye and lachrymal duct, by broken lines. Inserts of partially dissected head of chicken gives an idea of extent of fossa in relation to external landmarks. Not drawn to scale. (From Bang and Bang, 1959.)

in songbirds. It is very possible that development (or the lack of it) is related to the foods and other resources exploited by various species.

Vultures are carrion eaters, and their conspicuous circling behavior in the area of a carcass led to much early speculation that they locate sources of food by olfaction. This has been confirmed for the turkey vulture, and strongly suggested for the king vulture by Stager (1967). By careful release of ethyl mercaptan fumes in the path of migrating turkey vultures, Stager was able to demonstrate that vultures are led to the general area of food by olfaction. Once in the general area, these birds seem to rely more heavily on vision to lead them to the exact location of a food source. Conversely, ravens rely on visual cues to identify foraging locations, but sometimes then use subtle odor cues to locate buried food stuffs (Synder and Peterson, 1979).

Several species of the Procellariformes may use olfaction in navigation and nest location (Grubb, 1972). The most compelling evidence is for Leach's storm petrel, which usually return to their island nesting locations at night by flying upwind (Grubb, 1974). Severing the olfactory nerves or plugging the nares in these birds interferes with their ability to return to the nest. Nesting material effectively serves as a lure for the birds in total darkness, and these birds consistently chose the arm of a Y-maze that contains their own nesting material. An excellent review of seabird olfaction has been provided by Wenzel (1980).

The flightless kiwi is nocturnal and feeds largely on earthworms and other hidden food. Its vision is poor;

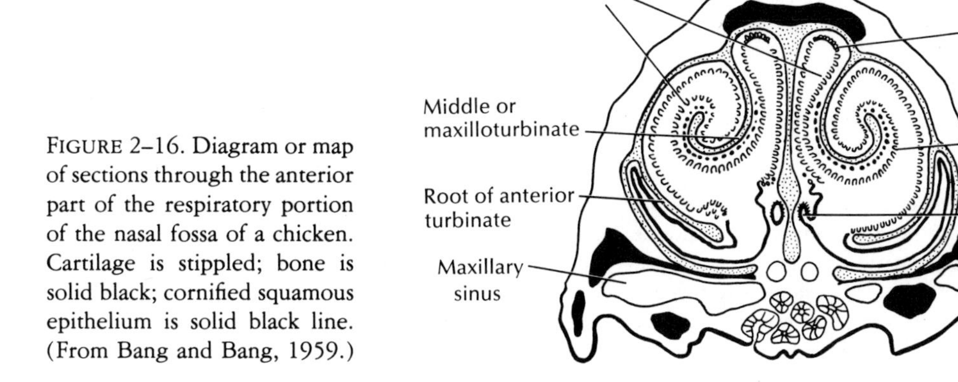

FIGURE 2–16. Diagram or map of sections through the anterior part of the respiratory portion of the nasal fossa of a chicken. Cartilage is stippled; bone is solid black; cornified squamous epithelium is solid black line. (From Bang and Bang, 1959.)

it is the only bird with nostrils at the tip of the beak, and it sniffs while foraging. It can detect food hidden or in the dark (Wenzel, 1968).

Stager (1967) suggested that African honeyguides locate beehives by means of olfaction and are attracted by burning beeswax candles. Honeyguides can locate abandoned hives presumably by smell, when other cues (e.g., vision, audition) are no longer available (Archer and Glen, 1969).

Operant conditioning techniques have been used to demonstrate that pigeons use olfactory cues to perform key-pecking and other tasks (e.g., Shumake et al., 1970). For example, homing pigeons with bilaterally sectioned olfactory nerves or nares plugged with cotton display an impaired ability to return to the home loft (Papi et al., 1973; Benvenuti et al., 1973). These and other experiments indicate that olfaction may play a role in the initial orientation of homing pigeons (Wallraff, 1979). Recent evidence suggests that olfactory cues are especially important for navigation by pigeons over long distances (e.g., 500 km). Like vultures, however, such cues appear relatively less important at short distances (e.g., 10 km), contrary to reports of homing by storm petrels (Grubb, 1974).

Gallinaceous birds have been the subjects of little olfactory research. Tucker (1965) presented electrophysiologic evidence that the bobwhite quail can perceive some odors. Stattleman et al. (1975) have determined that chickens are most sensitive to pentane and hexane, while pigeons are most sensitive to heptane. As in work with other species, it is unclear whether reagent-grade chemicals have any biological relevance to gallinaceous birds, and whether different results would obtain if biologically relevant stimuli were used.

The olfactory system in a number of aquatic species, e.g., penguins, geese, and terns, is well developed (Bang, 1971). Neuhaus (1963) reported that greylag geese respond to skatol, and odors of plants on which adult greylag geese will not feed (e.g., violet, lavender) are also aversive to goslings (Wurdinger, 1979).

The reports of olfactory behavior in many songbirds are predominantly negative, but the olfactory systems in such birds may be important for other purposes. Robinzon et al. (1979) reported that the surgical removal of olfactory bulbs in male red-wing blackbirds caused hyperphagia, weight gains, increased thyroid follicular activity, and increased testicular development, suggesting that the olfactory bulbs in red-wings may be involved in the photoperiodic regulation of activity of the gonads and thyroids.

Summary. The evidence available on the sense of smell in birds does not permit many generalizations. The failure of birds to react to odors as they are presented in the laboratory does not necessarily reflect a deficit in acuity, but that the odor may not have reached olfactory receptors. The lack of sniffing behavior in all birds but the kiwi may indicate that moving air is required to effect contact between the odor stimuli and receptors. On the other hand, the perfumes and reagent-grade chemicals that are most often used in the laboratory may not be biologically relevant to birds (Wurdinger, 1979) and/or may not be presented at levels that correspond to the birds' spectrum of chemical sensitivity.

Taste

The function of taste is to encourage the ingestion of nutrients, to discriminate among foods that are available, and possibly to avoid those that are toxic. The taste system in a particular species can be expected to complement digestion, metabolism, and the dietary requirements of that species. While relationships, if any exist, have yet to be clearly defined between number of taste buds and taste behavior, the relatively poor taste acuity of avian species may be related to the relatively small number of sensory cells (Table 2–4).

Taste Receptors (Buds). Receptors are largely found at the base of the tongue and the floor of the pharynx, commonly in close association with the salivary glands (Gentle, 1971; Wenzel, 1973; Dmitrieva, 1981). However, taste buds can be found in other areas (Saito, 1966; Warner et al., 1967; Berkhoudt, 1977), and the number of buds and their distribution may change over time (Duncan, 1960). Whether or not such changes have any ecological importance to these animals is not clear.

Little use has been made of birds in neuroanatomic research on taste, although both the chorda tympani (Gentle, 1983) and the glossopharyngeal nerve (Berkhoudt, 1983) carry taste information. The chorda tympani innervates taste buds adjacent to the anterior mandibular salivary glands, situated in the buccal epithelium of the lower jaw. The glossopharyngeal nerve innervates the posterior buccal and pharyngeal areas. Cutaneous (as well as taste) information is carried by both nerves. The details of their distribution at the level of the first relay and at all higher points in the brain are unclear (Wenzel, 1980). The reader is referred to Berkhoudt (1983) for a review of the structure and function of avian taste receptors. Beidler (1971) also provides a discussion, based largely on mammalian research, of current knowledge and theories concerning central functions and peripheral mechanisms in taste.

Methods of Study. Early studies of taste in birds involved observation of individuals as they consumed foods or fluids. Preference testing is now the most common laboratory method used to measure the sensitivity of birds to taste stimuli. Usually the material to be tested is placed in aqueous solution, and the animal is given a choice between the mixture and distilled

TABLE 2 4. Numbers of taste buds in birds and various other vertebrates

Species	n	Reference
Blue tit	24	Gentle (1975)
Pigeon	37–75	Moore and Elliott (1946) Van Kan (1979)
Bullfinch	46	Duncan (1960)
Barbary dove	54	Gentle (1975)
Japanese quail	62	Warner et al. (1967)
Starling	200	Bath (1906)
Chicken	250–350	Van Prooije (1978) Saito (1966)
Duck	375	Berkhoudt (1977)
Parrot	300–400	Bath (1906)
Snake	0	Payne (1945)
Kitten	473	Elliot (1937)
Bat	800	Moncrieff (1951)
Human	9,000	Cole (1941)
Pig and goat	15,000	Moncrieff (1951)
Rabbit	17,000	Moncrieff (1951)
Catfish	100,000	Hyman (1942)

water, the two being presented simultaneously. However, single-stimulus methods, in which choices are presented singly at different times, and three-choice methods, in which a tastant and two control solutions are offered simultaneously, have been used to eliminate confounding by position bias. Cafeteria-type tests in which more than three taste stimuli are presented simultaneously seem to overwhelm the chicken's discriminatory ability.

Chickens show a characteristic response to aversive oral stimulation, as produced by quinine hydrochloride, typified by persistent tongue and beak movements, and headshaking and beak-wiping behavior. No characteristic responses to presentations of neutral or appetitive oral stimuli such as sucrose have been observed (Gentle, 1978; Gentle and Harkin, 1979).

Neurophysiologic studies of taste in birds have been few and have involved chickens or pigeons as experimental animals (Halpern, 1963; Landolt, 1970). Such studies usually involve the application of substances to the tongue of the subject and measurement of multiunit or single-fiber activity in the glossopharyngeal nerve. Using these techniques, Kitchell et al. (1959) demonstrated the water taste phenomenon in birds. That is, water has been shown not to be a neutral carrier of taste stimuli, but to act as a taste stimulus itself under certain conditions. For example, adaptation to NaCl in humans will cause water to have a bitter or bitter-sour taste. Adaptation to the concentration of NaCl present in saliva is sufficient to produce this effect (McBurney, 1978). Electrophysiologically, water taste might be reflected in the response of sour- or bitter-sensitive neurons to water following adaptation. However, such results merely indicate whether a chemical can evoke a peripheral neural response. They do not indicate whether the chemical has an appealing or offensive taste to the animal, and while there are examples of positive correlations between behavioral and electrophysiological response, there are also contradictions (Halpern, 1963). Operant techniques, which often have been used successfully in studies on vision or olfaction (see above), are not often used in taste research with birds.

Research on taste in birds has been handicapped by the general assumption that they live in the human sensory world. The taste sensations experienced by man cannot be assumed to be the same as for birds. For example, dimethyl anthranilate, a flavoring used in human foods, has been used to reduce food intake in growing chicks and turkey poults, and has been suggested as a bird repellent livestock-feed additive. This compound is offensive to starlings, Japanese quail, pigeons, red-wing blackbirds, jungle fowl, herring gulls, and finches at dilutions as low as 1 part in 10,000 in two-choice tests (Kare and Pick, 1960; Mason et al., 1985). Nevertheless, in order to compare results obtained from birds with results from other species, the classical categories of sweet, sour, bitter, and salty are frequently used.

Ability to Taste. *Sweet.* Many avian species evidence little or no interest in the common sugars, although parrots, budgerigars, hummingbirds, and other nectar feeders actively select sugar solutions. Kare and Medway (1959) observed that fowl on an *ad libitum* diet failed to perceive, or were indifferent to, dextrose and sucrose in food when tested at concentrations ranging from 2.5 to 25%. The findings are different, however, when tastants are presented in aqueous solution. Gentle (1972) reported that chickens exhibit rejection of 30% sucrose, fructose, or glucose solutions, and that glucose is rejected at concentrations as low as 5%. Several investigators observed modest preferences for sugar solutions over water. Jacobs and Scott (1957) showed that chickens perferred a 12% sucrose solution to water. Japanese and bobwhite quail prefer some concentrations of sucrose and glucose (Brindley, 1965), but red-wing blackbirds select pure water over sucrose (Rogers and Maller, 1973). There is unanimity in the literature that birds reject such synthetic sweeteners as saccharin or dulcin. Curiously, even though chickens reject saccharin in behavioral tests, electrophysiologic techniques have failed to uncover neural activity when taste buds are rinsed with the substance (Welty, 1975). Collectively, the data suggest that nectar- or fruit-eating species are more likely to respond positively to sugars than are insectivorous or granivorous birds, which respond negatively or not at all.

The discovery that three tropical fruits contain intensely sweet proteins (Cagan, 1973) led to speculation on the role (if any) of the sweet principles in the plant. It is possible (though not demonstrated) that the sweet taste-active proteins may aid in seed dispersal by some frugivorous birds or other animals (Davison, 1962). No avian species have yet been tested with any of the sweet proteins.

A number of factors other than taste may be involved, individually or collectively, in the response of a bird to a sugar solution, e.g., osmotic pressure, viscosity, melting point, nutritive value, toxicity, and optical characteristics. Some have suggested that visual properties and surface texture sometimes take precedence over all other qualities in the birds' selection of food (Mason and Reidinger, 1983b). Across species, no physical or chemical characteristic can be used to reliably predict how a bird on an adequate diet will respond to the taste of a solution (Kare and Medway, 1959).

Salt. Birds kept on a salt-free diet will eagerly consume pure salt when it is made available to them. Numerous finches of the family Carduelidae have notorious appetites for salt, and cross-bills, for example, may be caught in traps baited with salt alone (Welty, 1975; Willoughby, 1971). Also, the domestic fowl maintained on a diet very low in sodium or calcium will exhibit a specific appetite and select, in a choice situation, the diet or solution that corrects its deficiency. However, the domestic chick delays drinking for extended periods to avoid consuming a sodium chloride solution whose concentration exceeds that which the chick's kidneys can handle (Kare and Biely, 1948). In fact, where no alternative is available, many chicks die of thirst rather than consume a toxic 2% salt solution. They accept sodium chloride solutions only up to about 0.9% (0.15 *M*). Various other birds without nasal salt glands that have been studied have similar taste-tolerance thresholds (Bartholomew and Cade, 1958). Mourning doves freely drink any solution that is hypotonic to their body fluids (Bartholomew and MacMillian, 1960).

Rensch and Neunzig (1925) investigated sodium chloride thresholds (i.e., the lowest concentration at which solutions are rejected) for 58 species, and found that the thresholds ranged from 0.35% in a parrot to 37.5% in the siskin. Unlike the rat, which avidly selects some hypotonic concentrations of sodium chloride, many birds are indifferent up to the concentration at which they reject the salt solution.

The common tern, which has a nasal salt gland, has a high threshold for salt that has been associated with the intake of brackish water with its food. However, when given a choice, the herring or laughing gull (with salt glands) selects pure water over saline solution (Harriman, 1967; Harriman and Kare, 1966). Similarly, penguins are said to prefer fresh water after having been at sea for extended periods (Warham, 1971). The role of the nasal salt gland in the handling of salt is discussed elsewhere in this volume.

The order of acceptability of ionic series by birds does not appear to fit into the lyotropic or sensitivity series reported for other animals. No physical or chemical theory has been offered to explain the responses to sodium salts and chlorides presented in Table 2–5.

Sour. Birds have a wide range of tolerance for acidity and alkalinity in their drinking water (Figure 2–17 and Table 2–6). Fuerst and Kare (1962) reported that over an 18-day period, chicks will tolerate strong mineral acid solutions, i.e., pH 2 (Table 2–6). Organic acids are less acceptable, and the tolerance for the hydrogen ion is not equivalent to that for the hydroxyl ion. The starling and the herring gull also readily accept hydrochloric acid solutions. The chick's aversion to acid (sour) solutions is reduced by the addition of glucose (Gentle, 1972). Brindley and Prior (1968) reported that bobwhites prefer 0.05% HCl to water.

Bitter. Many tastants are offensive at low concentrations. These include compounds that are bitter to man but quite acceptable to birds, some that are offensive to both man and birds, and a third category of those quite acceptable to man but rejected by some birds.

Sucrose octacetate at a concentration bitter to man is readily accepted by the herring gull and the chicken. Bobwhite quail, which do not respond to sucrose octacetate as very young birds, gradually develop the ability to discriminate this compound (Cane and Vince, 1968). Quinine hydrochloride or sulfate, both of which are used extensively as standard bitter stimuli for man and rats, are also rejected by some species of birds, although bread mixed with quinine is readily eaten by some parrots, and grain dipped in picric acid is readily consumed by seed eaters and titmice (Heinroth, 1938). Among those birds that reject quinine however,

TABLE 2–5. Preference for sodium and chloride metallic solutions at various concentrations over distilled water (chicks)[a]

Solution	Concentration (g/100 ml) 0.1	0.2	0.4	0.8	1.0
Na acetate	55[a]	52	56	52	51
Na sulfate	54	52	52	53	50
Na phosphate (monobasic)	52	53	52	52	54
Na succinate	49	52	54	50	56
Na citric	54	52	54	47	35
Na phosphate (diabasic)	51	49	47	44	14
Na tungstate	50	46	48	—	—
Na bicarbonate	52	43	38	20	14
Na benzoate	49	41	23	15	10
Na bisulfate	38	23	35	17	23
Na pyrophosphate	46	37	20	3	4
Na perborate	42	29	10	9	4
Na carbonate	42	30	10	4	2
Na phosphate (tribasic)	46	20	4	1	2
Na cholate	4	20	3	—	3
Sodium Cl	50[a]	50	55	50	45
Magnesium Cl	49	51	51	53	45
Choline Cl	51	48	49	50	51
Manganese Cl	49	51	46	16	—
Strontium Cl	50	38	44	18	9
Ammonium Cl	49	46	35	12	6
Barium Cl	36	48	41	—	15
Calcium Cl	43	45	27	15	5
Zinc Cl	33	24	10	2	2
Cobalt Cl	26	12	6	5	6
Tin Cl	30	7	1	1	2
Copper Cl	6	11	3	8	4
Iron Cl	2	4	2	3	4

[a] Preference = (salt solution consumed × 100)/total fluid intake.

TABLE 2–6. Percent intake in chickens of acids and bases at different pH levels[a]

pH	1.0	1.5	2.0	2.5	3.0	4.0
Acids						
HCl	4	19	50		59	
H_2SO_4	15	35	54		56	
HNO_3	8		62		52	
Acetic					16	53
Lactic				15	61	

pH	10.0	11.0	12.0	13.0
Bases				
NaOH	45	47	33	2
KOH		48	36	3

[a] Tabled values are the mean of replicate lots. The percent intake = (volume of tested fluid/total fluid intake) × 100 (18 daily values were averaged). The position of the numbers is an indication of the pH of the test solution. For example, at pH 1.5 the average daily consumption of HCl was 19% of the total fluid intake. Distilled water was the alternative in every instance.

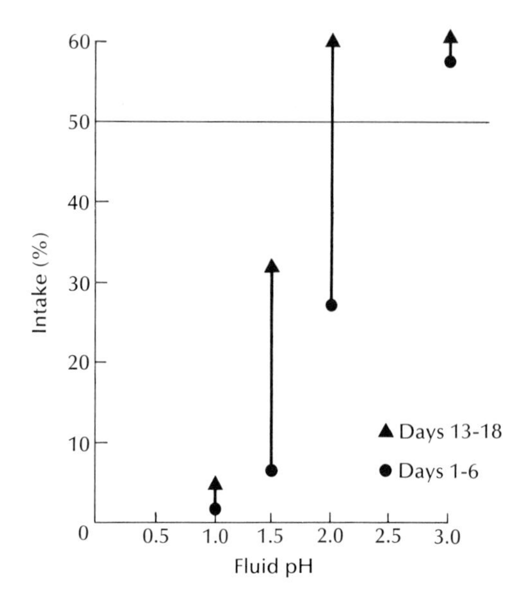

FIGURE 2–17. Daily consumption of HCl solutions for days 1–6 (●) compared with daily consumption for days 13–18 (▲), expressed as percent of total fluid intake at four pH levels. (From Fuerst and Kare, 1962.)

responding is influenced by deprivation, and as in the rat (Johnson and Fisher, 1973), there is increasing acceptance with increasing deprivation (Gentle, 1975). Both quinine and sucrose octaacetate evoke strong neural responses in the chicken.

The offensive secretions of some insects serve as a protective device against avian predators. Some of these have a caustic action on the eyes and possess offensive and possibly bitter tastes (Yang and Kare, 1968). Taste may serve as an important cue to adverse effects that may occur subsequent to ingestion of certain prey (Rozin and Kalat, 1971). For example, Wilcoxon *et al.* (1971) found that bobwhites could learn to associate adverse postingestinal effects with particular tastes, but they were more responsive to visual cues. Similar results have been obtained for red-wing blackbirds (Mason and Reidinger, 1983b). Little is known about the chemistry of offensive tastes in relation to the human senses.

Nutrition and Taste. The function of taste in nutrition is an enigma. In some instances, the birds' preference compliments its nutritional needs. For example, when caloric intake is restricted, a chick selects a sucrose solution to which it is normally indifferent and increases its fluid intake to make up the deficiency (Kare and Ficken, 1963). However, a similarly correct nutritional choice is not made when the sugar is replaced with an isocaloric solution of fat or protein. In a comparison of the responsiveness of domestic and wild jungle fowl to chemical stimuli in caloric regulation, Kare and Maller (1967) found that the wild strain was much more sensitive than the domestic. As such, the preference behavior of laboratory animals may not be a reliable guide to the nutritional adequacy of a diet (Kare and Scott, 1962). Presumably, this is a result of domestication, insofar as important traits may have been bred out of the population. Feed consumption is also discussed elsewhere in this volume.

Temperature and Taste. The domestic fowl is acutely sensitive to the temperature of water. Acceptability decreases as the temperature of the water increases above the ambient. Fowl can discriminate a temperature difference of only 5°F, rejecting the higher temperature. Chickens suffer from acute thirst rather than drink water 10°F above their body temperature.

At the other extreme, the chicken readily accepts water down to freezing temperatures. This pattern of sensitivity to temperature has also been observed in electrophysiologic studies. A sizable minority of chickens lacks this sensitivity, however. Because the response to temperature may take precedence over all chemical stimulants, temperature should be eliminated as a variable in taste studies of the fowl.

Saliva and Taste. Saliva is involved in the normal phenomena of taste. Birds have been described as having a limited salivary flow. Using a technique that permitted continuous collection, Belman and Kare (1961) observed that the flow of saliva in the chicken was greater than that of man in terms of body weight but less in terms of food consumed.

Individual Variation in Taste. Japanese quail and domestic chickens have been tested to measure their reactions to a variety of chemicals, including ferric ammonium and calcium chloride. Individuals show markedly different thresholds. The distribution of thresholds is continuous, with reactions among birds to a single concentration of one chemical varying from preference to rejection. Chemical specificity is involved, because an individual that can taste one chloride at either unusually low or only very high concentrations is likely to respond in an average manner to the others. It has been possible to select and breed for taste sensitivity to a specific chemical. This individual variation is not limited to birds (Kare, 1961).

That birds differ in their taste preference as individuals, strains, or species has obvious ecologic advantages. For example, it may permit a population composed of different species to utilize more of the potential food in an environment than would be possible if all were to compete for a limited group of foods, and it contributes to an adaptive plasticity of food habits, making the invasion of new habitats and utilization of new foods possible.

Variation in response to taste is made more complex by seasonal changes in sensitivity. It is interesting to consider whether taste directs or follows the abrupt changes in feeding pattern of birds that are insectivorous for part of the year and granivorous for the rest. A possible role for taste in the intensive feeding prior to migration is therefore to be considered.

Summary. Kare and Beauchamp (1984), in discussing the comparative aspects of the sense of taste in birds and mammals, pointed out that most of the work on the basic mechanism of taste stimulation has been conducted with mammals. This mammalian work has suggested that the initial interaction of a taste stimulus and a receptor cell may occur on the microvilli of the taste receptor cells, but this has not been demonstrated in Aves.

Birds have a sense of taste. However, no pattern, whether chemical, physical, nutrition, or physiologic, can be correlated consistently with the bird's taste behavior. The behavioral, ecologic, and chemical context of a taste stimulant can influence the bird's response. The observed response, particularly to sweet and bitter, indicates that the bird does not share human taste experiences. However, the supposition that there is a difference in degree between individual birds and an absolute difference between some species appears warranted.

References

Eye and Vision

Agarwal, P.K., K. Singh, and Y. Dayal. (1966). Retinal vascular patterns. I. Lower animals. Orient. Arch. Ophthalmol., 4, 35.

Amemiya, T. (1975). Electron microscopic and cytochemical study on paraboloid glycogen of the accessory cone of the chick retina. Histochemistry, 43, 185.

Barlow, H. B. (1982). What causes trichromasy? A theoretical analysis using comb-filtered spectra. Vision Res., 22, 635.

Barlow, H.B., and T.J. Ostwald. (1972). Pecten of the pigeon's eye as an inter-ocular shade. Nature (London), 236, 88.

Bellhorn, R.W., and M.S. Bellhorn. (1975). The avian pecten. 1. Fluorescein permeability. Ophthalmic Res., 7, 1.

Bliss, A.F. (1946). The chemistry of daylight vision. J. Gen. Physiol., 29, 277.

Bowmaker, J.K. (1977). The visual pigments, oil droplets and spectral sensitivity of the pigeon. Vision Res., 17, 1129.

Bowmaker, J.K., and A. Knowles. (1977). The visual pigments and oil droplets of the chicken retina. Vision Res., 17, 755.

Bowmaker, J.K., and G.R. Martin. (1978). Visual pigments and colour vision in a nocturnal bird, *Strix aluco* (tawny owl). Vision Res., 18, 1125.

Brach, V. (1975). The effect of intraocular ablation of the pecten oculi of the chicken. Invest. Ophthalmol., 14, 166.

Cohen, A.I. (1963). The fine structure of the visual receptors of the pigeon. Exp. Eye Res., 2, 88.

Crescitelli, F., B.W. Wilson and A.L. Lilyblade. (1964). The visual pigments of birds. I. The turkey. Vision Res. 4, 275.

Crozier, W.J., and E. Wolf. (1944a). Theory and measurement of visual mechanisms. X. Modifications of the flicker response contour, and the significance of the avian pecten. J. Gen. Physiol., 27, 287.

Crozier, W. J., and E. Wolf. (1944b). Flicker response contours for the sparrow, and the theory of the avian pecten. J. Gen. Physiol., 27, 315.

Danilov. V., G. Demirochaglyan, Z. Avetysan, M. Aelakhnyerdyan, S. Grigoryan, and G. Saribekhyan. (1970). Possible mechanisms of magnetic sensitivity in birds (in Russian). Biol. Zh. Arm., 23, 26. (Cited by Southern et al., 1982).

Delage, Y. (1920). Suggestion sur la raison d'être de la double fovea des rapaces diurnes. C.R. Hebd. Seances Acad. Sci., I Semestre, 170, 425.

Dodt, E., and A. Wirth. (1953). Differentiation between rods and cones by flicker electroretinography in pigeon and guinea pig. Acta Physiol. Scand., 30, 80.

Donner, K.O. (1951). The visual acuity of some passerine birds. Acta Zool. Fenn., 66, 2.

Dowling, J.E. (1968). Synaptic organization of the frog retina: An electronmicroscopic analysis comparing the retinas of frogs and primates. Proc. R. Soc. London, B, 170, 205.

Duecker, G., and I. Schulze. (1977). Color vision and color preference in Japanese quail with colorless droplets. J. Comp. Physiol. Psychol., 91, 1110.

Duijm, M. (1958). On the position of a ribbon-like central area in the eyes of some birds. Arch. Neerl. Zool., 13 (Suppl.), 128.

Duke-Elder, S. (1958). "System of Ophthalmology," Vol. I (The Eye in Evolution). St. Louis: C.V. Mosby.

Duke-Elder, S., and K.C. Wybar. (1961). "System of Ophthalmology," Vol. II (The Anatomy of the Visual System). St. Louis: C.V. Mosby.

Eichner, C., and H. Themann. (1963). Zur Frage der Glykogenlokalisation in der Netzhaut der Ratte, des Huhnes und des Goldfisches. Z. Mikrosk. Anat. Forsch., 70, 330.

Emmerton, J. (1983). Vision. In "Physiology and Behaviour of the Pigeon," (M. Abs, Ed.). London: Academic Press.

Emmerton, J., and M. Remy. (1983). The pigeon's sensitivity to ultraviolet and 'visible' light. Experientia, 39, 1161.

Engström, K. (1958). On the cone mosaic in the retina of *Parus major.* Acta Zool., 39, 65.

Evans, H.E. (1979). Organa sensoria. In "Nomina Anatomica Avium" (J.J. Baumel, A.S. King, A.M. Lucas, J.E. Breazile, and H.E. Evans (Eds.). London: Academic Press.

Ferens, N.B. (1947). On the ability of colour-discrimination of the tawny owl (*Strix aluco*). Bull. Int. Acad. Sci. Cracovie, B2, 309.

Fischer, A.B. (1969). Laboruntersuchungen und Freilandbeobachtungen zum Sehvermögen und Verhalten von Altweltgeiern. Zool. Jahrb. Abt. Syst., 96, 81.

Fischlschweiger, W., and R. O'Rahilly. (1968). The ultrastructure of the pecten oculi in the chick. II. Observations on the bridge and its relation to the vitreous body. Z. Zellforsch. Mikrosk Anat., 92, 313.

Fite, K.V. (1973). Anatomical and behavioral correlates of visual acuity in the Great Horned Owl. Vision Res., 13, 219.

Fite, K.V. (1974). The vertebrate fovea: a comparative study. Anat. Rec., 178, 355.

Francois, J., and A. Neetens. (1974). Comparative anatomy of the vascular supply of the eye in vertebrates. In "The Eye," Vol. 5 (H. Davson and L.T. Graham, Eds.). New York: Academic Press.

Fritsch, G. (1911). Der Ort des deutlichen Sehens in der Netzhaut der Vögel. Arch. Mikrosk. Anat., 78, 245.

Gallego, A., M. Baron, and M. Gayoso. (1975). Organization of the outer plexiform layer of the diurnal and nocturnal bird retinae. Vision Res., 15, 1027.

Giebel, C. (1857). Über den Sklerotikalring, den Fächer und die Hardersche Drüse im Auge der Vögel. Z. Gesamte Naturwiss., 9, 388.

Goldsmith, T.H. (1980). Hummingbirds see near ultraviolet light. Science, 207, 786.

Goodge, W.R. (1960). Adaptations for amphibious vision in the dipper (*Cinclus mexicanus*). J. Morphol., 107, 79.
Govardovskii, V.I. (1983). On the role of oil drops in colour vision. Vision Res., 23, 1739.
Gundlach, R.H., R.D. Chard, and J.R. Skahen. (1945). The mechanism of accommodation in pigeons. J. Comp. Psychol., 38, 27.
Hamilton, W.F., and T.B. Coleman. (1933). Trichromatic vision in the pigeon as illustrated by the spectral hue discrimination curve. J. Comp. Psychol., 15, 183.
Hanzely, L., W.E. Southern, and D. Molsen. (1975). Ultrastructure of the ring-billed gull eye pecten. Cytobios, 12, 191.
Harkness, L., and H.C. Bennett-Clark. (1978). The deep fovea as a focus indicator. Nature (London), 272, 814.
Hazlett, L.D., D.B. Meyer, and S.R. Susan. (1974). Visual cell ultrastructure in the Japanese quail (*Coturnix coturnix japonica*). Anat. Rec., 178, 371.
Hess, C. (1909–1910). Die Akkommodation bei Tauchervögeln. Arch. Vgl. Ophthalmol., 1, 153.
Hoffmann, C.K. (1877). Zur Anatomie der Retina. III. Über den Bau der Retina bei den Vögeln. Arch. Niederl. Zool., 3, 217.
Hogan, M.J., J.A. Alvarado, and J.E. Weddell. (1971). "Histology of the Human Eye." Philadelphia: W.B. Saunders.
Hudson, R.A., D. Johnston, and D.B. Meyer. (1971). The chemical composition of retinal oil droplets. Invest. Opthalmol. Res., 2, 217.
Kajikawa, J. (1923). Beiträge zur Anatomie und Physiologie des Vogelauges. Albrecht von Graefes Arch. Ophthalmol., 112, 260.
Kolmer, W. (1924). Über das Auge des Eisvogels (*Alcedo attis attis*). Pfleugers Arch. Gesamte Physiol. Menschen Tiere, 204, 266.
Kovach, J.K., G. Wilson, and T. O'Connor. (1976). On the retinal mediation of genetic influences in color preferences of Japanese quail. J. Comp. Physiol. Psychol., 90, 1144.
Kreithen, M.L., and W.T. Keeton. (1974). Detection of polarized light by the homing pigeon, *Columba livia*. J. Comp. Physiol., 89, 83.
Kreithen, M.L., and T. Eisner. (1978). Ultraviolet light detection by the homing pigeon. Nature (London), 272, 347.
Lashley, K.S. (1916). The color vision of birds. I. The spectrum of the domestic fowl. J. Anim. Behav., 6, 1.
Leuckart, R. (1876). Organologie des Auges. Vergleichende Anatomie. Handb. Gesamte Augenheilk., 2, 145.
Levy, B. (1979). Master of Science Dissertation, University of Waterloo, Waterloo, Ontario, Canada. (Cited by Sivak, 1980.)
Lindsay-Johnson, G. (1901). Contributions to the comparative anatomy of the mammalian eye, chiefly based on ophthalmoscopic examination. Philos. Trans. R. Soc. London, B, 194, 1.
Lord, J.D. (1956). A comparative study of the eyes of some falconiform and passeriform birds. Am. Midl. Nat., 56, 325.
Mann, I.C. (1924a). The function of the pecten. Br. J. Ophthalmol., 8, 209.
Mann, I.C. (1924b). The pecten of *Gallus domesticus*. Q. J. Microsc. Sci., 68, 413.
Mariani, A.P., and A.E. Leure-duPree. (1978). Photoreceptors and oil droplets in the red area of the pigeon retina. J. Comp. Neurol., 182, 821.
Martin, G.R. (1974). Colour vision in the tawny owl (*Strix aluco*). J. Comp. Physiol. Psychol., 86, 133.
Matsusaka, T.F. (1963a). Electron-microscopic observations on cytology and cytochemistry of the paraboloid glycogen of chick retina. Jpn. J. Ophthalmol., 7, 238.
Matsusaka, T.F. (1963b). Observations on the cytology and histochemistry of paraboloid glycogen of the chick retina. Acta Soc. Ophthalmol. Jpn., 67, 1151.
Maturana, H.R. (1962). Functional organization of the pigeon retina. In "Information Processing in the Nervous System." Amsterdam: Excerpta Medica.
Maturana, H.R., and S. Frenk. (1963). Directional movement and horizontal edge detectors in the pigeon retina. Science, 142, 977.
Mayr, I. (1972a). Das Farbunterscheidungsvermögen einiger Ploceidae (Passeriformes, Aves) im kurzwelligen Bereich des Spektrums. Vision Res., 12, 509.
Mayr, I. (1972b). Verteilung, Lokalisation und Absorption der Zapfenölkugeln bei Vögeln (Ploceidae). Vision Res., 12, 1477.
Menner, E. (1938). Die Bedeutung des Pecten im Auge für die Wahrnehmung von Bewegung nebst Bemerkungen über seine Ontogenie und Histologie. Zool. Jahrb. Abt. Allg. Zool. Physiol. Tiere, 58, 481.
Meyer, D.B. (1971). The effect of dietary carotenoid deprivation on avian retinal oil droplets. Ophthalmol. Res., 2, 104.
Meyer, D.B. (1977). The avian eye and its adaptations. In "Handbook of Sensory Physiology, The Visual System in Vertebrates," Vol. VII/5 (F. Crescitelli, Ed.). Berlin: Springer.
Meyer, D.B., and T.G. Cooper. (1966). The visual cells of the chicken as revealed by phase contrast microscopy. Am. J. Anat., 118, 723.
Meyer, D.B., T. G. Cooper, and C. Gernez. (1965). Retinal oil droplets. In "The Structure of the Eye" (J.W. Rohen, Ed.). Stuttgart: Schattauer.
Meyer, D.B., S.R. Stuckey, and R.A. Hudson. (1971). Oil droplet carotenoids of avian cones. I. Dietary exclusion: Models for biochemical and physiological studies. Comp. Biochem. Physiol. B: Comp. Biochem., 40, 61.
Meyer, D.B., L.D. Hazlett, and S.R. Susan. (1973). The fine structure of the retina in the Japanese quail (*Coturnix coturnix japonica*). I. Pigment epithelium and its vascular barrier. Tissue & Cell, 5, 489.
Meyer, D.B., and H.C. May, Jr. (1973). The topographical distribution of rods and cones in the adult chicken retina. Exp. Eye Res., 17, 347.
Meyer, M.E. (1964). Discriminative basis for astronavigation in birds. J. Comp. Physiol. Psychol., 58, 403.
Meyknecht, J. (1941). Farbensehen und Helligkeitsunterscheidung beim Steinkauz (*Athene noctua vidalii* A.E. Brehm). Ardea, 30, 129.
Morris, V.B. (1982). An afoveate area centralis in the chick retina. *J. Comp. Neurol.*, 210, 198.
Morris, V.B., and C.D. Shorey. (1967). An electron microscopic study of types of receptor in the chick retina. J. Comp. Neurol., 129, 313.
Munk, O. (1970). On the occurrence and significance of horizontal band-shaped retinal areae in teleosts. Videnskabsselsk. Meddr. Dansk. Naturuidensk. Foren., 133, 85.
Neumann, G.H. (1962). Das visuelle Lernvermögen eines Emus. J. Ornithol., 103, 153.
Oehme, H. (1961). Vergleichende histologische Untersuchungen an den Retina von Eulen. Zool. Jahrb. Abt. Anat. Ontog. Tiere, 79, 439.
Oehme, H. (1962). Das Auge von Mauersegler, Star und Amsel. J. Ornithol., 103, 187.
Oehme, H. (1964). Vergleichende Untersuchungen an Griefvogelaugen. Z. Morphol. Oekol. Tiere, 53, 618.
O'Rahilly, R., and D.B. Meyer. (1963). Etude histologique et histochimique des cellules visuelles de la rétine du poulet. Ann. Histochim., 8, 281.
Pennycuick, C.J. (1960). The physical basis of astro-navigation in birds: theoretical considerations. J. Exp. Biol., 37, 573.

Poley, D. (1968). Experimentelle Untersuchungen zur Nahrungssuche und Nahrungsaufnahme des Kolibris. Bonner Zool. Beitr., 1, 111.
Polyak, S. (1957). "The Vertebrate Visual System." Chicago: University of Chicago Press.
Pumphrey, R.J. (1948a). The sense organs of birds. Ibis, 90, 171.
Pumphrey, R.J. (1948b). The theory of the fovea. J. Exp. Biol., 25, 299.
Pumphrey, R.J. (1961). Sensory organs: vision. In "Biology and Comparative Physiology of Birds," Vol. 2 (A.J. Marshall, Ed.). New York: Academic Press.
Riggs, L.A., P.M. Blough, and K.L. Schafer. (1972). Electrical response of the pigeon eye to changes in wavelength of the stimulating light, Vision Res., 12, 981.
Rochon-Duvigneaud, A. (1919). Quelques données sur la fovea des Oiseaux. Ann. Ocul., 156, 717.
Rochon-Duvigneaud, A. (1920). La situation de foveae simples et doubles dans la rétine des Oiseaux et la problème de leurs relations fonctionelles. Ann. Ocul., 157, 673.
Rochon-Duvigneaud, A. (1943). "Les Yeux et la Vision des Vertébrés." Paris: Masson et Cie.
Schultze, M. (1866). Zur Anatomie und Physiologie der Retina. Arch. Mikrosk. Anat., 2, 175.
Schultze, M. (1867). Über Stäbchen und Zapfen der Retina. Arch. Mikrosk. Anat., 3, 215.
Sillman, A.J. (1973). Avian Vision. In "Avian Biology," Vol. III (D.S. Farner, J.R. King, and K.C. Parkes, Eds.). New York: Academic Press.
Sillman, A.J., D.A. Bolnick, L.W. Haynes, A.E. Walter, and E.R. Loew. (1981). Microspectrophotometry of the photoreceptors of palaeognathous birds—the emu and the tinamou. J. Comp. Physiol., 144, 271.
Slonaker, J.R. (1918). A physiological study of the anatomy of the eye and its accessory parts of the English sparrow (*Passer domesticus*). J. Morphol., 31, 351.
Snyder, A.W., and W.H. Miller. (1978). Telephoto lens system of falconiform eyes. Nature (London), 275, 127.
Soemmerring, D.W. (1818). De oculorum hominis animaliumque sectione horizontale commetatio. Gottingen: Vandenhoeck and Ruprecht.
Thompson, G. (1971). The photopic spectral sensitivity of gulls measured by electroretinographic and pupillometric methods. Vision Res., 11, 719.
Thomson, A. (1928). The riddle of the pecten with suggestions as to its use. Trans. Ophthalmol. Soc. U.K., 48, 293.
Thomson, A. (1929). The pecten, considered from an environmental point of view. Ibis, 12, 608.
van Genderen Stort, A.G.H. (1887). Über Form- und Ortsveränderungen der Netzhautelemente unter Einfluss von Licht und Dunkel. Albrecht von Graefes Arch. Ophthalmol. 33, 229.
Vrabec, F. (1958). Anatomical study of the pecten in *Dromaeus novaehollandiae.* Acta Soc. Zool. Bohemoslov., 22, 31.
Wald, G. (1937). Photo-labile pigments of the chicken retina. Nature (London), 140, 545.
Wald, G., P.K. Brown, and P.H. Smith. (1955). Iodopsin. J. Gen. Physiol., 38, 623.
Wald, G., and H. Zussman. (1938). Carotenoids of the chicken retina. J. Biol. Chem., 122, 449.
Walls, G.L. (1937). Significance of the foveal depression. Arch. Ophthalmol., 18, 912.
Walls, G.L. (1940). Postscript on image expansion by the foveal clivus. Arch. Ophthalmol., 23, 831.
Walls, G.L. (1942). "The Vertebrate Eye and Its Adaptive Radiation." Bloomfield Hills, Michigan: Cranbrook Inst. Sci. (Bull. 19).
Walls, G.L., and H.D. Judd. (1933). The intraocular filters of vertebrates. Br. J. Ophthalmol., 17, 641.
Watson, J.B. (1915). Studies on the spectral sensitivity of birds. Carnegie Inst. Wash. Publ. (Marine Biol.), 7, 87.
Wessels, R.H.A. (1974). Tetrachromatic vision in the daw (*Corvus monedula L.*). Thesis, Utrecht.
Wingstrand, K.G., and O. Munk. (1965). The pecten oculi of the pigeon with particular regard to its function. Biol. Skr. K. Dan. Vidensk. Selsk., 14, 1.
Winkel, W. (1969). Experimentelle Untersuchungen an Zuckervögeln (Coerebidae) im Funktionskreis der Nahrungssuche: Über die Bedeutung von Farben, Formen und Zuckerkonzentrationen. Z. Tierpsychol., 26, 573.
Wood, C.A. (1907). The eyes and eyesight of birds, with special reference to the appearances of the fundus oculi. Ophthalmology, 3, 377.
Wood, C.A. (1917). "The Fundus Oculi of Birds especially as viewed by the Ophthalmoscope." Chicago: Lakeside Press.

Ear and Hearing

Ashmore, J.F., and I.J. Russell. (1983). Sensory and effector functions of vertebrate hair cells. J. Submicrosc. Cytol., 15, 163.
Baumel, J.J., A.J. King, A.M. Lucas, J.E. Breazile, H.E. Evans (Eds.). (1979). "Nomina Anatomica Avium." New York: Academic Press.
Benjamins, C.E., and E. Huizinga. (1927). Untersuchungen über die Funktion des Vestibularapparates bei der Taube. Pfleuger's Arch. Gesamte Physiol., 217, 105.
Boord, R.L. (1960). The efferent cochlear bundle in the caiman and pigeon. Exp. Neurol., 3, 225.
Boord, R.L. (1969). The anatomy of the avian auditory system. Ann. N.Y. Acad. Sci., 167, 186.
Boord, R.L., and G.L. Rasmussen. (1963). Projection of the cochlear and lagenar nerves on the cochlear nuclei of the pigeon. J. Comp. Neurol., 120, 463.
Brand, A.R., and P.P. Kellogg. (1939a). Auditory responses of starlings, English sparrows and domestic pigeons. Wilson Bull., 51, 38.
Brand, A.R., and P.P. Kellogg. (1939b). The range of hearing of canaries. Science, 90, 354.
Bremond, J.C. (1963). Acoustic behaviour of birds. In "Acoustic Behaviour of Animals" (R.G. Busnel, Ed.). Amsterdam: Elsevier, p. 709.
Carlström, D., A. Engstrom, and S. Hjorth. (1953). Electron microscopic and x-ray diffraction studies of statoconia. Laryngoscope, 63, 1052.
Correia, M.J., J.P. Landolt, and E.R. Young. (1974). The sensura neglecta in the pigeon: A scanning electron and light microscopic study. J. Comp. Neurol., 154, 303.
de Burlet, H.M. (1934). Vergleichende Anatomie des statoakustischen Organs. (b) Die mittlere Ohrsphäre. In "Handbuch der vergleichenden Anatomie der Wirbeltiere" Vol. 2/2 (L. Bolk et al., Eds.). Urban and Schwarzenberg, p. 1381.
de Vincentiis, M., and F. Marmo. (1966). The ^{45}Ca turnover in the membranous labyrinth of chick embryos during development. J. Embryol. Exp. Morphol., 15, 349.
Dohlman, G. (1959). Modern views on vestibular physiology. J. Laryngol. Otol. Rhinol., 73, 154.
Dohlman, G.F. (1964). Secretion and absorption of endolymph. Ann. Otol. Rhinol. Laryngol., 73, 708.
Dohlman, G. (1965). The mechanism of secretion and absorption of endolymph in the vestibular apparatus. Acta Otolaryngol., 59, 275.
Dohlman, G.F. (1966). Excretion and absorption of endolymph in the vestibular apparatus. In "Myotactic, Kinesthetic and Vestibular Mechanisms" (A.V.S. DeReuck and J. Knight, Eds.). Ciba Foundation Symposium. Boston: Little, Brown.
Dohlman, G.F., and R.L. Boord. (1964). The effects of cupu-

lar removal on the activity of ampullary structures in the pigeon. Acta Otolaryngol., 57, 507.

Dooling, R.J., J.A. Mulligan, and J.D. Miller. (1971). Auditory sensitivity and song spectrum of the common canary (Serinus canarius). J. Acoust. Soc. Am., 50, 700.

Duijm, M. (1951). On the head posture in birds and its relation to some anatomical features. Proc. K. Ned. Akad. Wet., Ser. C, 54, 202.

Edwards, E.P. (1943). Hearing ranges of four species of birds. Auk, 60, 239.

Evans, H.E. (1979). Organa sensoria. In "Nomina Anatomica Avium" (J.J. Baumel et al., Eds.). London: Academic Press, p. 505.

Firbas, W., and G. Müller, G. (1983). The efferent innervation of the avian cochlea. Hear. Res., 10, 109.

Flock, A. (1971). Sensory transduction in hair cells. In "Handbook of Sensory Physiology," Vol. 1 (W. Lowenstein, Ed.). Berlin: Springer-Verlag, p. 396.

Freye, H.A. (1952–1953). Das Gehörorgan der Vögel. Wiss. Z. Martin-Luther Univ. Halle-Wittenberg, Math. Naturwissen. Reihe, 2, 5, 267.

Freye, H.A. (1953). Die Asymmetrie des Ohres der Waldohreule (*Asio otus* L.) Beitr. Vogelk. (Leipzig), 3, 231.

Freye-Zumpfe, H. (1952–1953). Befunde im Mittelohr der Vögel. Wiss. Z. Martin-Luther Univ. Halle-Wittenberg, Math. Naturwissen. Reihe, 2, 8, 445.

Friedmann, I. (1962). The cytology of the ear. Br. Med. Bull, 18, 209.

Frings, H., and B. Slocum (1958). Hearing ranges for several species of birds. Auk, 75, 99.

Gardner, E.D., D.J. Gray, and R. O'Rahilly. (1975). "Anatomy," 4th Ed. Philadelphia: W.B. Saunders.

Goodley, L.B., and R.L. Boord. (1966). Quantitative analysis of the hair cells of the auditory papilla of the pigeon. Am. Zool., 16, 542.

Granit, O. (1941). Beiträge zur Kenntnis des Gehörsinns der Vögel. Ornis Fenn., 18, 49.

Greenewalt, C.H. (1968). "Bird Song: Acoustics and Physiology." New York: Random House (Smithsonian Institute Press).

Griffin, D.A. (1953). Acoustic orientation in the oil bird, *Steatornis*. Proc. Natl. Acad. Sci. U.S.A., 39, 884.

Gulick, W.L. (1971). "Hearing. Physiology and Psychophysics." New York: Oxford University Press.

Heise, G.A. (1953). Auditory thresholds in the pigeon. Am. J. Psychol., 66, 1.

Hillman, D.E. (1969). New ultrastructural findings regarding a vestibular ciliary apparatus and its possible functional significance. Brain Res., 13, 407.

Hodges, R.D. (1974). "Histology of the Fowl." New York: Academic Press.

Hudspeth, A.J. (1982). Extracellular current flow and the site of transduction by vertebrate hair cells. J. Neurosci., 2, 1.

Hudspeth, A.J., and R. Jacobs. (1979). Stereocilia mediate transduction in vertebrate hair cells. Proc. Natl. Acad. Sci. U.S.A., 76, 1506.

Hunter-Duvar, I.M. (1983). An electron microscopic study of the vestibular sensory epithelium. Acta Otolaryngol., 95, 494.

Ishiyama, E., R.A. Cutt, and E.W. Keels. (1970). Ultrastructure of the tegmentum vasculosum and transitional zone. Ann. Otol. Rhinol. Laryngol., 79, 998.

Jahnke, V., P.-G. Lundquist, and J. Wersäll. (1969). Some morphological aspects of sound perception in birds. Acta Otolaryngol., 67, 583.

Jørgensen, J.M. (1970). On the structure of the macula lagenae in birds with some notes on the avian maculae utriculi and sacculi. Vidensk. Medd. Dan. Naturh. Foren., 133, 121.

Katsuki, Y. (1965). Comparative neurophysiology of hearing. Physiol. Rev., 45, 380.

Knecht, S. (1940). Uber den Gehörsinn und die Musikalität der Vögel. Z. Vgl. Physiol., 27, 169.

Kreithen, M. L. (1979). The sensory world of the homing pigeon. In "Neural Mechanisms of Behavior in the Pigeon" (A.M. Granda and J.H. Maxwell, Eds.). New York: Plenum Press, Chapter 3.

Kreithen, M.L., and D.B. Quine. (1979). Infrasound detection by the homing pigeon: A behavioral audiogram. J. Comp. Neurol., 129A, 1.

Kuwamoto, K., and F. Altmann. (1967). The atypical epithelial formations of the utricle. Arch. Otolaryngol., 85, 561.

Landolt, J.P., M.J. Correia, E.R. Young, R.P.S. Cardin, and R.C. Sweet. (1975). A scanning electron microscopic study of the morphology and geometry of neural surfaces and structures associated with the vestibular apparatus of the pigeon. J. Comp. Neurol., 159, 257.

Lim, D.J. (1974). The statoconia of the non-mammalian species. Brain Behav. Evol., 10, 37.

Lowenstein, O., and J. Wersäll. (1959). A functional interpretation of the electron-microscopic structure of the sensory hairs in the cristae of the elasmobranch *Raja clavata* in terms of directional sensitivity. Nature (London), 184, 1807.

Lüdtke, H. and H. Schölzel (1966). Die Morphogenese der Cristae im Labyrinth des Hühnchens. Zool. Jahrb. Abt. allg. Zool. Physiol. 72, 291.

Money, K.E., L. Bonen, J.D. Beatty, L.A. Kuehn, M. Sokoloff, and R.S. Weaver. (1971). Physical properties of fluids and structures of vestibular apparatus of the pigeon. Am. J. Physiol., 220, 140.

Money, K.E., and M.J. Correia. (1972). The vestibular system of the owl. Comp. Biochem. Physiol. A, 42, 353.

Mygind, S.H. (1965). Physiological interpretation of the anatomy of the labyrinth. Acta Otolaryngol., 59, 264.

Novick, A. (1959). Acoustic orientation in the cave swiftlet. Biol. Bull. (Woods Hole, Mass.), 117, 497.

Okajima, K. (1913). Macula and Pars acustica neglecta. Ergeb. Anat. Entwicklungsgesch., 21, 143.

Payne, R.S. (1971). Acoustic location of prey by barn owls (*Tyto alba*). J. Exp. Biol., 54, 535.

Pohlman, A.G. (1921). The position and functional interpretation of the elastic ligaments in the middle-ear region of *Gallus*. J. Morphol., 35, 229.

Portmann, A. (1950). Les Organes des Sens. In "Traité de Zoologie," Vol. 15 (P.P. Grassé, Ed.). Paris: Masson. Pp. 213–220.

Pumphrey, R.J. (1948). The sense organs of birds. Ibis, 90, 171.

Pumphrey, R.J. (1961). Sense Organs: Hearing. In "Biology and Comparative Physiology of Birds," Vol. II (A.J. Marshall, Ed.). New York: Academic Press, p. 69.

Rosenhall, U. (1970). Some morphological principles of the vestibular maculae in birds. Arch. Klin. Exp. Ohren.- Nasen Kehlkopfheilkd., 197, 154.

Rosenhall, U. (1971). Morphological patterns of the organ of Corti in birds. Arch. klin. Exp. Ohren.- Nasen Kehlkopfheilkd., 200, 42.

Sachs, M.B., J.M. Sinnott, and R.D. Hienz. (1978). Behavioral and physiological studies of hearing in birds. Fed. Proc. Fed. Am. Soc. Exp. Biol., 37, 2329.

Schwartzkopff, J. (1949). Uber Sitz und Leistung von Gehor und Vibrationssinn bei Vögeln. Z. Vgl. Physiol., 31, 527.

Schwartzkopff, J. (1952). Uber den Gehörsinn der Vögel. J. Ornithol., 93, 91.

Schwartzkopff, J. (1955a). On the hearing of birds. *Auk, 72,* 340.
Schwartzkopff, J. (1955b). Schallsinnesorgane, ihre funktion und biologische Bedeutung bei Vögeln. Proc. Int. Ornithol. Congr. XI, 1954, p. 189.
Schwartzkopff, J. (1957). Die Grössenverhältnisse von Trommelfell Columellafussplatte und Schnecke bei Vögeln verschiedenen Gewichts. Z. Morphol. Oekol. Tiere, 45, 365.
Schwartzkopff, J. (1962). Zur Frage des Richtungshören von Eulen (Striges). Z. Vgl. Physiol., 45, 570.
Schwartzkopff, J. (1968). Structure and function of the ear and of the auditory brain areas in birds. In "Hearing Mechanisms in Vertebrates" (A.V.S. DeReuck and J. Knight, Eds.). Boston: Little, Brown.
Schwartzkopff, J. (1973). Mechanoreception. In "Avian Biology," Vol. III (D.S. Farmer and J.R. King, Eds.). New York: Academic Press, Chapter 7.
Smith, G. (1904). The middle ear and columella of birds. Q. J. Microsc. Sci., 48, 11.
Stellbogen, E. (1930). Uber das äussere und mittlere Ohr des Waldkauzes (*Syrnium aluco* L.). Z. Morphol. Oekol. Tiere, 19, 686.
Stewart, P.A. (1955). An audibility curve for two ring-necked pheasants. Ohio J. Sci., 55, 122.
Stresemann, E. (1934). Aves. In "Kukenthal's Handbuch der Zoologie," Vol. 7 (2). Berlin: DeGruyter.
Takasaka, T., and C.A. Smith. (1969). Further observations on the pigeon's basilar papilla. Anat. Rec., 163, 343.
Takasaka, T., and C.A. Smith. (1971). The structure of the pigeon's basilar papilla. J. Ultrastruct. Res., 35, 20.
Tanaka, K., and C.A. Smith. (1978). Structure of the chicken's inner ear: SEM and TEM study. Am. J. Anat., 153, 251.
Thorpe, W.H. (1961). "Bird Song." London: Cambridge University Press.
Tiedemann, F. (1910). Anatomie und Naturgeschichte der Vögel, Teil 1. Landshut.
Tilney, L.G., and J.C. Saunders. (1983). Actin filaments, stereocilia and hair cells of the bird cochlea. 1. Length, number, width and distribution of stereocilia of each hair cell are related to the position of the hair cell on the cochlea. J. Cell Biol., 96, 807.
Trainer, J.E. (1946). "The Auditory Acuity of Certain Birds." Ph.D. Thesis, Cornell University.
Turkewitsch, B.G. (1934). Zur Anatomie des Gehörorgans der Vögel. Z. Anat. Entwicklungsgesch., 103, 551.
Vasquez, C.S. (1955). Calcareous formation in the endolymph sac of chicken embryos. Ann. Otol. Rhinol. Laryngol, 64, 1019.
Vigh, B., B. Aros, T. Wenger, I. Tork, and P. Zarand. (1963). Ependymosecretion (ependymal neurosecretion) III. Gomori-positive material in the choroid plexus and in the membranous labyrinth in different vertebrates. Acta Biol. Hung., 13, 347.
Vinnikov, Y.A., I. Osipova, L.K. Titova, and V.I. Govardovskii. (1965). Electron microscopy of the Corti's organ of birds. Zh. Obshch. Biol., 26, 138.
Vitali, G. (1911). Di un interessante derivato della prima fessura branchiale nel passero. *Anat. Anz., 39,* 219.
von Békésy, G. (1960). "Experiments in Hearing" (E.G. Wever, Trans.). New York: McGraw-Hill.
Wada, Y. (1924). Beiträge zur vergleichenden Physiologie des Gehörorganes. Pfleuger's Arch., 202, 46.
Wassiljew, M.P. (1933). Über das Tonunterscheidungsvermögen der Vögel für die hohen Töne. Z. Vgl. Physiol., 19, 424.
Werner, C.F. (1939). Die Otolithen im Labyrinth der Vögel, besonders beim Star und der Taube. J. Ornithol., 87, 10.
Werner, C.F. (1960). "Das Gehörorgan der Wirbeltiere und des Menschen." Leipzig:VEB Georg Thieme.
Werner, C.F. (1963). Schädel-, Gehirn- und Labyrinthtypen bei den Vögeln. Gegenbaur's Morphol. Jahrb., 104, 54.
Wersäll, J. (1956). Studies on the structure and innervation of the sensory epithelium of the cristae ampullares in the guinea pig. Acta Otolaryngol. [Suppl.] (Stockh.), 126, 1.
Wersäll, J. (1968). Efferent innervation of the inner ear. In "Structure and Function of Inhibitory Neuronal Mechanisms" (C. von Euler, Ed.). Oxford: Pergamon Press.
Wersäll, J., L. Gleisner, and P.-G. Lundquist. (1966). Ultrastructure of the vestibular end organs. In "Myotactic, Kinesthetic and Vestibular Mechanisms" (A.V.S. DeReuck and J. Knight, Eds.). Ciba Foundation Symposium, Boston: Little, Brown.
Wever, E.G. (1971). The mechanics of hair cell stimulation. Ann. Otol. Rhinol. Laryngol., 80, 786.
Wever, E.G., and C.W. Bray. (1936). Hearing in the pigeon as studied by the electrical responses of the inner ear. J. Comp. Physiol. Psychol., 22, 353.
Wever, E.G., P.N. Herman, J.A. Simmons, and D.R. Hertzler. (1969). Hearing in the black-footed penguin, *Spheniscus demersus,* as represented by the cochlear potentials. Proc. Natl. Acad. Sci. U.S.A., 63, 676.
Wever, E.G., M. Lawrence, and K.R. Smith. (1948). The middle ear in sound conduction. Arch. Otolaryngol., 48, 19.
Yashiki, K. (1968). Fine structure of the sensory epithelium of the lagena of birds. Yonago Acta Med., 12, 47.

Chemical Senses

Archer, A.T., and R.M. Glen. (1969). Observations on the behavior of two species of honey-guides *Indicator variegatus* (Lesson) and *Indicator exilis* (Cassin). Los Angeles County Mus. Contrib. Sci., 160, 1.
Audubon, J.J. (1826). Account of the habits of the turkey buzzard (*Vultura aura*), particularly with the view of exploding the opinion generally entertained of its extraordinary power of smelling. Edinburgh New. Phrlos. J., 2, 172.
Bang, B.G. (1971). Functional anatomy of the olfactory system in 23 orders of birds. Acta Anat., 58 (Suppl.), 1.
Bang, B.G., and F.B. Bang. (1959). A comparative study of the vertebrate nasal chamber in relation to upper respiratory infections. Bull. Johns Hopkins Hosp., 104, 107.
Bartholomew, G.A., and T.J. Cade. (1958). Effects of sodium chloride on the water consumption of house finches. Physiol. Zool., 31, 304.
Bartholomew, G.A., and R.E. MacMillan. (1960). The water requirements of mourning doves and their use of sea water and NaCl solutions. Physiol. Zool., 33, 171.
Bath, W. (1906). Die Geschmaksorgane der Vogel und Krokodile. *Arch. Biontol. (Berlin),* 1, 1.
Beidler, L.M. (Ed.). (1971). "Handbook of Sensory Physiology," Vol. 4, Pt. 1. Berlin and New York: Springer-Verlag.
Belman, A.L., and M.R. Kare. (1961). Character of salivary flow in the chicken. Poult. Sci., 40, 1377.
Benvenuti, S., V. Fiaschi, L. Fiore, and P. Papi. (1973). Homing performances of inexperienced and directionally trained pigeons subjected to olfactory nerve section. J. Comp. Physiol., 83, 81.
Berkhoudt, H. (1977). Taste buds in the bill of the Mallard (*Anas platyrhynchos* L.). Their morphology, distribution and functional significance. Neth. J. Zool., 27, 310.
Berkhoudt, H. (1983). Special sense organs: Structure and function of avian taste receptors. In "Form and Function in Birds," Vol. III. (A.S. Levy and J. McLelland, Eds.). Academic Press: New York, in press.
Brindley, L.D. (1965). Taste discrimination in bobwhite and Japanese quail. Anim. Behav., 13, 507.

Brindley, L.D., and S. Prior. (1968). Effects of age on taste discrimination in the bobwhite quail. Anim. Behav., 16, 304.

Cagan, R.H. (1973). Chemostimulatory protein: A new type of taste stimulus. Science, 181, 32.

Cane, V.R., and M.A. Vince. (1968). Age and learning in quail. Br. J. Psychol., 59, 37.

Cole, E.C. (1941). "Comparative Histology." Philadelphia: Blakiston.

Davison, V.E. (1962). Taste, not color, draws birds to berries and seeds. Audubon, 64, 346.

Dmitrieva, N.A. (1981). Fine structural pecularities of the taste buds of the pigeon *Columba livia.* Akad. Nauk USSR, Tr., Tom. XXIII, No. 8, 874.

Dubbledam, J.L. and H.J. Karten. (1978). The trigeminal system in the pigeon (*Columbia livia*). I. Projections of the Gasserian ganglion. J. Comp. Neurol., 180, 661.

Duncan, C.J. (1960). Preference tests and the sense of taste in the feral pigeon. Anim. Behav., 8, 54.

Elliot, R. (1937). Total distribution of taste buds on the tongue of the kitten at birth. J. Comp. Neurol., 66, 361.

Fuerst, F.F., and M.R. Kare. (1962). The influence of pH on fluid tolerance and preferences. Poult. Sci., 41, 71.

Gentle, M.J. (1971). The lingual taste buds of *Gallus domesticus* L. Br. Poult. Sci., 12, 245.

Gentle, M.J. (1972). Taste preferences in the chicken (*Gallus domesticus* L.). Br. Poult. Sci., 13, 141.

Gentle, M.J. (1975). Gustatory behavior of the chicken and other birds. In "Neural and Endocrine Aspects of Behaviors in Birds." (P. Wright, P.E. Caryl, and D.M. Vowles, Eds.). Amsterdam: Elsevier Scientific.

Gentle, M.J. (1978). Extra-longual chemoreceptors in the chicken (*Gallus domesticus*). Chem. Senses & Flavour, 3, 325.

Gentle, M.J. (1983). The chorda tympani nerve and taste in the chicken. Experientia, 39, 1002.

Gentle, M.J., and C. Harkin. (1979). The effect of sweet stimuli on oral behavior in the chicken. Chem. Senses & Flavour, 4, 183.

Grubb, T.C., Jr. (1972). Smell and foraging in shearwaters and petrels. Nature (London), 237, 404.

Grubb, T.C., Jr. (1974). Olfactory navigation to the nesting burrow in Leach's petrel (*Oceanodroma leucorrhoa*). Anim. Behav., 22, 192.

Halpern, B.P. (1963). Gustatory nerve responses in the chicken. Am. J. Physiol., 203, 541.

Harriman, G.E. (1967). Laughing gull offered saline in preference and survival tests. Physiol. Zool., 40, 273.

Harriman, G.E., and M.R. Kare. (1966). Tolerance for hypertonic saline solutions in herring gulls, starlings and purple grackles. Physiol. Zool., 39, 117.

Heinroth, O. (1938). "Aus dem Leben der Vögel." Berlin: Springer-Verlag.

Hyman, L.H. (1942). "Comparative Vertebrate Anatomy." Chicago: University of Chicago Press.

Jacobs, H.L. and M.L. Scott (1957). Factors mediating food and liquid intake in chickens. I. Studies on the preference for sucrose or saccharin solutions. Poult. Sci., 36, 8.

Johnson, A.K., and A.E. Fisher. (1973). Tolerance for quinine under cholinergic versus deprivation induced thirst. Physiol. Behav., 10, 613.

Kare, M.R. (1961). "Physiological and Behavioral Aspects of Taste." Chicago: University of Chicago Press, p. 13.

Kare, M.R., and J. Beily. (1948). The toxicity of sodium chloride and its relation to water intake in baby chicks. Poult. Sci., 27, 751.

Kare, M.R., and G.K. Beauchamp. (1976). Taste, smell and hearing. In "Duke's Physiology of Domestic Animals," 10th ed. (M.J. Swenson, Ed.). Ithaca: Comstock, Chapter 47.

Kare, M.R., and W. Medway. (1959). Discrimination between carbohydrates by the fowl. Poult. Sci., 38, 1119.

Kare, M.R., and H.L. Pick. (1960). The influence of the sense of taste on feed and fluid consumption. Poult. Sci., 39, 697.

Kare, M.R., and M.L. Scott. (1962). Nutritional value and feed acceptability. Poult. Sci., 44, 276.

Kare, M.R., and M.S. Ficken. (1963). Comparative studies on the sense of taste. In "Olfaction and Taste" (Y. Zotterman, Ed.). New York: Pergamon Press.

Kare, M.R., and O. Maller. (1967). Taste and food intake in domesticated and jungle fowl. J. Nutr., 92, 191.

Kitchell, R.L., L. Strom, and Y. Zotterman. (1959). Electrophysiological studies of thermal and taste reception in chickens and pigeons. Acta Physiol. Scand., 46, 133.

Landolt, J.P. (1970). Neural properties of pigeon lingual chemoreceptors. Physiol. Behav., 5, 1151.

Mangold, O. (1946). Die Nase der segelnden Vögel—ein Organ des Strömungssinnes? Die Naturwissenschaften, 33, 19.

Mason, J.R., and J.A. Maruniak. (1983). Behavioral and physiological effects of capsaicin in red-winged blackbirds. Pharmacol. Biochem. Behav., 19, 857.

Mason, J.R., and R.F. Reidinger. (1983a). Exploitable characteristics of neophobia and food aversions for improvements in rodent and bird control. In "Test Methods for Vertebrate Pest Control and Management Materials." Philadelphia: American Society for Testing and Materials. pp. 20–39.

Mason, J.R., and R.F. Reidinger. (1983b). Importance of color for methiocarb-induced taste aversions in red-winged blackbirds. J. Wildl. Manage., 47, 383.

Mason, J.R., and W.L. Silver (1983). Trigeminally mediated odor aversions in starlings. Brain Res., 269, 196.

Mason, J.R., J.F. Glahn, R.A. Dolbeer, and R.F. Reidinger, (1985). Field evaluation of dimethyl anthranilate as a bird repellent livestock feed additive. J. Wildl. Manage., 49: 636.

Matthes, E. (1934). "Geruchsorgan, Lubosch Handbuch der Vergleichenden Anatomie der Wirbeltiere, Groppert, Kallius," Vol. 11. Urban and Schwarzenberg.

McBurney, D.H. (1978). Psychological dimensions and perceptual analyses of taste. In "Handbook of Perception, VI-A: Tasting and Smelling" (E.C. Carterette and M.P. Friedman, Eds.). Academic Press: New York, p. 125.

Moncrieff, R.W. (1951). "The Chemical Senses." London: Hill, p. 172.

Neuhaus, W. (1963). On the olfactory sense of birds. In "Olfaction and Taste" (Y. Zotterman, Ed.). New York: Macmillan, p. 111.

Papi, F., L. Fiore, V. Fiaschi, and S. Benvenuti. (1973). An experiment for testing the hypothesis of olfactory navigation in homing pigeons. J. Comp. Physiol., 83, 93.

Parker, G.H. (1922). "Smell, Taste and Allied Senses in Vertebrates." Philadelphia: Lippincott, p. 192.

Payne, A. (1945). The sense of smell in snakes. J. Bomb. Nat. Hist. Soc., 45, 507.

Pfaffmann, C. (1978). The vertebrate phylogeny, neural code and integrative processes of taste. In "Handbook of Perception, VI-A: Tasting and Smelling" (E.C. Carterette and M.P. Friedman, Eds.). Academic Press: New York, p. 51.

Rensch, B., and R. Neunzig. (1925). Experimentelle Untersuchungen über den Geschmackssinn der Vögel, II. J. Ornithol., 73, 633.

Robinzon, B., Y. Katz, and J.G. Rogers. (1979). The involvement of the olfactory bulbs in the regulation of gonadal and thyroid activities of male red-winged blackbirds exposed to short-day light regime. Brain Res. Bull., 4, 339.

Rogers, J.G., and O. Maller. (1973). Effect of salt on the response of birds to sucrose. Physiol. Psychol., 1, 199.

Rozin, P., and J.W. Kalat. (1971). Specific hungers and poison avoidance as adaptive specializations of learning. Psychol. Rev., 78, 459.

Saito, I. (1966). Comparative anatomical studies of the oral organs of the poultry, V. Structure and distribution of taste buds of the fowl. Bull. Fac. Agric. Miyazaki Univ., 13, 95.

Shallenberger, R.J. (1973). Breeding biology, homing behavior and communication patterns of the Wedge-Tailed Shearwater, *Puffinus pacificus.* Ph.D. Dissertation, University of California, Los Angeles.

Shallenberger, R.J. (1975). Olfactory use in the Wedge-tailed Shearwater (*Puffinus pacificus*) on Manana Island, Hawaii. In "Olfaction and Taste, Vol. 5," (D.A. Deuter and J.P. Coghlan, Eds.). New York: Academic Press, p. 355.

Shumake, S.A., J.C. Smith, and D. Tucker. (1970). Olfactory intensity difference thresholds in the pigeon. J. Comp. Physiol. Psychol., 67, 64.

Silver, W.L., and J.A. Maruniak. (1980). Trigeminal chemoreception in the nasal and oral cavities. Chem. Senses, 6, 295.

Snyder, G.K., and T.T. Peterson. (1979). Olfactory sensitivity in the black-billed magpie and in the pigeon. Comp. Biochem. Physiol. A, 62, 921.

Soudek, S. (1929). The sense of smell in the birds. Proc. Int. Congr. Zool., 10th, 755.

Stager, K.E. (1967). Avian olfaction. *Am. Zool.,* 7, 415.

Stattleman, A.J., R.B. Talbot, and D.B. Coulter. (1975). Olfactory thresholds of pigeons (*Columba livia*), quail (*Colinus virginianus*) and chickens (*Gallus gallus*). Comp. Biochem. Physiol. A, 50, 807.

Tucker, D. (1965). Electrophysiological evidence for olfactory function in birds. Nature (London), 207, 304.

Van Kan, S. (1979). Touch and taste in the pigeon (*Columba livia domestica*). Internal report, Zool. Lab., Leiden.

Walker, J.C., D. Tucker, and J.C. Smith. (1979). Odor sensitivity mediated by trigeminal nerve in the pigeon. Chem. Senses Flavour, 4, 107.

Wallraff, H.G. (1979). Olfaction and homing in pigeons. A problem of navigation or of motivation. Naturwissenschaften, 66, 269.

Wallraff, H.G. (1980). Olfaction and homing in pigeons: Nerve-section experiments, critique, hypotheses. J. Comp. Physiol., 139, 209.

Warham, J. (1971). Aspects of breeding behaviors in the Royal Penquin, *Endyptes chrysolophus schlegeli.* Notornis, 18, 91.

Warner, R.L., L.Z. McFarland, and W.O. Wilson. (1967). Microanatomy of the upper digestive tract of the Japanese quail. Am. J. Vet. Res., 28, 1537.

Welty, J.C. (1975). "The Life of Birds." Philadelphia: W.B. Saunders.

Wenzel, B.M. (1968). The olfactory prowess of the Kiwi. Nature (London), 220, 1133.

Wenzel, B.M. (1973). Chemoreception. In "Avian Biology," Vol. III (D.S. Farner and J.R. King, Eds.). New York and London: Academic Press, Chapter 6.

Wenzel, B.M. (1980). Chemoreception in seabirds. In "Behavior of Marine Animals," Vol. 4 (J. Burger, B.L. Olla, and H.E. Winn, Eds.). New York: Plenum Press, pp. 41.

Wenzel, B.M., and M. Sieck. (1972). Olfactory perception and bulbar electrical activity in several avian species. Physiol. Behav., 9, 287.

Wilcoxon, H.C., W.B. Dragoin, and P.A. Kral. (1971). Illness-induced aversions in rat and quail: Relative salience of visual and gustatory cues. Science, 171, 826.

Willoughby, E.J. (1971). Drinking responses of the red crossbill (*Loxia curvirostra*) to solutions of NaCl, $MgCl_2$ and $CaCl_2$. Auk, 84, 828.

Wurdinger, I. (1979). Olfaction and feeding behavior in juvenile geese (*Anser a. anser* and *Anser domesticus*). Z. Tierpsychol., 49, 132.

Yang, R.S.H., and M.R. Kare. (1968). Taste response of a bird to constituents of arthropod defense secretions. Ann. Entomol. Soc. Am., 61, 781.

Ziegler, H.P. (1977). Trigeminal deafferentiation and feeding behavior in the pigeon: Dissociation of tonic and phasic effects. Ann. N.Y. Acad. Sci., 290, 331.

3
Muscle

A.L. Harvey and I.G. Marshall

General Structural Characteristics of Skeletal Muscle

Avian skeletal muscle is similar to mammalian skeletal muscle in that it contains the same types of contractile proteins arranged in the familiar striated pattern (Figure 3–1). As it is impossible to give a detailed account of all the properties of avian muscle, we have concentrated on reviewing the development and properties of the functionally important different fiber types, some aspects of the electrical properties of avian muscle membranes, innervation, and neuromuscular transmission. Because many of the basic properties of avian muscle structure and function have been reviewed comprehensively (Berger, 1960; George and Berger, 1966; Bowman and Marshall, 1972; van den Berge, 1975), we have concentrated on more recent work.

Development of Avian Muscle

Morphological studies of avian myogenesis have generally been restricted to chick muscle, and development has been examined in vivo and in tissue culture. Because of the difficulties associated with defining in situ which cells are destined to form skeletal muscle fibers, much of the knowledge of the early stages of myogenesis has been derived from tissue culture studies. By removing single cells from developing muscle and growing them as clones in culture, cells with the ability to form muscle can be identified. With this technique, the developmental change in numbers of muscle precursor cells (myoblasts) has been studied in chick limb muscle (Bonner and Hauschka, 1974). Clonable myoblasts appear on about day 3 in ovo and increase in numbers about six- to sevenfold in the next 6 days. Both in vivo and in vitro, skeletal muscle myogenesis involves fusion of spindle-shaped, uninucleated myoblasts (Figure 3–2) to form multinucleated cells (myotubes) that eventually grow into mature muscle fibers.

The early stages of myogenesis have been studied by transmission and scanning electron microscopy of chick embryo muscle cells in culture (Shimada, 1972a, b). Scanning electron micrographs have shown that myogenic cells grow close together; small projections extend from some myotubes, appearing to attach the cell to adjacent myoblasts and to the surface of the culture dish. Transmission electron micrographs reveal a cell-surface material possibly involved in cellular adhesion. Between pairs of myogenic cells are seen numerous close junctions with an intercellular distance of 2.5–10 nm and some focal tight junctions with no discernible gap. The fusion process probably initially involves the formation of close contact between cells, followed by the appearance of vesicles and tubules between the adjacent cytoplasms. Finally, remnants of cell membranes disappear, and a confluent cytoplasm is formed.

Both the thick myosin (15–16 nm in diameter) and

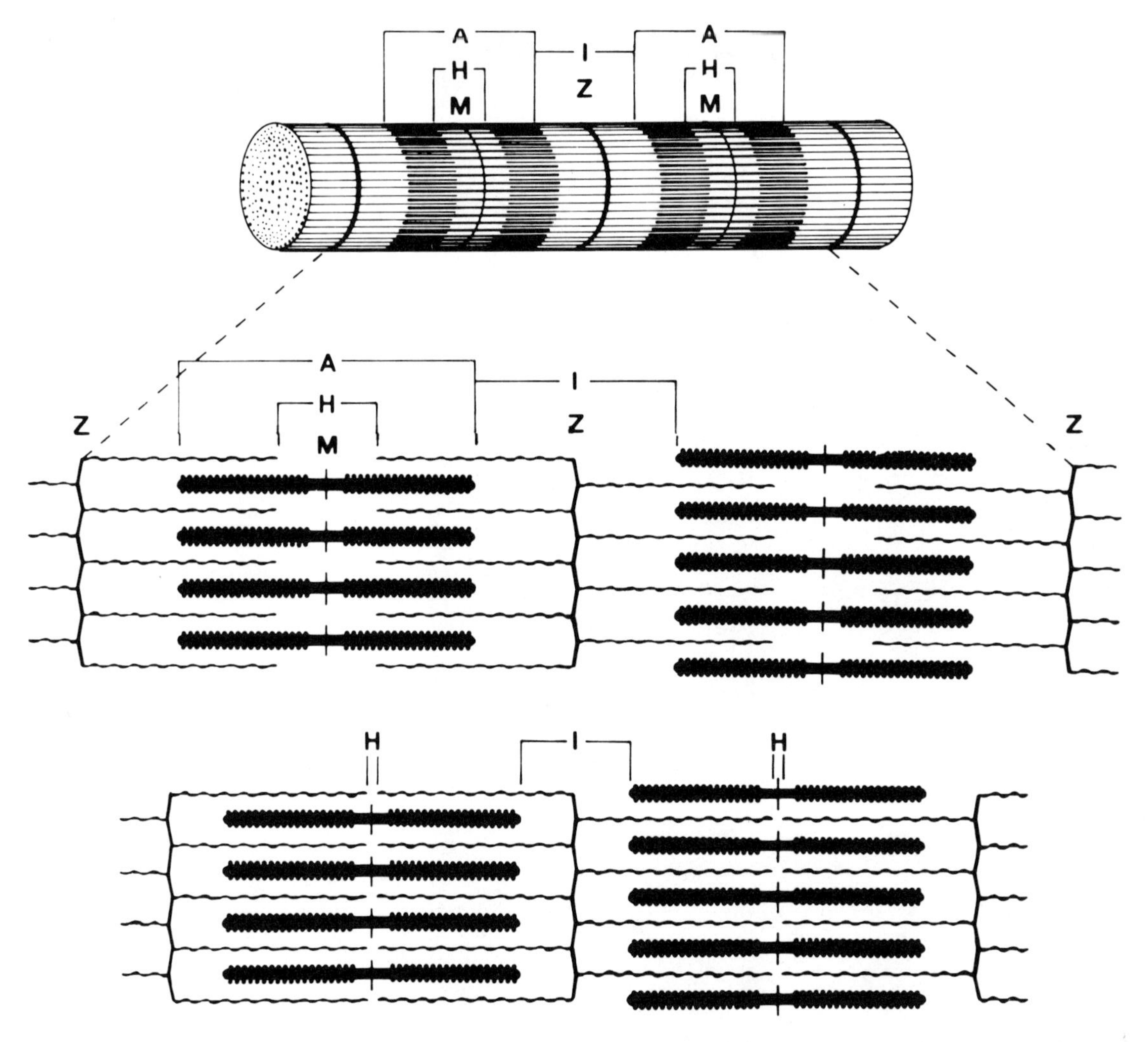

FIGURE 3–1. Schematic representation of the band pattern of a striated muscle myofibril related to the arrangement of the actin (thin) and myosin (thick) filaments. The area between two adjacent *z* lines is called a sarcomere, two of which are shown. *M* line is center of sarcomere (see text). The myofibrils within the muscle fiber are aligned in parallel and the band pattern stretches right across the muscle fiber as a whole. In the contracted state (bottom) the thick and thin myofilaments slide over one another but neither changes in length. This causes a narrowing of the H and I bands, but no change in the width of the A band, which reflects the constant length of the myosin filaments. (Bowman, 1980.)

the thin actin (5–6 nm in diameter) filaments are synthesized by clusters of ribosomes (polysomes) in the cytoplasm of muscle cells. The extent of synthesis greatly increases following fusion of myoblasts. During growth of myotubes, there is a progressive organization of myofibrils until most of the cell has the cross-striated pattern similar to that of adult skeletal muscle (Shimada *et al.,* 1967; Askanas et al., 1972).

Sarcoplasmic reticulum develops in isolated portions from rough-surfaced endoplasmic reticulum during the earliest myotube stage in embryonic chick muscle in vivo and in vitro (Ezerman and Ishikawa, 1967; Shimada et al., 1967, Larson et al., 1970). During subsequent development, these isolated portions of sarcoplasmic reticulum connect and form a network around the contractile filaments. The other specialized tubular system, the transverse tubule, develops more slowly than the sarcoplasmic reticulum. The T tubules are formed by invagination of the surface membrane of myotubes; at first they are little more than shallow vesicles connected to the sarcolemma, but they gradually project deeper into the myotube to make contact with the sarcoplasmic reticulum.

The next stages in the development of muscles are the growth of the muscle and the differentiation of different fiber types. As reviewed by Burleigh (1974),

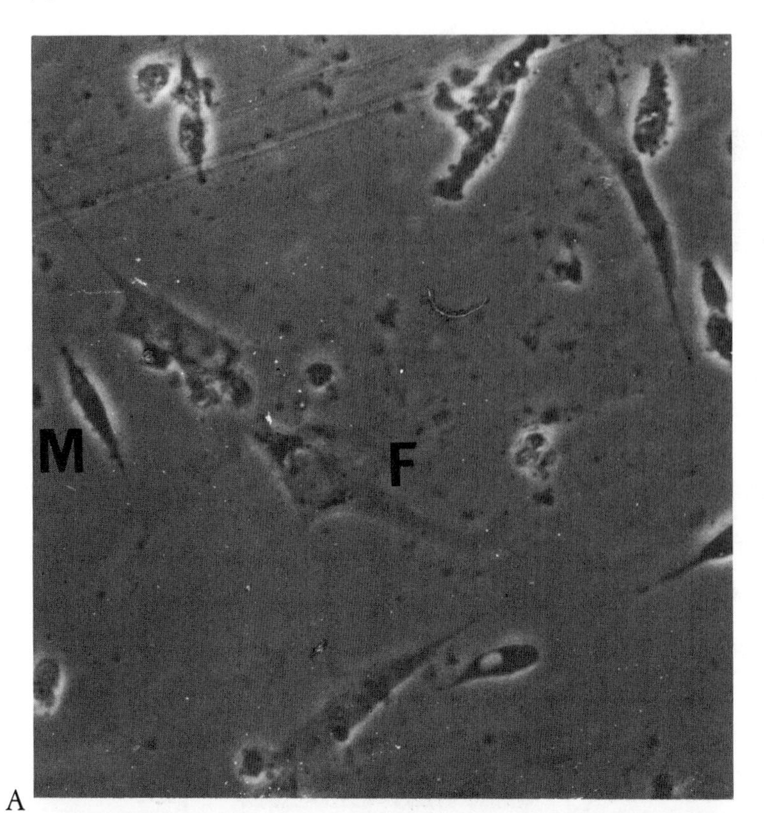

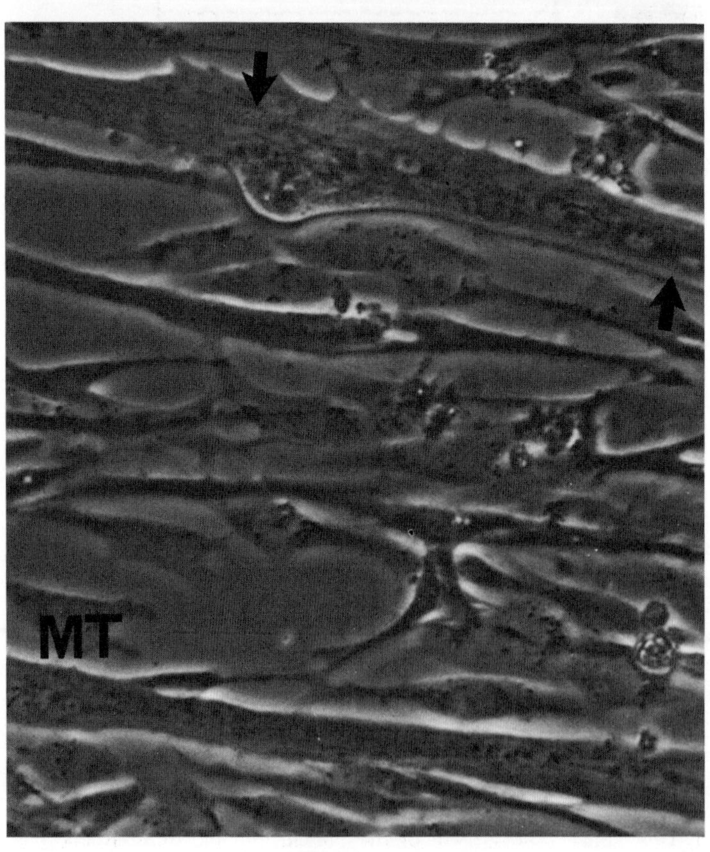

FIGURE 3–2. Phase-contrast micrographs of chick embryo skeletal muscle cells in culture. **(A)** 24-hr cell culture showing myoblasts (*M*) and fibroblasts (*F*). **(B)** 7-day culture showing several long myotubes (*MT*), one of which shows extensive cross-striations (*arrows*). Scale bar for both micrographs is 50 μm.

Goldspink (1974), and Vrbová et al. (1978), the increase in muscle mass is dramatic. Individual fibers increase in length by addition of new sarcomeres and in breadth as the number of myofibrils per fiber increases. For example, in the latissimus dorsi muscles of the chicken, the mean fiber diameter increases from 4 to 6 μm in 18-day-old embryos to 40–60 μm in 8-month-old chickens (Shear and Goldspink, 1971).

The growth of muscles after hatching is dependent on the presence of normal contractile activity. When posterior latissimus dorsi (PLD) muscles of chickens are immobilized for periods up to 11 months, there is a dramatic reduction in muscle fiber size (Shear, 1978, 1981). This atrophy is completely reversible in mature fowls, but is only partially reversible when the immobilization begins immediately after hatching (see also Chapter 14).

Muscle Fiber Types

Some fibers adapt for rapid, intermittent contraction, whereas others adapt for more continuous contraction. These specializations involve changes in the structure and biochemistry of the muscle fibers. Muscles can, therefore, be described as slow or fast contracting. However, the classification of muscle fiber types is more complicated than this suggests. The color of the muscles (red or white) does not adequately describe the variety of fiber types that exists, for most muscles contain a mixture of different types of fibers, and more than two types can be distinguished (e.g., Khan, 1976; Toutant et al., 1980). In a detailed examination of the fiber composition of several muscles of the chick (Barnard et al., 1982), five major fiber types were recognized on the basis of biochemical and morphological criteria (Table 3–1). With the exception of the multiply-innervated slow fibers (common in avian but not mammalian muscle), the fiber types in the chicken muscle are similar to those found in mammalian muscle.

In addition to the morphological and enzymatic differences indicated in Table 3–1, avian white and red fibers differ in their ultrastructure. Generally, white fibers have a very definite fibrillar appearance (*Fibrillenstruktur*) when viewed with a microscope, whereas red fibers have a more granular and indefinite appearance (*Felderstruktur*).

In fibers with a *Fibrillenstruktur* (Figure 3–3A), the myofibrils are polygonal in cross section and uniform in diameter. The fibers are regularly separated from each other by a granular sarcoplasm. Cross-striations in the form of dark A (anisotropic) and light I (isotropic) bands are evident, each I band being bisected by a smooth Z line that runs straight across the width of the fibril (see Figure 3–1). The H zone, where only thick filaments are present, is evident in the middle region of the A band. To either side of the H zone, the actin and myosin filaments interdigitate and cause

TABLE 3–1. Comparison of different fiber types in chicken muscle[a]

	Twitch fibers			Tonic fibers	
	I	IIA	IIB	IIIA	IIIB
Histochemical criteria					
ATPase (pH 9.4)	No staining	Strong	Strong	Medium	Strong
ATPase (pH 4.6)	Strong	No or weak	Weak	Weak	Medium
ATPase (pH 4.3)	Strong	No staining	No staining	Weak	Medium
NADH-TR[b]	Medium	Weak or medium	No staining	Medium	Medium or strong
Phosphorylase	None or weak	Strong	Strong	Weak	Medium
Fiber innervation	Multiple	Focal	Focal	Multiple	Multiple
Histological characteristics					
Fiber shape	Polygonal	Polygonal	Polygonal	Rounded	Rounded
Fascicle shape	Polygonal	Polygonal	Polygonal	Rounded	Rounded
Mitochondrial density	Very high	High	Low	Very high	Very high
Fiber lipid droplets	No	Yes	No	No	No
Relative fiber size	Small/medium	Medium	Medium	Large	Medium
Myonuclei distribution	Peripheral	Usually peripheral	Usually central	Peripheral	Peripheral
Fiber type composition (%)					
Pectoral	0	<1	>99	0	0
PLD	<3	5–20	80–95	0	0
ALD	0	0	0	65–80	20–35
Sartorius (red)	30–45	35–50	15–25	0	0
Sartorius (white)	0	10–20	80–90	0	0
Plantaris	0	0	0	65–75	25–35

[a] Adapted from Barnard *et al.* (1982).
[b] NADH-tetrazolium reductase.

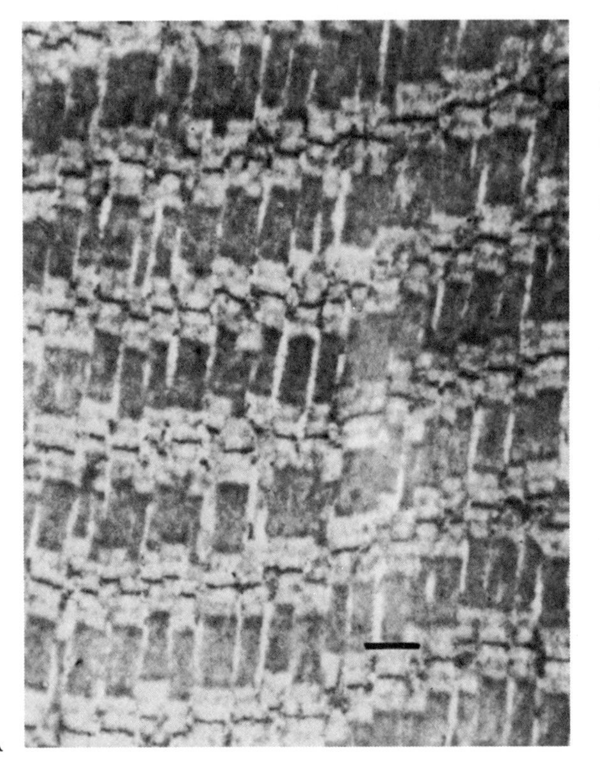

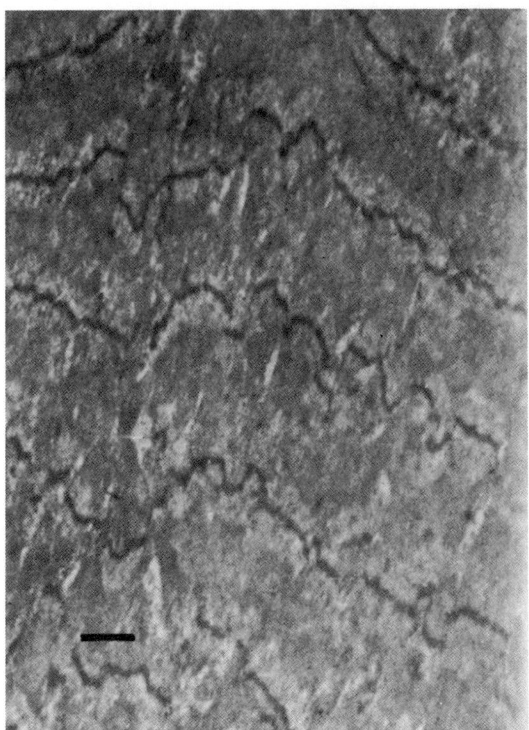

FIGURE 3–3. Electron micrographs of longitudinal sections of muscles from 2- to 3-month-old chickens. (**A**) White fibers have a definite fibrillar appearance (*Fibrillenstrukter*). This is a section from the posterior latissimus dorsi. (**B**) Red fibers have a more granular and indefinite appearance (*Felderstruktur*); a section from the anterior latissimus dorsi. Scale mark, 1 μm. See text for more details. (From Hess, 1961.)

the greatest optical density. An M band is seen bisecting every H zone in every sarcomere. In the chicken posterior latissimus dorsi (PLD) muscle, the myofibrils are 0.5–1 μm in diameter; the A filaments are 1.55–1.6 μm and the I filaments 1.05 μm long. The I filaments thicken as they approach the Z line and overlap 30–40 nm with those of the next sarcomere.

In fibers with a *Felderstruktur* (Figure 3–3B), the fibrils are irregular in size and distribution and appear to join with each other at points along the length of the fibrils. In the chicken anterior latissimus dorsi (ALD) muscle, they are often ribbon shaped in cross section, having diameters of 0.5–1 μm by 2–5 μm. The fibrils are not regularly surrounded by sarcoplasm and granules, and the Z lines take a zigzag course across the width of the fibril. A, I, H, and M bands are evident as in fibers with *Fibrillenstruktur,* although the M line is somewhat less distinct in *Felderstruktur* fibers. The A and I filaments of the ALD muscle are similar in length to those of the PLD muscle, but the I filaments do not show any regular arrangement at the level of the Z line, where they overlap those of the next sarcomere by about 100 nm. An unusual feature is a network of filaments encircling the fibrils at the Z line level.

The T tubules and sarcoplasmic reticulum are less well developed and less regular in arrangements in *Felderstruktur* than in *Fibrillenstruktur* fibers. (Carbohydrate metabolism of muscle is discussed in Chapter 13.)

Innervation

Final maturation and long-term survival of skeletal muscle depends on innervation by the motor neurones. The development of the neuromuscular junction thus allows individual muscle fibers to assume their adult role. In vivo, nerve fibers growing alongside developing muscle tissue can be revealed by silver staining. Nerve–muscle contacts have been localized by staining for acetylcholinesterase, and the development of these connections has been followed in many muscles of many species, including the chick (Hirano, 1967; Atsumi, 1971; Landmesser and Morris, 1975; Kikuchi and Ashmore, 1976; Atsumi, 1977; Burrage and Lentz, 1981; Bourgeois and Toutant, 1982; Adachi, 1983). Nerve fibers can be seen in muscle when myoblasts are the predominant cell type, but at this stage no evidence of specialization of the nerve ending or of localization of acetylcholinesterase activity is seen. Following the more widespread development of myotubes, occasional nerve–muscle contacts are seen. For example, in the intercostal muscles of chick embryos, a few neuromuscular contacts can be seen to occur by day 6 of incubation. Development of the first primitive neuromuscular junctions takes place between days 7 and 10, and by days 15–16 fully developed muscle fibers with mature neuromuscular junctions can be found. Subsequently, neuromuscular junction size increases with growth, but the morphology is essentially unchanged (Atsumi, 1971). The morphological development of neuromuscular junctions of embryonic chick ALD muscles has been related to the onset of transmitter release in a combined morphological and electrophysiological study by Bennett and Pettigrew (1974), and the sequence of innervation was studied in chick PLD muscles by Bourgeois and Toutant (1982). Adachi (1983) pointed out that the neuromuscular junctions of different muscles mature at different times, with proximal muscles preceding distal ones.

In the chicken, the red and white muscle fibers have different innervation patterns. White fibers containing a *Fibrillenstruktur* are focally innervated by one or only a few nerve terminals, whereas the *Felderstruktur*-containing red fibers are multiply innervated by many nerve terminals (Ginsborg and Mackay, 1961; Hess, 1961). Endplate structure in focally- and multiply-innervated fibers differs; endplates in focally-innervated fibers are elevated on the fiber and postjunctional folds are seen as in mammalian muscle, whereas the endplate in multiply-innervated muscle is not elevated and postjunctional folds are absent (Hess, 1967). Intracellular microelectrode studies have shown that the maximum distance between adjacent junctions is of the same order of magnitude as the space constant of the fiber in multiply-innervated fibers (see later section). Ginsborg and Mackay (1961) found that the average distance between neuromuscular junctions in the ALD of a 2-week-old chick was 225 μm. At 15 weeks this distance was increased to 790 μm, and Hess (1961) found a distance of 1000 μm in adult chickens. The distance appears to be directly proportional to the length of the muscle fiber suggesting that the number of junctions remains constant during growth. Ginsborg and Mackay (1961) estimated that there were about 80 junctions on each of the fibers of the ALD that extend the length of the muscle.

Electrical Properties of Muscle Fibers

The resting membrane potential of mature fibers in avian muscle is similar to that found in mammalian and amphibian muscles, i.e., around -70 to -90 mV. The mechanisms of the development of the mature resting potential are not fully understood. Adult muscle fiber membranes are much more permeable to K^+ than to Na^+, and this differential permeability develops during growth.

The response of a muscle fiber to an electrical stimulus is determined by its passive electrical properties. For example, if there is a high input resistance, the voltage response to a given current pulse will be large, and if there are long space and time constants, the response will spread over a large area of the membrane. In birds there are marked differences between the characteristics of the membrane passive properties of multiply—and focally-innervated fibers (Fedde, 1969; Gordon et al., 1977b). These differences relate to the fact that propagating action potentials are generated in focally-innervated fibers, but not necessarily in multiply-innervated fibers. Thus, in focally-innervated fibers of the chick PLD, propagated action potentials are elicited by a single nerve impulse or by direct muscle stimulation (Ginsborg, 1960b; Hník *et al.,* 1967). In multiply innervated chick muscle, the situation is less clear cut. In vivo, the chick ALD muscle responds to single-shock nerve stimulation with only local endplate potentials. Propagated action potentials can only be elicited in vivo by closely spaced twin pulses applied to the motor nerve, or by single shocks after a period of tetanic stimulation (Jirmanová and Vyklický, 1965; Hník et al., 1967). However, in vitro either nerve stimulation or direct muscle stimulation can elicit propagated action potentials in the ALD muscle (Ginsborg, 1960b).

In focally-innervated twitch fibers, propagated action potentials are initiated at the endplate. These fibers have short spaces (<1 mm) and time constants (3–4 msec) so that the potential change evoked by acetylcholine decays rapidly within a short distance of the endplate (Fedde, 1969; Entrikin and Bryant, 1975; Lebeda and Albuquerque, 1975; Gordon et al., 1977b). In multiply-innervated fibers, contraction is usually stimulated directly by the spread of the acetylcholine-induced potential change. To facilitate this, these fibers have long spaces (~2 mm) and time constants (~30 msec). To allow the greatest possible voltage change in response to the current flow produced by the transmitter, multiply-innervated fibers have high membrane resistances (3000–4000 Ω-cm^2) compared to those of focally innervated fibers (500–600 Ω-cm^2). At 14 days in ovo, the electrical properties of ALD and PLD fibers are similar. The properties of the PLD change within the first two weeks of hatching. Some of the changes are associated with the membrane becoming permeable to Cl^- ions (Poznansky and Steele, 1984).

The action potentials of the PLD muscles are larger and have a faster maximum rate of rise than those of ALD muscles (Cullen et al., 1975). In addition, the action-potential conduction velocity in the two muscles is different. Thus in the isolated PLD muscle at 31–36°C, the conduction velocity was 2.3–2.8 m/sec, compared to a value of 0.41–0.7 m/sec in isolated ALD muscle at 28–34°C (Ginsborg, 1960b). The difference in conduction velocity matches the difference in contraction speed of the two muscles.

Contractile Properties

Since avian skeletal muscle contains actin and myosin filaments, arranged in the classical interdigitated pattern, and also contains the regulatory contractile proteins troponin, tropomyosin, and α-actinin (Allen et al., 1979; Devlin and Emerson, 1978, 1979), there is no reason to assume that excitation–contraction coupling in avian muscle differs from that in mammalian muscle. In focally innervated avian muscle fibers, it is assumed that muscle action potentials spread down the T tubules to activate contraction. Although multiply-innervated fibers can conduct action potentials (Ginsborg, 1960b), the necessary depolarization stimulus for contraction is probably supplied directly by the action of acetylcholine (see later section).

The contraction velocities of multiply-innervated muscles with a *Felderstruktur* are five to 10 times slower than those of singly-innervated muscles with a *Fibrillenstruktur.* This difference is seen both in vivo (Hník et al., 1967) and in vitro (Ginsborg, 1960b; Gordon and Vrbová, 1975; Gordon et al., 1977b). For example, the time to reach one-half maximum response to a 40-Hz tetanus was about 400–500 msec in the chicken ALD, but only about 50 msec in the chicken PLD. Two other multiply-innervated muscles of the chicken, the adductor profundus and the plantaris, contracted with velocities similar to that of the ALD (Barnard *et al.,* 1982).

The development of mature contraction properties in chicken ALD and PLD muscles has been studied by Gordon *et al.* (1977a, b). The contraction speeds of both muscles were similar in embryos of 14–16 days incubation (time to half-maximal tension response to 40-Hz stimulation was ~400–500 msec). During development there was little change in the contraction speed of ALD muscles, but there was a progressive increase in the speed of contraction of PLD muscles: At hatching, the time to half-maximal tentanic response was only about 100 msec. During late embryonic growth, there was a marked increase in the amount of tension that could be generated by both ALD and PLD muscles.

Neuromuscular Transmission

There is no evidence against the assumption that acetylcholine is the neurotransmitter at avian skeletal muscle neuromuscular junctions. Choline acetyltransferase, the enzyme that synthesizes acetylcholine, is present in chicken ALD and PLD muscles, and its activity increases during development (Betz et al., 1980). Drugs such as hemicholinium, which inhibits choline uptake, and β-bungarotoxin, which blocks acetylcholine release, block neuromuscular transmission in chicken muscle (Marshall, 1969; Dryden et al., 1974). Langley (1905) and Gasser and Dale (1926) demonstrated that nicotine caused contracture of avian muscle, and subsequently Brown and Harvey (1938) showed that close-arterial injection of acetylcholine into the gastrocnemius muscle of anesthetized chickens caused a biphasic response, a rapid contraction of the focally-innervated fibers followed by a slower contracture of the multiply-innervated fibers. More recently, the presence of acetylcholine receptors in avian muscle has been demonstrated by the binding of α-bungarotoxin, an extremely potent and almost irreversible binding component from the venom of the Taiwan krait (Chang *et al.,* 1975).

The distribution of acetylcholine receptors is different on multiply- and focally-innervated avian muscle. In focally-innervated fibers such as those found in the chicken PLD, acetylcholine sensitivity is restricted to one area, at the neuromuscular junction, whereas there are several peaks of sensitivity in multiply-innervated fibers (Fedde, 1969). Corresponding to this difference, there is a difference in the pattern of spontaneous electrical activity, i.e., miniature endplate potentials (Figure 3–4). This was first described by Ginsborg (1960a), who observed two types of spontaneous activity in different muscles of the chicken. "Focal" activity was similar to that seen in focally-innervated mammalian and amphibian preparations. The rise time of the miniature endplate potentials recorded from any one fiber did not vary greatly (Figure 3–4A). In contrast, the "diffuse" activity included miniature endplate potentials with widely varying rise times. Usually the smaller potentials had longer rise times, suggesting that they originated some distance from the recording site (Figure 3–4B). The diffuse pattern of activity was, therefore, taken to correspond to multiple innervation and focal activity to focal or single innervation.

More recently, miniature endplate currents from chick ALD and PLD muscles have been recorded (Harvey and van Helden, 1981) (Figure 3–5). consistent with the focal innervation of PLD fibers, miniature endplate currents with fast growth times could only be recorded from a single site on each fiber, whereas miniature endplate currents could be recorded from many locations on fibers in ALD muscles. The miniature endplate currents in ALD muscles had a much wider range of amplitudes and growth times, consistent with multiple innervation. However, Harvey and van Helden (1981) pointed out that they frequently observed, in ALD fibers, very slowly rising miniature endplate currents of amplitudes much greater than would be predicted from cable theory for the propagation of a transient signal along a muscle fiber. The significance of these giant, slow spontaneous events

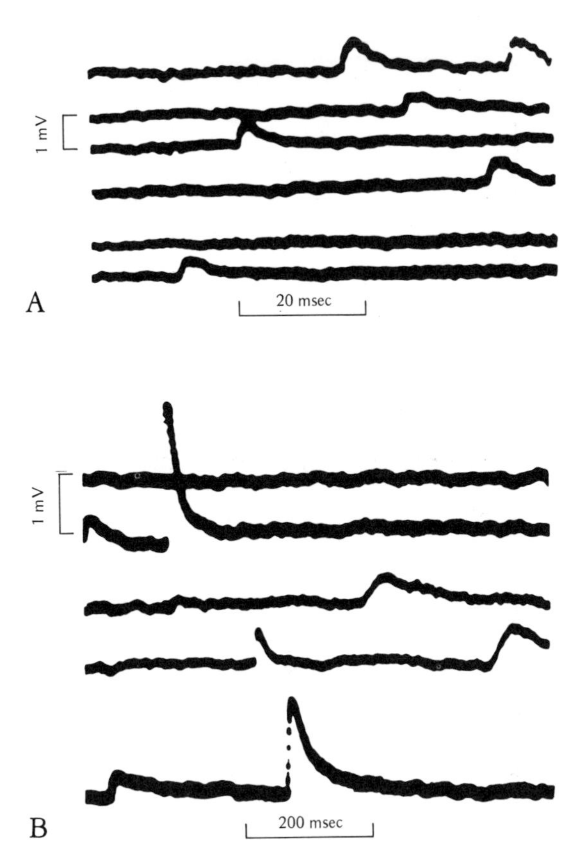

FIGURE 3–4. (A) Focal and (B) diffuse spontaneous activity (mepps) recorded with intracellular electrodes from two different fibers of the same chick biventer cervicis muscle at 37°C. The relatively constant shape of the potentials in A indicates that they originated at a focal endplate close to which the fiber was impaled. The variations in the shapes of the potentials in B indicate that they originated at junctions at a variety of distances from the point at which the fiber was impaled. (From Ginsborg, 1960a).

in unknown, although similar phenomena have been described in mammalian preparations treated with certain drugs (Molgó and Thesleff, 1982).

Although the distribution of acetylcholine receptors on focally- and multiply-innervated chicken muscles is different, there is no evidence for major differences in the functional (Harvey and van Helden, 1981) or biochemical (Sumikawa *et al.*, 1982a, b) properties of receptors from different types of chicken muscle. In contrast to findings on multiply- and singly-innervated fibers of the snake (Dionne and Parsons, 1981) and frog (Miledi and Uchitel, 1981), receptors in both types of chicken muscle have similar channel properties (Harvey and van Helden, 1981). That is, both types of muscle have the same acetylcholine equilibrium potential, miniature endplate currents of similar amplitude (~2 nA), similar single channel conductances (20–40 pS) and single channel lifetimes (4–8 msec at resting potential of −60 to −80 mV at room temperature), and similar temperature and voltage sensitivities. The channel lifetimes of chicken acetylcholine receptors are consistently longer than those of mammalian or amphibian cholinoceptors under the same conditions. Longer channel lifetimes are found in cholinoceptors of denervated muscle and in immature mammalian muscle, and there is evidence for an immunological similarity between such receptors and those on mature ALD muscles (Hall et al., 1985). During development of mammalian muscle there is a change to short open times (Sakmann and Brenner, 1978), but a corresponding change does not occur during the development of chicken ALD or PLD from 16-day-old embryos to 14-week-old animals (Harvey and van Helden, 1981). There do appear to be developmental changes in the molecular forms of acetylcholinesterase at ALD and PLD neuromuscular junctions (Jedrzejczyk et al., 1984).

Acetylcholine receptor turnover has also been estimated for chicken muscle developing in vivo (Betz et al., 1980; Burden, 1977a, b). The technique depends on measuring the rate of release of bound radioactivity from radiolabeled α-bungarotoxin (for review, see Fambrough, 1979). In chicken muscle, it was found that both junctional and extrajunctional receptors had similar half-lives (~30 hr) until about 3 weeks after hatching, when the junctional receptors had the adult half-life of about 5 days (Burden, 1977b; Betz et al., 1980). The mechanisms responsible for the change in rate of metabolism are not known, but it occurs much more slowly than in mammalian muscle. Close packing of receptors does not by itself explain the difference, as the half-life of receptors in "hot spots" is similar to that of the diffusely distributed receptors (Schuetze et al., 1978). Radiolabeled α-bungarotoxin can also be used to study the localization of surface and internal cholinoceptors during development (Atsumi, 1981).

Uses of Avian Muscle in Neuromuscular Pharmacology

It has long been known that avian and amphibian muscles respond differently from mammalian muscle to the addition of acetylcholine and endplate-depolarizing drugs such as nicotine and decamethonium. Mimicking of the effects of nerve stimulation by injection of putative chemical transmitters was the cornerstone of the theory of chemical synaptic transmission. However, innervated mammalian muscle does not respond to intravenously injected acetylcholine or to acetylcholine

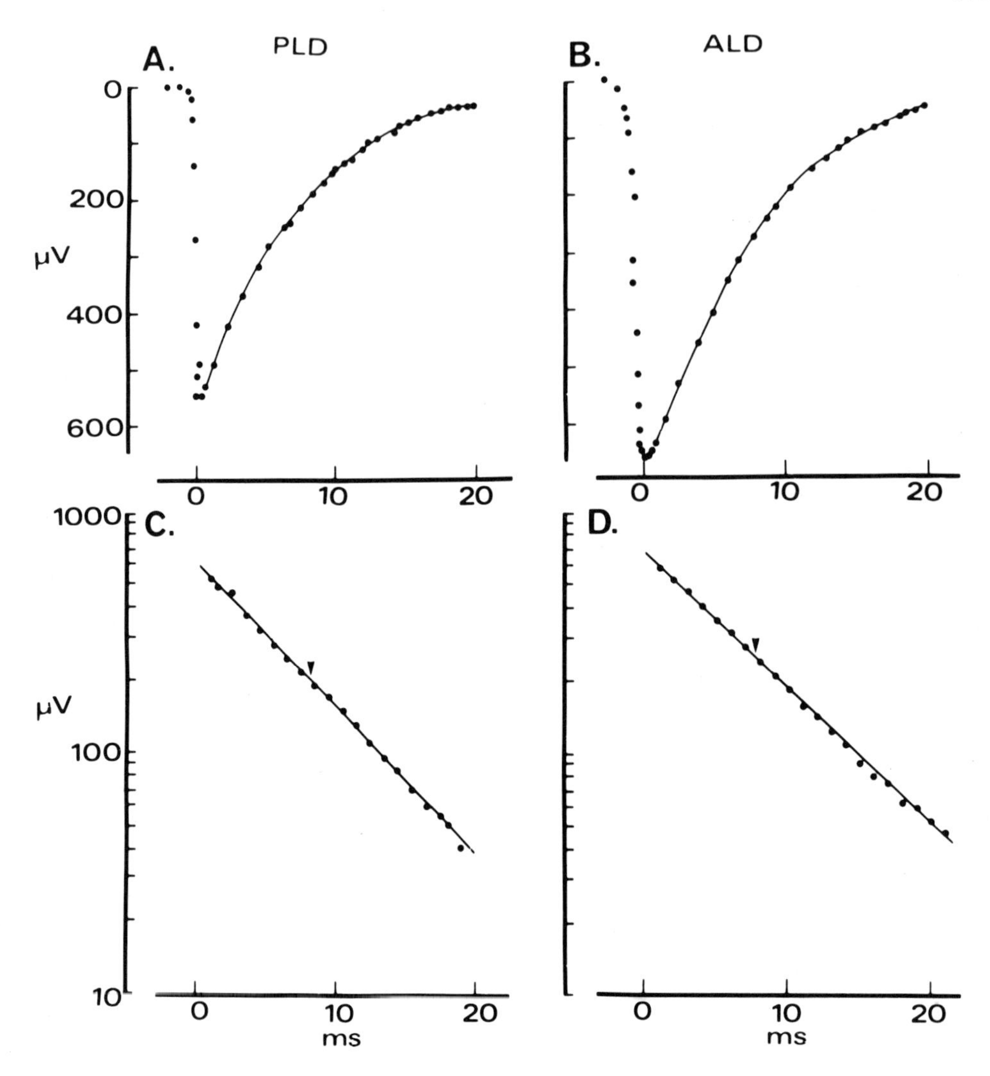

FIGURE 3–5. The decay of mepcs in PLD and ALD muscles is exponential. The averages of 30 extracellularly recorded mepcs from a PLD endplate (**A**) and from an ALD endplate (**B**) are shown. The amplitude of the decay phase of these averaged mepcs is plotted semilogarithmically as a function of time after peak in **C** (for the PLD) and **D** (for the ALD). The time constants of decay (τ_D) are shown by the arrowheads in **C** and **D.** (From an unpublished experiment of van Helden and Harvey.)

injected into the solution bathing an isolated muscle. After the discovery of the release of acetylcholine from mammalian motor nerves (Dale *et al.,* 1936), Brown (1938) showed that close intraarterial injection of acetylcholine was required to produce contraction responses in mammals similar to those produced by nerve stimulation. In contrast, nicotine and acetylcholine had been known for some time to produce contracture responses when injected intravenously into chickens (Langley, 1905; Gasser and Dale, 1926).

The difference between the responses of avian and mammalian muscle is related to the innervation and excitation–contraction coupling mechanisms of multiply- and focally-innervated muscles. In focally-innervated muscles, it is necessary to activate the endplate receptors on many individual muscle fibers simultaneously in order to obtain a synchronized contraction of the muscle. In multiply-innervated muscles, the local depolarizations directly excite the contractile mechanism, and the muscle will remain in contracture for as long as the depolarizing agent remains in contact with the receptors. The response is related to the number of receptors occupied by the drug, and this phenomenon can be exploited in the study of the action of drugs acting at the neuromuscular junction.

The contracture response of avian multiply-innervated muscle can thus be used to study the actions of nicotinic agonist drugs in the same way as other multiply-innervated muscles, such as the frog rectus abdominis and the leech dorsal muscle. Isolated muscles that have been used for this purpose are the anterior latissimus dorsi (Ginsborg and Mackay, 1960), the semispinalis cervicis (Child and Zaimus, 1960), the biventer cervicis (Ginsborg and Warriner, 1960), and the tibialis anterior (van Reizen, 1968) muscles. The presence of muscles containing larger numbers of multiply-innervated fibers in the necks and legs of birds has been utilized in the differentiation between neuromuscular blocking drugs of the depolarizing and nondepolarizing types (Buttle and Zaimis, 1949; Zaimis, 1953, 1959; Bowman, 1964). Such drugs are used clinically as skeletal muscle relaxants, but depolarizing drugs possess several undesirable effects that necessitate their identification and removal from testing programs designed to discover new agents. Depolarizing agents such as decamethonium, suxamethonium, and carbolonium produce, in birds, a characteristic spastic paralysis with the neck pulled back and the legs rigidly extended. In contrast, nondepolarizing agents such as tubocurarine produce a flaccid paralysis.

It is possible to construct concentration–response curves to nicotinic agonists on isolated avian multiply-innervated muscle and hence to study the effects of nicotinic antagonists. Thus it is possible to show classical competitive and noncompetitive blockade of nicotinic receptors by examination of the shape and position of concentration–response lines in the presence of the antagonist.

In our own laboratory and in others, the isolated chick biventer cervicis muscle has been widely used as a simple, inexpensive preparation for the initial screening of drugs thought to act at the neuromuscular junction. The muscle can be isolated together with its motor nerve, which is encapsulated in its tendon. Stimulation of the motor nerve results in twitch responses mainly of the focally-innervated fibers, whereas addition of agonists results in contracture responses of the multiply-innervated fibers (Figure 3–6). Drugs that act to change postjunctional acetylcholine receptor responsiveness affect responses to both nerve stimulation and added agonists. In contrast, drugs that act by changing the release of acetylcholine from the nerve terminals reduce responses to nerve stimulation, but the responses to added agonists remain unchanged. This is shown with 3,4-diaminopyridine (Figure 3–6).

Actions of drugs directly on muscle contractility can be assessed on preparations stimulated directly by high concentrations of KCl or by electrical stimulation after abolition of neuromuscular transmission. We have used these differences in the study of various groups of drugs affecting neuromuscular transmission including competitive neuromuscular blocking agents (Gandiha et al., 1975; Durant et al., 1979; Marshall et al., 1981), irreversible postjunctionally acting snake toxins (Dryden et al., 1974; Harvey et al., 1978, 1982; Harvey and Tamiya, 1980), postjunctionally active receptor-associated channel-blocking drugs (Harvey et al., 1984), prejunctionally active blocking drugs interfering with acetylcholine metabolism (Marshall, 1969, 1970a, b), snake toxins and antibiotics that reduce acetylcholine release (Dryden et al., 1974; Singh et al., 1978), aminopyridines (Bowman et al., 1977; Harvey and Marshall, 1977a, b, c), snake toxins that increase transmitter release (Barrett and Harvey, 1979; Harvey and Karlsson, 1980, 1982), and anticholinesterase agents (Gandiha et al., 1972; Green et al., 1978). In the case of the last class of drugs, it should be noted that chicken muscle acetylcholinesterase may be insensitive to some types of inhibitors (e.g., Anderson et al., 1985).

References

Adachi, E. (1983). Fluctuation in the development of various skeletal muscles in the chick embryo, with special reference to AChE activity and the formation of neuromuscular junctions. Dev. Biol., 95, 46.

Allen, R.E., M.H. Stromer, D.E. Goll, and R.M. Robson. (1979). Accumulation of myosin, actin, tropomyosin, and α-actinin in cultured muscle cells. Dev. Biol., 69, 655.

Anderson, A.J., A.L. Harvey and P.M. Mbugua (1985). Effects of fasciculin 2, an anticholinesterase polypeptide from green mamba venom, on neuromuscular transmission in mouse diaphragm preparations. Neurosci. Lett., 54, 123.

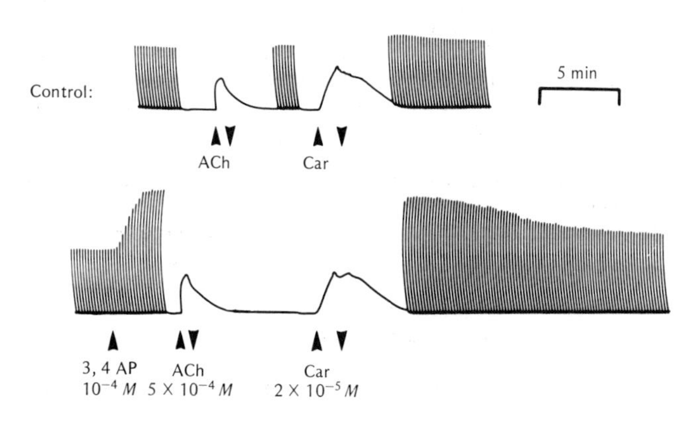

FIGURE 3–6. Indirectly stimulated chick biventer cervicis nerve–muscle preparation. The top panel shows responses to single-shock nerve stimulation and to exogenously applied acetylcholine (*ACh*) and carbachol (*Car*). The bottom panel shows the augmentation of twitch amplitude produced by 3,4-diaminopyridine (*3,4 AP*). Note that acetylcholine and carbachol responses are unaffected by the 3,4-diaminopyridine, indicating that the drug is acting to increase the output of transmitter acetylcholine on nerve stimulation. (From Harvey and Marshall, 1977a.)

Askanas, V., S.A. Shafiq, and A.T. Milhorat. (1972). Histochemistry of cultured aneural chick muscle. Morphological maturation of fibre types. Exp. Neurol., 37, 218.
Atsumi, S. (1971). The histogenesis of motor neurones with special reference to the correlation of their endplate formation. I. The development of endplates in the intercostal muscle in the chick embryo. Acta Anat., 81, 161.
Atsumi, S. (1977). Development of neuromuscular junctions of fast and slow muscles in chick embryo—light and electron microscopic study. J. Neurocytol., 6, 691.
Atsumi, S. (1981). Localisation of surface and internal acetylcholine receptors in developing fast and slow muscles of the chick embryo. Dev. Biol., 86, 122.
Barnard, E.A., J.M. Lyles, and J.A. Pizzey. (1982). Fibre types in chicken skeletal muscles and their changes in muscular dystrophy. J. Physiol., 331, 333.
Barrett, J.C., and A.L. Harvey. (1979). Effects of venom of the green mamba, *Dendroaspis angusticeps,* on skeletal muscle and neuromuscular transmission. Br. J. Pharmacol., 67, 199.
Bennett, M.R., and A.G. Pettigrew. (1974). The formation of synapses in striated muscle during development. J. Physiol., 241, 515.
Berger, A.J. (1960). The Musculature. In "Biology and Comparative Physiology of Birds," Vol. 1 (A.J. Marshall, Ed). London and New York: Academic Press, p. 301.
Betz, H., J.-P. Bourgeois, and J.-P. Changeux. (1980). Evolution of cholinergic proteins in developing slow and fast skeletal muscles in chick embryo. J. Physiol., 302, 197.
Bonner, P.H. and S.D. Hauschka. (1974). Clonal analysis of vertebrate myogenesis. I. Early developmental events in the chick limb. Dev. Biol., 37, 317.
Bourgeois, J.-P., and M. Toutant. (1982). Innervation of avian latissimus dorsi muscles and axonal outgrowth pattern in the posterior latissimus dorsi motor nerve during embryonic development. J. Comp. Neurol., 208, 1
Bowman, W.C. (1964). Neuromuscular blocking agents. In "Evaluation of Drug Activities: Pharmacometrics" (D.R. Laurence and A.L. Bacharach, Eds.). London and New York: Academic Press, p. 325.
Bowman, W.C. (1980). "Pharmacology of Neuromuscular Function." Bristol: John Wright & Sons, Ltd.
Bowman, W.C., and I.G. Marshall. (1972). Muscle. In "Physiology and Biochemistry of the Domestic Fowl," Vol. 2 (D.J. Bell and B.M. Freeman, Eds.). London: Academic Press, p. 707.
Bowman, W.C., A.L. Harvey, and I.G. Marshall. (1977). The actions of aminopyridines on avian muscle. Naunyn-Schmiedeberg's Arch. Pharmacol., 297, 99.
Brown, G.L. (1938). The preparation of the tibialis anterior (cat) for close-arterial injections. J. Physiol., 92, 22.
Brown, G.L., and A.M. Harvey. (1938). Neuromuscular conduction in the fowl. J. Physiol., 93, 285.
Burden, S. (1977a). Development of the neuromuscular junction in the chick embryo: the number, distribution and stability of acetylcholine receptors. Dev. Biol., 57, 317.
Burden, S. (1977b). Acetylcholine receptors at the neuromuscular junction: developmental change in receptor turnover. Dev. Biol., 61, 79.
Burleigh, I.G. (1974). On the cellular regulation of growth and development in skeletal muscle. Biol. Rev. Cambridge Philos. Soc., 49, 267.
Burrage, T. G., and T. L. Lentz. (1981). Ultrastructural characterization of surface specializations containing high-density acetylcholine receptors on embryonic chick myotubes *in vivo* and *in vitro.* Dev. Biol., 85, 267.
Buttle, G.A.H., and E.J. Zaimis. (1949). The action of decamethonium iodide in birds. J. Pharm. Pharmacol., 1, 991.
Chang, C.C., M.J. Su, and M.-C. Lee. (1975). A quantification of acetylcholine receptors of the chick biventer cervicis muscle. J. Pharm. Pharmacol., 27, 454.
Child, K.J., and E. Zaimis. (1960). A new biological method for the assay of depolarizing substances using the isolated semispinalis cervicis muscle of the chick. Br. J. Pharmacol., 15, 412.
Cullen, M.J., J.B. Harris, M.W. Marshall, and M.R. Ward. (1975). An electrophysiological and morphological study of normal and denervated chicken latissimus dorsi muscles. J. Physiol., 245, 371.
Dale, H.H., W. Feldberg, and M. Vogt. (1936). Release of acetylcholine at voluntary motor nerve endings. J. Physiol., 86, 353.
Devlin, R.B., and C.P. Emerson. (1978). Coordinate regulation of contractile protein synthesis during myoblast differentiation. Cell, 13, 599.
Devlin, R.B., and C.P. Emerson. (1979). Coordinate accumulation of contractile protein mRNAs during myoblast differentiation. Dev. Biol., 69, 202.
Dionne, V.E., and R.L. Parsons. (1981). Characteristics of the acetycholine-operated channel at twitch and slow fibre neuromuscular junctions of the garter snake. J. Physiol., 310, 145.
Dryden, W.F., A.L. Harvey, and I.G. Marshall. (1974). Pharmacological studies on the bungarotoxins: Separation of the fractions and their neuromuscular activity. Eur. J. Pharmacol., 26, 256.
Durant, N.N., I.G. Marshall, D.S. Savage, T. Sleigh, and I.C. Carlyle. (1979). The neuromuscular and autonomic blocking actions of pancuronium, Org NC45, and other pancuronium analogues in the cat. J. Pharm. Pharmacol., 31, 831.
Entrikin, R.K., and S.H. Bryant. (1975). Electrophysiological properties of biventer cervicis muscle fibers of normal and Roller pigeons. J. Neurobiol., 6, 201.
Ezerman, E.B., and H. Ishikawa. (1967). Differentiation of the sarcoplasmic reticulum and T system in developing chick skeletal muscle in vitro. J. Cell Biol., 35, 405.
Fambrough, D.M. (1979). Control of acetylcholine receptors in skeletal muscle. Physiol. Rev., 59, 165.
Fedde, M.R. (1969). Electrical properties and acetylcholine sensitivity of singly and multiply innervated avian muscle fibers. J. Gen. Physiol., 53, 624.
Gandiha, A., I.G. Marshall, D. Paul, I.W. Rodger, W. Scott, and H. Singh. (1975). Some actions of chandonium iodide, a new short-acting muscle relaxant. Clin. Exp. Pharmacol. Physiol., 2, 159.
Gasser, H.S., and H. Dale, (1926). The pharmacology of denervated muscle. II. Some phenomena of antagonism and the formation of lactic acid in chemical contracture. J. Pharmacol. Exp. Ther., 28, 290.
George, J.C., and A.J. Berger, (1966). "Avian Myology." London and New York: Academic Press.
Ginsborg, B.L. (1960a). Spontaneous activity in muscle fibres of the chick. J. Physiol., 150, 707.
Ginsborg, B.L. (1960b). Some properties of avian skeletal muscle fibres with multiple neuromuscular junctions. J. Physiol., 154, 581.
Ginsborg, B.L., and B. Mackay. (1960). The latissimus dorsi muscles of the chick. J. Physiol., 153, 19.
Ginsborg, B.L., and B. Mackay. (1961). A histochemical demonstration of two types of motor innervation in avian skeletal muscle. Bibl. Anat., 2, 174.
Ginsborg, B.L., and J. Warriner. (1960). The isolated chick biventer cervicis nerve–muscle preparation. Br. J. Pharmacol., 15, 410.
Goldspink, G. (1974). Development of muscle. In "Differentiation and Growth of Cells in Vertebrate Tissues," (G. Goldspink, Ed.). London: Chapman and Hall, p. 69.

Gordon, T., and G. Vrbová. (1975). The influence of innervation on the differentiation of contractile speeds of developing chick muscles. Pfleugers Arch., 360, 199.

Gordon, T., R. Perry, Srihari, T. and G. Vrobvá. (1977a). Differentiation of slow and fast muscles in chickens. Cell Tissue Res., 180, 211.

Gordon, T., R.D. Purves, and G. Vrbová. (1977b). Differentiation of electrical and contractile properties of slow and fast muscle fibers. J. Physiol., 269, 535.

Green, A.L., J.A.M. Lord, and I.G. Marshall. (1978). The relationship between cholinesterase inhibition in the chick biventer cervicis muscle and its sensitivity to exogenous acetylcholine. J. Pharm. Pharmacol., 30, 426.

Hall, Z.W., P.D. Gorin, L. Silberstein and C. Bennett. (1985). A postnatal change in the immunological properties of the acetylcholine receptor at rat missile endplates. J. Neurosci., 5, 730.

Harvey, A.L., and E. Karlsson. (1980). Dendrotoxin from the venom of the green mamba, *Dendroaspis angusticeps*. A neurotoxin that enhances acetylcholine release at neuromuscular junctions. Naunyn-Schmiedeberg's Arch. Pharmacol., 312, 1.

Harvey, A.L., and E. Karlsson. (1982). Protease inhibitor homologues from mamba venoms: Facilitation of acetylcholine release and interactions with prejunctional blocking toxins. Br. J. Pharmacol., 77, 153.

Harvey, A.L., and I.G. Marshall. (1977a). The actions of three diaminopyridines on the chick biventer cervicis muscle. Eur. J. Pharmacol., 44, 303.

Harvey, A.L., and I.G. Marshall. (1977b). The facilitatory actions of aminopyridines and tetraethylammonium on neuromuscular transmission and muscle contractility in avian muscle. Naunyn-Schmiedeberg's Arch. Pharmacol., 299, 53.

Harvey, A.L., and I.G. Marshall. (1977c). A comparison of the effects of amino-pyridines on isolated chicken and rat skeletal muscle preparations. Comp. Biochem. Physiol., C: Comp. Pharmacol., 58, 161.

Harvey, A.L., and N. Tamiya. (1980). Role of phospholipase activity in the neuromuscular paralysis produced by some components isolated from the venom of the sea snake, *Laticauda semifasciata*. Toxicon, 18, 65.

Harvey, A.L., and van Helden, D. (1981). Acetylcholine receptors in singly and multiply innervated skeletal muscle fibres of the chicken during development. J. Physiol., 317, 397.

Harvey, A.L., I.W. Rodger, and N. Tamiya. (1978). Neuromuscular blocking activity of two fractions isolated from the venom of the sea snake, *Laticauda semifasciata*. Toxicon, 16, 45.

Harvey, A.L., R.J. Marshall, and E. Karlsson. (1982). Effects of purified cardiotoxins from the Thailand cobra (*Naja naja siamensis*) on isolated skeletal and cardiac muscle preparations. Toxicon, 20, 379.

Harvey, A.L., S.V.P. Jones, and I.G. Marshall. (1984). Disopyramide produces non-competitive, voltage-dependent block at the neuromuscular junction. Br. J. Pharmacol., 81, 169P.

Hess, A. (1961). Structural differences of fast and slow extrafusal muscle fibres and their nerve endings in chickens. J. Physiol., 157, 221.

Hess, A. (1967). The structure of vertebrate slow and twitch muscle fibres. Invest. Opthalmol., 6, 217.

Hník, P., I. Jirmanová, L. Vyklický and J. Zelená. (1967). Fast and slow muscles of the chick after nerve cross-union. J. Physiol., 193, 309.

Hirano, H. (1976). Ultrastructural study on the morphogenesis of the neuromuscular junction in the skeletal muscle of the chick. Z. Zellforsch. Mikrosk. Anat., 79, 198.

Jedrzejczyk J., I. Silman, J. Lai and E.A. Barnard. (1984). Molecular forms of acetylcholinesterase in synaptic and extrasynaptic regions of avian tonic muscle. Neurosci. Lett., 46, 283.

Jirmanova, I., and L. Vyklický. (1965). Post-tetanic potentiation in multiply innervated muscle fibres of the chick (in Czech). Cslká Fysiol., 14, 351.

Khan, M.A. (1976) Histochemical sub-types of three fibre types of avian skeletal muscle. Histochemistry, 50, 9.

Kikuchi, T., and C.R. Ashmore. (1976). Developmental aspects of the innervation of skeletal muscle fibers in the chick embryo. Cell Tissue Res., 171, 233.

Landmesser, L., and D.G. Morris. (1975). The developmental of functional innervation in the hind limb of the chick embryo. J. Physiol., 249, 301.

Langley, J.N. (1905). On the reaction of cells and of nerve endings to certain poisons, chiefly as regards the reaction of striated muscle to nicotine and to curari. J. Physiol., 33, 374.

Larson, P.F., M. Jenkinson, and P. Hudgson. (1970). The morphological development of chick embryo skeletal muscle grown in tissue culture as studied by electron microscopy. J. Neurol. Sci., 10, 385.

Lebeda, F.J., and E.X. Albuquerque. (1975). Membrane cable properties of normal and dystrophic chicken muscle fibers. Exp. Neurol., 47, 544.

Marshall, I.G. (1969). The effects of some hemicholinium-like compounds on the chick biventer cervicis muscle preparation. Eur. J. Pharmacol., 8, 204.

Marshall, I.G. (1970a). Studies on the blocking action of 2-(4-phenylpiperidino)-cyclohexanol (AH 5183). Br. J. Pharmacol., 38, 503.

Marshall, I.G. (1970b). A comparison between the blocking actions of 2-(4-phenylipiperidino)-cyclohexanol (AH 5183) and its *N*-methyl quaternary analogue (AH 5954). Br. J. Pharmacol., 40, 68.

Marshall, I.G., A.L. Harvey, H. Singh, T.R. Bhardwaj, and D. Paul. (1981). The neuromuscular and autonomic blocking effects of azasteroids containing choline or acetylcholine fragments. J. Pharm. Pharmacol., 33, 451.

Miledi, R., and O.D. Uchitel. (1981). Properties of postsynaptic channels induced by acetylcholine in different frog muscle fibres. Nature (London), 291, 162.

Molgo, J., and S. Thesleff. (1982). 4-Aminoquinoline-induced 'giant' miniature endplate potentials at mammalian neuromuscular junctions. Proc. R. Soc. London, Ser. B, 214, 229.

Poznansky, M.J., and J.A. Steele. (1984). Membrane electrical properties of developing fast-twitch and slow-tonic muscle fibres of the chick. J. Physiol., 347, 633.

Sakmann, B., and H.R. Brenner. (1978). Change in synaptic channel gating during neuromuscular development. Nature (London), 276, 401.

Schuetze, S.M., E.F. Frank, and G.D. Fischbach, (1978). Channel open time and metabolic stability of synaptic and extrasynaptic acetylcholine receptors on cultured chick myotubes. Proc. Nat. Acad. Sci. U.S.A., 75, 520.

Shear, C.R. (1978). Cross-sectional myofibre and myofibril growth in immobilized developing skeletal muscle. J. Cell. Sci., 29, 297.

Shear, C.R. (1981). Effects of disuse on growing and adult chick skeletal muscle. J. Cell Sci., 48, 35.

Shear, C.R., and Goldspink, G. (1971). Structural and physiological changes associated with the growth of avian fast and slow muscle. J. Morphol., 135, 351.

Shimada, Y. (1972a). Scanning electron microscopy of myogenesis in monolayer culture: A preliminary study. Dev. Biol., 29, 227.

Shimada, Y. (1972b). Early stages in the reorganization of

dissociated embryonic chick skeletal muscle cells. Z. Anat. Entwicklungsgesch., 138, 255.

Shimada, Y., D.A. Fischman, and A.A. Moscona. (1967). The fine structure of embryonic chick skeletal muscle cells differentiated in vitro. J. Cell Biol., 35, 445.

Singh, Y.N., I.G. Marshall, and A.L. Harvey. (1978). Some effects of the amino-glycoside antibiotic amikacin on neuromuscular and autonomic transmission. Br. J. Anaesth., 50, 109.

Sumikawa, K., E.A. Barnard, and J.O. Dolly. (1982a). Similarity of acetylcholine receptors of denervated, innervated and embryonic chicken muscles. 2. Subunit compositions. Eur. J. Biochem., 126, 473.

Sumikawa, K., F. Mehraban, J.O. Dolly, and E.A. Barnard. (1982b). Similarity of acetylcholine receptors of denervated, innervated and embryonic chicken muscles. 1. Molecular species and their purification. Eur. J. Biochem., 126, 465.

Toutant, J.P., M.N. Toutant, D. Renaud, and G.H. Le Douarin. (1980). Histochemical differentiation of extrafusal muscle fibres of the *anterior latissimus dorsi* in the chick. Cell Differ., 9, 305.

Van Den Berge, J.C. (1975). Aves myology. In "Sisson and Grossman's The Anatomy of the Domestic Animals" R. Getty, W.B. Saunders, Philadelphia: p. 1802.

Van Reizen, H. (1968). Classification of neuromuscular blocking agents in a new neuromuscular preparation of the chick *in vitro.* Eur. J. Pharmacol., 5, 29.

Vrbová, G., T. Gordon, and R. Jones. (1978). "Nerve–Muscle Interaction." London: Chapman and Hall.

Zaimis, E.J. (1953). Motor endplate differences as a determining factor in the mode of action of neuromuscular blocking substances. J. Physiol., 122, 238.

Zaimis, E.J. (1959). Mechanisms of neuromuscular blockade. In "Curare and Curare-like Agents" (D. Bovet, F. Bovet-Nitti, and G.B. Marini-Bettolo, Eds.). Amsterdam: Elsevier, p. 191.

4 Immunophysiology

B. Glick

Introduction

Immunophysiology is dependent on the presence of immunocompetent and accessory cells. Immunocompetent cells are thymic derived (T) or bursal derived (B). Stem cells enter the thymus between 6.5 and 8.0 days of embryonic development. The thymus supplies the necessary microenvironment for the differentiation of the stem cell into three distinct subpopulations of thymic-derived (T) lymphocytes: T-helper (T_H), T-suppressor (T_S), and T-cytotoxic (T_c). These subpopulations of T cells function within both the humoral and cell-mediated expressions of the immunophysiology response. The humoral response includes the release of immunoglobulin (Ig) (antibody) and is dependent on T_H cells to initiate the response and T_s cells to modulate the response. The B cell differentiates in the bursa or its equivalent, and is the cell responsible for the synthesis of Ig and antibody. Accessory cells are phagocytic (e.g., macrophages) or adherent (ellipsoid-associated cells of the spleen) cells. Interaction of T or B cells with accessory cells is usually necessary for an optimum immunophysiology response.

Humoral Immunity

A foreign substance or antigen (e.g., a group of bacteria) may directly activate a B cell or initially bind an accessory cell that leads to the activation of T and then B cells. An antigen is generally a complex particle (e.g., a cell or a soluble protein) that will stimulate antibody synthesis when injected parenterally. A hapten (H) is a small molecule (e.g., dinitrophenyl) that is not antigenic and therefore will not stimulate antibody synthesis when injected parenterally. However, conjugating the hapten with a protein will confer antigenicity on the hapten. The protein portion of the hapten–protein moiety combines with a receptor that is an integral part of the T_H cell membrane (Figure 4–1). The T_H cell reacts with the major histocompatibility (MHC) product of the macrophage and the protein portion of the hapten-protein complex on the macrophage. This activates the T-cell to release two soluble products. One reacts with the resting B-cell to induce blast formation and the other reacts with the blast cell. The B

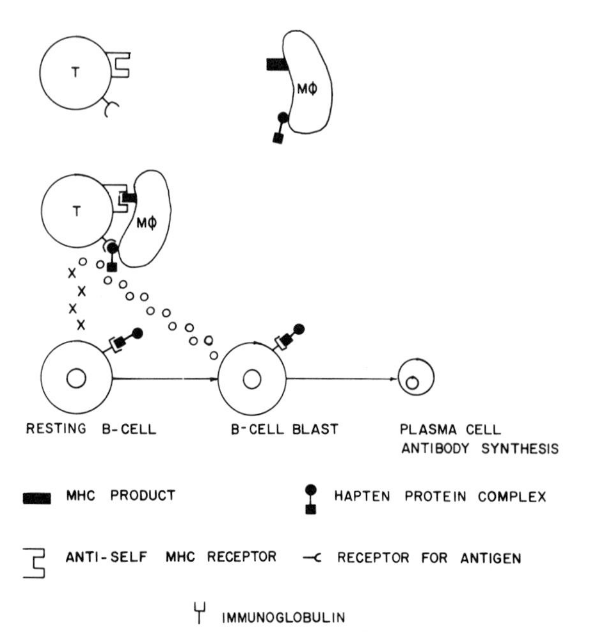

FIGURE 4–1. A model depicting soluble T-cell factors. The T-cell possesses an immunoglobulin-like molecule (tuning-fork symbol) that binds the protein (*solid circle*) part of the hapten (*solid square*)-protein moiety. [See Clark (1983) for genetic restraints on T-cell-macrophage (Mϕ) interaction.] The major histocompatibility complex (MHC) product of the Mϕ identifies the MHC receptor of the T-cell. The T-cell releases two soluble factors. (Reprinted with permission from Golub, E., 1981, Cellular Basis of the Immune Response. Sinauer Associates, Publishers, Sunderland, Massachusetts.)

cell then differentiates into the antibody-producing plasma cell. For genetic restraints and other models, the reader is referred to Golub (1981), Clark (1983) and Ewert et al. (1980). If this is the chicken's initial exposure to the antigen, a primary response occurs. Subsequent exposure to the specific antigen will elicit a secondary response (Figure 4–2). Within the primary response, one may identify a latent phase in which the immunocompetent cells are activated; there is also an exponential phase characterized by a progressive increase in circulating antibody, a peak antibody phase, and a phase in which circulating antibody declines. These phases are modified in the secondary response in that the latent period is shortened, the slope of the exponential phase is greater, the peak occurs earlier and is greater, and the regression phase is subdued. Also, there are isotypic (classes of Igs) differences in the antibodies produced during the primary (mainly a heavy Ig, termed IgM) and the secondary (mainly a lighter Ig, termed IgG) responses.

Immunoglobulin Synthesis

Immunoglobulin is a glycoprotein synthesized by B cells and plasma cells. Antibody is structurally identical to Ig, which appears in five distinct isotypes or classes: IgM, IgG, IgA, IgD, and IgE. Each isotype consists of polypeptide chains, two heavy (H) and two light (L) chains held together by disulfide linkages. Each H and L chain has a variable (N-terminal) and constant (C-terminal) end. The variable ends of each chain form the antigen-binding sites (F_{ab}) (Figure 4–3). The F_{ab} of an IgM, IgG, or IgA molecule may bind to the same antigenic determinant but still be distinguished as IgM, IgG, or IgA by differences that exist in the constant region of their heavy chains. Heavy chain no-

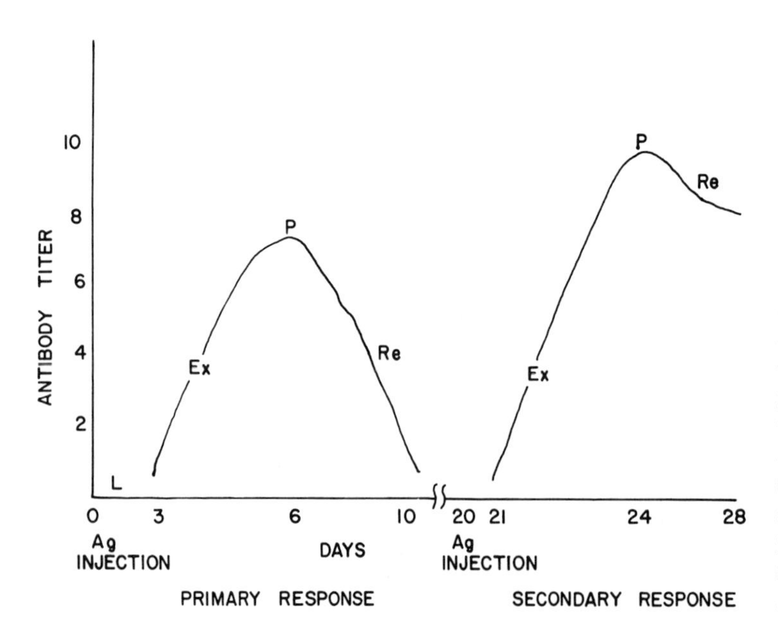

FIGURE 4–2. Primary and secondary responses. Antigen (*Ag*) injected at day 0. The latent (*L*) period in the primary response is 3 days: exponential (*Ex*) phase, peak (*P*) phase, and regression (*Re*) phase. Secondary response: Ag injected at day 20. There is a shorter *L* phase, more rapid *Ex* phase, more intense *P* phase, and slower *Re* phase than in the primary response.

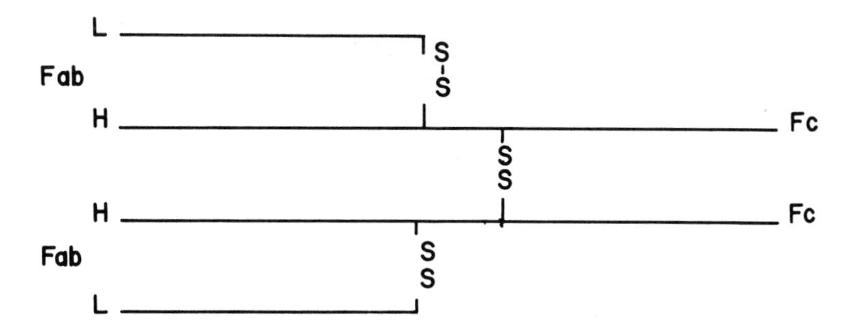

FIGURE 4–3. The polypeptide nature of an immunoglobulin molecule. Two light (L) chains and two heavy (H) chains held together by disulfide (S) linkages. F_{ab} identifies the two antigen-binding sites of the immunoglobulin molecule; F_c identifies the constant region of the molecule and its C-terminal end.

menclature for IgM, IgG, and IgA is μ, γ, and α, respectively.

Immunoglobulin Isotypes. The major Ig isotypes have been separated from the serum of birds. On the basis of physiocochemical and antigenic characteristics, the heavy Ig, with a sedimentation coefficient (S) of 16.7–16.9 and 890,000 daltons appears to resemble mammalian IgM (Leslie and Clem, 1969). On the other hand, there is less agreement that the other Ig (7.1–7.8 S) is homologous to mammalian IgG. Leslie and Clem (1969) first suggested an apparent difference between mammalian IgG and chicken IgG, in that the latter possesses more hexose and its γ chain is considerably larger (67,500 daltons) than the mammalian γ chain (50,000 daltons). Their proposal to call the putative IgG of chickens IgY was reinforced by the earlier data of Tenenhouse and Deutsch (1966), who on the basis of a lower isolectric point of 5.2, higher sedimentation coefficient of 7.73, and a higher hexose percentage of 3.1, compared to mammalian IgG, suggested that this Ig more closely resembled mammalian IgA than IgG. Ambrosius and Hadge (1982) reinforced this thesis when they found that human IgA was cross-reactive with chicken IgY. Functionally, however, chicken IgY is similar to mammalian IgG (Ivanyi, 1981). Apparently, on this basis most investigators have chosen the name IgG for the major serum 7 S Ig of birds.

Sequence data of heavy chains is necessary to determine homology. The carbohydrate peptide appears to be invariant in the evolution of mammalian γ chain (Howell et al., 1967). The carbohydrate peptide for avian 7.8 S Ig heavy chains exhibits little homology with the carbohydrate peptide from human γ chain (Travis and Sanders, 1973). On the other hand, the sequencing of six cysteine-containing peptides having 36 amino acids and including 108 base positions from avian 7.85 S Ig γ chains revealed 70% protein and 87% gene homology with the human γ chain (Travis and Sanders, 1975).

A third immunoglobulin secretory Ig has been identified in the serum and body secretions of birds (Lebacq-Verheyden et al., 1972; Bienenstock et al., 1972; Orlans and Rose, 1972; Watanabe et al., 1975). While this Ig has been referred to as IgA, its analogy and homology to mammalian IgA remains to be determined. Avian immunoglobulin A is concentrated in the bile (Lebacq-Verheyden et al., 1972; Bienestock et al., 1972) and may be the only bile Ig (Katz et al., 1974; Sanders and Case, 1977). Avian IgA exists in polymeric form (Leslie and Martin, 1973). Bile IgA has been reported to have a sedimentation coefficient of 15 S (Porter and Parry, 1976) and 17–19 S (Watanabe et al., 1975; Sanders and Case, 1977).

Immunogloblin A is highly concentrated (3–12 mg/ml) in bile (Lebacq-Verheyden et al., 1974; Leslie et al., 1976; Rose et al., 1981). Bile immunoglobulin flows from the cystic and hepatic ducts into the duodenum at the rate of 1.7 mg/ml (Rose et al., 1981). These same authors noted that ligation of the bile ductus increased IgA concentration in 3.5- to 4.5-week-old birds from 0.15–0.22 mg/ml to 0.29–0.49 mg/ml.

Secretory component is associated with chicken bile IgA (Bienenstock et al., 1973; Porter and Parry, 1976). Secretory component is a polypeptide, 60×10^3 daltons, that is synthesized by epithelial cells and attaches to IgA before secretion.

Bile immunoglobulin A may be synthesized in situ (Leslie et al., 1976) or may have a blood-borne origin (Rose et al., 1981). The latter authors injected intravenously human polymeric or monomeric IgA lacking secretory component into chickens. The protein-bound radio activity in bile was 10–20 times greater over a 7-hr period in those birds receiving polymeric than monomeric IgA. The authors concluded that polymeric IgA is transported from blood to bile. Apparently, polymeric IgA, which lacks secretory component, is transported from blood to bile by first combining with membrane secretory components of hepatocytes.

Bursa of Fabricius

Morphology. The bursa develops as a dorsal diverticulum of the proctadael region of the cloaca. The round or oval bursa of the chicken (Figure 4–4) is in marked contrast to the elongated bursa of the duck (Glick, 1963) and starling (*Sturnus vulgaris;* Glick and Olah,

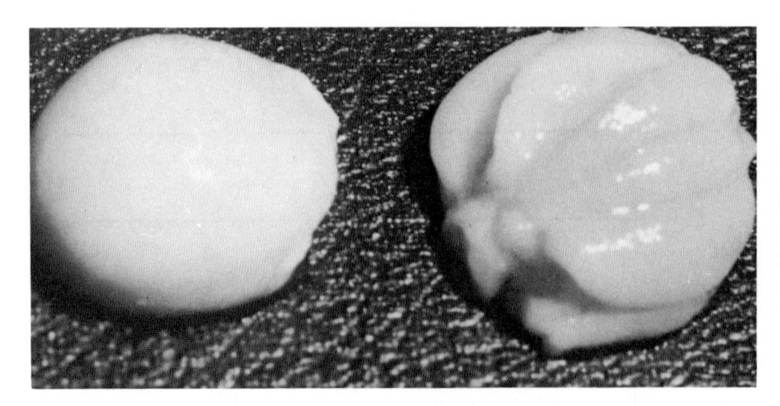

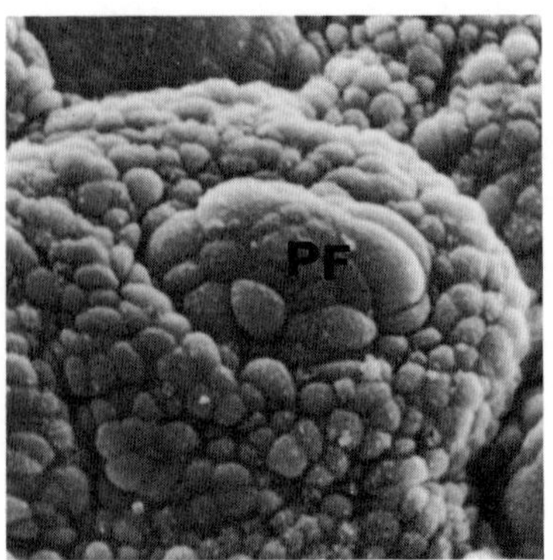

FIGURE 4–4. Bursa of Fabricius from a 2-week-old chicken. The bursa on the right has been turned inside out to reveal the numerous plicae. The projecting follicle (*PF*) is shown in middle insert. (Reprinted with permission from Glick, B. 1964. In Good, R.A. and A. Gabrielson, Eds., "Thymus in Immunobiology." New York, Hoeber Medical Division, Harper and Row, p. 344.)

1983), and to the ostrich (*Struthio camelus australis*) and the emu (*Dromaius novaehollandiae*) bursa, which is an integral part of the proctadael mucosa (von Rautenfeld and Budras, 1982). The luminal surface of the bursa in chickens, ducks, turkeys, and quail is plicated with as many as 15 primary and 7 secondary plicae or folds (Jolly, 1915). Plicae are not evident in the starling (Glick and Olah, 1983).

A reorganization of the surface epithelial cells between 10 and 15 days of embryonic development leads to the formation of epithelial buds, which develop into the bursal follicles. The bursal follicle is covered by follicle-associated epithelial (FAE) cells and possesses a medulla with lymphocytes, lymphoblasts, reticulocytes, macrophages, a few plasma cells and secretory cells, and a cortex with lymphocytes, lymphoblasts, macrophages, and plasma cells. The pinocytotic (engulfing liquid) activity of the FAE (Bockman and Cooper, 1973) allows a variety of soluble luminal substances entry to the bursa medulla (Glick, 1983). The FAE medullary unit may perform a sentinel role at the end of the alimentary canal (Glick and Olah, 1981), especially since antigens intoduced into the bursal lumen stimulate the cells of FAE medullary unit to produce antibody (Van Alten and Meuwissen, 1972). Carbon applied to the vent will enter the bursa, where the FAE selectively pinocytose it (Schaffner et al., 1974; Olah and Glick, 1978). Utilizing this technique with geometry, we have calculated 800 bursal follicles per fold or 8,000–12,000 FAE areas per bursa (Olah and Glick, 1978). Since the bursal surface area was calculated to be 10 cm^2 and 10% of this area was FAE, we concluded that 1 cm^2 (or 10%) of the bursal luminal surface was immunologically oriented.

Viewing the bursal plicae with scanning electron microscopy, one may identify projecting follicles (Figure 4–4), which occur embryonically (Glick *et al.,* 1977), and button-like follicles, which may be age related.

Ontogeny. Under normal conditions, the foundation for humoral immunity is established during the ontogeny of the bursa of Fabricius (Glick, 1977a). The anlage of the bursa appears between 3 and 5 days of embryonic development (DE) (Romanoff, 1960; Edwards et al., 1975; Olah and Glick unpublished observation). At this time lymphocytes are absent from the bursa, but they first appear between 12 and 15 DE (Ackerman, 1962; Leene et al., 1973). These bursal lymphocytes may be distinguished from other cells by the presence of monomeric IgM as an integral part of their cell membrane. Therefore, by raising antibodies to IgM one may identify the Ig-positive (Ig+) B cells at any location in the chicken. An Ig+ cell is identified by tagging an antibody against IgM with ^{125}I and incubating cells with this reagent (Glick *et al.,* 1977). Cell smears are made and coated with a thin emulsion of NTB-2 (Kodak). Following incubation in the dark, the slides are developed and stained. Any cell possessing surface IgM will bind the tagged anti-IgM, which releases γ rays from the ^{125}I, exposing the emulsion above the cell (Figure 4–5).

It is now apparent that bursal lymphocytes originate from a blood-borne progenitor cell (Le Douarin et al., 1975) that was elegantly identified by Nicole Le Douarin when she utilized the differences in heterochromatin pattern between the quail and chicken nuclei to construct chimeric embryos (Le Douarin and Jotereau, 1973). The nucleus from a quail reveals a large clump of chromatin, which is absent from the nucleus of the chicken.

The bursal anlage from a 5- to 6-day-old quail embryo, when transplanted into the somatopleure of a

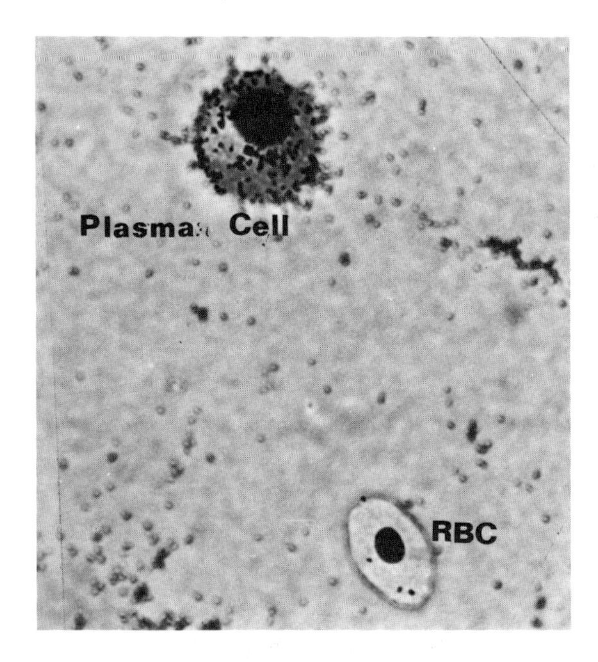

FIGURE 4–5. An Ig-positive *plasma* cell from the Harderian gland. Cell was incubated with ^{125}I-labeled antichicken Ig serum. From Glick et al. (1977).

3-day-old chick embryo for 13–14 days, maintained the quail chromatin pattern in all of its lymphocytes. If the transplants were delayed until 7 or 10 days, the embryos were chimeric for their lymphocytes in that they possessed both quail and chick chromatin patterns. Delaying the embryonic transplant until the quail was 11 days old led to the differentiation of only lymphocytes with the quail chromatin pattern. Therefore, the flow of progenitor cells into the bursal anlage of quail occurs between 7 and 11 days of embryonic development. Reciprocal experiments with the chicken reveal stem-cell migration into the chick bursa between 8 and 15 DE.

Control of Humoral Immunity. The role of the bursa in humoral immunity was identified in 1954 (Glick, 1983). Earlier studies had revealed that the bursa grew most rapidly during the first 3 weeks after hatching, plateaued, and then regressed as early as 8 weeks of age (Glick, 1956, 1960). Therefore, to study the bursa's function, it was logical to remove it within the first 3 weeks after hatching. Birds bursectomized at this time were shown to be deficient in their ability to produce antibodies to *Salmonella typhimurium* (Glick, 1983).

The bursa has been suggested as a primary site of immunoglobulin synthesis by various researchers (Glick, 1977a). Kincade and Cooper (1971) identified IgM in the bursa, the caecal tonsil, and spleen, by 13, 17, and 19 days DE, respectively. At 21 DE, occasional cells in the bursa medulla were observed to be $\mu+$ and $\gamma+$, suggesting a switch from IgM to IgG within the bursal microenvironment. Furthermore, the injection of anti-μ at 12 DE followed by bursectomy at hatching produced agammaglobulinemic chicks (Kincade et al., 1970). While these experiments indicated that the bursa is the primary site for Ig synthesis, other experiments suggested an independent site for IgM synthesis at least (Glick, 1983) (Figure 4–6). The embryonic bursa was eliminated in approximately 80% of embryos from fertile eggs dipped into 1–2 g% of testosterone propionate on the third day of incubation (Glick and Sadler, 1961; Glick, 1961). The hatched chick produced increased levels of IgM, suggesting a nonbursal site for Ig synthesis (Lerner et al., 1971; Subba Rao et al., 1978). Support for this thesis comes from a variety of laboratories (e.g., Fitzimmons et al., 1973; Jankovic et al., 1975, 1977). These authors removed the caudal portion of the tail buds before 70 hr of incubation, and although this eliminated the development of the bursa, it did not prevent the formation of B cells or Ig.

T-Helper Cells. Thymic involvement in humoral immunity is best observed by injecting a hapten–protein complex (see earlier section above and Figure 4–1). In mammals, helper cells from the thymus bind the protein part of the molecule, and with the assistance of an accessory cell, stimulate B cell differentiation (Miller and Mitchison, 1968). Similar experiments with the chicken have identified the presence of T-helper cells (McArthur et al., 1972; Weinbaum et al., 1973).

Lymphocyte Life Span. The bursa possesses 8,000–12,000 bursal follicles (Olah and Glick, 1978), which are packed with lymphocytes. The majority of embryonic B cells are in the resting or postmitotic phase of the cell cycle, with the DNA-synthesizing phase accounting for the next largest percentage (Glick et al., 1985).

Bursal lymphocytes have been shown to migrate to peripheral lymphoid tissue, e.g., spleen (Hemmingsson and Linna, 1972), while others may be activated in situ and differentiate into antibody-producing cells (Van Alten and Meuwissen, 1972). Most of the bursal lymphocytes have a short life span (Glick, 1976, 1977b). Application of tritiated thymidine ([^{3}H]tdR) to the air cell of fertile eggs will label all cells in DNA synthesis (S phase). These cells are large and medium lymphocytes, which on division become small lymphocytes. Within 48 to 72 hr after [^{3}H]tdR application, the entire small lymphocyte population will be labeled. The absence of nonlabeled small lymphocytes in this time period indicates the rapid turnover of this population of cells. Further evidence for the rapid turnover of bursal small lymphocytes is the observation that within 15 days after the application of [^{3}H]tdR, 94%

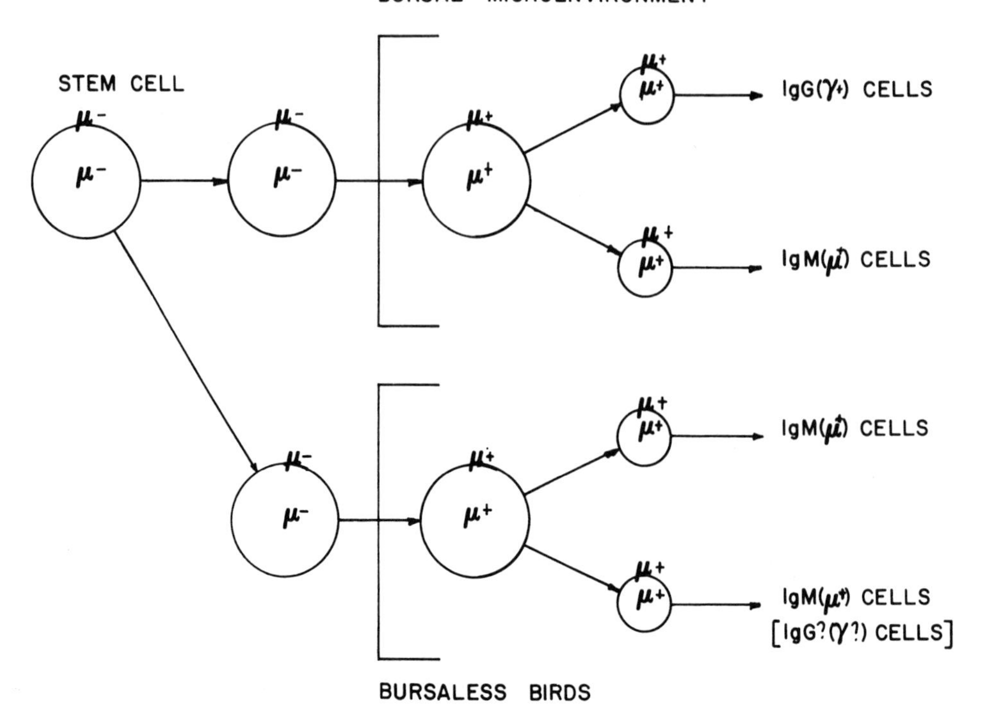

FIGURE 4–6. The large stem cell which is mu (μ) negative in its cytoplasm and on its membrane enters the bursal microenvironment, where it differentiates into a large cytoplasmic and membrane μ-positive cell. It then divides into small μ-positive cells. A portion of these cells experience a switch from μ-positive to gamma (γ) -positive. The IgM (μ-positive) and IgG (γ-positive) cells are then released into the circulation. In the bursa-less birds, the stem cells are located at a nonbursal site where they primarily differentiate into IgM.

of the labeled lymphocytes are replaced by nonlabeled lymphocytes. Long-lived populations of lymphocytes have been identified in the bone marrow (Glick, 1976, 1977b).

Cell-Mediated Immunity

Cell-mediated immunity includes immunity that is not dependent on the synthesis of antibody. The expansion of cell-mediated immunity is dependent on the normal development of the thymus. Several classic examples of cell-mediated immunity, graft-versus-host (GvH), allograft and delayed-type hypersensitivity, and cytotoxicity will be described in this section.

Thymus

The thymus develops from the third and fourth pharyngeal pouches. It appears as seven separate lobes on each side of the neck extending from the lower jaw to the thorax, with a portion of the caudal lobe occasionally embedded in the thyroid (Payne, 1971). Histological observations reveal the division of thymus into a cortex and medulla. Lymphocytes appear in the thymus on day 11 DE (Lucas and Jamroz, 1961). The initially large lymphocyte (11 μM) declines progressively in size with age: 11–13 DE, 8 μM; 13–15 DE, 8–7.5 μM; and 16 DE, 5.5 μM (Sugimoto et al., 1977). The appearance in the thymus, at 12 DE, of terminal deoxynucleotidyl transferase (Penit and Chaperille, 1977; Sugimoto and Bollum, 1979), which directs the synthesis of regions without benefit of template (Bollum, 1974), suggests that this enzyme may be important in cellular differentiation of the thymus (Sugimoto and Bollum, 1979).

Subpopulations of Thymic Cells. Cellular differentiation in the thymus leads to the release of a well-defined subpopulation of thymic or T cells, which participate in (1) promoting antibody synthesis (T-helper cells, discussed earlier), (2) suppression of the immune response (T-suppressor cells), (3) cytotoxicity (T-cytotoxic cells), (4) graft-versus-host and allograft responses, and (5) delayed-type hypersensitivity.

Suppressor T Cells. Bone marrow or spleen cells of agammaglobulinemic birds will suppress humoral immunity when transferred to normal birds (Blaese et al., 1974; Palladino et al., 1976). The transferred cell is a T cell, since agammaglobulinemic birds lack B cells but possess T cells. Incubating spleen or bone marrow cells of agammaglobulinemic birds with monospecific antiserum prepared against T cells will prevent their suppressive action. T cells from agammaglobulinemic birds may develop into suppressor cells after exposure to B cell antigen, provided the birds are older than 8 weeks (Grebenau et al., 1979). Spontaneously arising T-suppressor cells have been reported in the thymus of embryos and early hatched chicks (Droege et al., 1973, 1974; Droege, 1976). The T-suppressor cell is, in part, dependent on the bursa, since neonatal bursectomy reduces the suppressor ability of thymic cells of 7- to 11-week-old donors (Droege, 1976).

Cytotoxic T Cells. Splenic lymphocytes from a chicken injected with keyhole lympet hemocyanin (KLH) and incubated with KLH and chicken red blood cells labeled with chromium will lyse the chicken red blood cells, as evidenced by the release of chromium (Granlund et al., 1974; Granlund and Loan, 1974). In the presence of other antigens or in the absence of KLH, the splenic lymphocytes from the KLH-sensitized birds will not lyse chicken red blood cells. Bursectomy did not interfere with the development of the cytotoxic effector lymphocyte (Kirchner and Blaese, 1973). These and other observations suggested that the cytotoxic response is, in part, dependent on a subpopulation of T cells.

In other experiments, repeated immunization of normal or agammaglobulinemic chickens with allogeneic cells stimulated the production of cytotoxic cells to the allogeneic but not syngeneic cells (Palladino et al., 1980; Chi and Thorbecke, 1981). The cytotoxic cells in these normal chickens were not T cells, since their activity was not depressed in the presence of anti-T serum + complement. Since the cytotoxic cell of normal chickens adhered to plastic, a characteristic of phagocytic cells, it appeared to be a macrophage. On the other hand, the major cytotoxic cell in agammaglobulinemic birds was inactivated by anti-T serum + complement and therefore appeared to be a T cell (Chi and Thorbecke, 1981).

Allograft and Graft-Versus-Host Responses. The allograft and graft-versus-host (GvH) responses are dependent on cells derived from the thymus, since both responses are markedly inhibited or eliminated by thymectomy but not bursectomy. An allograft is a transplant (e.g., skin) that possesses different antigens from those of the host (recipient). The T cells of the host react to the foreign antigen displayed by the graft (host-versus-graft response) and will reject the graft within approximately 10 days (Sato and Glick, 1970). Skin transplants are affected by removing feathers and applying alcohol and then collodion to the sacral portion of an anesthesized donor and host. The skin secured from the donor should be about 1.5 times larger than the host bed, to allow for shrinkage of the transplant and enlargement of the bed. The graft may be secured by a drop of Eastman 910 or Band-Aid spots. The allograft response is useful, not only in evaluating an animal's cell-mediated immunity, but also in cell typing within the major histocompatible complex (MHC). The MHC defines a restricted region of a chromosome with an array of polymorphic genes that control the presence of membrane antigens associated with lymphocytes and other cells. The chicken MHC includes three loci—*B-F, B-G,* and *B-L*—which control the expression of B-F, B-G, and B-L antigens, respectively (Briles et al., 1982; Longenecker and Mosmann, 1981). The rejection of a skin graft does not appear to be related to differences at the *B-L* locus, but rather at either or both *B-F* and *B-G* loci (Hala, 1977). Therefore, compatible skin grafting between two chickens would identify homozygosity at these loci.

Graft-Versus-Host. Like the allograft response, the GvH response is governed by differences at the *B-F* and/or *B-L* loci (Longnecker and Mosmann, 1981). In the GvH response, it is the T cell within the graft reacting against the antigen differences of the host. The GvH response may be identified in chick embryos that are immunologically immature (Watabe and Glick, 1983a). White Leghorn eggs are incubated for 12 days, candled, and the major vein, immediately below the air cell, identified. A portion of the shell surrounding the vein is removed, and a drop of mineral oil is applied to the membrane. A 30-gauge needle is introduced into the vein that is made visible by mineral oil, and 0.1 ml of cell suspension is injected. The cut shell is then scotch-taped into place. Six days later, the embryo and spleen are weighed and a splenic index is calculated:

$$\text{SI} = \frac{\dfrac{\text{Cell-injected spleen weight}}{\text{Cell-injected embryo weight}}}{\dfrac{\text{Saline-injected spleen weight}}{\text{Saline-injected embryo weight}}}$$

A splenic index greater than 1.3 is indicative of a splenomegaly caused by T-effector cells in the graft reacting against antigenic differences of the host (Simonsen, 1957). Cells from the spleen, bone marrow, and thymus are capable of eliciting a GvH response in embryos (Danchakoff, 1918; Seto, 1968; Watabe and Glick, 1983a). While age will influence the intensity of the GvH response, all the tissues mentioned will produce a positive response by 4 weeks of age (Seto, 1968; Watabe and Glick, 1983a). A synergism occurs between chick bone marrow and thymic cells in inducing a greater GvH response than is observed with either bone marrow or thymic cells (Watabe and Glick, 1983a; Figure 4–7. The ability of chick bone marrow cells to produce a GvH response and to interact synergistically with thymic cells contrasts with mice in which neither event occurs (Stuttman and Good, 1969; Cantor et al., 1970).

Thymic mononuclear cells from 11-month-old chickens showed a significantly higher GvH capability than those of earlier ages (Watabe and Glick, 1983a). The lack of T-suppressor cells at this time (Droege, 1976) may explain the enhanced GvH activity.

A biphasic GvH response is produced by spleen cells from female chickens (McCorkle et al., 1979a). The SI is significantly higher for 6-month-old females than for 1, 3, 12, or 18-month-old females. The SI for 24-month old-females returned to the high value recorded at 6 months. Fluctuations in thymic growth, for example, thymus regeneration during a natural molt (Brake *et al.,* 1981), might explain these results. There appears

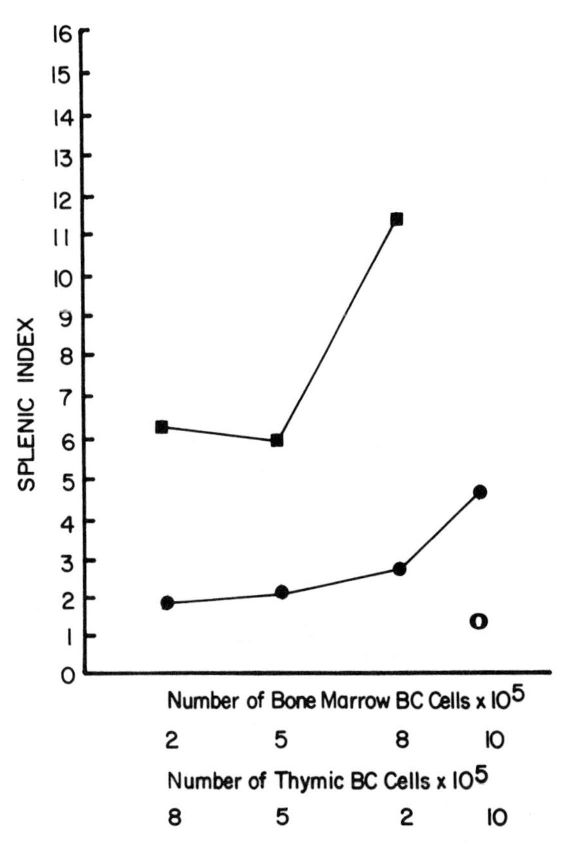

FIGURE 4–7. A graphic change in graft versus host response produced by bone marrow (*solid circle*) buffy coat (*BC*) cells, thymic BC cells (*open circle*), or the combination of bone marrow and thymic *BC* (*solid square*) from 4-week-old chickens. (From Watabe and Glick, 1983a.)

to be a slight decline in the GvH potential of male chickens with age (McCorkle et al., 1979a).

Delayed-type Hypersensitivity. The classic example of a cell-mediated response is delayed-type hypersensitivity (DTH). Animals are injected with a specific antigen, and then, several weeks later, receive a subcutaneous injection of the specific antigen. A visible or palpable swelling at the site of injection occurring 24–48 hr after the subcutaneous injection is characteristic of a DTH. Chicks produce DTH response to tuberculin (Jankovic et al., 1963; Warner et al., 1971; Fiedler et al., 1981), diptheria toxoid (Cooper et al., 1966; Klesius et al., 1977), *Salmonella adelaide* and dinitrophenylated chicken serum albumen (Warner et al., 1971), and human γ globulin (Palladino et al., 1978; Watabe and Glick, 1983b). In our laboratory, chicks 3, 6, 9, and 12 weeks of age have been injected at four subcutaneous sites over the pectoral muscle with a total of 1.0 ml of human γ globulin (HGG) in complete Freund's adjuvant. One, 2, or 3 weeks later, birds were intradermally injected in the right wattle with 0.1 ml of a saline solution of HGG (200 mg) (Watabe and Glick, 1983b). The left wattle received 0.1 ml saline. A thickness index (TI) was calculated 24 hr after challenge.

$$\mathrm{TI} = \frac{\text{Wattle diameter after HGG injection}}{\text{Wattle diameter after saline injection}}$$

In general, the optimum time of challenge after sensitization was 3 weeks. The TI of birds sensitized at 6 (4.02), 9 (3.58), and 12 (4.39) weeks of age, respectively, was significantly greater than that of 3-week-old birds (2.54). That T cells are important in mediating DTH is seen in the significantly reduced DTH response of thymectomized birds (McCorkle et al., 1979a).

Peripheral Lymphoid Tissue

It was an early suggestion that humoral immunity in chickens was controlled by the bursa and cell-mediated immunity by the thymus (Warner and Szenberg, 1962; Cooper et al., 1966). While this belief proved to be too narrow a view of immunity, the concept revealed that the bursa and thymus did contribute significantly to the normal development of certain glands in which the major cellular type was lymphoid. These glands have been designated peripheral lymphoid tissue (PLT). This section includes a brief description of selected PLT with emphasis on their contribution to the immune response.

Harderian Gland

The Harderian gland (HG) is located ventral and posteromedial to the eyeball (Wight et al., 1971). On the basis of lobular structure, three types of Harderian glands have been identified; Type I, compound tubuloacinar; Type II, compound tubular; and Type III, mixed compound tubuloacinar and tubular (Burns, 1975; Aitken and Survashe, 1977). Chickens are represented by Type I. Bang and Bang (1968) were the first to identify plasma cells in the Harderian gland. The plasma cell develops from B cells and produces antibodies. In studies to quantitate B cells, the plasma cell was identified as the major immunocompetent cell (Glick et al., 1977). Immunoglobulin, primarily monomeric IgM, is an integral part of the B-cell membrane. Ig+ cells are identified by using ^{125}I-labeled anti-Ig serum (see Figure 4–5). Unlike mouse plasma cells, chicken plasma cells have a high density of Ig receptors (Glick et al., 1977). At least 80% of the Harderian gland plasma cells are Ig+ small lymphocytes (i.e., B cells), which decline from 12% at 14 weeks to less than 2% at 8 months. Bursectomy will markedly reduce or eliminate the small lymphocyte and plasma cell populations of the Harderian gland (Mueller et al., 1971). The functional importance of the Harderian gland in the immune response was demonstrated by its ability to produce antibody to sheep red blood cells (Mueller et al., 1971). The presence of antibody-producing cells was revealed by plaques. The central cell of the plaque, the plaque-forming cell (PFC), is most likely a plasma cell. In this experiment, the antigenic determinants of the sheep red blood cells are bound by antibody released by the PFC. The antigen–antibody complex allows complement to bind to the red cells, resulting in lysis of the cell and thus accounting for the empty area around the PFC. The presence of T cells in the Harderian gland (Albini and Wick, 1974) would explain this gland's ability to make antibody to sheep red blood cells, since T- and B-cell interaction is necessary for the formation of the antibody.

Cecal Tonsil and Peyer's Patches

The cecal tonsil is an enlarged area (4–18 mm) of the proximal region of each caecum (Muthmann, 1913;

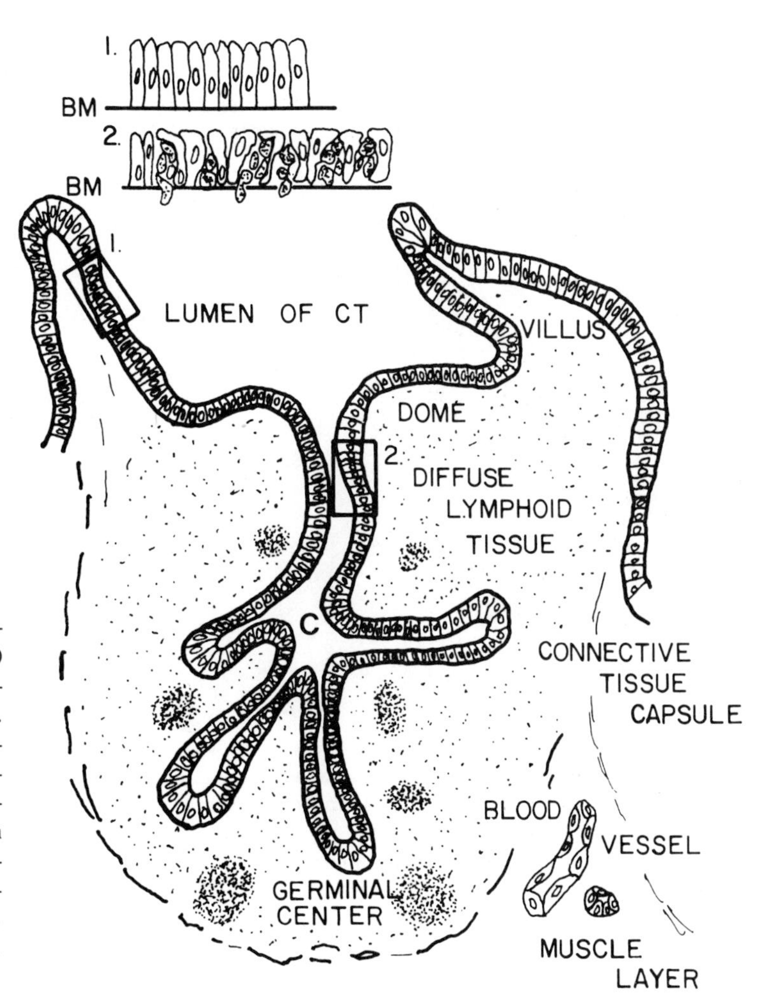

FIGURE 4–8. The tonsillar unit consists of a central branching crypt (*C*) surrounded by diffuse lymphoid tissue and numerous germinal centers. High epithelial cells (*1*), which secrete a mucous substance, line the distal portion of the villi while the central crypt is covered by epithelium invaded by lymphocytes (*2*). *BM,* basal membrane; *CT,* cecal tonsil. [Reprinted with permission from *Dev. Comp. Immunol., 5,* Glick *et al.* (1978b), Pergamon Press, Ltd.]

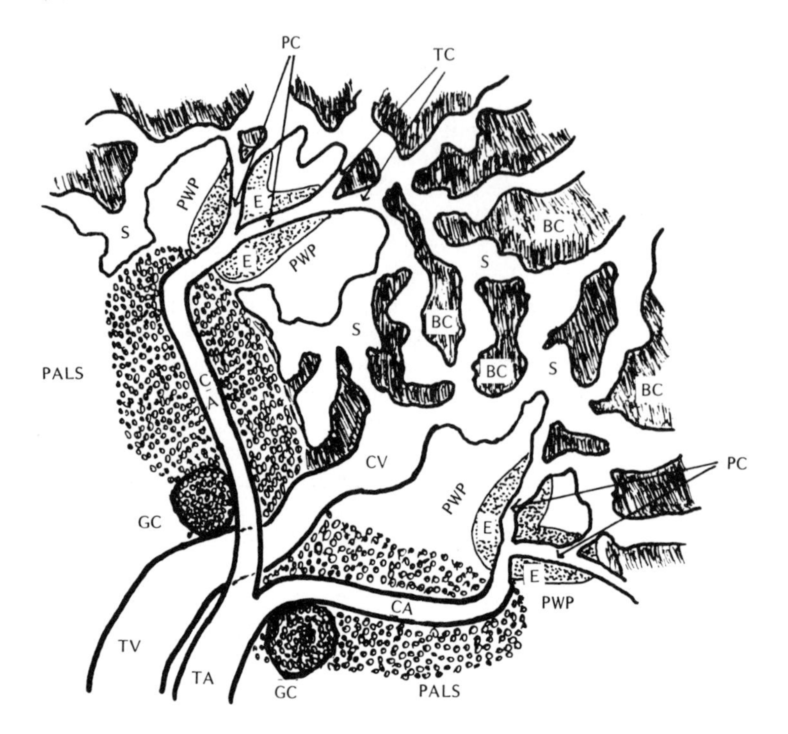

FIGURE 4–9. Diagram showing the organization of the chicken spleen and its circulation. The trabecular artery (*TA*) continues as a central artery (*CA*) surrounded by the T-dependent periarteriolar lymphatic sheath (*PALS*). The germinal centers (*GC*) are located at the beginning of the *CA* and the *PALS.* The midportion of the penicilliform (*PC*) is encompassed by the ellipsoid (*E*), which is surrounded by the periellipsoid white pulp (*PWP*). The distal portion of the *PC* or terminal capillary (*TC*) continues as the sinuses of the red pulp. Among the sinuses (*S*) the pulp cord or Billroth cord (*BC*) can be seen. The sinuses are drained by the collecting veins (*CV*), which join to form the trabecular vein (*TV*). (Reprinted with permission from Olah and Glick, 1982.)

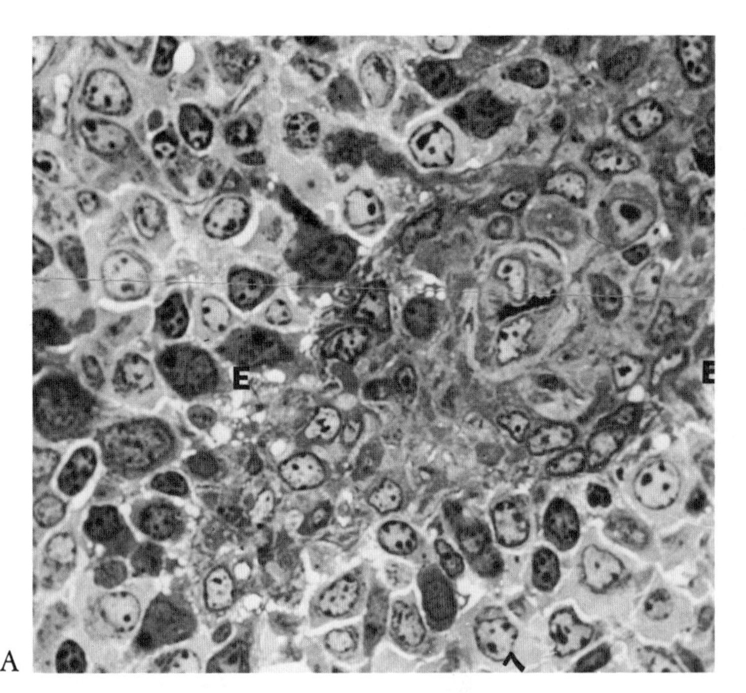

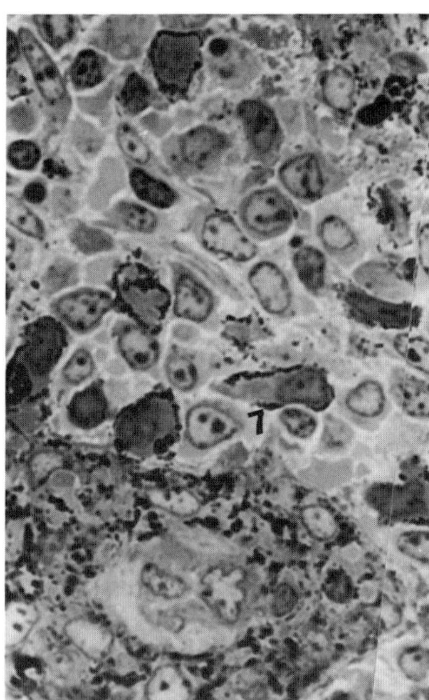

FIGURE 4–10. **(A)** The arrows indicate periellipsoid white pulp cells among the darkly staining ellipsoid-associated cells (*E*) that surround the penicilliform capillaries. ×1000 **(B)** Intravenously injected carbon is picked up by the ellipsoid-associated cells (7), which migrate into the adjacent regions. ×1100 (From Olah and Glick, 1982.)

Looper and Looper, 1929; Glick et al., 1978). The villi of the cecal tonsil are longer and not as broad as those of the remainder of the caecum's proximal region (Whitlock et al., 1975; Glick et al., 1978a). The cecal tonsil is polycrypic and similar to the mammalian palatine tonsil (Glick et al., 1978b) (Figure 4–8). The location of the cecal tonsil and the constant exposure of its villi to fecal content suggest a sentinel role for this peripheral lymphoid tissue (Glick and Olah, 1981). The cecal tonsil possesses T and B cells (Albini and Wick, 1974), and produces antibody to solubles antigens (Jankovic and Mitrovic, 1967; Orlans and Rose, 1970).

Peyer's patches appear along the intestine cranial to the ileocecal junction of 10-day-old chickens (Payne, 1971; Befus et al., 1980). They are similar to the cecal tonsil in that they possess lymphocytes beneath the epithelium but are not polycryptic like the tonsil (Befus et al., 1980). Hormonal bursectomy depopulates the lymphoid area of Peyer's patches and interferes with the ability to concentrate intravenous injection of colloidal carbon (Befus et al., 1980).

Spleen

The reddish-brown oval spleen is located adjacent to the dorsal surface of the right lobe of the liver and dorsal to the proventriculus (Nickel et al., 1977). Accessory spleens are located cranial, adjacent, and caudal to the spleen (Glick and Sato, 1969). The cranial accessory hypertrophied subsequent to splenectomy (Glick, 1970). The spleen grows most rapidly during the first 6 weeks after hatching, and attains maximum size (spleen-to-body weight ratio) by 10 weeks of age (Norton and Wolfe, 1949; Wolfe et al., 1962). The numerous cells that are critical for an immune response have been described in cytoarchitectural studies of the spleen (Olah and Glick, 1982). In this section, the location of these cells in the spleen and their interaction in the immune response will be emphasized.

The splenic artery enters the spleen and gives rise to the trabecular and central arteries (Figure 4–9). The central artery is surrounded by the periarteriolar lymphatic sheath (PALS), which contains large numbers of small lymphocytes; the PALS is considered to be thymic dependent (Sugimura and Hashimoto, 1980; Cooper et al., 1966; Hoffman-Fezer et al., 1977). These same authors have identified the germinal centers, located at the edge of the PALS (Olah and Glick, 1982), as a bursal-dependent region. The central artery becomes the penicilliform capillary, the midregion of which is surrounded by the ellipsoid (Figure 4–9). The ellipsoid is embroidered by darkly stained cells, the ellipsoid-associated cell (EAC) (Figure 4–10A). The EAC bind substances as diverse as carbon and *Brucella abortus* that have entered the area through channels formed by the endothelial cells of the midregion of the penicilliform capillaries (White and Gordon, 1970; Olah and Glick, 1982). The EAC are activated by the binding process, and within hours migrate into the periellipsoid white pulp, a bursal-dependent area (Figure 4–10B). After a few days, the EAC associate with plasma cells in the

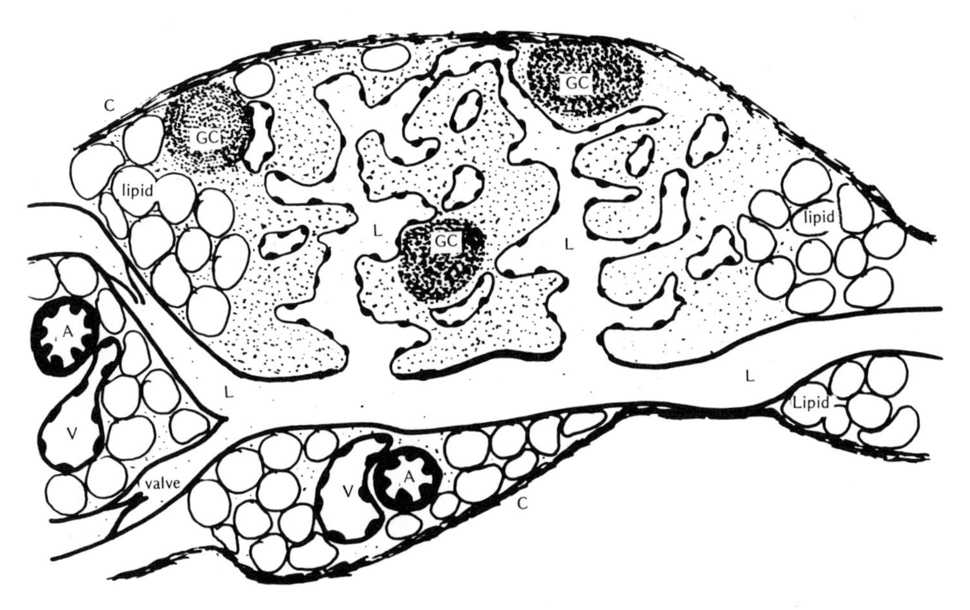

FIGURE 4–11. Lymph node with a larger amount of lymphoid tissue than adjacent tissue. Because of the large amount of lymphoid tissue, the capsule firmly adheres to the node. Abbreviations: *C*, capsule; *GC*, germinal center; *L*, lymphocyte; *V*, vein; *A*, artery. (From Olah and Glick, 1983.)

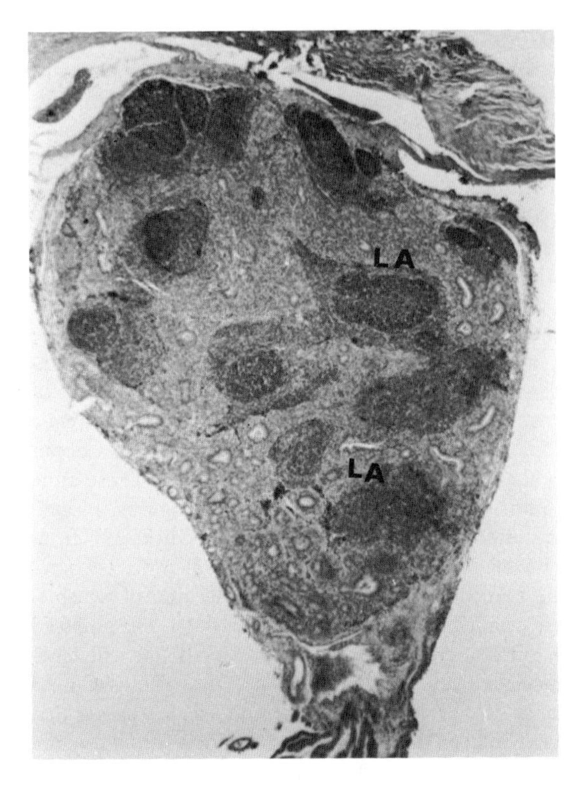

FIGURE 4–12. Sagittal section of a pineal showing lymphoid accumulation (*LA*) at age 32 days. (From Cogburn and Glick, 1981.)

red pulp, the PALS, and germinal centers. The EAC may transmit the antigenic signal that activates the T and B cells of the thymic- and bursal-dependent regions, respectively (Olah and Glick, 1982);

Lymph Node

Lymphoid accumulations have been identified in the chicken along the posterior tibio-popliteal and lower femoral veins (Olah and Glick, 1983). The most developed lymphoid accumulations are true lymph nodes possessing afferent and efferent lymphatic trunks, germinal centers, and an intricate lymphatic sinus system (Figure 4–11). These nodes are similar to the cervicothoracic lymph node of waterfowl (Kampmeier, 1969). The lymph nodes enlarge subsequent to a footpad injection of antigen (Good and Finstand, 1967; McCorkle et al., 1979b) or phytohemagglutinin, a mitogenic substance (Olah and Glick, 1983). They possess more B than T cells. The lymph node generated more plaque-forming cells (20×10^3 PFC/10^6 lymphoid cells) than the spleen (4×10^3 PFC/10^6 lymphoid cells) subsequent to a footpad injection of sheep red blood cells (McCorkle et al., 1979b). These data emphasize the lymph node role in responding to local stimulation.

Pineal Gland

The pineal has only recently been demonstrated to be a part of the peripheral lymphoid tissue of the chicken (Cogburn and Glick, 1981). Lymphocytes appear in the pineal after 9 days of age and reach maximal concentration by 32 days (Figure 4–12). The B and T cells in the pineal (Cogburn and Glick, 1979) are significantly reduced or eliminated in neonatally bursectomized/thymectomized animals (Cogburn and Glick, 1981). Plasma cells that synthesize antibody to bovine serum albumen were identified in the pineal 3–5 days after a carotid injection of bovine serum albumen (Cogburn and Glick, 1983). It is tempting to speculate that the pineal, in addition to its physiological role in diurnal periodicity, may supply immunocompetent cells important for the surveillance of the central nervous system.

References

Ackerman, G.A. (1962). Electron microscopy of the bursa of Fabricius of the embryonic chick with particular reference to the lymphoepithelial nodules. J. Cell Biol., 13, 127.

Aitken, I.D., and B.D. Survashe. (1977). Lymphoid cells in avian paraocular glands and paranasal tissue. Comp. Biochem. Physiol. A, 58, 235.

Albini, B., and G. Wick. (1974). Delineation of B and T lymphoid cells in the chicken. J. Immunol., 112, 444.

Ambrosius, H., and D. Hadge. (1982). A phylogenetic view of avian immunology. Folia Biol., 28, 1.

Bang, B.G., and F.B. Bang. (1968). Localized lymphoid tissues and plasma cells in paraocular and paranasal organ systems in chickens. Am. J. Pathol., 53, 735.

Befus, A.D., N. Johnson, G.A. Leslie, and J. Bienenstock, (1980). Gut-associated lymphoid tissue in the chicken. I. Morphology, ontogeny, and some functional characteristics of Peyer's patches. J. Immunol., 125, 2626.

Bienenstock, J., D.Y.F. Perey, J. Gauldie, and B.J. Underdown. (1972). Chicken immunoglobulin resembling γA. J. Immunol., 109, 403.

Bienenstock, J., B.Y.F. Perey, J. Gauldie, and B. J. Underdown. (1973). Chicken γA: Physiochemical and immunochemical characteristics. J. Immunol., 110, 50.

Blaese, R.M., P.C. Weiden, I. Koski, and N. Dooley. (1974). Infectious agammaglobulinemia: Transmission of immunodeficiency with grafts of agammaglobulinemic cells. J. Exp. Med., 140, 1097.

Bockman, D.C., and M.D. Cooper. (1973). Pinocytosis by epithelium associated with lymphoid follicles in the bursa of Fabricius, appendix Peyer's patch: An electron microscopic study. Am. J. Anat., 136, 455.

Bollum, F.J. (1974). Terminal deoxynucleotidyl transferase. In "The Enzymes," Vol. 10 (P.D. Boyer, Ed.). New York: Academic Press, p. 145.

Burns, R.B. (1975). Plasma cells in the avian Harderian gland and the morphology of the gland in the rook. Can. J. Zool., 53, 1258.

Brake, J., G.W. Morgan, and P. Thaxton. (1981). Recrudescence of the thymus and repopulation of lymphocytes during an artificially induced molt in the domestic chickens, proposed model system. Dev. Comp. Immunol., 5, 105.

Briles, W.C., N. Bumstead, D.L. Ewert, D.G. Gilmour, J. Gogusev, K. Hala, C. Koch, B.M. Longenecker, A.W. Nordskog, J.R.L. Pink, W.L. Schierman, M. Simonsen, A.

Toivanen, P. Toivanen, O. Vainio, and G. Wick. (1982). Nomenclature for chicken major histocompatibility (B) complex. Immunogenetics, *15*, 441.

Calhoun, M.L. (1932). The microscopic anatomy of the digestive tract of *Gallus domesticus*. Iowa State College J. Sci. 7, 261.

Cantor, H., M. Mandel, and R. Asofsky. (1970). Studies of thoracic duct lymphocytes of mice. II. A quantitative comparison of the capacity of thoracic duct lymphocytes and other lymphoid cells to induce graft versus host reactions. J. Immunol., 104, 409.

Chi, D.S., and G.J. Thorbecke, (1981). Cytotoxicity of allogeneic cells in the chicken. III. Antibody-dependent cell-mediated cytotoxicity in normal and agammaglobulinemic chickens. Cell. Immunol., 64, 158.

Clark, W.R. (1983). The experimental foundations of modern immunology. New York: John Wiley and Sons.

Cogburn, L.A., and B. Glick. (1979). The pineal gland: A prominent lymphoid tissue in young chickens. Physiologist, 22, 20.

Cogburn, L.A., and B. Glick. (1981). Lymphopoiesis in the chicken pineal gland. Am. J. Anat., 162, 131.

Cogburn, L.A., and B. Glick. (1983). Functional lymphocytes in the chicken pineal gland. J. Immunol., 130, 2109.

Cooper, M.D., R.D.A. Peterson, M.A. South, and R.A. Good. (1966). The function of the thymus system and bursa system in the chicken. J. Exp. Med., 123, 75.

Danchakoff, V. (1918). Equivalence of different hematopoietic anlages (by method of stimulation of their stem cells). II. Grafts of adult spleen on the allantois and response of the allantoic tissues. Am. J. Anat., 24, 127.

Droege, W. (1976). The antigen-inexperienced thymic suppressor cells: Class of lymphocytes in the chicken thymus that inhibits antibody production and cell-mediated responses. Eur. J. Immunol., 6, 279.

Droege, W., D. Malchow, J.L. Strominger, and T.J. Linna. (1973). Cellular heterogeneity in the thymus: Graft-vs-host activity of fractionated thymus cells in the chicken. Proc. Soc. Exp. Biol. Med., 143, 249.

Droege, W.R., R. Zucker, and K. Hanning. (1974). Developmental changes in the cellular composition of the chicken thymus. Cell. Immunol., 12, 186.

Edwards, J.R., C. Murphy, and R. Cho. (1975). On the development of the lymphoid follicles of the bursa of Fabricius. Anat. Rec., 181, 735.

Ewert, D.L., D.G. Gilmour, W.E. Briles, and M.D. Cooper. (1980). Genetics of Ia-like alloantigens in chickens and linkage with B major histocompatibility complex. Immunogenetics, 10, 169.

Fiedler, H., M. Scheu, I. Kuhlmann-Rabens, and U. Losch. (1981). Kinetics of delayed-type hypersensitivity (DHT) reaction and Ig levels in FCA-injected dyagammaglobulinemic B12 chickens. Immunology, 158, 293.

Fitzimmons, R.C., E.M.F. Garrod, and L. Garnett. (1973). Immunological responses following early embryonic surgical bursectomy. Cell. Immunol., 9, 377.

Glick, B. (1956). Normal growth of the bursa of Fabricius in chickens. Poult. Sci., 35, 843.

Glick, B. (1960). Growth of the bursa of Fabricius and its relationship to the adrenal gland in the White Pekin duck, White Leghorn, outbred and inbred New Hampshire. Poult. Sci., 39, 130.

Glick, B (1961). The influence of dipping eggs in hormonal solutions on the lymphatic tissue and antibody response of chickens. Endocrinology, 69, 984.

Glick, B. (1963). The effect of surgical and chemical bursectomy in the White Pekin duck. Poult. Sci., 42, 1106.

Glick, B. (1964). The bursa of Fabricius and development of immunologic competence. In "Thymus on Immunology" (R.A. Good and Anne Gabrielsen, Eds). New York: p. 343. Harper and Row, Publ. NY.

Glick, B. (1970). Hypertrophy of accessory spleen in splenectomized chickens. Folia Biol., 16, 74.

Glick, B. (1976). Lymphocyte lifespan in chickens. In "Phylogeny of Thymus and Bone marrow—Bursa cells." (R.K. Wright and E.L. Cooper, Eds.) New York: North Holland, p. 237.

Glick, B. (1977a). The bursa of Fabricius and immunoglobulin synthesis. Int. Rev. Cytol., 48, 345.

Glick, B. (1977b). Lymphocyte life span and migration of bursal lymphocytes. In "Developmental Immunology" (J.B. Solomon and J.D. Horton, Eds.). Amsterdam: Elsevier/North Holland Press, p. 371.

Glick, B. (1983). The bursa of Fabricius. In "Avian Biology," Vol. 7 (D.S. Farner, J.R. King, and K.C. Parkes, Eds.). New York: Academic Press.

Glick, B., and C.R. Sadler. (1961). The elimination of the bursa of Fabricius and reduction of antibody production in birds from eggs dipped in hormone solutions. Poult. Sci., 40, 185.

Glick, B., and K. Sato. (1964). Accessory spleens in the chicken. Poult. Sci., 43, 1610.

Glick, B., and I. Olah. (1981). Gut-associated lymphoid tissue of the chicken. In "Int. J. Scanning Electron Micros." Chicago: SEM, Inc., p. 99.

Glick, B., and I. Olah. (1983). The morphology of the starling (*Sturnus vulgaris*) bursa of Fabricius: A scanning and light microscope study. Anat. Rec., 204, 341.

Glick, B., D.S.V. Subba Rao, R. Stinson, and F.C. McDuffie. (1977). Immunoglobulin positive cells from the gland of Harder and bone marrow of the chicken. Cell. Immunol., 31, 177.

Glick B., K.A. Holbrook, I. Olah, W.D. Perkins, and R. Stinson. (1978a). A scanning electron microscope study of the caecal tonsil: The identification of a bacterial attachment to the villi of the caecal tonsil and possible presence of lymphatics in the caecal tonsil. Poult. Sci., 57, 1408.

Glick, B., K.A. Holbrook, I. Olah, W.D. Perkins, and R. Stinson. (1978b). A scanning electron microscope study of the caecal tonsil: The basic unit of the caecal tonsil. Dev. Comp. Immunol., 5, 95.

Glick, B., M. LaVia, and B. Koger. (1985). Flow cytometric analysis of embryonic bursal, thymic, and splenic cells from normal and cyclophosphamide-treated embryos. Poult. Sci., 64, 723.

Golub, E.S. (1981). The cellular basis of the immune response. Sinauer Associates Inc., Sunderland, MA.

Good, R.A., and J. Finstad. (1967). The phylogenetic development of immune responses and the germinal center system. In "Germinal Centers in Immune Response" (H. Cottier, N. Odartchenko, R. Schindler, and C.C. Congdon, Eds.). Berlin: Springer-Verlag, p. 4.

Granlund, R.D., G.M. Buening, and R.W. Loan. (1974). Cell-mediated immunity in the chicken. I. Development of a target cell destruction assay. Cell. Immunol., 11, 99.

Granlund, D.J., and R.W. Loan. (1974). Cell-mediated immunity in the chicken: Cytotoxicity induced by specific soluble antigens. J. Reticuloendothel. Soc., 15, 503.

Grebenau, M.D., D.S. Chi, and G.J. Thorbecke. (1979). T-cell tolerance in the chicken. II. Lack of evidence for suppressor cells in tolerant agammaglobulinemic and normal chickens. Eur. J. Immunol., 9, 477.

Hala, K. (1977). The structure of the major histocompatibility (B) complex of the chicken. Folia Biol., 23, 400.

Hemmingsson, E.J., and T.J. Linna. (1972). Ontogenetic studies on lymphoid cell traffic in the chicken. II. Cell traffic from the bursa of Fabricius to the thymus and spleen in the embryo. Int. Arch. Allergy Appl. Immunol., 42, 764.

Hoffman-Fezer, G., H. Rodt, D. Götze, and S. Thierfelder. (1977). Anatomical distribution of T and B lymphocytes identified by immunohistochemistry in the chicken spleen. Int. Arch. Allergy Appl. Immunol., 55, 86.

Howell, J., W. Hood, and B.G. Sanders. (1967). Carbohydrate analysis of the IgG heavy chain carbohydrate peptides. J. Mol. Biol., 30, 555.

Ivanyi, J. (1981). Functions of the B-lymphoid system in chickens. In "Avian Immunology" (M.E. Rose, L.M. Payne, and B.M. Freeman, Eds.). Edinburgh: British Poultry Science Ltd., p. 63.

Jankovic, B.D., and K. Mitrovic. (1967). Antibody-producing cells in the chicken, as observed by fluorescent antibody technique. Folia Biol. (Praha), 13, 406.

Jankovic, B.D., M. Isvaneski, D. Milosevic, and L. Popeskovic. (1963). Delayed hypersensitivity reactions in bursectomized chickens. Nature, (London), 198, 298.

Jankovic, B.D., Z. Knezevic, K. Isakovic, K. Mitrovic, B.M. Markovic, and M. Rajcevic. (1975). Bursa lymphocytes and IgM-containing cells in chicken embryos bursectomized at 52–64 hours of incubation. Eur. J. Immunol., 5, 656.

Jankovic, B.D., K. Isakovic, B.M. Markovic, and M. Rajcevic. (1977). Immunological capacity of the chicken embryo. II. Humoral immune responses in embryos and young chickens bursectomized and sham-bursectomized at 52–64 h of incubation. Immunology, 32, 689.

Jolly, J. (1915). La Bourse de Fabricius et les organes lymphoepitheliaux. Arch. Anat. Microsc. Morphol. Exp., 16, 363.

Kampmeier, O.F. (1969). "Evolution and Comparative Morphology of the Lymphatic System." Springfield: Thomas, p. 376.

Katz, D., A. Kohn, and R. Arnon. (1974). Immunoglobulins in the aviary washings and bile secretions of chickens. Eur. J. Immunol., 4, 494.

Kincade, P., A.R. Lawton, D.E. Bockman, and M.D. Cooper. (1970). Suppression of immunoglobulin G synthesis as a result of antibody-mediated suppression of immunoglobulin M synthesis in chickens. Proc. Natl. Acad. Sci. U.S.A., 67, 1918.

Kincade, P., and M.D. Cooper. (1971). Development and distribution of immunoglobulin-containing cells in the chicken: An immunofluorescent analysis using purified antibodies to μ, γ, and light chains. J. Immunol., 105, 371.

Kirchner, H., and R.M. Blaese. (1973). Pokeweed mitogen-, concanavalin A-, and phytohemagglutinin-induced development of cytotoxic effector lymphocytes. J. Exp. Med., 138, 812.

Klesius, P., W. Johnson, and T. Kramer. (1977). Delayed wattle reaction as a measure of cell-mediated immunity in the chicken. Poult. Sci., 56, 249.

Lebacq-Verheyden, A.M., J.P. Vaermans, and J.F. Heremans. (1972). Immunohistologic distribution of the chicken immunoglobulins. J. Immunol., 652, 109.

Lebacq-Verheyden, A.M., J.P. Vaermans, and J.F. Heremans. (1974). Quantification and distribution of chicken immunoglobulins IgA, IgM, and IgG in serum secretions. Immunology, 27, 683.

Le Douarin, N.M., and F.V. Jotereau. (1973). Recherches sur l'origine embryologique des lymphocytes du thymus chez l'embryon d'orseau. C.R. Hebd. Seances. Acad. Sci., Ser. D, 276, 629.

Le Douarin, N.M., E. Houssaint, F.V. Jotereau, and M. Belo. (1975). Origin of hemopoetic stem cells in embryonic bursa of Fabricius and bone marrow studied through interspecific chimaeras. Proc. Nat. Acad. Sci. U.S.A., 72, 2701.

Leene, W., M.J.M. Duyzings, and C. van Steeg. (1973). Lymphoid stem cell identification in the developing thymus and bursa of Fabricius of the chick. Z. Zellforsch. Mikrosk. Anat., 136, 526.

Lerner, K.G., B. Glick, and F.C. McDuffie. (1971). Role of the bursa of Fabricius in IgG and IgM production in the chicken: Evidence for the role of a nonbursal site in the development of humoral immunity. J. Immunol., 107, 493.

Leslie, G.A., L.W. Clem. (1969). Phylogeny of immunoglobulin structure and function. III. Immunoglobulins of the chicken. J. Exp. Med., 130, 1337.

Leslie, G.A., and L.N. Martin. (1973). Studies on the secretory immunologic system of the fowl. III. Serum and secretory IgA of the chicken. J. Immunol., 110, 1.

Leslie, G.A., R.P. Stankus, and L.N. Martin. (1976). Secretory immunological system of the fowl. V. The gall bladder: an integral part of the secretory immunological system of fowl. Int. Arch. Allergy Appl. Immunol., 51, 175.

Longnecker, B.M., and T.R. Mosmann. (1981). Structure and properties of the major histocompatability complex of the chicken. Speculations on the advantages and solution of polymorphism. Immunogenetics, 13, 1.

Looper, J.B., and M.H. Looper. (1929). The histologic study of the colic caeca in the bantam fowl. J. Morphol. Physiol., 48, 589.

Lucas, A.M., and C. Jamroz. (1961). "Atlas of Avian Hematology." Agric. Monogr. 25, U.S. Department of Agriculture, Superintendent of Documents, Washington, D.C.

McArthur, W.P., D.G. Gilmour, and G.J. Thorbecke. (1972). Immunocompetent cells in the chicken. Cell. Immunol., 8, 103.

McCorkle, F., R. Stinson, and B. Glick. (1979a). A biphasic graft vs. host response in aging chickens. Cell. Immunol., 46, 208.

McCorkle, F., R.S. Stinson, I. Olah, and B. Glick. (1979b). The chicken's femoral-lymph nodules: T and B cells and the immune response. J. Immunol., 123, 667.

Miller, J.F.A.P., and G.F. Mitchell. (1968). Cell-to-cell interaction in the immune response. I. Hemolysin-forming cells in neonatally thymectomized mice reconstituted with thymus or thoracic duct lymphocytes. J. Ex. Med., 128, 801.

Mueller, A.P., K. Sato, and B. Glick. (1971). The chicken lacrimal gland, gland of Harder, caecal tonsil and accessory spleens as a source of antibody-producing cells. Cell. Immunol., 2, 140.

Muthmann, E. (1913). Bietrage zur Vergleichenden Anatomic des Bunddarmes und der Lymphoiden Organe Saugetieren und Volgeln. Anat. Hefte. Abt, 48, 167.

Nickel, R., A. Schummer, E. Seiferle, W.G. Seller, and P.A.L. Wight. (1977). Anatomy of the domestic birds. New York: Springer-Verlag.

Norton, S., and H.R. Wolfe. (1949). The growth of the spleen in the chicken. Anat. Rec., 105, 83.

Olah, I., and B. Glick. (1978). The number and size of the follicular epithelium (FE) and follicles in the bursa of Fabricius. Poult. Sci., 57, 1445.

Olah, I., and B. Glick. (1982). Splenic white pulp and associated vascular channels in chicken spleen. Am. J. Anat., 165, 445.

Olah, I., and B. Glick. (1983). Avian lymph node: Light and electron microscope study. Anat. Rec., 205, 287.

Orlans, E., and M.E. Rose. (1970). Antibody formation by transferred cells in inbred fowls. Immunology, 18, 473.

Orlans, E., and M.E. Rose. (1972). An IgA-like immunoglobulin in the fowl. Immunochemistry, 9, 833.

Palladino, M.A., S.P. Lerman, and G.J. Thorbecke. (1976). Transfer of hypogammaglobulinemia in two inbred chicken strains by spleen cells from bursectomized donors. J. Immunol., 166, 1673.

Palladino, M.A., M.D. Grebenau, and G.J. Thorbecke. (1978). Requirements for induction of delayed hypersensitivity in the chicken. Dev. Comp. Immunol., 2, 121.

Palladino. M.A., D.S. Chi, N Blyznak, A.M. Paolino, and G.J. Thorbecke. (1980). Cytotoxicity to allogeneic cells in the chicken. I. Role of macrophages in the cytotoxic effect on 51Cr-labeled red blood cells by immune spleen cells. Dev. Comp. Immunol., 4, 309.

Payne, L.N. (1971). The lymphoid system. In "Physiology and Biochemistry of the Domestic Fowl" (D.J. Bell and B.M. Freeman, Eds.). New York: Academic Press, p. 985.

Penit, C., and F. Chaperille. (1977). Developmental changes in terminal deoxynucleotidyl transferase of the chicken thymus. Biochem. Biophys. Res. Commun., 74, 1046.

Porter, P., and S.J. Parry. (1976). Further characterization of IgA in chicken serum and secretions with evidence of a possible analogue of mammalian secretory component. Immunology, 31, 407.

Romanoff, A. (1960). "The Avian Embryo." New York: Macmillan Co.

Rose, M.E., E. Orlans, A.W.R. Payne, and P. Hesketh. (1981). The origin of IgA on chicken bile: Its rapid active transport from blood. Eur. J. Immunol., 11, 561.

Sanders, B.G., and W.L. Case. (1977). Chicken secretory immunoglobulin: chemical and immunological characterization of chicken IgA. Comp. Biochem. Physiol., 56B, 273.

Sato, K., and B. Glick. (1970). Antibody and cell mediated immunity in corticosteroid-treated chicks. Poult. Sci., 49, 982.

Schaffner, T., J. Mueller, M.W. Hess, C. Cottier, B. Sordat, and C. Ropke. (1974). The bursa of Fabricius: A central organ providing for contact between the lymphoid system and intestinal content. Cell. Immunol., 13, 304.

Seto, F. (1968). Quantitative study of the splenomegaly assay system of the graft vs. host reaction in chick embryos. Proc. Okla. Acad. Sci., 47, 134.

Simonsen, M. (1957). The impact on the developing embryo and newborn animal of adult homologous cells. Acta Pathol. Microbiol. Scand., 40, 480.

Stuttman, O., and R.A. Good. (1969). Absence of synergism between thymus and bone marrow in graft vs. host reactions. Proc. Soc. Exp. Biol. Med., 130, 848.

Subba Rao, D.S.V., F.C. McDuffie, and B. Glick. (1978). The regulation of IgM production in the chick: Roles of the bursa of Fabricius, environmental antigens, and plasma IgG. J. Immunol., 120, 783.

Sugimira, M., and Y. Hashimoto. (1980). Quantitative and histological studies on the spleen of ducks after neonatal thymectomy and bursectomy. J. Anat., 131, 441.

Sugimoto, M., and F. J. Bollum. (1979). Terminal deoxynucleotidyl transferase (TdT) in chick embryo lymphoid tissue. J. Immunol., 122, 392.

Sugimoto, M., T. Yasuda, and Y. Egashira. (1977). Development of the embryonic chick thymus. I. Characteristic synchronous morphogenesis of lymphocytes accompanied by the appearance of an embryonic thymus-specific antigen. Dev. Biol., 56, 281.

Tenenhouse, H.S., and H.F. Deutsch. (1966). Some physical-chemical properties of chicken γ-globulins and their pepsin and papain digestion products. Immunochemistry, 3, 11.

Travis, J.C., and B.G. Sanders. (1973). Structural comparisons between chicken low molecular weight immunoglobulin heavy chains and human gamma chains. Biochem. Genet., 8, 391.

Travis, J.C., and B.G. Sanders. (1975). Sequence of the cysteine peptides from the chicken. 7.8 S Ig heavy chain C-region. A comparison with the gamma chain of humans. Int. J. Biochem., 6, 719.

Van Alten, P.J., and H.J. Meuwissen. (1972). Production of specific antibody by lymphocytes of the bursa of Fabricius. Science, 176, 45.

von Rautenfeld, D.B., and K.D. Budras. (1982). The bursa cloacal (Fabricii) of struthioniforms in comparison with the bursa of other birds. J. Morphol., 172, 123.

Warner, N.L., and A. Szenberg. (1962). Effect of neonatal thymectomy on the immune response in the chicken. Nature (London), 96, 784.

Warner, N.L., Z. Ovary, and F.S. Kantor. (1971). Delayed hypersensitivity reactions in normal and bursectomized chickens. Int. Arch. Allergy Applied Immunol., 40, 719.

Watabe, M., and B. Glick. (1983a). Graft vs. host response as influenced by the origin of the cell, age of chicken and cellular interactions. Poult. Sci., 62, 1317.

Watabe, M., and B. Glick. (1983b). Influence of age in the induction of delayed hypersensitivity to human gamma globulin. Poult. Sci., 62, 563.

Watanabe, H., K. Kobayashi, and R. Isayama. (1975). Peculiar secretory IgA system identified in chickens. II. Identification and distribution of free secretory component and immunoglobulins of IgA, IgM, and IgG in chicken external secretions. J. Immunol., 115, 998.

Weinbaum, F.I., D.G. Gilmour, and G.J. Thorbecke. (1973). Immunocompetent cells of the chicken. III. Cooperation of carrier-sensitized T cells from agammaglobulinemic donors with hapten immune B cells. J. Immunol., 110, 1434.

White, R.G. and J. Gordon (1970). Macrophage reception and recognition mechanisms in the chicken spleen. In van Furth, R. (Ed.). "Mononuclear Phagocytes." Edinburgh, Blackwell Scientific, p. 511.

Whitlock, D.R., W.B. Lushbaugh, H.D. Danforth, and M.D. Ruff. (1975). Scanning electron microscopy of the caecal mucosa in *Eimeria tennella*-infected and uninfected chickens. Avian Dis., 19, 293.

Wight, P.A.L., R.B. Burns, B. Rothwell, and G.M. MacKenzie. (1971). The Harderian gland of the domestic fowl. I. Histology with reference to the genesis of plasma cells and Russell bodies. J. Anat., 110, 307.

Wolfe, H.R., S.A. Sheridan, N.P. Bilstad, and M.H. Johnston. (1962). The growth of lymphoidal organs and the testes of chickens. Anat. Rec., 142, 485.

5
Body Fluids: Blood

P.D. Sturkie with P. Griminger

Body Fluids

The water of the body is distributed into intracellular and extracellular compartments. The latter may be partitioned into plasma and interstitial fluids.

Estimation Methods

Body Water. Total body water may be estimated directly by desiccation and weighing and indirectly by measuring the dilution of some substance that can come into equilibrium with all water compartments. The volume distribution of an injected substance (x) equals the amount injected minus the amount excreted, divided by the concentration in the diluting fluid. Therefore, volume (V) = q/c, where q is the quantity injected minus the amount excreted, and c is the concentration in diluting fluid.

Antipyrine is a substance that has been used by a number of investigators in mammals and birds (chickens) e.g., Weiss, 1958; Medway and Kare (1959) to measure total body water. Isotope dilution (tritiated water) has also been used. Extracellular fluid volume is also determined by the dilution principle; a substance is injected that enters the extracellular but not intracellular compartment. Thiocyanate ion has been most commonly used, but other agents such as mannitol, Br 82, insulin, sucrose and thiosulfate have been used.

Intracellular space or fluid is total body water minus extracellular fluid. Interstitial fluid is calculated, and is equal to extracellular space minus plasma volume.

Blood volume may be determined directly by bleeding out and measuring the residual volume, or indirectly by the dilution principle. The latter may involve labeling red blood cells by injection of radioactive iron, chromium, or phosphorus, or by labeling the plasma with a dye such as Evans blue (T-1824). A known amount of dye is injected intravenously and allowed to mix completely with blood (within 2–3 min for chickens). The rate of disappearance of the dye from the blood of birds is much more rapid than in mammals, averaging about 1%/min in chickens.

$$\text{Total blood volume} = \frac{\text{Plasma volume} \times 100}{\text{Percent plasma}}$$

The most critical factor in determining blood volume is mixing time, which must be established accurately in each species involved. For example, if the mixing time in the chicken were estimated to be 10 min instead of 3 min, the calculations of blood volume would be too high. The accurate mixing time is based on the disappearance time or the time when the plasma dye concentration begins to decrease.

Total Body Water. The distribution of total body water in a number of selected species is presented in Table 5–1. The figures shown in Table 5–1 are based on the use of Antipyrine dilution, isotope dilution (tritiated water), and desiccation and weighing. Most of these data reveal a total body water of approximately 60% of body weight of adults; some figures are higher and some lower. Those based on isotope dilution may be more variable than those obtained by other methods. Skadhauge showed that there is an inverse relationship between body weight and water turnover, which reflects the higher relative rate of metabolism, drinking rate, and evaporation rate of small birds (Skadhauge, 1981). The body water of adult males appears to be higher than that of adult females. This may mean that males have less fat in their bodies than females; fat contains less water than lean body tissue. Weiss (1958) studied the body water of White Leghorn female chickens and found that as the females became older and fatter, their body water decreased (Table 5–2).

Growing birds, which have relatively more lean body tissue than adults, have more total body water than at later stages (see Table 5–3). The distribution of intracellular and extracellular (ECV) compartments also varies with age (Table 5–3), with the highest values in 1-week-old chickens and lowest values in those approaching maturity (Medway and Kare, 1959).

ECV figures tend to be higher than the same values for mammals, and some experimenters believe that this is due to methodology. Danby et al. (1980) compared markers for the determination of ECV in 8-week-old chickens. They used radioactive sucrose (6,6 n3H) thiosulfate (^{35}S S) and thiocyanate. They found that the value for sucrose (17.8 ml/100 g body weight) was significantly lower than those obtained with thiosulfate (22.3 ml/100 g) and thiocyanate (23.9 ml/100 g). The difference between thiosulfate and thiocyanate was not significant. These values compare favorably with those obtained by Harris and Kioke (1975)

TABLE 5–1. Total body water of some selected avian species

Species	Body weight (kg)	Body weight (%)	Reference
Hens (chicken)	1.81	61	Weiss (1958)
Chicken, growing	—	57–62	Medway and Kare (1959)
Cocks	2.6	64	Chapman and Black (1967)
Hens	1.7	62	Chapman and Black (1967)
Ducks, domestic	3.06	62	Thomas and Phillips (1975b)
On freshwater	3.09	69	Ruch and Hughes (1975)
On seawater	2.35	64	Ruch and Hughes (1975)
Emu, hydrated	32.7	63	Skadhauge et al. (1980)
Dehydrated	39.4	53	Skadhauge et al. (1980)
Pigeon	0.36	64	Le Febvre (1968)
Zebra finch			
Hydrated	0.013	63	Skadhauge and Bradshaw (1974)
Dehydrated	0.013	63	Skadhauge and Bradshaw (1974)
Japanese Quail			
Male	0.105	67	Chapman and McFarlend (1971)
Female	0.117	62	Chapman and McFarlend (1971)

on functionally nephrectomized chickens, in which renal loss of marker was prevented. The authors concluded that the agreement between their results and those of these workers suggests that their method did not suffer from significant renal loss of marker. The authors (Danby et al.), however, believe that sucrose gave the most reliable estimates, because it appears to be confined to extracellular space, whereas the other two markers have been reported to penetrate red blood cells to some extent.

Species with high total body water tend to have high ECVs. ECVs for some other species include: ducks, on freshwater, 24.9% of body weight, on saltwater, 26.4%; chickens (cocks) 28.8%, glaucous-winged gulls, 38.2% (Ruch and Hughes, 1975). Mongin and Carter (1976), using radiosulfate and correcting for marker loss in urine of laying hens, reported ECF volumes of 24.9 ml/100 g figures in the same range as those of Danby et al., who also used radiosulfate. Harris and Koike (1977) reported that chickens on a low sodium diet decreased their ECV from 28.8% to 20.5%. Douglas (1968) reported thiocyanate space in Adélie penguins of 29.1%, which is high compared to mammals.

Blood Volume

Total blood volume includes the cells and the plasma. After labeling the plasma with a dye and complete mixing (about 3 min in chickens), blood samples are taken; the blood is centrifuged and the plasma volume and packed cells are determined. The packed cell volume (PCV) is corrected for the amount of plasma trapped with the cells (about 4%). Red cell volume may be obtained directly by injecting tagged cells with radioactive chromium, phosphorus, iron, or some other suitable agent.

Normal Values. Values for growing chickens are shown in Table 5–4. It is apparent that total blood volume and plasma volumes are higher in younger birds as compared to adults. These values range from 6.6 to 8.3% of body weight (for TBV) to 4.4–5.2% for adult birds. Plasma volumes for most of the other avian species (Table 5–5) range from 3.2 to 6.5, with most values within 4–6% of body weight. Figures for TBV are more variable. High altitude and hypoxia tend to increase TBV. The value of adult chickens and other species are shown in Table 5–5.

TABLE 5–2. Body water (percent of body weight) of chicken hens of various ages[a]

	Ages (weeks)					
	26	30	36	42	55	61
Body weight (kg)	1.77	1.996	1.919	2.032	2.054	2.035
Body water (%)	66.0	61.2	56.7	53.3	52.9	53.4

[a] After Weiss (1958).

TABLE 5–3. Distribution of body fluids in White Leghorn female chickens[a]

Age (weeks)	Weight (g)	Percent of body weight: Total body water	Intracellular water	Extracellular water: Interstitial	Plasma	Total
1	55.1	72.4	11.4	52.3	8.7	61.0
2	108.4	71.6	21.0	42.3	7.3	50.6
3	175.3	70.5	24.6	39.1	6.8	45.9
4	241.8	68.4	24.1	38.3	6.0	44.3
6	372.3	—	—	36.8	5.9	42.7
8	527.3	68.7	26.6	36.1	6.1	42.2
16	1137.3	64.8	34.8	24.8	5.2	30.0
32	1759.5	57.3	31.1	21.7	4.6	26.2

[a] Medway and Kare (1959b).

Chronic acceleration causes an increase in plasma volume with a 1-*g* exposure causing an increase of as much as 25%. There is a shift in plasma volume from the thoracic area to the limbs (Smith, 1978).

Effects of Hormones. Earlier workers reported that estrogen administration increased the crude blood volume, but Sturkie (1951), using the dye technique, found no change in blood volume following massive doses of estrogen or thyroxine to hens. Estrogen produces hyperlipemia in birds, and consequently turbidity of the plasma, which interferes with the colorimetric determination of dye concentration unless the blood is diluted sufficiently or the fat is extracted.

Sturkie and Eiel (1966) determined the effects of estrogen on blood volume of males and females after extraction of fat from plasma, and found no change in blood volume of females, but an increase in that of males, as was reported by Gilbert (1963), and a decrease in hematocrit. Estrogen also increases blood volume in male and poulard geese significantly; but not in normal females (naturally estrogenized) (Hunsaker, 1968). Androgen had little effect on blood volume of male and female geese but increased volume in castrate males and females.

Dehydration for 3 or more days decreased blood volume in chickens, and increased plasma sodium and proteins but caused no change in potassium (Koike et al., 1983). Dehydration decreased blood pressure, but increased hematocrit and heart rates moderately.

Functions of Blood

Blood has many functions, among them (1) absorption and transport of nutrients from the alimentary canal to the tissues, (2) transport of the blood gases to and from the tissues, (3) removal of waste products of metabolism, (4) transportation of hormones produced by the endocrine glands, and (5) regulation of the water content of the body tissues. Blood is also important in the regulation and maintenance of body temperature.

Blood contains a fluid portion (the plasma), salts and other chemical constituents, and certain formed elements, the corpuscles. The corpuscles include the

TABLE 5–4. Blood volume of growing female White Leghorn chickens[a]

Age (weeks)	Body weight (g)	Mean blood volume (percent of body weight)	Mean plasma volume (percent of body weight)
1	61.8	12.0	8.7
2	115.0	10.4	7.3
3	163.3	9.7	6.8
4	249.9	8.7	6.0
6	398.7	8.3	5.9
8	571.6	8.4	6.1
16	1310.0	7.6	5.2
32	1789.0	6.5	4.6

[a] Medway and Kare (1959a).

TABLE 5–5. Blood volume of avian species

Species	Body weight (g)	Sex	Notes	Total blood volume (ml/100 g)	Plasma volume (ml/100 g)	Reference
Chicken W1	2000	F		6.6	4.7	Sturkie and Newman (1951)
Chicken W1	1789	—		7.6	5.2	Medway and Kare (1959)
Chicken BrL	1851	M		7.8	4.7	Gilbert (1963)
Chicken	2506	—		—	4.4	Ruch and Hughes (1975)
Chicken	800	—	Control	7.8	4.8	Harris and Koike (1977)
	800	—	Salt loaded	8.3	5.2	Harris and Koike (1977)
Turkey	664			7.8	3.8	Augustine (1982)
Duck	2924	—		7.6	6.4	Stewart (1972)
Duck	3000	—	Freshwater	7.8	6.5	Bradley and Holmes 1971 (Chapter 16)
Ducks						
Anas platyrhynchos	—	—		10.2	6.55	Portman et al. (1952)
(mallard and dabbling)	980	—		11.3	6.4	Bond and Gilbert (1958)
Tufted	632	M		—	114.2 (ml/kg)	Keijer and Butler (1982)
Mallard	1026	M		—	91.2 (ml/kg)	Keijer and Butler (1982)
Geese, adult	—	Male		5.99	3.36	Hunsaker (1968)
		Female		6.75	4.15	Hunsaker (1968)
		Capon		5.83	3.42	Hunsaker (1968)
		Poulard		6.07	3.69	Hunsaker (1968)
Geese	4240		Freshwater	6.0	—	Hanwell et al. (1971)
Geese	4720		Salt loaded	7.2	—	Hanwell et al. (1971)
Coot	550	—		9.5	5.1	Bond and Gilbert (1958)
Pigeon	310	—		9.2	4.4	Bond and Gilbert (1958)
Pheasant	1190	Male		6.7	4.5	Bond and Gilbert (1958)
	1110	Female		4.8	3.2	Bond and Gilbert (1958)
Red-tailed hawk	925	—		6.2	3.5	Bond and Gilbert (1958)
Great horned owl	1495	—		6.4	3.4	Bond and Gilbert (1958)
	—	—		—	—	
Coturnix	98	Male		7.4	4.73	Nirmalin and Robinson (1972)
	117	Female		6.76	4.29	Nirmalin and Robinson (1972)
Coturnix	(at sea level)	Male		10.3	5.5	Jaeger and McGrath (1974)
	(at 6100 m)	Male		14.3	5.5	Jaeger and McGrath (1974)
	(at sea level)	Female		10.8	5.7	Jaeger and McGrath (1974)
	(at 6100 m)	Female		15.3	5.3	Jaeger and McGrath (1974)

erythrocytes (red cells) and leukocytes (white cells).

Certain physical properties of blood are discussed here but such chemical constituents as proteins and lipids are treated in Chapters 14 and 15, carbohydrates in Chapter 13, calcium in Chapters 21 and 18, and other electrolytes in Chapter 16.

Physical Properties of Blood

Viscosity, specific gravity, and osmotic pressure of blood have important physical effects on the circulation and flow of blood and the exchange of fluid between blood and the tissues. Plasma proteins, which exert the greatest effect on plasma osmotic pressure, are discussed in Chapter 14 and the role plasma electrolytes play is discussed in Chapter 16 and elsewhere.

Viscosity

The viscosity of whole blood is most influenced by the number of cells and, as expected, it is higher in males, because males have more red cells than females (Table 5–6). The viscosity of plasma is also influenced considerably by plasma proteins and is significantly higher in females or estrogenized males, because in each of the latter, plasma proteins are higher than in normal males or capons.

Usami et al. (1970) studied the viscometric behavior of turkey blood over a wide range of hematocrits and shear rates. They have shown that the viscosity of the turkey red cell at high hematocrits is higher than in most mammalian species, which suggests a lower deformability of the nucleated cell. This lower deformability has also been demonstrated in the duck red cell (Gachtgens et al., 1981a).

TABLE 5–6. Physical properties of avian blood

Viscosity of blood in relation to water at given temperatures (°C)

Species	Whole blood	Plasma	Temperature (°C)	Reference
Chicken, male	3.67	1.42	42	Vogel (1961)
capon	2.47	1.28	42	Vogel (1961)
female	3.08	1.51	42	Vogel (1961)
Duck	4.0	1.5	14–20	Spector (1956)
Goose	4.6	1.5	14–20	Spector (1956)
Ostrich	4.5	—	—	
Turkey	—	1.12	—	Usami et al. (1970)

Specific gravity

Species	Whole blood	Plasma	Serum	Reference
Chicken, female	1.050	1.099		Sturkie and Textor (1960)
	—	1.0180 (very lipemic)		Sturkie and Textor (1960)
	1.0439	1.0177		Medway and Kare (1959)
male	—	1.0210		Sturkie and Textor (1960)
male	1.054	—	1.023	Sturkie and Textor (1960)
Goose, male	1.061	1.020	—	Hunsaker et al. (1964)
female	1.052	1.022	—	Hunsaker et al. (1964)
Goose	1.050	1.021	—	Wirth (1931) (Sturkie, 1976)
Duck	1.056	1.020	—	Wirth (1931) (Sturkie, 1976)
Guinea fowl	1.057	1.021	—	Wirth (1931) (Sturkie, 1976)
Ostrich	1.063	1.022	—	Devilliers (Sturkie, 1976)

Apparent Viscosity. There is a difference in the viscosity of blood measured in vivo and in vitro. As the diameter of the capillary decreases below 200 μm, the viscosity of the blood flowing through the capillaries decreases progressively, provided the shear rate is reasonably high (Fahraeus–Lindquist effect). It is theorized that the red cells in small tubes line up axially so there is little shear between the cells, and this causes less internal friction (the Sigma phenomenon). The lower viscosity of in vivo measurement is the *apparent* or effective viscosity. This value may be as low as one-half the in vitro viscosity in certain species with nonnucleated erythrocytes. Birds, however, have a much higher apparent viscosity than mammals at the same capillary diameter and hematocrit, mainly because of the nucleated red cells (Gachtgens et al., 1981b). For example, at a tube diameter of about 5 μm and the same hematocrit, the AP in duck blood is about 1.5 times that of humans, and at higher hematocrits, it ranges from 2.3 to approximately 3.5 times that in humans. At higher tube diameters (8–12 μm), the differences are much greater. The difference is least at a tube diameter of 7 μm, but still significantly higher for avian blood, particularly at higher hematocrits, where the apparent viscosity is nearly twice that of humans.

Avian red cells have a lower stability of orientation during capillary flow and greater hemodynamic disturbance. This is partially compensated for by the lower hematocrit of bird blood and by the greater capillary density of avian (duck) muscle (Gachtgens et al., 1981b).

Specific Gravity

The specific gravity of blood of various avian species is shown in Table 5–6. Whole blood has a higher specific gravity than plasma because of the erythrocytes. The figure for the plasma of female chickens is significantly lower than for that of males. This is surprising, because the plasma proteins, which supposedly influence specific gravity most, are significantly higher in females. However, female plasma is lipemic and this condition tends to depress specific gravity. The plasma of laying female chickens made hyperlipemic by nicarbazine administration had a lower specific gravity than did the plasma of untreated females (Sturkie and Textor, 1960), demonstrating that plasma specific gravity is not a good measure of plasma protein in birds having lipemic plasma.

Osmotic Pressure

The colloid osmotic pressure of avian plasma is considerably lower than that of most mammals because plasma albumins, which are relatively lower in birds, has more influence on colloid osmotic pressure than do globulins. Values for colloid osmotic pressure in chickens and doves are 150 and 100 mm H_2O, respectively (Albritton, 1952).

Erythrocytes

The erythrocytes of birds are oval shaped and, unlike those of mammals, are nucleated; they are also larger (Figure 5–1). The sizes of the erythrocytes of some avian species are presented in Table 5–7.

Hartman and Lessler (1963) made measurements on erythrocytes of 124 species in 46 families of wild birds. The range in size extended from 10.7×6.1 μm to 15.8×10.2 μm. The lower forms tended to have the largest erythrocytes; however, in a few families the cells were smaller in the smaller species (passerines and trochilids). Most of the domestic species have sizes intermediate between these ranges. The sizes are much larger than mammalian erythrocytes for many species but smaller than the cells of reptiles. For a comparison of human and avian red cell dimensions, see Gachtgens et al. (1981a), and Fourie and Hattingh (1983) for hematology of South African wild birds.

Numbers

The numbers of erythrocytes and corpuscular volume are influenced by age, sex, hormones, hypoxia, and other factors. Erythrocyte numbers for several species are shown in Table 5–8. In most species in which both sex hormones have been studied, there is a difference in numbers and packed cell volume (Tables 5–8 and 5–9), with higher levels in males. The goose and pheasant appear to be exceptions in which there is little or no difference in hematocrit (Hunsaker et al., 1964; Bond and Gilbert, 1958), although more studies are needed. Gilbert (1963) reported that estrogen administration to adult cocks depressed erythrocyte volumes and presumably numbers, although the latter was not determined; thyroxine administration tended to prevent the estrogen effect.

Estrogen administration to sexually immature *Coturnix* (quail) depressed erythrocyte numbers from 3.2 million to 1.6 million in males and from 3.19 million to 1.44 million in females (Nirmalan and Robinson, 1972) and had a similar effect on hematocrit; androgen, in contrast, increased the number significantly in immature males and females, suggesting a erythropoietic effect. Prolonged androgen administration increased hematocrit in chickens at sea level approximately 45% (Burton and Smith, 1972) and about the same degree at high altitude (12,500 ft). Domm and Taber (1946) and others have reported that red cell numbers in the castrate male are nearly the same as in the female; when androgen was administered to castrates, the numbers approached the normal male level. Sturkie and Textor (1960), however, found the number in male

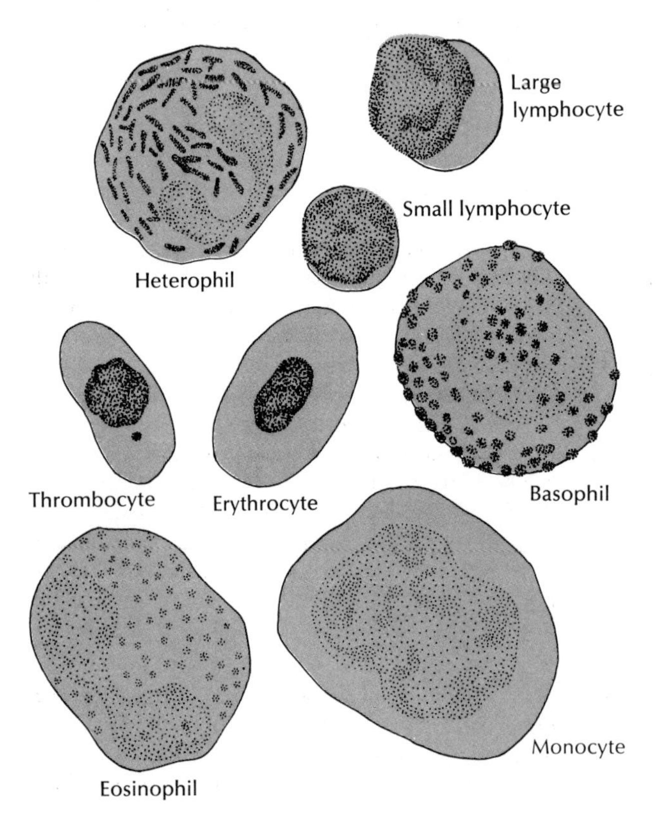

FIGURE 5–1. Drawing of mature blood cells of the chicken.

TABLE 5–7. Dimensions of avian erythrocytes

Species	Long diameter (μm)	Short diameter (μm)	Thickness (μm)	Thickness
Chicken (breed not given)				
3 days old	12.5	7.0	3.8	Groebbels (1932) (See Sturkie, 1965)
25 days old	13.0	7.2	3.5	Groebbels (1932) (See Sturkie, 1965)
70 days old	13.0	6.5	3.5	Groebbels (1932) (See Sturkie, 1965)
Adult	12.8	6.9	3.6	Groebbels (1932) (See Sturkie, 1965)
Chickens, White Leghorn	10.7	6.8	—	Lucas and Jamroz (1961)
Chickens, Brown Leghorn, male	12.8	7.9	2.41	Gilbert (1965)
Brown Leghorn, female	12.2	7.4	2.53	Gilbert (1965)
Chicken	10.7	7.06	—	Balasch et al. (1973) (See Sturkie, 1976)
Duck[a]	12.8	6.6	—	Groebbels (1932)
Duck	13.2	6.8	3.8	Gachtgens, et al. (1981a)
Turkey, male	15.5	7.5	—	Groebbels (1932) (See Sturkie, 1965)
Turkey, female	15.5	7.0	— —	Groebbels (1932) (See Sturkie, 1965)
(sex not given)	15.0	7.2	—	Usami et al. (1970)
Pigeon	12.7	7.5	—	Usami et al. (1970)
Numida meleagris, male (guinea)	11.96	5.96	—	Balasch et al. (1973) (See Sturkie, 1976)
Pavo cristatus (peacock)	12.55	6.97	—	Balasch et al. (1973) (See Sturkie, 1976)
Phasianus colchicus, male (pheasant)	10.64	6.77	—	Balasch et al. (1973) (See Sturkie, 1976)
Alectoris graeca, male (rock partridge)	11.26	6.40	—	Balasch et al. (1973) (See Sturkie, 1976)
Falcon	11–18	7.3–8.7		Hartman and Lessler (1963)

[a] Length of nucleus, 7.76 μm; surface area of nucleus, 271.4 μm^2; width of nucleus, 3.85 μm; volume of nucleus, 35.00 μm^3.

castrate chickens to be intermediate between those for normal males and females. Female castrates (poulards) have the same number of red cells as normal females, indicating that estrogen has no positive effect on erythropoiesis but that androgen does. Thyroxine has a erythropoietic effect; this tends to counteract the negative effect of estrogen, as later demonstrated by Gilbert (1963) and by Sturkie (1951).

Although Hunsaker (1968) reported that estrogen depressed erythropoiesis in geese, androgen had little or no effect on red cell counts of normal male and females, and castrate male and female geese. Starvation (forced molting) appears to increase cell numbers (packed cell volume) of erythrocytes and leukocytes of chickens (Brake et al., 1982).

Corpuscular Volume (Hematocrit)

The corpuscles may be separated from the plasma by centrifugation at speeds of 3000 rpm or higher for 15–60 mim. Hunsaker (1969) centrifuged blood of chickens, geese, and turkeys at 3215 × *g* for 30 min in macrotubes and determined hematocrits (see Table 5–9) and the amount of plasma trapped, with the erythrocytes and leukocytes making up the packed cell volume. The percentage of trapped plasma averaged from 2.35 in goose blood to 3.21 for chicken blood, with turkey blood intermediate. Cohen (1967) employed a microhematocrit method and centrifugation speeds of 11,650 rpm and 12,800 × *g* for varying periods up to 9.5 min. Almost complete packing was accomplished after 5 min centrifugation time. Values for several species are shown in Table 5–9.

Factors that affect cell numbers affect hematocrit, but cell size also influences the latter. Most species (adult males) have higher corpuscular volumes as their erythrocyte numbers increase. Changes in cell volume may occur or be induced, without influencing the absolute number of cells, by an increase in plasma water (hemodilution) or a decrease (hemoconcentration). Blood sampling may cause hemodilution (see Sturkie, 1965) and epinephrine administration and hypothermia may cause hemoconcentration.

Androgen is erythropoietic in chickens and quails but it does not account solely for the sex differences in red cell numbers because Sturkie and Textor (1960) found the erythrocyte numbers and hematocrits of chicken male castrates to be intermediate between those in normal males and females. Female castrates have the same number of red cells as normal females.

Anemias

An inherited type of macrocytic–normochromic anemia that has been reported in chickens (Washburn and Smyth, 1968) is characterized by a reduction of 15% in hematocrat, 14% in hemoglobin, and 23% in erythrocyte numbers. Different strains of chickens carrying the genetic defect behaved differently in response to a diet low in copper and iron (Washburn, 1969).

TABLE 5–8. Erythrocyte numbers in birds[a]

Species	Age	Sex: Male	Sex: Female	Sex not given	Reference
Chicken	Adult	3.8	3.0		Lucas and Jamroz (1961)
	Adult	3.32	2.72		Olson (1937)
	Adult	3.26	2.72		Lange (Sturkie, 1965)
	3 hr			1.84	Lange (Sturkie, 1965)
	3 days			2.23	Lange (Sturkie, 1965)
	12 days			2.65	Lange (Sturkie, 1965)
	26 days			2.77	Cook (1937)
	32–47 days			2.83	Twisselmann (Sturkie, 1976)
	50 days			2.34	Lange (Sturkie, 1965)
	70 days			2.39	Lange (Sturkie, 1965)
	82 days			2.79	Cook (1937)
Chicken, White Leghorn	42 days	—	3.02		Lucas and Jamroz (1961)
	84 days	—	3.02		Lucas and Jamroz (1961)
	Adult	3.91	2.95		Sturkie and Textor (1960)
	Adult castrate	3.50	—		Sturkie and Textor (1960)
Turkey (domestic)	Adult	2.38	2.24		Groebbels (See Sturkie, 1965)
Turkey	Embryo 24 days			1.95	
	Embryo 27–28 days			2.35–2.27	Christensen et al. (1982)
Ostrich	Adult	1.84 (mixed)			DeVilliers (Sturkie, 1976)
Rhenish goose	8 months	—	—	3.35	Kaleta and Bernhardt (Sturkie, 1976)
	2 years	—	—	3.49	Kaleta and Bernhardt (Sturkie, 1976)
Italian goose	5 months	2.53 (mixed)			Sova et al. (Sturkie, 1976)
Czech goose	5 months	2.22	—	—	Sova et al. (Sturkie, 1976)
Domestic goose	Adult	—	—	2.71	Groebbels (Sturkie, 1976)
Indian duck, adult, native		2.92	—	2.42	Surendranathon et al. (Sturkie, 1976)
Peking duck, adult		2.71	—	2.46	Halaj (Sturkie, 1976)
Peking duck, adult	—	—	—	2.58	Gachtgens, et al. (1981a)
Coturnix (quail)	Young	—	—	2.86	Nirmalan and Robinson (1971)
	Adult	4.14	3.86	Not laying	Nirmalan and Robinson (1971)
	Adult	—	3.81	Laying	Atwal et al. (1964)
Bobwhite (quail)	Adult	5.41	4.1		Bond and Gilbert, (1958)
Rock partridge		—	—	2.62	Balasch et al. (Sturkie, 1976)
Numida meleagris	Adult	2.82	—	—	Balasch et al. (Sturkie, 1976)
Phasianus colchicus	Adult	3.26	—	—	Balasch et al. (Sturkie, 1976)
Pavo cristatus	Adult	2.07 (mixed)		—	Balasch et al. (Sturkie, 1976)
Pigeon	Adult	4.00	3.07	—	Wastl and Leiner (Sturkie, 1965)
	Adult	3.23	3.09	—	Riddle and Braucher (Sturkie, 1965)
Dove	Adult	3.04	2.99	—	Riddle and Braucher (Sturkie, 1965)
Red-tailed hawk	Immature			3.2	Bond and Gilbert (1958)
Great horned owl	Adult			2.2	Bond and Gilbert (1958)
Diving duck	Adult			3.2	Bond and Gilbert (1958)
Dabbling duck	Adult			3.6	Bond and Gilbert (1958)
Red-throated loon	Adult			3.1	Bond and Gilbert (1958)

[a] In millions per cubic millimeter (milliliter).

Chickens raised at ambient temperatures of 10°, 21.1°, and 32.2°C had higher hematocrits and hemoglobins than those raised at lower temperature, regardless of dietary regime; those raised at 32.2°C were more susceptible to deficiencies of iron and copper than were those raised at other temperatures. A significant increase in hematocrit and hemoglobin levels occurred with increasing levels of copper and iron within limits at temperatures of 21.0° and 32.2° (Kubena et al., 1971).

TABLE 5–9. Erythrocyte volumes (%) of Aves

Species, age, and condition		No sex	Male	Female	Reference
Chicken, sexually immature			29	29	Newell and Shaffner (Sturkie, 1965)
Sexually mature			45	29	Newell and Shaffner (Sturkie, 1965)
Sexually mature			40	31	Lucas and Jamroz (1961)
6 weeks			—	31	Lucas and Jamroz (1961)
12 weeks			—	30	Lucas and Jamroz (1961)
Sexually mature, White Leghorn			48	31	Sturkie and Textor (1960)
Sexually mature, Capon			38	—	Sturkie and Textor (1960)
Adult			40.8	25.5	Hunsaker (1969)
Pilgrim goose, 34 weeks			48	47	Hunsaker et al. (1964)
50 weeks			49	36	Hunsaker et al. (1964)
59 weeks			42	41	Hunsaker et al. (1964)
European goose, 32 weeks		48	—	—	Kaleta and Bernhardt } Sturkie, 1976
156 weeks		46	—	—	Kaleta and Bernhardt } Sturkie, 1976
Bronze turkey, 9 months			45.1	36.4	Ringer (unpublished)
20 weeks			38.5 (mixed)		Sullivan, 1965 (Sturkie, 1976)
Adult			—	35.9	McCartney (Sturkie, 1965)
Adult			38.5	33.5	Hunsaker (1969)
Turkey, adult			—	41.0	Lewis et al. (1979)
Turkey, 2 weeks (664g)		26	—	—	Augustine (1982)
Turkey (embryos)					
24 days		33	—	—	Christensen et al. (1982)
28 days		33.6			
Indian native duck, adult			40.7	38.10, non-laying	Surendranathan et al. (Sturkie, 1976)
Peking duck, adult			46.7	44.2	Halaj (Sturkie, 1976)
Peking duck, adult		46	—	—	Gachtgens, Schmidt & Will (1981a)
Numida meleagris, 8 weeks			34.3	32.7	Fallaw et al. (1976)
8 months			43.5	37.5	Fallaw et al. (1976)
Numida meleagris (guinea)	adult	—	43	—	Balasch et al. } Sturkie, 1976
Pavo cristatus (peacock)	adult		37 (mixed)		Balasch et al. } Sturkie, 1976
Phasianus colchius	adult pheasant		41.5		Balasch et al. } Sturkie, 1976
Phasianus colchius	adult "		33	34	Bond and Gilbert (1958)
Coturnix (quail)	adult	—	53.1	48.7, not laying	Nirmalan and Robinson (1971)
	Adult (50 days)	—	42.0	37.0	Atwal et al. (1964)
Quail			46	—	Ernest et al. (1971)
	Adult		45 at sea level	45	Jaeger and McGrath (1974)
	Adult		61 high altitude	63	Jaeger and McGrath (1974)
Bobwhite	Adult		38		Bond and Gilbert (1958)
Alectoris g. (partridge)			37		Balasch et al. (1973)
		No sex			Bond and Gilbert (1958)
Red-tailed hawk, immature		43			Bond and Gilbert (1958)
Great horned owl		32			Bond and Gilbert (1958)
Red-throated loon		54			Bond and Gilbert (1958)
Diving ducks		37			Bond and Gilbert (1958)
Dabbling ducks (mallard)		43			Bond and Gilbert (1958)
Coot		46			Bond and Gilbert (1958)
Redwinged blackbird, adult		40	—	—	Ronald et al., 1968 (Sturkie, 1976)
Pigeon, adult		—	58.5	56.4	Kaplan (Sturkie, 1965)
Pigeon, adult			56.95 (Sea level)		McGrath (1970)
			75.6 (High Altitude)		McGrath (1970)

Effects of Hypoxia and High Altitude

Chickens, pigeons, and quail, similarly to mammals, respond to hypoxia by an increase in hematocrit and erythrocyte numbers, which in the acute stages results in a very high hematocrits (60–80% in chickens). When such birds become adapted to high altitude (12,500 ft), the increases in hematocrit, hemoglobin, and erythrocyte counts are 27, 36, and 33%, respectively, in male chickens (Burton et al., 1971). Three months after such birds were returned to near sea level, these parameters were not significantly different from birds raised at sea level.

Chickens of some strains adapted to high altitude were tested 2 weeks later at sea levels and their hematocrits at both locations were not significantly different. There was also no difference in their hemoglobin levels (Abati and McGrath, 1973). Chickens raised at high altitudes for 7 weeks (3300 m) had increased hematocrits and hemoglobins (Sillau et al., 1980). Japanese quail (males and females), held at simulated altitudes of 20,000 feet for 4 weeks and presumably acclimated, exhibited increases in hematocrits of approximately 35% in both sexes (Jaeger and McGrath, 1974) (see Table 5–9). Increases in hemoglobin were also reported. High altitude also produced high hematocrits in pigeons (McGrath, 1970).

Effects of Acceleration. Chickens respond to acceleration of 3–4 *g* by an increase in red cell numbers, and the increase is proportional to the *g* strength (Burton and Smith, 1969).

Formation of Corpuscles

There are two theories on the origin of the blood cells. The proponents of the monophyletic theory of evolution maintain that there is one specific stem cell, developing from the mesenchyme, that gives rise to both of two main types of blood cells, the white and the red. The proponents of the polyphyletic theory believe that erythrocytes and leukocytes have developed from two originally distinct cell types, the erythrocytes from the vascular endothelium of the bone marrow, and the leukocytes from reticular connective tissue cells. Lucas (1959) and Lucas and Jamroz (1961), who reviewed the early literature, stated that most avian hematologists subscribed to the polyphyletic view. Later work by Campbell (1967) suggested a monophyletic origin of blood cells.

Erythrocyte Series. Erythropoiesis in mammalian bone marrow is confined to the extravascular spaces, but in birds it occurs in the lumen of the medullary sinuses (Campbell, 1967). The bone marrow of birds, unlike that of mammals, contains large amounts of lymphatic tissue.

The immature erythroid cells of birds are located in the lumen of the medullary sinuses. The earliest progenitors of erythroid cells are large, 10–12 μm in diameter, with large nuclei, large nucleoli, and a basophilic cytoplasm. They are situated within the bone marrow sinuses near the sinus wall. These stem cells are called *hemocytoblasts.* Near these cells are smaller cells, with smaller nuclei and more basophilic cytoplasm, known as *basophil erythroblasts;* these arise from the hemocytoblasts.

More centrally located in the sinuses are smaller cells with nuclear chromatin, which are evenly distributed in small masses; these are described as *polychromatophil erythroblasts* and contain hemoglobin. The latter are believed to give rise to *mature erythrocytes,* which are more centrally located in the vascular lumen. Avian erythrocytes (Figure 5–1) are elongated and contain a dense oval nucleus. Nuclear hemoglobin is dispersed among the coarse chromatin particles and is continuous with cytoplasmic hemoglobin at the nuclear pores.

Barrett and Scheinberg (1972), showed that mature avian cells arise from spherical precursor cells that contain hemoglobin and become flat when conditions favor deoxygenated hemoglobin. The flat cells revert to spherical shape on reoxygenation. The authors suggest that the flat shape increases the survival of the cells because it gives them increased flexibility.

Control of Erythropoiesis. It is generally agreed that erythropoiesis is under humoral control and that hypoxia, the fundamental erythropoietic stimulus, operates through the production of a circulating substance, termed erythropoietin or an erythropoiesis stimulating factor (ESF) in mammals. It is a glycoprotein that is formed in the body tissues (kidney) and acts directly on bone marrow, increasing the production rate of erythroid cells and their conversion to mature erythrocytes.

In mammals, this increased production may cause the release of a large number of immature red cells (reticulocytes) into the circulation but in birds the amount is very small (< 2%); (Lucas and Jamroz, 1961), although some have reported a higher level (6–7%) in avian blood (Nirmalan and Robinson, 1971) and some a level as high as 25–35%. Much of this variation may be attributed to errors in distinguishing these cells. Coates and March (1966) reported that 35% of the differential count in young chickens was reticulocytes; this decreased to 10% at 15 weeks of age or older. Androgen and cortical hormones increase erythropoiesis, whereas estrogens depress it in mammals and birds, as previously pointed out.

Avian erythropoietin, which is stimulated and released following hypoxia and suppressed by induced polycythemia (Rosse and Waldmann, 1966), is different from the mammalian type because the latter does not stimulate erythropoiesis in birds, and the avian type is ineffective in mammals.

Life Span of Erythrocytes

The average life span of human erythrocytes is 50–60 days; that of chickens averages 28–35 days (Hevesy and Ottesen, 1945; Brace and Atland, 1956; Rodnan et al., 1957), pigeon cells, 35–45 days; and ducks, 42 days (Rodnan et al., 1957). The figure for *Coturnix* (quails) is 33–35 days (Nirmalin and Robinson, 1973). The short life span of avian cells is related to the higher body temperature and metabolic rates.

Resistance of Erythrocytes

Hemolysis is the act of discharging hemoglobin from the corpuscles into the plasma. A number of factors, such as freezing, thawing, and changes in osmotic pressure of the blood, produce hemolysis. Solutions that have the same osmotic pressure as blood and that do not cause hemolysis are isotonic. Solutions with lower osmotic pressures than blood are hypotonic, and those with higher pressures are hypertonic. Hypotonic solutions cause hemolysis and bursting by increasing the water content of the cells; hypertonic solutions cause a shrinking of the corpuscles, because water is lost from the cells. The fragility of the red cells may be measured by their resistance to solutions of known concentrations and osmotic pressures, usually NaCl solutions. Therefore, resistance may be expressed as percent hemolysis to a solution of known concentrations. Earlier workers on chicken blood and ostrich blood (see Sturkie, 1965) indicated that blood began to hemolyze in NaCl solutions of 0.40–0.48% and all cells were hemolyzed at 0.27–28% NaCl.

Data by March et al. (1966) on chickens also show maximum hemolysis (90–97%) at NaCl solutions of 0.2–0.30% and about 50% hemolysis at 0.4% solution. At 0.5% solution hemolysis dropped to 10% or less. They also reported that the cells of New Hampshire chickens were more resistant than those of White Leghorns and the cells of females were more resistant than those of males; estrogen administered to males increased the resistance to hemolysis but androgen had no significant effect.

Hunter (1953) has studied the permeability of chicken erythrocytes to hypotonic and hypertonic solutions. His results indicated that erythrocytes in chicken, as in man, behave as perfect osmometers in slightly hypotonic and slightly hypertonic solutions. In markedly hypertonic solutions, however, they shrank less than would be expected if they behaved as perfect osmometers. The blood of birds contains relatively more potassium and less sodium than mammalian blood (McManus, 1967; Lerner et al., 1982). The latter authors also studied the amino acid transport of erythrocytes.

Facilitated duffusion of starling red cells to urea, thiourea, ethylene glycol, and glycerine has been reported by Hunter and Lim (1978). Details concerning the exchange of Na and K in bird erythrocytes have been reported by Hunter et al. (1956) and by Tosteson and Robertson (1956).

Sedimentation Rate

Important Factors. Sedimentation rate of erythrocytes (ESR) is dependent on two main forces: (1) the force of gravity, causing cells to settle, and (2) the frictional resistance of the surrounding plasma, which holds the cells in suspension. The role played by each of these forces is related to a number of factors, such as cell size and shape, the specific gravity of cells and plasma, and the chemical composition of plasma. The sedimentation rate in human beings may be increased during infections and in diseases associated with tissue injury.

Less is known about the sedimentation rate in birds, because few reports are available, and in most cases the effects of disease, age, and sex have not been ascertained. Mean sedimentation rates ranging from 0.5 to 9 mm/hr, with most values falling between 1.5 and 4, have been reported (see Sturkie, 1965). These values were determined in the usual way, with the sedimentation tube held vertically, and are considerably lower than those reported for man. The vertical position is unsatisfactory for avian blood, because avian blood cells settle very slowly.

The sedimentation rate of human blood can be increased considerably by positioning the tube at 45° (Washburn and Meyers, 1957), and similar results have been obtained with chicken blood (Sturkie and Textor, 1958; Gilbert, 1968), goose blood (Hunsaket et al., 1964), turkey blood (Bierer et al., 1964), and pheasant blood (Dobsinska and Dobsinska, 1972). The results are shown in Table 5–10.

It is apparent that the sedimentation rate is lowest for males, intermediate for capons, and highest for females. Erythrocyte numbers (Sturkie and Textor, 1960) were highest in the male, intermediate in the capon, and lowest in the female. The same investigators (1958) showed that the sedimentation rate in the three groups was linear with time from 10 to 120 min, and this has been confirmed by Gilbert (1962, 1968).

Other Factors. Hyperlipemia produced by one drug nicarbazine (Sturkie and Textor, 1960) or by estrogen (Gilbert, 1962) increases the sedimentation rate significantly.

The number of cells has the greatest influence on the sedimentation rate, and cell size affects it to a much lesser degree (Sturkie and Textor, 1960; Gilbert, 1968). Differences in specific gravity of plasma and plasma proteins have little effect on the ESR.

It is apparent that the ESR alone has little physiologic

TABLE 5–10. Sedimentation rate in mm/min (tube slanted)

	Minutes				
	10	30	60	120	
Chicken, adults					Sturkie and Textor (1958)
Male	0.80	2.06	3.86	7.0	
Capon	0.73	2.87	6.45	12.9	
Female	1.35	5.30	10.5	18.05	
Geese					Hunsaker et al. (1964)
Male, 50 weeks	—	—	—	10.1	
Female, 50 weeks	—	—	—	21.9	
Pheasant					
Mixed sex	—	—	17.2	32.6	Dobsinska and Dobsinska (1972)

or pathologic significance. It must be related to cell number and hematocrit.

β-Adrenoreceptors. Adrenoreceptors of the beta 1 type have been reported in avian red cells, but they do not correspond strictly to the mammalian type (Dickinson and Nahorski, 1981).

Phosphate Compounds in Erythrocytes. Certain phosphate compounds influence the affinity of hemoglobin for O_2. In birds, this is myoinositol pentaphosphate, (IP_5) but 2,3 diphosglycerate in mammals. IP_5 decreases the affinity of Hb for O_2 and shifts the dissociation curve (see Bauer, 1978; Weingarten et al., 1978; and Chapter 8 for more details).

Other phosphate compounds in avian blood include ATP, inorganic phosphate (Pi), and, in some species (the ostrich), inositol tetraphosphate (IP_4) and inositol hexaphosphate (IP_6), Jaeger and McGrath (1974).

In early avian embryos, there is a high level of DPG and a low level of IP_5, but with an increase in embryonic age, DPG decreases and IP_5 increases. Fifty days after hatching there is no DPG in the blood (Bartlett and Borgese, 1976). Isaacs et al. (1976a) reported on the phosphate compounds of turkey embryos and poults and the role of IP_5 on O_2 binding. These authors (1982) reported that myoinositol appears to be the primary precursor of IP_5, and that the initial step in its synthesis is phosphorylation of myoinositol by a kinase. See Table 5–11 for phosphate compound in avian blood.

TABLE 5–11. Phosphate compounds in micromoles of P per ml of red blood cells[a]

	Pi	ATP	DPG	IP_5	References
Pigeon					
Embryo	3.7	14.0	4.0	0	Bartlett (1978)
2-week	2.4	25.8	0	5.6	Bartlett (1978)
4-week	0.93	13.6	0	14.1	Bartlett (1978)
Adult	1.2	11.0	0	17.0	Bartlett (1978)
Western gull					
Late embryo	13.1	12.5	8.3	1.2	Bartlett (1978)
3-week	13.5	20.2	0	5.3	Bartlett (1978)
Juvenile	1.6	12.7	0	16.5	Bartlett (1978)
Adult	1.6	11.5	0	18.4	Bartlett (1978)
Ostrich	3.9	11.5	0	10.3[a]	Bartlett (1978)
				5.3	Bartlett (1978)
Man	0.68	2.8	8.9	0	Bartlett (1978)
Duck					
23-day embryo	11.0	4.2	7.0	1.9	Bartlett (1978)
6-day (hatched)	2.1	11.0	0.32	10.0	Bartlett (1967)
49-day (hatched)	1.1	7.3	0.0	13.0	Bartlett (1978)
Chicken					
15-day embryo	7.4	7.6	6.5	0.9	Bartlett and Borgese (1976)
2-day (hatched)	1.4	10.0	0.3	4.6	Bartlett and Borgese (1976)
14-day (hatched)	1.0	7.5	0.0	8.5	Bartlett and Borgese (1976)

[a] Includes IP_4 and IP_6.

Thrombocytes

The thrombocyte or platelet of mammalian blood has its origin from the giant cells, megakaryocytes, of the lungs and bone marrow. The megakaryocyte is lacking in avian bone marrow and the thrombocytes arise from antecedent, mononucleated cells that have a blast stage like other cells (Lucas and Jamroz, 1961). However, Archer (1971) threw some doubt on this view. He stated that although large cells can be found in the bone marrow, they are not comparable to mammalian megakaryocytes; they appear to be multinucleated and may be presumed to be thrombocyte precursors. However, proof is lacking.

The thrombocytes show considerable variation in size, and their shape may vary from oval to round (Figure 5–1). The typical thrombocyte is oval, with a round nucleus in the center of a clear cytoplasm. A constant feature is the one or more brightly red-stained granules present at the poles of the cell when stained with Wright's stain. The chromatin of the nucleus is dense and is clumped into relatively large masses that are distinctly separated by the parachromatin.

Numbers of thrombocytes in avian blood are presented in Table 5–13. The number ranges in most species between 20,000 and 30,000 per cubic millimeter but may be as low as 10,000 in the ostrich and as high as 132,000 in the quail (Nirmalan and Robinson, 1971). The latter number seems inordinately high and should be checked.

When blood is centrifuged, the thrombocytes separate out with the leukocytes and are found in the buffy coat. The actual number of thrombocytes appears to be about the same as the total leukocytes (Table 5–13). There is no evidence of a sex difference in thrombocytes. The avian thrombocyte contains extremely high quantities of serotonin (Meyer, 1973).

Evidence from other sources suggests that the thrombocytes are cells belonging to the erythrocyte series. All stages of erythroblasts–thrombocyte series can be detected in smears from bone marrow.

Hemoglobin

Structure

Avian hemoglobins (Hb) have four heme subunits containing iron, as do mammals, but their protein moietes are different. They have been studied in many species, (see Sturkie, 1976, for older literature). The heme combines with the globins to form one molecule of Hb containing two alpha (α) and two beta (β) chains of polypeptides. The α chains are designated as HbA or α-A and HbD or α-D (Oberthür et al., 1983). Type A makes up about 70% and type D, 30%, of the alpha types. Both α chains have the same number (141) of amino acids, but they are arranged in different sequences within and between species. The pair of β chains contains the same number (146) and sequence of amino acids for the same species, but vary in sequence between species.

Aberrant or mutant types of Hb have been reported by many investigators, and some of the types are inherited (Washburn, 1968, and later). Oberthür et al. (1983) studied the structure of Hb in chickens, starlings, rheas, ostriches, and pheasants, and have reported substitutions or changes (mutations) in the α and β chains. The rate of mutation was low in the β chains, higher in the α-A chain, and highest in the α-D chain.

In embryos, at least four specific types have been identified: two major and two minor (Brown and Ingram, 1974, Chapman et al., 1981, 1982). The major types are HbP and HbP^1; the minor types are HbE and HbM, according to Baumann et al. (1982). These types undergo changes and develop into the adult types before hatching time, and this changes the affinity of the Hbs for oxygen. See Chapter 8 for more details on factors affecting binding of oxygen.

Levels of Hemoglobin

The amount of hemoglobin in avian blood, as given in the literature, is highly variable. Recent work has demonstrated that much of this variation may be attributed to the methods of determination. (See Sturkie, 1976, for details.) Wels and Horn (1965) and Pilaski (1972) reviewed and tested a number of methods. They reported that cyanmethemoglobin methods are very reliable and practicable for avian blood, and more recent data are based on this method (Table 5–12). Hemoglobin determinations have been made on a number of avian species other than chickens, but in many cases, because of unreliable methods, the values are too high. There is a higher level of hemoglobin in males than females. This is correlated with the usual higher number of erythrocytes in the male in most species and the fact that androgen tends to increase red cell numbers (see section on erythrocytes), whereas estrogen tends to decrease numbers (in quail) mainly by depressing erythropoiesis initially (Nirmalan and Robinson, 1972). However, with continued estrogen administration erythropoiesis is not depressed in chickens (Gilbert, 1963). Although estrogen depressed erythrocyte volume in geese, androgen had no significant effect (Hunsaker, 1968). Some have reported levels of hemoglobin lower in laying than in nonlaying chickens but others have reported no difference (see Bell and Sturkie, 1965); however, no attempts have been made to relate these alleged differences to estrogen levels.

Factors that affect erythropoiesis and red cell number also affect hemoglobin level. Exposure to high altitude

TABLE 5–12. Hemoglobin values (g/100 ml) of whole blood of Aves

Species	Age	Male	Female	No sex	Method	
Chicken	21 days	—	—	9.7	Cyanmethemoglobin (CMH)	Pilaski (1972)
	46 days	—	—	9.8	Cyanmethemoglobin	Pilaski (1972)
	71 days	11.1	11.0	—	Cyanmethemoglobin	Pilaski (1972)
	126 days	12.5	11.7	—	Cyanmethemoglobin	Pilaski (1972)
	180 days	11.3	8.9	—	Cyanmethemoglobin	Pilaski (1972)
	210 days	11.4	8.6	—	Cyanmethemoglobin	Pilaski (1972)
	Adult	—	—	8.9–9.2	Cyanmethemoglobin	Wels and Horn (1965)
	Adult	—	9.71	—	Modified	Bankowski (1942)
	Adult	—	8.90	—	Newcomer	Sturkie (1976)
	Adult	—	8.90		Newcomer	Schultze and Elvehjem (Sturkie, 1976)
White Holland turkey	28 days	10.8	10.3	—	Cyanmethemoglobin	Pilaski (1972)
	77 days	11.1	11.5	—	Cyanmethemoglobin	Pilaski (1972)
	149 days	12.7	12.7	—	Cyanmethemoglobin	Pilaski (1972)
	217, adult	15.2	13.4	—	Cyanmethemoglobin	Pilaski (1972)
Turkey (no breed)	Adult	—	12.7	—	Not given	Lewis et al. (1979)
Turkey (embryos)	24 days	—	—	9.4	CMH	Christensen et al. (1982)
	27 days			9.4	CMH	Christensen et al. (1982)
	28 (hatched)			11.1	CMH	Christensen et al. (1982)
Goose	56 days	—	—	11.3	Cyanmethemoglobin	Hunsaker et al. (1964)
	140 days	—	—	14.4	Cyanmethemoglobin	Hunsaker et al. (1964)
	Adult	15.7	12.7	—	Cyanmethemoglobin	Hunsaker et al. (1964)
Mallard duck	Adult	Mixed sex	17.1		CMH	Keijer and Butler (1982)
Tufted duck	Mixed	Mixed sex	18.4		CMH	Keijer and Butler (1982)
Peking duck	Adult	—		14.20	Cyanmethemoglobin	Gachtgens et al. (1981a)
Peking duck	Adult	14.2	12.7	—	Sahli	Halaj (Sturkie, 1976)
Indian native (duck)	Adult	13.3	12.7	—	Wong (iron)	Balasch et al. (Sturkie, 1976)
Diving duck	Adult	15.2	13.3	—	Sahli	Halaj (Sturkie, 1976)
Diving duck	Adult	—	—	10.3	Alkali hematin	Soliman et al. (1966)
Numida meleagris (guinea)	Adult	14.9	—	—	Drabkin	Balasch et al. (Sturkie, 1976)
Numida meleagris	8 weeks	11.5	11.4		Drabkin	Fallaw et al. (1976)
	Adult	15.8	13.3		Drabkin	Fallaw et al. (1976)
Phasianus colchicus (pheasant)	Adult	18.9	—	—	Drabkin	Balasch et al. } Sturkie (1976)
Oavocristatus (peacock)	Adult	Mixed sex	12.0	—	Drabkin	Balasch et al. } Sturkie (1976)
Coturnix japonica	14 days	Mixed sex	11.6	—	Cyanmethemoglobin	Nirmalin and Robinson (1971)
	Adult	15.8	14.6	—	Cyanmethemoglobin	Nirmalin and Robinson 91971)
	22 days	9.2	10.7	—	Cyanmethemoglobin	Atwal et al. (1964)
	29 days	12.0	11.0	—	Cyanmethemoglobin	Atwal et al. (1964)
	50 days	15.3	12.3	—	Cyanmethemoglobin	Atwal et al. (1964)
Coturnix japonica	Adult	14.5 (Sea Level)	14.5		CMH	Jaeger & McGrath (1974)
		20.0 (at 6100 m)	20.0		CMH	Jaeger & McGrath (1974)
Agelaius phoeniceus (blackbird)	Adult	Mixed	ca. 12.5	—	Drabkin	Ronald et al., 1968 (Sturkie, 1976)

(hypoxia) increases Hb in chickens (Burton et al., 1971) and quail (Jaeger and McGrath, 1974), but after chickens are returned to sea level conditions, hematocrit and hemoglobin returns to near normal levels.

The mean corpuscular hemoglobin concentration (MCHC) expresses the mean content of hemoglobin in grams per 100 ml of erythrocytes; it is calculated from the packed cell volume (PVC) and the hemoglobin measured in the whole blood. Its primary importance is in the diagnosis of anemic conditions and it reflects the capacity of the bone marrow to produce erythrocytes of normal size, metabolic capacity, and hemoglobin content.

Leukocytes

Description of Cell Types

The following description of cell types is that of Olson (1937) (see Figure 5–1), based on studies by light microscope.

Heterophils. This type of leukocyte is sometimes designated a polymorphonuclear–pseudoesinophilic granulocyte, but for the sake of brevity it is usually designated "heterophil." In man and in such other mammals as the dog, these leukocytes possess neutral-staining granules (neutrophils). In rabbits and birds, the granules of these leukocytes are acid in reaction.

The heterophils of the chicken are usually round and have a diameter of approximately 10–15 μm. The characteristic feature of these cells is the presence of many rod- or spindle-shaped acidophilic crystalline bodies in the cytoplasm. In routinely stained smears, these cytoplasmic bodies are frequently distorted, and they may then be variable in shape. In cases of such distortion the color reaction must be used as a criterion for distinguishing them. The bodies are of a distinct and sometimes brilliant red against a background of colorless cytoplasm. The nucleus is polymorphic with varying degrees of lobulation.

Eosinophils. Polymorphonuclear eosinophilic granulocytes are about the same size as the heterophils. The granules are spherical and relatively large. Their color is dull red, as compared to the brilliant red of the heterophil, when stained with Wright's stain. The cytoplasm has a faint yet distinct bluish-gray tint. The nucleus is often bilobed and is of a richer blue than that of the heterophil, giving the impression of a sharper differentiation between chromatin and parachromatin than in the nucleus of the latter.

The eosinophils of the Japanese quail exhibit distinctive vacuoles (Witkowski and Thaxton, 1981). The question has arisen as to the correctness of distinguishing between the heterophil and the eosinophil in avian blood. Some workers believe that the two cell types represent modified forms of the same group, but others think that the heterophil and eosinophil have different lineages.

Basophils. Polymorphonuclear basophilic granulocytes are of about the same size and shape as the heterophils. The nucleus is weakly basophilic in reaction and round or oval in shape; at times it may be lobulated. The cytoplasm is abundant and devoid of color. Deeply basophilic granules abound in the cytoplasm. Electron microscopy reveals that granuler of basophils are variable in size and are fibrillar in nature (Dhingra et al., 1969).

Lymphocytes. The lymphocytes constitute the majority of the leukocytes in the blood of the fowl. There is a wide range in the size and shape of these cells. The cytoplasm is usually weakly basophilic. It may consist of a narrow rim bordering on one side of the nucleus, as in the small lymphocytes, or it may constitute the major portion of the cell, as in the larger lymphocytes. The nucleus is usually round and may have a small indentation. There is usually a fairly coarse pattern of chromatin. In some instances, however, the chromatin is fine and is not distinctly separated by the parachromatin. Sometimes a few nonspecific azure granules are noted in the cytoplasm.

Monocytes. The monocytes of avian blood are sometimes difficult to identify or to distinguish from large lymphocytes because there are transitional forms between the two. In general, the monocytes are large cells with relatively more cytoplasm than the large lymphocytes. The cytoplasm of these cells has a blue-gray tint. The nucleus is usually irregular in outline. The nuclear pattern in the monocyte is of a more delicate composition than is that in the lymphocyte.

Counting Methods

Leukocyte counts in the blood of chickens and other avian species have been made by many investigators. There is considerable variation in the numbers of the various cell types. In part, these discrepancies may be attributed to the method of making the count and, in many cases, to the small numbers of birds used (see Lucas and Jamroz, 1961, for details of methods).

Total leukocytes include all of the white cells, and these are counted in a special chamber. The blood is usually diluted 1:100 instead of 1:200 before it is placed in the counting chamber. Determining the total white cell count is attended with difficulty because the red cells are nucleated. Diluting fluids containing acetic acid that are ordinarily used to dissolve the red cells of mammalian blood are unsatisfactory for bird blood because the stroma of the red cells contract about the nuclei, making it impossible to distinguish these from some of the leukocytes. If a fluid such as Toisson's solution, which preserves the red and white cells, is used then it is again difficult to distinguish with certainty between the thrombocytes and the small lymphocytes under the powers of magnification that can be used in conjunction with the counting chamber. Natt and Herrick (1952) described a new leukocyte diluent that contained methyl violet and that, according to Chubb and Rowell (1959), was the best of the several diluents tried. For the differential count, a number of stains are available (see Lucas and Jamroz, 1961; Sturkie, 1976).

Olson (1937) showed that in differential counts, the error in terms of coefficients of variability are lymphocytes, 8.6%; heterophils, 27.9%; eosinophils, 58.8%; basophils, 62.6%; and monocytes, 22.2%. The coefficient of variability for total leukocytes was 34.2%, using phyloxine as the stain.

Number of Leukocytes

The number of leukocytes changes under various conditions, such as stress, estrogen administration, disease, and certain drugs. Little is known, however, concerning the exact role played by these cell types in combating stresses and diseases, although numbers of lymphocytes and heterophils appear to change most under these conditions. Both of these cell types are believed to be active in phagocytosing or combating infective or foreign material. Topp and Carlson (1972) have demonstrated in vitro the phagocytosis of *Staphylococcus* organisms by heterophils.

Numbers of leukocytes for different species of birds are shown in Table 5–13. In most cases, the smears for the differential counts were stained with Wright's stain. The supravital technique may be more reliable, and the number of lymphocytes is higher and that of heterophils lower with Wright's stain. In most species, the percentage of lymphocytes is higher than for any other cell type, comprising 40–70% of the total count, and the heterophils are the second most numerous group. In the ostrich and pheasant the opposite is true, with the heterophils comprising over half of the total count. The significance of this difference is not known.

Sex and Age Differences. Most workers have not found a consistent sex difference in leukocytes. Olson (1937) did report a sex difference in adult chickens, but not in young chickens. Cook (1937) reported little variation in the count attributable to sex in chickens from 26 to 183 days of age. Lucas and Jamroz (1961), however, reported a significantly higher level of leukocytes in some adult female chickens than males. Slightly higher counts also were observed in *Coturnix* (quail) by Atwal et al., (1964) and by Nirmalan and Robinson (1971) and in ducks by Halaj (see Table 5–13). Some investigators have paid little attention to possible sex differences, nor to whether the females were laying or not.

Estrogen administration definitely increased leukocyte counts in male chickens (Meyer, 1973) and in *Coturnix* (Nirmalan and Robinson, 1972). They reported a significant increase in heterophils and a decrease in lymphocytes.

Glick (1960) has reported a diurnal variation in the leukocyte count of 3-week-old New Hampshire chicks. The leukocyte numbers were higher from 2 PM to 4 PM; the relative number of heterophils was lowest and that of lymphocytes highest at this time.

Young chicks and quails show slightly lower counts than adults. According to Burton and Harrison (1969), the blood of the neonate chick is low in leukocytes but relatively high in heterophils and basophils. The picture changes rapidly and by 3 weeks of age, the cell numbers increase and reach essentially the adult level.

Effects of Diet. Although the work of Cook (1937) suggested that diet might influence the leukocyte count, the data were not conclusive. Work by Goff et al. (1953) demonstrated conclusively that a deficiency of riboflavin significantly increases the heterophils and decreases the lymphocytes. Similar results were obtained with vitamin B_1 deficiency.

Effects of Environment. Very little experimental work has been conducted on the effects of changes in environment, exclusive of ration, on leukocyte count. Olson's (1937) studies indicate that more work should be conducted along these lines. It is known that changes in environment may induce stress, with the consequent release of adrenal corticoids and with changes in leukocyte numbers in mammals and birds. Newcomer (1958) and Besch et al. (1967) have shown that physical restraint and the use of ACTH, cortical hormones, and other stressing agents produce a relative increase in number of heterophils in chickens. *Eosinophila* associated with dermatitis and edema has been reported in chickens (Maxwell et al., 1979).

Effects of Hormones, Drugs, and Other Factors. There is a prominent increase in heterophils and a decrease in lymphocytes following injection of cortical hormones (Siegal, 1968). (See also Chapter 22.) The sensitivity of the heterophil is such that Wolford and Ringer (1962) have suggested that heterophil counts are a good means of assessing stress in birds.

Cortisone acetate increased the total count from 13,000 to 18,000 cells in chickens 3.5 hr after injection. Desoxycorticosterone acetate had a similar effect (Glick, 1961). Growth hormone alone did not influence the differential or the total leukocyte count.

Glick and Sato (1964) reported that ACTH increased heterophils in sham operated and bursectomized chicks and decreased lymphocytes only in bursec-

TABLE 5–13. Number of leukocytes and thrombocytes in bird blood and differential counts

Species, age, and sex	Number (× 10³/mm³)		Differential count (%)					Reference
	Leukocytes	Thrombocytes	Lymphocytes	Heterophils	Eosinophils	Basophils	Monocytes	
Chicken, adult male	19.8	25.4	59.1	27.2	1.9	1.7	10.2	Olson (1937)
Nonlaying female	19.8	26.5	64.6	22.8	1.9	1.7	—	Olson (1937)
Young, 2–21 weeks, males and females	29.4	32.7	66.0	20.9	1.9	3.1	8.1	Olson (1937)
Chicken, White Leghorn, 6 weeks to Maturity, males and females								
Supravital stain	32.6	—	40.9	35.6	2.7	4.3	16.5	Twisselmann (Sturkie, 1976)
Wright's stain	—	—	54.0	27.8	1.5	2.7	13.7	Twisselmann (Sturkie, 1976)
Chicken, average, all ages	30.4	—	73.3	15.1	—	2.7	6.3	Cook (1937)
Chicken, 5–10 weeks, males	—	—	69.5	20.4	1.3	3.3	3.7	Goff et al. (Sturkie, 1965)
6 weeks, female White Leghorn	28.6	30.4	81.5	10.1	1.5	2.3	4.5	Lucas and Jamroz (1961)
12 weeks, female White Leghorn	30.6	26.2	77.8	11.7	3.0	1.7	4.9	Lucas and Jamroz (1961)
Adult female White Leghorn	29.4	30.8	76.1	13.3	2.5	2.4	5.7	Lucas and Jamroz (1961)
Adult male White Leghorn	16.6	27.6	64.0	25.8	1.4	2.4	6.4	Lucas and Jamroz (1961)
Adult female farmstock White Leghorn	28.8	37.2	71.7	23.7	1.4	2.1	1.1	Lucas and Jamroz (1961)
Adult female Rhode Island Red	35.8	60.3	58.1	35.1	1.2	3.1	2.5	Lucas and Jamroz (1961)
Chicken, White Leghorn, Adult	28.9	—	70.1	22.6	2.4	2.2	5.5	Brake et al. (1982)
(Force Molted)	30.9	—	—	25.9	4.0	—	—	Brake et al. (1982)
Canada goose, male	—	—	46.0	39.0	7.0	2.0	6.0	Lucas and Jamroz (1961)
Crossbred European goose, no sex								
8 months	16.8	—	38.0	44.2	5.1	3.1	10.0	Kaleta and Bernhard (Sturkie, 1976)
3 years	18.2	—	36.2	50.0	4.0	2.2	8.0	Kaleta and Bernhardt (Sturkie, 1976)
Czech goose, no sex	27.0	—	48.5	34–44	—	—	—	Sova et al. (Sturkie, 1976)
Turkey, 20 weeks, mixed sex	26.8	—	—	—	—	—	—	Sullivan, 1965 (Sturkie, 1976)
No sex	—	—	50.6	43.4	0.9	3.2	1.9	Johnson and Lange (Sturkie, 1965)
Indian native duck, male	31.5		68.0	22	1.4	0.6	8.0	Surendranathan et al. (Sturkie, 1976)
Female	28.9	62.6	26.6	2.4	0.4	8.0	—	Surendranathan et al. (Sturkie, 1976)
Peking duck, male	24.0		31	52	9.9	3.1	3.7	Halaj (Sturkie, 1976)
Female	26.0		47	32	10.2	3.3	6.9	Halaj (Sturkie, 1976)
Duck	23.4	30.7	61.7	24.3	2.1	1.5	10.8	Magath and Higgins (Sturkie, 1965)
Ringneck, male pheasant	—	—	34.0	48.0	1.0	10.0	8.0	Lucas and Jamroz (1961)
Pigeon (no sex)	13.0	—	65.6	23.0	2.2	2.6	6.6	Shaw (Sturkie, 1965)
Ostrich, males and females	21.0	10.5	26.8	59.1	6.3	4.7	3.0	DeVilliers, 1938 (Sturkie, 1965)
Coturnix (Quail), adult male	19.7	117	73.6	20.8	2.5	0.4	2.7	Nirmalan and Robinson (1971)
Adult female	23.1	132	71.6	21.8	4.3	0.2	2.1	Nirmalan and Robinson (1971)
10 days	16.0	—	67.0	25.0	4.0	2.0	2.0	Atwal et al. (1964)
Male adult	24	—	46.	50.0	1.0	1.0	2.0	Atwal et al. (1964)
Female adult	25	—	40.0	52.0	4.0	3.0	1.0	Atwal et al. (1964)
Agelaius phoeniceus (blackbird), mixed	—	—	55.0	30.0	3.0	2.5	8.0	Ronald et al., 1968 (Sturkie, 1976)

tomized ones, suggesting that the bursa of Fabricius is involved in lymphocyte formation. Chemical bursectomy, however, did not significantly influence lymphocytes or heretophils (Glick, 1969).

Administration of ACTH to egrets and unilateral adrenalectomized crows and pigeons produced a decrease in heterophils and an increase in lymphocytes. In myna birds, however, both treatments resulted in an opposite effect (heterophila and lymphopenia; Bhattacharyya and Sarka (1968) also revealed that certain inhibitors of cortical hormone release caused lymphopenia and heterophilia.

A detailed study of the effects of x rays on blood cells of the chicken has been conducted by Lucas and Denington (1957). Total body irradiation with dosages of 50–300 roentgen units in chicks and hens decreased the total leukocyte count significantly, and the low level persisted for about 12 days after the treatment. There was likewise a decrease in lymphocytes but an increase in heterophils.

Effects of Changes in Gravity. Lymphocyte counts are a good indication or measurement of stress, since lymphocyte production is inhibited by adrenocorticoids (Burton and Smith, 1972). Birds adapted to 2 *g* after being on the centrifuge for 162 days were removed and then periodically exposed to a 2-*g* field for 24 hr with unadapted controls. After 24 hr at 2 *g* the previously adapted birds exhibited a 20% decrease in lymphocytes. The decrease was much greater in the nonadapted birds.

Leukocytosis and Disease. There is considerable variation in the blood picture of normal birds, and caution should be exercised before changes in the blood picture are attributed to disease. Olson (1965) has reviewed the hematologic changes associated with certain diseases.

Blood Cells Containing Biogenic Amines

The principal biogenic amines in avian blood are catecholamines, histamine, and serotonin (5HT). Practically all of the catecholamines are found in the plasma, but most of the histamine and serotonin is carried by the blood cells, mainly the leukocytes and thrombocytes, found in the buffy coat fraction. The serotonin content of whole blood of chickens is distributed mainly among the following cell types: thrombocytes, 50%; lymphocytes, 25–29%; polymorphs, mainly heterophils and monocytes, 19–25% (Sturkie et al., 1972, Meyer and Sturkie, 1974).

Females have higher blood levels of 5HT than males, mainly because the concentration of 5HT per cell, found in the buffy coat, is much higher in females. Most blood 5HT is derived from the synthesis and release of 5HT from the enterochromaffin cells of the intestines (Gross and Sturkie, 1975). It is released at a fairly constant rate in in vitro preparations (10 ng/min/g), and is increased significantly by the neurotransmitters acetylcholine and catecholamines (Schaible, 1977).

Avian plasma contains very little histamine, and most of it is in the cellular fractions (El Ackad and Sturkie, 1972), and particularly the buffy coat fraction, thrombocytes and leukocytes, which contain approximately 88% of whole blood histamine. About 10% is found in the red blood cells. The female chichen, as well as the female pigeon and duck, have higher levels of histamine in the blood than males (Meyer and Sturkie, 1974).

Histofluorescense studies indicate that the thrombocytes, lymphocytes, and heterophils carry high concentrations of histamine, and these are the most numerous cells in the buffy coat fraction. Histamine is also released by intestinal tissue in vitro by 5HT and catecholamines (Schaible, 1977).

Whether or not the biogenic amines have a regulating effect on the cardiovascular system is discussed in Chapter 7.

Hemorrhage and Replacement of Blood

Birds are able to tolerate severe blood loss better than mammals. The comparative mortality of several mammalian and avian species subjected to bleeding is presented in Figure 5–2 (Kovach et al., 1969). The mortality is much higher in mammals receiving the same treatment, and in land and nonflying birds like the chicken and pheasant, when compared to flying species like the pigeon and the duck. Each species was bled at the rate of 1% of body weight per hour for 4–5 hr (for most) and as long as 9 hr (pigeon). This means that nearly all of the blood of the pigeon was removed. Obviously blood was being replaced over the bleeding period, through absorption of tissue fluid. Mean arterial blood pressure dropped in all avian species to about 50–60 mm Hg at termination of bleeding periods. Stimulation of baroreceptors by compressing the carotid arteries increased blood pressure significantly in all species except the chicken. This and subsequent studies demonstrated that blood loss in the chicken (Wyse and Nickerson, 1971; Ploucha et al., 1981, 1982) resulting in a mean blood pressure of 50 mm Hg does not evoke reflex increases in precapillary resistance and blood pressure. If blood loss is increased so that the mean blood pressure is maintained at 25 mm Hg (in the chicken), then there is a significant increase in precapillary resistance in a perfused hind leg preparation (Ploucha et al., 1982). The increased resistance was associated with an enhanced release of circulating catecholamines whose effect could be blocked with alpha blockers. Since the chicken does not have a functional carotid sinus baroreceptor, the increased resistance was not attributable to such reflex,

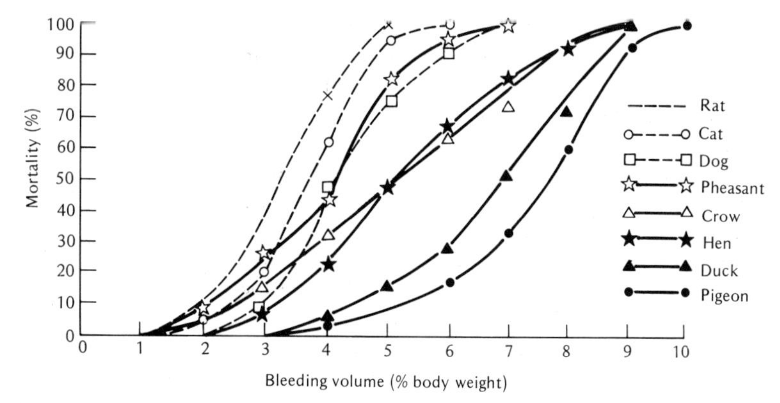

FIGURE 5–2. Mortality after blood losses in avian and mammalian species. Every hour, 1% of body weight of blood was withdrawn. (After Kovach et al., 1969.)

but to the release of adrenal medullary hormones, and is believed to be caused mainly by cerebral ischemia and low brain perfusion pressure.

Diving and Flying Birds. The pigeon and duck are most resistant to blood loss, and they respond apparently by evoking reflex vasoconstriction to raise blood pressure (Kovachs et al., 1969; Djojosugito et al., 1968). A rapid blood loss of 12% of the initial blood volume in ducks (Djojosugito et al., 1968) lowered mean blood pressure only 10–20 mm Hg; the pigeon was similarly resistant to a drop in blood pressure following bleeding (Kovach et al., 1969).

The chicken's blood pressure drops appreciably (20 mm Hg) when only small amounts of blood are lost (Wyse and Nickerson: Ploucha et al., 1981), but tends to return to normal after a few hours. The chicken responds to blood loss by increasing its blood volume, but the duck and pigeon try to maintain normal blood pressure by evoking baroreceptor reflexes, although the duck also has a very large capillary surface area that facilitates increased rate of absorption of tissue fluids (Djojosugito et al., 1968).

In chickens, the amount or volume of blood lost on first bleeding (FBV) required to drop blood pressure to 50 mm Hg is about one-half the amount that is required on the second bleeding (SBV) to maintain this pressure. The ratio of SBV to FVB is about 2, and indicates that in the chicken restoration of lost blood volume is the principal means of combating hemorrhage. This ratio is much less in mammals (about 0.10 in the dog), and less in ducks and pigeons.

The mobilization and restoration of fluid during the first 90 min after bleeding (in the chicken) amounted to 13–17% of the initial blood volume per hour, which is about twice that in the dog (Wyse and Nickerson, 1971).

Other Effects of Hemorrhage. Hemorrhage produces hemodilution, resulting in decreased hematocrit, hemoglobin, and plasma proteins, but may cause hyperglycemia and hyperpotassemia in chichens (Ploucha et al., 1981; Wyse and Nickerson 1971).

Induced bleeding, resulting in a loss of 10% of blood volume in ducks, increased the release of arginine vasotocin (AVT) in the blood significantly (Simon-Oppermann et al., 1984). AVT is the antidiuretic hormone of birds whose action conserves body water. Continuous bleeding of chickens to maintain a mean blood pressure of 50 mm Hg causes a decrease in cardiac output (CO) from 209 ml/min/kg (0.734) to 146 after 3 hr, and a drop in stroke volume. Heart rate, however, increased significantly, and total peripheral resistance decreased (Ploucha et al., 1981).

Shock may result from severe hemorrhage in mammals, but it does not occur in birds. Species that experience shock from hemorrhage exhibit acidosis, which results mainly from inadequate perfusion of tissues. The chicken does not become acidotic after sustained hemorrhage (Ploucha et al., 1981). Some have suggested that prostaglandins play a role in producing shock in mammals, but they have no effect in chickens, because inhibition of prostaglandin synthesis with indomethacin had no effect on the chicken suffering from severe blood loss (Zambraski and Schuler, 1980).

Blood Coagulation by P. Griminger

Hemostasis

Hemostasis, the arrest of bleeding, can be accomplished by vasoconstriction and blood coagulation, or artificially by surgical means. Blood coagulation requires an interplay between the tissue of the blood vessels, certain cellular elements of the blood, and a number of plasma proteins. On injury, the adjacent walls of the blood vessel contract and a hemostatic plug is formed through adhesion and aggregation of blood platelets. Avian blood lacks platelets, and their role in plug formation is assumed by thrombocytes (Grant and Zucker, 1973). A second and final hemostatic plug is then formed through blood coagulation.

The decisive step in blood coagulation is the transformation of the soluble plasma protein, fibrinogen, into the insoluble fibrin, followed by polymerization and stabilization of the fibrin monomers (Figure 5–3); it is mediated by the enzymatic activity of thrombin, which also activates factor XIII. This factor is a transglutaminase that cross-links fibrin monomers to form a stable clot. Thrombin is derived from an inactive precursor, prothrombin, a single-chain glycoportein with a molecular weight of approximately 72,000. The reaction is catalyzed by calcium ions, thromboplastin, and several accessory factors. The active element of thromboplastin (factor III) appears to be phospholipid in character. In the more carefully studied human clotting system, at least seven additional accessory factors have been described (factors V and VII through XII; see Table 5–14). They are mainly protein in character and, in electrophoresis, migrate with the plasma globulin fractions. While great strides have been made in recent years in the elucidation of their interaction, it is not perfectly understood; for more details, recent review papers should be consulted (Murano, 1980; Spurling, 1981).

Vitamin K

In 1929, Henrik Dam, at the University of Copenhagen, studied the biosynthesis of cholesterol in chickens, feeding them purified diets. He noticed subcutaneous and intramuscular hemorrhages that resembled those of scurvy, but were not prevented by lemon juice, a known remedy (McCollum, 1957). His observations eventually led to the discovery of vitamin K and its indispensable role in the synthesis of prothrombin and some of the accessory factors mentioned previously,

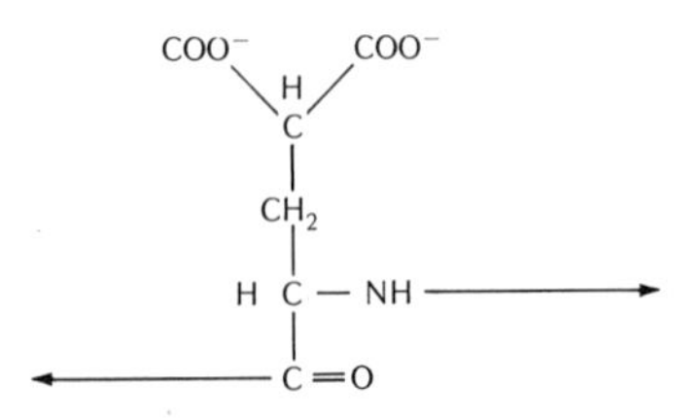

FIGURE 5–4. Gamma-carboxyglutamic acid in the activated prothrombin peptide chain.

namely factors VII, IX, and X. More recent investigations have shown vitamin K to be a required cofactor in an enzyme system that carboxylates glutamyl residues in precursor proteins to γ-carboxyglutamyl residues in completed proteins (Figure 5–4). The vitamin thus acts as a cofactor in a postribosomal protein modification, perhaps by activation of the γ-proton of the glutamic acid residues for reaction with CO_2. The function of these changes is to allow the modified proteins to participate in a specific protein–Ca^{2+}–phospholipid interaction that is necessary for their biologic role (Suttie, 1980).

Clotting Factors

While most clotting factors apparently are not absolutely species specific, there are pronounced differences between the clotting mechanisms of different classes of animals, and some differences, perhaps, even between species of the same class. The major factors involved in avian blood coagulation appear to be similar to those found in mammalian blood (see Table 5–14). However, according to Didisheim et al. (1959), avian plasma lacks thromboplastin component (factor IX)

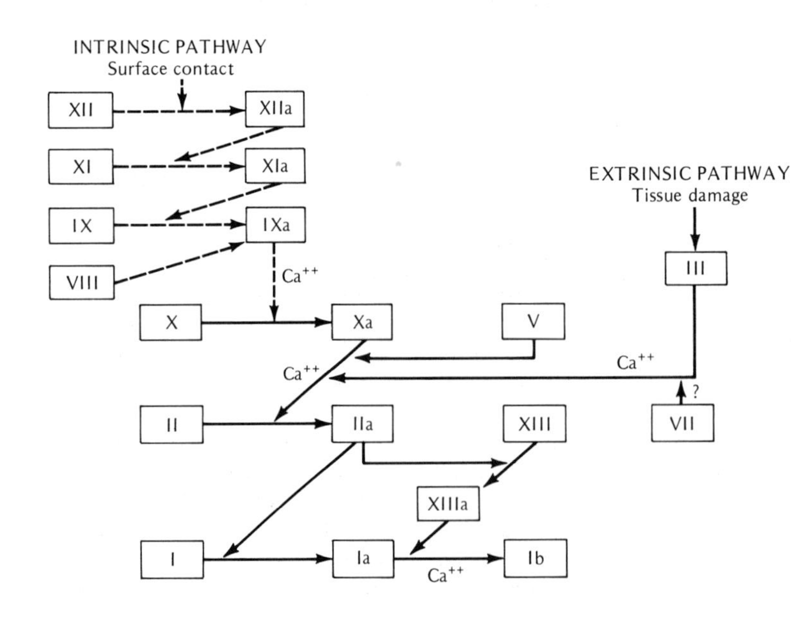

5–3. A scheme of blood coagulation. The broken lines represent the intrinsic pathway, which appears to play a minor role in *Aves*. The need for factor VII for avian blood coagulation has also been questioned. For names of factors, see Table 5–14.

TABLE 5–14. Synonyms for blood coagulation factors

Factor	Synonyms
I	Fibrinogen
Ia	Fibrin
Ib	Stabilized fibrin
II	Prothrombin
IIa	Thrombin
III	Thromboplastin
V	Labile factor, accelerator globulin
VII	Stable factor, proconvertin
VIII	Antihemophilic factor A, antihemophilic globulin
IX	Antihemophilic factor B, Christmas factor
X	Stuart–Prower factor
XI	Antihemophilic factor C, plasma thromboplastin antecedent
XII	Hageman factor
XIII	Fibrin stabilizing factor

and the Hageman factor (XII). According to Wartelle (1957), these factors are present, but in very small amounts. Stopforth (1970) thought that compared with mammalian systems, factors V and VII might be low or even absent in chickens, while factor X activity was definitely present. According to Walz et al. (1975), 2-week old chickens fed adequate levels of Vitamin K had lower levels of prothrombin (50%), factor IX (8%), and factor X (6%) than rats. Factor VII was not detectable in the plasma. The case made for the absence of factor VII is considered unconvincing by others, since the prothrombin time of avian plasma, when homologous tissue factor is used, is no longer than that of mammalian species known to contain high levels of factor VII activity (Kase et al., 1980). More recently, Belleville et al. (1982), using homologous reagents (thromboplastins and partial thromboplastins), demonstrated the presence of factors V, VII, VIII, IX, and X in Japanese quail plasma.

While there is evidence of a functioning intrinsic clotting mechanism in quail and in chickens (Doerr and Hamilton, 1981a), the coagulation of avian blood seems to be especially dependent on an extrinsic clotting system, involving the release of tissue thromboplastin.

Thromboplastin

A certain degree of species specificity exists for brain tissue thromboplastin, and probably for thromboplastin extracted from other tissues. Therefore, thromboplastic activity for avian prothrombin time determinations should be provided by bird tissue. The use of rabbit brain suspension (Simplastin) to accelerate the clotting of turkey blood resulted in prothrombin times in excess of 60 sec, while average prothrombin times of 13.6 were obtained with the same birds with a turkey brain-derived source of thromboplastin. Conversely, Simplastin resulted in human prothrombin times of 11.5–13 sec while they were prolonged to over 60 sec by the use of turkey brain (Lewis et al., 1979). Experimentation with ducks and turkeys (Didisheim et al., 1959; Griminger, 1957), and with chickens, pigeons, and quail (Griminger, 1971) has indicated that in assay work it might even be advantageous to use the brain of the same rather than of another avian species as a source of tissue thromboplastin.

A dry and relatively stable source of *thromboplastic activity* can be prepared by repeated extraction of brain tissue with acetone. The Vitamin K nutrition of the donor animal will influence the thromboplastic activity of the product. The shorter prothrombin times associated with acetone-extracted brain powder from vitamin K-fed, as compared to vitamin K-deficient, donors appear to be due to a vitamin K-dependent plasma clotting factor in the residual blood of the brain (Griminger et al., 1970). The brain powder is "activated" in saline before use in the prothrombin time determination. It appears to contain an inhibitor, since discarding the first and using the second extract enhances its thromboplastic activity (Shum and Griminger, 1972a).

Prothrombin Time

Vitamin K, essential for normal blood coagulation, can be supplied in the food as well as by bacterial synthesis in the lower part of the intestinal tract. Nutrients synthesized in this location do not supply the needs of the animal; therefore, birds are particularly dependent on an exogenous source for this vitamin, and growing chickens tend to become hypoprothrombinemic after a relatively short time on a vitamin K-deficient diet, if coprophagy is prevented. This fact has led to the widespread use of this species in vitamin K bioassay work.

In this assay, varying amounts of the material to be assayed are fed to groups of vitamin K-deficient chicks and known amounts of the vitamin to others. A comparison of plasma prothrombin times, usually by the one-stage method of Quick (1966), indicates the amount of vitamin K-active material in the assayed samples. In the determination of prothrombin times, blood is drawn and blood calcium bound chemically to prevent coagulation. This is accomplished by drawing blood with syringes prefilled with a measured amount of sodium oxalate or sodium citrate solution. While both solutions appear equally useful for mammalian blood, sodium citrate is the anticoagulant of choice for avian blood samples (Shum and Griminger, 1972b). The blood is then centrifuged, and calcium and a thromboplastic extract are added to the plasma. The time required for a clot to form is called the prothrombin time.

Prothrombin times of 12 sec or less can be obtained in growing chicks with the one-stage method of Quick (1966) if a carefully prepared chick brain thromboplastin extract is used. Comparable results can also be obtained with other extracts that promote slower clotting (Griminger, 1962). Prothrombin times even *shorter* than 12 sec have been observed in this laboratory for adult male birds, whereas those for adult females tend to be slightly longer. The adult male chicken appears to be relatively resistant to hypoprothrombinemia. A degree of vitamin K deprivation that will produce severe intramuscular and subcutaneous hemorrhage in growing chicks and will initiate a more moderate hypoprothrombinemia in laying hens may have little effect on adult males. Since prothrombin times do not follow a normal distribution, they have to be converted for statistical evaluation. In comparative studies, best results are obtained by using the reciprocals of the prothrombin times (Bliss and Griminger, 1969).

Whole blood clotting times are much more variable than prothrombin times. From 2 to 10 min appears to be the normal range for nonhypoprothrombinemic birds, although longer clotting times are not infrequently recorded (Bigland, 1964). Other coagulation measurements recorded in birds are plasma recalcification time, "Stypven" time, and the thromboplastin generation test (Stopforth, 1970).

A functional clotting system in the *chicken embryo* can be detected during the second half of the incubation period. Kane and Sizer (1953) found clotting material after 12–13 days of embryonic development.

Anticoagulants

While chickens are more resistant to oral anticoagulants of the coumarin type than are mammals, they will succumb to large doses. The relative resistance may be related to the low levels of factor VII (Archer, 1971). Coumarin-type anticoagulants reduce plasma coagulation factors much in the way as in Vitamin K deficiency (Losito, 1965). In our laboratory, 0.16% dicumarol in the feed given to 1-day-old chicks caused 70% mortality within 2 weeks, and 50% in chicks that were 1 or 2 weeks older at onset of medication. The reaction to lower doses is shown in Table 5–15. The anticoagulant effect of dicumarol, as well as of warfarin, another compound of this type, can be overcome by feeding an increased amount of vitamin K: 800 mg of K_1 per kilogram diet were required to obtain normal prothrombin times in chickens fed 100 mg warfarin per kilogram diet (Griminger, 1965), whereas less than 1 mg K_1 per kilogram is required in the absence of antivitamins or other stress factors. Menadione (K_3) is not effective in counteracting high doses of dicumarol-type anticoagulants. There seem to be quantitative limitations to the conversion of menadione to menaquinone (MK-4), which appears to be the active form of Vitamin K when menadione is the dietary source (Griminger and Brubacher, 1966).

The use of anticoagulants as rodenticides may lead to secondary poisonings of predatory birds, such as owls, when they consume poisoned rodents (Mendenhall and Pank, 1980).

Capillary Fragility

The question has been raised whether the hemorrhages that seem to be the cause of death in hypoprothrombinemia are precipitated by a simultaneous decrease in capillary strength. Increased capillary fragility in hypoprothrombinemia has indeed been demonstrated in the rat by the use of the negative-pressure method (Pastorova, 1957). In this method, pinpoint hemorrhages are produced by the application of negative pressure to a specific area of skin with the aid of a suction cup, varying either the pressure or the application time. Used on the chicken, this method produces bruises in the underlying flesh ("vascular faults"), which obscure the usual observation of petechiae in the skin. Observations made in this laboratory indicate that in the chicken dicumarol will also reduce capillary strength.

TABLE 5–15. Response of chicks to feeding of dicumoral for 1 week

Dicumarol added (mg/kg diet)	Average prothrombin times (sec)[a]
0	13.7
50	14.3
100	14.7
200	17.8
400	27.0
800	59.8

[a] Averages on the basis of reciprocals of individual plasma prothrombin times.

Heparin

Blood also contains natural anticoagulants in the form of antithrombin, heparin, and perhaps an antithromboplastin. Heparin is a mixture of sulfated polysaccharides; it is strongly acidic and highly charged and appears to be synthesized by basophilic mast cells throughout the body. It prevents the activation of factor IX and activates a plasma factor that inhibits the action of thrombin (Ganong, 1982), thus balancing the coagulation system in the blood, which is poised in favor of clotting (Engelberg, 1977). It is also a useful in vitro anticoagulant for studies not concerned with coagulation phenomena. Addition of 0.2 mg of heparin (2 drops of a 2 mg/ml solution) will prevent the coagulation of 1 ml of blood. On a dry basis, 1 mg heparin contains not less than 120 U.S.P. units, 1 unit being equal to approximately 0.01 mg of heparin sodium. Injected intravenously, 500 units/kg body weight will prevent coagulation. It is possible that heparin acts in conjunction with antithrombin. When a clot is formed, thrombin is also inactivated by the absorptive action of fibrin, thus localizing clot formation.

The shrinking of a blood coagulum, with the simultaneous expression of a serum, is called clot retraction. Clot retraction occurs in chicken blood appreciably slower than in human blood according to Bigland (1960), and is negligible according to Didisheim et al. (1959).

Hemorrhagic Syndrome

The so-called "hemorrhagic syndrome" observed in growing chickens is characterized by subcutaneous hemorrhages and occasionally by anemia and a pale bone marrow. The etiology of the syndrome has not been satisfactorily explained. Some early field cases probably resulted from a deficiency of Vitamin K or the injudicious use of coccidiostatic sulfonamides such as sulfaquinoxaline. Sulfonamides can act as Vitamin K antagonists in addition to their bacteriostatic effect (Griminger, 1957). Sauter et al. (1975) have used both maternal vitamin K depletion and sulfonamides to develop constant hemorrhagic symptoms in chicks. For further details on the history of this disorder, see Griminger (1976).

In the so-called fatty liver hemorrhagic syndrome in laying hens, the characteristic hemorrhages appear only in hens having enlarged livers of a high fat content (Wolford and Murphy, 1972).

References

Abati, A.L., and J.J. McGrath. (1973). Physiological responses to acute hypoxia in altitude-acclimatized chickens. J. Appl. Physiol., 34, 804.

Albritton, E.C. (1952). "Standard Values of Blood." Philadelphia: Saunders.

Archer, R.K. (1971). Blood coagulation. In "Physiology and Biochemistry of Domestic Fowl," Vol. II (D.J. Bell and B.M. Freeman, Eds.) London: Academic Press, Chapter 38.

Atwal, O.S., L.Z. McFarland, and W.O. Wilson. (1964). Hematology of *Coturnix* from birth to maturity. Poult. Sci., 43, 1392.

Augustine, P.C. (1982). Effects of feed and water deprivation on organ and blood characteristics of young turkeys. Poult. Sci., 61:796.

Bang, N.U., F.K. Beller, E. Deutsch, and E.F. Mammen. (1971). "Thrombosis and Bleeding Disorders." New York: Academic Press.

Bankowski, R.A. (1942). Studies of the hemoglobin content of chickens blood and evaluation of methods for its determination. Am. J. Vet. Res., 3, 363.

Barrett, L.A., and S.L. Scheinberg. (1972). The development of avian red cell shape. J. Exp. Zool., 182, 1.

Bartlett, G.R. Phosphate compounds in reptilian and avian red blood cells. (1978). Developmental changes. Comp. Biochem. Physiol. A, 61, 191.

Bartlett, G.R., and T.A. Borgese. (1976). Phosphate compounds in red cells of the chicken and duck embryo and hatching. Comp. Biochem. Physiol. A, 55, 207.

Bauer, C., W. Jelkmann, and H. Rollema. (1978). Mechanisms controlling the oxygen affinity of bird and reptile blood. A comparison between the functional properties of chicken and crocodile hemoglobin. Respiratory function in birds, adult and embryonic. New York: Springer-Verlag.

Baumann, R., S. Padeken, and E.A. Haller. (1982). Functional properties of embryonic chicken hemoglobins. J. Appl. Physiol. Respir. Environ. Exercise Physiol., 53, 1439.

Bell, D.J., and P.D. Sturkie. (1965). Chemical constituents of blood. In "Avian Physiology" (2d ed.), (P.D. Sturkie, Ed.). Ithaca: Cornell University Press.

Belleville, J., B. Cornillon, J. Paul, J. Baguet, G. Clendinnen, and R. Eloy. (1982). Haemostasis, blood coagulation and fibrinolysis in the Japanese quail. Comp. Biochem. Physiol. A, 71, 219.

Besch, E.L., A.H. Smith, R.R. Burton, and S.J. Sluka. (1967). Physiological limitations of animal restraint. Aerospace Med., 38, 1130.

Bhattacharyya, T.K., and A.K. Sarkar. (1968). Avian leucocytic responses induced by stress and corticoid inhibitors. Indian J. Exp. Biol., 6, 26.

Bierer, B.W., T.H. Eleazer, and D.E. Roebuck. (1963). Hematocrit and sedimentation rate values as an acid in poultry disease diagnosis. J. Am. Vet. Med. Assoc., 143, 1096.

Bierer, B.W., T.H. Eleazer, and D.E. Roebuck. (1964). Sedimentation rate, packed cell volume, buffy coat value, and rectal temperature of chickens and turkeys at various ages. J. Am. Vet. Med. Assoc., 144, 727.

Biggs, R., and R.G. Macfarlane. (1962). "Human Blood Coagulation." Philadelphia: F.A. Davis.

Bigland, C.H. (1960). Studies on avian blood coagulation. Ph.D. Thesis, University of Alberta, Edmonton, Canada.

Bigland, C.H. (1964). Blood clotting time of five avian species. Poult. Sci., 43, 1035.

Bliss, C.I., and P. Griminger. (1969). Response criteria for the bioassay of vitamin K. Biometrics, 24, 735.

Bond, C.F., and P.W. Gilbert. (1958). Comparative study of blood volume in representative aquatic and nonaquatic birds. Am. J. Physiol., 194, 519.

Brace, K., and P.D. Atland. (1956). Lifespan of the duck and chicken erythrocyte as determined with C-14. Proc. Soc. Exp. Biol. Med., 92, 615.

Brake, J. and M. Baker. (1982). Physiological changes in caged layers during a forced molt. 4. Leukocytes and packed cell volume. Poultry Sci., 61:790.

Brinkhous, K.M. (1959). Blood clotting; the plasma procoagulants. Am. Rev. Physiol., 21, 271.

Brown, J.L., and V.M. Ingram. (1974). Structural studies on chick embryo hemoglobins. J. Biol. Chem., 249, 3960.

Burton, R.R., and J.S. Harrison. (1969). The relative differential leukocyte count of the newly hatched chick. Poult. Sci., 48, 451.

Burton, R.R., and A.H. Smith. (1969). Hematological findings associated with chronic acceleration. Space Life Sci., 1, 503.

Burton, R.R., and A.H. Smith. (1972). The effect of chronic erythrocyte polycythemia and high altitude upon plasma and blood volumes. Proc. Soc. Exp. Biol. Med., 140, 920.

Burton, R.R., and A.H. Smith. (1972). Stress and adaptation responses to repeated acute acceleration. Am. J. Physiol., 222, 1501.

Burton, R.R., R. Sahara, and A.H. Smith. (1971). The hematology of domestic fowl native to high altitude. Environ. Physiol., 1, 155.

Campbell, F. (1967). Fine structure of bone marrow of chicken and pigeon. J. Morphol., 123, 405.

Campbell, J.W., (1970). "Comparative Biochemistry of Nitrogen Metabolism," Vol. 2, New York: Academic Press.

Chapman, T.E., and A.L. Black. (1967). Water turnover in chickens. Poult. Sci., 46, 761.

Chapman, T.E., and L.Z. McFarland. (1971). Water turnover in *Coturnix* quail, with individual observations on a burrowing owl, petz conure, and vulturine fish eagle. Comp. Biochem. Physiol. A, 39, 653.

Champan, B.S., A.J. Tobin, and L.E. Hood. (1981). Complete amino acid sequence of the major early embryonic beta-like globin in chickens. J. Biol. Chem., 256, 5524. (This gives earlier references on alpha globins.)

Chapman, B.S., L.E. Hood, and A.J. Tobin. (1982). Minor early embryonic chick hemoglobin M, amino acid sequences of the E and a-D chains. J. Biol. Chem., 257, 651.

Christensen, V.L., H.V. Bieller, and J.F. Forward. (1982). Physiology of turkey embryos during pipping, and hatching. I. Hematology. Poultry Sci., 61, 135.

Chubb, L.G., and J.G. Rowell. (1959). Counting blood cells of chickens. J. Agric. Sci., 42, 263.

Coates, V., and B.E. March. (1966). Reticulocyte counts in the chicken. Poult. Sci., 45, 1302.

Cohen, R.R. (1967). Anticoagulation centrifugation time and sample replicate number in the microhematocrit method for avian blood. Poult. Sci., 46, 214.

Cook, S.F. (1937). A study of blood picture of poultry and its diagnostic significance. Poult. Sci., 16, 291.

Dabrowski, Z. (1967). The absorption spectrum in the ultraviolet light of the hemoglobin of birds of the crow family. Comp. Biochem. Physiol., 21, 703.

Danby, R., F.E. Curry, and W.R. Gibson. (1980). Extracellular space in chickens, using sucrose, thiosulfate, and thiocyanate injected concurrently. Proc. Soc. Exp. Biol. Med., 164, 336.

Denmark, C.R., and K.W. Washburn. (1969). Hemoglobin types in chicks embryos with different adult hemoglobin genotypes. Poult. Sci., 48, 464

Dhingra, L.D., W.B. Parrish, and W.G. Venzke. (1969). Electron microscopy of granular leukocytes of chicken. Am. J. Vet. Res., 30, 637.

Dickinson, K.E.J., and S.R. Nahorski. (1981). Atypical characteristics of frog and chick erythrocyte beta adrenoceptors. Eur. J. Pharmacol., 94, 43.

Didisheim, P.K. Hattori, and J.H. Lewis. (1959). Hematologic and coagulation studies in various animal species. J. Lab. Clin. Med., 53, 866.

Djojosugito, A.M., B. Folkow, and A.G.B. Kovach. (1968). The mechanism behind the rapid blood volume restoration after hemorrhage in birds. Acta Physiol., Scand., 74, 114.

Dobsinska, E., and O. Dobsinska. (1972). Erythrocyte sedimentation rate in pheasant and its practical importance. Paper presented at the 5th State Conference on Poultry Physiology, Bratislava, September, 1972.

Douglas, D.S. (1968). Salt and water metabolism in the Adélie penguin. Antarct. Res., Ser. 12, 167.

Doerr, J.A., and P.B. Hamilton. (1981a). New evidence for intrinsic blood coagulation in chickens. Poultry Sci., 60, 237.

Domm, L.V., and E. Taber. (1946). Endocrine factors controlling erythrocyte concentration in the blood of the domestic fowl. Physiol. Zool., 19, 258.

Dunlap, J.S., V.L. Johnson, and D.S. Farner. (1956). Multiple hemoglobins in birds. Experientia, 12, 352.

El Ackad, T.M., and P.D., Sturkie. (1972). Histamine in blood and tissues of Aves. Proc. Soc. Exp. Biol. Med., 141, 448.

Engleberg, H. (1977). Probable physiologic functions of heparin. Fed, Proc. Fed. Am. Soc. Exp. Biol., 36, 70.

Ernst, R.A., T.H. Coleman, A.W. Kulenkamp, R.K. Ringer, and S. Pangborn. (1971). The packed cell volume and differential leukocyte count of bobwhite quail. Poult. Sci., 50, 389.

Fallaw, S.A., J.E. Jones, and B.L. Hughes. (1976). Hematocrit, erythrocyte and hemoglobin values for male and female guineas at various ages. Poult. Sci., 55, 814.

Fourie, F.Le R., and J. Hattingh. (1983). Comparative hematology of some South African birds. Comp. Biochem. Physiol. A, 74, 443.

Fraser, R., B. Horton, D. DuPorque, and A. Chernoii. (1972). The multiple hemoglobins of the chick embryo. J. Cell. Physiol., 60, 79.

Gachtgens, P., F. Schmidt, and G. Will. (1981a). Comparative rheology of nucleated and non-nucleated red blood cells. 1. Microrheology of avian erythrocytes during capillary flow. Pfleugers Arch., 390, 278.

Gachtgens, P., F. Schmidt, and G. Will. (1981b). II. Rheological properties of avian red cells suspensions in narrow capillaries. Pfleugers Arch., 390, 283.

Ganong, W.F. (1981). "Review of Medical Physiology" (10th ed.). Los Altos: Lange Medical Publications.

Gilbert, A.B. (1962). Sedimentation rate of erythrocytes in the domesticated cock. Poult. Sci., 41, 784.

Gilbert, A.B. (1963). The effect of estrogen and thyroxine on blood volume of the domestic cock. J. Endocrinol., 26, 41.

Gilbert, A.B. (1965). Sex differences in the erythrocytes in the adult domestic fowl. Vet. Sci., 6, 114.

Gilbert, A.B. (1968). The relationship between the erythrocyte sedimentation rate and packed cell volume in the domestic fowl. Br. Poult. Sci., 9, 297.

Glick, B. (1960). Leucocyte count variation in young chicks during an 18-hour period. J. Appl. Physiol., 15, 965.

Glick, B. (1961). The effect of bovine growth hormone, DCA and cortisone acetate on white cell counts of 2-week-old chickens. Poult. Sci., 40, 1537.

Glick, B. (1969). Hematology of chemically bursectomized birds. Avian Dis., 13, 142.

Glick, B., and K. Sato. (1964). White blood cell counts in bursectomized birds. Am. J. Physiol., 207, 1371.

Goff, S., W.C. Russell, and M.W. Taylor. (1953). Hematology of the chick in vitamin deficiencies. I. Riboflavin. Poult. Sci., 32, 54.

Grant, R.A., and M.B. Zucker. (1973). Avian thrombocyte aggregation and shape change in vitro. Am. J. Physiol., 225, 340.

Gratzer, W., and A. Allison. (1960). Multiple hemoglobins. Biol. Rev., 35, 459.

Griminger, P. (1957). On the vitamin K requirements of turkey poults. Poult. Sci., 36, 1227.
Griminger, P. (1962). Prothrombin bioassay for vitamin K with different thromboplastin preparations. Int. J. Vitam. Nutr. Res., 32, 405.
Griminger, P. (1976). Blood coagulation. (R 70) In "Avian Physiology" (3d ed.) Chapter 3. (P.D. Sturkie, Ed.). New York: Springer-Verlag.
Griminger, P. (1965). Vitamin K activity in chickens: phylloquinone and menadione in the presence of stress agents. J. Nutr., 87, 337.
Griminger, P., and G. Brubacher. (1966). The transfer of vitamin K and menadione from the hen to the egg. Poultry Sci. 45, 512.
Griminger, P., Y.S. Shum, and P. Budowski. (1970). Effect of dietary vitamin K on avian brain thromboplastin activity. Poult. Sci., 49, 1681.
Griminger, P. (1971). Nutritional requirement for vitamin K—animal studies. Proc. Symp. Assoc. Vitam. Chem., Chicago 1971, p. 39.
Gross, K., and P.D. Sturkie. (1975). Concentration of serotonin in intestines and factors affecting its release. Proc. Soc. Exp. Biol. Med., 141, 1261.
Hanwell, A., U. Linzell, and M. Peaker. (1971). Cardiovascular response to salt loading in conscious domestic geese. J. Physiol., 213, 389.
Harris, K.M., and T.I. Koike. (1977). The effects of dietary sodium restriction on fluid and electrolyte metabolism of chickens. Comp. Biochem. Physiol. A, 58, 311.
Hartman, F.A., and M.A. Lessler. (1963). Erythrocyte measurements in birds. Auk, 80, 476.
Hevesy, G., and J. Ottesen. (1945). Life cycle of red corpuscles of the hen. Nature (London), 156, 534.
Hunsaker, W.G. (1968). Blood volume of geese treated with androgen and estrogen. Poult. Sci., 47, 371.
Hunsaker, W.G. (1969). Species and sex differences in the percentage of plasma trapped in packed cell volume determinations on avian blood. Poult. Sci., 48, 907.
Hunsaker, W.G., J.R. Hunt, and J. Ranken. (1964). Physical characteristics of blood. Poult. Sci., 5, 249.
Hunter, F.R. (1953). An analysis of the photoelectric method for studying osmotic changes in chicken erythrocytes. J. Cell. Comp. Physiol., 41, 387.
Hunter, F.R., D. Chaltin, F.J. Finamore, and M.L. Sweetland. (1956). Sodium and potassium exchange in chicken erythrocytes. J. Cell Comp. Physiol., 47, 37.
Hunter, F.R., and J.O. Lim. (1978). Facilitated diffusion in starling erythrocytes. Comp. Biochem. Physiol. A, 59, 355.
Isaacks, R.E., D.R. Harkness, J.L. Adler, C.Y. Kim, P.H. Goldman, and S. Roth. (1976a). Studies on avian erythrocyte metabolism. 5. Relationship between major phosphorylated metabolic intermediates and whole blood O_2 affinity in embryos and poults of turkeys. Poult. Sci., 55, 1788.
Isaacks, R.E., D.R. Harkness, J.L. Adler, and P.H. Goldman. (1976b). Effects of organic phosphates on oxygen affinity on embryonic and adult type hemoglobins of the chick embryo. Arch. Biochem. Biophys., 173, 114.
Isaacks, R.E., D.R. Harkness, R.N. Sampsell, J.L. Adler, C.Y. Kim, and P.H. Goldman. (1977). Studies on avian erythrocyte metabolism. VI. Inositol tetraphosphate: the major phosphate compound in the erythrocyte of the Ostrich (*Struthio camelus camelus*). Eur. J. Biochem. 77:567.
Isaacks, R.E., C.Y. Kim, A.E. Johnson, Jr., P.H. Goldman, and D.R. Harkness. (1982). Studies on avian erythrocyte metabolism. XIII. The synthesis and degradation of inositol pentakis. Poult. Sci., 61, 2271.
Jaeger, J.J., and J.J. McGrath. (1974). Hematologic and biochemical effects of simulated high altitude on Japanese quail. J. Appl. Physiol., 37, 357.
Johnson, L.F., and M.E. Tate. (1969). Structure of phytic acid. Can. J. Chem., 47, 63.
Jordan, H.E. (1939). The lymphocytes in relation to erythrocyte production. Anat. Rec., 73, 227.
Kane, R.E., and I.W. Sizer. (1953). Some studies on the developing blood clotting system of chick embryo. Anat. Rec., 117, 614.
Keijer, E., and P.J. Butler. (1982). Volumes of the respiratory and circulatory systems in tufted and mallard ducks. J. Exp. Biol., 101, 213.
Kase, R., J. Butchers, and N.W. Spurling. (1980). Comparison of the rate-limiting influence of factor VII on mammalian plasma coagulation following extrinsic activation by avian and mammalian tissue thromboplastins. Comp. Biochem. Physiol. A, 65, 421.
Klein, R.E., and I.W. Sizer. (1963). Some studies on the developing blood clotting system of chick embryo. Anat. Rec., 117, 614.
Koike, T.I., L.R. Pryor, and H.L. Neldon. (1983). Plasma volume and electrolytes during progressive water deprivation in chickens. Comp. Biochem. Physiol. A, 74, 83.
Kovach, A.G.B., and T. Balint. (1968). Hemodilution in pigeon and rat. Acta Physiol. Acad. Sci. Hung., 35, 231.
Kovach, A.G.B., E. Szasz, and N.P.L. Mayer. (1969). Mortality of various avian and mammalian species following blood loss. Acta Physiol. Acad. Sci. Hung., 35, 109.
Kubena, L.F., J.D. May, F.N. Reece and W. Deaton. (1971). Hematocrit and hemoglobin levels of broilers as influenced by environmental temperature and dietary iron level. Poultry Sci. 51, 759.
Le Febvre, E.A. (1964). The use of D_2O for measuring energy metabolism of *Columbia livia* at rest and in flight. Auk, 81, 403. (Quoted by Skadhauge, (1981.)
Lerner, J., R.M. Smaglia, S.E. Hilchey, and R.G. Somes, Jr. (1982). Amino acid transport and intracellular Na and K content of chicken erythrocytes genetically selected for high and low leucine transport activity. Comp. Biochem. Physiol. A, 73, 243.
Lewis, J.H., U. Hasiba, and J.A. Spero. (1979). Comparative hematology: studies on class Aves, domestic turkey (*Meleagris gallopavo*). Comp. Biochem. Physiol. A, 62, 735.
Losito, R. (1965). Investigations into the presence of a competitive inhibitor (preprothrombin) in the plasma of chicks. Acta Chem. Scand. [B], 19, 2229.
Lucas, A.M. (1959). A discussion of synonymy in avian and mammalian hematological nomenclature. Am. J. Vet. Res., 20, 887.
Lucas, A.M., and E.M. Denington. (1957). Effect on total body x-ray irradiation on the blood of female single comb White Leghorn chickens. Poult. Sci., 36, 1290.
Lucas, A.M., and C. Jamroz. (1961). "Atlas of Avian Hematology." U.S. Department of Agriculture, Monograph 25.
March, B.E., V. Coates, and J. Biely. (1966). The effects of estrogen and androgen on osmotic fragility and fatty acid composition of erythrocytes in the chicken. Can. J. Physiol. Pharmacol., 44, 379.
Maxwell, M.H., W.G. Siller, and G.M. McKenzie. (1979). Eosinophiles associated with facial edema in fowls. Vet. Res., 105, 232.
McCollum, E.V. (1957). "A History of Nutrition." Boston: Houghton-Mifflin.
McGrath, J.J. (1970). Effects of chronic hypoxia on hematocrit ratios and heart size of pigeons. Life Sci., 9, 451.
McManus, T.J. (1967). Comparative biology of red cells. Fed. Proc. Fed. Am. Soc. Exp. Biol., 26, 1821.
Medway, W., and M.R. Kare. (1959a). Thiocyanate space in growing domestic fowl. Am. J. Physiol., 196, 873.
Medway, W., and M.R. Kare. (1959b). Blood and plasma volume hematocrit blood specific gravity and serum protein electrophoresis of the chicken. Poult. Sci., 38, 624.

Mendenhall, V.M., and L.F. Pank. (1980). Secondary poisoning of owls by anticoagulant rodenticides. Wild. Soc. Bull., 8. 311.

Meyer, D. (1973). Studies on factors regulating blood serotonin levels in the domestic fowl. Ph.D. Thesis, Rutgers University, New Jersey.

Meyer, D.C., P.D. Sturkie, and K. Gross. (1973). Diurnal rhythm in serotonin in blood and pineals of chickens. Comp. Biochem. Physiol. A, 46, 619.

Meyer, D.C., and P.D. Sturkie. (1974). Diurnal rhythm of histamine in blood and oviduct of the domestic fowl. Comp. Gen. Pharmacol., 5, 225.

Meyer, D.C., and P.D. Sturkie. (1974). Distribution of 5-HT among blood cells of the domestic fowl. Proc. Soc. Exp. Biol. Med., 147, 382.

Mongin, P. and N.W. Carter. (1976). Ann. Biol. Anim. Bioch. Biophys., 16, 649. (Quoted by Danby et al., 1980.)

Murano, G. (1980). A basic outline of blood coagulation. Sem. Thromb. Hemost. 6, 140.

Natt, M.P., and C.A. Herrick. (1952). A new blood diluent for counting the erythrocytes and leukocytes of the chicken. Poult. Sci., 31, 735.

Newcomer, W.S. (1958). Physiologic factors which influence acidophelia induced by stressors in the chicken. Am. J. Physiol., 194, 251.

Nirmalan, G.P., and G.A. Robinson. (1971). Hematology of the Japanese quail., *Coturnix japonica.* Br. Poult. Sci., 12, 475.

Nirmalan, G.P., and G.A. Robinson. (1972). Hematology of Japanese quail treated with exogenous stilbestrol disproprionate and testosterone propionate. Poult. Sci., 51, 920.

Nirmalan, G.P., and G.A. Robinson. (1973). The survival time of erythrocytes (DF^{32} I-label) in the Japanese quail. Poult. Sci., 52, 355.

Oberthür, W., J. Godovac-Zimmerman, and G. Braunitzer. (1983). The different evolution of bird hemoglobins. Faculte des. Sciences, Editions de l'Universite de Brussels, Hemoglobin symposium, 23 to 24 July.

Olson, C. (1937). Variation in cells and hemoglobin content in blood of normal domestic chickens. Cornell Vet., 27, 235.

Olson, C. (1965). Avian hematology. In "Diseases of Poultry" (H.E. Biester, and L.H. Schwarte, Eds.). Ames: Iowa State Press, p. 100.

Pastorova, V.E. (1957). Tzmenenie protshnosti kapilljornych sosidov pris K-avitaminose i spetsif tschnost diestvija vitina K na protschnost kapilljarov. Dokl. Akad. Nauk Arm. SSR, 113, 1379.

Paulsen, T.M., A.L. Moxon, and W.O. Wilson. (1950). Blood composition of broad-breasted bronze turkeys. Poult. Sci., 29, 15.

Pilaski, J. (1972). Vergleichende Untersuchungen über den Hämoglobingehalt des Hühner-und Putenblutes in Abhängigkeit von Alter und Geschlecht. Arch. Geflüegelkd. 36, 70.

Ploucha, J.M., J.B. Scott, and R.K. Ringer. (1981). Vascular and hematologic effects of hemorrhage in the chicken. Am. J. Physiol., 240, H9.

Ploucha, J.M., S.J. Bursian, R.K. Ringer, and J.B. Scott. (1982). Effects of severe hemorrhagic hypotension on the vasculature of the chicken. Proc. Soc. Exp. Biol. Med., 170, 160.

Portman, O.W., K.P. McConnell, and R.H. Rigdon. (1952). Blood volume of ducks using human albumin labelled with radioiodine. Proc. Soc. Exp. Biol. Med., 81, 599.

Quick, A.J. (1966). "Hemorrhagic Diseases and Thrombosis." Philadelphia: Lea & Febiger.

Rodnan, G.P., F.G. Ebaugh, Jr., and M.R.S. Fox. (1957). Life span of red blood cell volume in the chicken, pigeon, and duck as estimated by the use of Na_2C, $^{51}O_4$ with observations on red cell turnover rate in mammal, bird and reptile blood. J. Hematol., 12, 355.

Rosse, W.F., and T.A. Waldmann. (1966). Factors controlling erythropoiesis in birds. Blood, 27, 654.

Ruch, F.E. and M.R. Hughes. (1975). The effects of hypertonic sodium chloride injection on body water distribution in ducks, gulls and roosters. Comp. Biochem. Physiol. A, 52, 21.

Sauter, E.A., C.F. Petersen and E.E. Steele. (1975). Dietary management procedures for development of consistent hemorrhagic symptoms in chickens. Poultry Sci. 54, 1433.

Schaible, T.S. (1977). Some aspects of the physiology and pharmacology of histamine in the mesenteric vasculature and tissues of *Gallus domesticus. Ph.D.* Thesis, Rutgers University.

Shum, Y.S., and P. Griminger. (1972a). Thromboplastic activity of acetone-dehydrated chicken brain powder extracts after repeated extraction and dilution. Poult. Sci., 51, 402.

Shum, Y.S., and P. Griminger. (1972b). Prothrombin time stability of avian and mammalian plasma: effect of anticoagulant. Lab. Anim. Sci., 22, 384.

Siegel, H.S. (1968). Blood cells and chemistry of young chickens during daily ACTH and cortisone administration. Poult. Sci., 47, 1811.

Sillau, A.H., Cueva, S., and P. Morales. (1980). Pulmonary arterial hypertension in male and female chickens at 3300 meters. Pfleugers Arch., 366, 275.

Simon-Oppermann, C., D. Gray, E. Szezepanska-Sadowska, and E. Simon. 91984). Blood volume changes and arginine vasotocin blood concentration in conscious fresh water- and salt water-adapted ducks. Pfleugers Arch., 400, 151.

Skadhauge, E. (1981). Osmoregulation in birds. New York: Springer-Verlag.

Skadhauge, E., and S.D. Bradshaw. (1974). Saline drinking and cloacal excretion of salt and water in zebra finch. Am. J. Physiol., 227, 1263.

Skadhauge, E., R.M. Herd, and T.J. Dawson. (1980). Renal excretion and cloacal absorption of several ions and water in the emu. Proc. Int. Union Physiol. Sci., 14, 707. (Quoted by Skadhauge, 1981.)

Smith, A.H. (1978). Gravitional physiology. Physiol. Teacher, 7, 4.

Soliman, M.K., Elamrousi, S., and A.A.S. Ahmed. (1966). Cytological and biochemical studies on the blood of normal and spirochaete-infected chicks. Zentralb. Veterinaermed., Reihe B, 13, 82.

Sorby, O. (1966). "Studies Related to Coagulation of Chicken Blood." Oslo: Universitetsforlaget.

Spector, W.S. (1956). "Handbook of Biologic Data." Philadelphia: Saunders.

Spurling, N.W. (1981). Comparative physiology of blood clotting. Comp. Biochem. Physiol. A, 68, 541.

Stewart, D.J. (1972). Secretion of salt gland during water deprivation in duck. Am. J. Physiol 223:344.

Stopforth, A. (1970). A study of coagulation mechanisms in domestic chickens. J. Comp. Pathol., 80, 525.

Sturkie, P.D. (1951). Effects of estrogen and thyroxine upon plasma proteins and blood volume in the fowl. Endocrinology, 49, 565.

Sturkie, P.D. and H.J. Newman. (1951). Plasma proteins of chickens as influenced by time of laying, ovulation, number of blood samples taken and plasma volume. Poultry Sci. 30, 240.

Sturkie, P.D. (1976). In "Avian Physiology" (3d ed.) (P.D. Sturkie, Ed.). Chapter 3. New York: Springer-Verlag.

Sturkie, P.D., and K. Textor. (1958). Sedimentation rate of erythrocytes in chickens as influenced by method and sex. Poult. Sci., 37, 60.

Sturkie, P.D., and K. Textor. (1960). Further studies on sedimentation rate of erythrocytes in chickens. Poult. Sci., 39, 444.

Sturkie, P.D., and J.M. Eiel. (1966). Effects of estrogen on cardiac output, blood volume and plasma lipids. J. Appl. Physiol., 21, 1927.

Sturkie, P.D., Woods, J.J., and D. Meyer. (1972). Serotonin levels in blood, heart, and spleen of chickens, ducks and pigeons. Proc. Soc. Exp. Biol. Med., 139, 364.

Sturkie, P.D. (1965). In "Avian Physiology" (2nd ed.). Ithaca, N.Y.: Cornell University Press.

Suttie, J.W. (1980). The metabolic role of vitamin K. Fed. Proc. Fed. Am. Soc. Exp. Biol., 39, 2730.

Thomas, D.H., and J.G. Phillips. (1975). Studies on avian adrenal steroids. II. Chronic adrenalectomy and turnover of H_2O in domestic ducks. Gen. Comp. Endocrinol., 26, 404.

Topp, R.C., and H.C. Carlson. (1972). Studies on avian heterophils. III. Phagocytic properties. Avian Dis., 16, 374.

Tosteson, D.C., and J.S. Robertson. (1956). Potassium transport in duck red cells. J. Cell. Comp. Physiol., 47, 147.

Usami, S., V. Magazinovic, S. Chen, and M.I. Gregersen. (1970). Viscosity of turkey blood: Rheology of nucleated erythrocytes. Microvasc. Res., 2, 489.

Vogel, J. (1961). Studies on cardiac output in chicken. Ph.D. Thesis, Rutgers University, New Brunswick, New Jersey.

Walz, D.A., R.K. Kipfer and R.E. Olson. (1975). Effect of vitamin K deficiency, warfarin, and inhibitors of protein synthesis upon plasma levels of vitamin K-dependent factors in the chick. J. Nutrition 105, 972.

Wartelle, O. (1957). Mechanisme de la coagulation chez la poule 1: Etude des elements du complexe prothomique et de la thromboplastino formation. Rev. Hematol., 12, 351.

Washburn, K.W. (1968). Affects of age of bird and hemoglobin type on the concentration of adult hemoglobin components of the domestic fowl. Poult. Sci., 47, 1083.

Washburn, K.W. (1969). Hematological response of different stocks of chickens in iron-copper deficient diet. Poult. Sci., 48, 204.

Washburn, A.H., and A.J. Meyers. (1957). The sedimentation of erythrocytes at an angle of 45 degrees. J. CLin. Lab. Med., 49, 318.

Washburn, K.W., and J.R. Smyth, Jr. (1968). Hematology of an inherited anemia in the domestic fowl. Poult. Sci., 47, 1408.

Washburn, K.W., and C.Y. Chang Yen. (1976). Evaluation of heterogenity in hemoglobins of *Gallus domesticus* and *Coturnix* quail using disc-gel electroiphoresis. Poult. Sci., 55:1646.

Weingarten, J.P., H.S. Rollema, C. Bauer, and P. Scheid. (1976). Effects of IP_6, on the Bohr effect induced by CO_2 and fixed acids in chicken hemoglobin. Pfleugers Arch., 377, 135.

Weiss, H.S. (1958). Application to the fowl of the antipyrine dilution technique for the estimation of body composition. Poult. Sci., 37, 484.

Wels, A., and V. Horn. (1965). Beitrag zur Hämoglobinbestimung im Blut des Geflügels. Zentralb. Veterinaermed., Reihe A, 663.

Witkowski, A., and J.P. Thaxton. (1981). Morphology of eosinophils in Japanese quail. Poult. Sci., 60, 1587.

Wolford, J.H., and D. Murphy. (1972). Effect of diet on fatty liver-hemorrhagic syndrome incidence in laying chickens. Poult. Sci., 51, 2087.

Wolford, J.H., and R.K. Ringer. (1962). Adrenal weight, adrenal ascorbic acid, adrenal cholesterol, and differential leucocyte counts as physiological indicators of stress or agents in laying hens. Poult. Sci., 41, 1521.

Wyse, D.G. and M. Nickerson (1971). Studies on hemorrhagic hypotension in domestic fowl. Can. J. Physiol. Pharmacol. 49:919.

Zambraski, E.J., and R. Schuler. (1980). Failure of prostaglandin inhibition to attenuate the tolerance to hemorrhage in the domestic fowl. Poult. Sci., 59, 2567.

6
Heart and Circulation: Anatomy, Hemodynamics, Blood Pressure, Blood Flow

P.D. Sturkie

Anatomy of the Circulatory System

The heart of most birds is located in the thorax slightly to the left of the median line and is almost parallel to the long axis of the body, except that the apex may be bent to the right (Figure 6–1). The heart is surrounded by the pericardial sac. The bird heart, like the mammal's, has four chambers, two atria and two ventricles. The right atrium is usually larger than the left. The left ventricle is usually three or more times larger than the right and is considerably thicker. The radii of curvature at the apex are less than at the base and therefore tension and pressure here are less. Avian cardiac muscle has no transverse tubules and the fibers are much smaller than mammalian ones (Sommer and Steele, 1969; Akester, 1971). The atria have openings into the ventricles, which are closed by the atrioventricular (AV) valves (Figure 6–2). The left valve is thin, membranous, and bicuspid, as it is in mammals. The right valve is simply a muscular flap. It originates at the right side of base of the pulmonary artery, where it is held up by trabecular muscle. Contraction of the right ventricle forces the free leaf portion of the valve into the atrioventricular opening, thus closing it. The pulmonary and aortic valves are membranous and tricuspid, as are those of mammals.

The entrance of veins into and exit of arteries from the heart are illustrated in Figures 6–3 and 6–4. The mode of entry of veins into the right atrium exhibits considerable variation. In the chicken, the posterior vena cava and right anterior vena cava open together

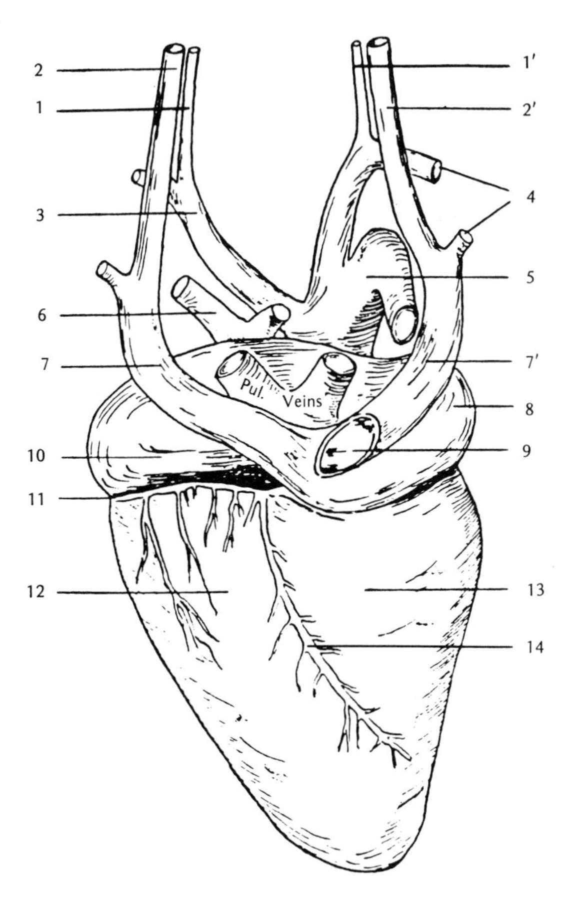

FIGURE 6–1. Abbreviations: *1* and *1′*, left and right carotid arteries; *2* and *2′*, left and right jugular veins; *3*, left brachiochephalic arteries; *4*, right brachial vessels; *5*, aorta; *6*, pulmonary arteries; *7* and *7′*, left and right vena cava; *8*, right artrium; *9*, posterior vena cava; *10*, left atrium; *11*, coronary groove; *12*, left ventricle; *13*, right ventricle; *14*, large median vein and branch. (After Sisson and Grossman, 1953.)

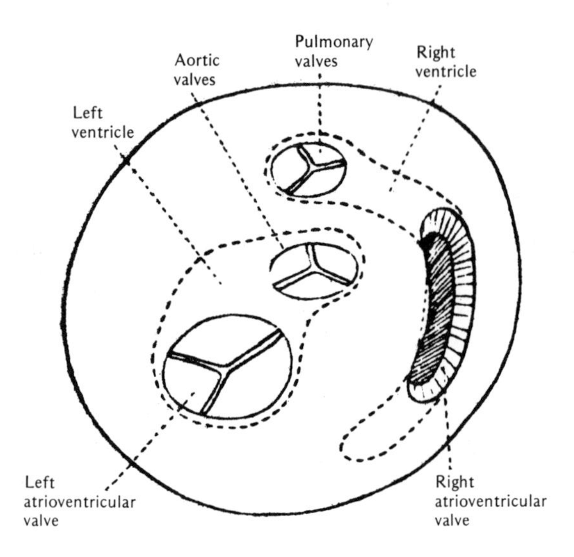

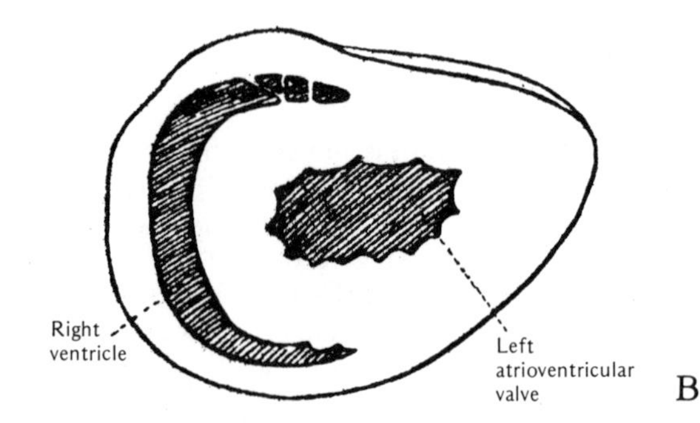

FIGURE 6–2. (**A**), Schematic drawing of bird heart at the level of and parallel to the atrioventricular junction. The dark area represents the AV orifice; the shaded area, the extent of the right AV valve. **B**, Transverse section of heart showing relative thickness of right and left ventricles. (After Jones and Johansen, 1972.)

into the sinus venous, whose slitlike entrance into the right atrium is demarcated by a pair of sinoatrial valves.

Size of Heart

The size of the heart varies considerably among birds according to body size (Table 6–1). The heart size appears to be relatively greater in small birds, although there is considerable variation (Grubb, 1983). For more details consult Hartman (1955), who has recorded heart weights of many species, and Brush (1966), who has plotted log heart weights against log body weight in many species. In birds weighing more than 100 g, the slope of the line (regression) is about 1, and for those weighing less than 100 g, the slope is about 0.6. Heart size tends to be greater in birds than mammals of equal weight, and this has some obvious advantages and disadvantages (see section on work of heart in Chapter 7).

Environmental factors such as hypoxia (high altitude) increase heart size (Abati and McGrath, 1973).

Innervation of Heart

The atria are innervated by sympathetic and parasympathetic fibers, and these influence heart rate considerably. The ventricles, unlike those of most mammals, also receive sympathetic and parasympathetic fibers. Based on pharmacologic data, Bolton (1967) reported vagal fibers in the ventricles. Abraham (1969) also reported that atria and ventricles are innervated by parasympathetic fibers, and Hirsch (1970) reported such fibers in the ventricles of owl, penguin, and goose. (See Chapter 7 for more details on heart nerves.)

Coronary Arteries and Veins

Most birds have two coronary arteries, but some have three (Petren, 1926). These arteries are illustrated in Figure 6–5. The right artery is larger than the left. It originates from the sinus of the semilunar valve of the aorta and divides into two branches, a superficial one and a large main branch (Figure 6–5; numbers 9 and 10). The large branch runs in the interventricular septum toward the apex of the heart, and branches into two; the small one of the right coronary runs in the coronary groove and goes to the wall of the right atrium and ventricle. The right artery supplies a large part of the heart, including the right atrium and ventricle, and the septum and wall of the left ventricle. The left coronary (Figure 6–5, 11a) arises from the sinus of the left semilunar valve, and it courses initially toward the base of the heart; it then divides into a superficial branch (12), and a main branch (13) that then branches as shown. The left coronary artery supplies the wall of the left and part of the right atrium and a large part of its septum and parts of the right and left ventricles.

There are two main coronary veins and their branches, which return blood to right atrium. These are the main vein (magna) shown in Figure 6–5, 5) and the median (4), which courses over the dorsal surface of the heart. It and its branches collect blood mainly from the right coronary artery. The main vein originates at the apex of heart, runs along the left ventricle into the coronary groove, and empties into the right atrium (Nickel et al., 1977). It collects blood from areas supplied by the left coronary artery, as well as branches of the right coronary.

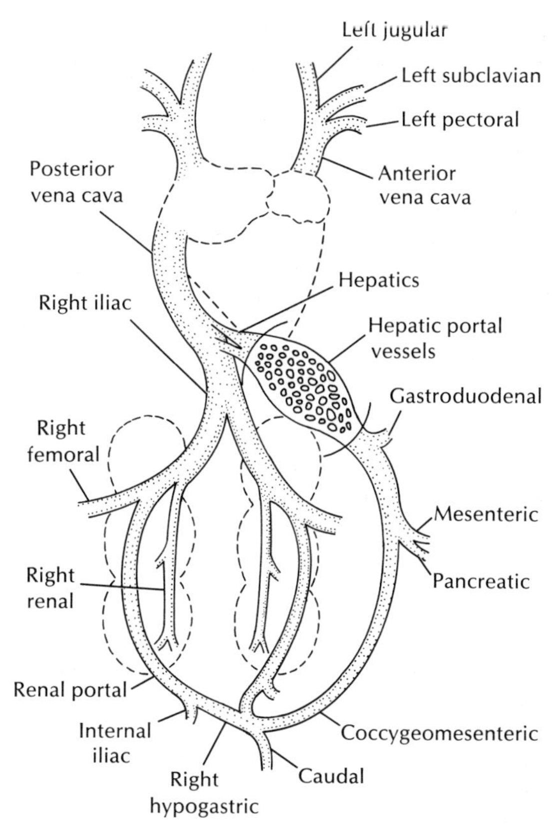

FIGURE 6–3. Venous system in ventral view. (After Portmann, 1950.)

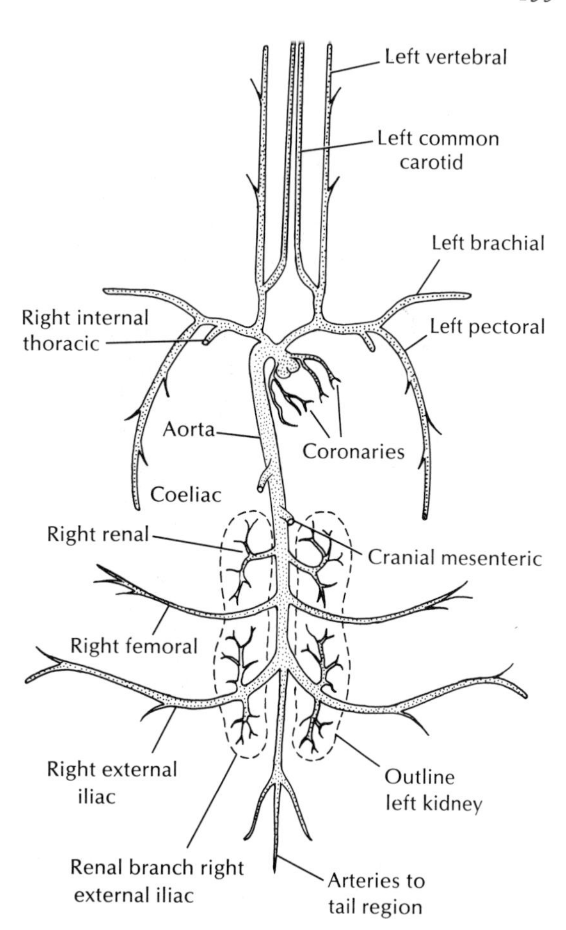

FIGURE 6–4. Principal arteries (ventrodorsal view). The outline of the heart is indicated by the coronary arteries. (After Akester, 1971.)

TABLE 6–1. Heart size and body weight

Species	Body weight (g)	Heart weight/body weight (%)	References
Goose	4405	0.8	Sturkie (1965)
Duck	1685	0.74	Sturkie (1965)
Chicken	3120	0.44	Sturkie (1965)
Ptarmigan	258	1.05	Johnson and Lockner (1968)
Hummingbird	—	2.4	Hartman (1955)
Coturnix (quail)	119	0.90	Sturkie (unpublished)
Pigeon	458	1.02	Sturkie (unpublished)
Emu	3750	0.85	Grubb (1983)
Goose	690	0.54	Grubb (1983)
Peking duck	290	0.81	Grubb (1983)
Vulture	204	0.81	Grubb (1983)
Turtle dove	15.3	0.85	Grubb (1983)
Monk parakeet	10.4	0.04	Grubb (1983)
Budgerigar	03.5	1.29	Grubb (1983)

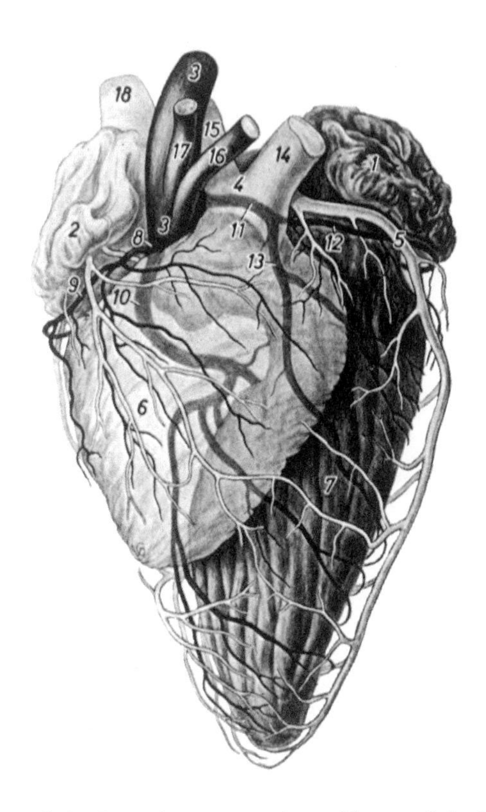

FIGURE 6–5. Corrosion preparation of heart of fowl showing coronary arteries and veins. Arteries in black; veins in white; *1,* left auricle; *2,* right auricle; *3,* aorta; *4,* truncus pulmonalis; *5,* v. cordis magna; *6,* right ventricle; *7,* left ventricle; *8,* a. coronary, right; with ramus supf. (*9*) and ramus prof. (*10*); *11,* a. coronary, left, with ramus supf. (*12*) and ramus prof. (*13*); *14, 15,* a. pulmonary, left and right; *16, 17,* a. brachiocephalics left and right; *18,* v. cava cran., right. (After Nickel et al., 1977.)

Blood Vessels

The principal arteries and veins are presented in Figures 6–3 and 6–4. The arterial supply to most organs is similar to that in mammals, except that the arrangement of carotid arteries in birds varies considerably among different species: (1) they may be paired; (2) they may be fused into a median vessel; (3) the right or left vessels may be unpaired; and (4) the right and left may be paired but with one or the other much smaller. The common carotid gives rise to the vertebral artery in some species and is often referred to as an internal carotid (Figure 6–6). The terminology is conflicting and confusing (Adams, 1958).

The branches of the internal carotids go to the brain. There is no circle of Willis, but there is a direct anastomosis between the two cerebral carotid arteries, immediately caudal to the hypophysis. The types of anastomes involve three arrangements or configurations: (1) an H type, (2) an I type, and (3) an X type (Baumel and Gerchman, 1968). Further details on arteries are treated in a later section on bloodflow to individual organs and systems.

Birds, unlike mammals, have a renal portal circulation. This system is treated in detail in Chapters 6 and 16, where it is shown that blood normally flowing to the liver from the coccygeomesenteric and hepatic portal veins may be shunted to the kidney via the renal portal vein, and vice versa. The coccygeomesenteric vein is the intervening link between the renal and hepatic portal systems.

Blood capillaries in the lungs are discussed in Chapter 8, where it is shown that they are much smaller than in mammals, but have a greatly increased blood volume and also present an increased and more efficient gas-exchange area compared to mammals.

Innervation of Arteries and Veins

Smooth muscle in the veins is innervated with adrenergic fibers (Bennett and Malmfors, 1970; Akester, 1971) and also cholinergic fibers (Akester, 1971), and so are those of arteries. The mesenteric arteries of chickens and turkeys contain longitudinal smooth muscle as well as circular, which is also innervated by adrenergic and cholinergic fibers (Bolton, 1968). The following arteries also receive a dual innervation, according to Akester (1971): renal, coeliac, common carotid, subclavian, and pulmonary, but not femoral, which receives adrenergic fibers only. Practically all veins, including the vena cava, pulmonary, jugular, hepatic, hepatic portal, renal portal vein and valve, and coccygeomesenteric, are dually innervated.

Hemodynamics

In each cardiac cycle (see Chapter 7) blood is ejected from the left ventricle through the aorta, forced through the arteries and arterioles to capillaries, and goes thence to veins, which carry blood back to the right atrium (systemic circulation). At the same time, blood from the right ventricles is ejected through pulmonary arteries to the lungs, where it is oxygenated and returned to the left atrium via the pulmonary veins (pulmonary circulation).

The heart pumps blood (a non-Newtonian fluid) through blood vessels of varying size, distensibility, and resistance. The volume of blood delivered to an organ (flow) is determined by the pressure difference (ΔP) exerted on it between two points and the resistance (R) to flow as follows: F (flow) $= P/R$. It follows therefore that $P = FR$ and $R = P/F$.

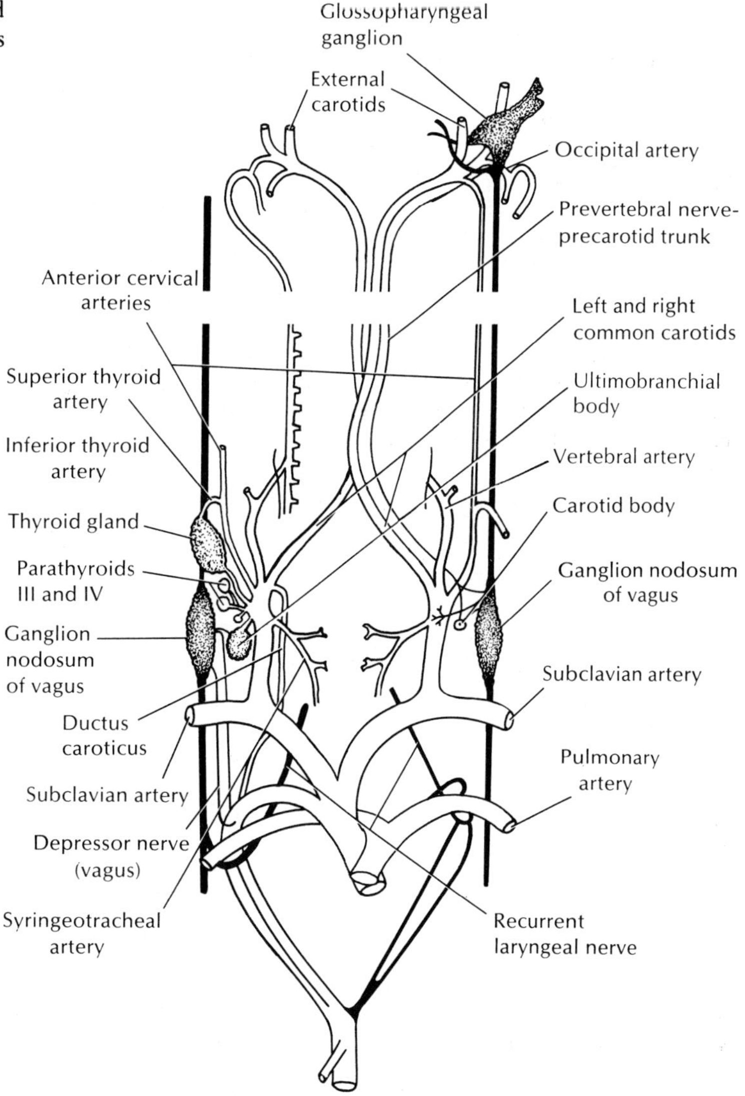

FIGURE 6–6. Innervation of carotid sinus complex and aortic arch of birds (according to Adams, 1958).

Poiseuille's equation describes in more detail the factors involved in resistance to flow:

$$R = \frac{\Delta P}{F} = \frac{8}{\pi} \times \eta \times \frac{L}{r^4}$$

where r^4 is the radius of the vessel, L is the length of vessel, η is the viscosity of blood, and 8 the factor of integration.

Bloodflow or cardiac output and the pressure drop (P) between points are easily measurable, but resistance is more difficult to measure and is usually calculated.

A slight change in radius has a great effect on resistance. Other factors affecting resistance are elasticity coefficient and distensibility, which vary in different blood vessels even of the same radius.

According to the law of LaPlace, tension in vessels varies directly with radius of vessel and transmural pressure; larger vessels with very high blood pressures, such as those of the turkey, are more susceptible to rupture (aortic aneurysm), as has been demonstrated (Speckman and Ringer, 1963).

The static elastic modulus of the thoracic and abdominal aortas of turkeys are 2.26 and 9.55×10^5 dynes/cm^2, respectively, suggesting a higher content of elastin (more distensible) and lower collagen in thoracic aorta than in the abdominal aorta (Speckman and Ringer, 1964).

Resistance may be calculated in relative units (RU) by dividing cardiac output in liters per minute or other desired units into blood pressure in millimeters Hg. This is a simple way of comparing resistance changes in some species but cannot be used to compare resistances in different species varying widely in cardiac output. In such cases the absolute unit (in dynes · sec/cm^5) is used. For example, pressure in millimeters Hg is converted to cgs units where 1 mm Hg P = 1332 dynes/cm^2. Blood flow or cardiac output is converted to cubic centimeters per second. For example, the mean blood pressure of a chicken before hyperthermia was 126 mm Hg and cardiac output was 388 ml/min. The peripheral resistance was 31,867 dynes-sec/cm^5. This figure is over 20 times that for an an average man. Following hyperthermia, blood pressure decreased, cardiac output increased, and resistance decreased to 22,298 dynes (Whittow et al., 1964). Chickens that have become acclimatized to heat, however, exhibit an increase in peripheral resistance, attributable mainly to a greater drop in cardiac output than in blood pressure (Sturkie, 1967). Therefore, peripheral resistance as thus determined is a mathematical expression and does not necessarily reflect differences in constrictor tone of blood vessels.

Peripheral resistance in hypertensive chickens is significantly higher than in hypotensive ones and probably reflects greater vasoconstrictor tone (Sturkie et al., 1962).

RUs in chickens range from 0.88 to 1.43 and in the duck from 0.567 to 0.62 (Sturkie, 1966). In the larger turkey, RUs ranges from 1 to 2 or higher (Speckman and Ringer, 1963).

Blood Pressure

The pressure in the heart and aorta reaches its peak on systolic ejection and its minimum during diastole (end diastolic pressure) (see also Chapter 7). The difference between the two pressures is pulse pressure (PP). Mean blood pressure (MBP) is equal to the area under pressure pulse wave × time. In the chicken, MBP = diastolic pressure (DBP) + 3/8 PP.

Pressure Pulses (Wave Forms)

In mammals, the pressure pulse increases in amplitude from the aorta to the more peripheral arteries such as femoral and saphenous, and the shape changes from a rectangular to a triangular form. In the goose and turkey, however, there is no appreciable change, according to Taylor (1964), who suggested a simple Windkessel model to characterize the avian circulation. The pulse waves of the chicken and duck aorta, contrarily, show significant changes in shape and amplitude. These changes in the chicken are apparently much less than in the dog (Strano et al., 1972) as revealed in Figures 6–7 and 6–8, but there is a slight change in amplitude toward the periphery. The changes in the pressure pulse of the Peking duck is greater than in the chicken, where the amplitude increase from the central to the more peripheral aorta was 29% (Langille and Jones, 1975). In the chicken there are significant differences in the amplitude of pressure transfer functions. This is shown in Figure 6–8 for the chicken and dog (Strano et al., 1972). The pressure transfer functions are based on frequency of heart rate and phase angle of waves, or the ratio of Fourier series component of output pressure of femoral data at each harmonic of the fundamental frequency of 4.67 Hz for the chicken.

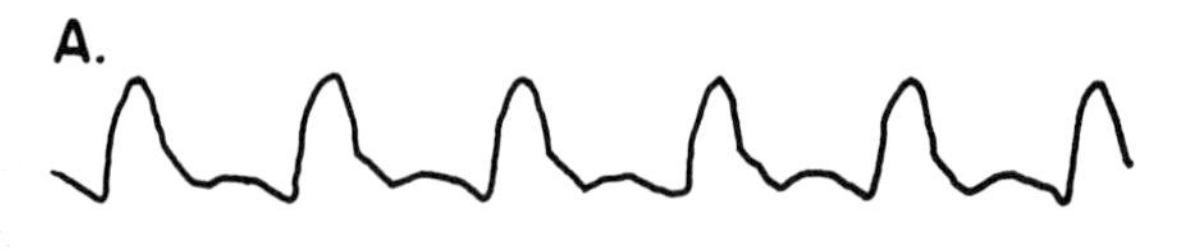

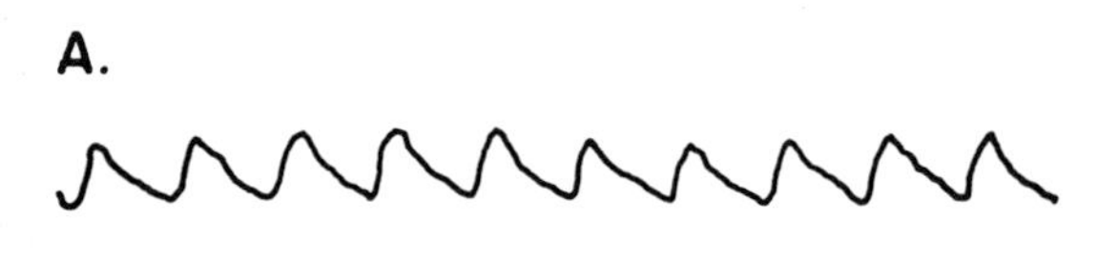

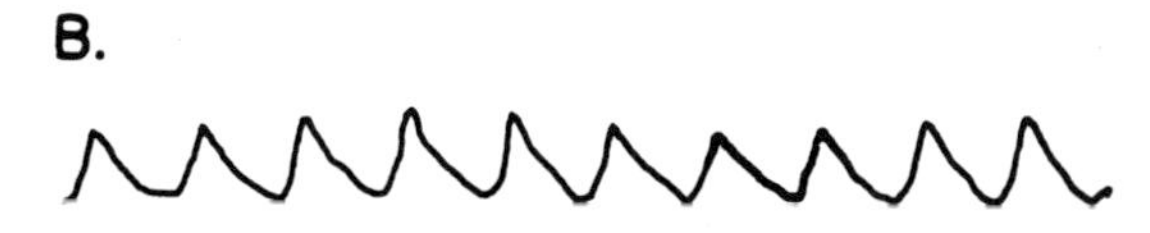

FIGURE 6–7. Measurement and utilization of in vivo blood pressure transfer functions of dog and chicken aortas. Upper. Simultaneous pressure-versus-time measurements taken in dog aorta. Curve A is pressure taken at femoral bifurcation, and B is pressure at root of aorta. Lower. Same for chicken aorta, with A and B curves at same locations as in dog. (Modified from Strano et al., 1972 ©, 1972, IEEE.)

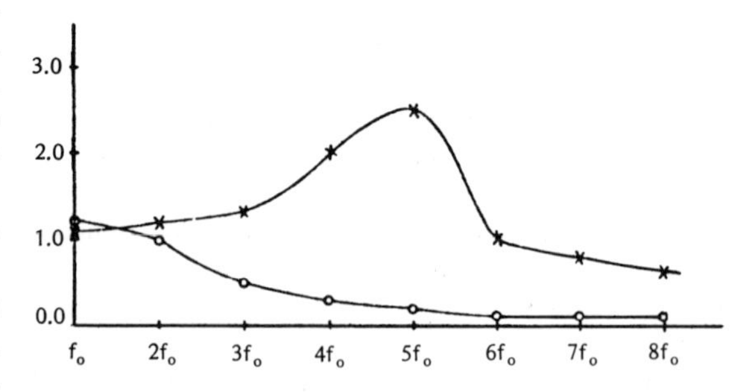

FIGURE 6–8. Amplitude of pressure transfer function (y axis) versus frequency (x axis) for dog (x) and chicken (o) aortas. (Modified from Strano et al., 1972 ©, 1972, IEEE.)

Pressure and flow oscillations at different sites in the mammalian arterial system are significantly out of phase, and these waves are believed to be altered by reflection (Taylor 1964). Reflection effects are considered to be maximal when the transit time between the heart and the reflecting site in the arterial bed constitutes a significant portion of the total cardiac cycle as in mammals, but minimal or absent when the transit time is low. Langille and Jones (1975) reported low transit times in the duck, amounting to 5–10% or less of the cardiac cycle. They concluded that the Windkessel model does not apply to the duck's systemic circulation nor, according to Strano et al. (1972) to that of the chicken.

Strano et al. and Welkowitz and Fich (1967) reported that the shape of pulse waves in mammals can be accounted for by the geometric taper of the vessel without regard to reflections from the periphery. The derived pulse wave forms based on their calculations showed close correspondence with actual measurements, as was the case in chickens (Fich and Welkowitz, 1973), which exhibit no reflected waves.

The thoracic aorta of birds contains elastin and is distensible, but the abdominal aorta contains little elastin and much collagen (see Figure 6–9 for pressure-volume curves), and it and other peripheral arterial branches have very high wave velocities in the goose and turkey, according to Taylor (1964).

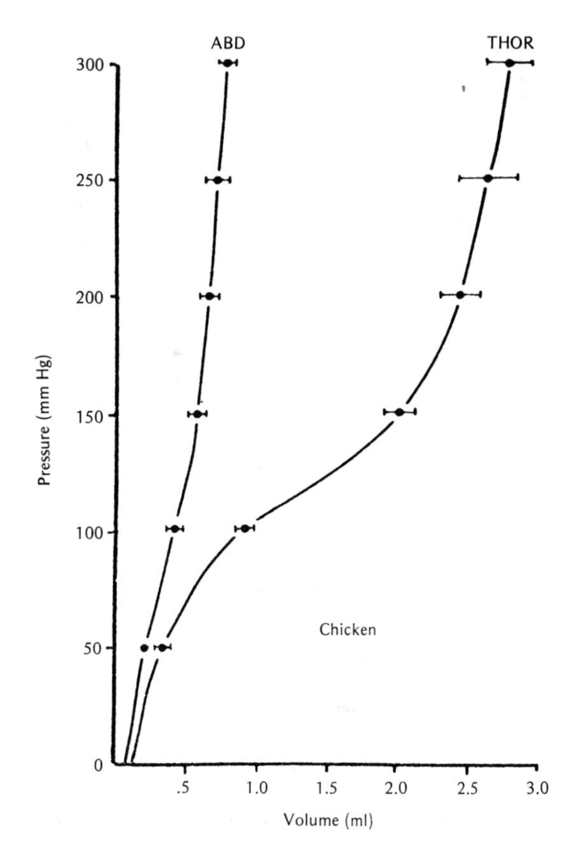

FIGURE 6–9. Pressure–volume curves for thoracic (*THOR*) and abdominal (*ABD*) aortic segments from chicken. (After Ploucha and Ringer, 1981.)

Velocity. The pulse wave velocity in the turkey aorta, based on data of Speckman and Ringer (1966) as calculated by Jones and Johansen (1972), is 4–6 m/sec for the thoracic aorta and 30–35 m/sec in the abdominal aorta; the latter value far exceeds such values in mammalian arteries. Actual determinations of pulse wave velocities in the duck by Langille and Jones (1975) revealed rates of 4.4 m/sec in the aortic arch to 11.7 m/sec in the abdominal aorta. The total time required for pulse waves to reach the distal end of aorta was about 20 msec, or 5–10% of the cardiac cycle. They also reported that dicrotic notches are rarely observed in the pulse waves of ducks and chickens.

Ploucha and Ringer (1981) determined the compliance of isolated segments of thoracic and abdominal aortas of avian species, and they compared the pulse wave velocities of these to man and dog. It is observed in Figure 6–9 that the abdominal aorta of chickens is almost nondistensible, but the thoracic aorta is fairly distensible. The velocities (Figure 6–10) are highest in the least distensible segments of the aorta, the abdominal aorta in the turkey, chicken, and duck, in that order.

Methods of Determining Blood Pressure

Direct. The methods may be classified as direct and indirect. In all of the commonly used direct methods, a cannula or needle is inserted into the artery or vein, and the pressure of the blood is exerted against a tube of liquid (containing an anticoagulant) that is attached to the manometer. Where a mercury manometer is used, the liquid is in contact with the mercury, which rises and falls in the U tube with heartbeat. Because of the inertia of mercury, systolic and diastolic pressures are not recorded accurately. There are other, more sensitive types of manometers, such as strain gages and capacitance manometers, in which the pressure is exerted against a relatively rigid membrane (Krista et al., 1981; Kuenzel et al., 1983).

Indirect. An apparatus that is commonly used clinically for measuring blood pressure is the sphygmomanometer, which consists of a compressing cuff, a manometer, and an air-inflating bulb. The cuff is applied to the upper arm (of man) and is inflated enough to obliterate the pulse. When the cuff is deflated slowly and when the pulse reappears, the reading of manometer pressure is taken. Indirect methods have been developed for a number of different animals, including the chicken. This method was first developed by Weiss and Sturkie (1951) and was later modified and im-

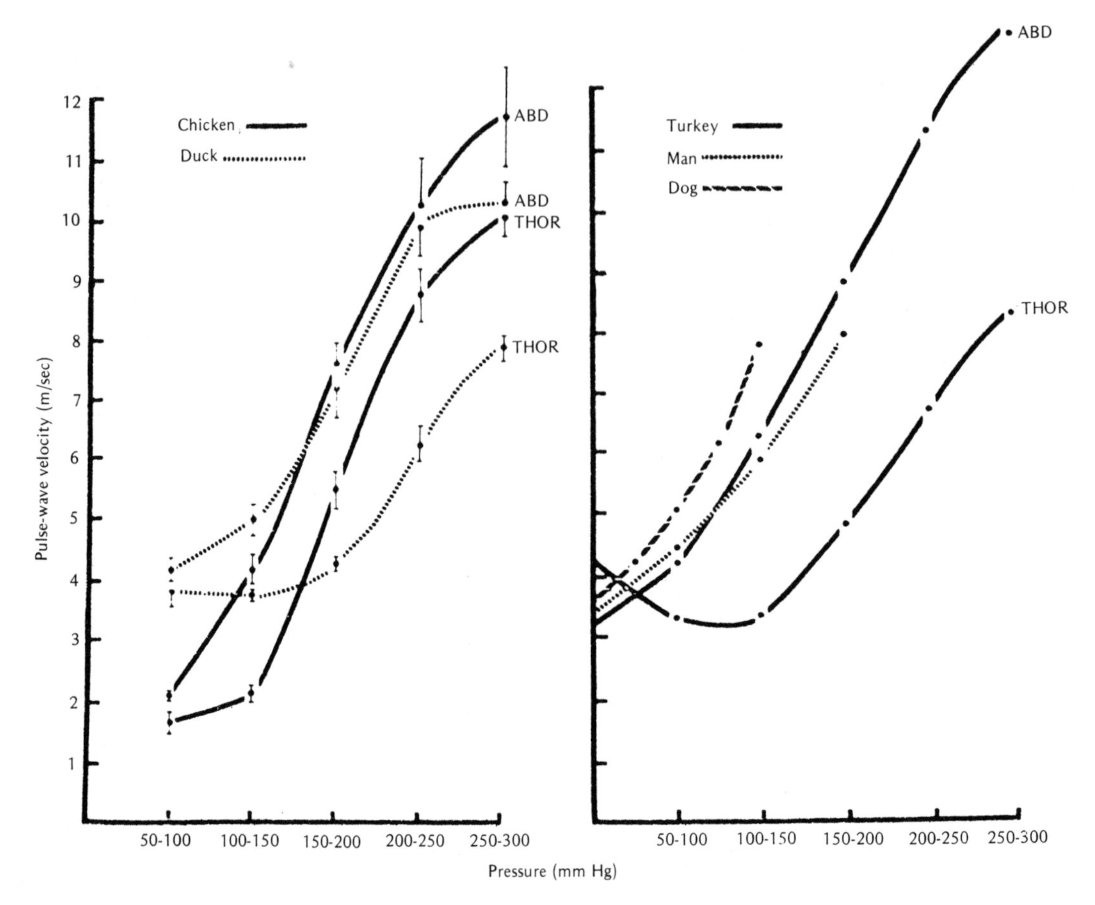

FIGURE 6–10. Calculated aortic wave velocities at various intraluminal pressures for the duck, chicken, turkey, dog, and man, in thoracic (*THOR*) and abdominal (*ABD*) segments for chicken and duck. Right curves for THOR and ABD for turkey and THOR only for dog and man. (After Ploucha and Ringer, 1981.)

proved by Sturkie et al. (1957). A cuff, 1 inch in diameter for adult chickens and made to fit the lower thigh, is inflated well above the point at which pulse disappears. It is then deflated, and the pressure at the point at which pulse reappears represents systolic pressure. The improved method involves the use of a capacitance pulse pickup, which is small and easily attached to the shank of the bird in an upright position, and a control box. The output from the latter is fed to the input of a suitable amplifier and recorded on a suitable oscillograph.

The appropriate cuff size is most important and must be determined for each age and species of bird, based on size of thigh. If the cuffs are too small or too large, the pressures recorded are too high or too low, respectively.

Changes in heart rate, output of the heart, and elasticity and resistance of the arteries all influence blood pressure. An increase in heart rate may increase pressure, provided the output of the heart is not decreased. It is possible to have a decreased cardiac output with an increased heart rate, resulting in no change in pressure.

Normal Values

Development of Blood Pressure in Embryos. Blood pressure can be recorded in chicks in the vitelline arteries as early as 46 hr of embryonic development and is very low. It then begins to rise and by the twentieth day measures 36/22 mm Hg, according to Van Mierop and Bertuch (1967; see also Paff et al., 1965).

Girard (1973), recorded pressures from the vitelline artery in 3- and 5-day-old embryos and from the chorioallantoic artery after these ages up to 20 days of age (Table 6–2). The pressures in embryos can be highly variable and are influenced greatly by temperature and hypoxia. Clark and Hu (1982) have recorded blood pressure, blood velocity, and heart rate in 3- to 5-day-old chick embryos (Table 6–2B).

Tazawa (1981a), employing the electromanometer and needle catheter in the allantoic artery and vein, measured blood pressure in chick embryos aged 13–17 days. Their arterial values were in the same range as those shown in Table 6–2; their heart rates were higher and constant from 13 to 16 days at a value of

TABLE 6–2 Blood pressures and heart rates in chick embryos and young hatched chicks[a,b]

Age (days)	Blood pressure (mm Hg) Systolic	Diastolic	Heart rate (beats/min)
From chorioallantoic artery			
3	0.61	0.43	138
5	1.19	0.83	191
From vitelline artery			
6	1.68	0.98	200
8	3.59	1.40	213
10	7.43	2.76	233
12	10.6	4.10	207
14	11.3	4.6	200
16	18.9	9.4	211
17	21.4	11.7	216
18	26.5	16.0	207
19	27.8	16.5	211
20	29.9	19.1	221
From carotid artery			
3 hr after hatching	43.7	23.6	108
3 hr + 1 day after hatching	60.2	37.2	156
1–2 days after hatching	61.6	35.3	116
3–4 weeks after hatching	114.6	72.9	345
5–6 weeks after hatching	130.9	95.3	376

[a] From Girard (1973).
[b] From Clark and Hu (1982):

Age (days)	Systolic	Diastolic	Heart rate (beats/min)	Aortic Blood Velocity (mm/sec)
3	0.82	0.52	165	2.63
4	1.29	0.81	183	6.11
5	1.46	0.94	188	11.32

Measurements from vitelline artery

about 250 b/min. *Venous* pressures, measured at 13–15 days, ranged from 2.5 to 3 torr.

Tazawa (1981b) also showed that hypoxia decreased embryo blood pressure and hyperoxia increased it. Hypoxia also increased heart rate.

Ventricular pressures have been measured, but such values are also highly variable depending on conditions. Paff et al. (1965) reported ventricular pressures in chick embryos as follows:

Age (days)	Systolic (mm H_2O)	Diastolic (mm H_2O)
3	29	10
5	63	17
6	74	5.45

Van Mierop and Bertuch (1967) reported ventricular pressures in 5-day-old embryos (chicks) and the diastolic pressures approached zero, indicating, according to them, the operation of an arterial valve mechanism at this stage.

Values in Adult Chickens. Values for normal arterial blood pressures of chickens are shown in Table 6–3. The values obtained with the more reliable direct methods indicate that the systolic pressure of the adult unanesthetized male is approximately 190 mm Hg and the diastolic pressure about 150 mm Hg, with a pulse pressure of 40 mm Hg. Systolic pressures for adult females range from about 140 to 160 mm Hg, with a pulse pressure of about 25 mm Hg. This important sex difference in blood pressure of chickens becomes evident at about 10–13 weeks of age (Ringer et al., 1957; Weiss et al., 1957).

The figures in Table 6–3 are based on birds restrained and in some instances anesthetized, both of which influence blood pressure.

Turkeys. The values indicate that the blood pressure level in turkeys is much higher than in chickens but it varies with species, sexes, and ages. Values obtained from use of the indirect technique are highly variable, depending on cuff size used (Weiss and Sheahan, 1958; Krista et al., 1963; and Weiss and Sturkie, 1951).

Earlier studies on blood pressures of turkeys were based on restrained birds, and this is known to increase blood pressure. Unrestrained turkeys allowed to move about, whose pressures and heart rates are determined by telemetry, are much lower (Krista et al., 1981) (see Table 6–4 and Figures 6–11 and 6–12).

TABLE 6–3. Blood pressure of chickens

Breed	Age		Sex	Blood pressure (mm Hg)			Method	Anesthetic	Artery	Reference
				Mean	Systolic	Diastolic				
—	Adult		Male	196	—	—	Hg manometer	None	Carotid	Stubel (1910)
	Adult		Female	164	—	—	Hg manometer	None	Carotid	Stubel (1910)
	Adult		Female	170	180	160	Membrane	None	Carotid	Stubel (1910)
Mixed	Adult		?	108	130	85	Hamilton manometer	Barbital, ether	Femoral or	Woodbury and Abreu (1944)
Mixed	6–10 weeks		?	128	125	120	Hamilton manometer	None	Ischiatic	Rodbard and Tolpin (1947)
White Leghorn	7 weeks		Male	—	151	128	Direct; strain gage	None	Carotid	Ringer et al. (1957)
	7 weeks		Capon	—	159	134	Direct; strain gage	None	Carotid	Ringer et al. (1957)
	7 weeks		Female	—	150	131	Direct; strain gage	None	Carotid	Ringer et al. (1957)
	7 weeks		Poulard	—	136	121	Direct; strain gage	None	Carotid	Ringer et al. (1957)
	13 weeks		Male	—	166	142	Direct; strain gage	None	Carotid	Ringer et al. (1957)
	13 weeks		Capon	—	157	135	Direct; strain gage	None	Carotid	Ringer et al. (1957)
	13 weeks		Female	—	156	131	Direct; strain gage	None	Carotid	Ringer et al. (1957)
	13 weeks		Poulard	—	162	135	Direct; strain gage	None	Carotid	Ringer et al. (1957)
	26 weeks	Adult	Male	—	191	154	Direct	None	Carotid	Ringer et al. (1957)
	26 weeks	Adult	Capon	—	180	149	Direct	None	Carotid	Ringer et al. (1957)
	26 weeks	Adult	Female	—	162	133	Direct	None	Carotid	Ringer et al. (1957)
	26 weeks	Adult	Poulard	—	189	152	Direct	None	Carotid	Ringer et al. (1957)
	5–7 months		Female		145		Indirect	None	1-in. cuff on femur	Hollands and Merritt (1973)
			Male		203					
	5–7 months	Cornell	Female		145		Indirect	None	1-in. cuff on femur	Sturkie (1970a)
		Random	Male		186		Indirect	None	1-in. cuff on femur	Sturkie (1970a)

TABLE 6–4. Blood pressure of birds other than the chicken[a]

Species	Age	Sex	Blood pressure: Mean	Blood pressure: Systolic	Blood pressure: Diastolic	Reference
Turkey, C		—	193	—	—	Stubel (1910)
Turkey, New Jersey Buff, C	6–7 weeks	Male	—	197	154	Weiss and Sheahan (1958)
	6–7 weeks	Female	—	190–	146	Weiss and Sheahan (1958)
	8–9 months	Male	—	226	152	Weiss and Sheahan (1958)
	8–9 months	Female	—	212	157	Weiss and Sheahan (1958)
Turkey, white, C	Hypertensive Adult	Male	165 aT	—	—	Krista et al. (1981)
	Hypotensive	Female	159 aT			Krista et al. (1981)
		Male	129 aT	—	—	Krista et al. (1981)
		Female	126 aT			Krista et al. (1981)
	6 weeks	Male	125	153	111	Simpson (1982)
		Female	127	153	113	Simpson (1982)
	14 weeks	Male	155	192	137	Simpson (1982)
		Female	148	177	133	Simpson (1982)
Turkey, bronze, C	8 weeks	Male		198	164	Ringer and Rood (1959)
	8 weeks	Female		189	158	Ringer and Rood (1959)
	12 weeks	Male		219	185	Ringer and Rood (1959)
	12 weeks	Female		226	185	Ringer and Rood (1959)
	16 weeks	Male		241	190	Ringer and Rood (1959)
	16 weeks	Female		248	194	Ringer and Rood (1959)
	22 weeks	Male		297	222	Ringer and Rood (1959)
	22 weeks	Female		257	200	Ringer and Rood (1959)
	18 months	Male		270	167	Ferguson et al. (1969)
	17–21 months	Female		223	170	Ferguson et al. (1969)
Turkey, Beltsville, C	21 months	Male		235	141	Ferguson et al. (1969)
Turkey, small white, C	17–18 months	Female		191	146	Ferguson et al. (1969)
Peking duck, C	4–5 months	Immature M		185	158	Ringer et al. (1955)
	4–5 months	Immature F		181	159	Ringer et al. (1955)
	12–13 months	Mature M		179	134	Ringer et al. (1955)
	12–13 months	Mature F		182	134	Ringer et al. (1955)
Peking duck, C	Adult	Male		179	142	Sturkie (1966)
	Adult	Female		168	134	Sturkie (1966)
White king, pigeon, C	Adult	Male		182	136	Ringer et al. (1955)
		Female		178	132	Ringer et al. (1955)
Pigeon, B	Adult	—		135	105	Woodbury and Hamilton (1937)

TABLE 6–4. (*Continued*)

Species	Age	Sex	Blood pressure			Reference
			Mean	Systolic	Diastolic	
Starling, C	Adult	—		180	130	Woodbury and Hamilton (1937)
Robin, V	Adult			118	80	Woodbury and Hamilton (1937)
Canary, C	Adult			130	—	Woodbury and Hamilton (1937)
Canary, V	Adult			220	154	Woodbury and Hamilton (1937)
Sparrow, C	Adult			180	140	Woodbury and Hamilton (1937)
Sparrow, V	Young			108	—	Woodbury and Hamilton (1937)
Coturnix (quail), C	Adult	Male		158	152	Ringer (1968)
	Adult	Female		156	147	Ringer (1968)
Bobwhite quail, C	Adult	Male		149	135	Ringer (1968)
	Adult	Female		145	129	Ringer (1968)

[a] Direct determinations were made from carotid artery (C), brachial artery (B), or from ventricle (V) with Hamilton manometer or strain gage. Local anesthesia was used.
aT, direct by telemetry.

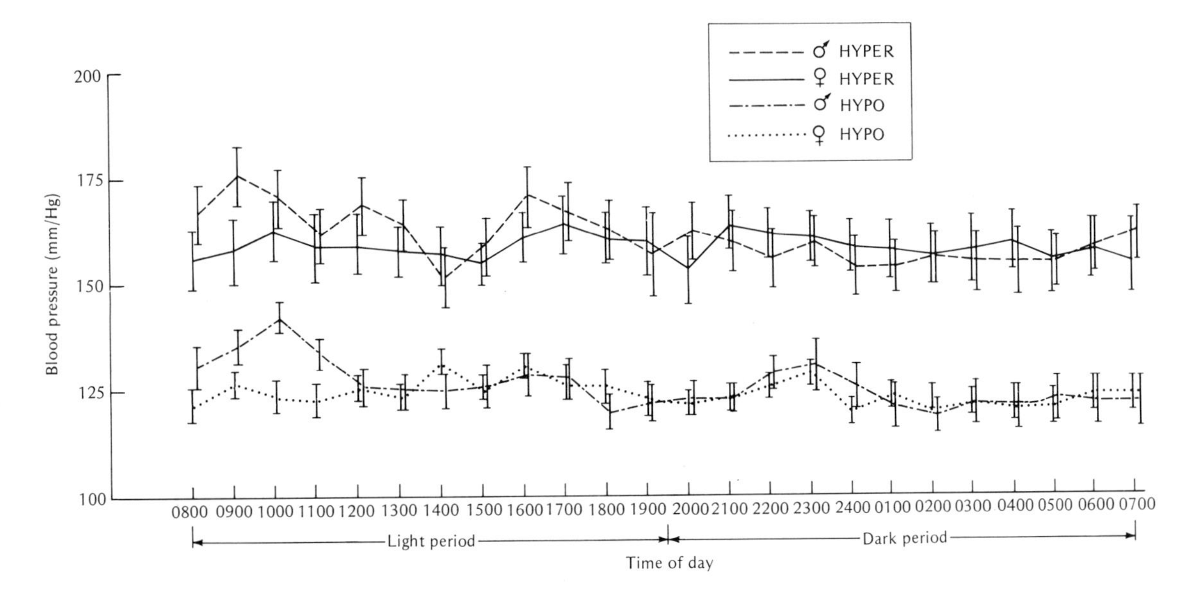

FIGURE 6–11. Mean blood pressure (mm Hg) with SEM for 24-hr period for male and female turkeys from hypertensive and hypotensive strains. (From Krista et al., 1981.)

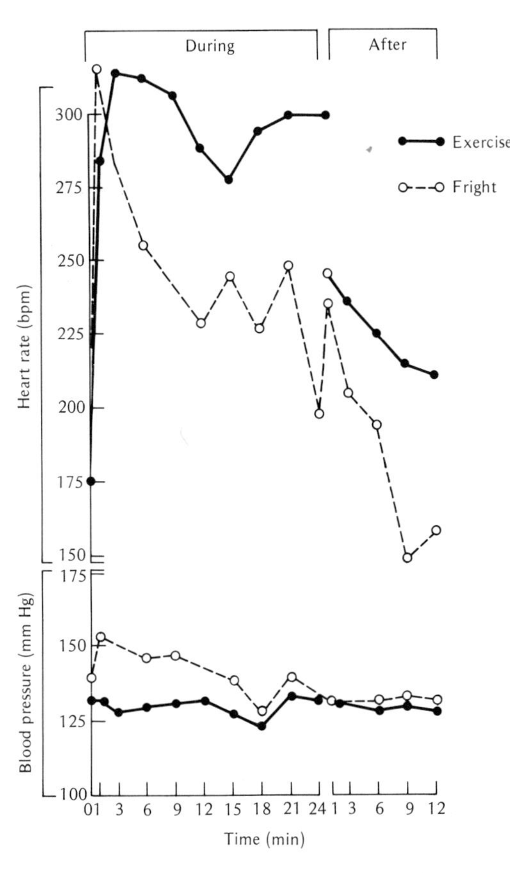

FIGURE 6–12. The effects of exercise and fright (sight of hawk) on the mean blood pressure (taken directly) and heart rate. (From Krista et al., 1981.)

Krista et al. (1981) also studied the effects of light and darkness, exercise, noise, fright, and mating behavior on heart rate and blood pressure of hypotensive and hypertensive strains of turkeys. Figure 6–11 shows the diurnal variation in blood pressure of these strains. The combined pressures of the two strains in light and darkness are shown in Table 6–5; the pressures for either strain is higher in the light period.

Heart rate was determined over an 18-hr period. The overall average was 161 beats/min in both sexes, and there was no difference between light and dark periods; however, there were significant sex and strain differences as follows (data are beats/min):

	Hypertensive strain	Hypotensive strain
Males	152	143
Females	186	163

Effects of Exercise and Fright. These are shown in Figure 6–12. Fright consisted of exposing birds to sight of a hawk. Both fright and exercise increased heart rate drastically, but had a lesser effect (fright) or no effect (exercise) on blood pressure (see later section for more details).

The birds were also subjected to various sounds, including a high-pitched dog whistle or a gun shot, which elicted only small responses in contrast to results by the same author (Krista) reported earlier.

Mating behavior of males increased both heart rate and blood pressure (see Table 6–6). Actually, the systolic blood pressure of the male reached 350 mm Hg during mating.

TABLE 6–5. Mean blood pressure (mm Hg) of hypertensive and hypotensive male and female turkeys during dark and light periods. Direct by telemetry.

	Hypertensive		Hypotensive		Combined strains	
Time[a]	Male	Female	Male	Female	Hyper	Hypo
Dark period	157 ± 1.96	158 ± 1.98	124 ± 1.29	123 ± 1.01	158[b] ± 1.40	123[b] ± .83
Light period	165 ± 1.92	159 ± 1.89	129 ± 1.30	126 ± .98	162[c] ± 1.36	128[c] ± .84

[a] Dark period: 2000–0700 hr; light period: 0800–1900 hr. From Krista et al. (1981).
[b,c] Mean ± SEM within a column with different letters differ significantly ($p < .05$).

Others. Blood pressures on other species, including ducks, pigeons, quails, robins, and sparrows, are given in Table 6–4. There is no significant sex difference in levels of most of these, unlike the turkey and chicken.

Effects of Age

Sturkie et al. (1953), using a limited number of birds, reported a significant increase in blood pressure in males and females from age 10–14 months to age 42–54 months. Muller and Carrol (1966) reported a decrease in blood pressure of females during an equivalent period of time but an increase in blood pressure of males. The reason for the discrepancies is not apparent, and further research is indicated.

There is no significant correlation between blood pressure level and plasma cholesterol, or between blood pressure and cholesterol in the thoracic aorta, but there is a significant correlation between systolic blood pressure and cholesterol level in the abdominal aorta (Weiss et al., 1957). A similar relation exists in the turkey (Speckman and Ringer, 1962). Cholesterol content of blood tends to increase with age.

Effects of Blood Pressure Level on Mortality

High blood pressure in chickens is not associated with higher mortality as it is in man, but low blood pressure may be (Sturkie et al., 1956). In more recent studies, Sturkie (1970a, b) reported no association between blood pressure level and mortality, and this was particularly true when total mortality was low.

Hollands and Merritt (1973) have shown higher mortality in some strains with lower than average blood pressure but no relationship in other strains. These results suggested that the differences in viability might be related to general level of mortality and that birds with higher than average blood pressure were better able to withstand greater disease exposure and physical stresses. This premise was tested (Sturkie and Textor, 1961) by exposing healthy birds with high and low blood pressures to physical exertion (walking) and low ambient temperature (hypothermia). The survival time to hypothermia was greater in hypertensive birds, and their ability to withstand physical exertion was likewise greater, but less so in turkeys (Bolden et al., 1983).

In turkeys, higher mortality is associated with hypertension, and an increased incidence of aortic ruptures or aneurysms (Ringer and Rood, 1959; Speckman and Ringer, 1962; Krista et al., 1970); there is little relationship between atherogenesis and blood pressure level in turkeys and chickens, however (Krista et al., 1978).

TABLE 6–6. The effects of presexual behavior (strutting) and mating (mounting) on blood pressure and heart rate of hypotensive and hypertensive male turkeys[a]

	Blood pressure change (%)		Heart rate change (%)	
	Hyper	Hypo	Hyper	Hypo
Strutting	5.5	9.2	6.3	7.6
Mating	33.0	42.0	89.0	122.0

[a] After Krista et al. (1981).

Heritability of Blood Pressure

Sturkie et al. reported in 1959 that heritability of blood pressure in chickens ranged from 25 to 28%, according to the method of calculation. In that study most of the change was in a hypertensive line, and no random line was available for comparison. In a later study Sturkie (1970a) selected for high and low blood pressure for seven generations and compared the changes to a random-bred line (Cornell). The results are shown graphically in Figure 6–13. The average realized heritabilities for all 7 years were 34.8 and 28% for the hypertensive and hypotensive males, respectively, compared to 45 and 33% for the females. Hollands and Merritt (1973), who selected chickens for blood pressure in three strains over a 5-year period, reported combined heritabilities of 26–27%. Krista et al. (1970) reported realized heritabilities in high- and low-blood pressure strains of turkeys of 34 and 23%, respectively. Levels of blood pressure of hypotensive and hyperten-

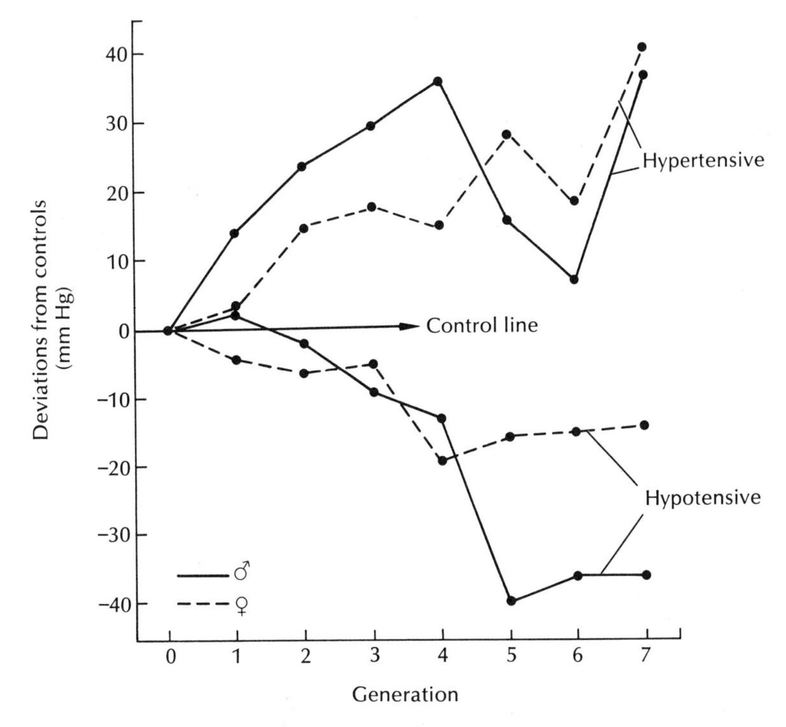

FIGURE 6–13. Effects of selection and breeding on blood pressure differences in hypertensive and hypotensive strains from a control line or strain. (After Sturkie, 1970a.)

sive strains of turkeys are presented in Tables 6–4, 6–5 and 6–6 and in Figure 6–11.

Respiratory Influence

When blood pressure is measured directly, waves are often seen in the trace, and the frequency of these waves is the same as that of the respiratory movements, although the phase relationship is variable. These waves have been attributed to changes in vasomotor tone, changes in cardiac output, or both. It is generally conceded that these pressure waves reflect changes in cardiac output, which is influenced by the gradient for venous return and by pulmonary vascular volume, which changes during lung inflation in mammals.

Inspiration in mammals with closed chests increases negative pressure in the thorax, venous return, and right heart output, and this is true also in birds that do not have a mammalian-type thorax and diaphragm (see Chapter 8). Aortic blood pressure, however, usually decreases on inspiration in the mammal (Manoach et al., 1971) and increases in birds. If both classes exhibit an increased right heart output on inspiration, differences in the pulmonary vascular volume, and therefore the subsequent return to and output of the left heart, may account for the difference in the change in blood pressure during respiration.

Based on work by Sturkie (1970b), it is concluded that there are minor changes in pulmonary vascular volume in the bird that have little influence on the cardiac output, and the latter is influenced most by changes in intrapulmonic pressure (IPP). Moreover, the outputs of the right and left hearts tend to be synchronous, although there is a phase lag of 58° between them, about half that observed in the rabbit. This can be demonstrated in the bird by unidirectional ventilation, whereby airflow rate and pressure can be kept constant or varied, when respiratory movements are minimized or eliminated by anesthesia or curare. In such instances there are no pressure wave variations. There are waves when positive-pressure breathing is induced by a reciprocating pump, but they are opposite in direction to those observed under normal breathing (Durfee and Sturkie, 1963) (Figure 6–14). This suggests an inverse relationship between intrapulmonic pressure and blood pressure and occurs even in animals with vagus and spinal cord transected. This relationship has been examined in detail by Magno and Sturkie (1969), who studied the effects of changes in intrapulmonic pressure, employing undirectional ventilation (with airflow constant but with changes in resistance) on blood pressure and right atrial pressure (RAP) and cardiac output by using electromagnetic flowmeters on the pulmonary artery and aorta. The lung of the bird does not collapse when the chest is opened; it is relatively fixed and moves mainly when the rib cage does.

Figure 6–15 shows that blood pressure and cardiac output are inversely related to IPP and that this relationship is linear. A similar relationship exists between IPP and RAP. Blood pressure waves in birds are caused mainly by changes in intrapulmonic and intrathoracic pressure.

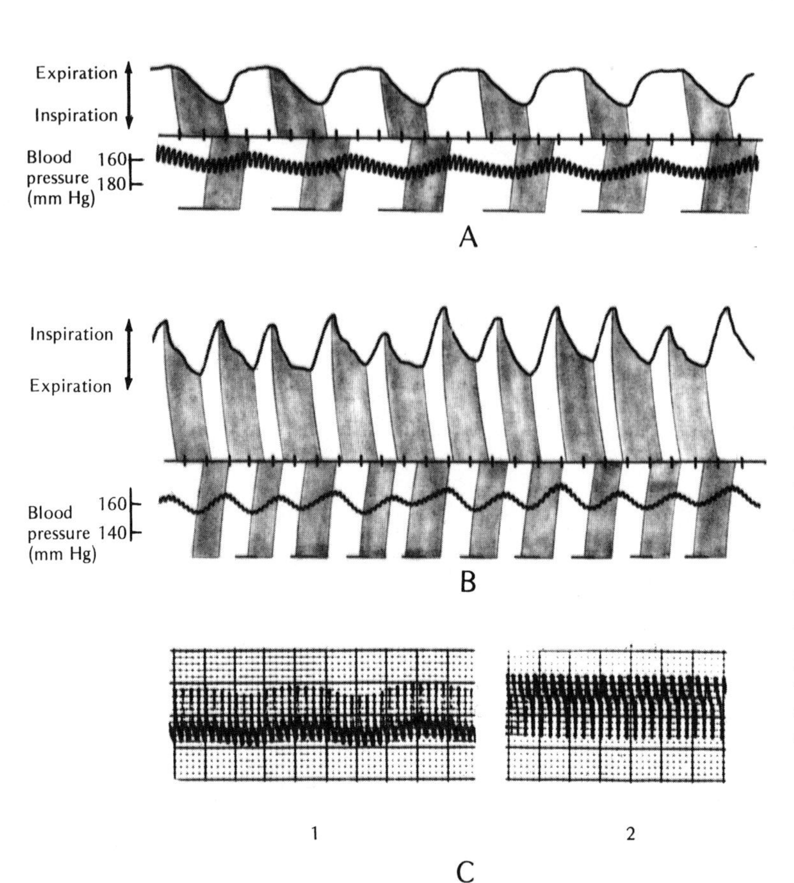

FIGURE 6–14. Relationship between rhythmic blood pressure changes and intrathoracic pressure changes. In **A**, there is natural respiration (slow) where peak blood pressure tends to be associated with low inspiratory pressure (negative) wave. Here there is some lag in the blood pressure wave. At higher respiratory rates, there is little or no lag. **B** shows changes occuring following artificial respiration with a reciprocating pump, where peak blood pressure coincides with peak inspiratory pressure wave (positive pressure). Note that blood pressure and respiratory pressure waves are in opposite direction to those in **A**. **C** shows blood pressure waves in natural respiration in 1. Artificial respiration by unidirectional air flow is shown in 2. Note absence of up-and-down changes in blood pressure wave. (After Durfee, 1963.)

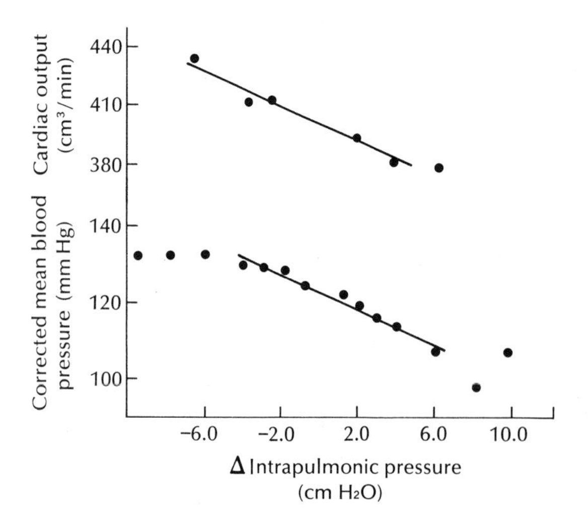

FIGURE 6–15. Effects of changes in intrapulmonic pressure on cardiac output and mean blood pressure. (After Sturkie, 1970b.)

Effects of Hypoxia

Durfee and Sturkie (1963) (see also Sturkie, 1970b), employing unidirectional artificial respiration, produced hypoxia and a slight degree of hypercapnia in chickens by decreasing the rate of ventilation from 500–350 ml/min to 230–150 ml/min for 1–8 min (slowly induced asphyxia). The average effects on systolic blood (SB) pressure and heart rate (HR) follow:

	Before asphyxia	For 1 min	After asphyxia 2	After asphyxia 8
SB (mm Hg)	191	184	178	151
HR (beats/min)	221	228	236	287

The resulting hypotension, which was maintained after vagotomy or spinal cord sectioning, apparently came from the direct effects of low oxygen on the blood vessels, and chemoreceptors were not involved.

The effects were reversible by hyperventilation. The increase in heart rate was reflexly induced. These results have been confirmed by Richards and Sykes (1967) and Ray and Fedde (1969). The increased heart rate could be abolished by vagotomy and beta blockade (Butler, 1967).

Blood pressures in ducks and pigeons exposed to hypoxia, $P_aO_2 < 50$ mm Hg, are only slightly reduced, and heart rate increases only after P_aO_2 is reduced to 35 mm Hg or less (Butler, 1970).

Right ventricular hypertrophy has been observed in pigeons, chickens, and ducks exposed to hypoxia and pulmonary arterial hypertension. Acute hypoxia resulting from exposure to low oxygen concentration (13%) decreased femoral artery pressure and increased pulmonary artery pressure in chickens (Besch and Kadono, 1977); see Figure 6–16. Heart rate increased reflexly but the increase did not persist, and there was a small but insignificant change in cardiac output.

Chickens raised for 7 weeks at 3300 m (hypoxia) developed pulmonary hypertension, compared to controls raised at sea level (Sillau et al., 1980). The levels of pulmonary pressure were as follows (in torr):

	Hypoxic	Normoxic
Systolic	50.7	20.4
Diastolic	40.1	12.5

Right ventricular heart weights were increased 90% over the control weights.

There is good evidence that the general hypotension resulting from hypoxia in chickens is caused by the lack of oxygen directly on the blood vessels, but what vascular beds are involved is not known. A few studies have been made on the effects of hypoxia on certain vascular beds, such as isolated mesenteric artery preparations (Gooden, 1980) and isolated vessels of hind limbs in the chicken (Ploucha et al., 1981). Other studies on the bloodflow in isolated systems are discussed in the section on bloodflow.

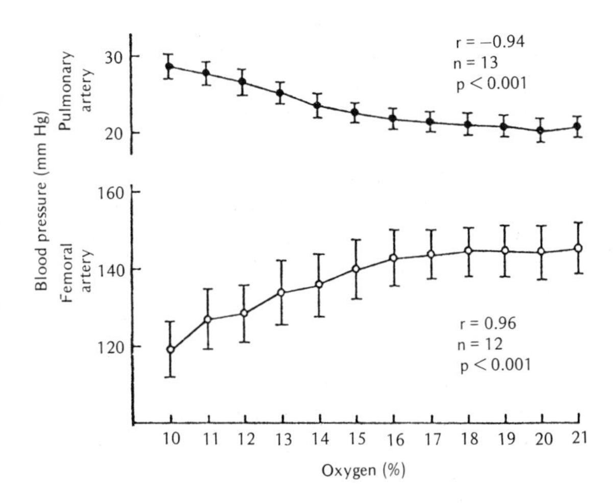

FIGURE 6–16. Relationship between ambient O_2 concentrations and pulmonary and femoral artery blood pressure (systolic). (After Besch and Kadono, 1977.)

Periarterial nerve stimulation of mesenteric arteries (in intact animals) plus hypoxia alone in the isolated mesenteric arterial segments with extravascular normoxia (Gooden, 1980), caused only a slight reduction in BP in the chicken but a great reduction in the duck (BP, 26% of normal). Under conditions of both intra- and extravascular hypoxia, periarterial stimulation reduced the pressure in both the duck and chicken preparations to about the same extent (38 and 35 torr, respectively). Norepinephrine administration gave the same response as nerve stimulation.

Gooden concluded that the duck blood vessels (mesenteric arteries) are less resistant to periarterial stimulation during hypoxia, and attributes this difference to a difference in diffusion distance for O_2 in the blood vessel that is greater in the duck because the adventitia of vessels is threefold thicker than in chickens. Therefore, a lack of O_2 would have a more drastic effect in the duck. Further studies, and studies on other blood vessels, are desired. Blood vessels in skeletal muscle of chicken (isolated preparation) constricted when exposed to extreme hypoxia (Ploucha et al., 1981). The intact duck is more resistant to hypoxia. It exhibited only a slight hypotension when exposed to a P_aO_2 less than 50 mm Hg and did not exhibit reflex tachycardia until P_aO_2 reached 35 torr (Butler, 1970). When P_aCO_2 is maintained at control levels in ducks, the heart rate responses are variable and do not exceed 14% even when the P_aO_2 is 37–45 mm Hg (Jones and Purves, 1970).

Hypercapnia and Hypoxia. Breathing high levels of CO_2 mixed with low levels of O_2 caused a reduction in blood pressure of chickens, according to Richards and Sykes (1967), but hypercapnia alone had little effect. Eiel and Sturkie (unpublished) administered mixtures of 20% O_2 and 5% CO_2, which had little effect, but when air containing 10% CO_2 was breathed, blood pressure was depressed significantly.

Hypoxia resulting from diving will be discussed later.

Effects of Drugs and Hormones

Effects of Anesthesia. Anesthesia may lower blood pressure in birds, as in mammals, depending on the type, the dose, and the time after administration. Sodium pentobarbital (25–30 mg/kg) injected intravenously into adult chickens depresses blood pressure 30 mm Hg within a few minutes after injection (Weiss and Sturkie, 1951). Harvey et al. (1954) induced anesthesia in chickens with urethane (1.4 g/kg body weight), sodium barbital (180 mg/kg), and sodium phenobarbital (200 mg/kg, in each instance administered intraperitoneally. The authors do not report blood pressure levels on unanesthetized birds, but it is apparent from their data that blood pressure levels were depressed, because the basal levels reported by them were quite low.

Dial, a commercial preparation containing urethane, has been used extensively by Durfee (1963) in dosages ranging from 0.50 to 0.75 ml/kg. At the lower level the birds were lightly anesthetized, but this dosage depressed blood pressure significantly.

Catecholamines. Most of the drugs and hormones that are pressor and depressor in mammals have the same effect in birds. Some of the pressor substances are epinephrine, benzedrine, ephedrine, and phenylephrine (Thompson and Coon, 1948). A detailed study of the effects of a number of drugs and adrenergic blockers on the blood pressure of chickens was made by Harvey et al. (1954). Isopropyl norepinephrine had a depressor effect, and considerable variation occurred in the responses to epinephrine and norepinephrine. The minimal pressor doses of both drugs vary from 0.02 to 1.5 μg/kg and the maximal doses range from 4 to 10 μg/kg. In most hens the two sympathomimetics were about equipotent, although occasionally norepinephrine was four times as potent (as a pressor) as epinephrine; however, the latter exhibited greater cardioaccelerator activity. Epinephrine and norepinephrine (1 μg/kg) increased diastolic pressures 39 and 20 mm Hg, respectively, in anesthetized chickens (Akers and Peiss, 1963). A number of ergot alkaloids (such as ergotamine and ergonovine) were tested, and most of them were consistently pressor at dosages lower than 200 μg/kg.

Sturkie (unpublished) has shown that norepinephrine (as free base) is effective in increasing blood pressure at dosages of 2–3 μg/kg. Similar results have been reported by Natoff and Lockett (1957). El-Halawani et al. (1973) studied the catecholamine levels in tissues of hypertensive and hypotensive turkeys and found that hypertensive birds had higher levels in the left ventricle and pulmonary aorta than hypotensive ones. They suggested that norepinephrine release was greater in the hypertensive birds, and there was less destruction of catecholamines because of lower monamine oxidase in the tissues.

In chickens, levels of catecholamines in blood or tissues do not appear to be related to blood pressure level. For example, female chickens have higher levels of catecholamines in blood and tissues but a significantly lower blood pressure (Sturkie and Lin, 1968, and unpublished data).

Adrenergic blockers were discussed by Harvey et al. (1954) These include the alpha and beta blockers. Beta blockers have been used in Aves, mainly for blocking heart rate, and alphas for blood pressure. Propranolol is a commonly used beta blocker, but recently a new irreversible blocker (BAMM) has been employed on turkeys; this lowers both heart rate and blood pressure (Kuenzel et al., 1983).

Ploucha et al. (1981) have studied the effects of norepinephrine and other constrictors and dilators on blood pressure in the hind limb preparation of chickens. In general, the effects were the same as in the intact animal.

Serotonin. (5-hydroxytryptamine) can be pressor or depressor in the chicken. It is normally a depressor, according to Bunag and Walaszek (1962), for they believe it releases histamine, which is depressor. When the tissues of chickens were depleted of histamine by continued administration of 48/80, serotinin and tryptamine were pressor.

Results from the author's laboratory also demonstrated that serotonin is mainly a depressor in the chicken. Dosages at levels of 0.5 to 10 μg/kg were effective in depressing blood pressure significantly; the maximum effect (20 mm Hg) was produced within 10 sec, and blood pressure level returned to normal within 20–30 sec (Woods, 1971). To determine whether serotonin released histamine, which might have depressed blood pressure, blood samples were taken at the same time as the peak drop in blood pressure occurred and analyzed for histamine with a highly sensitive fluorescent chemical method. There was no evidence of histamine release. The mechanism of the depression effect of serotonin is unexplained; moreover, when the substance is infused into the mesenteric artery of the isolated chicken intestine, it has a pressor effect in both mesenteric arteries and veins (Gross, 1974, and Gross and Sturkie, 1975).

Histamine. The depressing effect of histamine has been reported by Natoff and Lockett (1957), Bunag and Walaszek (1962) and El-Ackad (1972). A dose of histamine acid phosphate of 3.23 μg/kg was required to depress blood pressure 20 mm Hg (Natoff and Lockett, 1957). Histamine (free base) at 5 μg/kg dropped blood pressure (mean) approximately 40 mm Hg within 15 sec, and the level returned to approximately normal within 1–2 min (El-Ackad, 1972). When the blocking agent (trypelennamine) was administered along with the histamine, the latter depressor effect was blocked (El-Ackad, 1972; Meyer et al., unpublished). Similar results were obtained with Mepyramine an histamine antagonist (El-Ackad, 1972; Chand and Eyre, 1975).

When histamine was infused into an isolated intestinal preparation (chicken), an increased resistance or blood pressure was observed, and the effect was attributable to the veins (Schaible, 1977). The effect was blocked by Mepyramine.

Acetylcholine. Acetylcholine, methacholine, and other parasympathomimetic agents depress blood pressure in birds as they do in mammals. An injection of 0.2 mg of acetylcholine into chicks 6–12 weeks old depresses blood pressure 40–60 mm Hg within 3 sec (Rodbard and Fink, 1948). Intramuscular injections of 5 mg into adult fowls similarly decreases blood pressure and heart rate, but the latter returns to normal shortly thereafter. Methacholine, which has an effect of longer duration than does acetylcholine, produces significant depressor effects at levels as low as 0.15 mg/kg subcutaneously (Durfee, unpublished) and 2 μg/kg intravenously (Sturkie, unpublished).

Sex Hormones. The existence of a sex difference in the blood pressure of chickens suggests the influence of sex hormones. It was shown (Sturkie and Ringer, 1955) that the higher blood pressure level of adult males and capons could be depressed by administration of estrogen. Later it was shown that the blood pressure level of the castrate female was about the same as the adult male, and that this level could likewise be depressed to near the level of a normal female by estrogen. The mechanism of estrogen's action is not known, but it may have a direct vasodilating effect on the blood vessels by increasing arterial O_2 consumption and it may also exert some effect on the medullary centers (Malinow and Moguilevsky, 1961).

Reserpine. Reserpine, when injected intramuscularly in doses of 0.10–0.75 mg/kg, decreased blood pressure in the chicken significantly within 4 hr after the injection (Sturkie et al., 1958). Similar depressing effects have been reported in turkeys by Speckmann and Ringer (1962) and many others since (see Krista et al., 1963). Simpson et al. (1965) reported a slight but significant decrease in systolic blood pressure of turkeys fed reserpine, but reserpine did not reduce the incidence of aortic ruptures.

Oxytocin and Vasopressin. Synthetic oxytocin and the oxytocic fraction of posterior pituitary preparations depress blood pressure in the chicken within a few seconds (Chen, 1970). The drop is attributable not to changes in cardiac weakness or in heart rate but to vasodilation. Lloyd and Pickford (1961) found oxytocin always to be a depressor in the chicken but not always in mammals.

The response of the chicken to vasotocin and vasopressin is variable. In some instances a pure pressor response is obtained; in others a transient depressor response occurs, which may or may not be followed by a pressor response. Ames and Skadhauge (1971) reported that arginine vasotocin (AVT), when administered in doses of less than 50 ng/kg had no effect on blood pressure, but that doses of 150–500 ng/kg decreased blood pressure for less than 5 min. Braun and Dantzler (1974) injected AVT into Gambel quails and observed three types of reactions: (1) an increase in blood pressure, (2) a decrease in blood pressure, and (3) a biphasic response. The increase was about 20 mm/Hg.

Angiotensin. Angiotensin II has both pressor and depressor effects in conscious chickens, and only a pressor effect in anesthetized ones (Nishimura et al., 1982; Nakamura et al., 1982). Normally the first response is depressor, followed by a pressor response (see also Moore, 1980). There is good evidence that the pressor effect is caused by released catecholamines and that the depressor effect is mediated by angiotensin receptors in vascular smooth muscle. The pressor response could be abolished by preventing the release of catecholamines from nerves and adrenals by administration of 6-hydroxydopamine and adrenalectomy in ducks (Wilson and Butler, 1983a, b, c). Angiotensin II is known to release catecholamines in chickens, turtles, frogs, and fish (Carroll and Opdyke, 1982).

The chemical structures of Ag I and II are known. The sequence of amino acids in the polypeptide chain is

Ag I: Asp-Arg-Val-Tyr-Val-His-Pro-Phe-Ser-Leu
Ag II: Asp-Arg-Val-Tyr-Val-His-Pro-Phe --------

The first and eighth acids (aspartine and phenylalanine), respectively, are important sites for binding and action of receptors. The depressor action of Ag II can be blocked by substituting sarcosine and isoleucine at positions 1 and 8 of the molecule, but it was not inhibited by atropine, prostaglandin synthetase inhibitor, methysergide (serotonin inhibitor), or propranolol. Whether or not Ag II caused the release of histamine was not determined (Nakamura et al., 1982).

Ag I is converted to Ag II, which has the main effect on blood pressure. Inhibition of the converting enzyme by Captopril (Sq 14225) in turkeys (Simpson, 1982) prevented the rise in blood pressure normally produced by administration of Ag I. The same results were obtained in chickens (Nishimura et al., 1982).

Prostaglandins. Little work has been conducted on the effects of these on the cardiovascular system of birds. Horton (1971), who reviewed the subject, reported that Pr E (a type of prostaglandin) is an arterial vasodilator, but the pressor action of F_2 (another type) is attributable to venoconstriction and increased venous return and cardiac output as in the dog and rat. In the intact anesthetized chicken, the effect of prostaglandin, $F_{2\alpha}$ was depression. Pr $F_{1\alpha}$ and $F_{2\alpha}$ were depressors in chicks anesthetized with ether, chloralose, urethane, or barbitone.

Pressor Substances Released by Intracranial Compression. Acute compression of the brain causes an increase in both mammals' and birds' blood pressure (Rodbard et al., 1954). The rise in pressure could be blocked or prevented by administration of adrenergic blockers but not by tetraethylammonium, a ganglionic blocker, suggesting the release of a humoral substance akin to norepinephrine.

Later work by Ploucha et al. (1982) (see Chapter 5) indicated that severe hemorrhage in chickens results in cerebral ischemia, which causes a release of circulating catecholamines (which was demonstrated) that is responsible for the pressor response.

Oviposition and Blood Pressure. Immediately preceding contraction of the uterus and expulsion of the egg, there is a release of vasotocin from the posterior pituitary that is associated with oviposition (see Chapter 17). Because this hormone has a depressor effect, blood pressure might be expected to drop just prior to oviposition. Hunsaker and Sturkie (see Sturkie, 1965) recorded blood pressure of hens before, during, and after oviposition. Blood pressure did not change until after the egg was laid; the slight drop was transient and was not believed to be attributable to oxytocin or vasotocin.

Salt hypertension has been produced in chickens by administering sodium chloride in the drinking water at a level of about 1% (Lenel et al., 1948).

Effects of Ambient and Body Temperature

Sturkie and coworkers observed seasonal changes in blood pressure in chickens, with the pressures tending to drop with the advent of warm weather, and it was proven conclusively that ambient temperature and not the amount of light was responsible for the changes observed. Chickens acclimated to high temperature had a significantly lower blood pressure than those exposed to cold (Sturkie, 1967, and earlier).

Chickens exposed to acute heating (40.0°C) developed hypotention and an increase in cardiac output (Sturkie, 1967; Whittow et al., 1964). Hypothermia likewise depresses blood pressure in the chicken (Whittow et al., 1965a). Rewarming the bird after hypothermia produced an immediate rise in blood pressure and body temperature. Further heating and hyperthermia caused a decrease in blood pressure.

Heart Rate and Blood Pressure

There is little correlation between blood pressure and heart rate in chickens and turkeys. Male chickens and turkeys have higher blood pressures and lower heart rates than do females.

Ventricular and Pulmonary Pressures

Changes in pressure in the heart chambers are discussed in Chapter 7. Bredick (1960) measured systolic and diastolic pressures in the ventricules of chickens as follows: (peak pressures in mm Hg)

Left ventricle	145 systolic 0 diastolic
Right ventricle	27 systolic −0.3 to −2.0 diastolic

Respiratory movements caused fluctations averaging 8.8 mm Hg for the left ventricle and 3.6 for the right ventricle. These figures are of the same magnitude as those reported for the pulmonary artery by Rodbard and Katz (1949).

Right and left ventricular pressures for the duck are of the same magnitude as for the chicken (Jones and Langille; see Jones and Johansen, 1972). Pressure in the pulmonary arteries of chicken and duck are as follows (mm Hg):

	Duck (Jones and Johansen)	Chicken (Rodbard and Katz)
Systolic	25	24
Diastolic	10	12

Pressures reported by Burton et al. (1968) were lower (systolic, 21.5; diastolic 9.8). Pressure in the turkey pulmonary artery averaged 20 mm Hg (Hamlin and Kondrich, 1969).

Injection of epinephrine had no appreciable effect on pulmonary arterial pressure, even though it increased systemic arterial pressure 40–80 mm Hg. The pressures in the pulmonary arteries of most vertebrates are of similar magnitude.

Left ventricular peak systolic pressure in ducks ranged from 160 mm Hg to near 200 mm Hg (Folkow and Yonce, 1967; Jones and Johansen 1972). These pressures vary with venous return and contractility, which are discussed in Chapter 7. Right atrial pressure in chickens ranges from slightly negative on inspiration to slightly positive on expiration (Sturkie, unpublished).

Control of Blood Pressure and Heart Rate

Neural. Very little is known concerning central control of blood pressure and heart rate in Aves, although there is presumptive evidence of medullary centers as in mammals.

There is some evidence that the thalamus influences blood pressure level; it comes from the (crude, by modern standards) experiments by Dijk (1932), who reported that stimulation of this area produced changes in blood pressure.

Stimulation of an area in the mesencephalon of ducks near the midline produced a significant increase in blood pressure and a decrease in heart rate, whereas stimulation in the diencephalon produced vasodilation in skeletal muscle (Feigl and Folkow, 1963). Because the latter response could be blocked by atropine, it was attributed to cholinergic fibers.

It appears that the center affecting blood pressure is in the medulla near the center affecting heart rate (see also Chapter 7). Stimulation of the dorsal medulla produced tachycardia and hypertension (Cohen and Schnall, 1970). The need for more work on the location and stimulation of these centers is evident. Johansson (1983) has reviewed some of the recent literature on central control of blood pressure in Aves.

Vasoconstrictors. Most blood vessels of Aves receive a dual innervation (Akester, 1971; Bolton, 1968). The effects of drugs, hypoxia, and other factors that increase sympathetic discharge and produce vasoconstriction have been discussed, but few studies have been conducted on the effects of directly stimulating such nerves on blood pressure.

The longitudinal smooth muscles of the anterior mesenteric artery of the chicken are innervated by excitatory cholinergic and inhibitory adrenergic nerves, and the circular muscles by adrenergic vasoconstrictor nerves but no cholinergic nerves (Bell, 1969).

Gooden (1980) studied the vasoconstrictor responses of isolated mesenteric arteries of chickens and ducks to periarterial nerve stimulation and administration of norepinephrine. The mean pressure response to nerve stimulation in the duck mesentery was about threefold that in the chicken. The response to norepinephrine was almost twice that of the chicken. The wall thickness of the duck arteries was much greater than that of the chicken, and the nervous innervation (density) was about twice as great in the duck as in the chicken. The mesenteric vasculature of the duck is well adapted to the powerful vasoconstriction required during diving (see later section).

Based on nerve stimulation and the use of appropriate drugs, it was shown that the innervation of the blood vessels of the feet of ducks and chickens consisted of adrenergic vasoconstrictors and a vasodilator whose transmitter was unknown; it was not acetylcholine, a catecholamine, or histamine (McGregor, 1979). Results of Bell and Rome (1984) suggest that the vasodilator fibers of the duck foot may be dopaminergic.

The *Rete tibiotarsale* of the duck foot is a functional heat exchanger consisting of a shunt artery and vein and the rete artery and vein. Bloodflow may be shunted away from the rete artery following stimulation of nerves, or administration of adrenergic agents. The vasoconstrictor effect of nerve stimulation or of norepinephrine was much greater in the shunt artery than in the rete arteries, and the response could be blocked with alpha blockers. The shunt veins and rete veins were unresponsive to nerve stimulation or adrenergic agents (Midgard and Bech, 1981).

Reflex Changes in Blood Pressure and Heart Rate

The *carotid sinus* of birds is not situated high in the neck where the external and internal carotids diverge, as in mammals (Adams, 1958). There is no comparable area in birds. The location of the carotid body, which contain chemoreceptors, is shown in Figure 6–6 (see also Dreyer et al., 1977). The dimensions of the body are 0.8 × 0.5 mm in the chicken, and smaller in other species. The carotid body receives its blood supply from the carotid artery or its branch, and is drained by several veins. The innervation of the body is from the vagus nerve by way of the ganglion nodosum. Apparently the glossopharyngeal nerve does not contribute fibers to the carotid body, at least not in the chicken, although Terni (see Adams, 1958) asserted that the body may receive fibers from the recurrent laryngeal nerve.

Attempts to elicit reflex effects on blood pressure by means of occlusion or stimulation in the region high in the neck that corresponds to the sinus area of mammals have been unsuccessful (Adams, 1958; Heymans and Neil, 1958). Stimulation of the carotid body or sinus area gave variable and unconvincing results. Durfee (1963) was unable to elicit reflex responses by occluding either the common carotid in the sinus region of the chicken or the brachiocephalic artery. Reducing blood pressure significantly in one carotid artery had no reflex effect (McGinnis and Ringer, 1966). Moreover, that there is no evidence for cephalic receptors, as has been postulated, has been demonstrated by these authors.

King and King (1978), using the electron microscope, revealed what they believed were aortic baroreceptor endings of the vagus nerve of chickens located in the heart area; this is similar to reports for the duck by Jones (see later section).

Magno (1973), who isolated and perfused the carotid body of chickens with sodium cyanide, reported no effect on blood pressure or heart rate; however, this produced a respiratory response, indicating respiratory receptors but no cardiovascular ones.

Further morphological evidence for the presence of baroreceptors in the wall of the common carotid artery of the chicken was reported by Abdel-Magied (1978), but stimulation in this area was ineffective.

Although the location of baroreceptors in chickens has not been proven, there is ample evidence for their operation. Durfee (1963) showed that hypoxia and methacholine or acetylcholine, which depress blood pressure, caused a reflex rise in heart rate that could be blocked by atropine or bilateral vagotomy; moreover, norepinephrine-induced increases in blood pressure evoked a reflex decrease in heart rate that could also be blocked by bilateral vagotomy. These results suggested that reflex changes in heart rate (increase or decrease) are mediated via the vagus and that the cardioaccelerator nerve is not involved in such reflexes.

Serotonin, which likewise has a depressor effect in chickens (Woods, 1971), produces a reflex increase in heart rate that can be blocked with atropine but not with propranolol (β blockade). Further evidence from Tummons (1970) indicated that the reflex changes in heart rate induced by hypotensive or hypertensive agents are mediated by the vagus nerve and not by the cardioaccelerator nerve. The effects of methacholine and norepinephrine on blood pressure (BP) and heart rate (HR) before and after cardioaccelerator nerve (CA) transection are shown as follows:

Treatment	Before denervation		After denervation	
	BP	HR	BP	HR
Control (no drug)	204	291	181	242
Methacholine	113	344	109	276
Norepinephrine	275	220	231	177

The reflex changes in heart rate after CA nerve transection are significant, but are apparently not as great as before transection (Tummons and Sturkie, 1969).

The location of *baroreceptors* in the duck has been established by Jones (1973). The evidence is based on sectioning of the depressor nerve (vagus) arising from the ganglion nodosum and terminating in the root of the aorta just before the branching of the brachiocephalic arteries. This operation abolished the changes in heart rate induced by hypotensive and hypertensive agents. Moreover, changes in blood pressure evoked electrical discharge of the depressor nerve.

Presumptive evidence for baroreceptors in ducks and pigeons was deduced from experiments involving clamping of the carotid arteries following hemorrhage, which resulted in a more rapid return of normal blood pressure level (Kovach et al.; see Chap. 5). This treatment was ineffective in chickens. Reflex vasoconstriction occurs in ducks and pigeons following hemorrhage and in diving of ducks, but not in chickens (Ploucha et al., 1981, 1982).

Role of Serotonin and Histamine in Cardiovascular Regulation

It was shown in Chapter 5 that certain cells of blood, mainly those in the buffy coat fraction (thrombocytes and leukocytes) have high concentrations of serotonin (5-HT) and histamine. 5-HT and histamine are released by chromaffin cells and mast cells of the intestines, picked up and bound by blood cells, and circulated throughout the body. Also, it is known that the arteries, heart, and veins contain appreciable amounts of histamine, and the amount varies, depending on the size and location of the vessels (Schaible, 1977). Moreover, it has been demonstrated that catecholamines will cause the release of 5-HT and histamine from the chicken intestine (Gross and Sturkie, 1975), and that 5-HT will cause the release of histamine (Schaible). Likewise, histamine is known to cause the release of catecholamines of certain organs in mammals and probably in birds.

Little is known about the uptake and release of 5-HT and histamine by the blood cells, but it is possible that certain conditions of circulation, such as stasis of flow, low bloodflow, deformation of cells, and others might cause the release of these agents, which could have a significant effect on circulation, particularly in local areas.

Bloodflow and Cardiac Output

Cardiac Output

Cardiac output is the amount of blood ejected from the ventricles in a given time, usually referred to as minute volume. The amount ejected from the right and left ventricles is the same over a period of time, but may vary with each heart beat, and is influenced by heart rate and stroke volume. The latter is influenced by venous return, and diastolic volume, stretching of heart muscle (Frank–Starling mechanism), and increased force of contraction without change in heart muscle fiber length (see Chapter 7).

Methods of Determining. Cardiac output has been determined by a number of methods, including the direct Fick and various dilution methods (indirect Fick) involving the use of dyes, radioactive substances, and microspheres. Cardiac output may also be determined directly by placing electromagnetic flowmeters (probes) on the pulmonary arteries or ascending aorta.

With the dye-dilution method, a known quantity of dye is injected into a vein or the right heart, where it is mixed; blood from an artery is then sampled. The concentration of dye in the blood passing the sampling point changes with time and forms a concentration curve. By calculating the area under the curve or average concentration (C) and knowing the concentration of dye injected, cardiac output (CO) can be determined.

$$\mathrm{CO} = \frac{A \text{ (amount of dye injected, mg)}}{\mathrm{C} \cdot \text{time under curve}}$$

or

$$\frac{A}{\int_0^\infty \mathrm{C}dt} = \frac{A}{\Sigma \mathrm{C}\Delta t}$$

The principle of the *microsphere method* is that those radioactive particles of a given size (about 15 μm in diameter) are not recirculated in the vascular system. The method is used extensively to determine the distribution or regional blood flow to the various organs that trap the particles; their concentration or dilution within an organ is used to calculate the percentage of total microspheres introduced into the left ventricle or root of aorta that are trapped by a given organ. Microspheres may recirculate if they evade the capillary bed by passing through arteriovenous anastomosis (AVA) larger than the microspheres, which are later trapped in the lungs. The radioactivity measured in the liver, kidneys, and lungs following injection of 15 μm microspheres may result, in part, from the particles passing through the AVA in the lower organs of the body (Wolfenson et al., 1981).

Odlind (1978) has made detailed studies of the distribution of microspheres in the renal portal system of chickens.

Factors Affecting Cardiac Output

Normal Values in Birds. Normal values of chickens, turkeys, ducks, and pigeons are presented in Table 6–7. The values vary considerably, even within the same species, depending on method, anesthesia, and various other factors. Values in chickens obtained by isotope dilution (Sapirstein and Hartman, 1959) appear to be higher than those obtained with dye dilution.

Values in chickens determined by radioactive microspheres are in the same range as those based on dye dilution (Table 6–7). Birds have a greater CO per given body mass than mammals, according to Grubb (1983), who has studied the allometric relations of cardiovascular function of birds.

The values on turkeys and ducks are highly variable also. Cardiac output changes drastically with demands for blood flow, as illustrated in diving ducks (see later section on cardiovascular effects of diving). There appear to be sex differences in cardiac output, as well as blood pressure, of chickens, with the higher levels in females per kilogram of body weight. However, cardiac output is higher in male ducks (per kilogram body weight) than in females. The sex difference in chickens might be attributable to estrogen in the female. Sturkie and Eiel (1966), however, employing the thermodilution technique of cardiac output, which is not influenced by lipemic plasma, found that estrogen has no significant effect on cardiac output except to depress blood pressure slightly, which tends to somewhat increase cardiac output.

Boelkins et al. (1973) reported that the cardiac output of female chickens was consistently higher when determined by Evans Blue dye (277 ml/kg) than when determined by Indocyanine Green (177 ml/kg). The latter figure agrees closely with those of Vogel and Sturkie (1963a, b) and Sturkie (1967).

Oxygen consumption increases cardiac output tremendously, whereas rest, inactivity, and particularly starvation decrease it, as indicated in Table 6–7.

Ambient Temperature and Acclimatization. Acute heat increases cardiac output and decreases blood pressure and peripheral resistance (Whittow et al., 1964). Acute cold also increases cardiac output and blood pressure in chickens (Whittow et al., 1965a).

Acute heat and hyperthermia not only increase cardiac output but also influence regional bloodflow (Wolfenson et al., 1981); these will be discussed later.

Sturkie (1967) demonstrated conclusively that when chickens become acclimated to heat (similar to summer conditions), there is a significant decrease in cardiac output instead of an increase (Table 6–7) and the changes (adaptation) occur within 3–4 weeks. Birds were kept at control temperatures (23°–25°C) and at high temperatures (32°C for 12 hr), and at 25°C for 12 hr. The differences between treated and controls (treated minus controls) are expressed in percentages as follows:

Weeks exposure	Difference in cardiac output
1	8% increase
4	23% decrease

There were slight decreases in blood pressure, which persisted. Exposure to heat first causes vasodilation and increased cardiac output and oxygen consumption. As birds become acclimated to the heat, the demand for oxygen consumption decreases and so does bloodflow. Low temperatures and continued exposure to such temperatures had little effect on cardiac output.

Pigeons in *flight,* whose oxygen consumption is greatly increased, had cardiac outputs 3.4 times that of birds not flying (Butler et al., 1977).

Effects of Hypoxia. Few studies have been conducted on the effects of hypoxia on cardiac output other than hypoxia during diving. Jones and Holeton (1972) studied changes in oxygen tension from a normal P_aO_2 of about 100 mm Hg and at 63, 47, and 38 mm Hg. Cardiac output increased almost linearly over this range, reaching a peak increase of about 50%. The oxygen extraction increased greatly at the lower oxygen tensions with AV oxygen differences of 9.5 mm Hg at a P_aO_2 of 38 mm Hg compared to 30 mm Hg at 100 mm Hg.

Artificial ventilation of chickens following administration of succinycholine increased cardiac output; the effect was attributable to the chemical, rather than to ventilation rate (uptake of O_2), which was not changed (Piipers et al., 1970).

The chicken is apparently more sensitive to changes in ventilation rate and responds by an increase in cardiac output (Butler, 1967; Sturkie, unpublished). The effects of diving on cardiac output are discussed under cardiovascular adjustments.

Regional Bloodflow

Distribution of Flow to Various Organs. Bloodflow to organs is closely related to oxygen extraction and

TABLE 6–7. Cardiac output, blood pressure, and total peripheral resistance in Aves (means and standard errors)

Breed	Sex, age, condition	Body weight (kg)	Cardiac output per minute		Mean blood pressure	Peripheral resistance (units/kg)	Heart rate	Method[a]	Reference
			Per bird	Per kg					
White Leghorn	Male 16 months, starved	2.39	340 ± 18	143 ± 7	166	1.23	307	DD	Sturkie and Vogel (1959)
	Female 16 months, starved	1.79	308 ± 17	173 ± 7	142	0.88	378	DD	Sturkie and Vogel (1959)
White Leghorn	Male 12–14 months, winter	2.59	444 ± 22	173 ± 9	181	1.11	303	DD	Vogel and Sturkie (1963b)
	Male 12–14 months, summer	2.95	359 ± 11	135 ± 7	177	1.41	289	DD	Vogel and Sturkie (1963b)
	Female 18 months, winter	1.95	345 ± 15	181 ± 12	153	0.91	336	DD	Vogel and Sturkie (1963b)
	Female 18 months, summer	1.96	234 ± 7	121 ± 5	147	1.25	347	DD	Vogel and Sturkie (1963b)
	Male	2.2	—	190	182	—	276	DD	Vogel and Sturkie (1963b)
	Mixed female	1.6	218	—	—	—	—	TD	Sturkie and Eiel (1966)
	Female	1.6	430	269	—	—	—	ID	Sapirstein and Hartman (1959)
	Male adult (control)	2.8	481	171	169	0.98	—	DD	Sturkie (1967)
	Male (4 weeks, heat-acclimated)	2.6	330	127	163	1.28	—	DD	Sturkie (1967)
	Male, adult	—	—	150	187	1.24	247	RM	Merrill et al. (1981)
Cross-breds	Females (controls)	1.9	—	178	132	0.74	355	RM	Wolfenson et al. (1981)
Pigeon	(*Columba*), on land	0.39–0.48	245	555	142	0.255	115	DF	Butler et al. (1977)
	in flight	—	1007	2411	147	0.06	670	DF	Butler et al. (1977)
	at rest	0.42	143	341	106	0.31	178	DF	Grubb (1982)
Duck									
Peking	Males	3.3	—	286.8	161.2	0.566	175	DD	Sturkie (1966b)
	Females	3.0	—	253.4	147.2	0.62	185	DD	Sturkie (1966b)
Anas platyrhynchos	Control	3.0		482.0	—	—	—	TD	Folkow et al. (1967)
	Diving	3.0		24.4[b]	—	—	—	TD	Folkow et al. (1967)
	Control	2.1	393.9	187.5	—	—	186.2	[b]	Jones and Holeton (1972)
	Diving	2.1	25.9		—	—	21.1	[b]	Jones and Holeton (1972)
Peking	Male, female (control)	2.95	143	304	134	0.44	199	DF	Grubb (1982)
Anas platyrhynchos		2.4	973	405	—	—	179.0	DD	Jones and Holeton (1972)
Muscovy duck		2.16	844	390	—	—	130		Jones and Holeton (1972)
Turkey	B. B. Bronze males	14.5	1615	111.3	204	—	149	ID	Speckman and Ringer (1963)
	—	14.5	1615	200.0				DD	Hamlin and Kondrich (1969)

[a] DD, dye dilution; TD, thermo dilution; ID, isotope dilution; DF, direct Fick; RM, radioactive microspheres.
[b] Based on flowmeter on one pulmonary artery.

metabolic activity. In cases of decreased oxygen intake (hypoxia) the organ usually compensates by increasing blood flow and also the amount of oxygen extracted. The distribution of blood flow to various organs in the chicken has been studied by Sapirstein and Hartman (1959) and Boelkins et al. (1973), and more recently by Odlind (1978), Wolfenson et al. (1981, 1982), and Merrill et al. (1981). All used radioactive microspheres except Sapirstein and Hartman (1959), who injected Rubidium.

Bloodflow in Heart, Lungs, Liver, Kidneys, and Digestive Organs

Comparative results on regional bloodflow in certain organs are shown in Tables 6–8 and 6–9, where flows in milliliters per min per gram and as percent of cardiac output are given. It is apparent that bloodflow is usually highest in spleen, heart, kidneys, intestines (Table 6–8) and shell gland and ovarian follicles (Table 6–9). The percent of cardiac output going to an organ depends mainly on its size. The values in the tables vary, and are influenced by conditions of the experiment and environment. The largest discrepancies are in values for the heart and kidneys as reported by Merrill et al. (1981), whose values appear inordinately high compared to others in the table and to results of Sapirstein and Hartman (1954). Table 6–8 also includes flow in muscle, comb, wattles, and other organs.

Effects of Ambient Temperature on Flow. Acute heat and hyperthermia increased bloodflow in skin, comb and wattles, abdominal muscle, brain, trachea, and larynx, and decreased flow in proventriculus, gizzard, duodenum, jejunum, and liver, with little or no effect on ileum and kidney (Wolfenson et al., 1981). Increased blood flow in combs and wattles and vasodilation following heating and hyperthermia were reported by Nolan et al. (1978), suggesting an important heat-dissipating role for these organs.

Increased spinal cord temperature in ducks increased carotid artery flow from 13.1 to 32.8 ml/min and increased blood pressure, heart rate, and respiratory rate (Bech et al., 1980). Acute heating and hyperthermia and panting in ducks increased carotid artery (one) flow from 9.3 to 26.0 ml/min, but had no appreciable effect on sciatic artery flow (Bech and Johansen, 1980).

Endotoxin shock induced in White Leghorn males (Merrill et al., 1981) reduced blood flow in kidney, spleen, proventriculus, duodenum, and cloaca. There was little change in flow to heart, and no organs experienced an increased flow after shock.

Endocrine and Reproductive Organs. Bloodflow is highest in the thyroids and parathyroids (Table 6–9). Flow is very high in the shell gland, particularly when there is a hard-shell egg in the uterus and during the time shell is being secreted (Boelkins, 1973; Wolfenson et al., 1978, 1982). Blood flow (ml/min/g) in-

TABLE 6–8. Distribution of bloodflow to various organs of the chicken determined by radioactive microspheres

	A[a]		B[b]		C[c]	Organ weights
Organs	ml/m/g	% CO[d]	ml/m/g	% CO[d]	ml/m/g	(g)
Cerebrum	—	—	0.80	0.582	—	1.7
Lungs	—	—	—	—	3.6	—
Kidney	2.54	9.96	2.65	10.80	9.0	15.03
Heart	2.79	5.78	—	—	9.5	5.93
Spleen	5.20	3.37	—	—	3.6	1.95
Liver	0.60	7.05	1.25	17.5	0.56	47.50
Gizzard	—	—	0.20	1.04	—	18.60
Proventriculus	—	—	0.95	1.52	0.86	5.70
Duodenum	3.14	4.81	2.75	3.93	1.30	5.06
Ileum	—	—	1.25	1.75	—	5.00
Jejunum	—	—	1.70	3.82	—	8.00
Colon	1.06	0.40	—	—	—	1.52
Tibia	0.42	2.62	0.60	—	—	19.66
Femora	0.73	4.08	—	—	—	16.80
M. pectoralis major	—	—	0.10	—	0.050	—
M. oblique abdominal	—	—	0.20	—	—	—
Comb	—	—	0.12	0.286	—	8.50
Wattles	—	—	0.43	0.229	—	1.90

[a] A, Female White Leghorn chickens; data from Boelkins (1973).
[b] B, Cross-breds; data from Wolfenson et al. (1981).
[c] C, Male White Leghorns; Merrill et al. (1981).
[d] Percent cardiac output.

TABLE 6–9. Bloodflow to chicken endocrine and reproductive organs

	Bloodflow (ml/min/g)				Bloodflow (ml/min/follicle[e])
Organs	A[a]	B[b]	C[c]	D[d]	D[d]
Adrenals	0.72	2.59	2.75	—	—
Thyroids	2.08	0.33	2.10	—	—
Parathyroids	15.33	—	—	—	—
Pancreas	0.86	0.31	0.62	—	—
Infundibulum	—	0.30	0.43	0.33	—
Magnum	0.71	0.70	0.70	0.66	—
Isthmus	0.59	0.51	0.50	0.41	—
Shell Gland	2.20	0.99	2.00	2.10	—
Ovarian					
follicles C_1[f]	—	3.12	3.57	3.25	1.46
C_2	—	2.27	3.60	3.08	1.21
C_3	—	1.83	3.70	3.04	0.93
C_4	—	2.71	3.85	3.48	0.69

[a] A, Boelkins (1973), based on White Leghorn hens with hard-shell egg in uterus.
[b] B, Wolfenson et al. (1978), based on cross-bred hens with hard-shell egg in uterus, 1–4 hr.
[c] C, Wolfenson et al. (1981), based on cross-bred female birds in evening hours (probably with hard-shell egg in uterus, although not stated).
[d] D, Wolfenson et al. (1982), with hard-shell egg in uterus during calcification.
[e] Bloodflow per follicle with weights for respective follicles of 0.49, 0.43, 0.34, and 0.28 grams, in descending order.
[f] Cardiac output, 356 ml/min/2-kg hen.

creased with time spent in uterus from 1–3 hr to 4–6 hr and changed little thereafter. Calcium deposition becomes maximal after the egg is in the gland 4 or more hr (Wolfenson et al., 1982), at least four times that when no egg is in gland. Bloodflow in the magnum and isthmus is three to four times that when no egg is present (Moynihan and Edwards, 1975; Wolfenson et al., 1982).

Bloodflow is also high in the ovarian follicles and is higher in the larger follicles (Wolfenson et al., 1982; Scanes et al., 1982). There were no differences on a weight basis. Acute heat and hyperthermia decreased blood flow in ovaries, ovarian follicles, thyroids, and pancreas (Wolfenson et al., 1981). Administration of prostaglandin (PG-2) decreased blood flow to major preovulatory follicles (Scanes et al., 1982). See Table 6–9 for values of bloodflow.

Bloodflow in the Brain

Normal Levels. Few studies have been made on the brain, and it is not always clear exactly what parts of the brain are involved. Some have used microspheres (Merrill et al., 1981; Wolfenson et al., 1978), and others have employed a xenon clearance technique (Grubb and coworkers, 1977, 1978, 1979) that measures relative flows but not actual flows.

Wolfenson et al. (1978) determined bloodflow in the cerebrum and cerebellum of cross-bred hens and reported values of 2.00 and 1.49 ml/min/g of tissue, respectively.

Effects of Hypoxia on Cerebral Blood Flow (CBF). Grubb et al. (1978) showed that decreasing O_2 tension (below 75 mm Hg) increased cerebral bloodflow significantly, and at a P_aO_2 of 30 mm Hg, it increased over 600%. This magnitude is considerably greater in the duck than in mammals.

Decreasing CO_2 with hypoxia (*hypocapnic hypoxia*) shifts the bloodflow curve so that bloodflow does not increase until P_aO_2 is below 50–60 mm Hg rather than 75, as in normotoxic hypoxia. This is because the blood of hypocapnic animals contains more O_2 than that of normotoxic ones at the same P_aO_2 level. When CBF is expressed as a function of blood O_2 content, it does not differ significantly from that in the normotoxic animal.

Grubb et al. (1977) revealed that CO_2 does not decrease CBF in ducks when P_aO_2 is normal, in marked contrast to mammals.

Route of Bloodflow. Fedde and Guffy (1983), employing a technique involving injection of mercury and occlusion of brachiocephalic arteries, revealed that a small amount (approximately 2%) of blood reaches the brain of duck from branches arising between the dorsal intercostal arteries and descending vertebral arteries.

Bloodflow in Muscle

Blood flow (BF) in breast and abdominal muscles is presented in Table 6–8. Hyperthermia increased BF significantly in the abdominal muscles (Wolfenson et al., 1981).

Grubb (1981) determined the effects of hypoxia on BF to the leg muscles of the duck, employing an electromagnetic probe around the ischiadic artery. Arterial O_2 tension varied from 40 to 110 mm Hg. Over this range there was little or no change in BF. Most of the values were 44–45 ml/min/g of tissue. As O_2 tension decreased to 30–40 mm Hg, there was a significant increase in O_2 extraction by the leg muscle, which decreased as tension increased up to 70 mm and leveled off at tensions above this.

Avian skeletal muscle is able to maintain its O_2 consumption without an increase in bloodflow when exposed to severe hypoxia, unlike mammalian muscle. The duck has a very high muscle BF, and oxygen delivery is much greater than necessary. Therefore, the demands for O_2 must be excessive, such as occurs after hypoxia and exercise, before BF will increase (B. Grubb, unpublished data).

Regional Bloodflow in Diving

Distribution of bloodflow in ducks before and after diving has been reported by Johansen (1964) and by Jones et al. (1979). Before diving, the bloodflow is high in the spleen, heart, and kidneys, as it is in chickens; the amount of blood flowing to the heart, kidneys, and brain before the dive represented 2.6, 2.7, and less than 1% of the cardiac output, respectively. After the dive, the percentages were 15.9, 0.4, and 10.9, respectively (Jones et al., 1979). While cardiac output decreased drastically after the dive, the actual amount of flow to the heart did not change, but it increased to the brain. Spleen bloodflow decreased markedly after the dive, and adrenal flow increased greatly.

Bloodflow in Gut and Kidneys

The anatomical arrangement of blood vessels in the intestines and kidneys is shown in Figure 6–3. Sturkie and Abati (1975) and Sturkie et al. (1977, 1978) have studied bloodflow in various blood vessels as follows: cranial mesenteric vein (CMV), coocygeomesenteric (COCMV), hepatic portal (HPV), femoral vein (FV), renal portal vein (RPV), mesenteric artery (MA), and others. The birds were anesthetized, restrained, and surgically opened; electromagnetic probes were placed around the vessels. Flow was then measured under various conditions, such as clamping certain vessels and observing the direction and magnitude of flow. Blood flow in certain of these vessels is shown in Table 6–10.

The hepatic portal vein receives blood from the COCMV, MV, P, and gastroduodenal (GD), and represents liver bloodflow. Flow in HPV in this study represented all venous flow to the liver except that

TABLE 6–10. Bloodflow in veins of intestine of nonfasted chickens (ml/min)[a]

Blood vessel[b]	Mean per bird	Mean per kg	Percent of total bloodflow[c]
CM	14.7	6.4	3.3
COCM	18.1	7.9	4.1
HP	32.8	14.2	7.4
MA	33	14.3	7.4
PVC	98.3	45.3	

[a] After Sturkie and Abati (1975).
[b] White Leghorn males, lightly anesthetized cranial mesenteric (CM), coccygeomesenteric (COCM), portal (HP), mesenteric artery (MA), and posterior vena cava (PVC).
[c] Based on estimated cardiac output of 190 ml/kg for White Leghorn males.

from GD (gastro-duodenal) (see Figure 6–3). Our figure for HPV flow was about 0.7 ml/m/g, based on an estimated liver weight, or 7.4% of total bloodflow. This figure compares favorably with that of Sapirstein and Hartman (1959) of 6.7%. Liver flows reported by Boelkins (1973), Merrill et al. 1981), and Wolfenson et al. (1981) were 0.6, 0.56, and 1.25 ml/m/g, respectively, based on determinations employing microspheres.

The fact that bloodflow in the COCMV is higher than in the mesenteric vein is surprising, because the latter vessel drains a larger area (all of the small intestines other than duodenum and pancreas) than the COCMV, which drains a smaller area of large intstines. This suggested that some of the blood in the COCMV represents blood shunted from the femoral and renal portal veins. In order to test this view, experiments were set up involving flow measurements in HPV before and after ligation of femoral, renal portal, mesenteric, and pancreatic veins. Flows before and after ligation are shown as follows:

	Hepatic portal flow (less GD flow) (ml/min)	
Before ligation of renal portals	32.6	Difference
After ligation of renal portals	19.8	12.8
After ligation of renal portal and mesenteric veins	8.0	11.8

After the ligation of renal portals bilaterally, which includes bloodflow from renal portals, external iliac vein, or femoral veins, flow in HPV decreased 39.2%, or 12.8 ml/m; after further ligation of MV and pancreatic veins, flow in HPV decreased to 8 ml/m or 36.2%. The remaining 8 ml of flow represents true flow of COCMV drained from large intestines uncomplicated by contributions from the renal portal and other veins. Actually COCMV flow was not determined directly in this experiment, but it is calculated to be 20.8 ml/m. This figure is less than that of later experiments (Sturkie et al., 1978) of 26.4, where COCMV flow was measured directly and was slightly more than that reported in their 1975 experiments (16–19 ml/m).

Contribution of Femoral Vein Bloodflow. Bloodflow in COCMV and posterior vena cava (PVC) were determined after bilateral ligation of femoral veins at points proximal to the juncture of these veins with renal portal veins. Ligation caused a decrease of COCMV flow from 26.4 ml/m to 22.9, or approximately 13%; the flow in the PVC decreased from 66.0 to 59.7%, or 9.6%. Therefore, the contribution of bloodflow of both femoral veins to COCMV and PVC flow was 10.8 ml/m or only 30% of the total bilateral femoral blood flow of 36 ml/m. This suggested that much of the blood flowing through the femoral and renal portal veins is shunted off to the kidney tubules (Sperber effect) and does not reach the COCM and PVC veins by the usual routes.

Studies by Sturkie and Abati (1975) based on placement of electromagnetic probes on COCMV revealed that in nearly all cases the blood flow was toward the liver and not toward the renal portal and PVC in the opposite direction, as has been reported to occur in some instances (Akester, 1967). Odlind (1978), using microspheres, came to the same conclusions as Sturkie and Abati.

Role of Renal Portal Valves. Akester (1967) studied the course of bloodflow in the renal portal system by injecting radiopaque fluid into the external iliac vein of anesthetized chickens and following its course with cineradiography. In most instances, the fluid was observed to flow into the posterior vena cava (PCV), renal portal vein, and kidney, evidence to him that the renal portal valves were open (see Figure 6–3 and 16–4). He also observed that in about one-third of the cases, blood was shunted into the COCM even though the valves appeared to be open. In only one-fourth of the instances, the valves were closed and blood flowed to the kidney and COCM. Thus, it is apparent from these results that bloodflow may be shunted to COCM with the valve open and that other factors are involved in the shunting of flow.

In the studies of Sturkie and coworkers (see earlier section), it was presumed that the valves were partially or completely open, because when the posterior vena cava was ligated (Sturkie and Abati, 1975), which is equivalent to complete closure of the valves, bloodflow was shunted to the COCM vein and was increased tremendously.

The extent to which the renal portal valves influence bloodflow in vivo was determined by infusing adrenergic and cholinergic agents in chickens (Sturkie et al., 1978) in amounts known to open and close the valves in vitro or isolated [Rennick and Gandia (see Chapter 16); Braun et al., 1982]. Opening or closure of the valves was determined by changes in flow in PVC or hepatic portal veins. Cholinergic agents in amounts large enough to depress blood pressure should have decreased bloodflow in these vessels if the renal portal valves were closed. Actually, the flow was not changed appreciably, indicating that the valves were still open. Adrenergic agents increased flow in these vessels, but for reasons probably not related to condition of the valves. Although the valve is known to be innervated by cholinergic and adrenergic nerves (histologic evidence), this does not mean that the valve is functionally important (see Chapter 16) and that other factors are more important in regulating bloodflow in the renal portal system. Sturkie and Abati (1975) showed that starvation and anesthesia decreased bloodflow significantly in the MV and COCMV. Any condition that affects muscle bloodflow, such as exercise and temperature changes, might well affect bloodflow in the renal portal and COCM system.

Odlind (1978) made an extensive study of bloodflow in the renal portal and COCM system of laying and nonlaying chickens, using microspheres and a clearance technique. He injected these into a leg vein (external iliac or femoral vein) and recorded the uptake of the spheres in the kidneys, livers, and lungs; from this, he calculated the flow to these organs as a percentage of total flow. When microspheres were injected into a leg vein, it was shown that on the average 44.0, 47.0, and 8% of the bloodflow from the leg perfused the kidneys, livers, and lungs, respectively. Moreover, the distribution of flow varied considerably and particularly between right and left legs. Flow to the kidneys was higher on the left side and lower to the liver than on the right side. Odlind thought that this difference might be attributed to differences in local vasoconstriction and other factors as well as changes in status of the renal portal valves.

Bloodflow to the COCMV was in nearly all instances directed toward the liver, in agreement with the results of Sturkie and Abati (1975) and Sturkie et al. (1977). According to Odlind, this reduces the role often ascribed to the renal portal valves as the dominant regulatory factor of bloodflow to the kidneys, in agreement with the results of Sturkie et al. (1978). (See Chapter 16.)

Cardiovascular Adjustments to Exercise, Flight, Starvation, Restraint, and Diving

Exercise. Studies on exercise in land birds consist mainly of forced walking or running, usually on a treadmill. Exercise in penguins (running), based on telemetry studies, increased femoral bloodflow four times that of resting birds (Millard et al., 1972). Mean blood pressure increased from 80 to 125 torr, and heart rate from 90 to 180 beats/min.

Exercise in turkeys (moving about) (effects recorded by telemetry) increased heart rate drastically but had

TABLE 6–11. Effects of exercise (running) and atropine and propranolol on blood pressure and heart rates of turkeys[a]

	Running speed (km/hr)			
	0	1.9	3.3	5.6
Heart rate	145	217	265	301
Blood pressure	146	164	183	204
Systolic				
Diastolic	102	118	127	140
Mean	123	139	155	169
After drugs				
Heart rate				
Propranolol	118	122	152	176
Atropine	211	242	272	297
Propranolol and atropine	140	141	154	203
Mean blood pressure				
Propranolol	127	129	128	142
Atropine	113	116	129	137
Propranolol and atropine	112	132	138	147

[a] Blood pressure in mm Hg; heart rate in beats per minute. From Beaudinette et al. (1982).

little effect on blood pressure (see Figure 6–11). In more controlled studies, Baudinette et al. (1982) subjected turkeys to three different speeds of running, and measured the effects on heart rate and blood pressure (Table 6–11). Both heart rate and blood pressure were elevated at all speeds of running. In order to determine the extent to which sympathetic and parasympathetic tone accounted for the changes in heart rate, propranolol and atropine were administered; these abolish sympathetic and parasympathetic tone, respectively. Heart rates increased after atropine, as expected at rest, but after exercise the increases were no greater than in those birds exercised but not receiving atropine; therefore vagal control of heart rate during exercise was not important, but mean blood pressure appeared lower after atropine.

After administration of propranolol, however, heart rates were lowered significantly but less so at the higher level of exercise, indicating that much of the increased heart rate following exercise results from increased sympathetic tone. Propranolol decreased blood pressure because it abolished the effects of catecholamines, not only of neuronal sources, but of adrenal as well. After administration of atropine and propranolol, heart rate changed little except at extreme exercise, where it was significantly elevated. Blood pressure was elevated to some extent by exercise, particularly at the higher level.

Lactate levels were increased by exercise, but injection of lactic acid had no effect on heart rate. Thus, it appears that much of the increase in heart rate after abolition of sympathetic and parasympathetic tone is attributable to some metabolite other than lactate.

Exercise in adult turkeys had no effect on the severity of atherosclerosis or serum lipid levels. Sexually mature hypertensive turkeys showed less endurance to exercise than hypotensive ones (Bolden et al., 1983), a result just opposite to those of Sturkie and Textor (1961) in chickens, where hypotensive ones showed less resistance to exercise. Exercise increased certain cardiac measurements, such as length of heart and coronal axis, but there was no effect on ventricular weight; however, the hypertensive strain of turkeys had larger hearts than those of the hypotensive strain.

Bamford and Maloiy (1980) reported a significant increase in heart rate and O_2 consumption in the Marabou stork exercised on the treadmill. Ducks and pigeons exercised on a treadmill (Grubb, 1982) likewise experienced an increase in cardiac output (CO), heart rate (HR), and blood pressure (BP); CO exhibited a positive linear relationship with O_2 consumption. The increase in HR in the duck was much greater than in the pigeon with exercise.

In the duck, HR and mean BP increased from 199 beats/min and 134 torr to over 300 and 152, respectively, after exercise, depending on the extent of O_2 consumption. There was also an increase in cardiac output (CO) from 304 to 400–500 ml/min/kg, depending on severity of exercise. There was little correlation between stroke volume and O_2 consumption after exercise.

Exercise of Peking ducks on a treadmill at 1.3 km/hr increased CO from 539 to 870 ml/m/kg and heart rate from 174 to 328. O_2 consumption increased 2.6 (Bech and Nomoto, 1982). Heart rate increased in tufted ducks swimming at different velocities (Woakes and Butler, 1982).

The flightless emu, walking on a treadmill at 1.33m/sec on a 6° incline, exhibited changes in the following parameters (Grubb et al., 1983):

	At rest	Exercise
CO (ml/kg/m)	67.9	494
Heart rate/m	45.8	180
VO_2 (ml/kg/m	4.18	48.2

Blood pressure was elevated only at the most extreme exercise. There was a linear relationship between O_2 consumption and CO after exercise.

Flight. Birds can fly long distances at high speeds and at high altitudes where O_2 tension is low. The physiologic adaptations required to sustain flight under these conditions are remarkable, and were mainly speculative until the advent of more recent and appropriate techniques, such as the wind tunnel, which allow for more controlled experimental conditions in flight (Tucker, 1968a,b; 1972). The O_2 demand of birds flying at 6100 m was estimated to be about eight times that at sea level, and that this would require a CO of 2.8 liter/kg/min (Tucker 1968a).

Flight, even for short periods, increases HR considerably. Berger et al. (1970) reported that the heart rate of flying birds even at rest is greater than of nonflying ones. The increase in HR of small birds in flight is about twice those at rest, but in larger birds, it is three or four times the resting rate. O_2 consumption during flight has been estimated to be 5–14 times that of nonflying birds (Tucker (1968a; 1972). Butler and Woakes (1980) reported that in free-flying geese (*Branta leucopsis*) heart rate increased from 72 to 512 beats/mm after flying at speeds of 18.7 m/sec for 14.4 min. Rest rate increased from 8.5 to 121. The ratio of wingbeat frequency to heart rate was near 3:1.

Jones and Holeton (1972), who simulated high altitude in ducks, reported significant increases in CO from a level of 1000 ml/min per bird at a P_aO_2 of 95–100 torr to a level of about 1400 ml at a P_aO_2 of 40–50 torr (hypocapnic hypoxia).

Baudinette et al. (1976) observed large increases in bloodflow of the feet of flying gulls in a wind tunnel, and reported that the heat loss from the feet amounted to about 80% of the total heat production during flight. This appears inordinately high.

Butler et al. (1977) reported the effects of flight of pigeons in a wind tunnel (see Table 6–7). At a speed of 10 m/sec there was a tremendous increase in CO (7 times resting level) and HR (nearly 4 times resting level), but little change in BP. The wind tunnel alone had some effect on the measured parameters. When the tunnel was turned on, but the birds were not flying, HR and O_2 consumption increased significantly with some increase in BP.

Restraint. Restraining chickens on their backs increased heart rate and blood pressure initially; later, pressure decreased appreciably, but heart rate and cardiac output increased, probably reflexly (Whittow et al., 1965b). The initial increases are no doubt influenced by the release of adrenal and cardiac catecholamines, which are increased with initial excitement. Anesthesia, which lowers heart rate and blood pressure, also increases blood and cardiac catecholamines (Sturkie et al., 1970). Chronic restraint (50 days or more) produces a type of stress that may depress blood pressure and may ultimately result in death; however, individuals may be trained to withstand certain types of restraint without deleterious cardiovascular effects (Burton and Beljan, 1970).

These investigators also showed no significant changes in trained birds in blood volume, hematocrit, body weight, or lymphocyte counts.

Chronic acceleration is a means of partial restraint and orientation of animals in space or weightlessness, and it definitely produces stress and lymphopenia and increases red cell numbers and blood volume. Blood volume is decreased in the thorax and pooled in the limbs, and the increase is proportional to the *g* strength, increasing as much as 25% for an increase of 1 *g*. Smith et al. (1979) showed that chickens developed tolerance (some) to 6 *g*s, but as the animals grew older they became less tolerant, and they developed a more pronounced bradycardia (decrease of 25% in heart rate) with the onset of acceleration. Moreover, the bradycardia serves as a fair end point to the tolerance. Most birds tolerated the 6 *g*s for 4–12 min.

Starvation. Complete restriction of food intake for 3 or more days causes a drastic drop in blood pressure, heart rate, and cardiac output (Vogel and Sturkie, 1963a). Starvation apparently increases vagal tone, because atropine administration increased heart rate. A partial restriction of food intake to approximately 70–80% of normal food intake for an extended period of time reduced blood pressure and heart rate, but not greatly (Hollands et al., 1965).

Diving. Earlier studies on the cardiovascular adjustments to diving were reviewed by Anderson (1966) and Jones and Johansen (1972), and more recently by Butler and Jones (1982). When diving mammals or birds submerge there is intense bradycardia, apnea, and a drop in cardiac output to about one-fifth normal (Folkow et al., 1967; Jones and Holeton, 1972; Lillo and Jones, 1982a, b). (See table 6–7 and later section.) Blood is pooled into the visceral organs. The hypoxia and hypercapnia induced by the dive stimulate vasoconstriction and sympathetic discharge, particularly in the limbs where bloodflow virtually ceases, although pressure in femoral arteries is near normal (Folkow et al., 1967); however, there is some evidence that with sustained diving the blood pressure may fall significantly (Anderson, 1966). In many peripheral vascular beds, the resistance may increase eightfold or more, but in the pulmonary bed it increases only 1.5 to 3-fold (Jones, 1981). The fall in cardiac output (CO) is attributable mainly to the bradycardia, but a slight decrease in cardiac contractility may also be involved (Folkow et al., 1967). Sturkie and Abati (1977) (see Chapter 7), however, found the heart muscle of the diving duck was most resistant to hypoxia and decreased contractility, based on comparative studies with chickens and pigeons.

Cardiac output (CO) in the duck decreased from 482 to 24.4 ml/kg per bird during the dive, based on thermodilution studies (Folkow et al., 1967). The predive figures appear high. Jones and Holeton (1972), employing electromagnetic probes around the pulmonary artery, reported pre- and postdive figures (Table 6–7). The CO at the end of the dive was only 5–7% of the predive figure. Similar figures were reported later by Jones et al. (1979). CO dropped from 227 to 95 ml/kg 20–72 sec after the dive and to 59 ml/kg after 144 sec. Before the dive, the highest proportion of CO went to the heart and kidneys, and a small amount went to the brain. After the dive there was an increase in proportion of CO going to the heart and brain. Bloodflow to the skin, muscle (leg), gastrointestinal tract, and kidney decreased after the dive.

Changes before, during and after diving are shown as follows (Lillo and Jones, 1982b):

	HR	MBP	CO/ml	V_El/min
Before dive	195	159	452	0.7
During dive	34	149	88	—
After dive	302	159	806	2.73

While cardiac output and heart rate decrease drastically during diving, there is an actual increase in stroke volume (Langille, 1983), which is produced by or associated with an increased venoconstriction and central venous pressure via the Frank–Starling mechanism. When central venous pressure was maintained constant on diving by withdrawing venous blood, there was a decrease in stroke volume and contractility, but the intense bradycardia was inhibited. Thus, it appears that the diving evokes reflexes causing the bradycardia and decreased cardiac output, and these depend, in part, on increased venous pressure elicited by ventricular receptors (Langille). The author concluded that the rise in venous pressure during diving ultimately reduces myocardial O_2 requirements during a period when its availability is low.

Langille also demonstrated that mean circulatory pressure (MCP) increased greatly (doubled) during diving and is a good measure of central venous pressure. MCP represents the average pressure in the body when arterial and venous pressures are the same, and it is produced by inducing a cardiac arrest and allowing time for equilibration of the pressures.

Blood pressure does not change appreciably after diving, according to Jones et al. (1979). *Heart rate* drops drastically, from a predive value of 186 to 21, 120 sec after the dive (Jones and Holeton, 1972). The extent of the drop depends on the breed and habitat of the diver (Catlett and Johnston, 1974; Cook et al., 1977). The deep divers tend to show more extreme bradycardia in the earlier stages of the dive, whereas the surface feeders heart rate took longer to slow (Catlett and Johnston, 1974). Similarly, in a rare diver (shoveller) heart rates did not decrease as much or as rapidly as those of the intermediate and deep divers (Cook et al., 1977). The heart rate of the intermediate diver dropped to 30.0 and that of the deep diver to 24.7% of the predive rates, compared to 58.7 for the rare diver.

There can be no doubt that the ability to survive submergence for a long time is dependent mainly on adjustments of the cardiovascular system, although anaerobic metabolism is a factor.

Mechanisms Involved. The motor responses to diving are mediated by the sympathetic and parasympathetic nervous systems; also, catecholamines that are released may contribute to the process, particularly the extreme vasoconstriction in the periphery (Butler and Jones, 1982). The bradycardia results from vagal stimulation by chemoreceptors, which can be abolished by vagal transection. This reflex is independent of higher cerebral activity because it occurs in decerebrate animals, but is dependent on central integration in the hypothalamus and medulary areas; stimulation in these areas often mimics the natural reflex response (Folkow and Rubinstein, 1965; Feigl and Folkow, 1963).

The cardiac sympathetic nerves are not believed to be involved significantly in diving, according to Jones (1981) and Butler and Jones (1982), but they are involved according to Folkow et al. (1967).

What initiates the bradycardia reflex? Does submergence in water initiate it or is it induced by apnea and hypoxia? Some investigators have reported that wetting of head and beak, or submergence induced the bradycardia reflex, and there is good evidence that artificial ventilation during diving, partially or completely abolishes the reflex (Jones and Johansen, 1972, Jones, 1981).

If there is a true immersion reflex independent of apnea, it should show up in young ducklings, which are more sensitive to immersion than are adults. Jones and Butler (1982) demonstrated that about one-third of the bradycardia in ducklings is attributable to water immersion alone, and the remainder is caused by stimulation of the carotid chemoreceptors by hypoxia and hypercapnia. Moreover, Jones and Lillo in 1980 and Lillo and Jones (1982a, b) showed that the baroreceptors have little or no influence on the diving reflex, but they may be essential for the maintenance of blood pressure (see also Jones et al., 1983).

Jones et al. (1982) studied the effects of stimulation of the central and peripheral chemoreceptors on diving bradycardia and peripheral resistance in the hind limb blood vessels of the duck (HLVR). They concluded that about 85% of the total bradycardia resulting from diving is caused by stimulation of the peripheral (carotid body) chemoreceptors (PC). In the early part of the dive, the carotid baroreceptors cause some of the bradycardia but cause practically none of it in the later stages where the PCs are almost wholly responsible. Stimulation of both central and peripheral receptors increased HLVR to about an equal extent in the later stages of the dive.

References

Abati, A., and J.J. McGrath (1973). Physiological responses to acute hypoxia in altitude acclimatized chicken. J. Appl. Physiol., 34, 804.

Abdel-Magied, E.T.M. (1978). Encapsulated sensory receptors in the wall of avian common carotid. J. Anat., 127, 198.

Abraham, A. (1969). "Microscopic Innervation of the Heart and Blood Vessels in Vertebrates Including Man." Oxford: Pergamon Press.

Adams, W.R. (1958). "Comparative Morphology of the Carotid Body." Springfield: Thomas.

Akers, T.K., and C.N. Peiss. (1963). Comparative study of effect of epinephrine and norepinephrine on cardiovascular system of turtle, alligator, chicken and opossum. Proc. Soc. Exp. Biol. Med., 112, 396.

Akester, A.K. (1971). The heart (Chapter 31) and Blood vascular system (Chapter 32). In "Physiology and Biochemistry of the Domestic Fowl," Vol 2 (D.J. Bell and B.M. Freeman, Eds.). New York: Academic Press.

Akester, A.R. (1967). Renal portal shunts in the kidney of domestic fowl. J. Anat., 101, 241.

Ames, E., and E. Skadhauge. (1971). Effects of arginine vasotocin on excretion of Na, K, Cl, and urea in hydrated chicken. Am. J. Physiol., 221, 1223.

Andersen, H.T. (1966). Physiological adaptations in diving vertebrates. Physiol. Rev., 46, 212.

Bamford, O.S., and G.M.O. Maloiy. (1980). Energy metabolism and heart rate during treadmill exercise in the Marabou stork. J. Appl. Physiol., 53, 211.

Baudinette, R.V., A.L. Tonkin, J. Orbach, R.S. Seymour, and J.F. Wheeldrake. (1982). Cardiovascular function during treadmill exercise in the turkey. Comp. Biochem. Physiol. A, 72, 327.

Baudinette, R.V., J.P. Loveridge, K.J. Wilson, C.D. Mills, and K. Schmidt-Nielsen. (1978). Heat loss from feet of herring gulls at rest and during flight. Am. J. Physiol., 230, 920.

Baumel, J., and L. Gerchman. (1968). The avian intercarotid anastomosis and its homologue in other vertebrates. Am. J. Anat., 122, 1.

Bech, C., W. Rautenberg, B. May, and K. Johansen. (1980). Effect of spinal cord temperature on carotid blood flow in the Peking duck. Pfleugers Arch., 385, 269.

Bech, C., and K. Johansen. (1980). Blood flow changes in the duck during thermal panting. Acta Physiol. Scand., 110, 351.

Bech, C., and S. Nomoto. (1982). Cardiovascular changes associated with treadmill running in the Peking duck. J. Exp. Biol., 97, 345.

Bell, C. (1969). Indirect cholingeric vasomotor control of intestinal blood flow in the domestic fowl. J. Physiol., 205, 317.

Bell, C., and C. Rome. (1984). Pharmacological investigations of the vasodilator nerves supplying the duck's foot. Br. J. Pharmacol., 82, 801.

Bennett, T., and T. Malmfors. (1970). The adrenergic nervous system of the domestic fowl. Z. Zelliorsch. Mikrosk. Anat., 106, 22.

Berger, M., J.S. Hart, and O.Z. Roy. (1970). Respiration oxygen consumption and heart rate in some birds during rest and flight. Z. Vgl. Physiol., 66, 201.

Besch, E.B., and H. Kadono. (1977). Hypoxia: International Satellite Symposium on Respiratory Function of Birds, Adults and Embryonic, 27th, Göttingen, Germany, 1977. Berlin and New York: Springer-Verlag.

Boelkins, J.N., W.J. Mueller, and K.L. Hall. (1973). Cardiac output distribution in the laying hen during shell formation. Comp. Biochem. Physiol. A, 46, 735.

Bolden, S.L., L.M. Krista, G.R. McDaniel, L.E. Miller, and E.C. Nora. (1983). Effect of aortic atherosclerosis and other cardiovascular variables among hyper- and hypotensive turkeys. Poult. Sci., 62, 1287.

Bolton, T.B. (1967). Intramural nerves in ventricular myocardium of domestic fowl and other animals. Br. J. Pharmacol. Chemother., 31, 253.

Bolton, T.B. (1968). Studies on the longitudinal muscle of the anterior mesenteric artery of domestic fowl. J. Physiol. (London), 196, 273.

Braun, E.J., and W.H. Dantzler. (1974). Effect of ADH on single nephron glomerular filtration rate in the avian kidney. Am. J. Physiol., 226, 1.

Braun, E.J., M.E. Burrows, and S.P. Duckles. (1982). Avian renal portal valve: a reexamination of its control. Fed. Proc. Fed. Am. Soc. Exp. Biol., 41, 1008.

Bredick, H.E. (1960). Intraventricular pressure in chickens. Am. J. Physiol., 198, 153.

Brush, A.H. (1966). Avian heart size and cardiovascular performance. Auk, 83, 266.

Bunag, R.D., and E.J. Walaszek. (1962). Blockade of depressor responses to serotonin and tryptamine by lipergic acid derivatives in the chicken. Arch. Int. Pharmacodyn., 135, 1.

Burton, R.R., and J.R. Beljan. (1970). Animal restraint: Application in space environment. Aerospace Med., 41, 1061.

Burton, R.R., E.L. Besch, and A.H. Smith. (1968). Effect of chronic hypoxia on the pulmonary arterial blood pressure of the chicken. Am. J. Physiol., 214, 1438.

Butler, P.J. (1967). The effect of progressive hypoxia on the respiratory and cardiovascular systems of chickens. J. Physiol., 191, 309.

Butler, P.J. (1970). The effect of progressive hypoxia on the respiratory systems of pigeon and duck. J. Physiol., 210, 527.

Butler, P.J., and D.R. Jones. (1968). Onset of, and recovery from diving bradycardia in ducks. J. Physiol. (London), 196, 255.

Butler, P.J., and D.R. Jones. (1971). Variations in heart rate and regional distribution of blood flow on the normal pressor response to diving in ducks. J. Physiol. (London), 214, 457.

Butler, P.J., and A.J. Woakes. (1980). Heart rate, respiratory frequency and wing beat frequency of freeflying barnacle geese (*Branta leucopsis*). J. Exp. Biol., 85, 213.

Butler, P.J., and D.R. Jones. (1982). Comparative physiology of diving in vertebrates. Adv. Comp. Physiol. Biochem., 8, 179.

Butler, P.J., N.H. West, and D.R. Jones. (1977). Respiratory and cardiovascular responses of the pigeon to sustained level flight in a wind tunnel. J. Exp. Biol., 71, 7.

Carroll, R.G., and D.F. Opdyke. (1982). Evolution of angiotensin II-induced catecholamine release. Am. J. Physiol., 243, R65.

Catlett, R.L., and B.L. Johnson. (1974). Cardiac responses to diving in wild ducks. Comp. Biochem. Physiol. A, 47, 925.

Chand, N., and P. Eyre. (1975). Cardiovascular histamine receptors in the domestic chicken. Arch. Mt. Pharmacodyn. Ther., 216, 197.

Chen, T.W. (1970). Effects of oxytocin and adrenaline on oviduct motility, blood pressure, heart rate, and respiration rate of domestic hen. Nanyang Univ. J., 4, 178.

Clark, E.B., and N. Hu. (1982). Developmental hemodynamic changes in the chick embryo from stages 18 to 27. Circ. Res., 51, 810.

Cohen, D.H. and A.M. Schnall. (1970). Medullary cells of origin of vagal cardioinhibitory fibers in pigeon. II: Electrical stimulation of the dorsal motor nucleus. J. Comp. Neurol. 140, 321.

Cook, P.A., W.R. Siegfried, and P.G.H. Frost. (1977). Some physiological and biochemical adaptations to diving in three species of ducks. Comp. Biochem. Physiol. A, 57, 277.

Dijk, J.A. (1932). Arch. Neerl. Physiol., 17, 495. Cited from Prosser. (1950).

Dreyer, M.V., H.P.A. De Boom, P.D. De Wet, J.M.W. Le Roux, D.J. Coetzer, and F. Eloff. (1977). Excision and localization of the avian carotid body. Acta Anat., 99, 192.

Durfee, W.K. (1963). Cardiovascular reflex mechanisms in the fowl. Ph.D. Thesis, Rutgers University, New Brunswick, New Jersey.

Durfee, W.K., and P.D. Sturkie. (1963). Some cardiovascular responses to anoxia in the fowl. Fed. Proc. Fed. Am. Soc. Exp. Biol., 22, 182.

El-Ackad, T.M. (1972). Histamine in the avian cardiovascular system. Thesis, Rutgers University, New Brunswick, New Jersey.

El-Ackad, T., M. and P.D. Sturkie. (1972). Histamine in blood and tissues of Aves. Proc. Soc. Exp. Biol. Med., 141, 448.

El-Halawani, M.E., P.E. Waibel, J.R. Appel, and A.L. Good. (1973). Catecholamines and monoamine oxidase activity in turkeys with high or low blood pressure. Trans. N. Y. Acad. Sci., 35, 463.

Fedde, M.R., and M.M. Guffy. (1983). Routes of blood supply to the head of the Peking duck. Poult. Sci., 62, 1660.

Feigl, E., and B. Folkow. (1963). Cardiovascular responses in diving and during brain stimulation in ducks. Acta Physiol. Scand., 57, 99.

Ferguson, T.M., D.H. Miller, J.W. Bradley, and R.L. Atkinson. (1969). Blood pressure and heart rate of turkeys, 17–21 months of age. Poult. Sci., 48, 1478.

Fich, S., and W. Welkowitz. (1973). Perspectives in cardiovascular analysis and assistance. In "Perspectives in Biomedical Engineering" (R. M. Kenedi, Ed.). Baltimore: University Park Press. Proc. of Symposium held at University of Strathclyde, Glasgow, Scotland, in 1972.

Folkow, B., and E.H. Rubinstein. (1965). Effect of brain stimulation on diving in ducks. Hralrad Skrifter NR. Videnck-Akad. (Oslo), 48, 30.

Folkow, B., K. Fuxe, and R.R. Sonnenschein. (1966). Responses of skeletal musculature and its vasculature during diving in the duck: Pecularities of adrenergic vasoconstriction innervation. Acta Physiol. Scand., 67, 327.

Folkow, B., N.J. Nilsson, and L.R. Yonce. (1967). Effects of diving on cardiac output in ducks. Acta Physiol. Scand., 70, 347.

Folkow, B., and R. Yonce. (1967). The negative inotropic effect of vagal stimulation on the heart ventricles of the duck. Acta. Physiol. Scand., 71, 77.

Girard. H. (1973). Arterial pressure in chick embryos. Am. J. Physiol., 224, 454.

Gooden, B.A. (1980). The effects of hypoxia on the vasoconstrictor response of isolated mesenteric arterial vasculature from chickens and ducklings. Comp. Biochem. Physiol., C: Comp. Pharmacol., 67, 249.

Glenny, F.H. (1940). A systematic study of the main arteries in the region of the heart of aves. Anat. Rec., 76, 371.

Gross, K. (1974). Studies on Physiology of serotonin. Ph.D. Thesis, Rutgers University, New Brunswick, New Jersey.

Gross, K., and P.D. Sturkie. (1975). Concentration of serotonin in intestine and factors affecting its release. Proc. Soc. Exp. Biol. Med., 148, 1261.

Grubb, B. (1981). Blood flow and O_2 consumption in avian skeletal muscle during hypoxia. Am. J. Physiol., 50, R450.

Grubb, B. (1982). Cardiac output and stroke volume in exercising ducks and pigeons. J. Appl. Physiol., 53, 211.

Grubb, B. (1983). Allometric relations of cardiovascular function in birds. Am J. Physiol., 245, H567.

Grubb, B., C.D. Mills, J.M. Colacino, and K. Schmidt-Nielsen. (1977). Effect of arterial CO_2 on cerebral blood flow in ducks. Am. J. Physiol., 232, H596.

Grubb, B., J.M. Colacino, and K. Schmidt-Nielsen (1978). Cerebral blood flow in birds. Effect of hypoxia. Am. J. Physiol., 234, H230.

Grubb, B., J.H. Jones, and K. Schmidt-Nielsen. (1979). Avian cerebral blood flow. Influence of the Bohr effect on oxygen supply. Am. J. Physiol., 236, H744.

Grubb, B., D.D. Jorgensen, and M. Conner. (1983). Cardiovascular changes in the exercising Emu. J. Exp. Biol., 104, 193.

Hamlin, R.L., and R.M. Kondrich. (1969). Hypertension regulation of heart rate and possible mechanism contributing to aortic rupture in turkeys. Fed. Am. Soc. Exp. Biol., 28, 329.

Hartman, F.A. (1955). Heart weight in birds. Condor, 57, 221.

Harvey, S.C., E.G. Copen, D.W. Eskelson, S.R. Graff, L.D. Poulsen, and D.L. Rasmussen. (1954). Autonomic pharmacology of the chicken with particular reference to adrenergic blockade. J. Pharmacol. Exp. Ther., 112, 8.

Heymans, C., and E. Neil. (1958). "Reflexogenic Areas of the Cardiovascular System." Boston: Little, Brown & Co.

Hirsch, E.F. (1970). "The Innervation of the Vertebrate Heart." Springfield: Thomas.

Hollands, K.G., R.S. Gowe, and P.M. Morse. (1965). Effects of food restriction on blood pressure, heart rate and certain organ weights of the chicken. Poult. Sci., 6, 297.

Hollands, K.G., and E.S. Merritt. (1973). Blood pressure and its genetic variation and co-variation with certain economic traits in egg type chickens. Poult. Sci., 52, 1722.

Horton, E.W. (1971). Prostaglandins. In "Physiology and Biochemistry of the Fowl" (D.G. Bell and B.M. Freeman, Eds.), New York: Academic Press. 1971.

Hunsaker, W.G. (1959). Blood flow and calcium transfer through uterus of hen. Ph.D. thesis, Rutgers University, New Brunswick, New Jersey.

Johansen, K. (1964). Regional distribution of circulating blood during submersion asphyxia in the duck. Acta Physiol. Scand., 62, 1.

Johansson, P. (1983). Comparative aspects of central cardiovascular control with special reference to adrenergic mechanisms. Comp. Biochem. Physiol., C: Comp. Pharmacol., 74, 239.

Johnson, R.E., and F.R. Lockner. (1968). Heart size and altitude in ptarmigan. Condor, 70, 185.

Jones, D.R. (1969). Avian afferent vagal activity related to respiratory and cardiac cycles. Comp. Biochem. Physiol., 28, 961.

Jones, D.R. (1973). Systemic arterial baroreceptors in ducks and the consequence of their denervation on some cardiovascular responses to diving. J Physiol. 234, 499, 518.

Jones, D.R. (1981). Control of cardiovascular adjustments to diving in birds and mammals. "Advances in Physiological Sciences," Vol 20, Advances in Animal and Comparative Physiology (G. Pethes and V. L. Frenyo, Eds.), p. 307.

Jones, D.R., and P.J. Butler. (1982). On the cardioinhibitory role of water immersion per se in ducklings during submersion. Can. J. Zool., 60, 830.

Jones, D.R., and K. Johansen. (1972). The blood vascular system of birds. In "Avian Biology," Vol III, (D.S. Farner and J.R. King, Eds.). New York: Academic Press.

Jones, D.R., and M.J. Purves. (1970). The effects of carotid denervation upon the respiratory response to hypoxia and hypercapnia in the duck. J. Physiol., 211, 295.

Jones, D.R., and G.F. Holeton. (1972). Cardiovascular and respiratory responses of ducks to progressive hypocapnic hypoxia. J. Exp. Biol., 56, 657.

Jones, D.R., R.M. Bryan, Jr., N.W. West, R.H. Lord, and B. Clark. (1979). Regional distribution of blood flow during diving in the Auk. Can. J. Zool., 57, 995.

Jones, D.R., W.K. Milson, and N.H. West. (1980). Cardiac receptors in ducks: The effect of their stimulation and blockade on diving bradycardia. Am. J. Physiol., 7, R50.

Jones, D.R., W.K. Milsom, and G.R.J. Gabbott. (1982). Role of central and peripheral chemoreceptors in diving responses of ducks.

Jones, D.R., W.K. Milsom, F.M. Smith, N.H. West, and O.S. Bamford. (1983). Diving responses in ducks after acute barodenervation. Am. J. Physiol., 245, R222.

King, D.Z., and A.S. King. (1978). A possible aortic vasoreceptor zone in the domestic fowl. J. Anat., 127, 195.

Kramer, K., W. Lochner and E. Wetterer. (1963). Methods of measuring bloodflow. In "Handbook of Physiology," Section 2, Vol 2. (W. F. Hamilton & P. Dow, Eds.), American Physiological Society. Chapter 38, p. 1277.

Krista, L.M., R.E. Burger, and P.E. Waibel. (1963). Blood pressure and heart rate in the turkey as measured by indirect methods and their modifications by pharmacological agents. Poult. Sci., 42, 646.

Krista, L.M., P.E. Waibel, R.N. Shofkner, and J.H. Souther. (1970). A study of aortic rupture and performance as influenced by selection for hypertension and hypotension in the turkey. Poult. Sci., 49, 405.

Krista, L.M., E.C. Mora, and G.R. McDaniel. (1978). A comparison between aortic lumen surfaces of hypertensive and hypotensive turkeys. Poult. Sci., 58, 738.

Krista, L.M., S.D. Beckett, C.E. Branch, G.R. McDaniel, and R.M. Patterson. (1981). Cardiovascular responses in turkeys as affected by diurnal variation and stressors. Poult. Sci., 60, 462.

Kuenzel, W., J.W. Kusiak, P.C. Augustine, and J. Pitha. (1983). Effect of a beta-adrenergic antagonist on blood pressure, heart rate and beta-adrenoceptors in turkey poults. Comp. Biochem. Physiol., C: Comp. Pharmacol., 76, 371.

Langille, B.L., and D.R. Jones. (1975). Central cardiovascular dynamics of ducks. Am. J. Physiol., 228, 1856.

Langille, B.L. (1983). Role of venoconstriction in the cardiovascular responses to head immersion. Am. J. Physiol., 244, R292.

Lenel, R., L.N. Katz, and S. Rodbard. (1948). Arterial hypertension in the chicken. Am. J. Physiol., 152, 557.

Lillo, R.S., and D.R. Jones (1982a). Control of diving responses by carotid bodies and baroreceptors in ducks. Am. J. Physiol., 11, R105.

Lillo, R.S., and D.R. Jones. (1982b). Effects of cardiovascular variables on hyperpnea during recovery from diving. J. Appl. Physiol., 52, 206.

Lin, Y.C., P.D. Sturkie, and J. Tummons. (1970). Effects of cardiac sympathectomy reserpine and environmental temperatures on catecholamines, in the chicken heart. Can. J. Physiol. Pharmacol., 48, 182.

Lindsay, F.E. (1967). The cardiac veins of *Gallus domesticus*. J. Anat., 101, 555.

Lloyd, S., and M. Pickford. (1961). The persistence of a depressor response to oxytocin after denervation and blocking agents. B. J. Pharmacol. Chemother., 16, 129.

Magno, M.G. (1973). Cardio-respiratory responses to carotid body stimulation with NaCN in the chicken. Respir. Physiol., 17, 220.

Magno, M.G., and P.D. Sturkie. (1969). The mechanism of the phase lag of respiratory blood pressure waves in chickens. Physiologist, 12, 290 (Abstr.).

Malinow, M.R., and J.A. Moguilevsky. (1961). Effects of estrogens on atherosclerosis. Nature (London), 190, 422.

Manoch, M.A., S. Gitter, I.M. Levinger, and S. Stricker. (1971). On the origin of respiratory waves in the circulation. Pfleugers Arch., 325, 50.

McGinnis, C.H., and R.K. Ringer. (1966). Carotid sinus reflex in the chicken. Poult. Sci., 45, 402.

McGinnis, C.H., and R.K. Ringer. (1967). Arterial occlusion and cephalic baroreceptors in the chicken. Am. J. Vet. Res., 28, 1117.

McGregor, D.D. (1979). Non-cholinergic vasodilator innervation in the feet of ducks and chickens. Am. J. Physiol., 237, H117.

Merrill, G., R.E. Russo, and J.M. Halper. (1981). Cardiac output distribution before and after endotoxin challenge in the rooster. Am. J. Physiol., 241, R67.

Midgard, U., and C. Bech. (1981). Responses to catecholamines and nerve stimulation of the perfused rate tibotarsale and associated blood vessels in the hind limb of the Mallard (*Anas platyrhynchos*). Acta Physiol. Scand., 112, 77.

Millard, R.W., K. Johansen, and G. Milsom (1972). Unpublished data; see D.R. Jones and K. Johansen. (1972). In "Blood Vascular System of Birds" (D.S. Farner, U.J.R. King, Eds.). New York: Academic Press, Chapter 4.

Moore, A.F. (1980). The actions of Angiotensin II in the chicken. In "The Renin Angiotensin System" (J.A. Johnson and R.R. Anderson, Eds.). New York: Plenum, p. 79.

Moynihan, J.B., and N.A. Edwards. (1975). Blood flow in the reproductive tract of the domestic hen. Comp. Biochem. Physiol. A, 51, 745.

Muller, H.D., and M.E. Caroll. (1966). The relationship of blood pressure, heart rate, and body weight to aging in the domestic fowl. Poult. Sci., 45, 1195.

Murillo-Perrol, N.L. (1967). The development of the carotid body in *Gallus domesticus*. Acta Anat., 68, 102.

Nakamura, Y., H. Nishimura, and M.C. Khosta. (1982). Vasodepressor action of Angiotensin in conscious chickens. Am. J. Physiol., 243, H456.

Natoff, I.L., and M. Lockett. (1957). The assay of histamine, 5-HT, adrenaline and noradrenaline on blood pressure of the fowl. J. Pharm. Pharmacol., 9, 467.

Nickel, R., A. Schummer, E. Seiferle, W.G. Siller, and P.A.L. Wright. (1977). "Anatomy of the Domestic Birds. New York: Springer-Verlag.

Nishimura, H., Y. Nakamura, R.P. Sumner, and M.C. Khosla. (1982). Vasopressor and depressor actions of angiotensin in the anesthetized fowl. Am. J. Physiol., 242, H314.

Nolan, W.F., W.W. Weathers, and P.D. Sturkie. (1978). Thermally induced peripheral blood flow changes in chickens. J. Appl. Physiol., 44, 81.

Odlind, B. (1978). Blood flow distribution in the renal portal system of intact hen. A study of the venous system using microspheres. Acta Physiol. Scand., 102, 342.

Paff, G.H., R.J. Boucek, and G.S. Gutten. (1965). Ventricular blood pressures and competency of valves in early embryonic chick heart. Anat. Rec., 151, 119.

Petren, T. (1926). Die coronararterien des vogelherzens. Morphol., Jahrb., 56, 239.

Piiper, J., F. Drees, and P. Scheid. (1970). Gas exchange

in the domestic fowl during spontaneous breathing and artificial ventilation. Respir. Physiol., 9, 234.
Ploucha, J.M., R.K. Ringer, and J.B. Scott. (1981). Vascular responses of the chicken hind limb to vasoactive agents, asphyxia, and exercise. Can. J. Physiol., 59, i228.
Ploucha, J.M., and R.K. Ringer. (1981). Aortic wave velocity in chickens and ducks. Poult. Sci., 60, 2337.
Portmann, A. (1950). Des organes de la circulation sanguine. In "Traite de Zoologie," Vol 15 (P. P. Grasse, Ed.). Paris: Masson, p. 243.
Ray, P.J., and M.R. Fedde. (1969). Responses to alterations in respiratory PO_2 and PCO_2 in the chicken. Respir. Physiol., 6, 135.
Richards, S.A., and A.H. Sykes. (1967). The effects of hypoxia, hypercapnia, and asphyxia in the domestic fowl. Comp. Biochem. Physiol., 21, 691.
Ringer, R.K. (1968). Blood pressure of Japanese and bobwhite quail. Poult. Sci., 47, 1602.
Ringer, R.K., and K. Rodd. (1959). Hemodynamic changes associated with aging in the bronze turkey. Poult. Sci., 38, 395.
Ringer, R.K., H.S. Weiss, and P.D. Sturkie. (1955). Effect of sex and age on blood pressure in the duck and pigeon. Am. J. Physiol., 183, 141.
Ringer, R.K., P.D. Sturkie, and H.S. Weiss. (1957). Role of gonads in the control of blood pressure in chickens. Am. J. Physiol., 190, 54.
Rodbard, S., and A. Fink. (1948). Effects of body temperature changes on the circulation time in the chicken. Am. J. Physiol., 152, 383.
Rodbard, S., and Katz, L.N. (1949). The pulmonary arterial pressure. Am. Heart J., 38, 863.
Rodbard, S., and M. Tolpin. (1947). A relationship between body temperature and the blood pressure in the chicken. Am. J. Physiol., 151, 509.
Rodbard, S., N. Reyes, G. Mininni, and H. Saiki. (1954). Neurohumoral transmission of the pressor response to intracranial compression. Am. J. Physiol., 176, 455.
Sapirstein, L.A., and F.A. Hartman. (1959). Cardiac output and its distribution in the chicken. Am. J. Physiol., 196, 751.
Schaible, T.F. (1977). Some aspects of the physiology and pharmacology of histamine in the mesenteric vasculature and tissues of the chicken. Thesis, Rutgers University, New Brunswick, New Jersey.
Scanes, C.G., H. Mozolic, E. Kavanagh, G. Merrill, and J. Rabii. (1982). Distribution of blood flow in the ovary of domestic fowl and changes after prostaglandin $F\text{-}2_\alpha$ treatment. J. Reprod. Fertil., 64, 227.
Sillau, A.H., S. Cueva, and P. Morales. (1980). Pulmonary arterial hypertension in male and female chickens at 3300 meters. Pfleugers Arch., 366, 269.
Simpson, C.G. (1982). Antihypertensive effect of Captopril in turkeys. Proc. Soc. Exp. Biol. Med., 169, 101.
Simpson, C.F., R.H. Harms, and B.L. Damron. (1965). Failure of reserpine to modify the incidence of aortic ruptures induced in turkeys by diethylstilbestrol. Proc. Soc. Exp. Biol. Med., 120, 321.
Sisson, S., and J.D. Grossman. (1953). "Anatomy of domestic animals" (4th ed.). Philadelphia: Saunders.
Smith, A.H., W.L. Spangler, J.M. Goldberg, and E.A. Rhode. (1979). Tolerance of domestic fowl to high sustained G's. Aviat. Space Environ. Med., 50, 120.
Sommer, J.R., and R.J. Steere. (1969). Transverse tubules in chicken cardiac muscle. Fed. Proc. Fed. Am. Soc. Exp. Biol., 28, 328.
Speckman, E.W., and R.K. Ringer. (1962). The influence of reserpine on plasma cholestrol, hemodynamics, and arteriosclerotic lesions in Bronze turkeys. Poult. Sci., 41, 40.
Speckman, E.W., and R.K. Ringer. (1963). The cardiac output and carotid and tibial blood pressure of the turkey. Can. J. Biochem. Physiol., 41, 2337.
Speckman, E.W., and R.K. Ringer. (1964). Static elastic modulus of the turkey aorta. Can. J. Physiol. Pharmacol., 42, 553.
Speckman, E.W., and R.K. Ringer. (1966). Volume pressure relationship in the turkey. Can. J. Physiol. Pharmacol., 44, 901.
Strano, J.J., W. Welkowitz, and S. Fish. (1972). Measurement and utilization of in vivo blood pressure transfer functions of dog and chicken aortas. IEEE Trans. Biomed. Eng., 19, 261.
Stubel, H.S. (1910). Beiträge zur kenntnis der physiologie des blutkreislautes der verschiedenen. Arch. Gesamte Physiol. Menschen Tiere, 135, 249.
Sturkie, P.D. (1965). Avian Physiology, 2nd edition, Cornell Press, Ithaca, N.Y.
Sturkie, P.D. (1966). Cardiac output in ducks. Proc. Soc. Exp. Biol. Med., 123, 487.
Sturkie, P.D. (1967). Cardiovascular effects of acclimatization of heat and cold in chickens. J. Appl. Physiol., 22, 13.
Sturkie, P.D. (1970a). Seven generations of selection for high and low blood pressure in chickens. Poult. Sci., 49, 953.
Sturkie, P.D. (1970b). Circulation in Aves. Fed. Proc. Fed. Am. Soc. Exp. Biol., 29, 1674.
Sturkie, P.D. (Ed.). (1976). Circulation. In "Avian Physiology," (3d ed.) New York: Springer-Verlag.
Sturkie, P.D., and R.K. Ringer. (1955). Effects of suppression of pituitary gonadotrophin on blood pressure in the fowl. Am. J. Physiol., 180, 53.
Sturkie, P.D., and J.A. Vogel. (1959). Cardiac output, central blood volume and peripheral resistance in chickens. Am. J. Physiol., 197, 1165.
Sturkie, P.D., and K. Textor. (1961). Relationship of blood pressure level in chickens to resistance to physical stresses. Am. J. Physiol., 200, 1155.
Sturkie, P.D., and J.M. Eiel. (1966). Effects of estrogen on cardiac output, blood volume, and plasma lipids of the cock. J. Appl. Physiol., 21, 1927.
Sturkie, P.D. and Yuchong, Lin. (1968). Sex difference in blood norepinephrine of chickens. Comp. Biochem. Physiol., 24, 1973.
Sturkie, P.D., and D.W. Poorvin. (1973). The avian neurotransmitter. Proc. Soc. Exp. Biol. Med., 143, 644.
Sturkie, P.D., H.S. Weiss, and R.K. Ringer. (1953). The effects of age on blood pressure in the fowl. Am. J. Physiol., 174, 405.
Sturkie, P.D., R.K. Ringer, and N.S. Weiss. (1956). Relationship of blood pressure to mortality in chickens. Proc. Soc. Exp. Biol. Med., 92, 301.
Sturkie, P.D., W. Durfee, and M. Sheahan. (1957). Demonstration of an improved method on taking blood pressure in chickens. Poult. Sci., 36, 1160.
Sturkie, P.D., W.K. Durfee, and M. Sheahan. (1958). Effects of reserpine on the fowl. Am. J. Physiol., 194, 184.
Sturkie, P.D., J.A. Vogel, and K. Textor. (1962). Cardiovascular differences between high and low blood pressure chickens. Poult. Sci., 41, 1619.
Sturkie, P.D., D. Poorvin, and N. Ossario. (1970). Levels of epinephrine, and norepinephrine in blood and tissues of duck, pigeon, turkey and chicken. Proc. Soc. Exp. Biol. Med., 135, 267.
Sturkie, P.D., J.J. Woods, and D. Meyer. (1972). Serotonin levels in blood, heart, and spleen of chickens, ducks, and pigeons. Proc. Soc. Exp. Biol. Med., 139, 364.
Sturkie, P.D., and A. Abati. (1975). Blood flow in mesenteric, hepatic portal and renal portal veins of chickens. Pfleugers Arch., 359, 127.
Sturkie, P.D., G. Dirner, and R. Gister. (1977). Shunting of blood flow from the renal portal to hepatic portal circula-

tion of chickens. Comp. Biochem. Physiol. A, 58, 213.

Sturkie, P.D., G. Dirner and R. Gister. (1978). Role of renal portal valve in the shunting of blood flow in renal and hepatic portal circulations of chickens. Comp. Biochem. Physiol., C: Comp. Pharmacol. 59, 95.

Taylor, M.G. (1964). Wave travel in arteries and the design of the cardiovascular system. In "Pulsatile Blood Flow" (E.O. Attinger, Ed.). New York: McGraw Hill, Chapter 21.

Tazawa, H. (1981a). Measurement of blood pressure of chick embryos with an implanted needle catheter. J. Appl. Physiol., 51, 1023.

Tazawa, H. (1981b). Effect of O_2 and CO_2 in N_2, He, and SF_6 on chick embryo blood pressure and heart rate. J. Appl. Physiol., 51, 1017.

Thompson, R.M., and J.M. Coon. (1948). Effects of adrenolytic agents on response to pressor substances in the domestic fowl. Fed. Proc. Fed. Am. Soc. Exp. Biol., 7, 259.

Tucker, V.A. (1968a). Respiratory exchange and evaporative water loss in the flying Budgerigar. J. Exp. Biol., 48, 67.

Tucker, V.A. (1968b). Respiratory physiology of house sparrows in relation to high altitude flight. J. Exp. Biol., 48, 55.

Tucker, V.A. (1972). Metabolism during flight in the laughing gull, *Larus atricilla.* Am. J. Physiol., 22, 237.

Tummons, J.L. (1970). Nervous control of heart rate in domestic fowl. Ph.D. thesis, Rutgers University, New Brunswick, New Jersey.

Tummons, J.L., and P.D. Sturkie. (1969). Nervous control of heart rate during excitement in the adult Leghorn cock. Am. J. Physiol., 216, 1437.

Van Mierop, L.H.S., and C.J. Bertuch. (1967). Development of arterial blood pressure in the chick embryo. Am. J. Physiol., 212, 43.

Vogel, J. (1961). Studies on cardiac output in the chicken. Ph. D. thesis, Rutgers University, New Brunswick, New Jersey.

Vogel, J.A., and P.D. Sturkie. (1963a). Effects of starvation in the cardiovascular system of chickens. Proc. Soc. Exp. Biol. Med., 112, 111.

Vogel, J.A., and P.D. Sturkie. (1963b). Cardiovascular responses on chicken to seasonal and induced temperature changes. Science, 140, 1404.

Weiss, H.S., and P.D. Sturkie. (1951). An indirect method for measuring blood pressure in chickens. Poult. Sci., 30, 587.

Weiss, H.S., and M. Sheahan. (1958). The influence of maturity and sex on blood pressure of turkeys. Am. J. Vet. Res., 19, 209.

Weiss, H.S., R.K. Ringer, and P.D. Sturkie. (1957). Development of the sex difference in blood pressure of chickens. Am. J. Physiol., 188, 383.

Weiss, H.S., H. Fisher, and P. Griminger. (1961). Seasonal variation in avian blood pressure. Fed. Proc. Fed. Am. Soc. Exp. Biol., 2, 115 (abstr.).

Welkowitz, W., and S. Fich. (1967). A non-uniform hybrid model of the aorta. Trans. N. Y. Acad. Sci., 29, 316.

Whittow, G.C., P.D. Sturkie, and G. Stein, Jr. (1964). Cardiovascular changes associated with thermal polypnea in the chicken. Am. J. Physiol., 207, 1349.

Whittow, G.C., P.D. Sturkie, and G. Stein, Jr. (1965a). Cardiovascular effects of hypothermia in the chicken. Nature (London), 206, 200.

Whittow, G.C., P.D. Sturkie, and G. Stein, Jr. (1965b). Cardiovascular changes in restrained chickens. Poult. Sci., 44, 1452.

Wilson, J.R., and D.G. Butler. (1983a). Neural mechanisms affecting pressor responses to angiotensins in the Peking duck. Comp. Biochem. Physiol. A, 74, 351.

Wilson, J.R., and D.G. Butler. (1983b). 6-Hydroxydopamine treatment diminishes noradrenergic and pressor responses to angiotensin II in adrenalectomized ducks. Endocrinology, 112, 653.

Wilson, J.R., and D.G. Butler. (1983c). Adrenalectomy inhibits its noradrenergic, adrenergic and vasopressor responses to Angiotensin II in Peking duck. Endocrinology, 112, 645.

Woakes, A.J., and P.J. Butler. (1983). Swimming and diving in tufted ducks with particular reference to heart rate and gas exchange. J. Exp. Biol., 107, 311.

Wolfenson, D., A. Berman, Y.F. Frei, and N. Snapir. (1978). Measurement of blood flow distribution by radioactive microspheres in the laying hen. Comp. Biochem. Physiol. A, 61, 549.

Wolfenson, D., Y.F. Frei, N. Snapir, and A. Berman. (1981). Heat stress effects of capillary blood flow and its redistribution in the laying hen. Pfleugers Arch., 390, 86.

Wolfenson, D., Y.F. Frei, and A. Berman. (1982). Responses of the reproductive vascular system during the egg formation cycle of unanesthetized laying hens. Br. Poult. Sci., 23, 425.

Woods, J.J. (1971). Studies on the distribution and action of serotonin in the avian cardiovascular system. Ph.D. Thesis, Rutgers University, New Brunswick, New Jersey.

Woodbury, R.A., and W.F. Hamilton. (1937). Blood pressure studies in small animals. Am. J. Physiol., 119, 663.

Woodbury, R.A., and B.E. Abreu. (1944). Influence of oxytocin (pitocin) upon heart and blood pressure. Am. J. Physiol., 142, 114.

Wooley, P. (1959). The effect of posterior lobe pituitary extracts on blood pressure in several vertebrate species. J. Exp. Biol., 36, 453.

7
Heart: Contraction, Conduction, and Electrocardiography

P.D. Sturkie

The Cardiac Cycle

The sequence of events occurring in a complete heartbeat, a cardiac cycle, includes mechanical contraction of the atria and ventricles (systole) and relaxation of the heart muscle, (diastole). This sequence is followed by filling of the ventricles (diastasis). Accompanying these events are changes in volume and pressure in the atria and ventricles.

The contraction phase in mammals is normally the shorter phase and varies little with heart rate, but ventricular relaxation varies greatly and inversely with heart rate. Very little work has been conducted on these events in the avian cardiac cycle, but there is no reason to expect major differences between birds and mammals. The pressure-flow curves are different from those of mammals, but the difference is mainly one of degree. These are discussed in Chapter 6.

Langille and Jones (1975) reported that in the Peking duck contraction occurred synchronously in the right and left ventricles. At a mean heart rate of 219

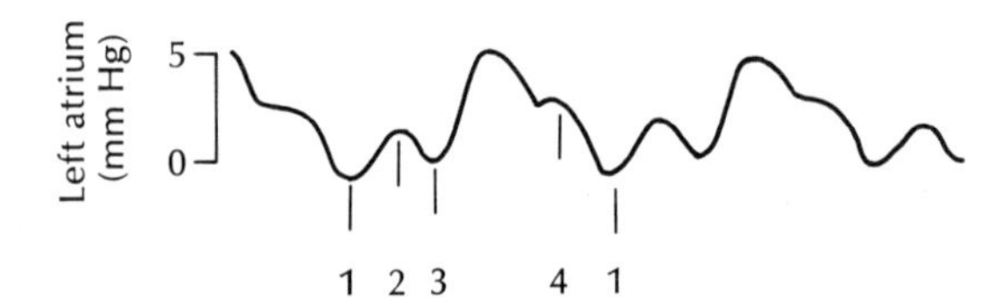

FIGURE 7–1. Pressure waves in the left atrium of the chicken (see text). (After Purton, © 1971 Cambridge University Press.)

beats/min, ventricular systole comprised 44% of the cardiac cycle, but the duration of systole in the right ventricular was 30% greater than in the left ventricle.

Purton (1971), who recorded atrial and ventricular pressures in chickens, reported no principal differences in the left ventricular pressures curves of birds and mammals, but he revealed significant differences in the atrial pressure waves which he divided into four stages (Figure 7–1). In stage 1 (points 1–2) pressure rises during systole. In stage 2 points (2–3), pressure drops during atrial relaxation, and there is a continued flow of blood into ventricules followed by closure of the AV (atrioventricular) valves and ventricular systole. Stage 3 (points 3–4) is atrial filling, in which the pressure shows two peaks; the first may be associated with the bulging of an AV valves into atrium and the second represents maximal atrial filling. In stage 4, (points 4–1), the AV valves open and atrial pressure falls to its lowest level as blood enters ventricle.

Factors Affecting Heart Rate

Small birds and mammals usually have higher heart rates than large ones, but there are exceptions.

Actually most heart rates previously determined (see Table 7–1) have been on birds restrained in different ways and degrees. It is now known that restraint influences heart rate considerably, probably in two ways. The initial excitement attending restraint increases heart rate and sympathetic discharge directly (Cain and Abbott, 1970; Cogger et al., 1974) and continued restraint up to 3 hr causes a progressive decrease in blood pressure and an increase in heart rate (reflexly; Whittow et al., 1965). When birds are allowed to move around (telemetry) or are only partially restrained, heart rates are much lower and probably represent the normal resting rates. The effect of a number of factors such as excitement, fright, exercise, flying, restraint, diving, and mating on heart rate, blood pressure, and cardiac output are discussed in Chapter 6.

Normal Heart Rates

The heart rates of several young and adult species are shown in Tables 7–1, 6–6, 6–7, and 6–11; those from embryos are given in Tables 6–2 and 7–2.

In Embryos. Embryonic heart rates have been determined by Girard (1973) (see Table 6–2), by Evans (1972), and by Soliman and Huston (1972), who have described methods of recording embryonic rates directly through the egg shell. These rates have been recorded from day 3 to hatching time and afterwards and range from 138 beats/min at 3 days to 221 beats/min on the 20th day (Girard, 1973). The figures found by Soliman and Huston (through the eggshell) range from 218 to 324/min during the same periods (Table 7–2). Differences in the handling temperature of egg and embryos may have accounted for much of this difference. The rates for young and growing chicks are shown in Tables 6–2 and 6–3 and Fig. 7–4 are discussed under age.

In Adults. It is apparent that adult heart rates vary considerably between and among species. Much of this variation is attributed to variation in tone or restraint of the cardioaccelerator (CA) and cardioinhibitor nerves (vagus) to the heart; the factors influencing the variation are discussed below.

Neural Control

Center for Heart Rate. There is evidence for a cardioinhibitory center in the medulla (Cohen et al., 1970). The cells involved have their greatest numbers in the ventral portion of the nucleus about 1 mm rostral to the obex. Stimulation in these areas produced bradycardia that could be blocked by atropine, indicative of a response from the cardioinhibitor center. Presumptive evidence for the operation of the cardioaccelerator and cardioinhibitor centers is derived from the reflex responses in heart rate resulting from factors that produce blood pressure changes (see Chapter 6).

Other afferent impulses arriving in the brain influence these centers. Stimulation of the central end of a sensory nerve such as vagus may affect these centers and thereby heart rate. Central stimulation of one cut vagus of chickens gives variable results (Sturkie, unpublished). In some instances, it decreased heart rate and in others there was no change, and in some there was an increase, apparently depending on the strength of the stimulus, and on whether one or both nerves was cut.

Central stimulation of the cut left vagus with the right one intact decreased heart rate and blood pressure in the duck, but opposite results were obtained when the central end of the right vagus was stimulated and the left vagus was intact (Johansen and Reite, 1964). Jones (1973) reported that central stimulation of one cut vagus in the duck caused an immediate fall in heart rate and blood pressure. Sectioning of the remaining vagus and central stimulation had no effect on heart rate or blood pressure even when the stimulus was 10 V or more.

TABLE 7–1. Heart rates of birds

Species	Age	Sex	Mean heart rate (beats/min)	Reference
White Leghorn chicken	1 day	—	286	Ringer et al. (1957)
	1 week	—	474	Ringer et al. (1957)
	7 weeks	Male	422	Ringer et al. (1957)
	7 weeks	Female	435	Ringer et al. (1957)
	7 weeks	Capon	425	Ringer et al. (1957)
	7 weeks	Poulard	452	Ringer et al. (1957)
	13 weeks	Male	367	Ringer et al. (1957)
	13 weeks	Female	391	Ringer et al. (1957)
	22 weeks	Male	302	Ringer et al. (1957)
	22 weeks	Female	357	Ringer et al. (1957)
	22 weeks	Capon	350	Ringer et al. (1957)
	22 weeks	Poulard	354	Ringer et al. (1957)
White Leghorn chicken				
Restrained	Adult	Male	304	Cain and Abbott (1970)
Unrestrained	Adult	Male	191	Cain and Abbott (1970)
Bronze turkey	8 weeks	Male	288	Ringer and Rood (1959; cited in Chapter 6)
	8 weeks	Female	283	Ringer and Rood (1959)
	12 weeks	Male	234	Ringer and Rood (1959)
	12 weeks	Female	230	Ringer and Rood (1959)
	18 weeks	Male	198	Ringer and Rood (1959)
	18 weeks	Female	212	Ringer and Rood (1959)
	22 weeks	Male	198	Ringer and Rood (1959)
	22 weeks	Female	232	Ringer and Rood (1959)
	22 weeks	Adult male	160	Ferguson et al. (1969)
	22 weeks	Adult female	219	Ferguson et al. (1969)
Turkey, white	6 weeks	Male	323	Simpson (1982)
	10 weeks	Male	307	Simpson (1982)
	14 weeks	Male	280	Simpson (1982)
	6 weeks	Female	340	Simpson (1982)
	10 weeks	Female	313	Simpson (1982)
	14weeks	Female	280	Simpson (1982)
				Simpson (1982)
				(Simpson entries: Cited in Chapter 6)
Turkey, white	Adult	Male	152	Krista et al. (1981)
Hypertensive		Female	186	Krista et al. (1981)
Hypotensive		Male	143	Krista et al. (1981)
		Female	163	Krista et al. (1981)
				(Krista entries: Cited in Chapter 6)
Peking duck	4 months	Male	194	Ringer et al. (1955)
	4 months	Female	190	Ringer et al. (1955)
	12–13 months	Male	189	Ringer et al. (1955)
	12–13 months	Female	175	Ringer et al. (1955)
White king pigeon	Adult	Male	202	Ringer et al. (1955)
	Adult	Female	208	Ringer et al. (1955)
Robin	Adult		570	Woodbury and Hamilton (1937; cited in Chapter 6)
	Adult		384	Lewis (1967)
Canary	Adult		795	Woodbury and Hamilton (1937; cited in Chapter 6)
Blue-winged teal	Adult		1000	Tigerstedt (1921)
Blue jay	Adult		165	Owen (1969)
Catbird	Adult		307	Lewis (1967)
Brown thrasher	Adult		465	Lewis (1967)
Wood thrush	Adult		303	Lewis (1967)
Coturnix (Japanese quail)			363	Lewis (1967)
Restrained	Adult	Male	445	Cogger et al. (1974)
Unrestrained (telemetry)	Adult	Male	368	Cogger et al. (1974)

TABLE 7–2. Average daily heart rate of Single Comb White Leghorn chick embryos[a]

Day	Mean beats/min ± SE[b]	Day	Mean beats/min ± SE
3	218 ± 4.3	12	252 ± 6.6
4	224 ± 6.5	13	261 ± 4.3
5	225 ± 3.6	14	272 ± 5.1
6	230 ± 5.9	15	276 ± 3.9
7	232 ± 6.7	16	274 ± 7.0
8	248 ± 5.0	17	273 ± 5.3
9	248 ± 3.9	18	265 ± 6.2
10	248 ± 6.2	19	263 ± 4.8
11	249 ± 6.8	20	324 ± 8.1[c]

[a] From Soliman and Huston (1972).
[b] Means of the heart beats per minute of 12 embryos, each measurement is the average value of three records at 5-min intervals (standard error is presented).
[c] Pipped eggs: heart beat rate during hatching.

Cardiac Nerves These are the vagus and cardioaccelerator nerves. Their origin and distribution to the heart are shown in Figure 7–2. The cardiac sympathetic nerve in the chicken arises from the first thoracic ganglion between the first and second ribs, near the most caudal branch of the brachial plexus. It runs as a single discrete nerve parallel to a small vertebral vein and thence with the vena cava to the heart, where it branches and joins with vagal fibers to form a cardiac plexus (Tummons and Sturkie, 1969). According to Ssinelnikow (1928) there are six plexuses; the right and left anterior cardiac plexuses, right and left posterior plexuses, and the anterior and posterior atrial plexuses. The atria and the sinoatrial and atrioventricular nodes receive the most extensive innervation (Bennett and Malmfors, 1970; Akester, 1971; Bolton, 1967; these references are cited in Chapter 6).

The origin and distribution of cardioaccelerator fibers in the pigeon appear to be more complex than in chickens, according to Macdonald and Cohen (1970), who reported that accelerator fibers arise from some of the cervical spinal segments as well as two and sometimes three thoracic segments. The relatively simple anatomic arrangement in the chicken, however, makes the operation of denervation much easier to perform in them; this was first done by Tummons and Sturkie (1968). After bilateral transection of the vagi and sympathetics, electrical stimulation of the peripheral end of right cardioaccelerator nerves (4 V, 40 Hz, 0.2 msec) increased heart rate from 235 to 345/min, an increase of 48%; the increase after stimulation of the left nerve was 32%. Stimulation of the right cardioaccelerator (CA) in the pigeon, but not stimulation of the left, consistently increases heart rate (Macdonald and Cohen, 1970).

To assess the degree of control or influence exerted by the intact accelerators and vagi nerves (tone), a series of experiments were conducted involving isolation, transection, and stimulation of nerves, employment of β-adrenergic and cholinergic blockade, various drugs known to deplete cardiac catecholamines, and the effects of various environmental factors.

Earlier work employing adrenergic blockade in ducks and seagulls (Johansen and Reite, 1964) and pigeons (Cohen and Pitts, 1968) emphasized the important role of accelerators in controlling heart rate. The report here relates to transection and stimulation studies on chickens. The accelerator and vagal nerves were isolated and transected as previously described (Tummons and Sturkie, 1968, 1969); the birds were allowed to recover from surgery, and the stimulation studies were carried out 6 days later. The pertinent results are shown in Figure 7–3. It is clear that sympa-

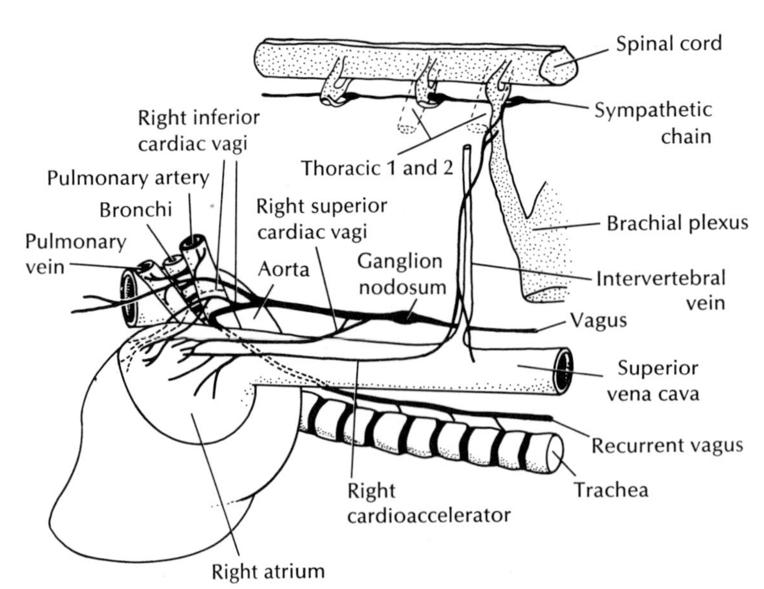

FIGURE 7–2. Diagram of cardiac nerves of chicken. (After Tummons and Sturkie, 1969.)

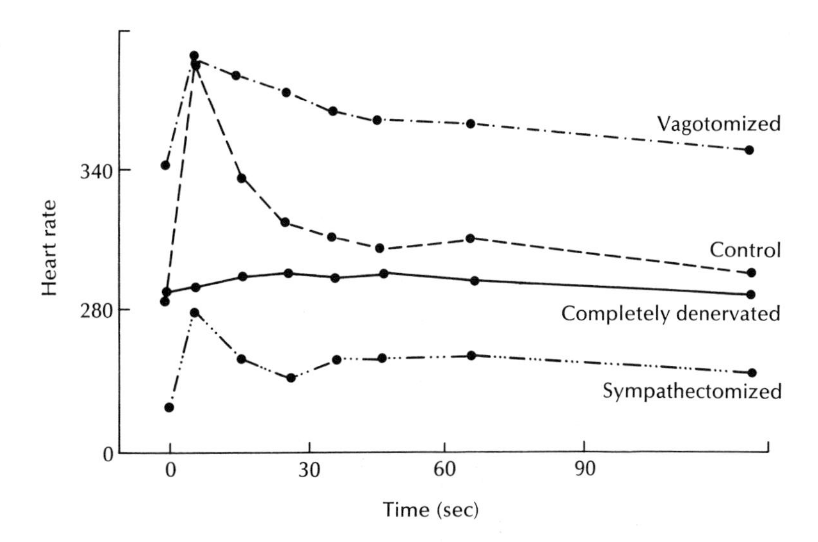

FIGURE 7–3. Heart rate response with time to excitement of control, vagotomized, sympathectomized, and completely denervated adult White leghorn cocks. (After Tummons and Sturkie, 1969.)

thectomy alone decreased heart rate about 20% from the basal resting rate (284/min). Vagotomy alone increased heart rate about 20% above the resting rate and atropinization gave the same results. When both nerves were cut, heart rate was almost the same as the basal resting rate (see Figure 7–3. This figure also illustrates the effect of excitement on relative tones of the vagus and accelerator nerves. Excitement increases heart rate most when the vagi are intact and less so when the vagi are cut. Even after sympathectomy, excitement increases heart rate some but not above the basal resting rate, indicating a decrease in vagal restraint. After complete denervation, heart rate does not change with excitement and remains near the resting rate level. Similar results have been obtained by complete adrenergic and cholinergic blockade except that basal heart rate is somewhat lower than in the completely denervated birds, presumably because β blockade affects circulating catecholamines as well as neuronal ones (Tummons and Sturkie, 1970a; Sturkie, 1970).

After abolition of vagal tone with atropine (A) or by vagotomy, the heart rate resulting from sympathetic tone (unrestrained by vagus) is evident and is increased. Abolition of sympathetic tone (ST) by propranolol (P) with the vagus intact depresses heart rate below the normal level. Abolition of vagal and sympathetic tone (PA) results in a heart rate about normal. This is also the intrinsic heart rate. In other words, vagal tone (VT) and sympathetic tone operate to about the same degree; therefore the sympathetic nerves increase heart rate to the same degree that the parasympathetic nerves (vagus) decrease heart rate. This technique has been used to determine the effects of age, sex, temperature, drugs, etc., on the degree of vagal and cardioaccelerator tone. This technique is illustrated in Figures 7–4, and 7–5, and 7–6.

Thus,

$$ST = \frac{\text{sympathetic rate (A} - \text{PA)}}{\text{intrinsic rate (PA)}} \times 100$$

and

$$VT = \frac{\text{parasympathetic rate (P} - \text{PA)}}{\text{intrinsic rate (PA)}} \times 100$$

Stimulation of the *right sympathetic nerve* of the isolated heart of chickens (20 Hz, 1 msec, 30 V) for 15 sec increased the rate from 140 to 194 beats/min. Infusion of tyramine for 30 min increased heart rate from 111 to 210 beats/min (Engel and Löffelholz, 1976). Some of the same authors (Desantis et al., 1975) who also stimulated right sympathetic nerve at 40 V, 1 msec, 20 HZ for 1 min reported a mean increase in heart rate of 119 beats/min.

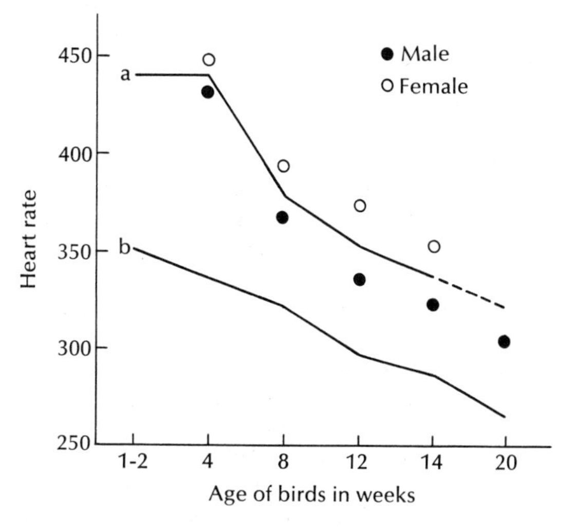

FIGURE 7–4. Normal resting heart rates (a) of chicken males and females at various ages; (b) intrinsic heart rates resulting from administration of propranolol and atropine (PA). (After Sturkie and Chillseyzn, 1972.)

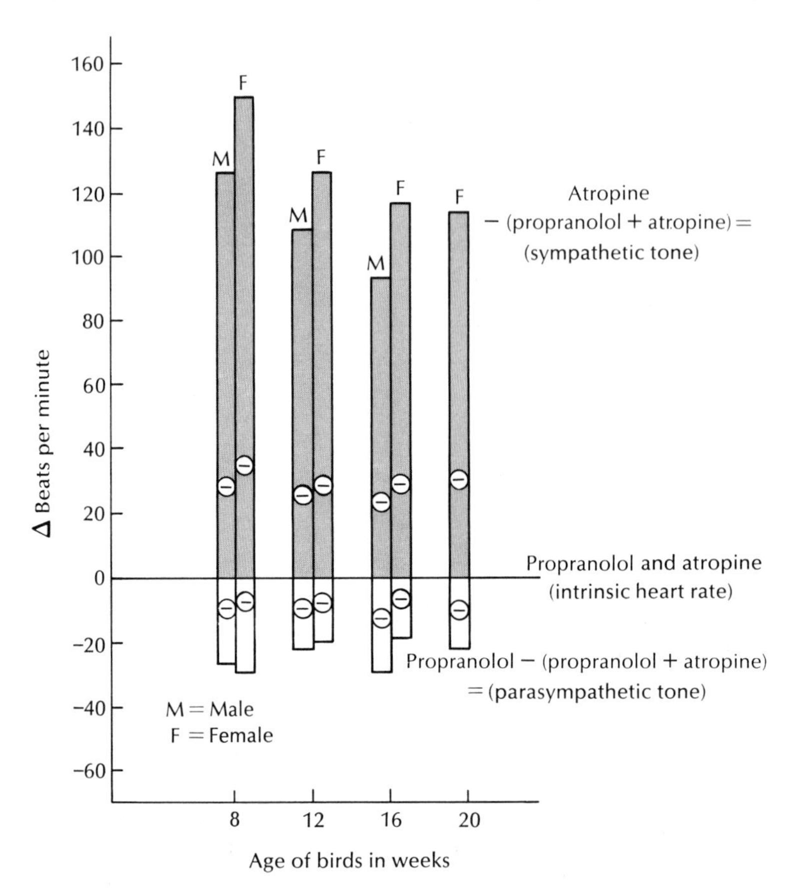

FIGURE 7–5. Differences in heart rate (actual, Δ, and percent, ⊖), from intrinsic rate after administration of atropine and propranolol for *M* (male) and *F* (female) birds. Change in sympathetic tone is greatest after 8 weeks and decreases thereafter with age, but with no significant change in parasympathetic tone. See Sturkie (1970) for further details of method.

Abati (1976) and Sturkie (1978) employed the isolated chicken heart for various studies and showed that to maintain a high basal heart rate and contractility, additional calcium must be added to the perfusion medium. Sturkie and Poorvin (1973) reported increases of heart rate after stimulation of CA nerve of chicken isolated hearts of the same magnitude as reported above.

Vagal restraint appears to vary considerably, depending on the species and investigator. Stubel (see Chapter 6) concluded that species with large hearts in relationship to body size, such as ducks, pigeons, and gulls have slower heart rates, and their vagi exert more control or tone than in species like the chicken. Bilateral vagotomy in the duck increases heart rate from 65 to 200% (Johansen and Reite, 1964; Butler and Jones, 1968), but in the chicken the increase is only 20–30% (Tummons and Sturkie, 1969). The tone exerted by each vagus is variable (Jones, 1973; Jones and Purves, 1970). In some instances, the left vagus was dominant, and in others, the right vagus; in others there was no difference. According to Butler and Jones (1968), control is exerted by only one vagus at a time, on the basis of increasing blockade of the right or left nerve with cold.

Stimulation of the peripheral end of a cut vagus with the other intact produced bradycardia and arrest in ducks (Johansen and Reite, 1964), and the right vagus had greater effect than the left. Langille (1983) produced cardiac arrest in ducks that was maintained for at least 8 sec before escape. The right vagus received a stimulus of 3–5 V, 5 msec duration, and 25 Hz. Blix et al. (1974) produced cardiac slowing in ducks with supramaximal stimulation of left vagus, but not arrest. Moore (1965) reported cardiac arrest with vagal escape in the turkey after vagal stimulation.

There seems to be some question about producing cardiac arrest in chickens by stimulating the vagus. Some of the older workers claimed to have produced arrest momentarily in chickens and pigeons (Jurgens, in 1909, and Paton, in 1912; see Sturkie, 1965). Recent work by Goldberg et al. (1983) indicated that cardiac arrest is not produced by stimulating either vagus of chickens at what they termed a supramaximal stimulus (20 Hz, 5 msec duration, and 4–8 V). Their results are (heart rate, in beats per minute):

Control	After stimulation of:	
	R. Vagus	L. Vagus
280	116	160

FIGURE 7–6. Degree of change in heart rate of Δ beats/min and Δ %) from resting rate (0) after administration of atropine alone, of propranolol, of or both atropine and propranolol for chickens acclimated to cold normal temperatures (control conditions) and heat. Birds acclimated to heat had slower heart rates. Their rates increased most after administration of atropine, evidence of increased parasympathetic tone. (After Sturkie et al., 1970.)

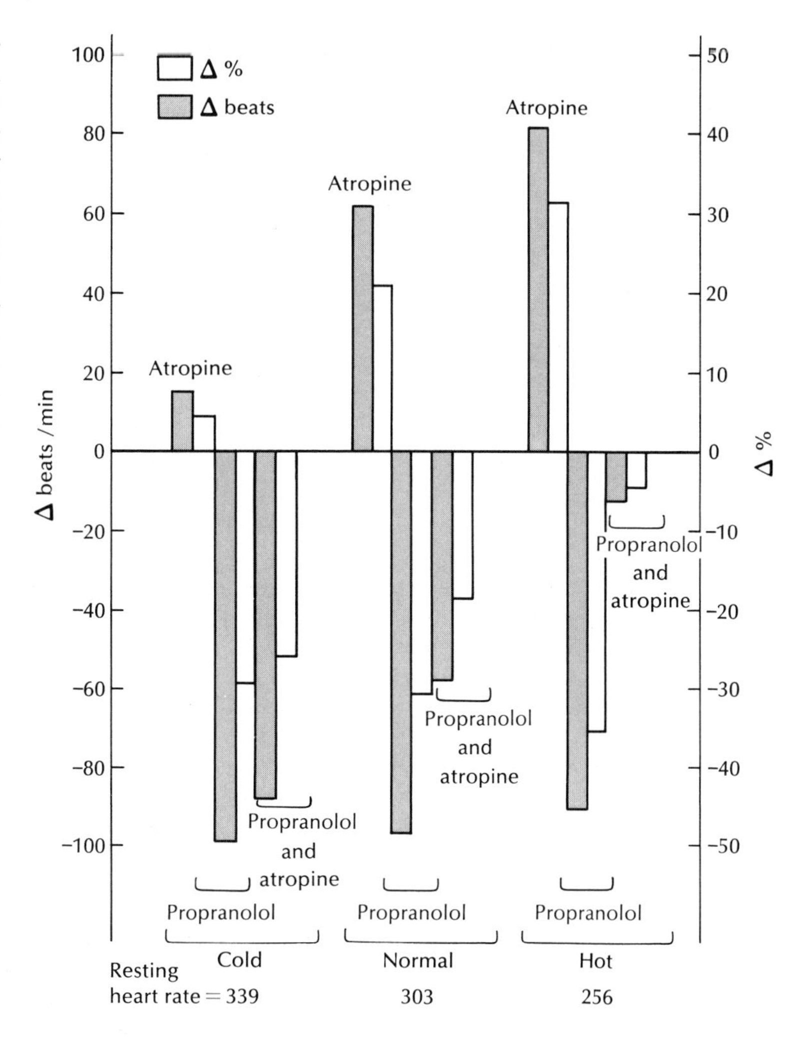

When the results were combined, the decrease averaged 42%.

Results from another quarter tend to support these results. Dieterich and Löffelholz (1977) and Dieterich et al. (1977) stimulated both vagi in the isolated chicken heart at 20 Hz, 40 V, for 15–20 min, and reported a decrease in heart rate of 50% in the chicken and 51% in the pigeon heart, compared to 90% in the cat heart. There was also a tremendous increase in the release of acetylcholine from the heart (see Table 7–3), at least 95 times that from the cat heart.

In another report, Dieterich et al. (1976) showed that heart rate following vagal stimulation of isolated heart decreased most (90%) during the first minute of stimulation, and thereafter increased so that after 5 min of stimulation, the rate was 54% of the control rate. The greatest decrease in rate coincided with the greatest output of acetylcholine (see later section). No evidence of cardiac arrest was reported. Sturkie (unpublished) infused acetylcholine in slug doses of 75–100 μg to chicken isolated hearts and produced cardiac arrest and mainly AV blocks in some of the specimens. Reflex changes in heart rate are discussed in Chapter 6.

Neurotransmitters

The *sympathetic* neurotransmitter (NT) in mammals is norepinephrine (Ne), and it is principally so in birds, although there is some evidence that epinephrine (E) is also a secondary NT. Details will be presented later. If E is considered a definite NT, more information is required on its origin. It is well known that Ne is synthesized in the nerve terminal in a series of steps beginning with tyrosine → dopa → dopamine → Ne. (See chapters 14 and 22 for more details.)

After its release from the nerve terminals, N may be acted on by phenylethanolamine (PNMT) to form epinephrine. The NH_2 group of Ne is now methylated to form N–H–CH_3: This is the only structural difference between Ne and E.

TABLE 7–3. Cardiac content of acetylcholine and acetylcholine overflow from isolated perfused hearts of various species. Both vagus nerves were stimulated at 20 Hz, 1 msec, 40 V for 15 min. The effuent was collected in 1- or 2-min samples immediately preceding the stimulation ("spontaneous overflow") and during the stimulation ("evoked overflow") for acetylcholine (ACh) assay[a]

Species	n	Cardiac weight (g w.w.)	Perfusion rate (ml/min)	Spontaneous ACh overflow (pmol/g/15 min)	Evoked ACh overflow (pmol/g/min)[b]	Ach in heart tissue	
						(nmol/g)	(nmol/heart)
Chicken	8	4.3 ± 0.2[c]	15	20.3 ± 2.7	1615 ± 380	8.3 ± 1.4	38.2 ± 6.9
Pigeon	3	3.6 ± 0.3	10	6.3 ± 0.9	949 ± 96	10.3 ± 1.2	37.8 ± 4.8
Duck	3	5.0 ± 0.2	20	21.1 ± 5.3	97 ± 34	11.5 ± 0.9	57.7 ± 3.3
Cat	4	6.5 ± 2.3	20	< Limit	< Limit	6.4 ± 0.9	38.8 ± 11.1
Rabbit	4	4.5 ± 0.3	20	< Limit	< Limit	2.2 ± 0.4	9.8 ± 2.7

[a] Dieterich et al. (1977).
[b] Note that these values indicate the evoked overflow after subtraction of the spontaneous overflow.
[c] Data are means ± SEM.

The critical test of a NT is its release following stimulation of the nerves involved. Stimulation of the cardioaccelerator (CA) nerve causes a release of Ne, which can be determined chemically. The Ne granules in the nerve terminals can be observed by a histofluorescent technique (Figure 7–7). There is presumptive evidence that E may be more important as a NT in birds than Ne, because the levels of E in the heart and blood of birds are higher than Ne (Sturkie and Poorvin, 1973; Sturkie et al., 1970; Lin et al. (1970). However, the higher levels of E in blood and heart may be derived from extraneuronal sources such as the adrenals or chromaffin cells in the heart. It has been demonstrated by the workers above that excitement, handling, and exsanguination increase Ne and E in heart and blood, particularly E; this was confirmed by Jurani et al. (1980).

Sturkie and Poorvin (1973) stimulated the CA nerve in the isolated heart of chickens and reported that the perfusate contained about 95% Ne and 5% E. They concluded that Ne was the NT, and that it was released from the nerve terminals where normally it is tightly bound; however, E is derived mainly from other sources and is loosely bound in heart muscle, except that found in the chromaffin cells. Further evidence for this view will be presented later.

Perfusion of the isolated heart (unstimulated) for 5 min decreased the level of E in the perfusate to almost unmeasurable levels. Collection of Ne and E from about 200 ml of perfusate following 7 min of stimulation resulted in 1888 ng/ml of Ne and 80 ng/ml of E, or about 95% Ne.

Desantis et al. (1975) and Engel and Löffelholz (1976) reported that after stimulation of the CA nerve in isolated chicken hearts, about 80% of catecholamines released in the perfusate were Ne and 20% were E.

The discrepances in the results of these workers and

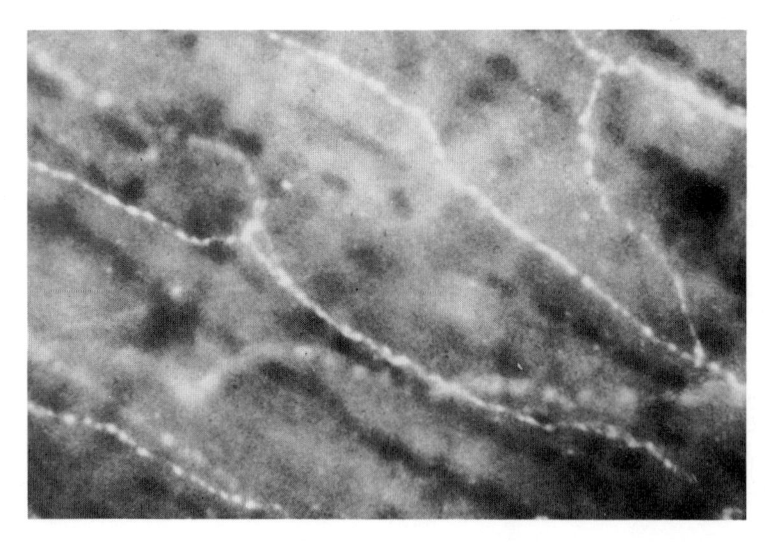

FIGURE 7–7. Granules of norepinephrine in sympathetic nerve terminals (*white beads in chain*) in heart of chicken (not showing fluorescense). Stretch preparation and histofluorescense. White chain of granules are greenish in fluorescence. (Poorvin, 1972.)

those of Sturkie and Poorvin are not evident. Both employed the trihydroxyindole fluorescent chemical method, but the levels of Ne and E in blood, heart, and perfusate were much higher in the results of Desantis et al. For example, their figures for Ne in the blood of chickens were 15–30 times, and for E, 10–15 times, those reported by Sturkie et al. (1970). Their figures were also much higher than those of Jurani et al. (1980) and other investigators. Further work is needed involving a more reliable and sensitive method of determination of Ne and E (see also Chapter 22). Likewise, more research is required on the source of E in the heart and whether or not it is loosely bound in heart muscle and not released from nerve terminals. Whether or not some of it is transformed from Ne by the enzyme PNMT, whose concentration is very high in chicken organs and hearts (Lindmar and Wolff, 1975), remains to be determined.

Sympathectomy results in the depletion of nerve terminals of Ne granules so that none are observed by histofluorescence. Likewise, reserpine administration depletes these granules in the chicken heart within 24 hr (Lin et al., 1970; Sturkie, 1970; Tummons and Sturkie, 1970a). After nerve depletion, there are still measurable amounts of Ne and E in the heart (Lin et al. 1970). The levels of Ne and E before CA nerve denervation were 848 Ne and 1879 E ng/g of tissue; after denervation, the levels were 58 Ne and 155 E ng/g. Similar results were obtained after reserpine administration.

After such depletion, the nerve terminals do not exhibit fluorescense, but the large chromaffin cells of heart do (Poorvin, 1972). CA nerve stimulation after reserpine depletion still causes an increase in heart rate (Tummons and Sturkie, 1970b), which is evidence of the effect of Ne or E, most probably the latter.

Types of Receptors in Sympathetic Nerves

Chicken hearts, unlike those of mammals, possess muscarinic inhibitor receptors but no nicotinic excitatory ones (Engel and Löffelholz, 1976). Employing the isolated chicken heart and strips, they showed that dimethylphenylpiperazinium (DMPP) caused the release of only Ne from the perfused heart, but that tyramine and high potassium solution released both Ne and E. Nicotine and acetylcholine (ACh) plus atropine were ineffective when added to the perfusate. Stimulation of the CA nerve of the isolated perfused chicken heart with ACh decreased the output of Ne to about 50%, but no E was detected. When ACh and atropine were perfused and the CA nerve stimulated, the output of Ne increased above control levels, presumably because previous inhibition of ACh had been eliminated.

On the basis of the use of α and β agonists and antagonists, St. Petery and Van Mierop (1974) suggested the presence of α and β receptors in the 6-day-old chick embryo. Bennett and Malmfors (1975) also reported details on sympathetic innvervation of chick atria, including a β receptor. However, catecholamine receptors, based on histofluorescense studies, are not observed until 12 and 14 days in the right atria and ventricles of chicken embryos (Higgins and Pappano, 1979).

Parasympathetic Neurotransmitter

Acetylcholine (ACh) is the substance released from vagal and other parasympathetic nerve endings, as was demonstrated in the frog in 1921 by Loewi. ACh is synthesized and released in postganglionic, cholinergic nerve fibers.

Release of ACh in isolated chicken hearts was studied by Dieterich and Löffelholz (1977) and in chicken, pigeon, and duck as well as in certain mammals (Dieterich et al., 1977). For an extensive review of ACh release, see Löffelholz (1981).

As soon as ACh is released from nerve endings it is acted on by acetylcholinesterase. The amount of unhydrolyzed ACh that escapes into the blood stream of intact animals or perfusate of isolated hearts varies, but is much higher in birds than mammals. When the vagus of the isolated chicken heart is stimulated at 20 Hz, 1 msec, and 40 V for 15 min, ACh is released. The unhydrolyzed portion is called ACh overflow. Actually, about 71% of that released is hydrolyzed. The inactivation of ACh is dependent on the rate of hydrolysis and rate of extracellular washout. The spontaneous and evoked overflow of ACh have been studied in a number of avian and mammalian species. (Dieterich et al., 1977) (see Table 7–3).

The rate of ACh overflow, both spontaneous and evoked by nerve stimulation, is very high in birds, particularly the chicken. When a cholinesterase inhibitor was employed (physostigmine), the content of ACh in the heart after perfusion and stimulation was increased 25% (Dieterich et al., 1976). The evoked maximum overflow in the chicken was 284 pmol/min, which is about 95 times that in the cat.

Cholinesterase as well as ACh content of the isolated heart have been determined by Dieterich et al., (1976) after vagal stimulation for 5 min. ACh content was 4.2 nmol/g of tissue after the stimulation period. Following the administration of a cholinesterase inhibitor, the ACh level was 6.8 nmol/g.

Cholinesterase activity in the perfused but nonstimulated heart averaged 551 units/g of wet tissue. When the cholinesterase inhibitor was given, cholinesterase activity was inhibited 85–96%.

The output of ACh of the nerve-stimulated heart was greatest during the first 0.5 min of stimulation (~250 pmol/g) and dropped to 175 within 1.5 min, and to about 100 in 2.5 min; it changed little during the remaining 2.5 min of stimulation. The authors also found that the greatest decrease in heart rate following vagal stimulation coincided with the highest output of ACh (see earlier section).

Hydrolysis of ACh in isolated chicken heart was influenced by rate of perfusion. A decrease in perfusion rate from 30 to 20 to 10 ml/min decreased the spontaneous and evoked output of ACh to one-half or one-third (Dieterich and Löffelholz, 1977).

The most likely explanation for the high ACh overflow from avian hearts seems to be a high release of the transmitter from the terminal nerves in both relative and absolute amounts relative to ACh stores. This would require a high rate of ACh synthesis (Dieterich et al., 1977), but this rate is no higher than in mammals (Dieterich et al. 1976) and suggests that the higher output of ACh in birds results from the greater vagal innervation of avian ventricles.

Stimulation of muscarinic receptors in the isolated chicken heart increased extracellular choline by mobilization of cellular choline, which is involved in acetylcholine synthesis (Corradetti et al., 1983).

Contractility

When heart muscle contracts, it shortens and moves a load (isotonic contraction); as the load increases, the muscle reaches a point where it no longer shortens; this is isometric contraction, but there is an increase in tension. When the heart contracts with all valves closed, it contracts under a constant load and volume (isometrically), but no blood is ejected (isometric phase). Contractility may be defined as the ability of the heart muscle to contract and do work at a definite fiber length. Contractile force is increased when the muscle fiber is stretched, as when the heart contracts and ejects blood with increased end diastolic volume and pressure (Frank–Starling mechanism).

When the heart's contractile force increases without a change in muscle fiber length, this represents a change (increase) in contractility. Contractility can be measured in heart muscle by recording the rate of change in ventricular pressure (dp/dt) from beginning pressure to peak pressure, or during the maximum rate of pressure rise (dp/dt max).

A number of factors, such as epinephrine, norepinephrine, and sympathetic nerve stimulation, are known to increase contractility. Histamine, vagal nerve stimulation, and others decrease contractility. Contractility changes are referred to as isotonic; chronotropic refers to changes in heart rate.

Embryos

Contractility studies on isolated avian embryonic hearts at different stages of development, usually before and after innervation, have been conducted. Lee et al. (1960) reported that nicotine had an inotropic effect on noninnervated as well as innervated embryonic heart, and so did acetylcholine on the atropinized heart. Reserpine, which depleted the embryonic hearts of catecholamines, decreased the positive inotropic response to nicotine. Norepinephrine had a positive inotropic effect on embryos at all stages (Michal et al., 1967), but tyramine had no effect until the embryo was at least 14 days old. Its effect may be attributable to a release of catecholamines at this stage.

Methylisobutylxanthine (MIX) has a positive inotropic effect in isolated ventricles of embryos (chick), and the effect may be due to the action of accumulated cAMP (Biegan et al., 1980); this effect is blocked by ACh, which may affect adenyl cyclase. Hypoxia decreases the contractility of cultured chick embryo ventricular cells (Barry et al., 1980).

(DMPP), which releases N from hearts, has a positive chronotropic effect in hatched chicks; it increased contractility in embryonic hearts, but nicotine did not (Pappano, 1976). As early as 4 days of embryonic life and later, isoproterenol has a positive inotropic effect, but sensitivity to it increases greatly at 16 days, and adrenergic nerves are capable of altering ventricular contraction at this time (Higgins and Pappano, 1981).

Adults

Effects of Sympathetic Transmitter. Stimulation of sympathetic nerves or administration of epinephrine or norepinephrine have positive chronotropic and inotropic effects in chickens and mammals. This has been demonstrated in the isolated chicken heart by Sturkie and Poorvin (1973), Poorvin (1972), Desantis et al., (1975), Abati (1976), and (in intact chickens) by Tummons and Sturkie (1968, 1970b).

Effects of Parasympathetic Transmitter. Folkow and Yonce (1967) paced electronically hearts of Peking ducks and stimulated the vagus, and reported a greatly reduced inotropic response (negative). They recorded a drop in ventricular pressure from a normal of about 160 to 110 mm Hg and a drop in cardiac output from 1312 to 950 ml/m.

Later data have proven conclusively that vagal stimulation or administration of ACh to isolated chicken hearts at a constant heart rate (paced) produce a negative inotropic effect (Sturkie, 1976, unpublished), and in unpaced chicken hearts (Dieterich et al., 1976; Dieterich and Löffelholz, 1977). The degree of decrease (21–74%) depended on the degree of stimulation and release of ACh. Pentobarbital decreased the negative effect and aminopyridine increased it (Weide and Löffelholz, 1980).

Effects of Histamine. The effect of histamine on contractility of mammalian hearts varies considerably. It

TABLE 7–4. Effects of 10μg of histamine (H) on contractility (dp/dt) (mm Hg/sec) of isolated chicken hearts 15, 30, 45, and 60 sec after administration[a]

	Time (sec)			
	15	30	45	60
Nonpaced	C-1719	2081	1990	1913
Mean	H-1090	1387	1382	1459
Difference (D)	−629	−694	−608	−454
D (%)	36.6	33.3	30.6	23.7
Paced	C-2158	2107	2107	2111
Mean	H-1929	1758	1777	1791
Difference (D)	−229	−349	−330	−320
D (%)	10.6	16.5	15.3	14.9

[a] After Sturkie (1978). Paired observations with controls (C).

has a positive inotropic effect in rabbits and guinea pigs, but a negative effect in the dog and sometimes in the rat (see Sturkie, 1978, for a review). Sturkie studied the effects of histamine on heart rate, contractility, and conduction in the isolated chicken heart under a constant load (end diastolic pressure) and under isovolumic conditions; he determined whether specific H_1 or H_2 receptors were involved and described the technique in detail. The results are shown in Table 7–4; it is obvious that histamine had a significant negative effect on contractility based on decreased dp/dt. This was true at a constant heart rate (paced) or unpaced. Since histamine caused sinus arrhythmia, heart rate was difficult to evaluate. In unpaced hearts with sinus rhythm, there was not a significant effect on heart rate. The effects of histamine on the ECG are discussed later. The negative effects on contractility, heart rhythm, and ECG could be blocked by the H_1 receptor antagonist, pyrilamine,

Effects of Hypoxia. The demand for and utilization of O_2 and the tolerance of O_2 deficiency (hypoxia) differ greatly in diving, flying, and land birds. Whether such differences may have had an adaptive influence on contractility of avian heart muscle, as revealed by the effects of hypoxia on isolated hearts of these species, was studied by Abati (1975) and by Sturkie and Abati (1977). Isolated hearts were perfused with a solution containing normal amounts of O_2, and then nitrogen (N_2) was then substituted for the oxygen, followed by a return to oxygen. The results are shown in Figure 7–8. The diving ducks, whose habits require that they operate without oxygen for some time, are most resistant to hypoxia, followed by the pigeon (a flyer), and last by the (land) chicken. On reoxygenation, the duck hearts returned to normal first, while the other hearts lagged.

Work of the Heart

The work involved in ejecting blood from the heart can be estimated as follows:

$$W = \bar{P}V$$

where $\bar{P}$ is the mean arterial pressure and V is the volume of blood ejected. To be strictly accurate,

$$W = \int P\,dV$$

over the ejection period.

However, calculations of stroke work using the first

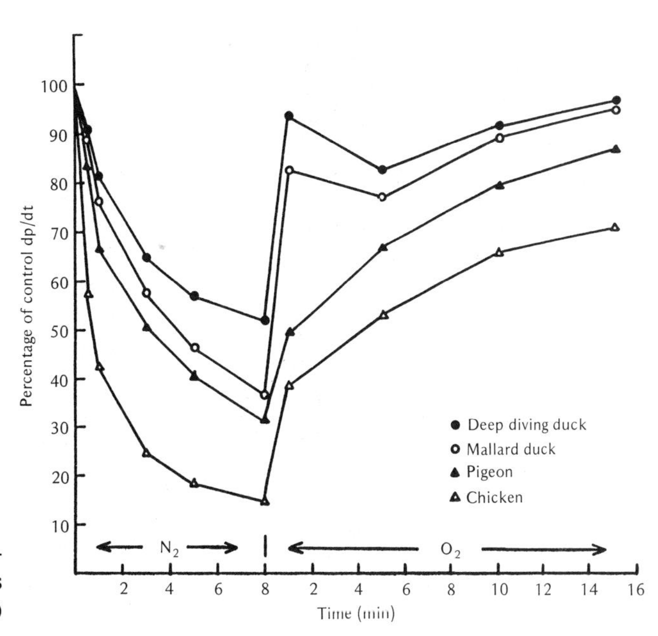

FIGURE 7–8. Effects of anoxia (N_2) and reoxygenation (O_2) on dp/dt of isolated avian hearts (see text). (After Sturkie and Abati, 1977.)

formula in mammals closely approximate those calculated from the latter formula.

The power of the heart is its rate of doing work. According to the calculations of Jones and Johansen (1972), the power of the left ventricle of chickens is (135 × 1330) dynes/cm² × 200 ml/min or 35.9 × 10^6 ergs/min; the right ventricle has a pressure of 18 instead of 135, giving a value for power of 4.79 × 10^6 ergs/min.

The higher the blood pressure (as in turkeys), the higher is the work of the left ventricle in relation to that of the right. The mechanical efficiency of the heart is found as follows:

$$\text{Me} = \frac{\text{Mechanical work done}}{\text{Mechanical work done} + \begin{array}{c}\text{maintenance heat}\\ \text{(tension energy)}\end{array}}$$

The tension energy is influenced mainly by O_2 consumed. It is much greater than the mechanical work energy because the heart's efficiency is low, ranging from 3 to 15%. The O_2 consumption of the chicken heart is estimated to be 7% of the total O_2 consumption of the bird (Jones and Johansen, 1972).

The systolic ejection period in birds seems long, occupying 25–33% of the resting cardiac cycle (Johansen and Aakus, 1963). As the heart rate increases, the time spent in ejection increases in relation to diastolic time. It is advantageous to have a high heart rate, with systole occupying a large part of the cycle (Jones and Johansen, 1972). This may also increase contractility.

The relatively larger heart of birds, compared to mammals, has advantages in that it need shorten less in ejection. However, it also has a disadvantage in that it uses more tension energy and is less efficient in accordance with the law of Laplace (see Chapter 6).

Electrophysiology and Electrocardiography

Anatomy of Specialized Conducting System

The existence of a specialized conducting system in bird hearts was doubted for a number of years. Later work by Davies (1930), Moore (1965), and Yousuf (1965) proves that birds indeed have a specialized conducting system (see Figure 7–9). This system represents the anatomic pathways by which electrical impulses spread to different parts of the heart; these impulses precede slightly the mechanical contraction that they induce. The system as described by Davies consists of (1) the sinoatrial node, (2) the atrioventricular node and branches, and (3) the right AV ring of Purkinje fibers.

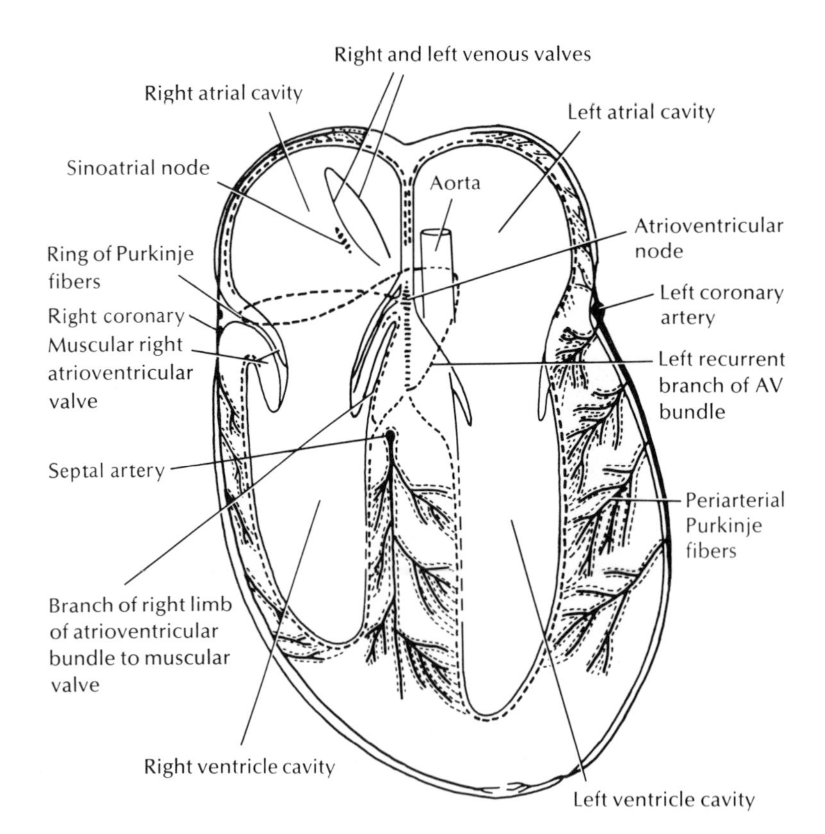

FIGURE 7–9. Diagram of the conduction system in the pigeon heart. (After Davies, 1930.)

The SA Node. The SA node is located near the entrance of the vena cava to the right atrium.

Hill and Goldberg (1980) reported that the chicken has no well-defined SA node, but has a shifting atrial pacemaker with preferential conduction pathways (more details are given later).

The AV Node. The AV node (pigeon) is embedded in connective tissue in the lower and posterior part of the atrial septum, a short distance in front and to the left of the opening of the left superior vena cava (this is a position similar to that of the mammalian node). It is ovoid in shape and its lower and anterior parts narrow into the commencement of the AV bundle. The lower part of the node consists of cells that are larger than the atrial myocardial cells proper and that are frequently multinucleated, the nuclei being rounded in shape and central in position.

Vassal-Adams (1978) has dissected out the AV node in the ostrich and turkey, thus confirming earlier work on other avian species.

The AV Bundle. Beginning as a narrow rounded bundle continuous with the lower and anterior end of the AV node, the AV bundle soon broadens out and runs forward and to the left in the ventricular septum (Davies, 1930). It then passes downward, forward, and to the left, to a point slightly below and to the right of the anterior septal attachment of the muscular right AV valve. This site is about one-quarter of the distance from the base to the apex of the ventricular system and is where the bundle divides into the right and left limbs. The right limb runs downward and slightly forward. It passes in front of and close to the main septal artery, but apparently no muscle fibers pass directly from the right limb to the collection of Purkinje fibers around the artery. The limb then reaches the subendocardial connective tissue on the right side of the septum, where it spreads out and becomes continuous with the subendocardial network of Purkinje fibers. The cells of the limbs are like those of the bundle.

The right limb also gives off a branch that runs up and around the right AV valve. The left limb branches at the point where the right one branches and passes posteriorly and to the left. It gives rise to another branch, the recurrent branch, which runs upward, forward, and to the left in the ventricular system. Finally, the left limb passes backward in the connective tissue on the left side of the root of the aorta and ends by joining the aortic end of the bundle of Purkinje fibers, which pass from the AV node around the right AV orifice, behind the root of the aorta.

The conducting system is poorly supplied with nerves and appears to be independent of the nervous system. Whereas Davies (1930) reported direct tracts of specialized conducting tissue between the SA and AV nodes of chickens, Moore (1965), and Hill and Goldberg (1980) did not find such tissue in the chicken and neither did Yousuf (1965) find it in sparrows. The connection between the SA and AV nodes, according to Yousuf, is made by ordinary muscle fibers of the interatrial septum.

According to Bogusch (1974), Purkinje fibers are richly innervated, contrary to the conclusions of Abraham (1969), who used a silver stain that stains both nerves and Purkinje fibers; this may account for his negative results, according to Bogusch.

Action Potentials and the Spread of Electrical Excitation

In Embryos. Action potentials in embryonic hearts have been studied by a number of workers, including Lieberman and Paes de Carvalho (1965), Pappano (1972), Krespi and Sleator (1966), and Sperelakis and Shigenobu (1972).

The electrophysiology of the embryonic heart is similar to that of the adult (Lieberman and Paes de Carvalho, 1965). The P–R interval defined later, for example, is established early in embryonic life (3–4 days) before the AV node is formed. Therefore, conduction delay is caused by tissue or cells other than the AV node, and these investigators found the delay was localized in a narrow band of tissue extending along the AV ring. These authors also suggested that the adult AV node may be a remnant of the embryonic AV ring.

Membrane potentials in embryonic chick atria change with age from 6 to 18 days, particularly the rate of rise (V_{max}) of action potential (Pappano, 1972). Reduction in sodium (Na^+) had no effect on resting membrane potentials but diminished overshoot and V_{max}. The atrial action potentials therefore underwent a transformation in their sodium electrode properties and in their susceptibility to blockade by tetrodotoxin, which was relatively ineffective in 6-day-old embryos but effective at 12 and 18 days. The embryonic atrial cells of all ages depended on Na^+, because the rate of rise in conductance varied in direct proportion to Na.

Embryonic ventricular cells react similarly to tetrodotoxin and Na (Ishima, 1968). Sodium ion is preferable, but not necessary, in early stages for generating the action potential. Further studies on ventricular embryonic action potentials by Sperelakis and Shigenobu (1972) demonstrated that the greatest changes in electrophysiologic properties occur between day 2 and day 8. The ratio of permeabilities of sodium to potassium (Na/K) is high in young embryos (0.2) and decreases to about 0.05 in later stage embryos. Further details on the role of Na conductance on the rise of V_{max} during embryonic development are given by Iijima and Pappano (1979).

In Adults. The electrical excitation wave precedes mechanical contraction slightly, and spreads from the SA node through the other branches of the conducting system. The paths and speed of conduction of this wave may be determined by placing electrodes at different areas of the heart and determining the change in potential at the electrodes (method of relative negativity).

Lewis (1915), Mangold (1919), and Kisch (1949, and 1951) used this method on the bird heart. Lewis and Mangold used bipolar leads. Lewis placed the exploring electrode on the heart and the other on the chest wall; Mangold placed both electrodes on the heart. Kisch used unipolar leads (chest, direct, and endocavity) for the most part. Mangold and Lewis reported that the impulse started in the region of the SA node and spread to the left side of the right atrium, thence to the left atrium, and then to the septum.

Hill and Goldberg (1980) determined the spread of the wave of depolarization in the atria of chickens by comparing the onset of the P wave of a number of local electrograms to the P wave of the ECG. The onset of activation begins in the right atrium at a point (indicated by hatched area) and spreads sequentially to other areas, as indicated in Figure 7–10. The time required for spread of the activation wave in the right atrium to its completion in the left atrium averaged 32 msec. There were certain preferential areas of conduction along the epicardium overlying the base of the left SA valve and along the rostral border of the right atrium.

The order of depolarization in the different areas of the ventricles (chicken) according to the three investigators is as follows:

Region of heart	Kish (1951)	Lewis (1915)	Mangold (1919)
Apex of right ventricle	1	1	1
Base of left ventricle	3	3	2
Base of right ventricle	2	4	3
Apex of left ventricle	4	2	4

The time required for the impulse to spread from the region of the septum to other parts of the ventricle's surface, according to Lewis (1915), is shown in Figure 7–11 from the distribution of surface potentials and studies of electrocardiograms, Lewis concluded that the impulse spread downward through the septum, then upward through the septum, and later upward through the free walls, almost in line with the latter rather than at right angles to them. The very rapid spread of the impulse downward (electrical axis, + 90°; see Figure 7–11) corresponds to the small upright R wave of the electrocardiogram. The depolarization wave then shifts abruptly upward (electrical axis, ≃90°), and its duration is relatively long. This produces the S wave of the electrocardiogram (Lewis, 1915).

Stimulation of the peripheral end of the cut vagus decreased AV conduction time 8–13 msec, and shifted the pacemaker in over 50% of those stimulated; the most frequent shift was to the lower AV node or ventricle (Goldberg et al., 1983).

Neither Mangold nor Lewis determined which of the endocardial and epicardial surfaces was activated first. Kisch (1951) made such studies on the hearts of chickens, pigeons, ducks, and seagulls. He showed that the epicardial surfaces of the ventricles were activated or depolarized before the endocardial surfaces. Depolarization on the surface of the right ventricles of the chicken heart begins about 0.02–0.03 sec earlier than it does inside the right ventricle, but on the left ventricle it starts about 0.01 sec earlier than on the inside. Depolarization in the interior of the left and right ventricles occurs at approximately the same time. Leads taken directly from the interior of the ventricle produce electrograms (EGs) with the configuration of waves in the opposite direction; these are similar to the normal ECGs of man and dog, in which the endocardial surfaces are depolarized before the epicardial surfaces.

Moore (1965) studied the activation and spread of action potentials in the turkey heart and the sequence according to him is somewhat different from that found by Kisch, Mangold, and Lewis. According to him, the apical third of the right ventricle is activated first, and activation spreads to the upper basilar part and thence on to the pulmonary conus region, which is the last part activated. The left ventricle is first activated in the anterior septal one-third and middle regions and in the basilar regions later.

The Electrocardiogram

The electrocardiogram (ECG) is a record of the electrical activity (depolarization and repolarization) of the heart picked up from electrodes attached to parts of the body other than the heart itself (leads). For details concerning the essentials and methods of taking and recording electrocardiograms, textbooks on the subject should be consulted.

Recording Methods. *Leads.* Leads taken directly from the heart produce records termed electrograms (EGs). Records from any two electrodes constitute a bipolar lead. The standard bipolar limb leads for man are: lead I, right arm and left arm; lead II, right arm and left leg; and lead III, left arm and left leg. The limb leads for the bird heart are the same as for man, except that the electrodes (usually needles) are attached to, or inserted in, the bases of the wings, and in legs. The three limb leads form roughly an equi-

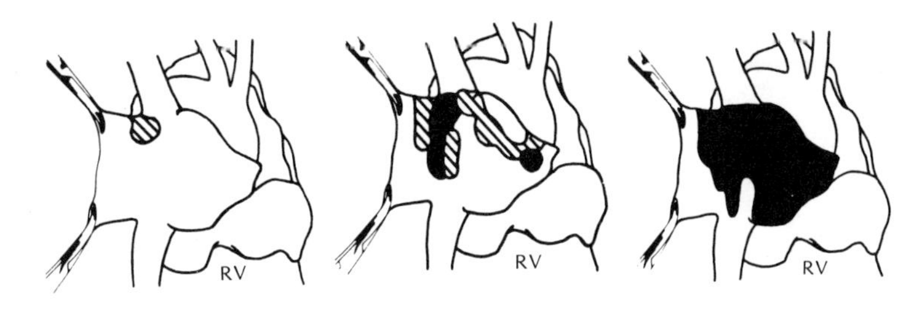

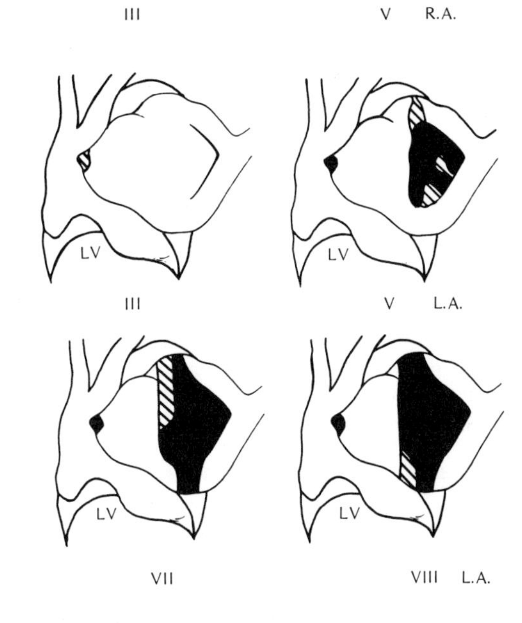

FIGURE 7–10. Spread of impulse in right (*RA*) and left atria, (*LA*) of chicken. Initial activation of right and left atria includes small islands of atrial tissue. Activation then spreads from these islands to form the broader wave fronts as follows: Time of spread is measured in eight intervals (not all are shown). Intervals, right atrium: I, −6.0 to −2.1 msec; III, +2.0 to +5.9 msec; V, +10.0 to +13.9. Left atrium: III and V, same as right atrium; VII, +18.0 to +21.9 msec; VIII, +22.0 to 25.9 msec. Hatched areas, beginning of activation; black areas, already activated; white areas, not activated. Right atrium was activated before left; right atrium was fully activated after interval V. See text for further details. (Modified from Hill and Goldberg, 1980.)

lateral triangle with the heart located near the center (Figure 7–12). This is more nearly true for the bird heart than for man.

The conventional chest leads (CR, CL, and CF) have also been recorded in birds (Sturkie, unpublished; Kisch, 1951; Douglas, 1960). The exploring electrodes are placed on the right arm (CR), left arm (CL) (of man, and the corresponding wings of birds), and the left leg or foot (CF); the chest electrode is the indifferent electrode.

It is impossible to record the true potential at any one point with bipolar leads because of the influence of the other electrode. Wilson devised a method of unipolar leads so that one of the electrodes was zero (see Lamb, 1957). The positive exploring electrode therefore records only the potential received at the point desired. Such electrodes placed on the right arm (VR), left arm (VL), and left leg or foot (VP) constitute the V leads. Because one lead is zero, the potential recorded in V leads is only half that of bipolar leads. Goldberger has augmented (amplified) the potential of unipolar leads by removing the negative electrode from VL when the positive electrode is on the left arm; this can be done in a like manner on the other

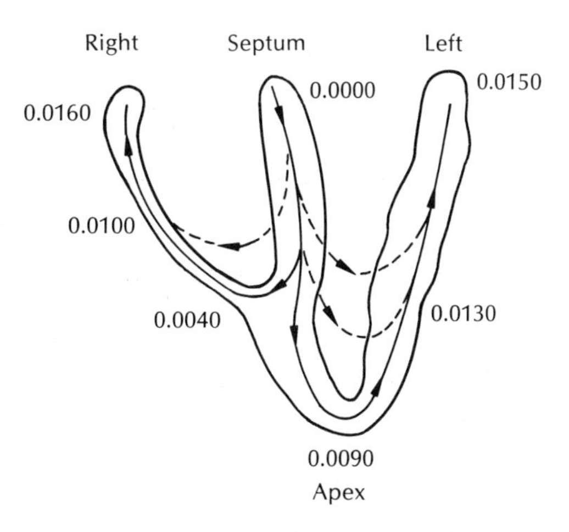

FIGURE 7–11. Diagram of the spread of impulse or excitation wave in the ventricles of bird heart. Coronal section; time in seconds. (After Lewis, 1915.)

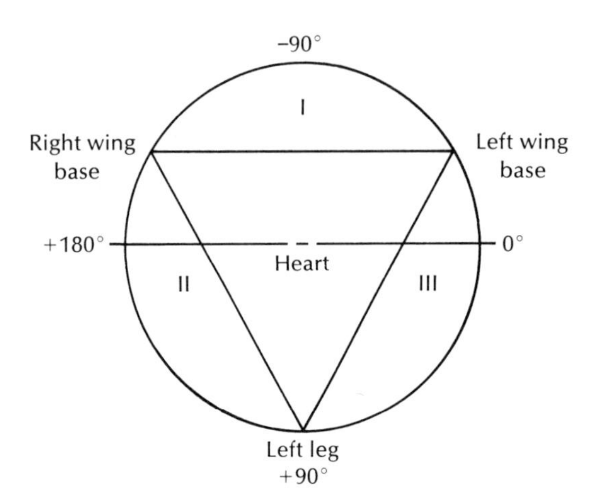

FIGURE 7–12. Limb leads for the ECG of bird, with heart in center.

extremities. Such leads are designated as aVR, aVL, and aVF, or augmented unipolar leads. Even with the augmentation, the amplitude obtained is only about 86% of that measured by bipolar leads (Lamb, 1957).

Chart Speed and Wave Designations. The standard speed for the paper of the electrocardiogram for human beings is 25 mm/sec. The vertical lines 1 mm apart on the paper represent units of time (0.04 sec) at standard speeds. The horizontal lines, also 1 mm apart, represent amplitude or voltage. For work with humans, the instrument is usually standardized at 1 mV. When 1 mV is impressed on the instrument, it causes a deflection of 10 mm. As the heart rate of the chicken is considerably faster than that of man, the standard chart speed in most cases is not fast enough to record all waves faithfully. Usually, if the heart rate is 300/min or more, the P (depolarization of atrial muscle) and T (repolarization of the ventricles) waves may be fused together, and the P wave is not always discernible (particularly in leads II and III) (see Figure 7–13). However, the

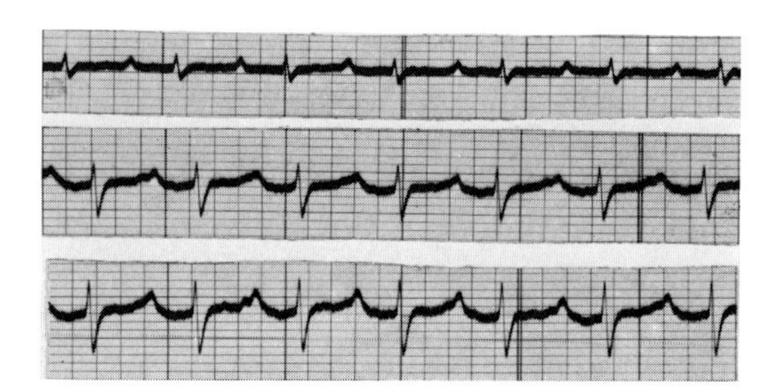

FIGURE 7–13. ECG of female chicken taken on a high-frequency instrument (500 Hz) showing limb leads (I, II, III) from top downward. Standardization, 1 mV = 1.5 cm; chart speed, 75 mm/sec. P and T waves are fused; compare with Figure 7–14, where they are not fused.

P wave of the chicken may not be discernible even when the chart speed is increased to 50 or 75 mm/sec or faster (Sturkie, 1948, 1949; Kisch, 1951) and when the frequency response of the instrument is increased to 500 Hz, which is more than adequate (Sturkie, unpublished; see Figure 7–14). It appears that the P and T waves are fused because the atria begin depolarizing before the ventricles are completely repolarized.

Characteristics of the Normal Bird Electrocardiogram

The normal ECG of man shows P, Q, R, S, and T waves in the limb leads. The bird ECG exhibits P, S, and T waves and usually a small abortive R wave in some leads, but no Q wave. The P wave, as in man (Figure 7–14), represents the depolarization of atrial muscle and slightly precedes atrial contraction. There is no recognizable wave of atrial repolarization. R and S waves represent the depolarization of ventricular muscle, which signals the onset of ventricular systole. The T wave represents repolarization of the ventricles.

Results from Different Leads. The P wave, when observed, is usually upright (positive) in all leads in the chicken, duck, and pigeon (Sturkie, 1948, 1949; Kisch 1951), although Douglas (1960), on the basis of studies of a small number of birds, claimed that the P wave is usually biphasic in leads I and II, and Kisch (1951) reported negative P III in the sea gull. The ECG of the turkey is essentially like that of chicken (Krista et al., 1970). The P wave is observed most frequently in lead I in the chicken, although its amplitude is less there than in leads II or III, because the T wave in lead I is usually flat or isoelectric. P and T may be fused in leads II and III.

Hill and Goldberg (1980) have described seven different P wave types in the chicken, based on P waves recorded from base of tail and neck, where the waves are more clearly discernible. Four of the seven types were upright or positive, two were inverted or negative, and one was biphasic. Often the P wave varied in the same bird over a period of time. Sturkie has never observed this in limb leads of chickens in the same bird. Hill and Goldberg ascribed these P wave changes to a shifting of the primary pacemaker. T is positive in leads II and III; a small upright R is usually (Sturkie, 1949, and unpublished) or always (Kisch, 1951) present in lead II of the chicken and pigeon, respectively, but it is of lower amplitude or absent in lead III. Occasionally the R wave is prominent in leads II and III of the duck (Kisch, 1951). The P wave is small and positive in lead I and in lead aVF but small and negative in lead V-10 (Szabuniewicz, 1967).

Limb Leads. Lead I is the most variable of the limb leads. The amplitude of all the complexes except the

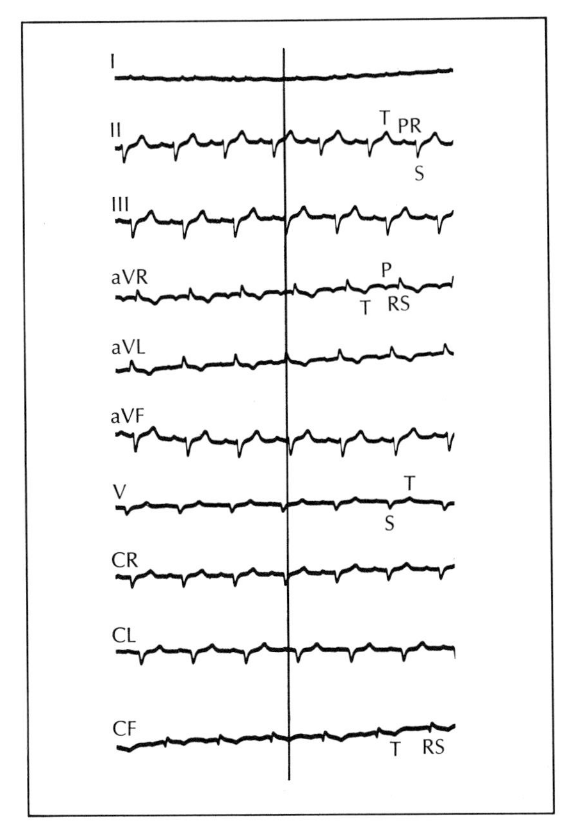

FIGURE 7–14. ECG of male chicken showing limb leads, augumented unipolar leads (aVR, aVL, aVF), V lead, and chest leads (CR, CL, CF). Standardization, 1 mv = 1.5 cm; chart speed, 50 mm/sec. Record taken on high-frequency instrument. P, R, S, and T waves are clearly observed. (Sturkie, unpublished).

P wave is usually low. With respect to the configuration of R, S, and T in lead I, the three main types are described as follows (Sturkie, 1948, 1949):

a. Relatively prominent S or no R or a small R. The T is usually isoelectric or slightly positive.
b. A relatively prominent R and no S or a small S. T is isoelectric or slightly negative.
c. Rs and Ss of about equal prominence. Type b is the most prevalent (see also Szabuniewicz, 1967).

The electrocardiogram of the chicken, showing limb leads, chest leads, and augmented unipolar leads, is reproduced in Figure 7–14. In leads CR and CL, the configurations of the waves are similar to those in leads II and III. In lead CF the P wave is usually biphasic, and the amplitude of all waves is low.

Augmented Unipolar Leads. In leads aVR and aVL the QRS complex is inverted when compared to lead II or lead III. The P wave is inverted in lead aVR and biphasic in lead aVL. The configuration of the waves in lead aVF is similar to those in leads II and III.

Lead V_{10}, used with lead aVF to determine the electrical axis (Z axis) in the dorsoventral position, is similar to lead aVR in the configuration of most waves, with P and T waves inverted. It is a unipolar lead and is obtained by placing the indifferent electrode over the third dorsal thoracic vertebrae (Szabuniewcz, 1967).

Unipolar Leads. Unipolar leads (VR, VL, and VF, not augmented) have been recorded in chickens and pigeons by Douglas (1960), Kisch (1951), Szabuniewicz (1967), and Sturkie (unpublished). In leads VR and VL of chickens and pigeons, there is a prominent R, no S, and an inverted T, but in the duck these leads show an S, a QS, or a small R and a prominent S. In lead VF of all birds studied by Kisch, there was a QS or a small R and prominent S.

Intervals. The duration in seconds of the various waves of the avian ECG cannot always be determined because of the fusion of P and T waves, particularly in females (see later section on sex differences); therefore the work that has been reported is on selected ECGs where P and T waves are clearly discernible (Sturkie, 1949; Kisch, 1951). Durations of P, S, and T waves and the time from the beginning of P to the start of S (P–S interval) and from the beginning of S to the end of T (ST interval) have been determined for leads II and III in the chicken. The R wave in leads II and III is usually small or absent; when it is present, its starting point is difficult to locate. For this reason the P–S, instead of the P–R, interval is determined. In many cases there is no ST segment. T usually begins where S ends, and that point is used as the starting point for T. ST represents the interval from the beginning of S to the end of T, or the time required for depolarization and repolarization of the ventricles.

The P–S interval is the time required for the impulse, beginning in the right atrium, to reach the ventricles. The wave intervals (durations) are shown in Table 7–5 for the chicken.

In general, as heart rate increases, the intervals for all complexes except S decrease; this has also been demonstrated by Kisch (1951) in the chicken, pigeon, duck, and seagull with direct leads. The intervals determined with direct leads on the chicken by Kisch (1951) are in close agreement with those determined with limb leads by Sturkie (1949), except for that of the ventricular complex. The duration of this interval is 0.037 sec according to Kisch and 0.024 sec according to Sturkie. The intervals determined by Rakalska (1964) on lead I and aVF leads agree in general with those of Sturkie for S and P–R but not for P; Rakalska's P values compare favorably with those of Hill and Goldberg.

Goldberg and Bolnick (1980) ran ECGs on the red-

TABLE 7–5. Intervals or durations and amplitudes of chicken electrocardiograms

	Intervals or durations (sec)			Amplitude (mV)			
	Female heart rates (Sturkie, 1949; with Hunsaker, 1957)		Rakalska (1964)	Szabuniewicz (1967)		Sturkie and Hunsaker (1957) Sturkie (unpublished)	
	260 to 280	300 to 341		Lead I	Lead aVF	Lead II	Lead III
P	0.0307 Male[a]						
	0.0421–0.0374		0.027	0.109	0.095	0.57 Male	0.48 Male
R				0.160	0.090	0.083 Female	0.030 Female
						0.265 Male	0.182 Male
S	0.0234	0.0234	0.0235	0.146	0.320	0.247 Female	0.250 Female
						0.795 Male	0.772 Male
RS	—	—	—			0.247 Female	0.250 Female
						0.530 Male	0.590 Male
T	0.1048	0.0925	—	0.074	0.120	—	—
						0.255 Male	0.268 Male
P–S	0.0849	0.0723	0.066	—	—	—	—
S–T	0.1281	0.1164	0.105	—	—	—	—

[a] Sturkie (unpublished).

tailed hawk, emu, Chilean tinamou, and the chicken that had been anesthetized. Unfortunately, their records were not good, because their instrumentation was not adequate, particularly to determine shape and amplitude of the waves. They determined interval and amplitudes on a few selected waves but not on the usual limb leads, but on a back-to-neck lead. Their intervals for the P wave ranged from 0.020 sec for the chicken to 0.012 for the hawk. Their P–R or P–S intervals ranged from 0.070 for the chicken, which compares favorably with those of Sturkie, to 0.050 for the hawk. The R–S interval averaged 0.030 sec for the female chicken. There was little difference in the appearance of the ECGs of these species.

Amplitude. The amplitude of all waves of the bird ECG is relatively low, and is considerably less than that of the human ECG. In lead I, it is so low that accurate measurements with the usual standardization are difficult or impossible. An estimate of average amplitude in lead I can be deduced from the differences in mean amplitude for the various waves in leads II and III, in accordance with Einthoven's law, which states that the amplitude of a given wave in leads I and III should equal that in lead II. This is true provided the three leads are run simultaneously or, if they are not, provided there is no appreciable change in heart rate and amplitude. For example, the amplitude in millivolts of S in lead I and S in lead III (added algebraically) should equal that in S II, as it does. Where S in lead I is the main ventricular wave (type a, lead I), S II is greater than that of S III. When R is predominant in lead I (type b), then the amplitude of S II is less than that of S III. If S II and S III are equal, then R or S waves are absent in lead I. The amplitudes of most of the waves are shown in Table 7–5 for leads II and III and also for leads I and AVF.

Electrical Axis. The electrical axis represents the mean or average electromotive force (magnitude) of depolarization and repolarization, acting in an average direction during the period of electrical activity of the heart. It is a vector quantity in that it has direction, magnitude, and sense. Axes for any of the waves may be determined by measuring the amplitude of complexes in any two of the limb leads and plotting these values. Lead I and II are usually used for human ECGs and leads II and III for the chicken, for in the latter the amplitude for all waves in lead I is too low for a accurate measurement.

Orthogonal Lead System. The previous lead systems have dealt with ECGs taken only in the frontal plane (*Y* axis). Leads may also be taken in the horizontal (*X* axis) and sagittal planes (*Z* axis). Each orthogonal lead system or plane is perpendicular to each of the others and these should intersect in the same plane in the heart. Szabuniewicz (1967) used the following orthogonal leads in the chicken: frontal (*Y*) plane leads, I (limb) and aVF (*Y*); horizontal plane (*X*), I and V-10 (unipolar lead attached at the base of the third dorsal vertebrae); sagittal or *Z* plane V-10 and aVF. Mean spatial vectors involving these leads have been determined in turkeys by Krista et al. (1970).

Electrical axes for chicken ECGs taken in the frontal

TABLE 7–6. Electrical axes (frontal plane) of chicken hearts (female) in relation to type of ECG in lead I[a]

Type of ECG in I	Number of birds	ECG		RS axes (°)	
		Number	%	Range	Average
S present	15	17	25.8	−91 to −120	−102.11
R present	30	46	74.2	−26 to −103[b]	− 74.04
Unknown	2	3		+10 to + 30	
Direction of T wave in I	**Number of birds**	**ECG**		**T axes (°)**	
		Number	%	Range	Average
Positive	34	43	70.49	+68 to + 89	+ 81.8
Negative	10	12	19.67	+95 to +115	+100.3
Isoelectric	6	6	9.83	+88 to + 91	+ 89.6

[a] From Sturkie (1949).
[b] Only one ECG above −90°.

plane are shown in Table 7–6. An RS axis of −90° (see Figure 7–12 means that the mean electromotive force is directed upward and parallel to the long axis of the body. Seventy-four percent of the chickens studied had RS axes averaging −74.04° and 26% had RS axes averaging −102.11° (between 11 and 12 o'clock). The normal RS axis in the human electricardiogram ranges from 0 to +90°.

The electrical axes for T are grouped according to the direction of T in lead I (Table 7–6). The more nearly the electrical axis parallels a given lead line, the higher the amplitude of the waves; when the axis runs almost perpendicular to the lead line (as in lead I of the chicken), the amplitude is low. This can be demonstrated experimentally by rotating the heart on its anterioposterior axis to the left or right (Sturkie, 1948). Goldberg and Bolnick (1980) reported frontal electrical axes in two emus of −117° and −102°, and −83° for one hawk.

Electrical Axes in Three Planes. Employing orthogonal leads I, aVF, and V-10, Szabuniewicz (1967) recorded ECGs taken in three planes for each of the waves in lead I and aVF, (Y axis); leads I and V_{10} (X axis), and aVF and V_{10} (Z axis), are as follows:

	QRS	P	T
Frontal plane (Y)	−77.1°	+43°	+59°
Sagittal plane (Z)	−55.4°	+130°	+132°
Horizontal plane (X)	+72.4°	−43°	−50°

Krista et al. (1970) reported differences in the spatial vectors of hypertensive and hypotensive turkeys.

Sex Differences in the ECG. There is a sex difference in the ECG of chickens characterized by a greater tendency toward fusion of P and T waves in leads II and III in females and by the greater amplitude of all waves in the male ECG. (Sturkie and Hunsaker, 1957). The female sex hormone, estrogen, plays a role in this difference. Administration of estrogen to males over a 2- or 3-week period depresses the amplitude of all waves approximately 35%. It is known that certain organs of the body are good conductors and others poor conductors of the cardioelectric force. If the heart of the bird is exposed by removing the sternum, sternal rubs, coracoids, clavicle, and the attached tissues, the amplitude of all waves in the limb leads is increased anywhere from two- to threefold (Sturkie, 1948). This suggests that the sternum and pectoral muscles influence the spread of cardioelectropotentials. Douglas (1960) has suggested that the air sacs of avian species likewise influence the spread.

Abnormal Electrocardiograms

Heart disorders as revealed by the electrocardiogram appear to cause few deaths in chickens during the first year of life. Serial ECGs were made on 72 adult female White Leghorns from 5 months (maturity) to 18 months of age (Sturkie, 1949). Although 37% of the birds died during this period, only 5% of these exhibited heart abnormalities that might have caused death.

Mineral Deficiencies. Acute potassium deficiency in growing chicks produces a high percentage of abnormal ECGs (70%) and 100% mortality within 2–4 weeks (Sturkie, 1950, 1952). Most of these ECG disorders are concerned with the rhythm and conduction of the heartbeat. They include partial and complete atrioventricular block, partial and complete sinoatrial block, sinus arrhythmia and marked sinus slowing, and premature nodal and ventricular systoles. Some examples are shown in Figures 7–15, 7–16, 7–17, 7–18, and 7–19.

Pathologic lesions in the AV node or bundle or in the SA node may cause AV and SA block. Gross lesions have not been observed in the hearts of potassium-

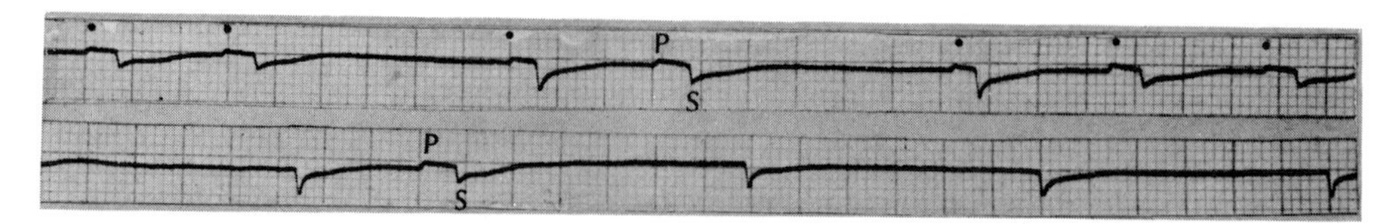

FIGURE 7–15. Partial (upper) and complete (lower) SA block with slow heart rate (100/min) and mild sinus arrhythmia. (Sturkie, 1952.)

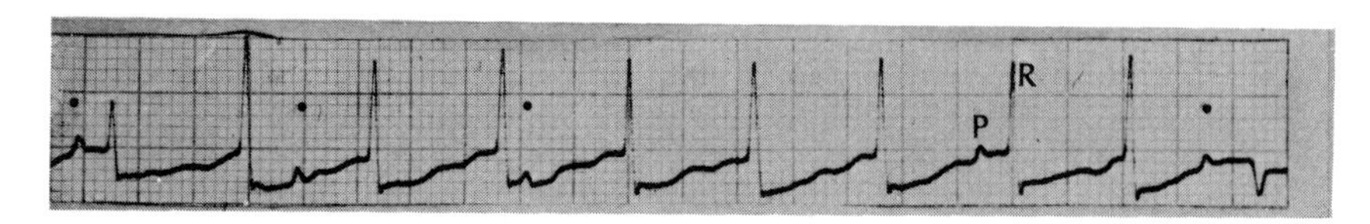

FIGURE 7–16. Nodal rhythm and AV dissociation in the chicken produced by potassium deficiency. See text. (After Sturkie, 1952.)

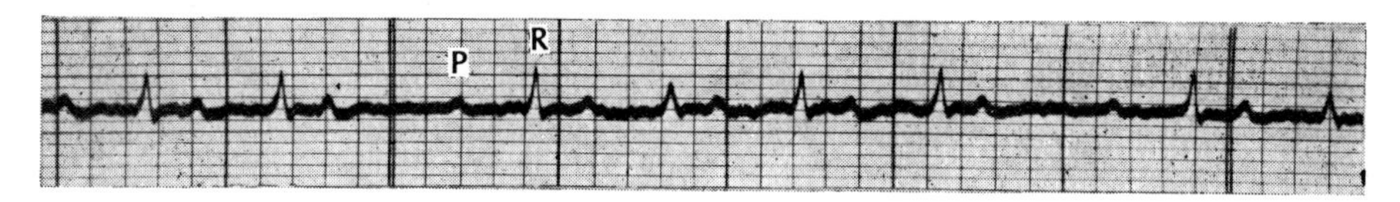

FIGURE 7–17. An example of partial AV block and prolonged conduction time in ECG of chicken (lead I). Every fifth ventricular beat (R wave) is blocked. The P–R interval following the dropped beat is shortened, and the interval increases on most successive beats (Wenckebach phenomenon). (From Sturkie et al., 1954.)

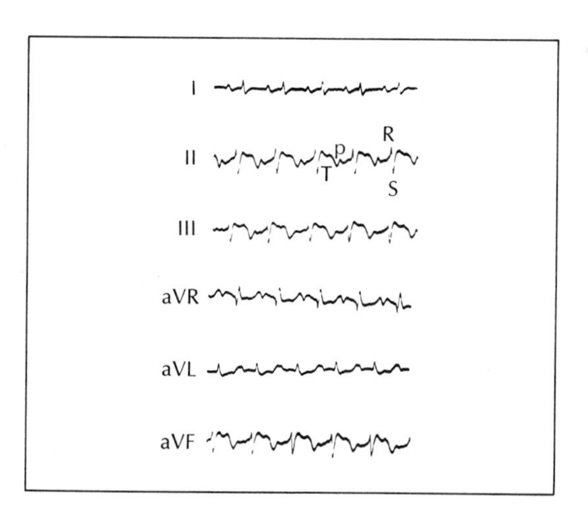

FIGURE 7–18. ECG of chicken administered cadmium sulfate. T wave is inverted in II, III, and aVF, instead of upright as in normal. (After Sturkie, 1973.)

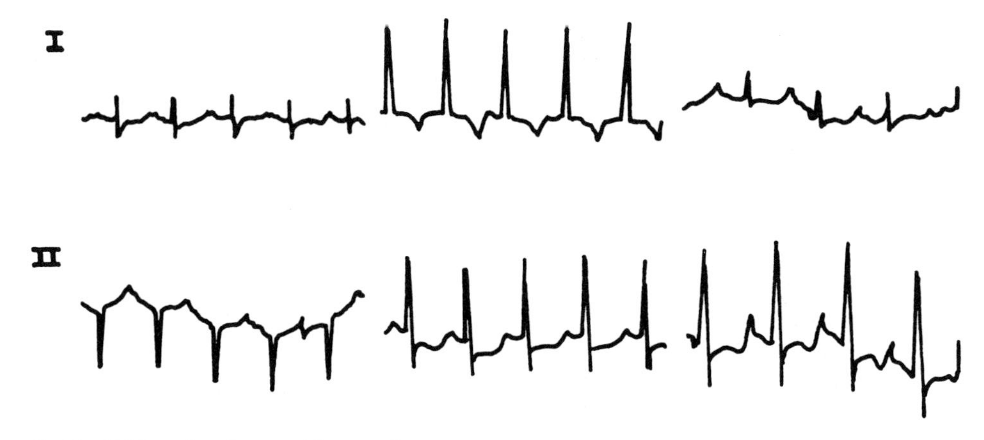

FIGURE 7–19. ECGs of turkeys with round heart disease at 3, 4, and 5 weeks of age, showing important changes at 4 and 5 weeks. Leads I and II. Normal record at 3 weeks of age. See text for details. (After Czarnecki and Good, 1980.)

deficient chickens, and it appears that the block results from a functional rather than a pathologic disturbance in conduction. In such cases it may be abolished by administration of atropine, a cholinergic blocker. In potassium-deficient chicks, the ECGs of which exhibited AV block and premature systoles, atropine and also diethylaminoethanol were effective in reverting the ECGs to normal. The requirement of potassium for the bird heart may be higher than for mammals, because the blood of birds contains more potassium than that of mammals.

Excess Potassium and Cadmium. Diving ducks, on emerging from the dive, exhibit changes in the ECG characterized by elevated and peaked T waves; Anderson (1965) has attributed these to a hyperpotassemia, which he demonstrated.

Cadmium is known to produce infarction and hypertrophy of the heart. After 15 daily injections of cadmium sulfate to chickens (Sturkie, 1973) the injections were discontinued and ECGs were run for 13–40 days thereafter. Deviations in the ECGs consisted mainly of T-wave changes. Normally the ECG exhibits an upright T in leads II, III, and aVF and inverted Ts or RS complexes in leads aVR and aVL (see Figure 7–18). The changes in comparison with normal ECGs are as follows, where + is an upright T wave and − is a negative one.

	Normal	Abnormal
II	+	−
III	+	−
aVR	−	+
aVL	−	+
aVF	+	−

Other abnormalities included AV block. Some of the abnormalities persisted for as long as 40 days. The changes are characteristic of varying degrees of myocardial infarction.

Isoproterenol Necrosis. Repeated and excessive dosages of isoproterenol (McGrath, 1972) produces myocardial infarction and enlarged hearts in rats and *Coturnix* (quails). Electrocardiograms run on quails (Sturkie, unpublished) revealed ECG changes almost identical to those reported for cadmium excess. These changes in T waves were progressive, and the first sign of myocardial damage was evident in the ECG. Histamine, when administered to the isolated chicken heart, produced arrhymias, sinus arrest, and complete AV blocks in about 30% of hearts treated (Sturkie 1978).

Vitamin Deficiencies. Acute thiamine deficiency in pigeons results in sinus arrhythmia, bradycardia, and AV block, but chronic deficiency of the vitamin rarely produces heart abnormalities (Swank and Bessey, 1941). Deficiencies of niacin, riboflavin, and Vitamins A and D have no effect on the ECG of the chicken (Sturkie, unpublished).

Further studies by Sturkie et al. (1954) show that deficiencies in the diet of Vitamin E alone, of E- and B-complex vitamins, or of B-complex vitamins (mainly thiamine) produce abnormal ECGs. The principal abnormalities of the vitamin E-deficient birds included right axis deviation, premature ventricular systoles, sinus arrhythmia, and elevated S–T segments. Right axis deviation was rare in thiamine-deficient birds, and the most prominent abnormalities observed were premature ventricular systoles and sinus arrhythmia.

Diseases and the ECG. Round heart disease is a condition first reported in chickens and later in turkeys, in which it causes many heart abnormalities and death

(Hunsaker et al., 1971; Jankus et al., 1971; Czarnecki and Good, 1980). The disease occurs in the first 4–6 weeks of life. The heart is extremely enlarged, particulary the right ventricle, giving the heart apex a rounded rather than a pointed appearance.

The characteristic change in the ECG is one indicating a marked shift in the electrical axis to the right such that the mean RS axis in the frontal plane averages +70° compared to −85° for the normal turkey. This represents a tremendous shift in the axes. Such shift produces significant changes in the configuration of most of the waves, particularly R–S complexes and T waves. The R–S wave is elevated or the original S wave has become an R wave, evidence of an axis shift (see Figure 7–19). The T wave is usually depressed or inverted in some of the leads. Sturkie (unpublished) also found right-axis deviation in turkeys suffering from round heart disease and also some evidence of myocardial infarction not unlike that previously reported and produced by cadmium and isoproterenol.

Jankus et al. (1971) reported a shift in axis in the spatial vectorelectrocardiogram (VCG) of turkeys with round heart disease and changes in amplitude of QRS complex. Czarnecki and Good (1980) described a technique of early screening and detection of round heart disease based on changes in amplitude of R wave in lead I.

Newcastle Disease. A virulent and neurotropic strain of Newcastle disease caused significant changes in the ECG of chickens, for example, in amplitude of R, R–S complex, and T waves, and in the duration of some of the intervals (Mitchell and Beard, 1976). Experimentally produced *Escherichia coli* infection in chickens causes myocarditis and changes in ECG, characterized by a decrease in the R wave and an increase in the S wave in leads I and a VF, and by changes in the electrical axes (Gross, 1966).

References

Abati, A. (1976). A comparison of the effect of anoxia on left ventricular function in the isolated perfused heart of the chicken (a non-flying bird), the pigeon (a flying bird), and the duck (a diving bird). Ph.D. Thesis, Rutgers University, New Brunswick, New Jersey.

Abraham, A. (1969). Microscopic innervation of the heart and blood vessels in vertebrates, including man. Oxford: Pergamon Press.

Andersen, H.T. (1965). Hyperpotassemia and electrocardiographic changes in the duck during prolonged diving. Acta Physiol. Scand., 63, 292.

Barry, W.H., J. Pober, J.D. Marsh, S.R. Frankel, and T.W. Smith. (1960). Effects of graded hypoxia on contraction of cultured chick embryo ventricular cells. Am. J. Physiol. 239, H651.

Bennett, T., and T. Malmfors. (1975). Characteristics of the noradrenergic innervation of left atrium in the chick. Comp. Biochem. Physiol., C: Comp. Pharmacol., 52, 47.

Biegan, RL, P.M. Epstein, and A.J. Pappano, (1980). Muscarinic antagonism of the effects of a phosphodiesterase inhibitor (methylisobutylxanthine) in embryonic chick ventricle. J. Pharmacol. Exp. Ther. 215, 348.

Blix, A.S., E.L. Gautvik, and H. Rebsum. (1974). Aspects of relative roles of vasoconstriction and vagal bradycardia in the establishment of the diving reflex in ducks. Acta. Physiol. Scand., 90, 289.

Bogusch, G. (1974). The innervation of Purkinje fibers in the atrium of the avian heart. Cell Tissue Res., 150, 57.

Butler, P.J., and D.R. Jones. (1968). Onset of, and recovery from diving bradycardia in ducks. J. Physiol. (London), 196, 255.

Cain, J.R., and U.K. Abbott. (1970). A system for diurnal heart rate measurement in chickens. Poult. Sci., 49, 1085.

Cogger, E.A., R.E. Otis, and R.K. Ringer. (1974). Heart rates in restrained and freely moving Japanese quail via radiotelemetry. Poult. Sci., 53, 430.

Cohen, D.N., and L.H. Pitts. (1968). Vagal and sympathetic components of conditioned cardio-acceleration in pigeons. Brain Res., 9, 15.

Cohen, D.N., A.M. Schnall, R.L. Macdonald, and L.H. Pitts. (1970). Medullary cells of origin of vagal cardioinhibitory fibers in the pigeon. Anatomical studies of peripheral vagus nerve and the dorsal motor nucleus. J. Comp. Neurol., 140, 299.

Corradetti, R., R. Lindmar, and K. Löffelholz. (1983). Mobilization of cellular choline by stimulation of muscarine receptors in isolated chicken heart and rat cortex in vivo. J. Pharmacol. Exp. Ther., 226, 826.

Czarnecki, C.M., and A.L. Good. (1980). Electrocardiographic technique for identifying developing cardiomyopathies in young poults. Poult. Science 59:1515.

Davies, F. (1930). The conducting system of the bird heart. J. Anat., 64, 9.

Desantis, V.P., W. Langsfeld, R. Lindmar, and K. Löffelholz. (1975). Evidence for noradrenaline and adrenaline as sympathetic transmitters in the chicken. Br. J. Pharmacol, 55, 343.

Dieterich, H.A., K. Löffelholz, and H. Pompetzki. (1977). Acetylcholine overflow from isolated perfused hearts of various species in the absence of cholinesterase inhibition. Naunyn-Schmiedeberg's Arch. Pharmacol., 296, 149.

Dieterich, H.A., H. Kaffel, H. Kilbinger, and K. Löffelholz. (1976). The effects of phostigmine on cholinesterase activity, storage and release of acetylcholine in the isolated chicken heart. J. Pharmacol. Exp. Ther., 199, 236.

Dieterich, H.A., and K. Löffelholz. (1977). Effects of coronary perfusion rate on the hydrolysis of exogenous and endogenous acetylcholine in the isolated chicken heart. Naunyn-Schmiedeberg's Arch. Pharmacol., 296, 143.

Douglas, S.D. (1960). Correlation between surface electrocardiogram and air sac morphology in the White Leghorn rooster. Am. J. Physiol., 199, 355.

Durfee, W.K. (1963). Cardiovascular reflex mechanisms in the fowl. Ph.D. Thesis, Rutgers University, New Brunswick, New Jersey.

El-Ackad, T., M.J. Meyer, and P.D. Sturkie. (1974). Inotropic and chronotropic actions of histamine in avian heart. Fed. Proc. Fed. Am. Soc. Exp. Biol., 33, 585.

Engel U., and K. Löffelholz. (1976). Presence of muscarinic inhibitory and absence of nicotine excitatory receptors at the terminal sympathetic nerves of chicken hearts. Naunyn-Schmiedeberg's Arch. Pharmacol., 295, 225.

Evans, J.H. (1972). A method of recording heart rate of chicken embryo. Physiol. Behav. 9, 131.

Ferguson, T.M., D.H. Miller, J.W. Bradley, and R.L. Atkinson. (1969). Blood pressure and heart rate of turkeys, 17–21 months of age. Poult. Sci., 48, 1478.

Flick, D.F. (1967). Effects of age and diet on heart rate of the developing cockerel. Poult. Sci., 46, 890.

Folkow, B., and L.R. Yonce. (1967). The negative inotropic effect of vagal stimulation on the heart ventricles of the duck. Acta Physiol. Scand., 71, 77.
Goldberg, J.M., and D.A. Bolnick. (1980). Electrocardiograms from the chicken, emu, red-tailed hawk and Chilean Tinamou. Comp. Biochem. Physiol., 67, 15.
Goldberg, J.M., M.H. Johnson, and K.D. Whitelaw. (1983). Effect of vagal stimulation on heart rate and atrioventricular conduction in the domestic fowl. Am. J. Physiol., 244, R235.
Girard, H. (1973). Arterial pressure in chick embryos. Am. J. Physiol., 224, 454.
Gross, W.B. (1966). Electrocardiographic changes of Escherichia coli infected birds. Am. J. Vet. Res. 27:1427.
Higgins, D., and A.J. Pappano. (1979). A histochemical study of the ontogeny of catecholamine containing axons in the chick embryonic heart. J. Mol. Cell. Cardiol., 11, 661.
Higgins, D., and A.J. Pappano. (1981). Developmental changes in the sensitivity of the chick embryo ventricle to β-adrenergic agonist during adrenergic innervation. Circ. Res., 48, 245.
Hill, J.R., and J.M. Goldberg. (1980). P-wave morphology and atrial activation in the domestic fowl. Am. J. Physiol., 239, R483.
Hunsaker, W.K., A Robertson, and S.E. Magwood. (1971). The effect of round heart disease on the electrocardiogram of turkey poults. Poult. Sci., 50, 1712.
Iijima, T., and A.J. Pappano. (1979). Ontogenetic increase of the maximal rate of rise of the chick embryonic heart action potential. Circ. Res., 44, 358.
Ishima, Y. (1968). The effect of tetrodotoxin and sodium substitution on the action of the embryonic chicken heart. Proc. Jpn. Acad., 44, 170.
Jankus, E.F., A.L. Good, K.A. Jordon, and S.K. Saxena. (1971). Electrocardiographic study of round heart disease in turkey poults. Poult. Sci., 50, 481.
Johansen, K., and T. Aakhus. (1963). Central cardiovascular responses to submersion asphyxia in the duck. Am. J. Physiol., 205, 1167.
Johansen, K., and O.B. Reite. (1964). Cardiovascular responses to vagal stimulation and cardioaccelerator nerve blockade in birds. Comp. Biochem. Physiol. 12, 474.
Jones, D.R., and K. Johansen. (1972). The blood vascular system of birds. In "Avian Biology," Vol. II (D.S. Farner and J.R. King, Eds.). New York: Academic Press.
Jones, D.R. (1973). Systemic arterial baroreceptors in ducks and the consequences of their denervation on some cardiovascular responses to diving. J. Physiol., 234, 499.
Jones, D.R., and J.J. Purves. (1970). The carotid body in the duck and consequences of its denervation upon the cardiac responses to immersion. J. Physiol, 211, 279.
Jurani, M., J. Nvota, P. Vyboh, and K. Boda. (1980). Effect of stress on plasma catecholamines in the domestic fowl. In: Catecholamines and Stress: USD In Vetriansky Kopin, Editors. North Holland: Elsevier.
Kisch, B. (1949). Electrocardiographic studies in sea-gulls. Exp. Med. Surg., 7, 345.
Kisch, B. (1951). The electrocardiogram of birds (chicken, duck, pigeon). Exp. Med. Surg., 9, 103.
Krespi, V., and W.W. Sleator, Jr. (1966). A study of the ontogeny of action potentials in chick embryo hearts. Life Sci., 5, 1441.
Krista, L.M., E.F. Jankus, B.E. Waibel, J.H. Souther, R.N. Shoffner, and G.J. Quarfoth. (1970). Comparison of electrocardiogram of hypertensive and hypotensive male turkeys. Poult. Sci., 49, 700.
Lamb, L.E. (1957). "Fundamentals of Electrocardiography and Vectorcardiography." Springfield: Thomas.
Langille, B.L., and D.R. Jones. (1975). Central cardiovascular dynamics of ducks. Am. J. Physiol., 228, 1856.
Langille, B.L. (1983). Role of venoconstriction in the cardiovascular responses of ducks to head immersion. Am. J. Physiol., 244, R292.
Lee, W.C., L.P. McCarty, W.W. Zodrow, and F.E. Shideman. (1960). Cardiostimulant action of certain ganglionic stimulants on embryonic chick heart. J. Pharmacol. Exp. Ther., 130, 30.
Lewis, A.R. (1967). Resting heart rate and respiratory rates of small birds. Auk, 84, 131.
Lewis, T. (1915). The spread of the excitatory process in the vertebrate heart. V: The bird's heart. Philos. Trans. Roy. Soc. London, 207, 298.
Lieberman, M., and A. Paes de Carvalho. (1965). The spread of excitation in the embryonic chick heart. J. Gen. Physiol., 49, 365.
Lin, Y.C., P.D. Sturkie, and J. Tummons. (1970). Effects of cardiac sympathectomy, reserpine and environmental temperatures on catecholamines, in the chicken heart. Can. J. Physiol. Pharmacol., 48, 182.
Lindmar, R., and U. Wolf. (1975). PNMT activity in various organs of the chicken. Naunyn-Schmiedeberg's Arch. Pharmacol., 287, 2.
Löffelholz, K. (1981) Release of acetylcholine in the isolated heart. Am. J. Physiol., 9, H431.
Macdonald, R.L., and D.H. Cohen, (1970). Cells of origin of sympathetic pre- and post-ganglionic cardio-accelerator fibers in the pigeon. J. Comp. Neurol., 140, 343.
McGrath, J.J. (1972). Experimental cardiac necrosis in the Japanese quail. Proc. Soc. Exp. Biol. Med., 139, 1334.
Mangold, E. (1919). Electrographische Untersuchungen des Erregungsverlaufes im Vogelherzen. Arch. Gesamte Physiol. (Pflügers), 175, 327.
Michal, F., F. Emmett, and R.H. Thorp. (1967). A study of drug action on the developing avian cardiac muscle. Comp. Biochem. Physiol., 22, 563.
Mitchell, B.W., and C.W. Beard. (1976). Electrocardiographic and respiratory responses to viscerotropic and neurotropic strains of Newcastle disease virus measured by radio-telemetry. Poult. Sci., 55, 874.
Moore, E.N. (1965). Experimental electrophysiological studies on Avian hearts. Ann. N.Y. Acad. Sci., 127, Art. 1, 127.
Owen, R.B., Jr., (1969). Heart rate, a measure of metabolism of blue winged teal. Comp. Biochem. Physiol., 31, 431.
Pappano, A.J. (1972). Increased susceptibility to blockade by tetrodotoxin during embryonic development. Circ. Res., 31, 379.
Pappano, A.J. (1976). Onset of chronotropic effects of nicotinic drugs and tyramine on the sino-atrial pacemaker in chick embryo heart. Relationship to the development of autonomic neuroeffector transmission. J. Pharmacol. Exp. Ther., 196, 676.
Poorvin, D.W. (1972). The avian cardiac nervous system. Thesis, Rutgers University, New Brunswick, New Jersey.
Purton, M.D. (1971). Pressure changes during the avian cardiac cycle. J. Anat., 108, 620 (Proc.).
Rakalska, Z. (1964). Electrocardiogram of young cockerels. Bull. Vet. Inst. Pulawy (Poland), 5, 145.
Ringer, R.K., H.S. Weiss, and P.D. Sturkie. (1955). Effect of sex and age on blood pressure in the duck and pigeon. Am. J. Physiol., 183, 141.
Ringer, R.K., H.S. Weiss, and P.D. Sturkie. (1957). Heart rate of chickens as influenced by age and gonadal hormones. Am. J. Physiol., 191, 145.
Soliman, F.F.A., and T.M. Huston. (1972). The photoelectric plethysmography technique for recording heart rate in chick embryos. Poult. Sci., 51, 651.
Speralakis, N., and K. Shigenobu. (1972). Changes in membrane properties of chick embryonic hearts during development. J. Gen. Physiol., 60, 430.

Ssinelnikow, R. (1928). Die herznerven der Vögel. Z. Anat. Entwichlungsgeschi, 86, 540.

St. Petery, L.B.,Jr., and L.H.S. Van Mierop. (1974). Evidence for the presence of adrenergic receptors in the 6-day-old chick embryo. Am. J. Physiol., 227, 1406.

Sturkie, P.D. (1948). Effects of changes in position of the heart of the chicken on the electrocardiogram. Am. J. Physiol., 154, 251.

Sturkie, P.D. (1949). The electrocardiogram of the chicken. Am. J. Vet. Res., 10, 168.

Sturkie, P.D. (1950). Abnormal electrocardiograms of chickens produced by potassium deficiency and effects of certain drugs on the abnormalities. Am. J. Physiol., 162, 538.

Sturkie, P.D. (1952). Further studies of potassium deficiency in the chicken. Poult. Sci., 31, 648.

Sturkie, P.D. (1963). Heart rate of chickens determined by radiotelemetry during light and dark periods. Poult. Sci., 72, 797.

Sturkie, P.D. (1965). "Avian Physiology" (2nd ed.). Ithaca: Cornell University Press.

Sturkie, P.D. (1970). Effects of reserpine on nervous control of heart rate in chickens. Comp. Gen. Pharmacol. 1, 336.

Sturkie, P.D. (1973). Effects of cadmium on ECG, blood pressure, and hematocrit of chickens. Avian Dis., 17, 106.

Sturkie, P.D. (1978). Effects of histamine on contractility and conduction in isolated chicken hearts. Comp. Biochem. Physiol., C: Comp. Pharmacol., 61, 55.

Sturkie, P.D., and J. Chillseyzn. (1972). Heart rate changes with age in chickens. Poult. Sci., 51, 906.

Sturkie, P.D., and W.G. Hunsaker. (1957). Role of estrogen in sex difference of the electrocardiogram of the chicken. Proc. Soc. Exp. Biol. Med., 94, 731.

Sturkie, P.D., and D.W. Poorvin. (1973). The avian neutrotransmitter. Proc. Soc. Exp. Biol. Med., 143, 644.

Sturkie, P.D., and A. Abati. (1977). Effects of hypoxia on heart activity in diving, flying and land birds. Satellite Symposium, 27th International Congress of Physiological Sciences, Paris, 1977. Held at Max Planck Institute for Experimental Medicine at Göttingen, Germany, July 1977 (p. 68).

Sturkie, P.D., E.P. Singsen, L.D. Matterson, A. Kozeff, and E.L. and Jungherr. (1954). The effects of dietary deficiencies of vitamin E and the B complex vitamins on the electrocardiogram of chickens. J. Vet. Res., 15, 457.

Sturkie, P.D., D. Poorvin, and N. Ossario. (1970). Levels of epinephrine and norepinephrine in blood and tissue of duck, pigeon, turkey and chicken. Proc. Soc. Exp. Biol. Med., 135, 267.

Sturkie, P.D., Yu-Chong-Lin, and N. Ossorio. (1970). Effects of acclimitization to heat and cold on heart rate in chickens. Am. J. Physiol., 219, 34.

Swank, R.L., and O.A. Bessey. (1941). Avian thiamine deficiency: Characteristic symptoms and their pathogenesis J. Nutr., 22, 77.

Szabuniewicz, M. (1967). The electrocardiogram of the chicken. Southwestern Vet., 20, (4).

Tigerstedt. R. (1921). "Physiologie des Kreislaufes," Vol. 2. Berlin and Leipzig.

Tummons, J.L. (1970). Nervous control of heart rate in the domestic fowl. Ph.D. Thesis, Rutgers University. New Brunswick, New Jersey.

Tummons, J.L., and P.D. Sturkie. (1968). Cardioaccelerator nerve stimulation in chickens. Life Sci., 7, 377.

Tummons, J.L., and P.D. Sturkie. (1969). Nervous control of heart rate during excitement in the adult White Leghorn cock. Am. J. Physiol., 216, 1437.

Tummons, J.L., and P.D. Sturkie. (1970a). Beta adrenergic and cholinergic stimulants from the cardioaccelerator nerve of the domestic fowl. Z. Vgl. Physiol., 68, 268.

Tummons, J.L., and P.D. Sturkie. (1970b). Chronotropic supersensitivity to norepinephrine and epinephrine induced by sympathectomy and reserpine pretreatment in the domestic fowl. Comp. Gen. Pharmacol., 1, 280.

Vassal-Adams, P.R. (1978). Dissection of the atrioventricular bundle and its branches in the avian heart. J. Anat., 127, 196.

Vogel, J.A., and P.D. Sturkie. (1963). Effects of starvation on the cardiovascular system of chickens. Proc. Soc. Exp. Biol. Med., 112, 111.

Weide, W., and K. Löffelholz. (1980). 4-Aminopyridine antagonizes the inhibitory effect of pentrobarbital on acetylcholine release in the heart. Naunyn-Schmiedeberg's Arch. Pharmacol., 312, 7.

West, N.H., B.L. Langille, and D.R. Jones. (1981). Cardiovascular system. In "Form and Function." (A.S. King and J. McLelland, Eds.). New York: Academic Press.

Whittow, G.C., P.D. Sturkie, and G. Stein, Jr. (1965). Cardiovascular changes in restrained chickens. Poult. Sci. 44, 1452.

Yousuf, N. (1965). The conducting system of the heart of the house sparrow (*Passer domesticus*). Anat. Rec., 152, 235.

8
Respiration

M.R. Fedde

The avian respiratory system is unique among vertebrates in its structure and in the manner by which it accomplishes its main function, that of providing oxygen (O_2) to and removing carbon dioxide (CO_2) from the blood. The model of the oxygen delivery system in Figure 8–1 shows the importance of this system in allowing normal cellular metabolism. This model illustrates that ventilation of the lung, along with gas exchange in the pulmonary capillaries, is coupled to the

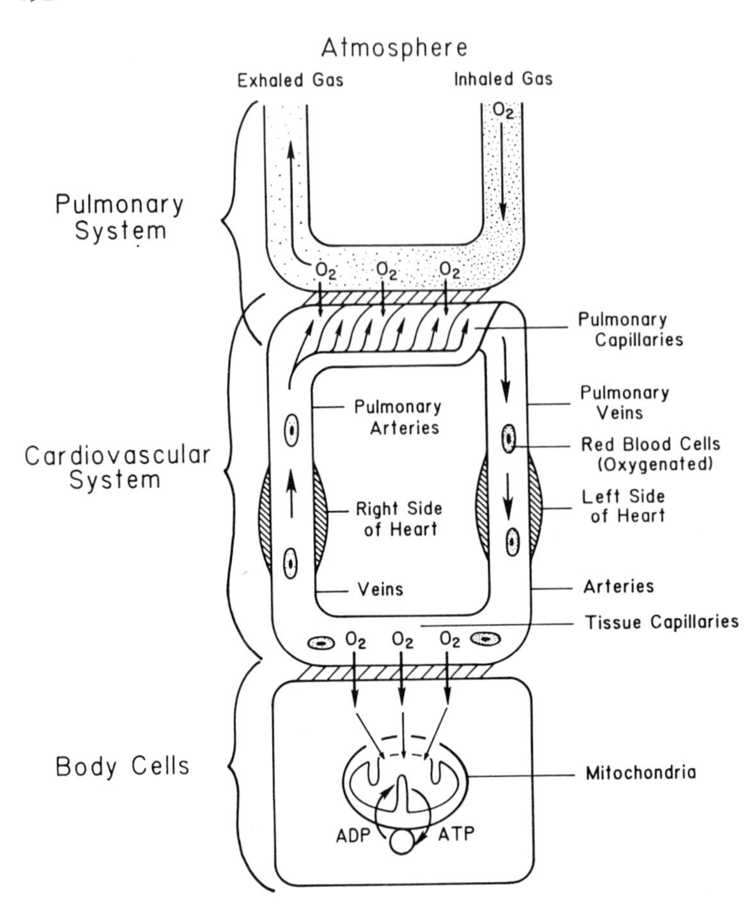

FIGURE 8–1. General model of the oxygen delivery system in birds. (Modified from Taylor and Weibel, 1981.)

cardiovascular function of transporting oxygen to the body cells where it moves across capillary walls. The mitochondria are the ultimate O_2 sinks and provide most of the high-energy phosphate compounds required for cellular function. For CO_2, the route is from mitochondria to lung gas and atmosphere.

The respiratory system also functions in eliminating heat from the body and has several nonrespiratory functions, such as detoxification of metabolic products, production of chemical messengers, and vocalization. In this chapter, the function of the respiratory system as a gas-exchange organ in the adult will be the primary concern, but consideration will also be given to the transport of gases in the blood to the body cells.

Structure of the Respiratory System

The morphology of the avian respiratory system, with its rigid lung and compliant air sac system, is considerably different from that of other vertebrates (Figure 8–2). The major features of the system will be briefly discussed here, but details for many species of birds and evolution of the system from that of ancient reptiles may be found in King (1966a, 1975), King and Molony (1971), and Duncker (1971, 1978a, b, 1981). The anatomic terminology used here is that agreed on by the International Committee on Avian Anatomical Nomenclature and described in detail by King (1979).

Anatomy of the Upper Airways

The gas spaces within the respiratory system are all in communication with the outside environment, which may contain potentially harmful agents. The upper part of the respiratory system is specialized to filter, heat, and humidify the inhaled gas.

The upper airways originate at the external nares and mouth, but it remains unclear how much of the inhaled gas passes through the nasal cavity versus the mouth during quiet breathing or during activity. The nasal cavity contains several mucosa-covered conchae, which provide an expanded surface over which the gas must pass on its way to the trachea. The nasal and buccal cavities communicate with the oropharynx and the larynx. The larynx is a complex structure with a cartilagenous frame, many ligaments, both intrinsic and extrinsic, and four paired skeletal muscles and one un-

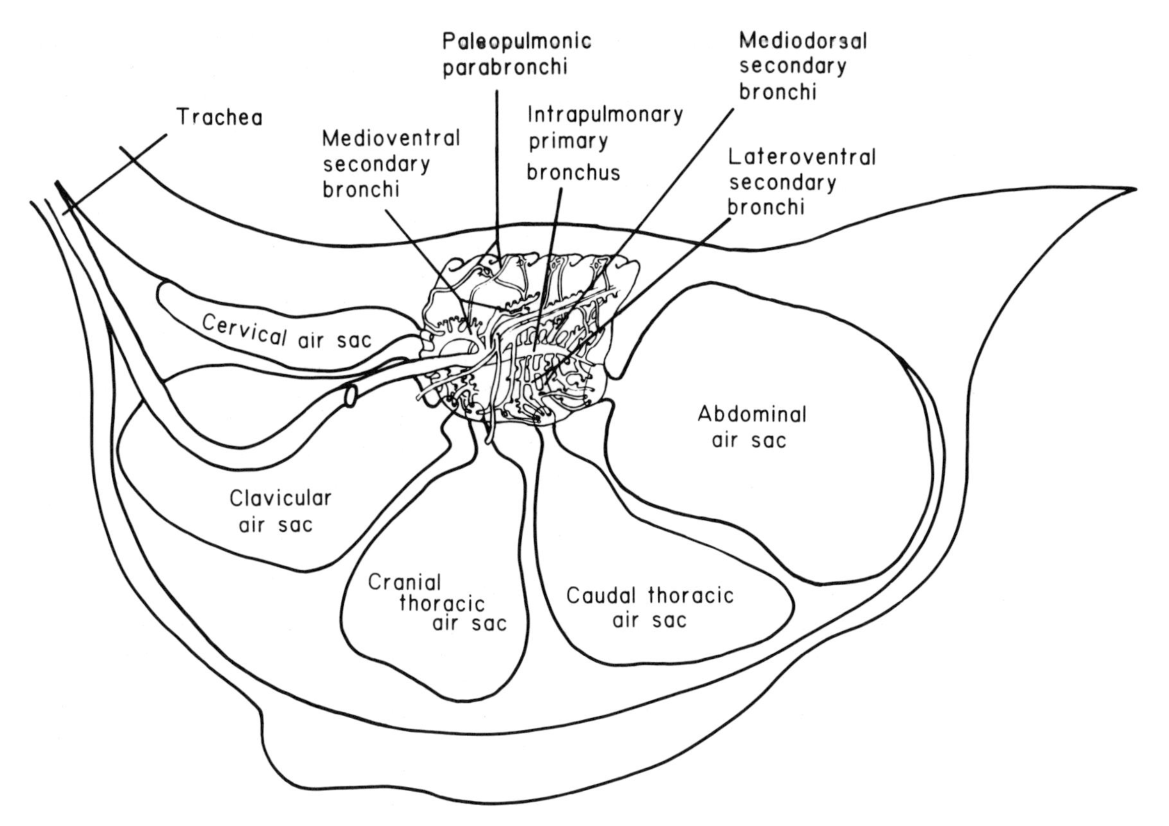

FIGURE 8–2. General arrangement of the lungs and air sacs in the chicken. (After King, 1966a, and Brackenbury, 1981.)

paired. This structure prevents entry of foreign bodies into the trachea (especially during swallowing). It may also alter airway resistance by changing the width of the inlet and act to modulate the pitch of voice.

The larynx connects with the trachea, a cartilaginous structure in some birds and ossified in others, which is composed of complete interlocking rings. The trachea takes the most direct route from the larynx to the syrinx in most species of birds; however, there are 55 species known to possess tracheal elongations varying from loops to complex convolutions (Niemeier, 1979). In the whooping crane (*Grus americana*), part of the coiled trachea lies in an infrasternal cavity; when uncoiled, the trachea is longer than the total length of the bird (Roberts, 1880). The elongated and coiled trachea is thought to influence the character of the voice and may allow production of lower frequency sounds that can be heard at greater distances by others of the same species (Prange et al., 1984). Whether this structure serves other important functions is not known.

The tracheal length and volume are characteristically greater in birds as compared to mammals; the tracheal dead-space volume is some 4.5-fold that of mammals of equal body mass (Hinds and Calder, 1971). Respiratory dead space is about 7.0 ml in 2.2-kg male chickens (Kuhlmann and Fedde, 1976). Birds appear to compensate for the large dead space volume by increasing their tidal volume and decreasing respiratory frequency as compared to equal-sized mammals.

The distal end of the trachea connects to the syrinx, the organ of voice production in birds. This complex structure lies just inside the coelomic inlet and is surrounded by the clavicular air sac. it is constructed of a cartilaginous skeleton, and contains tympanic membrances on its medial and lateral walls. Air movement past these membranes during expiration causes their vibration and hence voice production. The two extrapulmonary primary bronchi arise at the syrinx. These structures are relatively short and enter the lungs on the ventromedial aspect.

Anatomy of the Lungs and Air Sacs

Unlike mammalian lungs, avian lungs are relatively rigid and do not move appreciably during the respiratory cycle. There is no functional diaphragm separating the thoracic from the abdominal cavity and the lungs do not collapse when the thoracic cavity is entered (pneumothorax), as occurs in mammals. Emanating from various bronchi within the lungs are thin-walled air sacs that fill most of the body cavity not occupied by other viscera. There are nine air sacs in most birds (Figure 8–2); some penetrate into the interior of many bones and subcutaneous regions outside the coelom. The gas volumes of the lungs and air sacs vary with the size and species of the bird and the method of determination. Maximum capacities for a 2.9-kg chicken are (King, 1966a)

Air spaces	Volume (ml)
Cervical (both)	20
Clavicular (single)	55
Cranial thoracic (both)	50
Caudal thoracic (both)	24
Abdominal (both)	110
Lungs (both)	35
Skeletal air space	4
Total respiratory volume	298

In the domestic duck, the caudal thoracic air sacs are the largest and the abdominal sacs are relatively small (Scheid et al., 1974b). Additional measurements of the volumes of various parts of the respiratory system of chickens and other birds using a variety of techniques have been made (Burton and Smith, 1968; Scheid and Piiper, 1969; Duncker, 1972; Dubach, 1981; Maina et al., 1982; Keijer and Butler, 1982).

As indicated above, the extrapulmonary primary bronchus enters the hilus of the lung on the ventromedial surface; after that point, it is called the intrapulmonary primary bronchus. The tubular system of the lung consists of three bronchial subdivisions; the intrapulmonary part of the primary bronchus, various secondary bronchi, and many parabronchi. The general arrangement of these bronchi and the air sacs that extend from them is shown in Figure 8–2 and in more detail in Figure 8–3.

The first set of secondary bronchi (medioventral bronchi, usually four in number) arises from the intrapulmonary primary bronchus near the hilus of the lung. These bronchi extend over the medioventral (lower and innermost) surface of the lung and accept gas from parabronchi (tertiary bronchi) that emanate from their surfaces. The cervical air sac arises from the first medioventral bronchus. The clavicular air sac connects to the first, second, and third medioventral bronchi, whereas the cranial thoracic air sac connects principally to the third medioventral bronchus. Large connections of the fourth medioventral bronchus to air sacs are not present, but there may be small connections to both the cranial and caudal thoracic air sacs; additional studies are required to clearly determine those connections in various avian species. The cervical, clavicular, and cranial thoracic air sacs form a *cranial group,* all of which have direct connection only to medioventral secondary bronchi.

Just after the medioventral bronchi arise, the intrapulmonary primary bronchus makes an S-shaped bend through the substance of the lung coursing toward the dorsolateral (upper and outermost) surface just beneath the ribs; here second and third sets of secondary bronchi (mediodorsal and lateroventral bronchi) have their origin. The mediodorsal bronchi branch over the costal surface of the lung, with parabronchi emanating from them. The parabronchi joining the medioventral and mediodorsal bronchi form a series of parallel tubes called "paleopulmo" (Figure 8–4A). The lateroventral secondary bronchi pass along the ventral part of the costal surface of the lung, and the second (first or third in some species) forms a large direct connection to the caudal thoracic air sac. The intrapulmonary primary bronchus continues to the caudal border of the lung and empties into the abdominal air sac. The caudal thoracic and abdominal air sacs, therefore, form a *cau-*

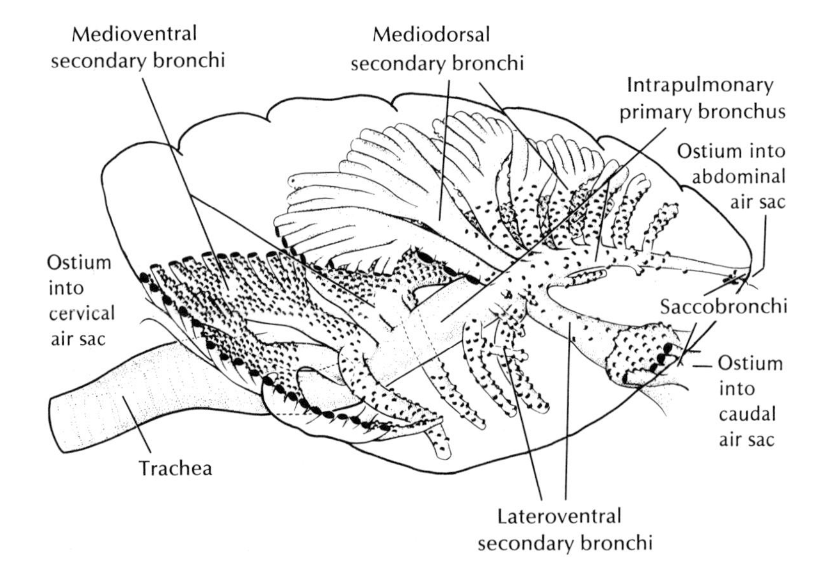

FIGURE 8–3. Drawing of the general arrangement of the avian bronchial system. Left lung of the mute swan. *Cygnus olor.* (After Duncker, 1971.)

FIGURE 8–4. Scheme of the organization of the parabronchial system in birds. (**A**) Only paleopulmic parabronchi are present in some birds (penguin and emu). (**B**) A highly developed neopulmonic parabronchial net, in addition to paleopulmonic parabronchi, is present in many birds (woodcocks, pheasants, quail, chicken, and songbirds). (After Duncker, 1972.)

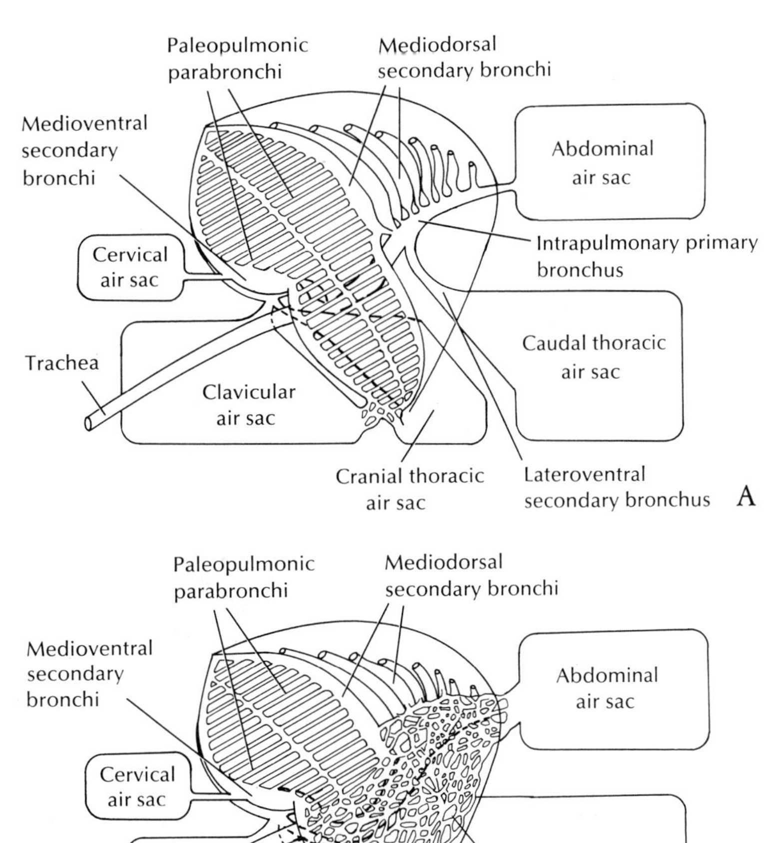

dal group that arises from the lateroventral bronchi and the intrapulmonary primary bronchus.

In most birds, an additional set of secondary bronchi (laterodorsal bronchi, not shown in Figure 8–3) and an additional network of parabronchi are present (Figure 8–4B). That additional network, which is absent in penguins and emus and only poorly developed in such birds as storks, cormorants, and cranes, is called the "neopulmo." It is more extensive in pigeons, ducks, geese, and gulls, and may originate from the initial segments of the mediodorsal bronchi as well as from the intrapulmonary primary bronchus and lateroventral bronchi. In these birds, the main trunks of the mediodorsal secondary bronchi are visible on the costal surface of the lung. In woodcocks, pheasants, quail, chickens, and songbirds, the neopulmo is so extensive that the secondary bronchi are displaced into the substance of the lung and are no longer visible from the outer surface. However, even in birds with a well-developed neopulmo it never exceeds 20–25% of the total lung volume.

At the entrances (ostia) of some of the air sacs, parabronchi empty into large tubes, called saccobronchi, which are similar in diameter to secondary bronchi. These parabronchial connections to the air sacs have been referred to as "recurrent bronchi" in earlier reports.

Parabronchi constitute the third level of bronchial division (tertiary bronchi) in the avian lung. There are 300–500 parabronchi in each lung of a chicken (King and Cowie, 1969), and these remain unchanged in position and number from a few days before hatching to adulthood. The neopulmonic parabronchi vary greatly in length, whereas the length of the paleopulmonic parabronchi is similar throughout the lung. The diameter of all parabronchi in the lung is similar in a given species and ranges from 0.5 mm in small birds to 2.0 mm in larger birds (Duncker, 1974; Dubach,

1981). Paleopulmonic and neopulmonic parabronchi cannot be differentiated in histological cross section.

Each parabronchus possesses large invaginations, called atria, from its lumen; these in turn lead to smaller funnel-shaped invaginations, termed infundibula (Figure 8–5). Extending from the infundibula are the air capillaries, which represent the smallest tubular network in the lung. The air capillaries are anastomosing cylindrical tubes with diameters ranging from 3 μm in songbirds to 10 μm in penguins, swans, and coots

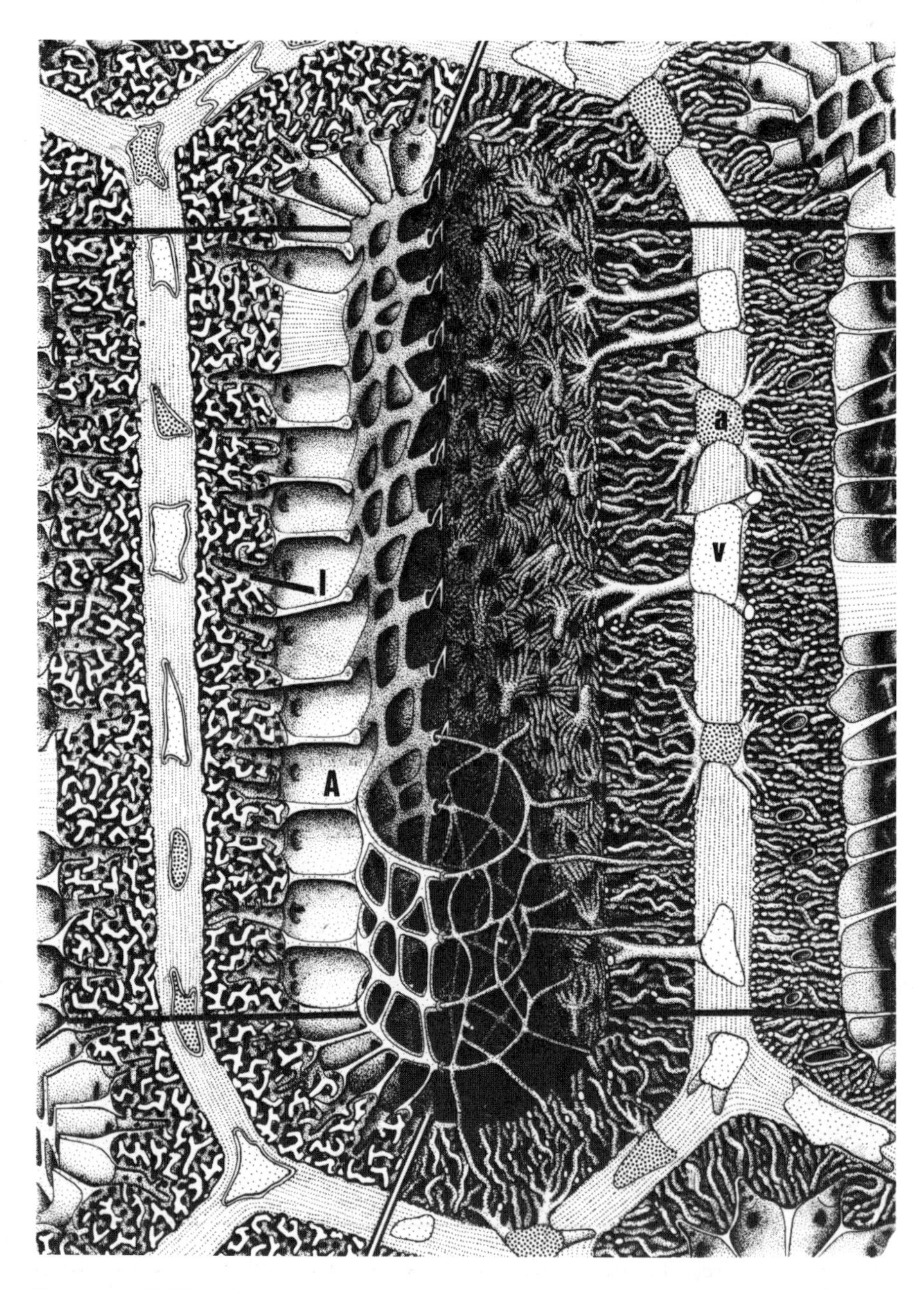

FIGURE 8–5. Drawing of an avian parabronchus showing, on the left, the pathway for gas diffusion from the lumen through the atria (*A*) and infundibula (*I*) to the air capillaries, and on the right, the pathway for blood flow from interparabronchial arteries (*a*) and intraparabronchial arterioles to the blood capillaries. Blood is collected in venules at the base of the atria and carried back to septal veins (*v*) between parabronchi and is finally deposited in the large pulmonary veins. (From Duncker, 1974.)

(Duncker, 1971; Brackenbury and Akester, 1978; Fujii et al., 1981; Maina, 1982; Powell and Mazzone, 1983). Gas exchange occurs across the walls of the air capillaries.

The avian lung contains an abundance of smooth muscle (King and Cowie, 1969). Circular, oblique, and longitudinal layers of smooth muscle are in the primary bronchus and initial segments of secondary bronchi. The musculature is organized in narrow, spiral bands further distally in the secondary bronchi, and these bands are especially thick where they surround the origin of the parabronchi. Smooth muscle forms the framework of the atrial openings with a network of bands just beneath the epithelium of the parabronchial lumen (Figure 8–5). The smooth muscle can reduce the diameter of the parabronchial lumen, or even close it completely, when activated by acetylcholine or stimulation of efferent nerve fibers in the vagi. Smooth muscle around the origin of the medioventral secondary bronchi may constrict these bronchi when intrapulmonary CO_2 concentration is lowered (Molony et al., 1976), but changes in intrapulmonary CO_2 concentration do not appear to induce parabronchial smooth muscle contraction (Barnas et al., 1978a, b).

Embryological Development of Lungs and Air Sacs

The first appearance of the lungs of the chick embryo occurs in the early part of the third day of development (Locy and Larsell, 1916a, b). The primitive lungs are paired and consist of two shallow pouches that open widely into the floor of the pharynx. The bronchial system develops from the lung pouches, which are lined by endoderm. At the end of the fourth day of incubation, the trachea becomes differentiated from the caudal portion of the laryngotracheal groove, and at the distal end of the lung tube there is an enlargement that later becomes the abdominal air sac.

The secondary bronchi begin to develop during the sixth day of incubation in the chick. On the ninth day of development, the lung has increased in size dorsoventrally and occupies a more lateral position in the thoracic cavity. The lung has begun to press against the ribs and exhibits shallow furrows where the lung substance has grown around the bodies of the ribs. The air sacs are formed and project beyond the surface of the lung. On the tenth day, outgrowths from the abdominal and caudal thoracic air sacs appear, which eventually ramify and anastomose with parabronchi in various parts of the lung. By 10.5 days of incubation, similar outgrowths appear from all air sacs except the cervical sacs.

At the beginning of the last third of embryonic development in the chick, the parabronchi have reached their final number and position (Duncker, 1978c). The beginning of the outgrowth of infundibula can be observed at this time. Many arterioles are present in the tissue between the parabronchi, while small venules lie directly at the base of the growing atria. Two or 3 days before hatching, blood capillary connections between arterioles and venules rapidly increase, and air-filled air capillaries sprout from the infundibula to surround the developing blood capillaries, forming a three-dimensional network of gas-exchange tissue. At this time, the amniotic cavity is aerated; respiratory movements become regular and serve to aerate the bronchial system, the air sacs, and the parabronchi. Thus, the lung gradually becomes used as an organ of gas exchange between the time of internal pipping and hatching as the bird prepares for a total air-breathing existence. This feature has also been recently demonstrated for the wedge-tailed shearwater (Pettit and Whittow, 1982). The lung, therefore, does not suddenly expand at the time of hatching, as is the case with the mammalian lung at birth, but rapidly grows as an air-filled structure during the latter stages of development. This method of air capillary development appears necessary because the high surface tension at the gas–liquid interface of the small-diameter air capillaries would not allow them to be expanded if they developed in a collapsed form.

During posthatch growth, the number of parabronchi in the lung does not increase; only their diameter and length increase. However, pneumatization of the skeleton is not present in the embryo and occurs totally after hatching (Duncker, 1978c).

Movement of Gases over the Respiratory Exchange Surface

Forces That Move Gas through the Lungs

The forces required to move gas through the lungs are derived from the action of respiratory muscles (Table 8–1). Electromyographic investigations of respiratory muscles have conclusively shown that inspiratory and expiratory muscles are alternately active during the respiratory cycle even in resting conditions in birds (Kadono and Okada, 1962; Kadono et al., 1963; Fedde et al., 1964a, b, c).

When inspiratory muscles contract, the body volume, and thereby that of the air sacs, increases, creating a subatmospheric pressure within these sacs. Because of the pressure differences, air then enters the mouth and nostrils of the bird and passes through the lungs into the air sacs. Conversely, when expiratory muscles contract, the volume of the air sacs is reduced increasing the pressure therein and forcing gas from the sacs back through the lungs and out the nostrils and mouth to the atmosphere (see later section on Pathway of

TABLE 8–1 Respiratory muscles of the chicken[a]

Inspiratory	Expiratory
M. scalenus	Mm. intercostales externi of fifth and sixth spaces
Mm. intercostales externi (except in fifth and sixth spaces)	Mm. intercostales interni of third to sixth spaces
M. intercostalis interni in second space	M. costosternalis pars minor
M. costosternalis pars major	M. obliquus externus abdominis
Mm. levatores costarum	M. obliquus internus abdominis
M. serratus profundus	M. transversus abdominis
	M. rectus abdominis
	M. serratus superficialis, pars cranialis and caudalis
	M. costoseptalis

[a] For innervation of the muscles, see deWet et al. (1967).

Gas through the Respiratory System for details). The air sacs thereby function as bellows, with their pressure and volume being modulated by the muscles of respiration, which force gas through the lungs during both inspiration and expiration. The avian lung is thus a "flow-through" system with a small gas volume and does not expand to accept the tidal volume, as is the case with mammalian lungs.

The anatomic arrangement in the avian respiratory system, which separates the gas-exchange function from the mechanical function of ventilation, allows the bird to be artificially ventilated by flowing a continuous stream of gas into the trachea, past gas-exchange surfaces in the lung, and out a surgically produced hole in the air sacs to the atmosphere. This procedure, called unidirectional artificial ventilation, permits the gas concentrations in the lung to be precisely controlled and quickly changed despite any changes in the breathing effort of the bird, and has been used extensively to study mechanisms of gas exchange and control of breathing (see Fedde, 1970, and Fedde et al., 1974a, for details of the procedure).

Sternum and Rib Cage Movement during Breathing

As the result of action of the respiratory muscles, the volume of the body cavity (and hence the air sacs) is altered by movement of the thoracoabdominal wall in both dorsoventral and lateral directions (Figure 8–6). The expansion of the cavity during inspiration results from the arrangement of the two articulations of each vertebral rib with the thoracic vertebra. When a rib is moved cranially by inspiratory muscles, it must also move ventrally and laterally. The sternum, coracoids, and furcula are simultaneously moved ventrally and cranially, pivoting at the shoulder joint. Because the sternum must move in a ventral-cranial direction for adequate body volume changes to sustain ventilation, birds must not be squeezed during restraint in such a way as to impede the sternal movement.

Air-Sac Pressures during Breathing

Although some studies in geese and chickens have found that pressures are identical in all the air sacs

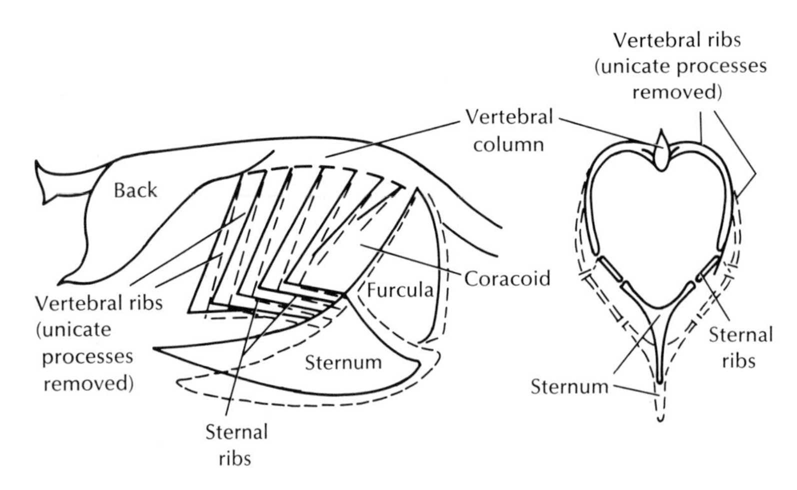

FIGURE 8–6. Changes in the position of the thoracic skeleton during breathing in a standing bird. Solid lines represent thoracic position at the end of expiration; dotted lines, at the end of inspiration. (After Zimmer, 1935.)

during a respiratory cycle (Cohn and Shannon, 1968; Baldwin, 1973), small differences of 0.5–2 mm H_2O have also been measured (Brackenbury, 1971a, b; 1972b). The air-sac pressures have the same general character as the airflow pattern in the trachea, and during quiet breathing exhibit a peak of approximately +1.0 cm H_2O (with respect to atmosphere) early in expiration that exponentially decays during the latter part of expiration; during inspiration, the peak pressure is approximately −1.0 cm H_2O, which does not decay appreciably until expiration begins (Cohn and Shannon, 1968; Brackenbury, 1973; Callanan et al., 1974). These small pressure differences provide the driving force for gas movement past the respiratory exchange surface during the breathing cycle. In addition, small pressure oscillations in the air sacs and in the gas spaces of the lung caused by the beating of the heart (cardiogenic oscillations) are often superimposed on these larger pressure fluctuations caused by the respiratory muscles (Powell et al., 1981). The cardiogenic oscillations may aid in gas mixing in the air sacs and lung.

Principal Impedances to Breathing

Of the principal impedances to breathing that must be overcome by the work of the respiratory muscles in mammals, two are of major importance in birds; impedance to movement of the thoracic cage and abdominal wall and impedance to gas flow.

Deformation of the Thoracic Cage and Abdominal Wall. The thoracic cage and abdominal wall are elastic structures that must be deformed by the respiratory muscles during the breathing cycle. This process requires a substantial amount of the work produced by these muscles. When all respiratory muscles are relaxed, the sterum rests approximately midway between its position at the peak of inspiration and the position at the peak of expiration. Thus, the elastic recoil of the thoracic cage aids the inspiratory muscles during the first half of inspiration, but resists these muscles during the latter part. Conversely, the elastic recoil aids the expiratory muscles during the first part of expiration but resists them during the last part. Both phases of respiration require active contraction of the respiratory muscles to produce the necessary change in volume of the thoracoabdominal cavity during each breathing cycle.

A measure of the deformability of the respiratory system is its compliance (the change in volume produced by a unit change in pressure). Because the lungs in birds are relatively stiff, the compliance is mainly determined by the air sacs and especially the surrounding body wall (Perry and Duncker, 1980). Compliance of the respiratory system has been measured in chickens whose skeletal muscles were relaxed by succinylcholine (Scheid and Piiper, 1969). Maximum values of 9.5 ml/cm H_2O were found in the transmural pressure range of −10 to 0 cm H_2O, and 6.5 ml/cm H_2O in the pressure range of 0 to 10 cm H_2O. Furthermore, because of changes in tonic activity of the muscles of the body wall, the compliance in the chicken is unstable and may spontaneously change over a wide range (Baldwin, 1973). The compliance of the respiratory system of the domestic pigeon (*Columba livia*) is 2.75 ml/cm H_2O over a transmural pressure range of 0 to 6 cm H_2O (Kampe and Crawford, 1973), and that of the duck is 7.7 ml/cm H_2O (Gillespie et al., 1982a).

Minimal work is required to move the thoracic cage when the frequency of movements is the same as the natural frequency of the respiratory system (the resonant frequency). The lower respiratory system of the duck has been shown to behave similarly to a series-mechanical network model consisting of resistive, inertial, and compliant elements (Gillespie et al., 1982a, b). The resonant frequency (f) can be determined by:

$$f = \frac{1}{2\pi}\sqrt{\frac{1}{CM}}$$

where C is the compliance of the system and M is the mass (tissue inertance). Although it appears that the pigeon utilizes these physical characteristics during panting by increasing its respiratory frequency to the resonant frequency of its system (and thereby minimizing the work of breathing, and, thus, the heat produced), the chicken and duck do not (Lacy, 1968; Crawford and Kampe, 1971; Gillespie et al., 1982a). The resonant frequency of the thoracoabdominal complex in the pigeon is 9.4 Hz, and pigeons pant at rates of about 10 breaths per second. The resonant frequency for the duck is 8 Hz, but they normally pant at only about four breaths per second.

Deformation of the Lungs and Air Sacs. It is generally believed that deformation of the avian lungs during breathing is minimal, thus requiring essentially no work for this process. However, the parabronchi are not totally rigid, and their diameter can be changed by altering intrapulmonary pressure (Macklem et al., 1979). The air capillaries would seem to be highly compliant structures because of their thin walls, but compressing the respiratory system does not cause them to collapse.

Although the thin-walled air sacs are highly compliant, intrapulmonary and coelomic pressures are not identical during inflation and compression of the respiratory system (Macklem et al., 1979; Brackenbury, 1973). Although some work must be utilized in deforming these sacs during breathing, it is likely to be only a small fraction of the total.

Overcoming the Surface Forces at the Gas–Liquid Interface in the Lung. In the mammalian lung, the majority of work expended in quiet breathing is used to overcome the surface forces that exist at the gas–liquid interfaces, especially in the alveoli. A surface-active material (called surfactant) is secreted onto the inner surface of the alveoli by the Type II pneumocytes in these animals. The surfactant reduces the surface tension and stabilizes the lung so that alveoli of different sizes can coexist.

In the avian lung, a surfactant (a complex dipalmitoyl lecithin–protein material) also exists in the parabronchi and air capillaries (see Pattle, 1978, for review). It is derived from laminated osmiophilic bodies formed in the epithelial cells of the atria, and is present by at least 18 days of incubation in the chick. In addition, a unique material called "trilaminar substance" is found in the same or similarly located cells and is composed of a sheet of two osmiophilic layers flanked by two nonosmiophilic layers. These sheets may occur singly or in packs.

As a result of the small volume changes in the avian parabronchial lung during the breathing cycle, it is likely that the principal function of surfactant is not to markedly alter the work of breathing but instead to prevent fluid transudation into the air spaces of the lung, especially the air capillaries. If the surface tension was high in these structures, there would be a tendency for the surface lining to pull away from the underlying epithelial cells, carrying water with it. This tendency would be greater the smaller the air capillary. Thus, a low surface tension is mandatory to keep these structures dry.

It is possible that the trilaminar substance also functions to keep the air capillaries dry. Because a liquid film lining a cylindrical space causes mechanical instability, the trilaminar substance may be able to absorb water without liquefying, thereby maintaining mechanical stability of the air capillaries.

As a consequence of the small amount of movement of the avian lung during the breathing cycle, the amount of work required to overcome the surface forces at the gas–liquid interface is likely to be small.

Overcoming the Resistances to Gas Flow. The tubular component of the respiratory system imposes a resistance to the flow of gas that requires a substantial amount of work to overcome. This resistance is defined as the ratio of driving pressure across the respiratory system (in cm H_2O) to gas flow through the system (in liters/sec). Three methods have been used to measure airway resistance in birds: (1) Flowing a constant stream of gas through the respiratory system (or part thereof) and measuring the pressure gradient across the resistance in question; (2) simultaneously measuring the flow of gas into the respiratory system and the pressure gradient across it in spontaneously breathing birds; and (3) measuring small induced sinusoidal flow and pressure changes at the tracheal cannula that are superimposed on the normal breathing movements of the bird (see Molony, 1978, and Gillespie et al., 1982a, for specifics). Airway resistance varies with the species from 4.8 cm H_2O/liter/sec in the duck to 41 cm H_2O/liter/sec in the pigeon (Gillespie et al., 1982a; Kampe and Crawford, 1973). Airway resistance increases with increasing flow and has been found to change suddenly during the course of a breath (Baldwin, 1973). The latter observation suggests active control of airway diameters. Several studies have indicated that airway resistance is greater during inspiration than during expiration, implying different gas pathways during the two parts of the respiratory cycle (Cohn and Shannon, 1968; Brackenbury, 1971b, 1972a; Baldwin, 1973); however, a recent study in the conscious duck indicates that resistances are about the same during inspiration and expiration (Gillespie et al., 1982a).

The respiratory time constant, an expression of how long it takes gas to move into or out of the respiratory system following a sudden muscular contraction, is given by the product of viscous resistance to airflow and tissue movement and compliance of the thoracoabdominal wall and internal septa. In birds, the time constant is long because of the large compliance and large airway resistance (Brackenbury, 1973). This feature highlights the necessity for both active expiration and active inspiration and may be responsible for the relatively slower, deeper breathing patterns in birds as compared with similar-sized mammals.

Pathway of Gas through the Respiratory System during Inspiration and Expiration

Early studies on the pattern and direction of gas flow through the avian lung were indirect and led to various conclusions (see Bretz and Schmidt-Nielsen, 1971, for review). However, direct measurements during spontaneous breathing have shown unequivocally that during both inspiration and expiration, gas moves *unidirectionally* from the intrapulmonary primary bronchus to the mediodorsal secondary bronchi and into the parabronchi of the paleopulmo (see Scheid, 1979a, b, 1982; Fedde, 1980; Powell, 1983a; and McLelland and Molony, 1983, for reviews).

During *inspiration* (Figure 8–7A), gas flow subdivides when it reaches the mediodorsal secondary bronchi, part going into these bronchi and part into and through the neopulmonic parabronchi to the caudal air sacs (caudal thoracic and abdominal air sacs). Gas entering the mediodorsal secondary bronchi moves into the paleopulmonic parabronchi toward, and perhaps into, the cranial group of air sacs (cranial thoracic, clavicular, and cervical air sacs) (James et al., 1976; Burns et al., 1978). As inspiration begins, gas that was contained within the mediodorsal secondary bronchi and paleopulmonic parabronchi at the end of the previous exhalation moves immediately into the cranial air sacs. Gas does not enter the medioventral secondary bronchi from the intrapulmonary primary bronchus during inspiration, nor does gas leave these bronchi

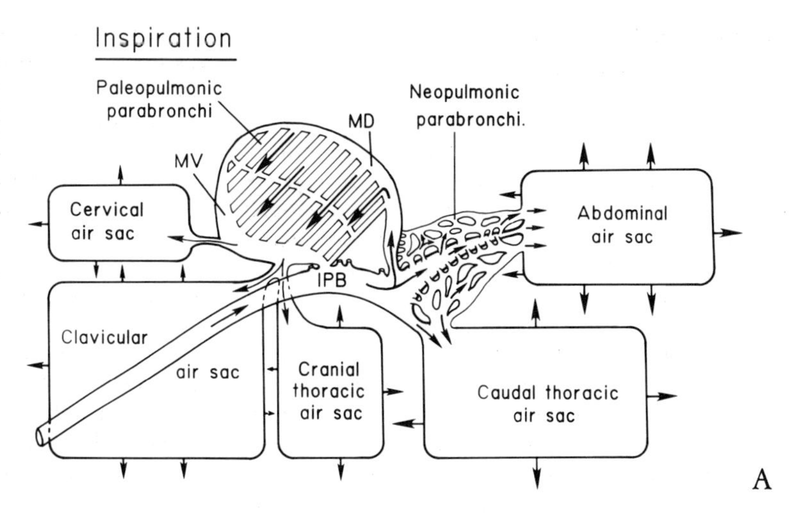

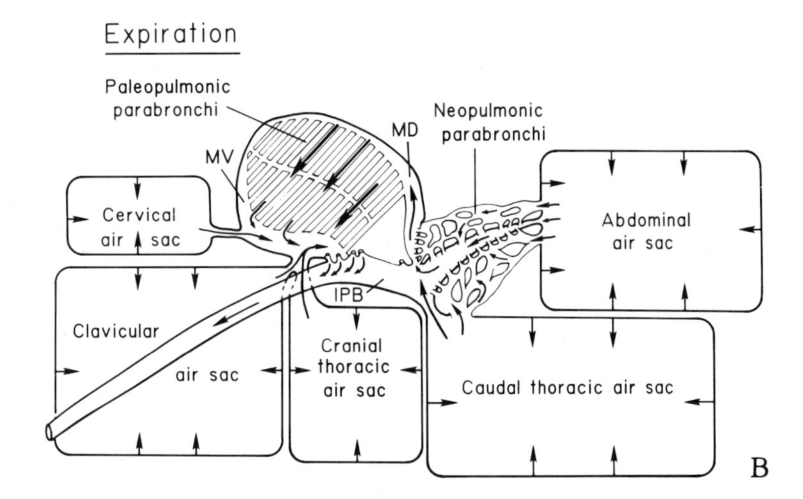

FIGURE 8–7. Schematic representation of the pathway of gasflow through the paleopulmonic and neopulmonic parabronchi during inspiration **(A)** and during expiration **(B)**. *MV,* Medioventral secondary bronchus; *MD,* mediodorsal secondary bronchus; *IPB,* intrapulmonary primary bronchus. (After Duncker, 1971.)

to reenter the intrapulmonary primary bronchus during this phase of the cycle (Powell et al., 1981).

During *expiration* (Figure 8–7B), gas moves from the caudal air sacs again through the neopulmo into the mediodorsal secondary bronchi and paleopulmonic parabronchi. At the same time, gas leaves the cranial air sacs through the medioventral secondary bronchi and flows into the primary bronchus and trachea to the atmosphere. At least during quiet breathing, little or no gas flows directly out of the lung without passing through the paleopulmonic parabronchi during expiration, thereby preventing an air shunt in the lung (Scheid, 1981; Powell et al., 1981). Thus, gas passes through the paleopulmo in a caudal-to-cranial direction during both phases of the respiratory cycle, but travels through the neopulmonic parabronchi bidirectionally.

The mechanisms responsible for the unidirectional gas pathway through the paleopulmo are thought to result from aerodynamic conditions within the lung, since no anatomic valves are present (see Jones et al., 1981, for review). These conditions are created by the shape and curvature of the intrapulmonary primary bronchus and the angle of departure and configuration of the secondary bronchi at their origin from the intrapulmonary primary bronchus. The gas flow rectification may be caused by the Bernoulli effect, detachment of gas flow from the wall thereby causing higher flow resistances in one direction, and gas jet formation at these sites. Recent studies indicate that the cranial air sacs play a predominant role in producing the caudal-to-cranial gas flow through the paleopulmo during inspiration and that the caudal air sacs are primarily responsible for gas flow in this direction during expiration (Brackenbury, 1979). Further studies are required to elucidate the exact mechanisms responsible for the aerodynamic valving.

Partial Pressures of Gases in the Respiratory System

At the beginning of inspiration, gas in the nasal passages, trachea, and extra- and intrapulmonary primary bronchi is the first to enter the gas-exchange region of the lung and the caudal air sacs. This dead-space gas has previously passed through the paleopulmonic parabronchi and has a high partial pressure of CO_2 (P_{CO_2}) and a low partial pressure of O_2 (P_{O_2}); the partial pressures of those gases are the same as that in end-expiratory gas (Table 8–2). The partitioning of the inspirate to the caudal air sacs and to other parts of the respiratory system occurs continually throughout inspiration, so that no preferential shunting of dead-space gas into these sacs occurs (Torre-Bueno et al., 1980). As inspiration continues, fresh gas from the atmosphere reaches the neopulmonic parabronchi, undergoes some gas exchange, and passes into the caudal air sacs (some gas may also pass directly into these sacs without passing over exchange surfaces) (Jammes and Bouverot, 1975). Thus, the last gas to enter these air sacs has a low P_{CO_2} and a high P_{O_2}. There appears to be incomplete mixing of the gas in the caudal air sacs within the time of one breath, resulting in gas layering or stratification (Torre-Bueno et al., 1980). Thus, the first gas to leave these air sacs at the beginning of expiration and then to pass over the exchange surfaces of the neopulmo and paleopulmo has a lower P_{CO_2} and a higher P_{O_2} than that leaving the sacs later. The gas-exchange surfaces are, therefore, exposed to gas containing a low P_{CO_2} and a high P_{O_2} during most parts of both inspiration and expiration.

The CO_2 in the caudal air sacs does not come from gas exchange across the air sac walls (Magnussen et al., 1976) or from recirculation of paleopulmonic gas (the so-called Hazelhoff loop) (Powell et al., 1981), but is derived from reinspiration of dead-space CO_2 and from gas exchange in the neopulmo (Jammes and Bouverot, 1975; Piiper, 1978; Fedde et al., 1981). The average P_{CO_2} is lower and the P_{O_2} higher in the caudal air sacs than in end-expired gas, as a result of dilution of dead-space gas with fresh gas and limited gas exchange by the neopulmo (Table 8–2).

There is a much higher P_{CO_2} and lower P_{O_2} in the gas in the cranial air sacs because of the greater degree of exchange with blood as gas passes through the paleopulmo (Table 8–2). Gas partial pressures in these sacs do not fluctuate much during a breath and represent end-parabronchial values; they are close to end-expired values (Powell et al., 1981). Because of the design of the gas-exchange system and the Haldane effect in the blood (see section on Transport of CO_2), the P_{CO_2} in the cranial air sacs and in end-expired gas can be higher than that in the blood entering the lung from the pulmonary artery (the mixed venous blood) (Meyer et al., 1976).

Mechanisms of Gas Exchange between Parabronchi and Blood

Several excellent reviews discussing parabronchial gas exchange have appeared (Scheid, 1979a, 1982; McLelland and Molony, 1983). The following sections present the highlights of the process.

TABLE 8–2 Partial pressure of O_2 and CO_2 in air sacs and in end-expiratory gas of indicated species of birds[a]

	Goose[b]	Goose[c]	Chicken[d]	Duck[e]
Clavicular				
P_{CO_2}, torr	35	39.4	44.0	39.2
P_{O_2}, torr	100	93.9	83.9	99.4
Cranial thoracic				
P_{CO_2}, torr	35	37.3	41.6	35.7
P_{O_2}, torr	100	95.3	99.1	104.3
Caudal thoracic				
P_{CO_2}, torr	28	20.0	24.2	18.9
P_{O_2}, torr	115	124.3	120.3	123.9
Abdominal				
P_{CO_2}, torr	28	17.3	14.7	17.5
P_{O_2}, torr	115	127.8	130.0	126.7
End expiratory				
P_{CO_2}, torr	35	35.9	36.7	35.7
P_{O_2}, torr	100	100.1	94.3	100.1

[a] All species were in an upright position and unanesthetized.
[b] Cohn and Shannon (1968), Figure 6.
[c] M.R. Fedde, R.E. Burger, J. Geiser, R.K. Gratz, J.A. Estavillo, and P. Scheid (unpublished data).
[d] Piiper et al. (1970).
[e] Vos (1934). Composition of gas was given as percentage; partial pressures were calculated by assuming a mean barometric pressure of 760 torr and a water vapor pressure of 60 torr.

Anatomical Arrangement of Gas and Blood Supply to a Parabronchus

From both anatomic and physiologic evidence, it is clear that gas flows through a parabronchus at right angles to the flow of blood in the blood capillaries (Figure 8–5) (Scheid and Piiper, 1970; Abdalla and King, 1975; West et al., 1977; Fujii et al., 1981). This arrangement can be modeled as a cross-current exchange system (Figure 8–8), and is inherently more efficient than the uniform pool system that exists in the mammalian lung (Piiper and Scheid, 1972, 1975, 1982). As gas flows through the parabronchus, CO_2 is continually added from the blood and O_2 is continually removed by the blood. More gas is exchanged in the region of the parabronchus where gas enters than where it leaves. On the other hand, blood of a similar composition (mixed venous blood) enters into gas exchange all along the parabronchus. The partial pressure of gases in the blood that leaves a parabronchus results from the mixing of capillary blood bathing its various segments.

Theoretically, the cross-current relationship of gas and blood permits the partial pressure of CO_2 in arterial blood leaving the lung (Pa_{CO_2}) to be lower than the end-expired P_{CO_2} ($P_{E'CO_2}$) and the partial pressure of O_2 in the arterial blood (Pa_{O_2}) to be higher than end-expired P_{O_2} ($P_{E'O_2}$). Direct measurements indicate that in resting normoxic birds Pa_{CO_2} is indeed less than $P_{E'CO_2}$, but Pa_{O_2} is usually not greater than $P_{E'O_2}$. The reason for the lower efficiency for O_2 exchange than for CO_2 exchange is not known, but may be related to several of the diffusion resistances that occur in the lung (Powell, 1982); it cannot be explained by ventilation–perfusion inequality. Of these resistances to gas exchange, the diffusion resistance of gas from the parabronchial lumen down the air capillaries should not be a limiting factor at rest (Crank and Gallagher, 1978; Scheid, 1978; Scheid et al., 1978a; Burger et al., 1979) nor should the thin tissue barrier between gas and blood in the parabronchial mantle. Although the rate of reaction between O_2 and hemoglobin in avian blood is not known, it is likely that blood remains in the pulmonary capillaries long enough (average, 0.9 sec) at rest so that equilibrium is established (Powell and Mazzone, 1983); this diffusion resistance should not limit gas exchange.

Factors Determining the Amount of Gas Exchanged

There is a cascade of conductances (defined as the ease with which a molecule can move from one place to another) from the atmosphere to the mitochondria of cells (Figure 8–1). The movement of O_2 from the atmosphere to the air capillaries in the lung is essentially determined by bulk flow of gas (ventilation of the lung) and thus is a convective conductance. Similarly, the movement of blood (cardiac output) from the lungs to the tissue depends on a convective conductance. However, the transfer of O_2 between the gas and the blood in the lung and between the blood and the cells

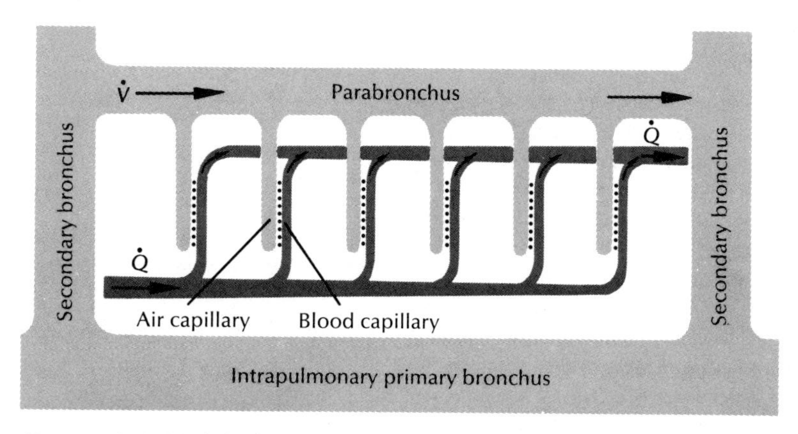

FIGURE 8–8. Model of the cross-current arrangement of blood capillaries and parabronchial gas flow in the avian lung. A paleopulmonic parabronchus lies parallel to the intrapulmonary primary bronchus. Air capillaries depart from the parabronchus (atria and infundibuli have been omitted) and are in gas-exchange contact with the blood capillaries. Respired gas flows through the parabronchus at a constant rate $\dot{V}$, and the composition of the gas changes along the parabronchus as CO_2 and O_2 are exchanged with the blood. The blood flow $\dot{Q}$ to the parabronchus is distributed over the parabronchial length and is directed at a 90° angle to the longitudinal axis of the parabronchus. (After Scheid and Piiper, 1970.)

in the tissues is determined by diffusive conductances. In the lung, the various components of the conductance, lumped together as the diffusing capacity of the lung ($D_{L_{O_2}}$), depend on the structural design of the exchange region (i.e., air capillary and blood capillary surface areas, blood capillary volume, and thickness of the gas–blood tissue barrier), diffusion distance within the red blood cell, and rate of O_2 binding with hemoglobin in the blood. Recent morphometric studies on the lungs of several species of birds have provided new insight into the reasons for the high gas-exchange efficiency exhibited by birds (Dubach, 1981; Maina and King, 1982; Maina et al., 1982; Abdalla et al., 1982; Powell and Mazzone, 1983; Maina, 1984).

Air Capillary–Blood Capillary Surface Area. Gas constitutes about 66% of the total lung volume in the chicken. The total volume of intrapulmonary gas in the lungs of a 2-kg chicken is about 16 cm^3, of which about 40% (6.7 cm^3) is in the air capillaries. The air capillaries are cylindrical, branching structures of 3–10 μm in diameter in different species (much smaller than mammalian alveoli), so the air–blood capillary network can be thought of as intertwining tubes (Figure 8–5). The gas–blood barrier in the chicken lung has a surface area of about 2 m^2. The surface area of the tissue barrier per unit volume of exchange tissue in the chicken lung is about 180 mm^2/mm^3, but is much higher in strong-flying birds (443 mm^2/mm^3 in the violet-eared hummingbird). The exchange surface of the avian lung related to the volume of the respiratory tissue is two to four times larger than that of the lung of a mammal of corresponding weight as a result of the very small diameters of the air capillaries. The human lung has only $\frac{1}{37}$ of the exchange surface per unit volume of the hummingbird.

Blood Capillary Volume. The total volume of blood (6.9 cm^3) within the two lungs of the chicken occupies about 28% of the total lung volume. Approximately 51% of the blood in the chicken lung resides in the pulmonary capillaries at any one time. This value, as high as 80% in the carrion crow, is much higher than in most mammals (i.e., only 20% in man).

The large capillary blood volume allows a relatively long time (0.90–0.92 sec) for red blood cells to equilibrate with air capillary gas. Furthermore, recruitment of blood capillaries may prevent a substantial decrease in transit time when cardiac output increases during exercise (Powell and Mazzone, 1983). Such a feature could maintain the P_{O_2} in a blood capillary near that of the gas in adjacent air capillaries, despite an increase in O_2 consumption during flight of as much as 10–20 times the basal level.

Thickness of the Gas-Blood Barrier. The remarkable thinness of the avian gas–blood barrier has been recognized for many years, but only recently have measurements of the harmonic mean thickness (a measure of the resistance to gas diffusion) and arithmetic mean thickness (a value that expresses the tissue density of the barrier and hence its oxygen consumption) been made on many avian species. The harmonic mean thickness varies from about 0.1 μm in the house sparrow and violet-eared hummingbird to about 0.3 μm in domestic fowl and domestic guinea fowl. The arithmetic mean thickness also varies from about 0.2 μm in the hummingbird to 1.7 μm in the common kestrel. The chicken tends to have a thicker barrier than most strong-flying birds. The harmonic mean thickness of the gas exchange barrier in mammals is three to four times greater than in birds of similar body weights. For example, the mammal with the thinnest barrier is the shrew (*Suncus etruscus*), with absolute minimum values of harmonic mean thickness of 0.23 μm. This value is similar to that of the chicken (0.30 μm). Thus, the mammalian lung with the thinnest barrier is comparable to the avian lung with the thickest barrier.

The gas–blood barrier is composed predominantly of the blood capillary endothelium (67% of the total in the chicken), the basal lamina (an intermediate layer comprising 21% of the total and thought to be secreted by the endothelium), and the epithelium plus osmiophilic surface lining (12% of the total). Collagen and fibrocytes are sparse in the barrier, but when present are found at the angles formed by the apposition of two blood capillaries. Despite the thinness of the barrier, there are undulations of thicker and thinner regions that suggest the barrier may be corrugated. Corrugation may be advantageous because it increases the degree of mechanical stability while at the same time lowering the overall resistance to gaseous diffusion, as compared to a barrier of uniform thickness.

The extremely thin barrier appears possible only because of the relatively rigid construction of the blood–air capillary net. Mammalian lungs, which expand and retract with each breath, must have more interstitial tissue to create strength of the barrier. Since the arithmetic mean thickness is considerably less than that of mammals, reptiles, or amphibians, the avian lung appears to have achieved mechanical stability of the barrier while keeping O_2 consumption of the lung tissue at a minimum. Furthermore, the small hamonic mean thickness of the gas–blood barrier in strong-flying birds allows increased efficiency of gas exchange, and this may be one of the factors that has permitted birds to achieve such high metabolic rates.

Rate of O_2 Binding with Hemoglobin. The rate of binding of O_2 with hemoglobin must be taken into account when considering the diffusing capacity of the lung for O_2. The binding rate is affected by the degree of saturation of hemoglobin and by the speed of the enzyme kinetics in various erythrocyte chemical reactions (see also Chapter 5). Recent experiments indicate that the reaction rate for binding O_2 with hemoglobin is similar in humans and in ducks and chickens, and that diffusion resistance, especially inside the erythrocytes, is the major determinant of the speed of O_2 uptake and release from these cells (Yamaguchi et al., 1985; D. Nguyen Phu, K. Yamaguchi, P. Scheid and J. Piiper, unpublished observations).

Lung and Membrane Oxygen-Diffusing Capacities

Because of limited data for birds, an estimated adjusted mammalian value for the rate of O_2 binding to hemoglobin has been used by Abdalla et al. (1982) to calculate the anatomic lung diffusing capacity for oxygen (D_{LO_2}) for the parabronchial lung of the chicken. The D_{LO_2} is a measure of the volume of oxygen that can move from the gas phase to the blood phase in the lung each minute per torr of driving pressure. The minimum D_{LO_2} calculated value (1.50 ml O_2/min/torr) is the same as the physiologic D_{LO_2} estimated in the chicken by Scheid and Piiper (1970). Furthermore, the mean anatomic D_{LO_2} value (2.53 ml O_2/min/torr) is near the physiologic value (2.28 ml O_2/min/torr or 102 μmol/min/torr) obtained by Burger et al. (1979) in the domestic muscovy duck after correcting for both shunts and inhomogeneities. Thus, even with the many assumptions, anatomic and physiologic estimates of D_{LO_2} are within reasonable agreement.

The morphometric estimation of specific membrane diffusing capacity in the hummingbird (0.271 ml O_2 moving through the membrane per torr of driving pressure per minute per gram of body weight) is twice as high as in the house sparrow or budgerigar. The specific membrane diffusing capacity in birds is 4.6–8.5 times greater than that of comparably sized mammals. The factors responsible for this high value are the larger effective exchange surface of the blood and air capillaries, and, to a lesser extent, the thin blood–gas barrier.

As indicated earlier, the O_2 flow rate from the air capillaries to the hemoglobin in the red cell of the blood capillary depends on the O_2 pressure gradient and the conductance to this gas. Abdalla et al. (1982) estimated that in this pathway 74% of the total resistance results from the erythrocyte resistance, and only 10% and 16% of the resistance can be attributed to the plasma layer and the tissue barrier, respectively. Thus, the rate of O_2 movement appears to be limited by intraerythrocyte diffusion.

In the resting, air-breathing bird, there is a substantial reduction in the P_{O_2} between the atmosphere and the arterial blood (typically, a difference of about 50 torr). Such a large difference would suggest a substantial degree of inefficiency in gas transfer at some point or points in the pathway. However, it has been shown that both Peking ducks and bar-headed geese (a high-altitude-adapted species) can dramatically narrow the P_{O_2} difference between moist inspired gas and arterial blood to within 1–5 torr when they are exposed to extreme hypoxia (Black and Tenney, 1980; Faraci et al., 1984). Under these conditions, it appears that the gas-exchange surfaces in the air capillaries are exposed to a P_{O_2} nearly the same as that in the inspired tracheal gas and that the diffusion resistance of the blood–gas barrier and erythrocyte as well as the reaction rate with hemoglobin are not limiting to gas exchange. The dynamic response of these conductances to changing environments and perhaps also to changing metabolic conditions may allow birds to meet the energy demands of flight, especially at high altitudes.

Gas Transport by Blood

At the same time O_2 moves from air capillaries into the blood, CO_2 moves from the blood into the air capillaries. Both of these gases are carried in the blood in a variety of forms, but their interaction with hemoglobin is critically important for transport in adequate quantities.

Transport of O_2

In Physical Solution. According to Henry's law, the concentration of a gas in solution is directly proportional to its partial pressure in the solution. Thus, $[O_2] = \alpha_{O_2} \cdot P_{O_2}$, where $[O_2]$ is the concentration of O_2 in aqueous solution, P_{O_2} is the partial pressure of oxygen in the solution, and α_{O_2} is the solubility coefficient for O_2 in the solution expressed as millimoles of O_2 per liter of solution per torr. If α_{O_2} in avian blood is similar to that for whole human blood (0.00124 mmol/liter/torr at 41°C) (Bartels et al., 1971), then avian blood would contain 0.102 mmol O_2/liter of blood in physically dissolved form at a P_{O_2} of 82 torr, the approximate P_{O_2} in arterial blood of unanesthetized, undisturbed chickens and ducks (Kawashiro and Scheid, 1975).

By Hemoglobin. Avian hemoglobin is composed of an iron-containing prosthetic group, heme, and a protein moiety, globin (see Chapter 5). The globin part of the molecule varies considerably among birds and among different types of hemoglobin in a given bird.

During embryonic development, avian red blood cells are formed in several different sites and exhibit successive changes in hemoglobin types with different O_2 affinities. Definitive erythrocytes containing adult hemoglobins type A and D appear at 6 days of development in the chick embryo, and by 14 to 16 days adult hemoglobin replacement is complete, causing a marked increase in O_2 affinity (P_{50} lowers to 35 torr) (Baumann and Baumann, 1978; Baumann et al., 1983). In the adult, hemoglobin A accounts for about 70% of the total and hemoglobin D about 30%. The structure of the α and β chains in these hemoglobins has now been determined in several birds (Schenk et al., 1978; Oberthür et al., 1983; see also Chapter 5).

As hemoglobin is exposed to increased partial pressures of O_2, more and more O_2 molecules bind to the heme subunits. Because of cooperativity among the heme subunits, the binding is not a linear function of P_{O_2}, but instead a sigmoid relationship exists (Figure 8–9). In the deoxygenated state, the iron atom of the heme is not very accessible to oxygen molecules because of its position in the protein matrix, and O_2 affinity is low. Hemoglobin in this state has a number of salt bridges and a tensed or T structure. When the P_{O_2} rises and the first O_2 molecules bind to an iron, salt bridges in the molecule are broken and the structure of hemoglobin suddenly changes to a relaxed or R structure. In this form, O_2 can very readily bind to the remaining iron atoms and the O_2 affinity rises dramatically. Hence, the hemoglobin saturation curve becomes steep as additional hemoglobin molecules become filled with four O_2 molecules each. At a P_{O_2} of 100 torr, there are still a few hemoglobin molecules that have not switched to an R structure and contain none or only one O_2 molecule (Perutz, 1978; Martin, 1981).

In vivo, as the shifts in saturation of hemoglobin occur in the pulmonary and tissue blood capillaries, the affinity of hemoglobin for O_2 does not simply move up and down one of the curves shown in Figure 8–9. At these capillary sites, there are simultaneous shifts in P_{CO_2} and pH of the blood. These chemical changes alter the affinity of hemoglobin for oxygen, as illustrated by the heavy line in Figure 8–10. Hence, when arterial blood has a P_{O_2} of 82 torr, hemoglobin is 91% saturated. If 100 ml of arterial blood has 12 g of hemoglobin, it would contain 0.663 mmol O_2 (0.653 mmol of O_2 bound to hemoglobin + 0.010 mmol of O_2 physically dissolved). As the blood passes through the tissue capillaries, O_2 diffuses from the blood and the P_{O_2} equilibrates with that in the tissues. Because the hemes give up their O_2 readily as the P_{O_2} drops between 70 and 20 torr, a large amount of O_2 can be delivered to the tissues with a rather small reduction in P_{O_2}. If the mixed venous blood, as represented by the blood in the pulmonary artery, has a P_{O_2} of 39 torr, the hemoglobin is only about 46% saturated and 100 ml of blood contains 0.337 mmol of O_2. Therefore, 100 ml of blood will deliver 0.326 mmol of O_2 to the tissues under these conditions.

Factors Influencing the Affinity of Hemoglobin for O_2. *Organic phosphates* inside the red blood cells (see Chapter 5), especially myoinositol 1,3,4,5,6-pentophosphate (IPP), interact with hemoglobin and alter its affinity for O_2 (Isaacks and Harkness, 1980; Isaacks et al., 1976, 1980, 1983). Oxygen affinity of hemoglobin varies in different avian species (see Lutz, 1980, for review) with P_{50} values of over 50 torr in chickens (Wells, 1976) to as low as 21 torr in the rhea (Lutz et al., 1974). Such a range also occurs among birds adapted to high-altitude life (such as bar-headed geese with a P_{50} of 29 torr) and those that normally live at or near sea level (P_{50} of 39 and 42 torr in greylag and Canada geese, respectively). The high O_2 affinity

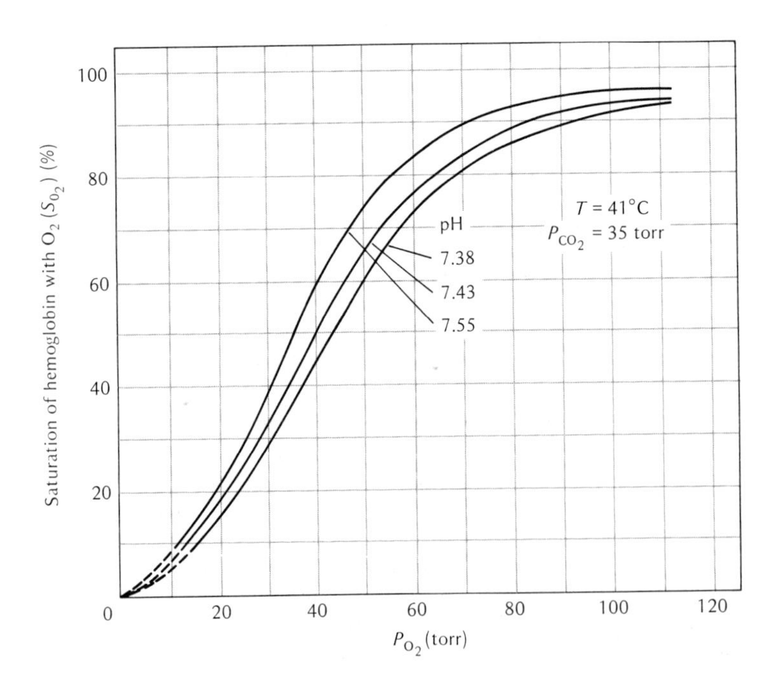

FIGURE 8–9. Oxyhemoglobin dissociation curves determined on duck blood. Bohr effect is demonstrated by the shift of the curve to the right as pH decreases. (After Scheipers et al., 1975.)

FIGURE 8–10. Effective or in vivo oxyhemoglobin dissociation curve for blood of an unanesthetized, undisturbed duck (heavy line). For reference, iso-pH dissociation curves derived as in Figure 8–9 are drawn through the arterial point of the same ducks (pH = 7.49) and the point of half-saturation, SO_2 = 50% (pH = 7.42). (After Scheipers et al., 1975.)

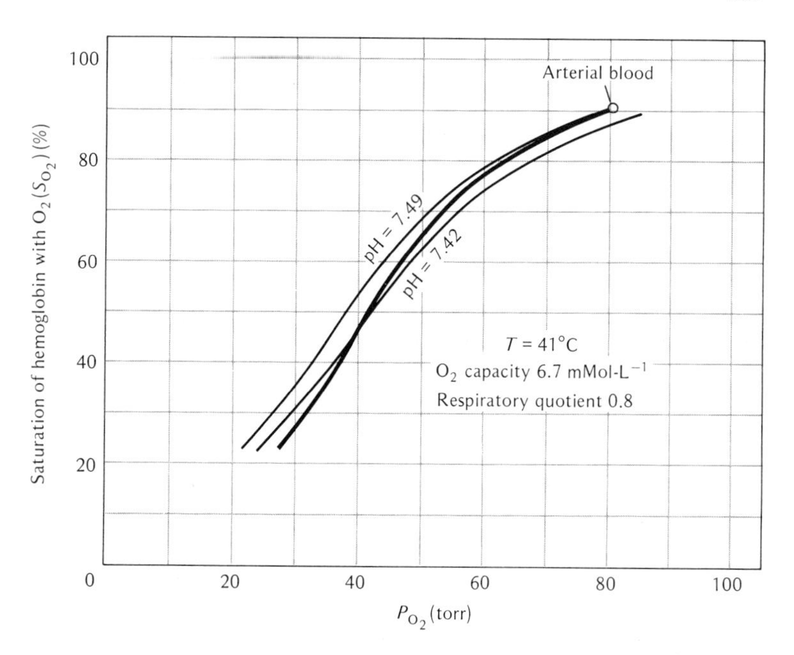

of hemoglobin in the high-altitude-adapted birds is not simply a function of the concentration of organic phosphates in the erythrocytes, which is similar in those birds that live at sea level (Petschow et al., 1977). The wide range in O_2 affinity of the hemoglobin of various avian species appears to result from different degrees of binding of IPP by their hemoglobins (Rollema and Bauer, 1979). As this binding increases, the O_2 affinity decreases, shifting the oxyhemoglobin dissociation curve to the right.

The *hydrogen ion concentration* in the blood also has an important influence on hemoglobin's affinity for oxygen. The effect of increasing H^+ concentration in the blood is reflected in a shift of the oxyhemoglobin dissociation curve to the right (Figure 8–9); hemoglobin then gives up its O_2 more easily. The Bohr factor is a measure of the influence of hydrogen ion concentration on O_2 affinity of hemoglobin. It represents the change in P_{O_2} at a constant O_2 saturation per change in pH and is commonly defined as $\Delta \log P_{50}/\Delta pH$. The change in O_2 affinity is the same in whole duck blood containing IPP in normal concentrations (Bohr factor of −0.44) irrespective of whether pH is changed by elevating P_{CO_2} (CO_2 Bohr effect) or by adding hydrogen ion (fixed-acid Bohr effect) (Meyer et al., 1978). However, removal of the organic phosphate from the hemoglobin of chicken blood induces differences in the O_2 affinity depending on the method used to change pH. Chicken hemoglobin is, therefore, capable of binding CO_2 as oxygen-linked carbamate, but the normal interaction of the hemoglobin with the organic phosphates abolishes that formation (Weingarten et al., 1978; Lapennas and Reeves, 1983). Thus, in avian blood CO_2 appears to exert no direct effect on the O_2 affinity of hemoglobin, and the increased P_{CO_2} in tissue capillaries favors O_2 release from hemoglobin only by H^+ formation from the hydration of CO_2.

Changes in temperature of the blood influence the O_2 affinity of hemoglobin; higher temperature decreases, and lower temperature increases, O_2 affinity. When blood passes through capillaries of organs that have a temperature higher than that in the lungs, such as exercising muscle, O_2 movement to the cells is enhanced.

Transport of CO_2

Carbon dioxide in avian blood, as in mammalian, is present principally in three forms: (1) physically dissolved CO_2; (2) CO_2 bound to proteins in the carbamate form; and (3) as bicarbonate ions, most of which are in the plasma. If the solubility coefficient of CO_2 in duck blood at 41°C is 0.0278 mmol CO_2/liter/torr (the values for bovine plasma determined by Bartels et al., 1971), and the arterial P_{CO_2} is 38 torr, there are 1.056 mmol of CO_2 physically dissolved in a liter of blood. This represents only a small fraction of the total CO_2 content in the blood.

The majority of the CO_2 in blood is in the form of HCO_3^-. When CO_2 diffuses into the capillary blood from cells of the various tissues, much of it enters the erythrocytes. There it quickly combines with water, catalyzed by the enzyme carbonic anhydrase, to form H_2CO_3, and in turn, HCO_3^- and H^+. The HCO_3^- diffuses into the plasma to be carried to the lungs where it moves back into the erythrocytes, again forming CO_2, which diffuses into the air capillary gas to be eliminated from the body.

As CO_2 enters and O_2 leaves the blood in the tissue capillaries, there is a change in the CO_2 content, as shown by the heavy line in Figure 8–11. That line represents the effective or in vivo CO_2 dissociation curve. The CO_2 dissociation curve of avian blood, unlike the oxyhemoglobin dissociation curve, is more linear in the physiological range and is much steeper. A small change in P_{CO_2} causes a large change in blood CO_2 content. The degree of oxygenation of hemoglobin greatly influences the quantity of CO_2 that blood contains, a property known as the Haldane effect. This property is especially important in the removal of CO_2 from the blood in the lung capillaries (Meyer et al., 1976).

Arterial Blood Gases and pH

The high metabolic rate of avian erythrocytes, especially embryonic erythrocytes (Grima and Girard, 1981), necessitates special caution in handling avian blood for blood gas measurements (Fedde and Kuhlmann, 1975; Scheid and Kawashiro, 1975). Blood gas and pH should be determined immediately after withdrawal from the bird; if this is impractical, the blood can be stored for a short time in chilled syringes placed in an ice bath. The blood gas electrodes should be set at the body temperature of the bird; however, if the electrode temperature differs from that of the bird, correction factors can be used (Kiley et al., 1979). As in mammalian blood, a blood gas correction factor must be applied to obtain the correct P_{O_2} value in the sample; this factor is obtained from P_{O_2} readings in gas and in blood equilibrated with the same gas (Nightingale et al., 1968).

A technique for collecting arterial blood from unanesthetized, undisturbed birds has been described by Scheid and Slama (1975). Values for arterial blood gases and pH in several species of birds are given in Table 8–3. Other blood gas values reported for unanesthetized, lightly restrained chickens (Piiper et al., 1970) are

	Partial pressure (torr)		Content (vol %)	
	O_2	CO_2	O_2	CO_2
Arterial blood	—	—	10.0	42.2
Mixed venous blood	40.8	39.3	4.3	45.8

Control of Breathing

Several excellent reviews on the control of breathing in birds are by Bouverot, 1978a; Scheid, 1979b, 1982; Burger, 1980; Jones and Milsom, 1982; and Powell, 1983a. It is generally recognized that both chemical and neural factors influence the inherent rhythm and neural output from the central respiratory controller.

Factors that Influence Ventilation

Carbon Dioxide. Inhaling this gas produces a powerful stimulus for increased ventilation. Tidal volume and ventilation are markedly increased in unanesthetized birds, even by inhaling low concentrations of CO_2, but the effects on respiratory frequency vary depending on the species studied (Bouverot et al., 1974; Fedde, 1976; Powell et al., 1978).

Carbon dioxide influences receptors in several locations, including carotid bodies and intrapulmonary chemoreceptors, and it also stimulates the central nervous system (Bouverot et al., 1974; Sébert, 1979; Fedde, 1981; Milsom et al., 1981). The sudden change in concentration of CO_2 within the lung with each breath

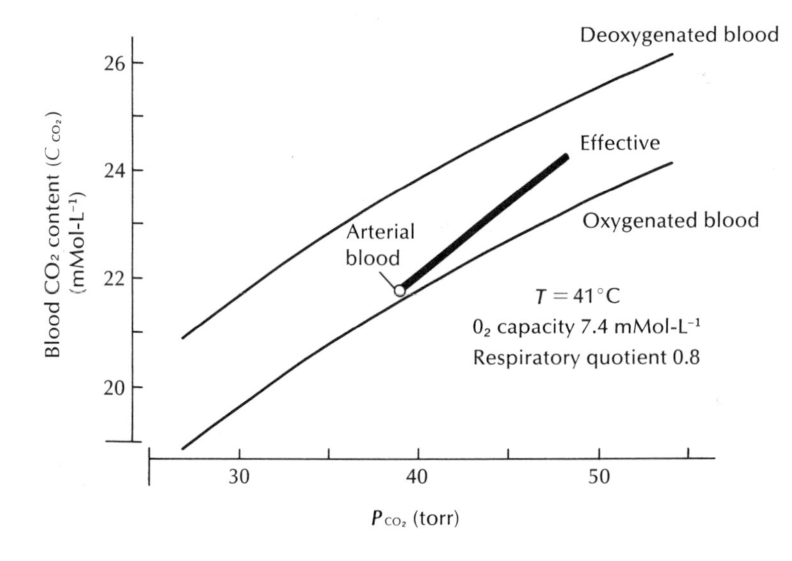

FIGURE 8–11. Carbon dioxide dissociation curves from the duck. Upper (deoxygenated blood) and lower (oxygenated blood) curves are derived from in vitro equilibration of blood samples. Heavy line is the effective or in vivo dissociation curve from unanesthetized, undisturbed birds, and illustrates the changes in CO_2 content (C_{CO_2}) in the blood as it changes from arterial blood to venous blood in the tissue capillaries. (After Scheipers et al., 1975.)

TABLE 8–3. Arterial blood gases and pH in unanesthetized birds breathing air

Bird	P_{O_2} (torr)	P_{CO_2} (torr)	pH
Female Black Bantam chicken[a]	—	29.9	7.48
Female White Leghorn chicken[d]	82	33.0	7.52
Male White Rock chicken[a]	—	29.2	7.53
Mallard duck[f]	81	30.8	7.46
Muscovy duck[d]	82	38.0	7.49
Muscovy duck[e]	96.1	35.9	7.46
Peking duck[c]	93.5	28.0	7.46
Peking duck[g]	100	33.8	7.48
Emu[m]	99.7	33.5	7.45
Bar-headed goose[c]	92.5	31.6	7.47
Domestic goose[b]	97.2	32.3	7.51
Domestic goose (in lay)[a]	—	24.5	7.53
Herring gull[a]	—	27.2	7.56
Red-tailed hawk[l]	108	27.0	7.49
Burrowing owl[n]	97.6	32.6	7.46
White pelican[a]	—	28.5	7.50
Adélie penguin[i]	83.8	36.9	7.51
Chinstrap penguin[i]	89.1	37.1	7.52
Gentoo penguin[i]	77.1	40.9	7.49
Pigeon[a]	—	28.5	7.52
Pigeon[k]	95	34.0	7.46
Roadrunner[a]	—	24.5	7.58
Abdim stork[j]	—	27.9	7.56
Mute swan[h]	91.3	27.1	7.50
Turkey vulture[a]	—	27.5	7.51

[a] Calder and Schmidt-Nielsen (1968).
[b] Fedde, M.R., R.E. Burger, J. Geiser, R.K. Gratz, J.A. Estavillo, and P. Scheid (unpublished data).
[c] Black and Tenney (1980).
[d] Kawashiro and Scheid (1975).
[e] Jones and Holeton (1972).
[f] Butler and Taylor (1983).
[g] Bouverot et al. (1979).
[h] Bech and Johansen (1980).
[i] Murrish (1982).
[j] Marder and Arad (1975).
[k] Bouverot et al. (1976).
[l] Kollias and McLeish (1978).
[m] Jones et al. (1983).
[n] Kilgore, D.L., F.M. Faraci, and M.R. Fedde (unpublished data).

appears to trigger the following inspiratory effort in the chicken (Miller and Kunz, 1977): Short pulses (0.2–0.4 sec) of CO_2 arriving in the lung near midexpiration induce the maximum change in the duration of expiration (Clanton et al., 1982). These studies indicate the importance of phasic changes in neural discharge in the vagus (presumably from intrapulmonary chemoreceptors) on the breath-to-breath control of breathing. However, the central response to the change in discharge of these receptors depends on the state of the bird (anesthetized or not), the method of restraint and degree of excitement, and the method by which ventilation is measured (i.e., disturbance of the respiratory system by a tracheal cannula, a face mask, or increased plethysmographic pressure). Attention to these problems would probably result in more comparable results among species.

Oxygen. Birds respond to reductions in ambient O_2 by increasing ventilation as do mammals, and in both classes there is a continuous tonic drive of ventilation originating from the carotid bodies during normoxia; when this drive is reduced by raising the inspired P_{O_2}, there is an accompanying hypoventilation and acidosis (Jones and Purves, 1970; Bouverot et al., 1979; Bouverot, 1981). However, all birds do not have the same threshold for ventilatory increases as ambient P_{O_2} decreases. Bar-headed geese and rheas do not increase ventilation until their arterial P_{O_2} has decreased to about 35 and 55 torr, respectively (van Nice et al.,

1980; Black and Tenney, 1980; Boggs and Birchard, 1983). The hemoglobin of these birds has a high affinity for O_2 (P_{50} less than 30 torr), and, hence, a relatively high saturation at these P_{O_2}s. Other species, such as the Pekin duck or pheasant whose hemoglobin has a lower affinity for O_2, begin to increase ventilation at a P_{O_2} 20–30 torr higher. However, burrowing owls, birds exposed to low P_{O_2} in burrows for long periods of time, are much more insensitive to both hypoxia and hypercapnia than comparably sized nonburrow dwellers, although their P_{50} is not reduced as in the bar-headed goose and rhea (Boggs and Kilgore, 1983; Boggs et al., 1984). These differences among species are beneficial; if the threshold arterial P_{O_2} for the hypoxic ventilatory response were constant among all species, those with high-affinity hemoglobin would be wasting ventilation when hemoglobin saturation was still high, whereas those species with low-affinity hemoglobins might exhibit severe arterial unsaturation before the ventilatory response was initiated.

The carotid bodies are primarily responsible for the ventilatory response to hypoxia, although recent evidence suggests other receptors may also be involved (Kunz et al., 1979; Lillo and Jones, 1983; Nye and Powell, 1984).

Stimulation of Peripheral Nerves. Stimulating almost any peripheral nerve will affect ventilation. Electrical stimulation of the skin as well as visual and auditory cues elicit changes in ventilation (Fowle and Weinstein, 1966). Recent experiments indicate that running exercise induces hyperventilation in ducks and chickens, and that the receptors involved may be in skeletal muscle (Kiley et al., 1979, 1982; Brackenbury et al., 1981; Kiley and Fedde, 1983a, b). However, changes in body temperature or arterial pH with exercise also lead to hyperventilation with subsequent fall in Pa_{CO_2} (Gleeson and Brackenbury, 1983, 1984). The possible function of muscle receptors in the control of breathing during exercise in birds suggests an important area for future research.

Electrical stimulation of the central end of the cut vagus nerve causes a variety of effects on ventilation depending on the frequency and strength of the stimulus (see King, 1966b, and McLelland, 1970, for reviews). A weak stimulus (1–3 volts, 5–25 Hz, 1-msec duration) applied to the central end of either the right or left vagus at any phase of the breathing cycle increases respiratory frequency and decreases respiratory depth. Stronger stimuli (higher voltages or frequencies) inhibit respiration.

Section of both vagi, or of pulmonary branches of the nerve (Jones and Bamford, 1978), markedly decreases respiratory frequency and increases respiratory amplitude in birds (Fedde et al., 1963). Stimulating the central end of one of the cut vagus nerves in the bilaterally vagotomized bird with a weak electrical stimulus prevents the changes in respiratory rate and amplitude. It also permits initiation and maintenance of thermal panting in the chicken in response to increasing body temperature (Richards, 1968, 1969; McLelland, 1970). Although the vagi contain afferent fibers, the neural activity of which is essential to maintain normal rhythmic respiration, it is not known exactly which receptors are responsible for the generation of this activity.

Temperature. Increases in body temperature produce thermal polypnea. However, despite marked increases in respiratory frequency, hyperthermia does not induce hyperventilation (and thus lower Pa_{CO_2}) in all birds, notably the ostrich and emu (Schmidt-Nielsen et al., 1969; Jones et al., 1983); other birds undergo a small degree of hyperventilation and mild respiratory alkalosis during moderate heat load (see Brackenbury et al., 1982, for review; Arad and Marder, 1983). Severe heat stress produces extreme hyperventilation and alkalosis in the chicken (Mather et al., 1980).

The pattern of breathing that occurs during panting (high frequency, low tidal volume) is markedly influenced by the reduction in Pa_{CO_2} and the alkalosis that results. If Pa_{CO_2} is prevented from decreasing during panting, respiratory frequency is substantially slowed and tidal volume greatly increased (Brackenbury, 1978; Mather et al., 1980; Barnas et al., 1981). In fact, if Pa_{CO_2} is held constant, the pattern of breathing is the same for the hyperpnea induced by exercise or by increases in body temperature (Brackenbury and Gleeson, 1983). Thus, Pa_{CO_2} influences the central respiratory controller and/or the interaction of the peripheral chemoreceptors and thermoreceptors and determines the resulting breathing pattern.

In pigeons, a compound pattern of ventilation involving two components occurs during elevated body temperature. One component entails the same frequency and tidal volume as during resting ventilation, the other being rapid and shallow (Bernstein and Samaniego, 1981). Thus, both normal parabronchial ventilation and enhanced dead-space ventilation can occur simultaneously to maintain near-normal arterial blood gases and pH and to enhance heat dissipation by evaporative water loss (see also Chapter 9).

Noxious Gases. Inhaling any one of several noxious gases (sulfur dioxide, ammonia vapor, acetic acid vapor) slows breathing or produces apnea (Eaton et al., 1971; Callanan et al., 1974). Simply exposing the upper respiratory system (upper trachea and larynx) to cold air or water also slows breathing or produces apnea (Eaton et al., 1971; Bamford and Jones, 1974). Irritant receptors in the upper respiratory system are strongly stimulated by these substances; anatomic evidence for the presence of such receptors exists (Walsh and McLelland, 1974a).

Acid-forming gases (SO_2 and H_2S) also alter the discharge of intrapulmonary chemoreceptors (Klentz and Fedde, 1978; Chiang et al., 1978). Changes in central input from these receptors may be partially responsible for the altered breathing pattern when these gases are inhaled.

Components of a Control System

The various components in a control system must be recognized and analyzed before the system can be thoroughly understood.

Controlled Variable. The most important consideration in defining the components of a control system is the identification of the controlled variable or variables, i.e., the variables to be kept at some long-term homeostatic level. Arguments have been made that the control of adequate oxygenation of the tissues is the paramount goal of the control of respiration (Dejours, 1981); indeed, changes in Pa_{O_2} that do not markedly alter blood oxygenation also do not markedly stimulate ventilation. On the other hand, small changes in Pa_{CO_2} or pHa evoke large changes in ventilation. However, decreases in Pa_{O_2} below 60 torr will stimulate ventilation and if the Pa_{O_2} is reduced below 30 torr, as might occur during flight at high altitude, severe hyperventilation and alkalosis result (Black and Tenney, 1980; Faraci et al., 1984). The system seems to be organized in such a way that the control of some variables can be overridden by disturbance of a variable that is one step higher in the hierarchy. Further definition of this hierarchy will substantially aid the understanding of the control system.

Receptor Systems. Disturbance of the controlled variable activates some receptor system, which detects the magnitude of the disturbance. Birds have several known receptor systems, but our understanding of their exact mechanisms of action in the breath-by-breath control of breathing is still inadequate.

Intrapulmonary Chemoreceptors (IPC). One of the recently discovered chemoreceptor systems involved in the control of breathing is located in the lungs (see Fedde and Kuhlmann, 1978; Fedde et al., 1980; Fedde, 1981; Powell 1983b; for reviews). This system has been directly demonstrated by neural recordings from vagal afferent fibers in both large and small species of birds (emu, goose, duck, chicken, pigeon, burrowing owl) (Fedde and Peterson, 1970; Fedde et al., 1974a; Barnas et al., 1978b, 1984; Burger et al., 1976a, b; Kilgore et al., 1984). These receptors decrease their discharge frequency logarithmically as the intrapulmonary CO_2 concentration increases (Figure 8–12) (Nye and Burger, 1978). They are not stimulated by stretching the respiratory system (Fedde et al., 1974b; Macklem et al., 1979) or by hypoxia (Fedde and Peterson, 1970; Tschorn and Fedde, 1974). They have some sensitivity to H^+ (Powell et al., 1978), but are most sensitive to changes in intrapulmonary CO_2 concentration (Barnas et al., 1978b). These receptors become less sensitive to changes in P_{CO_2} as the bird's body temperature increases above normal (Barnas et al., 1983). Although they have not been anatomically identified, they appear to reside in the gas-exchange region of the lung (Scheid et al., 1974a; Osborne et al., 1977a; Crank et al., 1980; Powell et al., 1980; Boon et al., 1982). The ultrastructure of afferent nerve endings and biogenic amine-containing cells in the avian lung and their possible relationship to IPC have been discussed by King et al. (1974) and Eaton and Fedde (1978).

Intrapulmonary chemoreceptors discharge rhythmically during spontaneous breathing and during ventilation with a reciprocating pump; most receptors have peak discharge frequencies during inspiration, but a few discharge maximally during expiration and some are biphasic with peak discharge frequencies during both phases of the respiratory cycle (Molony, 1974; Fedde and Scheid, 1976; Berger et al., 1980). These receptors respond to CO_2 added to the inspired gas or added to the mixed venous blood returning to the lung (Banzett and Burger, 1977; Boon et al., 1980). However, when CO_2 is added to the mixed venous blood in ducks where all the feedback mechanisms are intact, these birds adjust their breathing so the level of IPC discharge remains unchanged (Tallman and Grodins, 1982a). That observation suggests that a zero-error signal type of control system may exist.

There have been many studies indicating the importance of IPC in the control of breathing (Osborne et al., 1977b; Scheid et al., 1978b; Osborne and Mitchell, 1978; Mitchell and Osborne, 1978, 1979, 1980; Clanton et al., 1982; Berger and Tallman, 1982; Tallman and Kunz, 1982; Barnas and Burger, 1983). In some cases, only IPC were stimulated while other peripheral and central chemoreceptors were held in a constant environment (Fedde et al., 1982). However, some studies suggest that they may function primarily in controlling the pattern of breathing but not the total ventilation (Milsom et al., 1981; Tallman and Grodins, 1982b). Increased neural discharge from IPC acts to decrease respiratory amplitude and increase respiratory frequency (Barker et al., 1981). However, the precise role played by these receptors in the breath-to-breath control of breathing remains to be discovered.

Carotid Bodies. Receptors in the carotid bodies play a definite role in controlling breathing in birds (Bouverot, 1978b; Bouverot and Sébert, 1979; Milsom et al., 1981). The location, innervation, and ultrastructure of these organs have been well defined in several avian species (Kobayashi, 1969, 1971a, b; Hodges et al., 1975; Butler and Osborne, 1975; King et al., 1975; Dreyer et al., 1977, 1978; Abdel-Magied and King,

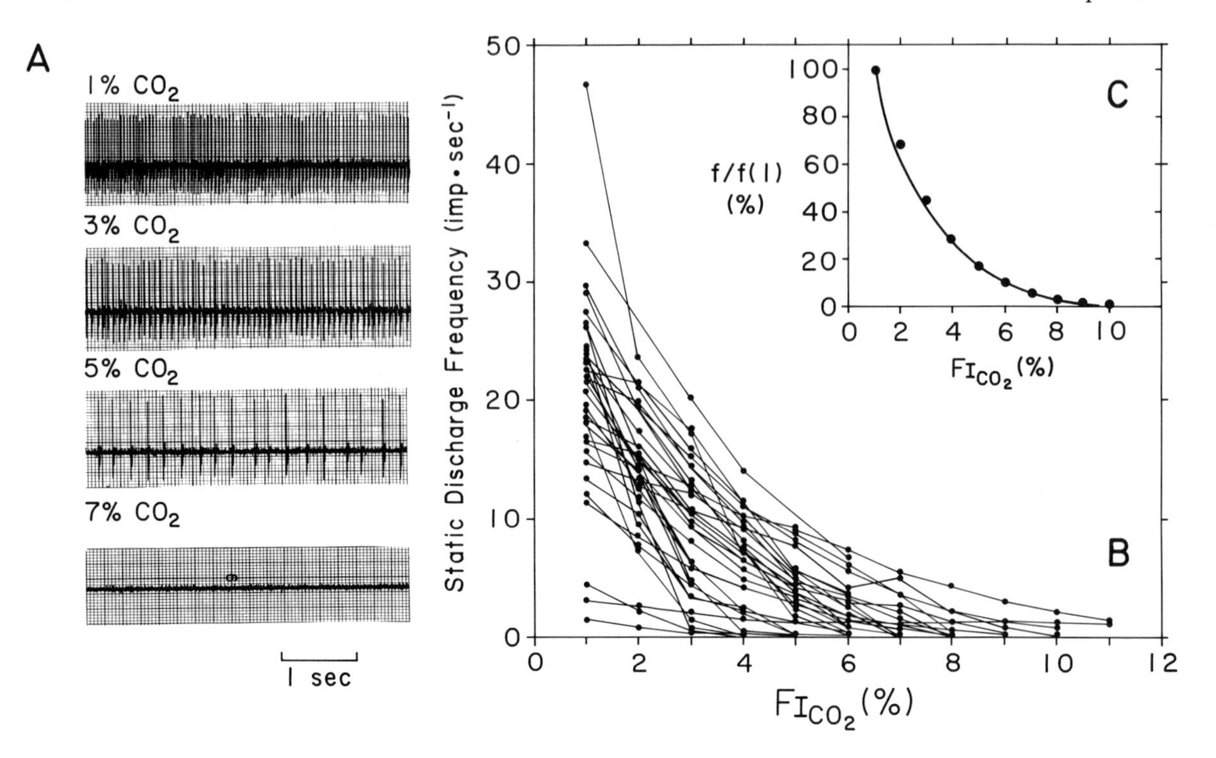

FIGURE 8–12. Response of intrapulmonary chemoreceptors in the pigeon to several static CO_2 concentrations ($F_{I_{CO_2}}$) in the unidirectional gas stream ventilating the lung. (A) Discharge frequencies of a single receptor in response to various CO_2 concentrations. (B) Discharge characteristics of a population of receptors from several pigeons. (C) Mean response for this population expressed as the fraction of the discharge frequency at a CO_2 concentration of 1%. (D.L. Kilgore, F.M. Faraci and M.R. Fedde, unpublished observations.)

1978, 1982; Taha and King, 1983). The carotid bodies are located bilaterally in the thoracic cavity near the nodose ganglia of the vagi, from which they receive innervation. Their blood supply is from the oesophago-tracheobronchial artery. Both Type I cells, postulated to have a chemoreceptor function, and Type II cells, believed to be supportive in function, are present, as in mammals, and innervation is primarily afferent with few efferent axonal endings.

The carotid bodies contain chemoreceptors responsible for detecting low arterial P_{O_2} and are also sensitive to elevated arterial P_{CO_2} (Nye and Powell, 1984). They provide a neural discharge to the central respiratory centers that is necessary for normal breathing. Surgical denervation of the carotid bodies usually eliminates the ventilatory response to hypoxia and hyperpoxia and results in hypoventilation and hypercapnia (Jones and Purves, 1970; Bouverot et al., 1974, 1979; although see Lillo and Jones, 1983). These organs can be stimulated by intravenous injections of sodium cyanide, similar to mammalian carotid bodies (Bouverot and Leitner, 1972; Magno, 1973; Tallman and Kunz, 1982).

Central Nervous System Receptors. Central chemoreceptors that respond to changes in P_{CO_2} in the blood entering the brain have been demonstrated in ducks (Sébert, 1978, 1979; Milsom et al., 1981). Ventilation and respiratory frequency both increase when these receptors are stimulated. Additional evidence for central chemoreceptors has been provided by experiments in which apnea was produced by reducing intrapulmonary CO_2 concentration below 3% in unidirectionally ventilated chickens following bilateral cervical vagotomy (Peterson and Fedde, 1968). The location of these receptors and their exact role in the control of breathing is unknown.

Pulmonary Stretch Receptors. When the avian respiratory system is inflated with air, birds exhibit an apneic response similar to that displayed by inflation of the lungs of mammals (Eaton et al., 1971). That response was initially interpreted to indicate the existence of stretch receptors in the pulmonary system and was equated with the Hering–Breuer reflex of mammals (see Fedde, 1970, for reviews). However, the intrapulmonary CO_2 receptors are strongly activated during inflation with air because air has a low CO_2 concentration. This induces high discharge from these receptors and leads to apnea. If the respiratory system is inflated with a gas containing 8% CO_2, most of the CO_2 receptors are inhibited from discharging and the apneic response is almost completely abolished. Hence, inflation of the respiratory system in the usual way cannot be used to test for pulmonary mechanoreceptors in birds.

Recent experiments have focused attention on pulmonary mechanoreceptor involvement in the control of breathing in birds (Miller, 1978; Ballam et al., 1982). Changing the thoracoabdominal volume in unidirectionally ventilated birds alters either inspiratory duration or expiratory duration, depending on when the volume is increased, which suggests the importance of mechanical feedback in the control of respiratory movements. However, it is still not certain that mechanoreceptors are located in the lung air-sac system, because inflation of that system also distorts the viscera and body wall. The air-sac walls are highly innervated and contain sensory endings (Groth, 1972), but mechanoreceptors have also been demonstrated in the gastrointestinal system (Duke et al., 1977). Afferent activity in some fibers in the avian vagus is phasic with respiration but not influenced by intrapulmonary CO_2 (Fedde et al., 1974b; Molony, 1974). These receptors may be mechanoreceptors, but their location and effect on the control of breathing are unknown.

Other Possible Receptors. Anatomic descriptions have been made of other possible peripheral chemoreceptors that lie (1) in the connective tissue between the ascending aorta and the pulmonary artery; (2) on the dorsal and lateral surfaces of the aorta, the brachiocephalic arteries, and the pulmonary arteries; (3) in the adventitia and media of the pulmonary trunk at the site of its junction with the right ventricle; and (4) in the wall of the pulmonary artery near the hilus of the lung (see Fedde, 1970, for review). These or other receptors may be responsible for the extracarotid body reception of hypoxia, but to date no information is available on the function of these receptors.

Biogenic amine-containing cells have been located in the epithelium of the trachea, syrinx, primary bronchus, and parabronchi of the chicken (Walsh and McLelland, 1974b; Eaton and Fedde, 1978). These cells contain high concentrations of either serotonin or catecholamines. It has been proposed that these cells function either as humoral or hormonal regulators of pulmonary ventilation and perfusion or as receptors involved in the control of breathing.

Afferent Pathway. The afferent pathway from the intrapulmonary CO_2 receptors to the brain is via the vagus nerves and, to a lesser degree, via sympathetic nerves (Burger et al., 1974). These nerves project centrally to the solitary complex, nucleus sulcalis dorsalis, parasolitary nuclei, and the nucleus commissuralis (Dubbeldam et al., 1979). The glossopharyngeal and vagus nerves both supply the carotid bodies. Many other nerves also contain afferent fibers that influence breathing.

Central Control Areas. Very few studies have dealt with the organization and function of the central respiratory control areas. Earlier studies have been reviewed previously (Fedde, 1976). These neurons appear to be in the medulla oblongata of the brainstem.

Neurophysiologic recordings have been made recently from medullary inspiratory and expiratory neurons in the brain (Peek et al., 1975; Michal et al., 1981). These neurons are sensitive to changes in intrapulmonary CO_2 concentration; some are also sensitive to changes in thoracic volume, and many are sensitive to stimulation of the midbrain call region (Peek and Phillips, 1971). Vocal and respiratory rhythms may be generated by the same neural oscillator composed of self-reexciting, reciprocally inhibiting neurons.

Models utilizing the interaction of chemoreceptors, volume receptors, and pacemaker cells in the brain have been developed to describe the behavior of the respiratory control system (Kunz and Miller, 1974a, b; Kunz and Tallman, 1978; Miller, 1980; Kunz et al., 1984). These models should aid in the interpretation of future neurophysiologic data from recordings of central respiratory neurons.

Respiratory Muscles. Respiratory muscles are involved in many actions in addition to respiratory movements. Expiratory muscles control airflow from the respiratory system during coughing, sneezing, and vocalization. Large volumes of gas are inhaled just prior to vocalization, and the air-sac gas concentrations of O_2 and CO_2 are transiently modified (Brackenbury, 1980).

Rhythmic contractions of the respiratory muscles are generated by a specific pattern of motoneuronal recruitment (Fedde et al., 1964a, 1969; Tschorn and Fedde, 1971). The small motor units are the first to be recruited followed by larger units. Electrical activity is either very low or ceases during the relaxation phase of the respiratory cycle for a particular muscle group. In birds made apneic by lowering intrapulmonary CO_2 concentration, there is a tonic discharge of small motor units in the respiratory muscles, and a small amount of tension is continuously generated.

Respiratory muscles are rhythmically active from before hatching to death of the bird; they must, therefore, possess muscle fibers that are fatigue resistant. Furthermore, some fibers in these muscles must be capable of contracting rapidly during vocalization. Fibers with the above capabilities have been demonstrated histochemically in chicken respiratory muscles (Fedde and Cardinet, 1977).

Normal Values of Some Respiratory Variables

Specific values for respiratory rate, tidal volume, and ventilation for various birds are shown in Table 8–4. Additional values for resting respiratory rate for many species have been compiled by Calder (1968). Because there are approximately 86,000 species of birds, ranging in weight from 2 to 150,000 g, the values of respiratory variables vary widely.

The relationship between respiratory variables and body weight has been useful in predicting the magnitude of a variable from a known body weight (Lasiewski and Calder, 1971; Lasiewski, 1972). The allometric relationship is based on a power function equation of the form $Y = aW^b$, where Y is some variable, W is the body weight in kilograms, and a and b are empirically derived constants. The key to success in deriving the constants is measuring respiratory variables accurately. Some respiratory variables, especially tidal volume, are extremely difficult to measure under truly undisturbed conditions. Tidal volume measurements used by Lasiewski and Calder (1971) to derive the constants for their allometric equations were obtained, in most cases, from restrained birds with a facemask or a cannulated trachea. Indeed, most of the values given in Table 8–4 suffer from the same problem. For example, tidal volume measured in unanesthetized ducks (2.0–2.6 kg) in a resting condition by plethysmography without facemasks or tracheal cannulation is considerably larger (55–63 ml) than values predicted from the allometric equations for birds of that size (28–37 ml) (Bouverot et al., 1974).

References

Abdalla, M.A., and A.S. King. (1975). The functional anatomy of the pulmonary circulation of the domestic fowl. Respir. Physiol., 23, 267.

Abdalla, M.A., J.N. Maina, A.S. King, D.Z. King, and J. Henry. (1982). Morphometrics of the avian lung. 1. The domestic fowl (*Gallus gallus* variant *domesticus*). Respir. Physiol., 47, 267.

Abdel-Magied, E.M., and A.S. King. (1978). The topographical anatomy and blood supply of the carotid body region of the domestic fowl. J. Anat., 126, 535.

Abdel-Magied, E.M., and A.S. King. (1982). Effects of distal vagal ganglionectomy and midcervical vagotomy on the ultrastructure of axonal elements in the carotid body of the domestic fowl. J. Anat., 134, 643.

Arad, Z., and J. Marder. (1983). Acid-base regulation during thermal panting in the fowl (*Gallus domesticus*): Comparison between breeds. Comp. Biochem. Physiol., 74A, 125.

Baldwin, J.K. (1973). Mechanics of respiration in euthermic and hyperthermic *Gallus domesticus*. Ph.D. Thesis, University of California, Davis.

Ballam, G.O., T.L. Clanton, and A.L. Kunz. (1982). Ventilatory phase duration in the chicken: role of mechanical and CO_2 feedback. J. Appl. Physiol., 53, 1378.

Bamford, O.S., and D.R. Jones. (1974). On the initiation of apnoea and some cardiovascular responses to submergence in ducks. Respir. Physiol., 22, 199.

Banzett, R.B., and R.E. Burger. (1977). Response of avian intrapulmonary chemoreceptors to venous CO_2 and ventilatory gas flow. Respir. Physiol., 29, 63.

TABLE 8–4 Values of various respiratory variables in resting, unanesthetized birds

Species	Weight (kg)	f^a (respirations/min)	V_T^a (ml)	$\dot{V}^a$ (ml/min)	Comments
Duck[b]					
Muscovy	2.16	10.5	69	700	Endotracheal tube,
Peking	2.4	8.2	98	807	pneumotachograph
Duck[c]					
Peking	2.01	11.9	55	650	Resting in plethysmograph
Pigeon[d]	0.38	26	4.6	118	Face mask
Chicken[e]					
Male	4.2	17	46	777	Tracheal cannula,
Female	3.4	27	31	766	valves, spirometer
Chicken[f]					
Female	1.6	23.0	33.0	760	Tracheal cannula, valves, spirometer
Goose[g]	4.5–9.0	12.6	112.5	1418	Face mask, pneumotachograph

[a] f, respiratory frequency; V_T, tidal volume; $\dot{V}$, ventilation.
[b] Jones and Holeton (1972).
[c] Bouverot et al. (1974).
[d] Hart and Roy (1966).
[e] King and Payne (1964).
[f] Piiper et al. (1970).
[g] Cohn and Shannon (1968, Fig. 3).

Barker, M.R., R.E. Burger, and P.C.G. Nye. (1981). Respiratory inhibition from chicken intrapulmonary chemoreceptors reduced by increasing rate of repetitive P_{CO_2} changes. Q. J. Exp. Physiol., 66, 367.

Barnas, G.M., and R.E. Burger. (1983). Interaction of temperature with extra- and intrapulmonary chemoreceptor control of ventilatory movements in the awake chicken. Respir. Physiol., 54, 223.

Barnas, G.M., F.B. Mather, and M.R. Fedde. (1978a). Response of avian intrapulmonary smooth muscle to changes in carbon dioxide concentration. Poult. Sci., 57, 1400.

Barnas, G.M., F.B. Mather, and M.R. Fedde. (1978b). Are avian intrapulmonary CO_2 receptors chemically modulated mechanoreceptors or chemoreceptors? Respir. Physiol., 35, 237.

Barnas, G.M., J.A. Estavillo, F.B. Mather, and R.E. Burger. (1981). The effect of CO_2 and temperature on respiratory movements in the chicken. Respir. Physiol., 43, 315.

Barnas, G.M., S.C. Hempleman, and R.E. Burger. (1983). Effect of temperature on the CO_2 sensitivity of avian intrapulmonary chemoreceptors. Respir. Physiol., 54, 233.

Barnas, G., K. Muckenhoff, and P. Scheid. (1984). Intrapulmonary chemoreceptors in the pigeon *Columba livia.* Pfleugers Arch., 400 (Suppl.), R58.

Bartels, H., C. Christoforides, J. Hedley-Whyte, and L. Laasberg. (1971). Solubility coefficients of gases. In "Biological Handbooks: Respiration and Circulation" (P.L. Altman and D.S. Dittman, Eds.). Bethesda: Federation of American Societies for Experimental Biology, p. 18.

Baumann, R., and F.H. Baumann. (1978). Respiratory function of embryonic chicken hemoglobin. In "Respiratory Function in Birds, Adult and Embryonic" (J. Piiper, Ed.). Berlin: Springer-Verlag, p. 292.

Baumann, R., S. Padeken, E.-A. Haller, and T. Brilmayer. (1983). Effects of hypoxia on oxygen affinity, hemoglobin pattern, and blood volume of early chicken embryos. Am. J. Physiol., 244, R733.

Bech, C., and K. Johansen. (1980). Ventilation and gas exchange in the mute swan (*Cygnus olor*). Respir. Physiol., 39, 285.

Berger, P.J., and R.D. Tallman, Jr. (1982). Lengthening of inspiration by intrapulmonary chemoreceptor discharge in ducks. J. Appl. Physiol., 53, 1392.

Berger, P.J., R.D. Tallman, Jr., and A.L. Kunz. (1980). Discharge of intrapulmonary chemoreceptors and its modulation by rapid $F_{I_{CO_2}}$ changes in decerebrate ducks. Respir. Physiol. 42, 123.

Bernstein, M.H., and F.C. Samaniego. (1981). Ventilation and acid-base status during thermal panting in pigeons (*Columba livia*). Physiol. Zool., 54, 308.

Black, C.P., and S.M. Tenney. (1980). Oxygen transport during progressive hypoxia in high-altitude and sea-level waterfowl. Respir. Physiol., 39, 217.

Boggs, D.F., and G.F. Birchard. (1983). Relationship between haemoglobin O_2 affinity and the ventilatory response to hypoxia in the rhea and pheasant. J. Exp. Biol., 102, 347.

Boggs, D.F., and D.L. Kilgore, Jr. (1983). Ventilatory responses of the burrowing owl and bobwhite to hypercarbia and hypoxia. J. Comp. Physiol., 149, 527.

Boggs, D.F., D.L. Kilgore, Jr., and G.F. Birchard. (1984). Respiratory physiology of burrowing mammals and birds. Comp. Biochem. Physiol. 77A, 1.

Boon, J.K., W.D. Kuhlmann, and M.R. Fedde. (1980). Control of respiration in the chicken: Effects of venous CO_2 loading. Respir. Physiol., 39, 169.

Boon, J.K., M.R. Fedde, and P. Scheid. (1982). A method for localizing intrapulmonary chemoreceptors in the parabronchial mantle of the duck. Comp. Biochem. Physiol. 72A, 463.

Bouverot, P. (1978a). Control of breathing in birds compared with mammals. Physiol. Rev., 58, 604.

Bouverot, P. (1978b). Role of arterial chemoreceptors in ventilatory acclimation to high altitude in unanesthetized Pekin ducks. In "Respiratory Function in Birds, Adult and Embryonic" (J. Piiper, Ed.). Berlin: Springer-Verlag, p. 84.

Bouverot, P. (1981). Hypoxia tolerance and ventilatory O_2-chemoreflexes. In "Advances in Physiological Sciences, Vol. 10, Respiration (I. Hutás and L.A. Debreczeni, Eds.). Elmsford, New York: Pergamon Press, p. 145.

Bouverot, P., and L.-M. Leitner. (1972). Arterial chemoreceptors in the domestic fowl. Respir. Physiol., 15, 310.

Bouverot, P., and P. Sébert. (1979). O_2-chemoreflex drive of ventilation in awake birds at rest. Respir. Physiol., 37, 201.

Bouverot, P., N. Hill, and Y. Jammes. (1974). Ventilatory responses to CO_2 in intact and chronically chemodenervated Pekin ducks. Respir. Physiol., 22, 1237.

Bouverot, P., G. Hildwein, and P. Oulhen. (1976). Ventilatory and circulatory O_2 convection at 4000 m in pigeon at neutral or cold temperature. Respir. Physiol., 28, 371.

Bouverot, P., D. Douguet, and P. Sébert. (1979). Role of the arterial chemoreceptors in ventilatory and circulatory adjustments to hypoxia in awake Pekin ducks. J. Comp. Physiol., 133, 177.

Brackenbury, J.H. (1971a). Pressure-flow phenomena within the avian respiratory system. J. Anat., 108, 609.

Brackenbury, J.H. (1971b). Airflow dynamics in the avian lung as determined by direct and indirect methods. Respir., Physiol., 13, 319.

Brackenbury, J.H. (1972a). Physical determinants of air flow pattern within the avian lung. Respir. Physiol., 15, 384.

Brackenbury, J.H. (1972b). Lung-air-sac anatomy and respiratory pressures in the bird. J. Exp. Biol., 57, 543.

Brackenbury, J.H. (1973). Respiratory mechanics in the bird. Comp. Biochem. Physiol. 44A, 599.

Brackenbury, J.H. (1978). Experimentally induced antagonism of chemical and thermal reflexes in the respiratory system of fully conscious chickens. Respir. Physiol., 34, 377.

Brackenbury, J. (1979). Corrections to the Hazelhoff model of airflow in the avian lung. Respir. Physiol., 36, 143.

Brackenbury, J.H. (1980). Respiration and production of sounds by birds. Biol. Rev., 55, 363.

Brackenbury, J.H. (1981). Airflow and respired gases within the lung-air-sac system of birds. Comp. Biochem. Physiol. 68A, 1.

Brackenbury, J.H., and A.R. Akester. (1978). A model of the capillary zone of the avian tertiary bronchus. In "Respiratory Function in Birds, Adult and Embryonic" (J. Piiper, Ed.). New York: Springer-Verlag, p. 125.

Brackenbury, J.H., and M. Gleeson (1983). Effects of P_{CO_2} on respiratory pattern during thermal and exercise hyperventilation in domestic fowl. Respir. Physiol., 54, 109.

Brackenbury, J.H., P. Avery, and M. Gleeson. (1981). Respiration in exercising fowl. I. Oxygen consumption, respiratory rate and respired gases. J. Exp. Biol., 93, 317.

Brackenbury, J.H., P. Avery, and M. Gleeson. (1982). Effects of temperature on the ventilatory response to inspired CO_2 in unanesthetized domestic fowl. Respir. Physiol., 49, 235.

Bretz, W.L., and K. Schmidt-Nielsen. (1971). Bird respiration: Flow patterns in the duck lung. J. Exp. Biol., 54, 103.

Burger, R.E. (1980). Respiratory gas exchange and control in the chicken. Poult. Sci., 59, 2654.

Burger, R.E., J.L. Osborne, and R.B. Banzett. (1974). Intrapulmonary chemoreceptors in *Gallus domesticus:* Adequate stimulus and functional localization. Respir. Physiol., 22, 87.

Burger, R.E., P.C.G. Nye, F.L. Powell, C. Ehlers, M. Barker, and M.R. Fedde. (1976a). Response to CO_2 of intrapulmonary chemoreceptors in the emu. Respir. Physiol., 28, 315.

Burger, R.E., J.C.G. Coleridge, H.M. Coleridge, P.C.G. Nye, F.L. Powell, C. Ehlers, and R.B. Banzett. (1976b). Chemoreceptors in the paleopulmonic lung of the emu: Discharge patterns during cyclic ventilation. Respir. Physiol., 28, 249.

Burger R.E., M. Meyer, W. Graf, and P. Scheid. (1979). Gas exchange in the parabronchial lung of birds: Experiments in unidirectionally ventilated ducks. Respir. Physiol., 36, 19.

Burns, B., A.E. James, G. Hutchins, G. Novak, and R.R. Price. (1978). Ventilatory 133-Xenon distribution studies in the duck (*Anas platyrhynchos*). In "Respiratory Function in Birds, Adult and Embryonic" (J. Piiper, Ed.). New York: Springer-Verlag, p. 129.

Burton, R.R., and A.H. Smith. (1968). Blood and air volumes in the avian lung. Poult. Sci., 47, 85.

Butler, P.J., and M.P. Osborne. (1975). The effect of cervical vagotomy (decentralization) on the ultrastructure of the carotid body of the duck, *Anas platyrhynchos*. Cell Tissue Res., 163, 491.

Butler, P.J., and E.W. Taylor. (1983). Factors affecting the respiratory and cardiovascular responses to hypercapnic hypoxia, in mallard ducks. Respir. Physiol., 53, 109.

Calder, W.A. (1968). Respiratory and heart rates of birds at rest. Condor, 70, 358.

Calder, W.A., and K. Schmidt-Nielsen. (1968). Panting and blood carbon dioxide in birds. Am. J. Physiol., 215, 477.

Callanan, D., M. Dixon, J.G. Widdicombe, and J.C.M. Wise. (1974). Responses of geese to inhalation of irritant gases and injections of phenyl diguanide. Respir. Physiol., 22, 157.

Chiang, M.J., P.J. Berger, and A.L. Kunz. (1978). A study of the effect of SO_2 on pacing and intrapulmonary chemoreceptor discharge in the domestic fowl. Respir. Physiol., 33, 229.

Clanton, T.L., G.O. Ballam, R.K. Moore, and A.L. Kunz. (1982). Rapid ventilatory responses to changes in insufflated CO_2 in awake roosters. J. Appl. Physiol., 53, 1371.

Cohn, J.E., and R. Shannon. (1968). Respiration in unanesthetized geese. Respir. Physiol., 5, 259.

Crank, W.D., and R.R. Gallagher. (1978). Theory of gas exchange in the avian parabronchus. Respir. Physiol., 35, 9.

Crank, W.D., W.D. Kuhlmann, and M.R. Fedde. (1980). Functional localization of avian intrapulmonary CO_2 receptors within the parabronchial mantle. Respir. Physiol., 41, 71.

Crawford, E.C., Jr., and G. Kampe. (1971). Resonant panting in pigeons. Comp. Biochem. Physiol. 40A, 549.

Dejours, P. (1981). "Principles of Comparative Respiratory Physiology," 2nd ed. Amsterdam: Elsevier/North-Holland Biomedical Press, p. 185.

deWet, P.D., M.R. Fedde, and R.L. Kitchell. (1967). Innervation of the respiratory muscles of *Gallus domesticus*. J. Morphol., 123, 17.

Dreyer, M.V., H.P.A. DeBoom, P.D. deWet, J.M.W. LeRoux, D.J. Coetzer, and F. Eloff. (1977). Excision and localization of the avian (*Gallus domesticus*) carotid body. Acta Anat., 99, 192.

Dreyer, M.V., H.P.A. DeBoom, P.D. deWet, N. Hugo, and F. Eloff. (1978). Histocytology of the avian (*Gallus domesticus*) carotid body. Acta Anat., 102, 217.

Dubach, M. (1981). Quantitative analysis of the respiratory system of the house sparrow, budgerigar and violet-eared hummingbird. Respir. Physiol., 46, 43.

Dubbeldam, J.L., E.R. Brus, S.B.J. Menken, and S. Zeilstra. (1979). The central projections of the glossopharyngeal and vagus ganglia in the mallard, *Anas platyrhynchos* L. J. Comp. Neurol., 183, 149.

Duke, G.E., W.D. Kuhlmann, and M.R. Fedde. (1977). Evidence of mechanoreceptors in the muscular stomach of the chicken. Poult. Sci., 56, 297.

Duncker, H.-R. (1971). The lung air sac system of birds. A contribution to the functional anatomy of the respiratory apparatus. Ergeb. Anat. Entwicklungsgesch., 45(6), 1.

Duncker, H.-R. (1972). Structure of avian lungs. Respir. Physiol., 14, 44.

Duncker, H.-R. (1974). Structure of the avian respiratory tract. Respir. Physiol., 22, 1.

Duncker, H.-R. (1978a). General morphological principles of amniotic lungs. In "Respiratory Function in Birds, Adult and Embryonic" (J. Piiper, Ed.). New York: Springer-Verlag, p. 2.

Duncker, H.-R. (1978b). Funktionsmorphologie des Atemapparates und Coelomgliederung bei Reptilien, Vögeln und Säugern. Verh. Dtsch. Zool. Ges., 71, 99.

Duncker, H.-R. (1978c). Development of the avian respiratory and circulatory systems. In "Respiratory Function in Birds, Adult and Embryonic" (J. Piiper, Ed.). New York: Springer-Verlag, p. 260.

Duncker, H.-R. (1981). Stammesgeschichte der Struktur- und Funktionsprinzipien der Wirbeltierlungen. Verh. Anat. Ges., 75, 279.

Eaton, J.A., Jr., and M.R. Fedde. (1978). Biogenic amine-containing cells in the chicken lung. Poult. Sci., 57, 793.

Eaton, J.A., Jr., M.R. Fedde, and R.E. Burger. (1971). Sensitivity to inflation of the respiratory system in the chicken. Respir. Physiol., 11, 167.

Faraci, F.M., D.L. Kilgore, Jr., and M.R. Fedde. (1984). Oxygen delivery to the heart and brain during hypoxia: Pekin duck vs. Bar-headed goose. Am. J. Physiol., 247, R69.

Fedde, M.R. (1970). Peripheral control of avian respiration. Fed. Proc. Fed. Am. Soc. Exp. Biol., 29, 1664.

Fedde, M.R. (1976). Respiration. In "Avian Physiology," 3rd ed. (P.D. Sturkie, Ed.). New York: Springer-Verlag, p. 122.

Fedde, M.R. (1980). Structure and gas-flow pattern in the avian respiratory system. Poult. Sci., 59, 2642.

Fedde, M.R. (1981). Intrapulmonary CO_2 receptors and their role in the control of avian respiration. In "Advances in Physiological Sciences, Vol. 10. Respiration" (I. Hutás and L.A. Debreczeni, Eds.). Elmsford, New York: Pergamon Press, p. 147.

Fedde, M.R., and D.F. Peterson. (1970). Intrapulmonary receptor response to changes in airway-gas composition in *Gallus domesticus*. J. Physiol. (London), 209, 609.

Fedde, M.R., and W.D. Kuhlmann. (1975). PO_2 changes during analysis of chicken arterial blood. Comp. Biochem. Physiol. 50A, 633.

Fedde, M.R., and W.D. Kuhlmann. (1978). Intrapulmonary carbon dioxide sensitive receptors: Amphibians to mammals. In "Respiratory Function in Birds, Adult and Embryonic" (J. Piiper, Ed.). Berlin: Springer-Verlag, p. 33.

Fedde, M.R., and P. Scheid. (1976). Intrapulmonary CO_2 receptors in the duck: IV. Discharge pattern of the population during a respiratory cycle. Respir. Physiol., 26, 223.

Fedde, M.R., and G.H. Cardinet, III. (1977). Histochemical studies of respiratory muscles of chicken. Am. J. Vet. Res., 38, 585.

Fedde, M.R., R.E. Burger, and R.L. Kitchell. (1963). The effect of anesthesia and age on respiration following bilateral, cervical vagotomy in the fowl. Poult. Sci., 42, 1212.

Fedde, M.R., R.E. Burger, and R.L. Kitchell. (1964a). Electromyographic studies of the effects of bodily position and anesthesia on the activity of the respiratory muscles of the domestic cock. Poult. Sci., 43, 839.

Fedde, M.R., R.E. Burger, and R.L. Kitchell. (1964b). Electromyographic studies of the effects of bilateral, cervical vagotomy on the action of the respiratory muscles of the domestic cock. Poult. Sci., 43, 1119.
Fedde, M.R., R.E. Burger, and R.L. Kitchell. (1964c). Anatomic and electromyographic studies of the costopulmonary muscles in the cock. Poult. Sci., 43, 1177.
Fedde, M.R., P.D. deWet, and R.L. Kitchell. (1969). Motor unit recruitment pattern and tonic activity in respiratory muscles of *Gallus domesticus*. J. Neurophysiol., 32, 995.
Fedde, M.R., R.N. Gatz, H. Slama, and P. Scheid. (1974a). Intrapulmonary CO_2 receptors in the duck: I. Stimulus specificity. Respir. Physiol., 22, 99.
Fedde, M.R., R.N. Gatz, H. Slama, and P. Scheid. (1974b). Intrapulmonary CO_2 receptors in the duck: II. Comparison with mechanoreceptors. Respir. Physiol., 22, 115.
Fedde, M.R., J.P. Kiley, and W.D. Kuhlmann. (1980). Are avian intrapulmonary chemoreceptors involved in the control of breathing? In "Acta XVII Congressus Internationalis Ornithologici" (R. Nohring, Ed.). Berlin: Deutsche Ornithologen-Gesellschaft, p. 360.
Fedde, M.R., R.E. Burger, J. Geiser, R.K. Gratz, J.A. Estavillo, and P. Scheid. (1981). Effects of dead space on caudal air sac gas composition in the goose. Physiologist, 24, 131.
Fedde, M.R., J.P. Kiley, F.L. Powell, and P. Scheid. (1982). Intrapulmonary CO_2 receptors and control of breathing in ducks: Effects of prolonged circulation time to carotid bodies and brain. Respir. Physiol., 47, 121.
Fowle, A.S.E., and S. Weinstein. (1966). Effect of cutaneous electric shock on ventilatory response of birds to carbon dioxide. Am. J. Physiol., 210, 293.
Fujii, S., T. Tamura, and T. Okamoto. (1981). Microarchitecture of air capillaries and blood capillaries in the respiratory area of the hen's lung examined by scanning electron microscopy. Jpn. J. Vet. Sci., 43, 83.
Gillespie, J.R., J.P. Gendner, J.C. Sagot, and P. Bouverot. (1982a). Impedance of the lower respiratory system in ducks measured by forced oscillations during normal breathing. Respir. Physiol., 47, 51.
Gillespie, J.R., J.P. Gendner, J.C. Sagot, and P. Bouverot. (1982b). Respiratory mechanics of Pekin ducks under four conditions: Pressure breathing, anesthesia, paralysis or breathing CO_2-enriched gas. Respir. Physiol., 47, 177.
Gleeson, M., and J.H. Brackenbury. (1983). Respiratory and blood gas responses in exercising birds. Comp. Biochem. Physiol. 76A, 211.
Gleeson, M., and J.H. Brackenbury. (1984). Effects of body temperature on ventilation, blood gases and acid-base balance in exercising fowl. Q. J. Exp. Physiol., 69, 61.
Grima, M., and H. Girard. (1981). Oxygen consumption by chick blood cells during embryonic and post-hatch growth. Comp. Biochem. Physiol. 69A, 437.
Groth, H.-P. (1972). Licht- und fluoreszenzmikroskopische Untersuchungen zur Innervation des Luftsacksystems der Vögel. Z. Zellforsch. Mikrosk. Anat., 127, 87.
Hart, J.S., and O.Z. Roy. (1966). Respiratory and cardiac responses to flight in pigeons. Physiol. Zool., 39, 291.
Hinds, D.S., and W.A. Calder. (1971). Tracheal dead space in the respiration of birds. Evolution, 25, 429.
Hodges, R.D., A.S. King, D.Z. King, and E.I. French. (1975). The general ultrastructure of the carotid body of the domestic fowl. Cell Tissue Res., 162, 483.
Isaacks, R.E., and D.R. Harkness. (1980). Erythrocyte organic phosphates and hemoglobin function in birds, reptiles, and fishes. Am. Zool., 20, 115
Isaacks, R.E., D.R. Harkness, J.L. Adler, and P.H. Goldman. (1976). Studies on avian erythrocyte metabolism. Effect of organic phosphates on oxygen affinity of embryonic and adult-type hemoglobins of the chick embryo. Arch. Biochem. Biophys., 173, 114.
Isaacks, R.E., C.Y. Kim, T.J. Legato, A.E. Johnson, P.H. Goldman, D.R. Harkness, and A. Costa. (1980). Studies on avian erythrocyte metabolism. IX. Relationship of changing organic phosphate composition to whole blood oxygen affinity during development of the ostrich (*Struthio camelus camelus*). Dev. Biol., 75, 485.
Isaacks, R., C.Y. Kim, H.L. Liu, P.H. Goldman, A. Johnson, Jr., and D.R. Harkness. (1983). Studies on avian erythrocyte metabolism. XIII. Changing organic phosphate composition in age-dependent density populations of chicken erythrocytes. Poult. Sci., 62, 1639.
James, A.E., G. Hutchins, M. Bush, T.K. Natarajan, and B. Burns. (1976). How birds breathe: Correlation of radiographic with anatomical and pathological studies. J. Am. Vet. Radiol. Soc., 17, 77.
Jammes, Y., and P. Bouverot. (1975). Direct P_{CO_2} measurements in the dorsobronchial gas of awake Pekin ducks: Evidence for a physiological role of the neopulmo in respiratory gas exchanges. Comp. Biochem. Physiol. 52A, 635.
Jones, D.R., and M.J. Purves. (1970). The effect of carotid body denervation upon the respiratory response to hypoxia and hypercapnia in the duck. J. Physiol. (London), 211, 295.
Jones, D.R., and G.F. Holeton. (1972). Cardiovascular and respiratory responses of ducks to progressive hypocapnic hypoxia. J. Exp. Biol., 56, 657.
Jones, D.R., and O.S. Bamford. (1978). The immediate effects of deafferentation of the lungs on heart and breathing frequencies in ducks. Can. J. Zool., 56, 149.
Jones, D.R., and W.K. Milsom. (1982). Peripheral receptors affecting breathing and cardiovascular function in nonmammalian vertebrates. J. Exp. Biol., 100, 59.
Jones, J.H., E.L. Effmann, and K. Schmidt-Nielsen. (1981). Control of air flow in bird lungs: Radiographic studies. Respir. Physiol., 45, 121.
Jones, J.H., B. Grubb, and K. Schmidt-Nielsen. (1983). Panting in the emu causes arterial hypoxemia. Respir. Physiol., 54, 189.
Kadono, H., and T. Okada. (1962). Electromyographic studies on the respiratory muscles of the domestic fowl. Jpn. J. Vet. Sci., 24, 215.
Kadono, H., T. Okada, and K. Ono. (1963). Electromyographic studies on the respiratory muscles of the chicken. Poult. Sci., 42, 121.
Kampe, G., and E.C. Crawford, Jr. (1973). Oscillatory mechanics of the respiratory system of pigeons. Respir. Physiol., 18, 188.
Kawashiro, T., and P. Scheid. (1975). Arterial blood gases in undisturbed resting birds: Measurements in chicken and duck. Respir. Physiol., 23, 337.
Keijer, E., and P.J. Butler. (1982). Volumes of the respiratory and circulatory systems in tufted and mallard ducks. J. Exp. Biol., 101, 213.
Kiley, J.P., and M.R. Fedde. (1983a). Cardiopulmonary control during exercise in the duck. J. Appl. Physiol., 55, 1574.
Kiley, J.P., and M.R. Fedde. (1983b). Exercise hyperpnea in the duck without intrapulmonary chemoreceptor involvement. Respir. Physiol., 53, 355.
Kiley, J.P., W.D. Kuhlmann, and M.R. Fedde. (1979). Respiratory and cardiovascular responses to exercise in the duck. J. Appl. Physiol., 47, 827.
Kiley, J.P., W.D. Kuhlmann, and M.R. Fedde. (1982). Ventilatory and blood gas adjustments in exercising isothermic ducks. J. Comp. Physiol., 147, 107.
Kilgore, D.L., Jr., F.M. Faraci, and M.R. Fedde. (1984). Static response characteristics of intrapulmonary chemoreceptors in the pigeon and the burrowing owl, a species with a

blunted ventilatory sensitivity to carbon dioxide. Fed. Proc. Fed. Am. Soc. Exp. Biol., 43, 638.

King, A.S. (1966a). Structural and functional aspects of the avian lungs and air sacs. In "International Review of General and Experimental Zoology," Vol. 2 (W.J.L. Felts and R.J. Harrison, Eds.). New York: Academic Press, p. 171.

King, A.S. (1966b). Afferent pathways in the vagus and their influence on avian breathing: A review. In "Physiology of the Domestic Fowl" (C. Horton-Smith and E.C. Amoroso, Eds.). London: Oliver and Boyd, p. 302.

King, A.S. (1975). Aves respiratory system. In "The Anatomy of the Domestic Animals," 5th ed., Vol. 2 (R. Getty, Ed.). Philadelphia: Saunders, Chapter 64, p. 1883.

King, A.S. (1979). Systema respiratorium. In "Nomina Anatomica Avium" (J.J. Baumel, A.S. King, A.M. Lucas, J.E. Breazile, and H.E. Evans, Eds.). London: Academic Press, p. 227.

King, A.S., and A.F. Cowie. (1969). The functional anatomy of the bronchial muscle of the bird. J. Anat., 105, 323.

King, A.S., and D.C. Payne. (1964). Normal breathing and the effects of posture in *Gallus domesticus*. J. Physiol. (London), 174, 340.

King, A.S., and V. Molony. (1971). The anatomy of respiration. In "Physiology and Biochemistry of the Domestic Fowl," Vol. 1 (O.J. Bell and B.M. Freeman, Eds.). New York: Academic Press, p. 93.

King, A.S., J. McLelland, R.D. Cook, D.Z. King, and C. Walsh. (1974). The ultrastructure of afferent nerve endings in the avian lung. Respir. Physiol., 22, 21.

King, A.S., D.Z. King, R.D. Hodges, and J. Henry. (1975). Synaptic morphology of the carotid body of the domestic fowl. Cell Tissue Res., 162, 459.

Klentz, R.D., and M.R. Fedde. (1978). Hydrogen sulfide: Effects on avian respiratory control and intrapulmonary CO_2 receptors. Respir. Physiol., 32, 355.

Kobayashi, S. (1969). On the fine structure of the carotid body of the bird, *Uroloncha domestica*. Arch. Histol. Jpn., 31, 9.

Kobayashi, S. (1971a). Comparative cytological studies of the carotid body. 1. Demonstration of monoamine-storing cells by correlated chromaffin reaction and fluorescence histochemistry. Arch. Histol. Jpn., 33, 319.

Kobayashi, S. (1971b). Comparative cytological studies of the carotid body. 2. Ultrastructure of the synapses on the chief cell. Arch. Histol. Jpn., 33, 397.

Kollias, G.V., Jr., and I. McLeish. (1978). Effects of ketamine hydrochloride in red-tailed hawks (*Buteo jamaicensis*) I.—Arterial blood gas and acid base. Comp. Biochem. Physiol. 60C, 57.

Kuhlmann, W.D., and M.R. Fedde, (1976). Upper respiratory dead space in the chicken: Its fraction of the tidal volume. Comp. Biochem. Physiol. 54A, 409.

Kunz, A.L., and D.A. Miller. (1974a). Pacing of avian respiration with CO_2 oscillation. Respir. Physiol., 22, 167.

Kunz, A.L., and D.A. Miller. (1974b). Effects of feedback delay upon the apparent damping ratio of the avian respiratory control system. Respir. Physiol., 22, 179.

Kunz, A.L., and R.D. Tallman, Jr. (1978). Effect of $F_{I_{CO_2}}$ dynamics on T_i and T_{tot} in spontaneously breathing birds. In "Respiratory Function in Birds, Adult and Embryonic" (J. Piiper, Ed.). Berlin: Springer-Verlag, p. 182.

Kunz, A.L., R.D. Tallman, Jr., E.K. Michal, and R.K. Moore. (1979). Effect of carotid body denervation on pacing in unidirectionally ventilated chickens. Physiologist, 22, 73.

Kunz, A.L., R.P. Kaminski, D.A. Rittinger, T.L. Clanton, and G.O. Ballam. (1984). Unified model explaining normal breathing, volume pacing and CO_2 pacing in birds. Fed. Proc. Fed. Am Soc. Exp. Biol., 43, 431.

Lacy, R.A., Jr. (1968). Mechanical determinants of panting frequency in the domestic fowl. M.S. Thesis, University of California, Davis.

Lapennas, G.N., and R.B. Reeves. (1983). Oxygen affinity of blood of adult domestic chicken and red jungle fowl. Respir. Physiol., 52, 27.

Lasiewski, R.C. (1972). Respiratory function in birds. In "Avian Biology," Vol. II (D.S. Farner and J.R. King, Eds.). New York: Academic Press, p. 287.

Lasiewski, R.C., and W.A. Calder, Jr. (1971). A preliminary allometric analysis of respiratory variables in resting birds. Respir. Physiol., 11, 152.

Lillo, R.S., and D.R. Jones. (1983). Influence of ischemia and hypoxia on breathing in ducks. J. Appl. Physiol., 55, 400.

Locy, W.A., and O. Larsell. (1916a). The embryology of the bird's lung based on observations of the domestic fowl. Part I. Am. J. Anat., 19, 447.

Locy, W.A., and O. Larsell. (1916b). The embryology of the bird's lung based on observations of the domestic fowl. Part II. Am. J. Anat., 20, 1.

Lutz, P.L. (1980). On the oxygen affinity of bird blood. Am. Zool., 20, 187.

Lutz, P.L., I.S. Longmuir, and K. Schmidt-Nielsen. (1974). Oxygen affinity of bird blood. Respir. Physiol., 20, 325.

Macklem, P.T., P. Bouverot, and P. Scheid. (1979). Measurement of the distensibility of the parabronchi in duck lungs. Respir. Physiol., 38, 23.

Magno, M. (1973). Cardio-respiratory responses to carotid body stimulation with NaCN in the chicken. Respir. Physiol., 17, 220.

Magnussen, H., H. Willmer, and P. Scheid. (1976). Gas exchange in air sacs: Contribution to respiratory gas exchange in ducks. Respir. Physiol., 26, 129.

Maina, J.N. (1982). A scanning electron microscopic study of the air and blood capillaries of the lung of the domestic fowl (*Gallus domesticus*). Experientia, 38, 614.

Maina, J.N. (1984). Morphometrics of the avian lung. 3. The structural design of the passerine lung. Respir. Physiol., 55, 291.

Maina, J.N., and A.S. King. (1982). The thickness of the avian blood–gas barrier: qualitative and quantitative observations. J. Anat., 134, 553.

Maina, J.N., M.A. Abdalla, and A.S. King. (1982). Light microscopic morphometry of the lung of 19 avian species. Acta Anat., 112, 264.

Marder, J., and Z. Arad. (1975). The acid base balance of abdim's stork (*Sphenorhynchus abdimii*) during thermal panting. Comp. Biochem. Physiol. 51A, 887.

Martin, D.W., Jr. (1981). Structure and function of a protein-hemoglobin. In "Harper's Review of Biochemistry," 18th ed. (D.W. Martin, Jr., P.A. Mayes, and V.W. Rodwell, Eds.). Los Altos, California: Lange Medical Publications, p. 40.

Mather, F.B., G.M. Barnas, and R.E. Burger. (1980). The influence of alkalosis on panting. Comp. Biochem. Physiol. 64A, 265.

McLelland, J. (1970). The innervation of the air passages of the avian lung and observations on afferent vagal pathways concerned in the regulation of breathing. Ph.D. Thesis, University of Liverpool, Liverpool.

McLelland, J., and V. Molony. (1983). Respiration. In "Physiology and Biochemistry of the Domestic Fowl," Vol. 4 (B.M. Freeman, Ed.). New York: Academic Press, p. 63.

Meyer, M., H. Worth, and P. Scheid. (1976). Gas–blood CO_2 equilibrium in parabronchial lungs of birds. J. Appl. Physiol., 41, 302.

Meyer, M., J.P. Holle, and P. Scheid. (1978). Bohr effect induced by CO_2 and fixed acid at various levels of O_2 saturation in duck blood. Pfleugers Arch., 376, 237.

Michal, E.K., G.O. Ballam, and A.L. Kunz. (1981). Effects of CO_2 and air sac volume on the activity of medullary respiratory neurons of the chicken. Physiologist, 24, 131.

Miller, D.A. (1978). Effect of stretch on the respiratory pattern of a chicken. In "Respiratory Function in Birds, Adult and Embryonic" (J. Piiper, Ed.). Berlin: Springer-Verlag, p. 188.

Miller, D.A. (1980). A CO_2 threshold mechanism in a closed-loop avian respiratory system. J. Appl. Physiol., 48, 1029.

Miller, D.A., and A.L. Kunz. (1977). Evidence that a cyclic rise in avian pulmonary CO_2 triggers the next inspiration. Respir. Physiol., 31, 193.

Milsom, W.K., D.R. Jones, and G.R.J. Gabbott. (1981). On chemoreceptor control of ventilatory responses to CO_2 in unanesthetized ducks. J. Appl. Physiol., 50, 1121.

Mitchell, G.S., and J.L. Osborne. (1978). Avian intrapulmonary chemoreceptors: Respiratory response to a step decrease in P_{CO_2}. Respir. Physiol., 33, 251.

Mitchell, G.S., and J.L. Osborne. (1979). Ventilatory responses to carbon dioxide inhalation after vagotomy in chickens. Respir. Physiol., 36, 81.

Mitchell, G.S., and J.L. Osborne. (1980). A comparison between carbon dioxide inhalation and increased dead space ventilation in chickens. Respir. Physiol., 40, 227.

Molony, V. (1974). Classification of vagal afferents firing in phase with breathing in *Gallus domesticus*. Respir. Physiol., 22, 57.

Molony, V. (1978). Airway resistance. In "Respiratory Function in Birds, Adult and Embryonic" (J. Piiper, Ed.). Berlin: Springer-Verlag, p. 142.

Molony, V., W. Graf, and P. Scheid. (1976). Effects of CO_2 on pulmonary air flow resistance in the duck. Respir. Physiol., 26, 333.

Murrish, D.E. (1982). Acid-base balance in three species of antarctic penguins exposed to thermal stress. Physiol. Zool., 55, 137.

Niemeier, M.M. (1979). Structural and functional aspects of vocal ontogeny in *Grus canadensis* (Gruidae: Aves). Doctoral Dissertation, University of Nebraska, Lincoln.

Nightingale, T.E., R.A. Boster, and M.R. Fedde. (1968). Use of the oxygen electrode in recording PO_2 in avian blood. J. Appl. Physiol., 25, 371.

Nye, P.C.G., and R.E. Burger. (1978). Chicken intrapulmonary chemoreceptors: Discharge at static levels of intrapulmonary carbon dioxide and their location. Respir. Physiol., 33, 299.

Nye, P.C.G., and F.L. Powell. (1984). Steady-state discharge and bursting of arterial chemoreceptors in the duck. Respir. Physiol., 56, 369.

Oberthür, W., G. Braunitzer, R. Baumann, and P.G. Wright. (1983). Die Primärstruktur der α- und β-Ketten der Hauptkomponenten der Hämoglobine des Strausses (*Struthio camelus*) und des Nandus (*Rhea americana*) (*Struthioformes*). Hoppe-Seyler's Z. Physiol. Chem., 364, 119.

Osborne, J.L., and G.S. Mitchell. (1978). Intrapulmonary and systemic CO_2-chemoreceptor interaction in the control of avian respiration. Respir. Physiol., 33, 349.

Osborne, J.L., R.E. Burger, and P.J. Stoll. (1977a). Dynamic responses of CO_2-sensitive avian intrapulmonary chemoreceptors. Am. J. Physiol., 233, R15.

Osborne, J.L., G.S. Mitchell, and F. Powell. (1977b). Ventilatory responses to CO_2 in the chicken: Intrapulmonary and systemic chemoreceptors. Respir. Physiol., 30, 369.

Pattle, R.E. (1978). Lung surfactant and lung lining in birds. In "Respiratory Function in Birds, Adult and Embryonic" (J. Piiper, Ed.). New York: Springer-Verlag, p. 23.

Peek, F.W., and R.E. Phillips. (1971). Repetitive vocalizations evoked by local electrical stimulation of avian brains. II. Anesthetized chickens (*Gallus gallus*). Brain Behav. Evol., 4, 417.

Peek, F.W., O.M. Youngren, and R.E. Phillips. (1975). Repetitive vocalizations evoked by electrical stimulation of avian brains. IV. Evoked and spontaneous activity in expiratory and inspiratory nerves and muscles of the chicken (*Gallus gallus*). Brain Behav. Evol., 12, 1.

Perry, S.F., and H.-R. Duncker. (1980). Interrelationship of static mechanical factors and anatomical structure in lung evolution. J. Comp. Physiol., 138, 321.

Perutz, M.F. (1978). Hemoglobin structure and respiratory transport. Sci. Am., 239(6), 92.

Peterson, D.F., and M.R. Fedde. (1968). Receptors sensitive to carbon dioxide in lungs of chicken. Science, 162, 1499.

Petschow, D., I. Würdinger, R. Baumann, J. Duhm, G. Braunitzer, and C. Bauer. (1977). Causes of high blood O_2 affinity of animals living at high altitude. J. Appl. Physiol., 42, 139.

Pettit, T.N., and G.C. Whittow. (1982). The initiation of pulmonary respiration in a bird embryo: Tidal volume and frequency. Respir. Physiol., 48, 209.

Piiper, J. (1978). Origin of carbon dioxide in caudal air sacs of birds. In "Respiratory Function in Birds, Adult and Embryonic" (J. Piiper, Ed.). New York: Springer-Verlag, p. 148.

Piiper, J., and P. Scheid. (1972). Maximum gas transfer efficacy of models for fish gills, avian lungs and mammalian lungs. Respir. Physiol., 14, 115.

Piiper, J., and P. Scheid. (1975). Gas transport efficacy of gills, lungs and skin: Theory and experimental data. Respir. Physiol., 23, 209.

Piiper, J., and P. Scheid. (1982). Models for a comparative functional analysis of gas exchange organs in vertebrates. J. Appl. Physiol., 53, 1321.

Piiper, J., F. Drees, and P. Scheid. (1970). Gas exchange in the domestic fowl during spontaneous breathing and artificial ventilation. Respir. Physiol., 9, 234.

Powell, F.L. (1982). Diffusion in avian lungs. Fed. Proc. Fed. Am. Soc. Exp. Biol., 41, 2131.

Powell, F.L. (1983a). Respiration. In "Physiology and Behavior of the Pigeon" (M. Abs., Ed.). New York: Academic Press, p. 73.

Powell, F.L. (1983b). Effects of acid-base balance on avian intrapulmonary chemoreceptors. In "Modeling and Control of Breathing" (B.J. Whipp and D.M. Wiberg, Eds.). Amsterdam: Elsevier Science Publishing Co., p. 70.

Powell, F.L., and R.W. Mazzone. (1983). Morphometrics of rapidly frozen goose lungs. Respir. Physiol., 51, 319.

Powell, F.L., R.K. Gratz, and P. Scheid. (1978). Response of intrapulmonary chemoreceptors in the duck to changes in PCO_2 and pH. Respir. Physiol., 35, 65.

Powell, F.L., M.R. Fedde, R.K. Gratz, and P. Scheid. (1978). Ventilatory response to CO_2 in birds. I. Measurements in the unanesthetized duck. Respir. Physiol., 35, 349.

Powell, F.L., M.R. Barker, and R.E. Burger. (1980). Ventilatory response to the P_{CO_2} profile in chicken lungs. Respir. Physiol., 41, 307.

Powell, F.L., J. Geiser, R.K. Gratz, and P. Scheid. (1981). Airflow in the avian respiratory tract: Variations of O_2 and CO_2 concentrations in the bronchi of the duck. Respir. Physiol., 44, 195.

Prange, H.D., J.S. Wasser, A.S. Gaunt, and S.L.L. Gaunt. (1984). Respiratory and thermoregulatory effects of tracheal coiling in cranes (Gruidae): The functions of a long trachea. Fed. Proc. Fed. Am. Soc. Exp. Biol., 43, 638.

Richards, S.A. (1968). Vagal control of thermal panting in mammals and birds. J. Physiol., (London), 199, 89.

Richards, S.A. (1969). Vagal function during respiration and

the effects of vagotomy in the domestic fowl (*Gallus domesticus*). Comp. Biochem. Physiol., 29, 955.
Roberts, T.S. (1880). The convolution of the trachea in the sandhill and whooping cranes. Am. Nat. 14, 108.
Rollema, H.S., and C. Bauer. (1979). The interaction of inositol pentaphosphate with the hemoglobins of highland and lowland geese. J. Biol. Chem., 254, 12038.
Scheid, P. (1978). Analysis of gas exchange between air capillaries and blood capillaries in avian lungs. Respir. Physiol., 32, 27.
Scheid, P. (1979a). Mechanisms of gas exchange in bird lungs. Rev. Physiol. Biochem. Pharmacol., 86, 137.
Scheid, P. (1979b). Respiration and control of breathing in birds. Physiologist, 22, 60.
Scheid, P. (1981). Significance of unidirectional ventilation for avian pulmonary gas exchange. Physiologist, 24, 131.
Scheid, P. (1982). Respiration and control of breathing. In "Avian Biology," Vol. 6 (D.S. Farner, J.R. King, and K.C. Parkes, Eds.). New York: Academic Press, p. 405.
Scheid, P., and J. Piiper. (1969). Volume, ventilation and compliance of the respiratory system in the domestic fowl. Respir. Physiol., 6, 298.
Scheid, P., and J. Piiper. (1970). Analysis of gas exchange in the avian lung: Theory and experiments in the domestic fowl. Respir. Physiol., 9, 246.
Scheid, P., and T. Kawashiro. (1975). Metabolic changes in avian blood and their effects on determination of blood gases and pH. Respir. Physiol., 23, 291.
Scheid, P., and H. Slama. (1975). Remote-controlled device for sampling arterial blood in unrestrained animals. Pfleugers Arch., 356, 373.
Scheid, P., H. Slama, R.N. Gatz, and M.R. Fedde. (1974a). Intrapulmonary CO_2 receptors in the duck: III. Functional localization. Respir. Physiol., 22, 123.
Scheid, P., H. Slama, and H. Willmer. (1974b). Volume and ventilation of air sacs in ducks studied by inert gas wash-out. Respir. Physiol., 21, 19.
Scheid, P., R.E. Burger, M. Meyer, and W. Graf. (1978a). Diffusion in avian pulmonary gas exchange: Role of the diffusion resistance of the blood–gas barrier and the air capillaries. In "Respiratory Function in Birds, Adult and Embryonic" (J. Piiper, Ed.). New York: Springer-Verlag, p. 136.
Scheid, P., R.K. Gratz, F.L. Powell, and M.R. Fedde. (1978b). Ventilatory response to CO_2 in birds. II. Contribution by intrapulmonary CO_2 receptors. Respir. Physiol., 35, 361.
Scheipers, G., T. Kawashiro, and P. Scheid. (1975). Oxygen and carbon dioxide dissociation of duck blood. Respir. Physiol., 24, 1.
Schenk, A.G., C. Paul, and C. Vandecasserie. (1978). Respiratory proteins in birds. In "Chemical Zoology, Vol. 10, Aves" (A.H. Brush, Ed.). New York: Academic Press, p. 359.
Schmidt-Nielsen, K., J. Kanwisher, R.C. Lasiewski, J.E. Cohn, and W.L. Bretz. (1969). Temperature regulation and respiration in the ostrich. Condor, 71, 341.
Sèbert, P. (1978). Do birds possess a central CO_2-H^+ ventilatory stimulus? IRCS Med. Sci., 6, 444.
Sèbert, P. (1979). Mise en évidence de l'action centrale du stimulus CO_2-[H^+] de la ventilation chez le Canard Pékin. J. Physiol. (Paris), 75, 901.
Taha, A.A.M., and A.S. King. (1983). Autoradiographic observations on the innervation of the carotid body of the domestic fowl. Brain Res., 266, 193.
Tallman, R.D., Jr., and A.L. Kunz. (1982). Changes in breathing pattern mediated by intrapulmonary CO_2 receptors in chickens. J. Appl. Physiol., 52, 162.
Tallman, R.D., Jr., and F.S. Grodins. (1982a). Intrapulmonary CO_2 receptor discharge at different levels of venous P_{CO_2}. J. Appl. Physiol., 53, 1386.
Tallman, R.D., Jr., and F.S. Grodins. (1982b). Intrapulmonary CO_2 receptors and ventilatory response to lung CO_2 loading. J. Appl. Physiol., 52, 1272.
Taylor, C.R., and E.R. Weibel. (1981). Design of the mammalian respiratory system. I. Problem and strategy. Respir. Physiol., 44, 1.
Torre-Bueno, J.R., J. Geiser, and P. Scheid. (1980). Incomplete gas mixing in air sacs of the ducks. Respir. Physiol., 42, 109.
Tschorn, R.R., and M.R. Fedde. (1971). Motor unit recruitment pattern in a respiratory muscle of unanesthetized chickens. Poult. Sci., 50, 266.
Tschorn, R.R., and M.R. Fedde. (1974). Effects of carbon monoxide on avian intrapulmonary carbon dioxide-sensitive receptors. Respir. Physiol., 20, 313.
van Nice, P., C.P. Black, and S.M. Tenney. (1980). A comparative study of ventilatory responses to hypoxia with reference to hemoglobin O_2-affinity in llama, cat, rat, duck, and goose. Comp. Biochem. Physiol. 66A, 347.
Vos, H.J. (1934). Über den Weg der Atemluft in der Entenlunge. Z. Wiss. Biol. Vgl. Physiol., 21, 552.
Walsh, C., and J. McLelland. (1974a). Intraepithelial axons in the avian trachea. Z. Zellforsch. Mikrosk. Anat., 147, 209.
Walsh, C., and J. McLelland. (1974b). Granular 'endocrine' cells in avian respiratory epithelia. Cell Tissue Res., 153, 269.
Weingarten, J.P., H.S. Rollema, C. Bauer, and P. Scheid. (1978). Effects of inositol hexaphosphate on the Bohr effect induced by CO_2 and fixed acids in chicken hemoglobin. Pfleugers Arch., 377, 135.
Wells, R.M.G. (1976). The oxygen affinity of chicken hemoglobin in whole blood and erythrocyte suspensions. Respir. Physiol., 27, 21.
West, N.H., O.S. Bamford, and D.R. Jones. (1977). A scanning electron microscope study of the microvasculature of the avian lung. Cell Tissue Res., 176, 553.
Yamaguchi, K., D. Nguyen-Phu, P. Scheid, and J. Piiper. (1985). Kinetics of O_2 uptake and release by human erythrocytes studied by a stopped-flow technique. J. Appl. Physiol., 58, 1215.
Zimmer, K. (1935). Beiträge zur Mechanik der Atmung bei den Vögeln in Stand und Flug. Zoologica, 33 (5 Heft 88), 1.

9
Regulation of Body Temperature

G.C. Whittow

The preparation of this chapter was supported by a grant (N1/R-14) from the Sea Grant College Program (NOAA).

Introduction

Birds are "homeotherms," which means that they maintain a relatively constant deep-body temperature (Bligh and Johnson, 1973). Birds are also "endotherms," a term indicating that they are able to increase their body temperature by generating a considerable amount of heat within their tissues instead of relying on heat gained directly from their surroundings.

In general, the features of thermoregulation are similar in birds and in the other large group of homeothermic animals, the mammals. However, there are some conspicuous differences between the two groups that have a direct bearing on the manner in which they regulate their body temperature. Thus, the plumage of birds has to subserve the dual functions of flight and the provision of thermal insulation. The disposition of fat tends to be different in birds and mammals, with implications as far as their tissue insulation is concerned. The salt glands of some birds enable them to avoid many of the consequences of dehydration as a result of evaporative heat loss, but the absence of sweat glands in birds places the onus of evaporative cooling on their respiratory mechanisms. In addition, the development of the embryo in an egg outside the body puts a different perspective on the ontogeny of thermoregulation in birds, while the incubation of the egg frequently exposes the parent birds to extremely demanding thermal conditions.

The purpose of this chapter is to present the general principles of physiological and behavioral temperature regulation, with special reference to birds. For older references, and for documentation of general statements made in this chapter, the reader is referred to two previous editions (Whittow, 1965, 1976).

Body Temperature

Deep-Body ("Core") Temperature

Reference to the body temperature invariably means to the temperature of deep-seated tissues and organs that constitute the "core" of the body, as opposed to the superficial tissues, such as the skin, that make up the "shell." It is the deep-body temperature that is regulated within narrow limits in homeotherms, and most of the body tissue is at deep-body temperature. The temperature of the peripheral tissues, on the other hand, may vary considerably in different parts of the body and under different conditions.

Measurement

In practice, the commonest method of measuring deep-body temperature is by means of a thermistor or thermocouple probe inserted into the large intestine (rectum) or proventriculus. The temperature of the brain or spinal cord is frequently measured under experimental conditions, because it provides information on an area important in the regulation of body temperature (see section on control mechanisms, below). In the domestic fowl (*Gallus domesticus*), the temperature of the hypothalamus is lower than that in the rectum by 0.7°C, and the brain temperature of all birds that have been studied is lower than that of other parts of the body core (Pinshow et al., 1982). Telemetry techniques permit the measurement of deep-body temperature of birds under natural conditions (e.g., Whittow et al., 1978).

Species Differences

In general, the deep-body temperature of birds is higher than that in mammals, although there are many exceptions to this generalization (Table 9–1). The core temperature of large flightless birds, diving birds, and of species that are capable of torpor tends to be low.

Circadian Rhythm

The deep-body temperature of birds fluctuates during the course of a day (Table 9–1). Circadian rhythms of body temperature are important because they reveal the extent to which the bird stores heat (see section on heat storage below) and the degree to which the thermoregulatory control mechanisms permit the body temperature to change. One of the factors that determines the extent of the fluctuation is the size of the bird; the variation in temperature ranges from 8°C in hummingbirds to less than 1°C in the ostrich. In general, the body temperature is highest when the bird is most active, and this is true for both diurnal and

TABLE 9–1. Deep-body temperature of selected birds, at rest, under thermoneutral conditions[a]

Species		Body mass (kg)	Deep-body temperature (°C)
Ostrich (*Struthio camelus*)		100.0	38.3
Emu (*Dromiceius novae-hollandiae*)		38.3	38.1
Rhea (*Rhea americana*)		21.7	39.7
Mute swan (*Cygnus olor*)		8.3	39.5
Domestic goose (*Anser anser*)		5.0	41.0
Gentoo penguin (*Pygoscelis papua*)		4.9	38.3
Peruvian penguin (*Spheniscus humboldti*)		3.9	39.0
Domestic turkey (*Meleagris gallopavo*)		3.7	41.2
Adélie penguin (*Pygoscelis adeliae*)		3.5	38.5
Chinstrap penguin (*Pygoscelis antarctica*)		3.1	39.4
Domestic pigeon (*Columba livia*)		3.0	42.2
Domestic fowl (*Gallus domesticus*)		2.4	41.5
Domestic duck (*Anas platyrhynchos*)		1.9	42.1
Double-crested cormorant (*Phalacrocorax auritus*)		1.33	
	Day		41.2
	Night		40.2
Black grouse (*Lyrurus tetrix*)			
	Summer	1.079	41.3
	Winter	0.931	40.2
Anhinga (*Anhinga anhinga*)		1.04	
	Day		39.9
	Night		39.1
Little penguin (*Eudyptula minor*)		0.9	38.4
Brown-necked raven (*Corvus corax ruficollis*)		0.610	39.9
Willow ptarmigan (*Lagopus lagopus*)		0.573	39.9
California quail (*Lophortyx californicus*)		0.139	41.3
American kestrel (*Falco sparverius*)		0.119	39.3
Evening grosbeak (*Hesperiphona vespertina*)		0.060	41.0
Speckled mousebird (*Colius striatus*)		0.053	39.0
Common redpoll (*Acanthis flammea*)		0.015	40.1
Zebra finch (*taeniopygia castanotis*)		0.012	40.3

[a] Adapted from Whittow (1976), with additional information from Marder (1973), Bech (1980), Shapiro and Weathers (1981), Murrish (1982), Stahel and Nicol (1982), Hennemann (1983), and Rintamaki et al. (1983).

nocturnal species. Because the highest body temperatures of nocturnal birds occur during the night when the environmental temperature is lowest, it may be argued that the rhythm of body temperature is independent of the environmental temperature; however, this is not the case because the fluctuation of body temperature in birds subjected to a constant high environmental temperature is reduced.

The daily cycle of body temperature is closely keyed to the photoperiod, and seasonal changes in the temperature cycle may be correlated with corresponding changes in day length. A circadian rhythm of body temperature was absent in a 5-month-old emperor penguin (*Aptenodytes forsteri*) during the 24-hr antarctic day in spite of the concurrent fluctuation of environmental temperature. Experimentally, the circadian rhythm of body temperatures may be reversed by changing the times of illumination. Nevertheless, the persistence of a daily temperature cycle in Inca doves (*Scardafella inca*) kept in complete darkness suggests that endogenous factors may be involved, at least in some species. In the house sparrow (*Passer domesticus*) and in several other species the circadian rhythm was eliminated by removal of the pineal organ (Ebihara and Kawamura, 1981). In the domestic fowl, the mean level of body temperature, its range of oscillation, and the activity of the bird are related to the light intensity. The oscillation in temperature persists, although with a reduced amplitude, under conditions of constant dim light.

The circadian rhythm of body temperature in domestic fowl was reduced in amplitude following the administration of thiouracil, suggesting that an underlying variation of thyroid activity was involved. Dehydration, on the other hand, results in a more pronounced circadian rhythm of deep-body temperature in birds. Clearly, a change in body temperature must be achieved by variations in either heat production, or heat loss, or both (see section on heat balance below). Thus there is a circadian rhythm of spinal cord temperature in the pigeon that correlates well with the heat production; both the spinal temperature and the heat production decrease at night and increase during the day (Graf and Necker, 1979).

Environmental Temperature

For most birds, there is a range of environmental temperature over which the deep-body temperature remains essentially constant. The extent of this range depends, among other things, on the size of the bird and the amount of plumage. At high air temperatures the body temperature increases, the air temperature at which this occurs depending on the degree to which evaporative cooling mechanisms are used (Figure 9–1). At low air temperatures, shivering may increase heat production so that body temperature is maintained, or to the extent that deep-body temperature increases. In small birds, Weathers (1981) has shown that the deep-body temperature is not constant within the thermoneutral zone (see section on heat production below), in contrast to the heat production which, by definition, does not vary significantly within the thermoneutral zone.

Acclimation, Acclimatization

The term "acclimation" generally refers to changes induced by prolonged exposure to a particular temperature, usually under laboratory conditions. "Acclimatization" refers to seasonally induced changes or to changes resulting from climatic variations, usually out-of-doors (Bligh and Johnson, 1973). The rectal temperatures of fowls acclimated to an air temperature of 31°C were significantly higher than those of birds kept at 0°C, both in males and females. The body temperature of black grouse (*Lyrurus tetrix*) was significantly higher in the summer than during the winter (Table 9–1), suggesting that the changes of body temperature during acclimation and acclimatization are similar.

Dehydration

The body temperature of dehydrated African ostrich (*Struthio camelus*) increased during exposure to heat (45°C), in contrast to the behavior of the normally hydrated bird. This phenomenon has been observed in the domestic turkey (*Meleagris gallopavo*), and it is related to reduced respiratory evaporative cooling (see later section on evaporative heat loss).

Food Deprivation

The body temperature of American kestrels (*Falco sparverius*) that were fasted for 79 hr decreased at a rate of 0.2–0.4°C, per day (Shapiro and Weathers, 1981).

Heat Balance

The body temperature is determined by the balance between the amount of heat produced in the body by metabolism and the quantity of heat lost from the body to the environment. If more heat is produced than is lost, heat is "stored" in the body and the body temperature rises. Conversely, if heat loss exceeds heat production, the body temperature decreases. This concept is embodied in the heat-balance equation:

$$H = E \pm R \pm C \pm K \pm S$$

where H = metabolic heat production, E = evaporative heat loss (always positive, implying heat loss from the body), and S = heat storage in the body (positive if the heat content of the body increases and body temperature rises). The nonevaporative heat-loss parameters are R = heat loss by radiation (positive if heat is lost to the environment), C = heat loss by convection (positive if heat is lost to the environment), and K = heat loss by conduction (positive if heat is lost to the environment).

Units

In the past, the kilocalorie (kcal) has been the unit of heat most frequently used, and the rate of heat loss, gain, or production was expressed as kcal/hr. In recent years, many authors have used the joule (J) instead of the kcal, and watts (W) in place of kcal/hr. This change in the use of units came about in response to an international attempt to standardize units according to the Système Internationale (SI). It is essential, therefore, to be able to convert one type of unit into another:

1 kcal = 4,187 J = 4.187 kJ
1 kcal/hr = 1.16 W

The present chapter is in keeping with the current trend to use SI units.

Heat Storage (S)

The quantity of heat stored in the body of a bird in a given time may be calculated with the following equation:

$$S = \Delta \overline{T}_b \times \text{body mass (kg)} \times 3.5$$

where S = heat stored (kJ) over the specified time period, $\Delta \overline{T}_b$ = change in *mean* body temperature (°C)

FIGURE 9–1. Deep-body temperature (T_b), oxygen consumption ($\dot{V}_{O_2}$), total thermal conductance (C_{total}), and total evaporative heat loss (E) of the sooty tern (*Sterna fuscata*) at various air temperatures (T_a). The solid line at the top of the figure represents equivalence between deep-body and air temperature; at the bottom it indicates that 100% of the heat production is lost by evaporative cooling. Symbols: ●, adults; *X*, fledglings; *EWL*, total evaporative water loss; *H*, heat production. (After MacMillen et al., 1977.)

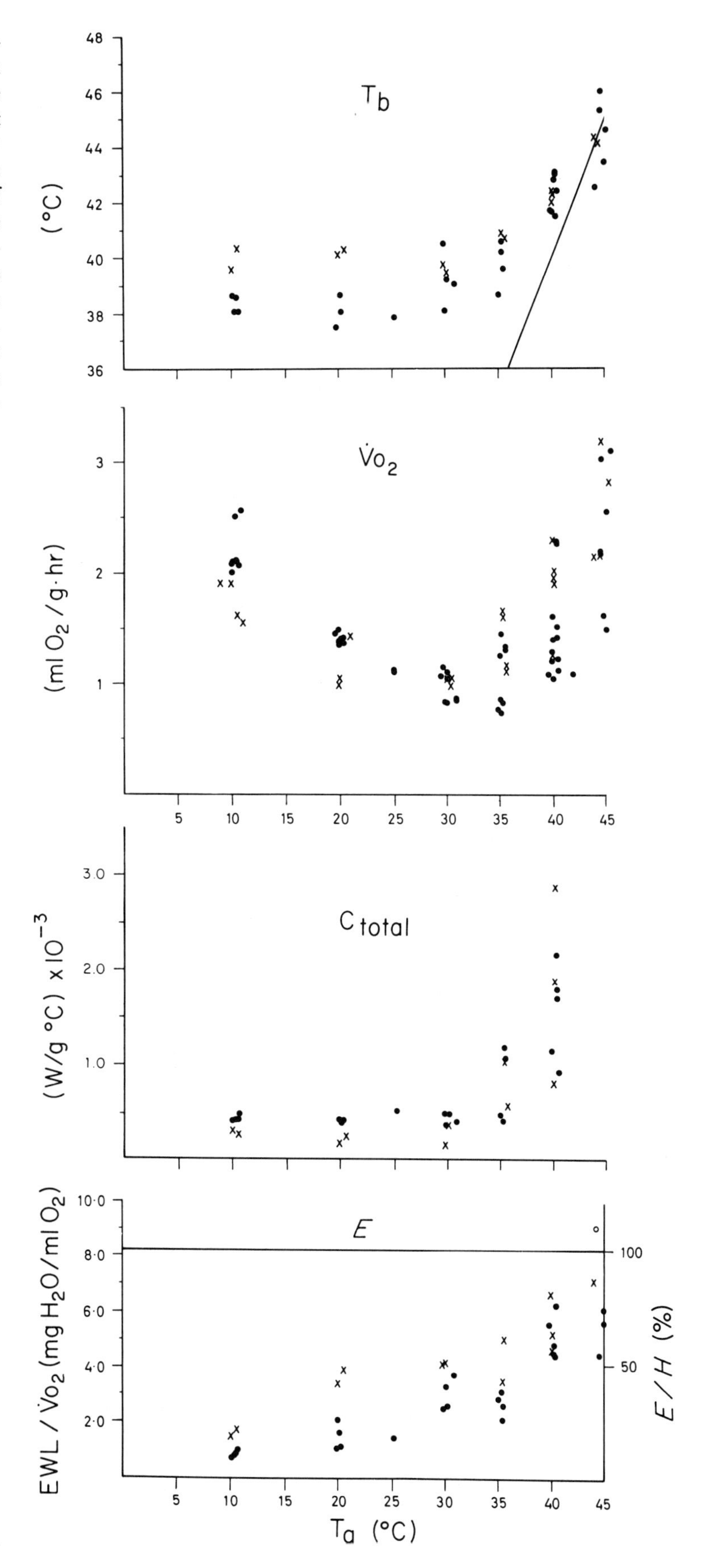

over the same period, and 3.5 = the mean specific heat of the body tissues (kJ/kg · °C).

The mean body temperature ($\overline{T}_b$) is the mean of a large number of temperatures measured both on the surface and deep within the body. This is obviously both a difficult and an impractical quantity to measure. It may be derived, with some loss of accuracy but with considerably greater facility, from the consideration that the body of a bird consists of an inner "core" at a relatively high temperature and an outer "shell" at a lower more variable temperature.

The core temperature may be measured by inserting a thermistor or thermocouple probe into the rectum or proventriculus. The shell temperature is estimated by averaging the temperatures taken at several different sites on the surface of the skin and applying weighting factors to the skin temperature of each site commensurate with the surface area of each site. The mean body temperature is then computed by taking the average of the core temperature and the mean skin temperature, the core and mean skin temperature being multiplied by other weighting factors appropriate to the amount of tissue at core and shell temperature, respectively. These factors have not been determined for birds.

Changes in the quantity of heat stored within the body constitute an important thermoregulatory mechanism, particularly during exposure to hot environments. The increased rate of heat storage results in a rise in the deep-body temperature; the temperature difference between the bird and the air increases, and so, therefore, does the nonevaporative heat loss from the bird to its environment. If the ambient temperature is higher than that of the bird itself, the bird gains heat from the environment. This happens when the bird is exposed to direct solar radiation, but it occurs particularly under desert conditions where the temperature of the sand and also the air may exceed the bird's temperature. In these circumstances, heat storage results in an increased body temperature, and this in turn leads to a reduction in the temperature difference between the bird and its environment. Consequently, heat gain from the environment diminishes.

The storage of heat within the body also implies a saving of water under hot conditions, simply because the alternative to storage of the heat is its dissipation by evaporative moisture loss. It is appropriate, therefore, that many desert birds become more hyperthermic when they are dehydrated and when their body water balance is more precarious.

The amount of heat stored cannot be such that the body temperature rises to lethal levels, which in many birds seems to be between 46° and 47°C. Heat storage is limited by the thermoregulatory heat-loss mechanisms, which are discussed later in this chapter. It is important to emphasize, however, that the curtailment of heat storage is usually achieved at the expense of an increased water loss and an enhanced risk of dehydration. Moreover, because the extent of heat storage is determined by tissue mass, small birds with a relatively high heat production are at a disadvantage in comparison with large birds.

The tissue most susceptible to heat damage, as the body temperature approaches the lethal temperature, is the brain. Some birds are able to keep their brains cool while the temperature of the rest of the body rises to high levels. This is accomplished by a special anatomic arrangement of the blood supply to the brain whereby the warm arterial blood going to the brain is cooled by venous blood from the head, eyes, and upper respiratory passages. The structure that permits the close apposition of small arteries and veins, the *rete mirabile ophthalmicum,* occurs in a variety of birds, including pigeons and tropical seabirds (Kilgore, 1976; Kilgore et al., 1979; Bernstein et al., 1979a, b; Pettit et al., 1981).

In the jackass penguin (*Spheniscus demersus*), the ophthalmic *rete* is believed to function to curtail excessive heat loss from the relatively poorly insulated heart (Frost et al., 1975). In this way, the warm arterial blood gives up heat to the cool venous blood in the *rete,* and this heat is returned to the heart and retained by the body (see section on countercurrent heat exchange below).

Heat Production (H)

Measurement

The heat production of birds is most commonly computed from the measured oxygen consumption. If the bird is at rest, and if it is not growing or laying down fat (energy storage; see Chapter 10), then almost all of the energy released in the tissues is in the form of heat. The energy is released by oxidative reactions; consequently, there is a direct connection between the amount of oxygen consumed by the bird and the amount of heat produced.

In order to measure its oxygen consumption, the bird is enclosed in a chamber through which fresh air is drawn at a constant and known rate. The air leaving the chamber is analyzed for its oxygen content. From the rate of airflow and the difference in the oxygen content of the air entering and leaving the chamber, the oxygen consumption may be calculated. The volume of oxygen consumed is corrected to standard conditions of temperature (0°C) and pressure (760 mm Hg) for the dry gas (STPD) and then multiplied by 20.1 (kJ/liter) to convert it to heat units. An assumption is made regarding the respiratory quotient (RQ) when this procedure is used, and there is a technical error when the RQ (the respiratory quotient is the ratio of carbon dioxide produced to oxygen consumed when energy-containing substrates are oxidized; it varies from 1.0 for carbohydrate to 0.70 for fat) is not actually measured. However, the errors of the procedure are small, and they may be calculated (Withers, 1977).

There are other methods of measuring or estimating the heat production of birds. Some of them are described later in this chapter and others in the following chapter. One method, which requires the measurement of skin temperature and the surface temperature of the plumage together with calculations of the heat transfer through the plumage, compared very favorably with direct measurements based on oxygen consumption (Hayes and Gessaman, 1982).

Basal Metabolic Rate (BMR)

If the heat production is measured in a bird that is rested, awake, fasting (postabsorptive), and in the thermoneutral zone (see below), the heat production is equal to the basal metabolic rate (Bligh and Johnson, 1973). Misson (1974) found that several training sessions were necessary to accustom the domestic fowl to the experimental situation before basal values were achieved, and 24–48 hr without food, depending on the size of the bird, were necessary to reach a postabsorptive state. The requisite conditions for the measurement of BMR are not always attainable in wild birds, but if the conditions under which the measurement is made are as close to those outlined above as possible, and if they are stated, the heat production is then equal to the standard metabolic rate (SMR; Bligh and Johnson, 1973). Some authors have used the term standard metabolic rate in a different context (Kendeigh et al., 1977). Other investigators have measured the heat production of birds that are at rest, under thermoneutral conditions, but not in a postabsorptive state, which according to Bligh and Johnson (1973) is a measure of resting metabolic rate. In some studies, the birds have been kept in the drak during the measurement of heat production, and the minimal values of heat production were recorded. In many instances, the birds were asleep during the measurements. Consequently, great care is needed in the interpretation of measured values of heat production, with particular attention to the specified conditions under which the measurements were made.

Species Differences

An extensive compilation of the measured basal metabolic rates of birds has been made by Kendeigh et al. (1977). The heat production (W) of large birds is greater than that of small ones. However, with occasional exceptions (MacMillen, 1981), the heat production per unit body mass (W/kg) is less in large birds. The relationship between basal metabolic rate and body mass in birds differs in passerines and other birds (Table 9–2), the basal metabolic rate being higher in passerines, which are mainly small species weighing less than 0.1 kg. Recently, the existence of a real difference between passerines and other birds has been questioned (Prinzinger and Hänssler, 1980). The basal metabolic rate is also higher if it is measured during the time of day that the birds are normally active than during that part of the day that they are ordinarily at rest (Table 9–2). Although the majority of birds conform to the relationships presented in Table 9–2, there are exceptions. For example, the basal metabolic rates of the flightless kiwis were found to be lower than predicted by the equations in Table 9–2 (Calder and Dawson, 1978). Other examples are given by Kendeigh et al. (1977).

Circadian Rhythm

There is a variation in the metabolic rate of birds during the course of the day that is independent of the effects of food intake, although food consumption contributes to the rhythm in birds that are fed. In the domestic fowl, the basal metabolic rate is highest in the forenoon and lowest at approximately 8:00 PM. The reduction in the basal metabolic rate at night amounts to 18–30% in the fowl but to as much as 49% in the English sparrow. The fluctuation is, in fact, greater in small than in large birds. Activity contributes to the circadian rhythm but the circadian rhythm in fasting fowls is not entirely the result of changes in activity. The cycle of light and darkness also plays a part; reversal of the daily photoperiod was followed by an inversion of the time relations of the circadian fluctuation in basal metabolic rate. Under conditions of constant light and temperature, an oscillation in metabolic rate is evident, but the rhythm was lost in dim light, suggesting that some light is essential for the generation of a circadian rhythm (Berman and Meltzer, 1978). There is some evidence for an underlying circadian rhythm in thyroid function in the domestic fowl (Klandorf et al., 1982),

TABLE 9–2. Equations for calculating the basal metabolic rate of birds (BMR, in W) in relation to body mass (M, kg) and the phase of the daily cycle[a]

	Lasiewski and Dawson (1967)	Aschoff and Pohl (1970a,b)
Active phase		
Passerines	$BMR = 6.3\ M^{0.73}$	$BMR = 6.8\ M^{0.704}$
Other birds	$BMR = 4.1\ M^{0.72}$	$BMR = 4.4\ M^{0.729}$
Rest phase		
Passerines	$BMR = 4.0\ M^{0.613}$	$BMR = 5.5\ M^{0.726}$
Other birds	$BMR = 3.6\ M^{0.76}$	$BMR = 3.5\ M^{0.734}$

[a] After Calder and King (1974).

which itself may be related to the light/dark cycle and to food intake.

Environmental Temperature

The oxygen uptake and heat production of many birds varies with environmental temperature in a manner similar to that depicted in Figure 9–1. The range of environmental temperature over which the oxygen uptake remains essentially constant and/or is at its lowest level is the thermoneutral zone. This is a range of air temperature defined by the upper and lower critical temperatures. The range of air temperatures over which birds are in a thermoneutral condition varies from a few degrees C in some small birds to 30°C in penguins (Le Maho, 1983). The heat production within the thermoneutral zone depends also on the level of food intake prior to the measurement. The higher the level of food intake, the higher the heat production, and this effect may persist for some time after the diet has changed. It is believed that these differences reflect changes in body composition. The heat production within the thermoneutral zone is, of course, equal to the basal metabolic rate (see earlier definition of basal metabolic rate). At air temperatures higher than the upper critical temperature, the heat production increases, usually as a result of an increased body temperature. The increased temperature of the tissues results in a generalized acceleration of chemical reactions and consequently in an increased oxygen requirement and heat production (van't Hoff–Arrhenius effect). Weathers (1981) has proposed the use of the slope of the increase in heat production with increasing air temperature, above the upper critical temperature ("coefficient of heat strain"), as a measure of the cost of thermoregulation in a hot environment. The coefficient of heat strain was greater for small than for large birds. The heat production also increases at air temperatures below the lower critical temperature, but the mechanism in this instance is quite different: The animal shivers and thereby produces heat.

The common redpoll (*Carduelis flammea*) is able to increase its heat production more than fivefold during exposure to cold. The heat produced by shivering is largely derived from the oxidation of fatty acids (Marsh and Dawson, 1982). The rate at which the heat production increases in response to exposure to low air temperatures depends on the insulation of the tissues and plumage: the greater the insulation, the lower the rate of increase in heat production. As the insulation of small birds is generally less effective than that of large ones (see section on the thermal conductance of the plumage below), the rate of increase in heat production below the lower critical temperature is usually greater in small birds. Nevertheless, the performance of some birds in the cold is extremely impressive. For instance, the lower critical temperature of a small arctic bird, the common redpoll, weighing approximately 14 g, is 9°C; yet the bird is able to survive exposure to −50°C for 3 hr.

The well-defined lower critical temperature in many species implies that the insulation of the tissues and plumage (see sections on thermal conductance below) has reached a maximum at the lower critical temperature, and that it does not increase further at air temperatures below the lower critical temperature. One consequence of this is that if a line is drawn through the values for heat production obtained below the lower critical temperature, and if this line is extrapolated to zero heat production, it transects the horizontal (temperature) axis at the deep-body temperature of the bird. This simply means that when the insulation is constant, the heat loss from the bird is proportional to the difference in temperature between the bird and the air surrounding it, an expression of Fourier's law. It follows from this that the heat flow is zero when the air is at the same temperature as the bird.

The change in heat production with temperature does not always follow the pattern illustrated in Figure 9–1. When the heat production of the evening grosbeak (*Hesperiphona vespertina*) was measured at night at different temperatures, there was no well-defined thermoneutral zone. Heat production and insulation changed simultaneously at ambient temperatures below 30°C. Two congeneric species of Hawaiian honeycreepers behaved quite differently at environmental temperatures below the lower critical temperature. In one species (*Loxops virens*), the thermal conductance varied below the lower critical temperature. In the other species (*L. parva*), the thermal conductance was constant but the body temperature diminished. Another potential anomaly was brought to light by Prinzinger (1982), who found that sinusoidal fluctuations of ambient temperature around a mean value resulted in a higher heat production in Japanese quail (*Coturnix c. japonica*) than did exposure to a constant temperature of the same value. As most experimental investigations involve exposure to constant temperatures, there may be quantitative errors in many such studies.

Acclimation, Acclimatization

Continuous experimental exposure to cold in the laboratory (acclimation) results in an increase in heat production that may amount to 85% in the siskin (*Chrysomitris spinus*). Earlier studies produced little evidence for "nonshivering thermogenesis," an increased heat production in the absence of shivering such as occurs in many mammals during acclimation to cold. In fact, birds do not appear to have brown fat, an important thermogenic tissue involved in nonshivering thermogenesis in mammals (see also Chapter 15). The evidence for nonshivering thermogenesis in mature, cold-acclimated birds is controversial (see Dawson, 1975).

If it does occur (El-Halawani et al., 1970; Arieli et al., 1979), catecholamines are not involved (Hissa and Palokangas, 1970) but corticosterone may be (El-Halawani et al., 1973).

In many birds tested during the winter (acclimatization), the lower critical temperature has shifted to the left (Kendeigh et al., 1977; Rintamaki et al., 1983) i.e., the lower critical temperature has diminished compared with birds tested under summer conditions. In addition, the slope of the relationship between heat production and air temperature is less (Hissa and Palokangas, 1970; Kendeigh et al., 1977). The differences reflect an underlying increase in insulation (see sections on thermal conductance below) in the birds tested in winter. These phenomena are illustrated in the willow ptarmigan, an arctic species in which the lower critical temperature during summer was 7.7°C, and in the winter, −6.3°C. There is evidence that smaller birds do not rely on seasonal changes in insulation to any great extent. Their primary adaptation to cold is an enhanced ability to sustain elevated heat production for long periods (Marsh and Dawson, 1982; Dawson et al., 1983). The common redpoll is apparently able to increase heat production during the day and to increase insulation at night during the winter (see also section on the thermal conductance of the plumage below). There is no evidence that brown fat develops in adult birds during acclimatization to cold. Nevertheless, there is some evidence for nonshivering thermogenesis in cold-acclimatized domestic fowl (Arieli et al., 1978) and also in the common redpoll (Koban and Feist, 1982), although the mechanism remains obscure. Catecholamines do not stimulate nonshivering thermogenesis in birds, as they do mammals (Hissa and Palokangas, 1970). One of the difficulties attending the attribution of nonshivering thermogenesis to birds is that propranolol, a drug that inhibits nonshivering thermogenesis in mammals, also inhibits shivering in birds (Hissa et al., 1980). There is some evidence that corticosterone might be involved (Hissa and Palokangas, 1970).

Aulie (1976b) detected differences in the shivering patterns of an arctic bird, the willow ptarmigan, and a domestic tropical species, the bantam hen. In the former, although shivering was intermittent, the oxygen consumption increased to higher levels than in the bantam hen, which shivered continuously. These differences were correlated with differences in the relative muscle mass in the two species: In the willow ptarmigan, muscle made up 41.3% of the body mass, as compared with 28.8% in the bantam hen.

Some birds experience an increased heat production within the thermoneutral zone during the cold season of the year (Kendeigh et al., 1977; Rintamaki et al., 1983), a feature of the common redpoll. There was an associated increase in thyroid activity, suggesting that the calorigenic effect of thyroid hormone was involved in the response.

Continued exposure to constant high temperatures in the laboratory (acclimation) results in a diminution in heat production in several species of birds. In the domestic fowl and Japanese quail, this has been attributed to a diminution in thyroid activity. The response to high temperatures is suppressed if the temperature is allowed to fluctuate.

The metabolic response of birds to high air temperatures in relationship to season has not been studied as extensively as have their reactions to low air temperatures. The basal heat production of yellow buntings (*Emberiza citrinella citrinella*) did not vary seasonally.

In contrast, the heat production of a population of house sparrows in Houston, Texas was lower than that of populations in colder climates of Michigan and Colorado, and a desert breed of the domestic fowl had a lower basal metabolic rate than in other breeds (Arad and Marder, 1982). Kendeigh et al. (1977) and Weathers (1979) have presented persuasive evidence that the basal metabolic rate varies with climate, being lower in the tropics than in cold climates, and Hails (1983) reported that the metabolic rates of 11 tropical passerines were only 47.6–85.1% of predicted values. Latitudinal variations in heat production are probably genetically determined, whereas seasonal changes are not.

The basal metabolic rate is adaptive to the prevailing climatic conditions, even within the tropics. For example, Weathers and Van Riper (1982) reported that the BMR of the Laysan finch (*Psittirostra cantans*), a species restricted to low treeless atolls in Hawaii, was only 80% of the value predicted from its body mass. No such depression was observed in the congeneric palila (*P. bailleui*), which is confined to the cool montane forests of Hawaii. Furthermore, different species may adopt different adaptive strategies. Thus, in two dark-plumaged herons nesting in a hot climate, the species with the higher metabolic rate nested in the shade and had a high thermal conductance (Ellis, 1980; see also sections on thermal conductance below).

Haim et al. (1979) showed that the heat production of birds may change, quite independently of changes in temperature, as they become acclimated to variations in photoperiod. Thus, acclimation to a short period of illumination during the day (long scotophase) had the same effect on the heat production of pigeons as did acclimation to a low ambient temperature.

Heat Transfer within the Body

Pathways

Heat produced in such deep-seated organs as the liver must be transported to the skin surface or the mucosa lining the upper respiratory tract before it can be lost to the environment. This transfer of heat occurs along two distinct pathways.

Conduction. Conduction through the tissues takes place by direct transfer of energy from molecule to molecule. The rate of conductive heat transfer depends on the thickness of the layer of tissue (fat, skin, muscle) that the heat must traverse and on the thermal conductivity of the tissue components.

	Fat	Skin	Muscle
Thermal conductivity (W/m·°C)	0.209	0.337	0.500

Although fat has the lowest thermal conductivity, its role as a barrier to heat flow is not as important in birds as it is in many mammals because some birds, e.g., the domestic fowl, do not have a substantial layer of fat beneath the skin. Instead, the fat deposition is localized in discrete areas in the abdominal cavity. Quite unlike the fowl in this respect are birds that have adapted to life in the water, notably penguins and ducks. Heat loss to water is considerably greater than that to air at the same temperature. Consequently, as in mammals, a layer of subcutaneous fat helps to limit heat loss in aquatic species, particularly in penguins, which encounter both very cold water and extremely low air temperatures. However, even in penguins, the major barrier to heat loss is provided by the plumage (see section on the thermal conductance of the plumage below). Some birds, such as pelicans, have a considerable subcutaneous pneumaticity that may provide additional insulation.

Vascular Convection. Heat transfer by conduction within the body is most important under cold conditions, when the blood flow through the skin is minimal. Within the thermoneutral zone, the more important avenue of heat transfer is by way of the bloodstream, a type of "internal convection." The blood acquires heat in the heat-producing tissues and conveys it to the skin. The skin temperature rises and heat loss to the environment increases.

Countercurrent Heat Exchange

Heat transfer between the core and the skin surface is also influenced by a process of countercurrent heat exchange in the limbs. The warm arterial blood entering the limb gives up some of its heat to the cool blood returning in the veins from the distal parts of the limbs. This heat is returned to the core. In order for this to occur, the veins must be in close apposition to the arteries. The temperature of the distal parts of the limbs diminishes, and so does the heat loss from the limb to its surroundings. It is easy to see that this process increases the amount of tissue making up the shell. In a warm environment, the venous return from the distal parts of the limbs occurs by way of superficial veins, so that the arterial blood entering the limb is not cooled until it reaches the skin. The advantage of the countercurrent system is that blood flow and oxygen supply to the distal parts of the extremities may be maintained without incurring a high level of heat loss. In other words, the supply of nutrients to the tissues of the extremities is separated from the delivery of heat to them.

Countercurrent heat exchange is strikingly illustrated in the wood stork (*Mycteria americana*), a species in which heat conservation is facilitated by the presence in the leg of a special vascular structure, the *rete mirabile.* In this structure the arteries and veins divide to form a network of intermingling vessels in which arteries and veins are in close apposition to each other. Such a complex structure is not essential for countercurrent heat exchange to occur, but it does increase its effectiveness. The wood stork spends long periods standing in cool water while feeding. The presence of the *rete* prevents excessive heat loss from the long uninsulated legs. In the heron (*Ardea cinerea*), which also has a *rete* in its legs, the heat loss from one leg immersed in water at 4–12°C was less than 10% of the heat production of the bird, when the air temperature to which the bird was exposed was less than 20°C. When the air temperature was 35°C, the heat loss from one leg to cold water was higher than the bird's heat production. In penguins, which face formidable problems in curtailing heat loss to water and cold air, the close apposition of veins and arteries supplying the flippers and legs may prevent excessive heat loss from these appendages (Frost et al., 1975).

Cold Vasodilatation

Curtailment of heat transfer from the deep tissues to the extremities by the mechanism described in the preceding paragraph obviously entails the risk of the peripheral tissues freezing if the environmental temperature is low enough. This possibility is circumvented by the intermittent flow of warm blood to the distal extremities, a phenomenon known as "cold vasodilatation," which was first described in the foot of the domestic fowl. Subsequently, it was reported in the feet of other birds including two antarctic species, the emperor penguin and the giant fulmar (*Marconectes giganteus*), which habitually expose their feet to cold sea water, ice, and snow (Johansen and Millard, 1973). Although cold vasodilatation may prevent freezing of the extremities, it does so at the expense of increased heat loss from the animal. In the mallard duck the increased heat loss is balanced by an increased heat production (Kilgore and Schmidt-Nielsen, 1975).

Thermal Conductance of the Tissues

In practice, it is difficult both to measure, and to distinguish between the avenues of heat transfer within the body. It is much simpler to consider the heat transfer (H) as a single process, as follows:

$$H = C_{tissues} (T_b - \overline{T}_{sk})$$

where H = heat transfer from the core to the skin surface (W/m^2), $C_{tissues}$ = thermal conductance of the tissues ($W/m^2 \cdot °C$), T_b = core temperature (°C), $\overline{T}_{sk}$ = mean skin temperature (°C; see section on heat storage above), and m^2 refers to the area of the skin surface. The thermal conductance ($C_{tissues} = H/T_b - \overline{T}_{sk}$) is a measure of the ease with which heat flows from the core to the skin surface. It incorporates factors for the thermal conductivity of the skin, fat, and muscle (tissues intervening between the deep-seated, heat-producing organs and the skin surface), the thickness of these tissues, the shape of the body, and the characteristics of blood flow to the skin. None of these factors can be readily measured in the intact bird, which is why they are amalgamated into a single factor, the thermal conductance. The reciprocal of the thermal conductance (1/C) of the tissues is the tissue insulation. Experimentally, C may be calculated from the measured heat production (H) after any heat stored (S) within the body, together with the heat loss by evaporation of moisture from the respiratory tract (E_{ex}; see section on evaporative heat loss below), are subtracted. The rationale for these corrections is that neither heat stored within the body nor that lost from the respiratory tract flows to the skin.

Short-term variations in the thermal conductance of the tissues are effected largely by changes in the blood flow to the skin. The thermal conductance is minimal when the blood flow through the skin is lowest and it is maximal when the cutaneous blood flow is high. Long-term changes or differences between species may be due to variations in the amount of insulating fat or the thickness of the skin.

In general, the thermal conductance of the tissues underlying the feathered areas changes little with variations in environmental temperature. The thermal conductance of small birds is greater than that of large ones, even in defeathered birds. This is probably due in part to differences in the thickness of the skin and fat layers, but it also reflects the fact that the physics of heat transfer is such that conductive and convective cooling is inversely proportional to the diameter of the body that is losing heat.

Marked changes in the tissue insulation of unfeathered skin sites, such as the comb and feet, do occur, however. The effectiveness of changes in tissue insulation in the unfeathered extremities is demonstrable in the great black-backed gull (*Larus marinus*), in which less than 3% of the heat production was lost from one leg immersed in water at 4–12°C when the air temperature was below 10°C. At an air temperature of 30°C, the heat loss could exceed the heat production of the bird (see also previous section on countercurrent heat exchange). In the Peking duck, heat loss from the bill may represent 18–117% of the resting heat production at 0°C, depending on the blood flow to the bill (Hagan and Heath, 1980). The blood flow to the bill is regulated by the preoptic region of the brain (see section on control mechanisms below).

Acclimation, Acclimatization. Information on changes in the thermal conductance of the tissues during long-term adaptations to cold or heat is both sparse and conflicting. Thus, the skin temperature of the trunk and extremities is higher in cold-acclimated than in heat-acclimated pigeons (*Columba livia*). This difference may be attributed to an increased peripheral blood flow in the cold-acclimated birds, which would lead to an increase in the thermal conductance of the tissues. However, the deposition of fat was greater in pigeons acclimated to an air temperature of 10°C than in those acclimated to 29°C. A decrease in the thermal conductance of the tissues of the legs of cold-acclimatized ring-necked pheasants (*Phasianus colchicus*) appeared to be brought about by the operation of the countercurrent heat-exchange mechanism (see previous section on countercurrent heat exchange). Winter fattening is a common phenomenon in small birds that winter in cold climates; however, see section on heat production above.

Heat Loss

Surface Area:Volume Ratio

Compared with large birds, small birds have a large area in relation to their volume. The same is true for a small appendage, such as a leg, as opposed to the torso of the bird. The larger the ratio of surface area to volume, the greater the heat loss, simply because there is a greater surface area from which to lose heat. This is why small birds in a very cold environment face an enormous task in preserving their body temperature.

Nonevaporative Heat Loss

Heat transported from the tissue where it is produced to the surface of the body may be lost from the surface to the environment by both evaporative and nonevaporative means. The latter, as its name implies, does not involve the evaporation of water. Nonevaporative heat loss, or "sensible" or "dry" heat loss as it is sometimes called, may occur by three distinct processes.

Radiation (*R*). Heat, in the form of electromagnetic waves, is transferred from the body to surfaces in the environment that are at temperatures lower than those of the skin or surface of the plumage. The intervening air is not involved. Of course, if the temperature of the surroundings is higher than that of the skin or feathers of the bird, the bird gains heat by radiation. This frequently occurs, particularly under desert conditions, and it should be borne in mind that whenever a bird is exposed to direct sunlight there is a radiant heat gain that has to be offset against any heat loss by radiation to other surfaces in the environment. The net radiant heat exchange (*R,* in W) between a bird and its environment may be expressed by the following equation:

$$R = \sigma \epsilon_s \epsilon_r (\overline{T}_s^4 - \overline{T}_r^4)A$$

where σ is a constant (the Stefan–Boltzmann constant) with a value of 5.67×10^{-8} W/m$^2 \cdot K^4$; ϵ_s is the emissivity of the surface of the bird, i.e., the ratio of the actual emission of heat from the surface to that from a "perfect black body," or "full radiator" as it is also known, at the same temperature; ϵ_r is the emissivity of the environmental surfaces; $\overline{T}_s$ is the mean surface temperature of the bird (*K*); $\overline{T}_r$ is the mean radiant temperature of the environment (*K*); and A is the effective radiating area (m^2).

The emissivity of the bare skin and feathers of most birds is very close to that of a perfect black body with respect to radiation of long wavelengths (>2.5 μm). This means that they are able to radiate heat efficiently to the environment, because the surface of the skin and plumage of birds at their usual temperatures in the biological range emits radiation with a wavelength of approximately 10 μm. However, a surface with a high emissivity is also a good absorber (it is said to have a high absorptance) of radiation. Consequently, the skin and plumage of birds absorb a great deal of the longwave radiation emitted by surfaces (ground, rocks, vegetation, etc.) in their environment. The net exchange between birds and their environment will depend on the relative temperatures of the skin/plumage surfaces and surfaces in the environment, as the above equation indicates. Birds vary a great deal in the emissivity of their plumage to *solar* radiation (wavelengths ~ 0.3–2.5 μm) depending on the color of the plumage, among other factors. The differences between birds in this respect are largely due to the extent to which their plumage reflects the visible part of the spectrum (0.3–0.7 μm) of solar radiation (Figure 9–2). The reflectance is equal to $1 - \epsilon$; in other words, the higher the emissivity of a surface, the lower the reflectance. A high reflectance to solar radiation reduces the amount of heat gained by the bird from its environment, and thus mitigates the effects of a hot environment. Thus, it has been shown by Lustick et al. (1970) that in a red-wing blackbird (*Aguelaius phoeniceus*) sitting in the sun at an air temperature above its lower critical temperature (30°C), there is a net heat flow into the bird. The reflectance of a blackbird's plumage to solar radiation is low.

Although it is intuitively obvious that, other things being equal, more solar radiation should be absorbed by a dark plumage than by a light-colored one, measurements of heat transfer reveal that the situation is considerably more complex. Thus, Walsberg et al. (1978) found that at low wind speeds, the absorptance of radiative heat by the plumage was greater for black than for white plumage. The temperatures in the superficial layers of the plumage were also much higher in

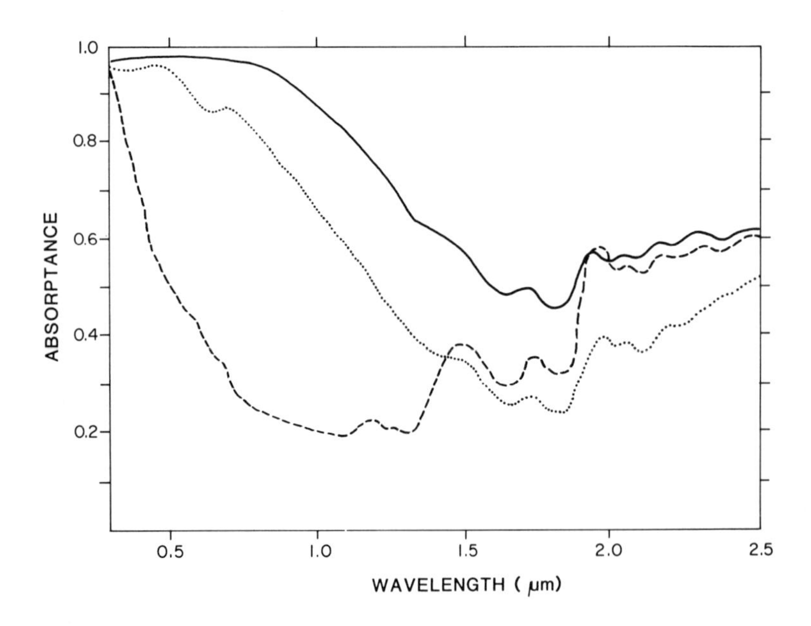

FIGURE 9–2. Absorptance of radiation of different wavelengths by the dorsal surface of the plumage of the black Muscovy duck (*Cairina moschata;* ———), the red cardinal (*Richmondena cardinalis;* ....) and the white whistling swan (*Cygnus columbianus;* ----). (After Gates, 1980.)

the dark plumage. However, at higher wind speeds the difference between the two plumages was greatly reduced, the heat absorbed by the dark plumage being dissipated by convection. This effect was not so pronounced in white plumage because the penetration of radiant heat was greater, probably as a result of greater reflection into the deeper layers of plumage. Consequently, the highest temperature occurred in a deeper layer of the plumage where it was less affected by convective cooling. When the plumage was erected, the amount of radiant heat reaching the skin was less because of the resistance offered by the thicker layer of plumage. Moreover, at wind speeds in excess of 3 m/sec, black plumage actually acquired less radiant heat than did white plumage. These observations may explain why the gray gull, nesting in the Atacama Desert of Chile, fluffs out its feathers during the hottest part of the day when there is a strong breeze (Howell et al., 1974). The price that a bird pays for having plumage of high absorptance is that, at high air temperatures combined with solar radiation, evaporative water loss is increased (Wunder, 1979).

On the other hand, in a cold environment, the heat gained from solar radiation reduces the amount of heat that the bird has to produce in order to maintain its body temperature (De Jong, 1976). Consequently, a bird with plumage that has a high absorptance (low reflectance) for solar radiation saves energy in a cold environment (Lustick, 1969; Lustick et al., 1979). In fact, it has been found that the emissivity (absorptance) of the owl's plumage was greater in the winter than during the summer.

Convection (C). Heat loss by convection involves the actual movement of molecules of the air. Air in contact with the skin warms, becomes less dense and rises, being replaced by cooler denser air. This is "natural convection." If the bird is exposed to moving air, or if the bird itself is actually moving through the air, considerably more heat may be lost. Provided that the air velocity is not too high, such "forced convection" is roughly proportional to the square root of the air velocity. Convective heat loss (C; W/m^2) may be expressed simply as follows:

$$C = h_c(\overline{T}_s - T_a)$$

where h_c is the convective heat transfer coefficient ($W/m^2 \cdot °C$), incorporating dimensionless numbers describing the flow and thermal properties of the air and the size and shape of the animal. $\overline{T}_s$ is the mean surface temperature and T_a is the air temperature.

Conduction (K). Heat loss by conduction involves the transfer of energy from molecule to molecule but, unlike convection, there is no actual gross translocation of molecules. The equation for heat transfer by conduction (K, in W/m^2) is

$$K = h_k(\overline{T}_s \quad T_a)$$

where h_k is the conductive heat transfer coefficient ($W/m^2 \cdot °C$). The conductive heat transfer coefficient depends on the thermal conductivity of the medium to which heat is being lost, i.e., air, nest materials, or, for some birds, water. It depends also on the thickness of the nest or of the "boundary" layer of air or water adjacent to the skin or feathers. The thickness of the boundary layer in turn varies with the roughness and shape of the animal's surface as well as with the velocity of the air or water movement adjacent to the animals and the size of the bird.

Respiratory Nonevaporative Heat Loss. A special instance of nonevaporative heat loss is the warming of inspired air to body temperature by a process of convection and conduction in the upper respiratory tract. The warming of the air as it enters the nasal passage removes heat from the nasal mucosa, the temperature of which falls. When the air is exhaled, cooling of the air occurs in the nasal passages so that some of the heat is regained by the mucosa. This is another example of a countercurrent heat exchange system, and it has important implications as far as respiratory evaporative heat loss is concerned (see section on respiratory evaporative heat loss below). The temperature of the exhaled air depends on the surface area and width of the nasal passages, and different birds vary significantly in this respect. In penguins kept at an air temperature of 5°C, the temperature of the exhaled air was 9°C, and 8.2% of the heat added to the inhaled air was reclaimed (Murrish, 1973). The following equation describes respiratory nonevaporative heat loss (H_{resp}, in W) in birds:

$$H_{resp} = \dot{V}_E \rho c_p (T_{ex} - T_{in})$$

where $\dot{V}_E$ is the respiratory minute volume (liters/sec), ρ is the density of air (g/liter), c_p is the specific heat of air ($J/g \cdot °C$), T_{ex} is the temperature of expired air (°C), and T_{in} is the temperature of inspired air (°C).

Thermal Conductance of the Plumage

The plumage of birds provides a very effective barrier to heat loss from the skin surface to the surrounding air. The down feathers trap air in which little convective movement occurs, and the distal parts of the contour feathers provide a windproof covering. The feathers are covered with a thin layer of oil secreted by the preen gland, and the spaces between the finest divisions of the feather structure are extremely small. Both factors render the plumage of birds relatively resistant to wetting. The thermal conductance of the plumage ($C_{plumage}$) may be calculated from the following equation:

$$C_{plumage} = H/(\overline{T}_{sk} - T_a) \quad (W/m^2 \cdot °C)$$

in which H is the nonevaporative heat transfer from the skin surface to the air surrounding the bird (W/m^2); $\overline{T}_{sk}$ = mean skin temperature under the plumage (°C) and T_a = air temperature (°C). This equation is analogous to that for the thermal conductance of the tissues presented above, and its converse (1/$C_{plumage}$) is the insulation of the feathers and entrapped air. The thermal conductance is, in fact, a heat transfer coefficient, but unlike the heat transfer coefficients for conduction and convection given above, it does not distinguish between the different pathways of heat transfer. Rather, it combines the heat transfer coefficients into a single measurable figure. The thermal conductance of the plumage is lower in large than in small birds. This reflects the fact that large birds are able to carry a heavier load of feathers than are small birds. In addition, the radial distribution of feathers around the torso and extremities of a small bird results in a less compact insulation than is the case in large birds. Furthermore, heat is transferred more rapidly through a structure with a small radius of curvature (see section on thermal conductance of the body tissues above).

The thickness of the plumage, and therefore its thermal conductance, may be varied by the activity of the ptilomotor nerves, which supply the arrectores plumorum muscles. These are smooth muscles and the available evidence suggests that their innervation may be cholinergic.

Birds fluff out their feathers during exposure to cold. This increases the insulation provided by the feathers and decreases the thermal conductance of the plumage; when eastern house wrens (*Trogolodytes aedon aedon*) were prevented from fluffing their feathers, their body temperature decreased more rapidly than did that of control birds. The ambient temperature at which the feathers of the Barbary dove (*Streptopelia risoria*) are fully raised is higher in birds deprived of food or water, presumably because their heat production is below normal. The lower critical temperature (see section on heat production above) of defeathered California quail is higher than that of birds with plumage, reflecting a diminution in insulation. The lower critical temperature of emperor penguins is also higher during molting (Le Maho, 1983). In fowls with poor or scant plumage, the level of heat production was higher than in controls; this effect could also be produced in normal fowls by removing the plumage. The higher heat production was probably related to the poor insulation in these birds. Hohtola et al. (1980) have shown that pigeons fluff out their feathers when they sleep at low air temperatures, rather than increase their heat production. Hill et al. (1980) obtained infrared radiographs from the plumage surface of black-capped chickadees (*Parus atricapillus*) exposed to air temperatures from −22°C to 27°C. The highest surface temperatures were recorded from the head and breast, and it was shown that the thermal conductance of the plumage of the breast was considerably less than that of the underlying tissues.

Penguins present a special case because they swim in cold water and often dive to considerable depths. They depend largely on the insulation of their plumage, rather than their subcutaneous fat, to prevent excessive heat loss to the water. During a dive, when the feathers are compressed, there is a substantial increase in the thermal conductance of the plumage (Kooyman et al., 1976). In air, also, the plumage provides the major part of the insulation of penguins (Stahel and Nicol, 1982; Le Maho, 1983).

Some birds elevate certain of their feathers during exposure to solar radiation (Figure 9–3). The raised scapulars of the great frigatebird (*Fregata minor*) illus-

FIGURE 9–3. A great frigatebird (*Fregata minor*) shades its chick and gular flutters at French Frigate Shoals in the Northwestern Hawaiian Islands. Note the raised scapulars.

trate this response; its purpose appears to be to permit a better circulation of air through the plumage and thus to enhance heat loss. Elevation of the plumage during exposure to heat is discussed above, in the section dealing with heat loss by radiation.

The weight of the plumage may be reduced by acclimating domestic fowl to a warm environment, and, conversely, the mass of the winter plumage may be about 30% greater than that of the summer plumage, in the English sparrow.

Total Thermal Conductance

The total thermal conductance (C_{total}) is a measure of the ease with which heat flows from the body core, through the tissues, plumage, and the boundary layer of air immediately contiguous to the outer surface of the plumage and skin, to the surrounding air. More information is available on the total thermal conductance of birds than on the separate thermal conductances of the tissues and plumage, because the former is simpler to compute. It may be derived from the following equation:

$$C_{total} = (H \pm S - E)/T_b - T_a \quad (W/m^2 \cdot {}^\circ C)$$

in which H = heat production (W/m^2); S = heat storage (W/m^2); E = total evaporative heat loss (W/m^2); T_b = deep-body temperature (°C), and T_a = air temperature (°C). Because the evaporative heat loss is subtracted from the metabolic heat production, the total thermal conductance is sometimes referred to as the "dry" conductance. The total thermal conductance is a combined heat transfer coefficient for the entire body of the bird, and its converse ($1/C_{total}$) is the total thermal insulation of the bird. The total thermal conductance may also be determined, with suitable precautions, from the slope of the increase in heat production below the lower critical temperature (McNab, 1980; see also section on heat production above). Calder and King (1974) have presented equations for the relationship between body mass and total thermal conductance; the thermal conductance of small birds is greater than that of large birds. Hennemann (1983) reported that the thermal conductance of two species of Pelecaniformes was less at night than during the day. Birds can vary their total thermal conductance two- to five-fold over a range of environmental temperatures.

The total thermal conductance of arctic birds is lower than that of tropical species, and the thermal conductance of the burrowing owl (*Speotyto cunicularia*) acclimatized to winter conditions is lower than that of summer-adapted birds. The owls appeared to have more extensive plumage in winter, and this probably contributed in large part to their lower thermal conductance. In the black grouse, the thermal conductance was actually higher in winter than in summer, at least within the thermoneutral zone. However, at lower ambient temperatures the thermal conductance decreased in winter but remained constant in summer (Rintamaki et al., 1983). The winter-acclimatized mute swan (*Cygnus olor*) responded to cold largely by reducing its thermal conductance, whereas summer-acclimatized birds relied more on an increase in heat production (Bech, 1980).

Changes in the photoperiod are believed to play an important part in seasonal variations in insulation in birds (Saarela and Vakkuri, 1982).

Evaporative Heat Loss (E)

Total (*E*). Evaporation of moisture occurs from the surface of the skin and from the respiratory tract. Total evaporative heat loss (E) may be represented as follows:

$$E = E_{ex} + E_{sw}$$

where E_{ex} is evaporative heat loss from the respiratory tract and E_{sw} is evaporative heat loss from the skin.

The partition of evaporative heat loss into its respiratory and cutaneous components has not been achieved in many birds. The technique is discussed in a later section. Total evaporative water loss may be measured simply by recording the mass loss of the animal over an accurately measured period of time, making appropriate corrections for the loss in mass contingent on respiratory gas exchange (oxygen, carbon dioxide). The total evaporative heat loss is calculated by multiplying the corrected mass loss by the latent heat of vaporization:

$$E = \dot{m}\,\lambda$$

where $\dot{m}$ is mass loss (g/hr) and λ is the latent heat of vaporization (J/g). The latent heat of vaporization is the heat that must be removed from the body in order to evaporate 1 g of water. The value varies with temperature. An approximate value is 2,427 J/g H_2O.

The importance of evaporative heat loss is that when the air temperature equals the body temperature of the bird, heat can be lost only by evaporation of moisture.

Cutaneous (E_{sw}). Cutaneous evaporative water loss may be measured by enclosing the bird in a chamber containing a flexible partition through which the bird's head is inserted. The purpose of this is to separate respiratory from cutaneous moisture loss. Both compartments within the chamber are ventilated with air at a known flow rate, and the humidity of the air entering and leaving the two compartments is recorded. This method also permits the quantitative evaluation of respiratory evaporative water loss.

The transfer of heat from the skin to the air by evaporation (E_{sw}, in W/m^2) is described by the following equation:

$$E_{sw} = h_e (\phi_{sk} P_{ws} - \phi_a P_{wa}) \lambda$$

where h_e is the evaporative heat transfer coefficient, a function of air movement, viscosity, density,the diffusivity of water vapor, and the dimensions of the body (g/sec · m² · mm Hg); ϕ_{sk} is the relative humidity at the skin surface (%); ϕ_a is the relative humidity of the air (%); P_{ws} is the saturated aqueous vapor pressure of skin at T_{sk} (mm Hg); P_{wa} is the saturated aqueous vapor pressure of air at T_a (mm Hg); and λ is the latent heat of vaporization (J/g; see above).

Neither sweat glands nor sebaceous glands are present in the skin of birds. The amount of moisture that may be lost from the skin under hot conditions is therefore limited. Nevertheless, under certain conditions the cutaneous water loss may exceed that from the respiratory tract, and evidence is available for a significant water loss from the skin of several species (Bouverot et al., 1974). Cutaneous evaporative water loss in the zebra finch (*Taeniopygia castanotis*) was reduced by depriving the birds of drinking water.

Respiratory (E_{ex}). Respiratory evaporative heat loss (E_{ex}) in W may be represented as follows:

$$E_{ex} = \dot{V}(\rho_{ex} - \phi \rho_{in}) \lambda$$

where $\dot{V}$ is respiratory minute volume (liter/sec), ρ_{ex} is grams of water per liter of air saturated with water vapor at the temperature of expired air, ϕ_a is the relative humidity of inspired (ambient) air, ρ_{in} is grams of water per liter of air saturated at ambient air temperature, and λ is the latent heat of vaporization of water (J/g). Respiratory evaporative cooling is extremely important in birds, and almost without exception, birds exhibit some form of panting.

Thermal Polypnea (Thermal Tachypnea). Thermal polypnea involves an increase in the respiratory minute volume; this leads to an increase in respiratory evaporative heat loss. The increased respiratory minute volume is brought about by an increase in respiratory frequency, while the tidal volume decreases. This particular pattern of respiration permits a maximal increase in respiratory minute volume and respiratory evaporative cooling with minimal disturbance of the blood gases, because the increased ventilation is limited to the respiratory dead space, in which gas exchange between blood and air does not occur. However, as the body temperature of the bird increases to high levels, the respiratory frequency reaches a maximal value and subsequently declines. This change in respiratory frequency is accompanied by an increase in tidal volume, while the minute volume increases further. At very high temperatures, the respiratory minute volume declines also. This change in the pattern of breathing is characteristic of many hyperthermic panting animals, including mammals. It represents the breakdown of thermal polypnea.

The extensive air-sac system of birds would seem to endow them with an unsurpassed means of directing the increased ventilation during panting to parts of the respiratory tract that do not participate in the exchange of gas between blood and air while at the same time permitting evaporative heat loss to occur. The ostrich appears to have taken advantage of this anatomic arrangement, and there is evidence that panting can occur in a number of species of birds without impairment of the blood gases and acid–base balance (see Bernstein and Samaniego, 1981). This is particularly true under conditions of mild or moderate heat stress. The mechanism by which this is achieved is not entirely understood. Bech and Johansen (1980) identified three different panting responses in birds, all resulting in a reduced parabronchial ventilation associated with an increase in the total ventilation. In many species (the mute swan studied by Bech and Johansen is an example), the tidal volume decreases to a level close to the respiratory dead space. In some birds, e.g., the pigeon, panting is superimposed on slower deeper breathing, which Ramirez and Bernstein (1976) called "compound ventilation." They presented evidence that this type of respiration is less likely to lead to a respiratory alkalosis than is simple panting. Thus, shallow rapid breathing, ventilating largely the respiratory dead space, accomplishes an increased loss of heat while gas exchange is achieved by the slower, deeper parabronchial component. A third pattern, illustrated by the greater flamingo (*Phoenicopterus ruber*), involves shallow breathing at tidal volumes less than the dead space. At regular intervals this rhythm is interrupted by a short sequence of deeper breaths ("flushouts"; Figure 9–4), which provide parabronchial ventilation (Bech et al., 1979). A different strategy is adopted by penguin chicks: panting results in a diminution in the partial pressure of carbon dioxide in the blood, but the alkalosis that might have resulted is compensated for by an increase in blood lactic acid so that blood pH remained unchanged (Murrish, 1983). However, the relationship between panting and the blood gases is complicated by the fact that in the domestic fowl there is evidence that maximal panting frequencies are attained only when some degree of alkalosis develops (Brackenbury, 1978; Mather et al., 1980; Barnas et al., 1981).

The initiation of thermal polynea varies in character in different birds. In some birds, e.g., the domestic fowl and double-crested cormorant (*Phalacracorax auritus*), the respiratory frequency increases steadily with increasing heat load. In other birds, e.g., the ostrich and the roadrunner (*Geococcyx californianus*), the respiratory frequency changes in response to heat from the normal low rate to a high value that does not change thereafter. This value matches the resonant frequency of the entire respiratory system in many birds. At this frequency the energy cost of panting is low, amounting to that required to keep the respiratory system oscillat-

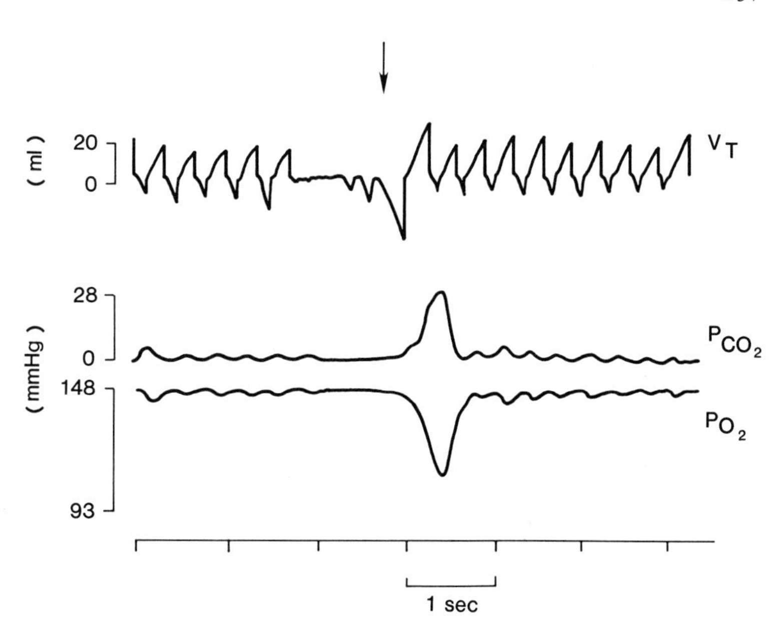

FIGURE 9–4. Tidal volume (V_T) and the partial pressure of carbon dioxide (P_{CO_2}) and oxygen (P_{O_2}) in the air from a flow-through respirometer, in the flamingo, during exposure to heat. The horizontal axis is a time scale. Rapid, shallow panting is interrupted by a single "flush-out" at the arrow. (Redrawn from Bech et al., 1979.)

ing at its natural frequency. If the frequency of thermal polypnea is constant in these birds, then they can vary evaporative cooling only by intermittent operation of the panting mechanism or by modulation of tidal volume. The energy cost of panting, however, can be substantial in some birds. It has been estimated that 40% of the heat dissipated by panting in the cardinal (*Richmondena cardinalis*) merely offsets the heat produced by the respiratory muscles.

The amount of heat that birds may lose by respiratory evaporative cooling is considerable, many birds being quite capable of dissipating all the heat that they produce by respiratory evaporation in a hot environment (Table 9–3; Figure 9–1; Dawson, 1982). Not all birds have effective mechanisms of evaporative cooling, however. Notable exceptions are two species of Hawaiian honeycreepers, which are comparatively heat intolerant.

Gular Flutter. Movements of the buccopharyngeal or gular area during exposure to heat occur in most, if not all, birds. In some species, however, ventilation

TABLE 9–3. Evaporative heat loss (E), expressed as a percentage of heat production, at high air temperatures (T_a)[a]

Species	Body mass (kg)	T_a (°C)	E (%)
Ostrich (*Struthio camelus*)	100.0	44.5	100
Bedouin fowl (*Gallus domesticus*)	1.427	48	159
Brown-necked raven (*Corvus corax ruficollis*)	0.610	50	167
Rock dove (*Columba livia*)	0.315	44.5	118
Roadrunner (*Geococcyx californianus*)	0.285	44.5	137
Galah (*Cacatua roseicapilla*)	0.271	>45	170
Burrowing owl (*Speotyto cunicularia*)	0.143	44.1	95
Japanese quail (*Coturnix coturnix*)	0.100	43	144
Common nighthawk (*Chordeiles minor*)	0.075	43.5	148
Speckled coly (*Colius striatus*)	0.044	44	99
Painted quail (*Excalfactoria chinensis*)	0.043	43.5	116
Inca dove (*Scardafella inca*)	0.042	43.5	108
Poorwill (*Phalaenoptilus nuttallii*)	0.040	43.5	175
House sparrow (*Passer domesticus*)	0.025	44.5	106
House finch (*Carpodacus mexicanus*)	0.020	44.5	130
Gouldian finch (*Poephila gouldiae*)	0.014	44.5	105
Zebra finch (*Taeniopygia castanotis*)	0.012	43.5	123
Costa's hummingbird (*Calypte costae*)	0.003	40	66

[a] From Whittow (1976), with additional information from Marder (1973), Weathers and Schoenbaechler (1976), and Dawson and Fisher (1982).

of the gular area is conspicuous and results in a significant loss of heat by evaporation from the gular mucous membranes. The gular flutter is actuated by flexure of the hyoid apparatus, and the gular area becomes conspicuously suffused with blood. The advantage of gular flutter over thermal polypnea is twofold. First, the air movement is restricted to surfaces that do not participate in gas exchange so that the dangers of hyperventilation are circumvented. Second, the energy cost of moving the gular area is considerably less than that of moving the larger thoracic cavity. Consequently, the heat loss contingent on gular flutter is not offset to any large extent by the heat produced in the muscles that move the hyoid apparatus.

Especially proficient at gular fluttering are birds such as the caprimulgids, which have a large gular area. The poorwill (*Phalaenoptilus nuttallii*) and the Australian spotted nightjar (*Eurostopodus guttatus*) are able to lose heat in excess of three times the amount of heat that they produce, at high air temperatures. This is facilitated in the poorwill by an unusually low heat production. Weathers and Schoenbaechler (1976) reported that gular flutter was responsible for 20% of the total evaporative water loss in the Japanese quail. In the domestic fowl, buccopharyngeal ventilation contributed up to 35% of the total water loss from the respiratory tract during exposure to heat (Brackenbury et al., 1981a).

In some species, the rate of gular flutter increases with increasing heat load; in others, the flutter frequency remains constant. In the latter instance, the contribution of gular flutter to evaporative cooling is augmented by an increase in the duration of episodes of gular flutter, by increasing the amplitude of the flutter movements or by increasing the area of the gular region involved.

In birds that demonstrate both overt gular flutter and thermal polypnea, the frequencies of the two movements are not necessarily the same. The frequency of gular flutter is usually higher than that of thermal polypnea, in keeping with the smaller mass of the gular–hyoid structure. It seems likely that gular flutter occurs at the resonant frequency of the gular region, and in several species gular flutter commences before thermal polypnea. However, this was not true for the pigeon or barn owl (*Tyto alba*).

Site of Evaporative Cooling. The site of evaporative cooling during gular flutter is clearly the buccopharyngeal region. Menaum and Richards (1975) have shown that evaporative cooling occurs mainly from the upper respiratory tract during thermal polypnea also, at least in the domestic fowl (Figure 9–5). In the Peking duck, evidence was obtained that the blood flow to the upper respiratory tract increased during thermal panting (Bech and Johansen, 1980).

Dehydration. Respiratory frequency and evaporation were less in the dehydrated ostrich during exposure to heat than in the hydrated ostrich. Pulmocutaneous water loss was also reduced in the zebra finch when the birds were exposed to heat in the dehydrated state.

Acclimation, Acclimatization. The respiratory frequency of hens acclimated to an air temperature of 31°C was significantly higher than that of birds acclimated to 0°C, as was the rectal temperature. The respiratory frequency may remain elevated in heat-acclimated birds after the deep-body temperature has returned to control levels, suggesting that the increased respiratory frequency was not necessarily related to an increased deep-body temperature. In pigeons, there was no difference in the threshold of mean body temperature for thermal polypnea between cold- and warm-acclimated birds. Evaporative water loss was higher in burrowing owls during the summer than in winter, and in a desert breed of domestic fowl the respiratory frequency was higher than in breeds adapted to cooler climates (Arad and Marder, 1982).

Respiratory Water and Heat Conservation. Murrish (1973) calculated that 81.7% of the heat required to humidify the air in the nasal passages of the penguin was recovered on exhalation at an air temperature of 5°C. The total amount of heat (evaporative and nonevaporative; see previous section on respiratory nonevaporative heat loss) recovered in the nasal passages represented 17% of the metabolic heat production. Brent et al. (1984) showed that the respiratory heat loss from the European coot (*Fulica atra*) at an air temperature of −25°C was only 9.6% of the theoretical maximum, largely as a result of cooling of the expired air.

Partition of Heat Loss

Using a gradient-layer calorimeter, which partitions heat loss from an animal into the separate components of convection, radiation, and evaporation, several investigators have measured the heat loss from domestic fowl at different environmental temperatures. At low environmental temperatures, most of the heat produced is lost by nonevaporative means; convection or radiation are the major pathways, depending on the rate of air movement. As the air temperature increases, the proportion of heat lost by evaporation increases; at an air temperature of 35°C, evaporative heat loss may amount to almost all the heat loss (see also Table 9–3). Attempts have been made also to partition heat exchange in birds in a natural situation out of doors. Thus, in the broad-tailed hummingbird (*Selasphorus platycercus*) sitting on its nest, it was estimated that radiative heat loss was 8.7–35% of the total heat loss, whereas convective heat loss amounted to 43.5–46%, at air temperatures of 0–4.6°C. Several other attempts have been made to partition the heat exchange between nesting hummingbirds and their natural envi-

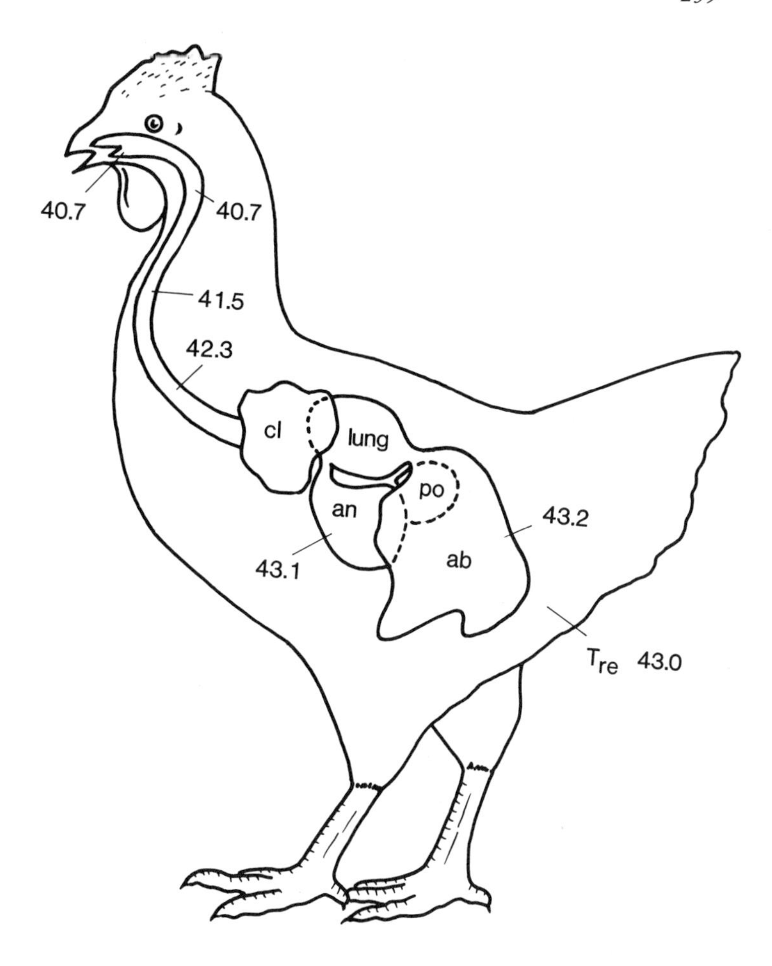

FIGURE 9–5. Respiratory surface temperatures (°C) of the domestic fowl during thermal panting. The air temperature was 43.0°C and the respiratory frequency was 141 breaths/min. Abbreviations: *cl*, clavicular; *an*, anterior thoracic; *po*, posterior thoracic; *ab*, abdominal air sacs; T_{re}, rectal temperature (°C). (Redrawn from Menaum and Richards, 1975.)

ronment (Smith et al., 1974; Southwick and Gates, 1975).

Heat Exchange under Natural Conditions

In recent years, the focus of attention of thermal physiologists has shifted from the indoor laboratory to the natural environment out-of-doors. The natural environment presents an immensely more complicated situation, including variations in air temperature, humidity, radiant heat, and air movement. While air temperature is the only variable that is usually changed in a laboratory experiment, changes in air temperature alone are clearly inadequate to describe the outdoor environment, because changes in the amount of solar radiation or air velocity may greatly exceed variations in air temperature. Recently, the "standard operative temperature" has been used as an index of sensible heat flow between the bird and its natural environment (Bakken, 1980). It may be measured by the use of a "taxidermic mount" (Bakken et al., 1981). This is a hollow model of the bird in question, covered with the bird's plumage. It has the same shape, size, color, and plumage characteristics of the species under study.

There is no doubt that periods of severe weather may increase the mortality of birds, and it is equally certain that birds may mitigate the effects of such episodes by the judicious selection of roosts with favorable microclimates (Lustick, 1983). Thus, the net radiative heat loss of three species of birds in a pine roost providing 75–95% cover was reduced 35–54% over that in the open (Lustick, 1983). On clear nights, at −10°C in the open, heat loss by radiation and convection exceeded the maximal capacities of the three species to produce heat. The birds switched from a deciduous roost to a pine roost in the winter and varied the height of their roost depending on weather conditions, roosting well below the canopy on cold windy nights.

Behavioral Thermoregulation

Thermoregulatory behavior usually involves movement of the entire bird, or part of the bird, such as a limb, in response to a change of either environmental or body temperature, and it often requires conscious effort. The most conspicuous thermoregulatory behavior of birds is migration to warmer or cooler areas.

In a cold environment the domestic fowl reduces its surface area, and hence its heat loss, by "hunching." An additional reduction in heat loss, amounting to 12% in the fowl, may be achieved by tucking the head under a wing. If the bird chooses to sit, it can reduce heat loss from the unfeathered legs and feet, a saving of 20–50% compared with the standing position. Penguins, in addition to squatting, rest only their tarsometatarsal joints on the ground, thus minimizing conductive cooling to the ice. Some birds—the anhinga is an example—spread their wings in order to dry their plumage and increase their heat gain by solar radiation, in this way supplementing their low rate of metabolic heat production (Hennemann, 1983). There are other examples of such "sunning" behavior, but, unfortunately, some observers have confused sunning with the spreading of the wings that accompanies heat stress in many birds.

Huddling is an effective means of reducing heat loss, and although it is not widespread among adult birds, it is extremely important in emperor penguins (Le Maho, 1983; Le Maho et al., 1976; Pinshow et al., 1976), which are exposed to a severe climate compounded by the fact that they fast when ashore. Birds make less use of burrows and shelters as a protection against cold than do mammals. One species that does so is the capercaillie (*Tetrao urogallus*), which utilizes burrows in the snow to protect itself from the rigors of the climate. The thermal conditions in the burrow were close to, or within, the thermoneutral zone of the birds (Marjakangas et al., 1984). Under laboratory conditions, it was possible to train Barbary doves to press a switch in order to turn on a heater. In this way, the birds were able to control their environmental temperature, and it is interesting that they maintained their environmental temperature at approximately 33°C, which is within the thermoneutral zone. However, the domestic fowl could not be trained to operate a source of heat in a cold environment (−5°C), relying rather on other manifestations of thermoregulatory behavior including fluffing of the feathers, the covering of unfeathered areas, and tucking the head under the wing, in addition to physiological measures such as an increase in heat production (Horowitz et al., 1978).

In response to a hot environment, birds may soar to avoid intense desert heat, seek shade if available, or reduce activity in the hottest part of the day. Many tropical seabirds face extremely demanding conditions during the nesting season (Figure 9–3). Air temperatures are high, solar radiation is intense, and there is no shade. The birds supplement their physiological responses by a number of behavioral adjustments. For example, the masked booby (*Sula dactylatra*) orients its body so that its back is to the sun, thereby placing its feet and gular area in the shade of its body. In addition, the wings are held away from the body and the scapular feathers are elevated, both responses facilitating convective heat loss to the air. The tropical albatross (*Diomedea nigripes* and *D. immutabilis*) pivot on their heels, raising their large webbed feet off the ground while orienting their bodies so that the feet are shaded (Howell and Bartholomew, 1961; Whittow, 1980). The herring gull (*Larus argentatus*) is able to reduce its solar heat gain by changing its orientation and posture with regard to the sun. In this way white reflective plumage is exposed to the sun and the area exposed is also reduced (Lustick et al., 1978; Hailman, 1982). The wood stork, when hot, resorts to the bizarre expedient of directing its liquid excrement onto its long legs, which are cooled by evaporation of the water in the excrement.

The domestic fowl achieves a similar end result, in a more socially acceptable way, by splashing water over its comb and wattles. In addition, the domestic fowl will actuate a source of cold air, in a hot environment, in contrast to its failure to operate a device for heat (see above). This operant behavior preceded physiological responses to heat (Horowitz et al., 1978) and suggests that, in this species, behavioral and physiological responses to heat and cold are evoked in different sequences. Some Charadriiformes, e.g., the black-necked stilt (*Himantopus mexicanus*), which nests under extremely hot conditions near the Salton Sea in California, achieve the dual purpose of cooling themselves and their eggs by wetting their ventral feathers ("belly soaking") (Grant, 1982). The Egyptian plover (*Pluvianus aegyptius*) buries its eggs—and its chicks—in the sand, and while this helps to conceal them, it also has thermoregulatory advantages. These are enhanced by the adult birds, which belly-soak and in this way wet the sand at the nest site, cooling the eggs or chicks in the process (Howell, 1979).

Control Mechanisms

There is evidence that both peripheral temperature receptors and temperature-sensitive neurons in the central nervous system are involved in the regulation of body temperature, heat production, and heat loss in birds.

When the domestic fowl is exposed to a cold environment, shivering can be detected before any change occurs in the deep-body temperature. This observation is consistent with the control of heat production by thermal receptors in the periphery. Peripheral cold receptors in birds have been identified in the tongue, the beak, and the bill (see Dawson, 1975), but there may well be others elsewhere. In fact, Necker (1977) has shown in the pigeon that feathered skin areas, particularly on the back, are very sensitive to changes of temperature. Their sensitivity was assessed in terms of the responses evoked by heating and cooling the skin areas. The beak and unfeathered areas of the feet were relatively insensitive.

There is also good evidence that the heat production may be influenced by changes in the temperature of thermoreceptors in the central nervous system, i.e., by changes of deep-body temperature. For example, localized heating of the anterior hypothalamus–preoptic region of the brain of the house sparrow results in a diminution of heat production, whereas cooling the same region has the opposite effect. Bilateral lesions in the anterior hypothalamus–preoptic region significantly impaired the maintenance of body temperature in a cold environment. In the domestic fowl, lesions of the anterior hypothalamus abolished the shivering response. However, shivering could not be elicited by localized cooling of the anterior hypothalamus of the pigeon, and there was no increase in oxygen consumption. On the other hand, localized cooling of the spinal cord resulted in both shivering and an increased oxygen consumption Figure 9–6).

There is evidence that heat production in the pigeon is regulated by a "proportional controller," a type of control system that implies that the magnitude of the effector response (in this instance, heat production) is proportional to the extent to which body temperature (spinal cord temperature, in the pigeon) deviates from a certain value, the "set point." In the pigeon, the set point varied with changes in skin temperature. There is also a circadian variation of spinal thermosensitivity that varies with the state of arousal, an increase in the heat production of the pigeon occurring at a lower spinal temperature when the bird is asleep and at night (Graf et al., 1983). There appear to be temperature-sensitive neurons in the spinal cord itself (Necker and Rautenberg, 1975). In the Adelie penguin (*Pygoscelis adeliae*) and in the Peking duck there did not appear to be any specific cold receptors in the hypothalamus, but there were warm receptors, albeit with rather different characteristics than those shown by mammalian hypothalamic warm receptors. Localized cooling of the lower mesencephalic and rostral rhombencephalic regions of the brainstem, however, elicited cutaneous vasoconstriction and an increase in metabolic heat production in the Peking duck (Simon et al., 1976; Simon-Oppermann et al., 1978; Simon-Oppermann and Martin, 1979; Eissel and Simon, 1980). Snapp et al. (1977) concluded that, in

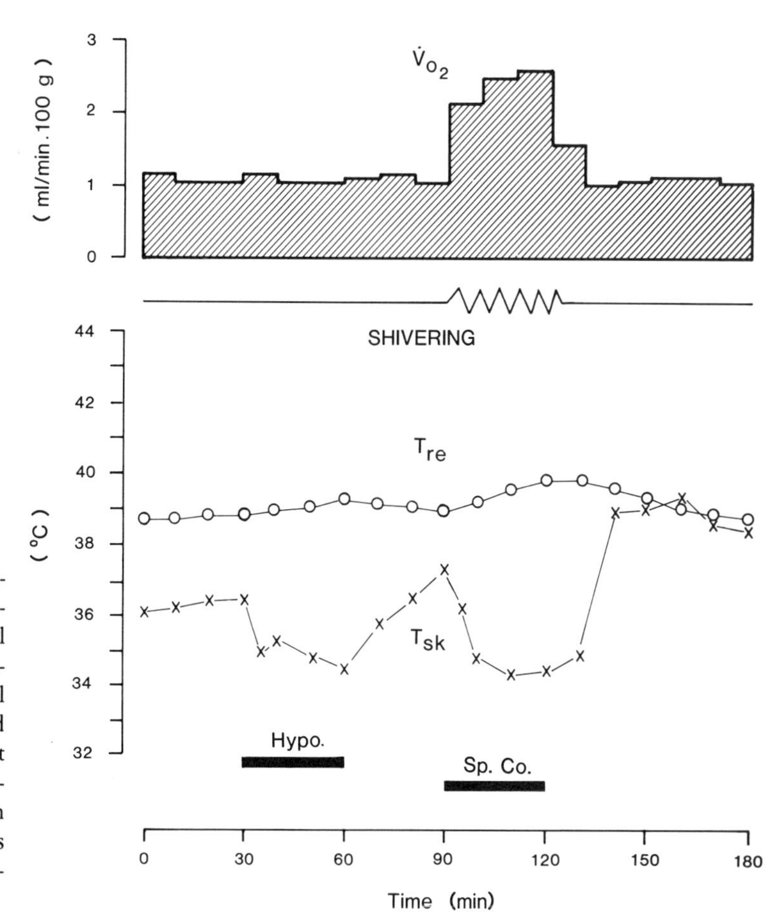

FIGURE 9–6. The effect of localized cooling of the anterior hypothalamus (*Hypo.*) and the spinal cord (*Sp. Co.*) on the oxygen consumption ($\dot{V}_{O_2}$), shivering, rectal temperature (T_{re}) (circles), and the skin temperature of the foot (T_{sk}) (x's) of the pigeon. The horizontal bars indicate the duration of cooling. Air temperature was 29.5°C. (Redrawn from Rautenberg et al., 1972.)

the California quail, ". . . hypothalamic thermosensitivity plays no significant role in thermoregulation . . . other than being a possible defense against severe brain hyperthermia."

Recent work has shown that in the domestic fowl also the thermal sensitivity of the spinal cord is more important than that of the hypothalamus (Avery and Richards, 1983). In view of this evidence in a number of species of birds, it is perhaps surprising that Helfmann et al. (1981) concluded that, in the goose, the thermal sensitivity of the spinal cord was relatively low. A further illustration of the complexity of the control mechanisms in birds is provided by Graf's (1980) demonstration that prolonged warming of the spinal cord in pigeons resulted in panting, which in turn evoked a decrease in extraspinal temperature that caused shivering so that the bird shivered and panted simultaneously! Clearly, the last word has not been written on the nature of the thermoregulatory control mechanisms in birds. There is also evidence for an interrelationship between peripheral and control temperatures in thermoregulation. Helfmann et al. (1981) reported that the heat production of the goose increased when the body temperature was lowered, but that the threshold body temperature at which this occurred was lower at higher air temperatures.

Intrahypothalamic injections of epinephrine, norepinephrine, and 5-hydroxytryptamine produced a decrease in core temperature in pigeons exposed to cold (Hissa and Rautenberg, 1974, 1975). The diminution in core temperature was the result of inhibition of shivering, a decrease in heat production, peripheral vasodilatation, and an attendant increase in heat loss (Figure 9–7). It remains to be elucidated whether the amines act as neurotransmitters in the hypothalamus.

There is clear evidence in the fowl that the tissue insulation of the unfeathered comb and feet may diminish in response to infrared irradiation of the thorax

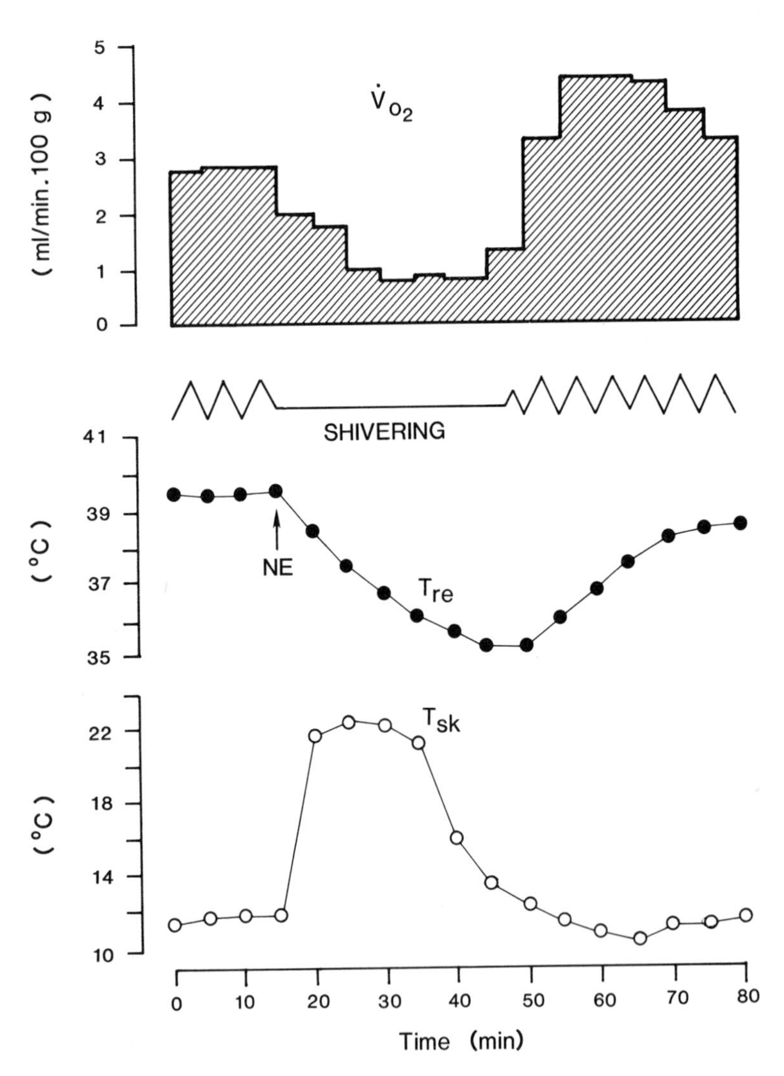

FIGURE 9–7. Effect of intrahypothalamic injection of norepinephrine (*NE*) on the oxygen consumption ($\dot{V}_{O_2}$), shivering, rectal temperature (T_{re}, *solid circle*) and the skin temperature of the foot (T_{sk}, *open circle*) of the pigeon. The norepinephrine was injected at the arrow. Air temperature was 1°C. (Reprinted with permission from Comp. Biochem. Physiol., A51, Hissa, R. and Rautenberg, W., © 1975, Pergamon Press.)

and abdomen, before the deep-body temperature increases, indicating that changes in the thermal conductance of the tissues may be effected by stimulation of peripheral thermal receptors. So far, peripheral warm receptors have been described only in the beak or bill in birds (see Dawson, 1975). However, cooling the spinal cord or the brainstem of the pigeon elicits a reduced blood flow and skin temperature in the foot (Figure 9–7) and similar responses were evident in the Peking duck, the effects of cooling the spinal cord being the more important in both species (Bech et al., 1982). Localized heating of the spinal cord evokes an increase in the skin temperature of the bare extremities of the pigeon and also of the Peking duck (Bech et al., 1982). In contrast, the effects of localized heating or cooling of the brainstem of the Peking duck were weak (Bech et al., 1982).

The control of ptilomotor activity seems to be invested in the anterior hypothalamus–preoptic region in the house sparrow. Localized heating of this region diminished the degree of fluffing of the feathers in birds exposed to a cold environment. Cooling the anterior hypothalamus of the pigeon resulted in erection of the feathers, and a similar effect was elicited by localized cooling of the spinal cord.

There is evidence that both thermal polypnea and gular flutter can occur in response to exposure to solar radiation before an increase in deep-body temperature takes place. Whether this is an effect specific to solar radiation is not known, but it is known that exposure of the domestic fowl to an increased air temperature in the absence of a radiant heat load results in thermal polypnea only after the deep-body temperature has increased. Although there is little evidence that an increase in skin temperature per se can provoke an increased respiratory frequency, there is an indication that skin temperature can influence thermal polypnea after it has become established.

Thermal polypnea has also been produced in the house sparrow by localized heating of the preoptic–anterior hypothalamic region of the brain alone and by exposing the sparrows to high air temperatures. Thermal polypnea was inhibited by localized cooling of the same region in the brain. Bilateral lesions in the hypothalamus abolished the polypneic response of the domestic fowl to heat. However, panting can occur after transection of the brain caudal to the hypothalamus. The role of the hypothalamus in the control of panting is clearly unsettled. In both the pigeon and the Peking duck, localized heating of the anterior hypothalamic region was relatively ineffective in producing thermal polypnea, but localized heating of the spinal cord induced panting in both species, which could be inhibited by localized cooling of the spinal cord (Bech et al., 1982). However, intrahypothalamic injections of 5-hydroxytryptamine abolished panting in pigeons (Hissa et al., 1980), while prostaglandins introduced directly into the hypothalamus resulted in hyperthermia (Nistico and Marley, 1976). The relative importance of amines and prostaglandins in the central control of body temperature of birds awaits further investigation.

Behavioral thermoregulatory responses in birds seem to be regulated by the anterior hypothalamus–preoptic region of the brain and by peripheral receptors. For example, in the fowl, drooping of the wings in response to heat was abolished by lesions in the hypothalamus. In Barbary doves, also, thermoregulatory behavior was controlled by hypothalamic rather than peripheral temperature, and the control mechanism appeared to operate as a simple "on-off" system. In the pigeon, there is evidence that appropriate behavioral thermoregulatory responses may be elicited from the hypothalamus while physiologic responses are engendered largely from the spinal cord (Schmidt, 1978a). In the pigeon, also, it appears that behavioral responses are related to the rate of change of skin temperature on exposed areas of the body (Schmidt, 1978b). In a further analysis of the effects of changing the temperature of the spinal cord, of the hypothalamus, and facial skin temperature in the pigeon, Schmidt (1983) concluded that displacements of spinal cord temperature mainly evoke physiologic thermoregulatory responses, while localized changes in the skin temperature of the facial region had a more powerful effect on behavioral thermoregulation. The relative effects of localized changes in temperature of the hypothalamus on behavioral and physiological responses were different for heating and cooling.

The role of higher levels of the central nervous system involved in conditioned responses, in thermoregulation, is suggested by work on the pigeon (Sieland et al., 1981). Pigeons were able to change their heart rates in order to interrupt an artificially induced aversive temperature of their spinal cord.

Birds develop a fever in response to some bacterial infections, and they respond to antipyretic drugs such as acetylsalicylic acid with an attenuation of fever (Kluger, 1979). These observations, together with the demonstrable hyperthermia induced by the intrahypothalamic injection of prostaglandin E_1 (see above), suggest that the characteristics of fever in birds are similar to those in mammals.

Thermoregulation during Flight

Flight is the most conspicuous activity of birds, but as it is difficult to study the physiology of flying birds, recourse has been made by some investigators to theoretical models of the energy cost of flight (see Rayner, 1982). Nevertheless, a number of direct measurements of the metabolic rate of flying birds has been made, and there is no doubt that a greatly increased heat production occurs during flight (Figure 9–8). In pigeons flying in a wind tunnel, the heat production increased to its highest value (12.5 × resting heat production) within 1 min of take-off and then declined to a steady level (10 × resting) 4 min after take-off (Butler et al., 1977). It was estimated that these values were at least 12% greater than they might be in a free-flying bird unencumbered by a mask and attached tubes (Figure 9–8). Hovering flight is the most expensive in terms of energy, although it is largely confined to hummingbirds. In the glittering-throated emerald hummingbird (*Amazalia fimbriata*), the cost of hovering flight was 14 times the basal metabolic rate (Berger and Hart, 1972). The energy expenditure during hovering flight increased with decreasing air temperature in two other species of hummingbirds (Schuchmann, 1979). Many seabirds glide for long periods of time when they are in the air. In the herring gull, the oxygen consumption during gliding was only twice the resting level (Baudinette and Schmidt-Nielsen, 1974).

The energy cost of flapping flight varies with a number of factors. Thus, in the fish crow (*Corvus ossifragus*), the flight metabolic rate was greater than that in the laughing gull (*Larus atricilla*), which has the same body mass. This difference was attributed to the long narrow wings and slower wingbeat of the gull (Bernstein et al., 1973). Hails (1979) obtained evidence that in house martins (*Delichon urbica*) and swallows (*Hirundo rustica*), which depend on continuous flight for their foraging activities, flight metabolic rates are lower than in other birds of similar size. The metabolic rate during flight varies also with the flight speed in some birds. In the budgerigar (*Melopsittacus undulatus*), the metabolic rate during level flight was lowest at a speed of 35 km/hr, increasing at lower and higher speeds. In the laughing gull the minimal metabolic rate occurred at a flight speed of 30 km/hr. However, in some species the metabolic rate does not change with flight speed (Torre-Bueno and Larochelle, 1978).

Some of the heat produced during flight is stored in the bird. In the starling (*Sturnus vulgaris*), the core temperature, measured by telemetry, was 42.7–44.0°C during flight, 2–4°C higher than the temperature of resting birds (Torre-Bueno, 1978), and body temperatures as high as 45°C were recorded in the flying white-necked raven (*Corvus cryptoleucus*) (Hudson and Bernstein, 1981). In the pigeon, body temperature rose steadily after take-off, reaching a stable value of

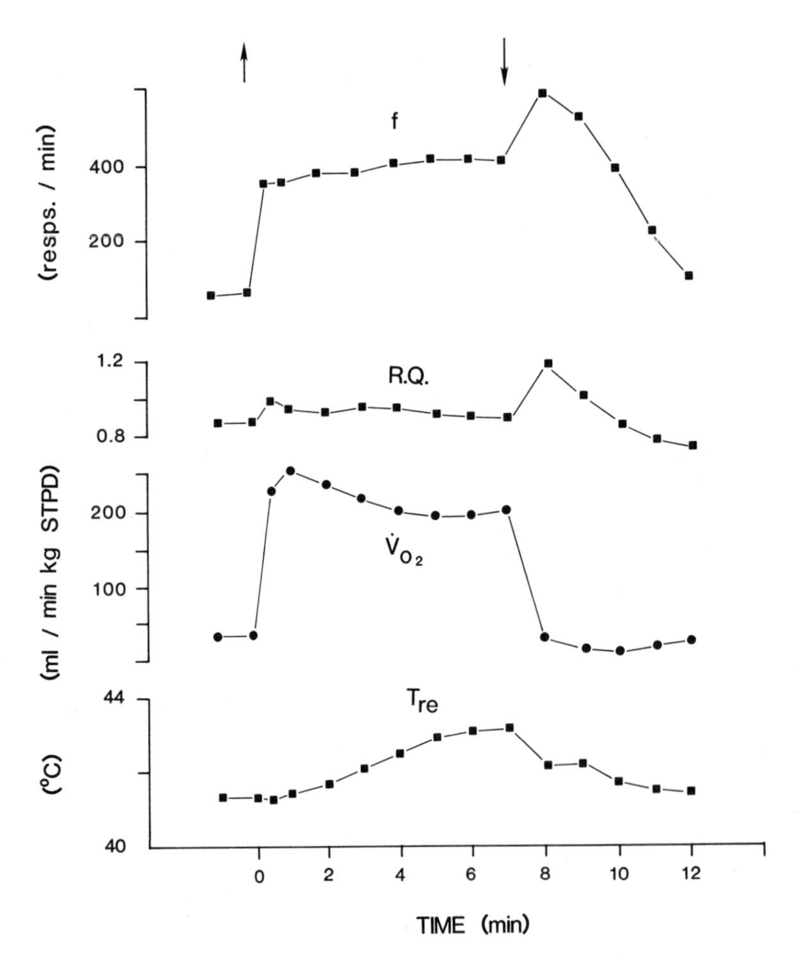

FIGURE 9–8. Respiratory frequency (f), respiratory quotient (RQ), oxygen consumption ($\dot{V}_{O_2}$) and deep-body temperature (T_{re}) of pigeons flying in a wind tunnel. The arrows indicate take-off and landing. (Redrawn from Butler et al., 1977.)

43.3°C, 2°C higher than the resting level, after 6 min of flight (Butler et al., 1977) (Figure 9–8). The heat is stored in regions other than the brain, because the increase in the temperature of the hypothalamus was very much less than that elsewhere, in the American kestrel (Bernstein et al., 1979a). Apparently the increase in brain temperature is attenuated by the operation of the *rete mirabile ophthalmicum* (see earlier section on heat storage). The arterial blood supply to the brain is cooled by contiguity with blood returning from respiratory and corneal evaporative surfaces.

The total thermal conductance of birds during flight appears to be more than five times that of birds at rest. A flying bird with its wings outstretched has a relatively large surface area. Moreover, it exposes poorly feathered areas under the wings, and the air movement past a flying bird greatly enhances convective cooling. It has been estimated that even at an air temperature of 30°C, nonevaporative heat loss accounted for 82% of the total heat loss in the budgerigar flying at a speed of 35 km/hr. In the herring gull, no less than 80% of the heat produced during flight was lost from the webbed feet (Baudinette et al., 1976).

At higher air temperatures, heat is lost by enhanced evaporative cooling. A considerable increase in respiratory minute volume occurs in rock doves during flight, and this subserves the increased oxygen requirements of the flying bird in addition to promoting heat loss. The respiratory minute volume during flight was 20 times that when the bird was at rest, the increase being brought about largely by an increase in respiratory frequency (Figure 9–8). In other species, the increased ventilation during flight was effected by an increased respiratory tidal volume in addition to an increased respiratory frequency. The increase in ventilation was also less in these species than in the rock dove, and proportional to the oxygen consumption. In the glittering-throated emerald hummingbird, respiratory evaporative water loss accounted for 40% of the heat produced during flight at an air temperature of 35°C (Berger and Hart, 1972). An augmented cutaneous evaporative heat loss is also likely to occur in flying birds, according to Hudson and Bernstein (1981). The cardiovascular effects of flight are discussed in Chapter 6.

Some birds are flightless, and it is illuminating to compare their temperature regulation during the only strenuous activitiy of which they are capable, i.e., running, with that of flying birds. The rhea (*Rhea americana*) dealt with the heat produced during running on a treadmill by storing a considerable proportion of it. This raised the bird's temperature, which in turn facilitated nonevaporative heat loss. Cutaneous evaporation did not increase, and although respiratory evaporation was augmented it played a minor role in the heat balance during running. Similar conclusions may be drawn from a study of the exercising domestic fowl (Brackenbury et al., 1981b), a species that spends most of its time on the ground. In Japanese quail trained to run on a treadmill, the increase in both deep-body temperature and respiratory frequency during exercise was greater at higher levels of exercise and heat production and at higher air temperatures (Nomoto et al., 1983).

Thermoregulation during Torpor

Representatives of several orders of birds have the extraordinary facility to allow their body temperature to diminish in certain circumstances, i.e., to become torpid. The particular circumstance that induces torpor seems to be curtailment of the bird's energy supply, and there is an indisputable saving of energy when a bird becomes torpid (see Calder and King, 1974). Many small birds that are able to become torpid feed on insects or nectar, both of which may become seasonally or temporarily unavailable. However, this does not mean that birds have depleted their energy reserves before becoming torpid; a dormant poorwill, found under natural conditions, was quite heavy and clearly in good nutritional condition. Furthermore, the speckled mousebird (*Colius striatus*), which does not feed on either insects or nectar, and which is not particularly small, may become torpid.

Some birds become torpid on a daily basis, during the inactive part of their circadian cycle. Nocturnal torpor has been reported in two small, tropical, frugivorous manakins, the body temperature falling to as low as 26.8°C (Bartholomew et al., 1983). The energy saving that accrued from torpor was calculated to be 58%, and the incidence of torpor was believed to be associated with occasional limited availability of fruit or insufficient opportunity to forage. Inca doves deprived of either water or food experience a pronounced nocturnal hypothermia. It is believed that the evolution of small nectar-feeding hummingbirds was related to their ability to become torpid (Brown et al., 1978). In the absence of torpor, the energy expenditure of a 2-g hummingbird would exceed its energy intake. Seasonal dormancy has been observed only in the Caprimulgidae.

Very little is known about the physiologic mechanisms by which a bird becomes torpid. There is evidence that torpor and sleep in birds are closely related phenomena (Walker et al., 1983). Temperature regulation is not in abeyance; rather there appears to be a diminution in the set point. Heat production decreases and cooling occurs passively. It is known that body cooling is facilitated in the giant hummingbird (*Patagonia gigas*) and in the poorwill by inhibition of the shivering mechanism. The rate of entry into torpor is inversely related to body mass. This is not surprising in view of the relatively greater thermal conductance and surface area of small birds. The metabolic rate of torpid hummingbirds may be only 1/50 of the basal rate at "normal" body temperature, and water loss is also reduced, amounting to 1/10–1/3 of that in homeothermic hummingbirds.

Birds vary in their tolerance of low body temperatures. For example, the poorwill and nightjar (*Caprimulgus europaeus*) survived body temperatures of 5–8°C, which are lethal to hummingbirds. However, it is important to draw a distinction between the body temperature that torpid birds can tolerate and the temperature from which they can arouse on their own. The latter temperature appears to be of the order of 13–20°C. The body temperature of many torpid birds tends to approximate that of the environment, and lowering the environmental temperature does not appear to stimulate thermoregulatory mechanisms that prevent the body temperature from falling further. For example, a poorwill could be subjected to an air temperature of 15°C, from which it would eventually arouse. If the temperature were lowered to and held at 5°C, however, the bird would eventually die simply becuase it was unable to arouse or to remain torpid indefinitely. There are exceptions to this generalization: the West Indian hummingbird (*Eulampis jugularis*) maintained its body temperature at 18–20°C in the face of a lowered air temperature by increasing its heat production. The thermal conductance of the tissues and plumage was the same in the torpid as in the nontorpid bird. Substantially similar results were obtained for two other species of hummingbirds, and it was shown that the regulated level of body temperature was related to the minimal environmental temperature encountered by the species under natural conditions. Clearly, torpidity can occur in tropical species as well as in those of colder regions.

Rewarming of a torpid bird is accomplished by vigorous shivering. The rate of rewarming is inversely related to body size, as is the rate of entry into torpor. This places a limitation on the capacity of birds of increasing size to indulge in short-term torpidity. Entry into torpor and rewarming would simply take too long. Daily torpor is therefore practicable only in small birds.

Ontogeny of Thermoregulation

Embryo

The responses of the embryo to changes of environmental temperature are a legitimate consideration in this chapter because the egg may be exposed to excessive heat or cold, depending on the attentiveness of the parent birds during the incubation period. The embryo of the domestic fowl is unable to regulate its own temperature during the early stages of incubation. From the 19th day of incubation, however, the embryo is able to respond to a small reduction in ambient temperature by a transient rise in metabolic rate. The metabolic response to cold appears to be associated with an increased thyroid activity. The respiratory quotient (RQ) increases, an indication of the metabolism of glycogen, and the temperature of the egg remains above that of the environment. The embryos of western gulls (*Larus occidentalis*) also appear to be able to regulate their body temperatures to some extent. In the week-old eggs of Heermann's gull (*L. herrmanni*), the incubation temperature may vary from 30° to 40°C, but the embryonic metabolic rate is independent of temperature within this range (Bennett and Dawson, 1979). The reader is referred to Chapter 10 for a further discussion of the oxygen consumption of the embryo.

Fortunately, the embryos of many species are able to withstand temporary periods of hypothermia when the adult is away from the nest. The 1-day-old embryo of the domestic fowl can survive an exposure of 76 hr to air at 0°C, although the normal incubation temperature for the chicken is 39–40°C. The lower lethal temperature of the fowl's egg is from −2.2° to −1.1°C. During the first 5 days of incubation of the fowl's egg, its upper lethal temperature is 42.2°C. By the eighth day, the upper lethal temperature has increased to 45.6–47.8°C, and it remains at this level for the remainder of the incubation period. The eggs of the eastern house wren appear to have a similar upper lethal temperature.

During the first 10 days of incubation, the evaporative heat loss from the artificially incubated egg of the domestic fowl exceeds the heat production, so that the temperature of the egg may be slightly below the air temperature. Toward the end of the incubation period the major channel of heat loss from the egg is nonevaporative.

Chick

Birds vary enormously in their thermoregulatory capacities immediately after hatching, and they have been classified according to these and other capabilities (Nice, 1962). Precocial species, e.g., the chick of the domestic fowl, are covered with down and are able to respond effectively to heat and to cold. In contrast, altricial birds, e.g., the Passeriformes, are naked when hatched and have little ability to regulate their body temperature at air temperatures below 35°C or above 40°C. Other birds (semiprecocial, semialtricial) are intermediate between precocial and altricial species in their thermoregulatory capacities.

Precocial Species. Immediately after hatching, the body temperature of the chick of the domestic fowl may diminish to below 30°C. The down feathers of the newly hatched chick are wet, and evaporative cooling probably explains the decrease in body temperature. Thereafter, the body temperature increases to reach the adult level at about 3 weeks. The increase results in part from an increase in the chick's metabolically active mass and metabolic rate, without a commensurate increase in area, and partly from an increase in set-point temperature (see section on control mechanisms above) with age (Myrhe, 1978). Given a choice of ambient temperatures in a thermal gradient box, the youngest chicks selected the highest ambient temperature. In the capercaillie (*Tetrao urogallus*), the increase in the body temperature of the chick coincided with an increase in basal metabolic rate (Hissa et al., 1983). The chicks of some arctic shorebirds permit their body temperature to vary widely (30–40°C) without impairing their ability to forage (Chappell, 1980).

The highly precocial hatchling of the mallee fowl (*Leipoa ocellata*) increased its heat production threefold during exposure to cold (Booth, 1984), and in the Xantus' murrelet (*Synthliboramphus hypoleucus*) the increase was 3.5 times the basal metabolic rate (Eppley, 1984). The chick of the domestic fowl also responds to cold exposure with an increase in heat production (Misson, 1977). Although the newly hatched chick is capable of shivering, it may also augment heat production without shivering, i.e., by nonshivering thermogenesis. Both carbohydrate and free fatty acids are mobilized as energy substrates (Freeman, 1977). However, the mechanism of nonshivering thermogenesis in young birds is not fully understood, and it may differ in some respects from that in mammals (see Dawson, 1975). Chicks do not have brown fat, an important site of nonshivering thermogenesis in young mammals. The involvement of thyroid hormones, norepinephrine, and glucagon in nonshivering thermogenesis has been postualted for different species (see Dawson, 1975; Freeman, 1977; Barré, 1983).

In the Japanese quail, another precocial species, the rate of cooling of the chick during exposure to an air temperature of 30°C decreased from the 3rd to the 11th day. This indicated an increase in thermoregulatory ability during this period that was correlated with a diminution of heat-seeking behavior and increased thyroid gland activity (Spiers et al., 1974). In the willow ptarmigan, the ability of the chick to increase its heat production by shivering increased *pari passu* with the development of the pectoral muscles (Aulie, 1976a). An increase in heat production rather than in thermal insulation is the key to the improved ability of the growing chick of the capercaillie to respond to cold (Hissa et al., 1983).

In three species of semiprecocial gulls, an elevation of heat production occurred in the newly hatched chick during exposure to cold. The increase varied from 50 to 100% of the basal metabolic rate (Palokangas and Hissa, 1971; Dawson et al., 1976; Dawson and Bennett, 1980).

The lower critical temperature of the fowl's chick decreases with age, and the rate of increase in heat production below the critical temperature diminishes. Both of these changes are consistent with a diminution in the relative surface area and in the thermal conductance with growth. The upper critical temperature is higher in the chick than in the adult. This conforms with the relatively greater surface area of the chick and its higher thermal conductance, both features promoting heat loss in a hot environment. In Xantus' murrelet chicks, the decrease in thermal conductance was not associated with concomitant changes in body mass or plumage, suggesting a vascular basis for the lower conductance (Eppley, 1984). The lower critical temperature for chicks in a group was lower than that of isolated chicks (Misson, 1982), reflecting the more favorable microclimate in the former instance.

The newly hatched chick responds to heat exposure with an identifiable thermal polypnea, which commences at a lower body temperature in the newly hatched chick than in the adult. The maximal respiratory frequency during exposure to heat is higher in the chick than in the adult. This is probably related to the size of the bird, and, in particular, to the mass of the thorax. The onset of peripheral vasodilatation and of evaporative cooling occurred at higher air temperatures in starved, as opposed to fed, 1-week-old chicks (Misson, 1982). Hatchling laughing gulls and ring-billed gulls (*Larus delawarensis*), and California gull chicks, were able to dissipate considerably more than their heat production at high air temperatures (Dawson et al., 1976, Chappell et al., 1984).

Little is known about the role of peripheral and central thermoreceptors in the regulation of body temperature in the chick. The nestling red-tailed tropicbird (*Phaethon rubricauda*) when placed in the sun pants before the deep-body temperature increases, indicating that stimulation of peripheral receptors alone is sufficient to elicit panting. Infusion of catecholamines into the hypothalamus of chicks of the domestic fowl resulted in a decrease in their body temperature and oxygen consumption. Intra-abdominal injection of 5-hydroxytryptamine resulted in a consistent decrease in body temperature in the newly hatched chick (Freeman, 1979). It is possible that the amines are involved as transmitter substances in the hypothalamus (see section on control mechanisms above).

Acclimation of newly hatched chicks of the domestic fowl to cold resulted in a higher oxygen consumption during exposure to cold than in nonacclimated controls (Aulie and Grav, 1979). The cytochrome oxidase activity of skeletal muscle and liver was elevated in the cold-acclimated birds, indicating a role for these tissues in the increased oxidative metabolism during cold acclimation. The leg muscles were considered to be a significant source of heat production in cold-acclimated chicks.

Altricial Species. Newly hatched altricial species are essentially naked; they are unable to maintain their body temperature or to increase their heat production in the face of a lowered environmental temperature (Figure 9–9). In the intervening time between hatching and the development of effective thermogenesis, altricial birds rely heavily on the protection of their parents and of the nest. The acquisition of thermoregulatory ability seems to be related to the rate of growth of heat-producing tissues, notably skeletal muscle (Dunn, 1975). Growth is accompanied by the development of the plumage and a reduction in the surface area relative to body mass. These factors help curtail heat loss and therefore facilitate the regulation of body temperature (O'Connor, 1975) (Figure 9–9). In nestling bank swallows (*Riparia riparia*), an effective thermoregulatory response to cold is more closely related to the increase in body mass than to the development of the plumage, and initially the increased resistance to cooling is due to the increase in body mass rather than an improvement in metabolic regulation (Marsh, 1979). Some species develop their thermoregulatory capability before others. Dunn (1975) has shown that the greater the rate of growth, the earlier the age at which altricial chicks are able to maintain their body temperature during exposure to cold. For birds with the same growth rate, those with the shorter nestling periods are also capable of maintaining their body temperature at an earlier age. There is some evidence that the nestlings of the larger altricial species acquire their thermoregulatory ability at an earlier stage of development than do smaller altricial species (Turner and McClanahan, 1981).

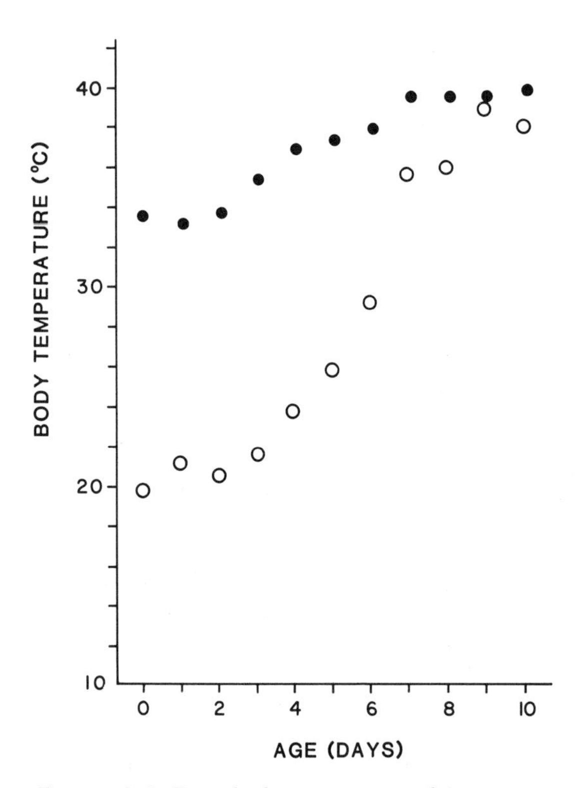

FIGURE 9–9. Deep-body temperature of the mountain white-crowned sparrow chick at different ages. Symbols: ●, while in the nest; ○, after removal from the nest and exposure to air at 5.8°C for 10 min. (After Morton and Carey, 1971.)

Hill and Beaver (1982) have drawn attention to the differences between the responses of a single chick to cold and that of a brood in the nest. The latter situation is the natural one, and as a result of the insulation of the nest and the effects of huddling with the other chicks, the body temperature of each chick was strikingly higher and the metabolic rate lower than in isolated chicks.

Many altricial birds are hatched under very hot conditions. There are very often differences between adult plumage and the downy plumage of the chick as far as their reflectance of solar radiation is concerned (Dunn, 1976). Some, such as the cattle egret (*Bubulcus ibis*), have evaporative cooling mechanisms operative on the first day of hatching (Hudson et al., 1974). Nestling cattle egrets were able to lose, by evaporative cooling, heat equivalent to more than double their metabolic heat production (Hudson et al., 1974). In general, altricial nestlings develop the ability to respond to heat stress before that to cold (O'Connor, 1975). Nevertheless, this potential is probably only realized under survival conditions, in the event that the parent must leave the nest. Ordinarily, altricial nestlings such as the great frigatebird that are hatched into a very hot environment are closely brooded by the parents (Figure 9–3).

Morton and Carey (1971) presented a detailed account of the development of thermoregulation in an altricial passerine, the mountain white-crowned sparrow. The nestlings were not able to respond to cooling by shivering before day 4, but they were able to pant on exposure to heat from day 1. The emergence of the feathers began on day 2 and by day 7 almost the entire body surface was covered (see also Chapter 10).

Insulation of the Nest

Not all birds build nests, but those that do evince a striking variety in the shape, size, and orientation of the nest (Skowron and Kern, 1980). Nests protect against excessive cooling of the eggs or chicks, and they may also mitigate the effects of solar radiation. In the Hawaiian honeycreeper amakihi (*Loxops v. virens*), Whittow and Berger (1977) presented evidence that the thermal conductance of the nest was similar to that of the tissues and plumage. A striking illustration of the amelioration of the external conditions by a nest is provided by the elaborate communal nests of the sociable weaver (*Philetairus socius*) (Bartholomew et al., 1976), the largest of all birds' nests.

Summary

The relatively high deep-body temperature, the absence of sweat glands, the effective insulation provided by the plumage, and the widespread incidence of gular flutter distinguish the thermoregulatory physiology of birds. Many birds become torpid on occasion, and almost all birds are able to fly. The physiology of temperature regulation during flight and during torpor are perhaps the most dramatic aspects of thermoregulation in birds. The avian egg provides unparalleled opportunities for the study of embryonic thermoregulation in a homeothermic animal. Recent studies on the heat exchange of birds out-of-doors promise to place the wealth of laboratory data in an ecological perspective.

References

Arad, Z., and J. Marder. (1982). Comparative thermoregulation of four breeds of fowls (*Gallus domesticus*) exposed to a gradual increase of ambient temperatures. Comp. Biochem. Physiol. A, 72, 179.

Arieli, A., A. Berman, and A. Meltzer. (1978). Indication for non-shivering thermogenesis in the adult fowl (*Gallus domesticus*). Comp. Biochem. Physiol., C: Comp. Pharmacol., 60, 33.

Arieli, A., A. Berman, and A. Meltzer. (1979). Cold thermogenesis in the summer-acclimatized and cold-acclimated domestic fowl. Comp. Biochem. Physiol., C: Comp. Pharmacol., 63, 7.

Aulie, A. (1976a). The pectoral muscles and the development of thermoregulation in chicks of Willow Ptarmigan (*Lagopus lagopus*). Comp. Biochem. Physiol. A, 53, 343.

Aulie, A. (1976b). The shivering pattern in an arctic (Willow Ptarmigan) and a tropical bird (bantam hen). Comp. Biochem. Physiol. A, 53, 347.

Aulie, A., and H.J. Grav. (1979). Effect of cold acclimation on the oxidative capacity of skeletal muscles and liver in young bantam chicks. Comp. Biochem. Physiol. A, 62, 335.

Avery, P. and S.A. Richards. (1983). Thermosensitivity of the hypothalamus and spinal cord in the domestic fowl. J. Therm. Biol., 8, 237.

Bakken, G.S. (1980). The use of standard operative temperature in the study of the thermal energetics of birds. Physiol. Zool., 53, 108.

Bakken, G.S., W.A. Buttemer, W.R. Dawson, and D.M. Gates. (1981). Heated taxidermic mounts: A means of measuring the standard operative temperature affecting small animals. Ecology, 62, 311.

Barnas, G.M., J.A. Estavillo, F.B. Mather, and R.E. Burger. (1981). The effect of CO_2 and temperature on respiratory movements in the chicken. Respir. Physiol., 43, 315.

Barré, H. (1983). Calorigenic action of glucagon in several species of chicks at neutral ambient temperatures. In "Environment, Drugs and Thermoregulation" (P. Lomax and E. Schönbaum, Eds.). Basel: Karger, p. 31.

Bartholomew, G.A., F.N. White, and T.R. Howell. (1976). The thermal significance of the nest of the Sociable Weaver *Philetairus socius:* Summer observations. Ibis, 118, 402.

Bartholomew, G.A., C.M. Vleck, and T.L. Bucher. (1983). Energy metabolism and nocturnal hypothermia in two tropical passerine frugivores, *Manacus vitellinus* and *Pipra mentalis.* Physiol. Zool., 56, 370.

Baudinette, R.V., and K. Schmidt-Nielsen. (1974). Energy cost of gliding flight in herring gulls. Nature (London), 248, 83.

Baudinette, R.V., J.P. Loveridge, K.J. Wilson, C.D. Mills, and K. Schmidt-Nielsen. (1976). Heat loss from feet of herring gulls at rest and during flight. Am. J. Physiol., 230, 920.

Bech, C. (1980). Body temperature, metabolic rate, and insulation in winter and summer acclimatized Mute Swans (*Cygnus olor*). J. Comp. Physiol., 136, 61.

Bech, C., and K. Johansen. (1980). Ventilatory and circulatory responses to hyperthermia in the Mute Swan (*Cygnus olor*). J. Exp. Biol., 88, 195.

Bech, C., K. Johansen, and G.M.O. Maloiy. (1979). Ventilation and expired gas composition in the flamingo, *Phoenicopterus ruber,* during normal respiration and panting. Physiol. Zool., 52, 313.

Bech, C., W. Rautenberg, B. May, and K. Johansen. (1982). Regional blood flow changes in response to thermal stimulation of the brain and spinal cord in the Pekin Duck. J. Comp. Physiol., 147, 71.

Bennett, A.F., and W.R. Dawson. (1979). Physiological responses of embryonic Heermann's Gulls to temperature. Physiol. Zool., 52, 413.

Berger, M., and J.S. Hart. (1972). Die Atmung beim Kolibri *Amizilia fimbriata* Während des Schwirrfluges bei verschiedenen Umgebungstemperaturen. J. Comp. Physiol., 81, 363.

Berger, M., J.S. Hart, and O.Z. Roy. (1970). Respiration, oxygen consumption and heart rate in some birds during rest and flight. Z. Vgl: Physiol., 66, 201.

Berman, A., and A. Meltzer. (1978). Metabolic rate: its circadian rhythmicity in the female domestic fowl. J. Physiol., 282, 419.

Bernstein, M.H., and F.C. Samaniego. (1981). Ventilation and acid-base status during thermal panting in pigeons (*Columba livia*). Physiol. Zool., 54, 308.

Bernstein, M.H., S.P. Thomas, and K. Schmidt-Nielsen. (1973). Power input during flight of the Fish Crow, *Corvus ossifragus.* J. Exp. Biol., 58, 401.

Bernstein, M.H., M.B. Curtis, and D.M. Hudson. (1979a). Independence of brain and body temperatures in flying American kestrels, *Falco sparverius.* Am. J. Physiol., 237, R58.

Bernstein, M.H., I. Sandoval, M.B. Curtis, and D.M. Hudson. (1979b). Brain temperature in pigeons: effects of anterior respiratory bypass. J. Comp. Physiol., 129, 115.

Bligh, J., and K.G. Johnson. (1973). Glossary of terms for thermal physiology. J. Appl. Physiol., 35, 941.

Booth, D.T. (1984). Thermoregulation in neonate Mallee Fowl *Leipoa ocellata.* Physiol. Zool., 57, 251.

Bouverot, P., G. Hildwein, and D. Le Goff. (1974). Evaporative water loss, respiratory pattern, gas exchange and acid-base balance during thermal panting in Pekin Ducks exposed to moderate heat. Respir. Physiol., 21, 255.

Brackenbury, J.H. (1978). Experimentally induced antagonism of chemical and thermal reflexes in the respiratory system of fully conscious chickens. Respir. Physiol., 34, 377.

Brackenbury, J., P. Avery, and M. Gleeson. (1981a). Respiratory evaporation in panting fowl: partition between the respiratory and buccopharyngeal pumps. J. Comp. Physiol., 145, 63.

Brackenbury, J.H., M. Gleeson, and P. Avery. (1981b). Respiration in exercising fowl. II. Respiratory water loss and heat balance. J. Exp. Biol., 93, 327.

Brent, R., P.F. Pedersen, C. Bech, and K. Johansen. (1984). Lung ventilation and temperature regulation in the European Coot *Fulica atra.* Physiol. Zool., 57, 19.

Brown, J.H., W.A. Calder, and A. Kodric-Brown. (1978). Correlates and consequences of body size in nectar-feeding birds. Am. Zool., 18, 687.

Butler, P.J., N.H. West, and D.R. Jones. (1977). Respiratory and cardiovascular responses of the pigeon to sustained level flight in a wind tunnel. J. Exp. Biol., 71, 7.

Calder, W.A., and J.R. King. (1974). Thermal and caloric relations of birds. In "Avian Biology," Vol. IV (D.S. Farner and J.R. King, Eds.). New York: Academic Press, p. 259.

Calder, W.A., and T.J. Dawson. (1978). Resting metabolic rates of ratite birds: the kiwis and the Emu. Comp. Biochem. Physiol. A, 60, 479.

Chappel, M.A. (1980). Thermal energetics of chicks of arctic-breeding shorebirds. Comp. Biochem. Physiol. A, 65, 311.

Chappell, M.A., D.L. Goldstein, and D.W. Winkler. (1984). Oxygen consumption, evaporative water loss, and temperature regulation of California Gull chicks (*Larus californicus*) in a desert rookery. Physiol. Zool., 57, 204.

Dawson, W.R. (1975). Avian physiology. Ann. Rev. Physiol., 37, 441.

Dawson, W.R. (1982). Evaporation losses of water by birds. Comp. Biochem. Physiol., 71A, 495.

Dawson, W.R., and A.F. Bennett. (1980). Metabolism and thermoregulation in hatchling Western Gulls. Condor, 82, 103.

Dawson, W.R., and C.D. Fisher. (1982). Observations on the temperature regulation and water economy of the Galah (*Cacatua roseicapilla*). Comp. Biochem. Physiol. A, 72, 1.

Dawson, W.R., A.F. Bennett, and J.W. Hudson. (1976). Metabolism and thermoregulation in hatchling Ring-billed Gulls. Condor, 78, 49.

Dawson, W.R., R.L. Marsh, W.A. Buttener, and C. Carey. (1983). Seasonal and geographic variation of cold resistance in House Finches *Carpodacus mexianus.* Physiol. Zool., 56, 353.

De Jong, A.A. (1976). The influence of simulated solar radiation on the metabolic rate of White-crowned Sparrows. Condor, 78, 1974.

Dunn, E.H. (1975). Growth, body components and energy content of nestling Double-crested Cormorants. Condor, 77, 431.

Dunn, E.H. (1976). The development of endothermy and existence energy expenditure in Herring Gull chicks. Condor, 78, 493.

Ebibara, S., and H. Kawamura. (1981). The role of the pineal organ and the supra-chiasmatic nucleus in the control of circadian locomotor rhythms in the Java Sparrow, *Padda oryzivora.* J. Comp. Physiol., 141, 207.

Eissel, K., and E. Simon. (1980). How are neuronal thermosensitivity and lack of thermoreception related in the duck's hypothalamus? A tentative answer. J. Therm. Biol., 5, 219.

El-Halawani, M. El-S., W.O. Wilson, and R.E. Burger. (1970). Cold acclimation and the role of catecholamines in body temperature regulation in male Leghorns. Poult. Sci., 49, 621.

El-Halawani, M.E., P.E. Waibel, J.R. Appel, and A.L. Good. (1973). Effects of temperature stress on catecholamines and corticosterone of male turkeys. Am. J. Physiol., 224, 384.

Ellis, H.I. (1980). Metabolism and solar radiation in dark and white herons in hot climates. Physiol. Zool., 53, 358.

Eppley, Z.A. (1984). Development of thermoregulatory abilities in Xantus' Murrelet chicks *Synthliboramphus hypoleucus.* Physiol. Zool., 57, 307.

Freeman, B.M. (1977). Lipolysis and its significance in the response to cold of the neonatal fowl, *Gallus domesticus.* J. Therm. Biol., 2, 145.

Freeman, B.M. (1979). Is 5-hydroxytryptamine concerned in avian thermoregulation? J. Therm. Biol., 4, 219.

Frost, P.G.H., W.R. Siegfried, and P.J. Greenwood. (1975). Arterio-venous heat exchange systems in the Jackass penguin *Spheniscus demersus.* J. Zool., 175, 231.

Gates, D.M. (1980). "Biophysical Ecology." New York: Springer-Verlag.

Graf, R. (1980). Simultaneously activated heat gain and heat loss mechanisms in pigeons. Proc. I.U.P.S., 14, 442.

Graf, R., and R. Necker. (1979). Cyclic and noncyclic variations of spinal cord temperature related with temperature regulation in pigeons. Pfleugers Arch., 380, 215.

Graf, R., H.C. Heller, and S. Sakaguchi. (1983). Slight warming of the spinal cord and the hypothalamus in the pigeon: effects on thermoregulation and sleep during the night. J. Therm. Biol., 8, 159.

Grant, G.S. (1982). Avian incubation: egg temperature, nest humidity, and behavioral thermoregulation in a hot environment. Ornithol. Monogr., No. 30.

Hagan, A.A., and J.E. Heath. (1980). Regulation of heat loss in the duck by vasomotion in bill. J. Therm. Biol., 5, 95.

Hailman, J.P. (1982). Radiation angle and heat transferred to a bird. Science, 219, 919.

Hails, C.J. (1979). A comparison of flight energetics in hirundines and other birds. Comp. Biochem. Physiol. A, 63, 581.

Hails, C.J. (1983). The metabolic rate of tropical birds. Condor, 8, 61.

Haim, A., S. Saarela, and R. Hissa. (1979). Heat production induced by photoperiodicity in the pigeon. Comp. Biochem. Physiol. A, 63, 547.

Hayes, S.R., and J.A. Gessaman. (1982). Prediction of raptor resting metabolism: comparison of measured values with statistical and biophysical estimates. J. Therm. Biol., 7, 45.

Helfmann, W., P. Jannes, and C. Jessen. (1981). Total body thermosensitivity and its spinal and supra-spinal fractions in the conscious goose. Pfleugers Arch., 391, 60.

Hennemann, W.W. (1983). Environmental influences on the energetics and behavior of Anhingas and Double-crested Cormorants. Physiol. Zool., 56, 201.

Hill, R.W., and D.L. Beaver. (1982). Inertial thermostability and thermoregulation in broods of Redwing Blackbirds. Physiol. Zool., 55, 250.

Hill, R.W., D.L. Beaver, and J.H. Veghte. (1980). Body surface temperatures and thermoregulation in the Black-capped Chickadee (*Parus atricapillus*). Physiol. Zool., 53, 305.

Hissa, R., and R. Palokangas. (1970). Thermoregulation in the Titmouse (*Parus major* L.). Comp. Biochem. Physiol., 33, 941.

Hissa, R., and W. Rautenberg. (1974). The influence of centrally applied noradrenaline on shivering and body temperature in the pigeon. J. Physiol., 238, 421.

Hissa, R., and W. Rautenberg. (1975). Thermoregulatory effects of intrahypothalamic injections of neurotransmitters and their inhibitors in the pigeon. Comp. Biochem. Physiol. A, 51, 319.

Hissa, R., A. Pyörnila, and J.C. George. (1980). The influence of intrahypothalamic injections of prostaglandins E_1 and $F_{2\alpha}$ and ambient temperature on thermoregulation in the pigeon. J. Therm. Biol., 5, 163.

Hissa, R., S. Saarela, H. Rintamäki, H. Linden, and E. Hohtola. (1983). Energetics and development of temperature regulation in Capercaillie *Tetrao urogallus.* Physiol. Zool., 56, 142.

Hohtola, E., H. Rintamäki, and R. Hissa. (1980). Shivering and ptiloerection as complementary cold defense responses in the pigeon during sleep and wakefulness. J. Comp. Physiol., 136, 77.

Horowitz, K.A., N.R. Scott, P.E. Hillman, and A. van Tienhoven. (1978). Effects of feathers on instrumental thermoregulatory behavior in chickens. Physiol. Behav., 21, 233.

Howell, T.R. (1979). Breeding biology of the Egyptian Plover, *Pluvianus aegyptius.* Univ. Calif. Publ. Zool., 113, 1.

Howell, T.R., and G.A. Bartholomew. (1961). Temperature regulation in Laysan and Black-footed albatrosses. Condor, 63, 185.

Howell, T.R., B. Araya, and W.R. Millie. (1974). Breeding biology of the Gray Gull, *Larus modestus.* Univ. Calif. Publ. Zool., 104, 1.

Hudson, D.M., and M.H. Bernstein. (1981). Temperature regulation and heat balance in flying White-necked Ravens, *Corvus cryptoleucus.* J. Exp. Biol., 90, 267.

Hudson, J.W., W.R. Dawson, and R.W. Hill. (1974). Growth and development of temperature regulation in nestling Cattle Egrets. Comp. Biochem. Physiol. A, 49, 717.

Johansen, K., and R.W. Millard. (1973). Vascular responses to temperature in the foot of the giant fulmar, *Macronectes giganteus.* J. Comp. Physiol., 85, 47.

Kendeigh, S.C., V.R. Dol'nick and V.M. Gavrilov. (1977). Avian energetics. In "Granivorous Birds in Ecosystems" (J. Pinowski and S.C. Kendeigh, Eds.). Cambridge: Cambridge University Press, p. 127.

Kilgore, D.L. (1976). Brain temperatures in birds. J. Comp. Physiol., 110, 209.

Kilgore, D.L., and K. Schmidt-Nielsen. (1975). Heat loss from duck's feet immersed in cold water. Condor, 77, 475.

Kilgore, D.L., D.F. Boggs, and G.F. Birchard. (1979). Role of the *rete mirabile ophthalmicum* in maintaining the body-to-brain temperature difference in pigeons. J. Comp. Physiol., 129, 119.

Klandorf, H., R.W. Lea, and P.J. Sharp. (1982). Thyroid function in laying, incubating and broody bantam hens. Gen. Comp. Endocrinol., 47, 492.

Kluger, M.J. (1979). Phylogeny of fever. Fed. Proc. Fed. Am. Soc. Exp. Biol., 38, 30.

Koban, M., and D.D. Feist. (1982). The effect of cold on norepinephrine turnover in tissues of seasonably acclimatized redpolls (*Carduelis flammea*). J. Comp. Physiol., 146, 137.

Kooyman, G.L., R.L. Gentry, W.P. Bergman, and H.T. Hammel. (1976). Heat loss in penguins during immersion and compression. Comp. Biochem. Physiol. A, 54, 75.

Le Maho, Y. (1983). Metabolic adaptations to long-term fasting in antarctic penguins and domestic geese. J. Therm. Biol., 8, 91.

Le Maho, Y., P. Delclitte, and J. Chatonnet. (1976). Thermoregulation in fasting emperor penguins under natural conditions. Am. J. Physiol., 231, 913.

Lustick, S. (1969). Bird energetics: Effects of artificial radiation. Science, 163, 387.

Lustick, S.I. (1983). Cost-benefit of thermoregulation in birds: influences of posture, microhabitat selection, and color. In "Behavioral Energetics: the Cost of Survival in Vertebrates" (W.P. Aspey and S.I. Lustick, Eds.). Columbus: Ohio State University Press, p. 265.

Lustick, S., S. Talbot, and E.L. Fox. (1970). Absorption of radiant energy in Red-winged Blackbirds (*Agelaius phoeniceus*). Condor, 72, 471.

Lustick, S., B. Battersby, and M. Kelty. (1978). Behavioral thermoregulation: orientation toward the sun in Herring Gulls. Science, 200, 81.

Lustick, S., B. Battersby, and M. Kelty. (1979). Effects of insolation on juvenile Herring Gull energetics and behavior. Ecology, 60, 673.

MacMillen, R.E. (1981). Nonconformance of standard metabolic rate with body mass in Hawaiian Honeycreepers. Oecologia, 49: 340.

MacMillen, R.E., G.C. Whittow, E.A. Christopher, and R.J. Ebisu. (1977). Oxygen consumption, evaporative water loss and body temperature in the Sooty Tern. Auk, 94, 72.

Marder, J. (1973). Body temperature regulation in the Brown-necked Raven (*Corvus corax ruficollis*) 1. Metabolic rate, evaporative water loss and body temperature of the raven exposed to heat stress. Comp. Biochem. Physiol. A, 45, 421.

Marjakangas, A., H. Rintamaki and R. Hissa. (1984). Thermal responses in the Capercaillie *Tetrao urogallus* and the Black Grouse *Lyrurus tetrix* roosting in the snow. Physiol. Zool., 57, 99.

Marsh, R.L. (1979). Development of endothermy in nestling Bank Swallows (*Riparia riparia*). Physiol. Zool., 52, 340.

Marsh, R.L., and W.R. Dawson. (1982). Substrate metabolism in seasonally acclimatized American goldfinches. Am. J. Physiol., 242, R 563.

Mather, F.B., G.M. Barnas, and R.E. Burger. (1980). The influence of alkalosis on panting. Comp. Biochem. Physiol. A, 67, 265.

McNab, B.K. (1980). On estimating thermal conductance in endotherms. Physiol. Zool., 53, 145.

Menaum, B., and S.A. Richards. (1975). Observations on the sites of respiratory evaporation in the fowl during thermal panting. Respir. Physiol., 25, 39.

Misson, B.H. (1974). An open circuit respirometer for metabolic studies on the domestic fowl: establishment of standard operating conditions. Br. Poult. Sci., 15, 287.

Misson, B.H. (1977). The relationships between age, mass, body temperature and metabolic rate in the neonatal fowl (*Gallus domesticus*). J. Therm. Biol., 2, 107.

Misson, B.H. (1982). The thermoregulatory responses of fed and starved 1-week-old chickens (*Gallus domesticus*). J. Therm. Biol., 7, 189.

Morton, M.L., and C. Carey. (1971). Growth and the development of endothermy in the Mountain White-crowned Sparrow (*Zonotrichia leucophrys oriantha*). Physiol. Zool., 44, 177.

Murrish, D.E. (1973). Respiratory heat and water exchange in penguins. Respir. Physiol., 19, 262.

Murrish, D.E. (1982). Acid-base balance in three species of antarctic penguins exposed to thermal stress. Physiol. Zool., 55, 137.

Murrish, D.E. (1983). Acid-base balance in penguin chicks exposed to thermal stress. Physiol. Zool., 56, 335.

Myrhe, K. (1978). Behavioral temperature regulation in neonate chick of bantam hen (*Gallus domesticus*). Poult. Sci., 57, 1369.

Necker, R. (1977). Thermal sensitivity of different skin areas in pigeons. J. Comp. Physiol., 116, 239.

Necker, R., and W. Rautenberg. (1975). Effect of spinal deafferentation on temperature regulation and spinal thermosensitivity in pigeons. Pfleugers Arch., 360, 287.

Nice, M.M. (1962). Development of behavior in precocial birds. Trans. Linn. Soc. (N.Y.), 8, 1.

Nistico, G., and E. Marley. (1976). Central effects of prostaglandins E_2, A_1 and $F_{2\alpha}$ in adult fowls. Neuropharmacology, 15, 737.

Nomoto, S., C. Bech, W. Rautenberg, and K. Johansen. (1983). Temperature regulation and cardiovascular responses during bipedal exercise in birds. J. Therm. Biol., 8, 175.

O'Connor, R.J. (1975). Growth and metabolism in nestling

passerines. In "Avian Physiology" (M. Peaker, Ed.). London: Academic Press, p. 277.

Palokangas, R., and R. Hissa. (1971). Thermoregulation in young Black-headed Gull (*Larus ridibundus* L.). Comp. Biochem. Physiol. A, 38, 743.

Pettit, T.N., G.C. Whittow, and G.S. Grant. (1981). Rete mirabile ophthalmicum in Hawaiian Seabirds. Auk, 98, 844.

Pinshow, B., M.A. Fedak, D.R. Battles, and K. Schmidt-Nielsen. (1976). Energy expenditure for thermoregulation and locomotion in emperor penguins. Am. J. Physiol., 231, 903.

Pinshow, B., M.H. Bernstein, G.E. Lopez, and S. Kleinhaus. (1982). Regulation of brain temperature in pigeons: effects of corneal convection. Am. J. Physiol., 242, R577.

Prinzinger, R. (1982). The energy costs of temperature regulation in birds: the influence of quick sinusoidal temperature fluctuations on the gaseous metabolism of the Japanese Quail (*Coturnix coturnix japonica*). Comp. Biochem. Physiol. A, 71, 469.

Prinzinger, R., and I. Hänssler. (1980). Metabolism-weight relationship in some small non-passerine birds. Experientia, 36, 1299.

Ramirez, J.M., and M.H. Bernstein. (1976). Compound ventilation during thermal panting in pigeons: a possible mechanism for minimizing hypocapnic alkalosis. Fed. Proc. Fed. Am. Soc. Exp. Biol., 35, 2562.

Rautenberg, W., R. Necker, and B. May. (1972). Thermoregulatory responses of the pigeon to changes of the brain and the spinal cord temperatures. Pfleugers Arch., 338, 31.

Rayner, J.M.V. (1982). Avian flight energetics. Annu. Rev. Physiol., 44, 109.

Rintamaki, H., S. Saarela, A. Marjakangas, and R. Hissa. (1983). Summer and winter temperature regulation in the Black Grouse *Lyrurus tetrix.* Physiol. Zool., 56, 152.

Saarela, S., and O. Vakkuri. (1982). Photoperiod-induced changes in temperature–metabolism curve, shivering threshold and body temperature in the pigeon. Experientia, 38, 373.

Schmidt, I. (1978a). Behavioral and autonomic thermoregulation in heat stressed pigeons modified by central thermal stimulation. J. Comp. Physiol., 127, 75.

Schmidt, I. (1978b). Interactions of behavioral and autonomic thermoregulation in heat-stressed pigeons. Pfleugers Arch., 374, 47.

Schmidt, I. (1983). Weighting regional thermal inputs to explain autonomic and behavioral thermoregulation in the pigeon. J. Therm. Biol., 8, 47.

Schuchmann, K.L. (1979). Metabolism of flying hummingbirds. Ibis, 121, 85.

Shapiro, C.J., and W.W. Weathers. (1981). Metabolic and behavioral responses of American Kestrels to food deprivation. Comp. Biochem. Physiol. A, 68, 111.

Sieland, M., J.D. Delius, W. Rautenberg, and B. May. (1981). Thermoregulation mediated by conditioned heart-rate changes in pigeons. J. Comp. Physiol., 144, 375.

Simon, E., C. Simon-Opperman, H.T. Hammel, R. Kaul, and J. Maggert. (1976). Effects of altering rostral brain stem temperature on temperature regulation in the Adelie Penguin, *Pygoscelis adeliae.* Pfleugers Arch., 362, 7.

Simon-Oppermann, C., and R. Martin. (1979). Mammalian-like thermosensitivity in the lower brainstem of the Pekin Duck. Pfleugers Arch., 379, 291.

Simon-Opperman, C., E. Simon, C. Jessen, and H.T. Hammel. (1978). Hypothalamic thermosensitivity in conscious Pekin ducks. Am. J. Physiol., 235, R130.

Skowron, C., and M. Kern. (1980). The insulation in nests of selected North American songbirds. Auk, 97, 816.

Smith, W.K., S.W. Roberts, and P.C. Miller. (1974). Calculating the nocturnal energy expenditure of an incubating Anna's Hummingbird. Condor, 76, 176.

Snapp, B.D., H.C. Heller, and S.M. Gospe, Jr. (1977). Hypothalamic sensitivity in California Quail (*Lophortyx californicus*). J. Comp. Physiol., 107, 345.

Southwick, E.E., and D.M. Gates. (1975). Energetics of occupied hummingbird nests. In "Perspectives of Biophysical Ecology" (D.M. Gates and R.B. Schmerl, Eds.). New York: Springer-Verlag, p. 417.

Spiers, D.E., R.A. McNabb, and F.M.A. McNabb. (1974). The development of thermoregulatory ability, heat-seeking activities, and thyroid function in hatchling Japanese Quail (*Coturnix coturnix japonica*). J. Comp. Physiol., 89, 159.

Stahel, C.D., and S.C. Nicol. (1982). Temperature regulation in the Little Penguin, *Eudyptula minor,* in air and water. J. Comp. Physiol., 148, 93.

Torre-Bueno, J.R. (1978). Evaporative cooling and water balance during flight in birds. J. Exp. Biol., 75, 231.

Torre-Bueno, J.R., and J. Larochelle. (1978). The metabolic cost of flight in unrestrained birds. J. Exp. Biol., 75, 223.

Turner, J.C., and L. McClanahan. (1981). Physiogenesis of endothermy and its relation to growth in the Great Horned Owl (*Bubo virginianus*). Comp. Biochem. Physiol. A, 68, 167.

Walker, L.E., J.M. Walker, J.W. Palca, and R.J. Berger. (1983). A continuum of sleep and shallow torpor in fasting doves. Science, 221, 194.

Walsberg, G.E., G.S. Campbell, and J.R. King (1978). Animal coat color and radiative heat gain: A re-evaluation. J. Comp. Physiol., 126, 211.

Weathers, W.W. (1979). Climatic adaptation in avian standard metabolic rate. Oecologia, 42, 81.

Weathers, W.W. (1981). Physiological thermoregulation in heat-stressed birds: consequences of body size. Physiol. Zool., 54, 345.

Weathers, W.W., and D.C. Schoenbaechler. (1976). Contribution of gular flutter to evaporative cooling in Japanese Quail. J. Appl. Physiol., 40, 521.

Weathers, W.W., and C. van Riper, III. (1982). Temperature regulation in two endangered Hawaiian honeycreepers: the Palila (*Psittirostra bailleui*) and the Laysan Finch (*Psittirostra cantans*). Auk, 99, 667.

Whittow, G.C. (1965). Regulation of body temperature. In "Avian Physiology," 2nd ed. (P.D. Sturkie, Ed.). Ithaca: Cornell University Press, p. 186.

Whittow, G.C. (1976). Regulation of body temperature. In "Avian Physiology," 3rd ed. (P.D. Sturkie, Ed.). New York: Springer-Verlag, p. 146.

Whittow, G.C. (1980). Thermoregulatory behavior of the Laysan and Black-footed Albatross. Elepaio, 40, 97.

Whittow, G.C., and A.J. Berger. (1977). Heat loss from the nest of the Hawaiian honeycreeper, "Amakihi." Wilson Bull., 89, 480.

Whittow, G.C., C.T. Araki, and R.L. Pepper. (1978). Body temperature of the Great Frigate-bird *Fregata minor.* Ibis, 120, 358.

Withers, PC. (1977). Measurement of $\dot{V}_{O_2}$, $\dot{V}_{CO_2}$ and evaporative water loss with a flow-through mask. J. Appl. Physiol., 42, 120.

Wunder, B.A. (1979). Evaporative water loss from birds: effects of artificial radiation. Comp. Biochem Physiol. A, 63, 493.

10
Energy Metabolism*

G.C. Whittow

Introduction

Birds derive all of their energy from the food that they eat. The regulation of the food intake (gross energy intake) would therefore be a logical starting point for this chapter. However, this aspect of energy metabolism is discussed in Chapter 11. Before proceeding to consider the fate of the energy contained in the food, a word should be said about the energy content of the food itself. For a discussion of the energy units used in this chapter, see Chapter 9. For older references, and for documentation of general statements made in this chapter, the reader is referred to two previous editions (Whittow, 1965, 1976).

Energy Content of the Food

Table 10–1 reveals that the three major types of food differ in their energy content. The values included in

* The preparation of this chapter was supported by a grant (N1/R-14) from the Sea Grant College Program (NOAA). The author is indebted to Dr. Hugh I. Ellis for his valuable comments on a draft of the chapter.

TABLE 10–1. Energy content of foods as determined by bomb calorimetry

Category	kJ/g
Carbohydrate	17.15
Protein	22.59 (17.99)
Fat	38.91

Table 10–1 are average values for each food, obtained by bomb calorimetry. In a bomb calorimeter, a small amount of the food under consideration is burned in an atmosphere of oxygen by passing an electrical current through a platinum wire in contact with the food material. All the energy contained in the food is liberated as heat, and it is the amount of heat liberated that is actually measured. The heat produced when fat and carbohydrate are completely oxidized in this way is a valid measure of the energy content of the food.

However, when protein is oxidized in the bird, it yields less than the figure (22.59 kJ/g) from bomb calorimetry, because the protein molecule is not completely oxidized by avian tissues. The end product of protein metabolism, uric acid, which is excreted in the urine, still contains some energy. Consequently, the energy yield of protein in the bird is 17.99 kJ/g. Therefore, carbohydrate and protein yield approximately the same amount of energy, while fat yields over twice as much. The diets of birds vary enormously, and so does the energy content of the various constituents of the diet. The energy content of dry seeds, for example, may be almost six times greater than that of fresh fruit. Kendeigh et al. (1977) cited average values of 18.0 kJ/g for grains and 23.0 kJ/g for insects, two common avian diets.

Disposition of the Food Energy

Although the disposition of the energy contained in the food is not in question, the terms used to describe the allocation of the energy to different uses differ among biologists. Thus, the energy contained in the food that is eaten is referred to as the "gross energy" by animal nutritionists and physiologists (Kendeigh et al., 1977), but variously as the "gross energy intake," "consumption," or "ingestion" by animal ecologists (Petrusewicz and Macfadyen, 1970; Krebs, 1978). The fate of this energy may be depicted as in Diagram [1] below (the letters used by ecologists to identify the quantity of energy are shown in parentheses).

Most of the energy in the food is absorbed from the gastrointestinal tract (the *digestible* energy). The digestible energy is contained in the molecules of amino acids, peptides, fatty acids, monoglycerides, and monosaccharides that are absorbed from the intestine. A very small proportion of these molecules is excreted by the kidney, along with much larger amounts of uric acid and other nitrogenous compounds derived from the breakdown of protein in the body tissues. Energy-containing molecules in the urine and feces make up the *excretory energy.* The energy retained in the body is the *metabolizable energy* (some authors refer to this as the *metabolized energy*). Shortly after the energy-containing molecules are absorbed from the intestine, there is an increased heat production, which is referred to variously as the *heat increment,* the *calorigenic effect,* or the *specific dynamic action* (SDA). Strictly speaking, the specific dynamic action of a particular constituent of the diet refers to the stimulation of metabolic rate that follows the absorption of the constituent from the digestive tract. Thus, the SDA is demonstrable if the constituent is injected into the blood stream.

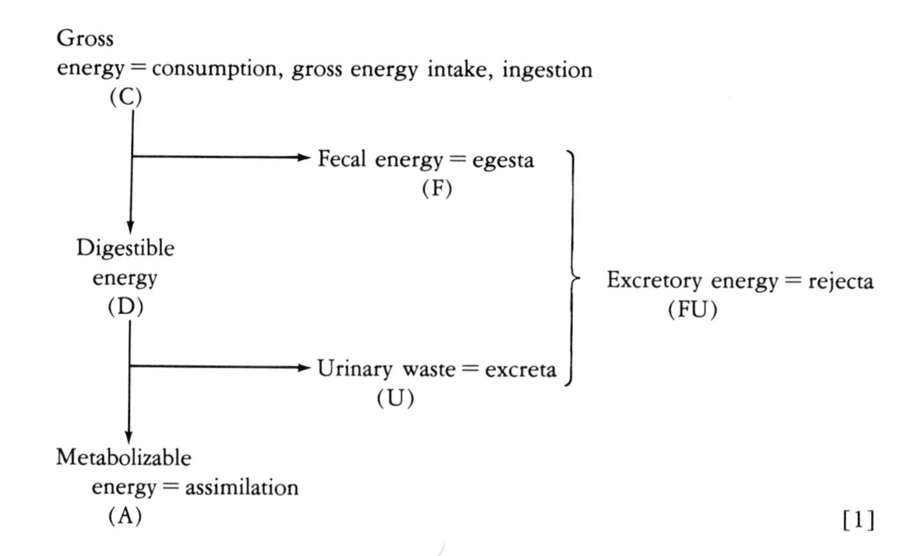

The elevation of basal metabolic rate attributable to the specific dynamic action is 45% for protein, 29% for starch, and 15% for fat (Ricklefs, 1974). When this amount of energy (SDA) is subtracted from the metabolizable energy, the remainder is the energy that is available to the bird for its use—to provide additional heat over and above that incidental to the specific dynamic action, to maintain body temperature, to provide the energy for flight and other activities, to be stored in the form of adipose tissue or (in growing chicks) as newly formed tissue, for maintenance activities such as breathing or cardiac activity, and for special functions such as the production of eggs. However, the performance of many of these functions entails its own energy cost. For example, if fat is deposited, some of the energy must be allocated to the cost of synthesizing fat from the fatty acids and monoglycerides absorbed from the gut. Animal nutritionists recognized the important distinction between energy released as heat (SDA) and energy (*net energy*) that may be used for other purposes:

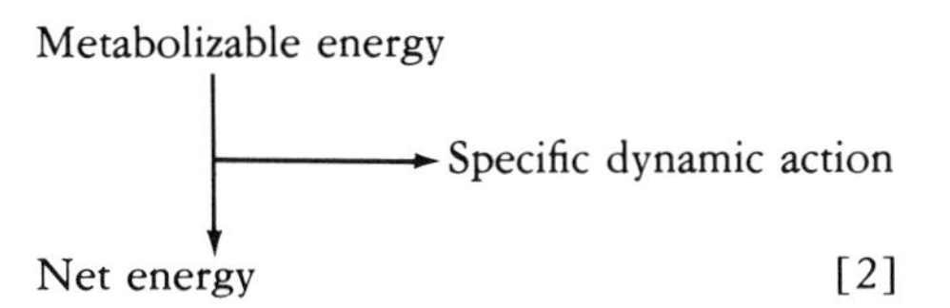

In the domestic fowl, the net energy is approximately 84% of the metabolizable energy. However, in practice the concept of net energy is not greatly used, and ecologists allocate the metabolizable energy to "production" (the energy contained in newly formed tissue as a result of growth or reproduction) or "respiration" as shown in Diagram [3] below.

Two points need to be emphasized, because they are not self-evident in the above diagram: although the energy actually incorporated in newly synthesized tissue would be included in "production," the cost of synthesizing the tissue would be measured as part of the "respiration" term. For this reason, the actual energy incorporated into tissue is often referred to as "net production." When the cost of synthesizing new tissue is subtracted from the "respiration" term, the remainder is a measure of the energy required for "maintenance," although strictly speaking "maintenance" refers to a situation in which there is no production and body mass is constant. The disposition of the food energy (gross) in a laying hen is as follows: 75% to metabolizable energy, and 25% to excretory energy; 18% of metabolizable energy is invested in egg laying (Polin and Wolford, 1973).

The energy costs of thermoregulation were dealt with in the preceding chapter, and they are touched upon in sections of this chapter. It remains to be pointed out that while some birds store energy (largely in the form of fat) to meet their energy requirements during winter, this is not true of all birds. The willow ptarmigan (*Lagopus lagopus*), for example, is an arctic species that has low fat reserves during the winter, relying on its immediate food supply for survival (Thomas, 1982).

Other terms have been used by different authors to refer to the allotment of energy for specific purposes. Thus, Kendeigh introduced *existence metabolism,* defining it as the energy expenditure of caged birds that maintained a constant body mass over a period of days when they were not undergoing reproduction, molting, migratory unrest, growth, or fat deposition. Therefore, this measures the basal metabolic rate, plus energy expended for thermoregulatory purposes, for specific dynamic action, and for the limited amount of activity permitted by the confines of the cage (Kendeigh et al., 1977). Kendeigh defined *productive energy* as the amount of energy that the caged bird mobilizes over and above that required for existence. It consists of the energy needed to produce eggs or feathers or to lay down fat or new tissue.

In the above discussion, energy units were not assigned to the terms used to describe the fate of the gross energy intake. In that way, the disposition of a

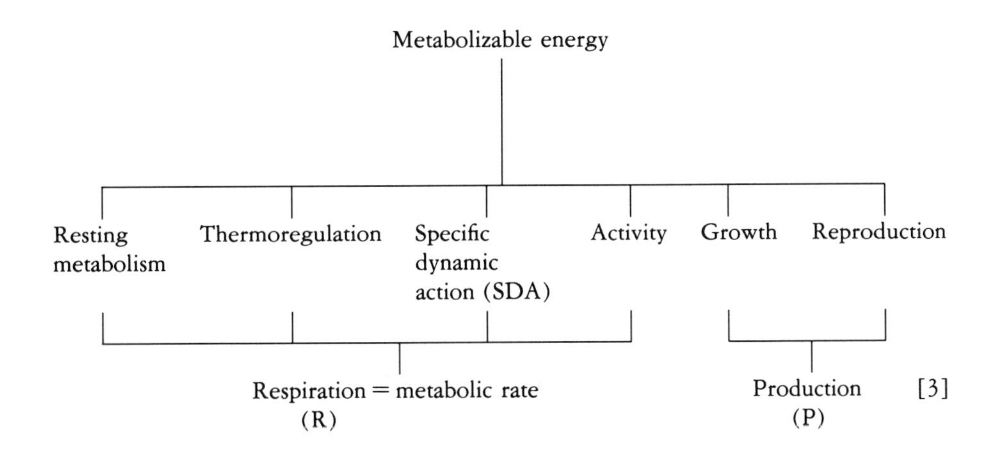

single meal could be envisaged. However, in practice, each of the quantities described by a term is measured as a rate process: e.g., kcal/hr, watts (W).

The Measurement of Energy Exchange

Energy balance studies are comparatively simple to perform on domestic birds and on wild birds kept under laboratory conditions. The amount of food consumed and of feces and urine produced, together with their respective energy content, may readily be determined, yielding the gross energy intake, digestible energy, excretory energy, and metabolizable energy. If the bird is growing, or producing eggs, its production may be measured from the mass of tissue produced and its energy content. Alternatively, if the oxygen consumption of the bird is measured (see Chapter 9), this provides a measure of its metabolic rate or respiration (see diagram [3] above). From diagrams [1–3] presented above, it is clear that, if some of the quantities are known, or can be measured, others may be obtained by addition or subtraction. Thus, if the respiration (metabolic rate) is determined from the oxygen consumption, and production is measured from the amount of tissue formed, then the metabolizable energy may be calculated simply by adding the two terms (using the symbols in diagram [1] and [3]):

$$A = R + P$$

This procedure is much used in studies in which it is difficult or impossible to measure one particular quantity.

There is only one satisfactory method of measuring the energy exchange of free-flying, unconfined birds under natural conditions. This method requires labeling the body water of the bird with stable isotopes of hydrogen and oxygen [deuterium (^{2}H) and ^{18}O, respectively; in current procedures, tritium (^{3}H) is used rather than deuterium]. Samples of body water are taken after a specified period of time, and the change in the concentration of the isotopes is determined. The turnover rate of oxygen is greater than that of hydrogen because the latter is lost only in water, but oxygen is lost both as water and as carbon dioxide. The difference between the two rates of loss is a measure of the carbon dioxide production, which in turn reflects the rate of energy exchange. The method requires that the bird be captured twice, at the beginning and end of the period of observation, and that a blood sample be taken on each occasion. Williams and Nagy (1984a) recently validated the method in savannah sparrows (*Passerculus sandwichensis*). It is important to point out that the doubly labeled water technique is a measure of the metabolic rate of the bird. In other words, it does not provide information on production. However, in birds that are not growing or laying eggs, the technique is a valid measure of the entire energy expenditure, as is evident from the diagram [3].

A more indirect approach has also been used to estimate the energy expenditure of free-living birds under natural conditions. The method depends on the compilation of "time–activity" budgets. A careful note is kept of the bird's activities, which are classified into three or more such categories as sleep, flight, and nonflying activity. From the percentage of time that the bird devotes to a particular activity and the known energy cost of that activity, it is possible to arrive at a figure for the energy expenditure of the bird without interfering with the bird's normal activity. Moreover, it is possible to partition the energy expenditure into the categories of activity used. The energy costs of the different activities are derived from laboratory determinations of the energy expenditure of birds. Clearly, the precision of the method is increased if such determinations are performed on the same species as are studied in the field. Utter and LeFebvre (1973) compared the method with the doubly labeled water ($^{2}H^{18}O$) technique, in the purple martin, and reported excellent agreement. Good agreement was noted also in the savannah sparrow (Williams and Nagy, 1984b). In *Phainopepla* (*Phainopepla nitens*), however, estimates of the daily energy expenditure derived from time–activity–energy budgets were 40% lower than those measured directly using the labeled water ($^{3}HH^{18}O$) technique (Weathers and Nagy, 1980). This discrepancy was attributed to a failure to take into account the effects of solar radiation in the time–activity–energy budgets.

An alternative to the time–activity–energy budget is the observation of feeding rates together with the measured caloric content of the food items (Wakeley, 1978). Another method for use in free-flying birds requires the measurement of heart rate by telemetry, and it is based on correlations between heart rate and either oxygen consumption or other methods of measuring energy expenditure in caged birds (Flynn and Gessaman, 1979). Although indirect in nature, and despite the fact that the heart rate may change in response to many factors, the method may permit an estimate of rapid changes in metabolic rate.

Gross Energy Intake

Among the environmental factors that affect energy intake, environmental temperature is important. For example, the gross energy intake of the zebra finch (*Taenopygia castanotis*) increased linearly with decreasing environmental temperature, a not unexpected phenomenon in view of the increased heat production required of birds below the lower critical temperature (see the preceding chapter). The feeding behavior of birds is affected by solar radiation; feeding by white-crowned sparrows (*Zonotrichia leucophrys gambelii*) was inhibited by insolation, particularly at low air temperatures (Morton, 1967). There are also seasonal differences in gross energy intake in some species. Thus, in the house sparrow (*Passer domesticus*), gross energy intake was greater in summer- than in winter-acclimatized birds (Blem, 1973). In addition to changes in the total amount of food consumed, the nature of the diet may change throughout the year (see following section on egg production). The gross energy intake is the product of the size of each meal, the number of meals, and the energy content of the diet. Hummingbirds appear to meet changes in their overnight energy requirements by varying meal size, but shorter term adjustments are made by variations in feeding frequency (Hainsworth, 1978).

Excretory Energy

Inverse relationships between the excretory energy and ambient temperature have been described for a number of species. This relationship was demonstrable in Alaskan Redpolls (*Acanthis flammea*) under conditions of both constant and fluctuating temperatures and at winter and summer photoperiods. Seasonal variations in the excretory energy occurred in the willow ptarmigan; the calorific value of the guano was lowest during the late-summer molt.

Metabolizable Energy

The metabolizable energy varies with the nature of the diet and in different species (Kendeigh et al., 1977). Thus, the metabolizable energy is only 32% of the energy contained in willow buds and twigs; it is 66% for insects, 76% for grains, and 82% for fish (Ricklefs, 1974). In view of the inverse relationships between gross energy intake and excretory energy on the one hand and environmental temperature on the other, it is not surprising that such a relationship also exists between metabolizable energy and ambient temperature. In the case of Alaskan Redpolls, this relationship is evident at constant and fluctuating temperatures and at winter and summer photoperiods. A diminution in the efficiency of assimilation of the food (metabolizable energy/gross energy intake) with decreasing air temperature was apparent in the zebra finch and in other species also (Kendeigh et al., 1977). Blem (1976) found that the assimilation efficiency in the house sparrow diminished with an increase in the protein content of the diet.

TABLE 10–2. Components (in percentages) of existence metabolism in the house sparrow[a]

Component	Air temperature	
	0°C	37°C
Basal metabolic rate	36	67
Thermoregulation	40	0
Specific dynamic action	14	14
Activity	11	19

[a] After Kendeigh et al. (1977).

Existence Metabolism

The existence metabolism has proved to be useful to physiological ecologists and others interested in estimating the daily energy budgets (DEB) of birds. Kendeigh et al. (1977) presented equations describing the relationship between existence metabolism and environmental temperature in a number of species of caged birds. No zone of thermal neutrality was evident, in contrast to the relationship between the basal metabolic rate and environmental temperature (see Chapter 9). Kendeigh et al. (1977) partitioned the existence metabolism of the house sparrow into its components at two different air temperatures (Table 10–2). It is clear that the basal metabolic rate and thermoregulation account for a major percentage of the total energy expenditure in caged adult birds.

Energy Cost of Activity

The existence metabolism includes a component for the limited activity permitted by the confines of the cage. Birds living under natural conditions indulge in a much greater variety of activity. This activity is influenced by both photoperiod and environmental temperature. For example, the house finch (*Carpodacus mexicanus*) awakened later when air temperature was low and day length was long. The circadian rhythm of activity was slowed by low environmental temperature, but the effect was obvious only at relatively high light intensities. Seasonal variation in the activity of the willow ptarmigan was dramatic. In the roadrunner (*Geococcyx californianus*), activity was less on overcast than on clear days, and the birds were most active in the evening.

Feeding

The energy that a bird expends to procure food must obviously offset, to a variable extent, the energy derived from the food. This is an important consideration in hummingbirds, which feed while hovering, because hovering is very expensive from an energetic point of view (see Chapter 9). The energy required to hover for a given time is greater in large than in small hummingbirds, but the latter have less storage capacity in their crops, relative to their metabolic rates. There may very well be an upper limit on body size for hummingbirds that hover in order to obtain food. The problem is exacerbated for a bird at high altitude, where the ambient temperature is lower and the energy requirements correspondingly greater. The production of nectar, on which hummingbirds feed, and its concentration are lower at high elevations, so that a bird must devote more time and energy to the acquisition of food than at low altitude. Time–activity–energy budgets constructed for hummingbirds revealed that a large part (26–64%) of the energy expenditure of these birds was accounted for by foraging. For additional information on the energy cost of foraging, see the following section on feeding the young.

Flight, Walking, and Swimming

The increase in metabolic rate during flight was discussed in the previous chapter. Many migrating birds perform prodigious feats of endurance flight. The golden plover (*Pluvialis dominica fulva*) migrates between the Aleutian Islands and Hawaii, a distance of 3,840 km, entirely over the ocean, with no opportunity to feed en route. The fuel for long-distance flight is mostly fat; before migration some birds may accumulate fat equivalent to 50% of their body mass, and most of this will have been metabolized by the end of the migratory flight (see following section on premigratory fattening). Indeed, fat is the only feasible source of energy for a migrating bird because of its high energy content per unit mass. As the bird flies and consumes fuel, its mass diminishes, and its rate of fuel consumption decreases also.

The energy cost of moving a unit mass of bird a given distance varies inversely with the size of the bird. This means that large birds can fly a greater distance for the expenditure of a given amount of fuel than is the case for small birds. This conforms with records of the distance over which birds migrate. For example, hummingbirds weighing 5 g are able to cross the Gulf of Mexico, whereas golden plovers (200 g) cross 3,840 km of ocean. Calculations of the energy cost of flight, in conjunction with the known energy reserves of birds, explain the ability of birds to fly great distances, but the margin of safety is apparently small for land birds flying over the ocean.

The most efficient speed of flight for a bird depends on the nature of the flight. If the bird needs to stay aloft for long periods of time while searching for food, it should fly at the speed that has the minimal energy cost. In the laughing gull (*Larus atricilla*), this is 8.5 m/sec. A migrating bird, in contrast, needs to cover the maximal distance on a given amount of fuel. This is accomplished at a flight speed at which the ratio of power expenditure to flight speed is minimal. In the laughing gull, this speed is 13.0 m/sec. Calculations of the flight performance of birds are complex, because of the variations in wind speed and direction that a bird encounters under natural conditions, the aerodynamic efficiency of formation flight in some birds (Rayner, 1982), vertical air movements, and probably other factors also. Kendeigh et al. (1977) presented equations for the total energy expenditure of migration, based on body mass, and they calculated the amount of energy saved by migration as opposed to overwintering in the birds' breeding grounds.

There is a linear relationship between the oxygen consumption of walking birds and the speed at which they walk (Bamford and Maloiy, 1980; Nomoto et al., 1983). However, the energy cost of walking does vary among species; in penguins, it is particularly high. Emperor penguins fast while they are at their breeding sites, which may be 120 km or more from the sea. As a result of the fast, they lose a considerable amount of weight and the energy cost of walking back to the sea after their stay at the rookery is much less than when they walked to the rookery at the beginning of the breeding season. It has been estimated that the total energy cost of walking to and from the rookery requires less than 15% of the initial energy reserves of the birds (Pinshow et al., 1976; Dewasmes et al., 1980; Le Maho, 1983). During locomotion on a treadmill, the oxygen consumption of the white-crowned sparrow was a linear function of treadmill speed, but the slope of the relationship depended on air temperature. It was concluded that at very low air temperatures the cost of foraging was no greater than that of shivering at rest (Paladino and King, 1984).

Ducks and penguins swim in quite different ways, the former paddling with their feet and the latter using their "wings" as flippers and diving to considerable depths. Nevertheless, estimates of the metabolic rate of swimming were remarkably similar in the two instances: 2.2 times the resting metabolic rate in ducks and 2.8 times the standard metabolic rate in penguins (Prange and Schmidt-Nielsen, 1970; Kooyman et al., 1982). Ducks have been shown to swim at speeds calculated to cost the minimal amount of energy.

Reproductive Activity

Many birds start the reproductive season with large energy reserves, but this is not true for all birds. The willow ptarmigan is an arctic species that enters its reproductive season with low levels of body fat, relying on its food supply to meet its energy requirements for reproductive purposes (Thomas, 1982).

Nest Construction. The energy expenditure of the zebra finch increases during nest building, and the increase is greater at lower air temperatures. The daily energy expenditure in the cliff swallow (*Petrochelidon pyrrhonota*) was greater during nest building than during incubation or the nestling period (Withers, 1977). The daily energy expenditure during construction of the nest was also greater than during incubation in the willow flycatcher (*Empidonax traillii;* Ettinger and King, 1980).

Incubation. The energy cost of incubation has long been a matter of controversy. This is partly because it is difficult to measure and partly because some investigators have considered the "waste" heat from normal metabolism to be adequate to warm the eggs (King, 1973; Vleck, 1981). However, this latter view is a fallacy that is inconsistent with the principles of heat transfer. Birds do not produce "excess" heat; they produce the requisite amount of heat to maintain their high body temperature, given the thermal conductance of their tissues and plumage and the temperature of their environment. If the thermal conductance changes, either the amount of heat stored or the heat production (or both) will have to change (see Chapter 9). Thus, the heat loss from a bird sitting on eggs will differ from that of the same bird before it has developed an incubation patch, unless the thermal conductance of the eggs is identical to that of the plumage that covered the area of the incubation patch before the development of the patch. If the heat loss is different in these two circumstances, then the heat production and/or the heat storage of the bird will also have to change, and there will be a definite energy cost of incubation. In view of these and other considerations, it is not surprising that models of the energy cost of incubation have not accurately predicted the measured cost (Vleck, 1981). A crucial comparison, which has rarely been made, is between (a) the energy expenditure of the bird without an incubation patch, sitting on the nest in the absence of eggs, and (b) the bird with an incubation patch (if it develops one), sitting on eggs in the nest.

In the starling (*Sturnus vulgaris*), the zebra finch, and the Laysan albatross (*Diomedea immutabilis*), the energy cost of incubating the eggs was considered to be negligible when the air temperature was within the thermoneutral zone (see Chapter 9) of the birds (Biebach, 1979; Vleck, 1981; Grant and Whittow, 1983). In the black-bellied magpie (*Pica pica hudsonia*), the energy expenditure during incubation was lower than that during any other phase of the bird's life cycle, 1.2 times the basal metabolic rate (Mugaas and King, 1981). Walsberg and King (1978) concluded that the energy expenditure of the incubating mountain white-crowned sparrow (*Zonotrichia leucophrys oriantha*) was 15% lower than that of a bird perching outside the nest but exposed to the same microclimate. However, in the great tit (*Parus major*), the wandering albatross (*Diomedia exulans*), and in both petrels and penguins, it was estimated that incubation involved a definite energy cost (Mertens, 1980; Croxall, 1982; Croxall and Ricketts, 1983).

Several studies have determined that the energy cost of incubation increases below the lower critical temperature (Vleck, 1981). As the thermal conductance of the egg does not change with temperature, this implies that the "tightness" with which the bird sits on the egg, or its attentiveness, changes with temperature. The goldcrest (*Regulus regulus*) is one of the smallest species to breed at high latitudes (Haftorn, 1978a, b). In spite of this, it is able to maintain egg temperature at 36.5°C, requiring the expenditure of a large percentage of its energy expenditure.

Feeding and Brooding of the Nestlings. Newly hatched chicks utilize their yolk sacs as a source of energy. Pettit et al. (1984) calculated how long some tropical seabird chicks may subsist on the energy contained in the yolk sac; in albatross chicks, this was as long as 6.2 days. In the domestic fowl, the yolk sac may provide substenance for 5 days (Freeman and Vince, 1974). Some species provide their nestlings with an energy-rich diet during the initial stages of the nestling period. Thus, petrels regurgitate stomach oils, rich in energy, to their nestlings (Warham et al., 1976), while pigeons receive only crop milk, also rich in energy, during the first 3 days after hatching (Vandeputte-Poma, 1980).

Altricial nestlings must be brooded by the parent birds initially, as they are unable to regulate their own body temperature (see Chapter 9). Kendeigh et al. (1977) presented an equation, based on the body mass of the brood, for estimating the energy cost of brooding. Using the doubly labeled water technique (see section on measurement of energy exchange), Hails and Bryant (1979) determined that the metabolic rate of house martins (*Delichon urbica*) feeding young averaged 3.89 times the basal metabolic rate. In the black-bellied magpie, the cost of feeding the young was also high, 2.08 times the basal metabolic rate (Mugaas and King, 1981). The daily energy expenditure of willow flycatchers was 28% greater during the nestling stage than while incubating the egg (Ettinger and King, 1980). In the mockingbird (*Mimus polyglottos*), the daily energy expenditure of the female birds caring for fledglings was 151% of that of incubating birds (Biedenweg, 1983). There is no doubt that gathering food for the nestlings is expensive in terms of both time and energy (Ricklefs, 1974).

Production (Energy Storage)

The storage of energy in the form of avian tissue is often important to the commercial poultryman (eggs, muscle), and it is always important biologically.

Growth

Embryo. The oxygen consumption of the egg increases throughout incubation (Figure 10–1) but the pattern of the increase differs depending on the mode of development of the species (Vleck et al., 1980). Thus, in precocial species (Nice, 1962) the oxygen consumption may slow prior to pipping, but in altricial species the oxygen consumption increases continuously. Pipping results in a further increase in oxygen consumption (Figure 10–1). The pattern of embryonic growth largely parallels that in oxygen consumption (Figure 10–1). The main source of energy for the embryo is the fat contained in the yolk. The total energy cost of development is higher in precocial than altricial embryos (Vleck et al., 1980), and this is reflected in the higher yolk content of precocial eggs (Carey et al., 1980; Pettit et al., 1984). Prolonged incubation also increases the energy cost of development, probably because of the increased maintenance costs of the embryo (Vleck et al., 1980, 1984; Whittow, 1980; Pettit et al., 1982). An extreme example of this trend is seen

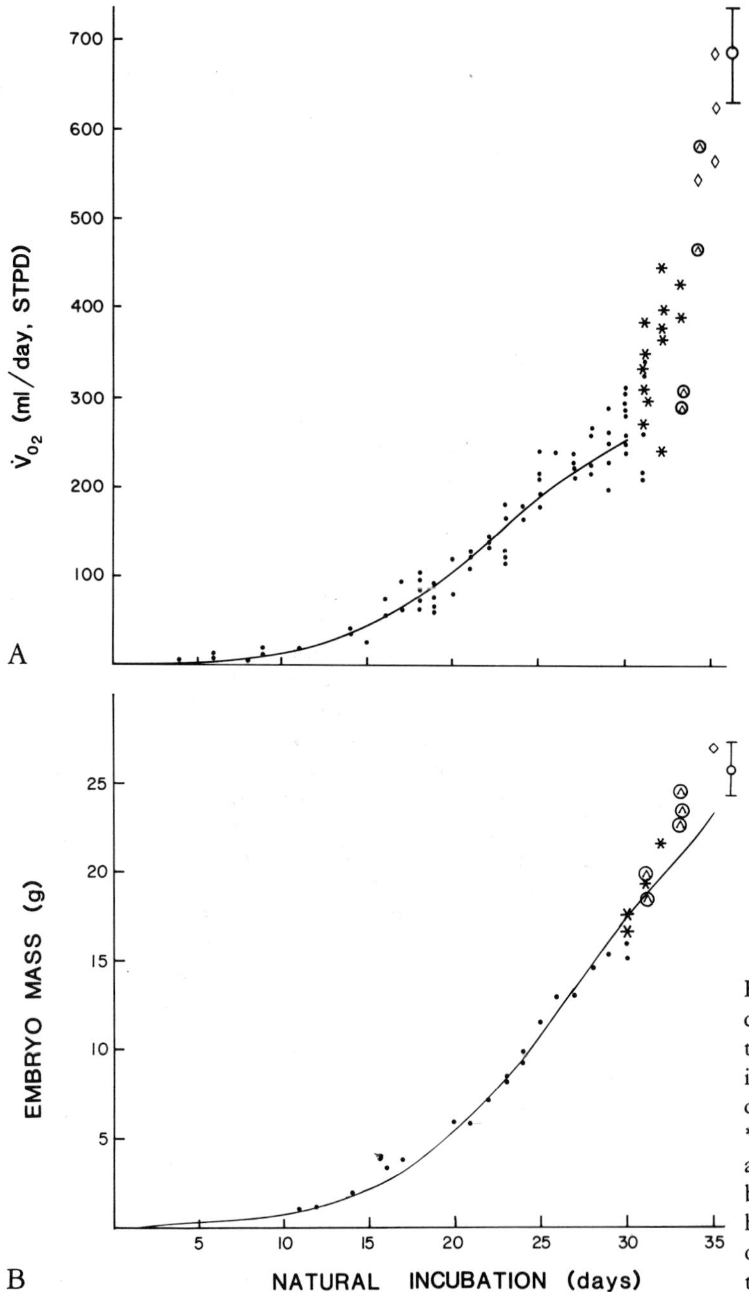

FIGURE 10–1. Embryonic oxygen consumption (**A**) and growth (**B**) of the brown noddy (*Anous stolidus*) during natural incubation. Symbols: ●, data prior to star-fracture of the shell; *, after star-fracture of the shell; Ⓐ after penetration of the air cell by the beak; ◇, after formation of shell pip-hole; O, mean oxygen consumption of newly hatched chicks (±SD). (After Pettit and Whittow, 1983.)

in the kiwi (*Apteryx australis*), which is both precocial and has a long incubation period. The kiwi's egg is one of the largest in relation to adult body mass, and it has one of the highest proportionate yolk contents known (Calder et al., 1978).

There is evidence that the porosity or "gas conductance" of the egg shell determines the rate at which oxygen may diffuse into the egg and therefore the rate at which the embryo is able to grow (Rahn et al., 1974). Avian eggs vary in the gas conductances of their shells, some species such as the small Procellariiformes having relatively low conductances and slow embryonic growth (Pettit et al., 1982). Even within species, individual eggs may have markedly different shell conductances, and in the domestic fowl, duck, and turkey it has been shown that these differences are reflected in the mass of the embryo (Burton and Tullett, 1983).

In some birds, the interval between pipping of the egg and hatching of the chick accounts for a disproportionate share of the total oxygen consumption of the egg. Thus, in the bonin petrel (*Pterodroma hypoleuca*) no less than 50% of the total oxygen consumption of the egg occurs during the pip-to-hatch interval (Pettit et al., 1982; Whittow, 1983). The increase in oxygen consumption that occurs between pipping and hatching may be attributed to the fracture of the shell and the penetration of the aircell by the embryo, both of which permit an increased oxygen transfer into the egg. In addition, the increased activity of the embryo, coincident with its attempts to fracture the shell and leave the egg, results in an increase in oxygen consumption. There is evidence for an increased level of thyroid activity, which might also be expected to increase the metabolic rate of the embryo, during pipping (Decuypere et al., 1979).

Chick. Within a few hours of hatching, an increase in heat production occurs as a correlate of the greatly increased heat loss of the chick of the domestic fowl (see Chapter 9). Subsequently, the oxygen uptake increases as the chick grows, the highest value per unit body mass (mass-specific) occurring at about 15 days. It then decreases until the adult level is reached. It is interesting that the thyroxine secretion rate is also highest at 2 weeks of age. The relationship between body mass and basal metabolic rate (BMR) is not the same in young chicks of the domestic fowl as it is in adults. Thus, in adults, the BMR is related to an exponent of body mass of approximately 0.75 (see Chapter 9) but in chicks the exponent is closer to 1.0 (Kuenzel and Kuenzel, 1977). In both precocial and altricial chicks, the basal metabolic rate per unit body mass increases to a peak and then declines as the chick increases in size (Ricklefs, 1974). In both cases, the maximal value exceeds that for an adult bird of the same size, but the peak in metabolic rate occurs earlier in the life of the chick in precocial than in altricial species. However, Bucher (1983) reported that the mass-specific metabolic rate of parrot nestlings never exceeds that of adults, and there may be other exceptions to Ricklefs' generalization. During the perinatal period, yolk lipid continues to be the major source of energy. The rather low metabolic rate of the highly precocial mallee fowl (*Leipoa ocellata*) hatchling is believed to be advantageous in times of food shortage (Booth, 1984).

Domestic fowl have been bred selectively for rapid growth (broilers) or egg production (layers). Kuenzel and Kuenzel (1977) found that the basal metabolic rates of the more rapidly growing broiler chicks were lower than those of layers when the BMR values were corrected for differences in body mass (Table 10–3). This resulted in a saving of energy to the broiler chicks of 582 kJ, which presumably was diverted, at least in part, to growth.

The rate of growth of cockerels is greater than that of hens, but the energy equivalent of the growth increments is greater in pullets than in cockerels. This may be a result of the greater ability of the hen to fatten during growth, because the energy equivalent of the gained body weight varies directly with the fat content of the increased body weight. The difference between the male and the female fowl in this respect is largely abolished if the males are caponized. If testosterone propionate is administered to pullets, their rate of growth increases to equal that of cockerels. Male turkeys also grow at a greater rate than do females.

The increase in body mass during growth mainly involves the synthesis of protein. The energy cost of protein deposition in fowls is 32.4 kJ of metabolizable energy per gram of protein. The deposition of 1 g of fat requires 65.5 kJ of metabolizable energy. The main factor determining the amount of fat deposited is the protein content of the diet in relation to the total energy; the greater the protein/energy ratio, the lower the fat content of the bird. Methionine-deficient

TABLE 10–3. Basal metabolic rates[a] of chicks of broilers and layers of domestic fowl[b]

Age (weeks)	Broilers	Layers
0	6.50	7.01
1	8.31	9.60
2	8.26	9.85
3	6.88	8.63
4	8.44	8.64
5	6.03	8.65
6	5.30	7.91
7	4.81	7.06
8	4.35	6.03

[a] Rates in W/g body mass.
[b] After Kuenzel and Kuenzel (1977).

chicks gain more fat but less protein than do chicks that receive supplementary methionine. The percentage utilization of the metabolizable energy of the diet for gains in tissue energy is significantly increased in fowls by the addition of supplementary methionine to the diet.

During growth, the organs and tissues of the chick may be considered to consist of those that require energy (e.g., muscle, skeleton) and those that not only consume energy but must also supply it to the other tissues (e.g., cardiovascular and digestive systems). Lilja (1981, 1982) found that in rapidly growing species, such as the goose (*Anser anser*), there is early development of the digestive tract and late development of the muscles and feathers. In more slowly growing Japanese quails (*Coturnix c. japonica*), the reverse was true. Rapid growth appears, therefore, to require early development of the energy-supplying tissues such as the gut.

The growth of young chicks is adversely affected by exposure to hot weather. The impaired growth is partly caused by diminished food intake but also by less efficient food conversion. The optimal environmental temperature for growth in chicks of the domestic fowl seems to be approximately 35°C. The composition of the gained body substance during growth is also influenced by the environmental temperature. For example, the amount of fat stored per gram body weight increase is maximal at an environmental temperature of 32°C and less at temperatures of 21° or 40°C. Growth rates of chicks are adversely affected by cold winds, probably because heat loss is accelerated under these conditions and energy otherwise available for growth is used to maintain the body temperature.

Ricklefs (1976) reported that the mean oxygen consumption of the nestlings of nine tropical species was 25% less than that of five temperate species at comparable stages of development, while the mean growth rate of the tropical species was 32% less. A recent study of three tropical insectivores revealed that the peak energy demands of the chicks were less than those of their counterparts in a temperate climate (Bryant and Hails, 1983). This was attributed to slow growth rates and reduced thermoregulatory requirements, among other factors, in the tropical species. Nestling activity levels were also low in the tropical chicks.

A reduction in thyroid activity as a result of thyroidectomy or the administration of thiouracil, which inhibits the thyroid, is associated with a reduced rate of growth (Davidson et al., 1980). The reduced rate of growth is probably related to a concomitantly reduced metabolic rate.

The energetics of growth are closely related to the development of temperature regulation. Morton and Carey (1971) identified three stages in the development of the mountain white-crowned Sparrow. During the first few days, growth was rapid because the chick was brooded by the parent and few calories were needed to keep warm. From the third to the sixth day, the chick was able to increase its heat production in response to exposure to cold, and the rate of increase in body mass was relatively slow because energy was diverted to thermoregulation and the growth of the feathers. From the seventh day to fledging (ninth day), the allocation of energy to activity, thermoregulation, and the further growth of the feathers meant that little increase in body mass occurred. In a comparison of the altricial double-crested cormorant (*Phalacrocorax auritus*) with the semiprecocial herring gull (*Larus orgentatus*), Dunn (1976) estimated that a much smaller percentage of the total energy expenditure was attributed to thermoregulation in the cormorant, which depended on its parents for protection from the climatic environment. Figure 10–2 shows that over the entire nestling period in the altricial house sparrow, by far the greatest proportion of the total energy expenditure is allocated to resting metabolism, which includes the cost of thermoregulation. The mutual insulation between members of a brood of nestlings was effective in reducing the energy requirements of the individual nestlings in house martins (Bryant and Gardiner, 1979).

The allocation of energy to growth is also affected by the rate of growth. Thus, in the more slowly grow-

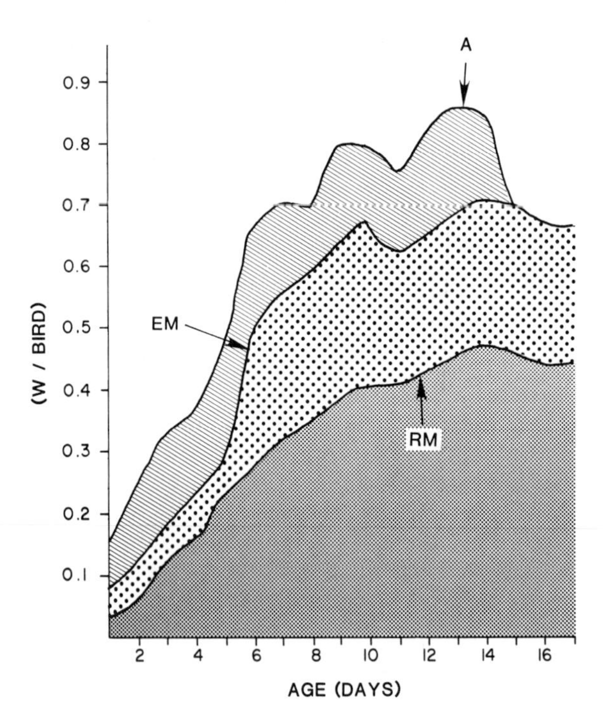

FIGURE 10–2. Resting metabolism (*RM*), existence metabolism (*EM*), and metabolizable energy (*A*) of the nestling house sparrow (*Passer domesticus*). The difference between *RM* and *EM* (stippled area) is an estimate of the cost of activity. The difference between *A* and *EM* (hatched area) is a measure of productive energy. (After Blem, 1975.)

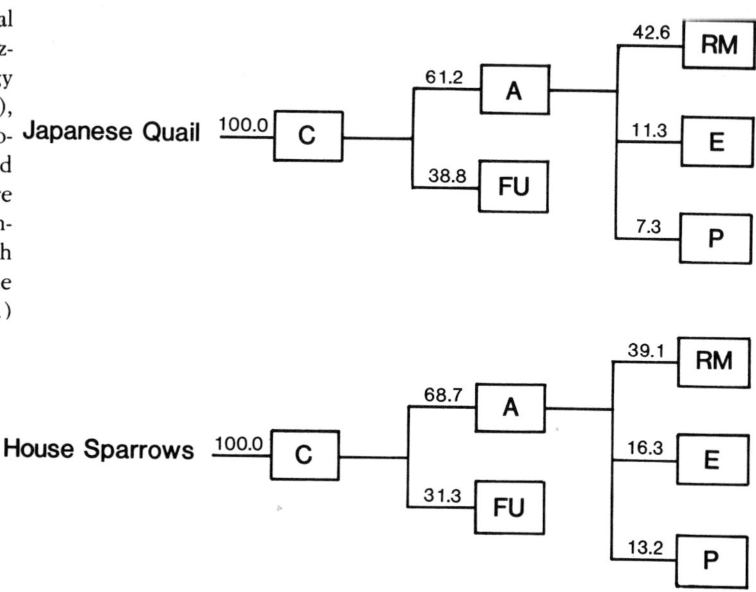

FIGURE 10–3. Allocation of the total gross energy intake (C) to metabolizable energy (A), excretory energy (FU), resting metabolic rate (RM), energy cost of activity (E), and production (P), in the Japanese quail and the house sparrow. The numbers are percentages. The total energy consumed by the nestling during growth from hatchling to adult is assigned the value of 100%. (After Blem, 1978.)

ing precocial dunlin (*Calidris alpina*), relatively less energy was accounted for by growth than in the altricial starling (Dunn, 1980). Ricklefs and White (1981) compared the energy budgets of two closely related terns with markedly different rates of growth. In the more rapidly growing common tern (*Sterna hirundo*), the allocation of energy to the synthesis of both fat and other tissues increased during the first half of the nestling period and then declined. In the more slowly growing sooty tern (*Sterna fuscus*), the allocation of energy to the formation of body tissues declined throughout the nestling period while the energy expended in maintenance (circulation of the blood, breathing, active transport of ions, thermoregulation, etc.) increased.

The assimilation efficiency (metabolizable energy/gross energy intake) changes during growth, even when the diet remains unchanged (Kendeigh et al., 1977). Thus, in sparrows, the efficiency increased to a maximum on the seventh day and then declined, the maximum coinciding with the period of most rapid growth. In ducks, on the other hand, the efficiency tended to increase with age.

Figure 10–3 presents the allocation of the total gross energy intake (represented as 100%) to metabolizable energy, excretory energy, resting metabolism, activity, and production in the altricial house sparrow and the precocial Japanese quail. It is clear that a smaller proportion of energy is lost (excreted) in the house sparrow and a correspondingly greater investment is made in tissue energy (production).

Energy Storage in the Adult Bird

In the adult bird, energy is stored partly as glycogen in liver and in muscle but mainly as fat. Not surprisingly, more fat is deposited in birds fed a high-energy diet than in those on a low-energy diet. The body mass and percentage fat are lower in fowls allowed access to food for only 2 hr per day (meal eaters) than in birds fed ad libitum (nibblers).

Premigratory Fattening. An increase in body mass has been observed in many species of birds prior to migration (Kendeigh et al., 1977; Blem, 1980). The increase in body mass is largely due to the deposition of fat, although hypertrophy of the flight muscles occurs also (Marsh, 1984). The main function of the accumulated fat, which may amount to 50% of the initial body mass (Blem, 1980), is to provide a reserve of energy for the migratory flight. The increase in the body mass of the ortolan bunting (*Emberiza hortulana*) before migration is associated with only a very slight increase in the metabolic rate. The metabolic rate per unit of lean body (fat-free) mass is unchanged during the fattening period, indicating that the deposition of fat has no detectable accelerator effect on the metabolic processes. In the brambling (*Fringilla montifringilla*), however, the oxygen consumption during the migratory season is lower, permitting the diversion of energy from heat production and into the deposition of fat. The fat is mainly deposited in the abdomen and beneath the skin (Blem, 1980); this involves an increase in the fat content of the cells and not an increased number of cells. The liver is the main site of fatty acid synthesis.

Although there is a reduction in the amount of energy used in normal activities, an increased food intake is the principal source of calories for lipogenesis preceding the spring migration (Kendeigh et al., 1977); the hyperphagia seems to be induced by the increasing vernal photoperiod. In the white-crowned sparrow, premigratory fattening may be simulated by increasing the duration of the daily photoperiod. The effect appears to be mediated by prolactin and gonadotropin.

The caloric value of the excreta decreases in the dickcissel (*Spiza americana*) during premigratory hyperphagia, a factor favoring the efficiency of deposition of fat. A diminished capacity for fatty acid oxidation in the premigratory rosy pastor (*Sturnus roseus*) is consistent with the bird's predilection to lay down fat rather than to metabolize it.

Egg Production. Egg production may impose considerable nutritional demands on female birds (Ricklefs, 1974). They meet these demands in various ways (see Ankney and Scott, 1980). Pintail ducks (*Anas acuta*) utilize body fat and eat a high-protein diet, while female cowbirds (*Molothrus ater*) rely entirely on their diet, which changes from seeds before egg laying to insects during egg formation. Species that breed in environments where food is scarce rely on stored reserves in the body (see Krapu, 1981). In the fiordland crested penguin (*Eudyptes pachyrhynchus*), which does not eat during egg formation, it was estimated that the quantity of nutrients transferred daily from tissues to eggs was not large compared with body reserves (Grau, 1982). Ricklefs (1974) assembled information on the energy cost of egg laying in several groups of birds; the daily energy requirement for production of the eggs was 40–50% of the BMR for altricial species and 125–180% for precocial species. The greater cost for precocial species correlates with the greater yolk and energy content of the eggs of precocial species. The basal metabolic rate of laying hens of the domestic fowl is higher than that of nonlayers.

The metabolizable energy is significantly increased during the egg-laying period in the zebra finch as is the efficiency of food utilization; these also increase in the willow ptarmigan during the egg-laying period. In the black-billed magpie (*Pica pica hudsonia*), ovogenesis accounted for 23% of the total daily energy expenditure (Mugaas and King, 1981). In fowls, the efficiency of conversion of metabolizable energy into production (eggs) is approximately 76%; the efficiency of converting dietary energy to eggs is not affected by the energy content of the diet. The percentage of the metabolizable energy that is diverted to egg production is increased by 20%, however, by feeding the birds only for two 2.5-hr periods each day as opposed to ad libitum feeding. This increased productive efficiency is achieved at the expense of body fat stores. The diminished egg production observed in a hot environment is often related to the lower energy intake. Birds have the highest productivity at environmental temperatures within the thermoneutral range, because their energy requirements for thermoregulatory purposes are at a minimum then.

Molting. The basal metabolic rate increases during molting. In the ortolan bunting, the increase is 26%, and in the fowl, (*Gallus domesticus*) 45%. The increase in metabolic rate during the molt is less in young birds than in adults. The greatest increase in metabolic rate coincides with the regeneration of the flight feathers. Molting is associated with an increase in thyroid activity, which is probably one of the causes of the increase in metabolic rate. Loss of feathers, naturally or artificially, causes an increase in heat production and heat loss (see Chapter 9). The frizzle fowl is a breed characterized by scanty plumage, and its heat production is greater than that of normal fowls; it also possesses a relatively high rate of thyroid activity.

During the early phase of molting, the metabolizable energy is high because the loss of feathers decreases the insulation and additional energy is needed to fuel the increased heat production and tissue synthesis. During the latter part of molting, the metabolizable energy falls below the premolt level, probably because the new plumage provides a more effective insulation (see Kendeigh et al., 1977). In the willow ptarmigan, the excretory energy is also low, and the efficiency of extracting energy from the food high, during molting. In the black-billed magpie, the molt was estimated to account for 8% of the daily energy expenditure (Mugaas and King, 1981).

Penguins fast when they are molting. This means that the materials and energy necessary for regeneration of the plumage must come from the breakdown of body tissues. The daily change in body mass was four times greater in fasting penguins that were molting than in those that were fasting but not molting (Le Maho, 1983). In molting macaroni (*Eudyptes chrysolophus*) and rockhopper (*E. chrysocome*) penguins, fat reserves declined at the rate of 38 and 33 g/day, respectively (Williams et al., 1977). The energy requirements of the molting birds were 1.6 and 2.1 times the standard metabolic rates of the two species, respectively. Croxall (1982) also calculated that the metabolic rates of molting penguins were approximately twice the BMR. In five species of penguins, 37–45% of the body mass was lost during the molt.

Effects of Energy Deprivation

When birds are deprived of food, their heat production diminishes. In the American kestrel (*Falco sparverius*), the basal metabolic rate had declined 23.4% by the third day of the fast (Shapiro and Weathers, 1981). Although the body temperature declined also, the observed reduction in BMR was greater than that contingent upon the decrease in temperature. The respiratory quotient (RQ) (see Chapter 9) decreases also, because fat is preferentially metabolized during starvation. Minimal RQ values are attained in the domestic fowl after a 24- to 48-hr period of starvation, depending on the size of the bird (Misson, 1974); the pigeon requires 28 hr and the ortolan and yellow bunting only 3 hr.

The glycogen reserves of pigeons are depleted within 24 hr of the start of a fast. Freeman et al. (1980) showed that the effects of fasting are evident within a few hours of the start of the fast in young domestic fowl. A subsequent rise in RQ after 72 hr of fasting in the domestic fowl was considered to reflect gluconeogenesis, the utilization of protein for glucose synthesis (Misson, 1974). Young swifts (*Micropus apus*) may lose 60% of their body weight during a fast, whereas adults may lose only 38% before death supervenes. The young birds were heavier initially and had greater fat deposits than adults. Next to adipose tissue, liver and muscle incur the greatest losses of weight during the fast. In the pigeon, adipose tissue loses 93%; spleen, 71%; pancreas, 64%; liver, 52%; heart, 45%; and muscle, 42%, of their respective initial weights during a fast.

Some birds voluntarily fast for long periods of time. Thus, the male emperor penguin (*Aptenodytes forsteri*) fasts for almost 4 months while incubating the egg (Le Maho et al., 1976). Over this period, the birds lose 40% of their initial body mass. In spite of the harsh conditions to which it is exposed, the fasting, incubating emperor penguin does not elevate its metabolic rate. Fasting, male Adélie penguins (*Pygoscelis adeliae*) lost 56 g of fat each day for 27 days during the breeding season (Johnson and West, 1973). Le Maho (1983) identified three phases in the loss of body mass in fasting penguins. Initially, there was a sharp decrease in both body mass and metabolic rate, as fat was mobilized as an energy source. Following this, the decline was more gradual, but toward the end of the fasting period, the rate of decrease in body mass increased, ostensibly due to the mobilization of body proteins to supply energy.

Daily Energy Budget

The total daily energy expenditure has been calculated for a number of species. One of the most extensive studies was carried out by Mugaas and King (1981) on the black-billed magpie. The observations, which covered the entire year, revealed that the periods of highest energy expenditure—on reproduction, molt, and thermoregulation—occurred at different times of the year. The total daily energy expenditure varied from 1.2 times the BMR during incubation to 2.08 times the BMR during the feeding of the nestlings. Differences in activity accounted for most of the variation in daily energy expenditure. A compilation of the daily energy expenditure by other species revealed that it exceeded the BMR by a factor of 1.3–7.2. Figure 10–4 illustrates the daily expenditure of energy by the house sparrow in Illinois throughout the year. In spite of the fact that reproduction, thermoregulation, and molt occurred sequentially throughout the year, the energy expenditure was clearly greatest during the winter, owing to the high cost of thermoregulation at that time of year (Kendeigh et al., 1977). In another passerine, the willow flycatcher, variations in air temperature accounted for 90% of the variation in daily energy expenditure (Ettinger and King, 1980).

Using the doubly labeled water technique Bryant

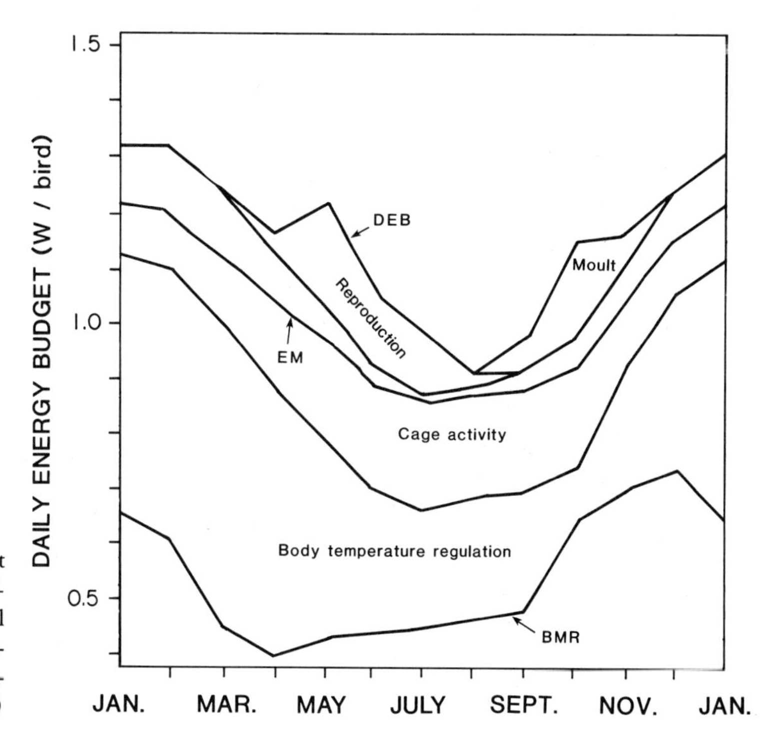

FIGURE 10–4. Daily energy budget of the house sparrow (*Passer domesticus*) throughout the year. *BMR*, basal metabolic rate; *EM*, existence metabolism; *DEB*, total daily energy budget. (After Kendeigh et al., 1977.)

and Westerterp (1980, 1983) found that the metabolic rate of house martins was greater when food (insects) was more abundant and feeding activity was increased. As might be expected, the average daily metabolic rate increased with the amount of time spent flying, particularly flapping flight. Larger birds had lower daily energy requirements per unit body mass. In the mockingbird, the basal metabolic rate and thermoregulatory requirements accounted for the bulk of the daily energy expenditure (Biedenweg, 1983).

Energy Requirements of an Avian Community

Several attempts have been made to estimate the total amount of energy required by an avian population. This is of considerable interest to ecologists, and it often has practical application also. This has been of particular concern to fishermen and fishery biologists, because dense colonies of nesting seabirds may have a considerable impact on the biological resources of the oceans. It is virtually impossible to measure directly the amount of food that a population of seabirds may consume. However, if the energy requirements of individual birds are determined (by the doubly labeled water technique, for instance) and the total numbers of the birds and their diet are known (they can both be estimated or determined), it is possible to calculate how much food the birds would have to consume in order to meet their energy requirements. Using this approach, Wiens and Scott (1975) concluded that four species of seabirds off the coast of Oregon consumed as much as 22% of the annual production of the fish; Furness (1978) estimated that the energy requirements of a seabird population in the Shetland Islands required the removal of 29% of the total annual fish production of the seas within a 45-km radius of the bird colony.

Summary

Recent work on the energetics of birds has attempted to define the total energy requirements of birds in their natural surroundings. The increasing use of the doubly labeled water technique promises further advances in our knowledge of the energy expenditure of free-flying birds. The use of computers has simplified the estimation of the energy requirements of entire populations of birds. The accessibility of the avian egg has prompted studies of the energy requirements of the embryo, and recent work has brought to light the diversity of avian growth patterns.

References

Ankney, C.D., and D.M. Scott. (1980). Changes in nutrient reserves and diet of breeding Brown-headed Cowbirds. Auk, 97, 684.

Bamford, O.S., and G.M.O. Maloiy. (1980. Energy metabolism and heart rate during treadmill exercise in the Marabou Stork. J. Appl. Physiol., 49, 491.

Biebach, H. (1979). Energetik des Brütens beim Star (*Sturnus vulgaris*). J. Ornithol., 120, 121.

Biedenweg, D.W. (1983). Time and energy budgets of the Mockingbird (*Mimus polyglottus*) during the breeding season. Auk, 100, 149.

Blem, C.R. (1973). Geographic variation in the bioenergetics of the House Sparrow. Ornithol. Monogr., 14, 96.

Blem, C.R. (1975). Energetics of nestling House Sparrows *Passer domesticus*. Comp. Biochem. Physiol. A, 52, 305.

Blem, C.R. (1976). Efficiency of energy utilization of the House Sparrow, *Passer domesticus*. Oecologia, 25, 257.

Blem, C.R. (1978). The energetics of young Japanese Quail, *Coturnix coturnix japonica*. Comp. Biochem. Physiol. A, 59, 219.

Blem, C.R. (1980). The energetics of migration. In "Animal Migration, Orientation and Navigation" (S.A. Gauthreaux, Ed.). New York: Academic Press, p. 175.

Booth, D.T. (1984). Thermoregulation in neonate Mallee Fowl (*Leipoa ocellata*). Physiol. Zool., 57, 251.

Bryant, D.M., and A. Gardiner. (1979). Energetics of growth in House Martins (*Delichon urbica*). J. Zool., 189, 275.

Bryant, D.M., and C.J. Hails. (1983). Energetics and growth patterns of three tropical bird species. Auk, 100, 425.

Bryant, D.M., and K.R. Westerterp. (1980). The energy budget of the House Martin (*Delichon urbica*). Ardea, 68, 91.

Bryant, D.M., and K.R. Westerterp. (1983). Short-term variability in energy turnover by breeding House Martins *Delichon urbica:* a study using doubly-labelled water ($D_2{}^{18}O$). J. Anim. Ecol., 52, 525.

Bucher, T.L. (1983). Parrot eggs, embryos, and nestlings: patterns and energetics of growth and development. Physiol. Zool., 56, 465.

Burton, F.G., and S.G. Tullett. (1983). A comparison of the effects of eggshell porosity on the respiration and growth of domestic fowl, duck and turkey embryos. Comp. Biochem. Physiol. A, 75, 167.

Calder, W.A., C.R. Parr, and D.P. Karl. (1978). Energy content of eggs of the Brown Kiwi *Apteryx australis;* an extreme in avian evolution. Comp. Biochem. Physiol. A, 60, 177.

Carey, C., H. Rahn, and P. Parisi. (1980). Calories, water, lipid and yolk in avian eggs. Condor, 82, 335.

Croxall, J.P. (1982). Energy costs of incubation and moult in petrels and penguins. J. Anim. Ecol., 51, 177.

Croxall, J.P., and C. Ricketts. (1983). Energy costs of incubation in the Wandering Albatross *Diomedea exulans*. Ibis, 125, 33.

Davidson, T.F., B.H. Misson, and B.M. Freeman. (1980). Some effects of thyroidectomy on growth, heat production and the thermoregulatory ability of the immature fowl (*Gallus domesticus*). J. Therm. Biol., 5, 197.

Dewasmes, G., Y. Le Maho, A Cornet, and R. Groscolas. (1980). Resting metabolic rate and cost of locomotion in long-term fasting emperor penguins. J. Appl. Physiol., 49, 888.

Decuypere, E., E.J. Nouwen, E.R. Kuhn, R. Greers, and H. Michels. (1979). Differences in serum iodohormone concentration between chick embryos with and without the bill in the air chamber at different incubation temperatures. Gen. Comp. Endocrinol., 37, 264.

Dunn, E.H. (1976). The development of endothermy and

existence energy expenditure in Herring Gull chicks. Condor, 78, 493.
Dunn, E.H. (1980). On the variability in energy allocation of nestling birds. Auk, 97, 19.
Ettinger, A.O., and J.R. King. (1980). Time and energy budgets of the Willow Flycatcher (*Empidonax traillii*) during the breeding season. Auk, 97, 533.
Flynn, R.K., and J.A. Gessaman. (1979). An evaluation of heart rate as a measure of daily metabolism in pigeons (*Columba livia*). Comp. Biochem. Physiol. A, 63, 511.
Freeman, B.M., and M.A. Vince. (1974). "Development of the Avian Embryo." London: Chapman and Hall.
Freeman, B.M., A.C.C. Manning, and I.H. Flack. (1980). Short-term stressor effects of food withdrawal on the immature fowl. Comp. Biochem. Physiol. A, 67, 569.
Furness, R.W. (1978). Energy requirements of seabird communities: a bioenergetics model. J. Anim. Ecol., 47, 39.
Grant, G.S., and G.C. Whittow. (1983). Metabolic cost of incubation in the Laysan Albatross and Bonin Petrel. Comp. Biochem. Physiol. A, 74, 77.
Grau, C.R. (1982). Egg formation in Fiordland Crested Penguins (*Eudyptes pachyrhynchus*). Condor, 84, 172.
Haftorn, S. (1978a). Egg laying and regulation of egg temperature during incubation in the Goldcrest *Regulus regulus*. Ornis Scand., 9, 2.
Haftorn, S. (1978b). Energetics of incubation by the Goldcrest *Regulus regulus* in relation to ambient air temperatures and the geographical distribution of the species. Ornis Scand., 9, 22.
Hails, C.J., and D.M. Bryant. (1979). Reproductive energetics of a free-living bird. J. Anim. Ecol., 48, 471.
Hainsworth, F.R. (1978). Feeding: models of costs and benefits in energy regulation. Am. Zool., 18, 701.
Johnson, S.R., and G.C. West. (1973). Fat content, fatty-acid composition and estimates of energy metabolism of Adélie Penguins (*Pygoscelis adeliae*) during the early breeding season fast. Comp. Biochem. Physiol. B: Comp. Biochem., 45, 709.
Kendeigh, S.C., V.R. Dol'nik, and V.M. Gavrilov. (1977). Avian energetics. In "Granivorous Birds in Ecosystems" (J. Pinowski and S.C. Kendeigh, Eds.). London: Cambridge University Press, p. 127.
King, J.R. (1973). Energetics of reproduction in birds. In "Breeding Biology of Birds" (D.S. Farner, Ed.). Washington, D.C.: National Academy of Science, p. 78.
Kooyman, G.L., R.W. Davis, J.P. Croxall, and D.P. Costa. (1982). Diving depths and energy requirements of King Penguins. Science, 217, 726.
Krapu, G.L. (1981). The role of nutrient reserves in mallard reproduction. Auk, 98, 29.
Krebs, C.J. (1978). "Ecology: The Experimental Analysis of Distribution and Abundance," 2nd ed. New York: Harper and Row.
Kuenzel, W.J., and N.T. Kuenzel. (1977). Basal metabolic rate of growing chicks *Gallus domesticus*. Poult. Sci., 56, 619.
Le Maho, Y. (1983). Metabolic adaptations to long-term fasting in antarctic penguins and domestic geese. J. Therm. Biol., 8, 91.
Le Maho, Y., P. Delclitte, and J. Chatonnet. (1976). Thermoregulation in fasting emperor penguins under natural conditions. Am. J. Physiol., 231, 913.
Lilja, C. (1981). Postnatal growth and organ development in the goose (*Anser anser*). Growth, 45, 329.
Lilja, C. (1982). Postnatal growth and organ development in the quail (*Coturnix coturnix japonica*). Growth, 46, 88.
Marsh, R.L. (1984). Adaptations of the Gray Catbird *Dumetella carolinensis* to long-distance migration: flight muscle hypertrophy associated with elevated body mass. Physiol. Zool., 57, 105.
Mertens, J.A.L. (1980). The energy requirements for incubation in Great Tits and other bird species. Ardea, 68, 185.
Misson, B.H. (1974). An open circuit respirometer for metabolic studies on the domestic fowl: establishment of standard operating conditions. Br. Poult. Sci., 15, 287.
Morton, M.L. (1967). Diurnal feeding patterns in White-crowned Sparrows, *Zonotrichia leucophrys gambelii*. Condor, 69, 491.
Morton, M.L., and C. Carey. (1971). Growth and the development of endothermy in the Mountain White-crowned Sparrow (*Zonotrichia leucophrys oriantha*). Physiol. Zool., 44, 177.
Mugaas, J.N., and J.R. King. (1981). Annual variation of daily energy expenditure by the Black-billed Magpie. Stud. Avian Biol., No. 5.
Nice, M.M. (1962). Development of behavior in precocial birds. Trans. Linn. Soc. (N.Y.), 8, 1.
Nomoto, S., C. Bech, W. Rautenberg, and K. Johansen. (1983). Temperature regulation and cardiovascular responses during bipedal exercise in birds. J. Therm. Biol., 8, 175.
Paladino, F.V., and J.R. King. (1984). Thermoregulation and oxygen consumption during terrestrial locomotion by White-crowned Sparrows *Zonotrichia leucophrys gambelii*. Physiol. Zool., 57, 226.
Pettit, T.N., and G.C. Whittow. (1983). Embryonic respiration and growth in two species of Noddy Terns. Physiol. Zool., 56, 455.
Pettit, T.N., G.S. Grant, G.C. Whittow, H. Rahn, and C.V. Paganelli. (1982). Respiratory gas exchange and growth of Bonin Petrel embryos. Physiol. Zool., 55, 162.
Pettit, T.N., G.C. Whittow, and G.S. Grant. (1984). Caloric content and energetic budget of tropical seabird eggs. In "Seabird Energetics" (G.C. Whittow and H. Rahn, Eds.). New York: Plenum.
Petrusewicz, K., and A. Macfadyen. (1970). "Productivity of Terrestrial Animals." IBP Handbook No. 13. Oxford: Blackwell.
Pinshow, B., M.A. Fedak, D.R. Battles, and K. Schmidt-Nielsen. (1976). Energy expenditure for thermoregulation and locomotion in emperor penguins. Am. J. Physiol., 231, 903.
Polin, D., and J.H. Wolford. (1973). Factors influencing food intake and caloric balance in chickens. Fed. Proc. Fed. Am. Soc. Exp. Biol., 32, 1720.
Prange, H.D., and K. Schmidt-Nielsen. (1970). The metabolic cost of swimming in ducks. J. Exp. Biol., 53, 763.
Rahn, H., C.V. Paganelli, and A. Ar. (1974). The avian egg: air-cell gas tension, metabolism and incubation time. Respir. Physiol., 22, 297.
Rayner, J.M.V. (1982). Avian flight energetics. Annu. Rev. Physiol., 44, 109.
Ricklefs, R.E. (1974). Energetics of reproduction in birds. In "Avian Energetics" (R.A. Paynter, Jr., Ed.). Cambridge: Nuttall Ornithological Club, p. 152.
Ricklefs, R.E. (1976). Growth rates of birds in the humid New World tropics. Ibis, 118, 179.
Ricklefs, R.E., and S.C. White. (1981). Growth and energetics of chicks of the Sooty Tern (*Sterna fuscata*) and Common Tern (*S. hirundo*). Auk, 98, 361.
Shapiro, C.J., and W.W. Weathers. (1981). Metabolic and behavioral responses of American Kestrels to food deprivation. Comp. Biochem. Physiol. A, 68, 111.
Thomas, V.G. (1982). Energetic reserves of Hudson Bay Willow Ptarmigan during winter and spring. Can. J. Zool., 60, 1618.
Utter, J.M., and E.A. LeFebvre. (1973). Daily energy expenditure of purple martins (*Progne subis*) during the breeding season: estimates using D_2O^{18} and time budget methods. Ecology, 54, 597.

Vandeputte-Poma, J. (1980). Feeding, growth and metabolism of the pigeon, *Columba livia domestica:* duration and role of crop milk feeding. J. Comp. Physiol., 135, 97.

Vleck, C.M. (1981). Energetic cost of incubation in the Zebra Finch. Condor, 83, 229.

Vleck, C.M., D. Vleck, and D. Hoyt. (1980). Patterns of metabolism and growth in avian embryos. Am. Zool., 20, 405.

Vleck, D., C.M. Vleck, and R.S. Seymour. (1984). Energetics of embryonic development in the megapode birds, Mallee Fowl *Leipoa ocellata* and Brush Turkey *Alectura lathami.* Physiol. Zool., 57, 444.

Wakeley, J.S. (1978). Activity budgets, energy expenditures, and energy intakes of nesting ferruginous hawks. Auk, 95, 667.

Walsberg, G.E., and J.R. King. (1978). The heat budget of incubating Mountain White-crowned Sparrows (*Zonotrichia leucophrys oriantha*) in Oregon. Physiol. Zool., 51, 92.

Warham, J., R. Watts, and R.J. Dainty. (1976). The composition, energy content and function of the stomach oils of petrels (order Procellariiformes). J. Exp. Mar. Biol. Ecol., 23, 1.

Weathers, W.W., and K.A. Nagy. (1980). Simultaneous doubly labeled water ($^{3}HH^{18}O$) and time-budget estimates of daily energy expenditures in *Phainopepla nitens.* Auk, 97, 861.

Whittow, G.C. (1965). Energy metabolism. In "Avian Physiology" 2nd ed. (P.D. Sturkie, Ed.). Ithaca: Comstock, Cornell University Press, p. 239.

Whittow, G.C. (1976). Energy metabolism. In "Avian Physiology," 3rd ed. (P.D. Sturkie, Ed.). New York: Springer-Verlag, p. 174.

Whittow, G.C. (1980). Physiological and ecological correlates of prolonged incubation in seabirds. Am. Zool., 20, 427.

Whittow, G.C. (1983). Physiological ecology of incubation in tropical seabirds. Stud. Avian Biol., No. 8. pp. 47–72.

Wiens, J.A., and J.M. Scott. (1975). Model estimation of energy flow in Oregon coastal seabird populations. Condor, 77, 439.

Williams, J.B., and K.A. Nagy. (1984a). Validation of the doubly labeled water technique for measuring energy metabolism in Savannah Sparrows. Physiol. Zool., 57, 325.

Williams, J.B., and K.A. Nagy. (1984b). Daily energy expenditure of Savannah Sparrows: comparison of time-energy budget and doubly-labeled water estimates. Auk, 101, 221.

Williams, A.J., W.R. Siegfried, A.E. Burger, and A. Berruti. (1977). Body composition and energy metabolism of moulting Eudyptid penguins. Comp. Biochem. Physiol. A, 56, 27.

Withers, P.C. (1977). Energetic aspects of reproduction by the Cliff Swallow. Auk, 94, 718.

11
Alimentary Canal: Anatomy, Regulation of Feeding, and Motility

G.E. Duke

Anatomy

As do other avian systems, the digestive system shows adaptations for flight (Farner, 1960). In the mouth area, the teeth and heavy jaw bones and muscles of reptiles and mammals have been replaced by a much lighter beak, jaw bones, and jaw muscles in birds. Since birds do not chew food, the esophagus is large in diameter to accommodate larger food items. The heavy muscular *gizzard,* or muscular stomach (for mechanical digestion), and the *proventriculus,* or glandular stomach (Figure 11–1), are located within the main mass of the bird's body. Less modification is evident in the avian small intestine and rectum; however, a cloaca is present as in reptiles.

The lengths of various parts of the tract vary with the size of the bird, type of food eaten, and other factors (Table 11–1). Birds eating coarse, fibrous food tend to have especially large digestive tracts, and granivorous birds have larger tracts than do carnivores. For details concerning anatomic and histologic variations and peculiarities of different wild species, see Ziswiler and Farner (1972) and McLelland (1979); for domestic species, see Calhoun (1954).

Mouth and Pharynx

The tongue and beak are important in food manipulation, and there are many adaptations of these organs in birds (McLelland, 1979). There is no sharp distinction between the mouth and pharynx, and there is no soft palate in most birds. The hard palate is pierced

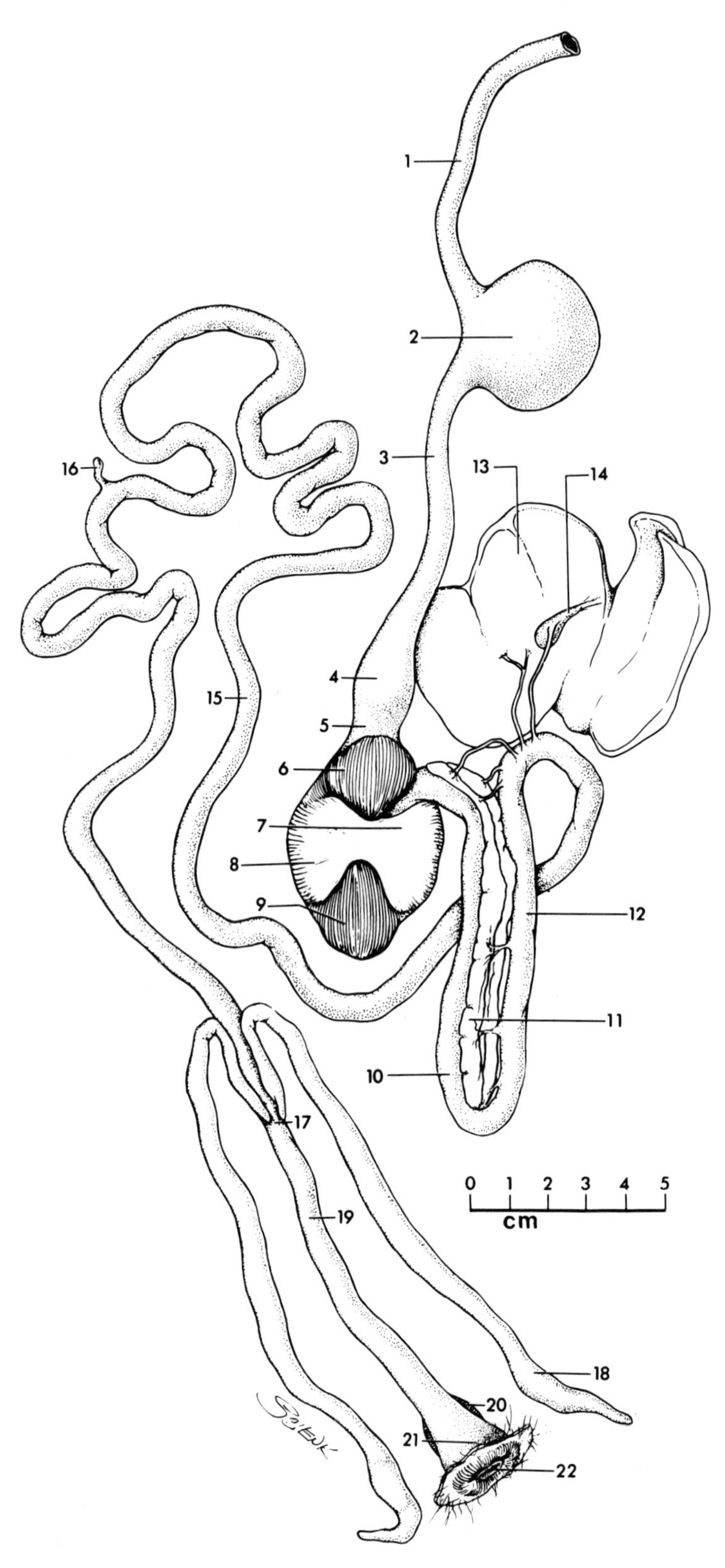

FIGURE 11–1. Digestive tract of a 12-week-old turkey. *1,* precrop esophagus; *2,* crop; *3,* postcrop esophagus; *4,* proventriculus; *5,* isthmus; *6,* thin craniodorsal muscle; *7,* thick cranioventral muscle; *8,* thick caudodorsal muscle; *9,* thin caudoventral muscle (6–9, gizzard); *10,* proximal duodenum; *11,* pancreas; *12,* distal duodenum; *13,* liver; *14,* gallbladder; *15,* ileum; *16,* Meckel's diverticulum; *17,* ileocecocolic junction; *18,* ceca; *19,* rectum; *20,* bursa of Fabricius; *21,* cloaca; *22,* vent. Scale is in centimeters. (Reprinted from Gary E. Duke, "Avian Digestion," in *Duke's Physiology of Domestic Animals,* Tenth Edition, edited by Melvin J. Swenson, p. 360. Copyright 1984 by Cornell University Press. Used by permission of the publisher.)

TABLE 11–1. Length of various parts of the digestive tract in young and adult chickens[a]

Part	Age, 20 days		Age, 1.5 years	
	cm	Total %	cm	Total %
Upper esophagus	7.5	8.3	20.0	9.8
Lower esophagus	4.0	4.4	15.0	7.4
Duodenum	12.0	13.3	20.0	9.8
Ileum	49.0	54.1	120.0	58.9
Ceca	5.0	5.5	17.5	8.6
Colon plus cloaca	4.0	4.4	11.3	5.5
Totals	90.5	100.0	203.8	100.0

[a] From Calhoun (1954).

by a median slit, which communicates with the nasal cavities. The cavity of the mouth is lined with stratified squamous epithelium.

Salivary glands are present and are usually tubular, but may be simple or branched. In general, species that ingest slippery aquatic food have poorly developed glands and those eating dry food have well-developed ones. The glands of some species (sparrow and others) contain appreciable amounts of amylase; others, such as the chicken and turkey, do not (Jerrett and Goodge, 1973). Secretion of saliva is increased by feeding and by parasympathetic stimulation (Chodnik, 1948).

The taste buds of birds vary in location and number between species (see Chapter 2). They are usually found on the back part of the palate, around the glottis and on the tongue and resemble those of other species (Gentle, 1971).

Esophagus and Crop

The esophagus of most birds is relatively long; it has outer longitudinal and inner circular muscles, and mucous glands are abundant to help lubricate the passage of food.

The size and shape of crops vary according to the eating habits of the species, and may be unilobular (Figure 11–1), bilobed, or spindle shaped. The crops of certain granivorous birds are bilobed and large, and in some species they are very large; in certain others, such as owls and insectivores, they are rudimentary or absent. Crop size also varies significantly between breeds of chickens and between the sexes within a breed (single Comb White Leghorns) (Wehner and Harrold, 1982). The crop serves principally for food storage.

The crops of several species, especially doves and pigeons, are adapted to produce crop "milk," and the proliferation of the crop epithelium and its sloughing are induced by prolactin. These birds regurgitate the crop milk to feed their young. The crop has essentially the same microstructure as the esophagus.

Proventriculus

The proventriculus is a fusiform organ (Figure 11–1) varying in size among species. It is relatively small in granivorous species, but may be quite large and distensible in aquatic carnivores (e.g., cormorants) that ingest large food items; in such species it serves a storage function.

In most species, food passes through the proventriculus and is held in the gizzard, where action of the gastric secretions occurs. The primary function of the proventriculus is the production and release of the gastric secretions, pepsin, hydrochloric acid, and mucus. The organ is lined with mucous membrane, which contains the gastric glands. In most species these glands contain only one cell, the *chief* cell, which secretes both acid and pepsin. The chief cells contain zymogen or pepsinogen granules (precursors of pepsin) in varying amounts, depending on the state of digestion. These granules increase during fasting and decrease immediately after feeding. Both inner circular and outer longitudinal smooth muscle layers are present in the proventriculus.

Gizzard

In most species, this organ is composed of two pairs of opposing muscles (Figure 11–1) termed the thin and thick muscle pairs (Dziuk and Duke, 1972). All four muscles consist almost entirely of circular smooth muscle arising from a central aponeurosis. A more simple, less muscular gizzard found in raptors and heron-like birds has both inner circular and outer longitudinal muscle layers. The main functions of both types of muscular stomachs are mechanical digestion (especially in granivores and frugivores) and serving as the site of preliminary proteolysis.

The dark lining of the gizzard, the koilin, is formed by secretion of protein (Webb and Colvin, 1964) from simple tubular glands and by entrapment of sloughed cells and cellular debris (Farner, 1960). Its greenish or brownish color is due to bile pigments refluxed from the duodenum.

There are a number of variations in the general anatomic scheme for the glandular and muscular stomachs described above (see McLelland, 1979). Most notable of these is the presence in several species of aquatic birds of a third chamber, aborad to the gizzard, called the *pyloric* stomach. This organ usually contains feathers or projecting mucosal processes that apparently serve as a filter to prevent larger portions of the digesta from leaving the muscular stomach.

The gizzard receives extrinsic fibers from the vagus and the sympathetic system, including both excitatory and inhibitory fibers (Nolf, 1937, 1938b; Burnstock, 1969). Anatomic, histochemical, and electrophysical studies have demonstrated the existence of pre- and postganglionic cholinergic excitatory fibers in the vagus and also in the perivascular sympathetic trunks. There are also nonadrenergic inhibitory fibers in the vagus that are apparently of sympathetic origin. The nonadrenergic nerve fibers serving the muscle of the muscular stomach appear to be associated only with ganglion cells or blood vessels (Bennett, 1969 a, b, c; Bennett and Cobb, 1969 a, b). A plexus comparable to Meissners' in mammals is absent in birds (Bennett and Cobb, 1969b). The excitatory fibers, which are cholinergic, are blocked by atropine or hyoscine (Bennett, 1969a). A more detailed description of morphologic and physiologic aspects of the gastric innervation is found in Bennett (1974).

Small Intestine

The avian small intestine consists of the duodenal loop and ileum (Figure 11–1); beyond the duodenum there are no histologically delimited areas. Some authors refer to the upper and lower ileum as corresponding to the jejunum and ileum in mammals, using the remnant of the attachment of the yolk stalk (Meckel's diverticulum) (Figure 11–1) as a boundary point. The intestines of birds are relatively shorter than those of mammals; however, there is considerable variation in the length. They are longer in herbivores and granivores and shorter in carnivores.

The mucosa of the small intestine is characterized by crypts of Lieberkühn of varying degrees of development. Brunner's glands are absent in the chicken (Calhoun, 1954), but in some species tubular glands that are similar to or homologous with Brunner's glands of mammals may be present (Ziswiler and Farner, 1972). The epithelium usually consists of simple columnar cells with many goblet cells. The layers, from the lumen outward, form the mucosa, submucosa, inner circular muscle, outer longitudinal muscle, and serosa, as in mammals. There is a submucosal nerve plexus, serving blood vessels and glands mostly, and a myenteric nerve plexus lying between the two muscle layers and serving those muscles. The myenteric system of the duodenum is intimately associated with that of the gizzard and proventriculus and is believed to be involved in coordination of gastric and duodenal motility (Duke et al., 1975b). Due to the near absence of a longitudinal muscle layer on the gizzard, the myenteric plexus is grossly visible just under the transparent serosa.

There are well-developed and finger-like villi in carnivorous birds, but in herbivores the villi are flattened and leaflike (Ziswiler and Farner, 1972). Lymphocytes, which are present in large numbers in the lamina propria, are distributed diffusely, in isolated lymph follicles, or in Peyer's patches.

The extrinsic innervation of the intestine is extensive, including pre- and postganglionic cholinergic fibers that may be vagal or sympathetic in origin. The perivascular sympathetic nerves contain cholinergic excitatory, and both adrenergic and nonadrenergic inhibitory, nerve fibers (Everett, 1968; see also Chapter 1). Stimulation of the perivascular nerves or nerve trunk excites mainly preganglionic nerves (Everett, 1968; Bolton, 1971). Catecholamines antagonize the effects of cholinergic nerve stimulation (Bolton, 1971), and histamine can overcome the relaxing effect of catecholamines if an adequate dosage is used (Everett and Mann, 1967).

In both adult chickens and in chicks, neuron density is greater in proximal than in distal parts of the intestines, being lowest in the rectum and distal ceca. Neuron density is two to three times higher in chicks than in corresponding gut regions in adults, but the total number of neurons is significantly higher in adults; i.e., density per square centimeter decreases with age, while total number of nerve cells increases. Both the changes in density and in number of neurons are probably a response to the increase in gut volume with age. Similarly, the differences in density proximally versus distally are related to the relative thickness of the intestine (Ali and McLelland, 1979).

The small intestine is the principal site of chemical digestion, involving enzymes of both intestinal and pancreatic origin. It also secretes hormones that are primarily involved in regulation of gastric and intestinal actions. Additionally, most nutrient absorption occurs in the small intestine.

Ceca, Rectum, and Cloaca

The ceca originate at the juncture of the small and large intestines (Figure 11–1). They are large, prominent, and paired in some species (e.g., herbivores, most granivores, and owls), and in others they may be single, rudimentary, or absent (e.g., parrots and hawks) (McLelland, 1979). Ileocecal valves separate the ceca from the intestines. The ceca may be histologically similar to the small intestine, except that the villi are not as tall, or they may be quite glandular and exhibit a high degree of secretory activity (owls). They also may be of the lymphoepithelial type, as present in many passerine species (Ziswiler and Farner, 1972). Several functions have been attributed to the ceca, but the principal one appears to be that of microbial fermentation of dietary fiber.

The rectum extends from the ileocecal junction to the cloaca; it is relatively short in most species except in the ostrich, in which it is long. Histologically, it is similar to the small intestine, except that the villi are shorter in the rectum. The *cloaca,* serving the digestive, urinary, and reproductive tracts, consists of three chambers. The large intestine empties into the *coprodaeum,* and the urinary and reproductive tracts terminate at the *urodaeum.* Finally, the *proctodaeum* opens externally through the anus. The bursa of Fabricius, a prominent lymphoid organ, is a dorsal projection of the urodaeum. Both the colon and cloaca are primarily involved in excretion and mineral and water balance (see Chapter 16).

According to Burnstock (1969), there is no evidence for a sacral parasympathetic nerve supply to the gut of birds, but Bolton (1971) suggested that parasympathetic nerves supply cholinergic nerves to all parts of the tract of birds. Whether these are exclusively vagal in origin is not stated.

Remak's nerve is a long nerve extending from the duodenum to the cloaca. The origin, homology, and function of this nerve are controversial (see Chapter 1). Some investigators consider the nerve sympathetic in origin, whereas others (Yntema and Hammond, 1952) claim that it is a derivative of the sacral parasympathetic system (see Bennett and Malmfors, 1970, for review). There is evidence of noncholinergic excitatory neural activity in the Remak's nerve of the chick (Bartlet and Hassan, 1971), which appears to involve a purinergic (e.g., ATP) neurotransmitter (Ahmad et al., 1978). Stimulation of Remak's nerve in an isolated rectum preparation caused the release of a nonadrenergic, noncholinergic (NANC) transmitter resulting in an initial fast contraction. This was followed by a slow contraction (produced by low-frequency stimulation) and caused by the release of acetylcholine, which acts mainly on prejunctional muscarine receptors and mediates an inhibitory effect on the release of the NANC transmitter (Komeri and Ohashi, 1984).

Liver and Pancreas

The liver is bilobed (Figure 11–1), and relatively large in most birds; the left hepatic duct communicates directly with the duodenum, whereas the right duct sends a branch to the gallbladder; alternatively, it may be enlarged locally as a gallbladder. Gallbladders are present in the chicken, duck, and goose, but not in some other species, such as the pigeon. The gallbladder gives rise to bile ducts, which empty into the duodenum near the distal loop. The pancreas lies within the duodenal loop (Figure 11–1). It consists of at least three lobes, and its secretions reach the duodenum via three ducts.

Circulation of Blood in the Alimentary Canal

The esophagus and crop receive arterial blood from branches of the external carotid arteries and are drained by branches of the jugular veins (see Chapter 6). The coeliac artery supplies the glandular stomach, muscular stomach (Nishida et al., 1969), liver, pancreas, duodenum, and spleen. The anterior mesenteric artery supplies the remainder of the small intestine, and the caudal mesenteric artery supplies the rectum and cloaca. Venous outflow from the glandular stomach, muscular stomach, duodenum, and spleen occurs via the gastroduodenal vein, which flows into the hepatic portal vein. The anterior mesenteric vein courses in the mesentery of the small intestines and also drains into the hepatic portal veins. The coccygeomesenteric vein drains the hindgut and then anastomoses with the hepatic portal and renal portal veins. Blood in the coccygeomesenteric vein may flow into the hepatic portal and to the liver, or may reverse its direction of flow via the renal portal to the kidney and thence to the vena cava (Akester, 1967; Sturkie et al., 1977). (For further details, see Chapters 6 and 16.)

Regulation of Feeding

The amount of feed consumed depends on many factors, including size and age, environmental temperature, day length, activity, stage in the reproductive cycle, appearance and taste of food, and availability of water. The quantity eaten may be expressed as a percentage of body weight per 24 hr. This figure averaged 6.4% of body weight for White Leghorn females (weighing 1.8 kg) and laying approximately 65% (Sturkie, 1965) and 6.5% of body weight (~1.8 kg) for Brown Leghorn hens laying at about 44% (Savory, 1978). The amount for males is about 50% less per unit of body weight, because a greater intake is required for egg production and also because small birds consume more per unit of body weight than large ones (roosters are larger than hens). Pullets that were 3–6 weeks old consumed an average of 10.1% of their body weight per day (Henken et al., 1982).

Hypothalamic Centers

There is general agreement that the satiety and appetite centers of mammals are located in the ventromedial and lateral nuclei of the hypothalamus, respectively.

Evidence regarding the function of these centers in birds, however, is not conclusive. Smith (1969), using chickens, and Kuenzel (1972), using white-crowned sparrows, reported that lesions in the ventromedial nucleus (VMN) caused hyperphagia whereas lesions in the lateral hypothalamus just slightly caudal to the region of the VMN resulted in aphagia. More recently, it has been concluded that lesions of the lateral hypothalamus that caused aphagia did so because neuronal pathways passing through the lateral hypothalamus were interrupted (Kuenzel, 1982). Also, Robinzon et al. (1982) reported that the structures involved in the inhibition of feeding are a dispersed neuronal system within the ventromedial brainstem; thus, it is difficult to establish a consistent relationship between hypothalamic lesion sites and their physiologic consequences.

Stimulation of these hypothalamic areas has not helped to clarify the situation. Equivocal results were obtained by stimulations in chickens (Phillips and Youngnen, 1971) and in mallard ducks (Maley, 1969). Ackerman et al. (1960) reported that stimulation of the lateral hypothalamus of pigeons caused hyperphagia; however, Goodman and Brown (1966) could not confirm these results in the same species. Tweeton et al. (1973) conducted an extensive study involving stimulation at 625 different sites in the anterior hypothalamus and supra- and preoptic areas of the brains of 68 chickens. Increased feed intake was elicited from only six sites, and none of these led to normal feeding during the stimulations.

Gold thioglucose, which produces lesions in the ventromedial nucleus of mammals, was ineffective in Japanese quail (Carpenter et al., 1969). Neural regulation of food and water intake has been reviewed by Wright (1973) and Smith (1979).

Adipose Tissue

Several mechanisms by which the hypothalamus or brain might regulate feed intake have been proposed. According to one mechanism, levels of *adipose tissue stores* act as set points for the hypothalamus. When the amount of fat in the tissues increases above a given level or set point, this causes a decrease in feed intake and promotes lipolysis (Mu et al., 1968). Lepkovsky (1973) reported that continued forced feeding of chickens twice normal nutrient requirements made the birds obese. When the force-feeding ceased, the birds refused to eat for 7–10 days until they lost weight and depleted their adipose stores to such level as to trigger the feeding center. Lepkovsky believed that testosterone was one of the adjustors of the set point. Conversely, when the fat level was below the set point, feed intake increased and lipogenesis and fat deposition were accelerated. Feed intake and body weight stabilized when the fat depots matched the level called for by the new set point (Lepkovsky, 1973). Hyperphagia resulting from lesions of portions of the ventromedial nucleus was produced because a new set point became operative (Lepkovsky and Yasuda, 1966; Lepkovsky, 1973). The above account is mainly from Sturkie (1976).

Some data apparently do not support this adipose tissue mechanism. Maurice (1983) found that partial lipectomy in chickens did not affect food intake or body weight gain, nor was there a compensatory hypertrophy in discrete fat depots.

Dietary

The work of Maurice (1983) also indicated that dietary fat level did not affect food intake in the partially lipectomized birds. Hill and Dansky (1954) and Gleaves et al. (1968) demonstrated, however, by varying caloric intake or nutrient density of feed, that *energy requirements* did influence feed intake. The latter study showed, in chickens, that

a. As dietary energy level was increased, food consumption decreased
b. As dietary protein level was increased, food consumption decreased
c. If dietary protein level was increased, but dietary energy level was decreased, then feed consumption increased. (This was apparently an effort to obtain more energy, so energy appeared to be a more important regulator of intake than protein.)

Chickens are also able to select diets on the basis of protein content. When given a choice of two or more isocaloric diets with different protein content, chickens chose 11 or 14% protein over 8%; they chose 16% over 23, 8 or 12% (Holcombe et al., 1976).

Blood Glucose Concentration

Mammals are believed to have glucoreceptors in the VMN of the hypothalamus by which food intake may be influenced. Injection (iv) or ingestion of glucose results in an increased blood glucose concentration and a suppression of feeding. This effect of glucose on food intake is known as the glucostatic theory. Several studies have shown that such a mechanism may not exist, or may be much less sensitive in birds (see Denbow et al., 1982a). The work of Matei-Vladescu et al. (1977), however, indicated that prolonged infusions of glucose into the lateral cerebral ventricles of chickens did result in decreased food intake, indicating that perhaps the receptors respond only to long-term changes in blood glucose levels rather than to acute changes.

The studies of Shurlock and Forbes (1981a) on young cockerels indicated that glucostatic regulation of feeding may operate at the *hepatic* level. They infused 5 ml of a glucose solution (40, 100, or 150 g/liter) into the hepatic portal vein of feeding birds and observed a depression of feeding, especially in previously fasted birds. A control solution of NaCl (9 g/liter) did not produce a similar effect, nor was this effect observed when glucose was infused into the jugular vein. Hepatic glucostatic regulation of feeding has been demonstrated in mammals (Campbell and Davis, 1974).

Osmolarity of Ingesta

In a companion study (Shurlock and Forbes, 1981b), infusion of hypertonic solutions of glucose or of a KCl/sorbitol mixture into the crops of feeding cockerels was followed by depression of food intake. Intraduodenal infusion of KCl/sorbitol also depressed feeding, but intraduodenal glucose infusions depressed feeding only in previously fasted birds. The relatively poor response to intraduodenal glucose was believed to be due to rapid absorption of glucose, thus decreasing its concentration in the duodenal lumen. KCl and sorbitol are both poorly absorbed. This osmotic control probably functions via a negative feedback mechanism to depress gut motility, which in turn acts to depress eating. Hypertonic duodenal ingesta have been shown to depress gastric motility in turkeys (Duke and Evanson, 1972) (see following sections).

Volume of Ingesta

The study of Gleaves et al. (1968) also showed that if dietary volume was increased by addition of polyethylene fluff to the feed, feed consumption increased in an effort to obtain more nutrients, but total nutrient intake was less than before addition of the fluff. So, distension of the tract may be another mechanism contributing to regulation of feed intake, and it may act quite independently of the energy or protein content of the feed. This may be due to distension-sensitive receptors in the gastrointestinal (GI) tract. By inflating a balloon in the crop and recording from single vagal fibers, Hodgkiss (1981) found two types of receptors in the crop that he described as slowly adapting and rapidly adapting. It was believed that the slow type detected prolonged periods of crop distension and thus may be involved in reflex suppression of feeding. Richardson (1970) found that artifical inflation of the crop by a permanently implanted balloon reduced food intake. Distension-sensitive receptors are also present in the gizzard (Duke et al., 1977).

Polin and Wolford (1973) believed that the volume of feed ingested, rather than energy requirements or fat deposits, was the main mechanism regulating feed intake. Using groups of chickens, fed ad libitum, fed ad libitum and force-fed, or force-fed 150% of the intake of the ad libitum group, they observed an increase in blood lipids and adipose tissue that correlated with the increased feed intake. When force-feeding was discontinued, they found no difference in the ad libitum feeding patterns (amount eaten) among the groups of birds, although there were considerable differences in their fat depots. They cite well-known observations that obese chickens do not lower their feed intake and that increasing nutrient density, but not volume of the diet, causes obesity as evidence that volume receptors in the gut are important in regulation of feeding. They propose that hormones or other factors may regulate the set point at which the receptors operate. That emptiness, fullness, or distension of the digestive tract influences feed intake in mammals is well known.

It is likely that all of the proposed mechanisms participate to some extent in the regulation of food intake.

Water Consumption

Water consumption is influenced by several factors, including size and age of bird, environmental temperature, and type and amounts of food consumed. As less feed is eaten, less water is consumed, except when the ambient temperature is high. Water consumption in a number of species of wild birds was found to be inversely related to body size, ranging from 5 to 30% of body weight per day (Bartholomew and Cade, 1963). Water consumption in normal adult male chickens amounted to 5.5% of body weight in 24 hr (Dunstan and Buss, 1969). In young, growing birds (3–4 weeks), the figure was 18–20%. Adult laying chickens not only eat more feed than males (see above), but also consume more water (13.6% of body weight), with a water to feed ratio of 2:1 (Sturkie, 1965). Daily water consumption in laying hens varied considerably, but increased significantly on days that hens laid, and increased most during the 12 hr preceding oviposition (Howard, 1968). For Brown Leghorns, the amounts consumed averaged 115 g/day on days when not laying and 225 g/day when laying (Howard, 1975). For this same breed, Savory (1978) showed that egg production declined (but not significantly) during restricted water intake.

Just as the hypothalamus is important in regulation of feeding, it also appears to be important in the regulation of water intake. An increase in concentration of NaCl in hypothalamic and juxtaventricular regions following water deprivation, hemorrhage, or administration of hypertonic saline can produce thirst, and angiotensin II (AII) is involved in this response (Kaufman and Peters, 1980; Skadhauge, 1981). This hormone causes dipsogenesis in domestic pigeons (Evered and Fitzsimons, 1976; Fitzsimons, 1978; Fitzsimons and Evered, 1978), white-crowned sparrows (Wada et al., 1975), and Japanese quail (Takei, 1977). AII is also involved in stimulation of normal physiologic drinking, and substances that inhibit AII (e.g., saralasin) inhibit naturally occurring drinking. Endogenous opiates appear to be involved in normal inhibition of drinking, and an injection (ip) of naloxone, a specific opiate receptor antagonist, caused an increase in water intake (Uemura et al., 1983). The preoptic area appears to be especially sensitive to angiotensin II, and injections into this area in Japanese quail produced a profound drinking response (Takei, 1977). Kobayashi (1978) studied the sensitivity of 17 avian species to angiotensin, and found that birds that normally drink little water (e.g., carnivores and those from arid regions) were much less sensitive to the hormone. Plasma concentration of AII increases during dehydration, and the concentrations of arginine vasotocin, aldosterone, and corticosterone increase as well. Since an injection of AII (ip) induces increases in plasma levels of these three hormones, AII must act as a "trigger" for their release during dehydration (Kobayashi and Takei, 1982).

Other neurohumoral factors also may play a role in hypothalamic regulation of both food and water intake. Denbow et al. (1982b) injected serotonin into the lateral ventricle of young chicks. In fasted birds, food intake was not affected, while data on water consumption were equivocal. In fed chicks, food intake was decreased while water intake was unaffected. Injections of epinephrine and norepinephrine into the lateral ventricle significantly decreased food intake in fed turkeys, whereas dopamine has no effect (Denbow, 1983). Denbow and Myers (1982) injected cholecystokinen (CCK) into the lateral ventricle of young cockerels and found that water consumption was unaffected but food intake was decreased. CCK may, therefore, be involved in satiety.

Lack of drinking (adipsia) has been produced in pigeons (Ackerman et al., 1960), white-crowned sparrows (Kuenzel, 1972), and chickens (Lepkovsky and Yasuda, 1966), by lesions in the lateral hypothalamus, i.e., the same area in which aphagia was produced in some studies (see above). Stimulation of this area in pigeons caused polydipsia (Ackerman et al., 1960). Lesions in the ventromedial hypothalamus (VMH) of white-crowned sparrows also caused polydipsia (Kuenzel, 1972).

Excessive drinking (polydipsia) has been observed in a few species of wild birds (Bartholomew and Cade, 1963). This may be due to a hypothalamic abnormality; such polydipsia is usually accompanied by polyuria, which is a secondary condition. Another cause of polyuria is a deficiency of antidiuretic hormone of the posterior pituitary. This also produces polydipsia but it is secondary to the primary polyuria or *diabetes insipidus*. Diabetes insipidus can be caused by lesioning or removal of the posterior lobe of chickens (Shirley and Nalbandov, 1956). Hereditary polydipsia (probably primary) has been reported in chickens (Dunston and Buss, 1969; Dunston et al., 1972) and in mockingbirds (Dunston, 1970). (For further details on water balance see Chapter 16.)

Motility

Motility was initially studied by fluoroscopy, direct observation in laparotomized birds, or by inflated balloons inserted into the tract, orally, rectally, or via a cannulated fistula. Balloons were connected to a pressure sensing device which, in turn, was connected to a recorder to provide a record of intraluminal pressure (ILP) changes detected by the balloons. The balloons themselves often stimulated motility (sometimes abnormal motility), and they were replaced by small fluid-filled, open-tipped tubes inserted into the tract. Most recently, image intensification radiography employing a photomultiplier system to reduce X-ray exposure to both the researcher and the subject has replaced fluoroscopy. Also, tiny implantable strain gauge transducers (SGT), to detect contractions, and bipolar electrodes, to detect electrical activity associated with muscular depolarization or electrical control potentials, have been employed. The latter permit much greater freedom for the subject than balloons or tubes, and can even be used telemetrically.

Prehension and Deglutition

The manner in which food is grasped, manipulated, or altered by beak activity, etc., before swallowing varies tremendously among species, depending on feeding habits and on the structure of the beak and tongue (Ziswiler and Farner, 1972; McLelland, 1979). In all species, however, it is likely that the presence of food in the mouth stimulates salivation and swallowing.

Detailed descriptions of swallowing by chickens are available from White (1970), using radiography, and from Suzuki and Nomura (1975), employing electromyography. Stimulation of the tongue by food causes rapid rostrocaudal tongue movements for 1–3 sec to move the bolus of food backward along the tongue to the pharynx (Figure 11–2). This is termed the *oral phase* of swallowing (Suzaki and Nomura, 1975). While this is occurring, the glottis closes, the hyoid apparatus is made concave ventrally, the tongue is moved backward and the esophagus forward, all by contraction of various muscles. The latter two movements shorten the distance between the oral cavity and the esophagus, and this is referred to as the *pharyngeal phase* of swallowing. Further tongue movements dislodge the bolus from the tongue into the esophagus. Then, muscular contractions return the glottis, hyoid apparatus, tongue, and esophagus to their starting positions, and esophageal peristalsis moves the bolus toward the stomach (*esophageal phase* of swallowing).

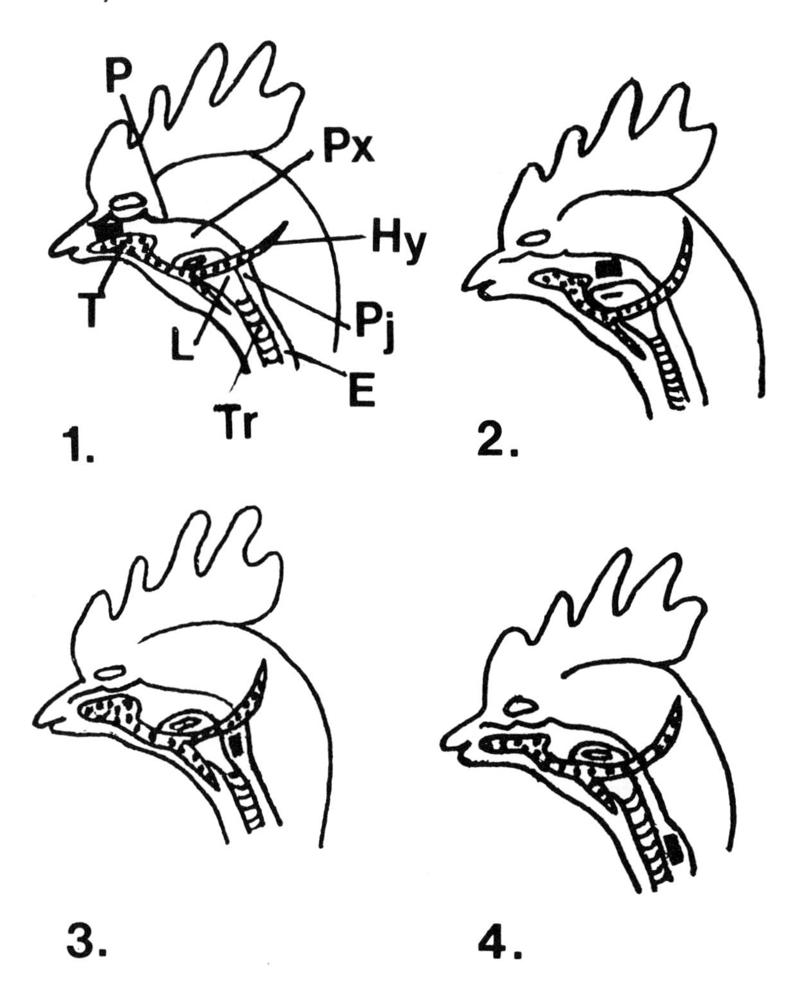

FIGURE 11–2. Passage of bolus into pharynx and esophagus. *T*, tongue; *Hy*, hyoid apparatus; *L*, larynx; *Tr*, trachea; *E*, esophagus; *Pj*, pharyngoesophageal junction; *Px*, pharynx; *P*, palate; bolus, black rectangle (see text for explanation). (From Suzuki and Nomura, 1975.)

Apparently, similar movements are employed in swallowing either food or fluid, except that fluid is more easily moved to the pharynx by gravity when the head is raised and the beak is pointed upward.

Esophagus and Crop

In a fasted chicken, food entering the esophagus is likely to be passed directly to the gizzard, bypassing the crop. According to most studies (Sturkie, 1976), subsequent ingesta are passed into the crop; however, it has been observed radiographically that while this may occur food may also move directly to the muscular stomach (Duke, unpublished). Further, a bolus that has already reached the opening to the glandular stomach may be returned to the crop. This reflux has also been observed by Vonk and Postma (1949). While further study is needed, it appears that the state of contraction of the gizzard at the time when a bolus nears the opening to the crop determines whether the bolus enters or bypasses the crop. If the gizzard is contracting, the bolus enters; if it is relaxed, the bolus bypasses the crop. Pastea et al. (1968) found that contractions of the crop and muscular stomach were coordinated and that the stomach may influence crop activity. The fullness of the crop may, in turn, influence gastric activity, according to Hill and Strachan (1975). The latter authors further postulated that inhibition of gizzard contractions following crop distension may be due to the increase in proventricular acid secretion also found to occur after crop distension (Rouff and Sewing, 1971). When the crop is full, its contractions and those of the opposing surface of the esophagus form boli and expel them into the esophagus for passage into the stomach (Macowan and Magee, 1932).

Other factors may influence crop activity. Extreme excitement, fear, or struggling may inhibit or retard crop contractions, as well as contractions throughout the GI tract, in the chicken and pigeon (Sturkie, 1976). Hunger produces restlessness and irregular crop activity in normal pigeons and chickens, but fear had no effect on crop activity of the decrebrate pigeon, and crop motility in hungry decerebrate pigeons was fairly regular. Groebbels (1932) found that the frequency of crop and esophageal contractions increased with length of fasting.

Innervation. The esophagus and crop receive parasympathetic excitatory fibers from the vagus and also both excitatory and inhibitory fibers from the sympathetic system (Sturkie, 1976). Ohashi (1971) found more inhibitory fibers in the esophagus distal than proximal to the crop and postulated that this may explain the slower rate of contraction of the distal portion of the esophagus as described by Pintea et al. (1957). Stimulation of the peripheral end of the left vagus causes contraction in the left side of the crop (cephalic and dorsal region), and stimulation of the right vagus causes contraction of the right side. Transection of the right vagus alone has little effect on crop motility, but ligation of the left vagus inhibits motility and particularly the ability of the crop to empty itself. The left vagus nerve apparently controls the peristaltic movements of the esophagus, because these movements are abolished after ligation of this nerve. Stimulation of the vagus produces contraction in the circular muscles of the crop, and stimulation of the sympathetics causes contraction of the longitudinal muscles. The latter are believed to have little influence on normal crop motility (see Sturkie, 1976).

Gastroduodenal

As indicated above, the gizzard of fowl consists of two pairs of muscles (thin and thick) (Figure 11–1), and because these pairs contract alternately, a biphasic ILP change or contraction pattern results (Figure 11–3, D)

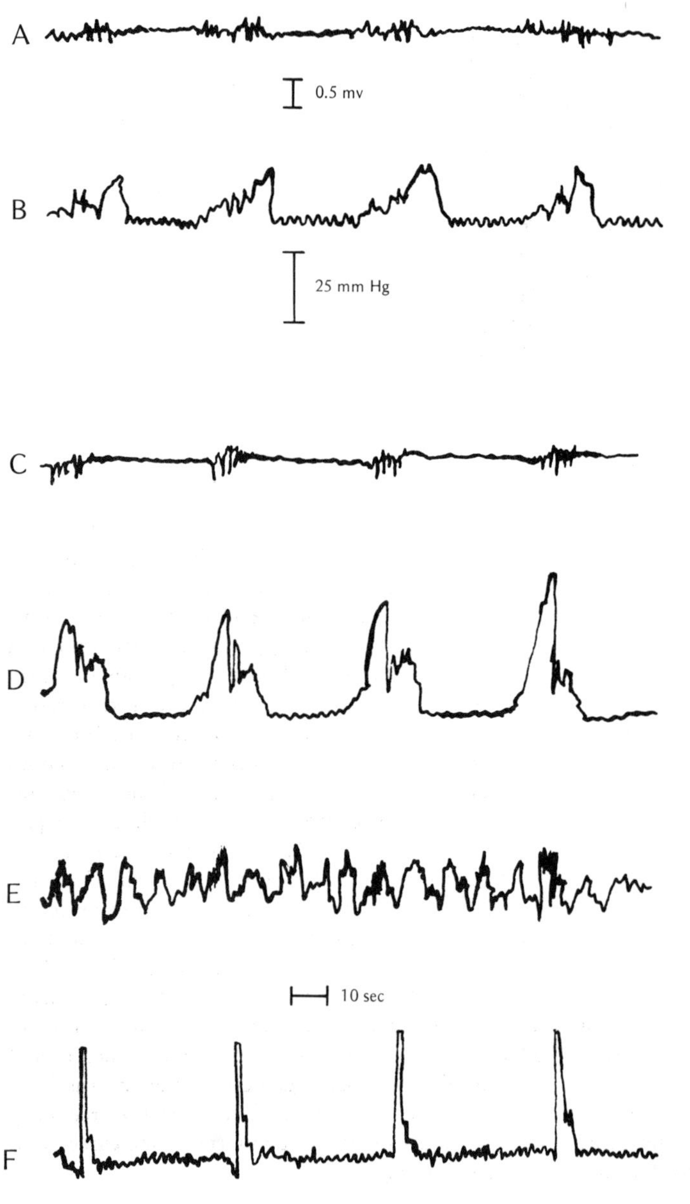

FIGURE 11–3. Tracings of typical records of electrical potential (EP) and intraluminal pressure (IP) changes from the proventriculus, gizzard, and duodenum of turkeys. (**A**), (**C**), and (**E**) are EP changes recorded from the proventriculus, thick cranioventral muscle of the gizzard, and proximal duodenum, respectively. Slow waves with spikes are evident in EPs from the duodenum (**E**); only electrical spike discharges associated with contractions are evident from the proventriculus (**A**) and gizzard (**C**). **B**, **D**, and **F** are IP changes recorded from the proventriculus, gizzard, and duodenum, respectively. Gizzard contractions cause small IP changes in the proventriculus prior to each proventricular contraction wave. Very small changes in thoracoabdominal pressure due to respiration are recorded between contractions of all three organs. Time constant for electrical recording was 1.1 sec (From Duke et al., 1975b.)

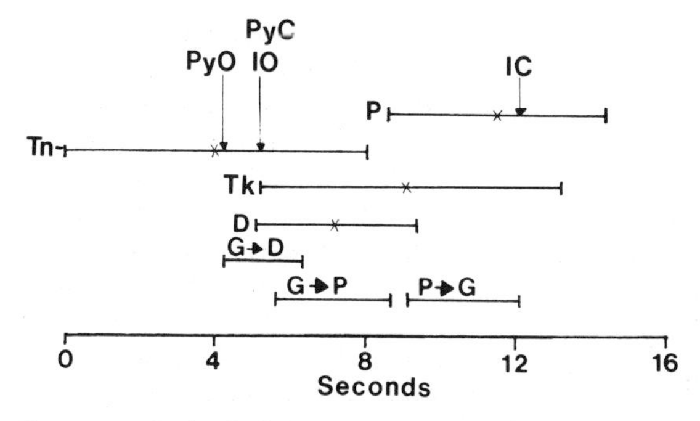

FIGURE 11–4. Relative sequence and duration of events in the gastroduodenal contraction cycle of turkeys. Horizontal lines indicate the relative sequence and duration of each event; the top four lines represent a contractile event; the bottom three represent ingesta flow; *x* is the end of contraction and beginning of relaxation. Vertical arrows indicate the point in this sequence at which events occur. Abbreviations: *Tn,* thin muscle pair; *Tk,* thick muscle pair; *D,* duodenum; *P,* proventriculus; *G,* gizzard; *I,* isthmus; *Py,* pylorus; *O,* open; *C,* closed. (From Dziuk and Duke, 1972.)

(Mangold, 1950; Duke et al., 1972a; Dziuk and Duke, 1972). Contractions of the gizzard, proventriculus, and duodenum are normally totally coordinated in a gastroduodenal sequence (Figure 11–3), which begins with contraction of the thin muscles followed by the duodenum, thick muscles, and proventriculus, respectively (Figure 11–4).

These gastroduodenal sequences occur at a mean frequency of 3.3 cycles/min in turkeys (Table 11–2). This frequency is similar to that reported by early investigators for fowl (see Sturkie, 1976). Akahori et al. (1971) found that contractile frequency decreased with age in quail and chickens and that frequency was greater in quail. The highest ILP changes are recorded during contraction of the pair of thick muscles (Duke et al., 1972a) (Table 11–2). The pressures reported by Duke et al. are considerably lower than those reported for chickens and other species by Mangold and coworkers, by Kato, and by others (see Sturkie, 1976), who reported pressures as follows (in mm Hg): buzzard, 8–26; duck, 180; hen, 100–150; and goose, 265–280. The pressure is much lower in carnivores, such as the buzzard, where the muscular stomach is poorly developed. Hard fibrous feeds such as barley produce higher gastric pressures and a larger, more powerful gizzard than do softer feeds such as wheat. The higher pressures reported by earlier workers may be due to the feeding of less refined diets and thus more muscular development of the muscular stomach. The amplitude of contractions is greater in males than females.

Another aspect of normal gastroduodenal motility is *intestinal refluxes* (Figure 11–5). They have been observed to occur about every 15–20 min while recording normal gastroduodenal motility in turkeys (Duke et al., 1972a; Dziuk and Duke, 1972). This function appears to be unique to birds. The reflux typically involves one or two very high-amplitude duodenal ILP waves. During the reflux period, apparently due to an intrinsic neural reflex, gastric motility is inhibited.

In many species (Rea, 1973), especially carnivores, one other unique gastric function occurs, viz, *egestion* (oral expulsion) of pellets of indigestible material. Usually bones and fur or feathers are consumed by carnivorous birds as prey is consumed. Since these items would be difficult to digest, they are compacted intragastrically and orally egested. The mechanism of this process has been described for great-horned owls, and as it is unlike either vomiting in mammals with a simple stomach or regurgitation in ruminants (Duke et al., 1976a), the term egestion is used. It proceeds as follows: About 12 min before egestion, gastric contractions increase in both frequency and amplitude. This results in the final compaction and shaping of the pellet as well as pushing it into the lower esophagus. During the last 8–10 sec before egestion, the pellet is moved orad by esophageal antiperistalsis. Contractions of the abdominal muscles are not involved, nor is duodenal motility.

Pellet egestion in owls is regulated primarily by intragastric factors so that the presence of amino acids or lipids inhibit egestion, while the presence of bulk (e.g., bones and fur) in the absence of nutrients stimulates egestion. Pellets are thus egested when gastric di-

TABLE 11–2. Mean frequency and amplitude of gastroduodenal intraluminal pressure (ILP) changes, electrical action potentials, and slow waves in turkeys.[a]

	ILP changes		Action potential			Slow waves	
	Frequency (cycles/min)	Amplitude (mm Hg)	Frequency (per sec)	Spikes per burst	Amplitude (mV)	Frequency (per min)	Amplitude (mV)
Proventriculus	3.3 ± 0.7	34.4 ± 12.9	5.3 ± 1.5	24.0 ± 9.0	0.18 ± 0.1	—	—
Gizzard, thin muscle	3.3 ± 0.7	42.5 ± 42.2	4.2 ± 1.4	27.2 ± 12.7	0.14 ± 0.1	—	—
Gizzard, thick muscle	3.3 ± 0.7	61.6 ± 62.8	3.8 ± 1.0	22.0 ± 7.3	0.42 ± 0.2	—	—
Duodenum	9.3 ± 1.2	25.0 ± 13.8	18.8 ± 3.2	11.1 ± 5.7	0.44 ± 0.3	6.0 ± 0.5	0.15 ± 0.1

[a] From Duke (1982).

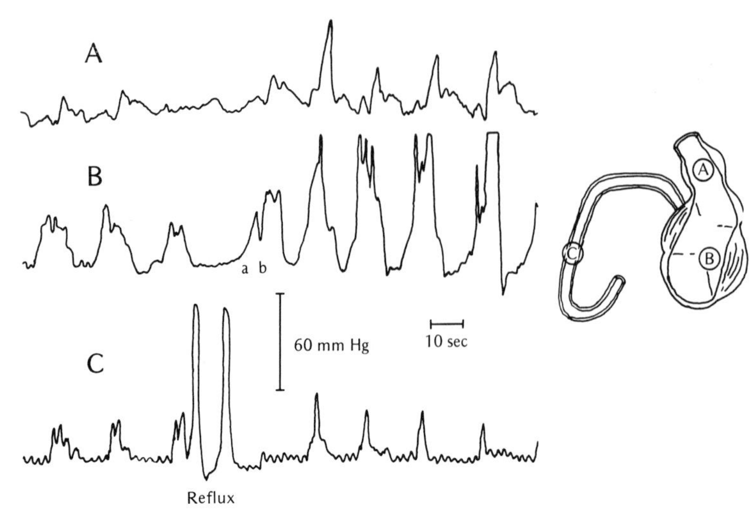

FIGURE 11–5. Tracings of typical records of pressure changes obtained from the proventriculus (**A**), gizzard (**B**), and the upper proximal duodenum (**C**) of a turkey, showing pressure events occurring during a duodenal reflux. Positions of open-tipped tubes within GI tract are indicated by letters *A, B,* and *C* (circled) on the diagram of a sagittal section of the stomach. The biphasic pattern (**B**) of the tracing representing the contraction of the gizzard is quite variable; two phases of one cycle are (*a*) pressure wave due to contraction of thin muscle pair; and (*b*) of thick muscle pair. (Duke et al., 1972a.)

gestion of a meal is completed (Duke and Rhoades, 1977). In hawks (all falconiforms, apparently), pellet egestion seems to be predominately regulated by an external factor, as dawn or "lights-on" in a research facility (Duke et al., 1975a). In fact, Durham (1983) found that gastric motility characteristic of egestion motility occurred even in fasted red-tailed hawks, indicating that egestion is not simply the end result of having eaten and digested a meal, but is probably an expression of an endogeneous circadian rhythm.

The gizzard of raptors (hawks, owls, etc.) lacks the two pairs of opposing muscles characteristic of the stomach of fowl (Figure 11–1) and most other birds. As a result, the gastroduodenal contraction cycle is simplified, consisting of a peristaltic wave arising in the proventriculus and passing aborad through the gizzard and on into the duodenum (Duke et al., 1976a; Duke and Evanson, 1976b). Three phases of gastric digestion, termed *mechanical, chemical,* and *pellet formation and egestion* based on gastric motility patterns, have been described for great horned owls (Kostuch and Duke, 1975), and similar phases have been described for red-tailed hawks (Durham, 1983).

A principal regulator of GI motility in mammals is intrinsic, myogenic electric "slow waves," also called 'pacesetter potentials" (Bortoff, 1972; Daniel, 1969; Ludwick and Bass, 1967). These are believed to arise in the longitudinal muscle layer and to be propagated through the circular muscle layer. They occur continuously whether contractions are occurring or not. As the smooth muscle layer of the gut depolarizes during a slow wave, the threshold for the depolarization required for contraction is approached, so that actual contractions generally occur only during slow waves and during the "peak" of the wave (representing the greatest depolarization). This results in one contraction per slow wave; thus, slow waves are "pacesetters" for maximal contraction frequency. Although contractions are primarily initiated and "paced" by slow waves, intrinsic and extrinsic nerves and hormones can alter contractile activity just as the vagus or epinephrine can alter the normally intrinsic cardiac activity.

Slow waves have not been recorded from either the proventriculus or gizzard of turkeys (Duke et al., 1975b), although they have been recorded from both the small (Figure 11–3) and large intestine and ceca. They apparently have no regulatory function in the duodenum or ceca, but may help regulate ileal and colonic motilities. In the duodenum of turkeys, there are usually 2–3 contractions per duodenal slow wave (Figure 11–6); therefore, the slow wave does not set the "pace" for duodenal motility. The gastroduodenal contraction sequence appears to be neurogenically initiated and coordinated via an intrinsic neural network, presumably the myenteric plexus (Duke et al., 1975b; Nolf, 1937; 1938a, b).

The absence of slow waves in electrical potential recordings from the avian gizzard may be due in part to the near absence of a longitudinal muscle layer (from which slow waves arise in mammals) in that organ. Additionally, regulation by slow waves propagated aborad through the stomach and duodenum, as in mammals, would be difficult in most species of birds in view of the complex gastroduodenal contraction sequence in which the most orad part, the proventriculus, contracts last, as stated above. This sequence is much less complex in birds of prey, being similar to that in the human stomach and duodenum, and these birds do have a longitudinal muscle layer in the gizzard; however, gastric electrical slow waves have not been recorded in these species either (Duke and Evanson, 1976b; Duke et al., 1976a).

Gastroduodenal motility in turkeys may be influenced by several factors. As in mammals, the chemical nature and volume of duodenal contents may regulate gastric motility; this is termed the *enterogastric reflex.* Intra-

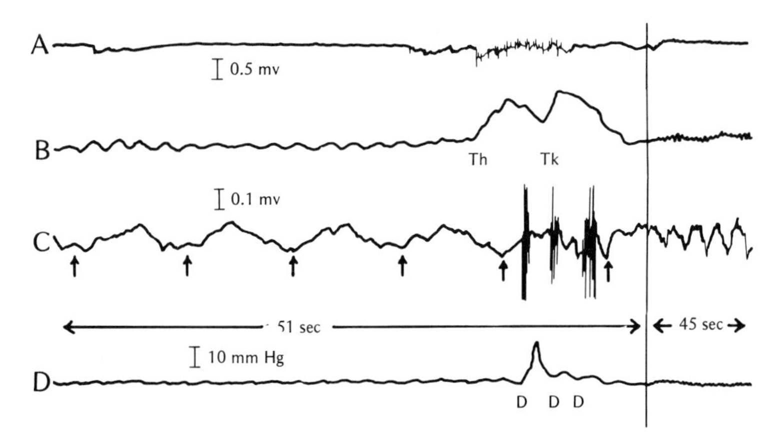

FIGURE 11–6. Tracings of typical records of electrical potential (EP) and intraluminal pressure (IP) changes from the gizzard and duodenum of a turkey. **A** and **C** are tracings of EP changes recorded from the thick cranioventral muscle of the gizzard and from the proximal duodenum, respectively. The burst of action potential spikes in **A** is associated with contraction of the thick cranioventral muscle. The arrows in **C** indicate the beginning of separate slow waves in the duodenum, and three bursts of spike potentials are associated with three contractions in **D. B** and **D** are IP changes recorded from the gizzard and duodenal flexure, respectively. *Th, Tk,* and *D* indicate the beginning of pressure changes associated with contractions of the thin and thick muscle pairs and three contractions in the duodenum, respectively. Time constant for electrical recording was 3.2 sec. Expanded time scale is in first portion of this record only. (From Duke et al., 1975b.)

duodenal injections of HCl, hypertonic NaCl, amino acid or lipid solutions, and distension via balloon inflation were all found to cause inhibition of gastric motility (Duke and Evanson, 1972; Duke et al., 1972b). The degree of inhibition produced was directly related to the concentration and volume introduced into the duodenum. Inhibition was produced within 3–30 sec, and persisted for 2–35 min (depending on dose), except for lipid solutions, which required 4–6 min to produce inhibition and 24 to more than 45 min to recover. The latter inhibitions were believed to be mediated primarily via a humoral mechanism, whereas the former were most likely induced by neural influence.

An aspect of this regulatory mechanism peculiar to turkeys, or perhaps to birds in general, was the occurrence of one or more intestinal *refluxes* during the period of inhibition. In most cases, a single reflux occurred immediately as inhibition began, but refluxes occurred repeatedly during inhibitions caused by hypertonic NaCl or lipid. Thus, not only do "offensive" stimuli in the duodenum inhibit further gastric emptying, as in mammals, but in birds, the reflux returns the offensive material to the stomach.

Humoral Regulation of Motility. Intravenous (iv) injections of the synthetic octapeptide of *cholecystokinin* (CCK) at 0.5, 5, and 15 μg/kg body weight strongly inhibited gizzard and duodenal motility in fed turkeys, but had little or no effect on ileal and cecal motility (Savory et al., 1981). In fasted turkeys, CCK had much less effect. These injections of CCK caused intestinal refluxes in the duodenum; whether CCK is involved in the gastroduodenal inhibition and intestinal refluxes of the enterogastric reflex is not known.

Avian pancreatic polypeptide (APP), a proposed GI hormone first discovered in fowl (Kimmel et al., 1971, 1975), also caused depression of gastric and duodenal motilities when injected iv (8, 10, 20, and 30 μg/kg body weight). Ileal, cecal, and colonic motility were depressed as well, but less than motility of the upper portion of the GI tract (Duke et al., 1979). Since both CCK and APP depress gastroduodenal motility and cause slowing of gastric emptying, both may be involved in satiety. Such a role has been proposed for CCK (Savory and Gentle, 1980; also see above).

The hormonal changes associated with sexual maturation and the ovulatory cycle of hens may also influence daily patterns of GI motility, particularly gastric motility (Roche and Decerprit, 1977).

It has been shown that motility of both the gizzard of turkeys (Duke and Evanson, 1976a) and the ceca of chickens (Oshima et al., 1974) are influenced by diurnal rhythms. When gastric motility was monitored,

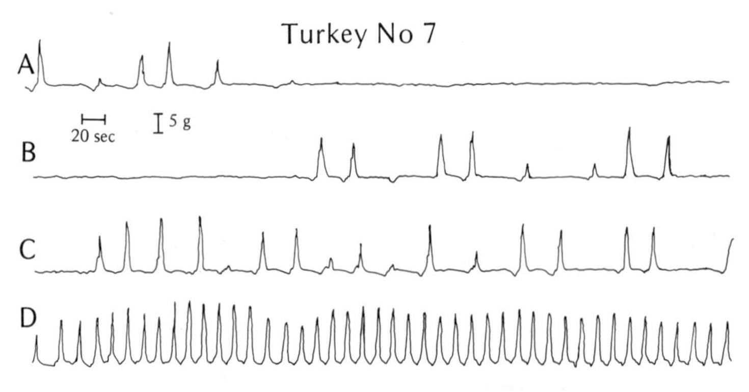

FIGURE 11–7. Tracings of records of contractions occurring in the gizzard of a fasted turkey before entry of an attendant (**A**), after entry (**B**), after seeing food (**C**), and after eating (**D**). Lines **A–D** are one continuous recording. Contractions were detected via implanted strain gauge transducers (SGT). (From Duke et al., 1976b.)

via implanted SGT, for several days before and after a period of 2 or 3 days of fasting (Duke and Evanson, 1976a), both frequency and amplitude of contractions were increased during daylight and depressed during darkness. In fact, the daily increases began slightly before "lights-on" in the holding room, and decreases likewise were initiated slightly before darkness; thus, the turkeys were somehow anticipating changes in illumination not simply responding to them. Although the diurnal patterns in frequency and amplitude of contractions were less pronounced during fasting, they were maintained, indicating that they were not simply due to food intake during the lighted period or lack of food intake in darkness. These findings provide evidence for the existence of an intrinsic biologic rhythm.

The sight of food is known to stimulate a "cephalic phase" of gastric motility in a previously fasted bird (Duke et al., 1976b). In turkeys, great horned owls, and red-tailed hawks fasted for 24 hr, the entry of an attendant into the holding room did not stimulate motility, whereas seeing feed (or in the case of the hawks, seeing a package of frozen mice) caused a significant increase in gastric contractile activity (Figure 11–7). If, after a brief delay, the birds were allowed to eat, gastric contractile activity was further increased. Stimulation of gastric motility following ingestion also demonstrates the existence of a "gastric phase" of motility.

Much less has been learned about the regulation of GI motility below the level of the duodenum, but descriptions of electrical potential changes and contractile activities are available for the ileum, ceca, and rectum of turkeys.

Ileum

Very prominent electric slow waves have been recorded from the ileum (Duke et al., 1975c). They had an average amplitude of 0.46 mV and a frequency of 6.1 waves/min (Table 11–3). Since these waves were very persistent, i.e., they seldom "faded" from the records (as occurred in records from the duodenum and cecum), and since maximal contraction frequency in the ileum was never greater than 6/min, it is believed that slow waves may be prominent in regulation of ileal motility. In general, ileal contractile and electrical activity are more similar to that of the corresponding portion of the mammalian gut than for any other part of the avian tract.

TABLE 11–3. Mean frequency and amplitude of ileal intraluminal pressure (ILP) changes and electric slow waves in turkeys[a]

ILP changes		Slow waves		
Frequency (cycles/min)	Amplitude (mm Hg)	Frequency (per min)	Amplitude (mV)	Persistence[b] (%)
40 ± 2.3	16.2 ± 13.0	6.7 ± 0.7	0.46 ± 0.3	97.9

[a] From Duke (1982).
[b] Persistence of slow waves is the proportion of time in which slow waves are evident in recordings of electrical potential changes.

TABLE 11–4. Mean frequency and amplitude of cecal contractile activity and electric slow waves in turkeys[a]

	Contractile activity		Slow waves		
	Frequency (cycles/min)	Amplitude (g)	Frequency (per min)	Amplitude (mV)	Persistence[b] (%)
Major contraction	1.2 ± 0.3	15.7 ± 3.8	—	—	—
Minor contraction	2.6 ± 0.9	2.8 ± 0.3	5.1 ± 1.2	.32 ± 0.3	6.6

[a] From Duke (1982).
[b] See Table 11–2.

Cecum

Two types of contractions have been recorded from the ceca of turkeys (Duke et al., 1980), those with a relatively low amplitude and those with a very high amplitude, called "minor" and "major" contractions, respectively (Table 11–4). About one-half the time, the minor contractions appeared to be coordinated with contractions occurring in the ileum and rectum at the same frequency but slightly out of phase. Such coordination would apparently be regulated via extrinsic nerves. Radiographic observations indicated that minor contractions were mixing rather than propulsive in nature, whereas major contractions were propulsive and were propagated both orad and aborad. Although aborad propagation occurred twice as often as orad, accumulation of ingesta in the cecal tip was apparently prevented by contractions arising in the distal cecum and moving orad, having a much greater amplitude than those moving aborad. Electric slow waves were rarely recorded in the ceca (Table 11–4) and are not believed to be involved significantly in regulation of cecal motility.

The contents of the ceca are homogeneous and pultaceous in consistency and are usually chocolate colored. Cecal contents or droppings can be readily distinguished from rectal feces, which has been used in determining when the ceca are evacuated. The ratio of cecal to rectal evacuations for the hen ranges from 1:7.3 after feeding of barley to 1:11.5 after the ingestion of corn (Röseler, 1929).

Rectum

The dominant and most unique aspect of rectal or colonic motility is the nearly continuous antiperistalsis. This can be readily observed radiographically following an enema of $BaSO_4$ solution. It is this motility that accounts for urinary "backflow" from the urodeum throughout the rectum and even into the ceca (Akester et al., 1967; Polin et al., 1967) and for cecal filling. Using a series of implanted SGT and electrodes, two types of electrical slow waves and of contactile activity have been recorded in the rectum (Lai and Duke, 1978). Small, short duration (sSW) and large, long duration (lSW) slow waves were recorded simultaneously by the two electrodes implanted in the proximal portion of the colon (Figure 11–8, A and C). Only sSW were recorded by the electrode on the distal colon (Figure 11–8E). The sSW were correlated with small contractions (A–E) and the lSW with the large contractions (A–C). Radiographic observations made while recording electrical and contractile activity disclosed that small contractions correlated with antiperistalsis. No movements correlated with large contractions could be observed radiographically.

A frequency gradient for recorded electric slow waves is characteristic in the mammalian gut (Christensen et al., 1974) and frequency gradients for both the sSW and lSW were observed (Table 11–5). The frequency of sSW was highest distally, while that of the lSW was highest proximally; in fact, lSW could not be recorded distally at all (nor could the large contractions) (Figure 11–8, D and E; Table 11–5). Thus, the sSWs were believed to arise in the distal colon and to be involved in regulation of antiperistalsis, while the lSW were believed to arise proximally in the rectum and be involved in regulation of large contractions. On the basis of the lSW frequency gradient, large contractions appeared to be peristaltic and to be primarily responsible for aborad movement of rectal digesta.

As indicated above, rectal antiperistalsis is nearly continuous. The only interruption in this activity observed so far occurs seconds before, during, and after defecation (rectal evacuation) (Dziuk, 1971; Lai and Duke, 1978). About 10 min prior to defecation, the amplitude of sSW became reduced and the frequency gradient of the lSW more pronounced. These changes would seem to favor depression of antiperistalsis and stimulation of peristalsis. During defecation, a strong contraction began in the proximal rectum and then appeared to be propagated aborally, moving all of the ingesta through the entire length of the rectum and through the cloaca in less than 4 sec. These contractions were extremely vigorous (Lai and Duke, 1978).

Rectal evacuation in turkeys usually precedes *cecal evacuation* (Duke et al., 1980); the latter typically occurred within 1–5 min after the lights came on daily in the animal holding room and again in late afternoon (at 1500–1700 hr on a photoperiod with illumination from

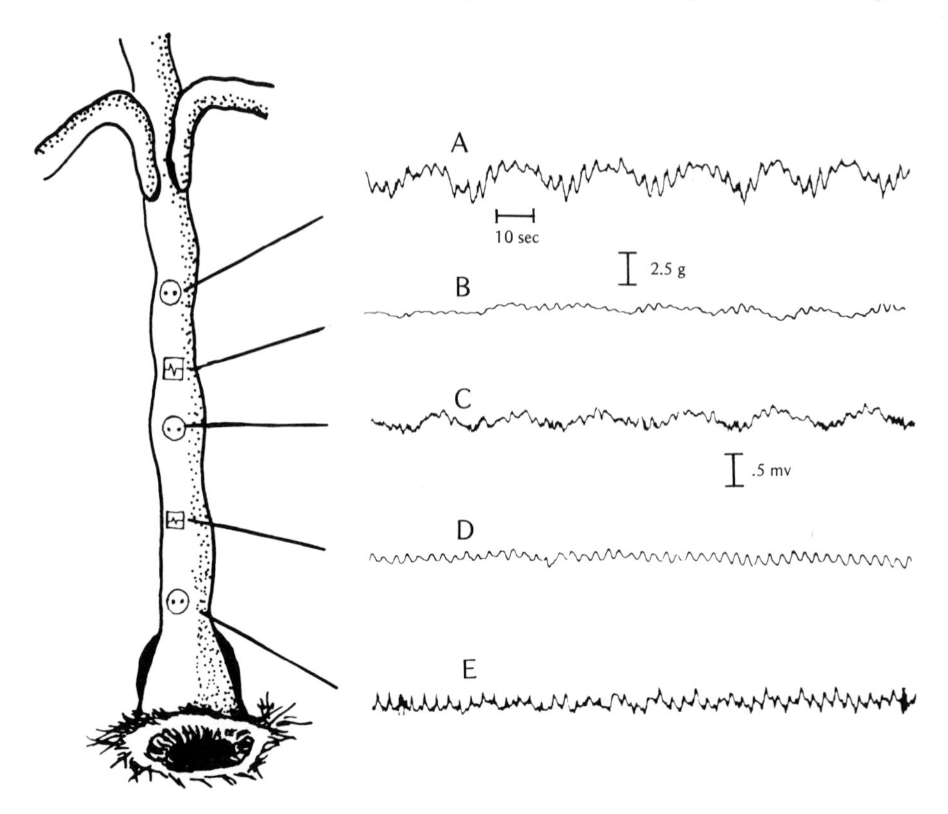

FIGURE 11–8. Electrical potential (EP) changes and contractile forces recorded from three bipolar electrodes (⊙) and two strain gauges (⊡) implanted on the colon of a turkey. EP changes are shown in **A**, **C**, and **E**, and intraluminal pressure changes in **B** and **D**. Both large and small electrical slow waves are evident in **A** and **C**, but only small slow waves in **E**. Small contractions are evident in **B** and **D**, but large contractions are only in **B**. (Lai and Duke, 1978.)

0700 to 2100 hr daily). As with rectal evacuation, contractile and electrical activity associated with cecal evacuation is very characteristic. An impressive increase in the frequency of major contractions is most apparent. Four to seven major contractions of much higher than normal amplitude occurred during the 2 min preceding cecal evacuation, and these contractions were primarily orally propagated at this time. Very dense, high-amplitude electrical spiking activity is associated with the high-amplitude contractile activity. A single, very high-amplitude contraction was recorded from both the ileum and rectum at the time of voiding

TABLE 11–5. Mean frequency and amplitude of small and large contractions and of small and large slow waves in the rectum of turkeys[a]

	Contractions				Slow waves			
	Small		Large		Small		Large	
	Frequency (cycles/min)	Amplitude (g)	Frequency (cycles/ min)	Amplitude (g)	Frequency (per min)	Amplitude (mV)	Frequency (per min)	Amplitude (mV)
Proximal[b]	14.6 ± 0.85	0.45 ± 0.24	2.66 ± 0.26	0.54 ± 0.20	15.4 ± 1.07	0.17 ± 0.08	2.83 ± 0.26	0.21 ± 0.09
Middle	—	—	—	—	15.8 ± 1.12	0.16 ± 0.09	2.76 ± 0.24	0.12 ± 0.06
Distal	15.4 ± 0.69	<0.70 ± 0.33	—	—	16.4 ± 2.16	0.25 ± 0.12	—	—

[a] From Lai and Duke (1978).
[b] Proximal, middle, and distal refer to electrode implant sites on the colon 10, 6, and 1 cm from the cloaca, respectively, or to strain gauge transducer implants at 8 (proximal) and 3.5 (distal) cm from the cloaca. (Large contractions were not recorded from the distal strain gauge nor were large slow waves recorded from the distal electrode.)

of cecal excreta; similar records were obtained also from the latter during the rectal evacuation.

Passage Rate

The rate of passage of ingesta through the digestive tract is usually measured by addition of nondigestible, nonabsorbable markers to the diet. A number of markers may be used, depending primarily on the dietary constituent under study. For example, iodinated cellulose can be used to determine the transit time of dietary fiber (Malagelada et al., 1980); the transit of all solids in the tract can be monitored via chromium- or cerium-mordanted plant cell walls (Uden et al., 1980) or radiopaque plastic pellets (Branch and Cummnings, 1978); passage rate of the liquid portion of the digesta can be determined using cobalt ethylenediamine tetracetic acid (EDTA) (Uden et al., 1980) or phenol red (Gonalons et al., 1982); and the passage of semisolids may be monitored by the gels of psyllium, guar gum, or polycarbopyil labeled with chromium-51 (Russell and Bass, 1983). Radiography may also be used, and via this technique, the rate of passage from the mouth through specific portions of the tract can be observed.

In chickens and turkeys, markers first appear in the excreta within about 2–2.5 hr, and most of the marker from a marked meal can be recovered within 24 hr (Dansky and Hill, 1952; Tuckey et al., 1958); however, a marker can be detected in the cecal excreta of turkeys for up to 72 hrs after feeding (Duke et al., 1968). Using radioactive barium, Imabayashi et al. (1956) found that about half of the label was excreted from chickens within 4–5 hr.

Passage rate through the avian alimentary canal may be affected by consistency, hardness, and water content of the food (Sturkie, 1976). Fluids pass more rapidly than solids, fiber passes more slowly than other solids, and pelleted diets pass faster than mash (Sibbald, 1979). Food passes faster through young chicks than through adults (Thornton et al., 1956), and the rate of passage through the adult turkey is similar to that of the chicken (Hillerman et al., 1953). Factors that affect the overall motility of the tract also influence the rate of food passage. Diseases that depress digestive function slow passage rate (Aylott et al., 1968; Duke et al., 1969) and the addition of antibiotics to the feed may also slow it (Hillerman et al., 1953). Increasing fat levels in a diet progressively decrease passage rate, and the slower passage may improve the digestibility of other nutrients in the diet (Larbier et al., 1977; Mateos, 1982). Higher protein content of the diet also tends to slow passage rate (Sibbald, 1979). Environmental conditions may also affect passage rate, e.g., high environmental temperature slows passage in ducks (Wilson et al., 1980). Lastly, fasting or overfeeding may slow passage, while passage is more rapid in underfed chickens (Sibbald, 1979).

Emesis

Little is known about emesis in avian species. Some researchers attempted to induce emesis as a means of collecting stomach contents to study food habits of wild birds (Prys-Jones et al., 1973; Radke and Frydendall, 1974). Apparently there are both species and individual differences in response to a variety of emetics.

Copper sulfate, given both orally and iv, and apomorphine, given iv, have been tested most commonly. Chaney and Kare (1966) tested these drugs plus hydergine and lanatoside in herring gulls, pigeons, chickens, and cowbirds. Chickens could not be induced to vomit with any of the emetics. Radke and Frydendall (1974) tested responses to copper sulfate, apomorphine, Ipecac (a mixture of the alkaloids cephaeline and emetine), and digitalis in house sparrows. Responses to the drugs varied, but digitalis produced emesis in all sparrows tested.

References

Ackerman, B., B. Anderson, E. Fabricius, and L. Svesson. (1960). Observations on central regulaion of body temperature and of food and water intake in the pigeon (*Columba livia*). Acta Physiol. Scand. 50, 328.

Ahmad, A., R.C.P. Singh, and B.D. Garg. (1978). Evidence of non-cholinergic excitatory nervous transmission in chick ileum. Life Sci., 22, 1049.

Akahori, F., M. Matsurra, and K. Arai. (1971). Studies on the movement of the alimentary canal. VI. Physiological values in growing female chicks and quails. Bull. Azabu Univ. Vet. Med., No. 22, p. 25 (English summary).

Akester, A.R. (1967). Renal portal shunts in the kidney of domestic fowl. J. Anat., 101, 569.

Akester, A.R., R.S. Anderson, K.J. Hill, and G.W. Osbaldiston. (1967). A radiographic study of urine flow in the domestic fowl. Br. Poult. Sci. 8, 209.

Ali, H.A., and J. McLelland. (1979). Neuron number in the intestinal myenteric plexus of the domestic fowl (*Gallus gallus*). Zentralbl. Veterinaermed. 8, 277.

Aylott, M.V., O.H. Vestad, J.F. Stephens, and D.E. Turk. (1968). Effect of coccidial infection upon passage rates of digestive tract contents of chicks. Poult. Sci. 46, 900.

Bartholmew, G.A., and T.J. Cade. (1963). The water economy of land birds. Auk, 80, 504.

Bartlet, A.L., and T. Hassen. (1971). Contraction of chicken rectum to nerve stimulation after blockade of sympathetic and parasympathetic transmission. Q. J. Exp. Physiol. 56, 178.

Bennett, T. (1969a). The effects of hyoscine and anticholinesterases on cholinergic transmission to the smooth muscle cells of the avian gizzard. Br. J. Pharmacol., 37, 585.

Bennett, T. (1969b). Studies on avian gizzard. Histochemical analysis of extrinsic and intrinsic innervation. Z. Zellforsch. Mikrosk. Anat., 98, 188.

Bennett, T. (1969c). Nerve-mediated excitation and inhibition of the smooth muscle cells of avian gizzard. J. Physiol. (London), 204, 669.

Bennett, T. (1974). Peripheral and autonomic nervous systems. In "Avian Biology" (D.S. Farner and J.R. King, Eds). New York: Academic Press.

Bennett, T., and J.L.S. Cobb. (1969a). Studies on avian gizzard morphology and innervation of smooth muscle. Z. Zellforsch. Mikrosk. Anat., 96, 173.

Bennett, T., and J.L.S. Cobb. (1969b). Studies on avian gizzard: Auerbach's plexus. Z. Zellforsch. Mikrosk. Anat., 99, 109.

Bennett, T., and J. Malmfors. (1970). The adrenergic nervous system of domestic fowl. Z. Zellforsch. Mikrosk. Anat., 106, 22.

Bolton, T.B. (1971). Physiology of nervous system. In "Physiology and Biochemistry of Fowl," Vol. 2 (D.J. Bell and B.M. Freeman, Eds.). London: Academic Press, Chapter 28, p. 675.

Bortoff, A. (1972). Digestion: Motility. Annu. Rev. Physiol., 34, 261.

Branch J., and J.H. Cummings. (1978). Comparison of radio-opaque pellets and chromium sesquioxide as inert markers in studies requiring accurate fecal collections. Gut, 19, 371.

Burnstock, C. (1969). Evolution of the autonomic innervation of visceral and cardiovascular systems in vertebrates. Pharmacol. Rev., 21, 247.

Calhoun, M. (1954). "Microscopic Anatomy of the Digestive System." Ames: Iowa State College Press.

Campbell, C.S., and J.D. Davies. (1974). Licking rate of rats reduced by intraduodenal and intraportal glucose infusion. Physiol. Behav., 12, 357.

Carpenter, J.W., C.M. Stein, A. Silverstein, and A. van Tienhoven. (1969). The effect of gold thioglucose on food consumption and reproduction of the Japanese quail. Poult. Sci., 48, 574.

Chaney, S.G., and M.R. Kare. (1966). Emesis in Birds. J. Am. Vet. Med. Assoc., 149, 938.

Chodnik, K.S. (1948). Cytology of the glands associated with the alimentary tract of the domestic fowl (*Gallus domesticus*). Q. J. Microsc. Sci., 89, 75.

Christensen, J., S. Anuras, and R.L. Hauser. (1974). Migrating spike bursts and electrical slow waves in the cat colon. Effect of sectioning. Gastroenterology, 66, 240.

Daniel, E.E. (1969). Digestion: Motor function. Annu. Rev. Physiol., 31, 203.

Dansky, L.M., and F.W. Hill. (1952). Application of the chromic oxide indicator method to balance studies with growing chickens. J. Nutr., 47, 449.

Denbow, D.M. (1983). Food intake and temperature response to injections of catecholamines into the lateral ventricle of the turkey brain. Poult. Sci., 62, 1088.

Denbow, D.M., and R.D. Meyers. (1982). Inhibition of food intake of chickens following injections of cholecystokinin into the lateral ventricle of the brain. Poult. Sci., 61, 1449.

Denbow, D.M., J.A. Cherry, H.P. VanKrey, and P.B. Siegel. (1982a). Food and water intake following injection of glucose into the lateral ventricle of the brain of broiler-type chicks. Poult. Sci., 61, 1713.

Denbow, D.M., H.P. VanKrey, and J.A. Cherry. (1982b). Feeding and drinking response of young chicks to injections of serotonin into the lateral ventricle of the brain. Poult. Sci., 61, 150.

Duke, G.E. (1982). Gastrointestinal motility and its regulation. Poult. Sci., 61, 1245.

Duke, G.E. (1984). Avian Digestion. In "Duke's Physiology of Domestic Animals," 10th ed. (M.J. Swenson, Ed.). Ithaca: Cornell University Press, p. 359.

Duke, G.E., and O.A. Evanson. (1972). Inhibition of gastric motility by duodenal contents in turkeys. Poult. Sci., 51, 1625.

Duke, G.E., and O.A. Evanson. (1976a). Diurnal cycles of gastric motility in normal and fasted turkeys. Poult. Sci., 55, 1082.

Duke, G.E., and O.A. Evanson. (1976b). Gastroduodenal electrical potential changes and contractile activity in birds of prey. (Abstr.) Fed. Proc. Fed. Am. Soc. Exp. Biol., 35, 303.

Duke, G.E., and D.D. Rhoades. (1977). Factors affecting meal to pellet intervals in great horned owls (*Bubo virginianus*). Comp. Biochem. Physiol. A, 56, 283.

Duke, G.E., G.A. Petrides, and R.K. Ringer. (1968). Chromium-51 in food metabolizability and passage rate studies with the ring-necked pheasant. Poult. Sci., 48, 1356.

Duke, G.E., H.E. Dziuk, and L. Hawkins. (1969). Gastrointestinal transit times in normal and bluecomb turkeys. Poult. Sci., 48, 835.

Duke, G.E., H.E. Dziuk, and O.A. Evanson. (1972a). Gastric pressure and smooth muscle electrical potential changes in turkeys. Am. J. Physiol., 222, 167.

Duke, G.E., O.A. Evanson, J.G. Ciganek, J.F. Miskowiec, and T.E. Kostuch. (1972b). Inhibition of gastric motility in turkeys by intraduodenal injections of ammino acid solutions. Poult. Sci., 51, 1749.

Duke, G.E., O.A. Evanson, and A.A. Jagers. (1975a). Meal to pellet intervals in 14 species of captive raptors. Comp. Biochem. Physiol. A, 53, 1.

Duke, G.E., T.E. Kostuch, and O.A. Evanson. (1975b). Gastroduodenal electrical activity in turkeys. Am. J. Dig. Dis., 20, 1047.

Duke, G.E., T.E. Kostuch, and O.A. Evanson. (1975c). Electrical activity and intraluminal pressure changes in the lower small intestine of turkeys. Am. J. Dig. Dis., 20, 1040.

Duke, G.E., O.A. Evanson, P.T. Redig, and D.D. Rhoades. (1976a). Mechanism of pellet egestion in great horned owls (*Bubo virginianus*). Am. J. Physiol., 213, 1824.

Duke, G.E., O.A. Evanson, and P.T. Redig. (1976b). A cephalic influence on gastric motility upon seeing food in domestic turkeys, great horned owls (*Bubo virginianus*) and red-tailed hawks (*Buteo jamaicensis*). Poult. Sci., 55, 2155.

Duke, G.E., W.D. Kuhlmann, and M.R. Fedde. (1977). Evidence for mechanoreceptors in the muscular stomach of the chicken. Poult. Sci., 56, 297.

Duke, G.E., J.R. Kimmel, P.T. Redig, and H.G. Pollack. (1979). Influence of exogenous avian pancreatic polypeptide on gastrointestinal motility of domestic turkeys. Poult. Sci., 58, 239.

Duke, G.E., O.A. Evanson, and B.J. Huberty. (1980). Electrical potential changes and contractile activity of the distal cecum of turkeys. Poult. Sci., 59, 1925.

Dunston, W.A. (1970). Excessive drinking (polydipsia) in a Galapagos mockingbird. Comp. Biochem. Physiol., 36, 143.

Dunston, W.A., and E.G. Buss. (1969). Abnormal water balance in a mutant strain of chickens. Science, 161, 167.

Dunston, W.A., E.G. Buss, W.H. Sawyer, and H.W. Sokol. (1972). Hereditary polydipsia and polyuria in chickens. Am. J. Physiol., 222, 1167.

Durham, K. (1983). The mechanism and regulation of pellet egestion in the Red-tailed hawk (*Bubeo jamaicensis*) and related gastrointestinal contractile activity. M. S. Thesis, University of Minnesota.

Dziuk, H.E. (1971). Reverse flow of gastrointestinal contents in turkeys. (Abstr.). Fed. Proc. Fed. Am. Soc. Exp. Biol., 30, 610.

Dziuk, H.E., and G.E. Duke. (1972). Cineradiographic studies of gastric motility in turkeys. Am. J. Physiol., 222, 159.

Evered, M.D., and J.T. Fitzsimons. (1976). Drinking induced by angiotensin in the pigeon (*Columba livia*). J. Physiol. (London), 263, 193.

Everett, S.D. (1968). Pharmacological responses of the isolated innervated intestine of the chick. Br. J. Pharmacol. Chemother., 33, 342.

Everett, S.D., and S.P. Mann. (1967). Catecholamine release by histamine from the isolated intestine of the chick. Eur. J. Pharmacol., 1, 310.

Farner, D.S. (1960). Digestion and the digestive system. In "Biology and Comparative Physiology of Birds," Vol. I (A.J. Marshall, Ed.). London: Academic Press, p. 411.

Fitzsimons, J.T. (1978). The role of the renin-angiotensin system in the regulation of extracellular fluid volume. In Skadhauge (1981).
Fitzsimons, J.D., and M.D. Evered. (1978). Eledoison, substance P, and related peptides: intracranial dipsogens in the pigeon and antidipsogens in the rat. Brain Res., 150, 533.
Gentle, M.J. (1971). The lingual taste buds of *Gallus domesticus. L.* Br. Poult. Sci., 12, 245.
Gleaves, E.W., L.V. Tonkinson, J.D. Wolf, C.K. Harman, R.H. Thayer, and R.D. Morrison. (1968). The action and interaction of physiological food intake regulators in the laying hen. Poult. Sci., 47, 38.
Goodman, I.J., and J.L. Brown. (1966). Stimulation of positively and negatively reinforcing sites in the avian brain. Life Sci., 5, 693.
Gonalons, E., R. Rial, and J.A. Tur. (1982). Phenol red as indicator of digestive tract motility in chickens. Poult. Sci., 61, 581.
Groebbels, F. (1932). "Der Vogel. Erster Band: Atmungswelt und Nahrungswelt." Berlin: Verlag von Gebrüder Borntraeger.
Henken, A.M., A.M.J. Groote Schaarsberg, and W. van der Hel. (1982). The effect of environmental temperature on immune response and metabolism of the young chicken. 4. Effect of environmental temperature on some aspects of energy and protein metabolism. Poult. Sci., 62, 59.
Hill, F.W., and L.M. Dansky. (1954). Studies of the energy requirements of chickens. I. The effects of dietary energy level in growth and feed consumption. Poult. Sci., 33, 112.
Hill, K.J., and P.J. Strachan. (1975). Recent advances in digestive physiology of the fowl. In "Symposium of the Zoological Society of London, No. 35" (M. Peaker, Ed.). London: Academic Press, p. 1.
Hillerman, J.P., F.H. Kratzer, and W.D. Wilson (1953). Food passage through chickens and turkeys and some regulating factors. Poult. Sci., 32, 332.
Hodgkiss, J.P. (1981). Distension-sensitive receptors in the crop of the domestic fowl (*Gallus domesticus*). Comp. Biochem. Physiol. (A), 70, 73.
Holcombe, D.J., D.A. Roland, Sr., and R.H. Harms. (1976). The ability of hens to regulate protein intake when offered a choice of diets containing different levels of protein. Poult. Sci., 55, 1731.
Howard, B.R. (1968). Drinking activity of hens in relation to egg laying. (Abstr.) Proc. Int. Congr. Physiol. Sci., 24th, 1968, 7, 202.
Howard, B.R. (1975). Water balance of the hen during egg formation. Poult. Sci., 54, 1046.
Imabayashi, K., M. Kametaka, and T. Hatano. (1956). Studies on digestion in the domestic fowl. Tokyo J. Agric. Res., 2, 99.
Jerrett, S.A., and W.R. Goodge. (1973). Evidence for amylase in avian salivary glands. J. Morphol., 139, 27.
Kaufman, S., and G. Peters. (1980). Regulatory drinking in the pigeon *Columba livia.* Am. J. Physiol., 239, R219.
Kimmel, J.R., H.G. Pollock, and R.L. Hazelwood. (1971). A new pancreatic polypeptide hormone. Fed. Proc. Fed. Am. Soc. Exp. Biol., 30, 1318.
Kimmel, J.R., L.J. Hayden, and H.G. Pollock. (1975). Isolation and characterization of a new pancreatic polypeptide hormone. J. Biol. Chem., 250, 9369.
Kobayashi, H. (1978). Evolution of the target organ. In "Comparative Endocrinology" (P.J. Gaillard and H.H. Boer, Eds.). Amsterdam: Elsevier/North Holland, p. 401.
Kobayashi, H., and Y. Takei. (1982). Mechanisms for induction of drinking with special reference to angiotensin II. Comp. Biochem. Physiol. A, 71, 485.
Komeri, S., and H. Ohashi (1984). Presynaptic muscarine inhibition of nonadrenergic, non-cholinergic neuromuscular transmission in the chicken rectum. Br. J. Pharmacol., 82, 73.
Kostuch, T.E., and G.E. Duke. (1975). Gastric motility in great horned owls. Comp. Biochem. Physiol. (A), 51, 201.
Kuenzel, W.J. (1972). Dual hypothalamic feeding system in a migratory bird, *Zonotrichia albicollis.* Am. J. Physiol. 223, 1138.
Kuenzel, W.J. (1982). Central neural structures affecting food intake in birds: the lateral and ventral hypothalamic areas. In "Aspects of Avian Endocrinology: Practical and Theoretical Implications" (C.G. Scanes, M.A. Ottinger, A.D. Kenny, J. Balthazart, J. Cronshaw, and I. Chester-Jones, Eds.). Graduate Studies, Texas Technical University, 26, 211.
Lai, H.C., and G.E. Duke. (1978). Colonic motility in domestic turkeys. Am. J. Dig. Dis., 23, 673.
Larbier, M., N.C. Baptista, and J.C. Blum. (1977). Effect of diet composition on digestive transit and amino acid intestinal absorption in chickens. Ann. Biol. Anim. Biochim. Biophys., 17, 597.
Lepkovsky, S. (1973). Hypothalamic adipose tissue interrelationships. Fed. Proc. Fed. Am. Soc. Exp. Biol., 31, 1705.
Lepkovsky, S. and M. Yasuda. (1966). Hypothalamic lesions, growth and body composition of male chickens. Poultry Sci., 45, 582.
Ludwick, J.R., and P. Bass. (1967). Contractile and electric activity of the extrahepatic biliary tract and duodenum. Surg. Obstet. Gynecol., 124, 536.
Macowan, M.M., and H.E. Magee. (1932). Observations on digestion and absorption in fowls. Q. J. Exp. Physiol., 21, 275.
Malagelada, J.R., S.E. Carter, M.L. Brown, and G.L. Carlson. (1980). Radiolabeled fiber, a physiologic marker for gastric emptying and intestinal transit of solids. Dig. Dis. Sci., 25, 81.
Maley, M.J. (1969). Electrical stimulation of agonistric behavior of the mallard. Behavior, 34, 138.
Mangold, E. (1950). "Die Verdauung bei den Nutztieren." Berlin: Akademie-Verlag, p. 87.
Matei-Vladescu, C., G. Apostol, and V. Popescu. (1977). Reduced food intake following cerebral intraventricular infusion of glucose in *Gallus domesticus.* Physiol. Behav., 19, 7.
Mateos, G.G., J.L. Sell, and J.A. Eastwood. (1982). Rate of food passage (transit time) as influenced by level of supplemental fat. Poult. Sci., 61, 94.
Maurice, D.V. (1983). Partial lipectomy and alterations in energy balance in chickens fed high and low fat diets. Fed. Proc. Fed. Am. Soc. Exp. Biol., 42, 668.
McLelland, J. (1979). Digestive system. In "Form and Function in Birds" (A.S. King and J. McLelland, Eds.). London: Academic Press, p. 69.
Mu, J.Y., T.H. Yin, C.L. Hamilton, and J.R. Brobeck. (1968). Variability of body fat in hyperphagic rats. Yale J. Biol. Med., 41, 133.
Nishida, T., Y.K. Paik, and M. Yasuda. (1969). LVIII. Blood vascular supply of the glandular stomach and muscular stomach. Jpn. J. Vet. Sci., 31, 51 (English summary).
Nolf, P. (1937). On the existence in the bird of a system of intrinsic fibers connecting the stomach to the small intestine. J. Physiol., 90, 53 p.
Nolf, P. (1938a). L'appareil nerveux de l'automatisme gastrique de l'oiseau. I. Essai d'analyse par la nicotine. Arch. Int. Physiol. Biochim., 46, 1.
Nolf, P. (1938b). L'appareil nerveux de l'automatisme gastrique de l'oiseau. II. Etude des effects causes par une ou plusieurs sections de l'anneau nerveux du gesier. Arch. Int. Physiol. Biochim., 46, 441.
Ohashi, H. (1971). An electrophysiological study of transmission from intramural excitory nerve to smooth muscle cells of the chicken oesophagus. Jpn. J. Pharmacol., 21, 585.

Oshima, S., K. Shimada, and T. Tonoue. (1974). Radiotelemetric observations of the durnal changes in respiration rate, heart rate and intestinal motility of domestic fowl. Poult. Sci., 53, 503.

Pastea, E., A. Nicolau, and J. Rosca. (1968). Dynamics of the digestive tract in hens and ducks. Acta Physiol. Hung., 33, 305.

Phillips, R.E., and O.M. Youngnen. (1971). Brain stimulation and species typical behavior. Activities evoked by electrical stimulation of the brains of chickens. Anim. Behav., 19, 757.

Pintea, V., V. Jarubescu, and M. Cotrut. (1957). Contributiuni la studiul esofagului de gaina. Lucr. Stiint. Inst. Agron., 1, 297.

Polin, D., and J.H. Wolford. (1973). Factors influencing food intake and caloric balance in chickens. Fed. Proc. Fed. Am. Soc. Exp. Biol., 32, 1720.

Polin, D., E.R. Wynosky, M. Loukides, and C.C. Porter. (1967). A possible urinary back flow to ceca revealed by studies on chicks with artificial anus and fed amprolium–C_{14} or thiamine–C_{14}. Poult. Sci., 46, 89.

Prys-Jones, R.P., L. Schifferli, and D.W. Macdonald. (1973). The use of an emetic in obtaining food samples from passerines. Ibis, 116, 60.

Radke, W.J., and M.J. Frydendall. (1974). A survey of emetics for use in stomach contents recovery in the house sparrow. Am. Midl. Nat., 92, 164.

Rea, A.M. (1973). Turkey vultures casting pellets. Auk, 90, 209.

Richardson, A.J. (1970). The role of the crop in the feeding behavior of the domestic chicken. Anim. Behav., 18, 633.

Robinzon, B., N. Snapir, and S. Lepkovsky. (1982). Hypothalamic hyperphagia, obesity, and gonadal dysfunction: absence of consistent relationship between lesion site and physiological consequences. In "Aspects of Avian Endocrinology: Practical and Theoretical Implications" (C.G. Scanes, M.A. Ottinger, A.D. Kenny, J. Balthazart, J. Gronshaw, and I. Chester-Jones, Eds.) Graduate Studies, Texas Technical University, 26, 201.

Roche, M., and J. Decerpit. (1977). Contrôles hormonal et nerveux de la motricité du tractue digestif de la pôule. Ann. Rech. Vet., 8, 25.

Röseler, M. (1929). Die Bedeutung der Blinddärme des Haushuhnes für die Resorption der Nahrung und Verdauung der Rohfaser. Z. Tierz. Zuechtungsbid., 13, 281.

Rouff, H.J., and K.F. Sewing. (1971). Die Rolle des Kropfs bei der Steuerung der Magensaftsekretion von Hühnern. Naunyn-Schmeidbergs Arch. Pharmakol. Exp. Pathol., 271, 142.

Russell, J., and P. Bass. (1983). Labeling and gastric emptying of gels in dogs. Fed. Proc. Fed. Am. Soc. Exp. Biol., 42, 759.

Savory, C.J. (1978). The relationship between food and water intake and the effects of water restriction on laying Brown leghorn hens. Br. Poult. Sci., 19, 631.

Savory, C.J., and M.J. Gentle. (1980). Intravenous injections of cholecystokenin and caerulin suppress food intake in domestic fowls. Experientia, 36, 1191.

Savory, C.J., G.E. Duke, and R.W. Bertoy. (1981). Influence of intravenous injections of cholecystokinin on gastrointestinal motility in turkeys and domestic fowls. Comp. Biochem. Physiol. A, 70, 179.

Shirley, H.V., and A.V. Nalbandov. (1956). Effects of neurohypohysectomy in domestic chickens. Endocrinology, 58, 477.

Shurlock, T.G.H., and J.M. Forbes. (1981a). Evidence for hepatic glucostatic regulation of food intake in the domestic chicken and its interaction with gastro-intestinal control. Br. Poult. Sci., 22, 333.

Shurlock, T.G.H., and J.M. Forbes. (1981b). Factors affecting food intake in the domestic chicken: the effect of infusions of nutritive and non-nutritive substances into the crop and duodenum. Br. Poult. Sci., 22, 323.

Sibbald, I.R. (1979). Passage of feed through the adult rooster. Poult. Sci., 58, 446.

Skadhauge, E. (1981). Osmoregulation in birds. New York: Springer-Verlag.

Smith, C.J.V. (1969). Alterations in the food intake of chickens as a result of hypothalamic lesions. Poult. Sci., 48, 475.

Smith, C.J.V. (1979). The hypothalamus and the regulation of feed intake. Poult. Sci., 58, 1619.

Sturkie, P.D. (Ed.). (1965). "Avian Physiology," 2nd ed. Ithaca: Cornell University Press.

Sturkie, P.D. (Ed.). (1976). "Avian Physiology," 3rd ed. New York: Springer-Verlag.

Sturkie, P.D., G. Dirner, and R. Gister. (1977). Shunting of blood from the renal portal to the hepatic portal circulation of chickens. Comp. Biochem. Physiol. A, 58, 213.

Suzuki, M., and S. Nomura. (1975). Electromyographic studies on the deglutition movement in the fowl. Jpn. J. Vet. Sci., 37, 289.

Takei, Y. (1977). Angiotensin and water intake in the Japanese quail (*Coturnix coturnix japonica*). Gen. Comp. Endocrinol., 31, 364.

Thornton, P.A., P.J. Schaible, and L.F. Wolterink. (1956). Intestinal transit and skeletal retention of radioactive strontium in the chick. Poult. Sci., 35, 1055.

Tuckey, R., B.E. March, and J. Biely. (1958). Diet and the rate of food passage in the growing chick. Poult. Sci., 37, 786.

Tweeton, J.R., R.E. Phillips, and F.W. Peek. (1973). Feeding behavior elicited by electrical stimulation of the brain of chickens, *Gallus gallus*. Poult. Sci., 52, 165.

Uden, P., P.E. Colucci, and P.J. Van Soest. (1980). Investigation of chromium, cerium and cobalt as markers in digesta. Rate of passage studies. J. Sci. Food Agric., 31, 625.

Uemura, H., H. Kobayashi, Y. Okawara, and K. Yamaguchi. (1983). Neuropeptides and drinking in birds. In "Avian Endogrinology: Environmental and Ecological Perspectives" (S. Mikami, K. Homma, and M. Wada, Eds.). Tokyo: Japan Scientific Society Press/Berlin: Springer-Verlag, p. 225.

Vonk, H.H., and N. Postma. (1949). X-ray studies on the movements of the hen's intestine. Physiol. Comp. Oecol., 1, 15.

Wada, M., H. Kobayashi, and D.S. Farner. (1975). Induction of drinking in the White-crowned sparrow, *Zonotrichia leucophrys gambelii*, by intracranial injection of angiotensin II. Gen. Comp. Endocrinol., 26, 192.

Webb, T.E., and J.R. Colvin. (1964). The composition, structure and mechanism of formation of the lining of the gizzard of the chicken. Can. J. Biochem., 42, 59.

Wehner, G.R., and R.L. Harrold. (1982). Crop volume of chickens as affected by body size, sex, and breed. Poult. Sci., 61, 598.

White, S.S. (1970). The larynx of *Gallus domesticus*. Ph.D. Thesis, University of Liverpool. In McLelland (1979).

Wilson, E.K., F.W. Pierson, P.Y. Hester, R.L. Adams, and W.J. Stadelman. (1980). The effects of high environmental temperature on feed passage time and performance traits of Pekin ducks. Poult. Sci., 59, 2322.

Wright, P. (1973). The neural basis of food and water intake in birds. Indian J. Physiol. Pharmacol., 17, 1.

Yntema, C.L., and W.S. Hammond. (1952). Experiments on the origin and development of the sacral autonomic nerves in chick embryo. J. Exp. Zool., 129, 375.

Ziswiler, V., and D.S. Farner. (1972). Digestion and digestive system. "Avian Biology," Vol. II (D.S. Farner and James R. King, Eds.). London: Academic Press, p. 343.

12
Alimentary Canal: Secretion and Digestion, Special Digestive Functions, and Absorption

G.E. Duke

Secretion and Digestion

Secretions: Their Functions and Regulation

The process of digestion involves all of the mechanical and chemical changes that ingested food must undergo before it can be absorbed in the intestines. Mechanical changes include swallowing, maceration, and grinding of food in the muscular stomach; chemical digestion consists of secretion of enzymes from the mouth, stomach, intestines, and pancreas, of bile from the liver, of hydrochloric acid from the stomach, and of bacterial action.

Before they can be absorbed from the small intestine ingested, carbohydrates must be converted into monosaccharides (see Chapter 13); fats will be hydrolyzed to fatty acids and glycerol (see Chapter 15); and proteins will be degraded to amino acids (see Chapter 14). In the latter process proteins, which are insoluble, are broken down to proteoses, peptones, and peptides. Although these are soluble, normally they are not absorbed but are converted finally into amino acids.

Mouth. The salivary glands in the oral area secrete mucus and may secrete amylase (Chapter 2). The latter has been found in the salivaries of the house sparrow (Jerrett and Goodge, 1973) and in several other avian species (Bhattacharya and Ghose, 1971), but not in those of the chicken or turkey (Chodnik, 1948; Jerrett and Goodge, 1973). The mucus is necessary as lubrication to aid in movement of ingesta through the pharynx and upper esophagus. From 7 to 25 ml of saliva are secreted per day by chickens (Leasure and Link, 1940). Amylase probably has its major activity in the crop, since food passes through the mouth too rapidly for much digestion to occur. Feeding results in increased saliva secretion due to parasympathetic stimulation (Chodnik, 1948).

Asian swifts produce an adhesive salivary secretion used to construct and attach their nests to a cave wall. These nests are eaten as "birds' nest soup" by wealthy Asians. The glands undergo seasonal regression and enlargement.

Esophagus and Crop. In response to feeding, these organs produce mucus for lubrication. Initial stages of carbohydrate digestion may occur in the crop of some species due to salivary amylase. Amylase is probably not secreted by the crop mucosa.

Proventriculus. The glandular stomach secretes both acid and pepsinogen from the chief cells and mucus from simple mucosal cells. Acid or pepsin can change pepsinogen to pepsin, the active protein enzyme, and an acid environment is best for pepsin activity. The pH of pure gastric secretions is about 2, but the pH of gastric contents is usually higher because the secretions are diluted by ingesta. HCl and pepsin have been found in the gastric secretions of all birds tested (Farner, 1960), and other enzymes (e.g., lipase) have been detected in the gastric contents, probably resulting from the duodenal reflux (Chapter 11). The concentration of pepsin is greater in carnivores than in grain eaters, and is considerably higher in the glandular stomach of pigeons than in chickens (Herpol, 1964).

Studies on the biochemical nature of pepsin and pepsinogen were reviewed by Sturkie (1976). There are at least five chicken pepsinogens. The molecular weight of fraction 4, pepsinogen, is 36,000; that of pure pepsin is 34,000. Chicken pepsin is unique in having a single thiol group, which is unusually stable for long periods at a high pH (7). However, the pH optimum for activity was near 2.0, at least for hemoglobin digestion. Assays of chicken pepsin, based on milk clotting and the hydrolysis of bisphenyl sulfite expressed as a percentage of specific activity of crystalline pig pepsin, were 58% for milk clotting, at a pH of 6.2, and 92% for hydrolysis of the sulfite, at pH 3.6.

Pepsin is always present in the contents of the muscular stomach, where most peptic digestion occurs. However, its removal has little effect on digestion if the food is soft (Fritz et al., 1936). It is likely that, following gizzardectomy, most protein digestion is accomplished by pancreatic and intestinal proteases. The optimum activity of these enzymes is at pH 6–8.

The gastric juice of birds may be collected by cannulation of the proventriculus (see Sturkie, 1976), or by aspiration of stomach contents via oral entubation (Duke et al., 1975), or by insertion of needles into the stomach (see Sturkie, 1976). Usually birds are starved for 18–24 hr or until the crop is empty before collection of gastric juice. This tends to decrease water consumption and leads to dehydration, which also decreases gastric secretion. To obtain fairly steady-state secretion, therefore, birds are usually hydrated (Long, 1967; Burhol and Hirschowitz, 1970, 1971a, b). Hcl production per kg body weight is higher in chickens than in humans and several laboratory mammals; pepsin, Cl^- and K^+ production are also higher (Table 12–1).

The regulation of gastric secretion in birds is complex, and although much is known, the process is not completely understood. In mammals, gastric secretion occurs in three phases. The *cephalic* phase occurs when an animal "senses" (visually, aurally, etc.) that it is about to eat or is eating; this phase operates via the vagal nerves. The *gastric* phase, initiated by food arriving in the stomach, involves both direct stimulation of the gastric mucosa by ingested nutrients and indirect stimulation via the autonomic nerves and release of gastric hormones. The *intestinal* phase, initiated by food arriving in the upper small intestine, also functions via the autonomic nervous system and by release of intestinal hormones. These three phases also appear to function in avian species.

Factors Affecting Gastric Secretion. Gastric secretion can be stimulated in ducks by auditory stimuli after conditioning (Walter, 1939), or by food or even empty food pans. The sight of food caused a decrease in the pH of gastric secretion in a barn owl, indicating increased secretion of Hcl (Smith and Richmond, 1972). Following sham-feeding (using an esophageal fistula so that ingesta did not reach the stomach), Karpov (1919) and Collip (1922) reported increased gastric secretion in geese and chickens. Farner (1960) found that sham-feeding of chickens with an esophageal fistula above the crop gave equivocal results, but this stimulated gastric secretion when the fistula was below the crop.

Evidence for a mechanism regulating the *cephalic* response was reviewed by Burhol (1982). In mammals, chemoreceptors in the lateral hypothalamus stimulate the vagus nerves to cause gastric secretion in response to a lack of metabolizable glucose. Lowering glucose level by injecting (i.v.) either insulin or 2-deoxy-D-glucose (2-DG), which competitively inhibits the intracellular oxidation of glucose, produced this result. In chickens, 2-DG stimulated gastric secretion of both

TABLE 12–1. Effects of histamine (dose response) on mean concentration of gastric juice of chickens[a]

Concentration (meq/liter)	Basal	Histamine (μg/kg/hr)[b] 75	150	300	600	1200
H^+	123.1	140.6	157.0	153.9	161.8	161.1
Cl^-	147.1	161.6	171.1	170.8	176.0	176.0
K^+	9.6	9.9	11.2	10.7	11.0	11.4
Na^+	16.9	13.3	5.9	8.8	4.6	5.7
Pepsin (Pu/ml)	1550	2200	1810	2070	2540	2130

[a] From Sturkie (1976), based on data of Burhol and Hirschowitz (1972).
[b] Subcutaneous infusion. Birds were starved for 18 hr and anesthetized with Penthrane.

H^+ and pepsin, but pepsin release was more strongly stimulated. Thus, it appears that central vagal stimulation can differentially affect H^+ and pepsin secretion in birds, even though they are produced by the same cell. Insulin injections inhibited H^+ secretion without affecting pepsin secretion in chickens. This, again, demonstrates a differential effect on the same cell. In mammals, it inhibits gastric water and electrolyte secretion, probably by a direct action on the gastric mucosa, and also indirectly stimulates gastric secretion via hypoglycemia. Thus, it appears that insulin may act only to inhibit gastric water and electrolyte secretion in birds without stimulating gastric secretion due to hypoglycemia; i.e., insulin probably does not act centrally in birds (Burhol, 1982).

Feeding increases gastric secretion, and foods having a higher protein content stimulate a proportionately higher rate of secretion. Further evidence of a gastric phase is that fasting greatly decreases gastric secretion (Sturkie, 1965).

Effects of Drugs and Hormones. Gastrin plays a central mediatory role in the gastric phase in mammals (Burhol, 1982). Gastrin-producing cells have been found in the glandular stomach, muscular stomach, and duodenum of birds (Polak et al., 1974; Larsson et al., 1974b); however, the relative significance of gastrin in regulation of gastric secretion in birds versus mammals is uncertain. Chickens are not particularly sensitive to mammalian gastrin (Kokas and Brunson, 1969; Burhol and Hirshowitz, 1970), but avian and mammalian gastrin differ structurally (Larsson et al., 1974a). The synthetic gastrin analog, pentagastrin, stimulates H^+ and pepsin secretion in chickens (Burhol, 1973), but contrary to the effect in mammals, pepsin release is more strongly stimulated than is H^+. It is believed that gastrin stimulates the secretion of H^+ and pepsin by acting on acetylcholine-releasing structures in the submucosal plexus of the proventriculus, rather than by direct action on the chief cells; acetylcholine (Ach) then stimulates secretion from the chief cells (Gibson et al., 1975). Also, gastrin is apparently not involved in vagally induced gastric secretion in birds. In mammals, vagal stimulation produces gastric secretion plus gastrin secretion. Since no gastrin is released in birds, vagal stimulation alone causes gastric secretion (Kokue and Hayama, 1975). Studies on the effect of avian gastrin on gastric secretion and studies confirming the postprandial release of gastrin are needed, but apparently none are available.

A second hormone that may be involved in the *gastric phase* of secretion in birds is *pancreatic polypeptide* (PP). Unlike other GI hormones, which were generally first discovered in mammals, PP was first discovered in chickens (Kimmel et al., 1968; Larsson et al., 1974a). Structurally, PP is distinct from insulin and glucagon, and apparently it does not normally influence carbohydrate metabolism (Hazelwood et al., 1973). PP originates only from the pancreas (Larsson, et al., 1974b).

Pancreatic polypeptide is not released by sham feeding, and apparently it is not involved in the cephalic phase (Kimmel and Pollock, 1975). It is, however, released postprandially with amino acids and HCl acting as secretogogues in the upper portion of the gut (Duke et al., 1982; Johnson and Hazelwood, 1982). Avian pancreatic polypeptide (APP) increases gastric secretion of pepsin and H^+, an effect not mediated via the vagus nerves (Hazelwood et al., 1973). The secretory response to APP within 1 hr was considerably greater than to pentagastrin when both were administered (i.v.) at 25 $\mu g/kg$ (Hazelwood et al., 1973; Table 12–2).

The hormones *cholecystokinin* (CCK) and *secretin*, both originating from the upper small intestine, are prominent regulators in the intestinal phase of gastric secretion in mammals. CCK stimulates H^+ secretion and is an agonist to gastrin in stimulation of H^+ (Burhol, 1982) in chickens also, but it does not seem to affect pepsin secretion (Burhol, 1974). Secretin inhibits H^+ secretion and stimulates pepsin secretion in mammals (Burhol, 1982), but stimulates both H^+ and pepsin secretion in birds (Burhol, 1974). Based on these differences in responses, the *intestinal phase* of gastric secretion may be more important in birds than in mammals (Burhol, 1982). These differences might be related to the greater role of the stomach in mechanical digestion versus chemical digestion in birds and the more thorough mixing of gastric and intestinal contents via the regularly recurring intestinal refluxes.

Other hormones and drugs may also affect gastric secretion. In chickens, *glucagon* (25 $\mu g/kg$, i.v.) decreased the volume of gastric secretion to one-third

TABLE 12–2. Gastric secretory responses[a] to avian pancreatic polypeptide (APP) and to pentragastrin (PG)[b]

	APP	PG
Gastric secretion (ml)	2.4 ± 0.37	0.22 ± 0.06
Free [H^+] (meq)	$9.52 \pm 0.49 \times 10^{-4}$)	$4.03 \pm 0.90 \times 10^{-6}$)
Pepsin (units)	$1179 \pm 95 \times 10^{4}$)	$1.06 \pm 0.35 \times 10^{4}$)
Total protein (mg)	16.01 ± 2.42	0.03 ± 0.01

[a] Within 1 hr after administration (iv) at 25 $\mu g/kg$ in chickens.
[b] From Hazelwood et al. (1973).

of the basal level (Kokas et al., 1971) and decreased pepsin output significantly. *Histamine* is a potent stimulant of gastric secretion (Table 12–1); however, when it was administered with glucagon the histamine effect was decreased. Similarly, glucagon administered with pentagastrin partially blocked the pentagastrin effect; no significant effect was observed on pepsin output.

Several cholinergic agents and vagal stimulation also stimulate gastric secretion (see Sturkie, 1976). In pigeons, *pilocarpine* (500 μg/kg) produced a secretion of moderate volume, and peptic activity and *acetylcholine* (100 μg/kg) induced the secretion of small amounts of gastric juice that was very rich in pepsin. Both pilocarpine and *mecholyl* also increased gastric secretory flow rate in chickens. *Urecholine* (200 μg/hr, subcutaneously) increased gastric secretion significantly, although not greatly, and it increased pepsin output considerably. Maximum response in volume was obtained with dosages of 1,600–3,200 μg/kg/hr, but urecholine was less effective than histamine (Burhol and Hirschowitz, 1971a). Rouff and Sewing (1970) reported that large doses of *carbachol* gave responses similar to histamine.

Vagal stimulation in pigeons and chickens is effective in increasing gastric secretion. Gibson et al. (1974) demonstrated that electrical stimulation of the chickens vagus at levels of 5–50 V at 10 pulses per second and duration of 1 msec produced a graded linear response in secretion of acid and pepsin. The threshold stimulus was 13.6 V for acid secretion and 13.9 V for pepsin. The response to vagal stimulation was greater than the response to cholinergic agents.

The cholinergic blocker, *atropine*, decreases gastric secretion in pigeons (Friedman, 1939) and chickens (Long, 1967).

Anesthetics generally affect gastric secretion. Nembutal significantly reduced volume of flow (Rouff and Sewing, 1970). *Methoxyflurane* reduced volume of gastric secretion, following histamine stimulation, by approximately one-half, and decreased acid and pepsin output significantly (Kessler et al., 1972). *Urethane*, however, did not significantly depress secretion (Kokas et al., 1971). Anesthetics also greatly reduce gastroduodenal motility (Duke et al., 1977), but have less of an effect on rectal contractions (Lai and Duke, 1978).

Intestines. Less is known about intestinal secretions than about gastric or pancreatic secretions because they are difficult to collect free from contamination by pancreatic, biliary and bacterial enzymes. Most of what we do know has been obtained from studies employing a Thiry-Vella loop. This is a segment of intestine separated from the rest of the gut and attached to the body wall to allow access to its contents; circulatory and neural connections to the loop are maintained. When a bird eats, intestinal secretions free from contamination may be collected from the lumen of the loop. The basal secretory rate, based on a 2- to 4-hr period, averaged 1.1 ml/hr per chicken (body weights, 2.5–3.5 kg), and the pH ranged from slightly acid to slightly alkaline (Kokas et al., 1967).

Intestinal secretions, or extracts from mucosal cells, are capable of digesting starch, sucrose, fats, and protein. *Amylase* is screted by the intestine as are *saccharidases, peptidases,* and *lipase* (Kokas et al., 1967; Lepkovsky and Furuta, 1970; Bhattacharya and Ghose, 1971; Radhakrishna and Nair, 1974). *Lactase* is not secreted from the avian intestine (Rutter et al., 1953; Zoppi and Shmerling, 1969), and although milk or milk products are frequently recommended for avian intestinal disorders, they usually cause diarrhea. *Trehalase,* which acts on the common plant carbohydrate trehalose, is also not secreted, but *maltase, isomaltase,* and *sucrase* are present (Zoppi and Shmerling, 1969). Sucrase concentrations are low; again, administration of sucrose in high concentration has been observed to cause diarrhea. *Sucrase* activity is present in the chick embryo intestines and rises after hatching (Brown, 1971). The upper ileum is found to have the greatest disaccharidase activity; the duodenum has less and the lower ileum almost none.

The intestine of chickens secretes *enterokinase,* which changes trypsinogen, secreted from the pancreas, to *trypsin* (Lepovsky et al., 1970). The presence of pancreatic secretions in the intestinal lumen apparently stimulates the intestine to secrete enterokinase. Raw soybeans in the diet inhibit trypsin by inhibition of the secretion of enterokinase. In addition to enzymes, the intestine also secretes the hormones *secretin, CCK* and *vasoactive intestinal peptide (VIP)*.

Intestinal secretion is increased by (1) mild duodenal distension (with food or by a balloon), which probably directly stimulates release of intestinal hormones, (2) vagal stimulation, and (3) *secretin* (Kokas et al., 1967). Secretion is not appreciably altered by administration of histamine and is inhibited by *glucagon* (Kokas et al., 1971). Glucagon stimulates intestinal secretion in mammals. Vagal stimulation has more affect on mucus secretion than on the secretion of digestive enzymes, so apparently cholinergic control does not have a major influence on Lieberkuhn's crypts. Although secretin stimulates duodenal secretion, if it is removed from a mucosal cell extract (by filtering through charcoal), the extract still stimulates duodenal secretion. Thus, some other intestinal hormone(s) is(are) also involved in regulation of intestinal secretion.

Pancreas. Pancreatic secretions and bile are emptied into the distal end of the duodenal loop. Pure secretions are obtained by cannulating one of the three ducts. When the main or larger duct is cannulated, one-third to one-half of the total pancreatic exocrine secretion may be collected. The cannulated main duct may remain patent for 30 days and in some instances for 90 days (Hulan et al., 1972). *Pancreatic juice* is pale yellow in color, and has a pH of 6.4–6.8 in chickens (Hulan et al., 1972) and 7.4–7.8 in turkeys (Duke, unpublished). White Leghorn chickens 14–20 weeks of age secreted 15–20 ml of pancreatic juice per chicken per 24 hr when the main pancreatic duct was cannulated.

Composition. There are two components to the pancreatic secretion: *aqueous,* containing water and bicarbonate ion, and *enzymic,* containing enzymes for degradation of carbohydrates, fats, and proteins. The slightly alkaline pH of the pancreatic secretion is determined by the aqueous component, and pancreatic enzymes have their highest activity at this pH.

Studies on the biochemical nature of the pancreatic enzymes were reviewed by Sturkie (1976). Highly purified chicken *chymotrypsin* and turkey *trypsin* have been extracted from the pancreas, as have less purified samples of chicken trypsin and turkey chymotrypsin. The molecular weight of turkey trypsin is 22,500, and the amino acid analysis indicated its similarity to mammalian trypsin. Chicken chymotrypsin (MW 20,000) exhibited esterase activity almost twice that of bovine α-chymotrypsin, but with only two-thirds of the protease activity. Turkey trypsin has many structural similarities with bovine and porcine trypsin. The amylolytic activity of pancreatic juice has been studied a great deal, but apparently such enzymes have not been purified or crystallized. Two types of amylases have been isolated from chicken pancreas (fast- and slow-acting types).

Bird (1971) determined the distribution of trypsin and amylase in different segments of the chicken duodenum (males 14–16 weeks old). The first three-quarters of the duodenum contained 45% of trypsin and 23% of the amylase, with 55% and 77%, respectively, in the remaining quarter. Most of the enzyme activity in the duodenum is, therefore, in the last quarter near the exit of the pancreatic ducts into the duodenum. Herpol (1967) determined the proteolytic activity of the pancreas of several species as related to diet.

Nitsan et al. (1973) determined levels of enzymes in pancreatic tissues of geese (Table 12–3); the concentration of trypsin was one-tenth that of chymotrypsin. Kokue and Hayama (1972) showed that pancreatic secretory rate per unit of body weight is higher in chickens than in mammals and that secretory rate is less affected by length of fasting in chickens (Table 12–4). Levels of pancreatic amylase and intestinal maltase as influenced by age, growth, and development have been studied by Laws and Moore (1963). Ovomucin from chicken egg albumen contains an inhibitor (MW 46,500 of trypsin and chymotrypsin (Tomimatsu et al., 1966) (see also Chapter 14).

TABLE 12–3. Concentration of pancreatic enzymes in the pancreatic tissue of geese fed a 16% protein diet[a]

Enzyme	Concentration (μg/g pancreas)
Trypsin	165
Chymotrypsin	1680
Amylase	368

[a] From Nitsan et al. (1973).

TABLE 12–4. Pancreatic secretory rate and influence of fasting in chicken, dog, rat, and sheep[a]

Species	Starvation time (hr)	Pancreatic secretory volume (ml/kg/hr)
Chicken[b]	24	0.70
	48	0.68
	72	0.65
Dog	24	0.1–0.3
	48	Negligible
Rat	24	0.6–0.7
Sheep	24	0.13
	48	0.07

[a] From Kokue and Hayama (1972).
[b] From ventral lobe only in chicken.

Regulation. An understanding of the regulation of pancreatic secretion may be aided if the regulatory scheme in mammals is first summarized. That scheme may be divided into *cephalic* and *intestinal* phases. When a meal is eaten, vagal stimulation causes the pancreas to secrete a low volume of the enzymic component. This secretion is short lived if the meal is prevented from reaching the duodenum (by sham-feeding). When food reaches the duodenum, gastric Hcl, (and the food to some extent) stimulates release of secretin from the intestine. Amino acids or peptides and fat from the food stimulate intestinal release of CCK. Secretin causes an initial secretion of the aqueous component from the pancreas, and CCK produces a prolonged flow of both the aqueous and enzymic components.

Kokue and Hayama (1972) showed that pancreatic secretion begins immediately when a fasted chicken is allowed to eat. If, however, the chickens are vagotomized, there is no immediate response, but pancreatic secretion eventually increases (Figure 12–1). Apparently the vagus mediates the immediate pancreatic secretory response. Ivanov and Gotev (1962) found that pancreatic secretion increased from 0.4–0.8 ml/hr to 3.0 ml/hr immediately after feeding. *Cholinergic* agents stimulate enzyme secretion from the pancreas of pigeons in vitro, and atropine blocks this effect (Hokin and Hokin, 1953).

Humoral regulation is apparently responsible for eventually causing pancreatic secretion. Secretin has been found in the intestinal mucosa of turkeys (Dockray, 1972) and injections (i.v.) of secretin produce an increased secretion of the pancreatic aqueous component

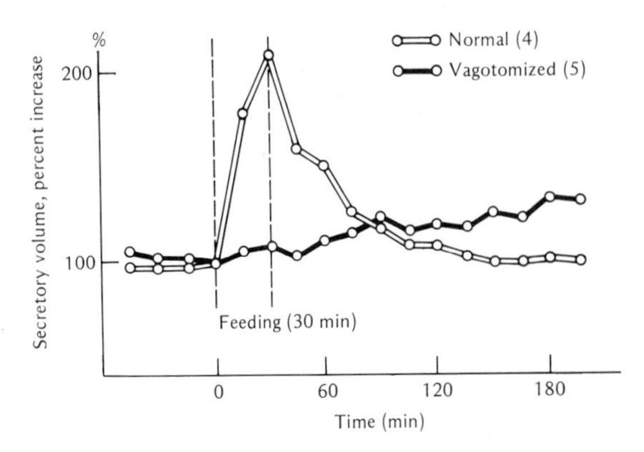

FIGURE 12–1. The effects of single feeding (for 30 min) on the pancreatic outflow of the normal (*open*) or vagotomized (*solid*) chickens after 24 hr of starvation. (From Kokue and Hayams, 1972.)

(Heatley et al., 1965). However, secretin is a less potent stimulant than is vasoactive intestinal peptide (VIP) (Vaillant et al., 1980). VIP is similar to secretin and has been isolated from the chicken intestine (Nilsson, 1974). Since the avian pancreas responds better to VIP than to secretin, VIP might control the avian pancreas in a manner similar to secretin in mammals. VIP has a predominately neural origin in the avian gut, so it may regulate digestive function more as a neurotransmitter or neurohormone than as a GI hormone (Vaillant et al., 1980).

The administration of CCK had no significant effect on pancreatic amylase activity in the chicken (Niess et al., 1972), or in the pigeon in most cases (Webster and Tyor, 1966). However, the latter authors did find that there was an increase in proteolytic enzyme activity. Sahba et al. (1970) found that administration of CCK increased pancreatic secretion in pigeons. Although secretion increased, the synthesis of protein by the pancreas did not increase. Porcine CCK was shown to increase the rate of flow of pancreatic secretions and to increase the rate of protein secretion from the pancreas of turkeys (Dockray, 1975). Thus, CCK apparently is involved in regulation of avian pancreatic secretion.

Dietary changes may influence the enzyme activity of pancreatic juice. Increased intake of carbohydrates and fat in the diet increased the amylase and lipase activity of pancreatic juice (Hulan and Bird, 1972). Increased protein (from 16 to 28%) had little effect, except that it increased the chymotrypsin activity in the duodenum and jejunum (Dal Borgo et al., 1968). Force-feeding (fattening) reduced the concentration of most of the enzymes. Unheated soybean meal in the diet decreased the specific activities of amylase, lipase, and chymotrypsin in pancreatic juice, but had no effect on trypsin activity (Dal Borgo et al., 1968), and secretory rate was greater (11.7 ml/kg/day) with unheated meal than with autoclaved meal (9.7 ml/kg/day). Little or no juice was secreted at night when birds were not eating.

Liver and Bile. The liver produces and secretes bile, which has several digestive functions. The principal function is emulsification of fats to aid in their absorption. Bile is also involved in the activation of pancreatic lipase (Chapter 15) and, because it contains amylase, in the digestion of carbohydrates. Amylase begins to appear in chicken bile at 4–8 weeks of age (Farner, 1943), and its activity in bile from the gallbladder is greater than in bile from the liver. Avian bile is slightly acid (Table 12–3).

Bile is secreted at a rate of 0.4–1 ml/kg/hr by Leghorns 10–18 weeks old (see Sturkie, 1976). In germfree domestic fowl only cholic, allocholic, and chenodeoxycholic acids were observed. In conventionally raised chickens, however, other bile acids were detected, which apparently originated from the intestinal flora. The common deoxycholic acid in mammalian bile has not been detected in the bile of chickens. Among 10 species of wild birds, *chenodeoxycholic acid* was the chief bile acid, with *cholic* and *allocholic acids* predominating in carnivorous species (see Sturkie, 1976).

The bile salts *glycocholate* and *taurocholate* are readily absorbed throughout the small intestine, with the rate of absorption being higher toward the distal end. This presumably aids recirculation of bile salts to the liver for their reuse (conservation of bile salts) (Lindsay and March, 1967).

Endogenous secretion rates for *biliverdin* and *bilirubin* in chickens were 14.7 and 0.9 μg/kg/min, respectively (Lin et al., 1974). Biliverdin levels were greater than those of bilirubin because the chicken has very low levels of liver glucuronyl transferase and little or no biliverdin reductase. Excretory rates for total endogenous bile pigements were found to be greater in chickens than in nonavian species, and bilirubin accounted for only 6% of the total.

Liver Function. In addition to its digestive functions, the liver is involved in the metabolism of protein, fats, and carbohydrates (Chapters 14, 15, and 13) and in the detoxification of metabolities. The principal method of assessing the functional state of the liver is a determination of its *rate of clearance* of *sodium bromsulphthalein* (BSP) from the blood and excretion of BSP in the bile. A review of efforts to apply this method to studies of the livers of chickens (Sturkie, 1976) disclosed disparate results. In one study, laying hens and estrogenized males or capons were found to clear BSP at a slower rate than normal males or capons. A second study, however, reported a greater clearance rate in laying females. Apparently, further studies are needed.

Bile secretion from the liver of Adélie penguins is stimulated by the presence of bile salts in the blood

TABLE 12–5. The pH of contents of the digestive tract of avian species[a]

	Chicken	Pigeon	Pheasant	Duck	Turkey
Crop	4.51	6.3[b] 4.28	5.8	4.9	6.0
Glandular stomach	4.80	1.4[b] 4.80	4.7	3.4	4.7
Muscular stomach	4.74[c] 2.50	2.00	2.0	2.3	2.2
Duodenum	5.7–6.0 6.4[c]	6.4[b] 5.2–5.4	5.6–6.0	6.0–6.2	5.8–6.5
Upper ileum	5.8–5.9 6.6[c]	5.3–5.9	6.2–6.8	6.1–6.7	6.7–6.9
Lower ileum	6.3–6.4 7.2[c]	6.8[b] 5.6	6.8	6.9	6.8
Rectum	6.3	5.4 6.6[b]	6.6	6.7	6.5
Ceca	5.7 6.9[c]	—	5.4	5.9	5.9
Bile	7.7[d] 6.6[c] 5.9	—	6.2	6.1	6.0

[a] From Sturkie (1976), based on the work of Farner (1942).
[b] From Herpol (1966).
[c] From Herpol and van Grembergen (1967).
[d] From Lin et al. (1974).

(Andrews, 1975). Bile flow is also stimulated by eating (Lavrenteva, 1963). Since this response occurs in mammals and is mediated by CCK, and since CCK is present in the avian intestine, CCK is probably involved in postprandial secretion of bile.

Hydrogen Ion Concentration of the Digestive Tract and Its Regulation

The hydrogen ion concentration or pH of the digestive tract is dependent mainly on the amount of HCl secreted in the glandular stomach and on the subsequent neutralizing action of bile and pancreatic juice in the small intestine. The pH for most parts of the tract has been determined for several species from the tracts of freshly killed specimens (Table 12–5). These data indicate that an acidic pH persists throughout the tract, although neutrality is approached in the lower small intestine and rectum. Data from studies in which pH was determined in situ in live birds indicated intestinal pH values to be somewhat higher than those reported for dead specimens. Winget et al. (1962) reported the following values for chickens: mouth, 6.7; crop, 6.4; ileum, 6.7; and rectum, 7.1; Hurwitz and Bar (1968) reported values of 7.0 and 7.9 for the upper and lower ileum, respectively.

Gastric pH values tend not to follow this pattern. Winget et al. found a value of 3.2 in the proventriculus, and Herpol (1966) reported a value of 1.4 for this organ. pH values of gastric juice collected from the stomach of live birds, via intubation, were similar to values from dead specimens, i.e., 2.3 and 2.1 for gizzard contents of domestic turkeys and ducks, respectively (Duke et al., 1975). In the latter study, values for two species of owls averaged 2.4, and for five species of hawks, 1.6. pH tends to be higher in fasted birds because fasting decreases bacterial populations and bacteria produce acid metabolites (Ford, 1974). Perhaps bacterial action accounts for lowered pH of the luminal contents of dead birds; i.e., in the absence of intestinal secretions, bacterial metabolites accumulate and lower the pH.

The in situ study of Winget et al. (1962) indicated that intraluminal pH was not static. Hurwitz and Bar (1968) demonstrated that the intestinal tract is quite capable of regulating its pH. Using ligated segments of jejunum or ileum, they introduced saline solutions at pH 4.3, 6.3, and 8.9, respectively; within 30 min the pH in the ligated segments had returned to approximately normal (Figure 12–2).

The age of chickens (1 day to adulthood) had no significant influence on the pH of digestive tract organs (Herpol, 1966).

Digestibility of Feedstuffs

Ingested energy may be partitioned into digestible, metabolizable, and net energy (Chapter 10). In order to determine the digestibile energy portion, i.e., digestibility of the diet, feces, and urine must be collected separately. This is difficult in birds, involving surgical procedures to exteriorize the ureters or to prepare a

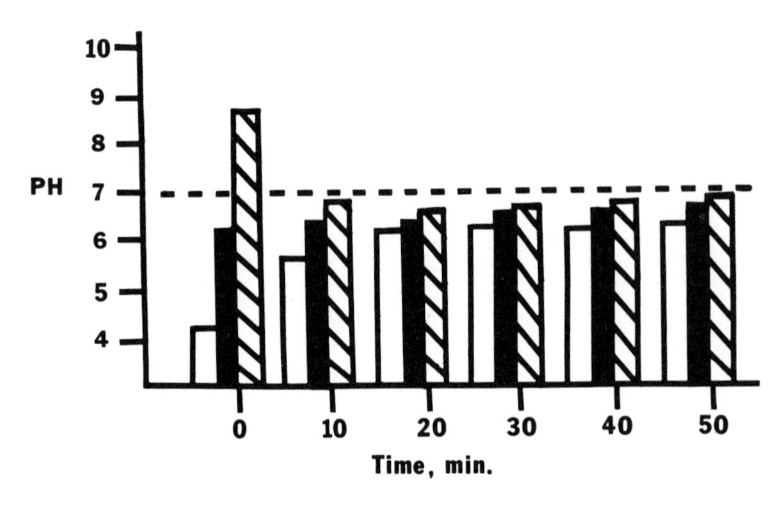

FIGURE 12–2. The intestinal pH in the jejunum following infusion of saline solutions at pH of 4.3 (open bars), 6.3 (solid bars), or 8.9 (hatched bars); dashed line indicates the in vivo pH for the jejunum. (From Hurwitz and Bar, 1968.)

colostomy. For this reason metabolizability of the diet is usually determined rather than digestibility; a correction for the amount of nitrogen in the urine must be made to convert metabolizability data to digestibility data. The digestibility of each nutrient in the feed may be determined by comparing the concentration of that nutrient in the diet to its concentration in the excreta. The metabolizability of the entire diet or the metabolizable energy in the diet is determined in the same way, i.e., the dry weight or energy value of the diet is compared to the dry weight and energy content in excreta.

In such a measurement, the test diet is fed for a period of time; then during this time period the total weight eaten and the weight of excreta are determined for a sample period (usually 3 days). Usually an indigestible marker such as chromic oxide is mixed with the feed, then the ratio of the concentration of the marker in the feed and excreta is a measure of the percent metabolizability. This allows sampling of the excreta rather than an accurate collection of all of it. This procedure, and variations on it, provide a measure of the *apparent metabolizability* (AM) of the diet. More recently, a procedure has been developed that appears to assess the *true metabolizability* (TM) of a diet (Sibbald, 1976). In this procedure, birds are fasted (usually for 24 hr) and force-fed known amounts of a test diet; excreta are then collected for 24–72 hr. The results of various tests of AM do not always agree, nor do results of tests of AM versus TM determinations. Such discrepancies are being examined (Schang et al., 1982).

Special Digestive Functions

Grit

Grit, or small bits of stone, seem to be very important to seed-eating birds. Wild birds select grit, and it is retained in the gizzard to aid that organ in macerating (grinding) hard food items. Grit is contained in commercially prepared feeds for poultry; without grit, chickens eat more feed per unit gain in body weight and the digestibility of the feed is decreased (Fritz, 1937). Digestibility of a diet may be increased by as much as 10% by addition of grit (Titus, 1955). The type of grit selected by birds is influenced by their diet and by the amount and types of grit available (Norris et al., 1975). Also, if coarse grit is available or the supply of grit is plentiful, birds tend to eat coarser food. The contraction frequency and amplitude of the gizzard is less without grit; this increases if grit is provided (Otani, 1966). The gizzard has a more elongated shape in birds raised without grit (Otani, 1967).

Cecal Function

Only 10–12% of the nutrients in standard commercial diets are digested in the ceca of pheasants (Duke et al., 1968) and turkeys (Duke et al., 1969), and the ingesta that enter the ceca are retained in the gastrointestinal (GI) tract 3–4 times longer than ingesta not in the ceca. The prolonged retention time is the result of infrequent emptying of the ceca, which occurs only once or twice per day in gallinaceous birds.

A number of possible functions have been ascribed to the ceca. Only the fluid portions of the digesta enter the ceca (Clemens et al., 1975; Gasaway et al., 1975) and several authors have shown that urinary water or other fluid in the rectum refluxes into the ceca (e.g., Polin et al., 1967; Clemens et al., 1975; Skadhauge, 1981). Thus the ceca appear to be important in water absorption. The moisture content of mixed cecal and rectal excreta was 1–2% higher following cecectomy than before (Thronburn and Willcox, 1965).

The ceca may play a role in nonprotein nitrogen metabolism and in protein utilization. Uric acid arriving in the ceca with urinary water is degraded in willow ptarmigans (Mortensen and Tindall, 1981), and the ammonia produced is incorporated into amino acids. Apparently only the bacteria use the amino acids, however, and none are absorbed from the ceca for use by the host. A similar function may occur in geese, since urea excretion was 5.7% greater following cecectomy (Gruhn et al., 1975), but this function is questionable in chickens. Even though chickens have been found to have cecal bacteria for decomposing uric acid (Barnes and Impey, 1974), uric acid excretion was unaffected by cecectomy in chickens (Kese and March, 1975). Unheated soybeans (Nitsan and Alumot, 1963) or fish meal (Nesheim and Carpenter, 1967) are less affected by intestinal proteolysis than are the heated foods. Although a significant portion of the proteinaceous materials escaping intestinal degradation may be broken down in the ceca, this probably provides little nutritional value to the host. Finally, Kessler et al. (1981) found that cecectomized chickens excreted 15–30% more amino acids than intact birds. Therefore, the ceca must be involved in protein degradation, but its significance is not clear. Since not all excreted amino acids are of dietary origin (McNab, 1973), it is likely that most proteins are digested and absorbed in the upper portion of the GI tract and that only those resistant to digestion (e.g., raw soybeans) and those of endogenous origin from the lower portion of the tract undergo cecal digestion.

Cecal bacteria synthesize vitamins. Coates et al. (1968) raised chicks in both germfree and conventional environments with diets devoid of one of eight selected vitamins. Negligible amounts of the omitted vitamins were detected in the cecal contents of germfree birds, and normal amounts were found in the ceca of conventionally riased birds. There was little evidence, however, that signs of dietary deficiency caused by the omitted vitamin were less severe in the latter than in the former chicks. Thus, while vitamins were synthesized by microbial action in the ceca of conventionally raised birds, they were—like amino acids (see above)—apparently not absorbed.

There is some evidence for a number of other possible cecal functions in fowl, such as cholesterol digestion and absorption (Tortuero et al., 1975) and protection from *Salmonella* infections (Schneitz et al., (1981). Bacteria capable of these functions have been identified (Bedbury and Duke, 1983).

Digestibility of Fiber. The most likely cecal function, and that receiving the most attention, is bacterial fermentation of dietary fiber. Early studies indicated that corn, oats, and wheat were degraded much more poorly by cecectomized than by intact birds (see Sturkie, 1976). However, subsequent work, usually involving commercially prepared diets, gave little evidence that domestic galliforms gain much of their daily energy needs via this route (Halnan, 1949; McNab, 1973; McBee, 1977). Cecectomy has little effect (Sunde et al., 1950; Beattie and Shrimpton, 1958), although it does decrease metabolizability of that small portion (1–10%; Sturkie, 1976) of dietary fiber usually digested by intact fowl. The cecal bacteria of domestic fowl appear to be able to digest little (Olsson et al., 1950) or no cellulose (Anderson et al., 1958; Barnes, 1972; Barnes and Impey, 1972). When birds were fed high-fiber or high-cellulose diets prior to experimentation, however, cellulose degradation was improved from 2.8% in turkeys not preconditioned to a high-fiber diet to 10.4% cellulose degradation in birds that were preconditioned (Duke et al., 1984). The glucose freed by this cellulose breakdown was absorbed from the ceca and used by the host birds. Apparently the preconditioning stimulated development of a cecal flora more capable of fiber digestion and cellulolysis (Bedbury and Duke, 1983). Perhaps the whole grain diets used in earlier studies provided this type of stimulation.

Most studies involving wild galliforms or their cecal contents indicate that they do seem to gain significant energy from fermentation via cecal bacteria (Thompson and Boag, 1975; Gasaway, 1976a, b) and do seem to be able to digest cellulose (Inman, 1973; Gasaway, 1976c; Moss, 1977; Hanssen, 1979a, b).

Several authors have shown that as fiber content of the diet is increased, the ceca increase in size (e.g., Gasaway, 1976a). This effect appears to be due to eating a greater quantity of food rather than to the quality of the food. Coturnix quail held at a low ambient temperature ate more than did those living in a high ambient temperature feeding on the same diet, and those in the low temperature had larger ceca (Fenna and Boag, 1974).

Crop Digestion

It is likely that some digestion occurs in the crop as a result of both enzymic and bacterial action. Amylase has been found in the crop of chickens, but it may not have originated there (see Farner, 1960; Sturkie, 1965). The source of the amylase could be the salivary glands, crop mucosa, bacteria in the crop, food in the crop, or intestine and pancreas (via reflux). The latter source probably accounts for proteases reported to be present in the crop. The pH in the crop is most favorable for plant amylase activity (Farner, 1960).

Jung and Pierre (1933) believed that little or no starch digestion occurred in the crop; however, Bolton (1965) found that as much as 25% of ingested starch is degraded to sugars in the crop and about 10% is lost from the crop. The sugars may have been degraded by bacteria, and sugar lost from the crop may either have been absorbed or used by crop bacteria. Pritchard (1972) found that sucrose was still digested in crop contents treated with chloroform (to kill bacteria) and then incubated. Probably both bacterial and nonbacterial amylase are involved. Glucose is absorbed from the crop but xylose is not (Soedarmo et al., 1961).

The crop also secretes mucus, which moistens ingesta and better prepares it for gastric and intestinal digestion. This may be more important to the digestive process overall than any digestion occuring in the crop. *Cropectomy* has little effect on digestion if normal diets are fed (Smith and Pilz, 1971). If a high-bulk ration is fed, then feeding frequency is increased to compensate for decreased food storage.

Absorption

The details of absorption of carbohydrates, proteins, and lipids are discussed in Chapters 13, 14, and 15. Most absorption takes place in the small intestine and occurs by *diffusion* along a concentration gradient or by some type of *active transport,* usually involving carrier-mediated transport, which is dependent on the active transport of sodium.

An impressive amount of absorption and secretion occurs in the small intestine of laying hens. Mongin (1976) found that hens drank approximately 2.3 g of water per gram of dry matter eaten. Their duodenal contents, however, had 7 g of water per gram of dry matter; i.e., for every gram of dry matter or water ingested, 2 g of pancreatic, biliary, and intestinal secretions had been added. Yet, by the time the digesta arrived in the proximal part of the rectum one-half of this fluid had been reabsorbed (the ratio was 3.5 g of water for every gram of dry matter). More fluid was absorbed in the rectum leaving the excreta with approximately 2 g of water per gram of dry matter. The rectum is involved primarily in water and electrolyte reabsorption (see Chapter 16).

Electrolytes

The rates and mechanisms of absorption of many electrolytes have been studied, e.g., iron (Edwards and Washburn, 1968; Featherston, et al., 1968), copper (Starcher, 1969), phosphate (Hurwitz et al., 1979), manganese (Cikrt and Vostal, 1969; Suso and Edwards, 1968), zinc and cobalt (Suso and Edwards, 1968), sodium (Medl and Scharrer, 1978), lead (Mykkanen and Wasserman, 1981), and many studies on calcium (e.g., Holdsworth, 1965; Hurwitz and Bar, 1969, 1971). Also, many factors have been found to influence absorption rates of electrolytes. The *site* of maximal phosphate absorption is the upper ileum (Blahos and Care, 1981). The *concentration* of an electrolyte may influence its absorption or the absorption of other electrolytes. Iron absorption decreased with increasing iron concentrations in the diet (Featherston et al., 1968). This is true for lead also (Mykkanen and Wasserman, 1981). Low concentrations of phosphate in the bathing medium stimulate in vitro intestinal calcium absorption (Bar and Hurwitz, 1969), but the absorption is independent of lead concentration and vice versa (Mykkanen and Wasserman, 1981).

Electrolyte absorption may be influenced by other substances. *Aldosterone* increases sodium absorption in the rectum (Thomas et al., 1980). *Prolactin* stimulates water absorption in the crop (Frantz and Rillema, 1968). *Vitamin D* increased phosphorous (Patrick and Schweitzer, 1952) and calcium (Hurwitz et al., 1967) absorption. *Bile salts* increase the absorption of calcium (Webling and Holdsworth, 1965, 1966a; Bar and Hurwitz, 1972) and influence the absorption of strontium and iron (Webling and Holdsworth, 1966b). Complexing zinc with EDTA preferentially increases zinc absorption over absorption of manganese; in the absence of EDTA manganese is absorbed more readily than zinc (Vohra and Gonzales, 1969). *Antibiotics* apparently do not alter electrolyte (zinc) or nutrient absorption (Turk, 1968), but diseases may. Intestinal parasitism by coccidia resulted in significantly decreased zinc but not calcium absorption (Turk and Stephens, 1970; Turk, 1973).

Most studies of electrolyte absorption have dealt with calcium absorption. This interest is because chicken eggs contain approximately 2 g of calcium each; calcium thus must have a high rate and volume of intestinal absorption to meet this demand. The rate of calcium absorption is greatest in the duodenum and jejunum and lowest in the ileum (Hurwitz and Bar, 1965). Absorption is higher when calcium is being deposited on the shell than when no calcium is being deposited, but there are no significant differences during periods of early and late calcification. The absorption of phosphorus follows a pattern similar to calcium.

Hurwitz and Bar (1969) studied the relationship of calcium absorption to changes in electrochemical potential difference (ECPD) between the intestinal lumen and the blood. When the calcium content of diet was high the ECPD was always positive in the duodenum and jejunum, where most absorption occurs indicating a mechanism of *simple diffusion.* When the calcium level in the diet of chicks was low (0.3%) the ECPD was negative, suggesting that absorption was against the concentration gradient and therefore involving *active* transport. There is a gradual increase in capacity for calcium absorption in female chickens at the onset of laying, and the increase is related to an increase in duodenal calcium-binding protein, which also occurs at the onset of laying (Hurwitz and Bar, 1971). When dietary calcium is restricted, calcium absorption reaches maximal rates in the duodenum but is much less in the jejunum. Changes in duodenal calcium absorption are correlated with mucosal calcium-binding activity. Hurwitz et al. (1973) have considered the factors involved in the regulation of calcium absorption.

TABLE 12–6. Osmotic pressures at several levels of the digestive tract in hens[a]

Level of tract	Osmotic pressure (mOsm)
Crop	537
Gizzard	312
Duodenum	571
Proximal jejunum	650
Distal jejunum	573
Proximal ileum	514
Distal ileum	451

[a] From Mongin et al. (1976).

Water

Water is absorbed throughout the tract (Skadhauge, 1981; Hill and Lumijarvi, 1968; Crocker and Holmes, 1971). It appears to be passively transported, with the net flux of water linked to net sodium and potassium movements (Mongin and deLaage, 1977). However, Mongin et al. (1976) found that water must be absorbed against a concentration gradient, and it is not clear how this is accomplished; a double-membrane model with solute-linked water movement was proposed. (See also Chapter 16) Mongin et al. (1976) measured osmotic pressures of the liquid phase of the contents of the tract in hens and rabbits. In rabbits, the osmolality was similar throughout the tract, averaging 331 mOsm, slightly hypertronic to plasma (297 mOsm). This osmotic pressure is regulated by either gastric dilution of a hypertonic solution or concentration of a hypotonic solution. In the hen, osmotic pressures varied between 312 and 650 mOsm at different levels of the tract (Table 12–6). Plasma osmolality in the hen was about 320 mOsm, so water absorption from the tract must occur against a concentration gradient.

References

Anderson, D.L., F.W. Hill, and R. Renner. (1958). Studies of metabolizable and productive energy of glucose for the growing chick. J. Nutr. 65, 561.

Andrews, C.J.H. (1975). Bile secretion in anesthetized Adélie penguin (*Pygoscelis adeliae*). J. Physiol., 246, 468.

Bar, A., and S. Hurwitz. (1969). *In vitro* calcium transport in laying fowl intestine: Characterization of the system and medium composition. Poult. Sci., 48, 1105.

Bar, A., and S. Hurwitz. (1972). *In vitro* calcium transport in laying fowl intestine: Effect of bile preparations. Comp. Biochem. Physiol. A, 41, 383.

Barnes, E.M. (1972). The avian intestinal flora with particular reference to the possible ecological significance of the cecal anerobic bacteria. Am. J. Clin. Nutr., 25, 1475.

Barnes, E.M., and C.S. Impey. (1972). Some properties of the non-sporing anerobes from poultry caeca. J. Appl. Bacteriol., 35, 241.

Barnes, E., and C. S. Impey. (1974). The occurrence and properties of uric acid decomposing anaerobic bacteria in the avian cecum. J. Appl. Bacteriol., 37, 393.

Beattie, J., and D.H. Shrimpton. (1958). Surgical and chemical techniques for *in vivo* studies of the metabolism of the intestinal microflora of domestic fowls. Q. J. Exp. Physiol., 43, 399.

Bedbury, H.P., and G.E. Duke. (1983). Cecal microflora of turkeys fed low or high fiber diets: Enumeration, identification and determination of cellulolytic activity. Poult. Sci., 62, 675.

Bhattacharya, S., and K.C. Ghose. (1971). Influence of food on amylase system in birds. Comp. Biochem. Physiol. B: Comp. Biochem., 40, 317.

Bird, F.H. (1971). Distribution of trypsin and amylase activities in the duodenum of the domestic fowl. Br. Poult. Sci., 12, 373.

Blahos, J., and A.D. Care. (1981). The jejunum is the site of maximal rate of intestinal absorption of phosphate in chicks. Physiol. Bohemoslov., 30, 157.

Bolton, W. (1965). Digestion in crop. Br. Poult. Sci., 6, 97.

Brown, K.M. (1971). Sucrose activity in the intestine of the chick; Normal development and influence of hydrocortisone, actinomycin D, cycloheximide and puromycin. J. Exp. Biol., 177, 493.

Burhol, P.G. (1973). Gastric secretory relationship between H^+ and pepsin in fistula chickens. Scand. J. Gastroenterol., 8, 283.

Burhol, P.G. (1974). Gastric stimulation by intravenous injection of cholecystokinin and secretin in fistula chickens. Scand. J. Gastroenterol., 9, 49.

Burhol, P.G. (1982). Regulation of gastric secretion in the chicken. Scand. J. Gastroenterol., 17, 321.

Burhol, P.G., and B.I. Hirschowitz. (1970). Single subcutaneous doses of histamine and pentagastrin in gastric fistula chickens. Am. J. Physiol., 218, 1671.

Burhol, P.G., and B.I. Hirschowitz. (1971a). Gastric stimulation by subcutaneous infusion of urecholine in fistula chickens. A comparison to histamine and pentagastrin. Scand. J. Gastroenterol., 6 (Suppl. 11), 1.

Burhol, P.G., and B.I. Hirschowitz. (1971b). Gastric stimulation by subcutaneous infusion of cholecystokinin-pancreozymin in fistula chickens. Scand. J. Gastroenterol., 6 (Suppl. 11), 41.

Burhol, P.G., and B.I. Hirschowitz. (1972). Dose responses with subcutaneous infusion of histamine in gastric fistula chickens. Am. J. Physiol., 222, 308.

Cheney, G. (1938). Gastric acidity in chicks with experimental gastric ulcers. Am. J. Dig. Dis., 5, 104.

Chodnik, K.S. (1948). Cytology of the glands associated with the alimentary tract of domestic fowl (*Gallus domesticus*). Q. J. Microsc. Sci., 89, 75.

Cikrt, M., and J. Vostal. (1969). Study of manganese resorption *in vitro* through intestinal wall. Int. Z. Klin, Pharmakol. Ther. Toxik., 2, 280.

Clemens, E.T., C.E. Stevens, and M. Southworth. (1975). Sites of organic acid production and pattern of digesta movement in the gastrointestinal tract of geese. J. Nutr., 105, 1341.

Coates, M.E., J.E. Ford, and G.F. Harrison. (1968). Intestinal synthesis of vitamins of the B complex in chicks. Br. J. Nutr., 22, 493.

Collip, J.B. (1922). The activation of the glandular stomach of the fowl. Am. J. Physiol., 59, 435.

Crocker, A.D., and W.N. Holmes. (1971). Intestinal absorption in ducklings maintained on fresh water and hypertonic saline. Comp. Biochem Physiol. A, 40, 203.

Dal Borgo, G.A., J. Salman, M.H. Pubols, and J. McGinnis. (1968). Exocrine function of the chick pancreas as affected by dietary soybean meal and carbohydrate. Proc. Soc. Exp. Biol. Med., 129, 877.

Dockray, G.J. (1972). Pancreatic secretion in the turkey. J. Physiol., 227, 49.
Dockray, G.J. (1975). Comparison of the actions of porcine secretin and extracts of chicken duodenum on pancreatic exocrine secretion in the cat and turkey. J. Physiol., 244, 625.
Duke, G.E., H.E. Dziuk, O.A. Evanson, and J.E. Miller. (1977). Studies of methods for *in situ* observations of gastric motility in domestic turkeys. Poult. Sci., 56, 1575.
Duke, G.E., G.A. Petrides, and R.K. Ringer. (1968). Chromium-51 in food metabolizability and passage rate studies with the Ring-necked pheasant. Poult. Sci., 47, 1356.
Duke, G.E., H.E. Dziuk, and L. Hawkins. (1969). Gastrointestinal transit-times in normal and Bluecomb diseased turkeys. Poult. Sci., 48, 835.
Duke, G.E., A.A. Jegers, G. Loff, and O.A. Evanson. (1975). Gastric digestion in some raptors. Comp. Biochem. Physiol., 50, 649.
Duke, G.E., J.R. Kimmel, K. Durham, H.G. Pollock, R. Bertoy, and D. Rains-Epstein. (1982). Release of avian pancreatic polypeptide by various intraluminal contents in the stomach, duodenum or ileum of turkeys. Dig. Dis. Sci., 27, 782.
Duke, G.E., E. Eckelstein, S. Kirkwood, C.F. Louis, and H.P. Bedbury. (1984). Cellulose digestion by domestic turkeys fed low or high fiber diets. J. Nutr., 114, 95.
Edwards, H.M., Jr., and K.W. Washburn. (1968). ^{59}Fe absorption by chickens. Poult. Sci., 47, 337.
Farner, D.S. (1942). The hydrogen ion concentration in avian digestive tracts. Poult. Sci., 21, 445.
Farner, D.S. (1943). Biliary amylase in the domestic fowl. Biol. Bull. (Woods Hole, Mass.), 84, 240.
Farner, D.S. (1960). Digestion and digestive system. In "Biology and Comparative Physiology of Birds" (A.J. Marshall, Ed.). New York: Academic Press, Chapter 11.
Featherston, W.R., T.J. Pockat, and J. Wallace. (1968). Radioactive iron absorption and retention by chicks fed different levels of dietary iron. Poult. Sci., 47, 946.
Fenna, L., and D.A. Boag. (1974). Adaptive significance of the ceca in Japanese quail and spruce grouse (Galliformes). Can. J. Zool., 52, 1577.
Ford, D.J. (1974). The effect of microflora on gastrointestinal pH in the chick. Br. Poult. Sci. 15, 131.
Frantz, W.L., and J.A. Rillema. (1968). Prolactin-stimulated uptake of amino acids–^{14}C and ^{3}HOH in pigeon crop mucosa. Am. J. Physiol., 215, 762.
Friedman, M.H.F. (1939). Gastric secretion in birds. J. Cell. Comp. Physiol., 13, 219.
Fritz, J.C. (1937). The effect of feeding grit on digestibility in the domestic fowl. Poult. Sci., 16, 75.
Fritz, J.C., W.H. Burrows, and W.H. Titus. (1936). Comparison of digestibility of gizzardectomized and normal fowls. Poult. Sci., 15, 289.
Gasaway, W.C., D.F. Holleman, and R.G. White. (1975). Flow of digesta in the intestine and cecum of the Rock ptarmigan. Condor, 77, 467.
Gasaway, W.C. (1976a). Seasonal variation in diet, volatile fatty acid production and size of the cecum of Rock ptarmigan. Comp. Biochem. Physiol. A, 53, 109.
Gasaway, W.C. (1976b). Volatile fatty acids and metabolizable energy derived from cecal fermentation in the Willow ptarmigan. Comp. Biochem. Physiol. A, 53, 115.
Gasaway, W.C. (1976c). Cellulose digestion and metabolism by captive Rock ptarmigan. Comp. Biochem. Physiol., 54, 179.
Gibson, R.G., H.W. Colvin, Jr., and B.I. Hirschowitz. (1974). Kinetics for gastric response in chickens to graded electrical vagal stimulation. Proc. Soc. Exp. Biol. Med., 145, 1058.
Gibson, R.G., H.W. Colvin, Jr., and R.E. Burger. (1975). Ganglion-mediated hydrogen ion and pepsin secretion in *Gallus domesticus*. Comp. Biochem. Physiol., 51, 633.
Gruhn, K., A. Hennig, D. Jamroz, and M. Zieger. (1975). Influence of the caecum in geese fed diets containing urea on nitrogen balance and digestibility of nutrients, also on the urea content of excreta and blood. Arch. Exp. Veterinaermed., 29, 199.
Halnan, E.T. (1949). The architecture of the avian gut and tolerance of crude fiber. Br. J. Nutr., 3, 245.
Hanssen, I. (1979a). Micromorphological studies on the small intestine and ceca in wild and captive willow grouse (*Lagopus lagopus lagopus*). Acta Vet. Scand., 20, 351.
Hanssen, I. (1979b). A comparison of the microbiological conditions in the small intestine and ceca of wild and captive willow grouse (*Lagopus lagopus lagopus*). Acta Vet. Scand., 20, 365.
Hazelwood, R.L., S.D. Turner, J.R. Kimmel, and H.G. Pollock. (1973). Spectrum effects of a new polypeptide (third hormone?) isolated from the chicken pancreas. Gen. Comp. Endrocrinol., 21, 485.
Heatley, N.G., F. McElheny, and L. Lepkovsky. (1965). Measurement of rate of flow of pancreatic secretion in anesthetized chicken. Comp. Biochem. Physiol., 16, 29.
Herpol, C. (1964). Activite proteolytique de l'appareil gastrique d'oiseaux granivores et carnivores. Ann. Biol. Anim. Biochim. Biophys., 4, 239.
Herpol, C. (1966). Influence de l'ago sur le pH dans le tube digestif de gallus doméstícus. Ann. Biol. Anim. Biochim. Biophys., 6, 495.
Herpol, C. (1967). Etude de l'activité proteolytique des divers organes du système digestif de quelques espèces d'oiseaux en rapport avec leur régime alimentaire. Z. Vgl. Physiol., 57, 209.
Herpol, C., and G. van Grembergen. (1967). La signification du pH dans le tube digestif de gullus domesticus. Ann. Biol. Anim. Biochim. Biophys., 7, 33.
Hill, F.W., and D.H. Lumijarvi. (1968) Evidence for an electrolyte-conserving function of the colon in chickens. Fed. Proc. Fed. Am. Soc. Exp. Biol., 27, 421 (Abstr. 1165).
Hokin, L.E., and M.R. Hokin. (1953). Enzyme secretion and the incorporation of ^{32}P into phospholipids of pancreas slices. J. Biol. Chem., 203, 967.
Holdsworth, E.S. (1965). Vitamin D_3 and calcium adsorption in the chick. Biochem. J., 96, 475.
Hulan, H.W., and F.H. Bird. (1972). Effect of fat level in isonitrogenous diets on composition of avian pancreatic juice. J. Nutr., 102, 459.
Hulan, H.W., G. Moreau, and F.H. Bird. (1972). A method for cannulating the main pancreatic duct of chickens: The continuous collection of avian pancreatic juice. Poult. Sci., 51, 531.
Hurwitz, S., and A. Bar. (1965). Absorption of calcium and phosphorus along the gastrointestinal tract of the laying fowl as influenced by dietary calcium and egg shell formation. J. Nutr., 86, 433.
Hurwitz, S., and A. Bar. (1968). Regulation of pH in the intestine of the laying fowl. Poult. Sci., 47, 1029.
Hurwitz, S., and A. Bar. (1969). Relation between the lumen blood-electrochemical potential difference of calcium, calcium absorption and calcium-binding protein in the intestine of the fowl. J. Nutr., 99, 217.
Hurwitz, S., and A. Bar. (1971). Relationship of duodenal Ca-binding protein to calcium absorption in the laying fowl. Comp. Biochem. Physiol. A, 41, 735.
Hurwitz, S., A. Bar, and I. Cohen. (1973). Regulation of calcium absorption by fowl intestine. Am. J. Physiol., 225, 150.
Hurwitz, S., H.C. Harrison, and H.E. Harrison. (1967). Effect of vitamin D_3 on the *in vitro* transport of calcium by the chick intestine. J. Nutr., 91, 319.

Hurwitz, S., U. Eisner, D. Dubrov, D. Skelan, G. Risenfeld, and A. Bar. (1979). Protein, fatty acids, calcium and phosphate absorption along the gastrointestinal tract of the young turkey. Comp. Biochem. Physiol. A, 62, 847.

Inman, D.L. (1973). Cellulose digestion in Ruffed grouse, Chukar partridge, and Bobwhite quail. J. Wildl. Manage., 37, 114.

Ivanov, N., and R. Gotev. (1962). Untersuchungen über die aussensekretorische Tätigkeit der Bauchspeicheldrüse bei Hühnern. Arch. Tierernaeh., 12, 65.

Jerret, S.A., and W.R. Goodge. (1973). Evidence for amylase in avian salivary glands. J. Morphol., 139, 27.

Johnson, E., and R.L. Hazelwood. (1982). Avian pancreatic polypeptide (APP) levels in fasted-refed chickens: Locus of post-prandial trigger. Proc. Soc. Exp. Biol. Med., 169, 175.

Jung, L., and M. Pierre. (1933). Sur le rôle de la salive chez les oiseaux granivores. C. R. Soc. Biol. 113, 115.

Karpov, L.V. (1919). O perevarivanii nekotorykh rastitelnykh i zhivotnykh belkov gusiiym zheludochnom sokom. Fiziol. Zh SSSR im I.M. Sechenova 2, 185. Russ. Physiol. J. 2, 185 (In Russian) (Physiol. Abstr. 5, 469, 1920). In Sturkie (1965).

Kese, A.G., and B.E. March. (1975). The role of the avian ceca in energy and protein metabolism (Abstr.). Poult. Sci., 54, 1781.

Kessler, C.A., B.I. Hirschowitz, P.G. Burhol, and G. Sachs. (1972). Methoxyflurane (penthrane) anesthesia effect on histamine stimulated gastric secretion in the chickens. Proc. Soc. Exp. Biol. Med., 139, 1340.

Kessler, J.W., T.H. Nguyen, and O.P. Thomas. (1981). The amino acid excretion values in intact and cecectomized negative control roosters used for determining metabolic plus endogenous urinary losses. Poult. Sci., 60, 1576.

Kimmel, J.R., and H.G. Pollock. (1975). Factors affecting blood levels of avian pancreatic polypeptide (APP), a new pancreatic hormone. Fed. Proc. Fed. Am. Soc. Exp. Biol., 34, 454.

Kimmel, J.R., H.G. Pollock, and R.L. Hazelwood. (1968). Isolation and characterization of chicken insulin. Endocrinology, 83, 1323.

Kokas, E., and W. Brunson, Jr. (1969). Gastric secretion inhibition in chickens. Physiologist, 12, 272.

Kokas, E., L. Phillips, Jr., and W.D. Brunson, Jr. (1967). The secretory activity of the duodenum in chickens. Comp. Biochem. Physiol., 22, 81.

Kokas, E., S.H. Kaufman, and J.C. Long. (1971). Effect of glucagon on gastric and duodenal secretion in chickens. Z. Vgl. Physiol., 74, 315.

Kokue, E., and T. Hayama. (1972). Effects of starvation and feeding in the endocrine pancreas of chicken. Poult. Sci., 51, 1366.

Kokue, E., and T. Hayama. (1975). Doubtful role of endogenous gastrin in chicken gastric secretion by vagal stimulation. Experientia, 31, 197.

Lai, H.C., and G.E. Duke. (1978). Colonic motility in domestic turkeys. Am. J. Dig. Dis., 23, 673.

Larsson, L.I., F. Sundler, R. Hakanson, H.G. Pollock, and J.R. Kimmel. (1974a). Localization of APP, a postulated new hormone to a pancreatic cell type. Histochemistry, 42, 377.

Larsson, L.I., F. Sundler, R. Hakanson, J.F. Rehfeld, and F. Stadil. (1974b). Distribution and properties of gastrin cells in the gastrointestinal tract of the chicken. Cell Tissue Res., 154, 409.

Lavrentera, G.F. (1963). (Bile secretion in chickens) (in Russian). Tr. Gor'k. Skho Inst., 13, 80. In Ziswiler and Farner (1972).

Laws, B.M., and J.H. Moore. (1963). Some observations on pancreatic amylase and intestinal maltase of the chick. Can. J. Biochem. Physiol., 41, 2107.

Leasure, E.E., and R.P. Link, (1940). Studies on the saliva of the hen. Poult. Sci., 19, 131.

Lepkovski, S., and F. Furuta. (1970). Lipase in pancreas and intestinal contents of chickens fed, heated and raw soybean diets. Poult. Sci., 49, 192.

Lin, G.L., J.A. Himes, and C.E. Cornelius. (1974). Bilirubin and biliverdin excretion by the chicken. Am. J. Physiol., 226, 881.

Lindsay, O.B., and B.E. March. (1967). Intestinal absorption of bile salts in the cockerel. Poult. Sci., 46, 164.

Long, J.F. (1967). Gastric secretion in unanesthetized chickens. Am. J. Physiol., 212, 1303.

McBee, R.H. (1977). Fermentation in the hindgut. In "Microbial Ecology of the Gut" (R.T.J. Clark and T. Bauchop, Eds.). New York: Academic Press, Chapter 4, p. 185.

McNab, J.M. (1973). The avian caeca: A review. World's Poult. Sci. J., 29, 251.

Medl, M., and E. Scharrer. (1978). Active trans-epithelial transport of sodium across chicken crop mucosa. Zentralbl. Veterinaermed., 25, 441.

Mongin, P. (1976). Ionic constituents and osmolarity of the small intestinal fluids of the laying hen. Br. Poult. Sci., 17, 383.

Mongin, P., and X. deLaage. (1977). A study of water and electrolyte movements through the duodenal mucosa in laying fowl by perfusion *in vivo*. C. R. Hebd. Seances Acad. Sci. Naturelles, 285, 225.

Mongin, P., M. Larbier, N.C. Baptista, D. Licois, and P. Coudert. (1976). A comparison of the osmotic pressure along the digestive tract of the domestic fowl and the rabbit. Br. Poult. Sci., 17, 379.

Mortensen, A., and A. Tindall. (1981). On caecal synthesis and absorption of amino acids and their importance for nitrogen recycling in Willow ptarmigan (*Lagopus lagopus lagopus*). Acta Physiol. Scand., 113, 465.

Moss, R. (1977). The digestion of heather by Red grouse during the spring. Condor. 79, 471.

Mykkanen, H.M., and R.H. Wasserman. (1981). Gastrointestinal absorption of lead (203-Pb) in chicks: Influence of lead, calcium and age. J. Nutr., 111, 1757.

Nesheim, M.C., and K.J. Carpenter. (1967). The digestion of heat-damaged protein. Br. J. Nutr., 21, 399.

Niess, E., C.A. Ivy, and M.C. Niesheim. (1972). Stimulation of gallbladder emptying and pancreatic secretion in chicks by soybean whey protein. Proc. Soc. Exp. Biol. Med., 140, 291.

Nilsson, A. (1974). Isolation, amino acid composition and terminal amino acid residue of the vasoactive octacosapeptide from chicken intestine. Partial purification of chicken secretin. FEBS Lett., 47, 284.

Nitsan, Z., and E. Alumot. (1963). Role of the caecum in the utilization of raw soybean in chicks. J. Nutr., 80, 299.

Nitsan, Z., I. Nir, Y. Dror, and I. Bruckental. (1973). The effect of forced feeding and dietary protein level on enzymes associated with digestion, protein and carbohydrate metabolism in geese. Poult. Sci., 52, 474.

Norris, E., C. Norris, and J.B. Sken. (1975). Regulation and grinding ability of grit in the gizzard of Norwegian Willow ptarmigan (*Lagopus lagopus*). Poult. Sci., 54, 1839.

Olsson, N., G. Kihlen, A. Ruudvere, C. Wadne, and G. Anstrand. (1950). Smaltbarhetsforsok med fjaderfa. Kgl. Lantbrukshoegsk. Statens Lantbruksfoer. Medd. 43., (In Ziswiler and Farner, 1972.)

Otani, I. (1966). Fundamental studies on the digestion in the domestic fowl. II. Effects of the grit on the movements of gizzard. J. Fac. Fish. Anim. Husb. Hiroshima Univ., 6, 457.

Otani, I. (1967). Fundamental studies on the digestion in

the domestic fowl. III. Effects of the grit on the developments of gizzard. J. Fac. Fish. Anim. Husb. Hiroshima Univ. 7, 119.

Patrick, H., and G.K. Schweitzer. (1952). Absorption and retention of radioactive phosphorus by chicks. Poult. Sci. 31, 888.

Polak, J.M., A.G.E. Pearce, C. Adams, and J.C. Garaud. (1974). Immunohistochemistry and ultrastructural studies on the endocrine polypeptide (APUD) cells of the avian GI tract. Experentia, 30, 564.

Polin, D., E.R. Wynosky, M. Loukides, and C.C. Porter. (1967). A possible urinary back flow to ceca revealed by studies on chicks with artificial anus fed amprolium-C^{14} or thiamine-C^{14}. Poult. Sci., 46, 89.

Pritchard, P.J. (1972). Digestion of sugars in the crop. Comp. Biochem. Physiol. A, 43, 195.

Radhakrishna, P.C.R., and S.G. Nair. (1974). Studies on some of the gastrointestinal mucosal enzymes in chicks and ducklings. Indian, Vet. J., 51, 683.

Rouff, H.J., and K.F. Sewing. (1970). Die Wirkung von Histamine, Carbacol, Pentagastrin und Hühnergastrinextrakten auf die Magensekretion von nicht narkotisierten Hühnern mit einer Magenfistel. Naunyn-Schmiedsberg's Arch. Pharmacol., 267, 170.

Rutter, W.J., P. Krichevsky, H.M. Scott, and R.H. Hansen. (1953). The metabolism of lactose and galactose in the chick. Poult. Sci., 32, 706.

Sahba, M.M., J.A. Morisset, and P.D. Webster. (1970). Synthetic and secretory effects of cholecystokinin-pancreozymin on the pigeon pancreas. Proc. Soc. Exp. Biol. Med., 134, 728.

Schang, M.J., I.R. Sibbald, and R.M.G. Hamilton. (1982). Comparison of two direct bioassays using young chicks and two internal indicators for estimating the metabolizable energy content of feeding stuffs. Poult. Sci., 62, 117.

Schneitz, C., E. Seuna, and A. Rizzo. (1981). The anaerobically cultured cecal flora of adult fowls that protects chickens from *Salmonella* infections. Acta Pathol. Microbiol. Scand., B, 89, 109.

Sibbald, I.R. (1976). A bioassay for true metabolizable energy in feeding stuffs. Poult. Sci., 55, 303.

Skadhauge, E. (1981). "Osmoregulation in Birds." New York: Springer-Verlag.

Smith, C.J.V., and D.R. Pilz. (1971). Feeding behavior of chickens: Effect of cropectomy. Poult. Sci., 50, 226.

Smith, C.R., and M.E. Richmond. (1972). Factors influencing pellet egestion and gastric pH in the barn owl. Wilson Bull., 84, 179.

Soedarmo, D., M.R. Kare, and R.H. Wasserman. (1961). Observations on the removal of sugar from the mouth and crop of the chicken. Poult. Sci., 40, 123.

Starcher, B.C. (1969). Studies on mechanism of copper absorption in chick. J. Nutr., 97, 321.

Sturkie, P.D. (Ed.) (1965). "Avian Physiology" (2nd ed.). Ithaca: Cornell University Press.

Sturkie, P.D. (Ed.) (1976). "Avian Physiology" (3rd ed.). New York: Springer-Verlag.

Sunde, M.L., W.W. Gravens, C.A. Elvehjem, and J.G. Halpin. (1950). The effect of diet and cecectomy on the intestinal synthesis of biotin. Poult. Sci., 29, 10.

Suso, F.A., and H.M. Edwards. (1968). Influence of various chelating agents on absorption of ^{60}Co, ^{59}Fe, ^{54}Mn and ^{65}Zn by chickens. Poult. Sci., 47, 1417.

Thomas, D.H., M. Jallageas, B.G. Munck, and E. Skadhauge. (1980). Aldosterone effects on electrolyte transport of the lower intestine (coprodeum and colon) of the fowl (*Gallus domesticus*) *in vitro*. Gen. Comp. Endocrinol., 40, 44.

Thompson, D.C., and D.A. Boag. (1975). Role of the caeca in Japanese quail energetics. Can. J. Zool., 53, 166.

Thornburn, C.C., and J.S. Willcox. (1965). The ceca of the domestic fowl and digestion of the crude fiber complex. I. Digestibility trials with normal and caecectomized birds. Br. Poult. Sci., 6, 23.

Titus, H.W. (1955). "The Scientific Feeding of Chickens" (3rd ed.). Danville, Illinois: Interstate Press.

Tomimatsu, Y., J.J. Clary, and J.J. Bartulovich. (1966). Physical characterization of oviinhibitor, a trypsin and chymotrypsin inhibitor from chicken egg white. Arch. Biochem. Biophys., 115, 36.

Tortuero, F., A. Brenas, and J. Riperez. (1975). The influence of intestinal (ceca) flora on serum and egg yolk cholesterol levels in laying hens. Poult. Sci., 54, 1935.

Turk, D.E. (1968). Dietary antibiotics and the absorption of zinc-65 and ^{131}I-labeled oleic acid. Poult. Sci., 47, 1768.

Turk, D.E. (1973). Intestinal parasitism and nutrient absorption. Fed. Proc. Fed. Am. Soc. Exp. Biol., 3, 106.

Turk, D.E., and J.F. Stephens. (1970). *Eimeria necatrix* and zinc absorption in the chick: Effect of sulfa-quinoxaline treatment of the infection. Poult. Sci., 49, 285.

Vaillant, C., R. Dimaline, and G.J. Dockray. (1980). The distribution and cellular origin of vasoactive intestinal polypeptide in the avian gastrointestinal tract and pancreas. Cell Tissue Res., 211, 571.

Vohra, P., and N. Gonzales. (1969). The effect of EDTA on the preferential intestinal absorption of zinc than manganese in turkey poults. Poult. Sci., 48, 1509.

Walter, W.G. (1939). Bedingte Magensaftsekretlon bei der Ente. Acta Brevia Neerl. Physiol. Pharmacol. Microbiol., 9, 56.

Webling, D.D., and E.S. Holdsworth. (1965). The effect of bile, bile acids and detergents on calcium absorption in the chick. Biochem. J., 97, 408.

Webling, D.D., and E.S. Holdworth. (1966a). Bile salts and calcium absorption. Biochem. J., 100, 652.

Webling, D.D., and E.S. Holdsworth. (1966b). Bile and the absorption of strontium and iron. Biochem. J., 100, 661.

Webster, P.D., and M.P. Tyor. (1966). Effect of intravenous pancreozymin on amino acid *in vitro* by pancreatic tissue. Am. J. Physiol., 211, 157.

Winget, C.M., G.C. Ashton, and A.J. Cawley. (1962). Changes in gastrointestinal pH associated with fasting in laying hen. Poult. Sci., 41, 115.

Ziswiler, V., and D.S. Farner. (1972). Digestion and digestive system. In "Avian Biology," Vol. 2 (D.S. Farner and J.R. King, Eds.). London: Academic Press, p. 343.

Zoppi, G., and D.H. Shmerling. (1969). Intestinal disaccharidase activities in some birds, reptiles and mammals. Comp. Biochem. Physiol., 29, 289.

13 Carbohydrate Metabolism

R.L. Hazelwood

General Introduction

Literature Resources

Since the appearance of the previous (third) edition of this text, several excellent monographs have appeared that impact and extend our knowledge of avian carbohydrate metabolism. While not restricted to avian forms, "The Evolution of Pancreatic Islets" (Grillo et al., 1976) contains sufficient chapter material devoted to birds to place these organisms properly in their phylogenetic strata with regard to embryogenesis, secretory regulation, and functional significance of the islet products in embryos as well as adult life. "Avian Endocrinology" (Epple and Stetson, 1980) is an overview of those endocrine mechanisms (many of which are unique) important in avian metabolism. "Recent Advances in Avian Endocrinology" (Pethes et al., 1981) extends and broadens the aforementioned sources and, in addition, adds a utilitarian flavor to the wealth of information contained therein as far as the relationship of didactic material to the physiology of production is concerned.

The reader, therefore, has a wide array of erudite information touching all aspects of avian life from early embryo to aged forms. The regulation of carbohydrate metabolism, so critical to the developing embryo but fleetingly important to the growing chick and perhaps of questionable "value" in avian adulthood, is a thoroughly treated topic of significance. The reader is referred to Romanoff (1967) and Freeman and Vince (1974) for quantitative treatment of carbohydrate metabolism in avian embryos; to Bell and Freeman's five volumes (1984) for a biochemical approach to embryonic and adult metabolism in birds; and to Farner and King's six holistic volumes (1971–1982), which have covered virtually every aspect of adult avian life and metabolism and the relationship of carbohydrates thereto.

Current Trends

At present, the current interest in avian carbohydrate metabolism appears to be slanted toward the relationship of such metabolism to growth and development of the young bird and the hormonal controls over such processes. Growth hormone, in particular, is receiving considerable attention in this regard. Much emphasis is being directed toward a better understanding of the availability of nutrients for production in fowl and to the timed endocrine requirements for such utilization. A third area of research appears to be the molecular mechanisms of hormone release in Aves, particularly those of the pancreas. Associated with the latter is the continued search for extrapancreatic sites of peptide production, such as those of insulin, glucagon, etc.

Future Needs

Better understanding of molecular events involved with the absorption and metabolism of carbohydrate in Aves prompts the need for research directed toward elucidation of the binding–receptor transduction of regulatory hormone messages. The role that the exocrine pancreas plays in "intrapancreatic" regulation of hormone release (and subsequent carbohydrate utilization) needs to be studied thoroughly. The cytologic and morphologic aspects of hormone action remain a void in our body of information. Hormone molar ratios may be very important both to the bird and to our understanding. Finally, a significant area "pregnant with possibilities" is the distribution and role of calmodulin, the calcium binding regulating protein, in avian metabolic processes.

Patterns of Avian Carbohydrate Metabolism

Carbohydrate Metabolic Pathways

Glucose (D form) increases in avian plasma from less than 100 mg/dl early in embryonic life to levels approaching 150–160 mg/dl at the time of hatching. After hatching, plasma glucose levels continue to increase for several weeks or months, reaching values of 190–220 mg/dl (depending on species) that are characteristic of adult birds fed ad libitum. Virtually no glucose resides in the erythrocyte. However, the sugars D-fructose (1–3 mg/dl) and D-galactose (<1 mg/dl) also reside in plasma.

Embryonic Pathways

The fundamental enzyme-controlled reactions characterizing anabolic and catabolic carbohydrate metabolism in mammals appears to prevail in Aves. Enzymes associated with the anaerobic degradation of glucose are apparent in most avian embryonic tissues from the earliest days in ovo. However, those associated with glucose phosphorylation (liver hexokinase, muscle glucokinase, etc.) exist at extremely low activities until a few days before hatching. Nonetheless, liver carbohydrate is available to the developing embryo from the action of phosphohexoisomerase, phosphofructokinase, fructose diphosphoaldolase, glyceraldehyde-3-phosphate dehydrogenase, enolase, and pyruvate kinase (Figure 13–1). All of the gluconeogenic enzymes tend to increase in activity throughout embryonation to reach a peak level about 10–20 days after hatching, only to decrease to relatively low levels from then on into adulthood. As most glycolytic enzyme levels decrease in avian tissue over the normal lifespan, hexokinase activity continues to increase greatly, reaching levels 300–400% higher than that of glucokinase during adulthood. Simultaneously, D-glycerophosphate dehydrogenase (liver) decreases rapidly at hatching and continues to decrease with an increase in age (Goodridge, 1968a). Apparently, the embryo is preparing for an abrupt change of diet from one high in lipid content (yolk) to one high in carbohydrate (growing mash or field grain).

Glycogen Synthesis and Degradation

Glycogen levels in liver (mainly) and muscle (to a lesser extent) wax and wane throughout embryonic life and, to a large degree, reflect the rather sudden onset of secretion of various glycogenic or glycogenolytic hormones. Despite a sudden decrease in liver glycogen levels on days 13–15 in ovo, the overall trend is one of gradually increasing glycogen levels that peak near hatching time. Concomitantly, glucose-6-phosphatase, β-glucuronidase, uridine diphosphoglucose (UDPG) synthase, and phosphorylase activities also increase greatly over the last week of incubation. One day after hatching, both heart and liver glycogen levels decrease dramatically to 40% and 16%, respectively, of prehatching values (Freeman, 1969). Liver glycogen levels remain relatively low for several posthatch weeks only to climb subsequently to adult levels by about 4 months. These polysaccharide depots are subject to the impact of circadian rhythms, to ovulatory stress, to rapidly changing ovarian steroid levels during laying time, to molting stress, etc. Plasma glucose approaches 150–160 mg/dl at hatching (in the chicken; lower in ducks) and gradually increases over the subsequent 4–6 weeks to reach levels considered "mature," i.e., 190–220 mg/dl (lower in ducks).

Anaerobic and Aerobic Glucose Degradation

Goodridge (1968b) presented evidence indicating that the citric acid (Krebs') cycle activity increases throughout the incubation period, particularly within the liver, and continues to increase into adulthood based on estimations of malate and succinic dehydrogenase during embryogenesis and early posthatch growth periods. As in mammals, anaerobic and aerobic glucose degradation appear to be the major routes by which energy is provided for morphogenesis and the growing phases of early avian life (Figure 13–1). Contributions by the hexosemonophosphate shunt to overall energy, substrates, and/or cofactors in birds after hatching appears to be minimal at best (see below).

Gluconeogenesis

The formation of substrate for de novo plasma glucose synthesis is accelerated through gluconeogenic pathways. Whenever a vertebrate organism is confronted with a diminishing supply of ATP (hence more AMP), glycolysis is enhanced and gluconeogenesis is diminished. In most animals, simply fasting overnight is an adequate stimulus to activate the rate-limiting reactions that reverse the anaerobic glycolytic sequence presented in Figure 13–1. In birds, the key enzymes for gluconeogenesis are the same as those in mammals, namely glucose-6-phosphatase, fructose-1,6-diphosphatase, pyruvate carboxylase, and phosphoenol-pyruvate carboxykinase (Figure 13–1).

Before hatching, the avian embryo draws heavily on the yolk lipid for energy substrate to support morphogenesis. Yolk is relatively low in carbohydrate content. Thus gluconeogenic reactions are important to support embryonic differentiation and maturation of tissues. After hatching, dietary composition switches from high lipid, low carbohydrate to low lipid, high carbohydrate, thereby reducing the demand on gluconeogenic pathways. Estimates indicate that glucose-6-phosphatase activity decreases more than 60% from hatchout time to that of adulthood, and fructose-1,6-diphosphatase decreases at least 50%.

Gluconeogenesis in isolated chicken liver cells appears to be greatest from lactate, followed by pyruvate, dihydroxyacetone, glyceraldehyde, and fructose (Brady et al., 1979). Conversion of alanine to glucose is relatively low, compared with mammals. It is suggested that the differences in gluconeogenic activity in pigeons and chickens as opposed to those in various experimental mammals (rat, dog, guinea pig, etc.) may well be due to the localization of rate-limiting gluconeogenic enzymes within the mitochrondria. Phosphoenolpyruvate carboxykinase (PEPCK) is thus located in chicken hepatocyte mitochondia, whereas in rats it is a cytosolic enzyme. Since there is a low utilization of citric acid cycle intermediates for de novo glucose formation, and that from pyruvate is modest at best, the lack of cytosolic reducing equivalents (due to the mitochondrial localization of PEPCK) may be responsible (Brady et al., 1979).

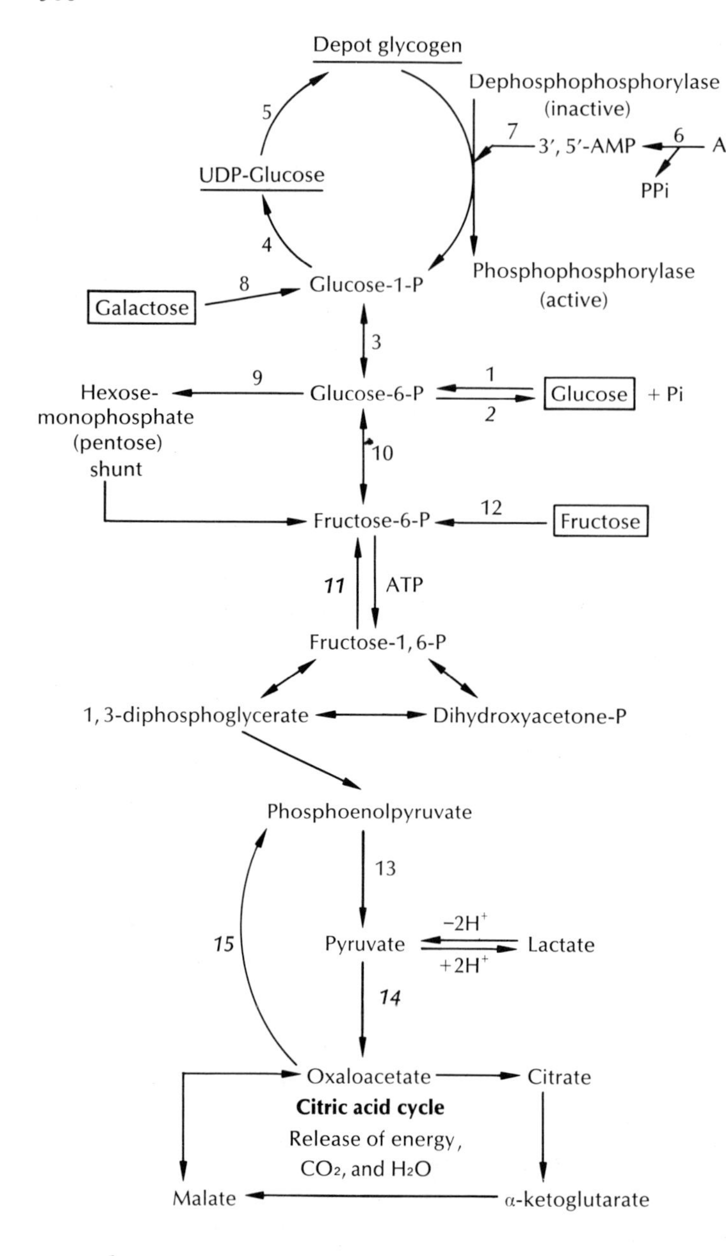

FIGURE 13–1. Metabolism of glycogen and simple sugars. Controlled reactions are by the number: (1) glucokinase, (2) glucose-6-phosphatase, (3) phosphoglucomutase, (4) UDPG-pyrophosphorylase, (5) branching enzyme + glycogen synthetase, (6) adenylcyclase, (7) Dephosphophosphorylase kinase, (8) galactokinase + UDPG, (9) glucose-6-phosphate dehydrogenase, (10) phosphohexosisomerase, (11) fructose-1,6-diphosphatase, (12) hexokinase (fructokinase), (13) pyruvate kinase, (14) pyruvate carboxylase, (15) Phosphoenolypyruvate carboxykinase. Reactions governed by enzymes with italicized numerical designations are reactions essential to gluconeogenesis (see text) and therefore circumvent certain critical energy barriers to a simple reversal of glycolysis.

Blood glucose levels of carnivorous avian species remain remarkably stable during prolonged periods of fasting, unlike the fluctuating glucose levels observed in granivorous birds. Thus, gluconeogenesis supports available carbohydrate supplies to carnivores and appropriately reflects the "meal-eating" nature of these birds as opposed to the "constant-nibbling" regimen of domestic fowl. Actually, the very nature of the black vulture's normal diet (carrion) suggests effective gluconeogenic mechanisms for providing glucose substrates. A comparison of liver enzyme activity from chickens and vultures indicates that the vulture has much higher gluconeogenic enzyme levels than does the chicken. During a prolonged fast, vulture muscle glycogen is depleted markedly and the activity of critical gluconeogenic enzymes (e.g., PEP-carboxykinase) is increased 300–500%. As a result, hepatic glycogen depots are spared to a large extent (Migliorini et al., 1973). Similar results have been described in the horned owl and other carnivorous avian forms.

Uronic Acid Pathways and Glycogen Synthesis

In many ways the embryonic avian liver is similar to the adult mammalian liver on the basis of the metabolic patterns of the two classes. Even in the absence of detectable amounts of insulin, the avian embryonic liver synthesizes significant amounts of glycogen, much of it from the sugar acid D-glucuronic acid. After oxidizing the primary hydroxyl group at position 6 on the D-glucose-1-phosphate molecule to that of a carboxyl, uridine diphosphate (UDP) acts as a carrier (specific) of the glucose molecules in the biosynthesis of glycogen. Glucuronic acid thus acts as an intermediate and also is important as a polysaccharide component in the chondroitin foundation in cartilage and corneal structures, as well as in steroid conjugation and other events in the rapidly growing embryo. Birds also use the UDP glucuronic path to synthesize a related sugar acid, namely ascorbic acid (Vitamin C), which is very quickly oxidized to dehydroascorbic acid. Mammals are incapable of this conversion and must rely on dietary intake of Vitamin C to prevent deficiency states evidenced by scurvy, connective tissue abnormalities, and capillary fragility. Other uronic acids of importance are mannuronic and galacturonic acids.

Differences between Avian and Mammalian Carbohydrate Metabolism

All metabolic pathways studied in mammals appear to be operative also in avian forms, although in several cases there are differences in the relative contribution of a given pathway to the overall energy requirements of a specific tissue or even to the organism as a whole. The difference in contribution of the direct oxidative pathway, also known as the hexosemonophosphate shunt, in birds as opposed to mammals (Figure 13–1) is probably best known. The production of pentoses, three-carbon phosphates, NADPH, and CO_2 by this shunt pathway makes available substrates for a wide variety of reactions, including pentoses for nucleotide and nucleic acid synthesis, three-carbon phosphates for recombination leading to glucose-6-phosphate, and reduced NADP for lipogenesis and steroidogenesis (Goodridge, 1968b). Activity of the pentose shunt in avian embryos varies from tissue to tissue, and usually peaks in activity between days 8 and 15. Thereafter, shunt enzyme activities decline, and after hatching the products of shunt activity contribute very little to the overall metabolic economy of the growing bird. While glucose degradation by the phosphate shunt is very important to the growing avian embryo, its existence and activity is virtually inoperative in most adult tissues, particularly in the liver. In contrast, the mammalian liver relies heavily on the products of this shunt pathway; also, mammalian erythrocytes metabolize glucose preferentially by this pathway. Avian adipocytes synthesize less than 5–8% of the bird's lipid; thus, the liver manufactures the vast majority, and does so without significant use of the hexose monophosphate shunt (O'Hea and Leveille, 1969). The importance of malic enzyme thus becomes obvious. Neural, cardiac and intestinal embryonic tissue appear to be the most dependent on pentose shunt activity.

Fatty acid and protein degradation products support erythrocytic structural integrity and function in birds. These nucleated cell types do not possess significant citric acid cycle enzyme machinery in adult species. Neither do they harbor a very large glucose moiety. Alternating "peaks and valleys" of energy release by citric acid cycle reactions in the embryonic chick have been associated with hemoglobin synthesis, during which the 5-day-old chick embryo may complete the ribosomal assembly of one of the hemoglobin polypeptide chains in less than 2 min! Mammalian red blood cells (RBCs) cannot synthesize hemoglobin after the reticulocyte stage due to the absence of a nucleus, mitochondria, RNA, and citric acid cycle enzymes. Also, the cytosolic anaerobic enzyme machinery is responsible for supplying energy to maintain the biconcave shape, Na–K pump activity, and required reduction of pyridine nucleotides. The hexose monophosphate shunt appears to be important in the adult mammalian RBC for the purpose of generating energy, and even more important for generating NADPH, the lack of which results in fragile RBC membranes and increased hemolysis.

Prolonged flight in migrating avian species places special metabolic demands on large skeletal muscles. Such muscles appear to be "well constructed" to meet these demands by having a rich vascular supply, high levels of myoglobin, many mitochondria, and a luxurious supply of lipid droplets. Glycogen levels are modest, if not actually low, in these red fibers. Fatty acid oxidases, hormone-sensitive lipases, and oxidative enzymes of the citric acid cycle are found in abundance in red fibers, indicating once again that this type of muscle is adapted for prolonged, strenuous and sustained flight. Such activity is strongly supported by oxidative reactions, largely of noncarbohydrate substrates. Preferential utilization of fatty acids (concomitant with elevated plasma glucagon levels) by flight muscles allows for a more efficient liberation of energy concomitant with a sparing of existing (somewhat limited) carbohydrate supplies during migration.

White ("light meat") muscle fibers appear best suited for short burst of "explosive" work such as liftoff for flight, the braking action of wings, or setdown for landing (see also Chapter 3). Structurally and enzymatically, these muscles appear best suited for anaerobic work, that is, there is an abundance of glycogen, few lipid droplets, high enzyme activity associated with the Embden–Myerhoff pathway, few mitochondria, and low myoglobin levels. Experimental studies have verified preferential use of various metabolic fuels in relation both to the type of muscle fiber and to the task normally performed by that muscle (Drummond, 1967; George and Berger, 1966).

Interrelationship of Carbohydrate, Lipid, and Protein Metabolism

Intestinal Absorption

The development of the everted gut-sac technique for the in vitro study of ^{14}C-labeled nutrients has played a significant role in elucidating the transport mechanisms that the avian intestine employs to move amino acids, fatty acids, and simple sugars from the gut lumen into the bloodstream. A considerable body of evidence indicates that active transport systems in the avian intestine exist that translocate amino acids and glucose (perhaps fatty acids) against chemical concentration gradients (Fearon and Bird, 1968; Bogner, 1966). This appears to be true as early as in the 18-day-old chick embryo, wherein the middle segment of the small intestine is most active in transporting sugars. Transport activity increases for an additional 2–3 weeks in the posthatch chick, the increase being due primarily to increased carrier availability rather than increased affinity for the sugar substrate (Ziswiler and Farner, 1972). Any interference with coexistent Na^+ transport reduces sugar transport as well, indicating that the "active" component of the transport system may well be that of Na^+ and that sugar transport may be a secondary, co–phenomenon (Alvarado and Monreal, 1967). Preferential sugar absorption in the avian intestine follows the pattern of D-galactose > D-glucose > D-xylose > D-fructose. More than 98% of all carbohydrates are carried in the plasma moiety of blood; in fact, galactose and fructose can be found only in plasma.

The intestine appears to play an important homeostatic role in modulating perturbations caused by ingestion and absorption of carbohydrates. Generally, plasma glucose level and body weight gain of chicks are not affected by glucose intake. The intestinal wall converts over 30% of ingested carbohydrate to lactate in chicks, thereby regulating glucose homeostasis (Riesenfeld et al., 1982).

Schema of Fuel Interconversion

Once absorbed, the basic nutrients are carried to the liver as amino acids (both glucogenic and ketogenic), pentoses, and hexoses (but very few disaccharides), and as free fatty acids (most of which were taken up by the gut lymphatic system) and glycerol where they enter the metabolic pool. The direction of activity that this pool takes is dictated largely by the immediate previous nutritional history of the avian species, as well as the endocrine background against which the specific metabolic (enzymatic) machinery is operating.

Carbohydrates are either polymerized to glycogen in the liver, or degradated to CO_2, H_2O, and energy release over the classical oxidative citric acid pathway, or, at the two- and three-carbon fragment stage, shunted into lipogenic paths as indicated in Figure 13–2. Hepatic glycogen is synthesized quickly even in fasted birds following meal (carbohydrate) intake and absorption. Lepkovsky et al. (1960) found [^{14}C]glucose given as a test meal to be recovered as [^{14}C]glycogen in less than 2 hr, only to be phosphorylized and released subsequently as [^{14}C]glucose. Fructose and galactose metabolism in birds has attracted thorough investigation due to reported differences between Mammalia and Aves in disposing of these sugars (see below).

Depending on local tissue enzyme kinetics, alanine, glycine, cysteine, and serine (or threonine) may form acetyl-CoA via pyruvate (route 4 in Figure 13–2) prior to entering into the Krebs' oxidative cycle to yield CO_2, H_2O, and energy. Alternatively, these amino acids (glucogenic types) can follow gluconeogenic paths (routes 4, 5, and 6 in Figure 13–2) to result in increased free plasma glucose or liver and/or skeletal muscle glycogen. Thus, fasting favors gluconeogenic activity to provide the plasma glucose absolutely required by certain tissues, such as brain, retina, and adrenal medulla. High-protein diets lead to formation of surplus quantities of the aforementioned L-amino acids, and thus sugar storage in a polymerized form commonly results. Glutamic acid can be formed from arginine, histidine, ornithine, and/or proline, and then, by route 6 in Figure 13–2, converted by deamination to the corresponding α-keto acid. L-asparagine is converted to aspartate and then contributes its entire four-carbon skeleton to form oxalacetate (route 5 in Figure 13–2) and thus enters the metabolic caldron. Other amino acids (tyrosine, phenylalanine, etc.) may be converted to acetoacetate and/or acetyl-CoA (via routes 9 and 10 of Figure 13–2), or they may enter gluconeogenic reactions. It is important to note that other than the release and recapture of energy, the majority of amino acid metabolism in Aves, as in mammals, is deployed over pathways that support the relatively high plasma glucose levels.

Very little (<4% total) lipid is formed de novo in peripheral depots; rather, most is of hepatic origin. After absorption, reesterification, and transport to the liver [either by portal vein blood (15%) or lymph (85%)], that which is not used for synthesizing complex lipids can be degradated, step by β-oxidative step, to acetyl-CoA and acyl-CoA (route 2 in Figure 13–2); see also Chapter 15). The latter compound is the activated free fatty acid preparatory to β–oxidation. Thus, metabolism of lipids leads either to the formation of three-carbon fragments (route 1 in Figure 13–2) or to activated two-carbon fragments, both resulting in a "push forward" on the Krebs' metabolic tricycle. The end result is fat utilization that spares carbohydrate and some proteinaceous substrates and also yields approximately double the energy per unit lipid weight

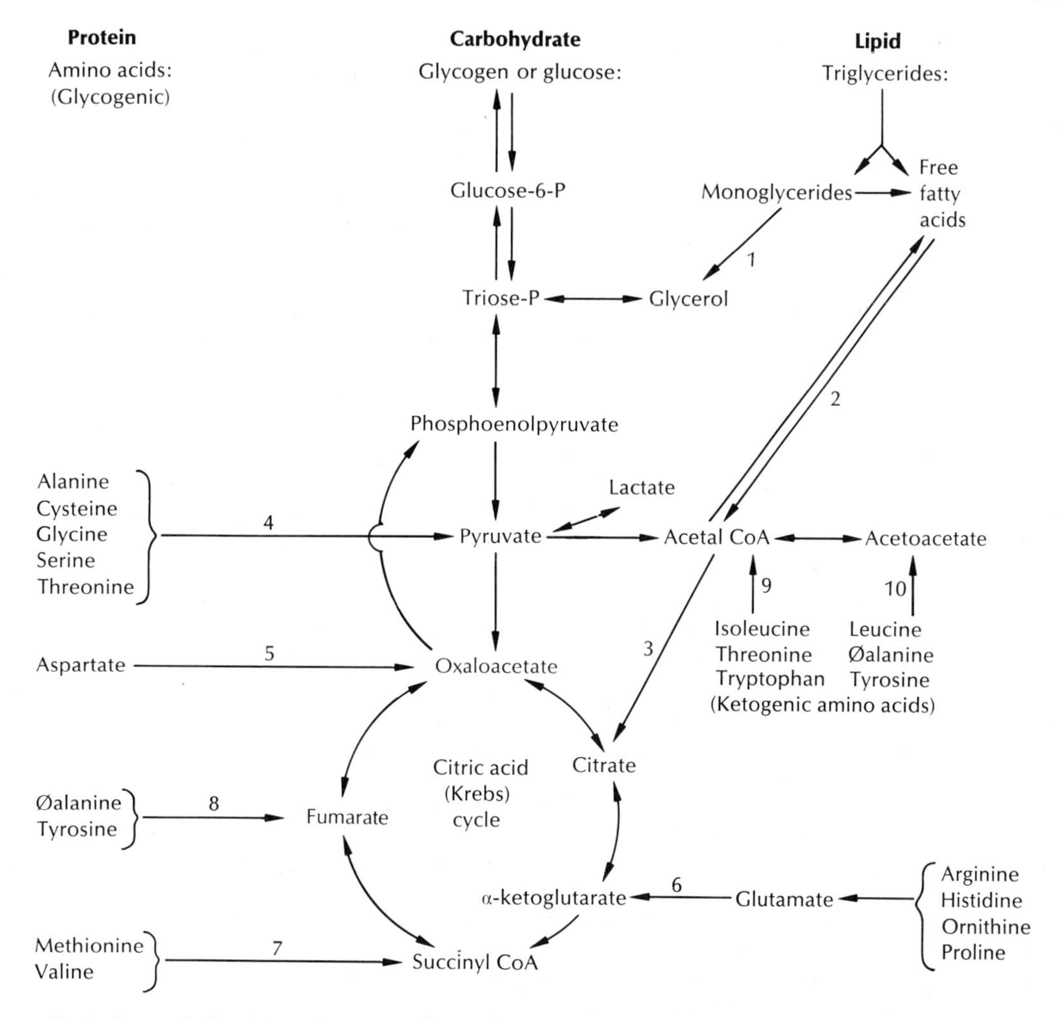

FIGURE 13–2. Interrelationships of protein, fat, and carbohydrate metabolism in Aves. Steps number 1, 2, and 3 indicate the principle gluconeogenic pathways for fat substrate (step 1) or for the release of energy within the avian cell (steps 2 and 3). Steps 4, 5, and 6 are the principle routes of entry of amino acids for gluconeogenesis. Steps 4, 5, 6, 7, and 8 also indicate entry of individual "glycogenic amino acids," whereas steps 9 and 10 indicate entry of "ketogenic amino acids," which also may be considered glycogenic in characteristic.

metabolized. The everchanging ambient environment that confronts most noncaptive birds requires considerable latitude in meeting the metabolic challenge thus placed. The interconversion of absorbed nutrients, as depicted in Figure 13–2, allows birds to meet this challenge. The prominent gluconeogenic pathway activity in carnivores such as vultures, owls, ravens, kestrels, etc., attests to the composition of the normal diets of these species and makes possible the maintenance of normal circulating glucose levels simultaneous with the oxidation of metabolites to release energy. Preferential use of noncarbohydrate substrates (e.g., fatty acid utilization by red skeletal muscle fibers) may allow for a more efficient release of energy concomitant with the sparing of existing carbohydrate supplies, as is seen during migratory flights. (Additional details of fuel conversion and metabolic pathways are presented in Chapters 14 and 15.)

Special Considerations: Starvation, Fructose, Lactose, and Galactose; Rhythms; Temperature Effects

Fasting and Starvation

Short-term starvation (fasting) causes almost immediate mobilization of hepatic carbohydrate reserves to liberate free glucose to the plasma in support of the metabolic needs of certain tissues (brain, retina, adrenal medulla, etc). Thus, the normal blood glucose level of 180–250 mg/dl (depending on species) is supported by hepatic glycogenolysis but not directly by muscle glyconeolysis. In this way, liver glycogen levels may decrease by 90% with a 24- to 36-hr fast. Skeletal muscle glycogen depots are much more stable, although they too will decrease somewhat. Curiously, cardiac glycogen levels increase dramatically with prolonged fasting, a response similar to that reported in mammals. This will occur as long as adequate circulating levels of growth hormone are available, and represents the phenomenon of "carbohydrate sparing" resulting from preferential use by the heart of fatty acids to sustain metabolism during periods of food deprivation. Cardiac glycogen levels may double or triple after 48- to 72-hr fast in chickens (Hazelwood and Lorenz, 1959).

As glycogen depots are reduced during a prolonged fast, the avian blood glucose levels off at 10–15% below prefasting levels and then begins to rise toward normal levels over a period of several days. Blood nonprotein nitrogen levels increase correspondingly, while cardiac glycogen levels level off at 200–300% higher than normal levels after 3 days of starvation (Hazelwood and Lorenz, 1959). Concomitant with these events, activities of glucose-6-phosphatase, pyruvate carboxylase, and phosphoenolpyruvate carboxykinase increase markedly, indicating compensatory gluconeogenesis (see Figure 13–1). Thus, both protein and fatty acid catabolism are increased greatly during periods of starvation (Langslow et al., 1970).

Short-term fasting (1–8 days) in chickens does not decrease glucose utilization per unit body weight as it does in fasted mammals. Rather it remains at about 10–13 mg/min/kg body weight over an 8-day fasting period, a level that is almost twice as high as that found in mammals (Brady et al., 1978). Thus, even though the bird loses body weight during a period of starvation, its greatest energy loss is due to fat depletion and to some extent body protein mobilization. Plasma glucose levels remain remarkably stable (Hazelwood and Lorenz, 1959; Brady et al., 1978), and lactate, pyruvate, and glycerol levels remain relatively constant. The constancy of the aforementioned metabolites, along with the elevated (but constant) levels of glycine, alanine, and serine during short-term fasting, is reinforced by the constant level of tricarbon units originally derived from glucose that are once again reincorporated into the glucose molecule (Brady et al., 1978). One may conclude that there exists little glucose-sparing adaptation during short-term starvation, in the chicken.

Normally, intracellular translocation of glucose is associated with depression of plasma free fatty acid and elevated insulin levels in most vertebrates. This is not true in Aves, in which insulin injections increase free fatty acid levels as well as intracellular movement of plasma (free) glucose (Lepkovsky et al., 1967; Langslow et al., 1970). It is important to realize that birds are unique in their resistance to the well-documented effect of insulin and epinephrine in mammals, namely the antilipolytic effects of the former hormone and the lipolytic effects of the latter. Thus, the stress associated with periods of starvation in birds elevates free fatty acids by mechanisms other than epinephrine action at the adipocyte or perturbations in plasma insulin levels. Fasting chickens over a 4-day period increases *glucagon* levels (Cieslak and Hazelwood, 1983), while having little effect, if any, on insulin levels (Johnson and Hazelwood, 1982). Glucagon is a powerful mobilizer of depot lipid (see below), and therefore may be regarded as a primary regulator of metabolism during prolonged periods of food deprivation, including those associated with 500- to 1800-mile nonstop migratory journeys.

Use of experimental diets has allowed for the evaluation of selective nutrient deprivation on residual metabolism in many avian species. A protein-free diet that is high in dextrose will depress the elevated lysine and threonine levels observed in fasted pullets. Insulin action mimics these events. Thus, intracellular availability of carbohydrate depresses gluconeogenesis associated with fasting. Or, stated in other terms, urinary carbohydrate excretion increases as dietary protein availability decreases.

Both intermittent protein deprivation and intermittent total starvation improve glucose tolerance in otherwise normal chickens; both regimens also decrease the insulin response. Thus, it appears that in chickens intermittent feeding increases the insulin sensitivity of target tissues while decreasing B-cell sensitivity to glucose (Simon and Rosselin, 1979).

Fructose, Lactose, and Galactose Metabolism

Although intestinal absorption of glucose far exceeds that of D-fructose, avian liver metabolism of the latter sugar is rapid and efficient and is not impeded by simultaneous high levels of D-glucose. This is not true in mammals. Observations that respiration quotients (RQs) greater than unity obtain when chick liver slices are incubated with fructose as a substrate, that the replacement of glucose with sucrose in the diet results in higher liver cholesterol levels, and that hepatic lipogenesis from fructose exceeds that of all other carbohydrate substrates collectively indicate that in birds the metabolism of fructose and its conversion to lipid holds high metabolic priority (Pearce and Brown, 1971).

Gut absorption of certain pentoses (e.g., L-arabinose, D-xylose), or of lactose, a disaccharide, occurs in most birds, but the ultimate metabolism of these sugars does not meet the energy needs of the organism. Thus, chickens fed diets high in any of the above listed sugars (but particularly lactose and pentoses) succumb to retarded growth patterns, decreased liver cellularity, diminished glycogen stores, severe diarrhea, and cessation of egg production. Intracellular transfer of glucose appears to be hindered simultaneously with accelerated protein and fatty acid utilization (Wagh and Waibel, 1966).

Galactose, the hexose with the fastest rate of intestinal absorption by chicken gut-sac preparations, is toxic when added to the diets of birds. Thus, when dietary galactose levels in chickens exceed 10%, epileptiform convulsions frequently occur. Female birds appear to be more sensitive to galactose toxicity than are males. If 50% of the diet contains galactose, diminished tissue glycogen depots, normal blood glucose, high blood galactose, severe renal damage, and central nervous system lesions are observed (Mayes et al., 1970). ATP, phosphocreatine, galactose-1-phosphate, and uridyltransferase activity are all markedly reduced in brain tissue of toxic chickens, while galactose-1-phosphate accumulates. The entire glycolytic scheme (Figure 13–1) appears to "jam-up" as galactose intermediates increase. Thus, once again glucose oxidation is depressed in the face of a rising galactose tide in neural tissue. Recall that both sugars compete for the same carrier (transport) molecule both in avian and mammalian tissues.

Effects of Circadian Rhythms and Environmental Temperature on Carbohydrate Metabolism

Circadian peaks and valleys in avian hepatic glycogen levels may occur twice or more daily. In the avian embryo, hepatic glycogen levels wax and wane predictably in two cycles each day. The adult chicken, however, exhibits a more true circadian rhythm of hepatic glycogen levels. Liver weight, liver fat, and liver glycogen decrease with onset of darkness in many species (blackbirds, starlings, chickens, etc.). On the other hand, Tweist and Smith (1970) reported that shifting the day–night light pattern by 12 hr also shifts the hepatic glycogen increment–decrement cycle accordingly. Chickens are known to demonstrate higher blood glucose levels (20–30 mg/dl) during the daytime than nighttime; but, then again, birds generally are not night feeders. Thus, photoperiodism, through its effect on the feeding regimen as well as its impact on appropriate diencephalic regulatory centers, probably affects circulating glucose and liver glycogen depots: Investigators should be alert to this environmental effect when quantitatively assessing these metabolic parameters.

Ambient temperature fluctuations also may alter plasma glucose and hepatic glycogen levels in birds. Freeman (1966) concluded from studies on 1-day-old chick hatchouts that were subjected to cold (20°C) that such exposure led to a nonthermogenic shivering response independent of adrenomedullary involvement. These cold-exposed chicks demonstrated increased Q_{O_2}, decreased plasma glucose (20%), decreased liver glycogen (50–70%), and increased plasma fatty acid levels (40%). Thus, in contrast to mammals, such responses are mediated by noncatecholamine paths, probably involving the liver (Freeman, 1967). Adult birds, when subjected to cold stress—but not when subjected to heat stress—responded with increased adrenal catecholamine biosynthesis and release (Lin and Sturkie, 1968). Pigeons exposed to 48°C for a few hours respond with a marked hepatic glycogenolysis and a modest hyperglycemia. Hepatic depolymerization of glycogen may occur with either form of deviation from normal temperature, and plasma glucose perturbations do not necessarily follow adrenomedullary secretory patterns. Since double vagotomy in chickens abolishes the plasma glucose response to heat stress, a cholinergic pathway via the liver is strongly implicated.

Pancreatic–Enteric Regulation of Avian Carbohydrate Metabolism

General

While insulin has been considered, for more than 50 years, to be the "main director" of mammalian carbohydrate metabolism, only recently has glucagon received recognition in Aves as the principal director of metabolism. This is not to overstate the case by implying that glucagon is unimportant in mammals (as attested by the frequently observed hyperglucagonemia in diabetic mammals) or that insulin is not important in birds (note that antiinsulin antibodies produce hyperglycemia in ducks). Rather, the predominant hormone regulator of metabolic forces appears to be different in the two classes of animals; namely, insulin plays the key role in mammals while glucagon appears to do so in birds. Recently, research has been directed at the significance of the glucagon/insulin molar ratio in birds in recognition that any/all hormone(s) act superimposed on a background of other hormone expressions. No hormone works in a vacuum!

Somatostatin (SRIF) from the pancreatic D cell probably has no direct effect on glycogenesis or glycolysis per se. Rather, any impact that this polypeptide has on carbohydrate metabolism is most probably exerted through its modulation of release of other pancreatic hormones. Thus, equal efficacy in altering both insulin and glucagon release rates could lead to quantitative alterations in metabolic reactions, whereas disparate perturbation of these two hormones by SRIF would lead to altered glucagon : insulin molar ratios and thus a qualitative modification of carbohydrate metabolism.

Avian pancreatic polypeptide (APP) has proven enigmatic to students of avian carbohydrate metabolism, because even though first reports indicated that the putative hormone was glycogenolytic, further work has not elucidated its precise role in the overall metabolic sphere. Additional work is necessary.

Glucagon

Glucagon is a single-chain, 29-amino-acid polypeptide (3485 daltons), which is secreted by the A cell of the vertebrate pancreas. Structures of various mammalian glucagons evidently are identical, whereas the chicken and turkey molecules differ from those of mammals by a single substitution at position 28 (serine for asparagine). The duck has an additional substitution at position 16 (threonine for serine), as shown in Figure 13–3. This highly conserved molecule is synthesized as a large preproglucagon molecule on the A-cell endoplasmic reticulum, is subsequently shortened by cleavage to proglucagon (biologically inactive) at the Golgi apparatus, and finally shortened even further to the final 29-residue form (biologically active) as the molecules are packaged and processed within secretory vesicles. Apparently, any modification/substitution of glucagon residues markedly reduce the hormone activity. The hormone is very susceptible to destruction by endogenous protease activity, and is mainly destroyed by kidney tissue.

Glucagon is secreted from pancreatic A cells as early as day 4 in chicken embryos (Table 23–1, Chapter 23), and circulates either free or very loosely bound to plasma albumin. Although the dorsal and ventral lobes have a paucity of A cells, the A-cell complement of the "third" and splenic lobes is so concentrated that the avian pancreas has been reported to have two to five times more glucagon per unit wet weight pancreas than that found in mammalian tissue. Actual values of avian pancreatic glucagon are presented in Chapter 23, Table 23–2), and in Weir et al., 1976. Plasma levels of avian glucagon are uniformly measured in nanograms per milliliter in contrast to picograms per milliliter in man and other mammals. Thus, circulating levels of glucagon of 1–4 ng/ml are common to domestic avian species and as such are 10–50 times higher than mammalian levels. Values of 3.70, 3.00, 2.74, and 1.25 ng/ml plasma have been reported for nonfasted adult chickens, ducks, pigeons, and geese, respectively. These plasma levels increase 100–200% during a 24- to 48-hr fasting period (Cieslak, 1984; Cieslak and Hazelwood, 1983; Sitbon and Mialhe, 1978).

Release of glucagon has been studied in ducks and chickens by use of hormone provocators as well as nutrients known to exist normally in the digesta of the small intestine of birds. Additionally, the paracrine modulation of hormone release by pancreatic SRIF has been evaluated. From these studies we know that cholecystokinin (CCK), insulin, and free fatty acids stimulate release of duck pancreatic glucagon, while glucose decreases it (Samols et al., 1969; Sitbon and Mialhe, 1978). Somatostatin has been reported to *increase* glucagon release in ducks, an event that is quite contrary to reports in mammalian systems (Strosser et al., 1980; also see below). Data collected from in vivo and in vitro studies in chickens (especially those wherein isolated pancreatic islets were perifused) have been extremely illuminating in identifying "triggers" of pancreatic glucagon release. Fasting of chickens moderately increases plasma glucagon levels by 16 hr and doubles control levels by 24–30 hr (Cieslak, 1984). Long-chain fatty acids appear to have an attenuating effect in vitro on glucagon release, while linoleate increases glucagon release (Colca and Hazelwood, 1981). Glucose profoundly suppresses glucagon release in chickens, as does SRIF (Cieslak and Hazelwood, 1983). Thus, the avian A cell responds to a mixture of intra- and extrapancreatic humoral agents, the glucagon release kinetics being the resultant of the algebraic sum of counteracting forces (see Figure 13–4). The efficacy of amino acids in releasing glucagon and insulin in diabetic ducks has been investigated, and a possible feedback regulation between plasma free fatty acids and glucagon release has been established (Gross and Mialhe, 1978; Laurent and Mialhe, 1978). No studies relative to α- or β-adrenergic control of the avian A cell have been reported, although it is known that cholinergic stimuli increase glucagon release (reviewed by Hazelwood, in Epple and Stetson, 1980). A glucagon-like peptide isolated from the avian gut is discussed below.

Insulin

The so-called light islets of the avian pancreas are the major source of insulin. No intestinal equivalent of "gut glucagon" appears to exist for insulin, although there is accumulating evidence that suggests that extrapancreatic sources of avian insulin do exist, at least

1 5 10
His–Ser–Gln–Gly–Thr–Phe–Thr–Ser–Asp–Tyr–Ser–Lys–Tyr–Leu

29 25 20 15
Thr–Ser–Met–Leu–Trp–Gln–Val–Phe–Asp–Gln–Ala–Arg–Arg–Ser–Asp

(Asn: All mammals) (Thr: Duck)

FIGURE 13–3. Structure of avian glucagon molecules. Reported structures for all mammals are invariant among themselves, but differ from avian glucagon as shown.

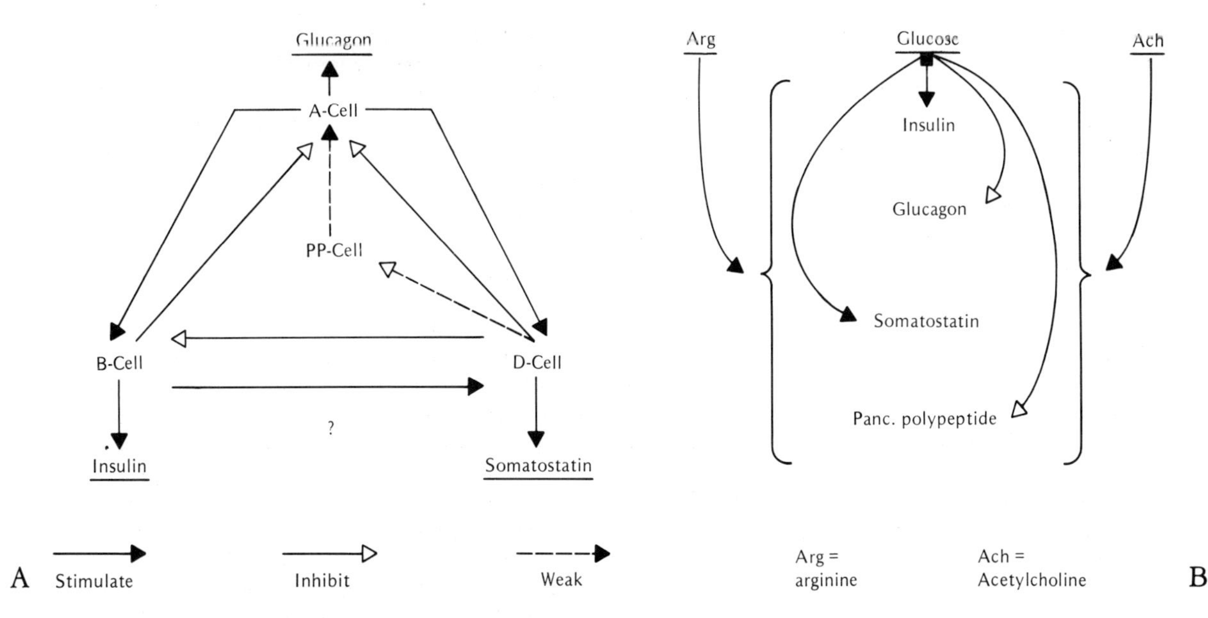

FIGURE 13–4. Functional relationships among the pancreatic endocrine cell types in birds (**A**) and their response to neural, nutrient, and humoral inputs (**B**). (Also see Hazelwood, in Epple and Stetson, 1980; and Cieslak, 1984.)

in the chicken (Colca and Hazelwood, 1976, 1982a). As in the biogenesis of the glucagon molecule, insulin is first formed as a "supermolecule" on the endoplasmic reticulum of the B cells and undergoes cleavage shortening enroute to the Golgi structures. The proinsulin that results is approximately 9,100 daltons and is inactive biologically, although it does cross-react (poorly; i.e., 14–16%) with antibodies to chicken insulin. Further processing (shortening) by removal of a 33-amino-acid residue (biologically inactive connecting or C-peptide) occurs within the nascent secretory vesicles, leaving equimolar concentrations of the active 51-residue insulin molecule (5,734 daltons). As shown in Figure 13–5, the avian insulin molecule is a double polypeptide chain connected by two disulfide bonds, position 7 to position 7 on both chains and position 20 on

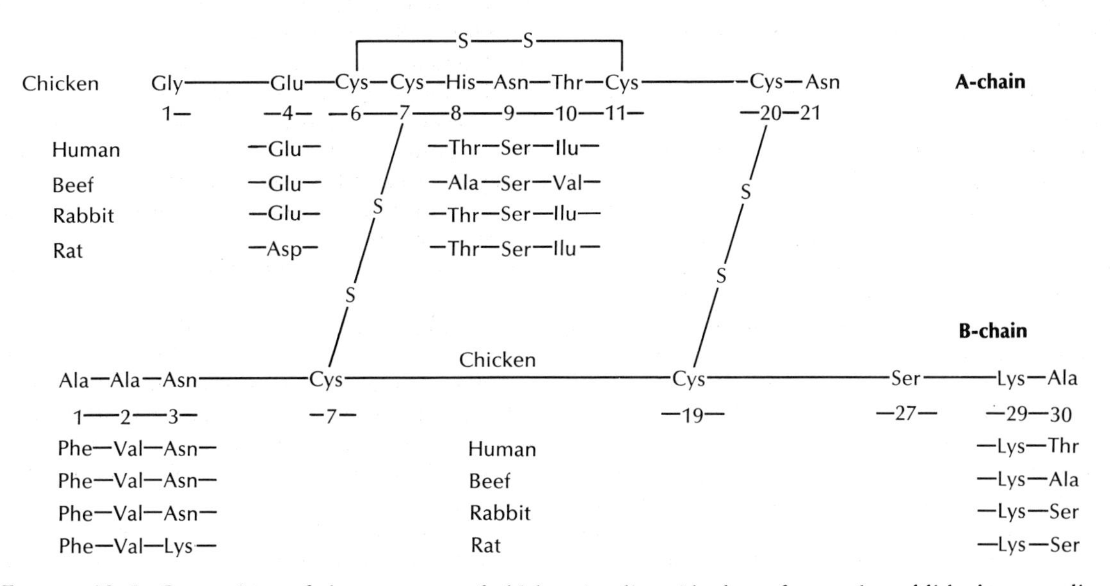

FIGURE 13–5. Comparison of the structure of chicken insulin with that of several established mammalian insulin structures. Turkey insulin is identical with the chicken molecule and duck insulin differs from chicken by substitution of Glu and Pro at position 8 and 10, respectively, on the A chain, and Thr at position 30 on the B chain.

the A chain to that of 19 on the B chain. These bonds must not be reduced, or activity will be lost at the receptor cell. Additionally, an intrachain disulfide bond exists between positions 6 and 11 on the A chain. Some workers feel that this area of the structure serves as a "handle" for the attachment to a tissue receptor. The isolation, homogeneity, amino acid sequence, and phylogenetic placement of chicken insulin have been established (Smith, 1966; Kimmel et al., 1968). Turkey insulin is identical with that of chicken, while duck insulin differs from the former two structures with three replacements; i.e., a GLU and PRO at positions 8 and 10, respectively, on the A chain, and a THR at position 30 on the B chain (Marhussen and Sunby, 1973). The differences in molecular structure presented in Figure 13–5 are modest substitutions, indeed, and convince us that avian insulins do not differ from mammalian forms to any greater degree than the latter differ among themselves (guinea pig insulin is an exception). Furthermore, the biological activity of avian insulin is equal to, if not greater than, that of all mammalian insulins evaluated (Hazelwood et al., 1968; Simon et al., 1974).

Insulin release from the pancreatic B cell is the result of a wide variety of inputs: exocrine, endocrine, paracrine, and nutrient. Glucose may not be the primary stimulator of avian insulin release. A marked increase in circulating glucose does indeed double or triple basal insulin release within minutes from perfused organ, perifused microdissected islets, or various in vivo chicken preparations (Smith and Hazelwood, 1981; Colca and Hazelwood, 1981; Stellenwerf and Hazelwood, 1979; Cieslak, 1984). Progressive fasting has no or a slight depressant effect on plasma insulin in ducks and chickens (3.58 versus 2.88 ng/ml) (Langslow and Hales, 1971; Johnson and Hazelwood, 1982; Cieslak, 1984). Hyperaminoacidemia (in vivo) or leucine, isoleucine, lysine, phenylalanine, and arginine perifusion of isolated pancreatic islets increases insulin release (Langslow and Hales, 1971; Colca and Hazelwood, 1982b). These observations suggest that the composition of normal digesta reinforces, on absorption, intestinal and intrapancreatic hormonal stimuli to the B cells. Actually, a certain synergism appears to exist between amino acids and glucose in regulating insulin release in birds (Simon and Rosselin, 1979). CCK (gut) and glucagon (pancreas) are powerful stimuli to insulin release in ducks and chickens; epinephrine, fatty acids, and secretin are for the most part without effect. Use of either α-adrenergic blockers or β-blockers is without effect in vivo on basal insulin levels, but the β-blocker propranolol prevents the rapid glucose-induced rise in plasma insulin in adult chickens (Smith and Hazelwood, 1981). The β-receptor site thus appears stimulatory to glucose-induced insulin secretion. Vagal stimulation and/or acetylcholine impact on avian insulin release has not been reported. The myriad factors affecting insulin release in Aves is, once again, the algebraic sum of stimulatory and inhibitory forces, as presented in Figure 13–4. Overall, with a few exceptions, B-cell function and regulation of insulin release are qualitatively similar in chickens and mammals.

Somatostatin (SRIF)

This tetradecapeptide emanates from the pancreatic D cell as well as from D-type cells scattered throughout the upper small intestine in birds (Figure 13–6). Possessing an essential disulfide bond between position 3 and 14, this polypeptide has been implicated in the absorption and distribution of recently digested nutrients (gut source) as well as in the paracrine regulation of glucagon, insulin, and APP secretion (pancreatic source). Somatostatin (SRIF) is synthesized as a large molecule, probably as an eicosoctapeptide. While SRIF has no direct effect on avian carbohydrate metabolism, its functional impact is observed by its efficacy in modifying glucagon and insulin release from the pancreas. [It is well to note that although avian SRIF circulates in plasma at levels 200–400% higher than those of mammals (nonfasted chickens, 0.90–1.30 ng/ml; 24-hr-fasted, 0.60–0.80 ng/ml), and although chicken and pigeon pancreata contain up to 20–21 and 200–300 times greater amounts, respectively, of the polypeptide than that found in mammals, very little seeps into the plasma. Plasma SRIF probably represents nonpancreatic SRIF.] Thus, SRIF depresses glucagon secretion (most sensitive), insulin secretion, and APP (least sensitive) secretion from the chicken pancreas (Cieslak, 1984; Honey and Weir, 1979). SRIF itself is released by direct actions of glucagon, insulin, acetylcholine, glucose, and many amino acids in chickens. Thus, when one considers that the avian D cell resides in juxtaposition with neighboring A cells and B cells within an islet, and that all pancreatic endocrine cells share the same extracellular fluid wherein endocrine secretions are 6–10 times more concentrated than in peripheral plasma, the ramifications of paracrine control of pancreatic hormone release are apparent (Hazelwood, in Epple and Stetson, 1980).

The fact that one endocrine cell (A cell) is more susceptible to the depressant effects of SRIF than another pancreatic cell type indicates that this diffusion polypeptide modulates the insulin–glucagon (I/G) molar ratio. This ratio, then, and not the individual hormones per se, may be *the* critical factor in regulating carbohydrate metabolism in birds such that metabolic demands are satisfied efficiently. The suggestion that injected SRIF *elevates* the rate of release of a particular hormone is best explained by a consideration of the immediate past history of the recipient cell, namely

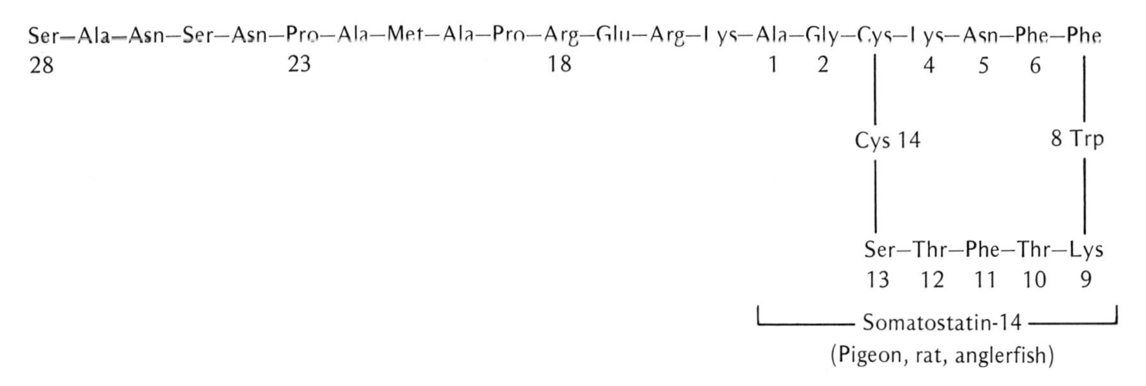

Ser—Ala—Asn—Ser—Asn—Pro—Ala—Met—Ala—Pro—Arg—Glu—Arg—Lys—Ala—Gly—Cys—Lys—Asn—Phe—Phe
28 23 18 1 2 4 5 6
Cys 14
8 Trp
Ser—Thr—Phe—Thr—Lys
13 12 11 10 9
Somatostatin-14
(Pigeon, rat, anglerfish)
Somatostatin-28
(May be tissue precursor (Prohormone) to the above SRIF-14 molecule)

FIGURE 13–6. Somatostatin structure as isolated from the pigeon. No species variations have been reported for the basic 1–14 SRIF molecule.

at what level of activity it was releasing hormone when exogenous SRIF acted on it. Thus, at certain levels of activity, SRIF may act directly on a neighbor cell to suppress the latter's release capabilities; at other levels, injected SRIF may suppress D-cell activity to a greater extent than A- or B-cell activity and in this manner actually increase either/both of the latter hormone release. Thus, data obtained from exogenous SRIF experiments may not truly reflect endogenous relationships (Cieslak, 1984). Also, exogenous SRIF can definitely suppress endogenous SRIF release and action. Interrelationships of D-cell activity and other pancreatic cell types are presented in Figure 13–4.

Avian Pancreatic Polypeptide (APP)

This putative hormone was discovered accidently during the isolation of chicken insulin (Kimmel et al., 1968, 1975). It is a 36-amino acid-residue molecule, synthesized in the PP cell with a nonactive connecting peptide of 23 additional residues, and is subject to wide fluctuations in the plasma dependent on the prandial state of the organism (Figure 13–7). Any impact that APP may have on avian carbohydrate metabolism is based largely on the early observation that the peptide mobilizes hepatic glycogen while plasma glucose levels remain unperturbed. Such an observation sug-

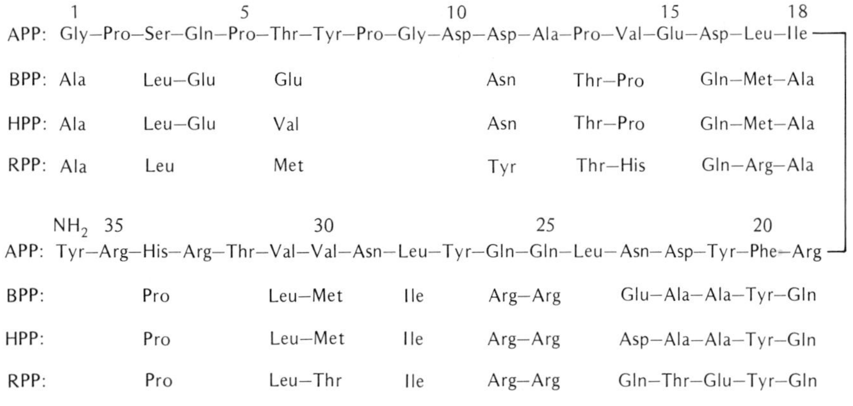

1 5 10 15 18
APP: Gly—Pro—Ser—Gln—Pro—Thr—Tyr—Pro—Gly—Asp—Asp—Ala—Pro—Val—Glu—Asp—Leu—Ile
BPP: Ala Leu—Glu Glu Asn Thr—Pro Gln—Met—Ala
HPP: Ala Leu—Glu Val Asn Thr—Pro Gln—Met—Ala
RPP: Ala Leu Met Tyr Thr—His Gln—Arg—Ala
NH2 35 30 25 20
APP: Tyr—Arg—His—Arg—Thr—Val—Val—Asn—Leu—Tyr—Gln—Gln—Leu—Asn—Asp—Tyr—Phe—Arg
BPP: Pro Leu—Met Ile Arg—Arg Glu—Ala—Ala—Tyr—Gln
HPP: Pro Leu—Met Ile Arg—Arg Asp—Ala—Ala—Tyr—Gln
RPP: Pro Leu—Thr Ile Arg—Arg Gln—Thr—Glu—Tyr—Gln

*Turkey and duck APP structures are identical with that of chicken.

FIGURE 13–7. Structure of APP and homology with pancreatic polypeptide of other vertebrates. Pancreatic polypeptide structures: BPP, bovine; HPP, human; RPP, rat; APP, chicken (avian). Turkey and duck APP structures are identical with that of the chicken.

gests an efficient shunting of glycolytic precursors into a lipogenic pathway, a suggestion for which some evidence exists in chickens (Hazelwood et al., 1973; Hayden, 1976). The fact that APP release from the avian pancreas is responsive to CCK, vagal input, certain fatty acids, and amino acids suggests strongly that an enteropancreatic functional axis exists (Colca and Hazelwood, 1982b). It implies further that APP may be intimately involved in the absorption and distribution of nutrients, including carbohydrate. Normal nonfasted plasma values for APP in the chicken, duck, and pigeon are 4–6, 2–3, and 1–2 ng/ml, respectively; fasting 24 hr depresses these values rapidly to 2–3, 0.90–1.25, and 0.80–1.50 ng/ml, respectively. Glucose loads depress plasma APP in chickens, the β-adrenergic receptor being inhibitory to this response (Smith and Hazelwood, 1981). APP secretion in the chicken is most probably under dual control by complementary α- and β-adrenergic receptor mechanisms (Meglasson and Hazelwood, 1983). Despite this growing evidence linking the presence of APP with the prandial state of vertebrates, the precise role that APP plays in the regulation of avian carbohydrate metabolism remains to be elucidated. Its interaction with other pancreatic hormones in Aves is presented in Figure 13–4.

"Gut Glucagon" (GLI)

Glucagon-like immunoreactivity (GLI) has been reported for extracts of certain parts of the small intestine in ducks and geese that react with nonspecific glucagon antibodies. It appears to be of small molecular weight and of peptide nature. While some evidence exists that this material may be glycogenolytic in nature, it definitely is associated with the prandial state of ducks and geese and is antilipolytic in vivo and in vitro. Thus, at least in two groups, GLI appears to reduce release of free fatty acids during the well-fed state (Sitbon and Mialhe, 1979).

Mechanism of Pancreatic Hormone Action: Molecular Events

Glucagon (Pancreatic)

All evidence at hand indicates that avian pancreatic glucagon exerts its action at the molecular level in the classic "second-messenger" fashion, namely, that the hormone binds to a fixed receptor on the exterior of the membrane and this complex (H·R) interacts with a membrane regulator protein, activating it. Activated, this protein increases the activity of adenylate cyclase, which most frequently is located on the inner surface of the plasma membrane and which cyclizes ATP into cyclic adenosine monophosphate (cAMP). The latter nucleotide now binds to a protein kinase, one that is specific for the hormone-sensitive cell type involved, and this activation of a kinase may initiate a cascade of events leading to a multiplication of the initial signal into the final products formed. Phosphorylation of proteins or enzymes is the usual end result, resulting in a conformational change, and in the case of enzymes this either decreases or increases the latter activities. Thus, glucagon, which decreases hepatic (but not muscle) glycogen levels within seconds, not only activates the enzyme phosphorylase (via cAMP action) to initiate glycogenolysis, but also deactivates glycogen synthase, thus decreasing glycogenesis. In contrast, neither insulin nor APP influences cAMP levels in isolated hepatocytes (Langslow and Siddle, 1979). Both actions are carried out by a phosphorylation of the appropriate enzymes, phosphorylase and synthase. Thus, the hormone remains (largely) outside the plasma membrane and its "message" is carried inside and is amplified by a sequence of events initiated by cAMP. Destruction of cAMP by phosphodiesterase is rapid and is necessary to prevent prolongation of the initial signal. Specificity exists because of the actual H·R complex formed, as is evidenced by the fact that both epinephrine and glucagon elevate cAMP in liver to cause glycogenolysis, but only epinephrine can do this in skeletal muscle. Glucagon receptor numbers appear to be far greater than insulin receptors on the same avian tissue (Cramb et al., 1982; Simon et al., 1977a). Although glucagon increases glycogenolysis in isolated chick embryo hepatocytes, it apparently does not increase lactate gluconeogenesis (Dickson, 1983). Glucagon-induced glycogenolysis is not influenced by APP or insulin (Langslow and Siddle, 1979).

Insulin

The diverse effects of insulin in avian tissues have not yet been totally elucidated. While most of the peptide interacts with exteriorly oriented membrane receptors, it is known in mammals that the hormone may become trapped by membrane vesicles and internalized. The significance of such a process in avian systems is presently under study.

The best evidence at hand indicates that in mammals insulin effects its metabolic action by interaction with one if not both subunits of the insulin receptor on the plasma membrane. Both subunits are glycoproteins, with the α unit 135 Kdaltons in size and externally oriented, and the smaller β unit (95 Kdaltons) bound to the α unit by —S—S— bonds and large enough to penetrate most of the membrane, exposed on the cytosol side. The binding site for insulin is the α subunit. The β subunit possesses insulin-stimulated tyrosine kinase activity. Phosphorylation of tyrosine (and possibly serine) initiates the insulin action, which includes such diverse activity in Aves as increased glucose and amino acid uptake, increased protein and glycogen synthesis, increased membrane potentials, and increased facilitative diffusion of glucose. Presumably, similar systems exist in Aves. Insulin is not antilipolytic in birds.

Other Hormones

Somatostatin (SRIF) is structurally similar in birds and mammals, but the mechanism of action of this tetradecapeptide is virtually unknown. No definite signal has been identified with SRIF effects, although it is known that cAMP is not involved in the gut action of the polypeptide. Sitbon (in Epple and Stetson, 1980) suggested that SRIF reduces cAMP levels in avian adipocytes, thus reducing glycerol release. How SRIF exerts its paracrine action in the pancreas remains to be established; this is of considerable importance because of the extremely high levels found in both chicken and pigeon pancreata. Although equivalent observations have not as yet been reported in birds, SRIF has been identified in several mammals as a probable neurotransmitter in the hypothalamus.

Of all the diverse effects attributed to APP, only the observation that the polypeptide acts as an antilipolytic agent can be traced to a probable molecular basis of action. Thus, APP has been shown in vitro to reduce cAMP levels significantly in isolated chicken adipocytes (but not hepatocytes) in the presence of glucagon (McCumbee and Hazelwood, 1978). Other than this, the mechanism of its action in promoting gastric secretion, reducing gut motility, or depressing plasma glycerol levels is totally unknown at this time.

The effects of gut GLI as an antilipolytic agent are similar to both APP and SRIF, namely, in the reduction of cAMP levels at the adipocyte. Thus, the nonactivity of insulin as an antilipolytic agent in birds is more than covered by such an activity of SRIF, APP, and GLI.

Peripheral Carbohydrate Effects of Pancreatic Hormones

Those tissues in the bird that are to a large extent metabolically regulated by pancreatic hormones include (in decreasing order of significance) skeletal muscle, liver, adipose tissue, cardiac muscle, and possibly the erythrocyte. Various tissue responses have been tabulated by Hazelwood (see Sturkie, 1976).

Skeletal Muscle

Of the carbohydrate-directed hormones, those of the pancreas are of major importance in vertebrates, and of these insulin is the most important as far as directing metabolic events in voluntary muscle. Skeletal muscle responds to the antiglycogenolytic effects of exogenous insulin regardless of whether or not the hormone is homologous with the avian species being tested. Muscle glycogen increases, membrane potentials increase, and restoration of previously depleted glycogen stores is hastened. Hexokinase activity in skeletal muscle appears to parallel glycogen levels. In vitro, chicken muscle strips are relatively resistant to these anabolic effects, although chicken insulin is far superior to various mammalian insulins (Hazelwood et al., 1968; Simon et al., 1974).

Glucagon apparently has little effect on the carbohydrate moiety of striated muscle, there being no phosphatase to release free glucose to the circulation even if a small glycogenolytic effect can be demonstrated. However, there appears to be a strong lipolytic effect of glucagon, and many workers believe that the hormone makes possible the protracted flight required in migration due to the mobilization of free fatty acids as (red) muscle substrates. Many of these neutral fatty acids find their way to the liver where they are deposited (see below). No effects of either SRIF or APP on skeletal muscle have been reported to date.

Liver

Hepatocytes are extremely sensitive to circulating glucagon, releasing free (nonphosphorylated) glucose to the plasma within minutes of injection/release of the hormone. Despite the fact that APP also causes a modest glycogenolysis, only glucagon action results in an hyperglycemia. Glucagon is a powerful lipolytic agent; however, hepatocytes remove the triglycerides resulting from the hormone's adipokinetic action and thereby block simultaneous release of lipids from the liver (Grande and Prigge, 1970). Actually, plasma neutral fatty acids are better correlated with plasma glucagon levels than are glucose levels, once again emphasizing the importance of this pancreatic hormone. Thus, existent plasma glucose levels may be seen somewhat as a "spared substrate" in the face of fatty acid utilization, which is largely under the control of glucagon or the G/I molar ratio. Mialhe and coworkers (Gross and Mialhe, 1978) have presented data supporting the concept that in ducks and geese, at least, there is a feedback regulation between glucagon and fatty acid levels (but not ketone bodies).

Insulin action on avian liver is largely reflected by its ability to reduce glycogenolysis simultaneously with a pronounced incorporation of glucose molecules into de novo glycogen. Although hepatic uptake of glucose is not dependent on insulin, the facilitative diffusion mechanisms within the hepatocyte are more efficient when insulin is present. Generally, hypoglycemia attends this insulin-induced glycogenesis as a result of the increased facilitative diffusion, which, in turn, creates a downhill gradient for glucose to enter the liver cell. Avian insulin appears to be more active in this regard than nonavian insulins, when evaluated in vitro and in vivo. Chicken insulin has been reported by many laboratories to be more potent in biological systems, both in vivo and in vitro, than equal amounts of various mammalian insulins. In preparations of rat liver plasma membranes, chicken insulin is up to 2.5 times more potent than the porcine hormone in stimulating glucose transport (Simon et al., 1974). This difference in potency can be attributed to an equally greater affinity of the chicken hormone for the liver binding site as well as to a slower dissociation rate of the chicken insulin–receptor complex (Simon et al., 1977b). Data on the effects of insulin in fed or fasted chickens are plentiful (Hazelwood, in Sturkie, 1976). The antigluconeogenic nature of insulin is found also in Aves, there being a decrease in activity of those enzymes that are rate limiting in the gluconeogenic pathways. Insulin probably potentiates the lipolytic activity of glucagon within the liver. There are no reports available on what significant role, if any, SRIF plays at the level of the hepatocyte.

Adipose Tissue

Adipose tissue in birds plays quite a different role than that documented in mammals. That is, the avian adipocyte appears largely to play a depository role for lipids that were synthesized previously in the liver and then transported, loosely complexed with apoprotein or albumin, to the peripheral fat pad. Here it is stored until mobilized and deesterified. Glucagon plays a major role in mobilizing glycerol and neutral fatty acids from these adipose deposits ("rests"), an action that is greatly potentiated by insulin (Langslow and Hales, 1969). The lipolytic sensitivity of isolated chicken adipocytes to glucagon increases 8- to 10-fold during the hatching period; an increase that probably results from an increased capacity for cAMP accumulation and possibly also an increased sensitivity of the lipase activation system, to cAMP (Langslow et al., 1979). Metabolism of glucose by fat cells is accelerated by chicken insulin to an extent much greater than that by mammalian insulin (Simon et al., 1974). However, this hormone is not antilipolytic (in marked contrast to a major effect observed in mammals). Thus, a major role of the pancreatic hormones in avian adipose tissue is that of glucagon activation of specific lipases, releasing free fatty acids and glycerol to the circulation to support various forms of tissue metabolic activity (particularly skeletal and cardiac muscle).

APP is known to reduce liver glycogen in chickens, to depress plasma glycerol levels, and to increase liver lipid in nonfasted birds (Hazelwood et al., 1973; Hayden, 1976). Also, in vitro the polypeptide is antilipolytic at the adipocyte level, although it has no effect on glucose metabolism (McCumbee and Hazelwood, 1978). Thus, along with GLI, which also exerts antilipolytic activity in ducks and geese, APP may well fill the void left by insulin's lack of antilipolytic effects, restoring metabolic regulation and counterregulation in adipose tissue. The role, if any, of SRIF at the adipose tissue level is yet to be elucidated.

Cardiac Muscle

Heart tissue of birds, metabolically similar to that of mammals, derives a disproportionately large amount of its metabolic fuel from plasma fatty acids and lactic acid. Also, as in mammals, avian hearts have relatively low levels (160–200 mg/100 gm wet tissue) of glycogen in the nonfasted state. At fasting, these levels may triple within 48–92 hours (see Hazelwood, in Sturkie, 1976). The pituitary gland is essential for this sparing of cardiac muscle carbohydrate. Insulin is of little value in increasing cardiac glycogen, but glucagon appears to have a modest glycogenic effect in fasting chickens, probably as a result of increased gluconeogenesis.

Embryonic hearts (chick) have very high glycogen levels relative to other tissues. At day 3, the myocardium may be 20 times richer than liver or skeletal muscle. This glycogen moiety increases until day 13, after which it gradually declines toward "adult levels" by hatching time. The decline of cardiac glycogen levels in ovo occurs concomitant with the onset of secretion of many hormones (e.g., glucagon, epinephrine, thyroxine, etc.) known to affect tissue glycogen levels. No effects of SRIF or APP on the heart have been reported.

Glycogen Body

The corpora sciatica, otherwise known as the glycogen body, is an opaque structure that lies dorsal to the spinal cord in all birds at the emergence of the sciatic plexus (see Chapter 2 for details). Its role in avian carbohydrate metabolism remains an enigma even though more than 80% of its lipid-free weight is glycogen. The peashaped, gelatinous structure contains glycolytic enzymes common to other avian tissues, yet its carbohydrate component is resistant to dietary, metabolic, or hormonal perturbation (Hazelwood, 1972). The suggestion that the glycogen body serves as a reserve tissue to supply the avian central nervous system with glucose (via cerebrospinal fluid) has not received experimental support, although glycogen in this structure is continuously synthesized and degraded (De Gennaro, 1982). The direct oxidative pathway (pentose shunt; Figure 13–1) is very active in the glycogen body, suggesting that production of reduced NADP for use by neural tissue may be a functionary attribute of this structure. Thus at present it appears that in some way the glycogen body contributes to the "wellbeing" (lipogenesis for myelin synthesis, fatty acid components of cerebrospinal fluid, etc.) of the avian central nervous tissue (see review by De Gennaro, 1982). Other than these suggestions, the significance of this glycogen-laden structure remains obscure. See also Chapter 1.

Other Tissues

Pancreatic hormones affect many other tissues in the bird in many diverse ways. Thus, while insulin encourages glucose uptake by most chicken tissues (except gut and kidney), glucagon is quite specific for releasing glucose to the plasma, etc. For the most part, most effects reported for these polypeptides at tissues other than those described above are noncarbohydrate-related effects.

Pancreatectomy, Pancreotoxins, and Pancreatropic Agents

Surgical Ablation

Removal of the pancreas in a variety of birds has been accomplished since 1891, the latter following quickly on the observation that such a procedure in dogs led to experimental diabetes. It was soon established that most, if not all, birds survived the operation quite well, exhibiting only a transitory hyperglycemia and glycosuria that disappeared within 7 days after the operation (reviewed in Hazelwood, 1973). Possible exceptions were some of the carnivorous species, such as certain hawks, owls, and kestrels. Mialhe's observation (1958) on the duck suggested for the first time that failure of earlier workers to produce diabetes in most birds may have been due to the presence of the splenic pancreatic lobe after surgery was completed. [Workers in this field usually employ ducks and geese as their experimental animal because the vascular components of the discrete pancreas are much less complex than those in chickens and other avian species. Thus, surgical extirpation of the duck pancreas is less traumatic and is more likely to be more complete (Sitbon et al., 1980).] This isthmus of pancreas (splenic lobe) (see Chapter 23, Figure 23–1) represents about 1–1.5% of the total pancreatic tissue, is highly endocrine, and has marked regenerative powers. Total pancreatectomy in ducks leads to a severe hypoglycemic crisis and death within 48 hr unless glucagon or glucose replacement therapy is initiated (Mialhe, 1958). Subsequently, chickens (Mikami and Ono, 1962) and geese (Sitbon, 1967) were found to respond to pancreatectomy in the same manner as ducks. Thus, once again, it appears that glucagon-supported carbohydrate and free fatty acid levels are extremely important to metabolic normalcy in birds, with insulin playing a secondary or supportive role. Decreased glucose tolerance, depressed liver glycogen levels, and disturbed I/G molar ratios in depancreatized ducks and geese are normalized by hypophysectomy; however, the procedure aggravates the fasting hypoglycemia (Sitborn et al., 1980).

The existence of an auxillary or secondary source of insulin from nonpancreatic sources has been postulated in both mammals and birds, but only in the latter has ample experimental evidence been presented to support the suggestion. Langslow and Freeman (1972) employed partial pancreatectomy in young chickens to reduce blood glucose and free fatty acid levels and to lower plasma insulin levels 75% over a 72-hr period. Other workers have observed in both 99%-depancreatized and totally depancreatized chickens a transitory hyperphagia and hyperglycemia (lasting 4–5 days), and a marked decrease in plasma insulin, which by day 3–4 returned to normal and then increased even further to reach supranormal levels in 8 days after the operation (Colca and Hazelwood, 1976). During 16 days of postsurger · observation, the splenic lobe remnant in the 99%-depancreatized birds hypertrophies greatly, which may explain in part the apparent normalcy of the chicken (Colca and Hazelwood, 1976; Cieslak, 1984). However, how does one explain the similar metabolic responses observed in totally depancreatized chickens?

Subtotal pancreatectomy in the chicken has been reported to encourage an increase in the SRIF cell population and activity in the gizzard-duodenal junctional tissue. This response in insulin-deprived chickens is similar to that reported in diabetic mammals and may be related in some way to the reduced release of insulin observed 5 weeks postoperatively in response to a meal (Simon and Dubois, 1983).

Persistence of immunoreactive insulin, glucagon, and APP has been reported for the adult chicken following total pancreatectomy. Plasma insulin and glucagon are readily detectable in the chicken 5 hr, 1 day, and 5 days after total pancreatectomy. These hormones respond to known stimulants of release such as arginine (Colca and Hazelwood, 1982a). APP levels, however, fall to undetectable levels 5 hr following pancreatectomy, only to rise modestly to subnormal levels during the next 5 days. This APP appears to be a very large molecular form (>100,000 daltons) and is not responsive to a release probe (Colca and Hazelwood, 1982a).

Thus, the earlier suggestion of the existence of an extrapancreatic source of insulin may also apply to glucagon in birds, but not to APP wherein a true deficiency state is created by total pancreatectomy. The course of plasma and pancreatic splenic lobe hormonal events following 99% pancreatectomy in the chicken is summarized in Table 13–1.

TABLE 13–1. Plasma and pancreatic hormone responses to partial (99%) pancreatectomy in fasted adult chickens (SCWL)[a]

Days after pancreatectomy	Observations (n)	Plasma parameters[b]				Splenic lobe characteristics[b]			
		GLUC	IRI	IRG	APP	Wt (mg)	IRI/mg	IRG/mg	APP/mg
0	36	233	2.70	1.65	3.04	19.0	244	452	3.5
1	30	276	1.48	1.72	2.13	21.6	314	400	3.8
2	24	328	1.77	1.82	2.27	30.8	342	411	3.7
4	27	292	2.34	2.56	2.28	36.0	133	263	3.3
8	19	236	1.75	3.64	2.53	80.3	88	279	3.6
16	15	229	1.09	3.52	2.28	97.9	118	263	3.5

[a] Data taken from Colca and Hazelwood (1976, 1982a); Cieslak (1984); Cieslak and Hazelwood (1983); on adult Single Comb White Leghorn chickens weighing 1.3–1.7 kg each.
[b] Plasma glucose in mg/dl; plasma hormones in ng/ml plasma; splenic lobe hormones in ng/mg wet tissue. GLUC, glucose; IRI, insulin; IRG, glucagon; APP, avian pancreatic polypeptide.

Alloxan and Streptozotocin

Two beta cytotoxic agents have been employed in efforts to cause "chemical diabetes" in birds by destruction of insulin-producing B cells. All attempts reported to date have failed, observations that contrast markedly with those in mammals. Glucose tolerance tests carried out in alloxanized chickens are indistinguishable from those done in normal chickens; both differ markedly from the typical diabetic tolerance curves seen in alloxanized rats (Figure 13–8).

Streptozotocin also is without effect on chicken plasma glucose and insulin levels, although the cytotoxin does hinder the insulin response to a glucose load (Stellenwerf and Hazelwood, 1979). Glucose uptake is unimpaired. Thus, different metabolic responses to alloxan and streptozotocin occur in chickens, but neither B-cytotoxin leads to a diabetic state. The suggestion has been made that the normal blood glucose level of birds (200–250 mg/dl, i.e., double that of mammals) protects the B cell from cytodamage because it is known that infusion of glucose into rats minutes before alloxan injection prevents alloxan diabetes. These interesting comparisons merit further investigation.

Other Cytotoxins

Insulin release and immediate B-cell depletion has been reported following injection of ascorbic acid into 6- to 10-week-old chickens. The action of ascorbate in chickens leads to a marked intolerance to a glucose load, and is not attributable to the oxidation product, dehydroascorbic acid, which is diabetogenic in rat, guinea pig, and man (Meglasson and Hazelwood, 1982). Repeated injections of ascorbic acid results in repeated release of APP in chickens, unlike the singular release of insulin.

Chemical agents of the biguanide family and metal salts such as cobaltous chloride, etc., have been used to selectively destroy the A cell and thus the glucagon-secreting potential of birds. Early reports of success have been replaced with those of guarded skepticism as more recent work has demonstrated that the hyperglycemia followed by hypoglycemia, and the hyperfatty-acidemia that ensue, are traceable to the general toxic effects of these agents and not to selective A-cell damage, release, and then exhaustion of glucagon reserves. Selective A-cell removal appears best achieved by discrete extirpation of the third *and* splenic pancreatic lobes in birds, as discussed above and in Chapter 23 (Assenmacher, 1973). Such operations lead to an hypoglycemic crisis and death.

Pancreatropic Agents

Substances such as tolbutamide, chlorpromamide, carbutamide, and other related sulfonylurea compounds have been shown to be effective triggers of insulin release in birds when injected intravenously or via intracardiac means. A profound decrease in blood glucose levels (triggered by insulin release) occurs in response to these agents; a decrease in free fatty acids also occurs, even before the effect on blood glucose is seen. Chronic feeding of tolbutamide increases chicken body weight as well as decreasing plasma insulin levels; however, plasma glucose levels remain unchanged (Simon, 1980). Apparently, there is no peripheral inhibition by the sulfonylurea of adipokinesis in depot fat. These observations have been made in ducks, pigeons, geese, and chickens, but for unknown reasons responses in owls differ. Tolbutamide actually decreases duck glucagon simultaneously with releasing insulin (Samols et al., 1969).

Again, it is important to realize that all evidence at hand indicates that the sulfonylureas act on the B-cell release mechanism, not on the biosynthetic pathway per se. Thus, to be effective as an hypoglycemic or antidiabetic agent, the sulfonylurea must act on

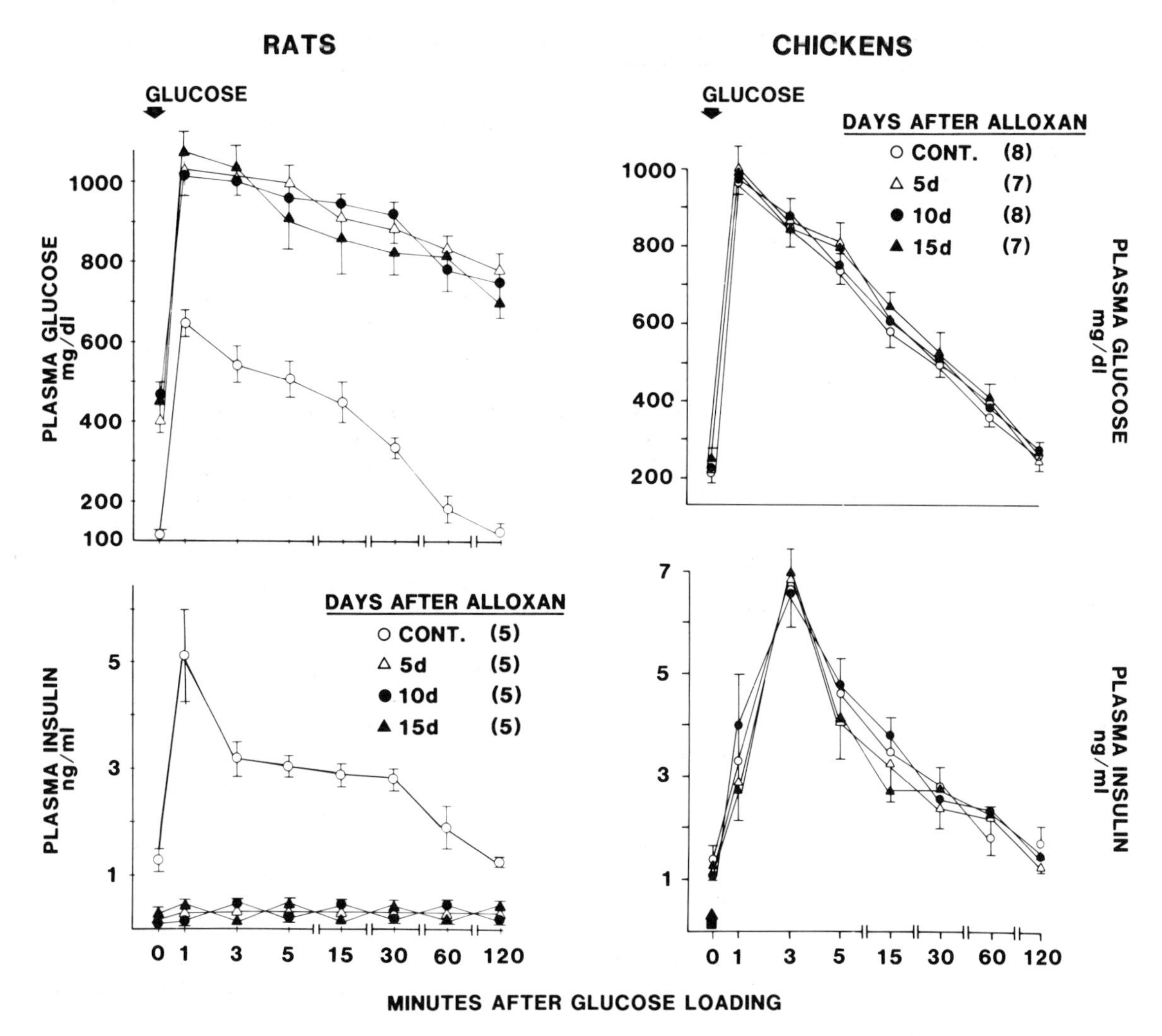

FIGURE 13–8. The effects of a B-cell cytotoxin (alloxan) on the response to glucose loading in rats and chickens. Rats (left side) are compared with chickens (right side) in their response to glucose loading (upper graphs) and the insulin secreted (lower graphs) in response to such a glucose load. Control (*cont*) animals were injected with subdiabetogenic doses of alloxan. Other animals were injected with alloxan either 5, 10, or 15 days (*d*) prior to the glucose test. Number of animals in each group is shown in brackets (–). (Modified from Stellenwerf and Hazelwood, 1979.)

healthy B cells containing preformed insulin. Tolbutamide is almost as effective as a blood glucose depressant in 99%-depancreatized and "totally" depancreatized chickens as it is in intact birds (Hazelwood, 1958; Colca and Hazelwood, 1976). Other agents known to decrease blood glucose in mammals by means other than releasing preformed insulin, e.g., by interference with glycolysis, have not been systematically studied in Aves.

Peripheral Endocrine Secretions and Carbohydrate Metabolism

Pituitary Hormones

Hypophysectomy in birds is difficult, indeed. In those studies in which it has been carried out successfully, reproductive function has been the main objective of the study. Other studies directed at the effects of various pituitary secretions on avian carbohydrate metabolism are few. However, the recent development of a sensitive immunoassay for growth hormone in chickens and improved methods of hypophysectomy in ducks have rekindled interest in this area of avian metabolism, and certain facets of growth hormone physiology have been elucidated. Hypophysectomy (hypox) of ducks with previously established "diabetes" is known to ameliorate the metabolic condition, although it does not improve glycemic levels during starvation (Sitbon et al., 1980). Growth hormone (GH) appears to be the main pituitary secretion affecting avian carbohydrate metabolism (see also Chapter 17). Injection of GH does not alter plasma glucose levels (chicks), but it does suppress glucagon (ducks) and increase insulin (ducks and geese) levels. As a result, it should not be surprising that it has been reported that GH injection in nonfasted chickens increases cardiac glycogen levels quickly (100% in 1 hr) while having no effect on liver glycogen levels (Hazelwood, in Sturkie, 1976). Whether its effect on pancreatic hormones or on the heart are direct effects of GH or, alternatively, effects of the hormone on metabolites, the levels of which alter other hormone secretion, is yet to be established.

While hypophysectomy in chickens reduces voluntary food intake, it also increases hepatic lipogenesis greatly. Pituiprivic birds become obese. The increased rate of hepatic lipogenesis is thought to result from either increased sensitivity of the hepatocyte to (increasing) endogenous insulin levels or from a release from pituitary GH (inhibition) influence (Kompiang and Gibson, 1976). Avian GH is far more effective than equivalent amounts of mammalian GH in promoting lipolysis and glycerol release in avian (turkey, pigeon, and chicken) adipocytes, even in the absence of supportive glucocorticoids (Harvey et al., 1977). Insulin and APP action at the adipocyte may be countered by an action of GH. Therefore, it would appear that avian GH spares carbohydrate by providing an alternate energy substrate, namely free fatty acids and glycerol. In this manner, GH in birds acts synergistically with glucagon, even though the latter hormone encourages glycogenolysis and hyperglycemia simultaneously.

One should recall that both fasting and insulin depress plasma glucose in mammals; GH release is an immediate response in both cases. Thus, it may be surprising to find that while fasting decreases plasma glucose in birds and increases circulating GH levels, insulin administration depresses the pituitary hormone, and does so even if hypoglycemia is prevented by simultaneous infusion of glucose (Harvey et al., 1978). Thus, the interaction of pancreatic insulin and pituitary GH would appear to be either a direct effect of the former hormone on the release of the latter, or, alternatively, to be secondary to the peripheral effects of insulin on uptake (by the pituitary?) of carbohydrate by metabolically active tissues. Hypophysectomy in young chickens invariably leads to greatly increased hepatic glycogen levels.

The role of prolactin (PRL) in birds has fascinated avian endocrinologists and physiologists for more than 50 years. Although glucose infusion into immature chickens appears to have little or no effect on plasma levels of GH or prolactin (PRL), hyperglycemia induced by glucagon elevates PRL release and simultaneously depresses GH release from the pituitary gland (Harvey et al., 1978). Once again, these effects are most probably the direct action of the pancreatic hormone on the pituitary gland. Fasting markedly decreases PRL levels in cockerels while increasing GH levels. When injected, PRL is an effective hyperglycemic agent both in birds and mammals. Hepatic lipogenesis from glucose substrates occurs in pigeons and chicks in response to PRL, while abdominal fat pads appear relatively uninfluenced. Early studies with prolactin preparations indicated that the exceedingly high hepatic glycogen levels in hypophysectomized pigeons are readily mobilized. At the same time, pancreatic endocrine cell types increased in size, number, and activity (reviewed by Hazelwood, in Sturkie, 1976). The cell most affected by PRL appears to be the pancreatic D cell.

Environmental cues associated with the onset of vernal premigratory or nesting periods are undoubtedly processed in the avian hypothalamus, and PRL, whose secretion is controlled by the hypothalamus, could play a very important regulator role in the metabolism of migratory species. Thus, its antiinsulin effects, its prominent role in avian lipid metabolism, and its ability to modulate pancreatic endocrine cell activity (increase SRIF?) suggest that PRL may be important to birds to shift "metabolic gears" favoring the provision of fuel to support prolonged flights.

The role that ACTH plays in avian carbohydrate metabolism is based on results obtained with use of mammalian ACTH preparations in various avian species. Generally, only pharmacologic doses of ACTH evoke an hyperglycemia in recipient birds, and this occurs 1–3 hr after administration as a result of ACTH-mediated corticosterone release from the adrenal cortex (see Chapter 22). Gluconeogenesis and hepatic glycogenesis are increased and glycogenolytic enzyme activity is decreased. Lipolysis provides noncarbohydrate fuel, which in turn spares carbohydrate substrate, thus reinforcing the hyperglycemia initiated by increased gluconeogenic enzyme activity. Considerably more research is needed to elucidate the physicochemical nature and biologic efficacy of avian ACTH.

Adrenal Hormones

Corticosterone is the primary glucocorticoid secreted from the avian adrenal cortex (see Chapter 22). All birds studied thus far have demonstrated a resistance to exogenous cortisone. Effective glucocorticoids increase hepatic glycogen deposition and cause a modest hyperglycemia (reactions that are driven largely by the strong gluconeogenic action of corticosterone and hydrocortisone). As protein catabolism increases, and glucogenic amino acids are converted to de novo carbohydrate substrates, nitrogen excretion increases, and, in chronic experiments, body weight decreases. Cholesterol levels increase in the plasma and true lipemic conditions may develop as a result of corticosterone injections, especially in 99%-depancreatized cockerels. "Steroid diabetes" has been produced in 8- to 33-week-old cockerels; even glycosuria can be produced. All of these effects are brought about by ACTH release, particularly over a prolonged period of time such as in chronic stressful situations. How important these metabolic effects of increased glucosteroid production are is yet to be established.

Both epinephrine and norepinephrine are secreted by the avian adrenal medulla, although the ratio of the concentration of one catecholamine to the other varies greatly depending on the species of bird (see Chapter 22). Both secretions are glycogenolytic in nature (in addition to their nonmetabolic actions), rapidly increasing hepatic cAMP levels and effecting a rapid delivery of free glucose to the plasma. Neural-based stress, exercise, and radical temperature changes act as potent stimuli to catecholamine release and subsequent hyperglycemia. Glucagon probably is a more potent glycogenolytic agent than epinephrine in birds, acting via the same "second-messenger" regulator scheme in liver, but it is not activated by neural mechanisms (sympathetic). Unlike epinephrine, glucagon secretion is most responsive to metabolic needs of the organism and thus plays a major role in maintaining glycemic homeostasis during fasting periods.

Thyroid and Gonadal Hormones

Chickens appear to be less sensitive than doves, pigeons, and ducks to the carbohydrate effects of thyroxine. Yet, in all birds studied, normal thyroxine levels appear essential to maintain a normoglycemia. Thyroidectomy in ducks and pigeons is without effect on liver and skeletal muscle glycogen deposits, but extirpation does induce a mild hypoglycemic condition (see Hazelwood, in Sturkie, 1976). Injection of T_4 or T_3 in fasted pigeons or chicks leads to a marked, although delayed, mobilization of liver glycogen; however, a glycogenic action occurs in cardiac and skeletal muscles. The glycemic effects of extreme ambient temperatures may well be indicated through thyroid–adrenal medulla mechanisms. Parathyroidal alteration or regulation of avian carbohydrate metabolism has not been reported.

Although differences in blood glucose levels have been reported for the two sexes, plasma levels of the two are approximately the same. Thus, the degree of packed red cell volume (which is greater in males) probably accounts for the differences reported in whole blood (Sturkie, 1976). Castration of males at an early age equates the blood glucose values without having an effect on plasma glucose levels. Androgen treatment of capons increases the hematocrit, thus reducing the blood glucose compartment. It is doubtful that the gonadal hormones play any significant role in avian carbohydrate regulation.

Summary

Avian carbohydrate metabolism is regulated, much like that in mammals, by a multifaceted force of neural, hormonal, humoral, and nutrient components. While neural (direct) input may be of relatively little importance in maintaining metabolic homeostasis, it appears that pancreatic hormones play a dominating role in this regard, with the I/G molar ratio being the single most important factor. Thus, avian carbohydrate metabolism regulation is qualitatively similar to that of most mammals studied, although differing in the set points for various reactions to occur. With the possible exception of GH, peripheral (extrapancreatic) endocrine secretions do not appear to play an important regulatory role in Aves, especially in nonstressful, hour-to-hour situations.

References

General Reviews

Bell, D.J., and B.M. Freeman (Eds.). (1984). "Physiology and Biochemistry of the Domestic Fowl," Vols. 1, 2, 3, 4, and 5. London: Academic Press.

Epple, A., and M.H. Stetson (Eds.). (1980). "Avian Endocrinology." New York: Academic Press.

Farner, D.S., and J.R. King (Eds.). (1971–1981). "Avian Biology," Vols, I–VI. New York: Academic Press.

Freeman, B.M., and M.A. Vince (Eds.). (1974). "Development of the Avian Embryo." New York: Wiley.

Grillo, T.A., L. Leibson, and A. Epple (Eds.). (1976). The Evolution of Pancreatic Islets. Oxford: Pergamon Press.

Pethes, G., P. Peczely, and P. Rudas (Eds.). (1981). Recent advances of avian endocrinology. In "Advances in Physiological Sciences," Vol. 33. Budapest: Pergamon Press.

Romanoff, A.L. (1967). "Biochemistry of the Avian Embryo." New York: Wiley.

Selected Papers

Alvarado, F., and J. Monreal. (1967). Na^+-dependent active transport of phenylglucosides in the chicken intestine. Comp. Biochem. Physiol., 20, 471.

Assenmacher, I. (1973). The peripheral endocrine glands. In "Avian Biology." Vol. III (D.S. Farner and J.R. King, Eds.). New York: Academic Press.

Bogner, P.H. (1966). Development of sugar transport in the chick intestine. Biol. Neonate, 9, 1.

Brady, L.J., D.R. Romsos, P.S. Brady, W.G. Bergen, and G.A. Leveille. (1978). The effects of fasting on body composition, glucose turnover, enzymes and metabolites in the chicken. J. Nutr., 108, 648.

Brady, L.J., D.R. Romsos, and G.A. Leveille. (1979). Gluconeogenesis in isolated chicken liver cells. Comp. Biochem. Physiol. B: Comp. Biochem., 63, 193.

Cieslak, S.R. (1984). Master's Thesis, University of Houston, Houston, Texas.

Cieslak, S.R., and R.L. Hazelwood. (1983). Does somatostatin (SRIF) stimulate glucagon release in Aves? Endocrinology, 112 (Suppl.; Abstr. 1271).

Colca, J.R., and R.L. Hazelwood. (1976). Pancreatectomy in the chicken: does an extrapancreatic source of insulin exist? Gen. Comp. Endocrinol., 28, 151.

Colca, J.R., and R.L. Hazelwood. (1981). Insulin, pancreatic polypeptide and glucagon release from the chicken pancreas *in vitro.* Gen. Comp. Endocrinol., 45, 482.

Colca, J.R., and R.L. Hazelwood. (1982a). Persistence of immunoreactive insulin, glucagon and pancreatic polypeptide in the plasma of depancreatized chickens. J. Endocrinol., 92, 317.

Colca, J.R., and R.L. Hazelwood. (1982b). Amino acids as *in vitro* secretogogues of avian pancreatic polypeptide (APP) and insulin from the chicken pancreas. Gen. Comp. Endocrinol., 47, 104.

Cramb, G., D.R. Langslow, and J.H. Phillips. (1982). The binding of pancreatic hormones to isolated chicken hepatocytes. Gen. Comp. Endocrinol., 46, 297.

DeGennaro, L.D. (1982). The glycogen body. In "Avian Biology," Vol. VI (D.S. Farner and J.R. King, Eds.). New York: Academic Press, Chapter 6.

Dickson, A.J. (1983). Gluconeogenesis in chick embryo isolated hepatocytes. Int. J. Biochem., 15, 861.

Drummond, G.I., (1967). Muscle metabolism. Fortschr. Zool., 18, 359.

Fearon, J.R., and F.H. Bird. (1968). Site and rate of active transport of D-glucose in the intestine of the fowl at various initial glucose concentrations. Poult., Sci., 47, 1412.

Freeman, B.M. (1966). The effects of cold, nor-adrenaline and adrenaline on oxygen consumption and carbohydrate metabolism of the young fowl. Comp. Biochem. Physiol., 18, 369.

Freeman, B.M. (1967). Some effects of cold on the metabolism of the fowl during the perinatal period. Comp. Biochem. Physiol., 20, 179.

Freeman, B.M. (1969). The mobilization of hepatic glycogen in *Gallus domesticus* at the end of incubation. Comp. Biochem. Physiol., 28, 1169.

George, J.C., and A.J. Berger. (1966). "Avian Myology." New York: Academic Press.

Goodridge, A.G. (1968a). Lipolysis *in vitro* in adipose tissue from embryonic and growing chicks. Am. J. Physiol., 214, 902.

Goodridge, A.G. (1968b). Metabolism of glucose-U-^{14}C *in vitro* in adipose tissue from embryonic and growing chicks. Am. J. Physiol., 214, 897.

Grande, F., and W.F. Prigge. (1970). Glucagon infusion, plasma FFA and triglycerides, blood sugar and liver lipids in birds. Am. J. Physiol., 218, 1406.

Gross, R., and P. Mialhe. (1978). Roles of insulin and glucose in the regulation of plasma FFA in the duck. Diabetologia, 14, 185.

Harvey, S., C.G. Scanes, and T. Howe. (1977). Growth hormone effects on *in vitro* metabolism of avian adipose and liver tissue. Gen. Comp. Endocrinol., 33, 322.

Harvey, S., C.G. Scanes, A. Chadwick, and N.J. Bolton. (1978). Influence of fasting, glucose and insulin on the levels of growth hormone and prolactin in the plasma of the domestic fowl. J. Endocrinol., 76, 501.

Hayden, L.J. (1976). An examination of the metabolic function of a new pancreatic polypeptide hormone in *Gallus domesticus.* Ph.D. Dissertation, University of Kansas Medical School. Kansas City.

Hazelwood, R.L. (1958). The peripheral action of tolbutamide in the fowl. Endocrinology, 63, 611.

Hazelwood, R.L. (1972). Intermediary metabolism. In "Avian Biology," Vol. II (D. Farner and J. King, Eds.). New York: Academic Press, Chapter 8.

Hazelwood, R.L. (1973). The avian endocrine pancreas. Am. Zool., 13, 699.

Hazelwood, R.L. (1976). Carbohydrate metabolism. In "Avian Physiology," (3d ed.) (P.D. Sturkie, Ed.). New York: Springer-Verlag, Chapter 11.

Hazelwood, R.L., and F.W. Lorenz. (1959). Effects of fasting and insulin on carbohydrate metabolism in the domestic fowl. Am. J. Physiol., 197, 47.

Hazelwood, R.L. J.R. Kimmel, and H.G. Pollock. (1968). Biological characterization of chicken insulin activity in rats and domestic fowl. Endocrinology, 83, 1331.

Hazelwood, R.L., S.D. Turner, J.R. Kimmel, and H.G. Pollock. (1973). Spectrum effects of a new polypeptide (third hormone?) isolated from the chicken pancreas. Gen. Comp. Endocrinol., 21, 485.

Honey, R.N., and G.C. Weir. (1979). Insulin stimulates somatostatin and inhibits glucagon secretion from the perfused chicken pancreas–duodenum. Life Sci., 24, 1747.

Johnson, E.M., and R.L. Hazelwood. (1982). Avian pancreatic polypeptide (APP) levels in fasted-refed chickens: locus of postprandial trigger? Proc. Soc. Exp. Biol. Med., 169, 175.

Kimmel, J.R., H.G. Pollock, and R.L. Hazelwood. (1968). Isolation and characterization of chicken insulin. Endocrinology, 83, 1323.

Kimmel, J.R., L.J. Hayden, and H.G. Pollock. (1975). Isolation and characterization of a new pancreatic polypeptide hormone. J. Biol. Chem., 250, 9369.

Kompiang, I.P., and W.R. Gibson. (1976). Effect of hypophysectomy and insulin on lipogenesis in cockerels. Horm. Metab. Res., 8, 340.

Langslow, D.R., and C.N. Hales. (1969). Lipolysis in chicken adipose tissue *in vitro.* J. Endocrinol., 43, 285.

Langslow, D.R., and B.M. Freeman. (1972). Partial pancreatectomy and the role of insulin in carbohydrate metabolism in *Gallus domesticus.* Diabetologia, 8, 206.

Langslow, D.R., and C.N. Hales. (1971). The role of the endocrine pancreas and catecholamines in the control of carbohydrate and lipid metabolism. In "Physiology and Biochemistry of the Domestic Fowl," Vol. I (D.J. Bell and B. Freeman, Eds.). London: Academic Press, Chapter 21.

Langslow, D.R., and K. Siddle. (1979). The action of pancreatic hormones on the cyclic AMP content of isolated chicken hepatocytes. Gen. Comp. Endocrinol., 39, 531.

Langslow, D.R., E.J. Butler, C.N. Hales, and A.W. Pearson. (1970). The response of plasma insulin, glucose and nonesterified fatty acids to various hormones, nutrients and drugs in the domestic fowl. J. Endocrinol., 46, 243.

Langslow, D.R., G. Cramb, and K. Siddle. (1979). Possible mechanisms for the increased sensitivity to glucagon and catecholamines of chicken adipose tissue during hatching. Gen. Comp. Endocrinol., 39, 527.

Laurent, F., and P. Mialhe. (1978). Effect of free fatty acids and amino acids on glucagon and insulin secretions in diabetic ducks. Diabetologia, 15, 313.

Lepkovsky, S., A. Chari-Bitron, R. Lemmon, R. Ostwald, and M. Dimick. (1960). Metabolic and anatomic adaptations in chickens "trained" to eat their daily food in two hours. Poult. Sci., 39, 385.

Lepkovsky, S., M.K. Dimick, F. Furuta, N. Sapir, R. Park, N. Narita, and K. Komatsu. (1967). Response of blood glucose and plasma free fatty acids to fasting and injection of insulin and testosterone in chickens. Endocrinology, 81, 1001.

Lin, Y.C., and P.D. Sturkie. (1968). Effect of environmental temperature on the catecholamines of chickens. Am. J. Physiol., 214, 237.

Marhussen, J., and F. Sundby. (1973). Duck insulin: isolation, crystallization and amino acid sequence. Int. J. Pept. Protein Res., 5, 37.

Mayes, J.S., L.R. Miller, and F.K. Myers. (1970). The relationship of galactose-1-phosphate accumulation and uridyl transferase activity to the differential galactose toxicity in male and female chicks. Biochem. Biophys. Res. Commun., 39, 661.

McCumbee, W.D., and R.L. Hazelwood. (1978). Sensitivity of chicken and rat adipocytes and hepatocytes to isologous and heterologous pancreatic hormones. Gen. Comp. Endocrinol., 34, 421.

Meglasson, M.D., and R.L. Hazelwood. (1982). Ascorbic acid diabetogenesis in the domestic fowl. Gen. Comp. Endocrinol., 47, 205.

Meglasson, M.D., and R.L. Hazelwood. (1983). Adrenergic regulation of avian pancreatic polypeptide secretion *in vitro*. Am. J. Physiol., 244, E408.

Mialhe, P. (1958). Glucagon, insuline et regulation endocrine de la glycemie chez le conard. Acta Endocrinol. (Suppl.), 36, 9.

Migliorini, R.H., C. Linder, J.L. Moara, and J.A. Veiga. (1973). Gluconeogenesis in a carniverous bird (black vulture). Am. J. Physiol., 225, 1389.

Mikami, S., and K. Ono. (1962). Glucagon deficiency induced by extirpation of alpha islets of the fowl pancreas. Endocrinology, 71, 464.

O'Hea, E.K., and G.A. Leveille. (1969). Lipid biosynthesis and transport in the domestic chick (*Gallus domesticus*). Comp. Biochem. Physiol., 30, 149.

Pearce, J., and W.O. Brown. (1971). Carbohydrate Metabolism. In "Physiology and Biochemistry of Domestic Fowl," Vol. 1 (D.J. Bell and B.M. Freeman, Eds.). London: Academic Press, Chapter 11.

Reisenfeld, G., A. Geva, and S. Hurwitz. (1982). Glucose homeostasis in the chicken. J. Nutr. 112, 2261.

Samols, E., J. Tyler, and P. Mialhe. (1969). Suppression of pancreatic glucagon release by the hypoglycemic sulfonylureas. Lancet, II, 174.

Simon, J. (1980). Effects of acute and chronic ingestion of tolbutamide in the chicken. Horm. Metab. Res., 12, 489.

Simon, J., and G. Rosselin. (1979). Effect of intermittent feeding on glucose–insulin relationships in the chicken. J. Nutr., 109, 631.

Simon, J., and M.P. Dubois. (1983). Subtotal pancreatectomy in the chicken: effect on the somatostatin cells of the digestive tract. Diabete Metab., 9, 75.

Simon, J., P. Freychet, and G. Rosselin. (1974). Chicken insulin: radioimmunological characterization and enhanced activity in rat cells and liver plasma membranes. Endocrinology, 95, 1439.

Simon, J., P. Freychet, and G. Rosselin. (1977a). A study of insulin binding sites in chicken tissues. Diabetologia, 13, 219.

Simon, J., P. Freychet, and G. Rosselin, and P. DeMeyts. (1977b). Enhanced binding affinity of chicken insulin in rat liver membranes and human lymphocytes: Hormone–receptor interaction. Endocrinology, 100, 115.

Smith, L. (1966). Species variation in the amino acid sequence of insulin. Am. J. Med., 40, 662.

Smith, M.K., and R.L. Hazelwood. (1981). Alpha- and beta-adrenergic control of pancreatic polypeptide and insulin secretion in adult chickens. Proc. Soc. Exp. Biol. Med., 168, 319.

Sitbon, G. (1967). La pancreatectomic totale chez l'oie. Diabetologia, 3, 427.

Sitbon, G., and P. Mialhe. (1978). Pancreatic hormones and plasma glucose: regulation mechanisms in the goose under physiological conditions. Horm. Metab. Res., 10, 12.

Sitbon, G., and P. Mialhe. (1979). Pancreatic hormones and plasma glucose: regulation meahanisms in the goose under physiological conditions. Horm. Metab. Res., 11, 85.

Sitbon, G., F. Laurent, A. Mialhe, et al. (1980). Diabetes in birds. Horm. Metab. Res., 12, 1.

Stellenwerf, Jr., W.A., and R.L. Hazelwood. (1979). Peripheral utilization of a glucose load after alloxan and streptozotocin in the rat and chicken: a comparison. Gen. Comp. Endocrinol., 39, 131.

Strosser, M.T., L. Cohen, S. Harvey, and P. Mialhe. (1980). Somatostatin stimulates glucagon secretion in ducks. Diabetologia, 18, 319.

Sturkie, P.D. (Ed.). (1976). "Avian Physiology" (3d ed.). New York: Springer-Verlag, (Chapter 11).

Tweist, G., and C.J. Smith. (1970). Circadian rhythm on blood glucose levels of chickens. Comp. Biochem. Physiol., 32, 371.

Wagh, P.V., and P.E. Waibel. (1966). Metabolizability and nutritional implications of L-arabinose and D-xylose for chicks. J. Nutr., 90, 207.

Weir, G.C., P.C. Goltsos, E.P. Steinberg, and Y.C. Patel. (1976). High concentration of somatostatin immunoreactivity in chicken pancreas. Diabetologia, 12, 129.

Ziswiler, V., and D.S. Farner. (1972). Digestion and the digestive system. In "Avian Biology," Vol. II (D.S. Farner and J.R. King, Eds.). New York: Academic Press, Chapter 6.

14
Protein Metabolism

P. Griminger and C.G. Scanes

Introduction

In 1817, Francois Magendie expounded the idea of a division of the "proximate principles of animals" into nitrogenous and nonnitrogenous matter. Twenty years later, Gerardus Mulder of Utrecht coined the term "protein" on the (albeit erroneous) supposition that such nitrogen-rich organic compounds as fibrin, egg albumen, and gluten contained a common basic component. The discovery of the real basic components of protein, the amino acids, stretched over more than a century, culminating with the discovery of threonine by Rose in 1935. For detailed historical accounts of these developments, see McCollum (1957) and Munro (1964).

One-fifth to one-quarter of the fat-free body of mammals and birds consists of protein (Mitchell et al., 1931; Griminger and Gamarsh, 1972). Because of variations in body fat content (mostly between 5 and 20%), which are only partially balanced by adjustments in water content, whole-body protein is somewhat more variable.

In birds, as much as 20–30% of the body protein may be found in the feathers. Other structural proteins are found in bone, muscle, and skin. The regulatory role of protein is exemplified by enzymes, plasma proteins, and such transport proteins as hemoglobin. Proteins can also play a functional role by supplying energy in the course of their degradation. There are clearly more similarities than differences in the protein metabolism of mammals, birds, and reptiles, and the major differences are in the nature of the end products. While the principal end product in mammals is urea, it is uric acid or related compounds in birds and reptiles (see Chapter 16).

Much of the available information on avian protein metabolism has been derived from experiments with the domestic chicken; pigeons and other domestic species have received their share of attention. Information on wild species, however, is extremely scanty.

Absorption

The digestion of dietary protein and the hydrolysis to amino acids have been discussed in Chapter 12. The amino acids enter the organism by way of the portal circulation almost entirely in the free form, but not all of the amino acids absorbed are derived from the proteins in the ingesta. As in mammals, amino acids are also contributed by the hydrolysis of desquamated intestinal mucosal epithelial cells and of digestive enzymes. It has been estimated that the mucosal cells of the jejunum of very young chicks are replaced in approximately 48 hr and somewhat more slowly in the more proximal and distal parts of the intestinal tract (Imondi and Bird, 1966).

The Absorptive Process

Kratzer (1944) concluded from a series of experiments that amino acid absorption in the chick was a function of the rate of diffusion and was not controlled by cellular metabolism. More recent work is at variance with this view, indicating an active absorption process. Major differences have been found in the absorptive ability for certain amino acids in different avian species. For example, methionine/lysine absorption ratios in *Eudromia elegans* (Martineta tinamou) and *Coturnix coturnix japonica* (Japanese quail) are 11.4 and 10.5, respectively, while in *Phasianus colchicus* (ring-necked pheasant) and *Meleagris gallopavo* (turkey) they are 3.17 and 1.46, respectively. Some differences have even been observed among strains of a single species (Lerner and Kratzer, 1976).

Absorption occurs by attachment to a specific site, presumably on the mucosal epithelial membrane, and when different sites appear to be involved, researchers speak of different "pathways." Learner (1971) concluded from his experiments and those of others that in the chicken there are at least three separate absorptive pathways for neutral amino acids: a system for methionine and related aliphatic compounds, one for glycine, and one for proline and related amino acids. There is also a distinct pathway for basic amino acids.

The existence of pathways for groups of amino acids entails competition among the members of a group in the absorptive process. For example, L-lysine absorption is inhibited by L-arginine, L-phenylalanine, or L-histidine, while L-leucine absorption is inhibited by L-valine, L-isoleucine, or L-methionine, with little or no competition observed for the glycine site (Gous et al., 1977a). Apparently, absorption sites exhibit preference rather than exclusiveness. The complexity of absorption mechanisms is highlighted by the findings of Burrill and Lerner (1972) that proline uptake occurs partly in a leucine-shared system and partly in one shared with other cyclic secondary amino acids. Additional and separate evidence on the heterogeneity of the neutral amino acid transport system across the avian intestinal wall was obtained by the same group (Miller et al., 1973) by the use of model amino acids specially synthesized for this purpose. After studying the uptake of glutamic acid and alanine by jejunal slices and by brush-border membrane vesicles of 2- and 21-day-old chicks, scientists from the same laboratory also concluded that each specific brush-border membrane transport mechanism follows separate developmental time tables (Shehata et al., 1984).

The natural isomers (L-amino acids) are generally absorbed more rapidly than the D forms. A common but L-preferring site exists for the transport of both isomers of methionine. Other neutral L-amino acids have a high affinity for this site, whereas the D-isomers, except for D-methionine, have a very low affinity (Lerner and Taylor, 1967). Both L- and D-amino acids therefore seem to be transported actively, albeit at different rates, and, to a degree, in competition with each other.

The sodium dependency of the uptake of L-lysine and L-histidine by chicken intestinal rings in vitro was demonstrated by Gous et al. (1977b).

Absorption Rates and Locations

Nitrogenous compounds disappear rapidly from the intestine of chickens. In 6-week-old birds, four-fifths of the nitrogen leaving the duodenum was removed

TABLE 14–1. Cumulative digestion and absorption of protein nitrogen in 3-week-old chickens on a low- (14%) and high- (20%) protein diet[a]

Section	Length (cm)	Cumulative digestion (%)		Cumulative absorption (%)	
		Low-protein	High-protein	Low-protein	High-protein
Duodenum	23	58.6	62.6	3.9	11.5
Upper jejunum	22.5	75.8	79.8	55.0	57.4
Lower jejunum	22.5	85.5	84.1	70.8	69.9
Upper ileum	23.5	90.9	90.5	79.5	78.5
Lower ileum	23.5	92.3	93.2	82.9	83.8

[a] Adapted from Hurwitz et al. (1972).

from the gut in the adjacent section (upper jejunum) of the intestinal tract (Bird, 1968).

Since endogenous protein is not digested as rapidly as exogenous protein, the more distal segments of the intestinal tract may play a more important role in the absorption of amino acids derived from endogenous protein (Crompton and Nesheim, 1969). These authors also noted that in Khaki Campbell ducks, the central portion of the intestinal tract contains the largest quantity of free amino acids. When segments of everted chicken intestine were incubated with lysine, the duodenum, jejunum, and ileum accumulated the amino acid at a similar rate, whereas ceca and colon-rectum slices accumulated appreciably less amino acid in the same period of time (Wakita et al., 1970). Within the jejunum, the proximal third accumulated the greatest amount of glutamic acid, methionine, or lysine. The absorption of amino acids can therefore take place in essentially all sections of the avian small intestine, albeit at varying rates.

Because of incomplete hydrolysis and a consequent relative scarcity of free amino acids, actual amino acid absorption in the duodenum is slight; it is again low in the distal parts of the small intestine, because the bulk of amino acids have been absorbed previously. This is demonstrated in Table 14–1, which shows the cumulative percent protein digested and absorbed by 3-week-old chicks receiving high-protein (20%) and low-protein (14.4%) diets.

If the natural flow of ingesta is impeded by artificial separation of sections of the gastrointestinal tract, absorption can occur even in the most proximal parts. Teekell et al. (1967) found that threonine was absorbed from the crop and gizzard as well as from the proventriculus and the duodenum when these sections were ligated and injected with [^{14}C]threonine. However, absorption at the latter two locations was appreciably greater than from the crop and gizzard.

The hydrogen ion concentration in the gut lumen influences the rate of absorption. Using Thiry-Vella fistulas and five pH levels from 2.6 to 10, Ivanov and Ivanov (1962) observed maximum absorption of a protein hydrolyzate at pH 6.8 from the small intestine of adult chickens.

Amino Acids

Essential Amino Acids

It is customary to classify amino acids as essential and nonessential. Clearly, all amino acids normally found in tissues are vital; the classification really pertains to their being essential in the diet. Nine amino acids are considered essential for all higher animals (Table 14–2). Chickens, turkeys, and probably other birds require arginine in addition to these nine. Histidine is known to be a dietary requirement for at least some species. However, since it was not until recently considered generally necessary for adult maintenance, it had been classified as semiessential. More recent work justifies the inclusion of this amino acid in the list of dietary essentials (Robbins et al., 1977; Cianciaruso et al., 1981). Apparently, nonprotein sources of histidine in muscle, such as carnosine, necessitate long-term tests to show the indispensability of this amino acid in the adult (Amend et al., 1979).

Cystine and tyrosine are synthesized from methionine and phenylalanine, respectively, and their adequacy therefore depends on the supply of these two essential acids. Glycine (or serine) as well as proline are considered essential for optimum growth of young chickens. Graber and Baker (1973) found that growing chickens could synthesize about 90% of the proline,

TABLE 14–2. Nutritionally essential amino acids

For maintenance and growth of all species:	
Histidine	Phenylalanine
Isoleucine	Threonine
Leucine	Tryptophan
Lysine	Valine
Methionine	
For growth of some species:	
Arginine	
Under certain conditions:	
Cystine	Tyrosine
Glycine	Proline

but only 60–70% of the glycine required for maximal growth. A mixture of glycine, proline, and glutamic acid that supplies one-third of the dietary nitrogen adequately meets the needs for all nutritionally dispensable amino acids (Stucki and Harper, 1961).

Metabolic Pathways

Ornithine Cycle. In mammalian liver, reversal of the reactions of ornithine catabolism can lead to ornithine synthesis. Ornithine, by way of the urea cycle, can serve as a source of arginine (Figure 14–1). The effectiveness of this pathway for the supply of arginine is very low (if it exists at all) in chickens and probably in other birds. Chickens cannot synthesize ornithine except from arginine and cannot convert ornithine to citrulline. This conversion requires carbamyl phosphate, which is synthesized in the liver of ureothelic animals from ammonia, biotin-activated carbon dioxide, and ATP in the presence of carbamylphosphate synthetase. In the chicken, this key enzyme is absent (Tamir and Ratner, 1963). When citrulline is provided, however, arginine can be synthesized in the avian kidney.

Young chicks can convert ornithine and therefore arginine to proline (Figure 14–2). However, the amount of proline synthesized in this manner is insufficient for adequate growth (Austic, 1976). Proline synthesis by this pathway is not stimulated by excess arginine or ornithine, and ornithine-δ-transaminase as well as the conversion of Δ^1-pyrroline-5-carboxylic acid may be the rate-limiting reaction.

Exogenous α-ketoglutaric acid greatly stimulates the

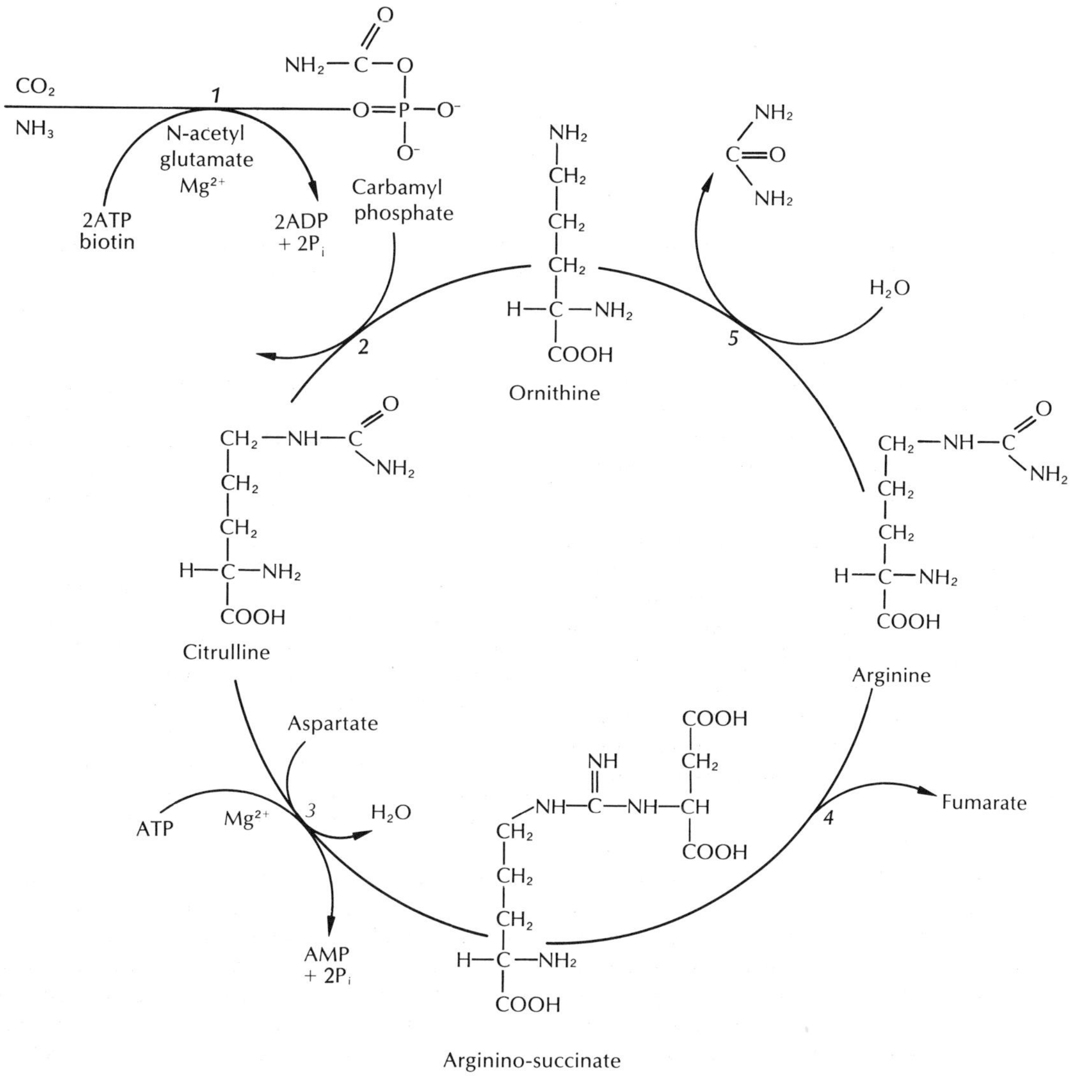

FIGURE 14–1. Ornithine cycle. Enzyme systems: (*1*) carbamyl phosphate synthetase, (*2*) ornithine transcarbamylase, (*3*) argininosuccinate synthetase, (*4*) succinase, (*5*) arginase.

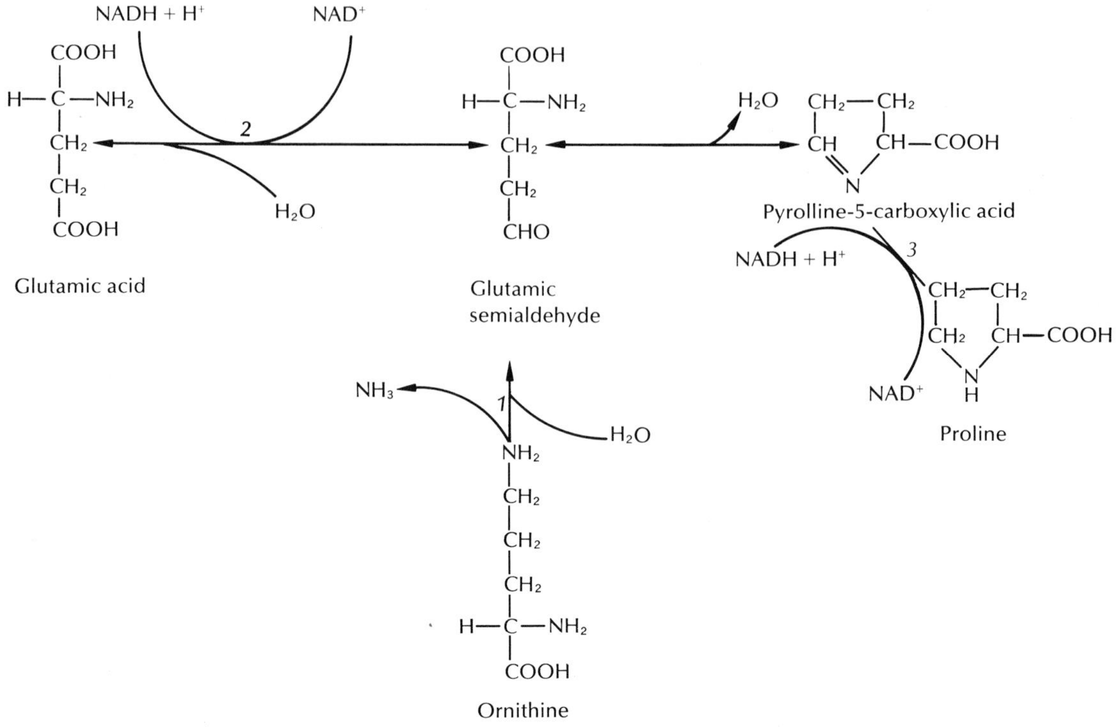

$NADH + H^+$ NAD^+

Glutamic acid

Glutamic semialdehyde

H_2O

Pyrolline-5-carboxylic acid

$NADH + H^+$ NAD^+

Proline

NH_3 H_2O

Ornithine

FIGURE 14–2. Proline synthesis from ornithine. Enzyme systems: (*1*) ornithine-δ-transaminase, (*2*) pyrolline-5-carboxylic acid dehydrogenase, and (*3*) pyrolline-5-carboxylic acid reductase.

in vitro conversion of ornithine to both Δ^1-pyrroline-5-carboxylic acid and proline. Employing ^{14}C-labeled glutamic acid, it was also demonstrated in vivo and in vitro that glutamic acid can be converted to proline (Shen et al., 1973a, b).

Glycine. There is no agreement about the ability of the chicken to satisfy its glycine needs by synthesis alone. Graber and Baker (1973) estimated from their experiments that 60–70% of the glycine needed for maximal growth of chicks could be synthesized. Ngo et al. (1977) observed that a deficiency of dietary glycine in young chicks limits the synthesis of uric acid, creatine, and protein necessary for optimum growth. However, adult birds can probably satisfy all their glycine needs by synthesis.

Cystine and Tyrosine. It is well established that tyrosine (hydroxyphenylalanine) has a sparing effect on phenylalanine, as does cystine on methionine, because tyrosine can be readily synthesized from phenylalanine and cystine from methionine. If there is ample dietary intake of tyrosine or cystine, the actual requirement for methionine and phenylalanine is reduced. This sparing effect is, of course, limited to the requirement for the two nonessential amino acids, and they cannot replace the basic needs for phenylalanine and methionine. An exception to this was claimed by Ishibashi (1972), who maintained adult White Leghorn males in positive nitrogen balance on a phenylalanine-free, tyrosine-containing diet. There was some body weight loss, however. Tracer studies indicated some conversion of tyrosine to phenylalanine, but the rate of conversion seemed quite insufficient to compensate for the lack of dietary phenylalanine.

The addition of an oxygen atom to phenylalanine in the synthesis of tryosine increases the molecular weight by approximately 10%. In theory this means that more tyrosine for protein synthesis can be provided per unit of weight by dietary phenylalanine than by dietary tyrosine, if a similar efficiency of absorption is assumed for the two amino acids. However, Sasse and Baker (1972) found equivalency in the efficiency with which these two amino acids supply tyrosine for protein synthesis.

In the conversion of methionine to cysteine in mammals, removal of a methyl group first leads to homocysteine. The addition of serine results in cystathione, which is then split into cysteine and α-ketobutyrate with the loss of one amino group. It is assumed that birds utilize the same pathway. During protein synthesis, cystine is created by the formation of disulfide bonds between two cysteine molecules. In the final analysis, two molecules of methionine are therefore required for each molecule of cystine in protein; therefore, 100 g of dietary methionine are required to produce 80.5 g of cystine. Indeed, experiments in our laboratory indicate that dietary methionine, on a molar basis, may actually be a better source of cystine for body maintenance of adult birds than dietary cystine. Graber and Baker (1971), in contrast, found that the molar efficiency of methionine, 1/2-cystine, and cysteine were alike and that, on a weight basis, less of a methionine–cystine combination was needed than of methionine alone. The problem is made somewhat more complex by the absorption rates of the two amino acids. Although there are no comparative measurements in birds, it is possible that the rate of absorption of L-cystine/cysteine is lower than that of L-methionine. Furthermore, cystine probably must be split into cysteine for incorporation into peptides, to be recombined later.

Dietary guanidoacetic acid accelerates the conversion of methionine to cysteine, whereas creatine, betaine, and choline have the opposite effect. The latter compounds accentuate the symptoms of muscular dystrophy in Vitamin E-deficient chicks, and the acceleration of the conversion of methionine reduces their severity. Hathcock and Scott (1966) considered this additional proof that cysteine, not methionine, was the metabolically active sulfur amino acid aiding in the prevention of nutritional muscular dystrophy in Vitamin E-deficient chicks.

The conversion of methionine to cystine is more efficient in White Leghorns than in Black Australorp chickens, but it appears adequate to meet the physiologic needs of both breeds on a methionine-supplemented, low-cystine diet (Miller et al., 1960).

The methionine hydroxy analog can replace methionine in the diet of chickens. The conversion of the hydroxy analog into methionine involves the formation of a keto acid as an intermediate (Gordon and Sizer, 1965). This keto acid also serves as an intermediate in the conversion of D- to L-methionine (Figure 14–3). The amino group of leucine, and probably also that of other amino acids, participates in the transamination reaction.

When most of the methionine is supplied by the L-methionine of protein-containing foods, as is the case with birds eating natural feed, equimolar levels of the OH analog are approximately equivalent to methionine in nutritional value. In amino acid diets, however, L-methionine appears superior to DL-methionine and to the OH analog (Smith, 1966). The efficiency of conversion may depend on the availability of suitable amino group donors.

Amino Acid Isomers

Although most ingested D-amino acids can probably be catabolized to yield energy, incorporation into structural or functional proteins requires prior conversion to the corresponding L-isomer. It is generally assumed that this involves conversion to an α-ketoacid by action of liver or kidney D-amino acid oxidase, followed by transamination. D-Amino acids may be absorbed less efficiently than the corresponding L-isomers; furthermore, they differ in their susceptibility to attack by D-amino oxidase; the conversion also depends on the total amount of D-isomers present. The ingestion of significant amounts of several D-amino acids causes a growth depression that may be a result of competition for the oxidase.

Sugahara and coworkers (1967) have developed a classification for the growth-promoting value of 16 D-amino acids in chickens. Among the essential amino acids, they assigned equal or nearly equal replacement values for the L-isomers to D-methionine, D-phenylalanine, D-leucine, and D-proline. They considered D-valine to have half the replacement value; D-tryptophan, D-isoleucine, and D-histidine to have little replacement value; and the D-isomers of lysine, threonine, and arginine to have none.

These categories are in part at variance with the experience of other researchers. For example, Fisher et al. (1957) observed relatively poor utilization of D-phenylalanine, listed above in the high replacement value category. Boebel and Baker (1982) also observed differences in the utilization of branched-chain

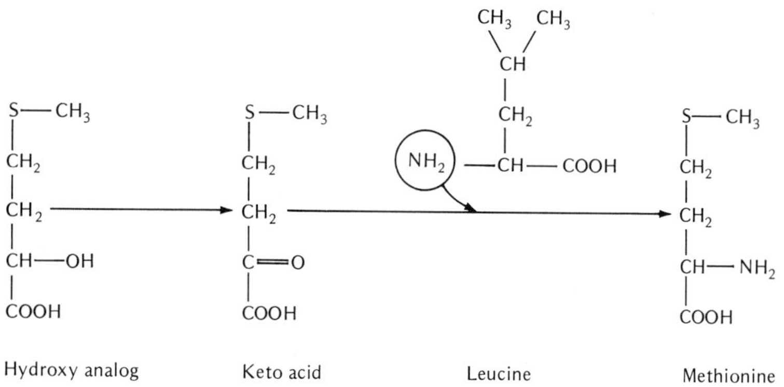

Hydroxy analog: $S{-}CH_3$ / CH_2 / CH_2 / $CH{-}OH$ / $COOH$ → Keto acid: $S{-}CH_3$ / CH_2 / CH_2 / $C{=}O$ / $COOH$ → (Leucine: CH_3 CH_3 / CH / CH_2 / $NH_2{-}CH{-}COOH$) → Methionine: $S{-}CH_3$ / CH_2 / CH_2 / $CH{-}NH_2$ / $COOH$

Hydroxy analog | Keto acid | Leucine | Methionine

FIGURE 14–3. The amination of methionine hydroxy analog.

amino acid analogs for chickens and rats. The DL- forms of isoleucine, valine, and leucine had approximately 80% of the effectivity of the L- forms for growing chicks. Parson and Potter (1981) have cataloged utilization data for various analogs of several amino acids for growing turkeys.

In general, the nitrogen from D-amino acids is used as efficiently, or only slightly less efficiently, than that from excess indispensable L-amino acids for the synthesis of dispensable amino acids, depending on the criteria used (Featherston et al., 1962). An excess of nonessential D-amino acids, however, may cause varying amounts of growth depression (Maruyama et al., 1972). D-Serine seems to be deaminated in chicken liver and kidney by serine dehydratase: even at the 2% level, it did not depress growth. The presence of 1.5% D-alanine in the diet depressed growth, while D-aspartic acid was deleterious even at only 1% of the diet. Accumulations of the latter two amino acids in the plasma pool indicated a limited capacity for their oxidation.

Amino Acid Imbalance and Toxicity

Harper et al. (1970) described an amino acid imbalance as the condition created by the addition, to a low-protein diet, of amino acids other than the one which is most growth limiting. The condition causes depression of food intake and growth. It can be readily prevented by supplementation with the limiting amino acid. Amino acid "antagonism," in contrast, can be alleviated by supplements of structurally similar amino acids but not by supplementation with the amino acid that is limiting in the basal diet. "Toxicity" applies to conditions where an adverse effect is caused by a large surplus of an individual amino acid. Fisher et al. (1960) defined amino acid imbalance differently. The addition of a protein or amino acid supplement may decrease the ratio of the most limiting amino acid to the sum total of available amino acids. If this decrease is such that the amount of the most limiting amino acid is below the minimum requirement for optimum performance, an "imbalance" is created.

Of special interest in birds is the antagonism between arginine and lysine. Excess dietary lysine decreases arginine efficacy by accentuating the arginine deficiency (Allen and Baker, 1972). Among the reasons given for the arginine–lysine antagonism are urinary losses of arginine caused by competition of lysine for renal tubular reabsorption and the depression of liver transamidinase activity caused by excess lysine (Austic and Nesheim, 1972). According to Harper et al. (1970) and D'Mello and Lewis (1971), lysine–arginine antagonism as well as the growth-depressing effect of excess leucine do not operate on the level of appetite regulation. Calvert et al., (1982), however, have demonstrated that in branched-chain amino acid antagonism in chickens, where the effect of excess leucine can be overcome by increased levels of isoleucine and valine, approximately 70% of the reduced growth rate caused by excess leucine can be ascribed to reduced food intake. A further portion may be due to increased branched-chain amino acid catabolism, which limits the availability of isoleucine and valine for growth.

Methionine has repeatedly been shown to be the most toxic amino acid when ingested in excess. The effect of high levels of methionine and related compounds on chickens and on turkeys have been described by Griminger and Fisher (1968) and by Hafez et al. (1978), respectively. At levels greater than three times the requirement, an increase in spleen iron levels and a decrease in blood hemoglobin levels have also been observed in chickens (Harter and Baker, 1978).

Inherited Response to Deficiency. The growth response of chickens on an arginine-deficient diet is more variable than that of chickens subjected to a comparable deficiency of other amino acids. An inherited potential to grow slower or faster on low-arginine diets was demonstrated in a random-bred flock of chickens by Griminger and Fisher (1962) and between different strains of chickens by Nesheim and Hutt (1962). The latter workers believed that the inherited differences were polygenic. Kidney arginase was found to increase in response to casein-based, arginine-deficient diets in high-requirement strains but not in those having a low requirement (Nesheim et al., 1971). Chickens with a high arginine requirement also had a much higher lysine-arginine plasma ratio, which is believed to depress feed intake. It was indeed shown that the low-requirement strain had a higher level of lysine-ketoglutarate reductase, the initial degradation enzyme of lysine (Wang et al., 1973).

Metabolic Transformations of Amino Acids

In the presence of adequate amounts of the essential amino acids and a source of nitrogen such as glutamic acid, optimal growth of birds can be obtained. Therefore, provisions must exist for the biosynthesis of serine, alanine, aspartic acid, asparagine, glutamic acid, glutamine, and, especially for collagen synthesis, for hydroxylysine and hyroxyproline. Not only is there similarity in the metabolic pathways of nitrogenous compounds in various vertebrate species, but avian tissues have been used as a basis for the observations of many investigators and have thus become the source of a body of generalized information on pathways of amino acid metabolism.

Transamination, the transfer of an amino group from one carbon skeleton to another, and deamination, culminating in the excretion of the amino group, are of special importance among the metabolic transformations. Amino group transfers are catalyzed by pyridoxal

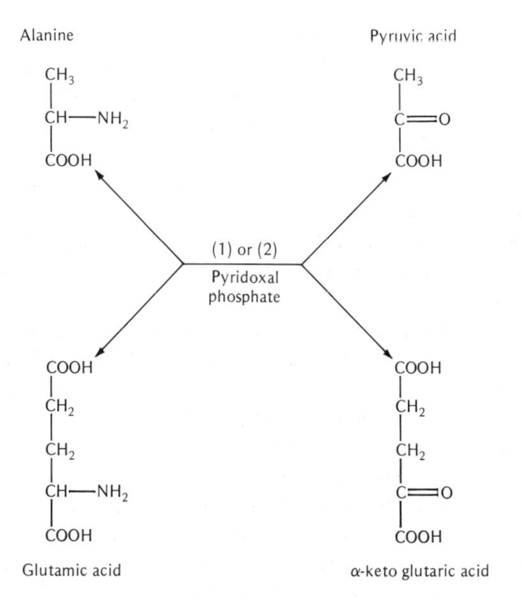

FIGURE 14–4. Pyridoxine-dependent transamination. Alanine pyruvate transaminase (*1*) transfers the α-amino group of alanine, and glutamate-α-keto glutarate transaminase (*2*) transfers the α-amino group of glutamate to an appropriate α-keto acid.

6-phosphate-containing enzyme systems. Examples are the transfer of an amino group from ornithine to α-ketoglutaric acid to form glutamate (Figure 14–2), the transfer of an amino group from alanine to α-ketoglutaric acid (Figure 14–4), or an exchange of amino groups between a five-carbon and a four-carbon amino acid. In the latter sequence, glutamic acid is deaminated to α-ketoglutarate, while oxaloacetate is aminated to aspartate, or vice versa (I). Glutamic acid can also be either deaminated or synthesized without involvement of a receptor or donor acid (II). Such deamination initiates the complete catabolism of glutamate and removal of ammonia and is mediated by glutamate dehydrogenase, an enzyme containing nicotinic acid. Glutamic acid therefore plays a central role in a complex system regulating amino nitrogen metabolism. Although oxidative deamination, mediated by L-amino oxidases, has also been observed in avian tissues, it is not clear whether it plays a significant role in avian nitrogen metabolism (Equations I, II, III, below).

Carbamylphosphate synthetase (CPS), which is found in mammals and in some other classes of animals, is apparently absent in chickens (Figure 14–1). The uptake of ammonia, however, can be catalyzed by other enzyme systems. One of these is glutamate dehydrogenase, in a reversal of the reaction (II) given above. Another is glutamine synthetase (III). In this reaction, ammonia combines with glutamate in the presence of Mg^{2+} and of adenosine triphosphate as a source of energy to yield glutamine.

Carbamyl phosphate is an essential precursor of uridylic acid, which in turn is the precursor for other pyrimidines utilized in the synthesis of ribo-and deoxyribonucleic acids. The absence of the enzyme synthesizing this compound has been given as the reason for the lack of arginine synthesis in birds. Clearly, however, birds do synthesize pyrimidines. Jones (1970) offers a possible explanation by assuming two types of CPS. CPS I is located in the mitochondria and is not found in Aves. It occurs in conjunction with ornithine transcarbamylase (Figure 14–1), which is also found in low concentration in birds. Avian livers, however, contain CPS II, which occurs with aspartate transcarbamylase in the soluble supernatant. Although not active in the synthesis of arginine and, eventually, urea, CPA II may provide carbamyl phosphate for pyrimidine synthesis. This hypothesis requires further confirmation.

Amino Acid Degradation

The deaminated carbon skeletons of amino acids can serve as a source of energy for the organism. They are converted to intermediates that ultimately form either glycogen or fat and are accordingly classified as glucogenic or ketogenic. Leucine belongs to the latter group, whereas isoleucine, lysine, phenylalanine, tyrosine, and threonine may be either ketogenic (by way of acetyl-CoA) or glycogenic. All other amino acids are considered glycogenic.

Part of the tryptophan ingested by chickens—and by many other species—is converted to nicotinic acid. Fisher et al. (1955) observed that 50 mg dietary L-tryptophan replaced approximately 2.5 mg nicotinic acid. Injected ring-labeled L-tryptophan is to a large extent recovered as CO_2, while the label of the D-isomer is primarily detected in the excreta (Nesheim et al., 1974).

Although threonine does not participate in transamination reactions, carbon-14 from labeled threonine can be found in the carbon chains of several amino acids within hours of administering it orally to chicks (Teekell et al., 1967). The effect of threonine-imbalanced diets on threonine catabolism has been investigated by Davis and Austic (1982).

Lysine is degraded in chicken liver by an L-amino acid oxidase and lysine-ketoglutarate reductase, leading

$$\text{Glutamate} + \alpha\text{-ketoglutarate} = \text{oxaloacetate} + \text{aspartate} \quad (I)$$
$$\text{Glutamate} + NAD^+ + H_2O = \alpha\text{-ketoglutarate} + NADH + H^+ + NH_4^+ \quad (II)$$
$$\text{Glutamate} + ATP + NH_4^+ = \text{glutamine} + ADP + P_i \quad (III)$$

to the formation of pipecolic acid and saccharopine, a tricarboxylic amino acid, respectively. Carbon-14 studies indicate that the saccharopine pathway may be the major one for the in vivo degradation of L-lysine in chicks (Wang and Nesheim, 1972). Both metabolites are subsequently converted to α-amino adipate and eventually to CO_2. D-Lysine is catabolized via pipecolate, which is further metabolized. In this the chicken differs from the rat, and probably other mammals, which excrete pipecolate in the urine without further catabolism (Grove and Roghair, 1971). These authors think that the rapid elimination of the catabolic products of both lysine isomers speaks against a conversion of D- to L-lysine; however, if the capacity to catabolize D-amino acids readily were common to birds, it would be useful for the utilization of D-amino acids found in insects and other lower animal forms ingested by birds. There seems to be a similar difference between birds and mammals in the degradation of D-hydroxylysine (Hiles et al., 1972).

Proline is probably oxidized to glutamate and then to α-ketoglutarate. Austic (1973) observed that in the absence of dietary proline, increased levels of liver pyrolline-5-carboxylic acid dehydrogenase were present in chickens (Figure 14–2). Although it is unclear why this increase occurs, it is possible that it reflects an increase in the activities of the enzyme complex for proline synthesis.

Brown (1970) has tabulated many of the enzyme systems involved in nitrogen metabolism in avian tissues.

Biogenic Amines. Most of these are derivatives of amino acids and act as neurotransmitters or hormones (see also Chapters 5 and 7). Histamine, produced by the decarboxylation of histidine, is a potent gastric secretagog. It stimulates the production of hydrochloric acid as well as of pepsin. Serotonin arises from the conversion of tryptophan by tryptophan hydroxylase to 5-hydroxytryptophan, followed by the action of L-aromatic amino acid decarboxylase. The serum, tissue, and intestinal levels of this amine in several species of birds have been tabulated by Brown (1970) and by Sturkie et al. (1972). Part of the serotonin is acetylated by serotonin-*N*-acetyltransferase to *N*-acetylserotonin. Hydroxyindole-*0*-methyltransferase, which is highly localized in the pineal glands of mammals and birds (Chap 23B) effects the methylation to melatonin, with *S*-adenosylmethionine serving as a methyl donor. Melatonin is a pineal "hormone" to which many different actions have been ascribed in various species (Axelrod, 1974).

Norepinephrine, a hormone produced by the adrenal medulla, results from the decarboxylation of dihydroxyphenylalanine (DOPA), which is an oxidation product of phenylalanine (Figure 14–5). The conversion of tyrosine to DOPA requires the presence of the enzyme tyrosine hydroxylase. Norepinephrine differs from epinephrine in that the primary amine of the latter is methylated by the enzyme phenylethanolamine *N*-methyltransferase; the methyl donor for this conversion is *S*-adenosylmethionine (Iversen, 1967). Chemically, these two hormones are classified as catecholamines, because they are derivatives of catechol (*o*-dihydroxybenzene). The effects of catecholamines are discussed elsewhere (Chapters 6, 7, and 22).

Taurine is a derivative of cysteine. It is found in bile of birds as a conjugate with cholic, chenodeoxycholic, and other bile acids (see Chapter 12). Examples of other amines or their derivatives are acetylcholine and γ-aminobutyrate. In the intestinal tract bacteria synthesize amines, including histamine, tyramine, putrescine, and cadaverine, by decarboxylation of histidine, tyrosine, ornithine and lysine, respectively.

Other Derivatives of Amino Acids. Thyroxine and triiodothyronine are iodinated derivatives of tyrosine that are synthesized in thyroglobulin, a protein elaborated by the thyroid gland, and are then released from the globulin by proteolytic action (see Chapter 20). Thyroid tissue also contains mono- and diiodotyrosine.

Melanin is a dark pigment of skin and feathers that is derived from tyrosine by way of phenylalanine-3,4-quinone (Figure 14–5). The early stages of synthesis

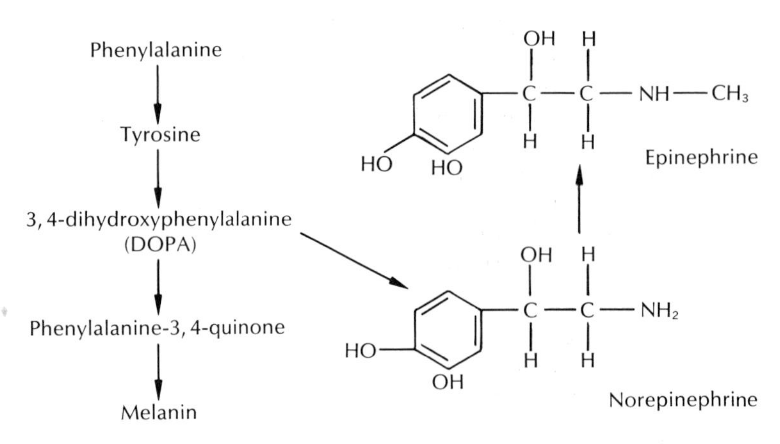

FIGURE 14–5. Abbreviated pathway of epinephrine and melanin synthesis.

require the copper-containing enzyme tyrosine (*o*-diphenol oxidase). The final steps of melanin synthesis are obscure. The term "melanin" has been used for the quinoid pigment as well as for an aggregate of the pigment and several enzyme systems in a protein matrix.

Nitrogen Excretion

Urea is the major nitrogenous end product of mammalian metabolism. In humans, this compound accounts for 80–90% of the urinary nitrogen. In chickens, uric acid is the excretory vehicle for four-fifths of the metabolized nitrogen, whereas ammonia N accounts for 10–15% of total N (Chapter 16). Chickens, and birds in general, are therefore said to be uricotelic, whereas man is ureotelic. Whereas blood levels of urea in mammals exceed uric acid levels by a factor of 10–20, levels of these two nitrogenous compounds are of approximately equal magnitude in birds. Urea, creatine, and amino acids are found in bird urine in lesser amounts, and creatinine may not be found. The major nitrogenous compounds in urine of chickens have been determined by a number of workers and details are presented in Chapter 16. The sites of synthesis of uric acid are the kidneys and liver (see Chapter 16).

The steps that lead to the synthesis of urea from amino acid nitrogen in mammals include the removal of ammonia by transamination and oxidative decarboxylation, the transport of ammonia, and the synthesis of carbamyl phosphate (Figure 14–1), which enters the ornithine–urea cycle. As has been mentioned previously, birds lack carbamylphosphate synthetase and therefore cannot use this route to dispose of the ammonia. A different route is the use of purines, such as uric acid, which derive their nitrogen from glutamine, glycine, and aspartate. This is the major pathway for the excretion of the end products of nitrogen metabolism in birds and some other classes of animals, which are, therefore, uricotelic. Another reason given for uricotely in birds is the adaptation to embryonic development in hard-shelled eggs, which are nearly impervious to liquid water. An accumulation of urea in the absence of sufficient water for its dissolution would be toxic to the embryo. The recent discovery of a ureotelic turtle laying hard-shelled eggs is an indication of the complexity of this problem (Packard and Packard, 1983).

There is one further difference between birds and saurian reptiles (i.e., lizards), on the one hand, and most species of other classes: uric acid is excreted as such and not broken down further. In mammals other than the primates, uricase decarboxylates uric acid to allantoin, and in some animals a further breakdown of allantoin occurs. The renal clearance of uric acid and of other nitrogenous substances is discussed in detail in Chapter 16.

Although many of the excretory products that are the result of detoxification reactions are similar to those found in mammals, birds also make use of a conjugation with ornithine (Nesheim and Garlich, 1963). Ornithuric acid (dibenzoyl ornithine) and dinicotinyl ornithine are examples of such conjugations. As in mammals, 3-methyl histidine is found in actin and myosin, which constitute approximately two-thirds of tissue protein. On degradation of tissue protein, 3-methyl histidine is released quantitatively by several mammals, including man and rat, thus permitting an estimate of muscle protein catabolism. Saunderson and Leslie (1983) have shown that 3-methyl histidine excretion is also a useful measure of muscle protein breakdown in chickens and Japanese quail, but not in turkeys.

Studies of the end products of digestion and metabolism in birds are complicated by the mixing of feces and urine prior to voiding. One may use chemical methods to separate the excreta, create an artificial anus, or canulate or exteriorize the ureters. The problem as well as the methods, none of which is without drawbacks, are discussed in Chapters 12 and 16. Of interest is the work of Krogdahl and Dalsgard (1981), who observed that 90% of all excreted ammonia and 86% of all urea were found in the urine. By means of these distribution data and measurements of the total excreted nitrogen compounds, nitrogen digestibility could be estimated without surgical intervention.

When an organism cannot, by dietary means, obtain sufficient amounts of an essential amino acid, it catabolizes body protein to obtain this amino acid. This process naturally leads to an increased excretion of nitrogen, in birds mostly as uric acid. With increasing dietary levels of the limiting amino acid, uric acid excretion decreases per unit of nitrogen (protein) consumed, until a plateau is reached when the birds' needs are fulfilled. In this manner, the level of uric acid excreted per unit of nitrogen consumed can serve as a measure of the avian amino acid requirement and an indicator of dietary protein quality (Miles and Featherston, 1974, 1976).

Protein Turnover

Protein turnover is the phenomenon by which the synthesis and the degradation of any protein type is occurring simultaneously. Therefore, in steady state conditions the rate of synthesis would be expected to be identical to the rate of degradation. Increases in tissue protein (accretion) would occur when the rate of synthesis exceeds the rate of degradation (catabolism).

$$\text{Accretion} = \text{Synthesis} - \text{Degradation} \quad [1]$$

TABLE 14–3. Breast muscle protein synthesis and degradation during growth in the domestic fowl

Strain	Age (weeks)	Protein synthesis (%/day)	Protein degradation (%/day)
White Leghorn[a]	1	42	26
	2	20	10
	4	24	17
	6	17	13
	7	17	13
New Hampshire × White Leghorn[b]	1	38	21
	2	26	10

[a] From MacDonald and Swick (1981).
[b] Adapted from Maruyama et al. (1978).

Conversely, there will be a decrease in tissue protein (breakdown, depletion, wasting) when the rate of degradation is greater than that of synthesis.

$$\text{Breakdown} = \text{Degradation} - \text{Synthesis} \quad [2]$$

Proteins in virtually all organs of the body are constantly turning over. In birds, protein turnover studies have been limited to only three muscles in the domestic fowl.

Muscle Protein Turnover

In vivo methods have been developed for the measurement of muscle protein synthesis, breakdown, and accretion. Accretion or breakdown can simply be estimated by determining the increase or decrease in muscle protein. In vivo protein synthesis is determined by examining the incorporation of radioactive amino acid into the protein(s). In vivo degradation is not estimated directly but as either the difference between synthesis and accretion (when muscle protein is increasing) or the sum of synthesis (when muscle protein is decreasing) and breakdown (see Equations 1 and 2).

$$\left\{\begin{array}{c}\text{Plasma}\\\text{Amino-}\\\text{Acid}\\\text{Pool}\end{array}\right\} \rightleftharpoons \left\{\begin{array}{c}\text{Muscle}\\\text{Free}\\\text{Amino Acid}\\\text{Pool}\end{array}\right\} \underset{\text{Degradation}}{\overset{\text{Synthesis}}{\rightleftarrows}} \left\{\begin{array}{c}\text{Muscle}\\\text{Protein}\\\text{Pool}\end{array}\right\}$$

The determination of the in vivo rate of protein synthesis for a particular muscle is complex and subject to possible artifacts, including changes in the size of the free amino acid pool and dilution of the radioactive amino acid resulting from degradation of proteins. The in vivo rate of muscle protein synthesis is determined from the ratio of the specific activity of radioactive amino acid (RAA) in the protein pool to the specific activity of the RAA in the muscle free amino acid pool. The amino acids used (tyrosine, valine, and proline) are not major energy sources and are not recycled. For details of methods see Laurent et al. (1978a, b). These methods have been employed to examine the changes of muscle protein synthesis and degradation during muscle growth, stretch-induced hypertrophy, atrophy following nutritional deprivation, and also the rapid growth of the muscle following restoration of a replete diet in a chick previously protein deprived.

The rapid growth of muscle in the young chick includes considerable muscle protein accretion. This might be explained by high synthetic rates or by low degradation rates. In an exhaustive study, MacDonald and Swick (1981) characterized muscle protein metabolism in young chicks as having a very high synthesis rate and a high degradation rate (Table 14–3), the decline in synthesis with age being more pronounced. Muscle growth also occurs when the muscle is stretched by the application of a weight. Stretch-induced hypertrophy increases muscle protein (both collagen and noncollagen), largely due to increases in protein synthesis. However, a paradoxical increase in the degradation rate was also observed (Laurent et al., 1978a, b).

The influence of diet on breast muscle protein synthesis and degradation is illustrated in Table 14–4. Chronic deprivation of energy or one essential amino

TABLE 14–4. Effect of nutritional deprivation on breast muscle protein synthesis and degradation

	Protein synthesis (as % of control)	Degradation (as % of control)
Control[a]	100	100
Energy-deficient[a] (50% energy of control)	95	256
Lysine-deficient[a] (50% requirement)	94	212
Control[b]	100	100
Protein-free diet[b]	124	164
Refed protein[b] (2 days)	241	171

[a] Adapted from Maruyama et al. (1978).
[b] Adapted from MacDonald and Swick (1981).

acid (lysine) or feeding a protein-free diet leads to some atrophy of the breast muscles, caused mainly by protein degradation. However, restoration of a normal diet results in a 2.4-fold increase in the rate of synthesis and no change in degradation.

It appears that muscle protein accretion or breakdown is due to the balance of synthesis and degradation; furthermore, they seem to have independent control mechanisms. These may include hormonal mechanisms, local and intrinsic factors, and direct effects of metabolites.

Blood Proteins and Amino Acids

Plasma Proteins

Most plasma proteins are synthesized in the liver from amino acids derived from the food or the catabolism of tissues. Their presence in the plasma is short lived. Replenishment of these proteins is essential, since they exert a colloidal osmotic pressure that aids in the preservation of blood volume and helps maintain the blood pH within a narrow range. The plasma also contains specialized proteins, such as prothrombin and fibrinogen, which act in the hemostatic mechanism. It is also essential that certain proteins be able to pass through the capillary walls to fulfill their extravascular tasks; these provide antibodies to control infections, transport protein-bound hormones and other compounds to their target organs, and are the major source of cellular metabolic protein.

Chickens can produce high levels of precipitating antibodies following injections of heterogenous serum proteins. The demonstrable passage of antibodies to the egg yolk offers a mechanism of transfer of maternal antibodies to the offspring. Antibodies disappear rapidly; the half-life of γ-globulin in newly hatched chicks is only about 3 days, and it is progressively shorter as the bird matures (Patterson et al., 1962). The level of antibody production is influenced by such nutritional factors as protein and amino acids, and optimal antibody production may not be obtained at the same nutrient level as optimum growth (Fisher et al., 1964; Bhargava et al., 1971). The chicken, a good producer of antibodies, also has a relatively high level of γ-globulin.

Measurements of plasma proteins are usually based on the determination of amino and imino nitrogen multiplied by 6.25. This factor is a good estimate of the ratio of the weighted average of the molecular weights of the amino acids of plasma proteins to the molecular weight of nitrogen and is therefore a good estimate of the polypeptide component of the plasma proteins. These proteins may, however, contain significant nonpeptide components, such as carbohydrates and lipids, which contain little or no nitrogen. This problem, and other assay methods, have been reviewed by Bell and Sturkie (1965) and by Martinek (1970).

Plasma Protein Fractions. Fractionations of plasma or serum proteins are frequently based on differences in ionic mobility of the various fractions, a process known as electrophoresis. This may be carried out in a buffered solution, as in the classic method devised by Tiselius ("free" electrophoresis), or on filter paper or starch gel ("zone" electrophoresis). If only the major components are to be determined, salting-out procedures may be satisfactory. Advantages and disadvantages of various methods have been discussed in the reviews by Bell and Sturkie (1965) and Martinek (1970), and a comparison of electrophoretic techniques for chicken serum protein fractionation has been published by Torres-Medina et al. (1971). Gel electrophoresis appears superior to the other electrophoretic methods of fractioning avian plasma or serum. Five main fractions of plasma proteins can be discerned in all systems. These fractions correspond to albumin and four globulin fractions of mammalian plasma: α_1, α_2, β, and γ. Immunophoresis can aid greatly in the identification of the proteins. By this method, Tureen et al. (1966) identified 12 different proteins in chickens 1–210 days of age. Also using the same method, a δ-globulin has been demonstrated; it was described in detail by Richards (1975).

Several workers have reported distinct prealbumin fractions. Tureen et al. (1966) found that these decrease after 8 days of age and only exist in traces after 18 days. Harris and Sweeney (1969) noted prealbumins in adult males, and Elliott and Bennett (1971) observed a prealbumin peak in all female samples, but none in males.

Albumins are believed to act as a protein reserve and a protein source at times of subnormal intake. In their normal role, however, albumins also act as carriers of many nutrients, including mineral elements, vitamins, and fatty acids. In the chicken and quail, albumins have also been shown to be carriers of thyroid hormones (McNabb and Hughes, 1983).

Transferrin, an iron-binding protein, behaves electrophoretically as a γ_1-globulin in the chicken, whereas mammalian transferrin exhibits the mobility of β_1-globulin (Torres-Medina et al., 1971). Most of the copper in serum is bound to ceruloplasmin, an α_2-globulin that is found in lower levels in chickens than in mammals.

Lipoproteins consist of lipids bound mostly by noncovalent forces to α- or β-globulins. The major function of plasma lipoproteins is the transport of lipids, predominantly triglycerides but including a variety of other compounds, such as cholesterol, fat-soluble vitamins, and phospholipids (Schumaker and Adams, 1969).

Immunoglobulins are synthesized by the cells of the

reticuloendothelial system in response to a variety of antigenic stimuli that are ever present in the interior and exterior environment of all living organisms. In the electrophoretic pattern of fractionation, part of the β_2-globulins and essentially all of the γ-globulins are immunoproteins.

In the chicken plasma electrophoretic pattern, fibrinogen appears as a separate peak in the γ-globulin range (Sturkie, 1965). Fibrinogen, prothrombin, and other proteins that are an integral part of the hemostatic mechanism are discussed in Chapter 5.

Plasma Protein Levels. On the tenth day of embryonic development, chick as well as *Phasianus colchicus* (ring-necked pheasant) serum protein is relatively low in albumin, which increases rapidly in subsequent days relative to globulin (Weller, 1966). Total serum proteins increased approximately two-fold in chickens from week 1 to week 12 (Morgan and Glick, 1972) and in Japanese quail from day 2 to day 50 (Atwal et al., 1964). Values for mature birds of several species are shown in Table 14–5. To permit comparison of the data from various sources, prealbumin and albumin are combined as "albumin" and all globulin fractions as "globulin." It is probably correct to assume that the "globulin" fraction includes fibrinogen in the determinations by Sturkie and Newman (1951) as well as those by Balasch et al. (1973). Using a salting-out method, Dabrowski (1966) determined plasma fibrinogen separately. The average fibrinogen values obtained [from 0.30 g/100 ml for *Corvus corone cornix* (crow) to 0.80 g/100 ml for *Pica pica pica* (magpie)] were not, for the purpose of this table, included in the "globulin" fraction. While total plasma protein levels in the species listed in Table 14–5 are remarkably similar, there is considerable variation in albumin and globulin

Table 14–5. Some measurements of total plasma (or serum) proteins, "albumins," "globulins," and "A/G" ratio in various species of birds[a]

Species	Sex	Total protein (g/100 ml)	"Alb" (g/100 ml)	"Glob" (g/100 ml)	A/G	References
Galliformes						
Gallus gallus domesticus (chicken)	M	4.00	1.66	2.33	0.71	Sturkie and Newman (1951)
	F	5.24	1.97	3.27	0.60	Sturkie and Newman (1951)
Meleagris gallopavo (S)[a] (turkey)	M	4.40	2.69	1.35	1.98	Lynch and Stafseth (1953)
Numida meleagris (guinea fowl)	M	3.52	1.45	1.98	0.73	Balasch et al. (1973)
Phasianus colchicus (pheasant)	M	4.90	2.29	2.62	0.87	Balasch et al. (1973)
Chrysolophus pictus (pheasant)	M	3.23	1.55	1.68	0.92	Schram and McNabb (1975)
Alectoris graeca (rock partridge)	M	4.66	1.66	2.98	0.56	Balasch et al. (1973)
Gallus gallus gallus (bankiva)	M and F	4.43	1.95	2.47	0.79	Balasch et al. (1973)
Pavo christatos (peacock)	M and F	4.36	2.41	1.94	1.24	Balasch et al. (1973)
Penelope waenieri (guan)	M and F	3.69	2.03	1.60	1.22	Balasch et al. (1973)
Passeriformes						
Corvus frugilegus (rook)	M and F	4.10	0.81	2.69	0.30	Dabrowski (1966)
Corvus corone cornix (crow)	M and F	4.40	1.30	2.80	0.46	Dabrowski (1966)
Coleus monedula spermologus (jackdaw)	M and F	4.60	1.20	2.80	0.43	Dabrowski (1966)
Pica pica pica (magpie)	M and F	4.30	1.00	2.50	0.40	Dabrowski (1966)
Garrulus glandarus (jay)	M and F	4.80	1.12	3.16	0.35	Dabrowski (1966)

[a] Plasma except for *Meleagris gallopavo* and *Chrysolophus pictus*, where serum (S) was used. In the Passeriformes samples (Dabrowski, 1966), fibrinogen (0.30–0.80 g/100 ml) was not included in the globulin fraction; it was, apparently, in the other plasma samples. A/G, albumin/globulin ratio.

and their ratio to each other. There appears to be appreciable variation within zoological classes. Whereas adult chickens and many other birds have an albumin/globulin ratio of less than 1, this is not true of some species shown in Table 14–5 or the pigeon and turkey. Additional data for various breeds of domestic chickens as well as a number of other species have been listed by Sturkie (1965).

Factors Affecting Plasma Proteins. The state of hydration or dehydration, or hemorrhage, influences the level of plasma proteins, as does the level of protein nutrition (Leveille and Sauberlich, 1961), sex (Grant and Anastassiadis, 1962), and the stage of development. Changes in plasma proteins induced by egg formation are discussed in Chapter 18.

Plasma Nonprotein Nitrogen (NPN)

Nitrogenous substances in protein-free plasma constitute only a small, but physiologically important, fraction of the total plasma nitrogen. About half of NPN consists of amino acids; the other half includes urea, urates, creatine, ammonia, and a host of lesser, mostly unidentified, components (see also Chapter 16). Whereas avian erythrocytes contain, on a volume basis, about six times as much NPN as the plasma, all of the urates are found in plasma (Bell et al., 1959).

Uric acid concentrations in plasma are greatly influenced by such factors as age, sex, and reproductive and nutritional status, whereas urea and ammonia appear relatively unaffected. Featherston (1969) found that increasing protein from 25 to 75% of the diet did not significantly change plasma ammonia but did increase plasma uric acid fourfold and free plasma amino acids by one-third (see also Chapter 16). It was shown in other experiments with chicks subjected to similar treatment that the plasma uric acid level increase was in agreement with changes in levels of liver xanthine dehydrogenase, an enzyme responsible for uric acid formation. Changes in dietary electrolytes can also affect plasma uric acid concentrations (Austic, 1980).

Free Plasma Amino Acids. Data on free plasma amino acids in very young chicks have been tabulated by Zimmerman and Scott (1967) and levels for adult cockerels by Ohno and Tasaki (1972) and Desmarais and Pare (1972). The latter authors indicated that heredity plays an important role in the free plasma amino acid levels of mature chickens. The response of free plasma amino acid levels of poults to various protein sources has also been tabulated (Dunkelgod and Winkleman, 1982). Methods for the determination of free amino acids in avian serum by high-performance liquid and ion-exchange chromatography have been compared by Elkin (1984).

Miscellaneous Proteins and Derivatives

Structural Proteins

It can be assumed that the structural proteins of birds, such as muscle and collagen, are similar to those of mammals. The comb and wattles of mature chickens are epidermal appendages; their high collagen content, exemplified by their hydroxyproline content, is indicative of their similarity with skin.

Feathers. Although feathers generally constitute only 5–10% of body weight, up to one-third of the total protein of a bird may be feather keratin. Harrap and Woods (1967) have listed the amino acid composition of feathers and feather parts of several species. Serine, glycine, and proline are the most abundant amino acids in feathers and methionine, histidine, lysine, and tryptophan occur at rather low levels. It should be noted, however, that tryptophan also occurs in relatively low concentrations in muscle tissue. According to Fisher et al. (1981), the methionine content of chicken feathers decreases with age, while that of threonine, isoleucine, and valine increases. Feathers are a unique biologic entity, and feather proteins differ among zoologic groups to various degrees. It is not surprising, therefore, that efforts have been made to use feather protein composition for taxonomic purposes (Brush, 1976).

Other Proteins and Peptides

Histones and other nuclear proteins may control protein biosynthesis by controlling messenger RNA formation through partial blocking of the DNA molecule. Information on the composition of avian histones has been reviewed by Brown (1970).

Polypeptide Hormones. Many hormones and neuropeptides are polypeptides or proteins. For example, adrenocorticotropic hormone (ACTH) consists of 39 amino acids, of which only the first 23 are necessary for activity. Other hormones secreted by the anterior pituitary, such as growth hormone and the tropic hormones, contain larger numbers of amino acids, and the hormones of the latter group, especially, are classified as proteins; they are discussed in detail in Chapters 17 and 22.

Another protein hormone is insulin, produced by the β-cells of the pancreas. Fowl insulin differs from porcine insulin (which is very similar to human insulin) in the sequence of only 6 of the 51 constituent amino acids. The α_2 cells of the pancreas produce glucagon, a polypeptide hormone consisting of 29 amino acids. These endocrine secretions of the pancreas are discussed in Chapter 13. Secretin, one of the hormones stimulating pancreatic secretion, consists of 27 amino acid residues, with about half in the same position as in glucagon. The activities of this hormone, secreted in the walls of the intestine, are described in Chapter 13. The digestive enzymes, as well as the metabolic enzymes, are protein in nature. There is reason to believe that the chemical properties of avian digestive enzymes closely resemble those of their mammalian counterparts (see Chapter 12).

Hormonal Control of Amino Acid and Protein Metabolism

Our knowledge of the endocrine control of protein metabolism in any avian species is very limited. Where data are available from studies of different species of birds, they will be cited. Otherwise, the mammalian situation will be tacitly assumed to exist in Aves. This is not necessarily true, as hormonal effects vary considerably from species to species.

The major hormones affecting protein metabolism are insulin, the thyroid hormones, growth hormones, androgens, glucocorticoids and estrogens. Other hormones may prove to have important actions in protein metabolism but as yet their effects tend to be limited to a single tissue or organ system [e.g., the effects of nerve growth factor (NGF) or epidermal growth factor (EGF)].

Insulin

In mammals, insulin has a distinct anabolic role, stimulating the uptake of amino acids and hence protein synthesis by muscle tissue. There is strong circumstantial evidence for an equivalent role in avian species. In the domestic fowl, insulin administration has been observed to reduce the plasma concentration of amino acids by approximately 40% (Langslow et al., 1970), probably because insulin increases tissue uptake of amino acids and perhaps decreases the release of amino acids by various tissues. Insulin appears to play a similar role in domestic ducks. This is supported by the increase in plasma amino acids following pancreatectomy (and hence loss of circulating insulin) (Samsel and Ledig, 1976). Furthermore, pancreatectomy affects the ability of ducks to respond to arginine load. In the normal duck, arginine stimulates insulin release (Laurent and Miahle, 1978), which would be expected to depress the circulating concentration of amino acids. However, arginine infusion results in a greater increase of the plasma concentration of amino acids and a slower clearance of amino acids in pancreatectomized or partially pancreatectomized ducks (Samsel and Ledig, 1976; Laurent and Miahle, 1978).

Thyroid Hormones

In birds, thyroid hormones are required for growth, particularly that of bone and muscle. King and King (1973) demonstrated that chemical thyroidectomy reduces the growth rate, and furthermore, that this can be overcome by the administration of thyroxine (T_4). It is probable that T_4 has to be monodeiodinated to triiodothyronine (T_3) in order to exert its effect on growth (for details, see Chapter 20). Indeed, T_3 stimulates growth in sex-linked dwarf chicks with an isolated T_3 deficiency (Scanes et al., 1984). The effect of T_4 and particularly T_3 on growth in birds is critically determined by the dose administered or endogenously present in the bird. While low-to-intermediate levels of T_3 and T_4 are required for growth, high doses of T_3 and to a lesser extent of T_4 will depress growth (May, 1980).

Thyroid hormones also have various other influences on nitrogen metabolism. For instance, T_4 administration to Japanese quail decreases the plasma protein concentration (Konecka and Majewska, 1981). Antithyroid compounds inhibit the oxidation of trimethylamine in chicks; however, this seems to be because of their direct effect on liver microsomes (Pearson et al., 1981).

Growth Hormone (GH)

On the basis of the mammalian situation, it is likely that growth requires the presence of both GH and thyroid hormones (T_3 and T_4). (See also Chapter 17.) The growth-promoting effects of these are mediated, at least in part, via the somatomedins (insulin-like growth factors IGF-I and IGF-II). It should be noted that IGF-I appears to be synonymous with somatomedin C. The importance of GH in protein growth (accretion) by muscle and bones is supported by the ability of hypophysectomy to depress growth (King, 1969), the reduction in growth rate in chicks following the administration of antisera against chicken GH, and the correlation between plasma GH concentrations and growth rate in birds (reviewed by Scanes and Harvey, 1982). It is likely that GH acts either directly or indirectly to stimulate amino acid uptake and protein synthesis in muscles and to reduce protein catabolism and the release of amino acids by muscle. The latter appears to be the more pronounced or quantitively important effect, as the circulating concentrations of amino acids fall following hypophysectomy in ducks and conversely are elevated by GH replacement therapy (Foltzer and Miahle, 1976).

Androgens

On the basis of mammalian studies, it would be reasonable to assume that androgens exert a general anabolic effect in birds. Androgens stimulate growth of the secondary sexual and accessory organs and secondary sexual characteristics (for further details, see Chapter 19). Investigation of the use of anabolic androgens to promote growth in poultry has met with little success. An important exception to this is the ability of testosterone and anabolic androgens (e.g., trienbolone acetate) to markedly stimulate growth in turkeys (Harvey et al., 1979; Wise and Ranaweera, 1981).

Glucocorticoids

Administration of glucocorticoids strongly suppresses growth. For instance, in chicks receiving daily cortisol injections for 8 days, body weight was 40% that of the control group; this was largely due to decreases in muscle and skeleton (heart weight, 25% of control; pectoralis, 4% of control) (Bellamy and Leonard, 1965). As in mammals, glucocorticoids appear to have a catabolic role in birds. Evidence for this comes from the decrease in nitrogen balance and increased nitrogen excretion and urinary uric acid in glucocorticoid-treated chickens and quail (Adams, 1968; de la Cruz et al., 1981). Corticosterone had no effect on the circulating concentration of amino acids (amino nitrogen) in hypophysectomized ducks (Foltzer and Miahle, 1976). It is possible either that catabolism is already maximal in the hypophysectomized birds or that the pituitary hormones are required for glucocorticoids to exert their effect on protein metabolism.

Estrogens

There is little to suggest that estrogens have a general role in protein metabolism in Aves. However, a specific role for estrogens on at least two major aspects of protein metabolism associated with egg production has been demonstrated. Estrogens act to cause the liver to produce the yolk proteins (and lipids) that are transported to the developing ova (or egg yolk). In addition, estrogens provoke the oviduct to synthesize the egg white proteins. Chapter 18 on female reproduction considers these aspects of egg protein metabolism in more detail.

References

Adams, B.M. (1968). Effect of cortisol on growth and uric acid excretion in the chick. J. Endocrinol., 40, 145.

Allen, N.K., and D.H. Baker. (1972). Effect of excess lysine on the utilization of and requirement for arginine by the chick. Poult. Sci., 51, 902.

Amend, J.F., D.H. Strumeyer, and H. Fisher. (1979). Effect of dietary histidine on tissue concentrations of histidine-containing dipeptides in adult cockerels. J. Nutr., 109, 1779.

Atwal. O.S., L.Z. McFarland, and W.O. Wilson. (1964). Hematology of *Coturnix* from birth to maturity. Poult. Sci., 43, 1392.

Austic, R.E. (1973). Influence of proline deficiency on enzymes of proline metabolism in the chick. Poult. Sci., 52, 801.

Austic, R.E. (1976). Nutritional and metabolic interrelationships of arginine, glutamic acid and proline in the chicken. Fed. Proc. Fed. Am. Soc. Exp. Biol., 35, 1914.

Austic, R.E. (1980). Dietary balance of sodium, potassium, and chloride influences plasma uric acid concentrations in chicks. Proc. Soc. Exp. Biol. Med., 170, 411.

Austic, R.E., and M.C. Nesheim. (1972). Arginine and creatine interrelationships in the chick. Poult. Sci., 51, 1099.

Axelrod, J. (1974). The pineal gland: a neurochemical transducer. Science, 184, 1341.

Balasch, J., L. Palacios, S. Musquera, J. Palomeque, M. Jimenez, and M. Alemany. (1973). Comparative hematological value of several galliformes. Poult. Sci., 52, 1531.

Bell, D.J., and P.D. Sturkie. (1965). Chemical constituents of blood. In "Avian Physiology" (2nd ed.) (P.D. Sturkie, Ed.). Ithaca: Cornell University Press, p. 32.

Bell, D.J., W.M. McIndoe, and D. Gross. (1959). Tissue components of the domestic fowl. 3. The nonprotein nitrogen of plasma and erythrocytes. Biochem. J., 71, 355.

Bellamy, D., and R.A. Leonard. (1965). Effect of cortisol on the growth of chicks. Gen. Comp. Endocrinol., 5, 402.

Bhargava, K.K., R.P. Hanson, and M.L. Sunde. (1971). Effects of threonine on growth and antibody production in chicks infected with Newcastle disease virus. Poult. Sci., 50, 710.

Bird, F.H. (1968). Role of the avian small intestine in amino acid metabolism. Fed. Proc. Fed. Am. Soc. Exp. Biol., 27, 1194.

Boebel, K.P., and D.H. Baker. (1982). Comparative utilization of the α-keto and D- and L-α-hydroxy analogs of leucine, isoleucine and valine by chicks and rats. J. Nutr., 112, 1929.

Brown, G.W., Jr. (1970). Nitrogen metabolism of birds. In "Comparative Biochemistry of Nitrogen Metabolism," Vol. 2 (J.W. Campbell, Ed.). New York: Academic Press, p. 711.

Brush, A.H. (1976). Waterfowl feather proteins: analysis of use in taxonomic studies. J. Zool., 179, 467.

Burrill, P., and J. Lerner. (1972). A distinct component of proline transport in chicken small intestine. Comp. Biochem. Physiol. A, 42, 437.

Calvert, C.C., K.C. Klasing, and R.E. Austic. (1982). Involvement of food intake and amino acid catabolism in the branched-chain amino acid antagonism in chicks. J. Nutr., 112, 627.

Cianciaruso, B., M.R. Jones, and J.D. Kopple. (1981). Histidine, an essential amino acid for adult dogs. J. Nutr., 111, 1074.

Crompton, D.W.T., and M.C. Nesheim. (1969). Amino acid patterns during digestion in the small intestine of ducks. J. Nutr., 99, 43.

Dabrowski, Z. (1966). Electrophoretic studies of blood serum proteins of birds of the crow family (Corvidae). Acta Biol. Cracov. Ser. Zool., 9, 259.

Davis, A.T., and R.E. Austic. (1982). Threonine metabolism of chicks fed threonine-imbalanced diets. J. Nutr., 112, 2177.

de la Cruz, L.F., F.J. Mataix, and M. Illera. (1981). Effects of glucocorticoids on protein metabolism in laying quails (*Coturnix coturnix japonica*). Comp. Biochem. Physiol. A, 70, 649.

Desmarais, M., and J.P. Pare. (1972). Genetic analysis of free plasma amino acids in pure strains of mature chicks. Poult. Sci., 51, 751.

D'Mello, J.P.F., and D. Lewis. (1971). Amino acid interactions in chick nutrition. 4. Growth, food intake and plasma amino acid patterns. Br. Poult. Sci., 12, 345.

Dunkelgod, K.E., and G.E. Winkleman. (1982). Free amino acids in the plasma of poults used as an indicator of possible limiting amino acids in presently available protein sources. Poult. Sci., 61, 1674.

Elkin, R.G. (1984). Measurement of free amino acids in avian blood serum by reverse-phase high-performance liquid chromatography as compared to ion-exchange chromatography. J. Agric. Food Chem., 32, 53.

Elliott, J.W., and J. Bennett. (1971). Genic determination of a protein in the immunoglobulin region in the chicken. Poult. Sci., 50, 1365.

Featherston, W.R. (1969). Nitrogenous metabolites in the plasma of chicks adapted to high protein diets. Poult. Sci., 48, 646.

Featherston, W.R., H.R. Bird, and A.E. Harper. (1962). Ability of the chick to utilize D- and excess L-indispensable amino acid nitrogen in the synthesis of dispensable amino acids. J. Nutr., 78, 95.

Fisher, H., and P. Griminger. (1963). Aging and food restriction: changes in body composition and hydroxyproline content of selected tissues. J. Nutr., 80, 350.

Fisher, H., H.M. Scott, and B.C. Johnson. (1955). Quantitative aspects of the nicotinic acid–tryptophan interrelationship in the chick. Br. J. Nutr., 9, 340.

Fisher, H., D. Johnson, Jr., and G.A. Leveille. (1957). The phenylalanine and tryosine requirement of the growing chick with special reference to the utilization of the D-isomer of phenylalanine. J. Nutr., 62, 349.

Fisher, H., P. Griminger, G.A. Leveille, and R. Shapiro. (1960). Quantitative aspects of lysine deficiency and amino acid imbalance. J. Nutr., 71, 213.

Fisher, H., J. Grun, R. Shapiro, and J. Ashley. (1964). Protein reserves: evidence for their utilization under nutritional and disease stress conditions. J. Nutr., 83, 165.

Fisher, M., S. Leeson, W.D. Morrison, and J.D. Summers. (1981). Feather growth and feather composition of broiler chickens. Can. J. Anim. Sci., 61, 769.

Foltzer, C., and P. Miahle. (1976). Pituitary and adrenal control of pancreatic endocrine function in the duck. Diabete Metab., 2, 101.

Gordon, R.S., and I.W. Sizer. (1965). Conversion of methionine hydroxy analogue to methionine in the chick. Poult. Sci., 44, 673.

Gous, R.M., W.A. Lindner, W.J. Stielau, and I.E. Dreosti. (1977a). Saturation kinetics and competition of amino acid uptake in domestic fowl (*Gallus domesticus*). Comp. Biochem. Physiol., A, 57, 77.

Gous, R.M., W.A. Lindner, W.J. Stielau, and I.E. Dreosti. (1977b). Sodium dependence and counterflow of some amino acids in chick intestine. Poult. Sci., 56, 793.

Graber, G., and D.H. Baker. (1971). Sulfur amino acid nutrition of the growing chick: quantitative aspects concerning the efficacy of dietary methionine, cysteine and cystine. J. Anim. Sci., 33, 1005.

Graber, G., and D.H. Baker. (1973). The essential nature of glycine and proline for growing chickens. Poult. Sci., 52, 892.

Grant, D.L., and P.A. Anastassiadis. (1962). Effects of reproductive stage, sex, and gonadal hormones on hexosamine and protein levels of avian and bovine sera. Can. J. Biochem. Physiol., 40, 639.

Griminger, P., and H. Fisher. (1962). Genetic differences in growth potential on amino acid deficient diets. Proc. Soc. Exp. Biol. Med., 111, 754.

Griminger, P., and H. Fisher. (1968). Methionine excess and chick growth. Poult. Sci., 47, 1271.

Griminger P., and J.L. Gamarsh. (1972). Body composition of pigeons. Poult. Sci., 51, 1464.

Grove, J.A., and H.G. Roghair. (1971). The metabolism of D- and L-lysine in the chicken. Arch. Biochem. Biophys., 144, 230.

Hafez, Y.S.M., E. Chavez, P. Vohra, and F.H. Kratzer. (1978). Methionine toxicity in chicks and poults. Poult. Sci., 57, 699.

Harper, A.E., N.J. Benevega, and R.M. Wohlhueter. (1970). Effects of ingestion of disproportionate amounts of amino acids. Physiol. Rev., 50, 428.

Harrap, B.S., and E.F. Woods. (1967). Species differences in the proteins of feathers. Comp. Biochem. Physiol., 20, 449.

Harris, G.C., and M.J. Sweeney. (1969). Electrophoretic evaluation of blood sera proteins of adult male chickens. Poult. Sci., 48, 1590.

Harter, J.M., and D.H. Baker. (1978). Factors affecting methionine toxicity and its alleviation in the chick. J. Nutr., 108, 1061.

Harvey, S., C.G. Scanes, and P.M.M. Godden. (1979). Plasma growth hormone levels in normal and testosterone implanted growing turkeys. Poult. Sci., 58, 745.

Hathcock, J.N., and M.L. Scott. (1966). Alterations of methionine to cysteine conversion rates and nutritional muscular dystrophy in chicks. Proc. Soc. Exp. Biol. Med., 121, 908.

Hiles, R.A., C.J. Willett, and L.M. Henderson. (1972). Hydroxylysine metabolism in rats, mice, and chickens. J. Nutr., 102, 195.

Hurwitz, S., N. Shamir, and A. Bar. (1972). Protein digestion and absorption in the chick: effect of *Ascaridia galli.* Am. J. Clin. Nutr., 25, 311.

Imondi, A.R., and F.H. Bird. (1966). The turnover of intestinal epithelium in the chick. Poult. Sci., 45, 142.

Ishibashi, T. (1972). Protein metabolism in the fowl. Part IV. Possibility of conversion of tyrosine to phenylalanine in the adult rooster. Agric. Biol. Chem., 36, 596.

Ivanov, N., and C. Ivanov. (1962). Die Eiweissresorption im Hühnerdarm in Abhängigkeit von der Wasserstoffionenkonzentration im Darm und dem Kohlendioxydgehalt des Blutes. Arch. Tierernaeh., 12, 109.

Iversen, L.L. (1967). "The Uptake and Storage of Noradrenaline in Sympathetic Nerves." London: Cambridge University Press.

Jones, M.E. (1970). Vertebrate carbamyl phosphate synthetase I and II. Separation of the arginine-urea and pyrimidine pathways. In "Urea and the Kidney" (B. Schmidt-Nielson, Ed.). Amsterdam: Excerpta Medica.

King, D.B. (1969). Effect of hypophysectomy of young cockerels, with particular reference to body growth, liver weight, and liver glycogen level. Gen. Comp. Endocrinol., 12, 242.

King, D.B., and C. King. (1973). Thyroidal influence on early muscle growth of chickens. Gen. Comp. Endocrinol., 21, 517.

Konecka, A.M., and H. Majewska. (1981). Effect of L-thyroxine on metabolism in Japanese quails (*Coturnix coturnix japonica*). I. Glycolytic enzymes activity in liver after multiple injections of L-thyroxine. Comp. Biochem. Physiol. B: Comp. Biochem., 69, 307.

Kratzer, F.H. (1944). Amino acid absorption and utilization in the chick. J. Biol. Chem., 153, 237.

Krogdahl, A., and B. Dalsgard. (1981). Estimation of nitrogen digestibility in poultry: content and distribution of major urinary nitrogen compounds in excreta. Poult. Sci., 60, 2480.

Langslow, D.R., E.J. Butler, C.N. Hales, and A.W. Pearson. (1970). The response of plasma insulin, glucose and nonesterified fatty acids to various hormones, nutrients and drugs in the domestic fowl. J. Endocrinol., 46, 243.

Laurent, F., and P. Miahle. (1978). Effect of free fatty acids and amino acids on glucagon and insulin secretions in normal and diabetic ducks. Diabetologia, 15, 313.

Laurent, G.J., M.P. Sparrow, and D.J. Millward. (1978a). Turnover of muscle protein in the fowl. Changes in rates of protein synthesis and breakdown during hypertrophy of the anterior and posterior latissimus dorsi muscles. Biochem. J., 176, 407.

Laurent, G.J., M.P. Sparrow, P.C. Bates, and D.J. Millward. (1978b). Turnover of muscle protein in the fowl. Collagen content and turnover in cardiac and skeletal muscles of the adult fowl and the changes during stretch-induced growth. Biochem. J., 176, 419.

Lerner, J. (1971). Intestinal absorption of amino acids *in vitro* with special reference to the chicken: a review of recent findings and methodological approaches in distinguishing transport systems. University of Maine, Orono, Technical Bulletin 50.

Lerner, J., and F.H. Kratzer. (1976). A comparison of intestinal amino acid absorption in various avian and mammalian species. Comp. Biochem Physiol. A, 53, 123.

Lerner, J., and M.W. Taylor. (1967). A common step in the intestinal absorption mechanisms of D- and L-methionine. Biochem. Biophys. Acta, 135, 991.

Leveille, G.A., and H.E. Sauberlich. (1961). Influence of dietary protein level on serum protein components and cholesterol in the growing chick. J. Nutr., 74, 500.

Lynch, J.E., and H.J. Stafseth. (1953). Electrophoretic studies on the serum proteins of turkeys. I. The composition of normal turkey serum. Poult. Sci., 32, 1068.

MacDonald, M.L., and R.W. Swick. (1981). The effect of protein depletion and repletion on muscle-protein turnover in the chick. Biochem. J., 194, 811.

Martinek, R.G. (1970). Review of methods for determining proteins in biologic fluids. J. Am. Med. Technol., 32, 177.

Maruyama, K., M.L. Sunde, and A.E. Harper. (1972). Effect of D-alanine and D-aspartic acid on the chick. J. Nutr., 102, 1441.

Maruyama, K., M.L. Sunde, and R.W. Swick. (1978). Growth and muscle protein turnover in the chick. Biochem. J., 176, 573.

May, J.D. (1980). Effect of dietary thyroid hormone on growth and feed efficiency of broilers. Poult. Sci., 59, 888.

McCollum, E.V. (1957). "A History of Nutrition." Boston: Houghton Mifflin.

McNabb, F.M.A., and T.E. Hughes. (1983). The role of serum binding proteins in determining free thyroid hormone concentrations during development in quail. Endocrinology, 113, 957.

Miles, R.D., and W.R. Featherston. (1974). Uric acid excretion as an indicator of the amino acid requirement of chicks. Proc. Soc. Exp. Biol. Med., 145, 686.

Miles, R.D., and W.R. Featherston. (1976). Uric acid excretion by the chick as an indicator of dietary protein quality. Poult. Sci., 55, 98.

Miller, E.C., J.S. O'Barr, and C.A. Denton. (1960). The metabolism of methionine by single comb White Leghorn and Black Australorp chicks. J. Nutr., 70, 42.

Miller, D.S., D. Houghten, P. Burrill, G.R. Herzberg, and J. Lerner. (1973). Specificity characteristics in the intestinal absorption of model amino acids in domestic fowl. Comp. Biochem. Physiol. A, 44, 17.

Mitchell, H.H., L.E. Card, and T.S. Hamilton. (1931). A technical study of the growth of White Leghorn chickens. Bull.-Agric. Exp. Stn. (Ill.), No. 367.

Morgan, G.W., Jr., and B. Glick. (1972). A quantitative study of serum proteins in bursectomized and irradiated chickens. Poult. Sci., 51, 771.

Munro, H.N. (1964). Historical introduction: the origin and growth of our present concepts of protein metabolism. In "Mammalian Protein Metabolism," Vol. 1 (H.N. Munro and J.B. Allison, Eds.). New York: Academic Press.

Nesheim, M.C., and F.B. Hutt. (1962). Genetic differences among White Leghorn chicks in requirements of arginine. Science, 137, 691.

Nesheim, M.C., and J.D. Garlich. (1963). Studies on ornithine synthesis in relation to benzoic acid excretion in the domestic fowl. J. Nutr., 79, 311.

Nesheim, M.C., R.E. Austic, and S.H. Wang. (1971). Genetic factors in lysine and arginine metabolism of chicks. Fed. Proc. Fed. Am. Soc. Exp. Biol., 30, 121.

Nesheim, M.C., L.O. Crosby, and W.J. Kuenzel. (1974). The oxidation and kidney clearance of D- and L-tryptophan by chickens. Proc. Soc. Exp. Biol. Med., 147, 850.

Ngo, A., C.N. Coon, and G.R. Beecher. (1977). Dietary glycine requirements for growth and cellular development in chicks. J. Nutr., 107, 1800.

Ohno, I., and I. Tasaki. (1972). Effect of dietary lysine level on plasma free amino acids in adult cockerels. J. Nutr., 102, 603.

Packard, G.C., and M.J. Packard. (1983). Patterns of nitrogen excretion by embryonic softshell turtles (*Trionyx spiniferus*) developing in cleidoic eggs. Science, 221, 1049.

Parsons, C.M., and L.M. Potter. (1981). Utilization of amino acid analogues in diets of young turkeys. Br. J. Nutr., 46, 77.

Patterson, R., J.S. Youngner, W.O. Weigle, and F.J. Dixon. (1962). Antibody production and transfer to egg yolk in chickens. J. Immunol., 89, 272.

Pearson, A.W., N.M. Greenwood, E.J. Butler, and G.R. Fenwick. (1981). The inhibition of trimethylamine oxidation in the domestic fowl (*Gallus domesticus*) by antithyroid compounds. Comp. Biochem. Physiol. C: Comp. Pharmacol., 69, 307.

Richards, C.B. (1975). Molecular weight and amino acid composition of fowl delta-globulin. Br. Poult. Sci., 16, 241.

Robbins, K.R., D.H. Baker, and H.W. Norton. (1977). Histidine status in the chick as measured by growth rate, plasma free histidine and breast muscle carnosine. J. Nutr., 107, 2055.

Samsel, J.F., and M. Ledig. (1976). Effets de l'arginine et de l'alanine sur la secretion du glucagon pancreatique et de l'enteroglucagon. J. Physiol. (Paris), 72, 841.

Sasse, C.E., and D.H. Baker. (1972). The phenylalanine and tyrosine requirements and their interrelationship for the young chick. Poult. Sci., 51, 1531.

Saunderson, C.L., and S. Leslie. (1983). N^{τ}-methyl histidine excretion by poultry: not all species excrete N^{τ}-methyl histidine quantitatively. Br. J. Nutr. 50, 691.

Scanes, C.G., and S. Harvey. (1982). Hormones, nutrition, and metabolism in birds. In "Aspects of Avian Endocrinology: Practical and Theoretical Implications" (C.G. Scanes, M.A. Ottinger, A.D. Kenny, O. Balthazart, J. Cronshaw, and I.C. Jones, Eds.). Lubbock: Texas Tech Press, p. 173.

Scanes, C.G., S. Harvey, J.A. Marsh, and D.B. King. (1984). Hormones and growth in poultry. Poult. Sci., 63, 2062.

Schram, A.C., and O.W. McNabb. (1975). Serum proteins of the Golden Pheasant (*Chrysolophus pictus*). Comp. Biochem. Physiol. B: Comp. Biochem., 52, 449.

Schumaker, V.N., and G.H. Adams. (1969). Circulating lipoproteins. Annu. Rev. Biochem., 38, 113.

Shehata, A.T., J. Lerner, and D.S. Miller. (1984). Development of nutrient transport systems in chick jejunum. Am. J. Physiol., 246, G101.

Shen, T.F., H.R. Bird, and M.L. Sunde. (1973a). Conversion of glutamic acid to proline in the chick. Poult. Sci., 52, 676.

Shen, T.F., H.R. Bird, and M.L. Sunde. (1973b). Relationship between ornithine and proline in chick nutrition. Poult. Sci., 52, 1161.

Smith, R.E. (1966). The utilization of L-methionine, DL-methionine and methionine hydroxy analogue by the growing chick. Poult. Sci., 45, 571.

Stucki, W.P., and A.E. Harper. (1961). Importance of dis-

pensable amino acids for normal growth of chicks. J. Nutr., 74, 377.
Sturkie, P.D. (Ed.). (1965). "Avian Physiology" (2nd ed.). Ithaca: Cornell University Press.
Sturkie, P.D., and H.J. Newman. (1951). Plasma proteins of chickens as influenced by time of laying, ovulation, number of blood samples taken and plasma volume. Poult. Sci., 30, 240.
Sturkie, P.D., J.J. Woods, and D. Meyer. (1972). Serotonin levels in blood, heart, and spleen of chickens, ducks, and pigeons. Proc. Soc. Exp. Biol. Med., 139, 364.
Sugahara, M., T. Morimoto, T. Kobayashi, and S. Ariyoshi. (1967). The nutritional value of D-amino acid in the chick nutrition. Agric. Biol. Chem., 31, 77.
Tamir, H., and S. Ratner. (1963). Enzymes of arginine metabolism in chicks. Arch. Biochem. Biophys., 102, 249.
Teekell, R.A., E.N. Knox, and A.B. Watts. (1967). Absorption and protein biosynthesis of threonine in the chick. Poult. Sci., 46, 1185.
Torres-Medina, A., M.B. Rhodes, and H.C. Mussman. (1971). Chicken serum proteins: a comparison of electrophoretic techniques and localization of transferrin. Poult. Sci., 50, 1115.
Tureen, L.L., K. Warecka, and P.A. Young. (1966). Immunophoretic evaluation of blood serum proteins in chickens. I. Changing protein patterns in chickens according to age. Proc. Soc. Exp. Biol. Med., 122, 729.
Wakita, M., S. Hoshino, and K. Morimoto. (1970). Factors affecting the accumulation of amino acid by the chick intestine. Poult. Sci., 49, 1046.
Wang, S.H., and M.C. Nesheim. (1972). Degradation of lysine in chicks. J. Nutr., 102, 583.
Wang, S.H., L.O. Crosby, and M.C. Nesheim. (1973). Effect of dietary excesses of lysine and arginine on the degradation of lysine by chicks. J. Nutr., 103, 384.
Weller, E.M. (1966). Comparative development of pheasant and chick embryo sera. Proc. Soc. Exp. Biol. Med., 122, 264.
Wise, D.R., and K.N.P. Ranaweera. (1981). The effects of trienbolone acetate and other anabolic agents in growing turkeys. Br. Poult. Sci., 22, 93.
Zimmerman, R.A., and H.M. Scott. (1967). Plasma amino acid pattern of chicks in relation to length of feeding period. J. Nutr., 91, 503.

15
Lipid Metabolism

P. Griminger

Introduction

"Lipids" include a variety of materials of differing chemical composition. Triglycerides or "neutral fats" are fatty acid esters of glycerol and serve as a source as well as a store of energy for all higher animals. Phospholipids, which are complex lipids found in all plant and animal tissues, are especially abundant in nervous tissue; they may constitute up to 30% of the dry matter of the brain. On hydrolysis, some phospholipids yield glycerol, fatty acids, phoshoric acid, and a base such as choline (in phosphatidyl choline or "lecithin") or ethanolamine (in phosphatidyl ethanolamine), or the amino acid serine (in phosphatidyl serine); the latter two groups of phospholipids were originally classified as "cephalins." Other phospholipids yield glycerol, fatty acids, phosphoric acid, and the cyclic polyalcohol inositol (phosphatidyl inositol). Sphingomyelins, another important group, consist of a fatty acid, phosphoric acid, choline, and the base sphingosine.

A number of other compounds, extractable in such nonpolar organic solvents as benzene, ether, petroleum ether, or chloroform, are generally included in the "lipid" group. Among these are the fatty acids with medium or long carbon chains, the glycolipids, long-chain alcohols esterified with fatty acids (waxes), steroids, including cholesterol and various hormones, and the fat-soluble vitamins.

Lipid Intake

The only dietary requirements for lipids by birds and other higher animals are for relatively small amounts of those few fatty acids that cannot be synthesized by the organism (essential fatty acids). At the same time, birds are capable of utilizing appreciable quantities of fats. Depending on eating habits, fat intake among nondomestic species varies widely. In carnivorous species and others feeding on seeds that contain high levels of oil, fat may constitute nearly half of the dry matter consumed. At the other extreme, in birds sustaining their needs from low-fat grains or nectar, it may be negligible. Under laboratory conditions, chickens can also adapt to considerable variation in fat intake. Fisher et al. (1961) gave an essentially fat-free ration to female chickens for 19 months; except for initial problems arising from the physical condition of the extremely dry feed, no difficulties were encountered, and eggs laid by these hens hatched normally. Vermeersch and Vanschoubroek (1968), in their excellent review, indicated that levels of fat in the diet as high as 30% had no deleterious effect on growth. Good chick growth was also obtained by Renner (1964) when essentially all the nonprotein calories in a diet were derived from lard, the latter exceeding 35% (by weight) of the diet.

Body Fat

To a large extent, body fat acts as an energy reserve. It is not surprising, therefore, that it is the most variable among the major body constituents. While it varies with species, sex, and age, it is also strongly affected—quantitatively as well as qualitatively—by nutrition. Table 15–1 illustrates the effect of changes in dietary protein level on the fat content of a 7-week-old fast-growing strain of chickens receiving an otherwise well-balanced diet. The influence of age and heredity is shown in F_4 generation measurements of chickens selected for abdominal fat content (Table 15–2). In a study of several strains of 6-month-old domestic geese, carcass fat varied from 33.25 to 35.71%, while that of modern chicken broilers is roughly half this level (Fortin et al., 1983). It has been noted that the capability of birds for storing triglycerides as an energy reserve exceeds that of other classes of vertebrates (Blem, 1976). The fatty acids of these triglycerides are predominantly of the 16- and 18-carbon variety and are generally more unsaturated than those of mammals. Following feeding, lipids are mostly stored in existing adipocyte vacuoles, rather than in newly formed cells.

TABLE 15–1. Fat, protein and water in carcasses of 7-week-old birds fed graded levels of protein[a]

	Carcass composition		
Dietary protein (%)	Fat (%)	Protein (%)	Water (%)
16	18.8	15.3	62.5
20	16.4	16.0	64.4
24	14.6	16.5	65.5
28	12.9	16.1	67.2
32	12.8	16.4	67.4
36	12.4	16.4	67.6

[a] Adapted from Jackson et al. (1982).

TABLE 15–2. Body fat content of male chickens bred for fatness and leanness (F_4 generation)[a]

Age (weeks)	Fat line[b]	Lean line[b]
At hatch	4.39 ± 0.31 (13)	3.87 ± 0.37 (13)
2	6.85 ± 0.56 (11)	6.60 ± 0.34 (11)
4	11.90 ± 0.45[c] (6)	9.60 ± 1.05 (6)
9	13.70 ± 1.10[d] (5)	9.49 ± 0.83 (5)
14	13.40 ± 1.07[c] (6)	9.34 ± 1.24 (5)
17	13.54 ± 2.18 (4)	9.71 ± 0.43 (4)

[a] Adapted from Simon and Leclercq (1982).
[b] Data are percentages of life body weights. Numbers in parentheses are n.
[c] Lines differ significantly at $p < .05$.
[d] Lines differ significantly at $p < .01$.

Even following prolonged starvation, body fat is never completely depleted and is not likely to drop below 4% of total body weight; the need to protect the integrity of tissues and organs, especially those containing phospholipids as a functional necessity, creates a threshold value for lipid content below which life would be endangered. At the opposite end of the scale, nearly half of the body weight of some species of birds just prior to migration may consist of fat, which will serve as a source of energy during long flights (Griminger and Gamarsh, 1972). The importance of depot fat is also demonstrated by male emperor penguins (*Aptenodytes forsteri*), who have been reported to incubate eggs for 6 weeks without interruption and to fast for up to 90 days before the females return to take care of the chicks (Oring, 1982). Macaroni (*Eudyptes chrysolophus*) and rockhopper penguins (*E. chrysocome*) lose essentially all of their depot fat during a 3- to 4-week molting period. Premolting fat levels in these two species were measured as 21 and 26%, respectively. Protein losses, on the other hand, were less than one-fifth by weight and less than one-tenth by caloric value during the same period (Williams et al., 1977). Presumably, low water temperatures require that these penguins finish their molt prior to returning to their aquatic feeding grounds.

Sources of Lipids

A hen fed a commercial laying diet will absorb little more than 3 g of fat per day. Since an average egg yolk contains approximately 6 g of fat, an appreciable part of the yolk lipid of a hen laying an egg almost every day must be synthesized from nonlipid constituents. Similarly, the massive build-up of depot fat prior to migration or other activities requiring a large energy reserve is not necessarily based on the consumption of lipid material. Under these circumstances, carbohydrates and, to a lesser extent, proteins, serve as source material for lipogenesis. Thus, absorption of fats from the intestinal tract and the synthesis of fats from nonlipid compounds become two distinct sources of depot, organ and egg lipids.

Lipid Absorption

Lipase activity in the small intestine and to a lesser extent, in the stomach, hydrolyzes triglycerides to diglycerides, monoglycerides, fatty acids, and glycerol. In mammals, long-chain fatty acids and their glycerol esters are mostly transported via the lymphatic system. Conrad and Scott indicated as early as 1942 that in the laying hen fatty acids may be absorbed by the portal system rather than by way of the lymphatic system. More recent evidence has been recorded by Bensadoun and Rothfeld (1972). These authors also reported that in the presence of a lipoprotein lipase inhibitor, in a conscious but functionally hepatectomized chicken, long-chain fatty acids were absorbed in the form of triglycerides as the major component of a lipoprotein. The authors suggest the term "protomicrons" for the lipoproteins absorbed via the portal system, by analogy to the term "chylomicrons" chosen half of a century ago for the fat-rich particles absorbed by mammals via the lymphatic capillaries.

In the human, dog, and rat, the pancreatic bile ducts enter the proximal part of the duodenum; in the chicken, they enter at the distal end. Lipids passing through the first part of the jejunum of a chicken may, therefore, be hydrolyzed and emulsified for absorption at a more distal location than in the above-mentioned mammals, where the digestive preparation for absorption may have already occurred in the duodenum. Indeed, Renner (1965) found a large part of the fat absorption to have taken place in the third fifth of the small intestine, which would correspond to the distal part of the jejunum, and a lesser amount in the fourth fifth (the proximal ileum). In mammals, the entire jejunum seems to play a prominent role in fat absorption. No general conclusions for the class Aves can be drawn from these observations, however, since there are significant anatomic differences among its species. The domestic pigeon (*Columba livia*), for example, has one bile duct entering the proximal part of the duodenal loop, while a second duct secretes into the distal leg of the loop (Griminger, 1983). Age may also affect the site of absorption, since the hen can absorb fat in more proximate sections than the chick (Whitehead, 1973).

Overall absorption of fatty acids given as triglycerides is higher than when they are given as free fatty acids; absorption was high for linolenic and linoleic acid (85–92%), but only 58–66% for stearic acid (Sklan, 1979). In growing chickens, soybean oil and lard were very well absorbed (96 and 92%, respectively), but tallow only to the extent of 67%. In hens, tallow absorption was much better, except if high levels were given. In general, absorption improves with age (Griminger, 1976). Pullen and Polin (1984) measured an average secretion of 21 g of bile containing 20% dry matter per day in chickens weighing 1850 g. Others have found 8 g of bile acids to be secreted into the duodenum of adult chickens; over 90% of these bile acids are reabsorbed. According to Sklan et al. (1975), most of this reabsorption is taking place in the proximal small intestine. Bile also serves as a vehicle for the excretion of lipids. In this connection it is of interest that the bile lipids of domestic fowl differ significantly from that of any other animal species yet studied. While mammalian bile lipids consist mostly of phospholipids and cholesterol, fowl bile contains significant amounts of cholesteryl esters and triglycerides. Noble and Connor (1984) think that the excretion of these lipids may play a regulatory role.

Sterols. Blood and tissue cholesterol is affected by endogenous cholesterol synthesis as well as by exogenous cholesterol. Chickens will absorb in excess of 50% of low levels of ingested cholesterol, but less when the sterol is given at higher levels (Janacek et al., 1959). Chickens, and the few other species of birds that have been investigated, respond to exogenous cholesterol with higher blood cholesterol levels, while some other vertebrates, such as the rat, appear to be able to balance intake, at least in part, by reduced cholesterol synthesis. The rate of cholesterol absorption, and, consequently, plasma cholesterol levels, can be reduced by various polysaccharides and other materials (Griminger and Fisher, 1966). The binding of bile salts by these compounds appears to play an important role in their antihypercholesterolemic effect.

When plant sterol mixtures are fed to pigeons, campesterol is well absorbed, while β-sitosterol is poorly absorbed (Subbiah et al., 1971). Similar observations have been made with chickens. Plant sterols appear to interact at some level with animal sterols, such as cholesterol, since feeding of plant sterols to laying hens reduces the cholesterol content of egg yolks (Godfrey et al., 1976).

Lipid Synthesis

Fatty Acids. The liver as well as adipose tissue contribute to the synthesis of fatty acids in the animal organism. In the mouse and the rat, adipose tissue is an important location for the synthetic steps. In chickens and pigeons, on the other hand, the liver is the major site of fatty acid synthesis (Leveille, 1969; Goodridge and Ball, 1967). In young chicks killed 15 min after injection of labeled acetate, 47% of the total fatty acid synthesis took place in the liver, 44% in the carcass, 7% in the skin, and 2% in the intestine (Yeh and Leveille, 1973a). When relative weights of tissues

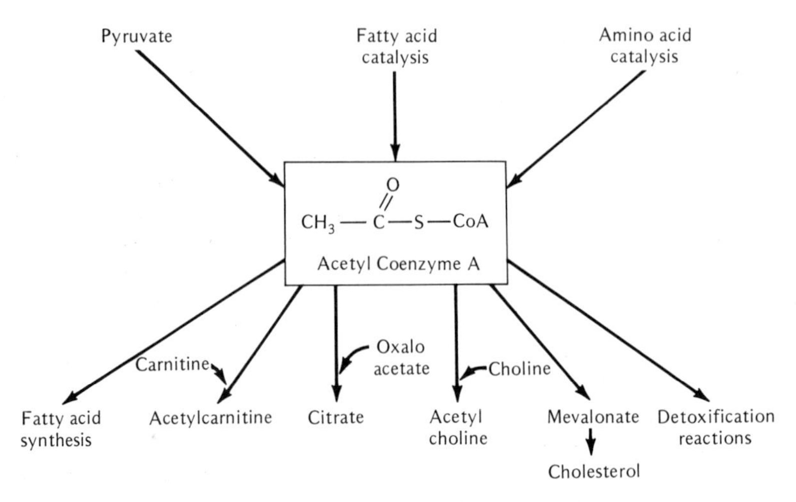

FIGURE 15–1. Metabolic roles of acetyl-CoA.

are taken into account, liver was approximately 20 times as active per unit of weight in fatty acid synthesis as was the carcass. According to Shrago et al. (1971), there is little fatty acid synthesis in the adipose tissue of humans; this increases the usefulness of the avian model for the experimental study of the pathways of fatty acid synthesis and the regulation of lipogenesis in humans.

Recent work also indicated that the avian skeleton was an important site of lipogenesis in the chick. While bone had one-tenth to one-third of the lipogenic activity of the liver, that of bone marrow was approximately two-thirds of hepatic activity (Nir and Lin, 1982).

The two major enzyme systems involved in fatty acid synthesis are acetyl-CoA carboxylase, a biotin-dependent enzyme, and the multienzyme system fatty acid synthetase. The starting material, acetyl-CoA, sometimes also designated "active acetate," may come from the oxidative decarboxylation of pyruvate, an end product of the oxidation of glucose (glycolysis); it can also result from the breakdown of ingested or previously synthesized fatty acids and from the catabolism of certain amino acids. Acetyl-CoA has many other metabolic roles, such as condensation with oxaloacetate to form citric acid in order to serve as a source of energy via the tricarboxylic acid cycle, and the formation of acetylcholine. Some of these roles are indicated in Figure 15–1.

By virtue of the activity of acetyl-CoA carboxylase, acetyl-CoA, in the presence of carbon dioxide (as bicarbonate) and ATP, is converted to malonyl-CoA (Figure 15–2).

The next step involves the combination of malonyl-CoA with the acyl-carrying enzyme to provide acetoacetate, a reaction that requires the decarboxylation of the malonyl component. Following reduction in the presence of NADPH, an acyl–enzyme complex is created (Figure 15–3). This complex can again combine with malonyl-CoA, followed by decarboxylation, until a saturated fatty acid, e.g., palmitate, has been formed (Fig. 15–4).

The system requires the presence of NADPH + H^+, and in the case of yeast also of $FMNH_2$, which is recharged by NADPH + H^+. The avian synthetase does not appear to contain flavin. For a kinetic analysis of the decarboxylation of malonyl coenzyme A and the subsequent condensation reactions, see Srinivasan and Kumar (1981). These authors also showed that high concentrations of substrate (malonyl-CoA) will irreversibly inactivate chicken liver fatty acid synthetase, thus stopping the condensation reaction (Kumar and Srinivasan, 1981).

In addition to the extramitochondrial system for de novo lipogenesis, other pathways for the elongation of existing fatty acids in birds have been observed (Stumpf, 1969). Cell-free extracts from the uropygial gland of the goose also catalyze the carboxylation of propionyl-CoA. A connection appears to exist between this reaction and the finding that 2,4,6,8-tetramethyldecanoic acid is the major fatty acid determined in

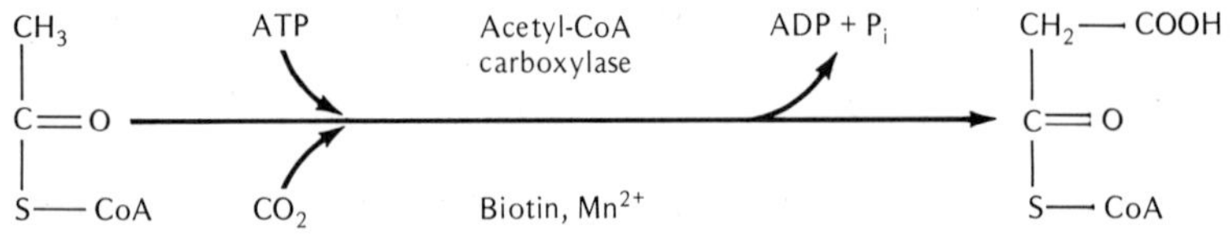

$$CH_3\text{–}C(=O)\text{–}S\text{–}CoA + ATP + CO_2 \xrightarrow[\text{Biotin, }Mn^{2+}]{\text{Acetyl-CoA carboxylase}} HOOC\text{–}CH_2\text{–}C(=O)\text{–}S\text{–}CoA + ADP + P_i$$

FIGURE 15–2. Conversion of acetyl-CoA to malonyl-CoA.

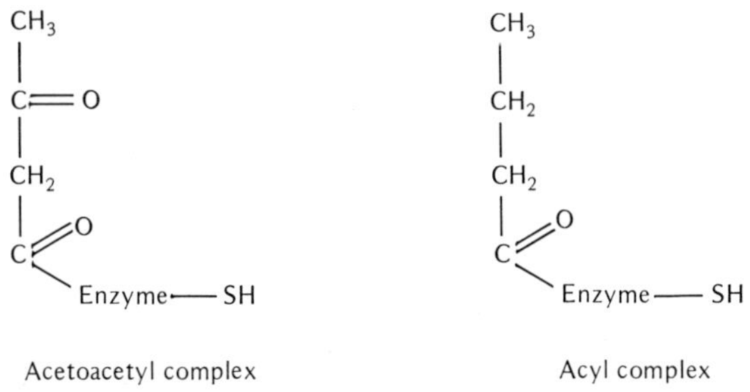

$$\begin{array}{c} CH_3 \\ | \\ C{=}O \\ | \\ CH_2 \\ | \\ C(=O)\text{—Enzyme—SH} \end{array} \qquad \begin{array}{c} CH_3 \\ | \\ CH_2 \\ | \\ CH_2 \\ | \\ C(=O)\text{—Enzyme—SH} \end{array}$$

Acetoacetyl complex — Acyl complex

FIGURE 15–3.

the urophygial gland of the goose (Bloch and Vance, 1977).

With the exception of the mitochondrial fatty acid elongation system, the enzymes mentioned here seem to be affected by nutritional factors. Donaldson (1979) and his associates have shown that in vivo lipogenesis can be affected by biotin status, dietary fat level, degree of unsaturation of dietary fat, starvation, and dietary levels of mercury and selenium. While acetyl-CoA carboxylase is considered the rate-limiting enzyme, fatty acid synthetase may be subject to short-term control, limiting the rate of lipogenesis under conditions of fasting or continued feeding of high-fat diets.

Meal eating, as compared to nibbling, increases lipogenesis. This can be observed in rats fed once daily for 2 hr rather than at times of their choice. This approach will not work in the chicken, which may respond with an enlargement of the crop and thus remain, for all practical purposes, a nibbler. Circumvention by surgical removal of the crop is possible, but young chickens find it difficult to consume sufficient feed for growth and fat deposition at the same rate as control birds fed ad libitum, even if offered several meals a day (Griminger et al., 1969).

Quantitative Effects of Fat Intake. High levels of dietary carbohydrate will increase, and high levels of dietary fat decrease, lipogenesis. For example, the inclusion of 10% corn oil in the ration of young chickens reduced the incorporation of labeled acetate into liver lipids by 38% and reduced significantly the specific activities of citrate cleavage and of "malic" enzymes (Pearce, 1971). In most tests in which the dietary fat content is increased, this is accomplished by replacing dietary carbohydrate. According to Hillard et al. (1980), it is the decrease in carbohydrate rather than the increase in fat which decreases chicken liver lipogenesis. When carbohydrate was held constant, no inhibitory effect of fat was noted. A decrease in lipogenesis in vitro in response to an increase of dietary protein was also noted in the livers of 3-week-old turkey poults, while dietary carbohydrate had the opposite effect (Rosebrough and Steele, 1982).

Qualitative Effects of Fat Intake. Since tissue lipids are derived not only from lipogenesis but also from dietary lipids, the nature of the fat consumed will influence the composition of the lipids deposited in various tissues. The triglycerides in each tissue, as well as the compound lipids, have a specific fatty acid composition. When liponeogenesis is the source of tissue triglycerides, they will correspond to the specific fatty acid composition of the tissue in which they are deposited. When significant amounts of fats are consumed, however, the fatty acids of the ingested lipids will change the fatty acid composition of the tissue lipids to varying degrees. Because of the dynamic nature of adipose tissue and the lipid moieties of other tissues, changes occur continuously so that the fatty acid composition of a tissue may ultimately return to its constant composition. Continuous desaturation and saturation and elongation of fatty acids are therefore taking place.

The effect of dietary fat on body and egg fatty acids was well demonstrated by Feigenbaum and Fisher (1959). While body fat became either more unsaturated or more saturated after the ingestion of the appropriate fats, egg fat was influenced by unsaturated fatty acids only. Edwards and Denman (1975) observed higher levels of linoleic and linolenic acids and lower levels of palmitic and palmitoleic acids in the adipose tissue of pullets of different breeds that were given 10% poultry fat and a high protein level rather than 2% poultry fat and a low protein level. In addition to nutrition, climate also appears to affect carcass fat. When young Japanese quail were acclimated to high environmental temperatures, they had significantly higher levels of myristic, stearic, and palmitic acid, while cold-acclimated birds had higher levels of palmitoleic and oleic acids. These five acids, plus linoleic,

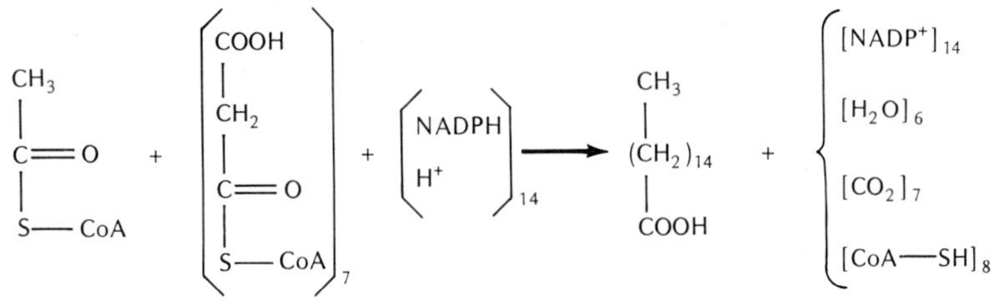

$$\begin{array}{c} CH_3 \\ | \\ C{=}O \\ | \\ S\text{—}CoA \end{array} + \left(\begin{array}{c} COOH \\ | \\ CH_2 \\ | \\ C{=}O \\ | \\ S\text{—}CoA \end{array}\right)_7 + \left(\begin{array}{c} NADPH \\ H^+ \end{array}\right)_{14} \longrightarrow \begin{array}{c} CH_3 \\ | \\ (CH_2)_{14} \\ | \\ COOH \end{array} + \left\{\begin{array}{l} [NADP^+]_{14} \\ [H_2O]_6 \\ [CO_2]_7 \\ [CoA\text{—}SH]_8 \end{array}\right.$$

FIGURE 15–4. Synthesis of palmitic acid.

eicosatrienoic, and arachidonic acid constituted more than 95% of all of the fatty acids in the quail (Durairaj and Martin, 1975).

The fatty acids in 22 species of nocturnal migrating Passeriformes, which flew to their death by hitting the lights of television towers when the skies were overcast, were determined by Caldwell (1973). Eighteen fatty acids were identified, with oleate plus palmitate averaging 57% of all fatty acids. The levels of 20-carbon acids were low, but were high in two specimens of hummingbirds (*Archilochus colubris*) that were analyzed at the same time. Differences between specimens of the same species point toward the importance of the source of dietary lipids for each bird, which may vary according to the birds' "home turf."

Cholesterol. Acetyl-CoA molecules, the building blocks of fatty acid synthesis, are also the original source of all carbon atoms in the in vivo synthesis of cholesterol, which occurs in several stages. The end product of the first stage is mevalonate, a six-carbon compound. When mevalonate is decarboxylated, isoprenoid units are formed. By way of lanosterol and several other sterol intermediates, cholesterol is formed (Figure 15–5).

Location and rate of cholesterol synthesis vary with species, age, and nutrition. Injection of carbon-labeled acetate into chickens helped determine the locus of synthesis. Nearly two-thirds of the synthesis over a 15-min period took place in the liver, one-fourth in the carcass, and about 6% in the intestine and skin (Yeh and Leveille, 1973a). Other loci are the thoracic as well as the abdominal aorta, especially in young birds.

While increased dietary protein decreased fatty acid synthesis in liver and carcass of young chickens, it increased cholesterol synthesis in liver and intestine. Plasma cholesterol was decreased, but carcass cholesterol was increased, pointing to the possibility of a shift in the location of cholesterol. By administering cholesterol labeled with ^{14}C in the C-4 position, Yeh and Leveille (1973b) showed that the hypocholesterolemic effect of a high-protein diet was at least partly mediated by an increase in the rate of cholesterol turnover. This resulted in more speedy removal of cholesterol from the bloodstream and excretion in the form of fecal cholesterol and bile acids. A similar conclusion was drawn by Wagner and Clarkson (1974), who bred two lines of Show Racer pigeons that differed greatly in their response to oral cholesterol. There were no indications of a feedback inhibition of cholesterol synthesis or a shift of cholesterol to other tissues, but greater removal from the cholesterol pool by the hyporesponders was observed. This indicates a genetic control of plasma cholesterol at the level of excretion rather than an adaptation of the rate of synthesis.

While it is clear from these examples that a lowering of blood cholesterol is not necessarily an indication of a lower rate of cholesterol synthesis, a connection was observed in a study by Qureshi et al. (1983a). Inclusion of 0.25 or 0.28% of Wisconsin or Chinese red ginseng root powder, respectively, in the diet of chickens significantly lowered serum cholesterol, β-hydroxy-β-methylglutaryl-CoA (HMG-CoA) reductase,

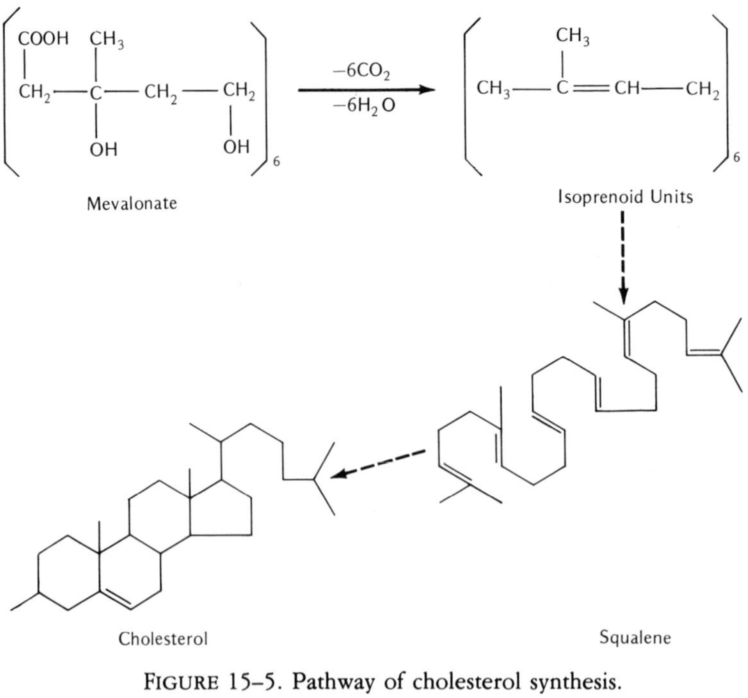

Mevalonate $\xrightarrow[-6H_2O]{-6CO_2}$ Isoprenoid Units → Squalene → Cholesterol

FIGURE 15–5. Pathway of cholesterol synthesis.

and cholesterol-7α-hydroxylase. HMG-CoA reductase is a rate-limiting enzyme in the synthesis of mevalonate and, therefore, of cholesterol. The active ingredients in ginseng root appear to be saponins (ginsenosides). The same authors also obtained a significant reduction of HMG-CoA reductase, cholesterol-7α-hydroxylase, and fatty acid synthetase by feeding growing chickens an essentially cholesterol-free diet supplemented with garlic extract. This treatment also reduced total serum cholesterol, low-density lipoprotein cholesterol, and triglycerides, without at the same time affecting growth (Qureshi et al., 1983b). Other aspects of nutrient and drug effects on cholesterol synthesis and metabolism have been reviewed by Naber (1983).

Essential Fatty Acids

The fatty acids that are necessary for normal physiological functions but cannot be synthesized by the organism are called essential fatty acids (EFA). Linoleic, linolenic, and arachidonic acid are commonly listed under this heading. Linoleic acid can be converted to arachidonic acid by way of

$$CH_3(CH_2)_4CH{=}CHCH_2CH{=}CH(CH_2)_7COOH$$

Linoleic acid (18:2)

$$CH_3CH_2(CH{=}CHCH_2)_3(CH_2)_6COOH$$

Linolenic acid (18:3)

$$CH_3(CH_2)_4(CH{=}CHCH_2)_4(CH_2)_2COOH$$

Arachidonic acid (20:4)

γ-linolenic (Δ-6,9,12) and Δ-5,8,11-eicosatrienoic acid. Therefore, a dietary supply of linoleic acid can satisfy the needs for arachidonic acid. The question of the essentiality of linolenic acid has been somewhat controversial.

Linoleic and arachidonic acid belong to the ω-6 family, since the first unsaturation occurs after C-6, counting from the CH_3 end of the compound. Linolenic is a ω-3 acid, and the nonessential oleic acid (18:1), a ω-9 acid. The intermediary γ-linolenic acid in the conversion of linoleate to arachidonate is also a ω-6 acid, and thus different from the α-linolenic acid (18:3, ω-3), which is not a precursor of arachidonic acid.

The essential fatty acids, especially linoleic acid, appear in various tissues as constituent fatty acids of phospholipids and cholesteryl esters. Their major biologic function, however, appears to be as precursors of prostaglandins and thromboxanes, which are 20-carbon unsaturated acids with a cyclopentane ring in the center of the molecule (Figure 15–6). There are three main series of prostaglandins, PG_1, PG_2, and PG_3. They are synthesized from linoleic, arachidonic, and linolenic acid, respectively, and have, in that order, one, two, or three double bonds in the carbon chain.

Japanese quail on a linoleic acid-deficient diet grown in a germ-free environment only gained a few grams over a 4-week period, while quail on the same ration but exposed to a normal environment grew somewhat better. Quail given linoleic acid gained incomparably better under either of these two environmental conditions (Müller et al., 1976). While this clearly shows the need of linoleic acid for normal development, the authors also contemplated the possibility that microorganisms in the gut may be synthesizing linoleic acid and may thus be contributing to the supply of ω-6 acids of the deficient birds. In laying hens, a linoleic acid deficiency reduces egg size. Balnave (1971a) has shown in studies with ^{14}C that about one-quarter of dietary linoleate was deposited rapidly as ovarian fatty acid. He also demonstrated an increase in liver size and lipid concentration and in the specific activity of hepatic lipogenic and glutamate-metabolizing enzymes in response to an EFA-deficient diet (Balnave, 1975).

As mentioned previously, the EFA status of α-linolenic acid is controversial. While certain fish species require linolenate, a requirement by mammals or birds has not been clearly established; linolenate appears to improve growth in fat-deficient rats, but it fails to cure the characteristic dermatitis and sterility associated with an EFA deficiency. It is clear, on the other hand, the linolenate is a precursor of certain prostaglandins. The subject, including existing information on distribution and metabolism of linolenate in bird tissues, has been exhaustively reviewed by Tinoco (1982).

The arachidonic acid-derived prostaglandins $PGF_{2\beta}$ and PGE_2 cause uterine contractions, resulting in the expulsion of the egg. PGE_2 also facilitates oviposition by relaxing the vaginal musculature (Claeys et al., 1981). The same group demonstrated the conversion of [^{14}C]arachidonate into thromboxanes and prostaglandins by different tissues of the hen's oviductal tract.

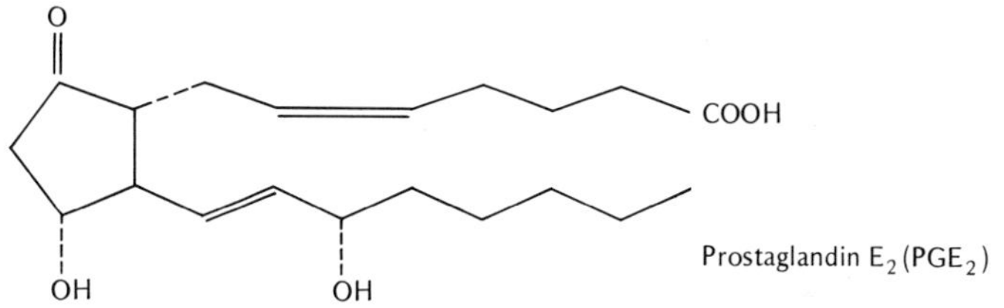

O
COOH
OH
OH
Prostaglandin E_2 (PGE_2)

FIGURE 15–6. Prostaglandin E_2 (PGE_2)

Blood Lipids

The lipids circulating in the blood are derived from intestinal absorption, synthesis, or mobilization from carcass fat. They may be classified into neutral lipids (triglycerides), phospholipids (phosphatides), cholesterol esters, free fatty acids, and lesser quantities of various fat-soluble compounds, such as the fat-soluble vitamins.

Few of these lipids occur in free form. To transport lipids, which are hydrophobic, in an aqueous medium such as blood, the relatively insoluble triglycerides must be associated with the more polar phospholipids, which combine with proteins to form hydrophilic lipoprotein complexes. Since the density of pure lipids is less than that of water, an increase in the lipid-to-protein ratio in lipoproteins decreases the overall density of these proteins. Table 15–3 illustrates the relative lipid composition of typical very-low-, low-, and high-density lipoproteins extracted from hen plasma.

TABLE 15–3. Percentage of lipid classes in lipids extracted from hen plasma lipoproteins

Lipoprotein fraction	VLDL[a]	LDL[a]	HDL[a]
Hydrocarbons	1.1	2.4	0.5
Sterol esters	2.9	9.2	24.9
Sterols	5.4	13.5	3.8
Mono- and diglycerides	1.1	1.5	0.6
Triglycerides	60.2	44.1	21.9
Cephalins	4.4	5.1	5.7
Lecithins	26.3	24.2	43.2

[a] Data are percentages of total lipid; fractions were determined by silicic acid chromatography. VLDL, LDL, HDL: very-low-, low-, and high-density lipoproteins, respectively. (Adapted from Evans et al., 1977.)

Blood lipids of birds are qualitatively and quantitatively similar to those of other vertebrates, except that very high levels are observed in mature females during periods of egg formation. This is caused, in part, by the lipoprotein complexes that transport the lipids synthesized in the liver to the ovary for deposition in the follicles (Christie and Moore, 1972).

As in other classes of animals, the concentration of avian plasma lipids is influenced by such factors as species, age, sex, reproductive stage, state of health and nutrition, and energy demands. Table 15–4 shows the proportions of the major lipid classes in the plasma of laying hens. As the following examples will illustrate, comparisons of different sets of data are difficult unless all the conditions under which they were obtained are known.

TABLE 15–4. Proportions of the major classes of lipids in the plasma of laying hens[a]

Lipid	% of Lipids (weight)	Phospholipids	% of Phospholipids (mole)
Triglycerides	59.7	Phosphatidyl ethanolamine	18.5
Diglycerides	4.3	Phosphatidyl choline	69.6
Free fatty acids	1.8	Phosphatidyl inositol	2.3
Cholesterol esters	2.4	Lysophosphatidyl choline	5.8
Phospholipids	31.8	Sphingomyelin	3.8

[a] Blood samples collected 2–3 hr after feeding (Christie and Moore, 1972).

Reproductive stage: In prebreeding female bobwhites (*Colinus virginianus*), triglycerides averaged 57 ± 19 mg/100 ml serum (mean ± SD); in breeding females, the readings were 524 ± 180 mg/100 ml. In males, on the other hand, triglyceride levels changed from prebreeding to breeding only from 64 ± 33 to 39 ± 24 mg/100 ml (McRae and Dimmick, 1982).

Timing with respect to food consumption: Broiler-type pullets, 10 weeks old and weighing 2200 g, had 1010 mg total lipid/100 ml serum 80 min after a meal containing 50 g dry matter, but only 572 and 578 mg after 2 and 4 hr, respectively (Leclercq et al., 1974). Force-feeding a meal of 130 g corn produced lipid levels that were approximately 30% higher, with most of the increase coming from a significant increase in the very-low-density lipoprotein fraction.

Energy demands: In response to the need for energy, fat is released from adipose tissue in the form of free fatty acids (FFA) and carried as a protein–FFA complex in an unesterified state. While FFA are only a small fraction of total plasma lipids, they are the metabolically most active fraction, and changes in the plasma FFA levels mirror changes in energy demands. Expressed in μeq/liter, the FFA level of adult pigeons held at room temperature was 479, that of pigeons exposed to −25°C for 30 min was 745, and that of partially defeathered pigeons under the same cold stress was 1296 (Parker, 1978). Chickens acclimated to heat (35°C for 4 weeks) had significantly lower plasma FFA than controls (343 versus 463 μeq/liter). When unacclimated chickens were exposed to 35°C for 4 hr, however, their plasma FFA increased significantly. In contrast to the effect of heat acclimation, sudden heat exposure acts as a stressor (Braganza et al., 1973). The same authors also noted a significant increase in plasma FFA in response to glucagon injection. For additional data on plasma lipids of various avian species, consult Bell and Sturkie (1965).

Cholesterol

If it is permissible to draw conclusions from the few avian species investigated to the entire class of Aves, then it becomes clear that serum (or plasma) cholesterol of birds is strongly affected by heredity and nutrition. Age, sex, and environmental conditions may also have an effect.

A group of Japanese quail 5–15 weeks of age maintained a relatively constant serum cholesterol level between 2.5 and 2.9 mg/ml on a control diet. Another group, receiving an "inducer" diet with 1% cholesterol, peaked at nearly 25 mg/ml after 4 weeks and then leveled off at approximately 16–17 mg/ml. Older quail reacted similarly: 22 weeks old at the start and given 1% cholesterol, they had increased their serum cholesterol fivefold at the end of a 12-week period, and returned to normal (2.4–2.9 mg/ml) after another 12 weeks on a cholesterol-free diet (Morrissey and Donaldson, 1977a, b). Chickens given 2% cholesterol from the time of hatching had, at that time, serum cholesterol levels of nearly 6 mg/ml. It decreased to 1 mg/ml at 2 weeks of age, followed by a steady increase over the next 4 weeks, and varied between 9 and 11 mg/ml from 6 to 14 weeks of age (Keely, 1979). The high level after hatching was obviously not related to the diet. It has also been observed in our laboratory, and may result from utilization of the yolk sac by the chick. Normal serum cholesterol levels for chickens from 2 to 14 weeks are 1–2 mg/ml.

The factors that can reduce diet-induced hypercholesterolemia in birds include various complex carbohydrates, Vitamin A, certain plant sterols, and a number of drugs. The interaction between dietary cholesterol, cholestyramine, a bile acid sequestrant, and candicidin, a heptaene macrolide antifungal antibiotic, is illustrated in Table 15–5. The table also shows a greater hypercholesterolemic effect of dietary cholesterol provided by egg powder than when it is provided in crystalline form. Resistance to diet-induced hypercholesterolemia and, subsequently, to atherosclerosis, can also be increased by stimulation of cholesterol metabolism in early life (Subbiah et al., 1983). Apparently, the increased activity of cholesterol 7α hydroxylase, stimulated by oral cholestyramine given to pigeons in early life, was still maintained 2 weeks later when the pigeons were challenged with oral cholesterol.

TABLE 15–5. Interaction of dietary cholesterol with antihypercholesterolemic substances affecting plasma cholesterol in chickens

	Dietary cholesterol (%)			
	0.0	0.25	0.5	1.0
Cholestyramine (%)[a]	Plasma cholesterol (mg/100 ml)			
0.00	122	—	177	178
0.25	116	—	108	151
0.50	93	—	96	94
1.00	72	—	84	97
2.00	76	—	74	74
Candicidin (%)[b]				
0.00	117	171	270	622
0.01	116	145	237	510
0.025	106	136	127	239
0.05	78	86	99	155

[a] White Leghorn chickens, given cholestyramine from 7 to 21 days of age; cholesterol added in crystalline form. (Adapted from Hwang et al., 1975.)
[b] Male white Leghorn chickens, given candicidin at 10 weeks of age for 31 days; cholesterol provided by graded levels of dried whole egg powder. (Adapted from Hausheer and Fisher, 1978.)

It has been estimated that approximately 30% of the variation in plasma cholesterol concentrations in chickens is hereditarily controlled. This is in line with estimates for certain small mammals (Clarkson and McMahan, 1980). The effect of genotype was also demonstrated by Sutton et al. (1983). Dietary cholesterol increased plasma cholesterol more in a strain of chickens selected for high oxygen consumption than in one selected for low oxygen consumption. Although hepatic cholesterol synthesis was significantly decreased in cholesterol-fed birds of both strains, this could not prevent an increase in both plasma and liver cholesterol. While most liver cholesterol in cholesterol-fed birds is found in free form, the esterified form predominates in blood (Teekell et al., 1975).

Endocrine Effects on Lipid Metabolism

As has been discussed previously, much of the lipid in birds, especially in species consuming seeds low in fat, is synthesized in the liver and extrahepatic tissue. Glucose, the end product of carbohydrate digestion, thus becomes a major precursor of lipids. Glucagon and insulin, which play important roles in glycogen mobilization and glucose transport, are therefore principal factors in the regulation of lipogenesis. Glucagon is a potent lipolytic and antilipogenic hormone, whereas insulin stimulates lipogenesis. Both glucagon and insulin have been chemically defined in several avian species (Yanaihara et al., 1983). Another pancreatic hormone, an "avian pancreatic polypeptide," which reduces the concentration of circulating free fatty acids and partially suppresses lipolysis by glucagon, has also been proposed (see also Chapter 13).

Insulin acts as an important antilipolytic hormone by inhibiting release of glycerol and free fatty acids in adipose tissue and by stimulating the conversion of glucose to fat. By stimulating lipoprotein lipase, it increases incorporation of circulating triglycerides into cells. These processes result in the sparing of existing triglycerides and an increased fat deposition. Secondarily, insulin influences triglyceride synthesis by regulating esterification through provision of substrate, α-glycerol phosphate.

In contrast to most mammalian species, where exogenous insulin causes a decrease in plasma free fatty acids, an increase in birds has been observed by several researchers (Nir and Levy, 1973). A single intravenous injection of 5 units/kg of bovine insulin raised plasma free fatty acids twofold in cockerels and threefold in domestic geese in a 3-hr period. In the latter species, plasma triglycerides also increased substantially (see also Chapter 13).

Catecholamines may be of less importance in the mobilization of free fatty acids in birds than in mammals, and glucagon may be the major lipolytic hormone in birds. It certainly displays a very potent action on lipolysis in chickens (LeClercq, 1984).

Braganza et al. (1973), in a study of the influence of heat stress, measured free fatty acids in 15-min intervals after intramuscular injection of glucagon. There was a considerable increase after 15 min and a peak was reached after 30 min. At a level of 100 μg/kg, plasma free fatty acids increased by a factor of more than three in birds kept at 21°C, and by a factor of more than five in birds acclimated to 35°C, both groups averaging more than 1500 μeq of free fatty acids per liter of plasma.

According to Neely and Oram (1973), at least 12 hormones are known to stimulate lipolysis in adipose tissue by affecting triglyceride lipase. These include, in addition to glucagon, the biogenic amines, epinephrine and norepinephrine (Chapters 6, 7, 14, and 22), the polypeptide hormones secretin and ACTH (Chapter 14), and two hormones secreted by the adenohypophysis: the mucoprotein TSH (thyroid-stimulating hormone) and the glycoprotein LH (luteinizing hormone, a pituitary gonadotropin). It is not surprising, therefore, that triglyceride lipase is also referred to as "hormone-sensitive lipase." The action of these hormones is believed to be mediated by 3,′ 5′-adenosine monophosphate (cAMP), and the lipolytic response reaches maximum within a very short time. An apparently different mechanism, involving a lag period of not less than an hour, is exerted by growth hormone and glucocorticoids, which also stimulate lipolysis. Growth hormone (GH) stimulates lipolysis and inhibits lipogenesis. This results in an elevation of plasma fatty acids as well as of glucose. As discussed by Scanes and Harvey (1982), the increase in adipose tissue in hypophysectomized chickens can be explained both by a decrease in lipolysis and an increase in lipogenesis. The administration of GH appears to have the opposite effect: in in vitro tests, GH stimulated lipolysis and inhibited lipogenesis in the tissues of domestic fowl. In vivo, GH administration overcame the depression of the plasma free fatty acid concentration caused by hypophysectomy in the duck.

Hypothyroidism in chickens results in increased fat deposition (Ringer, 1976). Thyroid hormones could, therefore, be expected to diminish fat deposition. However, not all thyroid hormones have the same effect. May (1982) observed that feeding 3,5,3′-triiodothyronine (T_3) and thyroxine (T_4) at 0.25 ppm of the diet over a 54-day period reduced abdominal fat significantly in male birds, but that only T_3 affected female birds in this manner. Likewise, carcass fat of female birds was significantly reduced and water content increased by T_3, but not by T_4. Carcass protein was not affected by either treatment.

Reproductive Hormones

Estrogen enhances lipid metabolism. The increase in blood lipids at the time of egg formation is mostly governed by estrogen (see also Chapter 18), and treatment with estrogen elevates blood lipids considerably

TABLE 15–6. Effects of estrogens on blood lipids

Species and treatment	Total lipids (g/100 ml) Control	Treated	Reference
Duck (Pekin), male; estradiol 1 mg/day	0.45	3.14	Laudauer et al. (1941)
Pigeon (White Carneaux), both sexes, immature; 0.25–0.5 mg estradiol/per day	0.47	1.78	McDonald and Riddle (1945)
Japanese quail, immature, stilbesterol propionate 1 mg/day			
Female	1.97	15.1	Nirmalan and George (1972)
Male	1.80	13.1	Nirmalan and George (1972)
Chickens (Shaver Starcross 288), female; estradiol dipropionate 5 mg/48 hr	0.65	7.53	Balnave (1971b)
Chickens, mature, 3 mg estradiol-17-β-dipropionate/day for 10 days			
Rhode Island Red			
Male	0.6	7.6	Bruce and Anastassiadis (1977)
Female	4.2	10.8	Bruce and Anastassiadis (1977)
Leghorn			
Male	0.5	4.7	Bruce and Anastassiadis (1977)

(Table 15–6). Young turkey hens given two doses of 17-β-estradiol and killed 1 week after the second dose had larger livers and significantly increased lipids per unit of wet or dry liver, although the treatment did not affect overall body weights (Rosebrough et al., 1982). De novo lipogenesis, measured in vitro with l-[^{14}C]sodium acetate as a lipid precursor, was shown to be increased in the treated birds by a factor of 2.5.

The increased blood lipids of the avian female during the reproductive stage appear necessary for the deposition of the relatively large amounts of fat needed for yolk formation (see Chapter 18). When blood lipids in male birds are increased by the injection of estrogens, some of the surplus lipids are stored in the tissues, making the bird fatter and more desirable as a source of human food. To this end, diethylstilbesterol or its dimethyl ether (dianisylhexane) can be injected or implanted; diethylstilbesterol is relatively ineffective when given orally (Sturkie, 1965). When implanted, about 2 mg/week must be absorbed to be effective, and treatment for 4–6 weeks was generally found to be optimum. For this purpose a pellet ranging in size from 15 to 25 mg (depending on the size of the bird) is implanted in the skin high on the neck or head. This treatment increases tenderness and plumpness in chickens and turkeys by virtue of increased subcutaneous fat deposits and favors a smoother and softer skin. At present, estrogen treatment of fowl grown for human consumption is not permitted in the United States and some other countries, since hormonal residues in the neck may have a carcinogenic effect on the consumer.

In contrast to estrogen, testosterone and progesterone do not affect the lipid composition of the liver. While they seem to enhance synthesis of liver lipids, they also stimulate degradation (Griminger, 1976).

Miscellaneous

Brown Fat

Normally, adipose tissue is pale yellow in color; the fat is present in one large vacuole within the distended cell and the nucleus is eccentric. In contrast to this "white" fat, some mammals have, in certain locations, brown adipose tissue cells that contain multiple small fat droplets and a central nucleus. This tissue, characterized by a high rate of substrate oxidation and, consequently, a large capacity to produce and distribute heat (Glick et al., 1981), is especially prominent in hibernating animals and those adapted to a cold environment. Since birds are relatively unstable thermoregulators, Johnston (1971) attempted to find brown adipose tissue in avian tissue. Preparing several hundred sections from 11 species of birds, ranging from *Archilochus colibris* (ruby-throated hummingbird) to *Coragyps atratus* (black vulture), he was unable to detect any brown adipose tissue.

Stomach Oils

Three of the four families of the order Procellariiformes, also called tube-nosed swimmers, may store appreciable quantities of stomach oils in their proventriculus. The composition of these oils varies not only from species to species, but among individual birds. Monoester waxes, triglycerides, and free fatty acids are prominent, but other species of lipids and hydrocarbons (pristane and squalene, especially) can be found. Originally, an endogenous origin of the stomach oils had been widely assumed, with proventricular glands mentioned as the source. However, intraspecies variation strongly hints of an exogenous source. Recent analyses of stomach oils and comparison with the food consumed by certain birds seem to confirm the nutritional origin of the stomach oils. It has been hypothesized that the monoester waxes act as lipase inhibitors, thus favoring the accumulation of stomach oils in the proventricula. Since most of the tube-nosed swimmers predigest the food given to their chicks, stomach oils may play a role in the nutrition of the offspring. The oils are also used in a defensive mode. The birds spit at their enemies, which may spoil the plumage of the affected birds. For more details, the excellent review of Jacob (1982) should be consulted.

Atherosclerosis

This degenerative disease is a major health problem of man; since fewer people fall victim to infectious diseases during early and middle life, atherosclerosis and related cardiovascular diseases have become a foremost cause of disability and death in older people. Adequate animal models for the experimental study of this disease can be found among mammals as well as birds. Several avian species develop atherosclerosis spontaneously and are also suitable for the experimental induction and manipulation of the disease. Those most studied are domestic chickens and turkeys, Japanese quail, and certain strains of pigeons.

Moderate food restriction lessens the degree of atherosclerotic involvement of the abdominal aorta (Griminger et al., 1963), as do various dietary supplements ranging from pectin to antibiotics (Fisher et al., 1964, 1974). More recently, a herpesvirus has been shown to induce atherosclerosis in chickens by causing alterations in the lipid metabolism of arterial smooth muscle cells of chickens. This is of great interest, since herpesviruses cause widespread and persistent infections in humans and other animals (Fabricant et al., 1983). Atherosclerosis in Japanese quail has been reviewed by Shih (1983), who also demonstrated the preventive effect of subcutaneously implanted and slowly released lipoic acid. Some domestic pigeon strains are very susceptible to spontaneous as well as to nutritionally induced atherosclerosis, while others are relatively resistant; those differences enhance their usefulness as experimental models (Griminger, 1983). The pathologic aspect of pigeon atherosclerosis and the interaction with lipid metabolism has been discussed by St. Clair (1983). Although it is not possible in this chapter to do justice to the wealth of experimental work accomplished in this field, the references cited here can serve as a starting point for further readings on this subject.

Fatty Livers

Some domestic birds are able to accumulate large amounts of fat in their livers. This is of value in the creation of fatty goose livers for the preparation of *pate de fois gras,* but is less welcome when it occurs in domestic laying hens. Force-fed geese may increase their liver weight sixfold while increasing their body weight by two-thirds only. About half of the weight of these livers is fat, and although RNA and DNA in the total liver increases, it is decreased per unit of liver weight (Nitsan et al., 1973).

The "fatty liver syndrome" of laying hens, sometimes accompanied by hemorrhages (see also Chapter 5) and/or reduced egg production, is essentially due to a positive energy balance and is usually observed in caged, exercise-deprived birds. Among the complicating factors is the steroid hormone balance of the birds. Harms et al. (1977) induced histologic changes in the liver simulating the syndrome by feeding 0.5% iodine as potassium iodide for 8 days and injecting the hens with 12 mg estradiol 3 days prior to termination of the experiment. The role of the steroid balance is underlined by the observation of Miles and Harms (1981) that elevated serum calcium and phosphorus levels and increased comb size have been observed in natural outbreaks of the syndrome. Akiba et al. (1983) reviewed the efforts made to date to elucidate the fatty liver syndrome and suggested a laboratory model with chicks, involving the injection of estradiol diproprionate, for the assay of nutritional factors affecting lipid accumulation in laying hens.

References

Akiba, Y., L.S. Jensen, and C.X. Mendonca. (1983). Laboratory model with chicks for assay of nutritional factors affecting hepatic lipid accumulation in laying hens. Poult. Sci., 62, 143.

Balnave, D. (1971a). The contribution of absorbed linoleic acid to the metabolism of the mature laying hen. Comp. Biochem. Physiol. A, 40, 1097.

Balnave, D. (1971b). The influence of exogenous estrogens and the attainment of sexual maturity on fatty acid metabolism in the immature pullet. Comp. Biochem. Physiol. B: Comp. Biochem., 40, 189.

Balnave, D. (1975). The influence of essential fatty acids and food restriction on the specific activities of hepatic lipogenic and glutamate-metabolizing enzymes in the laying hen. Br. J. Nutr., 33, 439.

Bell, D.J., and P.D. Sturkie. (1965). Chemical constituents of blood. In "Avian Physiology" (2nd ed.) (P.D. Sturkie, Ed.). Ithaca: Cornell University Press.

Bensadoun, A., and A. Rothfeld. (1972). The form of absorption of lipids in the chicken, *Gallus domesticus.* Proc. Soc. Exp. Biol. Med., 141, 814.

Blem, C.R. (1976). Patterns of lipid storage and utilization in birds. Am. Zool., 16, 671.

Bloch, K., and D. Vance. (1977). Control mechanisms in the synthesis of saturated fatty acids. Annu. Rev. Biochem., 46, 263.

Braganza, A.F., R.A. Peterson, and R.J. Cenedella. (1973). The effects of heat and glucagon on the plasma glucose and free fatty acids of the domestic fowl. Poult. Sci., 52, 58.

Bruce, K.R., and P.A. Anastassiadis. (1977). Connective tissue constituents of the fowl. Effects of exogenous estrogen. Poult. Sci., 56, 1073.

Caldwell, L.D. (1973). Fatty acids of migrating birds. Comp. Biochem. Physiol. B: Comp. Biochem., 44, 493.

Christie, W.W., and J.H. Moore. (1972). The lipid components of the plasma, liver and ovarian follicles in the domestic chicken (*Gallus gallus*). Comp Biochem. Physiol. B: Comp Biochem., 41, 287.

Claeys, M., E. Wechsung, A.G. Herman, and A. Houvenaghel. (1981). Metabolism of arachidonic acid by the reproductive tract of the hen. Prog. Lipid Res., 20, 269.

Clarkson, T.B., and M.R. McMahan. (1980). Individual differences in the response of serum cholesterol to change in diet: animal studies. In "Childhood Prevention of Atherosclerosis and Hypertension" (R.M. Lauer and R.B. Shekelle, Eds.). New York: Raven Press.

Conrad, R.M., and H.M. Scott. (1942). Fat absorption in the laying hen. Poult. Sci., 21, 407.

Donaldson, W.E. (1979). Regulation of fatty acid synthesis. Fed. Proc. Fed. Am. Soc. Exp. Biol., 38, 2617.

Durairaj, G., and E.W. Martin. (1975). The effect of temperature and diet on the fatty acid composition of Japanese quail. Comp. Biochem. Physiol. B: Comp. Biochem., 50, 237.

Edwards, H.M., Jr., and F. Denman. (1975). Carcass composition studies. 2. Influences of breed, sex and diet on gross composition of the carcass and fatty acid composition of the adipose tissue. Poult. Sci., 54, 1230.

Evans, R.J., C.J. Flegal, C.A. Foerder, D.H. Bauer, and M. LaVigne. (1977). The influence of crude cottonseed oil in the feed on the blood and egg yolk lipoproteins of laying hens. Poult. Sci., 56, 468.

Fabricant, C.G., J. Fabricant, C.R. Minick, and M.M. Litrenta. (1983). Herpesvirus-induced atherosclerosis in chickens. Fed. Proc. Fed. Am. Soc. Exp. Biol., 42, 2476.

Feigenbaum, A.S., and H. Fisher. (1959). The influence of dietary fat on the incorporation of fatty acids into body and egg fat of the hen. Arch. Biochem. Biophys., 79, 302.

Fisher, H., A.S. Feigenbaum, and H.S. Weiss. (1961). Requirement of essential fatty acids and avian atherosclerosis. Nature (London), 192, 1310.

Fisher, H., P. Griminger, H.S. Weiss, and W.G. Siller. (1964). Avian atherosclerosis: retardation by pectin. Science, 146, 1063.

Fisher, H., P. Griminger, and W. Siller. (1974). Effect of candicidin on plasma cholesterol and avian atherosclerosis. Proc. Soc. Exp. Biol. Med., 145, 836.

Fortin, A., A.A. Grunder, J.R. Chambers, and R.M.G. Hamilton. (1983). Liver and carcass characteristics of four strains

of male and female geese slaughtered at 173, 180, and 194 days of age. Poult. Sci., 62, 1217.
Glick, Z., R.J. Teague, and G.A. Bray. (1981). Brown adipose tissue: thermic response increased by a single low protein, high carbohydrate meal. Science, 213, 1125.
Godfrey, J.C., J.R. Luttinger, H.D. Taylor, and G.M. Sanhueza. (1976). Dietary plant sterol-induced reduction of egg yolk cholesterol in the chicken. Nutr. Rep. Int., 13, 263.
Goodridge, A.G., and E.G. Ball. (1967). Lipogenesis in the pigeon: in vivo studies. Am. J. Physiol., 213, 245.
Griminger, P. (1976). Lipid metabolism. In "Avian Physiology" (P.D. Sturkie, Ed.). New York: Springer-Verlag.
Griminger, P. (1983). Digestive system and nutrition. In "Physiology and Behaviour of the Pigeon" (M. Abs, Ed.). London: Academic Press.
Griminger, P., and H. Fisher. (1966). Anti-hypercholesterolemic action of scleroglucan and pectin in chickens. Proc. Soc. Exp. Biol. Med., 122, 551.
Griminger, P., and J.L. Gamarsh. (1972). Body composition of pigeons. Poult. Sci., 51, 1464.
Griminger, P., H. Fisher, and H.S. Weiss. (1963). Food restriction and spontaneous avian atherosclerosis. Life Sci., 6, 410.
Griminger, P., V. Villamil, and H. Fisher. (1969). The meal eating response of the chicken—species differences, and the role of partial starvation. J. Nutr., 99, 368.
Harms, R.H., D.A. Roland, Sr., and C.F. Simpson. (1977). Experimentally induced "fatty liver syndrome" condition in laying hens. Poult. Sci., 56, 517.
Hausheer, W.C., and H. Fisher. (1978). The antihypercholesterolemic activity of candicidin as a function of dietary cholesterol in cockerels. J. Nutr., 108, 712.
Hillard, B.L., P. Lundin, and S.D. Clarke. (1980). Essentiality of dietary carbohydrate for maintenance of liver lipogenesis in the chick. J. Nutr., 110, 1533.
Hwang, E.C., P. Griminger, and H. Fisher. (1975). Effect of varying levels of cholestyramine and sephadex on liver vitamin A concentration of chickens fed three levels of cholesterol. Nutr. Rep. Int., 11, 193.
Jackson, S., J.D. Summers, and S. Leeson. (1982). Effect of dietary protein and energy on broiler carcass composition and efficiency of nutrient utilization. Poult. Sci., 61, 2224.
Jacob, J. (1982). Stomach oils. In "Avian Biology," Vol. VI (D.S. Farner, J.R. King, and K.C. Parkes, Eds.). New York: Academic Press.
Janacek, H.M., R. Suzuki, and A.C. Ivy. (1959). Endogenous excretion and capacity to absorb dietary cholesterol in the chicken. Am. J. Physiol., 197, 1341.
Johnston, D.W. (1971). The absence of brown adipose tissue in birds. Comp. Biochem. Physiol. A, 40, 1107.
Keeley, F.W. (1979). The synthesis of soluble and insoluble elastin in chicken aorta as a function of development and age. Effect of a high cholesterol diet. Can. J. Biochem., 57, 1273.
Kumar, S., and K.R. Srinivasan. (1981). Inactivation of chicken liver fatty acid synthetase by malonyl coenzyme A. Effects of acetyl coenzyme A and nicotinamide adenine dinucleotide phosphate. Biochemistry, 20, 3393.
Landauer, W., C.A. Pfeiffer, W.U. Gardner, and J.C. Shaw. (1941). Blood serum and skeletal changes in two breeds of ducks receiving estrogens. Endocrinology, 28, 458.
LeClercq, B. (1984). Adipose tissue metabolism and its control in birds. Poult. Sci. 63, 2044.
LeClercq, B., I. Hassan, and J.C. Blum. (1974). The influence of force-feeding on the transport of plasma lipids in the chicken (*Gallus gallus* L.). Comp. Biochem. Physiol. B: Comp. Biochem., 47, 289.
Leveille, G.A. (1969). *In vitro* hepatic lipogenesis in the hen and chick. Comp. Biochem. Physiol., 28, 431.
May, J.D. (1982). Effect of dietary thyroid hormones on serum hormone concentration, growth, and body composition of chickens. In "Aspects of Avian Endocrinology: Practical and Theoretical Implications" (C.G. Scanes et al., Eds.). Graduate Studies. Lubbock, Texas Tech University.
McRae, W.A., and R.W. Dimmick. (1982). Body fat and blood-serum values of breeding wild bobwhites. J. Wildl. Manage., 46, 268.
McDonald, M.R., and O. Riddle. (1945). The effect of reproduction and estrogen administration on the partition of calcium, phosphorus, and nitrogen in pigeon plasma. J. Biol. Chem., 159, 445.
Miles, R.D., and R.H. Harms. (1981). An observation of abnormally high calcium and phosphorus levels in laying hens with fatty liver syndrome. Poult. Sci., 60, 485.
Morrissey, R.B., and W.E. Donaldson. (1977a). Rapid accumulation of cholesterol in serum, liver and aorta of Japanese quail. Poult. Sci., 56, 2003.
Morrissey, R.B., and W.E. Donaldson. (1977b). Diet Composition and cholesterolemia in Japanese quail. Poult. Sci., 56, 2108.
Muller, M., V. Girard, A. L. Prabucki, and A. Schurch. (1976). Linoleic acid deficiency in young animals. Bibl. Nutr. Dieta, 23, 27.
Naber, E.C. (1983). Nutrient and drug effects on cholesterol metabolism in the laying hen. Fed. Proc. Fed. Am. Soc. Exp. Biol., 42, 2486.
Neely, J.R., and J.F. Oram. (1973). Control of fatty acid metabolism in adipose tissue. In "Best and Taylor's Physiological Basis of Medical Practice" (9th ed.) (J.R. Brobeck, Ed.). Baltimore: Williams and Wilkins.
Nir, I., and V. Levy. (1973). Response of blood plasma glucose, free fatty acids, triglycerides, insulin and food intake to bovine insulin in geese and cockerels. Poult. Sci., 52, 886.
Nir, I., and H. Lin. (1982). The skeleton, an important site of lipogenesis in the chick. Ann. Nutr. Metab., 26, 100.
Nirmalan, G.P., and J.C. George. (1972). The influence of exogenous oestrogens and androgen on respiratory activity and total lipid of the whole blood of the Japanese quail. Comp. Biochem. Physiol. B: Comp. Biochem., 42, 237.
Nitsan, Z., I. Nir, Y. Dror, and I. Bruckental. (1973). The effect of forced feeding and of dietary protein level on enzymes associated with digestion, protein and carbohydrate metabolism in geese. Poult. Sci., 52, 474.
Noble, R.C., and K. Connor. (1984). Unique lipid patterns associated with the bile of the domestic fowl (*Gallus domesticus*). Proc. Nutr. Soc., 43, 52A.
Oring, L.W. (1982). Avian mating systems. In "Avian Biology," Vol. VI (D.S. Farner, J.R. King, and K.C. Parkes, Eds.). New York: Academic Press.
Parker, G.H. (1978). Changes in muscle, liver, and plasma free fatty acid levels in the pigeon on acute exposure to cold. Arch. Int. Physiol. Biochim., 86, 771.
Pearce, J. (1971). An investigation of the effects of dietary lipid on hepatic lipogenesis in the growing chick. Comp. Biochem. Physiol. B: Comp. Biochem., 40, 215.
Pullen, D.L., and D. Polin. (1984). Effect of bile acids and diet composition on lipid absorption in chickens with cannulated bile ducts. Poult. Sci. 63, 2020.
Qureshi, A.A., N. Abuirmeileh, Z.Z. Din, Y. Ahmad, W.C. Burger, and C.E. Elson. (1983a). Suppression of cholesterogenesis and reduction of LDL cholesterol by dietary ginseng and its fractions in chicken liver. Atherosclerosis, 48, 81.
Qureshi, A.A., Z.Z. Din, N. Abuirmeileh, W.C. Burger, Y. Ahmad, and C.E. Elson. (1983b). Suppression of avian hepatic lipid metabolism by solvent extracts of garlic: impact on serum lipids. J. Nutr., 113, 1746.
Renner, R. (1964). Factors affecting the utilization of "car-

bohydrate-free" diets by the chick. I. Level of protein. J. Nutr., 84, 322.
Renner, R. (1965). Site of fat absorption in the chick. Poult. Sci., 44, 861.
Ringer, R.K. (1976). Thyroids. In "Avian Physiology" (P.D. Sturkie, Ed.). New York: Springer-Verlag.
Rosebrough, R.W., J.P. McMurtry, and N.C. Steele. (1982). Effect of estradiol on the lipid metabolism of young turkey hens. Nutr. Rep. Int., 26, 373.
Rosebrough, R.W., and N.C. Steele. (1982). The role of carbohydrate and protein level in the regulation of lipogenesis by the turkey poult. Poult. Sci., 61, 2212.
Scanes, C.G., and S. Harvey. (1982). Hormones, nutrition, and metabolism in birds. In "Aspect of Avian Endocrinology: Practical and Theoretical Implications" (C.G. Scanes et al., Eds.). Graduate Studies. Lubbock: Texas Tech University.
Shih, J.C.H. (1983). Atherosclerosis in Japanese quail and the effect of lipoic acid. Fed. Proc. Fed. Am. Soc. Exp. Biol., 42, 2494.
Shrago, E., J.A. Glennon, and E.S. Gordon. (1971). Comparative aspects of lipogenesis in mammalian tissues. Metabolism, 20, 54.
Simon, J., and B. Leclercq. (1982). Longitudonal study of adiposity in chickens selected for high or low abdominal fat content: further evidence of a glucose-insulin imbalance in the fat line. J. Nutr., 112, 1961.
Sklan, D. (1979). Digestion and absorption of lipids in chicks fed triglycerides or free fatty acids: synthesis of monoglycerides in the intestine. Poult. Sci., 58, 885.
Sklan, D., S. Hurwitz, P. Budowski, and I. Ascarelli. (1975). Fat digestion and absorption in chicks fed raw or heated soybean meal. J. Nutr., 105, 57.
Srinivasan, K.R., and S. Kumar. (1981). Kinetic analysis of the malonyl coenzyme A decarboxylation and the condensation reaction of fatty acid synthesis. Application to the study of malonyl coenzyme A inactivated chicken liver fatty acid synthetase. Biochemistry, 20, 3400.
St. Clair, R.W. (1983). Metabolic changes in the arterial wall associated with atherosclerosis in the pigeon. Fed. Proc. Fed. Am. Soc. Exp. Biol., 42, 2480.
Stumpf, P.K. (1969). Metabolism of fatty acids. Annu. Rev. Biochem., 38, 159.
Sturkie, P.D. (1965). Gonadal hormones. In "Avian Physiology" (2nd ed.) (P.D. Sturkie, Ed.). Ithaca: Cornell University Press.
Subbiah, M.T.R., B.A. Kottke, and I.A. Carlo. (1971). Uptake of campesterol in pigeon intestine. Biochim. Biophys. Acta, 249, 643.
Subbiah, M.T.R., D. Deitemeyer, and R.L. Yunker. (1983). Decreased atherogenic response to dietary cholesterol in pigeons after stimulation of cholesterol catabolism in early life. J. Clin. Invest., 71, 1509.
Sutton, C.D., M.M. Muir, and G.E. Mitchell, Jr. (1983). Effect of dietary cholesterol and genotype on cholesterol metabolism in roosters. Poult. Sci., 62, 1606.
Teekell, R.A., C.P. Breidenstein, and A.B. Watts. (1975). Choleserol metabolism in the chicken. Poult. Sci., 54, 1036.
Tinoco, J. (1982). Dietary requirements and functions of α-linoleic acid in mammals. Prog. Lipid Res., 21, 1.
Vermeersh, G., and F. Vanschoubroek. (1968). The quantification of the effect of increasing levels of various fats on body weight gain, efficiency of food conversion and food intake of growing chicks. Br. Poult. Sci., 9, 13.
Wagner, W.D., and T.B. Clarkson. (1974). Mechanism of the genetic control of plasma cholesterol in selected lines of Show Racer pigeons. Proc. Soc. Exp. Biol. Med., 145, 1050.
Whitehead, C.C. (1973). The site of fat absorption in the hen. Proc. Nutr. Soc., 32, 16A.
Williams, A.J., W.R. Siegfried, A.E. Burger, and A. Berruti. (1977). Body composition and energy metabolism of moulting eudyptid penguins. Comp. Biochem. Physiol. A, 56, 27.
Yanaihara, N., T. Mochizuki, and C. Yanaihara. (1983). Chemistry of avian peptide hormones. In "Avian Endocrinology: Environmental and Ecological Perspectives" (S. Mikami et al., Eds.). Tokyo: Japan Scientific Society Press/ Berlin: Springer-Verlag.
Yeh, S-J.C., and G.A. Leveille. (1973a). Significance of skin as a site of fatty acid and cholesterol synthesis in the chick. Proc. Soc. Exp. Biol. Med., 142, 115.
Yeh, S-J.C., and G.A. Leveille. (1973b). Influence of dietary protein level on plasma cholesterol turnover and fecal steroid excretion in the chick. J. Nutr., 103, 407.

16
Kidneys, Extrarenal Salt Excretion, and Urine

P.D. Sturkie

Introduction

The ions of Na and Cl, the major electrolytes of plasma and extracellular fluids, constitute with K and NH_4 the major fraction of osmolality of ureteral urine; the osmotic and volume regulation (osmoregulation) is defined as the turnover and homeostasis of these ions (Skadhauge, 1981). This chapter is concerned primarily with the involvement of the kidney, urine, and salt glands in osmoregulation and other functions. Also, body fluids are considered in Chapter 5, water consumption in Chapter 11, and water absorption in the gut in Chapter 12; water loss (evaporation) is discussed in Chapter 9. For those readers interested in greater detail in these areas and in osmoregulation in general, consult the reviews of Skadhauge (1981) and Willoughby and Peaker (1979). The reviews of Dantzler and Braun (1980) and Braun (1982) are concerned more specifically with the type of nephrons involved in osmoregulation and with the handling of uric acid, a nonelectrolyte.

A recent publication "Scaling of Osmotic Regulation in Mammals and Birds" (Calder and Braun, 1983) deals with allometric equations that provide a means of summarizing and relating the components of renal structure and function in water balance and osmotic regulation. The authors conclude that "there are now sufficient allometric descriptions of single variables in renal morphology and physiology that general patterns can be seen; patterns that can be rationalized. The accumulation and synthesis of these correlations can contribute significantly to our understanding of renal structure and function."

Anatomy

The urinary organs consist of paired symmetrical kidneys and the ureters, which transport urine to the urodeum of the cloaca. A urinary bladder and a renal pelvis are absent in birds. The kidneys are located in the bony depressions of the fused pelvis. Each kidney is made up of three divisions, often called lobes, designated as cranial, middle, and caudal (Siller and Hindle, 1969). Each division is made up of lobules, composed of a large cortical mass of tissue and a smaller medullary component (Figures 16–1, 16–2, and 16–3). It is more correct to consider the avian lobe as the total complement of lobules drained by a single uretral branch (Siller, 1971).

Nephrons and Glomeruli

The description that follows is from Braun (1982) and Dantzler and Braun (1980), and is based mainly on quail and starling kidneys. The kidney contains cortical and medullary areas, but the line of demarcation between them is indistinct. The cortex contains simple nephrons, without loops of Henle, of the reptilian type (RT) and located on the surface of the kidney, which empty at right angles into collecting ducts. Each tubule is straight and is folded upon itself three or four times. These nephrons are arranged in radial fashion about the central efferent vein to form cylindrical units or lobules (Figures 16–1 and 16–2).

Deeper into the cortex are found nephrons with loops of Henle of the mammalian type (MT). The loops of Henle, which arise from the efferent arterioles of MT glomeruli, and the collecting ducts, which drain both types of nephrons, are bound in parallel arrangement by a connective tissue sheath into a tapering unit, the medullary cone. The nephrons, collecting ducts, and ureters are contiguous. Each cone terminates as a branch of a ureter. The number of medullary cones per kidney is consistent for a given species, but varies considerably among species. The arrangement of tubules in the cone permits the operation of a countercurrent multiplier system (Braun, 1982).

Siller and Hindle (1969) reported a one-to-one relationship between cortical and medullary lobules based on sectioned material of chickens. However, different conclusions were reached by other investigators, including Poulson (1965), and Johnson and Mugas

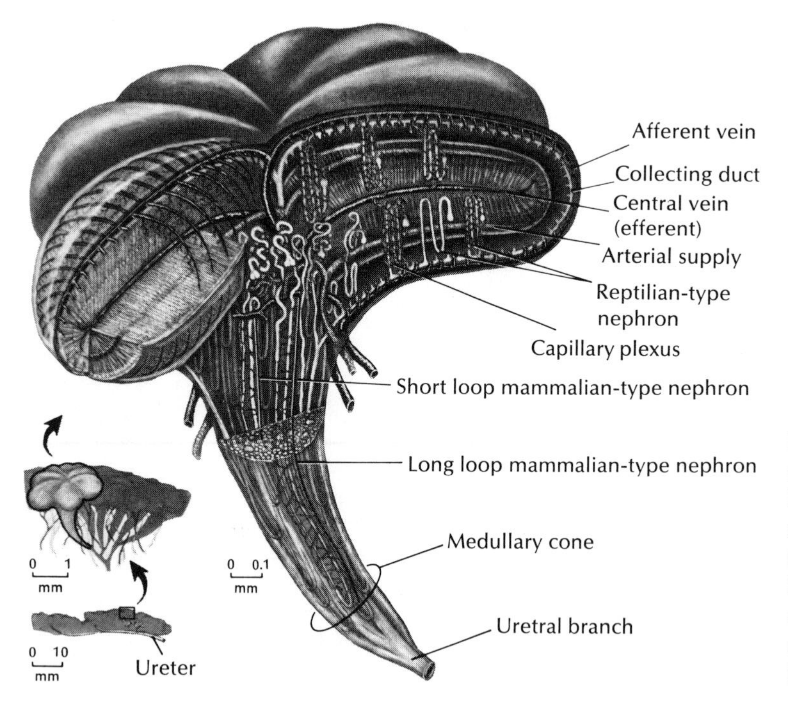

FIGURE 16–1. A three-dimensional drawing of a section of avian kidney, showing the types of nephrons present and their relationship to other renal structures. Interlobular veins are designated as afferent veins, intralobular veins as central vein. (After Braun and Dantzler, 1972.)

FIGURE 16–2. Diagram of the bird kidney showing the arrangement of the glomeruli and tubules in two lobules and blood vessels supplying them. (After Spanner, 1925.)

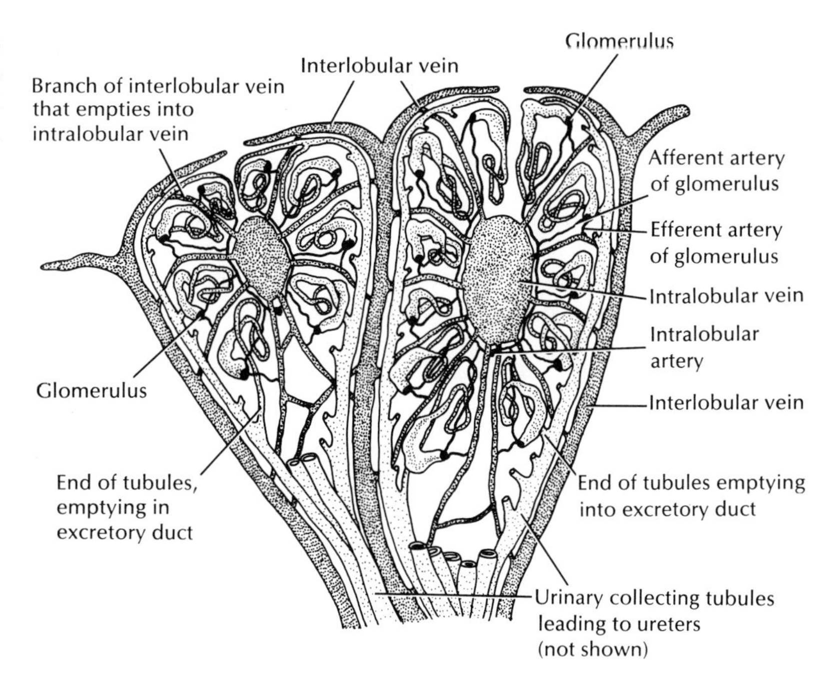

(1970a,b), who proposed that each medullary lobule drains several cortical ones. Their studies, which were based on injected specimens, indicated that a typical cortical lobule is associated with several medullary lobules. There is considerable overlap in the organizational pattern: One medullary lobule may be associated with parts of separate cortical lobules.

Capsule size or diameter of RT nephrons is small (28–35 μm) (for MT nephrons, 90–120 μm) (Braun, unpublished). Birds have a juztaglomerular apparatus, but a rudimentary macula densa (Sokabe and Ogawa, 1974); whether or not there is a feedback mechanism regulating single GFR (glomerular filtration rate) in RT or MT nephrons is not known (Braun, 1982). The arrangement of tubules in the cone permits the operation of a countercurrent multiplier system (Braun, 1982).

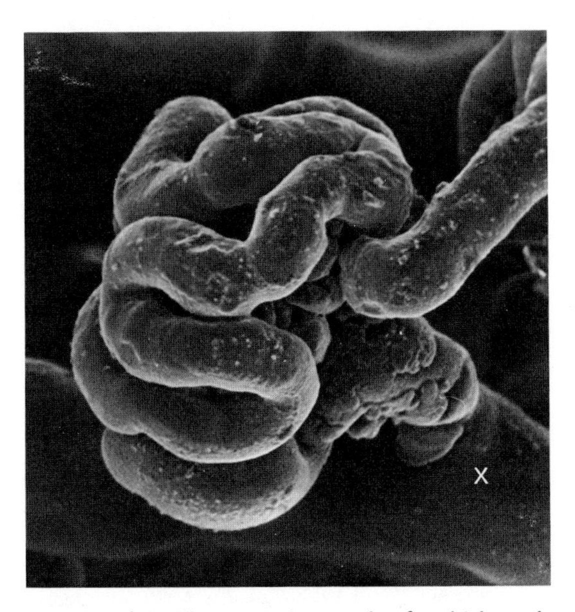

FIGURE 16–3. Electron micrograph of a chicken glomerulus showing afferent arteriole arising from intralobular artery (X) and the arrangement of capillary loops and efferent arteriole. (After Siller, 1971.)

Estimates for numbers of nephrons are as follows (for both kidneys):

	Body weight, g	Number of glomeruli (×1000)	Author
Chicken	2500	840	Marshall (1934)
Duck	3670	1989	Marshall (1934)
Goose	5400	1659	Marshall (1934)
Pigeon	232–420	274–535	Marshall (1934)

The number ranges from 30,000 in small passerines to 200,000 in the chicken, according to Benoit (1950). Dantzler and Braun (1980) estimated that there are about 74,000 nephrons in the starling and 47,000 in the Gambel quail kidneys.

Size of Kidney

The kidneys of birds are relatively larger than those of mammals, ranging from 2 to 2.6% of body weight, depending on the species (Benoit, 1950). Based on a study of 181 species, Johnson (1968) reported that species weighing less than 100 g, those weighing 100–1,000 g, and those weighing 1000 g had kidney weights higher than 1%, exactly 1%, and less than 1%, respectively. The size of certain kidney components has been reported in 17 species of wild birds, as follows, by Johnson and Mugaas (1970b):

Diameter of cortical lobules, 0.36–1.72 mm
Diameter of medullary lobules, 0.12–0.97 mm
Number of collecting ducts entering lobules, 9–85
Diameter of thick limb of tubule, 18–35 μm
Diameter of thin limb of tubule, 11–22 μm

Certain measurements on individual nephrons of *Lophortyx gambel* (Gambel's quail) were made by Braun and Dantzler (1972) as follows:

	Mammalian		
Type	Long loop	Short loop	Reptilian
Length of proximal convoluted tubule (mm)	3.6	3.4	1.6
Length of Henles loop (mm)	2.7	1.5	—
Volume of glomeruli (nl)	0.247	0.237	0.032

Measurements of chicken nephrons (see Sturkie, 1965) reveal dimensions similar to those for quail. The diameters for certain hen nephrons are as follows: glomerulus, 0.086 mm; proximal tubule, 0.062 mm; thick part of loop, 0.0349 mm; thin part of loop, 0.0186 mm. Umflat et al. (1985) recorded the diameter (D) and circumference (C) of glomeruli of female chickens (993 g). The D's ranged from 0.02–0.14 mm. The C's in most glomeruli ranged from 0.15–0.26 mm.

Circulation

The main blood vessels that supply and drain the lobule, and the arrangement of these vessels within the lobules, are shown in Figures 16–3 and 16–4 and also in Figure 6–1 in Chapter 6. There are three pairs of renal arteries; the anterior, arising directly from the aorta, and the middle and posterior ones, arising from the sciatic artery (Siller and Hindle, 1969) or external iliac artery (Chapter 6). The latter two branches supply the middle and posterior divisions (lobes) of the kidney. The femoral artery sends no branches to the kidney. Branches from the renal arteries become the intralobular arteries, and those ramify to form the afferent arteriole to the glomeruli and the efferent artery leaving the glomeruli (Figure 16–3).

Renal Portal. Birds, like reptiles, amphibia, and fish, have a renal portal circulation, but it is only since 1946 that the physiologic significance of this system has become evident. Sperber (1948, 1960) has demonstrated that the renal portal veins carry blood to the kidney tubules and behave as arteries. When a substance normally secreted by tubules is injected into the leg vein (femoral vein or external iliac vein), it is excreted first by the tubules on the injected side before it enters the general circulation. Apparently the tubules do not receive an afferent arterial blood supply directly, but the efferent arterioles from the glomeruli anastomose with the capillary network of afferent veins; the peritubular blood is thus mixed with arterial and venous (renal portal) blood (Sperber, 1960). However, it is not possible from injection preparations to determine where the arteriole ends and the venous network begins (Siller and Hindle, 1969). These authors have reported that arterioles branching from the intralobular arteries may bypass the glomeruli and run directly to the peritubular capillary network. Ultimately, the venous network empties into the central renal vein.

The renal portal circulation enters the peritubular spaces of RT nephrons and spaces about the proximal and distal tubules of MT nephrons, but does not enter the circulation of the medullary cones (Wideman et al., 1981). The efferent arterioles of the RT nephrons course unbranched toward the periphery of the cortical lobules, where they merge with the large sinus-like peritubular vessels of the renal portal system.

Located at the juncture of the renal vein and the iliac vein is a prominent valve (Figure 16–4) that may govern the flow of blood into the renal vein. This valve varies in shape and size among different species (Spanner, 1925, 1939; Burrows et al., 1983). Until recently, little was known about the physiology of this valve, but it was presumed that pressure and flow relations in the renal portal veins might influence its opening and closing, and therefore the relative amounts of afferent venous blood supplying the kidney by way of interlobular veins and extrarenal blood, which bypass the kidney (see also Chapter 6). However, new data on the role of the valve have been presented (Sturkie; see Chapter 6).

Renal Portal Valves. It is known that the renal portal valve can be made to open and close in vitro by cholinergic and adrenergic agents, respectively (Rennick and Gandia, 1954), and that the valve is innervated by sympathetic and parasympathetic nerves (Akester and Mann, 1969) (see also Chapter 6) and Burrows et al (1983). The latter authors also stimulated the nerves to the isolated valve. Their results suggested that the opening and closing of the valves as regulated

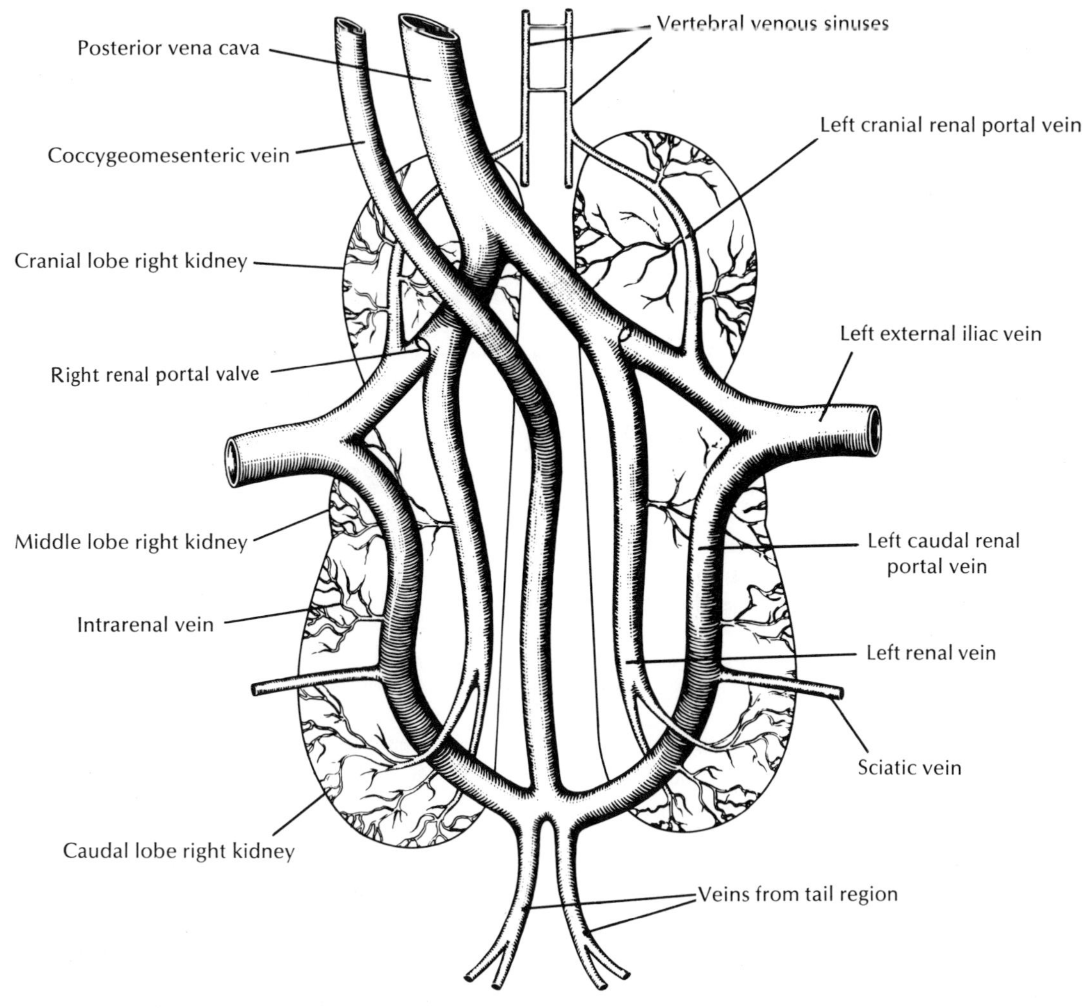

FIGURE 16–4. The principal veins concerned in the renal portal system of the chicken. Cranial and caudal renal portals also referred to as external and internal iliac veins respectively. (After Akester, © 1967, Cambridge University Press.)

by nerves influences bloodflow in the renal portal system. Sturkie and coworkers (Chapter 6) studied the role of cholinergic and adrenergic drugs on the behavior of the valves and subsequent bloodflow in the renal portal system by infusing these agents in physiologic and pharmacologic amounts. Bloodflow was determined by placing bloodflow probes around appropriate blood vessels (Chapter 6). Cholinergic agents, which in vitro closed the valves, had no effect, based on changes in bloodflow. Norepinephrine increased bloodflow slightly, presumably by relaxing and opening the valves more widely or by dilating the renal portal blood vessels. The authors concluded that the valves were open most of the time and that other factors were more important in regulating bloodflow in the renal portal system; these results are supported by the experiments of Odlind (1978) (see Chapter 6).

Kidney Function

The kidney performs three main functions: filtration, excretion or secretion, and absorption. It filters water and some substances normally used by the body from the blood, along with waste products of metabolism, which are voided in the urine. It conserves needed body water, glucose, sodium, and other substances by reabsorption. These processes make the kidney an important homeostatic mechanism whereby the body water and solutes are maintained at fairly constant levels.

Filtration takes place in the glomeruli, where crystalloids and substances with molecules of medium to small size pass through the capillary walls of the glomeruli into the capsule. The plasma proteins, which are composed of large molecules, do not normally pass through the capillary walls and are not filterable. Some of the filterable substances of the blood are sodium, potassium, chloride, inorganic phosphate, glucose, urea, creatinine, and uric acid. These substances have the same concentration in the capsular fluid as in the blood plasma and this is evidence of filtration.

The concentration of some of the substances in the urine may be higher or lower than that in the blood plasma. A lower concentration usually indicates that the substance is being reabsorbed by the kidney tubules. Glucose normally does not appear in the urine, but it is completely filterable and therefore must be reabsorbed. When the kidney tubule suffers impairment, such as follows administration of the glucoside, phlorizin, or when the blood sugar level is inordinately high, not all of the glucose that is filtered is reabsorbed and some appears in the urine.

The kidney tubules also reabsorb water, which aids in maintaining normal blood volume.

The concentration of certain substances in the tubules and urine may be higher than that in the plasma or glomerular filtrate. When the increased concentration cannot be accounted for by reabsorption of water from the tubules, it is evidence of tubule secretion.

In the aglomerular kidney of some marine teleosts, in which there is no filtration, the kidney tubules secrete the urine. Certain metabolites are excreted by the tubules, including uric acid in birds and reptiles and creatinine in fishes, chicken, anthropoid apes, and man. A number of foreign substances, such as phenol red, diodrast, and hippuran, when administered to mammals and birds, have a higher renal clearance than does inulin, indicating tubular secretion. Secretory activity is relatively a more important function in birds than mammals.

Energy is expended by the tubular cells in secretion and reabsorption. Reabsorption occurs even when the substance reabsorbed is many times more concentrated in blood than in the urine and is therefore not accomplished by simple diffusion.

Changes in arterial pressure do not appreciably affect secretion or reabsorption but do affect rate of filtration. The pressure required to drive the fluid through the glomerular blood vessels must be sufficient to overcome the pressure exerted by the capsular membrane and the osmotic pressure of blood colloids. The effective filtration pressure (P_f) therefore equals the pressure in the glomerular blood vessels (P_b) minus the capsular pressure (P_c) and the osmotic pressure of the blood colloids (P_o):

$$P_f = P_b - (P_o + P_c)$$

P_b and P_f can be estimated by determining the pressure required in the kidneys or ureters to stop urine flow. P_b in the dog approximates 60% of mean arterial pressure. Direct determinations of P_b in rat kidney indicate levels (45 mm Hg) considerably lower than indirect levels. The pressure required to stop urine flow in the chicken ureter may be as high as 32 mm Hg or as low as 7.5–15 mm Hg, according to Gibbs (1929a). Other factors affecting filtration are considered below.

Renal Clearance

Renal clearance may be defined as the volume of blood that can be cleared of a given substance by excretion of urine for 1 min or the minimum volume of blood required to furnish the quantity of substances excreted in the urine in 1 min (Smith, 1951). The volume of blood cleared of a particular substance by the kidney in 1 min may be expressed as follows:

U_x = concentration of x in each milliliter of urine
V = rate of urine formation (ml/min)
P_x = concentration of x in each milliliter of plasma
U_xV = rate of excretion of x (mg/min)

Then U_xV/P_x is the volume of plasma required to supply the quantity of x excreted in each minute's time. Therefore UV/P equals the clearance of a given substance x in the plasma.

The substance to be tested is usually infused at a constant rate, and urine samples are collected at regular and frequent intervals (usually at intervals of 2–7 min in the chicken). Blood samples are usually taken near the middle of the collection periods. The plasma concentrations are plotted against time and the exact values at the middle of each urine collection period are determined by interpolation. The technique for determining GFR in single nephrons is discussed later.

Clearance therefore involves filtration (C_F), secretion (T_S), and reabsorption (T_R):

$$T_R = C_F P_x - U_x V$$

The point at which rate of reabsorption becomes constant and does not increase with increased plasma concentration (saturated) is known as the transport maximum (T_m). Tubular secretion is

$$T_S = U_x V - C_F P_x$$

Filtration

Inulin Clearance and Glomerular Filtration Rate (GFR). A substance suitable for measuring glomerular filtration must be completely filterable and physiologically inert and must not be reabsorbed, excreted, or synthesized by the kidney tubules. Such a substance is inulin, a starchlike polymer containing 32 hexose molecules with a molecular weight of 5200. Inulin is neither reabsorbed nor excreted by the kidney tubules of mammals, amphibians, reptiles, or birds, and varies directly with the concentration of inulin in the plasma. Because UV/P is constant and independent of plasma concentration, the clearance of inulin is a measure of glomerular filtration.

The ratio of the inulin clearance to the simultaneous clearance of other substances, such as urea, uric acid, creatinine, and phenol red, indicates whether the substance in question is reabsorbed or secreted by the kidney tubules. For example, a ratio of less than 1 indicates reabsorption and a ratio greater than 1 indicates excretion or secretion by the tubules. Figures on filtration (inulin clearance) and on a number of avian species are shown in Table 16–1.

Glomerular Filtration Rate. Glomerular filtration rate (GFR) is more variable in birds than mammals, probably because there is intermittent filtration of glomeruli in birds but not in mammals. Comparisons of GFR of various avian species (Table 16–1) reveal no consistent figures. Results from birds with salt glands, such as the duck, goose, and gull, are not greatly different from those without such glands. Data on chickens from earlier workers such as Korr, Pitts and Korr, and Shannon (see Sturkie, 1976, for references) were somewhat lower than those of more recent investigators. Differences in technique, handling, sampling, etc., may account for these discrepancies.

The state of hydration, dehydration, and salt loading do influence GFR. Most studies indicate that dehydration decreases GFR, but there are some exceptions, notably those on the goose and glaucous-winged gull reported by Robinson-Hughes (1980); likewise salt loading decreased filtration, depending on the degree of loading (see Table 16–1).

Arginine vasotocin (AVT) decreased GFR in chickens (Ames et al., 1971) and in Gambel quails (Braun and Dantzler, 1974). The release of AVT appeared to be related to increases in plasma osmolality caused by salt loading (Koike et al., 1979) or dehydration (Koike et al., 1977, 1983). Exogenous AVT apparently decreased GFR by constriction of the afferent arteriole of the RT nephron (Braun, 1976) and by increasing

TABLE 16–1. Glomerular filtration (inulin clearance) in Aves[a]

Species	Condition	Mean GFR (ml/kg/min)	Reference
Chicken	Control	1.23	Dantzler, 1966
	Salt loaded	0.35	Dantzler, 1966
	Water loaded	3.18	Dantzler (1966)
Chicken	Dehydrated	1.73	Skadhauge and Schmidt–Nielson (1967a)
	Hydrated	2.12	Skadhauge and Schmidt–Nielsen (1967a)
	Salt loaded (mild)	2.06	Skadhauge and Schmidt–Nielsen (1967a)
Chicken	Fasted, mild hydration	1.81	Berger et al. (1960)
Chicken	Laying, not fasted	3.0	Sperber (1960)
Chicken	AVT Antidiuresis	2.71	Ames et al. (1971)
Chicken		3.03	Svendsen and Skadhauge (1976)
Duck (domestic)	Seawater	2.10	Holmes et al. (1968)
Duck (domestic)	Freshwater	2.5	Holmes et al. (1968)
Duck (Mallard)	Freshwater	3.3	Robinson-Hughes (1980)[b]
Duck (Mallard)	Seawater	2.2	Robinson-Hughes (1980)
Sea Gull	Freshwater	4.6	Douglas (1966)
Sea Gull	Salt loaded	4.2	Douglas (1966)
Glaucous winged Gull	Freshwater	1.9	Robinson-Hughes (1980)
Glaucous winged Gull	Seawater	4.0	Robinson-Hughes (1980)
Canadian Goose	Freshwater	1.2	Robinson-Hughes (1980)
Canadian Goose	Seawater	2.2	Robinson-Hughes (1980)
Quail (Gambels)	Control	0.88	Braun and Dantzler (1972)
	Salt loaded	0.015	Braun and Dantzler (1972)
	Water loaded	1.39	Braun and Dantzler (1975)
Starling	Control	2.83	Braun and Dantzler (1972)
	Salt loaded	2.81	Braun and Dantzler (1972)
Doves	Mild hydration	2.6	Shoemaker (1967)
Budgerigar	Hydrated	4.4	Krag and Skadhauge (1972)
	Dehydrated	3.2	Krag and Skadhauge (1972)

[a] See Sturkie (1976) for earlier figures and references.
[b] [^{14}C]polyethylene glycol injected intramuscularly.

the permeability of the collecting ducts of MT nephrons (Skadhaüge, 1981).

Infusion of hypertonic saline into the renal portal system produced local effects and caused a prolonged antidiuresis and presumed release of AVT, which was not measured, and a decreased output of plasma renin activity (Bailey and Nishimura, 1984).

Filtration in Single Nephrons. The study of filtration in single nephrons of RT and MT types was made possible by the technique of Hanssen, which involves measurement of GFR with a nonabsorbable market, Na ferrocyanide, labeled with an isotope (Braun and Dantzler, 1972). An arterial injection of unlabeled filtration marker is administered as a bolus, and after a few seconds, when the bolus is in the proximal convoluted tubule, the blood supply and flow are stopped by rapid freezing of the kidney. The tissue is then thawed and individual nephrons are isolated and viewed under a stereomicroscope. The radioactive material in the glomerulus is collected and determined, and the filtration is calculated from the concentration in the glomerulus and the plasma. This technique has been checked by studies on micropuncture of surface RT nephrons in the starling kidney (Laverty and Dantzler, 1982). The results obtained are in the same range as those determined by the ferrocyanide technique.

Heavy salt loading decreased filtration in MT nephrons slightly and completely inhibited filtration in RT nephrons (see Table 16–2). The decreased GFR following osmotic stress and dehydration is related to the decrease in the number of nephrons filtering rather than rate of filtering of individual nephrons (Johnson, 1974; Johnson and Skadhäuge, 1975; Braun and Dantzler, 1972; Braun, 1975, 1982). Large doses of arginine vasotocin (AVT) likewise inhibited filtration in single RT nephrons, and apparently increased filtering in some MT nephrons. However, under these conditions whole body filtration would have been significantly decreased, because RT nephrons in the Gambel's quail comprise about 90% of total nephrons (Braun and Dantzler, 1972).

TABLE 16–2. Glomerular filtration in single nephrons[a]

Condition	MT[b]	RT[b]	RT nephrons filtering (%)
2.5% Mannitol diuresis	14.6	6.4	71
Salt loading	12.7	0.0	0
AVT (ng/kg)[b]			
10	11.3	4.7	52
50	16.5	6.9	26
200	24.7	0.0	0
Neurohypophysectomy	11.3	4.6	100

[a] Data are in nl/min. After Braun and Dantzler (1972, 1974) for AVT and Braun (1975) for neurohypophysectomy. (Quoted by Braun, 1982.)

[b] Abbreviations: MT, mammalian-type nephrons; RT, reptilian-type nephrons; AVT, arginine vasotocin.

When the posterior lobe of the pituitary, which produces AVT, was removed (see Table 16–2) all RT nephrons became active and filtered. Further details on the concentrating ability of nephrons will be considered later.

Apparent Tubular Excretion Fraction (ATEF). Sperber (1948, 1960) demonstrated that the renal portal vein carries blood to the avian renal tubules. When a substance was injected into a leg vein or iliac vein, the substance was first excreted by the tubules on the injected or ipsilateral side before it got into the general circulation. This was determined by cannulating each ureter and collecting urine samples for determining the concentration of injected material. This technique has been used to demonstrate the tubular handling of various substances (x) and for determining the apparent tubular excretion fraction (ATEF), which represents the difference between the excretion rate (E_x) (clearance) of the ipsilateral (I, injected) and contralateral (C) kidney. Thus,

$$\text{ATEF} = \frac{E_x - E_{xc}\ (100)}{\text{Rate of infusion}}$$

The ATEF varies for different compounds at different concentrations. The ATEF for histamine, for example, ranges from 46 to 62%; for phenol red, it is a bit lower. Figures for some other substances are choline (23%), PAH (65°b), and *d,l* epinephrine (38%).

Renal Clearance of Uric Acid. *Secretion.* Uric acid is the end product of protein metabolism in Aves, whereas urea is the end product in mammals. Clearance of uric acid has been studied in whole kidney by a number of workers and in single nephrons by Laverty and Dantzler (1983). The preponderant evidence indicates that uric acid is filtered and excreted by the tubules into the urine, with uric acid/inulin ratios of 6–15, depending on varying conditions and species. Thus, uric acid is highly concentrated in the urine, and constitutes 52–88% of total nitrogen of the urine of chickens and ducks (Skadhäuge, 1981). Earlier studies by Shannon (1938b) indicated that 83–90% of the chicken's uric acid is secreted by the tubules. The uric acid/inulin clearance ratios at moderate plasma uric acid concentrations ranged from 7.5 to 15.8. As plasma levels of uric acid increase, the ratio is depressed such that at high levels (100 mg%) the U/I ratios range from 1.8 to 3.4. These figures are similar to those for phenol red. As the plasma level rises, the amount filtered is increased and the amount secreted is decreased.

Not all of uric acid is in solution. The ratio of uric acid in solution to that in colloidal state to that as a precipate is 1:2:12 (Skadhäuge, 1981). The transport maximum T_m for uric acid (in chickens) is about 5 mg/min (Sykes, 1960a). Absolute clearances of uric acid in chickens average about 30 ml/kg/min, and are of the same order as reported by Sykes (1960a) and by Nechay and Nechay (1959). For details on clearance methods, see Sturkie (1976).

Uric acid clearance in doves is about 25 ml/kg/min and of the same magnitude in ducks (Stewart et al., 1969). These authors reported a U/I ratio of 6.6 in birds administered both freshwater and hypertonic (saline) water.

The excretion of uric acid depends on tubular secretion and is largely independent of tubular water resorption; ureteral concentration is very high on a low urine flow rate.

Urates have a low aqueous solubility and do not contribute to osmotic activity of urine as urea does in mammals; and excretion of uric acid permits the excretion of inorganic cations in excess of amounts expected by the osmolality of urine (Dantzler and Braun, 1980), and may actually decrease osmotic activity of urine (McNabb, 1974). The extent to which cations are trapped with uric acid precipitates depends on the content of the diet particularly protein and salts, and degree of hydration, and ranges from 3–75% for Na to 8–84% for K and less for Ca and Mg (quoted by Skadhauge, 1981).

The principal site of secretion of uric acid, based on micropuncture studies of RT tubules in starlings is the proximal part of tubule but some secretion may occur more distally (Laverty and Dantzler, 1983), and possibly in MT nephrons, although this has not been determined.

Pertinent data on the handling of uric acid by RT nephrons are as follows (Laverty and Dantzler, 1983):

GFR (ml/kg/min	6.74
U/P	81.8
C_{UA} ml/kg/min)	27.3
F/Fr	4.7

Starlings excrete nearly five times (4.7) the amount of uric acid filtered. Since the starlings were fed ad libitum, it is likely that synthesis may have accounted for a significant portion of the apparent secretion, as high as 29 percent (Skadhauge, 1981).

Impaired clearance of uric acid of hereditary origin in chickens was reported by Austic and Cole (1972). The abnormal strain had a blood uric acid level of 15 mg/100 ml, compared to 8 mg in a normal strain on a standard diet. As the protein level in the diet increased, plasma levels of uric acid increased to 55 mg in the abnormal strain and much less in the normal strain. Clearance of uric acid was markedly lower in the abnormal strain, and tubular secretion was 40% lower than normal.

Employing the micropuncture technique, Laverty and Dantzler (1983) studied the transport of a number of substances in RT tubules of the starling. Sodium and chloride and water were transported and at equivalent rates, with about 24% of the filtered load absorbed at point of tubule puncture. Calcium was absorbed continuously along the tubules in excess of sodium and water. Magnesium appeared to undergo net secretion early and net absorption late in the tubules. Potassium transport was highly variable, exhibiting net absorption or no net transport.

Individual tubules exhibited either net absorption or net secretion of phosphorus.

Phosphate Transport. Earlier workers reported that phosphate is mainly reabsorbed in intact chickens with a P/I ratio of 0.1. After the administration of parathyroid hormone, the P/I was 3.00, indicating a high rate of secretion and possible inhibition of reabsorption (Levinsky and Davidson, 1957; Ferguson and Wolbach, 1967).

Wideman et al. (1980) made an extensive and important study of the renal handling of phosphate in the intact starlings, those parathyroidectomized, and those parathyroidectomized (PTX) and those receiving parathyroid hormone (PTH). Briefly, their results showed that in intact starlings there was no correlation between filtered phosphate load and the amount secreted. At low filtered loads secretion predominated, but at higher loads (20–35 μmol/kg), net reabsorption prevailed. PTX birds reabsorb phosphate over a wide range of filtered loads. In phosphate-loaded PTX birds, infused with PTE, secretion was maximal and reabsorption minimal at all levels of phosphate infusion and parathyroid hormone. Their data also show that both the secretion and reabsorption are limited by maximal tubular transport capacities. PTH is necessary for the appearance of net phosphate secretion during phosphate loading, and secretion is not an inherent tubular response to hyperphosphatemia (Wideman et al., 1980). While PTH inhibits phosphate reabsorption, as in mammals, the factors regulating the release of PTH are discussed in Chapter 21; one of these is plasma calcium level.

The renal clearance of calcium in the laying hen was decreased slightly during the period of shell deposition onto the egg, and there was a significant increase in the clearance of phosphorus (Prashad and Edwards, 1973).

Potassium. Potassium is an inorganic constituent that is filtered, reabsorbed, and secreted. Its level in the plasma is low and fairly constant (4–5 meq/liter); it is higher in urine (about 5–25 meq/liter), depending on conditions of hydration and loading. Potassium is secreted in the fowl, but at a relatively low rate (Orloff and Davidson, 1956). The maximal rate of transport for one kidney (the one perfused) ranged from 60 to 85 μmol/min for a 2-kg chicken. Reducing P_{CO_2} locally increases K excretion, and increasing P_{CO_2} decreases it. A mercurial diuretic inhibited the excretion of K, and its virtual disappearance from the urine may indicate that the filtered K is entirely reabsorbed. Dehydration and salt loading increase its level in the urine.

Hydrochlorothiazide belongs to the sulphonamyl group of compounds and has been reported to have a diuretic action in turkeys. Sykes (1961) confirmed this and also reported that it increased the excretion of sodium and potassium in the urine.

Creatinine and Creatine Clearance. Creatinine is a normal constituent of the urine in mammals, but in birds the amount formed is negligible in relation to the amount of creatine. Clearance studies on birds, man, apes, and certain fishes indicate that the tubules of these species secrete creatinine. The creatinine/inulin ratio in the chicken at plasma concentrations of 9–12 mg% average 1.54 (Shannon, 1938a). At higher concentrations the ratio is depressed, and at concentrations of 200–230 mg%, the ratios are 1.09 and 1.07, respectively. Sykes (1960b) reported that the absolute clearance of exogenous creatinine in hens is 3.90 ml/min at plasma concentrations ranging from 3.2 to 57.0 mg%, with a creatinine/inulin ratio of 1.46. The clearance of endogenous creatinine averaged 2.25 ml/min, with a creatinine/inulin ratio of 0.78. At normal or endogenous plasma levels (0.2–0.5 mg%), therefore, creatinine is not secreted but is reabsorbed. The ability of the kidney to clear the plasma of creatinine is therefore considerably lower than it is for uric acid, PAH, Phenol Red, and other substances.

The clearance of exogenous creatinine in hens averaged 6.70 ml/min, with a creatinine/inulin ratio of 2.2, at plasma concentrations of 2.8–13.3% (Sykes, 1960b). The mean clearance of endogenous creatine in hens is 3.90 ml, with a creatine/inulin ratio of 1.41.

The tubular secretion of creatinine can be inhibited by substances that inhibit the transport of either organic acids or bases, such as probenecid and priscoline, respectively, indicating transport by two systems (Rennick, 1967). Water diuresis increases creatinine clearance and urine flow (Dicker and Haslam, 1966).

Phenol Red Clearance. The dye phenol red is filtered and is secreted by the tubules of mammalian and avian species. Some of the dye is bound by the plasma proteins, and only the free dye is filterable. The amounts bound and free depend on the concentration of the dye in the plasma. At very low concentrations in chicken plasma (1 mg%), only 15–20% of the dye is free or filterable; at concentrations of 15 mg%, 60% is filterable, however, and at higher concentrations more is filterable because more is free (Pitts, 1938).

The ratio of free phenol red in the plasma to that in the urine is not constant and decreases after the concentration in the plasma has reached a certain point. At low plasma levels of the dye (1–3 mg%) the clearance ratio of phenol red to inulin ranges from 10:1 to 17:1 in the chicken (Pitts, 1938), as contrasted with much lower ratios in man and dog.

At high concentrations of the dye, the clearance ratio of phenol red to inulin approaches 1. This means that as the plasma concentration reaches a certain point, proportionately less of the dye is excreted by the tubules and that the tubules have a limited capacity (T_m) to excrete the dye. The absolute clearances of phenol red and inulin in the adult chicken at low plasma concentrations are 28 and 1.8 ml/min, and at high concentrations (100 mg), they are 2.59 and 2.57, respectively (Pitts, 1938). A number of substances inhibit or depress the secretion of phenol red (see Sturkie, 1976).

Clearance of Other Substances. *Urea.* Urea is handled by the avian kidney similarly to the mammalian kidney; it is completely filterable, but is partially reabsorbed by the tubules independently of the plasma concentration (Pitts and Korr, 1938).

The average urea clearance in the chicken is 1.5 ml/kg/min, and the urea/inulin ratio is 0.74 (Pitts and Korr, 1938). Similar results were obtained by Owen and Robinson (1964), who reported the U/I ratio of 0.70. However this ratio is highly variable, depending on the state of hydration. In the hydrated state, Skadhauge and Schmidt-Nielsen (1967b) showed that 100% of the filtered urea was excreted; in the dehydrated state, only 1% was excreted, and 99% was resorbed in the tubules. Urea does not influence medullary hypertonicity in birds.

A number of other substances are known to be secreted by the kidney tubules of chickens, including glucuronides of menthol and phenol, sulphuric esters of phenol, resorcinol, hydroquinone, tetraethylammonium (TEA), thiamine, choline, epinephrine, potassium, histamine, serotonin, acetylcholine, and atropine. See Sturkie (1976) for detailed references.

Amiloride, a potassium-sparing mild diuretic that is actively secreted, could be blocked by the infusion of certain organic cations (guanidine, quinine) but not the organic anion probenecid. Amiloride also exerted a natiuretic and kaliuretic effect; it was concluded that amiloride is secreted by proximal tubules (Besseghir and Rennick, 1981).

Renal Bloodflow

At low plasma concentrations certain substances are almost completely excreted by the tubules. The volume of the substance cleared per minute (see section on renal clearance for details) is therefore theoretically equal to the volume of plasma flowing through the kidney during the same time. In the bird, paraminohippuric acid (PAH), diodrast, and uric acid are very efficiently cleared from the plasma, and the clearance of these has been used to estimate renal plasma flow. Figures reported in the literature vary considerably, depending on the method used and on other factors, such as the variation caused by the presence of a renal portal circulation (see below). Sykes (1960a) reported mean clearances of uric acid and PAH of 68.5 and 67.6 ml/min, respectively, for Light Sussex chickens ranging in weight from 2.3 to 2.9 kg; this gives a renal plasma flow of 23–30 ml/kg/min compared to 40 ml/kg/min reported by Skadhauge (1964). Other values include 27 ml/kg/min for ducks on fresh water (Holmes et al., 1968), and 25–30 ml/kg/min for mourning doves (Shoemaker, 1967). According to Sykes (1960a) there is less variation in the clearance of uric acid than there is for PAH, and uric acid may therefore be preferable for estimating plasma flow. Clearance of PAH decreases as the concentration increases above a given level (2 mg/100 ml) and bloodflow may therefore be underestimated. The transport maximum (T_m) of PAH is reached in chickens when the blood level of PAH reaches 10–20 mg/100 ml (Dantzler, 1966) and the amount secreted becomes constant. This value ranges from 1.08 to 1.93 mg/kg/min.

The ratio of inulin clearance to PAH clearance, the filtration fraction, is about 13% in doves (Shoemaker, 1967), 15% in chickens, and 11–15% in ducks (Holmes et al., 1968). This avian fraction is somewhat lower than that in mammals (20–25%), most likely because the arterial supply to the avian kidney is augmented with inflow of blood from the renal portal venous system.

The relative contribution of this blood to the kidney circulation is not taken into account by the usual methods involving percentage of total cardiac output by Boelkins, Wolfenson, and others (see Chapter 6), by which kidney flow is estimated to be about 10% of total cardiac output. Actually, the contribution of renal portal blood flow (venous afferent blood) to total kidney flow has been determined by Sturkie and coworkers (see Chapter 6). Their results show that bilateral ligation of renal portal veins decreased venous flow 39% or approximately 12 ml/min in a chicken weighing 2.5–2.7 kg. Shideman et al. (1981) estimated that the renal venous circulation accounted for approximately 50 percent of the clearance of PAH or uric acid in Aves.

Renal Handling of Sodium Chloride

In birds without nasal glands, sodium is handled by the kidney, but in those with salt glands, 60–88% of the salt is eliminated by the nasal glands (see section on extrarenal salt excretion).

Sodium is the principal osmotically active electrolyte in the plasma and urine. It is actively absorbed in the intestine (see Chapter 12) and carried to the kidney, where it may be resorbed into the plasma or secreted by kidney tubules and excreted.

The renal concentrating ability depends on a countercurrent multiplier system in the medulla, or the arrangement of loops of Henle in the medulla, vasa recta, and collecting ducts in parallel fashion within medullary cones (Skadhauge and Schmidt-Nielsen, 1967b; Dantzler and Braun, 1980). The effects of salt loading, hydration, and dehydration on GFR have been discussed previously. Tubular resorption of water increases to 80% in the overhydrated state to 95% during mild hydration, and to at least 99% in the dehydrated state (Krag and Skadhauge, 1972).

The osmolalities (mOsm/liter) of the urine of several avian species are presented in Table 16–3. There is a wide range of variation among the different species, ranging from 362 in the salt-loaded chicken to 2,000 in the salt-loaded savannah sparrow.

Skadhauge (1981), in commenting on this variation, stated that the results are influenced by sampling. Most of the studies on urine osmolalities were based not on ureteral urine but on anal droppings or the liquid, milky part of the droppings from which a supernatant fluid was obtained. Another source of error, according to Skadhauge, occurs after large salt loads where fluid may have passed rapidly through the intestinal tract, and samples thus would not represent truly renal urine. Even so, it is apparent from Table 16–3 that salt loading and dehydration influence significantly the urine/plasma osmolalities. These ratios in the chicken were highest in the dehydrated state, next highest in the salt-loaded state, and lowest in the hydrated condition (Skadhauge and Schmidt-Nielsen, 1967a). Similar results were obtained in other species under like conditions. During water loading and diuresis, birds excrete more of the filtered load than do mammals, according to these authors. Sodium excretion in most birds is about 1% of the filtered load, unless an external load is given (Skadhauge, 1981). A higher rate of Na excretion has been observed in the desert quail after mannitol diuresis, because mannitol causes osmotic diuresis (Dantzler, 1966; Braun and Dantzler, 1972). In the domestic duck (with salt glands), salt loading does not result in high Na excretion by the kidney, because 88% of the filtered load is excreted by the salt glands (Holmes et al., 1968).

Koike et al. (1983) reported that water deprivation in chickens for 24 hr increased plasma osmolality from a control level of 315 to 325 mOsm/liter, and after 72 hr of dehydration, to 340 mOsm/liter. The principal changes were in Na, Cl, and K, all of which, except K, were increased after dehydration; K was decreased. Plasma volume decreased, and red cell volume increased (hemoconcentration). Most of the changes were reversed after rehydration.

The amount of filtered solute resorbed or excreted is influenced by the loading of the solute and its trans-

TABLE 16–3. Osmolality of urine of avian species under conditions mainly of dehydration and salt loading

Species	Treatment	Urine/plasma ratio	Osmolality (mOs)	Ions meq/liter			References
				Na	Cl	K	
Chicken	Dehydrated	1.6	538	134	70	19	Skadhauge and Schmidt-Nielsen (1967a)
	Salt loaded	1.06	362	161	141	25	Skadhauge and Schmidt-Nielsen (1967a)
	Hydrated	0.37	115	38	27	5.6	Skadhauge and Schmidt-Nielsen (1967a)
Turkey	Salt loaded	1.6	553	—	—	—	Skadhauge and Schmidt-Nielsen (1967a)
	Dehydrated	1.4	492	—	—	105	Skadhauge and Schmidt-Nielsen (1967a)
Ostrich	Dehydrated	2.7	800	—	—	—	Louw et al. (1969)
Duck (dom.)[a]	Salt loaded	1.4	444	—	124	—	Scothorne (1959)
Duck (dom.)[a]	Salt loaded	1.3	462	133	—	—	Skadhauge and Schmidt-Nielsen (1967a)
Pelican[a]	Dehydrated	1.7	580	—	—	—	Calder and Bentley (1967)
Gambel's quail	Dehydrated	2.0	669	470	—	—	McNabb (1969a)
	Salt loaded	2.5	884	—	—	—	McNabb (1969b)
	Salt loaded	—	962	493	—	—	Carey and Morton (1971)
Savannah sparrow	Salt loaded	5.8	2000	—	—	—	Poulson and Bartholomew (1962)
	Salt loaded	3.2	1000	—	—	—	Poulson and Bartholomew (1962)
Budgerigar	Dehydrated	2.3	848	—	—	—	Krag and Skadhauge (1972)
Senegal dove	Dehydrated	1.74	661	—	—	121	Skadhauge (1974)
Zebra finch	Dehydrated	2.78	1005	—	—	135	Skadhauge (1974)
	Salt loaded	2.8	1027	—	—	—	Skadhauge and Bradshaw (1974)

[a] Species with salt glands.

port maximum, T_m. The habitat of avian species influences considerably their handling of water and electrolytes. Desert species, subjected to long periods of insufficient water, and species inhabiting salt marshes tend to conserve body fluids by increasing the rate of absorption of filtered water and concentration of solutes.

The ability of the bird kidney to conserve and excrete water has a wider range than that of mammals. Skadhauge (1981) concluded that the avian kidney permits a high fractional excretion of water (as high as 33% during hydration) to as low as 1% during dehydration. The water shutdown is aided by the shutdown of RT type nephrons. The excretion of strong electrolytes may be regulated independently of the excretion of water.

Extrarenal Salt Excretion

It had long been suspected that marine birds might be able to consume and handle salt water, but only in relatively recent years has the mechanism responsible for this ability been discovered (Schmidt-Nielsen et al., 1958). Such birds have a specialized nasal gland that is able to secrete large quantities of NaCl. This gland has been reported and studied in several species, including cormorants (Schmidt-Nielsen et al., 1958), the brown pelican (Schmidt-Nielsen and Fange, 1958), the domestic duck (see Holmes et al., 1968; Holmes, 1972; Scothorne, 1959), the herring gull (Fange et al., 1958; Douglas, 1970), the Humboldt penguin (Schmidt-Nielsen and Sladen, 1958), roadrunner (Ohmart et al., 1970a,b), and in several other species reported by Shoemaker (1972), including ostriches, geese, flamingoes, and desert partridges.

Anatomy of the Nasal Glands

The account that follows is mainly from Schmidt-Nielsen (1960). The existence of a nasal gland, so called, has been known for centuries. It is not, however, always located in the nose. In most marine birds it is located on the top of the head, above the orbit of the eye; it has also been called the supraorbital gland. On top of the skull of the gull are two flat, crescent-shaped "nasal" glands, located in the shallow depressions in the bone. Two ducts run from each side down to the nose, where they open into the vestibular concha. From the anterior nasal cavity, the secretion flows out through the nares and drips off from the tip of the beak. The gland consists of longitudinal lobes, which in cross section show tubular glands radiating from a central canal (see Figure 16–5). The gland receives its main arterial supply from the arteria ophthalmica interna. For further details on histology, see Ernest and Ellis (1969). The gland is innervated from the ganglion ethmoidale, which is supplied by a relatively large branch of the ophthalmic nerve and a small one from the facial nerve, as well as by sympathetic fibers. Stimulation of the parasympathetic nerve or branches going to the nasal gland produces an increase in volume of secretion rich in sodium (see Ash et al., 1969;

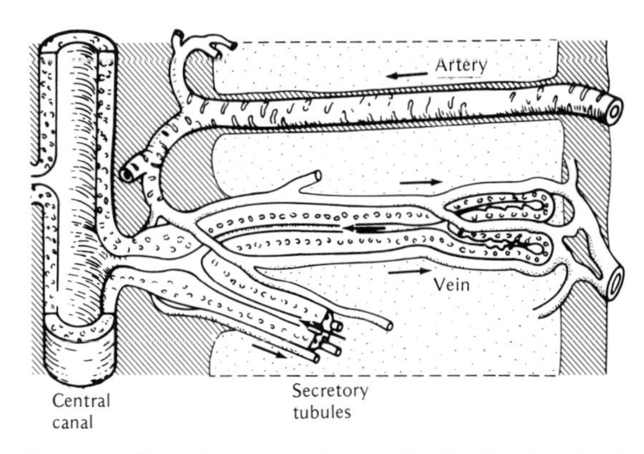

FIGURE 16–5. Anatomy of the salt gland. The gland consists of longitudinal lobes with a central canal from which radiate secretory tubules with blind ends. [After Fänge, R., K. Schmidt-Nielsen, and H. Osaki (1958).]

Hakansson and Malcus, 1969). Acetylcholine also increases gland secretion and atropine blocks or depresses it (Fange et al., 1958; Ash et al., 1969). Acetylcholinesterase fibers are prominent and these disappear after denervation and nerve degeneration. Denervation abolished nasal secretion and changes in blood flow of geese following a salt load (Hanwell et al., 1971a,b). Stimulation of sympathetic nerves to the gland does not influence secretion directly but causes vasoconstriction (Fange et al., 1963).

Although a cholinergic mechanism is essential for normal nasal secretion, the latter is also dependent on an intact adenohypophysical–adrenocortical axis (Holmes, 1972). Removal of adenohypophysis reduces secretion of nasal gland of duck to about 5% of normal rate. Elucidation of the secretory mechanisms of salt glands has been hampered by the absence of methods whereby secretory cells can be studied under rigorously defined in vitro conditions (Hootman and Ernst, 1980). These authors reported a method of separation of salt gland cells involved in secretion. These cell suspensions provide a promising system for in vitro study of secreting salt gland cells, according to these authors.

Function of the Nasal Gland

Details of the anatomy and physiology of the nasal gland of a number of species have been reviewed by Shoemaker (1972) and Holmes (1972), Peaker and Linzell (1975) and Skadhauge (1981).

Composition and Rate of Secretion. The average ionic composition of the nasal secretion of a seagull is shown as follows, based on data of Schmidt-Nielsen (1960) in meq/liter.

Na^+	K^+	Ca^{2+} and Mg^{2+}	Cl^-	HCO_3^-
718	24	2.0	720	13

The variation in composition depends on the species involved, but the concentration of NaCl usually falls between one and two times that of sea water (Shoemaker, 1972). The concentrations are usually higher in marine species. Shoemaker (1972) has tabulated the concentration of ions in the nasal secretions and the maximum secretory rate following mild salt loading in some 19 species.

The usual concentration of Na in meq/liter for most species ranges from 450 to 700 but in some approaches 1,000. Nasal secretion is not restricted to marine species but occurs also in certain terrestrial species such as *Ammoperdix heji* (desert partridge) and the ostrich. In these species nasal secretion occurred without osmotic stimulation but in response to high temperature (Schmidt-Nielsen et al., 1963). The chicken, pigeon, and many other species do not have nasal glands and handle all of their NaCl through the kidneys and cloaca.

The amount or percentage of the intake of water and electrolytes excreted in the urine of ducks given fresh and salt water are as follows (Fletcher and Holmes, 1968):

	Maintained on	
	Fresh water (%)	Salt water (%)
Water	60.3	36.5
Na^+	59.5	10.0
K^+	71.3	71.3

The rate of secretion is more variable among species than is the concentration of fluid and is influenced by degree of salt loading and hydration. Data on rates of secretion are subject to greater variation because of difficulties of determining accurately flow rates. Secretion is usually collected by runoff from the beak into attached containers. Continuous secretion from moment to moment has been recorded by a new improved flowmeter (Hakansson and Malcus, 1969). The maximal secretory capacity of the nasal gland of herring gulls following initial electrical nerve stimulation for 3 min ranged from 0.280 to 0.375 ml/g/min of the tissue. These figures are about the same as those reported by Douglas (1970). Equal volumes were secreted from the two ducts of the glands. With continued secretion for periods of 1–2 hr, flow rate decreased to one-third to one-fourth the maximum level following electrical stimulation. These values are based on young herring gulls weighing approximately 1 kg, with salt glands weighing about 0.6 g per bird. The secretion therefore amounted to about 0.148–0.225 ml/min/kg of body weight, a figure similar to values reported for gulls by others (see the tabulation by Shoemaker, 1972) but lower than the value reported by Schmidt-Nielsen (1960) for herring gulls. Salt loading tends to increase the amount of secretion and the electrolyte concentration but to a lesser extent (Hanwell et al., 1971a; Holmes, 1972, and others). The values reported by different investigators for different species vary from 0.074 to 0.50 ml/kg/min (Shoemaker, 1972). Hanwell et al. (1971a) reported secretion rates of 0.07–1.33 ml/g nasal tissue per minute in geese following different salt loads.

Factors Affecting Nasal Secretion

The obvious stimulus for nasal secretion is osmolality of the plasma. Within minutes after a salt load, nasal secretion begins, and a solution two to six times hyperosmotic to plasma is produced.

A seagull that had ingested 134 ml of seawater containing 54 meq/liter of sodium excreted most of the sodium through the salt gland, but some of it in the urine and feces, within 175 min (as follows) (Schmidt-Nielsen, 1960).

Nasal secretion		Cloacal excretion	
Volume (ml)	Sodium (meq/liter)	Volume (ml)	Sodium (meq/liter)
56.3	43.7	75.2	4.41

Douglas (1970) reported that herring gulls receiving 80 ml of seawater containing 32.4 meq of Na excreted 66–74% of the Na load in the nasal fluid within 3.5–4 hr. The volume of nasal fluid ranged from 25.9 to 27.8 ml, or roughly 6 ml/hr. The volume of fluid excreted in the cloacal fluid was 80.1 ml in the unfasted bird and 38.2 ml in birds fasted for 24 hr before receiving the salt load. Fifteen to 18% of salt load was in the cloacal fluid.

Willoughby and Peaker (1979) and Skadhauge (1981) have tabulated the nasal secretion rates of a number of species under different conditions. These values ranged from over 1500 mOsm/liter to a low of 765, depending on species and degree of salt loading. The Na content ranged from 776 to as low as 263 meq/liter. Na and Cl are present in nearly equal proportions and constitute the major ions contributing to osmolality of nasal secretion.

The flow rates tabulated by Skadhauge (1981) are highly variable and depend on many factors; they range from about 20 to 200 μ liter/kg/min.

The flow rate of nasal glands of the cormorant and the amounts of water and electrolytes excreted from the nasal gland and cloaca are presented in Table 16–4 (Willoughby and Peaker, 1979).

By secreting salt, the nasal glands offer a mechanism for gaining osmotically free water. Estimates indicate that in normally fed ducks, approximately 20% of the water loss occurs through the nasal gland; during fasting for 24 hr with no water consumption, nasal water loss may increase to 40% (Willoughby and Peaker, 1979).

While osmolality of the plasma is the main stimulus for nasal secretion, plasma volume expansion may also influence it under certain conditions. When blood volume is increased greatly nasal secretion may be induced (Zucker et al., 1977), but when plasma volume is increased much less (9–16%), there was no change in nasal secretion (Hanwell et al., 1972; Peaker, 1978). There is some evidence for a nonosmotic volume expansion factor (see discussion by Skadhauge, 1981). Other factors influencing nasal secretion include intake of food, water, and NaCl, neural factors, and adrenal cortical hormones. Stimuli such as physical stress of temperature, light, and sound may induce nasal secretion. Excitement decreases nasal secretion in penguins (Douglas, 1968), and dehydration decreases it (Douglas and Neely, 1969).

TABLE 16–4. Amounts of electrolytes and water excreted by cloaca (a) and salt glands (b) of cormorant given NaCl and water[a]

	NA^+ (meq)	K^+ (meq)	Cl^- (meq)	Water (ml)
Amount given	54	4.0	54	50
Cloacal excretion (a)	25.6	2.66	27.5	108.9
Nasal excretion (b)	23.8	0.31	26.1	51.4
Percent (a)	52.0	90.0	51.0	68.0
(b)	48.0	10.0	49.0	32.0

[a] Modified from Willoughby and Peaker (1979).

Endocrines. The stimulating effect of stresses may be exerted through the release of adrenal cortical steroids, a subject studied extensively in ducks by Holmes and coworkers and reviewed by him (Holmes and Wright, 1968; Holmes, 1972).

Although the normal triggering mechanism for nasal secretion may be osmoregulatory in nature, the final response is influenced by adrenocortical hormones. In the adrenalectomized salt-loaded duck, there is no nasal secretion of NaCl (Phillips et al., 1961), but it can be restored after administration of cortisol. Administration of cortisol, corticosterone, or ACTH to salt-loaded ducks increases nasal secretion (see Holmes, 1972). Moreover, adrenocortical hormones influence not only nasal secretion but also intestinal absorption of NaCl. Crocker and Holmes (1971) showed that ducks maintained on seawater as compared to freshwater had higher rates of Na transport and absorption (see Chapter 12) in the proximal parts of the small intestine and higher rates of nasal salt secretion. When saltwater-adapted ducks received spironolactone, which inhibits aldosterone action, these ducks absorbed Na at the same rate as those in freshwater but did not secrete NaCl through the nasal gland. Aldosterone therefore not only influences intestinal absorption of Na, which is required in salt loading, but also the increased nasal secretion of salt. Nasal secretion is influenced also by thyroid hormones, because it is decreased in the thyroidectomized duck given a salt load orally. This does not occur when salt is injected intravenously, indicating that the main effect is through a decreased absorptive capacity of the intestine (Ensor et al., 1970; see Holmes, 1972). Hypophysectomy reduces nasal secretion drastically; it can be restored with ACTH (Bradley and Holmes, 1971). Holmes (1975) has reviewed osmoregulation in marine birds. Adrenalectomized ducks, however, can adapt to a mild salt load and secrete salt nasally. Spontaneous secretion was also observed in adrenalectomized ducks (Butler, 1980).

Following a salt load in ducks or gulls, there is an initial diuresis (renal phase), which is followed by nasal secretion and a progressive decrease in renal Na output as nasal secretion of Na progresses. As the Na is eliminated from the nasal gland this provides for an increase in osmotically free water and is a means of decreasing Na concentration in blood and urine. There is some evidence that separate mechanisms are involved in the handling of water and sodium by the salt gland (Inoue, 1963).

Initiation and Control of Nasal Secretion

At the cellular level of nasal gland activity, Na^{+}–K^{+}, ATPase is most likely involved in increased nasal secretion, because it has been amply demonstrated that adaptation to salt loading results in a three- to fourfold increase in ATPase activity (Fletcher et al., 1967 and others; see Shoemaker, 1967; Holmes, 1972). When this activity is inhibited by administration of oaubain to the nasal gland, nasal secretion is also inhibited (Shoemaker, 1967). The level of potassium available in the media for ATPase activity is critical because deficiencies of potassium in chicken kidney slices depress ATPase activity (Dantzler, 1972). The evidence for volume receptors is not conclusive. If there are osmoreceptors, they appear not to be located in the brain or centrally, because intracarotid injections of saline (Hanwell et al., 1971b) failed to elicit nasal secretion in the goose. Moreover the results of infusions of NaCl into the renal portal vein on nasal secretion were negative; however, because of the complicated circulation in the kidney and viscera of Aves, these authors do not rule out the possibility of receptors in the gut or renal portal blood vessels. Further studies are indicated.

Regardless of the nature and location of the receptors, the sensory (afferent) input goes to the central nervous system, which then relays the stimulus to the hypothalamus, pituitary, and adrenals and directly to the nasal gland via the seventh cranial nerve (efferent), to the ganglion ethmoidale, and thence to postganglionic (cholinergic) fibers that run to the gland to produce the final stimulus (secretion; see Holmes, 1972).

Recent investigations have confirmed that there is a selective binding of corticosterone to the cytosol of salt gland cells (Sandor et al., 1977). Studies on dispersed cells of avian salt glands suggest that cholinergic stimulation of these cells by metacholine triggers a Cl^{-}-dependent uptake of Na^{+}, which elicits a compensatory increase in Na^{+} pump turnover. The decrease in cellular Cl content caused by metacholine suggests that the agonist either directly or indirectly mediates an efflux of CL from the cells (Hootman and Ernst, 1981).

Urine

Collection and Amount of Urine Voided

The urine and feces of the bird are voided into the cloaca. In order to obtain urine free from fecal matter, therefore, it is necessary to cannulate the ureters or to separate the openings of the ureters or the rectum by surgery. Improved surgical techniques for exteriorizing the ureters (Dixon and Wilkinson, 1957) or the anus (Imabayashi et al., 1956; Fussell, 1969; Paulson, 1969) have been reported. Uncontaminated urine in the cloaca may also be obtained by plugging the anal opening.

Most of the work relating to quantities of urine voided is based on short-period collections involving clearance studies in which calculations were made on the output for 24 hr. Most of those who cannulated the ureters and collected urine for short periods (usually 30–100 min) estimated the output of the adult chicken (weight about 2 kg) at 500–1000 ml/day (Davis, 1927; Coulson and Hughes, 1930; Hester et al., 1940). Short-period collections from exteriorized ureters gave similar estimates (Hester et al., 1940).

The estimates of urine output based on short-period collections are obviously unduly high. An estimated output of 1000 ml in 24 hr is considerably higher than water consumption for the same period (50–250 ml). It was demonstrated by Hester et al. (1940) and by Hart and Essex (1942) that such estimates are in error; the work of these authors showed the urine output from cannulated ureters to average 13.9 ml for the first 30 min but only 2.5 ml during the second 30 min. Estimates for 24 hr based on first collections therefore gave an output of 667 ml, and on the second period, 120 ml. These investigators and others showed that cannulation and handling of the bird causes diuresis, which persists for at least 30 min. In later experiments urine was collected in bags from exteriorized ureters for 24 hr; here the average output was 86.8 ml, with a range of 61–123.4 ml. In another experiment the output was as high as 180 ml. Urine collections made by Dixon (1958) from exteriorized ureters and after the rectum had been exteriorized show urine flows of 132 and 155 ml/24 hr. Dixon's results are not in agreement with those of Dicker and Haslam (1966), who reported that birds with exteriorized ureters (Dixon's method) drank almost twice as much water after the operation as before (~100 ml/kg per 24 hr). Urine collected by Ariyoshe and Morimoto (1956) from chickens with artificial anuses (cocks, weighing 2.5–3.0 kg) averaged 115 ml/bird/24 hr.

Urine flows of selected avian species under certain specified conditions are presented in Table 16–5. Most of the collections of urine were not based on ureteral cannulations, which may influence flow rates considerably.

Factors Affecting Urine Flow and Output

Role of Ureters. The ureters tend to force or milk the urine along by peristaltic action. The peristaltic waves move caudally. The pressure they exert is considerable, and the urine may be forced along against a pressure as high as 30 mm Hg (Gibbs, 1929c). The ureters, according to Gibbs (1929c), appear to be under sympathetic control.

Resorption of Water. Urine production is influenced significantly by the tubular resportion of water. As little as 1% of filtered water may be resorbed during water diuresis and as much as 99% at low rates of urine flow (Skadhauge, 1981). Mammals never resorb less than 90% of the filtered water, even when they are made diuretic (Skadhauge and Schmidt-Nielsen, 1967a).

Diuresis. Administration of water alone or of hypertonic solutions increases urine output, but dehydration decreases it (see also glomerular filtration and renal handling of electrolytes). Dicker and Haslam (1966) determined the effects of 50 ml/kg of water given orally on urine flow in chickens with exteriorized ureters. They reported that urine flow increased considerably (from <0.05 ml/kg/min to >0.2 ml) within 15–20 min and reached a peak in 80 min; it then declined to almost the control level at 120 min after water administration. Changes in osmolality followed those of urine flow, decreasing from a control level of 490–520 to 60 mOsm/liter at high urine flows.

Dilation of the crop by filling it with liquid paraffin caused diuresis similar to that produced by water; the mechanism for this effect is not known. Insertion of ureteral cannulas are known to produce diuresis; in pigeons this lasts from 30 to 70 min, after which urine flow is stable at 0.4 ml/kg/hr (McNabb et al., 1970).

Salt Loading. The effects are illustrated in Table 16–5. Some workers have shown a decreased flow (see Skadhauge, 1981, and others) or an increased flow (apparently depending on conditions of loading and hydration), and diuresis.

Drugs and Hormones. A number of drugs and hormones affect urine flow and water consumption (see Chapter 12), some of which are discussed here. Urine flow is decreased by pitressin or pituitrin (Hester et al., 1940) but increased by ether anesthesia and caffeine (Hester et al., 1940). Atropine and pilocarpine have no appreciable effect on urine flow, according to Gibbs (1929b). The antidiuretic hormone arginine vasopressin, in mammals, and arginine vasotocin (AVT), its counterpart in birds, both produce antidiuresis in chickens, but the latter is considerably more potent (Munsick et al., 1960). Unilateral antidiuresis can be produced by unilateral infusion of AVT into the portal vein (Skadhauge, 1964) demonstrating that AVT increases reabsorption of water in the collecting ducts and tubules.

A dose of 30 ng/kg of AVT was found to yield a maximum antidiuretic response in hydrated chickens (Ames et al., 1971). The duration of effect ranged from 6 to 30 min, depending on dosage. Removal of the neural lobe of the hypophysis produces polyuria and polydipsia in chickens and in ducks (Bradley et

TABLE 16–5. Urine flows of selected species under certain specified conditions. Most urine samples were not taken by ureter cannulation.

Species	Conditions	Urine flow (ml/kg/hr)[a]	Reference
Chicken	Dehydrated	1.08	Skadhauge and Schmidt-Nielsen (1967a)
	Hydrated	17.9	Skadhauge and Schmidt-Nielsen (1967a)
	Salt loaded	10.9	Skadhauge and Schmidt-Nielsen (1967a)
Chicken	Control	86.8 (mg/hr/bird)	Karasawa and Sakae (1981)
	Diuresis	148.7 (mg/hr/bird)	Karasawa and Sakae (1981)
Turkey	Control	3.30	Vogel et al. (1965)
	Hydrated	30.5	Vogel et al. (1965)
	Control	32 (μl/kg/min)	Palmore et al. (1981)
	Diuresis (epinephrine)	55.0 (mg/hr/bird)	Palmore et al. (1981)
Duck	Hydrated	7.8	Holmes (1965)
	Hydrated	2.4	Holmes et al. (1968)
	Salt loaded	0.57	Holmes (1965)
Gambel's quail	Mannitol diuresis	11.2	Braun and Dantzler (1972)
	Hydration	6.2	Braun and Dantzler (1975)
Budgerigar	Control	6.30	Krag and Skadhauge (1972)
	Dehydration	1.68	Krag and Skadhauge (1972)

[a] Unless otherwise stated.

al., 1971). Administration of arginine vasotocin restored urine flow and excretion of NaCl to the normal levels.

Arecoline, pilocarpine, and propionylcholine (cholinomimetric agents) increase urine flow in chickens. The diuretic effect is belived to be independent of the vasodilating effect of these agents, because other vasodilating noncholinomimetic agents, papaverine and isoproterenol, had no significant effect (May and Carter, 1970). Acetylcholine (2.5–7.5 mg/kg/min), when infused unilaterally into the chicken's renal portal system, increased excretion of water (urine flow) and NaCl, which could be blocked by atropine at dosages of 10 μg/kg/min (Parmlee and Carter, 1968).

Angiotensin has a diuretic effect on the pigeon and its site of action appears to be the tubules (Langford and Fallis, 1966). Hydrochlorothiazide (HCZ) belongs to the sulphonamyl group of compounds and has been reported to have a diuretic action in turkeys (Sykes, 1961), increasing urine flow fivefold in these birds after a 25-mg dose. Reserpine (1 mg/kg) greatly reduced the diuretic effect of HCZ in chickens (Nechay and Sanner, 1961), as did theophylline, but had no effect on a mercurial diuretic, which is quite effective in chickens (Campbell, 1957).

Catecholamines. Palmore et al. (1981) demonstrated that a number of catecholamines, including epinephrine, norepinephrine, isoproterenol, and dopamine all had a diuretic effect in turkeys, but the birds were most sensitive to epinephrine and isoproterenol and least sensitive to dopamine. Diuresis occurred following administration of either epinephrine or epinephrine at doses that produced few or any cardiovascular effects. The authors concluded that the catecholamines appear to induce a renal response opposite from that produced by AVT. Catecholamines have an effect like water diuresis, that is, diuresis and increased glomerular filtration. This suggests an interplay between AVT and endogenous catecholamines in body fluid homeostasis, according to these authors, and probably increases the number of poorly concentrating RT nephrons that are operating.

Submersion Anuria. Submersion of the head of the Aylesbury duck for 3–4 min resulted in an immediate cessation of urine flow under normal conditions, and water diuresis was resumed 2 min after emersion. Asphyxia alone also resulted in anuria (Sykes, 1966).

Laying Cycle and Urine Flow. There is an increase in urine flow a few hours before the first egg is laid, reaching a peak 1 hr before laying, which appears to be associated with an increased excretion of Na and P (Mongin and Lacassagne, 1967).

Other Factors. *Calcitonin* administered to starlings significantly increased urine flow and Na clearance (Clark and Wideman, 1980). In parathyroidectomized starlings, calcitonin also increased urine flow.

Cooling of the hypothalamus of Peking ducks caused diuresis, which suggests an inhibition of endogenous AVT; hypothalamic cooling also reduced salt gland secretion (Simon-Oppermann et al., 1979).

Role of the Cloaca in the Absorption of Urine Water. Many of the early investigators believed that water from ureteral urine was absorbed in the cloaca and rectum and that the cloaca served an important function in the conservation of water (for review of some of the early work, see Korr, 1939; Hester et al., 1940; Hart and Essex, 1942; Sturkie, 1965). Some of these workers observed an increase in the flow of urine collected from cannulated ureters or from the cloaca after the rectum was plugged. See also review by Skadhauge (1981).

Experiments by Hart and Essex (1942), based on long-term urine collections, suggested that some water is reabsorbed into the cloaca, but not appreciable amounts; the evidence for this conclusion is indirect (see Sturkie, 1965). Experiments by Dixon (1958), involving improved surgical techniques for exteriorizing the ureters and anus, indicated that little or no water was reabsorbed in the cloaca of the chicken. His data also indicated that surgical treatment had no significant effect on water consumption and urine flow. This is contrary to the work of Dicker and Haslam, (1966), Scheiber and Dziuk (1969), and others, who have reported increased consumption of water and diuresis following surgery and cannulation and manipulation of ureters.

Weyrauch and Roland (1957) attempted to measure water reabsorption in the cloaca of the chicken by introducing a solution containing tracer isotopes into the cloaca and measuring the content of isotope in the bloodstream. The figures indicate that 7.6% of the isotope was absorbed within 4 hr. However, these studies really measure not cloacal absorption alone, but absorption into the anus and into the gut at a distance 10 cm above the anus.

Later data by Akester et al. (1967), Skadhauge (1968), and Nechay et al., (1968) on chickens; by Peaker et al., (1968) on ducks; and by Ohmart et al. (1970b) on the roadrunner, employing x rays or tracer isotopes, definitely showed that water is absorbed in the cloaca. The urine accumulated in the coprodaeum of the cloaca and was forced by antiperistaltic waves into the colon (Akester et al., 1967). The urine enters the bowel from the cloaca within 4 min after intravenous administration of a urographic contrast medium (Nechay et al., 1968). Absorption from the cloaca into the ceca of chickens with artificial anuses has been reported by Polin et al. (1967). The absorption of water and electrolytes can occur during hydration, dehydration, and salt loading in chickens and ducks (Peaker et al., 1968). Although less is absorbed during dehydration (Skadhauge, 1968), there appears to be no means of keeping uretral urine from exposure to cloacal membranes and large intestine, across which the exchange of water and solutes may occur (Shoemaker, 1972). However, these rates of exchange are sufficiently low so that they can have only minor effects when urine is produced at moderate to high rates (Skadhauge, Schmidt-Nielsen, 1967a).

Skadhauge (1968) did not instill fluid into the lower intestines, as did most other investigators, but employed an unabsorbable water marker, [^{14}C]inulin, which came with the urine after it had been filtered by the kidney of dehydrated cocks. He determined the concentrations of inulin, Na, K, and Cl, and the osmolality of the cloacal fluid supernatant. The results obtained support the concept that there is considerable absorption of water in the cloaca (Skadhauge, 1981). The study indicated that 50% of the volume of cloacal urine water and salts are absorbed even from the hyperosmotic contents present in the cloaca in the dehydrated state (Skadhauge, 1981).

Skadhauge and Bradshaw (1974) studied water consumption, urine flow, and ability to handle high levels of salt in *Taeniophygia castanotis* (zebra finch), which inhabits the semiarid zones of western Australia. These finches were able to consume greater quantities of salt water than domesticated finches, and their urine osmolality was 40% greater than in domesticated finches (1475 mOsm).

Physical Characteristics of Urine

Bird urine is usually cream colored and contains thick mucoid material; the latter contains much uric acid, which is usually in the insoluble form. Under most conditions the urine is hyperosmotic (see Table 16–3), but less so than mammalian urine. Under extreme conditions of hydration, the urine may be thin, watery, and hypoosmotic. The hyperosmolality ranges from a low of 115 to a high of 2000, depending on conditions.

Specific gravity of avian urine normally ranges from 1.0018 to 1.015, but increases or decreases, depending on conditions of hydration and osmolality.

Acid–Base Balance. The pH of avian urine varies from 4.70 to 8.00, depending on stage of laying in females, hypoxia, and other factors. The pH of male urine is 6.4 (Ariyoshi and Morimoto, 1956). During the period when calcium is being deposited on the shell, the pH of the urine is acid (5.3), and when the egg is laid or no Ca is being deposited, it is alkaline (7.6) (Prashad and Edwards, 1973; see also Anderson, 1967). This is because the production of Ca for the shell results in an excess of hydrogen ions, which are buffered by blood, and there is a fall in blood bicarbonate and pH. This fall also occurs in the urine, and most of the bicarbonate is absorbed; there is also an increase in ammonia, phosphate, and titratable acidity (Anderson, 1967; Prashad and Edwards, 1973). Hypoxia occurring in the diving duck may lower pH of urine to 4.7 (Sykes, 1966). Although the chicken excretes 1–2 g of uric acid per day, the uric acid and its salts are poorly dissociated and act as efficient buffers; however, the proportion of uric acid that can act as an acid or a base is not known (Sykes, 1971).

Long and Skadhauge (1983) determined the effects

TABLE 16–6. Distribution of nitrogen in urine of chickens and ducks[a]

Species	Uric acid	Urea	NH_4	Purines	Creatine and creatnine	Amino acid	References
Chicken	82.0	—	5.6	—	—	—	Katayama (1924)
	66.0	—	—	—	6.0	—	Mayrs (1923)
	63.0	10.4	17.3	—	8.0	—	Davis (1927)
	66.0	6.5	7.6	9.6	4.6	—	Coulson and Hughes (1930)
	60.0	—	—	20.0	—	10[b]	Edwards and Wilson (1954)
Chicken, fed	84.0	5.2	6.8	—	0.5	1.7	Sykes (1960)
Chicken, starved	58.0	2.9	23.0	—	4.3	2.8	Sykes (1960)
Chicken, practical diet	81.0	4.5	10.5	—	0.9	2.2	O'Dell et al. (1960)
Chicken, purified diet	76.0	5.6	15.4	—	0.2	1.7	O'Dell et al. (1960)
Chicken, normal diet	60.0	6.0	23.0	—	4.0	2.0	Teekel et al. (1968)
Chicken, fed	82.0	2.4	11.0	—	2.4	—	Sell and Balloun (1961)
Chicken, fasted	74.6	2.9	18.6	—	4.0	—	Sell and Balloun (1961)
Duck, freshwater	52.0	—	30.3	—	1.3	—	Stewart et al. (1969)
Duck, hypertonic saline	57.0	—	27.5	—	2.5	—	Stewart et al. (1969)
Duck, normal	78.0	4.2	3.2	0.5	—	2.7	Szalagyi and Kriwuscha (1914)

[a] Data are % of total nitrogen.
[b] Aminonitrogen

of different diets on the production of renal acid in chickens. Birds on a commercial diet had urinary pHs ranging from 4.74 to 7.25. At an average pH of 6.1, uric acid accounted for 52% of total acid excreted in the urine; H_2PO_4 was 26% and NH_4 was 28%. Net acid secretion in ureteral urine was 5–10 times that in ureotelic vetebrates (amphibians and mammals), according to these investigators. On diet B (Na poor and protein rich), average pH fell to 5.12, and net renal acid excretion rate doubled.

Nitrogenous Constituents. The principal differences in the chemical constituents of the urine in birds, as compared with that of mammals, are the preponderance of uric acid over urea and of creatine over creatinine in birds. Creatinine exists only in minute amounts in bird urine (Davis, 1927). O'Dell et al. (1960) reviewed the literature and presented new data on growing male chicks 5–6 weeks of age. Their results, and those of others, are presented in Table 16–6. In general, most of the investigators reported a lower percentage of uric acid nitrogen in the urine than did O'Dell et al. (1960). The reason for the discrepancy is not clear, but certain differences did exist in the experiments performed. The results of O'Dell et al. were from growing chickens that were well adapted to the conditions of urine collection, whereas most of the previous work had been conducted on mature birds, some of which were under physiologic stress during the collection periods. The determination of uric acid by O'Dell et al. was by the uricase method, which is highly specific and which was not used by most of the other investigators.

Starvation significantly decreases the amount of uric acid N in the urine (84–58%) and blood, and increases the level of ammonia (from 6.8 to 23.0%), according to Sykes (1971). Feeding as compared to starvation increased the percentage of uric acid excreted in the urine (Sykes, 1960; Sell and Balloun, 1961). Uric acid is only slightly soluble in water, but its salts (urates) are considerably more soluble (Sykes, 1971).

The ratio of uric acid in solution to that in colloidal state to that in form of a precipitate is 1:2:12 (Skadhauge, 1981). Urates, because of their low aqueous solubility, do not contribute to osmotic activity of avian urine as urea does to mammalian urine (see earlier section).

Intravenous administration of purine compounds to chickens changed the production of uric acid in the urine (Karasawa and Sakae, 1981), in plasma, and certain other tissues. These increases ranged from 8 to 103% above control levels for most of the compounds. The most important of these were IMP, inosine, GMP, guanosine, xanthine, and hypoxanthine. AMP, however, caused a marked decrease in uric acid production.

Uric acid was first believed to be synthesized mainly in the liver of birds, as was indicated by Minkowski in 1886 and by Edson et al. (1936). It is now generally agreed that the kidney is also concerned in the synthesis of uric acid in the pigeon. Chou (1972) believed that the kidney produces nearly twice as much uric acid as the liver. The livers of chickens and pigeons produce hypoxanthine, which is then oxidized to xanthine and then to uric acid, in the kidney of the pigeon and in the liver of chickens, by xanthine oxidase. The metabolism and active excretion of hypoxanthine by the renal tubules of chickens was reported by Cacini and Quebbemann (1978).

The uric acid content of blood, liver, and kidneys of birds is given in Table 16–7. These values and those of more recent data are highly variable (Wideman et al. 1983). Ligation of the ureters of birds caused

TABLE 16–7. Uric acid content of avian tissues

Species	Blood (whole or from plasma	muscle)	Liver	Kidney	Reference
Duck, normal	6.7	1.8	22.2	70.5	Folin et al. (1924)
Duck with ligated ureters	224.0	30.2	101.0	354.0	Folin et al. (1924)
Chicken, normal	5.8	—	—	—	Levine et al. (1947)
Chicken with ligated ureters	304.0	—	—	—	Levine et al. (1947)
Chicken, normal	9.8	Plasma	4.8	14.81	Karasawa and Sakae (1981)
Chicken given certain purines (IMP)	37.9	—	7.7	23.24	Karasawa and Sakae (1981)

Data are in milligrams per 100 ml for plasma or per mg of wet tissue.

marked increases in the concentration of uric acid in the blood and tissues but not in the urine. Birds with ligated ureters usually died of uricemia within 12–24 hr (Folin et al., 1924; Levine et al., 1947).

Intravenous administration of certain purines increased the uric acid content of blood, liver, and kidneys of chickens, as well as that of the urine (Karasawa and Sakae, 1981). Uricase from *C. utilis* and uricase to which polyethylene glycol had been attached were injected into Leghorn cockerels in an attempt to lower plasma urate levels. Twenty units of either substance reduced urate levels to zero for 24 hr. Smaller doses were less effective (Abuchowski et al., 1981).

References

Abuchowski, A., D. Karp, and F.F. Davis. (1981). Reduction of plasma urate levels in cockerels with polyethylene glycol-uricase. J. Pharmacol. Exp. Ther., 219, 352.

Akester, A.R. (1967). Renal portal shunts in the kidney of the domestic fowl. J. Anat., 101, 569.

Akester, A.R., and S.P. Mann. (1969). Adrenergic and cholinergic innervation of the renal portal valve in the domestic fowl. J. Anat., 104, 241.

Akester, A.R., R.S. Anderson, K.J. Hill, and G.W. Osbaldiston. (1967). A radiographic study of urine flow in the domestic fowl. Br. Poult. Sci., 8, 209.

Ames, E., K. Steven, and E. Skadhauge. (1971). Effects of arginine vasotocin on renal excretion of Na^+, K^+, Cl^- and urea in the hydrated chicken. Am. J. Physiol., 221, 1223.

Anderson, R.S. (1967). Acid base changes in the excreta of the laying hen. Vet. Rec., 80, 314.

Ariyoshi, S., and H. Morimoto. (1956). Studies on nitrogen metabolism in the fowl. I. Separation of urine for nutritional balance studies. Bull. Natl. Inst. Agric. Sci., Series G, No. 18, 37.

Ash, R.W., J.W. Pearce, and A. Silver (1969). An investigation into the nerve supply to the salt gland of the duck. J. Exp. Physiol., 54, 284.

Austic, R.E., and R.K. Cole. (1972). Impaired renal clearance of uric acid in chickens having hyper-uricemia and articular gout. Am. J. Physiol., 223, 525.

Bailey, J.R., and H. Nishimura. (1984). Renal response of fowl to hypertonic saline infusion into renal portal system. Am. J. Physiol., 245, H628.

Ballantyne, B., and W.G. Wood. (1969). Mass and the function of the avian nasal gland. Cytobios, 4, 337.

Benoit, J. (1950). "Traité de zoologie," Tome XV. "Oiseaux" (P.P. Grasse, Ed.). Paris: Masson & Co., p. 341.

Berger, L., T.F. Yu, and A.B. Gutman. (1960). Effects of drugs that alter uric acid excretion in man on uric acid clearance in the chicken. Am. J. Physiol., 198, 575.

Besseghir, K., and B. Rennick. (1981). Renal tubule transport and electrolyte effects of Amiloride in the chicken. J. Pharmacol. Exp. Ther., 219, 435.

Bradley, E.L., and W.N. Holmes. (1971). The effects of hypophysectomy on adrenocorticol function in the duck (Anas platyrhynchos). J. Endocrinol., 49, 437.

Bradley, E.L., W.N. Holmes, and A. Wright. (1971). The effects of neurohypophysectomy on the pattern of renal excretion of the duck. J. Endocrinol., 51, 57.

Braun, E.J., and W.H. Dantzler. (1972). Function of mammalian-type and reptilian type nephrons in kidney of desert quail. Am. J. Physiol., 222, 617.

Braun, E.J., and W.H. Dantzler. (1974). Effects of ADH on single nephron glomerular filtration rates and blood pressure on birds. Am. J. Physiol., 226, 1.

Braun, E.J., and W.H. Dantzler. (1975). Effects of water load on renal glomerular and tubular functions in desert quail. Am. J. Physiol., 229, 222.

Braun, E.J. (1975). Effects of neurohypophysectomy on single nephron glomerular filtration rate in desert quail. Physiologist, 18, 150.

Braun, E.J. (1976). Intrarenal blood flow distribution in the desert quail following salt loading. Am. J. Physiol., 231, 1111.

Braun, E.J. (1978). Renal response of starling (*Sturnus vulgaris*) to an intravenous salt load. Am. J. Physiol., 234, F270.

Braun, E.J. (1982). Renal function. Comp. Biochem. Physiol., A, 70, 511.

Burrows, M.E., E.J. Braun, and S.P. Duckles. (1983). Avian renal portal valve: a reexamination of its innervation. Am. J. Physiol., 245, H628.

Butler, D.G. (1980). Functional nasal salt glands in adrenalectomized domestic ducks. Gen. and Comp. Endocrinol., 40:15.

Cacine, W., and A.J. Quebbemann. (1978). The metabolism and active excretion of hypoxanthine and urine by the renal tubules in the chicken. J. Pharmacol. Exp. Ther., 207, 574.

Calder, W.A., and J.P. Bentley. (1967). Urine concentrations of two carnivorus birds, the white pelican and the roadrunner. Comp. Biochem. Physiol., 22, 607.

Campbell, D. (1957). Excretion and diuretic action of mercurial diuretics, Experientia, 13, 327.

Carey, C., and M.L. Morton. (1971). A comparison of salt and water regulation in California quail and Gambel's quail. Comp. Biochem. Physiol. A, 39, 75.

Chou, S.T. (1972). Relative importance of liver and kidney in synthesis of uric acid in chickens. Can. J. Physiol. Pharmacol., 50, 936.

Clark, N.B., and R. Wideman, Jr. (1980). Calcitonin stimulation of urine flow and Na excretion in the starling. Am. J. Physiol., 71, R406.

Coulson, E.J., and J.H. Hughes. (1930). Collection and analysis of chicken urine. Poult. Sci., 10, 53.

Crocker, A.D., and W.N. Holmes. (1971). Intestinal absorption in ducklings maintained on fresh water and hypertonic saline. Comp. Biochem. Physiol. A, 40, 203.

Dantzler, W.H. (1966). Renal response of chickens to infusion of hyperosmotic sodium chloride solution. Am. J. Physiol., 210, 640.

Dantzler, W.H. (1972). Effects of incubations in low potassium and low sodium media on $Na^{+}-K^{-}$ ATPase activity in snake and chicken kidney slices. Comp. Biochem. Physiol. B, 41, 79.

Dantzler, W.H., and E.J. Braun. (1980). Comparative nephron function in reptiles, birds, and mammals. Am. J. Physiol., 8, R197.

Davis, R.E. (1927). The nitrogenous constituents of hens urine. J. Biol. Chem., 74, 509.

Dicker, S.E., and J. Haslam. (1966). Water diuresis in the domestic fowl. J. Physiol., 183, 225.

Dixon, J.M. (1958). Investigation of urinary water reabsorption in the cloaca and rectum of hen. Poult. Sci., 37, 410.

Dixon, J.M., and W.S. Wilkinson. (1957). Surgical technique for exterioraization of the ureters of the chicken. Am. J. Vet. Res., 18, 665.

Douglas, D.S. (1966). Low urine salt concentrations in salt loaded gulls. Physiologist, 9, 171.

Douglas, D.S. (1968). Salt and water metabolism of the Adelie penguin. Antarctic Res. Ser., 12, 167.

Douglas, D.S. (1970). Electrolyte excretion in sea water loaded herring gulls. Am. J. Physiol., 219, 534.

Douglas, D.S., and S.M. Neely. (1969). The effect of dehydration on salt gland performance. Am. Zool., 9, 1095.

Edson, N.L., A. Krebs, and A. Motel. (1936). The synthesis of uric acid in the avian organism: Biochem. J., 36, 1380.

Edwards, W.H., and W.O. Wilson. (1954). Relationship of hyperthermy to nitrogen excretion in chickens. Am. J. Physiol., 179, 76.

Ensor, D.M., Thomas, D.H., and Phillips, J.G. (1970). The possible role of the thyroid in extrarenal secretion following a hypertonic saline load in the duck (*Anas platyrhynchos*). Proceedings of Society of Endocrinology. J. Endocrinol., 46, x.

Ernst, S.A., and R.A. Ellis. (1969). Development of surface specialization in secretory epithelium of avian salt gland in response to osmotic stress. J. Cell. Biol., 40, 305.

Fange, R., K. Schmidt-Nielsen, and H. Osaki. (1958). The salt gland of the herring gull. Biol. Bull. (Woods Hole, Mass.), 115, 162.

Fange, R., J. Krog, and O. Reite. (1963). Blood flow in avian salt gland studied by polarographic oxygen electrodes. Acta. Physiol. Scand., 58, 40.

Fange, R., K. Schmidt-Nielsen, and M. Robinson. (1958). Control of secretion from the avian salt gland. Am. J. Physiol., 195, 321.

Ferguson, R.K., and R.A. Wolbach. (1967). Effects of glucose, phlorizin, and parathyroid extract on renal phosphate transport in chickens. Am. J. Physiol., 212, 1123.

Fletcher, G.L., and W.N. Holmes. (1968). Observations on the intake of water and electrolytes by the duck maintained on fresh water and on hypertonic saline. J. Exp. Biol., 49, 325.

Fletcher, G.L., I.M. Stainer, and W.N. Holmes. (1967). Sequential changes in the adenosinetriphosphatase activity and the electrolyte excretory capacity of the nasal glands of the duck (*Anas platyrhynchos*) during the period of adaptation to hypertonic saline. J. Exp. Biol., 47, 375.

Folin, O., H. Berglund, and C. Derick. (1924). The uric acid problem: an experimental study on animals and man, including gouty subjects. J. Biol. Chem., 60, 361.

Fussell, M.H. (1969). A method for the separation and collection of urine and feces in the fowl. Res. Vet. Sci., 10, 332.

Gibbs, O.S. (1929a). The secretion of uric acid by the fowl. Am. J. Physiol., 88, 87.

Gibbs, O.S. (1929b). The effects of drugs on the secretion of uric acid in the fowl. J. Pharmacol. Exp. Ther., 35, 49.

Gibbs, O.S. (1929c). The function of the fowl's ureters. Am. J. Physiol., 87, 594.

Hakansson, C.H., and B. Malcus. (1969). Secretive response of electrically stimulated nasal salt gland in herring gull. Acta Physiol. Scand., 76, 385.

Hanwell, A., J.L. Linzell, and M. Peaker. (1971a). Salt gland secretion and blood flow in geese. J. Physiol. (London), 210, 373.

Hanwell, A., J.L. Linzell, and M. Peaker. (1971b). Cardiovascular responses to salt loading in conscious domestic geese. J. Physiol., 213, 389.

Hanwell, A., J.L. Linzell, and M. Peaker. (1972). Nature and location of receptors for salt gland secretion in the goose. J. Physiol., 226, 453.

Hart, W.M., and H.E. Essex. (1942). Water metabolism of the chicken with special reference to the role of the cloaca. Am. J. Physiol., 336, 657.

Hester, H.R., H.E. Essex, and F.C. Mann. (1940). Secretion of urine in the chicken. Am. J. Physiol., 128, 592.

Holmes, W.N. (1965). Some aspects of osmoregulation in reptiles and birds. Arch. Anat. Microsc. Morphol. Exp., 54, 491.

Holmes, W.N. (1972). Regulation of electrolyte balance in marine birds with special reference to the role of the pituitary adrenal axis in the duck. Fed. Proc., 31, 1587.

Holmes, W.N. (1975). Hormones and osmoregulation in marine birds. Gen. Comp. Endocrinol., 25, 249.

Holmes, W.N., and A. Wright. (1968). Some aspects of the control of osmoregulation and homeostasis in birds. Progress in Endocrinology. Proc. Int. Cong. Endocrinol., 3rd, p. 237.

Holmes, W.N., G.L. Fletcher, and D.J. Steward. (1968). The patterns of renal electrolyte excretion in ducks, maintained on freshwater and on hypertenic saline. J. Exp. Biol., 48, 487.

Hootman, S.R., and S.A. Ernst. (1980). Dissociation of avian salt glands: suspension procedures and characterization of dissociated cells. Am. J. Physiol., 238, C184.

Hootman, S.R., and S.A. Ernst. (1981). Effect of methacholine on Na pump activity and ion content of dispersed avian salt gland cells. Am. J. Physiol., 241, R77.

Hughes, M.R. (1970). Relative kidney size in nonpasserine birds with functional salt glands. Condor, 72, 164.

Imbayashi, K., M. Kametaka, and T. Hatano. (1956). Studies on the digestion in the domestic fowl. I: "Artificial anus operation" for the domestic fowl and the passage of the indicator throughout the digestive tract. Tohoku J. Agric. Res., 6, 99.

Inoue, T. (1963). Nasal salt gland: Independence of salt and water transport. Science, 142, 1299.

Johnson, O.W. (1968). Some morphological features of avian kidneys. Auk, 85, 216.

Johnson, O.W., and J.N. Mugaas. (1970a). Some histological features of avian kidneys. Am. J. Anat., 127, 423.

Johnson, O.W., and J.N. Mugaas. (1970b). Quantitative and organizational features of avian renal medulla. Condor, 72, 288.

Johnson, O.W., and R.D. Ohmart. (1973). Some features of water economy and kidney microstructure in the large billed savannah sparrow. Physiol. Zool., 46, 276.

Johnson, O.W., G.L. Phipps, and J.N. Mugaas. (1972). Injection studies of cortical and medullary organization in the avian kidney. J. Morphol., 136, 181.

Johnson, O.W., and E. Skadhauge. (1975). Structural and functional correlations in the kidney of observations on colon and cloaca morphology in certain Australian birds. J. Morphol., 136, 181.

Karasawa, Y., and A. Sakae. (1981). Comparative effect of intravenously administered purine compounds on uric acid production in chickens. Comp. Biochem. Physiol. A, 70, 591.

Katayami, T. (1924). Uber die verdaulichkeit der Futtermittel bei Huhnern. Bull. Imp. Agric. Exp. Sta. Jpn. (Quoted by Skadhauge, 1981.)

Koike, T.I., L.R. Pryor, H.L. Neldon, and R.S. Venable. (1977). Effect of water deprivation on plasma radio immunoassayable AVT in conscious chickens. Gen. Comp. Endocrinol., 33, 259.

Koike, T.I., L.R. Pryor, and H.L. Neldon. (1979). Effect of saline infusion on plasma immunoreactive vasotocin in conscious chickens. Gen. Comp. Endocrinol., 37, 451.

Koike, T.I., L.R. Pryor, and H.L. Neldon. (1983). Plasma volume and electrolytes during progressive water deprivation in chickens. Comp. Biochem. Physiol. A, 74, 83.

Korr, I.M. (1939). The osmotic function of the chicken kidney. J. Cell. Comp. Physiol., 13, 175.

Krag, B., and E. Skadhauge. (1972). Renal salt and water excretion in the budgerïgar. Comp. Biochem. Physiol. A, 41, 667.

Langford, H.G., and N. Fallis. (1966). Diuretic effect of angiotension in the chicken. Proc. Soc. Exp. Biol. Med., 123, 317.

Laverty, G., and W.H. Dantzler. (1982). Micropuncture of superficial nephrons in avian (*Sturnus vulgaris*) kidney. Am. J. Physiol., 243, F561.

Laverty, G.N., and W.H. Dantzler. (1983). Micropuncture study of urate transport by superficial nephrons in avian (*Sturnus vulgaris*) kidney. Pfleügers Arch., 397, 232.

Levine, R., W.Q. Wolfson, and R. Lenel. (1947). Concentration and transport of true urate in the plasma of the azotemic chicken. Am. J. Physiol., 151, 186.

Levinsky, M.G., and D.G. Davidson. (1957). Renal action of parathyroid extract in the chicken. Am. J. Physiol., 191, 530.

Long, S., and E. Skadhauge. (1983). Renal acid excretion in the domestic fowl. J. Exp. Biol., 104, 51.

Louw, G.N., P.C. Belonde, and H.J. Coetzee. (1969). Renal function, respiration heart rate and thermoregulation in the ostrich. Sci Pap. Namib Desert Res. Stn., 42, 43. Quoted by Skadhauge, (1981).

Marshall, E.K. (1934). Comparative physiology of vertebrate kidney. Physiol. Rev., 14, 133.

May, D.G., and M.K. Carter. (1970). Effect of vasoactive agents on urine and electrolyte excretion in the chicken. Am. J. Physiol., 218, 417.

Mayrs, E.B. (1923). Secretion as a factor in elimination by bird kidney. J. Physiol. (London), 58, 276.

McNabb, F., M. Anne, and T.L. Poulson. (1970). Uric acid excretion in pigeons. Comp. Biochem. Physiol., 33, 933.

McNabb, F.M.A. (1969a). A comparative study of water balance in three species of quail. I. Water turnover in the absence of temperature stress. Comp. Biochem. Physiol., 28, 1045.

McNabb, F.M.A. (1969b). A comparative study of water balance in three species of quail. II. Utilization of saline drinking solutions. Comp. Biochem. Physiol., 28, 1059.

McNabb, R.A. (1974). Urate and cation interactions in the liquid and precipitated fractions of avian urine and speculations on their physicochemical state. Comp. Biochem. Physiol. A, 48, 45.

Mongin, P., and L. Lacassagne. (1967). Excretion urinaire chez la poule au moment de la ponte de son premier oeuf. C. R. Seances Acad. Sci. Paris, 264, 2479.

Munsick, R.A., W.H. Sawyer, and H.B. Van Dyke. (1960). Avian neurohypophyseal hormones: pharmacological properties and tentative identification. Endocrinology, 66, 860.

Nechay, B.R., and L. Nechay. (1959). Effects of probenecid, sodium salicylate, 2,4 dinitrophenol and pyrazinamide on renal secretion of uric acid in chickens. J. Pharmacol. Exp. Ther., 126, 291.

Nechay, B.R., and E. Sanner. (1961). Interference of reserpine with diuretic action of theophylline and hydrochlorothiazide on the chicken. Acta. Pharmacol. Toxicol., 18, 339.

Nechay, B.R., S. Boyarsky, and P. Catacutan-Labay. (1968). Rapid migration of urine into intestine of chickens. Comp. Biochem. Physiol., 26, 369.

O'Dell, G.L., W.D. Woods, O.A. Laerdal, A.M. Jeffay, and J.E. Savage. (1960). Distribution of the major nitrogenous compounds and amino acids in chicken urine. Poult. Sci., 39, 426.

Ohmart, R.D., T.E. Chapman, and L.Z. McFarland. (1970a). Water turnover in Roadrunners under different environmental conditions. Auk, 87, 787.

Ohmart, R.D., L.Z. McFarland, and J.P. Morgan. (1970b). Urographic evidence that urine enters the rectum and ceca of the roadrunner. Comp. Biochem. Physiol., 35, 487.

Orloff, J., and D. Davidson. (1956). Mechanism of potassium excretion in the chicken. Fed. Proc. Fed. Am. Soc. Exp. Biol., 15, 452.

Owen, R.E., and R.R. Robinson. (1964). Urea production and excretion by the chicken kidney. Am. J. Physiol., 206, 1321.

Palmore, W.P., M.J. Fregly, and C.E. Simpson. (1981). Catecholamine-induced diuresis in turkeys. Proc. Soc. Exp. Biol. Med., 167, 1.

Parmalee, M.L., and M.K. Carter. (1968). The diuretic effect of acetylcholine in the chicken. Arch. Int. Pharmacodyn. Ther., 174, 108.

Paulson, G.D. (1969). An improved method for separate collection of urine, feces, and expiratory gases from the mature chicken. Poult. Sci., 48, 1331.

Peaker, M., A. Wright, S.J. Peaker, and J.G. Phillips. (1968). Absorption of tritiated water by cloaca of the domestic duck. Physiol. Zool., 41, 461.

Peaker, M., J.L. Linzell. (1975). Salt glands in birds and reptiles. Cambridge, England: Cambridge University Press, p. 1.

Peaker, M. (1978). Do osmoreceptors or blood volume receptors initiate salt gland secretion in birds. J. Physiol. (London), 276, 66P.

Phillips, J.G., W.N. Holmes, and D.G. Butler. (1961). The effect of total and subtotal adrenalectomy on the renal and extrarenal response of the domestic duck (*Anas platyrhynchos*) to saline loading. Endocrinology, 69, 958.

Pitts, R.F. (1938). The excretion of phenol red by chickens. J. Cell. Comp. Physiol. 11, 99.

Pitts, R.F., and I.M. Korr. (1938). The excretion of urea by the bird. J. Cell. Comp. Physiol., 11, 117.

Polin, D., E.R. Wynosky, M. Loukides, and C.C. Porter. (1967). A possible urinary backflow to ceca revealed by studies on chicks with artificial anus and fed amprolium ^{14}C. Poult. Sci., 46, 88.

Poulson, T.L. (1965). Counter current multipliers in avian kidneys. Science, 148, 389.

Poulson, T.L., and G.A. Bartholomew. (1962). Salt balance in the savannah sparrows. Physiol. Zool., 35, 109.

Prashad, D.N., and N.E. Edwards. (1973). Phosphate excretion in the laying fowl. Comp. Biochem. Physiol. A, 46, 131.

Rennick, B.R. (1967). Transport mechanisms for renal tubular excretion of creatinine in the chicken. Am. J. Physiol., 212, 1131.

Rennick, B.R., and H. Gandia. (1954). Autonomic pharmacology of the smooth muscle valve in renal portal venous circulation in birds. Fed. Proc. Fed. Am. Soc. Exp. Biol., 13, 396.

Robinson-Hughes, M. (1980). Glomerular filtration rate in saline acclimated ducks, gulls and geese. Comp. Biochem. Physiol. A, 65, 211.

Sandor, T., A.Z. Mehdi, and A.G. Fazekas. (1977). Corticosteroid binding macromolecules in the salt-activated nasal gland of the domestic duck. Gen. Comp. Endocrinol., 32, 348.

Scheiber, A.R., and H.E. Dziuk. (1969). Water ingestion and excretion in turkeys with a rectal fistula. J. Appl. Physiol., 26, 277.

Schmidt-Nielsen, K. (1960). The salt secreting gland of marine birds. Circulation, 21, 955.

Schmidt-Nielsen, K., and R. Fange. (1958). The function of the salt gland in the brown pelican. Auk, 75, 282.

Schmidt-Nielsen, K., and W.J. Sladen. (1958). Nasal salt secretion in the Humboldt penguin. Nature (London), 181, 1217.

Schmidt-Nielsen, K., A. Borut, P. Lee, and E. Crawford, J. (1963). Nasal salt excretion and possible function of the cloaca in water conservation. Science, 142, 1300.

Schmidt-Nielsen, K., C. E. Jorgensen, and H. Osaki. (1958). Extrarenal salt excretion in birds. Am. J. Physiol., 193, 101.

Scothorne, J.J. (1959a). The nasal glands of birds: a histological and histochemical study of the inactive gland in the domestic duck. J. Anat., 93, 246.

Scothorne, R.J. (1959). On the response of the duck and pigeon to intravenous hypertonic saline solutions. Q. J. Exp. Physiol., 44, 200.

Sell, J.L., and S.L. Balloun. (1961). Nitrogen retention and nitrogenous urine components of growing cockerels as influenced by diethylstilbestrol, methyl testosterone and porcine growth hormone. Poult. Sci., 40, 1117.

Shannon, J.A. (1938a). The excretion of exogenous creatinine by the chicken. J. Cell. Comp. Physiol., 11, 123.

Shannon, J.A. (1938b). The excretion of uric acid by the chicken. J. Cell. Comp. Physiol., 11, 135.

Shideman, J.R., R.L. Evans, D.W. Bierer, and A.J. Quebbeman. (1981). Renal venous portal contribution to PAH and uric acid clearance in the chicken. Am. J. Physiol., F46.

Shoemaker, V.H. (1967). Renal function in mourning dove. Am. Zool., 7, 736.

Shoemaker, V.H. (1972). Osmoregulation and excretion in birds. In "Avian Biology," Vol. II (D.S. Farner and J.R. King, Eds.). Chapter 9. New York: Academic Press.

Siller, W.G., and R.N. Hindle. (1969). The artificial blood supply to the kidney of the fowl. J. Anat., 194, 117.

Siller, W.G. (1971). Structure of the kidney. In "Physiology and Biochemistry of Domestic Fowl" (D.G. Bell and B.M. Freeman, Eds.). New York: Academic Press, Chapter 8.

Simon-Opperman, C., H.I. Hammel, and E. Simon. (1979). Hypothalamic temperature and osmoregulation in Peking duck. Pfleugers Arch., 378, 213.

Skadhauge, E. (1964). The effect of unilateral infusion of arginine vasotocin into the portal circulation of the avian kidney. Acta Endocrinol., 47, 321.

Skadhauge, E. (1968). The cloacal storage of urine in the rooster. Comp. Biochem. Physiol., 24, 7.

Skadhauge, E., and B. Schmidt-Nielsen. (1967a). Renal function in the domestic fowl. Am. J. Physiol., 212, 793.

Skadhauge, E., and B. Schmidt-Nielsen. (1967b). Renal medullary electrolyte and urea gradient in chickens and turkeys. Am. J. Physiol., 212, 1313.

Skadhauge, E., and S.D. Bradshaw. (1974). Saline drinking and cloacal excretion of salt and water in the zebra finch. Am. J. Physiol., 227, 1263.

Skadhauge, E. (1974). Renal concentrating ability in selected West Australian birds. J. Exp. Biol. Physiol., 61, 269.

Skadhauge, E. (1981). Osmoregulation in birds. Heidelberg: Springer-Verlag.

Sokabe, H., and M. Ogawa. (1974). Comparative studies of the juxtaglomerular apparatus. Int. Rev. Cytol. 37, 271.

Smith, H.W. (1951). "The kidney: Structure and function in health and disease." London: Oxford University Press.

Spanner, R. (1925). Der Pfortaderkreislauf in der Vogelniere. Morphol. Jahrb., 54, 560.

Spanner, R. (1939). Die Drosselklappe der veno-venosen Anastomose und ihre Bedeutung für den Abkürzungskreislauf in portocavalen. System des Vogels; zugleich ein Beitrag der epitheloiden Zellen. Z. Anat. Entwicklungsmech., 109, 443.

Sperber, I. (1948). Investigations on the circulatory system of the avian kidney. Zool. Bidrag (Uppsala), 27, 429.

Sperber, I. (1960). Excretion. In "Biology and Comparative Physiology of Birds" (A.J. Marshall, Ed). New York: Academic Press, Chapter 12.

Stewart, D.D., W.N. Holmes, and G. Fletcher. (1969). The renal excretion of nitrogenous compounds by the duck maintained on freshwater and hypertonic saline. J. Exp. Biol., 359, 127.

Stewart, D.J. (1972). Secretion by salt gland during water deprivation in the duck. Am. J. Physiol., 223, 384.

Sturkie, P.D. (1965). In "Avian Physiology" (2nd ed.) (P.D. Sturkie, Ed.). Ithaca: Cornell University Press, Chapter 13.

Sturkie, P.D. (1976). In "Avian Physiology" (3rd ed) (P.D. Sturkie, Ed.). New York: Springer-Verlag. Berlin.

Svendsen, C., and E. Skadhauge. (1976). Renal function in hens fed graded dietary levels of Ochratoxin A. Acta Pharmacol. Toxicol., 38, 186.

Sykes, A.H. (1960a). The renal clearance of uric acid and *p*-amino hippurate in the fowl. Res. Vet. Sci., 1, 308.

Sykes, A.H. (1960b). The excretion of inulin, creatinine and ferrocyanide by the fowl. Res. Vet. Sci., 1, 315.

Sykes, A.H. (1961). The action of hydrochlorothiazide in the turkey. Vet. Rec., 73, 396.

Sykes, A.H. (1962). The excretion of urea following its infusion through the renal portal system of the fowl. Res. Vet. Sci., 3, 183.

Sykes, A.H. (1966). Submersion anuria in the duck. J. Physiol., 184, 16. (Proceedings).

Sykes, A.H. (1971). Formation and composition of urine. "Physiology and Biochemistry of the Fowl," Vol. 1 (D.J. Bell and B.M. Freeman, Eds.). New York: Academic Press, Chapter 9.

Szalagyi, K., and A. Kriswuscha. (1914). Untersuchungen uber die chemische Zusammensetzung und die physikalischen Eigenschaften des Enten und Huhnerharnes. Biochem. Z., 66, 122. (Quoted by Skadhauge, 1981.)

Teekell, R.A., C.E. Richardson, and A.B. Watts. (1968). Dietary protein effects on urinary nitrogen components of the hen. Poult. Sci., 47, 1260.

Umflat, J.G., R.E. Kissell, R.F. Wideman Jr., and F.V. Muir. (1985). A comparison of two techniques for determining glomerular size distributions in the domestic fowl. Poult. Sci., 64, 1210.

Vogel, G., I. Stoeckert, W. Kroger, and I. Dobberstein. (1965). Harn und berertung bei terrestrisch lebenden vögeln. Untersuchungen am Truthuhn. Z. Veterinarmed., 12, 132. (Quoted by Skadhauge, 1981.)

Weyrauch. H.M., and S.I. Roland. (1957). Electrolyte absorption from fowl's cloaca. Trans. Am. Assoc. Genito-Urinary Surgeons, 49, 117.

Wideman, R.F., Jr., N.B. Clark, and E.J. Braun. (1980). Effects of phosphate loading and parathyroid hormone on starling renal phosphate excretion. Am. J. Physiol., 239, F2333.

Wideman, R.E., Jr., E.J. Braun, and G.L. Anderson. (1981). Microanatomy of the domestic fowl renal cortex. J. Morph. 168, 249.

Wideman, R.F., E.T. Mallinson, and H. Rothenbacher. (1983). Kidney function of pullets during outbreaks of urolithiasis. Poult. Sci., 62, 1954.

Willoughby, E.J., and M. Peaker. (1979). In "Comparative Physiology of Osmoregulation in Animals," Vol. 2, Birds. (G.M.O. Maloiy, Ed.). New York: Academic Press.

Zucker, I.H., C. Gilmore, J. Dietz, and J.P. Gilmore. (1979). Effects of volume expansion and Veratrine on salt gland secretion. Am. J. Physiol., 232, R185.

17
Pituitary Gland

C.G. Scanes

Introduction

Birds have a complement of endocrine organs similar to those of mammals and lower vertebrates; these are (1) the pituitary–hypothalamus complex, (2) gonads, (3) pancreatic islets, (4) adrenal glands, (5) thyroid glands, (6) parathyroid gland, (7) ultimobranchial gland, and (8) the endocrine cells of the gut. These ductless or endocrine glands release hormones into the bloodstream. The hormone(s) then acts on specific tissues, cells, or organs (referred to as the target organ, etc.). The hormone acts by interacting with receptors on the surface of the cells (proteins and polypeptides) or within the cytoplasm and/or nucleus (steroids). In addition to the "classical" endocrine glands, hormone-like substances are produced by other organs including the pineal (melatonin), thymus (thymosin), liver (somatomedins), and kidney (renin, Vitamin D metabolites).

Anatomy of the Hypothalamic–Hypophyseal Complex

The pituitary gland (hypophysis) is intimately connected to the hypothalamus at the base of the brain. The pituitary gland has a complex structure and an interesting embryonic development. Pituitary tissue can be classified as either adenohypophysis or neurohypophysis; each has a distinct embryonic origin. The adenohypophysis is derived from Rathke's pouch (probably ectoderm from the roof of the mouth), and the neurohypophysis is derived from the infundibulum (an outgrowth of the brain). The adenohypophysis in mammals goes to form the pars distalis (anterior pituitary gland), the pars intermedia, and the pars tuberalis. However, in birds there is no pars intermedia, and hence the adenohypophysis forms the anterior pituitary gland and the pars tuberalis. The cells of the pars tuberalis are found at the base of the brain (hypothalamus). The neurohypophysis forms the pars nervosa (the equivalent of the posterior pituitary gland in birds), the infundibular stalk, and the median eminence. The structure of the hypothalamic–hypophyseal complex is shown in Figures 17–1 and 17–2.

The anterior lobe of the pituitary gland is well supplied with blood vessels, including the hypophyseal portal vessels. These latter provide a route from the neurosecretory nerve terminals in the median eminence to the anterior pituitary gland. Indeed it is by way of the portal blood vessels that the anterior pituitary gland is controlled, by *releasing factors* from the median eminence. It should be stressed that there is little or no nervous innervation of the anterior pituitary gland. The size of the anterior lobe is relatively small, depending on body size, and weighs 7–10 mg in adult chickens and 16–23 mg in adult turkeys.

The posterior pituitary gland consists of the neurosecretory terminals, which release mesotocin or arginine vasotocin. These hormones are synthesized in cell bodies in nuclei in the hypothalamus and are transported to the posterior pituitary gland through modified axons.

Adenohypophyseal Hormones

The avian pars distalis produces a full complement of hormones. These will be considered under the follow-

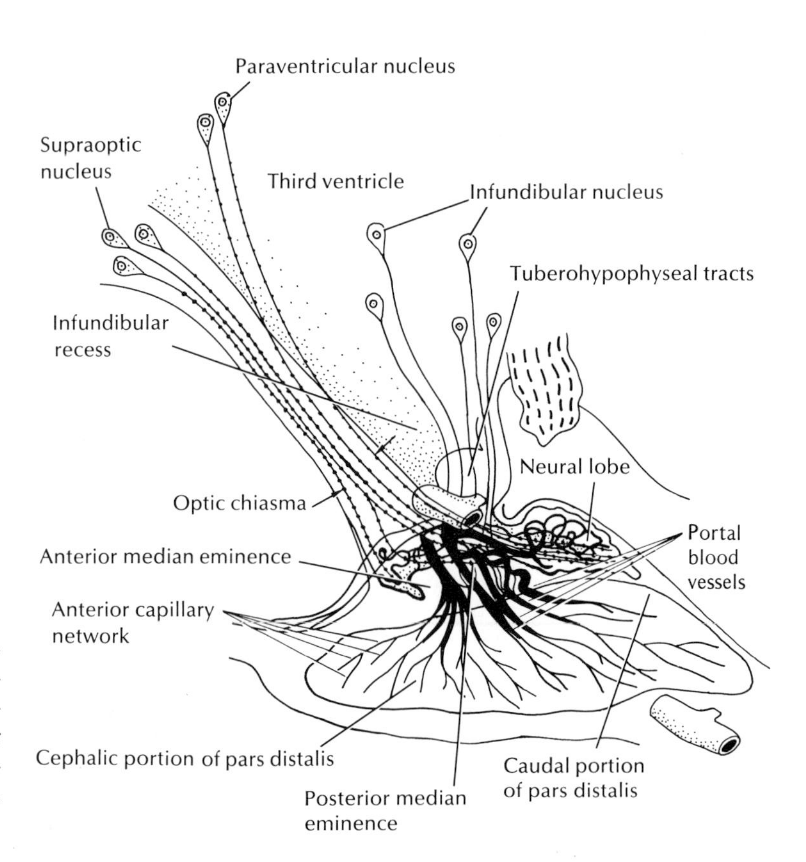

FIGURE 17–1. Schematic parasagittal section through the infundibulum and hypophysis of white-crowned sparrow, showing fiber tracts from hypothalamus and portal blood vessels in the hypophysis. (After Oksche, 1965.)

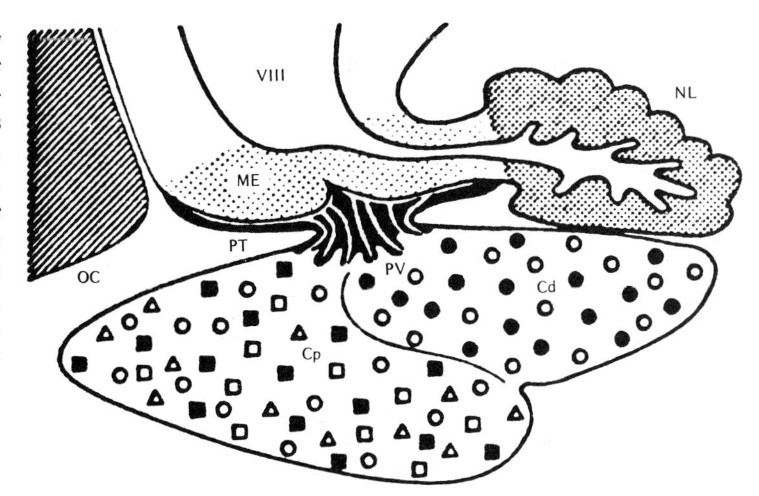

FIGURE 17–2. Diagrammatic midsagittal section of the hypophysis of the Japanese quail to show the distribution of cell types of the pars distalis in Japanese quail. Abbreviations: *Cd*, caudal lobe; *Cp*, cephalic lobe; *ME*, mediam eminence; *NL*, neural lobe (pars nervosa); *OC*, optic chiasma; *PT*, pars tuberalis; *PV*, portal vessels; *VIII*, third ventricle. Symbols: ●, GH; ○, LH/FSH; ■, ACTH; □, PRL; △, TSH. (Adapted from Mikami, 1960.)

ing topics: gonadotropins, thyrotropin, prolactin, growth hormone, and adrenocorticotropin and related hormones.

Gonadotropins

Chemistry. Luteinizing hormone (LH) and follicle-stimulating hormone (FSH) have been isolated from chicken, turkey, and ostrich pituitary tissue. The purification of avian LH and FSH has been achieved using techniques developed for the isolation of mammalian gonadotropins. These methods include differential extraction of pituitary tissue, salt precipitation, and separation of proteins on the basis of their charge, by ion-exchange chromatography, and of their size, by gel-filtration chromatography.

Both avian LH and FSH are glycoprotein hormones with molecular weights of approximately 30,000 (Burke et al., 1979a; Papkoff et al., 1982). Ostrich FSH has a high carbohydrate content (25.4 g/100 g glycoprotein) and a relatively high sialic acid content (5%). Ostrich LH has a lower content of both carbohydrate (17.0%) and sialic acid (2%) (Papkoff et al., 1982). Turkey LH and FSH have even lower contents of carbohydrate, 14% and 7.6%, respectively (Burke et al., 1979a).

Avian LH and FSH, like their mammalian homologs, consist of two glycoprotein subunits (molecular weight, 15,000 each); an α subunit [common to LH, FSH, and thyroid-stimulating hormone (TSH)] and a β subunit (hormone specific). The subunits of turkey LH have been separated and isolated (Burke et al., 1979b); the subunits of avian LH, and presumably FSH and TSH also, are biologically inactive when separate but regain activity when α and β are recombined.

Action. The role of LH and FSH in controlling testicular functioning in the male bird is now well established (see Chapter 19). As in mammals, LH acts primarily to stimulate the Leydig cells to differentiate and produce testosterone, while FSH promotes spermatogenesis. The administration of chicken LH to hypophysectomized Japanese quail greatly increased the number of mature Leydig cells and decreased the number of fibroblasts and transitional cell types in the interstitium (Brown et al., 1975). The injection of preparations of avian or mammalian LH into Japanese quail elevated the plasma concentration of testosterone; for instance, 10 μg of chicken LH increases the plasma concentration of testosterone from 4.1 to 13.5 ng/ml (Maung and Follett, 1978). Similarly, the in vitro synthesis of testosterone is stimulated by preparations of avian or mammalian LH using tissue from quail (Maung and Follett, 1977; Jenkins et al., 1978), chickens, ducks, and turkeys (Chase, 1982). Neither avian nor mammalian FSH preparations have any effect on the in vivo or in vitro production of testosterone by bird testes, except at very high concentrations. In this case, the activity can be explained by LH contamination in the FSH preparations (Jenkins et al., 1978; Maung and Follett, 1978; Chase, 1982). On the other hand, avian FSH, when injected into hypophysectomized quail, increases testicular size (8.8-fold) and seminiferous tubule diameter (2.1-fold), and promotes Sertoli cell differentiation and spermatogenesis (Brown et al., 1975).

Tsutsui and Ishii (1978) observed increases in the concentration of FSH receptors in quail following testosterone administration. In a more detailed investigation, the amount of FSH specifically bound per Sertoli cell in quail testes was decreased by hypophysectomy to 12.2% of that in the intact bird. Conversely, both FSH and testosterone increase FSH binding by Sertoli cells, and together exert a synergistic effect on FSH binding (Tsuisui and Ishii, 1980). Thus, FSH appears to affect the number of its own receptors, whereas LH may influence FSH action by increasing the production of testosterone production and hence FSH receptors.

Gonadotropins are obviously essential for ovarian functioning in birds. However, their exact roles have not been established. Both LH and FSH are necessary for normal ovarian physiology (see also Chapters 18 and 19). Ovulation is induced by LH (Imai, 1973) and prevented by the administration of antisera against chicken LH (Sharp et al., 1978). Although both LH and FSH stimulate ovarian steroidogenesis, LH appears to have the major effect, at least in the largest follicles. In vivo, mammalian LH has been demonstrated to increase progesterone and testosterone production by the domestic hen (Shahabi et al., 1975). In vitro, LH stimulates progesterone synthesis by granulosa cells (Hammond et al., 1978; Scanes and Fagioli, 1980; Wells et al., 1980). Huang et al. (1979) showed that high concentrations of either mammalian LH or FSH increased in vitro estrogen and testosterone production by theca cells and progesterone production by granulosa cells. The relative importance of the gonadotropins in affecting testosterone and estrogen biosynthesis is not established. LH is by far the predominant gonadotropin affecting progesterone production by granulosa cells of the largest follicle (Huang et al., 1979; Scanes and Fagioli, 1980; Hammond et al., 1981). However, FSH appears to have some—perhaps the major—influence on granulosa cells from the smaller follicles (Huang et al., 1979; Hammond et al., 1981).

Mechanism of Gonadotropin Action. It appears that LH exerts its effect on the testis (Leydig cells) and the granulosa cells of the ovary by binding to a receptor, activating adenyl cyclase, and increasing the intracellular concentration of cyclic adenosine 3′, 5′-monophosphate (cAMP) (see also Chapter 19). This is supported in the male by the high-affinity (K_a, $16.7 \times 10^9\ M^{-1}$) receptors for LH in the turkey testis (Bona Gallo and Licht, 1979); by the stimulation of androgen synthesis by cAMP in vitro, and by the increase in intracellular cAMP concentration induced by LH in quail testes cells (Maung and Follett, 1977). Similarly LH binds to avian ovarian tissue (Bona Gallo and Licht, 1979) and stimulates the granulosa adenyl cyclase system (Calvo and Bahr, 1982; Tákats and Hertelendy, 1982). Moreover, progesterone production from granulosa cells can be increased by dibutyryl cAMP, while the effect of LH is potentiated if cAMP degradation is depressed by inhibiting phosphodiesterase (Hammond et al., 1980; Zakar and Hertlendy, 1980). It is likely that FSH acts in a similar manner.

Assay. The purification of avian gonadotropins requires the availability of biological assays. The fractionation of LH and FSH has been followed by in vivo and in vitro biological assays using various responses in gonadal tissue obtained from rodents, birds, and even reptiles. In addition, radioreceptor assays for FSH and LH have been employed to monitor gonadotropin purification (Follett et al., 1978; Sakai and Ishii, 1980). These "biological" systems are largely unsuitable for the measurement of the concentration of LH or FSH in body fluids because of their lack of sensitivity and problems of poor precision, high cost, and lack of specificity (Follett et al., 1972). However, the application of the technique of radioimmunoassay to avian gonadotropins has led to tremendous advances in our knowledge of their physiology. The purification of LH from chicken and turkey pituitary tissue allowed the development of homologus radioimmunoassays for these hormones (Follett et al., 1972; Wentworth et al., 1976); these assays are specific and make possible the determination of LH concentrations in small quantities of plasma. The chicken LH radioimmunoassay (Follett et al., 1972) cross-reacts with LH throughout the Aves, and has been applied to a variety of problems of LH secretion in many species of birds (see below for examples). A heterologous radioimmunoassay for mammalian FSH was validated by Follett (1976) for determining plasma concentrations of FSH in birds, and this has been used extensively in investigating the photoperiodic control of reproduction in birds, particularly in the Japanese quail and white-crowned sparrow (see below for details). The purification of chicken FSH led to the development of a homologous radioimmunoassay (Scanes et al., 1977a) that is both specific and sensitive. However, the low titer of the antisera precludes its extensive application to problems of avian endocrinology.

Pituitary Location. The LH-producing cells are found throughout the pars distalis, but their concentration is higher in the caudal section of the anterior pituitary gland (Follett et al., 1978). The distribution of the FSH-producing cells within the anterior pituitary gland is not known.

Hypothalamic Control. The secretion of avian gonadotropins is controlled by a hypothalamic-hypophysiotropic principle, the luteinizing-hormone-releasing hormone (LHRH). Recently King and Millar (1982a, b) determined the structure of chicken LHRH to be as follows:

p Glu-His-Trp-Ser-Tyr-Gly-Leu-GlNH$_2$-Pro-Gly-NH$_2$

This decapeptide differs from its mammalian homolog by the substitution of a glutamine residue for arginine at position 8.

The location of the LHRH cell bodies and neurons within the hypothalamus has been examined by immunohistochemistry (Bons, 1980). LHRH cell bodies are located in the periventricular preoptic nucleus and in the tuberoinfundibular neuron. The neurons extend to the median eminence, where they terminate. LHRH is released from neurosecretory terminals into the portal blood vessels and hence to the anterior pituitary gland.

The hypothalamic control of LH secretion has been extensively examined by Davies and Follett (1975, 1980). The electrical stimulation of the infundibular nuclear complex caused large increases in the plasma concentration of Japanese quail. Conversely, if these loci are destroyed by lesioning, low concentrations of LH are observed.

Mammalian LHRH stimulates the secretion of LH from the avian pituitary gland in vivo (Bonney et al., 1974) and in vitro (Bonney and Cunningham, 1977a, b). The action of LHRH on the gonadotroph in the pituitary gland requires calcium, and can be accentuated by either inhibiting cAMP degradation or adding a cAMP analog (Bonney and Cunningham, 1977a, b).

Studies using pharmacologic neurotransmitter synthesis blockers, agonists, and antagonists suggest that the principal neurotransmitter affecting LHRH, and hence LH release, is norepinephrine (El Halawani et al., 1980b; Buonomo et al., 1981; Knight et al., 1982).

Negative and Positive Feedback. Sex steriods affect gonadotropin secretion in birds, primarily by influencing the release of LHRH from the median eminence in the portal blood vessels. This is predominantly an inhibitory effect, as gonadectomy leads to large increases in the circulating concentration of both LH and FSH. Studies in the castrated Japanese quail by Davies et al. (1980) demonstrated that LH secretion in the male is inhibited by estrogens (estradiol) and androgens. The effectiveness of the androgens was similar whether they could be aromatized to estrogens (such as testosterone or androstenedione) or were nonaromatizable (5α-dihydrotesterone, DHT). On the other hand, 5β- reduced androgens were largely inactive in affecting LH secretion. FSH secretion in quail is similarly depressed by testosterone and estradiol, but is not affected by DHT. This suggests that the pituitary is one of the sites of negative feedback, as changes in LHRH release would probably affect both LH and FSH secretion. This is supported by the observation that implanting testosterone in castrated Japanese quail decreases the in vivo response to LHRH (Davies and Bicknell, 1976). However, reduced basal LHRH release from the hypothalamus, due to lesioning, decreases, the responsiveness of the pituitary gland to LHRH (Davies and Follett, 1980).

In the domestic hen, there is evidence that estradiol exerts a *negative feedback* effect on LH secretion, but progesterone can either stimulate or inhibit LH release. In the ovariectomized hen, plasma concentrations of LH are decreased by a single injection of either progesterone or estradiol (Wilson and Sharp, 1976a, b). Progesterone has a *positive feedback* effect on LH release in the intact or ovariectomized hen that has been primed with progesterone and estradiol (Etches and Cunningham, 1976; Wilson and Sharp, 1976a, b). Perhaps the best evidence that the locus of progesterone feedback is within the hypothalamus is the ability of passive immunization with antisera against LHRH to block induction of LH release by progesterone (Frazer and Sharp, 1978). It is probable that the preovulatory LH surge in the hen is induced by progesterone.

Female Reproductive Cycle. The hormonal control of the ovulation cycle will be considered elsewhere in this volume (Chapter 18). However, it is pertinent to briefly outline the secretion of gonadotropins during the ovulation cycle of the domestic fowl. There is a preovulatory surge in plasma LH concentrations approximately 4–6 hr before ovulation (Furr et al., 1973). In addition to this LH peak, there is evidence for a spike of LH secretion at the time of artificial dusk (Wilson and Sharp, 1973). Similar patterns of plasma LH concentrations occur during the ovulation cycles of other birds.

There are only small differences in the circulatory concentrations of FSH, as determined by radioimmunoassay, during the ovulatory cycle of the hen (Scanes et al., 1977a). Some increase in plasma FSH concentrations was observed 14 hr before ovulation. This corresponds with a peak in FSH biological activity in the plasma of the hen (Imai and Nalbandov, 1971) and increases the ability of ovarian tissue to bind FSH (Etches and Cheng, 1981).

During egg laying, incubation, and care of the chicks, there are changes in the plasma concentrations of LH that decrease dramatically at the beginning of the period of incubating eggs, remain low during incubation, and rise gradually while the hens are caring for the chicks, in preparation for the resumption of egg laying (Sharp et al., 1979a). A similar decline in plasma LH concentration has been observed in broody turkeys (Burke and Dennison, 1980).

Annual Reproductive Cycles (Males). Reproductive activity in many species of birds is limited to the specific season that is optimal for the successful production and survival of chicks (Chapter 19). In white-crowned sparrows, for example, reproduction occurs in May and June after the spring migration to the summer breeding grounds. The seasonal pattern of reproduction, gonadal functioning, and circulating hormone concentrations has been characterized in a wild population of white-crowned sparrows (Wingfield and Farner, 1978) (see Figure 17–3). It is apparent that plasma concentrations of LH and testosterone, together with the testes weight and the length of the cloacal protuberance, increase in May following migration. All reproductive

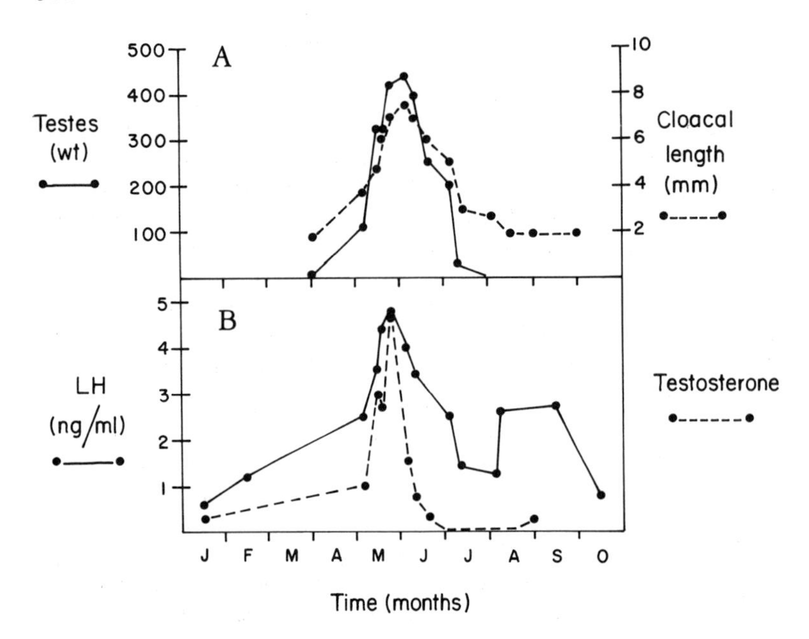

FIGURE 17–3. Variations of reproductive parameters during annual cycle of white-crowned sparrow. (**A**) Testes weight in mg (solid line), cloacal length in mm (dotted line). (**B**) LH in ng/ml (solid line), testosterone in ng/mg (dotted line).

indexes including plasma concentrations of LH and testosterone decline at the beginning of the period of incubating eggs and remain low during feeding of the young, the postnuptial molt, and the migration to the wintering grounds.

In both quail and sparrows, plasma concentrations of LH rise in the late spring at a time when the ambient day length is increasing. Photoperiod is obviously the best environmental parameter to indicate time of year and predict the optimal breeding season. The effect of daylength on reproduction has been extensively investigated in the Japanese quail by Follett. When quail are transferred from a short to a long daylength, plasma concentrations of LH and FSH rise. This leads to the rapid growth of the gonads and to the production of gonadal steroids (Follett, 1976) (see Figure 17–4). This photostimulation of reproductive development is very rapid. In Japanese quail exposed to a long daylength, plasma concentrations of both LH and FSH rise between 18 and 22 hr after the onset of dawn on the first day of imposing long daylength (Nicolls and Follett, 1974; Follett et al., 1977). The mechanism by which long photoperiods affect gonadotropin secretion is mediated by the light acting on photoreceptors within the brain, probably the hypothalamus, wherein there appears to be a circadian rhythm of photosensitivity. If light coincides with the time of sensitivity, LHRH will be released and, hence, gonadotropin secretion increased. If, however, light coincides with the insensitive phase of the rhythm, then there is no response (Follett et al., 1974).

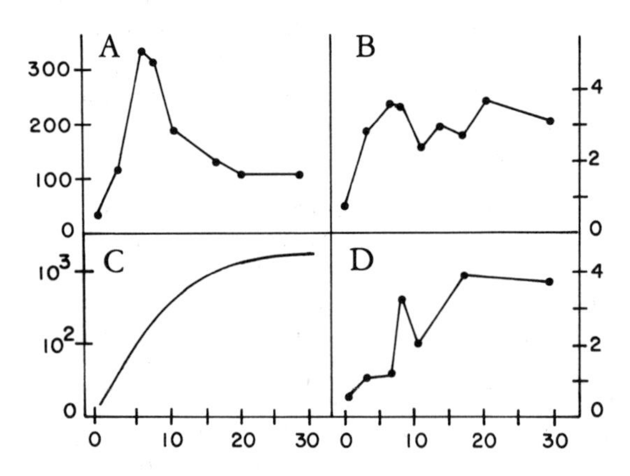

FIGURE 17–4. Effects of transferring Japanese quail from short to long daylength (from 0 to 30 days on horizontal axis). (**A**) Blood FSH, ng/ml. (**B**) Blood LH, ng/ml. (**C**) Testes weight (*log* mg). (**D**) Testosterone, ng/ml. (Follett and Maung, 1978).

The annual cycle of LH secretion in the white-crowned Sparrow is such that the long photoperiods of late June and July do not stimulate gonadotropin secretion (Figure 17–3). Thus, the sparrows are refractory to the long daylengths. A similar refractoriness develops with sparrows that are exposed to long daylengths. Increasing the daylength from 8 to 20 hr of light per 24-hr period (day) is followed by increases in the plasma concentrations of LH (from 0.6 to 2.0 ng/ml) and FSH (from 10 to 160 ng/ml) in white-crowned sparrows (Wingfield et al., 1980). However, after 50 days on long daylength lighting, the birds became refractory to the photostimulation, and plasma concentrations of both LH and FSH began to fall. Similarly, in castrated sparrows, long photoperiod increases the plasma concentrations of LH and FSH prior to refractoriness.

Other factors influencing gonadotropin secretion include behavior and nutritional and other stresses. For instance, plasma LH concentrations are reduced in blackbirds during an aggressive encounter (Harding and Follett, 1979). Similarly, in the domestic fowl plasma concentrations of LH are depressed by nutritional deprivation, e.g., prolonged protein deprivation (Buonomo et al., 1982), while stress handling suppresses the episodic secretion of LH (Wilson and Sharp, 1975).

Thyrotropin (TSH)

Chemistry. There have been numerous attempts to isolate TSH from chicken, ostrich, and turkey pituitary tissue. Conventional purification techniques (salt precipitation and ion-exchange chromatography) have resulted in TSH preparations with considerable LH contamination (e.g., Godden and Scanes, 1975). Papkoff et al. (1982) purified ostrich TSH by salt fractionation, ion-exchange chromatography, and gel filtration coupled with countercurrent distribution, which dissociated LH into subunits. The resulting TSH preparation was both potent (76% as active as NIH-TSH) with low contaminating LH activity (3% that of ostrich LH).

Ostrich TSH is a glycoprotein containing 9.2% carbohydrate (including sialic acid, hexose, and hexosamines) (Papkoff et al., 1982). Assuming avian TSH is similar to its mammalian homolog, it consists of α and β glycoprotein subunits; the α subunit is common to LH and TSH.

Release and Action. TSH affects many aspects of thyroid functioning, stimulating growth (Robinson et al., 1976) together with the production and release of thyroid hormones (Mackenzie, 1981) (see Chapter 20). It is likely that TSH acts on the avian thyroid by binding to specific receptors on the cell surface. The observation that TSH increases the release (production) of cyclic adenosine monophosphate (cAMP) from chicken thyroid tissue in vitro (Tonoue and Kitoh, 1978) suggests that TSH may act by activating adenyl cyclase and thus elevate cellular cAMP (as second messenger).

Thyroid-stimulating hormone has a number of effects on the thyroid cell that ultimately lead to thyroxine (T_4) secretion. In vitro, TSH increases the number of colloid droplets in the thyroid cells (Tonoue and Kitoh, 1978); in vivo in chicks, TSH stimulates the thyroid uptake of both radioactive phosphate (Lamberg, 1953) and iodide (Shellabarger, 1954). TSH has been demonstrated to stimulate T_4 and tri-iodothyronine (T_3) release in the snow goose (Campbell and Leatherland, 1979) and T_4 release in ducks (Campbell and Leatherland, 1979) and domestic fowl (Mackenzie, 1981). The increases in the plasma concentration of T_3 and T_4 following TRH administration presumably reflect the influence of peak plasma TSH concentrations on the avian thyroid. In the domestic fowl (Klandorf et al., 1978) and Japanese quail (Kamis and Robinson, 1978), TRH stimulates (via TSH) approximately synchronous increases in the circulating concentrations of T_4 and T_3. However, the magnitude of the peaks for T_4 and T_3 differed in the two species (in the chicken, 32% for T_4 and 230% for T_3; in the quail, 60% for T_4 and 330% for T_3). The reason for this difference in T_3 and T_4 is not apparent. It is possible that T_3 may be secreted preferentially by the thyroid in response to TSH. Alternatively, the newly released T_4 is very rapidly deiodinated to T_3 in peripheral sites (Kuhn and Nouwen, 1978). It is interesting to note that TSH is capable of stimulating T_4 secretion in the chicken embryo as early as day 6.5 of incubation (Thommes and Hylka, 1978).

Assay. The most widely used biological assay for avian TSH depends on the ability of TSH to stimulate the uptake of [^{32}P]-phosphate by the thyroids of 1-day-old chicks (Lamberg, 1953) (see Table 17–1). Other bioassays also employ chicks but use as end points the uptake or depletion of radioactive iodide by the thyroid. More recently, Mackenzie (1981) developed a sensitive bioassay (see Table 17–1) that depends on the increase in the plasma T_4 concentration. This system has the disadvantage that the response to TSH is affected by the presence of GH or prolactin. None of the bioassays available is sufficiently sensitive for the measurement of TSH in small quantities of plasma.

There are no homologous radioimmunoassays for avian TSH. Almeida and Thomas (1981), however, reported that a human TSH radioimmunoassay detects immunoreactive (IR) TSH in the plasma of Japanese quail. Plasma concentrations of IR-TSH were found to be elevated by the administration of TRH or a goitrogen and depressed by T_3 or T_4 injection, and to

TABLE 17–1. Effect of mammalian thyrotropin on plasma concentrations of thyroxine and on thyroidal uptake of [^{32}P]-phosphate in young chicks

Treatment	Plasma T_4[a] (ng/ml ± SEM)	Thyroidal [^{32}P]-phosphate uptake[b] (cmp/mg ± SEM)
Vehicle	2.9 ± 0.2	28.7 ± 0.8
TSH (2 μg)	13.9 ± 1.8	64.5 ± 2.0
TSH (8 μg)	44.1 ± 6.2	109.8 ± 5.9

[a] One-day-old male chicks were injected twice with TSH; 20 and 5 hr before taking blood samples for T_4 determination. (Data from MacKenzie, 1981.)
[b] Chicks were injected with TSH 6 hr before estimating thyroidal activity by the incorporation of [^{32}P]-phosphate in 1 hr. (Data from Scanes and Follett, 1972.)

be lower in quail on a long daylength compared to a short daylength.

Control. The cells producing TSH, the thyrotrophs, are predominantly found in the cephalic lobe of the avian anterior pituitary gland. Evidence for this comes from light microscopy (reviewed by Tixier-Vidal and Follett, 1973), from immunocytochemistry using antisera against bovine TSH β subunit, and biologic assay of the separate lobes (Sharp et al., 1979b).

Thyrotropin-Releasing Hormone (TRH). In mammals, the secretion of TSH is under the control of the hypothalamohypophysiotropic factor, thyrotropin-releasing hormone (TRH). This tripeptide (L-pyro-glutamyl-L-histidyl-L-proline amide) (Schally et al., 1969) is released from terminals in the median eminence into the portal blood vessels and hence to the pituitary gland. There is evidence for a similar hypothalamic control mechanism for TSH in birds. The hypothalamus of the domestic fowl contains IR—TRH (Jackson and Reichlin, 1974) while *in vitro* TRH release for the chicken pars distalis is increased by synthetic TRH and extracts of avian hypothalami (Scanes, 1974). Furthermore, TRH increases avian TSH secretion in vivo, as indicated by elevated circulating concentrations of T_4 (Klandorf et al., 1978; Kuhn and Nouwen, 1978; Campbell and Leatherland, 1979).

There is evidence that thyroid hormones exert a negative feedback effect on TSH secretion in birds (see also Chapter 20). Inhibiting thyroid function by goitrogen administration increases the number of thyrotrophs, the pituitary TSH content, and the plasma TSH concentration (as indicated by thyroid growth) in ducks (Sharp et al., 1979b). Moreover, the addition of T_3 to the diet of chickens is followed by a dramatic decrease in the plasma concentrations of T_4 (May, 1980). This presumably indicates that T_3 has a negative feedback effect on TSH secretion. It is not well established whether the pituitary or the hypothalamus is the principal sites of T_4/T_3 feedback in birds.

Prolactin

Chemistry. Prolactin has been purified from adenohypophyseal tissue from chickens (Scanes et al., 1975), turkeys (Burke and Papkoff, 1980; Proudman and Corcoran, 1981), and ostrichs (Papkoff et al., 1982). The isolation of avian prolactin has employed conventional fractionation techniques (salt precipitation, ion-exchange chromatography, and gel filtration) and also preparative isotachophoresis. Avian prolactin is a protein (Papkoff et al., 1982) with an isoelectric point (pI) of 5.6 (Scanes et al., 1975) and a molecular weight between 21,700 and 26,000 (Burke and Papkoff, 1980). Although mammalian prolactin is characterized by six half-cystine residues (three disulfide bridges), there appears to be less cysteine in avian prolactin (Papkoff et al., 1982). These avian prolactin preparations are active in stimulating proliferating mucosa in the pigeon crop sac gland.

Action. Many effects have been ascribed to prolactin in different species of vertebrates. In birds, the actions of prolactin are related either to reproduction or to osmoregulation. (The role of prolactin in carbohydrate metabolism is discussed in Chapter 13.) Prolactin is also thought to be involved in the control of metabolism and growth in some birds (de Vlaming, 1979).

Prolactin stimulates the production of "crop milk" and the proliferation of mucosal cells of the crop sac gland in pigeons and doves (Riddle et al., 1933). This unique feature of the Columbidae has been used as the basis of the biologic assay for prolactin. During the latter part of the period when both male and female doves are incubating eggs, the crop increases in weight, and there is a concomitant increase in the plasma prolactin concentrations (Goldsmith et al., 1981) (see Figure 17–5). There is an increase not only in the plasma concentration of prolactin, but also in the specific binding of prolactin to crop membrane preparations in "lactating" pigeon compared with nonlactating adults (Kledzik et al., 1975). Thus increases in both the circulating prolactin and number of prolactin receptors induce crop sac growth.

In other avian species, prolactin is related to incubation behavior (broodiness) and to inhibition of ovarian functioning, but the precise relationship between prolactin and broodiness remains to be elucidated (see Chapter 18). Circulating concentrations of prolactin are elevated during the period when the birds are incubating their eggs; in many species including turkeys (see Figure 17–5) (Burke and Dennison, 1980; Proudman and Opel, 1981a), bantam chickens (Sharp et al., 1979a; Lea et al., 1981), and ducks (Goldsmith and williams, 1980). Furthermore, the plasma concentrations of prolactin fall rapidly if incubation behavior or resting is interrupted (El Halawani et al., 1980a). There is good evidence that prolactin secretion is stimulated by resting and incubation but no conclusive evidence that it induces incubation behavior (see also Chapter 18). Impure preparations of prolactin have been shown to induce broodiness in the domestic fowl (Riddle et al., 1935). However, more recent studies administering ovine prolactin to chickens or turkeys failed to substantiate this (Opel and Proudman, 1980). It is worth noting that prolactin implanted into the hypothalamus appears to stimulate broody behavior in the chicken (Opel, 1971).

Prolactin tends to inhibit gonadal function in female birds. In both chickens and turkeys, prolactin administration decreases ovarian weight and the number of

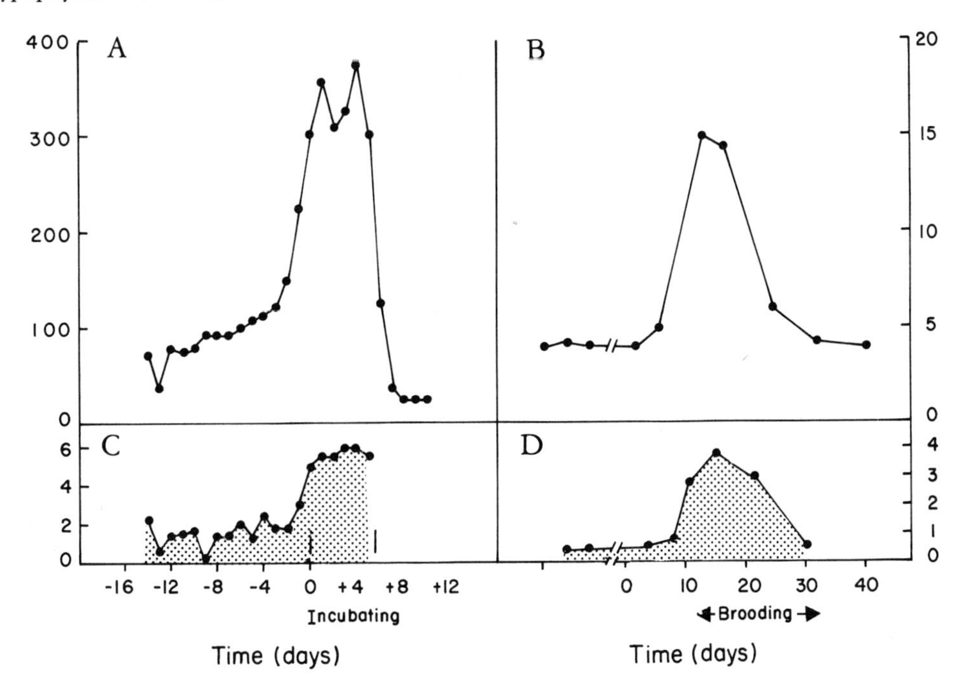

FIGURE 17–5. Plasma prolactin concentrations in ng/ml in female turkeys (**A**) and doves (**B**). (**C**) Nesting frequency and days incubating; (**D**) crop weight (g) during brooding (days).

normal follicles and increases the number atretic follicles (Opel and Proudman, 1980). Thus, prolactin either depresses gonadotropin secretion (directly, or indirectly via the hypothalamus) or inhibits gonadal functioning. Actually, prolactin exerts both effects. The observation that plasma LH is elevated in broody chickens by the administration of antisera against chicken prolactin (Lea et al., 1981) suggests a pituitary–hypothalamic site of action. On the other hand, prolactin inhibits LH-stimulated steroidogenesis in turkeys (Camper and Burke, 1977) and LH-induced ovulation in chickens (Tanaka et al., 1971). However, prolactin does not appear to affect progesterone synthesis by chicken granulosa cells in vitro (Hammond et al., 1982).

Phillips and Harvey (1982) concluded that prolactin probably affects food and water consumption and intestinal/cloacal transport of water and salt.

Assay. The pigeon crop bioassay (Nicoll, 1967) was used extensively during the isolation of mammalian and avian prolactin and other pituitary hormones. Prolactin levels have been estimated in the pituitary gland of birds of different physiologic states (Saeki and Tanabe, 1954), and prolactin secretion and its control were determined in vitro (Hall et al., 1975). Pituitary prolactin concentrations have also been estimated by densitometry following polyacrylamide electrophoresis (Hall et al., 1975). The insensitivity of these systems precludes their use for the determinations of prolactin in small quantities of avian plasma.

In recent years, it has become possible to examine prolactin secretion and physiology by employing radioimmunoassays. Homologous radioimmunoassays have been developed, using purified avian prolactin preparations for chicken (Scanes et al., 1976a) and turkey (Burke and Papkoff, 1980; Proudman and Opel, 1981a). In addition, a heterologous prolactin radioimmunoassay (McNeilly et al., 1978) has been employed to estimate the variations of circulating prolactin concentration during annual and reproductive cycles in avian species (see below) (also see Figure 17–5). A radioreceptor assay for prolactin has been developed using the membrane fractions from pigeon crop-sac glands and ^{125}I-labelled ovine prolactin (Kledzik et al., 1975). However, this system has been employed in few physiologic studies.

Control. The prolactin-producing cells are largely restricted to the cephalic lobe of the anterior pituitary gland (Burke and Papkoff, 1980; Mikami, 1980). In mammals, the hypothalamus exerts a predominantly inhibitory control of prolactin secretion. Evidence is accumulating that dopamine is a major contributor to hypothalamic prolactin inhibitory factor (PIF) activity and may be the PIF (Macleod and Lehmeyer, 1974) in mammals.

Prolactin-Releasing Factor (PRF). In birds, prolactin release is under stimulatory control by hypothalamic-hypophysiotropic releasing factor(s) (PRF). This is supported by in vitro studies in which extracts of avian hypothalami increased prolactin secretion from pituitary tissue from a number of birds; prolactin was estimated by bioassay (Kragt and Meites, 1965; Nicoll, 1965), by densitometry (Hall et al., 1975), and by radioimmunoassay (Harvey et al., 1979a). Furthermore, chicken hypothalamic extract increases the circulating concentrations of prolactin in the domestic fowl (Harvey et al., 1979b).

The chemical nature of avian PRF is not established. One possibility is that thyrotropin-releasing hormone (TRH) is the PRF in birds. Synthetic TRH increases prolactin secretion from chicken and pigeon adenohypophyseal tissue in vitro (Hall et al., 1975; Harvey et al., 1978b) and from the turkey pituitary gland in vivo (El Halawani et al., 1980b). The TRH content in the avian hypothalamus does not account for all its prolactin-releasing activity (Harvey et al., 1979b). Recently, Harvey et al. (1982a) determined the extent to which neurotransmitters affect prolactin release from the chicken pituitary gland in vitro and PRF activity. Prolactin secretion was increased by norepinephrine, serotonin, and histamine. Moreover, extracts of chicken hypothalami, which were alumina treated to remove biogenic amines, lost most of their PRF activity. It is interesting in view of its supposed role of the mammalian PIF that dopamine inhibited in vitro prolactin release from chicken pituitary tissue.

That serotonin plays a role in the physiologic control of prolactin secretion in birds is supported also by in vivo studies. In broody turkeys, inhibition of serotonin biosynthesis by *p*-chlorophenylalanine is followed by a dramatic decrease in circulating prolactin, while the administration of the serotonin precursor, 5-hydroxytryptophan, increased prolactin secretion (El Halawani et al., 1980). Similarly, pharmacologic studies demonstrated that serotonin is required for normal (tonic) prolactin secretion in the domestic fowl (Rabii et al., 1981). It is not clear whether serotonin exerts its effect solely at the level of the pituitary or whether actions within the hypothalamus are also found.

There is evidence that prolactin secretion is influenced by the breeding season (see above) and prostaglandins of the E series (reviewed by Harvey et al., 1982a).

Growth Hormone

Chemistry. Growth hormone (GH) (somatotropin) has been isolated from pituitary tissue obtained from turkeys, ducks, pigeons, chickens, and ostrichs (Farmer et al., 1974; Harvey and Scanes, 1977; Papkoff et al., 1982). The techniques employed to purify avian GH, developed for mammalian GH and then applied to birds, included salt precipitation, ion-exchange chromatography, and gel filtration. The avian GH preparations are active in increasing the width of the epiphyseal cartilage in hypophysectomized young rats.

The purification of avian GH has enabled some chemical characterization of the hormone. In birds and mammals, GH is a protein hormone with four half-cysteine residues, and, presumably, two disulfide bridges (Farmer et al., 1974; Papkoff et al., 1982). The amino acid composition of the avian GH preparations has been determined. In addition, the N-terminal amino acid is phenylalanine in ducks, pigeons (NH_2-Phe-Pro-Ala-Met-) (Farmer et al., 1974), and ostrichs (Papkoff et al., 1982). Chicken GH has an isolectric point of 7.5 and a molecular weight of approximately 23,000 (Harvey and Scanes, 1977).

Assay. During the purification of avian GH, the biologic activity was determined by the rat tibia assay; there is no in vitro or in vivo biologic assay for GH that uses birds or avian tissue for the end point. The rat tibia test has not been used in physiologic studies in birds. The pituitary content of GH has been estimated during growth in control and dwarf strains of chickens by densitometry following separation of GH by disc electrophoresis on polyacrylamide gel (Hishino and Yamamoto, 1977). This assay method has also been employed to examine the control of GH secretion. For increase, the effects of TRH and somatostatin (SRIF) on GH release and synthesis have been demonstrated by the incubation of chicken anterior pituitary gland in vitro with the GH released and stored and estimated by the densitometry method (Hall et al., 1975). The lack of sensitivity of these methods has precluded their use for detailed examination of the physiology of GH secretion.

The purification of chicken and turkey GH has allowed the development of homologous radioimmunoassays (Harvey and Scanes, 1977; Proudman and Wentworth, 1978). These systems appear very similar in terms of sensitivity, specificity, and precision. The assays are capable of determining GH concentrations in small quantities of plasma and have been extensively employed to examine the physiology of GH secretion in vivo and in vitro (see below). The radioimmunoassay for chicken GH appears capable of detecting GH in many species of birds including doves, ducks, and turkeys (Scanes and Balthazart, 1981; Harvey and Phillips, 1980; Harvey et al., 1977a).

GH and Growth. Lack of avian GH resulting from hypophysectomy is followed by a dramatic decrease in the growth rate of chickens (Nalbandov and Card,

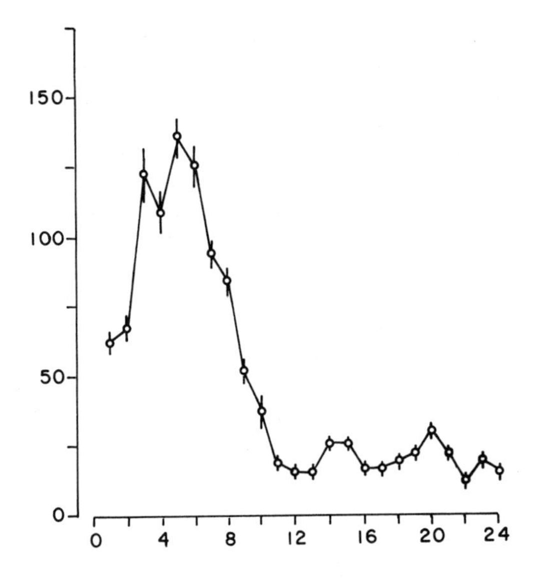

FIGURE 17–6. Changes in plasma concentrations of GH in ng/ml (ordinate) during development in the domestic fowl (age in weeks on abcissa). (Reprinted with permission from *Life Sciences* 28, C.G. Scanes and S. Harvey, Growth hormone and prolactin in avian species, © 1981, Pergamen Press, Ltd.)

1943). Moreover, the administration of antisera against chicken GH significantly reduces chicken growth (Scanes et al., 1977b). Data on the effect of mammalian GH on growth in Aves are inconsistent. Bovine GH failed to stimulate growth rate in intact domestic fowl (Libby et al., 1955), but did increase body weight in hypophysectomized pigeons (Bates et al., 1962). On the other hand, bovine GH that had been trypsin modified stimulated the growth of "broiler type" chicks (Myers and Peterson, 1974).

High plasma concentrations of GH are observed during the period of rapid posthatching growth; low GH concentrations are seen in older and adult chickens (Harvey et al., 1979a, d), turkeys (Harvey et al., 1977a; Proudman and Wentworth, 1980), ducks (Harvey and Phillips, 1980; Foltzer et al., 1981), and doves (Scanes and Balthazart, 1981) (Figure 17–6). Studies in the domestic fowl indicate that GH is secreted in pulses approximately once an hour. This episodic secretion pattern is described as circhoral (Scanes et al., 1983a).

GH and Metabolism. GH exerts short-term effects on metabolism in birds, and has a lipolytic effect. Chicken GH stimulates lipolysis in vitro by adipose explants of chickens, turkeys, and hypophysectomized pigeons (Harvey et al., 1977b). Moreover, bovine GH increases the circulating concentration of free fatty acids in pigeons (John et al., 1973) and in hypophysectomized ducks (Foltzer and Mialhe, 1976). GH also affects lipogenesis. Bovine or chicken GH prevented insulin-stimulated synthesis of fatty acids from ^{14}C-labeled acetate by chick hepatocytes in vitro (Harvey et al., 1977b).

The effect of GH on carbohydrate and lipid metabolism may be characterized as (1) increasing the free fatty acids in the circulation, which are then available as an energy source, and (2) decreasing lipogenesis and reducing glucose utilization. Bovine GH injections reversed the decrease in circulating concentration of glucose in hypophysectomized ducks (Foltzer et al., 1975). These actions of GH may be related to the increased plasma concentrations of GH in birds following nutritional deprivation (see below).

Control of GH Secretion. The somatotroph cells producing GH are in the caudal lobe of the anterior pituitary gland (Mikami, 1980). The release of GH from the avian pituitary gland is under hypothalamic control. This appears to involve at least three hypophysiotropic factors, of which two are stimulatory: (1) thyrotropin-releasing hormone (TRH), (2) a GH-releasing factor (GRF), and (3) somatostatin, which inhibits GH secretion. In the domestic fowl, TRH stimulates GH secretion from the pituitary gland, both in vivo and in vitro (Harvey et al., 1978b) (see Table 17–2). Although TRH increases GH secretion in vivo in chickens between 1 day and 10 weeks old (Harvey et al., 1978b; Scanes et al., 1981b), it does not exert an effect in the conscious adult birds (Scanes et al., 1981b) or in the chick embryo, at least until the process of hatching is underway (Decuypere and Scanes, 1983). Extracts of chicken hypothalami stimulate GH release in vivo but not in vitro (Harvey et al., 1979b, c). It is possible that this lack of an in vitro effect may reflect differences in the clearance rates for the releasing and release-inhibiting activities present in the hypothalamic extract. Although the avian hypothalamus contains TRH (Jackson and Reichlin, 1974), it is generally agreed that the TRH content is insufficient to account

TABLE 17–2. Effect of TRH, prostaglandins, and epinephrine on the plasma concentration of GH in young male domestic fowl[a]

Treatment	Plasma GH[b]
Vehicle (control)	89.5 ± (13) 7.6
TRH (10 μg/kg)	307.6 ± (6) 33.6[c]
PGE_1 (200 μg/kg)	35.0 ± (7) 11.4[c]
PGE_2 (200μg/kg)	15.0 ± (7) 6.2[c]
$PGF_{2\alpha}$ (200 μg/kg)	61.3 ± (7) 13.8
Epinephrine (1 mg/kg)	11.7 ± (6) 5.3[c]

[a] From Scanes and Harvey (1980).
[b] At 20 min after treatment; data are in ng/ml ± (*n*) SEM.
[c] $p < .05$ compared to vehicle (control).

for GH-releasing activity of chicken hypothalamic extract (Harvey et al., 1979a).

GRF. The observation that synthetic human pancreatic GRF (Guillemin et al., 1982; Rivier et al., 1982) increases the plasma concentration of GH in both young and adult chickens (Scanes et al., 1984) provides additional evidence for an avian GRF. *Somatostatin* (SRIF) appears to play a negative role in the hypothalamic control of GH release in birds. Not only are there somatostatin neurons present in the hypothalamus of the duck (Blahser et al., 1978), but also synthetic (SRIF) has been demonstrated to depress both the in vivo and in vitro GH response to TRH in the domestic fowl (Hall and Chadwick, 1976; Harvey et al., 1978b).

Factors Affecting GH Secretion. Neurotransmitters are involved (or may be) in the control of the circulating GH, and exert their effect both at a peripheral level (for instance, by increasing the clearance of GH) and at a central nervous locus, presumably in the hypothalamus. Injections of either epinephrine or norepinephrine can depress the plasma concentration of GH in young domestic fowl (Harvey and Scanes, 1978b) (also see Table 17–2). This effect may be mediated via peripheral β-adrenergic receptors (Harvey and Scanes, 1978b). Moreover, the release of the hypothalamo-hypophysiotropic factors controlling GH secretion is under adrenergic control via α_1-adrenergic receptors (Buonomo et al., 1984). Serotonin, which is involved in the control of GH release, exerts a negative role and probably acts within the hypothalamus (Rabii et al., 1981).

A number of *chemical* and *hormonal factors* influence GH secretion in birds. Plasma concentrations of GH are depressed by prostaglandins E_1, E_2, and $F_{2\alpha}$ in young domestic fowl (Scanes and Harvey, 1980; Harvey and Scanes, 1981) and by insulin and glucagon in young chickens and ducks (Harvey et al., 1978a; Foltzer et al., 1981). Nutritional deprivation increases the plasma concentration of GH in birds. Plasma concentrations of GH are elevated by short periods (1–2 days) of starvation in chickens (Harvey et al., 1978a) and by chronic feed restriction in young chickens and turkeys (Engster et al., 1979; Proudman and Opel, 1981b). Similarly, long-term protein deficiency is accompanied by elevated plasma concentrations of GH in young chickens (Scanes et al., 1981a). The mechanism by which nutritional deprivation affects the plasma concentration of GH is not established, but it is likely that it increases the pituitary pool of readily releasable GH; Proudman and Opel (1981b) found that the in vivo GH response to TRH was increased in turkeys following restricting food intake.

Hormones produced by the other target organs may inhibit GH secretion in birds. Elevated plasma concentrations of GH are found in domestic fowl with aberrantly low plasma levels of thyroid hormones resulting from goitrogen administration (Chaisson et al., 1979), autoimmune thyroiditis (Scanes et al., 1976b), and triiodothyronine-deficient dwarfism (Scanes et al., 1983b). Administration of androgens (Harvey and Scanes, 1978a) decreases plasma concentrations of GH in chickens.

A distinct annual cycle of plasma GH concentrations is characteristic of a number of temperate zone birds. In captive red grouse exposed to natural daylength and weather conditions, levels of GH are found to be high in the spring and summer (maximum, June) and low in the fall and winter (minimum, November) (Harvey et al., 1982b). Similar patterns of GH with summer maxima and winter maxima are observed in Peking ducks (Scanes et al., 1980a) and snow geese (John et al., 1983).

Adrenocorticotropic Hormone and Related Polpeptides

Chemistry. Evidence from peptide sequencing and gene mapping suggests that adrenocorticotropic hormone (ACTH) is synthesized as part of a large protein (proopiomelanocortin). proopiomelanocortin also contains the sequences of β-endorphin (β-EP), which itself is part of β-lipotropin (β-LPH), together with α- and β-melanophore-stimulating hormone (MSH).

Probable Sequence of Proopiomelanocortin

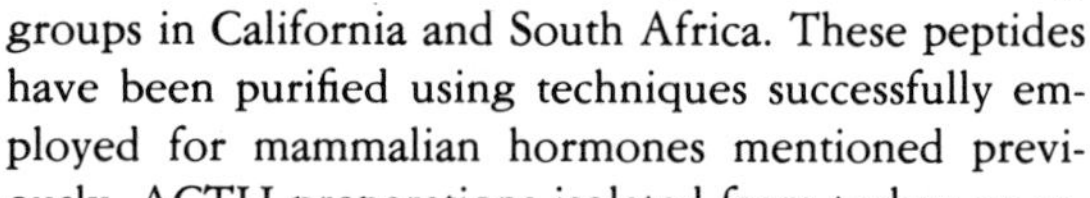

ACTH
β-LPH
α-MSH
α-MSH
β EP

ACTH, β-EP, and β-LPH have been purified from adrenohypophyseal tissue from several avian species. The amino acid sequences of ACTH, β-EP, and β-LPH have been determined by two collaborating groups in California and South Africa. These peptides have been purified using techniques successfully employed for mammalian hormones mentioned previously. ACTH preparations isolated from turkey or ostrich pituitary glands are very potent in stimulating steroidogenesis by rat adrenal cells in vitro; for example, turkey ACTH has 82% of the activity of sheep ACTH (Chang et al., 1980). Avian ACTH, like its mammalian homolog, is a simple polypeptide containing 39 amino acids (Li et al., 1978; Naude et al., 1977; Chang et al., 1980). Ostrich ACTH (see Figure 17–7) differs from mammalian ACTH at positions 15, 27, 28, 29, 31, and 32 (Li et al., 1978). Similar techniques have been used to purify β-LPH and β-EP from ostrich and turkey adenohypophyseal tissue (Naudé and Oelofen, 1981; Naudé et al., 1981a, b; Chang et al., 1980; Naudé et al., 1980). As in mammals, the 31 amino acid sequence of ostrich β-EP is identical to the C

ACTH (αMSH sequence underlined)

H Ser–Tyr–Ser–Met–Glu–His–Phe–Arg–Try–Gly–
Lys–Pro–Val–Gly–Arg–Lys–Arg–Arg–Pro–Val–
Lys–Val–Tyr–Pro–Asn–Gly–Val–Gln–Glu–Glu–
Thr–Ser–Glu–Phe–Pro–Leu–Glu–Phe–OH

β-LPH (β-EP underlined)

H– Ala–Leu–Pro–Pro–Ala–Ala–Met–Leu–Pro–Ala–
Ala–Ala–Glu–Glu–Glu–Glu–Gly–Glu–Glu–Glu–
Glu–Glu–Gly–Glu–Ala–Glu–Lys–Glu–Asp–Gly–
Gly–Ser–Tyr–Arg–Met–Arg–His–Phe–Arg–Trp–
Gln–Ala–Pro–Leu–Lys–Asp–Lys–Arg–Tyr–Gly–
Gly–Gly–Phe–Met–Ser–Ser–Arg–Gly–Arg–Ala–
Pro–Leu–Val–Thr–Leu–Phe–Lys–Asn–Ala–Ile–
Val–Lys–Ser–Ala–Tyr–Lys–Lys–Gly–Gln–OH

FIGURE 17–7. Structures of ostrich ACTH and β-LPH. (From Li et al., 1978; Naudé et al., 1981 a, b.)

terminal of ostrich β-LPH (see Figure 17–7). Avian β-EP is very potent in stimulating mammalian opioid receptors (Naudé et al., 1980; Yamashiro et al., 1982; Chang et al., 1980), while avian β-LPH has poor lipolytic activity, at least with rabbit adipose tissue (Naude et al., 1981a; Chang et al., 1980.

Source of ACTH. ACTH is produced by acidophilic cells in the cephalic lobe of the anterior pituitary gland (Yamashiro et al., 1982). The corticotrophs have been first observed on day 7 in embryonic development in the domestic fowl by use of immunocytochemistry (Jozsa et al., 1979). It is interesting that MSH activity is also found in the cephalic lobe at approximately the same embryonic stage (Betz et al., 1976).

Action. ACTH acts to stimulate corticosterone production by the avian adrenal cortical cells (Holmes and Cronshaw, 1980) (see Chapter 22). The effect of ACTH is observed in vivo (Etches, 1976) and in vitro (Carsia, unpublished observations). ACTH appears to act on the adrenal cells by a mechanism that depends on both cAMP and intracellular ionized calcium (Carsia et al., unpublished observations). The physiology and actions of β-LPH and β-EP in birds are poorly understood.

Control. The secretion of ACTH is under the control of hypothalamic-hypophysiotropic factors. These are released from the median eminence and travel to the anterior pituitary gland via the portal blood vessels. In mammals, and presumably in birds, ACTH release is controlled by corticotropin-releasing factor (CRF). This factor has been demonstrated to be a 41-amino-acid-containing polypeptide in mammals. The possibility of other hypothalamic factors affecting ACTH release cannot be precluded.

The absence of either a homologous RIA for avian ACTH or the validation of a mammalian ACTH RIA for bird plasma has limited studies on the control of ACTH release in birds. At present, using the circulating concentration of glucocorticoid(s) as an indicator of plasma ACTH concentration, it is possible to estimate times when ACTH is likely to be released. ACTH secretion is increased by various stresses, including handling, immobilization, repeated bleeding (Beuving and Vander, 1978; Harvey et al., 1980), fasting (Nir et al., 1976), cold stress (Etches, 1976), ether anesthesia, and *Escherchia coli* endotoxin challenge (Scanes et al., 1980b).

The control of β-EP and β-LPH release has received little attention.

Neurohypophysis

Introduction

The anatomy of the hypothalamic neurosecretory tracts (neurons) leading to their secretory terminals in the pars nervosa is shown in Figures 17–1 and 17–2. It is apparent that the two hormones of the neurohypophysis, arginine vasotocin (AVT) and mesotocin (MT), are produced by and secreted from separate neurosecretory neurons. However, the AVT and MT cell bodies are located in both separate and overlapping areas within the hypothalamus. It is likely that hormones of avian pars nervosa are synthesized in the hypothalamic cell bodies. AVT and MT are transported bound to carrier proteins (neurophysins) by axoplasmic transport. The hormones are then stored in pars nervosa prior to release. Peek and Watkins (1979) identified three neurophysins in the domestic fowl and designated these B_1, B_2, and C_1. All bind radiolabeled oxytocin; B_1 and B_2 bind to a lysine vasopressin and are depleted by water deprivation. It might be suggested that neurophysins B_1 and B_2 are the carrier proteins for AVT, while mesotocin is bound to neurophysin C_1.

In view of the similarities of their chemical structures and (to some extent) of their actions also, the hormones of the pars nervosa will be considered together.

Arginine Vasotocin and Mesotocin

Chemistry. The hormones of the avian neurohypophysis are similar to AVT and oxytocin, based on partial purification and potencies in a number of biological assay systems (Munsick et al., 1960; Munswick, 1964). The isolation and chemical characterization of these hormones from chicken, turkey, and goose pituitary glands confirmed that the avian antidiuretic hormone is AVT and that the oxytocic principle is mesotocin (Acher et al., 1970).

Arginine vasotocin (8-arginine oxytocin):

CyS-Tyr-Ile-Glu(NH_2)-Asp(NH_2)-CyS-Pro-Arg-Gly (NH_2)

Mesotocin (8-isoleucine oxytocin):

CyS-Tyr-Ile-Glu(NH_2)-Asp (NH_2)-CyS-Pro-Ile-Gly (NH_2)

AVT differs from arginine vasopressin, the mammalian antidiuretic hormone, by one amino-acid residue (isoleucine at position 3) in place of phenylalanine, while MT differs from the mammalian homolog, oxytocin, by the substitution of isoleucine for leucine (position 8).

Action. The peptide hormones of the pars nervosa have two major effects, acting on the uterus and on the kidney.

There is no doubt that AVT is the major antidiuretic hormone in birds. The surgical removal of the pars nervosa of the domestic fowl is followed by large increases in the volume of water consumed and urine produced (Shirley and Nalbandov, 1956). Similarly, polydipsia and polyuria were found in chickens with a deficiency of AVT (Dunson et al., 1972). Conversely, AVT has an antidiuretic effect in birds. In hydrated (hypotonic glucose solution-infused) domestic fowl, AVT depresses urine production by 86% with a concomitant 3.35-fold increase in urine osmolality (Ames et al., 1971). Similar results have been obtained also in ducks (Möhring et al., 1980). The mechanism by which AVT affects kidney function includes decreasing the kidney glomerular filtration rate; this is discussed in Chapter 16 on the kidney.

Oviposition, the physical process of laying eggs, is probably controlled in part by oxytocic hormones of the pars nervosa. The hen uterus contracts in vitro in response to AVT and also oxytocin (Munsick et al., 1960). In addition, administration of AVT to a domestic hen is followed by a prolonged increase in intrauterine pressure (Rzasa and Ewy, 1971). Moreover, in the chicken at oviposition a large (133-fold) increase in blood vasotocin-like activity has been reported (Sturkie and Lin, 1966) (see also Chapter 18). The release of AVT at the time of oviposition might be expected to have consequences for the physiology of the bird, in view of its effects on water balance and its long half-life (~20 min) (Hasan and heller, 1968).

Assay. Many biologic assays have been used during the purification and characterization of the avian neurohypophyseal hormones. These include the hen uterus (oxytocic), the rat uterus (oxytocic), the chicken antidiuretic, the rat pressor, and the chicken depressor assays (Munsick et al., 1960; Acher et al., 1970), and frog bladder (Sturkie and Lin, 1966). In addition, biologic assays have been employed in studies of the control of AVT and MT secretion. Follett and Farner (1966) demonstrated a differential release of arginine vasotocic and oxytocic activities from the quail pars nervosa following redehydration of Japanese quail.

Recently, RIAs have been validated for the determination of AVT in the plasma of birds. Rosenbloom and Fisher (1974) developed an assay for AVT using antisera to arginine vasopressin and ^{125}I-labeled AVT as the tracer. Alternatively, an arginine vasopressin RIA has been validated for AVT in avian plasma (Möhring et al., 1980). In these systems, MT has very low cross-reactivity ($<1\%$ that of AVT). Skadhauge (1981) has compiled blood AVT values of several species of Aves.

Control. Arginine vasotocin is released when blood osmolality is high and consequently when water loss should be prevented. In the domestic fowl, withholding drinking water for more than 1 day produced increases in both blood osmolality and AVT concentrations (Koike et al., 1977). Similar infusion of sodium chloride increases both the plasma AVT concentration and osmolality (Koike et al., 1979). Similarly, Peking ducks maintained on saltwater had higher plasma concentrations of AVT than those on freshwater [22.7 ± 3.0 ($n = 7$) fmol/ml versus 5.1 ± 1.4 ($n = 7$) fmol/ml]. Moreover, infusion of hypoosmotic glucose depressed plasma AVT concentrations to below the detection limit for the assay (<0.5 fmol/ml) (Möhring et al., 1980). The control of AVT release in ducks appears to involve peripheral and central osmolality receptors. Plasma concentrations of AVT are elevated by the administration of hypertonic saline into the cerebroventricular system. Similarly, AVT secretion is stimulated in hydrated ducks if the inhibitory influence of the vagus is temporarily removed by procaine blockade (Simon-Oppenmann et al., 1980).

References

Acher, R., J. Chauvet, and M.T. Chauvet (1970). Phylogeny of the neurohypophyseal hormones; the avian active peptides. Eur. J. Biochem., 17, 509.

Almeida, O.F.X., and D.H. Thomas (1981). Effects of feeding pattern on the pituitary thyroid axis in the Japanese quail. Gen. Comp. Endocrinol., 44, 508.

Ames, E., K. Steven, and E. Skadhauge. (1971). Effects of arginine vasotocin on renal excretion of Na^+, K^+, Cl^-, and urea in the hydrated chicken. Am. J. Physiol., 221, 1223.

Bates, R.W., R.A. Miller, and M.M. Garrison. (1962). Evidence in the hypophysectomized pigeon of a synergism among prolactin, growth hormone, thyroxine and pregnisone upon weight of the body, digestive tract, kidney and fat stores. Endocrinology, 71, 345.

Betz, T.W., A. Enemar, and M. Rohss. (1976). On the development of chicken adenohypophyseal melanophore-expanding activity Can. J. Zool., 54, 1554.

Beuving, G., and G.M.A. Vander. (1978). Effect of stressing factors on corticosterone levels in the plasma of laying hens. Gen. Comp. Endocrinol., 35, 153.

Blahser, S., D. Fellman, and C. Bugnon. (1978). Immunocytochemical demonstration of somatostatin-containing neurons in the hypothalamus of the domestic mallard. Cell Tissue Res., 195, 183.

Bona Gallo, A., and P. Licht. (1979). Differences in the properties of FSH and LH binding sites in the avian gonad revealed by homologous radioligands. Gen. Comp. Endocrinol., 37, 521.

Bonney, R.C., and F.J. Cunningham. (1977a). A role for cyclic AMP as a mediator of the action of LH-RH on cyclic anterior pituitary cells. Mol. Cell. Endocrinol., 7, 233.

Bonney, R.C., and F.J. Cunningham. (1977b). Effect of ionic environment on the release of LH from chicken anterior pituitary cells. Mol. Cell. Endocrinol., 7, 245.

Bonney, R.C., F.J. Cunningham, and B.J.A. Furr. (1974). Effect of synthetic luteinizing hormone releasing hormone on plasma luteinizing hormone in the female domestic fowl (*Gallus domesticus*). J. Endocrinol., 63, 539.

Bons, N. (1980). Immunocytochemical localization of neurohormones in bird hypothalamus. In "Recent Advances in Avian Endocrinology" (G. Pethes, P. Peczely, and P. Rudas, Eds.). London: Pergamon Press, p. 15.

Brown, N.L., J.D. Baylé, C.G. Scanes, and B.K. Follett. (1975). The actions of avian LH and FSH on the testes of hypophysectomized quail. Cell Tissue Res., 156, 499.

Buonomo, F.C., J. Rabii, and C.G. Scanes. (1981). Aminergic involvement in the control of luteinizing hormone secretion in the domestic fowl. Gen. Comp. Endocrinol., 45, 162.

Buonomo, F.C., P. Griminger, and C.G. Scanes. (1982). Effects of gradation in protein-calorie restriction on the hypothalamic pituitary-gonadal axis in young domestic fowl. Poult. Sci., 61, 800.

Buonomo, F.C., T.J. Lauterio, N. Zimmermann, and C.G. Scanes. (1984). Catecholamine involvement in the control of growth hormone secretion in the domestic fowl. Gen. Comp. Endocrinol., 54, 360.

Burke, W.H., and D.T. Dennison. (1980). Prolactin and luteinizing hormone levels in female turkeys (*Meleagris gallopavo*) during a photoinduced reproductive cycle and broodiness. Gen. Comp. Endocrinol., 41, 92.

Burke, W.H., and H. Papkoff. (1980). Purification of turkey prolactin and the development of a homologous radioimmunoassay for its measurement. Gen. Comp. Endocrinol., 40, 297.

Burke, W.H., P. Licht, H. Papkoff, and A. Bona Gallo. (1979a). Isolation and characterization of luteinizing hormone and follicle-stimulating hormone for pituitary glands of the turkey (*Meleagris gallopavo*). Gen. Comp. Endocrinol., 37, 521.

Burke, W.H., H. Papkoff, P. Licht, and A. Bona Gallo. (1979b). Preparation and properties of luteinizing hormone (LH) subunits from the turkey (*Meleagris gallopavo*) and their recombination with subunits of ovine LH. Gen. Comp. Endocrinol., 37, 508.

Calvo, F.O., and J.M. Bahr. (1982). Hormone-stimulatable adenylate cyclase system in the granulosa cells of the domestic hen (*Gallus domesticus*). Gen. Comp. Endocrinol., 48, 452.

Campbell, R.R., and J.F. Leatherland. (1979). Effect of TRH, TSH and LHRH on plasma thyroxine and triiodothyronine in the lesser snow goose (*Anser caerulescens caerulescens*) and plasma thyroxine in the Rouen duck (*Anas platyrhynchos*). Can. J. Zool., 57, 271.

Camper, P.M., and W.H. Burke. (1977). The effects of prolactin on the gonadotropin-induced rise in serum estradiol and progesterone of the laying turkey. Gen. Comp. Endocrinol., 32, 72.

Chaisson, R.B., P.J. Sharp, H. Klandorf, C.G. Scanes, and S. Harvey. (1979). The effect of rapeseed meal and methimazole on levels of plasma hormones in growing broiler cockerels. Poult. Sci., 58, 1575.

Chang, W.C., D. Chung, and C.H. Li (1980). Isolation and characterization of β-lipotropin and adrenocorticotropin from turkey pituitary glands. Int. J. Pept. Protein Res., 15, 261.

Chase, D.J. (1982). Gonadotropin specificity of acute testicular androgen secretion in birds. Gen. Comp. Endocrinol., 46, 486.

Davies, D.T., and R.J. Bicknell. (1976). The effect of testosterone on the responsiveness of the quail's pituitary to luteinizing hormone-releasing hormone (LHRH) during photoperiodically induced testicular growth. Gen. Comp. Endocrinol., 30, 487.

Davies, D.T., and B.K. Follett. (1975). Electrical stimulation of the hypothalamus and luteinizing hormone secretion in Japanese quail. J. Endocrinol., 67, 431.

Davies, D.T., and B.K. Follett. (1980). Neuroendocrine regulation of gonadotrophin-releasing hormone secretion in the Japanese quail. Gen. Comp. Endocrinol., 40, 220.

Davies, D.T., R. Massa, and R. James. (1980). Role of testosterone and of its metabolites in regulating gonadotrophin secretion in the Japanese quail. J. Endocrinol., 84, 211.

Decuypere, E. and C.G. Scanes. (1983). Variation in the release of thyroxine, triiodothyronine and growth hormone in response to thyrotropin releasing hormone during development of the domestic fowl. *Acta Endocrinol,* 102, 220.

de Vlaming, V.L. (1979). Actions of prolactin among the vertebrates. In "Hormones and Evolution" (E.J.W. Barrington, Ed.). New York: Academic Press, Chapter 12.

Dunson, W.A., E.G. Buss, W.H. Sawyer, and H. Sokol. (1972). Hereditary polydipsia and polyuria in chickens. Am. J. Physiol., 222, 1167.

El Halawani, M.E., W.H. Burke, and P.T. Dennison. (1980a). Effect of nest deprivation on serum prolactin level in resting female turkeys. Biol. Reprod., 23, 118.

El Halawani, M.E., W.H. Burke, and L.A. Ogren. (1980b). Involvement of catecholaminergic mechanisms in the photoperiodically induced rise in serum luteinizing hormone in Japanese quail (*Coturnix coturnix japonica*). Gen. Comp. Endocrinol., 41, 14.

Engster, H.M., L.B. Carew, S. Harvey, and C.G. Scanes. (1979). Growth hormone metabolism in essential fatty acid-deficient and pair-fed non-deficient chicks. J. Nutr., 109, 330.

Etches, R.J. (1976). A radioimmunoassay for corticosterone and its application to the measurement of stress in poultry. Steroids, 28, 763.

Etches, R.J., and F.J. Cunningham. (1976). The interrelationship between progesterone and luteinizing hormone during the ovulation cycle of the hen (*Gallus domesticus*). J. Endocrinol., 71, 51.

Etches, R.J., and K.W. Cheng. (1981). Changes in the plasma concentrations of luteinizing hormone, progesterone, oestradiol and testosterone and the binding of follicle stimulatory hormone to the theca of follicles during the ovulatory cycle of the hen (*Gallus domesticus*). J. Endocrinol., 91, 11.

Farmer, S.W., H. Papkoff, and T. Hayashida. (1974). Purification and properties of avian growth hormones. Endocrinology, 95, 1560.

Farmer, S.W., H. Papkoff, and P. Licht. (1975). Purification of turkey gonadotropins. Biol. Reprod., 12, 415.

Follett, B.K. (1976). Plasma follicle-stimulating hormone during photoperiodically induced sexual maturation in male Japanese quail. J. Endocrinol., 69, 117.

Follett, B.K., and D.S. Farner. (1966). The effects of the daily photoperiod on gonadal growth, neurohypophysial hormone content and neurosecretion in the hypothalamo-hypophyseal system of the Japanese quail (*Coturnix coturnix japonica*). Gen. Comp. Endocrinol., 7, 111.

Follett, B.K., and S.L. Maung. (1978). Rate of testicular maturation, in relation to gonadotrophin and testosterone levels, in quail exposed to various artificial photoperiods and to natural day lengths. J. Endocrinol., 78, 267.

Follett, B.K., C.G. Scanes, and F.J. Cunningham. (1972). A radioimmunoassay for avian luteinizing hormone. J. Endocrinol., 52, 359.

Follett, B.K., P.W. Mattocks, and D.S. Farner. (1974). Circadian function in the photoperiodic induction of gonadotropin secretion in the White-crowned sparrow, *Zonotrichia leucophrys gambelii.* Proc. Natl. Acad. Sci. U.S.A., 71, 1666.

Follett, B.K., D.T. Davies, and B. Gledhill. (1977). Photoperiodic control of reproduction in Japanese quail: changes in gonadotrophin secretion on the first day of induction and their pharmacological blockade. J. Endocrinol., 74, 449.

Follett, B.K., D.T. Davies, R. Gibson, K.J. Hodges, N. Jenkins, S.L. Maung, Z.W. Maung, M.R. Redshaw, and J.P. Sumpter. (1978). Avian gonadotropins—their purification and assay. Pavo, 16, 34.

Foltzer, C., and P. Mialhe. (1976). Pituitary and adrenal control of pancreatic endocrine function in the duck. II. Plasma free fatty acids and insulin variations following hypophysectomy and replacement therapy with growth hormone and corticosterone. Diabete Metab., 2, 101.

Foltzer, C., V. Leclercq-Meyer, and P. Mialhe. (1975). Pituitary and adrenal control of pancreatic control of pancreatic endocrine function in the duck. I. Plasma glucose and pancreatic glucagon variations following hypophysectomy and corticosterone. Diabete Metab., 1, 39.

Foltzer, C., S. Harvey, M.T. Strosser, and P. Mialhe. (1981). Influence of insulin and glucagon on secretion of growth hormone in growing ducks (*Anas platyrhynchos*). J. Endocrinol., 91, 189.

Fraser, H.M., and P.J. Sharp. (1978). Prevention of positive feedback in the hen (*Gallus domesticus*) by antibodies to luteinizing hormone releasing hormone. J. Endocrinol., 76, 181.

Furr, B.J.A., R.C. Bonney, R.J. England, and F.J. Cunningham. (1973). Luteinizing hormone and progesterone in peripheral blood during the ovulatory cycle of the hen (*Gallus domesticus*). J. Endocrinol., 57, 159.

Godden, P.M.M., and C.G. Scanes. (1975). Studies on the purification and properties of avian gonadotrophins. Gen. Comp. Endocrinol., 27, 538.

Goldsmith, A.R., and D.M. Williams. (1980). Incubation in mallards (*Anas platyrhynchos*) changes in plasma levels of prolactin and luteinizing hormone. J. Endocrinol., 86, 371.

Goldsmith, A.R., C. Edwards, M. Koprucu, and R. Silver. (1981). Concentrations of prolactin and luteinizing hormone in plasma of doves in relation to incubation and development of the crop. J. Endocrinol., 90, 437.

Guillemin, R., P. Brazeau, P. Bohlen, F. Esch, N. Ling, and W.B. Wehrenberg. (1982). Growth hormone-releasing factor from a human pancreatic tumor that caused acromegaly. Science, 218, 585.

Hall, T.R., and A. Chadwick. (1976). Effect of growth hormone inhibiting factor (somatostatin) on the release of growth hormone and prolactin from pituitaries of the domestic fowl *in vitro*. J. Endocrinol., 68, 163.

Hall, T.R., A. Chadwick, N.J. Bolton, and C.G. Scanes. (1975). Prolactin release *in vitro* and *in vivo* in the pigeon and the domestic fowl following administration of synthetic thyrotrophin-releasing factor (TRF). Gen. Comp. Endocrinol., 25, 298.

Hammond, R.W., H. Todd, and F. Hertelendy. (1978). Effect of bovine LH in steroidogenesis in avian granulosa cells. IRCS. Med. Sci., 6, 452.

Hammond, R.W., H. Todd, and F. Hertelendy. (1980). Effects of mammalian gonadotropins on progesterone release and cyclic nucleotide production by isolated avian granulosa cells. Gen. Comp. Endocrinol., 41, 467.

Hammond, R.W., W.H. Burke, and F. Hertelendy. (1981). Influence of follicular maturation on progesterone release in chicken granulosa cells in response to turkey and ovine gonadotropins. Biol. Reprod., 24, 1048.

Hammond, R.W., W.H. Burke, and F. Hertelendy. (1982). Prolactin does not affect steroidogenesis in isolated chicken granulosa cells. Gen. Comp. Endocrinol., 48, 285.

Harding, C.F., and B.K. Follett. (1979). Hormonal changes triggered by aggression in a native population of blackbirds. Science, 203, 918.

Harvey, S., and C.G. Scanes. (1977). Purification and radioimmunoassay of chicken growth hormone. J. Endocrinol., 73, 321.

Harvey, S., and C.G. Scanes. (1978a). Plasma concentrations of growth hormone during growth in normal and testosterone-treated chickens. J. Endocrinol., 79, 145.

Harvey, S., and C.G. Scanes. (1978b). Effect of adrenaline and adrenergic active drugs on growth hormone secretion in immature cockerels. Experientia, 34, 1096.

Harvey, S., and J.G. Phillips. (1980). Growth, growth hormone and corticosterone secretion in freshwater and saline-adapted ducklings (*Anas platyrhynchos*). Gen. Comp. Endocrinol., 42, 334.

Harvey, S., and C.G. Scanes. (1981). Inhibition of growth hormone release by prostaglandins in immature domestic fowl (*Gallus domesticus*). J. Endocrinol., 91, 69.

Harvey, S., P.M.M. Godden, and C.G. Scanes. (1977a). Plasma growth hormone concentrations during growth in turkeys. Br. Poult. Sci., 18, 547.

Harvey, S., C.G. Scanes, and T. Howe. (1977b). Growth hormone effects on *in vitro* metabolism of avian adipose and liver tissue. Gen. Comp. Endocrinol., 33, 322.

Harvey, S., C.G. Scanes, A. Chadwick, and N.J. Bolton. (1978a). Influence of fasting, glucose and insulin on the levels of growth hormone and prolactin in the plasma of the domestic fowl (*Gallus domesticus*). J. Endocrinol., 78, 501.

Harvey, S., C.G. Scanes, A. Chadwick, and N.J. Bolton. (1978b). The effect of thyrotropin-releasing hormone (TRH) and somatostatin (GHRIH) on growth hormone and prolactin secretion *in vitro* and *in vivo* in the domestic fowl (*Gallus domesticus*). Neuroendocrinology, 26, 249.

Harvey, S., T.F. Davison, and A. Chadwick. (1979a). Ontogeny of growth hormone and prolactin secretion in the domestic fowl (*Gallus domesticus*). Gen. Comp. Endocrinol., 39, 270.

Harvey, S., C.G. Scanes, A. Chadwick, G. Border, and N.J. Bolton. (1979b). Effect of chicken hypothalamus on prolactin and growth hormone secretion in male chickens. J. Endocrinol., 82, 193.

Harvey, S., C.G. Scanes, A. Chadwick, and N.J. Bolton. (1979c). *In vitro* stimulation of chicken pituitary growth hormone and prolactin secretion by chicken hypothalamic extract. Experientia, 15, 694.

Harvey, S., C.G. Scanes, A. Chadwick, and N.J. Bolton (1979d). Growth hormone and prolactin secretion growing domestic fowl; influence of sex and breed. Br. Poult. Sci., 20, 9.

Harvey, S., B.J. Merry, and J.G. Phillips. (1980). Influence of stress on the secretion of corticosterone in the duck (*Anas platyrhynchos*). J. Endocrinol., 87, 161.

Harvey, S., A. Chadwick, G. Border, C.G. Scanes, and J.G. Phillips. (1982a). Neuroendocrine control of prolactin secretion. In "Aspects of Avian Endocrinology—Practical and Theoretic Implications" (C.G. Scanes, M.A. Ottinger, A.D. Kenny, J. Balthazart, J. Cronshaw, and I. Chester-Jones, Eds.). Lubbock: Texas Tech Press, p. 41.

Harvey, S., C.G. Scanes, and P.J. Sharp. (1982b). Annual cycle of plasma concentrations of growth hormone in red grouse (*Lagopus lagopus scoticus*). Gen. Comp. Endocrinol., 48, 411.

Hasan, S.H., and H. Heller. (1968). The clearance of neurohypophyseal hormones from the circulation of non-mammalian vertebrates. Br. J. Pharmacol. Chemother., 33, 523.

Hishino, S., and K. Yamamoto. (1977). Synthesis and release of growth hormone, prolactin and other proteins from the anterior pituitary of normal and dwarf chickens. Gen. Comp. Endocrinol., 31, 7.

Holmes, W.N., and J. Cronshaw (1980). Adrenal cortex: structure and function. In "Avian Endocrinology" (A. Epple and M. Stetson, Eds). New York: Academic Press, p. 271.

Huang, E.S.R., K.J. Kao., and A.V. Nalbandov. (1979). Synthesis of sex steroids by cellular components of chicken follicles. Biol. Reprod., 20, 454.

Imai, K. (1973). Effects of avian and mammalian pituitary preparations on induction of ovulation in the domestic fowl, *Gallus domesticus*. J. Reprod. Fertil., 33, 91.

Imai, K., and A.V. Nalbandov. (1971). Changes in FSH activity of anterior pituitary glands and of blood plasma during the laying cycle of the hen. Endocrinology, 88, 1465.

Jackson, I.M.D., and S. Reichlin. (1974). Thyrotropin-releasing hormone (TRH); distribution in hypothalamic and extra-hypothalamic brain tissues of mammalian and submammalian chordates. Endocrinology, 95, 854.

Jenkins, N., J.P. Sumpter, and B.K. Follett. (1978). The effects of vertebrate gonadotrophins on androgen release *in vitro* from testicular cells of Japanese quail and a comparison with the radioimmunoassay activities. Gen. Comp. Endocrinol., 35, 309.

John, J.M., B.A. McKeown, and J.C. George. (1973). Influence of exogenous growth hormone and its antiserum on plasma free fatty acid level in the pigeon. Comp. Biochem. Physiol. A, 46, 497.

John, T.M., J.C. George, and C.G. Scanes. (1983). Circulating levels of luteinizing hormone and growth hormone during the annual cycle of the migratory Canada goose. Gen. Comp. Endocrinol., 51, 44.

Jozsa, R., C.G. Scanes, S. Vigh, and B. Mess. (1979). Functional differentiation of the embryonic chicken pituitary gland studied by immunohistological approach. Gen. Comp. Endocrinol., 39, 158.

Kamis, A.B., and G.A. Robinson. (1978). Serum T_3 and T_4 concentrations of Japanese quail treated with thyrotropin-releasing hormone. Gen. Comp. Endocrinol., 36, 636.

King, J.A., and R.P. Millar. (1982a). Structure of chicken hypothalamic luteinizing hormone-releasing hormone. I. Structural determination on partially purified material. J. Biol. Chem., 257, 10722.

King, J.A., and R.P. Millar. (1982b). Structure of chicken hypothalamic luteinizing hormone-releasing hormone. II. Isolation and Characterization. J. Biol. Chem., 257, 10729.

Klandorf, H., P.J. Sharp, and R. Sterling. (1978) Induction of thyroxine and triiodothyronine release by thyrotrophin-releasing hormone in the hen. Gen. Comp. Endocrinol., 34, 377.

Kledzik, G., S. Marshall, M. Gelato, G. Campbell, and J. Meites. (1975). Prolactin binding activity in the crop sacs of juvenile, mature, parent and prolactin-injected pigeons. Endocr. Res. Commun., 2, 345.

Knight, P.G., S.C. Wilson, R.T. Gladwell, and F.J. Cunningham. (1982). Evidence for the involvement of central catecholaminergic mechanisms in mediating the preovulatory surge of luteinizing hormone in the domestic hen. J. Endocrinol., 94, 295.

Koike, T.K., L.R. Pryor, H.L. Neldon, and R.S. Venable. (1977). Effect of water deprivation on plasma radioimmunoassayable arginine vasotocin in conscious chickens (*Gallus domesticus*). Gen. Comp. Endocrinol., 33, 359.

Koike, T.I., L.R. Pryor, and H.L. Neldon. (1979). Effect of saline infusion on plasma immunoreactive vasotocin in conscious chickens (*Gallus domesticus*). Gen. Comp. Endocrinol., 37, 451.

Kragt, G.L., and J. Meites. (1965). Stimulation of pigeon pituitary release by pigeon hypothalamic extract *in vitro*. Endocrinology, 76, 1169.

Kuhn, E.R., and E.J. Nouwen. (1978). Serum levels of triiodothyronine and thyroxine in the domestic fowl following mild cold exposure and injection of synthetic thyrotropin-releasing hormone. Gen. Comp. Endocrinol., 34, 336.

Lamberg, B.A. (1953). Radioactive phosphorus as indicator in a chick assay of thyrotropic hormone. Acta Med. Scand. (Suppl.), 279.

Lea, R.W., A.S.M. Dods, P.J. Sharp, and A. Chadwick. (1981). The possible role of prolactin in the regulation of nesting behavior and the secretion of luteinizing hormone in broody bantams. J. Endocrinol., 91, 89.

Lea, R.W., P.J. Sharp, and A. Chadwick. (1982). Daily variations in the concentrations of plasma prolactin in broody bantams. Gen. Comp. Endocrinol., 48, 275.

Li, C.H., D. Chung, W. Oelofsen, and R.J. Naude. (1978). Adrenocorticotropin 53. The amino acid sequence of the hormone from the ostrich pituitary gland. Biochem. Biophys. Res. Commun., 83, 900.

Libby, D.A., J. Meites, and J. Schaible. (1955). Growth hormone in chickens. Poult. Sci., 34, 1329.

MacKenzie, D.S. (1981). *In vivo* thyroxine release in day-old cockerels in response to acute stimulation by mammalian and avian pituitary hormones. Poult. Sci., 60, 2136.

MacLeod, R.M., and J.E. Lehmeyer. (1974). Studies on the mechanisms of the dopamine-mediated inhibition of prolactin secretion. Endocrinology, 94, 1077.

Maung, Z.W., and B.K. Follett. (1977). Effects of chicken and ovine luteinizing hormone on androgen release and cyclic AMP production by isolated cells from the quail testis. Gen. Comp. Endocrinol., 33, 242.

Maung, S.L. and B.K. Follett. (1978). The endocrine control by luteinizing hormone of testosterone secretion from the testis of the Japanese quail. Gen. Comp. Endocrinol., 36, 79.

May, J.D. (1980). Effect of dietary thyroid hormone on growth and feed efficiency of broilers. Poult. Sci. 59, 888.

McNeilly, A.S., R.J. Etches, and H.G. Friesen. (1978). A heterologous radioimmunoassay for avian prolactin: application to the measurement of prolactin in the turkey. Acta Endocrinol. (Copenh.), 89, 60.

Mikami, S. (1980). Hypothalamic control of the avian adenohypophysis. In "Biological Rhythms in Birds: Neural and Endocrine Aspects" (Y. Tanabe, K. Tanaka, and T.

Ookawa, Eds.) Tokyo: Japan Scientific Society Press Berlin: Springer-Verlag, p. 17.

Möhring, J., J. Schoun, C. Simon-Oppermann, and E. Simon. (1980). Radioimmunoassay for arginine-vasotocin (AVT) in serum of Pekin ducks: AVT concentrations after adaptation to fresh water and sea water. Pfleugers Arch., 387, 91.

Munsick, R.A. (1964). Neurohypophyseal hormones of chickens and turkeys. Endocrinology, 75, 104.

Munsick, R.A., W.H. Sawyer, and H.B. Van Dyke. (1960). Avian neurohypophyseal hormones: Pharmacological properties and tentative identification. Endocrinology, 66, 860.

Myers, W.R., and R.A. Peterson. (1974). Responses of six and ten-week old broilers to a tryptic digest of bovine growth hormone. Poult. Sci., 53, 508.

Nalbandov, A.V., and L.E. Card. (1943). Effect of hypophysectomy on growing chicks. J. Exp. Zool., 94, 387.

Naude, R.J., and W. Oelofsen. (1977). The isolation and characterization of corticotropin from the pituitary gland of the ostrich, *Struthio camelus.* Biochem. J., 165, 519.

Naudé, R.J., and W. Oelofsen. (1981). Isolation and characterization of β-lipotropin from the pituitary gland of the ostrich, *Struthio camelus.* Int. J. Pept. Protein Res., 18, 135.

Naudé, R.J., W. Oelofsen, and R. Maske. (1980). Isolation, characterization and opiate activity of β-endorphin from the pituitary gland of the ostrich, *Struthio camelus.* Biochem. J., 187, 245.

Naudé, R.J., D. Chung, C.H. Li, and W. Oelofsen. (1981a). β-lipotropin: primary structure of the hormone from the ostrich pituitary gland. Int. J. Pept. Protein Res., 18, 138.

Naudé, R.J., D. Chung, C.H. Li, and W. Oelofsen. (1981b). β-endorphin. Primary structure of the hormone from the ostrich pituitary gland. Biochem. Biophys. Res. Commun., 98, 108.

Nicoll, C.S. (1965). Neural regulation of adenohypophysial prolactin secretion in tetrapods: indications for *in vitro* studies. J. Exp. Zool., 158, 203.

Nicoll, C.S. (1967). Bio-assay of prolactin. Analysis of the pigeon crop-sac response to local prolactin injection by an objective and quantitative method. Endocrinology, 80, 641.

Nicholls, T.J., and B.K. Follett. (1974). The photoperiodic control of reproduction in *Coturnix* quail. The temporal pattern of LH secretion. J. Comp. Physiol., 93, 301.

Nir, I., D. Yamand, and M. Perek (1976). Effects of stress on the corticosterone content of the blood plasma and adrenal gland of intact and bursectomized *Gallus domesticus.* Poult. Sci., 54, 2101.

Okschje, A. (1965). The fine structure of the neurosecretory system of birds in relation to its functional aspects. Proc. Int. Congr. Endocrinol. 2nd, 1964. (London). Int. Congr. Ser., Excerpta Med. 83, 167.

Opel, H. (1971). Induction of incubation behavior in the hen by brain implants of prolactin. Poult. Sci., 50, 1613.

Opel, H., and J.A. Proudman. (1980). Failure of mammalian prolactin to induce incubation behavior in chickens and turkeys. Poult. Sci., 59, 2550.

Papkoff, H., P. Licht, A. Bona-Gallo, D.S. Mackenzie, W. Oelofsen, and M.M.J. Oosthuizen. (1982). Biochemical and immunological characterization of pituitary hormones from the ostrich (*Struthio camelus*). Gen. Comp. Endocrinol., 48, 181.

Peek, J.C., and W.B. Watkins. (1979). Identification of neurophysin-like proteins in the posterior pituitary gland of the chicken (*Gallus domesticus*). Neuroendocrinology, 28, 52.

Phillips, J.G., and S. Harvey. (1982). A reappraisal of the role of prolactin in osmoregulation. In "Aspects of Avian Endocrinology: Practical and Theoretical Implication" (C.G. Scanes, M.A. Ottinger, A.D. Kenny, J. Balthazart, J. Cronshaw, and I. Chester-Jones, Eds.). Lubbock: Texas Tech Press, p. 329.

Proudman, J.A., and B.C. Wentworth. (1978). Radioimmunoassay of turkey growth hormone. Gen. Comp. Endocrinol., 36, 194.

Proudman, J.A., and B.C. Wentworth. (1980). Ontogenesis of growth hormone in large and midget white strains of turkeys. Poult. Sci., 59, 906.

Proudman, J.A., and Opel, H. (1981a). Turkey prolactin: validation of a radioimmunoassay and measurement of changes associated with broodiness. Biol. Reprod., 25, 575.

Proudman, J.A., and H. Opel. (1981b). Effect of feed or water restriction on basal and TRH-stimulated growth hormone secretion in the growing turkey poult. Poult. Sci., 60, 659.

Proudman, J.A., and D.H. Corcoran. (1981). Turkey prolactin: purification by isotachophoresis and partial characterization. Biol. Reprod., 25, 375.

Rabii, J., F.C. Buonomo, and C.G. Scanes. (1981). Role of serotonin in the regulation of growth hormone and prolactin secretion in the domestic fowl. J. Endocrinol., 90, 355.

Robinson, G.A., D.C. Wasnidge, F. Floto, and S.E. Downie. (1976). Ovarian ^{125}I transference in the laying Japanese quail; apparent stimulation by FSH and lack of stimulation by TSH. Poult. Sci., 55, 1665.

Riddle, O., R.W. Bates, and S.W. Dykshorn. (1933). The preparation, identification and assay of prolactin—a hormone of the anterior pituitary gland. Am. J. Physiol., 105, 191.

Riddle, O., R.W. Bates, and E.L. Lahr. (1935). Prolactin induces broodiness in fowl. Am. J. Physiol., 11, 352.

Rivier, J., J. Spiess, M. Thorner, and W. Vale. (1982). Characterization of a growth hormone-releasing factor from a human pancreatic islet tumour. Nature (London), 300, 276.

Rosenbloom, A.A., and D.A. Fisher. (1974). Radioimmunoassay of arginine vasotocin. Endocrinology, 95, 1726.

Rzasa, J., and Z. Ewy. (1971). Effects of vasotocin and oxytocin on intrauterine pressure in the hen. J. Reprod. Fertil., 25, 115.

Saeki, Y., and Y. Tanabe. (1954). Changes in prolactin content of fowl pituitary during broody periods and some experiments on the induction of broodiness. Poult. Sci., 34, 900.

Sakai, H., and S. Ishii. (1980). Isolation and characterization of chicken follicle-stimulating hormone. Gen. Comp. Endocrinol., 42, 1.

Scanes, C.G. (1974). Some *in vitro* effects of synthetic thyrotrophin releasing factor on the secretion of thyroid stimulating hormone from the anterior pituitary gland of the domestic fowl. Neuroendocrinology, 15, 1.

Scanes, C.G., and B.K. Follett. (1972). Fractionation and assay of chicken pituitary hormones. Br. Poult. Sci., 13, 603.

Scanes, C.G., and J.H. Fagioli. (1980). Effects of mammalian and avian gonadotropins on *in vitro* progesterone production by avian ovarian granulosa cells. Gen. Comp. Endocrinol., 41, 1.

Scanes, C.G., and S. Harvey. (1980). Inhibition of thyrotropin-releasing hormone-induced growth hormone secretion in domestic fowl by adrenaline and prostaglandin E_1 and E_2. Horm. Metab. Res., 12, 634.

Scanes, C.G., and J. Balthazart. (1981). Circulating concentrations of growth hormone during growth, maturation, and reproductive cycles in ring doves (*Streptopelia risoria*). Gen. Comp. Endocrinol., 45, 381.

Scanes, C.G., and S. Harvey. (1981). Growth hormone and prolactin in avian species. *Life Sci.,* 28, 2895.

Scanes, C.G., N.J. Bolton, and A. Chadwick. (1975). Purification and properties of an avian prolactin. Gen. Comp. Endocrinol., 27, 371.

Scanes, C.G., A. Chadwick, and N.J. Bolton. (1976a). Radioimmunoassay of prolactin in the plasma of domestic fowl. Gen. Comp. Endocrinol., 30, 12.

Scanes, C.G., L. Gales, S. Harvey, A. Chadwick, and W.S. Newcomer. (1976b). Endocrine studies in young chickens of the obese strain. Gen. Comp. Endocrinol., 30, 419.

Scanes, C.G., P.M.M. Godden, and P.J. Sharp. (1977a). An homologous radioimmunoassay for chicken follicle-stimulating hormone: observations on the ovulatory cycle. J. Endocrinol., 73, 473.

Scanes, C.G., S. Harvey, and A. Chadwick. (1977b). Hormone and growth in poultry. "Growth and Poultry Meat Production" (K.N. Boorman and B.J. Wilson, Eds.). Edinburgh: British Poultry Science, p. 79.

Scanes, C.G., M. Jallageas, and I. Assenmacher. (1980a). Seasonal variations in the circulating concentrations of growth hormone in male Pekin duck (*Anas platyrhynchos*) and teal (*Anas crecca*); correlations with thyroidal function. Gen. Comp. Endocrinol., 41, 76.

Scanes, C.G., G.F. Merrill, R. Ford, P. Mauser, and C. Horowitz. (1980b). Effects of stress (hypoglycaemia, endotoxin and ether) on the peripheral circulating concentration of corticosterone in the domestic fowl (*Gallus domesticus*). Comp. Biochem. Physiol. C: Comp. Pharmacol., 66, 183.

Scanes, C.G., P. Griminger, and F.C. Buonomo. (1981a). Effects of dietary protein restriction on circulating concentrations of growth hormone in growing domestic fowl (*Gallus domesticus*). Proc. Soc. Exp. Biol., 168, 334.

Scanes, C.G., S. Harvey, B.A. Morgan, and M. Hayes. (1981b). Effect of synthetic thyrotrophin-releasing hormone and its analogues on growth hormone secretion in the domestic fowl (*Gallus domesticus*). Acta Endocrinol., 97, 448.

Scanes, C.G., T.J. Lauterio, and F.C. Buonomo. (1983a). Annual, developmental and diurnal cycles of pituitary hormone secretion. In "Avian Endocrinology: Environmental and Ecological Perspectives" (S. Mikami, K. Homma, and M. Wada, Eds.). Tokyo: Japan Scientific Society Press Berlin: Springer-Verlag, p. 307.

Scanes, C.G., J. Marsh, E. Decuypere, and P. Rudas. (1983b). Abnormalities in the plasma concentrations of thyroxine, triiodothyronine and growth hormone in sex-linked dwarf and autosomal dwarf White Leghorn domestic fowl (*Gallus domesticus*). J. Endocrinol., 97, 127.

Scanes, C.G., T.J. Lauterio, L. Huybrechts, J. Rivier, and W. Vale. (1984). Human pancreatic growth hormone-releasing factor (HPGRF) stimulates growth hormone secretion in the domestic fowl. *Life Sci.,* 34, 1127.

Schally, A.V., T.W. Redding, C.Y. Bowers, and J.F. Barrett. (1969). Isolation and properties of porcine thyrotropin-releasing hormone. J. Biol. Chem., 244, 4077.

Shahabi, N.A., J.M. Bahr, and A.V. Nalbandov. (1975). Effect of LH injection on plasma and follicular steroids in the chicken. Endocrinology, 96, 969.

Sharp, P.J., C.G. Scanes, and A.B. Gilbert. (1978). *In vivo* effects of an antiserum to partially purified chicken luteinizing hormone (CM2) in laying hens. Gen. Comp. Endocrinol., 34, 296.

Sharp, P.J., C.G. Scanes, J.B. Williams, S. Harvey, and A. Chadwick. (1979a). Variations in concentrations of prolactin, luteinizing hormone, growth hormone and progesterone in the plasma of broody bantams (*Gallus domesticus*). J. Endocrinol., 80, 51.

Sharp, P.J., R.B. Chiasson, M.M. El Tounsy, H. Klandorf, and W.J. Radke. (1979b). Localization of cells producing thyroid-stimulating hormone in the pituitary gland of the domestic fowl. Cell Tissue Res., 198, 53.

Shellabarger, C.J. (1954). Detection of thyroid stimulating hormones by I^{131} uptake in chicks. J. Appl. Physiol., 6, 721.

Simon-Oppenmann, C., E. Simon, H. Deutch, L. Mohring, and J. Schoun. (1980). Serum arginine-vasotocin (AVT) and afferent and central control of osmoregulation in conscious Pekin ducks. Pfleugers Arch., 387, 99.

Shirley, H.V., and A.V. Nalbandov. (1956). Effects of neurohypophysectomy in domestic chickens. Endocrinology, 58, 477.

Skadhauge, E., 1981. "Osmoregulation in Birds." Berlin: Springer-Verlag.

Stockell-Hartree, A., and F.J. Cunningham. (1969). Purification of chicken pituitary follicle-stimulating hormone and luteinizing hormone. J. Endocrinol., 43, 609.

Sturkie, P.D., and Y.C. Lin. (1966). Release of vasotocin and oviposition in the hen. J. Endocrinol., 33, 325.

Takáts, A., and F. Hertelendy. (1982). Adenylate cyclase activity of avian granulosa: effect of gonadotropin-releasing hormone. Gen. Comp. Endocrinol., 48, 515.

Tanaka, K., M. Kamiyoshi, and Y. Tanabe. (1971). Inhibition of premature ovulation by prolactin in the hen. Poult. Sci., 50, 63.

Thommes, R.C., and V.W. Hylka. (1978). Hypothalamo-adenohypophyseal–thyroid interactions in the chick embryo. I. TRH and TSH sensitivity. Gen. Comp. Endocrinol., 34, 193.

Tixier-Vidal, A., and B.K. Follett. (1973). The adenohypophysis. In "Avian Biology," Vol. III (D.S. Farner and J.R. King, Eds.). New York: Academic Press, p. 110.

Tonoue, T., and J. Kitch. (1978). Release of cyclic AMP from the chicken thyroid stimulated with TSH *in vitro.* Endocrinol. Jpn., 25, 105.

Tsutsui, K., and S. Ishii. (1978). Effects of follicle stimulating hormone and testosterone on receptors of follicle-stimulating hormone in the testis of the immature Japanese quail. Gen. Comp. Endocrinol., 36, 297.

Tsutsui, K., and S. Ishii. (1980). Hormonal regulation of follicle-stimulating hormone receptors in the testes of Japanese quail. J. Endocrinol., 85, 511.

Wells, J.W., A.B. Gilbert, and J. Culbert. (1980). Effect of luteinizing hormone on progesterone secretion *in vitro* by the granulosa cells of the domestic fowl (*Gallus domesticus*). J. Endocrinol., 84, 249.

Wentworth, B.C., W.H. Burke, and G.P. Birrenkott. (1976). A radioimmunoassay for turkey luteinizing hormone. Gen. Comp. Endocrinol., 29, 119.

Wilson, S.C., and P.J. Sharp. (1973). Variations in plasma LH levels during the ovulatory cycle of the hen, *Gallus domesticus.* J. Reprod. Fertil., 35, 561.

Wilson, S.C., and P.J. Sharp. (1975). Episodic release of luteinizing hormone in the domestic fowl. J. Endocrinol., 64, 77.

Wilson, S.C., and P.J. Sharp. (1976a). Induction of luteinizing hormone release by gonadal steroids in the ovariectomized domestic hen. J. Endocrinol., 71, 87.

Wilson, S.C., and P.J. Sharp. (1976b). Effects of androgens, oestrogens and deoxycorticosterone acetate on plasma concentrations of luteinizing hormone in laying hens. J. Endocrinol., 69, 93.

Wingfield, J.C., and D.S. Farner. (1978). The annual cycle of plasma in LH and steroid hormone in feral population of the White-crowned sparrow, *Zonotrichia leucophrys gambelii.* Biol. Reprod., 19, 1046.

Wingfield, J.C., B.K. Follett, K.S. Matt, and D.S. Farner. (1980). Effect of daylength on plasma FSH and LH in castrated and intact White-crowned sparrows. Gen. Comp. Endocrinol., 42, 464.

Yamashiro, D., R.G. Hammonds, and C.H. Li. (1982). β-endorphin synthesis and receptor-binding activity of the ostrich hormone. Int. J. Pept. Protein Res., 19, 251.

Zakar, T., and f. Hertelendy. (1980). Effects of mammalian LH, cyclic AMP and phosphodiesterase inhibitors on steroidogenesis, lactate productive, glucose uptake and utilization by avian granulosa cells. Biol. Reprod., 22, 810.

18
Reproduction in the Female

A.L. Johnson

Shell Matrix
Cuticle

Layers of Crystallization

Mammillary Knob Layer
Palisade Layer
Surface Crystal Layer
Respiration via the Eggshell

Calcium Metabolism

Sources of Eggshell Calcium
Vitamin D Metabolism
Calcium Mobilization from Medullary Bone
Calcium Absorption and Secretion by the Shell Gland
Carbonate Formation and Deposition

References

Anatomy of the Female Reproductive System

The right ovary and oviduct are present in embryonic stages of all birds, but the distribution of primordial germ cells to the ovaries of the chicken becomes asymmetrical by day 4 of incubation, and by day 10 regression of the right oviduct begins. The reproductive system of birds (Galliformes) consists of a single left ovary and its oviduct, although on occasion a functional right ovary and oviduct may be present. Among the falconiformes and in the brown kiwi, both left and right gonads and associated oviducts are commonly functional, although the ovaries may be asymmetrical in size; in sparrows and pigeons, about 5% of specimens have two developed ovaries (see Romanoff and Romanoff, 1949; Kinsky, 1971).

Ovary

The left ovary is attached by the mesovarian ligament at the cephalic end of the left kidney. The number of oocytes of the chick embryo increases from approximately 28,000 on the 9th day of development to 680,000 on the 17th day, and subsequently decreases to 480,000 by the time of hatching, when oogenesis is terminated (Benoit, 1950). The ovary of the immature bird consists of a mass of small ova, of which at least 2,000 are visible to the naked eye. Only a relatively few of these (200–500) reach maturity and are ovulated within the life span of most domesticated species, and considerably fewer mature in wild species. The functionally mature ovary of the hen is arranged with an obvious hierarchy of follicles, and in its entirety weighs 20–30 g. Commonly, there are four to six large yolk-filled follicles 2–4 cm in diameter, accompanied by a greater number of 2- to 10-mm yellow follicles and numerous small white follicles (Figure 18–1).

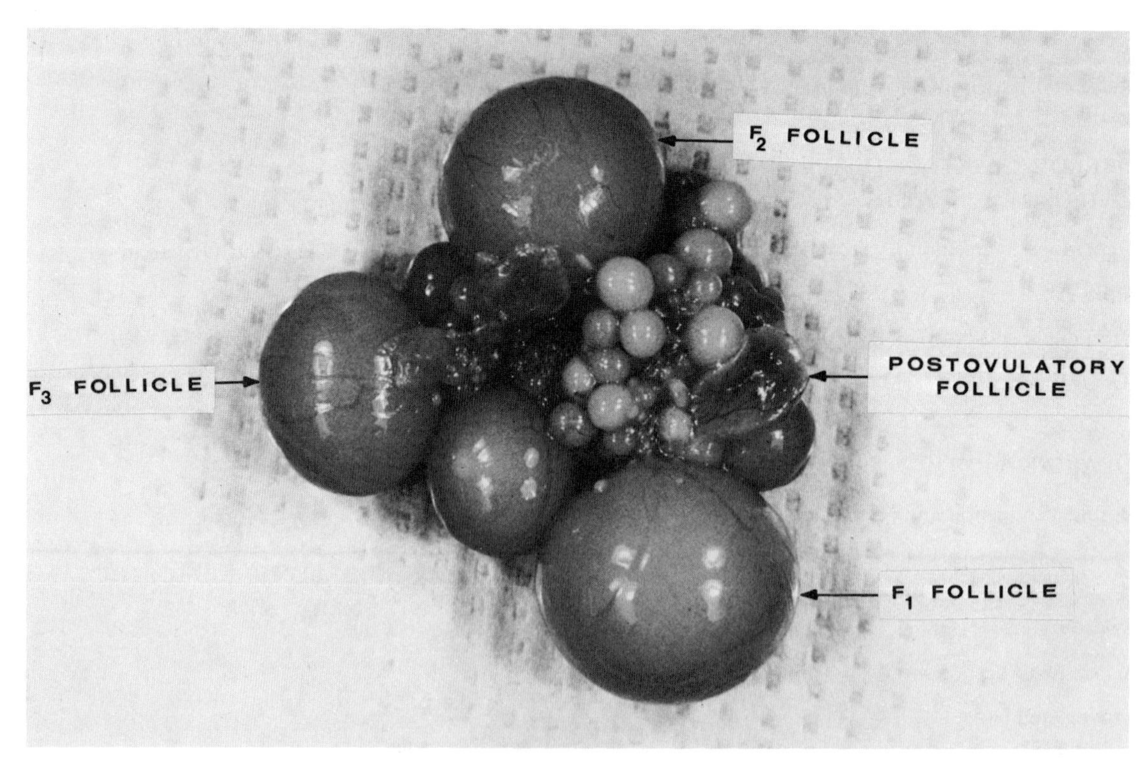

FIGURE 18–1. Follicular hierarchy in the laying hen.

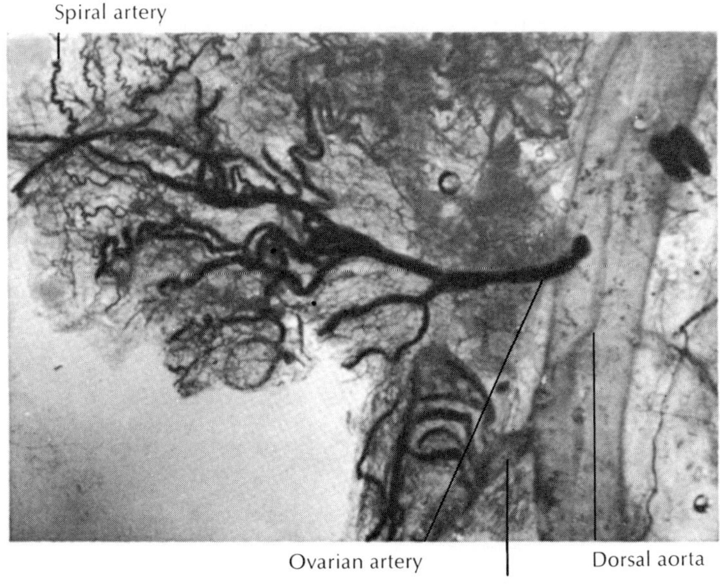

FIGURE 18–2. Arterial supply to the ovary of the hen. (From Nalbandov and James, 1949).

The ovary receives its blood supply from the ovarian artery, which usually arises from the left renolumbar artery but may branch directly from the dorsal aorta (Figure 18–2) (see also Nalbandov and James, 1949; Hodges, 1965). Within the ovary, bloodflow is greatest to the five largest preovulatory follicles (Wolfenson et al., 1981; Scanes et al., 1982; see Chapter 4 for further details). The ovarian artery divides into many branches and leads to a single follicular stalk, usually from two to four separate arterial branches. All veins from the ovary unite into the two main arterior and posterior veins, which subsequently drain into the posterior vena cava.

Several workers have shown that the ovary is well innervated by both adrenergic and cholinergic fibers (Gilbert, 1969; Dahl, 1970b; Unsicker et al., 1983) and that a greater number of neurons are present as a follicle progressively matures.

Ovarian Follicle. The follicle consists of concentric layers of tissue that surround the oocyte and yolk, including: (1) the oocyte plasma membrane, (2) the perivitelline membrane, (3) granulosa cells, (4) basal lamina, and (5) the theca (interna and externa) (Figure 18–3). The follicle is highly vascularized except for the stigma (the point of rupture during ovulation), which contains a lesser number of underlying small veins and arteries (Nalbandov and James, 1949) (Figure 18–4). The main arteries from the stalk are directed toward the fastest growing follicles, branch into arterioles, and pass through the theca to the basal lamina to form arterial capillaries (Dahl, 1970a). The venous system of the follicle is more elaborate than the arterial system and forms three layers: (1) the inner venous capillary layer, lying proximate to the basal lamina; (2) the middle venous layer, lying outside the theca interna; and (3) the outer or peripheral venous layer, which consists of numerous large veins that encircle the follicle and drain via the stalk.

Postovulatory and Atretic Follicles. The postovulatory follicle (POF) consists of the remaining follicular tissue subsequent to ovulation (Figure 18–1). The POF of the hen is metabolically active, as indicated by the presence of enzymatic activity in granulosa cells for 3 days after ovulation and for at least 7 days in thecal cells (Chalana and Guraya, 1978). Total resorption of the POF ranges from 8–10 days in the chicken to a period of several months in the mallard (Lofts and Murton, 1973). The POF may influence nesting behavior (Wood-Gush and Gilbert, 1975), and has been demonstrated to be functional in timing oviposition (Rothchild and Fraps, 1944; Gilbert et al., 1978). There is no structure in birds analagous to the corpus luteum of mammals.

Follicles that have become enlarged and yolk filled, but fail to reach the stage at which they are ovulated, become atretic. The process of atresia consists of the reabsorption of the oocyte and yolk either via the vascular system or by the rupture of the thecal layers and slow loss of the yolk into the peritoneal cavity. Natural instances of follicular atresia occur subsequent to a change from the egg-laying state to incubation and broody behavior and with the onset of molt (Gilbert, 1979). Atresia of follicles can also be induced by prolonged daily administration of pregnant mare serum gonadotropin (PMSG) (Johnson, unpublished observation), hypophysectomy (Opel and Nalbandov, 1961), or by the administration of ovine luteinizing hormone (LH) (Gilbert et al., 1981).

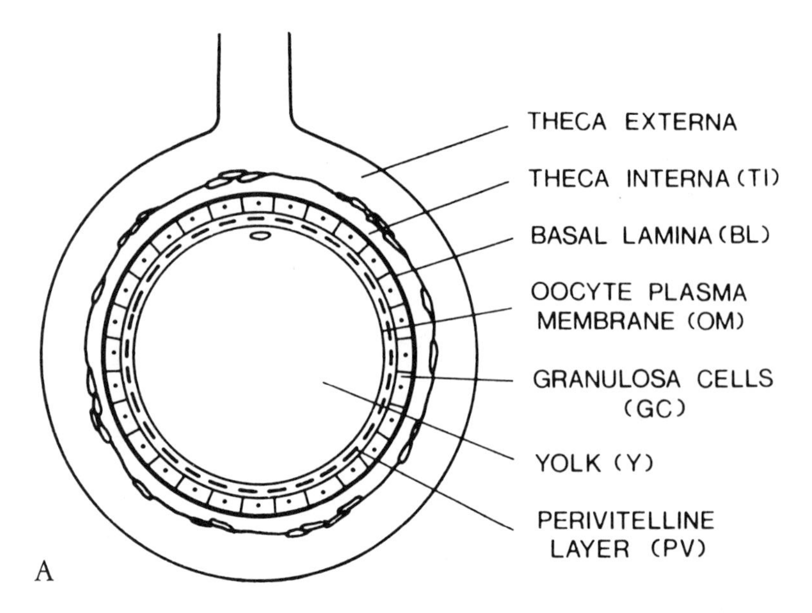

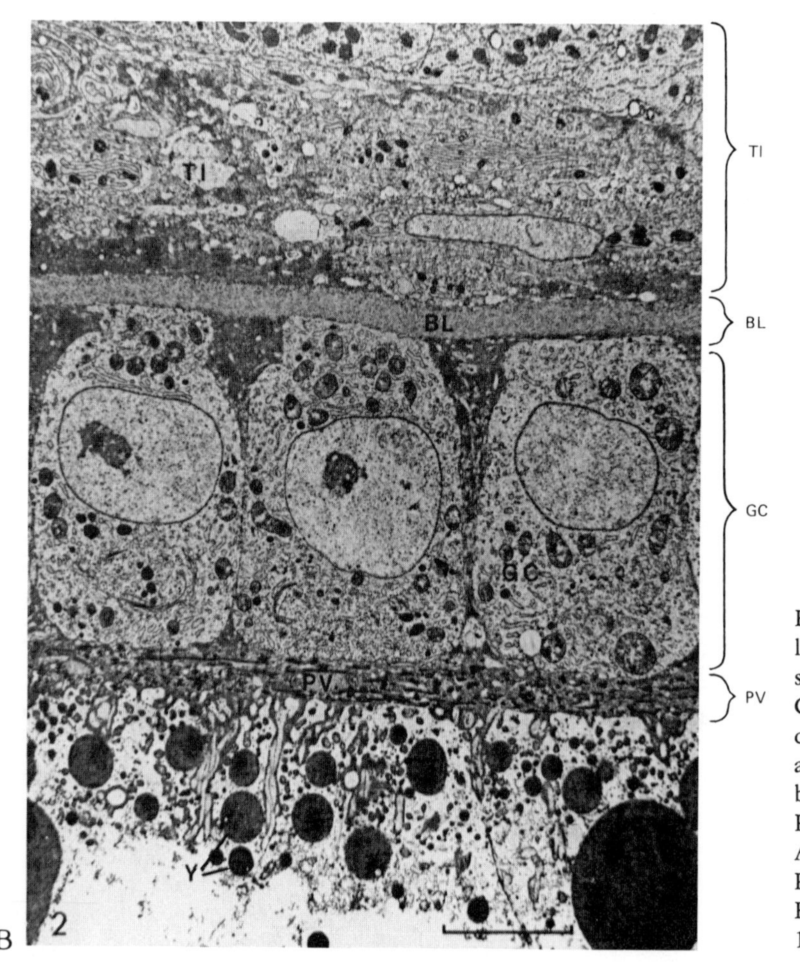

FIGURE 18–3. (**A**) Diagram of relationship of follicle tissue layers surrounding the oocyte. (From Gilbert, 1979.) (**B**) Electron micrograph of tissue layers depicted above. Abbreviations are as labeled in Figure 18–3**A**. (From Perry and Gilbert, 1979, and from Agricultural Research Council Poultry Research Centre Report, Ruslin, Midlothian, Scotland, 1980.)

FIGURE 18–4. Mature ovarian follicle of the hen. (From Nalbandov and James, 1949.)

Growth of the Follicle. Growth of the follicle can be divided into three phases: (1) slow growth of follicles 0.05–1.0 mm in diameter, lasting months to years; (2) a 2-month period of increasingly rapid growth that consists mainly of deposition of yolk protein; and (3) a rapid growth phase during the final 7–11 days before ovulation when the majority of yolk and lipids are deposited (Marza and Marza, 1935). In this final phase of development, the follicle will increase, on the average, from 8 to 37 mm in diameter and from 0.08 to 15–18 g in weight. Yolk formation occurs in the liver and is regulated by gonadotropin and steroid hormones.

A schematic representation of vitellogenesis in the hen is presented in Figure 18–5. The yolk protein precursor, vitelloginin, is transported via the blood to the ovary, where it is cleaved into the two yolk proteins, lipovitellin and phosvitin (Deely et al., 1975). Triglycerides are transported to the yolk in the form of β-lipoproteins and subsequently are incorporated into the yolk as lipid globules. Lipids and protein are deposited into the growing follicle at about the same ratio for most of the growth phase, but during the final rapid growth phase, relatively more lipid is incorporated. The final composition of yolk in the hen's egg consists of a greater percentage of lipid (33% of wet weight) compared to protein (16% wet weight). It has been suggested that deposition of yolk into the maturing follicle is terminated by 24 hr before ovulation. The ultrastructure of developing follicles has been described by Wyburn et al. (1965), Rothwell and Solomon (1977), Perry et al. (1978a, b), and Gilbert et al. (1980).

Observations indicate that in chickens the decrease in egg production with age is in part caused by a reduction in the number of follicles reaching the phase of rapid growth; that is, fewer follicles receive proportionately greater quantity of yolk, resulting in larger sized eggs (Williams and Sharp, 1978).

The mechanisms controlling the development and maintenance of the follicular hierarchy are not well understood. Pituitary gonadotropin involvement is indicated by the finding that hypophysectomy rapidly leads to follicular atresia, whereas ablation and treatment with a chicken pituitary extract results in the maintenance of the normal follicular hierarchy (Opel and Nalbandov, 1961; Mitchell, 1967). On the other hand, repeated injections of follicle-stimulating hormone

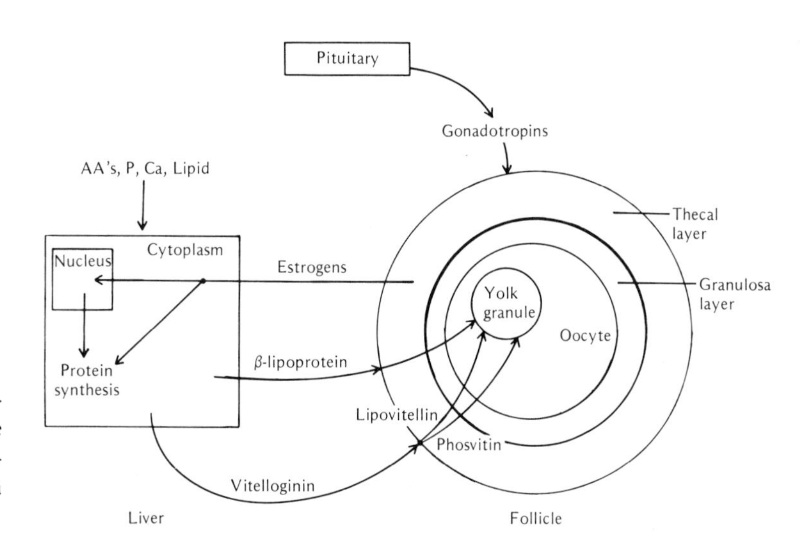

FIGURE 18–5. Diagrammatic representation of vitellogenesis in the hen. *AA,* amino acids; *P,* phosphorus; *Ca,* calcium. (Redrawn from Redshaw and Follett, 1972.)

(FSH) or pregnant mare serum gonadotropin (PMSG) to intact laying hens stimulates numerous small follicles to enter the rapid growth phase and results in the elimination of the follicular hierarchy and the cessation of ovulation.

Oviduct and Sperm Storage Glands

The oviduct of the hen is derived from the left Mullerian duct. Generally, the right Mullerian duct remains vestigial, although certain strains of chickens may develop two functional oviducts (Morgan and Kohlmeyer, 1957). The oviduct consists of five distinguishable regions: infundibulum, magnum, isthmus, shell gland, and vagina (Figure 18–6). [For details on general oviduct histology, see Aitken (1971) and King (1975); on the infundibulum and magnum, see Wyburn et al. (1970); for the isthmus, see Hoffer (1971) and Solomon (1975); and for the shell gland, see Breen and DeBruyn, 1969, Nevalainen (1969), and Wyburn et al. (1973).] Cilia are found along the entire length of the oviduct (Fujii, 1975); and one proposed function of these cilia is that of sperm transport. A dorsal and a ventral ligament suspend the oviduct within the peritoneal cavity.

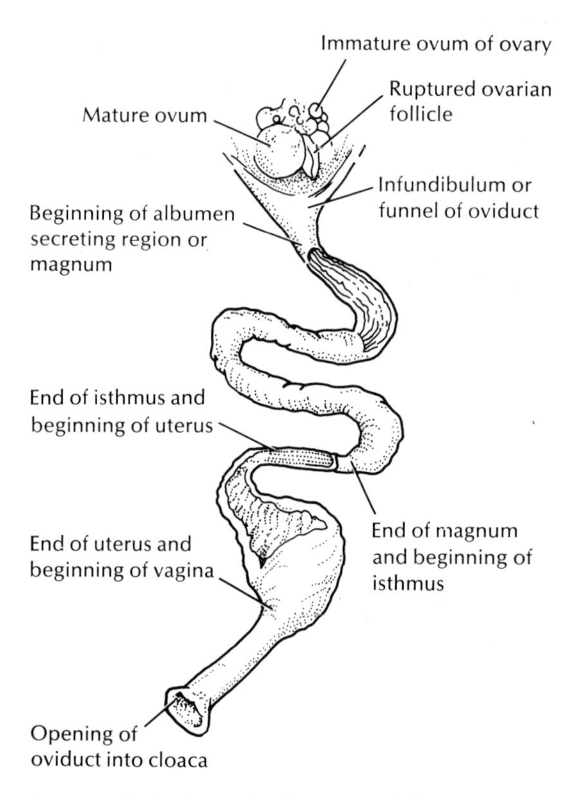

FIGURE 18–6. Reproductive tract of the laying hen. (From Sturkie and Mueller, 1976.)

Infundibulum. Subsequent to ovulation, the ovum is engulfed by the infundibulum (which is not directly connected to the ovary) where it resides for approximately 18 min (range, 15–30 min) (Warren and Scott, 1935). Occasionally, the ovum fails to be picked up by the infundibulum (an "internal ovulation") and is reabsorbed, possibly via the ovary itself (Callebaut, 1979), in 24 hr or less (Sturkie, 1955). Infundibular activity appears not to be controlled by ovulation per se, as foreign objects placed into the abdominal cavity prior to ovulation will also be taken up, and reports indicate that entire unovulated follicles may be engulfed and later laid as fully developed eggs. Fertilization of the ovum occurs in the infundibulum, and it is here that in the chicken the first layer of albumin is produced (Gilbert, 1979).

Magnum. The ovum next passes to the largest portion of the oviduct, the magnum (a length of 33 cm in the chicken), where the majority of albumen is formed. The magnum is highly glandular, two types of glands (tubular and unicellular or epithelial) being distinguishable (Aitken, 1971). Tubular glands are responsible for the production of ovotransferrin (conalbumin) and ovomucoid (Schimke et al., 1977), and epithelial cells synthesize avidin (Tuohimaa, 1975). A measurable amount of calcium secretion occurs within the magnum, but at no time during the laying cycle is calcium secretion greater than the basal rate of secretion found in the shell gland (Eastin and Spaziani, 1978a). Movement of the ovum through the magnum is primarily via peristaltic action of longitudinal circular layers of muscle. The ovum remains in the magnum approximately 2–3 hr.

Isthmus. The isthmus is clearly demarcated from the magnum. It has a thick circular layer of muscle, and glandular tissue is less developed compared to the magnum. This tissue is characterized by a layer of epithelial cells with underlying tubular gland cells. Both inner and outer shell membranes are formed during the 1- to 2-hr (mean time, 1 hr and 14 min) passage through the isthmus.

Shell Gland. The shell gland (uterus) is characterized by a prominent longitudinal muscle layer lined medially with both tubular and unicellular goblet cells. Prior to calcification, the egg takes up salts and about 15 g water into the albumen from the tubular glands, a process termed "plumping" (Wyburn et al., 1973). The ovum remains in the shell gland for 20–26 hr.

The initiation of calcification within the shell gland appears to be associated with stimuli initiated by ovulation or by neuroendocrine factors that control and coordinate both ovulation and calcium secretion. Additional evidence suggests that distension of the shell gland by the egg is not the principal stimulus for the initiation of a high rate of calcium secretion, which is characteristic of calcification, nor is autonomic innervation involved (Eastin and Spaziani, 1978a). Calcification of the egg occurs over a period of 15 hr, while the pigment of the shell is formed during the last 5 hr (Warren and Conrad, 1942). The high rate of calcium secretion returns to low basal levels approximately 2 hr before expulsion of the egg (Eastin and Spaziani, 1978a).

Vagina. The vagina is separated from the shell gland by the uterovaginal sphincter muscle and terminates at the cloaca. There are numerous folds of mucosa, which are lined by ciliated and nonciliated cells, but secretory glands are absent. The vagina has no role in the formation of the egg, but, in coordination with the shell gland, participates in the expulsion of the egg.

Bloodflow and Innervation to the Oviduct. The blood supply to the oviduct and shell gland of the domestic hen has been described by Freedman and Sturkie (1963a) and Hodges (1965) (Figures 18–7 and 18–8). For a review of the vasculature to the oviduct of additional avian species, see Gilbert (1979). Bloodflow to the shell gland of the hen is increased during the presence of a calcifying egg compared to times when no egg is present (Boelkins et al., 1973; Wolfenson et al., 1981; Scanes et al., 1982). For further details on bloodflow, see Chapter 6.

The oviduct is innervated by both sympathetic and parasympathetic nerves. Innervation of the infundibulum is via the aortic plexus and the magnum by the aortic and renal plexuses (Hodges, 1974). Sympathetic innervation of the shell gland is via the hypogastric nerve, which is the direct continuation of the aortic plexus. Parasympathetic pelvic nerves, which constitute the left pelvic plexus, arise from the pelvic visceral rami of spinal nerves 30–33 (Freedman and Sturkie, 1963b).

Oviduct Motility and Expulsion of the Egg. Egg transport is primarily accomplished by contractions of the oviduct; the oviduct probably functions as a stretch receptor and the mechanical stimulus is produced by the ovum itself (Arjamaa and Talo, 1983). Changes in electrical activity and oviduct motility have been recorded in the magnum, isthmus, and shell gland during the ovulatory cycle, with the greatest frequency of electrical activity and contractions occurring in the shell gland at the time of oviposition (Shimada, 1978; Shimada and Asai, 1978). Both α- and β-adrenergic receptors are present throughout the length of the oviduct and have been shown to affect oviduct motility (Verma and Walker, 1974; Crossley et al., 1980).

Administration of arachidonic acid (a prostaglandin precursor) or prostaglandin $F_{2\alpha}$ ($PGF_{2\alpha}$) increases intraluminal pressure of the infundibulum, magnum, isthmus, and shell gland (Wechsung and Houvenghel, 1978, 1981). Both $PGF_{2\alpha}$ and arginine vasotocin (see later section) stimulate shell gland contractility in vitro (Olson et al., 1978; Rzasa, 1978), while $PGF_{2\alpha}$ has

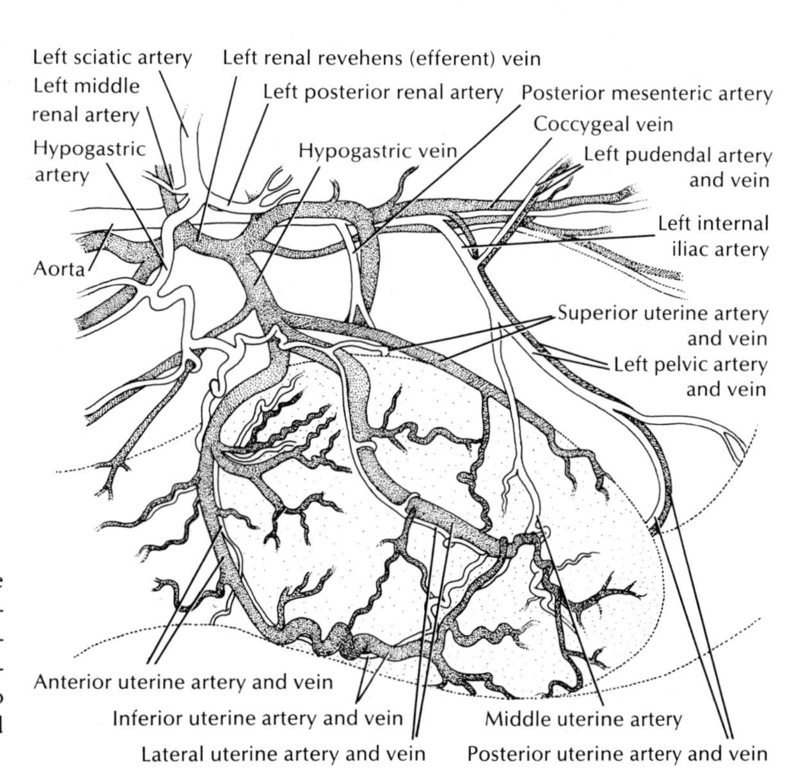

FIGURE 18–7. Laterial view of the blood vessels to the uterine portion of the hen's oviduct. The arterial system is in white and the venous system is shaded. Drawn to scale. (From Freedman and Sturkie, 1963a.)

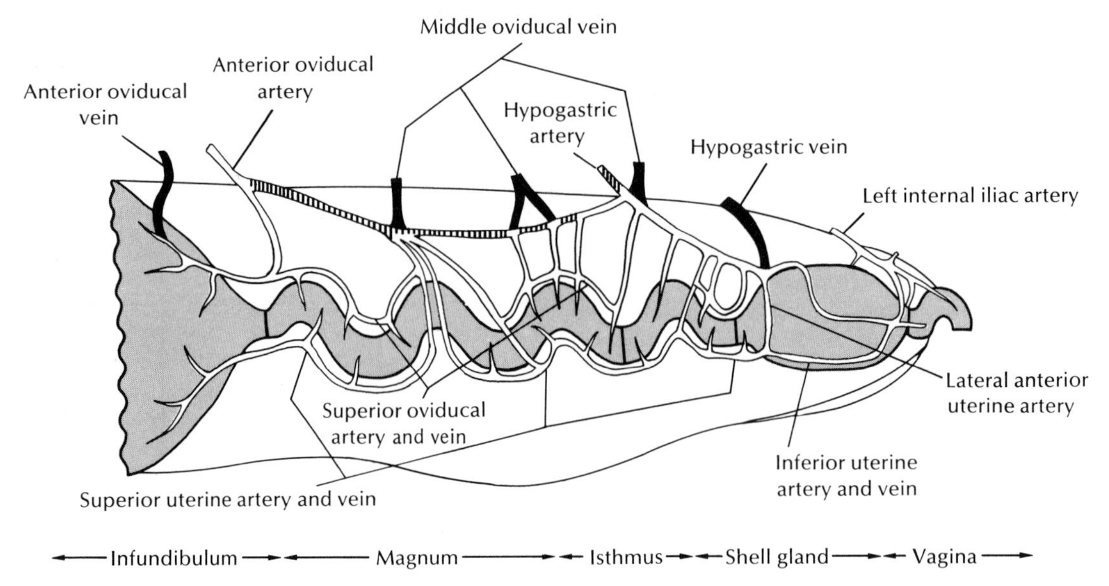

FIGURE 18–8. Blood vessels in the oviduct. Arteries are crosshatched and white, veins are black. (From Hodges, 1965.)

been demonstrated to be effective in vivo (Shimada and Asai, 1979). Specific binding sites for $PGF_{2\alpha}$ have been identified and characterized in the membranes of shell gland muscle (Toth et al., 1979). In contrast, prostaglandins of the E series (PGE, PGE_2) suppress spontaneous motility and decrease intraluminal pressure in the vagina (Wechsung and Houvenaghel, 1978; Verma et al., 1976). These observations suggested that ovum transport through the oviduct and expulsion of the egg from the shell gland may involve prostaglandin-stimulated contractility of oviduct smooth muscle.

Sperm-Host Glands. Spermatozoa are stored and retain their fertilizing capacity within the oviduct for a period of 7–14 days in the chicken and for 40–50 days in the turkey hen. Two sets of sperm-host glands have been identified in the domestic hen; infundibular glands and uterovaginal junction glands. The existence of infundibular glands in the turkey is in question (for contrasting views, see Verma and Cherms, 1965; and Bakst, 1981). In addition, the functional significance of sperm stored within the infundibular glands of the chicken is uncertain (Lake, 1975).

Uterovaginal junction glands are apparently devoid of innervation and contractile tissue, but possess a well-developed vascular system (Gilbert et al., 1968; Tingari and Lake, 1973). The ultrastructure of these glands has been studied by Burke et al. (1972). The mechanism by which sperm are released from the glands is unknown, but Bakst (1981) suggested that the pattern of release is either continuous or episodic and not associated with either ovulation or oviposition. Evidence suggests that spermatozoa fill the uterovaginal glands in a sequential fashion without mixing, so that with successive inseminations, sperm from the latest insemination is most likely to fertilize an ovum (Compton et al., 1978).

Breeding and Ovulation–Oviposition Cycles

Breeding cycles between species of birds can be classified according to the length of the cycle and the time of the year at which each species become reproductively active. Continuous breeders, such as the domestic hen or khaki Campbell duck, are, under optimal conditions, reproductively active throughout the year. Most nondomestic species that breed in temperate, subarctic, and arctic zones display yearly cycles, while birds adapted to tropical or desert climates may breed with cycles less than a year, at 6-month intervals, or when favorable conditions exist (opportunistic breeders) (for further details, see Lofts and Murton, 1973; van Tienhoven, 1983).

Wild birds usually lay one or more eggs in a clutch and then terminate laying to incubate the eggs. The number of eggs per clutch and total number of clutches vary with the species and season. For example, some birds, such as the auk or sooty term (*Sterna fuscata*) lay a single egg to incubate. The king penguin (*Aptenodytes patogonica*) will lay but one egg, and breeding will not necessarily occur on a yearly basis. In contrast, the European partridge will lay a single clutch per year consisting of as many as 12–20 eggs (mean, 14.6). The pigeon usually lays two eggs per clutch and averages eight clutches per year. The interval between clutches is approximately 45 days in the fall and winter and from 30 to 32 days in the spring and early summer. Finally, clutch size of the bobwhite quail declines from a mean high of 19.2 eggs in early May to 11.3 eggs in late July (see Sturkie and Mueller, 1976).

When eggs are removed from the nest of some species, the birds then lay more than the normal number of eggs in a clutch (indeterminate layers; an example is the duck, *Anas platyrhynchos*) (Donham et al., 1976). Determinate layers fail to lay additional eggs on the removal or destruction of eggs in the nest. In seasonal (or noncontinuous) breeders, the ovary undergoes periods of growth and regression. The weight of the European starling ovary may vary from 8 mg during the regression phase to 1400 mg at the height of the breeding season (Lofts and Murton, 1973).

Ovulation–Oviposition Cycle and Rate of Lay

The ovulation–oviposition cycle (the time from ovulation of an ovum to the oviposition of the egg) of the domestic hen generally ranges from somewhat longer than 24 hr (24 plus) to 28 hours in length, and ovulations proceed uninterrupted for several days, or as long as 1 year in extreme instances. The number of eggs laid on successive days is called the sequence, and each sequence is separated by 1 or more pause days on which no egg is laid. The term clutch is sometimes used synonomously with sequence, although the former term is generally considered more appropriate to describe the group of eggs laid by nondomesticated species prior to incubation behavior. The 24-plus-hr ovulation–oviposition cycle is characteristic of the chicken (Fraps, 1955), turkey (Kosin and Abplanalp, 1951), bobwhite quail (Wilson et al., 1973), and perhaps several other species, including the Japanese quail (Planck and Johnson, 1975). By contrast, the interval between successively laid eggs in the pigeon has been reported to be 40–44 hr (Sturkie and Mueller, 1976) and in the Khaki Campbell duck is 23.5 to 24.5h (Simmons and Hetzel, 1983).

In the domestic hen, the longer the sequence, the shorter the duration of the ovulation–oviposition cycle. The delay, or "lag," in hours between the oviposition of successive eggs in a sequence is not constant (Figure 18–9). The differences in lag time within a sequence represent mainly differences in the amount of time between oviposition and the subsequent ovulation; this interval ranges from 15 to 75 min (Warren and Scott, 1935). It is clear that even when the lag between the oviposition of successive eggs approaches 24 hr (i.e., sequence length 13, Figure 18–9) there remains a progressive shift towards laying the egg later and later in the day. As the sequence becomes shorter, the lag becomes greater and the length of the ovulation–oviposition cycle deviates more and more from 24 hr. The total lag time between the first ovulation of a sequence (C_1 ovulation) and last ovulation of a sequence (C_t ovulation) is greater in chickens (4–8 hr, depending on sequence length) than in Japanese quail [1.5–2.0 hr (Opel, 1966); 4–5 hr (Tanabe and Nakamura, 1980)]. The normal release of the ovulation-inducing hormone [luteinizing hormone (LH)] in the hen is restricted to a 4- to 11-hr period (the "open period") beginning at the onset of the scotophase (dark phase). The timing and regulation of the "open period" is as yet incompletely understood. For further details concerning the ovulation–oviposition cycle and the sequence, see Fraps (1955), Gilbert and Wood-Gush, (1971), and van Tienhoven (1981).

A chicken typically ovulates the first egg of a se-

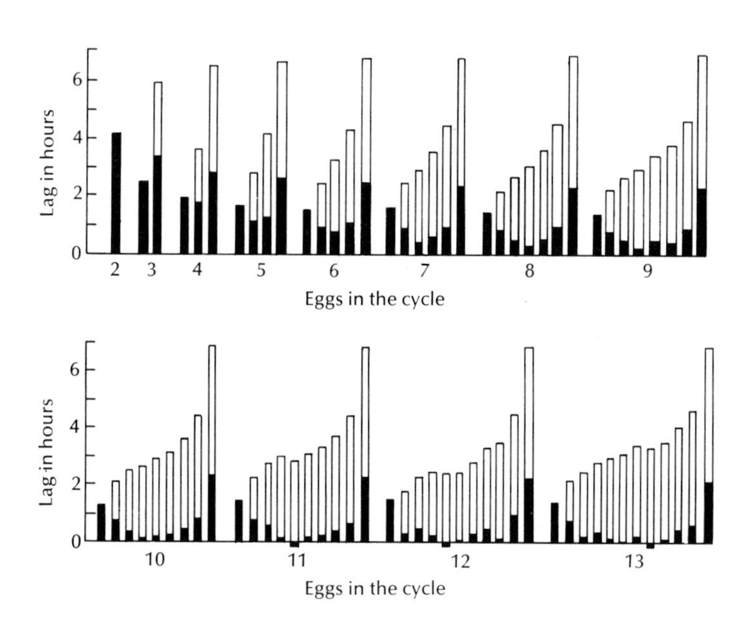

FIGURE 18–9. The delay or lag in hours between the laying of successive eggs in a cycle. A two-egg cycle is represented by one bar; a three-egg cycle by two bars, etc. A lag of 0 hr means an interval of 24 hr between eggs. A lag of 4.2 hr occurs between the laying of the first and second egg of a two-egg cycle (black bar). For the three-egg cycle, the lag between the first and second eggs is about 2.5 hr, and between the second and third, about 3.4 hr (second black bar). The cumulative lag (white bar) is about 5.9 hr. In a sequence, the lag between the last egg and the preceding one is always greatest. (Courtesy of Dr. R.M. Fraps, based on data of Heywang, 1938.)

quence early in the photophase. The timing of ovulation is apparently synchronized by the onset of the scotophase (Tanabe and Nakamura, 1980). In comparison, Japanese quail ovulate the C_1 egg 8–9 hr after the onset of the photophase (Opel, 1966), and this appears to be synchronized by the timing of the light phase (Tanabe and Nakamura, 1980).

The rate of lay, or laying frequency, is the number of eggs laid in a given period of time, irrespective of the regularity or pattern of laying. For example, a chicken laying at the rate of 50% for 60 days produces 30 eggs, the same number of eggs as another that lays at a rate of 75% for 40 days. On reaching sexual maturity (at ~20–22 weeks of age), the domestic hen lays with an erratic pattern (sequences with 2 or more pause days or occasionally more than 1 egg per day), and with a high incidence of abnormal (soft-shelled and double-yolked) eggs for 1–2 weeks. Six to 10 weeks after the onset of egg laying, production peaks (frequently at a rate of 80–90%) and gradually decreases over a period of 40–50 weeks, depending on whether the chicken is an egg- or meat-type breed. Subsequently, egg production rapidly declines and the hen begins a period of molt. A typical laying strain of chicken will produce 280 or more eggs in a 50-week production period.

Molt

The phenomenon of molt consists of the orderly replacement of feathers, and is accompanied by the total regression of the reproductive organs and the cessation of lay. In nondomesticated species, a prenuptial molt may occur prior to the breeding season, while the postnuptial molt occurs between the end of the reproductive season and the onset of autumn or the autumn migration. In the domestic fowl, molt occurs after approximately 1 year of egg production. The physiologic mechanisms that cause molt remain obscure; however, the pituitary, thyroid, ovary, and adrenal have been suggested as mediators (see Payne, 1972). Forced molt in domestic fowl can be induced by restriction of feed and water, decreased daylength, or by administration of thyroxine, prolactin, or large doses of progesterone.

Broodiness and Nesting Behavior

Broody behavior consists of termination of egg production, the incubation of eggs, and care of the young. In domesticated birds, broodiness is virtually nonexistent in present-day commercial egg-producing hens, but is still common in the bantam hen, feral domestic fowl, and broiler-breeder, and in turkeys. The onset of incubation is coincident with complete regression of the ovary and accessory reproductive tissues such as the oviduct and comb. Broodiness is often accompanied by the development of an incubation (or brood) patch, and this increased vascularity, edema, and thickening of the epidermis is probably under the control of estrogen and prolactin. Induction of incubation behavior can be initiated by injection of prolactin (for review, see de Vlaming, 1979). Incubation activity stimulates the development of the crop sac in the pigeon. Proliferation of this gland is thought to be under the direct control of prolactin. The endogenous endocrine changes during broodiness will be discussed later in this chapter.

In the ring dove, endogenous estrogen production, stimulated by FSH secretion, is thought to induce the female to build nests. In turn, the initiation of nest-building activity in the male is dependent on the hormonal environment and behavior of the female (Cheng, 1979). In the domestic fowl, initiation of nesting behavior is more closely related to ovulation than to the presence of the egg in the oviduct or oviposition (Wood-Gush and Gilbert, 1970); 100% of ovulations are followed 24 hr later by nesting behavior, and 99% of nestings are preceded by ovulation, yet only 60–95% of ovulations result in a subsequent oviposition.

Parthenogenesis

Parthenogenesis (development from an unfertilized oocyte) has been documented in turkeys (Olsen and Marsden, 1954), and occurs less frequently in chickens (Poole and Olsen, 1958). Genetic selection can increase the incidence of parthenogenesis in turkeys, and viable poults are usually homozygous diploid males. Cytologic studies indicate that parthenogenesis in turkeys is initiated from a haploid oocyte and proceeds to the diploid state after a meiosis that is unaccompanied by cytokinesis (mitosis) (see van Tienhoven, 1983).

Ovarian Hormones

Much work has recently been published concerning the production, metabolism, and secretion of ovarian steroids in the domestic fowl (see Figures 18–10 and 18–11). In addition, other hormones (i.e., prostaglandins, neurohormones) have been identified in ovarian tissue. The present discussion summarizes research related to follicular synthesis, content and secretion of hormones, and the influence of gonadotropins.

Steroid Production and Secretion

Preovulatory Follicle. Shahabi et al. (1975b) reported that 6–8 hr prior to ovulation, highest concentrations of progesterone and estrogen occur in the largest and smaller preovulatory follicles, respectively, while concentrations of testosterone decrease in the

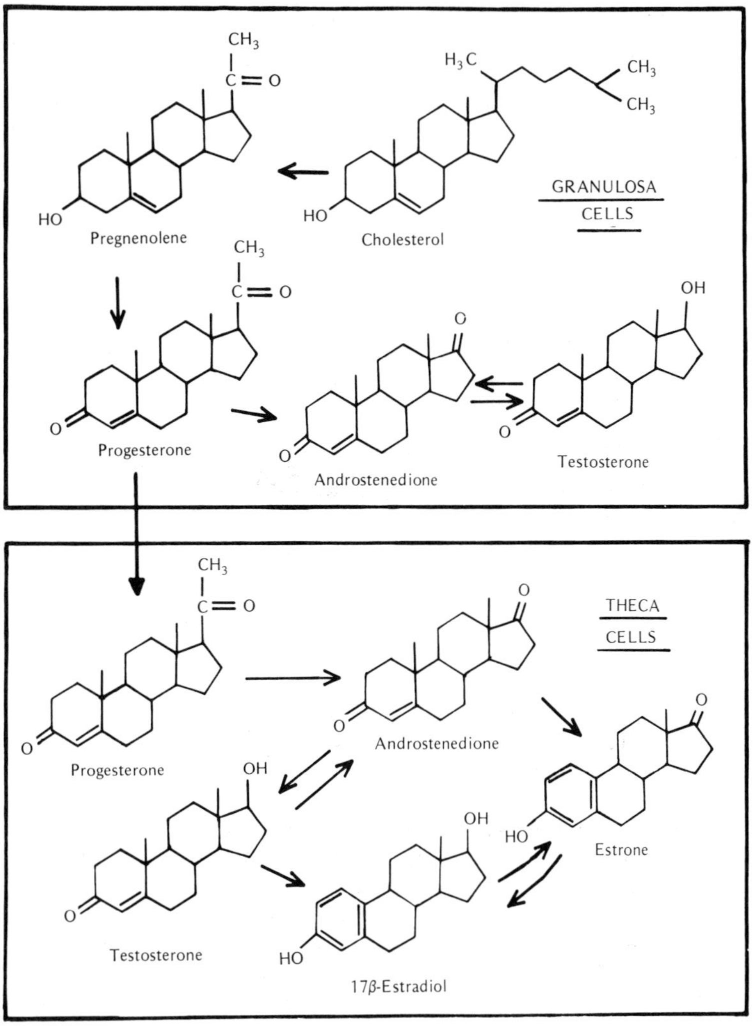

FIGURE 18–10. Structural formulas and pathways of steroid hormone synthesis by theca and granulosa tissues of the domestic fowl.

largest preovulatory follicle as the time of ovulation approaches (Figure 18–11). The principal cellular source of progesterone appears to be the granulosa cells (Huang and Nalbandov, 1979). In contrast, granulosa cells incubated in vitro spontaneously produce little to no estrogen; instead, this steroid is produced by thecal tissue. Huang et al. (1979) proposed a two-cell model for steroidogenesis within the avian follicle, which suggests that the granulosa cells are required to produce the precursor, progesterone, which is further metabolized by the thecal tissue to androgens and estrogens (Figure 18–10). Additional metabolites have been isolated and identified after incubation of [^{14}C]-progesterone with whole ovary homogenates (Nakamura et al., 1974), but the biologic activity of these metabolites has yet to be established.

Production of progesterone by granulosa cells is greatly enhanced by exposure to mammalian LH (Shahabi et al., 1975a; Huang et al., 1979; Culbert et al., 1980); this effect is greatest in granulosa cells from the largest preovulatory follicle (Hammond et al., 1981). FSH stimulates granulosa cell progesterone production of all follicles to a much lesser extent (Scanes and Fagioli, 1980; Hammond et al., 1981), with the greatest stimulatory effect produced by less mature follicles (Hammond et al., 1981). Prolactin (Hammond et al., 1982), prostaglandins (Hertelendy and Hammond, 1980), and luteinizing hormone-releasing hormone (LHRH) (Hertelendy et al., 1982) appear to have no direct effect on progesterone production by granulosa cells.

The effect of LH on progesterone production appears to be mediated by the adenylyl cyclase system (Calvo et al., 1981; Zakar and Hertelendy, 1980). Adenylyl cyclase activity in response to LH is greatest in the largest follicle, compared to the second and third

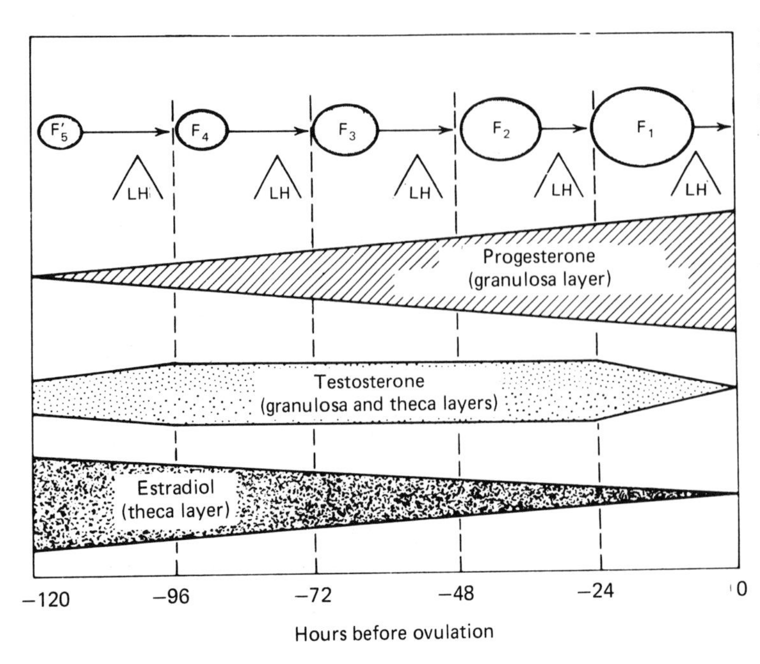

FIGURE 18–11. Schematic representation of steroid concentrations in theca and granulosa tissues during follicular maturation. F_1–F_5 represent preovulatory follicles within the hierarchy; *LH* designates the time of the preovulatory LH surge that occurs 4–6 hr before ovulation of the F_1 follicle. Hours before ovulation represents time to ovulation of that follicle (F_1–F_5) within the hierarchy. (From Bahr et al., 1983.)

largest follicle, and in both the largest and second largest follicles, basal and LH-stimulated adenylyl cyclase activity is highest approximately 8 hr before ovulation (Calvo et al., 1981).

Injection of LH increases the content of testosterone within the three largest follicles (Shahabi et al., 1975a); the increased follicular content is attributed to production by both thecal and granulosa tissue (Huang et al., 1979). Both FSH and LH stimulate in vitro estrogen production by thecal cells from the second and third largest, but not largest, follicles (Huang et al., 1979); estrogen synthesis is potentiated by the addition of the precursor, testosterone (Wang and Bahr, 1983).

Postovulatory and Atretic Follicles. There is much evidence to suggest that the postovulatory follicle is steroidogenically active. Progesterone content of the most recent postovulatory follicle declines over the first 15 hr after ovulation (Dick et al., 1978), and this decrease parallels that of the steroid-synthesizing enzyme, 3β-hydroxysteroid dehydrogenase (Armstrong et al., 1977). There is a transient increase, however, in tissue estrogens between 20 and 27 hr after ovulation (Dick et al., 1978). LH-stimulable adenylyl cyclase activity is present for at least 24 hr after ovulation (Calvo et al., 1981).

Both the enzymes 3β-hydroxysteroid dehydrogenase and 17β-hydroxysteroid dehydrogenase have been identified in follicles during the early stages of atresia; however, as atresia progresses this activity is lost. The biologic significance of steroidogenesis in atretic follicles, if any, has yet to be determined (Lofts and Murton, 1973).

Nonsteroidal Hormones

Both prostaglandins E and F have been identified in preovulatory and postovulatory follicles (Hammond et al., 1980; Shimada et al., 1984), however, prostaglandin production appears not to be stimulated by gonadotropins in vitro (Hertelendy and Hammond, 1980). High concentrations of epinephrine and norepinephrine, but not dopamine, have been identified in the largest, most mature follicles (Moudgal and Razdan, 1983). Finally, concentrations of histamine in tissue of preovulatory follicles increase as the follicle matures, decrease at the time of ovulation, and increase again for at least 2 days in tissue of postovulatory follicles (Schaible and Sturkie, 1980).

Action of Ovarian Steroid Hormones

The ovarian steroids that have been studied most extensively to date include progesterone, the estrogens, and androgens. The role of these steroids in relation to ovulation and oviposition is discussed in subsequent sections.

Progesterone. Progesterone circulates in the blood of birds bound to a high-affinity corticoid-binding globulin (CBG) or to albumin and other γ-globulins (Wingfield et al., 1984). There would appear to be no specific, high-affinity, progesterone-binding globulin in Aves such as occurs in some mammalian species. The metabolic clearance rate (MCR) of progesterone is lower in laying (122.4 ml/min) than in nonlaying birds (pullets and molting hens, 163.4 ml/min) (Johnson and van Tienhoven, 1981a), and may in part account for the lower basal concentrations of plasma progesterone found in nonlaying birds. The biologically active half-life of progesterone in the circulation of the hen is approximately 11 min.

When administered in large doses, progesterone may inhibit ovulation or induce molt (see Sturkie, 1965), but the mechanism of these actions is as yet unknown. Progesterone receptors (cytosolic and nuclear) have been demonstrated in the hypothalamus, pituitary, and oviduct of the hen (Kawashima et al., 1978, 1979; Spelsberg and Halberg, 1980) and the number of receptors is influenced by the reproductive state. Progesterone priming of immature or estrogen-primed birds induces the production of the secretory protein avidin via oviductal progesterone receptors and enhances epithelial differentiation in the oviduct (for review, see O'Malley et al., 1975).

Estrogens. There has been no specific, high-affinity, estrogen-binding protein found in any avian species studied to date. Rather, estrogens appear to circulate loosely associated with albumin and γ-globulins. The MCR for estradiol-17β is comparable to that of progesterone, and the difference in basal circulating concentrations between the two steroids results from the substantially lower production rate of estradiol. The half-life of estradiol in the circulation is calculated to be approximately 28 min (Johnson and van Tienhoven, 1981b).

Treatment of immature Japanese quail and young female chickens with estradiol enhances growth of the oviduct and promotes the formation of tubular secretory glands and epithelial differentiation. Estradiol treatment also induces the synthesis of the specific oviductal proteins, ovalbumin, conalbumin, and lysozyme (Schimke et al., 1975; Boogaard and Finnegan, 1976), and the yolk protein precursor, vitellogenin (Deely et al., 1975), and can modify the concentration of cytoplasmic progesterone receptors in the reproductive tract (Pageaux et al., 1983).

Estrogens stimulate vitellogenesis (via its action on the liver) (Figure 18–5), food intake, and the deposition of calcium within the medullary portion of long bones (to serve as a calcium reserve during periods of high egg production) (see also Chapter 21).

Secondary sex characteristics, such as color and shape of plumage, and sexual behavior are also under the control of estrogens. Finally, sexual differentiation of the female brain may be under the influence of estrogen, as indicated by the finding that treatment of female quail embryo with an antiestrogen results in behavioral masculinization (Adkins, 1978).

Androgens. Despite the fact that several studies (discussed above) have convincingly shown that the ovary of the chicken both produces and secretes androgens, much less is known about the physiologic role of androgens. As in the male, androgens are responsible for the growth and coloring of the comb and wattles at the time of sexual maturation, and synergize with estrogen to induce medullary ossification (Bloom et al., 1941). In addition, androgens induce protein synthesis in the oviduct of estrogen-primed birds; however, the mechanism has been less intensively studied than that for progesterone.

Hormonal and Physiologic Factors Affecting Ovulation

The release of the ovum from the largest ovarian follicle (ovulation) occurs from the stigma by a mechanism as yet unknown. Ovulation in most domesticated birds usually follows an oviposition within 15–75 min, except for the first ovulation of a sequence, which is not associated with oviposition. Neither the premature expulsion of the egg (which can be induced with prostaglandins or other agents) nor a delayed oviposition (affected by epinephrine or progesterone) influences the time of ovulation.

Hormones and Ovulation

Luteinizing Hormone. Plasma concentrations of LH in the domestic hen (and most other birds studied to date) peak 4–6 hr prior to ovulation (Figure 18–12) (Laguë et al., 1975; Johnson and van Tienhoven, 1980a; Etches and Cheng, 1981). It is generally agreed that this surge is required for ovulation to occur. Some workers have reported an additional peak of LH at 14–16 hr prior to ovulation (Etches and Cheng, 1981); however, the significance of this second peak has yet to be determined. Finally, in addition to these ovulatory–oviposition cycle-related peaks, there occurs a crepuscular (occurring at the onset of darkness) peak of LH, which has a periodicity of 24 hr and has been proposed to act as a timing cue for the subsequent preovulatory LH surge (Johnson and van Tienhoven, 1980a; Wilson et al., 1983; see also Chapter 17).

Action. Injection of mammalian LH into laying hens increases plasma concentrations of progesterone, estrogens, and testosterone (Shahabi et al., 1975a; Imai and Nalbandov, 1978). However, the ovulatory response subsequent to administration of mammalian LH is dependent on the stage within the sequence; treatment 14–11 hr prior to the first ovulation of a sequence results in premature ovulation, whereas the same treatment before a midsequence ovulation commonly results in follicular atresia and blocked ovulation (Gilbert et al., 1981). Finally, passive immunization of laying hens with antiserum generated against partially purfied chicken LH results in the cessation of ovulation for

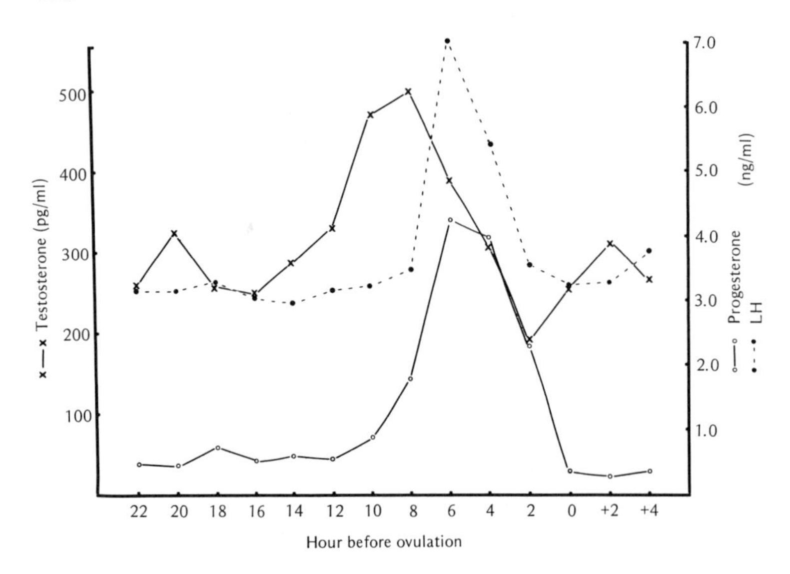

FIGURE 18–12. Plasma concentrations of progesterone, testosterone, and LH relative to time of ovulation. (From Johnson and van Tienhoven, 1980a.)

approximately 5 days and extensive atresia of existing follicles (Sharp et al., 1978).

Progesterone. Highest concentrations of plasma progesterone are found 6–4 hr before ovulation and coincide with the final LH peak (Figure 18–12). This increase is almost entirely the result of progesterone secretion by the largest preovulatory follicle (Etches et al., 1981).

Action. Systemic and intraventricular injection of progesterone can induce both a preovulatory surge of LH and premature ovulation (Etches and Cunningham, 1976; Johnson and van Tienhoven, 1980b). In addition, administration of progesterone antiserum prior to the preovulatory surge of progesterone has been reported to block ovulation (Furr and Smith, 1975). To date, it has not been possible to determine from serial samplings whether the preovulatory increase in progesterone precedes that of LH, or vice versa. Indirect evidence, however, suggests that the preovulatory surge of LH is initiated by an increase in progesterone, in view of the findings that: (1) in the absence of the normal preovulatory rise in progesterone (blocked by the steroidogenesis inhibitor, aminoglutethimide), there is no preovulatory increase in plasma LH; and (2) intramuscular injection of 500 μg progesterone in aminoglutethimide-treated hens induces a normal preovulatory LH surge in the absence of any increase in plasma testosterone or estrogen (Johnson and van Tienhoven, 1984).

The mechanism by which the preovulatory LH surge is potentiated and ovulation is induced, whether by a stimulatory action or by a positive feedback mechanism, has also been investigated. A stimulatory action implies simply that either progesterone secretion stimulates the release of LH, which in turn directly initiates ovulation, or that LH stimulates progesterone secretion and ovulation. By contrast, a true positive feedback mechanism is defined as the generation of an initial stimulus (i.e., progesterone secretion) that acts on another signal generator (i.e., the hypothalamus and pituitary), whose signal (LH) amplifies the initial stimulus. By definition, this spiraling increase in progesterone and LH is necessary for ovulation to occur. Several lines of evidence suggest that a true positive feedback mechanism initiates and potentiates the progesterone and LH preovulatory surges in the hen (for details, see Johnson, 1984; and Johnson and van Tienhoven, 1984).

Androgens. Peak preovulatory concentrations of testosterone occur 10–6 hr prior to ovulation (Figure 18–12), whereas highest levels of 5α-dihydrotestosterone (DHT) occur approximately 6 hr before ovulation (Figure 18–13). The increase in testosterone at this time is the result of secretion from at least the four largest follicles (Etches et al., 1981).

Action. As the preovulatory increase in plasma testosterone clearly precedes that of LH, one might suggest that testosterone is responsible for initiating the preovulatory LH surge and ovulation. However, intraventricular injections of testosterone fail to release LH or induce premature ovulation, and peripheral injections of testosterone that result in LH secretion produce unphysiologically high circulating concentrations of plasma testosterone (Johnson and van Tienhoven, 1981c). Finally, ovulation can be demonstrated to occur in the absence of any preovulatory increase in plasma testosterone (Johnson, 1984). Similarly, the preovulatory surge of DHT is in all probability not directly involved in ovulation.

Estrogens. Both estradiol-17β and estrone show highest plasma concentrations 6–4 hr before ovulation (Fig-

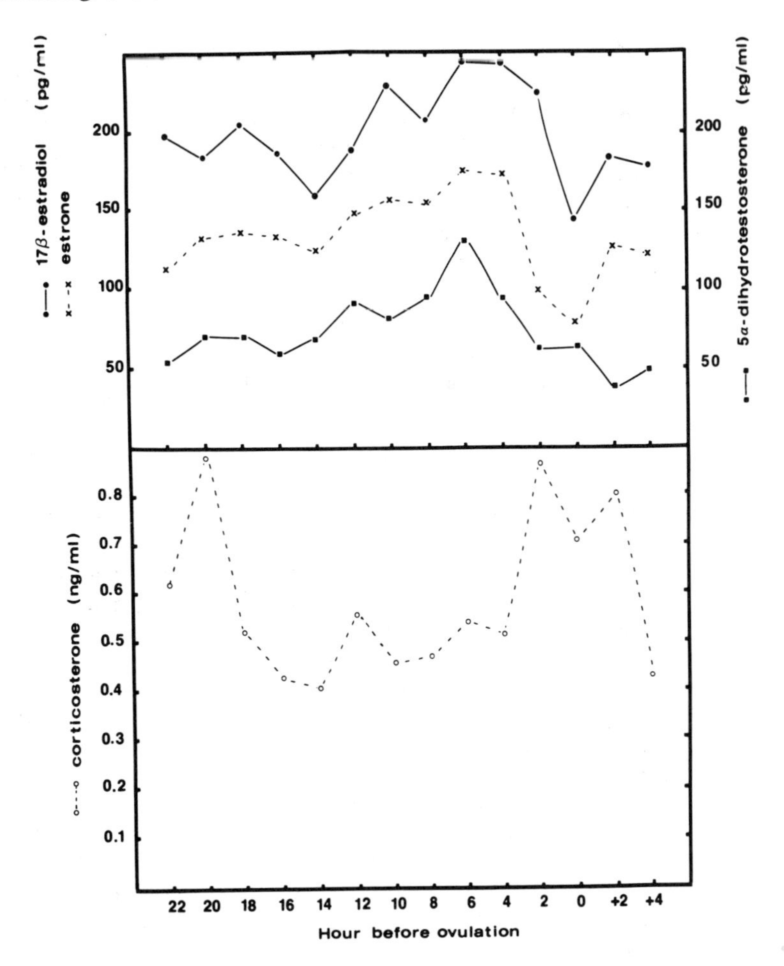

FIGURE 18–13. Plasma concentrations of estradiol-17β, estrone, corticosterone, and 5α-dihydrotestosterone relative to time of ovulation. (From Johnson and van Tienhoven, 1980a.)

ure 18–13). It has also been reported that there is a smaller rise in estrogens 23–18 hr before ovulation (Shodono et al., 1975); however, this rise appears to be less consistent among hens. Follicular secretion of estradiol 6–3 hr prior to ovulation increases in each of the four largest follicles; however, the increase is greatest in the third and fourth largest follicles (Etches et al., 1981).

Action. Like testosterone, estrogens are unlikely to be involved in the direct induction of LH secretion or ovulation, as indicated by the finding that injection of estradiol in laying hens either inhibited, or had no effect on, ovulation or LH secretion (Laguë et al., 1975); moreover, ovulation can occur in the absence of an increase in plasma estrogens (Johnson and van Tienhoven, 1984). Instead, estrogens, together with progesterone, are required for priming of the hypothalamus and pituitary in order that progesterone can induce LH release (Wilson and Sharp, 1976).

Corticosterone. Plasma concentrations of corticosterone are highest 20 and 2 hr prior to an ovulation (Figure 18–13). However, several investigations have demonstrated that corticosterone displays both a daily (photoperiod-related) rhythm and a peak coincident with oviposition, but no ovulation-related increase (Johnson and van Tienhoven, 1981d). Injection of corticosterone or ACTH has been reported to induce premature ovulation; however, it is unlikely to act at the level of the hypothalamus and/or pituitary to directly induce LH release. It remains possible that the adrenal gland, via corticosterone secretion, regulates the timing of the preovulatory LH surge (Wilson and Cunningham, 1980).

Prolactin. Circulating prolactin has been reported to increase at the onset of egg laying in the turkey hen (Etches et al., 1979b; Proudman and Opel, 1981), Japanese quail (Goldsmith and Hall, 1980), and duck (Hall and Goldsmith, 1983). During the ovulation–oviposition cycle of the domestic hen, prolactin is highest approximately 10 hr and lowest 6 hr prior to ovulation. The nadir in plasma prolactin occurs at a time of peak plasma progesterone and LH concentrations (Scanes et al., 1977b). These authors suggested that the decrease in prolactin prior to ovulation may facilitate steroid secretion and initiate the preovulatory LH surge.

Follicle-Stimulating Hormone. A rise in plasma follicle-stimulating hormone (FSH) has been observed 15 hr prior to ovulation in the domestic hen (Scanes et al., 1977a), and it occurs coincident with an increase in FSH binding in ovarian tissue (Etches and Cheng, 1981). Other workers, however, have observed additional peaks of pituitary and plasma FSH at times corresponding to 11 hr before and immediately after ovulation of a midsequence follicle. Additional study of the role of FSH in follicular maturation and the ovulatory process is necessary.

Other Factors. The role of neurotransmitters and prostaglandins in the ovulatory process has been investigated. Administration of serotonin (5-hydroxytryptamine) has been reported to block ovulation, but this effect is probably mediated via its inhibitory effect on LH release (Soliman and Huston, 1978). The catecholamine-synthesis inhibitor, α-methylmetatyrosine, and the α-adrenergic blocking agents phentolamine and dibenzyline block ovulation when injected intrafollicularly, whereas anti-β-adrenergic agents are ineffective. Subsequent administration of epinephrine or norepinephrine overcomes the blocking effect of phentolamine, while only epinephrine overcomes the effect of α-methylmetatyrosine, and neither catecholamine could reverse the blocking effect of dibenzyline (Kao and Nalbandov, 1972). It remains to be determined whether catacholamines play a physiologic role in the process of ovulation.

There is little evidence to suggest that prostaglandins are involved in the ovulatory process, as neither the administration of prostaglandins into the follicle nor injection of indomethacin, a prostaglandin-synthesis inhibitor, affects the time of ovulation.

Finally, injection of synthetic mammalian luteinizing-hormone-releasing hormone (LHRH) induces premature ovulation (van Tienhoven and Schally, 1972), whereas in vivo administration of LHRH antiserum has been shown to block ovulation (Fraser and Sharp, 1978). These effects, however, are mediated via the hypothalamus and pituitary, and there is no evidence of a direct effect of LHRH on the ovary.

Ovulation in Vitro

Excised mature ova may be ovulated in vitro by placing the follicle in Ringers solution (41.7°C) or by incubation with proteolytic enzymes (Neher et al., 1950; Nakajo et al., 1973; Dukelow and Maatman, 1978). Treatment with human chorionic gonadotropin (HCG) plus blood plasma will induce early ovulation in vitro, whereas HCG alone is ineffective (Ogawa and Nishiyama, 1969). Whereas administration of antiadrenergic drugs in vivo inhibits ovulation, incubation of follicles with the catecholamines, norepinephrine and epinephrine, or LH, but not dopamine, induces ovulation in vitro. Epinephrine and norepinephrine induce ovulation in the absence of LH via α-adrenergic receptors (Moudgal and Razdan, 1981).

$PGF_{2\alpha}$, PGE_2, acetylcholine, and oxytocin increase the contraction of smooth muscles in the connective tissue wall of the follicle in vitro, and may in combination with proteolytic enzymes, play a role in the rupture of the stigma at ovulation (Yoshimura et al., 1983).

Hormones during Sexual Maturation, Molt, and Broodiness

Sexual Maturation. Investigations of the changes in gonadotropin concentrations between the time of hatching and the onset of lay in the domestic hen show that there is an early LH peak at about 1 week of age, and a prepubertal LH rise beginning approximately 15 weeks posthatch that is highest about 3 weeks prior to sexual maturity (Sharp, 1975). Ovarian steroidogenesis can be demonstrated as early as day 3.5 of incubation. Plasma estradiol increases from less than 100 pg/ml 6 weeks before lay to a peak of about 350 pg/ml 2–3 weeks before lay; it then decreases to basal concentrations (100–150 pg/ml) at the time of first lay (Senior, 1974). Plasma progesterone concentrations are low (0.1–0.5 ng/ml) until about 1 week before laying, at which time baseline levels increase to 0.4–0.6 ng/ml (Williams and Sharp, 1977). The hen's ovary is unresponsive to stimulation by either mammalian gonadotropins or avian pituitary extracts until 16–18 weeks of age. Contrarily, the pituitary becomes less responsive to exogenous LHRH as the onset of puberty approaches (Wilson and Sharp, 1975); the change is probably the result of negative feedback by increasing steroid secretion by the ovary.

Molt. Basal concentrations of plasma prolactin, growth hormone, and LH are lower in molting females than in laying hens (Scanes et al., 1979). In turkey hens, plasma prolactin levels during molt were less than half those found in laying hens (Proudman and Opel, 1981). However, pituitary content of LH as measured by bioassay was reported not to be different, while FSH activity was about twofold higher in molting versus laying hens (Imai et al., 1972). Ovarian progesterone and estrogen secretion decreases with the onset of molt, and is correlated with a decrease in Δ^5, 3β-hydroxysteroid dehydrogenase enzyme activity (Boucek and Savard, 1970), but testosterone and thyroxine secretion increases during the molting period (Scanes et al., 1979). The elevation in plasma thyroxine may not only play a role in the induction of molt, but may also be required for its thermogenetic effect during the period of heavy feather loss.

Broodiness. In general, behavioral broodiness is preceded by decreases in circulating concentrations of LH, progesterone, testosterone, and estradiol, while growth hormone (GH) tends to be lower only while hens are caring for the young (Harvey et al., 1981). Concentrations of prolactin are significantly elevated in broody and incubating hens compared to laying birds [i.e., bantam fowl (Sharp et al., 1979); turkeys (Etches et al., 1979b; Proudman and Opel, 1981); ruffed grouse (Etches et al., 1979a); ducks (Goldsmith and Williams, 1980); black swans and ring doves (Goldsmith and Follett, 1980); and canaries (Goldsmith, 1982)]. Despite those findings and the fact that prolactin administration can induce broody behavior, it is generally believed that broodiness is not directly initiated or maintained by prolactin secretion (see Chapter 17 for further details).

Plasma Calcium and Egg Laying

Blood calcium circulates in two forms, as nondiffusible protein-bound calcium (un-ionized) and as diffusible ionized calcium. Nondiffusible calcium is bound by the plasma calcium-binding proteins vitellogenin and albumin (Guyer et al., 1980). Estrogen treatment increases total plasma calcium, primarily by stimulating the production of blood calcium-binding proteins (Bacon et al., 1980). Similarly, total plasma calcium increases several weeks before laying, and the increase is attributable to an increase in the protein-bound, but not in the diffusible, fraction. During the ovulation–oviposition cycle, concentrations of ionized calcium peak (2.83 meq/liter or 5.67 mg%) 4 hr after oviposition, then decrease significantly during the period of shell calcification (minimum of 2.42 meq/liter or 4.85 mg%) (Parsons and Combs, 1981). By contrast, there are no detectable differences in total (ionized plus un-ionized) plasma calcium at comparable periods during the ovulation–oviposition cycle (Table 18–1).

Feeding a calcium-deficient diet to laying hens causes a significant decrease in plasma ionized calcium concentrations, a significant decrease or complete cessation of lay, and regression of the ovary within 6 to 9 days. However, birds on the same diet injected with crude chicken pituitary extracts continued to lay for a more prolonged period, despite a comparable decrease in ionized calcium (Taylor et al., 1962; Luck and Scanes, 1979a). Basal concentrations of LH in birds fed a calcium-deficient diet (0.2% Ca) are significantly lower compared to birds on a calcium-rich diet (3.0% Ca), but between the two groups the LH-releasing activity of exogenously administered LHRH is similar (Luck and Scanes, 1979b). These results suggest that pituitary LH synthesis and release, and the ability of the ovary to respond to gonadotropins are unchanged with a calcium deficiency.

Finally, Mongin and Sauveur (1974) showed that voluntary calcium consumption by laying hens increases at the time the egg enters the shell gland, and given a choice between a calcium-deficient diet (1.0% Ca) and a diet supplemented with calcium-rich oyster shell, hens will preferentially consume the calcium-rich diet during the period of eggshell calcification. For further details see Gilbert, 1983 and Chapter 21.

Effects of Light on the Ovary and Ovulation

Photoperiod can influence the reproductive activity of birds relative to both seasonal changes (signaling the onset and termination of breeding seasons in temperate latitudes) and daily changes (entraining the time of ovulation and/or oviposition). Ovarian development appears to be most stimulated by increasing photope-

TABLE 18–1. Variation in total and ionized calcium relative to the ovulation–oviposition cycle in 18-month-old laying hens[a]

Hours following previous oviposition[b]	Total calcium (meq/liter[c]	Ionized calcium (meq/liter[c]
0	12.18 ± 1.26[d]	2.77 ± .05[e]
+4	12.83 ± .80	2.83 ± .05[e]
+8	11.56 ± .56	2.59 ± .04[f]
+12	11.11 ± 1.08	2.46 ± .06[gh]
+16	10.04 ± .98	2.48 ± .03[gh]
+20	10.09 ± .85	2.42 ± .03[g]
+24	10.08 ± .95	2.48 ± .03[gh]
Oviposition	10.20 ± 1.37	2.79 ± .07[e]

[a] From Parsons and Combs (1981).
[b] Average time of entry of egg into shell gland was 5.4 hr following oviposition. Average interval between ovipositions was 27.2 hr.
[c] 1 meq/liter calcium = 2.004 mg/dl.
[d] No differences among means (not significant).
[e-h] Mean ± SE for 7 to 10 birds per sampling period. Values in column with same superscripts do not significantly differ ($p > .05$).

riod, as normally occurs during the spring in the northern hemisphere. Migratory species such as the junco or white-crowned sparrow as well as the domestic turkey fail to show sexual development when held under constant short-day conditions. In contrast, domestic fowl raised under a continuous photoperiod of either 6 hr of light *or* 22 hr of light will reach sexual maturity at about 21 weeks of age. Even in this species, however, an increasing photoperiod can be demonstrated to be important, as pullets raised under a continuously increasing photoperiod from hatch to 18 weeks of age initiate egg laying 2–3 weeks earlier than birds raised under a constant photoperiod (see Morris, 1978).

Maximal photostimulation is generally considered to occur with a photoperiod of 12–14 hr of light; however, normal egg production can be obtained using any photoperiod between 12 and 18 hr, and hens will continue to lay (although less frequently) even if kept in continuous darkness (Wilson and Woodward, 1958; Morris, 1968). The latter observation suggests that light is not the only factor involved in the maintenance of the ovary and the release of LH. Intermittent photoperiods have also proven effective in maintaining egg production. Van Tienhoven and Ostrander (1973) showed that egg production in domestic fowl was not affected when a total of 10 hr of light in a photoperiod (hr) of 8 L:10 D:2 L:4 D was provided compared to a lighting regimen of 14 L:10 D. Darkness influences the time at which ovulation occurs; this was demonstrated by Lanson (1959), who showed that as little as 1¼ hr of darkness in 24 hr was sufficient to alter time of ovulation and oviposition. Subsequent experiments indicated that neither egg production, egg weight, shell-breaking strength, nor feed efficiency was different in photoperiods of 2 L:10 D:2 L:10 D or 2 L:12 D:2 L:8 D compared to 16 L:8 D (van Tienhoven and Ostrander, 1976).

Ahemeral photoperiods (e.g., light cycles of greater or less than 24 hr) most affect the rate of lay and the size of the egg. A light/dark cycle of 27 hr reduces the rate of lay but increases the mean egg weight. Light cycles shorter than 24 hr often reduce the frequency of ovulation due to the inability of the ovarian follicle to adequately mature (Morris, 1973).

A photoperiodic response in birds may be elicited in the absence of the eyes and pineal gland (Turek, 1978). Recent evidence suggests that the extraretinal photoreceptor is located within the hypothalamus (Yokoyama et al., 1978). For further details, see Oishi and Lauber (1973), van Tienhoven and Planck (1973), Morris (1978), and van Tienhoven (1983).

Oviposition

Expulsion of the egg (oviposition) involves the relaxation of abdominal muscles and sphincter between the shell gland and vagina and the muscular contraction of the shell gland. The majority of work to elucidate the mechanisms controlling oviposition has been performed in the quail and domestic hen, and has implicated neurohypophyseal hormones, prostaglandins, and hormones of the pre- and postovulatory follicles.

Neurohypophyseal Hormones

Both oxytocin and arginine vasotocin have been shown to induce premature oviposition in hens (see Sturkie and Mueller, 1976). In addition, stimulation of the preoptic area of the hen's brain results in premature expulsion of the egg, presumably through the release of neurohypophyseal hormones (Opel, 1964), whereas stimulation of the telencephalon delays oviposition (Juhasz and van Tienhoven, 1964). Arginine vasotocin activity in the blood of the laying hen was reported to be highest during oviposition (Niezgoda et al., 1973); however, removal of the neurohypophysis failed to affect the pattern of timing of oviposition (Sturkie and Mueller, 1976). It is possible that increased vasotocin release is not the cause, but rather the result, of oviposition. Finally, there is evidence that the effect of oxytocin may be mediated via prostaglandins, as administration of indomethacin, a prostaglandin-synthesis inhibitor, blocks oxytocin-induced premature oviposition (Hertelendy, 1973).

Prostaglandins

There is much experimental data suggesting that prostaglandins are directly involved in oviposition. Exogenous administration of $PGF_{2\alpha}$ stimulates shell-gland contractility, relaxes the vagina, and induces premature oviposition (Shimada and Asai, 1979), while treatment with indomethacin or aspirin depresses the peak of prostaglandins in the plasma and in pre- and postovulatory follicles, suppresses uterine contractility, and delays oviposition (Day and Nalbandov, 1977; Hertelendy and Biellier, 1978; Shimada and Asai, 1979). In addition, injection of PGE_1 induces premature oviposition in chickens and Japanese quail, while passive immunization with PGE_1 antiserum delays oviposition in the hen (Hertelendy and Biellier, 1978). Further evidence to suggest that prostaglandins mediate the oviposition-inducing activity of arginine vasotocin is supplied by the finding that plasma prostaglandins are significantly elevated at the time of arginine vasotocin-induced premature oviposition, and that arginine vasotocin stimulates the biosynthesis and release of prostaglandin from the uterus (Rzasa, 1984). Arginine vasotocin administered to longitudinal shell-gland muscle strips from the hen increased tissue content of prostaglandins and stimulated shell-gland muscle contractility *in vitro,* while cotreatment of muscle strips with indomethacin blocked the production of prostaglandins and decreased the rate of muscle contraction (Olson et al., 1981).

Endogenous concentrations of prostaglandins E and F increase in the largest preovulatory follicle beginning 6–4 hr prior to the expected time of ovulation (or oviposition). In contrast, postovulatory levels of prostaglandin F increase about 100-fold approximately 24 hr after that ovulation or 2 hr before the next expected oviposition (Day and Nalbandov, 1977). There is also evidence to suggest that increased prostaglandin production occurs in the shell gland at the time of spontaneous oviposition and of oviposition induced by phosphates (Ogasawana and Koga, 1978). Increased prostaglandin production and secretion from one or more of these sources is reflected by significantly elevated plasma concentrations of the prostaglandin metabolite 13,14-dihydro-15-keto prostaglandin $F_{2\alpha}$ coincident with the time of oviposition (Olson et al., 1981).

Postovulatory Follicle

Removal of the most recent postovulatory follicle results in a 1- to 7-day delay in oviposition (Rothchild and Fraps, 1944). More recently, this effect has been attributed to the removal of the granulosa cell layer (Gilbert et al., 1978). The hormone(s) or hormone-like factor(s) produced by the granulosa cells of the postovulatory follicle may be a small peptide, possibly analagous or identical to the neurohypophyseal hormones (Tanaka and Goto, 1976) or prostaglandins.

Preovulatory Follicle

A role for the preovulatory follicle in the process of oviposition was first suggested by Fraps (1942), who found that a premature midsequence ovulation induced by the injection of LH was accompanied by premature expulsion of the egg within the oviduct. Soon after, Rothchild and Fraps (1944) demonstrated that removal of the most mature preovulatory follicle resulted in the delayed oviposition of the egg in the shell gland, although the delay was not as long as that found after removal of the postovulatory follicle.

Other Factors

As previously discussed, plasma concentrations of corticosterone increase dramatically at the time of oviposition. Premature oviposition (induced with vasopressin) or delayed oviposition (resulting from the removal of the postovulatory follicle) failed to alter the timing of the corticosterone peak from when the egg would normally have been laid (Beuving and Vonder, 1981). Additionally, there is an increase in corticosterone at the time of vasopressin-induced oviposition. These results, however, fail to indicate that corticosterone plays a direct role in inducing oviposition.

Despite the fact that the shell gland is innervated by both sympathetic and parasympathetic nerves, there is no evidence that these nerves influence the timing or process of oviposition (Sturkie and Mueller, 1976). Many other factors, including acetylcholine, histamine, pentobarbital, (Sturkie and Mueller, 1976), and lithium induce premature oviposition, but the mechanisms of their action are unclear.

Formation of Albumen, Organic Matrix, and Shell

Components of the egg include the yolk, albumen, the organic matrix, and the crystalline shell. These components have been reviewed by a number of workers, and the reader is referred to the following literature for more details: Gilbert (1971a, b), Simkiss and Taylor (1971), Parsons (1982), and Sturkie and Mueller (1976). There appears to be a remarkable uniformity among species of birds as to the density of yolk and albumen, as well as the proportion of egg yolk with respect to egg weight. Variations do occur, however, in the amino acid and carbohydrate content of both albumen and yolk (Roca et al., 1984).

Albumen

There are four distinct layers of albumen in the fully formed egg: (1) the chalaziferous layer, attached to the yolk; (2) the inner liquid layer; (3) the dense or thick layer; and (4) the outer thin or fluid layer. Approximately one-fourth of the total albumen is found in the outer layer and one-half in the dense, thick layer. The inner layer is 16.8% and the chalaziferous layer 2.7% of the total (Table 18–2).

In the chicken, the initial albumen layer is deposited by the caudal region of the infundibulum (see Gilbert, 1979). However, the majority of albumen is deposited by secretory cells of both epithelial cells and tubular gland cells of the magnum. The discrete layers of albumen are the result of either its consecutive deposition by different regions of the magnum (see Gilbert, 1979) or changes that occur during plumping within the shell gland and movement of the egg through the oviduct (see Sturkie and Mueller, 1976). When the egg leaves the magnum, it contains approximately one-half the final volume of albumen compared to when the egg is laid.

The major proteins found in albumen include: (1) ovalbumin, 54%; (2) ovotransferrin (conalbumin), 13%; (3) ovomucoid, 11%; (4) ovoglobulins (G_2 and G_3), 8%; (5) lysozyme, 3.5%; and (6) α- and β-ovomucin, 1.5–3.0%. There are also several characteristic proteins present in lesser concentrations, including avidin, flavoprotein, ovomacroglobulin, and ovoinhibitors (Robinson, 1972).

Ovomucoid is a protease inhibitor, which in the chicken specifically inhibits the action of trypsin. Alpha-ovomucin (a glycoprotein) and β-ovomucin (a carbohydrate-rich protein) are insoluble, fibrous proteins

TABLE 18–2. Composition of the hen's egg[a]

	Yolk	Outer	Middle	Inner	Chalaziferous	Shell
Weight (g)	18.7	7.6	18.9	5.5	0.9	6.2
Water (%)	48.7	88.8	87.6	86.4	84.3	1.6
Solids (%)	51.3	11.2	12.4	13.6	15.7	98.4

	All layers		
	Yolk	Albumen	Shell
Proteins (%)	16.6	10.6	3.3
Carbohydrates (%)	1.0	0.9	—
Fats (%)	32.6	Trace	0.03
Minerals (%)	1.1	0.6	95.10

[a] From Romanoff and Romanoff (1949).

that probably are responsible for the gel-like qualities of egg white, particularly in the thick or dense layer. The enzymatic action of β-N-acetylglucosaminidase may in part be responsible for the thinning of the thick white, via its action on β-ovomucin (Solomon, 1979). The main biologic properties of the lysozyme of egg albumin is its lytic activity against bacterial cell walls and the hydrolyzation of polysaccharides. Several of the proteins are known for their ability to bind specific substrates (e.g., ovotransferrin binds iron; flavoprotein, riboflavin; and avidin, biotin). Avidin is a progesterone-dependent secretory protein. Although the biologic role of avidin in the hen egg is unknown, its presence has provided the molecular biologist with an excellent model for the study of steroid action on the cellular metabolism of the oviduct (see O'Malley et al., 1975). For further information on albumin synthesis and properties, see Sturkie and Mueller (1976) and Gilbert (1971a, b).

Organic Matrix

The organic fraction of the eggshell consists of shell membranes, the mammillary cores, the shell matrix, and the cuticle. Although these components constitute only a small fraction of the entire eggshell, their integrity is critical to its formation and strength.

Shell Membranes. Shell membranes, organized into an inner and outer membrane, are produced by the isthmus region of the oviduct. There is some question as to whether epithelial cells or tubular gland cells of the isthmus are responsible for the secretion of the membranes (see Leach, 1982). The membranes consist of a meshwork of protein fibers, cross-linked by disulfide and lysine-derived bonds, with small fibrous protuberances of unknown function (Figure 18–14).

The membranes are composed in part of collagen (10%), inasmuch as both hydroxyproline and hydroxylysine can be identified. The remainder of the fibrous component is composed of as yet uncharacterized protein (70–75%) and glycoprotein (Candlish, 1972). The membranes are semipermeable and permit the passage of water and crystalloids; there is no relationship between thickness of the membranes and thickness of the shell, but membrane thickness does decrease with the age of the hen.

Mammillary Cores. The mammillary cores are embedded in the outer shell membrane (Figure 18–15) and are proposed as the initial sites of calcification (Stemberger et al., 1977). They are composed of mucopolysaccharides and sulfated protein and are thought to be formed by the epithelial cells of the isthmus. The mammillary cores represent the greatest proportion of organic material in the eggshell.

Shell Matrix. The organic shell matrix is a series of layers of protein plus acid mucopolysaccharide on which calcification takes place. It represents approxi-

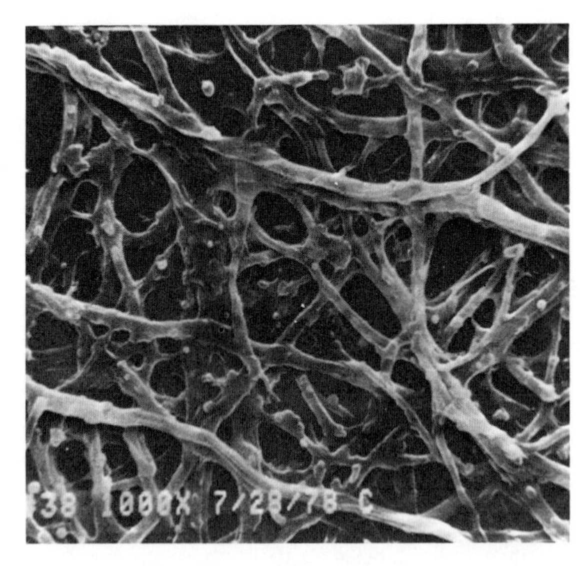

18–14. Scanning electron micrograph of eggshell membranes (×1000). (From Leach, 1982.)

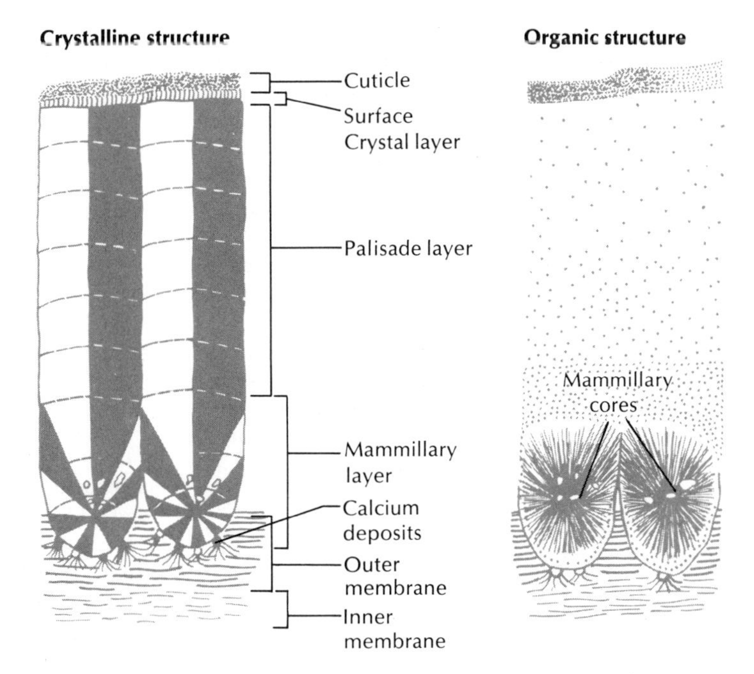

FIGURE 18–15. Diagram of section through the shell and membranes of a hen's egg, showing crystalline structure and organic material that remains after calcium carbonate has been dissolved. (From Simons, 1971.)

mately 2% of the total organic composition of the eggshell. The matrix plus calcified crystals make up the palisade layer of the shell. The innermost region of the shell has a greater density of matrix compared to the outer regions. Both calcium-binding protein (Abatangelo et al., 1978) and carbonic anhydrase (Krampitz et al., 1974) have been identified within the matrix. Deposition of the matrix occurs soon after the egg reaches the shell gland.

Cuticle. The outermost surface of the egg is covered by a thin waxy cuticle composed of protein, polysaccharide, and lipid. The cuticle is unevenly distributed over the entire surface, ranging in depth from 0.5 to 12.8 μm (Parsons, 1982). Its function is believed to be to protect the egg from water evaporation and microbial invasion, although it adds little or nothing to the structural integrity of the shell. The source of the cuticle is uncertain.

Layers of Crystallization

The calcified portion of the shell can be arbitrarily divided into the mammillary knob layer, the palisade layer, and the outer surface crystal layer (see Figure 18–15) (see below).

Mammillary Knob Layer. Outward crystallization of the mammillary cores results in the formation of the mammillary knob layer. In addition, calcium deposits radiate from the bottom of the mammillary cores to penetrate the outer shell membrane. Formation of the mammillary knobs occurs in the shell gland during the first 5 hr of calcification. The passage of water during plumping stretches the shell membranes and increases the distance between the tips of the mammillae. Crystals that form laterally grow and eventually abut with crystals from other mammillae, whereas those that grow outward may extend to the shell surface (Figure 18–15). At some points the crystals do not grow completely together, leaving pores with a diameter of 0.3–0.9 μm (Simons, 1971). There is some disagreement as to whether or not the density of mammillary knobs is related to the structural integrity of the eggshell (See Simons, 1971, and van Toledo et al., 1982).

Palisade Layer. The crystallized palisade (or spongy) layer (approximately 200 μm thick) is composed mainly of crystalline calcium carbonate in the form of calcite and represents the greatest proportion of the shell. Calcification of this layer is initiated 5–6 hr after the entrance of the egg into the shell gland, during the process of plumping. The palisade layer is arranged as columns situated directly over mammillary knobs that are perpendicular to the surface. It has been suggested that the termination of palisade layer calcification is regulated by the shell matrix (Fujii, 1974).

Surface Crystal Layer. The outermost layer of calcification is designated the surface crystal layer. The crystalline structure is more dense than that of the palisade region (Simons, 1971) and lies perpendicular to the shell surface. The overall thickness of this layer ranges from 3 to 8 μm (Parsons, 1982).

Respiration via the Eggshell. Shell pores of the hen egg are simple funnel-shaped openings that arise at the shell surface and protrude through to the mammillary knob layer. In most species, however, pores traverse the shell radially and branch longitudinally along the axis of the egg (Board et al., 1977). The pores are the result of areas of incomplete crystallization. The number of pores is generally related to egg weight, the number of pores per unit area decreasing with increasing egg weight (Tullett, 1978). The function of eggshell pores is to serve as a mechanism of chemical communication between the air cell of the egg and the external environment (see review by Tullett, 1984). The exchange of oxygen, carbon dioxide, and water occurs via passive diffusion. Inner and outer shell membranes play little or no role in limiting the flow of oxygen, carbon dioxide, and water vapor, and the major resistance to passage of these substances comes from the shell itself. Therefore, air-cell gas tensions are determined almost exclusively by the number and size of the pores (total functional pore area) and the thickness of the shell (pore length), relative to the metabolic rate of the embryo (Paganelli et al., 1978). Furthermore, experimental data suggest that a change in environmental conditions may be reflected by an alteration in shell conductance (see Tullett, 1978).

Calcium Metabolism

The shell gland transports 2–2.5 g of calcium within 15 hr for the calcification of a single egg. A hen that lays 280 eggs in a production year will use a quantity of calcium for the purpose of shell formation corresponding to 30 times the calcium content of her entire body. A detailed description concerning mechanisms of absorption and changes in uptake by the intestine relative to age and egg production are beyond the scope of the present discussion, and the reader is urged to consult the following references: Sturkie and Mueller (1976); Bar et al. (1977); Gilbert, 1983; (see also Chapter 21).

Sources of Eggshell Calcium. Total blood calcium concentrations of a laying hen range, on a daily basis, from approximately 20 to 30 mg%. During the final 15 hr of shell formation, calcium is secreted by the shell gland at approximately 100–150 mg/hr. At this rate, circulating blood calcium would be depleted in 8–18 min if not continually replenished by increased intestinal absorption and bone mobilization. The relative importance of these two organs as sources of calcium depends on the concentration of dietary calcium. Hens consume approximately 25% more feed on days when eggshell formation occurs than on days when it does not. If concentrations of calcium in the food are 3.56% or higher, most of the eggshell calcium is derived directly from the intestine (Hurwitz and Bar, 1969). If the concentration is 1.95%, bone supplies 30–40% of the shell calcium, and on calcium-free diets, the skeleton is the obvious principal source (Mueller et al., 1969). However, it is likely that these relationships vary depending on the time of day. When hens are provided constant free access to feed, much of the daily intake occurs early in the photoperiod; the remainder is consumed at the end of the photoperiod. However, much of the shell is formed during the night, when generally no calcium is consumed (Mongin and Sauveur, 1974) and when the calcium content of the digestive tract is gradually decreasing. Therefore, bone may be a particularly important source of shell calcium during the early morning hours (Sturkie and Mueller, 1976).

Vitamin D Metabolism. Vitamin D plays an important role in the regulation of calcium metabolism via the metabolically active metabolite 1,25-dihydroxyvitamin D_3 (1,25 $(OH)_2D_3$) (see Chapter 21). Conversion to this form occurs by the 1-hydroxylation of 25-hydroxyvitamin D_3 (25OHD_3) in the kidney. Sex steroids and prolactin have been implicated in the control of 25OHD_3-1-hydroxylase activity in the chicken and quail (Tanaka et al., 1976; Spanos et al., 1976).

Renal 25-dihydroxy-D_3-1-hydroxylase activity increases just prior to the initiation of egg laying, at a time corresponding with an observed increase in total plasma calcium. In addition, 25OHD_3-1-hydroxylase activity during the ovulation–oviposition cycle increases at the time of ovulation. The increase in activity is followed by an increase in circulating concentrations of 1,25$(OH)_2D_3$ 4 hr after ovulation, and elevated levels persist until 10 hr after ovulation or after the initiation of eggshell formation (Castillo et al., 1979). It has been suggested that 1,25$(OH)_2D_3$ is involved in increasing calcium absorption from the intestine during the ovulation–oviposition cycle.

Calcium Mobilization from Medullary Bone. Earlier work has been reviewed by Simkiss (1967) and Sturkie and Mueller (1976). In the long bones (e.g., femur and tibia), medullary bone lines the endosteal surface and grows with a system of interconnecting spicules that may completely fill the narrow spaces. Medullary bone develops under the influence of the ovarian hormones estrogen and testosterone (Bloom et al., 1941), and it can be readily induced in intact male birds by administration of estrogen (Landauer and Zondek, 1944) or in castrated males by estrogen and testosterone (Common et al., 1948). Gonadal steroids apparently act directly on the cells in the medullary cavity (Benoit and Clavert, 1945), and independently of calcium intake.

Medullary bone forms in female birds during the final 10 days before egg laying. It is unlikely that 1,25$(OH)_2D_3$ mediates the transfer of calcium to medullary bone at this time, as the appearance of this bone occurs 1–2 weeks prior to the increase in renal 1-hydroxylase activity and the elevation of total plasma calcium (Castillo et al., 1979). The role for parathyroid hormone (PTH) on calcium secretion is discussed in Chapter 21.

During the ovulation–oviposition cycle, periods of intense medullary bone formation alternate with periods of severe bone depletion. Hens fed a high-calcium diet are generally able to replenish the calcium lost from medullary bone during shell calcification when shell formation is not taking place, but on a low-calcium diet the cortical bone of the femur is eroded, while medullary bone is maintained in a fairly constant amount (Taylor and Moore, 1954). Under these conditions the new medullary bone that forms is only partially calcified, and an increase in the number of osteoblasts is indicative of a more rapid turnover rate (Zambonin and Mueller, 1969).

Calcium Absorption and Secretion by the Shell Gland. Calcium-binding protein (CaBP) has been identified in the shell gland and acts to sequester and transport calcium (Bar and Hurwitz, 1973). Shell gland CaBP synthesis is under the control of Vitamin D_3 (Corradino et al., 1968) and the ovarian steroids, estrogen and progesterone (Navickis et al., 1979).

Calcium secretion from the shell gland increases approximately 7 hr after ovulation, reaches a maximum as the shell is being formed, and decreases to the basal secretion rate after shell formation is complete but before the expulsion of the egg (Eastin and Spaziani, 1978a). The presence of an egg within the shell gland is not likely to be the major stimulus for the initiation of calcium secretion but appears to be more closely related to ovulation. The hormonal signal(s) that affect changes in the rate of calcium secretion are unknown, although estrogen involvement has been suggested (Eastin and Spaziani, 1978a). Similarly, termination of a high rate of calcium secretion is not related to egg removal, as calcium deposition decreases fully 2 hr before eggs are laid. Movement of calcium across the shell gland, which may occur by both diffusion and active transport (Eastin and Spaziani, 1978b), involves expenditure of metabolic energy.

Carbonate Formation and Deposition. The mineral content of the eggshell consists of about 60% carbonate and 97% calcium carbonate. The formation and precipitation of the carbonate radical of the shell involve equilibria among CO_2, H_2CO_3, HCO_3^-, CO_3^{2-}, H^+, OH^-, and H_2O.

Bicarbonate synthesis is thought to occur through synthesis from tissue CO_2 (Lorcher et al., 1970). In addition, luminal carbonate ion content is supplemented by bicarbonate in fluid flowing from the magnum to the shell gland (Beadle et al., 1938). The secretion from the shell gland of HCO_3^-, like that of calcium, is accomplished by active transport (Eastin and Spaziani, 1978b), and may be influenced by luminal HCO_3^- concentrations. Finally, the net amount of calcium secretion is functionally linked to HCO_3^- production and to luminal HCO_3^- concentrations. For excellent reviews of carbonate/bicarbonate production and calcium–bicarbonate interactions and secretion, see Sturkie and Mueller (1976) and Eastin and Spaziani (1978b).

References

Abatangelo, G., D. Daga-Gordini, I. Castellani, and R. Cortivo. (1978). Some observations on the calcium ion binding to the eggshell matrix. Calcif. Tissue Res., 26, 247.

Adkins, E.K. (1978). Sex steroids and the differentiation of avian reproductive behavior. Am. Zool., 18, 501.

Aitken, R.N.C. (1971). The oviduct. In "Physiology and Biochemistry of the Domestic Fowl" Vol. 3 (J. Bell and B.M. Freeman, Eds.). London and New York: Academic Press, Chapter 53.

Armstrong D.G., M.F. Davidson, A.B. Gilbert, and J.W. Wells. (1977). Activity of 3β-hydroxysteroid dehydrogenase in the postovulatory follicle of the domestic fowl (*Gallus domesticus*). J. Reprod. Fert., 49, 253.

Arjamaa, O., and A. Talo. (1983). The membrane potential of the Japanese quail's oviductal smooth muscle during ovum transport. Acta Physiol. Scand., 118, 349.

Bacon, W.L., K.I. Brown, and M.A. Musser. (1980). Changes in plasma calcium, phosphorus, lipids, and estrogens in turkey hens with reproductive status. Poult. Sci., 59, 444.

Bahr, J.M., S.-C. Wang, M.Y. Huang, and F.O. Calvo. (1983). Steroid concentrations in isolated theca and granulosa layers of preovulatory follicles during the ovulatory cycle of the domestic hen. Biol. Reprod., 29, 326.

Bakst, M.R. (1981). Sperm recovery from oviducts of turkeys at known intervals after insemination and oviposition. J. Reprod. Fert., 62, 159.

Bar, A., and S. Hurwitz. (1973). Calcium restriction and intestinal calcium-binding protein in the laying fowl. Comp. Biochem. Physiol. A, 45, 571.

Bar, A., A. Cohen, G. Montecuccoli, S. Edelstein, and S. Hurwitz. (1977). Relationship of intestinal calcium and phosphorus absorption to vitamin D metabolism during reproductive activity of birds. In "Vitamin D Biochemical, Chemical and Clinical Aspects Related to Calcium Metabolism" (A.W. Norman, et al., Eds.). Berlin: de Gruyter, p. 93.

Beadle, B.W., R.M. Conrad, and H.H. Scott. (1938). Composition of the uterine secretion of the domestic fowl. Poult. Sci., 17, 498.

Benoit, J. (1950). Organes uro-genitaux. In "Traite de Zoologie," Vol. 15 (P.P. Grasse, Ed.). Paris: Mason.

Benoit, J., and J. Clavert. (1945). Action osteogenetique directe et locale de le folliculine, demontree, chez le canard et le pigeon, par son introduction localisee dans un os long. C. R. Seances Soc. Biol. Ses Fil., 139, 728.

Beuving, G., and G.M.A. Vonder. (1981). The influence of ovulation and oviposition on corticosterone levels in the plasma of laying hens. Gen. Comp. Endocrinol., 44, 382.

Bloom, M.A., W. Bloom, and F.C. McLean. (1941). The role of androgen in the production of medullary bone in pigeons by the administration of sex hormones. Am. J. Physiol., 133, 216.

Board, R.G., S.G. Tullett, and H.R. Perrott. (1977). An arbitrary classification of the pore systems in avian eggshells. J. Zool., 182, 251.

Boelkins, J.N., W.J. Mueller, and K.L. Hall. (1973). Cardiac output distribution in the laying hen during shell formation. Comp. Biochem. Physiol. A, 46, 735.

Boogaard, C.L., and C.V. Finnegan. (1976). The effects of estradiol and progesterone on the growth and differentiation of the quail oviduct. Can. J. Zool., 54, 324.

Boucek, R.J., and K. Savard. (1970). Steroid formation by the avian ovary *in vitro* (*Gallus domesticus*). Gen. Comp. Endocrinol., 15, 6.

Breen, P.C., and P.P.H. DeBruyn. (1969). The fine structure of the secretory cells of the uterus (shell gland) of the chicken. J. Morphol., 128, 35.

Burke, W.H., F.X. Ogasawara, and C.L. Fuqua. (1972). A study of the ultrastructure of the uterovaginal sperm-storage glands of the hen, *Gallus domesticus*, in relation to a mechanism for the release of spermatozoa. J. Reprod. Fert., 29, 29.

Callebaut, M. (1979). The avian ovary is an open organ. A study of the lacunar system. Anat. Embryol., 158, 103.

Calvo, F.O., S.-C. Wang, and J.M. Bahr. (1981). LH-stimulable adenylyl cyclase activity during the ovulatory cycle in granulosa cells of the three largest follicles and the postovulatory follicle of the domestic hen (*Gallus domesticus*). Biol. Reprod., 25, 805.

Candlish, J.K. (1972). The role of the shell membranes in the functional integrity of the egg. In "Egg Formation and Production" (B.M. Freeman and P.E. Lake, Eds.). Endinburgh: British Poultry Science Ltd., Chapter 6.

Castillo, L., Y. Tanaka, M.J. Wineland, J.O. Jowsey, and H.F. DeLuca. (1979). Production of 1,25-dihydroxyvitamin D_3 and formation of medullary bone in the egg-laying hen. Endocrinology, 104, 1598.

Chalana, R.K., and S.S. Guraya. (1978). Histophysiolgical studies on the postovulatory follicles of the fowl ovary. Poult. Sci., 57, 148.

Cheng, M.-F. (1979). Progress and prospects in ring dove research: A personal view. Adv. Study Behavior, 9, 97.

Common, R.H., N.A. Rutledge, and R.W. Hale. (1948). Observations on mineral metabolism of pullets. VIII. The influence of gonadal hormones on the nutrition of calcium and phosphorus. J. Agric. Sci., 38, 64.

Compton, M.M., H.P. Van Krey, and P.B. Siegel. (1978). The filling and emptying of the uterovaginal sperm-host glands in the domestic hen. Poult. Sci., 57, 1696.

Corradino, R.A., R.H. Wasserman, M.H. Pubols, and S.I. Chang. (1968). Vitamin D_3 induction of a calcium-binding protein in the uterus of the laying hen. Arch. Biochem. Biophys., 125, 378.

Crossley, J., G. Ferrando, and H. Eiler. (1980). Distribution of adrenergic receptors in the domestic fowl oviduct. Poult. Sci., 59, 2331.

Culbert, J., M.A. Hardie, J.W. Wells, and A.B. Gilbert. (1980). Effect of ovine LH on the progesterone content of the granulosa cells in preovulatory follicles of the domestic fowl (*Gallus domesticus*). J. Reprod. Fert., 58, 449.

Dahl, E. (1970a). Studies on the fine structure of ovarian interstitial tissue. 2. The ultrastructure of the thecal gland of the domestic fowl. Z. Zellforsch. Mikrosk. Anat., 109, 195.

Dahl, E. (1970b). Studies of the fine structure of ovarian interstitial tissue. 3. The innervation of the thecal gland of the domestic fowl. Z. Zellforsch. Mikrosk. Anat., 109, 212.

Day, S.L., and A.V. Nalbandov. (1977). Presence of prostaglandin F (PGF) in hen follicles and its physiological role in ovulation and oviposition. Biol. Reprod., 16, 486.

de Vlaming, V.L. (1979). Actions of prolactin among the vertebrates. In "Hormones and Evolution" (E.J.W. Barrington, Ed.). London and New York: Academic Press.

Deely, R.G., K.P. Mullinix, W. Wetekam, H.M. Kronenberg, M. Meyers, J.D. Eldridge, and R.F. Goldberger. (1975). Vitellogenin synthesis in the avian liver. Vitellogenin is the precursor of the egg yolk phosphoproteins. J. Biol. Chem., 250, 9060.

Dick, H.R., J. Culbert, J.W. Wells, A.B. Gilbert, and M.F. Davidson. (1978). Steroid hormones in the postovulatory follicle of the domestic fowl (*Gallus domesticus*). J. Reprod. Fert., 53, 103.

Donham, R.S., C.W. Dane, and D.S. Farner. (1976). Plasma luteinizing hormone and the development of ovarian follicles after loss of clutch in female mallards (*Anas platyrhynchos*). Gen. Comp. Endocrinol., 29, 152.

Dukelow, W.R., and T.J. Maatman. (1978). Enzymatic degradation of avian follicular tissue by proteases and lipase. Anim. Reprod. Sci., 1, 75.

Eastin, W.C., and E. Spaziani. (1978a). On the control of calcium secretion in the avian shell gland (uterus). Biol. Reprod., 19, 493.

Eastin, W.C., and E. Spaziani. (1978b). On the mechanism of calcium secretion in the avian shell gland (uterus). Biol. Reprod., 19, 505.

Etches, R.J., and F.J. Cunningham. (1976). The interrelationship between progesterone and luteinizing hormone during the ovulation cycle of the hen (*Gallus domesticus*). J. Endocrinol., 71, 51.

Etches, R.J., and K.W. Cheng. (1981). Changes in the plasma concentrations of luteinizing hormone, progesterone, oestradiol and testosterone and in the binding of follicle-stimulating hormone to the theca of follicles during the ovulation cycle of the hen (*Gallus domesticus*). J. Endocrinol., 91, 11.

Etches, R.J., A. Garbutt, and A.L. Middleton. (1979a). Plasma concentrations of prolactin during egg laying and incubation in the ruffed grouse (*Bonasa umbellus*). Can. J. Zool., 57, 1624.

Etches, R.J., A.S. McNeilly, and C.E. Duke. (1979b). Plasma concentration of prolactin during the reproductive cycle of domestic turkey (*Meleagris gallopavo*). Poult. Sci., 58, 963.

Etches, R.J., F. Croze, and C.E. Duke. (1981). Plasma concentrations of luteinizing hormone, progesterone, testosterone and estradiol in follicular and peripheral venous plasma during the ovulation cycle of the hen. Recent Adv. Avian Endocrinol., 33, 89.

Fraps, R.M. (1942). Synchronized induction of ovulation and premature oviposition in the domestic fowl. Anat. Rec., 84, 521.

Fraps, R.M. (1955). Egg production and fertility in poultry. In "Progress in the Physiology of Farm Animals," Vol. 2 (J. Hammond, Ed.). London: Butterworths, Chapter 15.

Fraser, H.M., and P.J. Sharp. (1978). Prevention of positive feedback in the hen (*Gallus domesticus*) by antibodies to luteinizing hormone-releasing hormone. J. Endocrinol., 76, 181.

Freedman, S.L., and P.D. Sturkie. (1963a). Blood vessels of the chicken's uterus (shell gland). Am. J. Anat., 113, 1.

Freedman, S.L., and P.D. Sturkie. (1963b). Extrinsic nerves of the chicken's uterus (shell gland). Anat. Rec., 147, 431.

Fujii, S. (1974). Further morphological studies on the formation and structure of hen's eggshell by scanning electron microscopy. J. Fac. Fish. Anim. Husb. Hiroshima Univ., 13, 29.

Fujii, S. (1975). Scanning electron microscopal observation on the mucosal epithelium of hen's oviduct with special reference to the transport mechanism of spermatozoa through the oviduct. J. Fac. Fish. Anim. Husb. Hiroshima Univ., 14, 1.

Furr, B.J.A., and G.K. Smith. (1975). Effects of antisera

against gonadal steroids on ovulation in the hen *Gallus domesticus*. J. Endocrinol., 66, 303.

Gilbert, A.B. (1969). Innervation of the ovary of the domestic hen. Q. J. Exp. Physiol., 54, 404.

Gilbert, A.B. (1971a). Egg albumin and its formation. In "Physiology and Biochemistry of the Domestic Fowl," Vol. 3 (D.J. Bell and B.M. Freeman, Eds.). London and New York: Academic Press, Chapter 54.

Gilbert, A.B. (1971b). The egg: its physical and chemical aspects. In "Physiology and Biochemistry of the Domestic Fowl," Vol. 3. (D.J. Bell and B.M. Freeman, Eds.). London and New York: Academic Press, Chapter 58.

Gilbert, A.B. (1979). Female genital organs. In "Form and Function in Birds," Vol. 1 (A.S. King and J. McLelland, Eds.). London and New York: Academic Press, Chapter 5.

Gilbert, A.B. (1983). Calcium and reproductive function in the hen. Proc. Nutr. Soc., 42, 195.

Gilbert, A.B., and D.G.M. Wood-Gush. (1971). Ovulatory and ovipository cycles. In "The Physiology and Biochemistry of the Domestic Fowl," Vol. 3 (D.J. Bell and B.M. Freeman, Eds.). New York and London: Academic Press, Chapter 57.

Gilbert, A.B., M.E. Reynolds, and F.W. Lorenz. (1968). Distribution of spermatozoa in the oviduct and fertility in domestic birds. VII. Innervation and vascular supply of the uterovaginal sperm-host glands of the domestic hen. J. Reprod. Fert., 17, 305.

Gilbert, A.B., M.F. Davidson, and J.W. Wells. (1978). Role of the granulosa cells of the postovulatory follicle of the domestic fowl in oviposition. J. Reprod. Fert., 52, 227.

Gilbert, A.B., M.A. Hardie, M.M. Perry, H.R. Dick, and J.W. Wells. (1980). Cellular changes in the granulosa layer of the maturing ovarian follicle of the domestic fowl. Br. Poult. Sci., 21, 257.

Gilbert, A.B., M.F. Davidson, M.A. Hardie, and J.W. Wells. (1981). The induction of atresia in the domestic fowl (*Gallus domesticus*) by ovine LH. Gen. Comp. Endocrinol., 44, 344.

Goldsmith, A.R. (1982). Plasma concentrations of prolactin during incubation and parental feeding throughout repeated breeding cycles in canaries (*Serinus canarius*). J. Endocrinol., 94, 51.

Goldsmith, A.R., and B.K. Follett. (1980). Anterior pituitary hormones. In "Avian Endocrinology" (A. Epple and M.H. Stetson, Eds.). London and New York: Academic Press, p. 147.

Goldsmith, A.R., and M. Hall. (1980). Prolactin concentrations in the pituitary gland and plasma of Japanese quail in relation to photoperiodically induced sexual maturation and egg laying. Gen. Comp. Endocrinol., 42, 449.

Goldsmith, A.R., and D.M. Williams. (1980). Incubation in mallards (*Anas platyrhynchos*): changes in plasma levels of prolactin and luteinizing hormone. J. Endocrinol., 86, 371.

Guyer, R.B., A.A. Grunder, E.G. Buss, and C.O. Clagett. (1980). Calcium-binding proteins in serum of chickens: Vitellogenin and albumin. Poult. Sci., 59, 874.

Hall, M.R., and A.R. Goldsmith. (1983). Factors affecting prolactin secretion during breeding and incubation in the domestic duck (*Anas platyrhynchos*). Gen. Comp. Endocrinol., 49, 270.

Hammond, R.W., D.M. Olson, R.B. Frenkel, H.V. Biellier, and F. Hertelendy. (1980). Prostaglandins and steroid hormones in plasma and ovarian follicles during the ovulation cycle of the domestic hen (*Gallus domesticus*). Gen. Comp. Endocrinol., 42, 195.

Hammond, R.W., W.H. Burke, and F. Hertelendy. (1981). Influence of follicular maturation on progesterone release in chicken granulosa cells in response to turkey and ovine gonadotropins. Biol. Reprod., 24, 1048.

Hammond, R.W., W.H. Burke, and F. Hertelendy. (1982). Prolactin does not affect steroidogenesis in isolated chicken granulosa cells. Gen. Comp. Endocrinol., 48, 285.

Harvey, S., E. Bedrak, and A. Chadwick. (1981). Serum concentrations of prolactin, luteinizing hormone, growth hormone, corticosterone, progesterone, testosterone and oestradiol in relation to broodiness in domestic turkeys (*Meleagris gallopavo*). J. Endocrinol., 89, 187.

Hertelendy, F. (1973). Block of oxytocin-induced parturition and oviposition by prostaglandin inhibitors. Life Sci., 13, 1581.

Hertelendy, F., and H.V. Biellier. (1978). Evidence for a physiological role of prostaglandins in oviposition by the hen. J. Reprod. Fert., 53, 71.

Hertelendy, F., and R.W. Hammond. (1980). Prostaglandins do not affect steroidogenesis and are not being produced in response to oLH in chicken granulosa cells. Biol. Reprod., 23, 918.

Hertelendy, F., F. Lintner, E.K. Asem, and B. Raab. (1982). Synergistic effect of gonadotropin releasing hormone on LH-stimulated progesterone production in granulosa cells of the domestic fowl (*Gallus domesticus*). Gen. Comp. Endocrinol., 48, 117.

Heywang, B.W. (1938). The time factor in egg production. Poult. Sci., 17, 240.

Hodges, R.D. (1965). The blood supply to the avian oviduct, with special reference to the shell gland. J. Anat., 99, 485.

Hodges, R.D. (1974). In "The Histology of the Fowl." New York and London: Academic Press.

Hoffer, A.P. (1971). The ultrastructure and cytochemistry of the shell membrane-secreting region of the Japanese quail oviduct. Am. J. Anat., 131, 253.

Huang, E.S.-R., and A.V. Nalbandov. (1979). Steroidogenesis of chicken granulosa and theca cells: *In vitro* incubation system. Biol. Reprod., 20, 442.

Huang, E.S.-R., K.J. Kao, and A.V. Nalbandov. (1979). Synthesis of sex steroids by cellular components of chicken follicles. Biol. Reprod., 20, 454.

Hurwitz, S., and A. Bar. (1969). Intestinal calcium absorption in the laying fowl and its importance in calcium homeostasis. Am. J. Clin. Nutr., 22, 391.

Imai, K., and A.V. Nalbandov. (1978). Plasma and follicular steroid levels of laying hens after the administration of gonadotropins. Biol. Reprod., 19, 779.

Imai, K., K. Tanaka, and S. Nakajo. (1972). Gonadotrophic activities of anterior pituitary and of blood plasma and ovarian response to exogenous gonadotrophin in moulting hens. J. Reprod. Fert., 30, 433.

Johnson, A.L. (1984). Interactions of progesterone and LH leading to ovulation in the domestic hen. In "Reproductive Biology of Poultry" (F.J. Cunningham, P.E. Lake, and D. Hewitt, Eds.). Poultry Science Symposium No. 17, British Poultry Science Ltd., p. 133.

Johnson, A.L., and A. van Tienhoven. (1980a). Plasma concentrations of six steroids and LH during the ovulatory cycle of the hen, *Gallus domesticus*. Biol. Reprod., 23, 386.

Johnson, A.L., and A. van Tienhoven. (1980b). Hypothalamo-hypophyseal sensitivity to hormones in the hen. I. Plasma concentrations of LH, progesterone, and testosterone in response to central injections of progesterone and R5020. Biol. Reprod., 23, 910.

Johnson, A.L., and A. van Tienhoven (1981a). Pharmacokinetics of progesterone in laying and non-laying hens (*Gallus domesticus*). J. Endocrinol., 89, 1.

Johnson, A.L., and A. van Tienhoven. (1981b). Pharmacokinetics of estradiol-17β in the laying hen. Poult. Sci., 60, 2720.

Johnson, A.L., and A. van Tienhoven. (1981c). Hypothalamo-hypophyseal sensitivity to hormones in the hen. II. Plasma concentrations of LH, progesterone, and testosterone in response to peripheral and central injections of LHRH or testosterone. Biol. Reprod., 25, 153.

Johnson, A.L., and A. van Tienhoven. (1981d). Plasma concentrations of corticosterone relative to photoperiod, oviposition, and ovulation in the domestic hen. Gen. Comp. Endocrinol., 43, 10.

Johnson, A.L., and A. van Tienhoven. (1984). Effects of aminoglutethimide on luteinizing hormone and steroid secretion, and ovulation in the hen, *Gallus domesticus*. Endocrinology, 114, 2276.

Juhasz, L.P., and A. van Tienhoven. (1964). Effect of electrical stimulation of telencephalon on ovulation and oviposition in the hen. Am. J. Physiol., 207, 286.

Kao, L.W.L., and A.V. Nalbandov. (1972). The effect of antiadrenergic drugs on ovulation in hens. Endocrinology, 90, 1343.

Kawashima, M., M. Kamiyoshi, and K. Tanaka. (1978). A cytoplasmic progesterone receptor in hen pituitary and hypothalamic tissues. Endocrinology, 102, 1207.

Kawashima, M., M. Kamiyoshi, and K. Tanaka. (1979). Nuclear progesterone receptor in the hen pituitary and hypothalamus. Endocrinol. Jpn., 26, 501.

King, A.S. (1975). Aves urogenital system. In "Sisson and Grossman's The Anatomy of Domestic Animals," Vol. 2, (5th ed.) (R. Getty, Ed.). Philadelphia: Saunders, p. 1919.

Kinsky, F.C. (1971). The consistent presence of paired ovaries in the Kiwi (*Aptryx*) with some discussion of this condition in other birds. J. Ornithol., 112, 334.

Kosin, I.L., and H. Abplanalp. (1951). The pattern of egg laying in turkeys. Poult. Sci., 30, 168.

Krampitz, G., J. Engels, and I. Helfgen. (1974). Uber das vorkommen von carboanhydratase in der eischale des huhnes. Experientia, 30, 228.

Laguë, P.C., A. van Tienhoven, and F.J. Cunningham. (1975). Concentrations of estrogens, progesterone and LH during the ovulatory cycle of the laying chicken (*Gallus domesticus*). Biol. Reprod., 12, 590.

Lake, P.E. (1975). Gamete production and the fertile period with particular reference to domesticated birds. Symp. Zool. Soc. London, 35, 225.

Landauer, W., and B. Zondek. (1944). Observations on the structure of bone in estrogen-treated cocks and drakes. Am. J. Pathol., 20, 179.

Lanson, R.K. (1959). The influence of light and darkness upon the reproductive performance of the fowl. Ph.D. Thesis, Rutgers University, New Brunswick, New Jersey.

Leach, R.M., Jr. (1982). Biochemistry of the organic matrix of the eggshell. Poult. Sci., 61, 2040.

Lofts, B., and R.K. Murton. (1973). Reproduction in Birds. In "Avian Biology," Vol. 3 (D.S. Farner and J.R. King, Eds.). London and New York: Academic Press, Chapter 1.

Lorcher, K., C. Zscheile, and K. Bronsch. (1970). Transfer of continuously i.v. infused $C^{14}O_3$ and Ca^4 Cl_2 to the hen's eggshell. Ann. Biol. Anim. Biochim. Biophys., 10, 193.

Luck, M.R., and C.G. Scanes. (1979a). Plasma levels of ionized calcium in the laying hen (*Gallus domesticus*). Comp. Biochem. Physiol. A, 63, 177.

Luck, M.R., and C.G. Scanes. (1979b). The relationship between reproductive activity and blood calcium in the calcium-deficient hen. Br. Poult. Sci., 20, 559.

Marza, V.D., and E.V. Marza. (1935). The formation of the hen's egg, I-IV. Q. J. Microsc. Sci., 78, 134.

Mitchell, M.E. (1967). Stimulation of the ovary in hypophysectomized hens by an avian pituitary preparation. J. Reprod. Fert., 14, 249.

Mongin, P., and B. Sauveur. (1974). Voluntary food and calcium intake by the laying hen. Br. Poult. Sci., 15, 349.

Morgan, W., and W. Kohlmeyer. (1957). Hens with bilateral oviducts. Nature, (London), 180, 98.

Morris, T.R. (1968). Light requirements of the fowl. In "Environmental Control in Poultry Production" (T.C. Carter, Ed.). Endinburgh: Oliver and Boyd, p. 15.

Morris, T.R. (1973). The effects of ahemeral light and dark cycles on egg production in the fowl. Poult. Sci., 52, 423.

Morris, T.R. (1978). The influence of light on ovulation in domestic birds. In "Animal Reproduction," BARC Symposium Number 3 (H. Hawk, Ed.). Montclair: Allenheld, Chapter 19, p. 307.

Moudgal, R.P., and M.N. Razdan. (1981). Induction of ovulation *in vitro* by LH and catecholamines in hens is mediated by α-adrenergic receptors. Nature (London), 293, 738.

Moudgal, R.P., and M.N. Razdan. (1983). Catecholamines in ovarian follicles during the ovulatory cycle of white Leghorn hens. Br. Poult. Sci., 24, 173.

Mueller, W.J., R.L. Brubaker, and M.D. Caplan. (1969). Egg shell formation and bone resorption in laying hens. Fed. Proc. Fed. Am. Soc. Exp. Biol., 28, 1851.

Nakajo, S., A.H. Zakaria, and K. Imai. (1973). Effect of the local administration of proteolytic enzymes on the rupture of the ovarian follicle in the domestic fowl, *Gallus domesticus*. J. Reprod. Fert., 34, 235.

Nakamura, T., Y. Tanabe, and H. Katukawa. (1974). Steroidogenesis *in vitro* by the ovarian tissue of the domestic fowl (*Gallus domesticus*). J. Endocrinol., 63, 507.

Nalbandov, A.V., and M.F. James. (1949). The blood–vascular system of the chicken ovary. Am. J. Anat., 85, 347.

Navickis, R.J., B.S. Katzenellenbogen, and A.V. Nalbandov. (1979). Effects of the sex steroid hormones and vitamin D_3 on calcium-binding proteins in the chick shell gland. Biol. Reprod., 21, 1153.

Neher, B.N., M.W. Olsen, and R.M. Fraps. (1950). Ovulation of the excised ovum of the hen. Poult. Sci., 29, 554.

Nevalainen, T.J. (1969). Electron microscope observations on the shell gland mucosa of calcium-deficient hens (*Gallus domesticus*). Anat. Rec., 164, 127.

Niezgoda, J., J. Rzasa, and Z. Ewy. (1973). Changes in blood vasotocin activity during oviposition in the hen. J. Reprod. Fert., 35, 505.

Ogasawana, T., and O. Koga. (1978). Prostaglandin production by the uterus of the hen in relation to spontaneous and phosphate-induced ovipositions. Jpn. J. Zootech. Sci., 49, 523.

Ogawa, K., and H. Nishiyama. (1969). Studies on the mechanism of ovulation in the fowl. Mem. Fac. Agric. Kagoshima Univ., 7.

Oishi, T., and J.K. Lauber. (1973). Photoreception in the photosexual response of the quail. II. Effects of intensity and wavelength. Am. J. Physiol., 225, 880.

Olsen, M.W., and S.J. Marsden. (1954). Development in unfertilized turkey eggs. J. Exp. Zool., 126, 337.

Olson, D.M., H.V. Biellier, and F. Hertelendy. (1978). Shell gland responsiveness to prostaglandins $F_{2\alpha}$ and E_1 and to arginine vasotocin during the laying cycle of the domestic hen (*Gallus domesticus*). Gen. Comp. Endocrinol., 36, 559.

Olson, D.M., H.V. Biellier, and F. Hertelendy. (1981). The role of prostaglandins in oviposition. Recent Adv. Avian Endocrinol., 33, 185.

O'Malley, B.W., S.L. Woo, S.E. Harris, J.M. Rosen, J.P. Comstuck, L. Chan, C.B. Bordelon, J.W. Holder, P. Sperry, and A.R. Means. (1975). Steroid hormone action in animal cells. Am. Zool., 15 (Suppl. 1), 215.

Opel, H. (1964). Premature oviposition following operative interference with the brain of the chicken. Endocrinology, 74, 193.

Opel, H. (1966). The timing of oviposition and ovulation in the quail (*Coturnix coturnix japonica*). Br. Poult. Sci., 7, 29.

Opel, H., and A.V. Nalbandov. (1961). Follicular growth and ovulation in hypophysectomized hens. Endocrinology, 69, 1016.

Paganelli, C.V., R.A. Ackerman, and H. Rahn. (1978). The avian egg: *In vivo* conductances to oxygen, carbon dioxide, and water vapor in late development. In "Respiratory Function in Birds, Adult and Embryonic," (J. Piper, Ed.). New York: Springer-Verlag, p. 212.

Pageaux, J.F., C. Laugier, D. Pal, and H. Pacheco. (1983). Analysis of progesterone receptor in the quail oviduct. Correlation between plasmatic estradiol and cytoplasmic progesterone receptor concentrations. J. Steroid Biochem., 18, 209.

Parsons, A.H. (1982). Structure of the eggshell. Poult. Sci., 61, 2013.

Parsons, A.H., and G.F. Combs. (1981). Blood ionized calcium cycles in the chicken. Poult. Sci., 60, 1520.

Payne, R.B. (1972). Mechanisms and control of molt. In "Avian Biology," Vol. 2 (D.S. Farner and J.R. King, Eds.). London and New York: Academic Press, Chapter 3.

Perry, M.M., A.B. Gilbert, and A.J. Evans. (1978a). Electron microscope observations on the ovarian follicle of the domestic fowl during the rapid growth phase. J. Anat., 125, 481.

Perry, M.M., A.B. Gilbert, and A.J. Evans. (1978b). The structure of the germinal disc region of the hen's ovarian follicle during the rapid growth phase. J. Anat., 127, 379.

Perry, M.M., and A.B. Gilbert. (1979). J. Cell Science, 39, 257–272.

Planck, R.J., and H.J. Johnson. (1975). Oviposition patterns and photoresponsivity in Japanese quail (*Coturnix coturnix japonica*). J. Interdiscip. Cycle Res., 6, 131.

Poole, H.K., and M.W. Olsen (1958). Incidence of parthenogenetic development in eggs laid by three strains of dark cornish chickens. Proc. Soc. Exp. Biol. Med., 97, 477.

Proudman, J.A., and H. Opel. (1981). Turkey prolactin: Validation of a radioimmunoassay and measurement of changes associated with broodiness. Biol. Reprod., 25, 573.

Redshaw, M.R., and B.K. Follett. (1972). The physiology of egg yolk production in the hen. In "Egg Formation and Production" (B.M. Freeman and P.E. Lake, Eds.). Edinburgh: British Poultry Science Ltd., Chapter 3.

Robinson, D.S. (1972). Egg white glycoproteins and the physical properties of egg white. In "Egg Formation and Production" (B.M. Freeman and P.E. Lake, Eds.). Edinburgh: British Poultry Science Ltd., Chapter 5.

Roca, P., F. Sainz, M. Gonzalez, and M. Alemany. (1984). Structure and composition of the eggs from several avian species. Comp. Biochem Physiol. A, 77, 307.

Romanoff, A.L., and A.J. Romanoff. (1949). "The Avian Egg." New York: Wiley.

Rothchild, I., and R.M. Fraps. (1944). On the function of the ruptured ovarian follicle of the domestic fowl. Proc. Soc. Exp. Biol. Med., 56, 79.

Rothwell, B., and S.E. Solomon. (1977). The ultrastructure of the follicle wall of the domestic fowl during the phase of rapid growth. Br. Poult. Sci., 18, 605.

Rzasa, J. (1978). Effects of arginine vasotocin and prostaglandin E_1 on the hen uterus. Prostaglandins, 16, 357.

Rzasa, J. (1984). The effect of arginine vasotocin on prostaglandin production of the hen uterus. Gen. Comp. Endocrinol., 53, 260.

Scanes, C.G., and J.H. Fagioli. (1980). Effects of mammalian and avian gonadotropins on *in vitro* progesterone production by avian ovarian granulosa cells. Gen. Comp. Endocrinol., 41, 1.

Scanes, C.G., P.M.M. Godden, and P.J. Sharp. (1977a). An homologous radioimmunoassay for chicken follicle-stimulating hormone: observations on the ovulatory cycle. J. Endocrinol., 73, 473.

Scanes, C.G., P.J. Sharp, and A. Chadwick. (1977b). Changes in plasma prolactin concentration during the ovulatory cycle of the chicken. J. Endocrinol., 72, 401.

Scanes, C.G., P.J. Sharp, S. Harvey, P.M.M. Godden, A. Chadwick, and W.S. Newcomer. (1979). Variations in plasma prolactin, thyroid hormones, gonadal steroids and growth hormone in turkeys during the induction of egg laying and moult by different photoperiods. Br. Poult. Sci., 20, 143.

Scanes, C.G., H. Mozelic, E. Kavanagh, G. Merrill, and J. Rabii. (1982). Distribution of blood flow in the ovary of domestic fowl (*Gallus domesticus*) and changes after prostaglandin F-2α treatment. J. Reprod. Fert., 64, 227.

Schaible, T.F., and P.D. Sturkie. (1980). Histamine levels of preovulatory and postovulatory ovarian follicles of hens. Poult. Sci., 59, 1658.

Schimke, R.T., G.S. McKnight, D.J. Shapiro, D. Sullivan, and R. Palacios. (1975). Hormonal regulation of ovalbumin synthesis in the chick oviduct. Recent Prog. Horm. Res., 31, 175.

Schimke, R.T., P. Pennequin, D. Robins, and G.S. McKnight. (1977). Hormonal regulation of egg white protein synthesis in chick oviduct. In "Hormones and Cell Regulation," Vol. 1 (J. Dumont and J. Nunez, Eds.). Amsterdam: North Holland.

Senior, B.E. (1974). Oestradiol concentration in the peripheral plasma of the domestic hen from 7 weeks of age until the time of sexual maturity. J. Reprod. Fert., 41, 107.

Shahabi, N.A., J.M. Bahr, and A.V. Nalbandov. (1975a). Effect of LH injection on plasma and follicular steroids in the chicken. Endocrinology, 96, 969.

Shahabi, N.A., H.W. Norton, and A.V. Nalbandov. (1975b). Steroid levels in follicles and the plasma of hens during the ovulatory cycle. Endocrinology, 96, 962.

Sharp, P.J. (1975). A comparison of variations in plasma luteinizing hormone concentrations in male and female domestic chickens (*Gallus domesticus*) from hatch to sexual maturity. J. Endocrinol., 67, 211.

Sharp, P.J., C.G. Scanes, and A.B. Gilbert. (1978). *In vivo* effects of an antiserum to partially purified chicken luteinizing hormone (CM2) in laying hens. Gen. Comp. Endocrinol., 34, 296.

Sharp, P.J., C.G. Scanes, J.B. Williams, S. Harvey, and A. Chadwick. (1979). Variations in concentrations of prolactin luteinizing hormone, growth hormone and progesterone in the plasma of broody bantams (*Gallus domesticus*). J. Endocrinol., 80, 51.

Shimada, K. (1978). Electrical activity of the oviduct of the laying hen during egg transport. J. Reprod. Fert., 53, 223.

Shimada, K., and I. Asai. (1978). Uterine contraction during the ovulatory cycle of the hen. Biol. Reprod., 19, 1057.

Shimada, K., and I. Asai. (1979). Effects of prostaglandin $F_{2\alpha}$ and indomethacin on uterine contraction in hens. Biol. Reprod., 21, 523.

Shimada, K., D.M. Olson, and R.J. Etches. (1984). Follicular and uterine prostaglandin levels in relation to uterine contraction and the first ovulation of a sequence in the hen. Biol. Reprod., 31, 76.

Shodono, M., T. Nakamura, Y. Tanabe, and K. Wakabayashi. (1975). Simultaneous determinations of oestradiol-17β, progesterone and luteinizing hormone in the plasma during the ovulatory cycle of the hen. Acta Endocrinol., 78, 565.

Simkiss, K. (1967). "Calcium in Reproductive Physiology." New York: Reinhold.

Simkiss, K., and T.G. Taylor. (1971). Shell formation. In

"Physiology and Biochemistry of the Domestic Fowl," Vol. 3 (D.J. Bell, and B.M. Freeman, Eds.). New York: Academic Press, Chapter 55.

Simmons, G.S., and D.J.S. Hetzel. (1983). Time relationships between oviposition, ovulation and egg formation in Khaki Campbell ducks. Br. Poult. Sci., 24, 21.

Simons, P.C.M. (1971). Ultrastructure of the hen eggshell and its physiological interpretation. Ph.D. Thesis, Landbouwhogeschool, Wageningen, Netherlands: Centre for Agricultural Publishing and Documentation.

Soliman, K.F.A., and T.M. Huston. (1978). Inhibitory effect of serotonin on ovulation in the domestic fowl, *Gallus domesticus.* Anim. Reprod. Sci., 1, 69.

Solomon, S.E. (1975). Studies on the isthmus region of the domestic fowl. Br. Poult. Sci., 16, 255.

Solomon, S.E. (1979). The localization of beta-*N*-acetylglucosaminidase in the oviduct of the domestic fowl. Br. Poult. Sci., 20, 139.

Spanos, E., J.W. Pike, M.R. Haussler, K.W. Colston, I.M.A. Evans, A.M. Goldner, T.A. McCain, and I. MacIntyre. (1976). Circulating 1α,25-dihydroxyvitamin D in the chicken: Enhancement by injection of prolactin and during egg laying. Life Sci., 19, 1751.

Spelsberg, T.C., and F. Halberg. (1980). Circannual rhythms in steroid receptor concentration and nuclear binding in the chick oviduct. Endocrinology, 107, 1234.

Stemberger, B.H., W.J. Mueller, and R.M. Leach. (1977). Microscopic study of the initial stages of eggshell calcification. Poult. Sci., 56, 537.

Sturkie, P.D. (1955). Absorption of egg yolk in body cavity of the hen. Poult. Sci., 34, 736.

Sturkie, P.D. (1965). Reproduction in the female and egg production. In "Avian Physiology" (2nd ed.). Ithaca: Cornell University Press, Chapter 15.

Sturkie, P.D., and W.J. Mueller. (1976). Reproduction in the female and egg production. In "Avian Physiology" (3rd ed.). New York: Springer-Verlag, Chapter 16.

Tanaka, K., and K. Goto. (1976). Partial purification of the ovarian oviposition-inducing factor and estimation of its chemical nature. Poult. Sci., 55, 1774.

Tanabe, Y., and T. Nakamura. (1980). Endocrine mechanism of ovulation in chickens (*Gallus domesticus*), quail (*Coturnix coturnix japonica*), and ducks (*Anas platyrhynchos domestica*). In "Biological Rhythms in Birds: Neural and Endocrine Aspects" (Y. Tanabe, K. Tanaka, and T. Ookawa, Eds.). Berlin: Springer-Verlag, p. 179

Tanaka, Y., L. Castillo, and H.F. DeLuca. (1976). Control of renal vitamin D hydroxylases in birds by sex hormones. Proc. Natl. Acad. Sci. U.S.A., 73, 2701.

Taylor, T.G., and J.H. Moore. (1954). Skeletal depletion in hens laying on a low-calcium diet. Br. J. Nutr., 8, 112.

Taylor, T.G., T.R. Morris, and F. Hertelendy. (1962). The effect of pituitary hormones on ovulation in calcium deficient pullets. Vet. Rec., 74, 123.

Tingari, M.D., and P.E. Lake. (1973). Ultrastructural studies on the uterovaginal sperm-host glands of the domestic hen, *Gallus domesticus.* J. Reprod. Fert., 34, 423.

Toth, M., D.M. Olson, and F. Hertelendy. (1979). Binding of prostaglandin $F_{2\alpha}$ to membranes of shell gland muscle of laying hens: Correlations with contractile activity. Biol. Reprod., 20, 390.

Tullett, S.G. (1978). Pore size versus pore number in avian eggshells. In "Respiratory Function in Birds, Adult and Embryonic" (J. Piper, Ed.). New York: Springer-Verlag, p. 219.

Tullett, S.G. (1984). The porosity of avian eggshells. Comp. Biochem. Physiol. A, 78, 5.

Tuohimaa, P. (1975). Immunofluorescence demonstration of avidin in the immature chick oviduct epithelium after progesterone. Histochemistry, 44, 95.

Turek, F. (1978). Diurnal rhythms and the seasonal reproductive cycle in birds. In "Environmental Endocrinology" (I. Assenmacher and D.S. Farner, Eds.). New York: Springer, p. 144.

Unsicker, K., F. Seidel, H.-D. Hofmann, T.H. Muller, R. Schmidt, and A. Wilson. (1983). Catecholaminergic innervation of the chicken ovary. Cell Tissue Res., 230, 431.

van Tienhoven, A. (1981). Neuroendocrinology of avian reproduction, with special emphasis on the reproductive cycle of the fowl (*Gallus domesticus*). World's Poult. Sci. J., 37, 156.

van Tienhoven, A. (1983). "Reproductive Physiology of Vertebrates" (2d ed.). Ithaca and London: Cornell University Press.

van Tienhoven, A., and C. Ostrander. (1973). The effect of interruption of the dark period at different intervals on egg production and shell breaking strength. Poult. Sci., 52, 998.

van Tienhoven, A., and C. Ostrander. (1976). Short total photoperiods and egg production of white Leghorns. Poult. Sci., 55, 1361.

van Tienhoven, A., and R.J. Planck. (1973). The effect of light on avian reproductive activity. In "Handbook of Physiology," Vol. II, Endocrinology. Part 1. Bethesda, Maryland: American Physiological Society, p. 79.

van Tienhoven, A., and A.V. Schally. (1972). Mammalian luteinizing hormone-releasing hormone induces ovulation in the domestic fowl. Gen. Comp. Endocrinol., 19, 594.

van Toledo, B., A.H. Parsons, and G.F. Combs, Jr. (1982). Role of ultrastructure in determining eggshell strength. Poult. Sci., 61, 569.

Verma, O.P., and F.L. Cherms. (1965). The appearance of sperm and their persistency in storage tubules of turkey hens after a single insemination. Poult. Sci., 44, 609.

Verma, O.P., and C.A. Walker. (1974). Adrenergic activity of the avian oviduct, *in vivo.* Theriogenology, 2, 47.

Verma, O.P., B.K. Prasad, and J. Slaughter. (1976). Avian oviduct motility induced by prostaglandin E_1. Prostaglandins, 12, 217.

Wang, S.-C., and J.M. Bahr. (1983). Estradiol secretion by theca cells of the domestic hen during the ovulatory cycle. Biol. Reprod., 28, 618.

Warren, D.C., and H.M. Scott. (1935). The time factor in egg formation. Poult. Sci., 14, 195.

Warren, D.C., and R.M. Conrad. (1942). Time of pigment deposition in brown-shelled hen eggs and in turkey eggs. Poult. Sci., 21, 515.

Wechsung, E., and A. Houvenaghel. (1978). Effect of prostaglandins on oviduct tone in the domestic hen *in vivo.* Prostaglandins, 15, 491.

Wechsung, E., and A. Houvenaghel. (1981). Effect of arachidonic acid on oviductal pressure in the domestic hen. Biol. Reprod., 24, 519.

Williams, J.B., and P.J. Sharp. (1977). A comparison of plasma progesterone and luteinizing hormone in growing hens from eight weeks of age to sexual maturity. J. Endocrinol., 75, 447.

Williams, J.B., and P.J. Sharp. (1978). Ovarian morphology and rates of ovarian follicular development in laying broiler breeders and commercial egg-producing hens. Br. Poult. Sci., 19, 387.

Wilson, S.C., and P.J. Sharp. (1975). Effects of progesterone and synthetic luteinizing hormone-releasing hormone on the release of luteinizing hormone during sexual maturation in the hen (*Gallus domesticus*). J. Endocrinol., 67, 359.

Wilson, S.C., and P.J. Sharp. (1976). Induction of luteinizing hormone release by gonadal steroids in the ovariectomized domestic hen. J. Endocrinol., 71, 87.

Wilson, S.C., and F.J. Cunningham. (1980). Modification by

metyrapone of the "open period" for pre-ovulatory LH release in the hen. Br. Poult. Sci., 21, 351.

Wilson, H.R., M.W. Holland, Jr., and R.L. Renner, Jr. (1973). Egg-laying cycle characteristics of the bobwhite (*Colinus virginianus*). Poult. Sci., 52, 1571.

Wilson, S.C., R.C. Jennings, and F.J. Cunningham. (1983). An investigation of diurnal and cyclic changes in the secretion of luteinizing hormone in the domestic hen. J. Endocrinol., 98, 137.

Wilson, W.O., and A.E. Woodward. (1958). Egg production of chickens kept in darkness. Poult. Sci., 37, 1054.

Wingfield, J.C., K.S. Matt, and D.S. Farner. (1984). Physiologic properties of steroid hormone-binding proteins in avian blood. Gen. Comp. Endocrinol., 53, 281.

Wolfenson, D., Y.F. Frei, N. Snapir, and A. Berman. (1981). Heat stress effects on capillary blood flow and its redistribution in the laying hen. Pfleugers Arch., 390, 86.

Wood-Gush, D.G.M. (1971). "The Behaviour of the Domestic Fowl." London: Heinemann.

Wood-Gush, D.G.M., and A.B. Gilbert. (1970). The rate of egg loss through internal laying. Br. Poult. Sci., 11, 161.

Wood-Gush, D.G.M., and A.B. Gilbert. (1975). The physiological basis of a behaviour pattern in the domestic hen. Symp. Zool. Soc. London, 35, 261.

Wyburn, G.M., R.N.C. Aitken, and H.S. Johnston. (1965). The ultrastructure of the zona radiata of the ovarian follicle of the domestic fowl. J. Anat., 99, 469.

Wyburn, G.M., H.S. Johnston, M.H. Draper, and M.F. Davidson. (1970). The fine structure of the infundibulum and magnum of the oviduct of *Gallus domesticus*. J. Exp. Physiol., 55, 213.

Wyburn, G.M., H.S. Johnston, M.H. Draper, and M.F. Davidson. (1973). The ultrastructure of the shell-forming region of the oviduct and the development of the shell of *Gallus domesticus*. J. Exp. Physiol., 58, 143.

Yokoyama, K., A. Oksche, T.R. Darden, and D.S. Farner. (1978). The sites of encephalic photoreception in photoperiodic induction of the growth of the testes in the white-crowned sparrow, *Zonotrichia leucophyrs gambelii*. Cell Tissue Res., 189, 441.

Yoshimura, Y., K. Tanaka, and O. Koga. (1983). Studies on the contractility of follicular wall with special reference to the mechanism of ovulation in hens. Br. Poult. Sci., 24, 213.

Zakar, T., and F. Hertelendy. (1980). Steroidogenesis in avian granulosa cells: Early and late kinetics of oLH- and dibutyryl cyclic AMP-promoted progesterone production. Biol. Reprod., 23, 974.

Zambonin A.A., and W.J. Mueller. (1969). Medullary bone of laying hens during calcium depletion and repletion. Calcif. Tissue Res., 4, 136.

19
Reproduction in the Male

A.L. Johnson

Anatomy of the Male Reproductive System: Passage and Storage of Sperm

Anatomy

Testes of the male bird are paired, and unlike those of most mammals are located within the body cavity, ventral and toward the cephalic border of the kidneys (Figure 19–1). Each testis is attached to the body wall by the mesorchium and is encapsulated by a fibrous inner coat, the tunica albuginea, and a thin outer layer, the tunica vaginalis. One or the other of the two testes may be larger, depending on the species (Lofts and Murton, 1973; Lake, 1981), but both are functional. The weight of the testes in chickens constitutes about

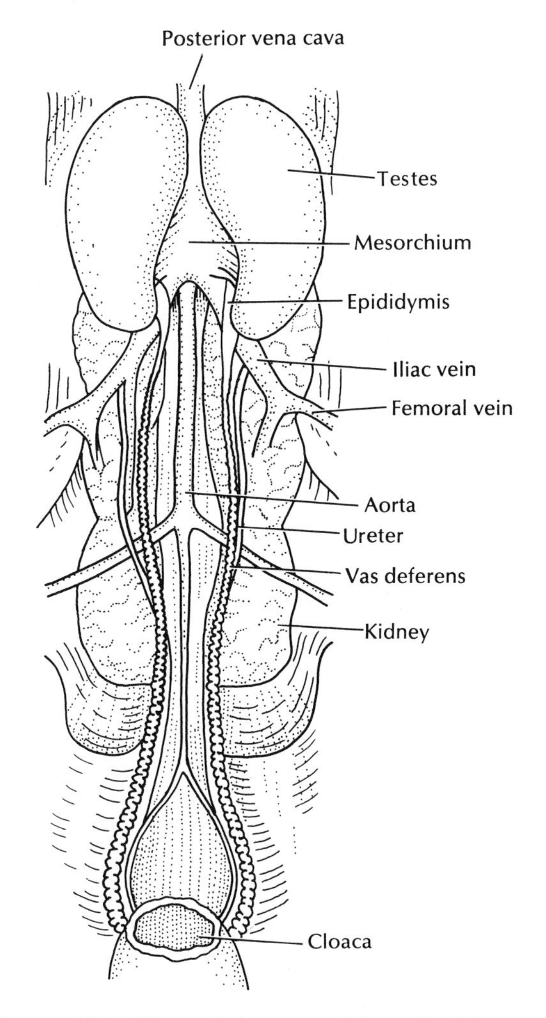

FIGURE 19–1. Urogenital system of the male chicken. (From Sturkie and Opel, 1976.)

1% of the total body weight, or about 9–30 g per testis at sexual maturity, depending on the breed (Sturkie and Opel, 1976). In seasonal breeders, testis size may increase by 300- to 500-fold during the reproductively active season as compared to the nonbreeding state.

Internally, the testes consist of a series of convoluted seminiferous tubules, connective tissue permeated by blood capillaries, and Leydig (or interstitial) cells. The seminiferous tubules consist of Sertoli cells and germinal epithelial cells (stem spermatogonia), whereas the Leydig cells are responsible for steroidogenesis and androgen secretion. The ultrastructure and differentiation of the Sertoli cell and seminiferous tubules have been reviewed by Cooksey and Rothwell (1973), Rothwell and Tingari (1973), and Osman (1980, 1981), and that of the Leydig cell by Nicholls and Graham (1972) and Rothwell (1973).

The testicular blood supply in avian species is less complex than that found in mammals (see Nishida, 1964). This is partly because birds have no pampiniform plexus, which in mammals is a countercurrent mechanism to maintain testes temperature below body temperature. Blood flows to each testis from the abdominal aorta, via a common trunk with the anterior renal artery, to the testicular artery (Figure 19–2). This testicular artery subsequently branches to form numerous smaller arteries that intertwine among the seminiferous tubules. There may also be an accessory testicular artery, which, when present, branches directly from the aorta. Venous flow occurs via superficial veins, which successively merge to form the testicular vein and empty directly into the posterior vena cava. Lymphatic vessels to the testis have not been well characterized, but are present within Leydig cells (see Lake, 1981).

The accessory sexual reproductive organs of the male include the vasa efferentia and epididymis, the ductus deferens, and the ejaculatory groove and phallus (penis). Birds have no organs comparable to the mammalian prostate gland, bulbourethral gland, or seminal vessicle, and seminal plasma is derived from the vasa efferentia and seminiferous tubules (Lake, 1981). The phallus of the rooster and many other birds is small and arises as a modification of the cloaca (Figure 19–3). Erection of the phallus occurs when folds within the cloaca become engorged with lymph (Nishiyama, 1955).

Intromission does not occur in the chicken or turkey; rather, semen is transferred to the vagina of the female by positioning the engorged phallus in contact with

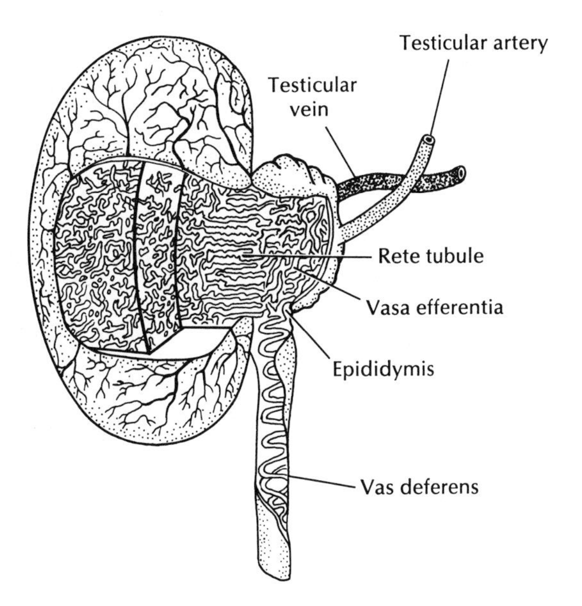

FIGURE 19–2. Internal structure of the chicken testes. (After Marshall, 1961.)

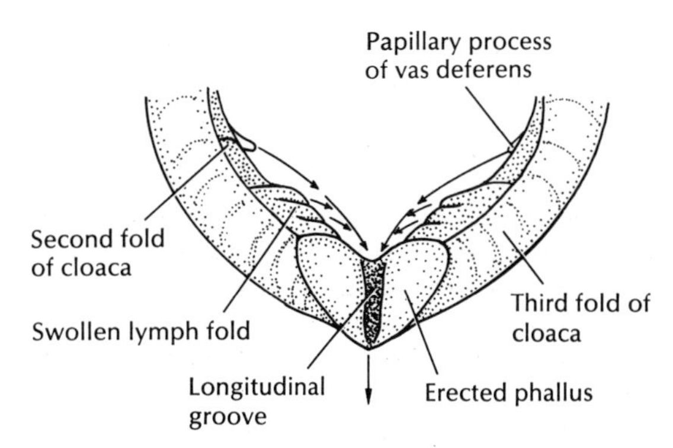

FIGURE 19–3. Ejaculatory groove of the male chicken. Shown is the ejection of semen from the papillary process of the vas deferens and outflow of transparent fluid from the swollen lymph fold, as well as the ejaculation of the semen (the mixture of vas deferens semen with transparent fluid) along the longitudinal groove of the erected phallus to the outside anus. (After Nishiyama, 1955.)

the everted cloaca of the female. By contrast, the phallus of the drake and several other species (King, 1981) is a vascularized, spiral-shaped sac with an ejaculatory groove that can be protruded by muscle and lymph infiltration and retracted by a cartilage-like ligament (Figure 19–4). The ejaculatory groove region of the drake is an androgen-dependent accessory organ, as demonstrated by the finding that administration of testosterone proprionate to juvenile male ducks causes the development of the ejaculatory groove to an extent comparable to the adult drake, while castration of the adult results in its involution (Fujihara and Nishiyama, 1976). Arterial blood to the cloacal region of the duck is supplied by the pudendal artery (Fujihara et al., 1976). Transfer of semen in these species occurs via intromission.

Nerves to the ductus deferens and penis include pelvic nerves (lumbosacrals 8–11), some of which are involved in erection of the penis, and probably sympathetic fibers (hypogastric), which are likely involved in ejaculation (Figure 19–5). The ductus deferens of the fowl has been demonstrated to be well innervated by adrenergic nerves (Bennett and Malmfors, 1970).

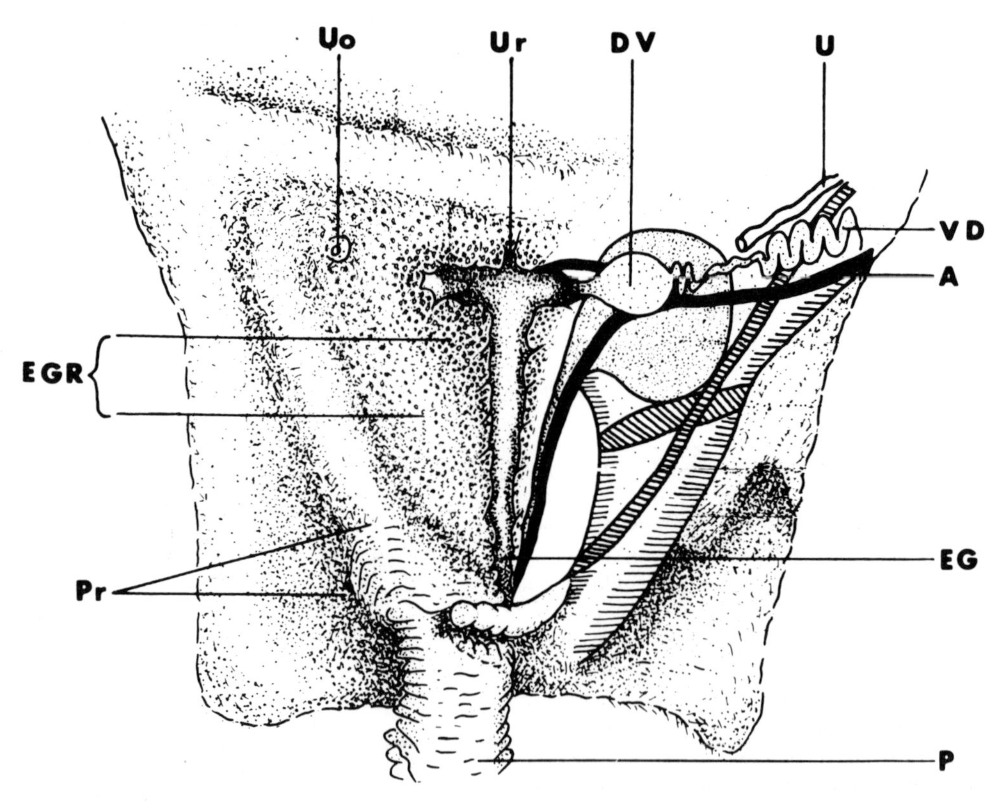

FIGURE 19–4. Cloacal region of the drake, with special reference to the ejaculatory groove; left, dorsal view of the cloaca; right, mucosa of ejaculatory groove region is removed to show blood supply to the ejaculatory groove region. Abbreviations: *A,* pudendal artery; *DV,* dilated portion of vas deferens; *EG,* ejaculatory groove; *EGR,* ejaculatory groove region; *P,* penis; *Pr,* proctodeum; *U,* ureter; *Uo,* orifice of ureter; *Ur,* urodeum; *VD,* ductus deferens. (From Fujihara et al., 1976.)

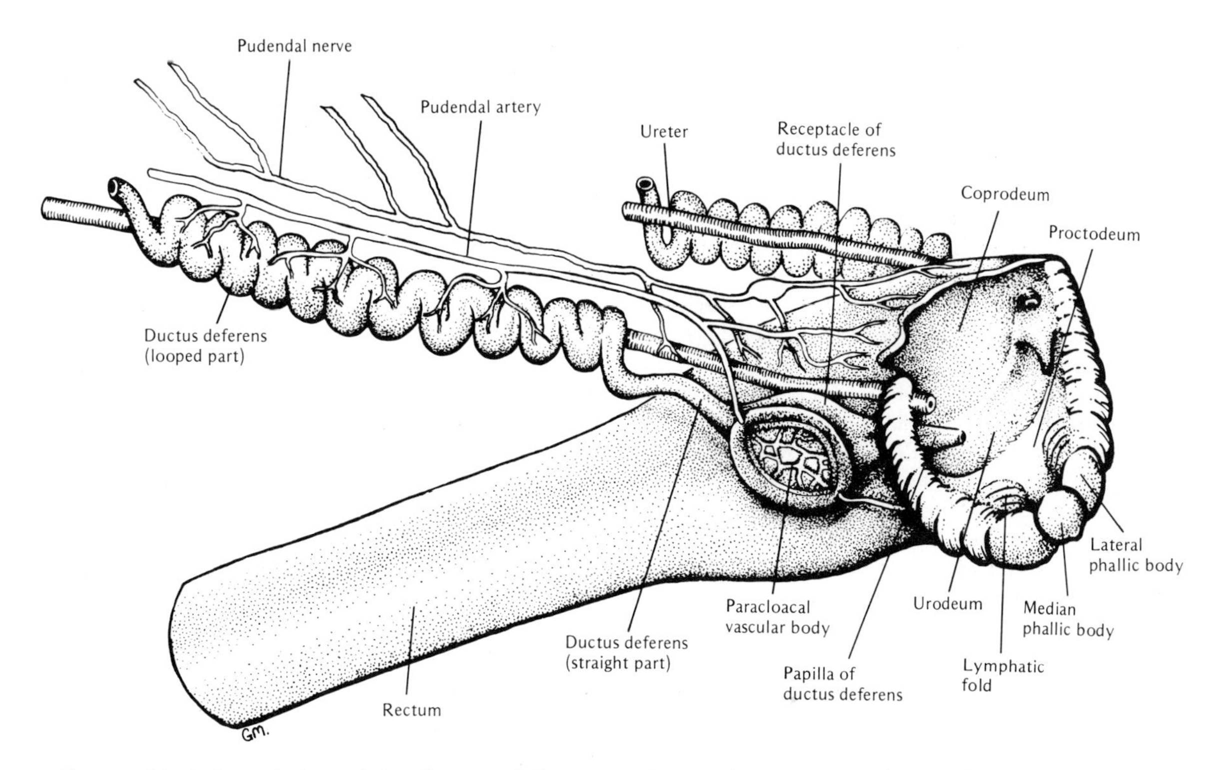

FIGURE 19–5. Lateral view of the cloaca and the terminal part of the ductus deferens of the domestic fowl. (From Lake, 1981.)

For further details concerning accessory organs of the rooster, see Tingari (1971); of the drake, see Fujihara and Nishiyama (1976) and Fujihara et al. (1976).

An accessory sexual organ unique to the Japanese quail is the cloacal (proctodeal) gland. The cloacal gland is an androgen-dependent gland of uncertain function that is well developed in the male and rudimentary in the female. In the mature male, there is a high correlation between size of the gland and weight of the testes. The morphology of this gland has been reviewed by Klemm et al. (1973).

Transport and Storage of Sperm

Mature spermatozoa from the seminiferous tubules pass to the rete tubules, then to the highly convoluted vasa efferentia, the epididymis, and ductus deferens (Figure 19–2). In seasonal breeders, all of these ducts decrease in size to a variable extent during the nonbreeding season. Sperm are stored in the ductus deferens of the rooster (Lake, 1981). In passerine birds, the terminal portion of the ductus deferens is enlarged into a seminal sac and serves as an additional area for sperm storage. Estimates of the time required for sperm to pass from the testes to the distal region of the ductus deferens range from 1 to 4 days (Munro, 1938). Structural differentiation of spermatozoa is thought to be complete before it leaves the rete tubules (Tingari, 1973a); however, motility was found to be minimal in the testis, increased in the epididymal region, and maximal in the ductus deferens (Munro, 1938). The fertilizing capacity of fowl spermatozoa is thought to increase during its passage through the epididymis and/or ductus deferens.

There is evidence that both Sertoli cells and the epithelial cells of the epididymis can resorb spermatozoa, and it is suggested that these are possible routes for the elimination of unejaculated sperm (Tingari and Lake, 1972; Breucker, 1978).

Development of the Testes, Spermatogenesis, and Semen Production

Development of the Testes, and Spermatogenesis

The testes are derived from the germinal epithelium of the mesonephros, and can be differentiated by day 4 of incubation. The epididymis, originating from the anterior portion of the mesonephros, and the ductus deferens, formed from the mesonephric duct, are evident by 18 days of incubation. For further details concerning the ontogenesis of the male genital organs, see Lofts and Murton (1973) and Budras and Sauer (1975).

The growth and development of the testes is largely dependent on hypothalamic releasing factors and the secretion of gonadotropic hormones. Hypothalamic neurosecretory activity, on the other hand, is influenced by the stage of sexual maturation (age) and environmental stimuli (to the greatest extent, photoperiod). It is generally believed that in avian species, as in mammals, follicle-stimulating hormone (FSH) stimulates growth, differentiation, and spermatogenic activity of the seminiferous tubules, and that luteinizing hormone (LH) affects steroidogenic activity of the Leydig cells (Brown et al., 1975; see also Chapter 17).

The mechanisms involved with the onset of sexual maturity are not well understood. In the cockerel, the increase in plasma levels of LH at the onset of puberty may be the result of a decrease in the sensitivity of the luteinizing hormone-releasing hormone (LHRH)-secreting neurons to the negative feedback effects of testicular steroids (Sharp and Gow, 1983). Parasagittal cuts in the preoptic to mammillary region of the hypothalamus of 2-week-old male chicks significantly increased plasma concentrations of LH and caused precocious puberty (Kuenzel and Sharp, 1982; Mass and Kuenzel, 1983). In addition, the pineal has been implicated in the development of sexual maturation. Pinealectomy at 8–9 days of age significantly delayed testes growth and the onset of spermatogenesis, and depressed plasma concentrations of testosterone and comb growth, compared to sham-pinealectomized and control cockerels. These effects were thought to be mediated via the inhibition of gonadotropin secretion (Cogburn and Harrison, 1977).

The testes of all avian species undergo marked changes during the development of spermatogenesis, and these are essentially similar in most species (for detailed studies, see Blivaiss, 1947; Kumaran and Turner, 1949a, b; and Budras and Sauer, 1975). During the first 5 weeks of age (in the chicken), the tubules become organized, and multiplication of the basal layer of cells (the spermatogonia) occurs. The primary spermatocytes begin to appear at about the sixth week of age (Figure 19–6A). During the next 2–3 weeks, growth of the primary spermatocytes takes precedence over the further multiplication of the spermatogonial layer.

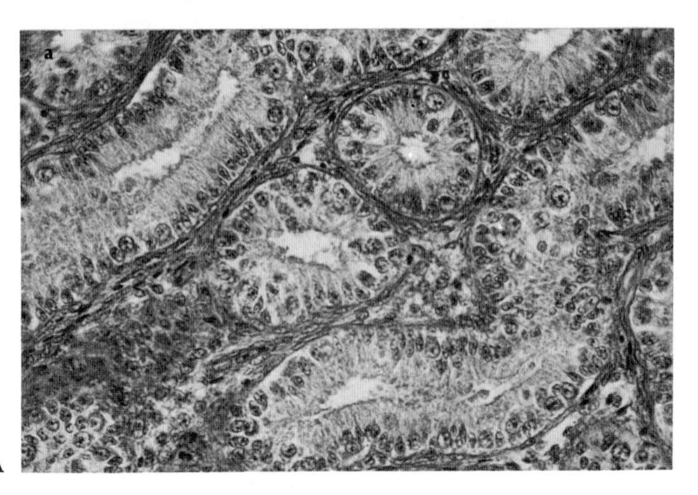

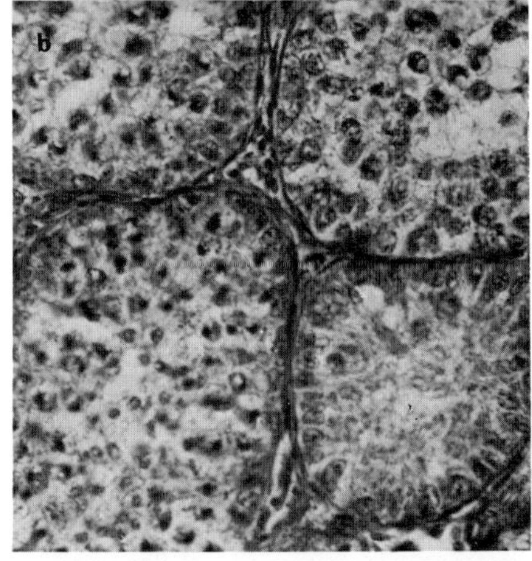

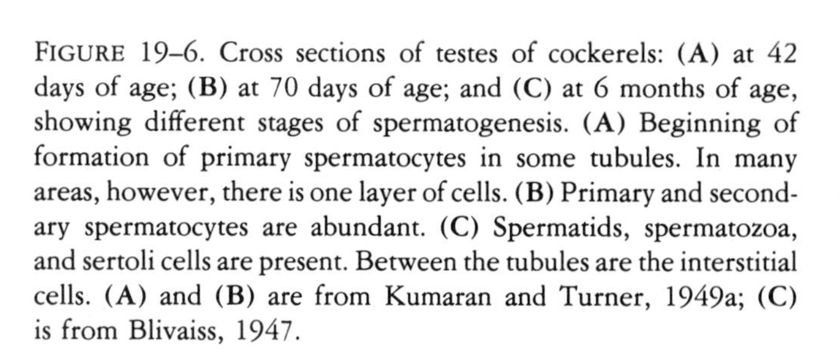

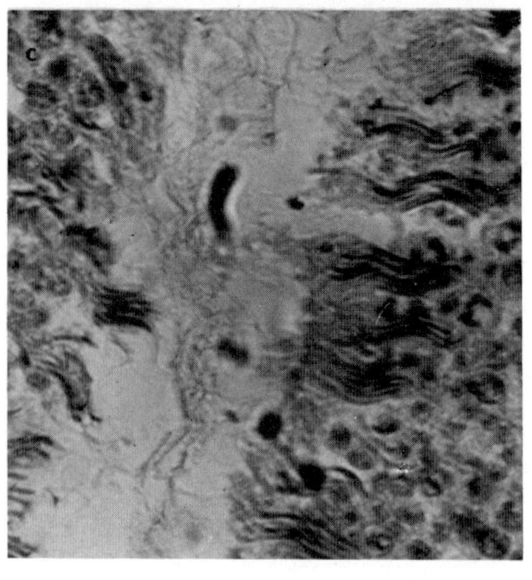

FIGURE 19–6. Cross sections of testes of cockerels: (**A**) at 42 days of age; (**B**) at 70 days of age; and (**C**) at 6 months of age, showing different stages of spermatogenesis. (**A**) Beginning of formation of primary spermatocytes in some tubules. In many areas, however, there is one layer of cells. (**B**) Primary and secondary spermatocytes are abundant. (**C**) Spermatids, spermatozoa, and sertoli cells are present. Between the tubules are the interstitial cells. (**A**) and (**B**) are from Kumaran and Turner, 1949a; (**C**) is from Blivaiss, 1947.

The secondary spermatocytes begin to appear at about 10 weeks of age as a result of the reduction division of the primary spermatocytes (Figure 19–6B). Spermatids (immature spermatozoa) begin to appear in the seminiferous tubules at about 12 weeks of age, and by week 20 are usually present in all of the tubules (Figure 19–6C). The onset of puberty in the cockerel is characterized by a phase of rapid testicular growth and the completion of spermatogenesis. This period generally occurs over a period of 8–10 weeks (from 16 to 24 weeks of age). By contrast, in the photostimulated Japanese quail, spermatozoa can be detected in the testes by day 26 and in the ductus deferens by day 30. By day 35, large concentrations of sperm are present in both the testes and ductus deferens (Ottinger and Brinkley, 1979).

The seminiferous tubules of prepuberal males are small and are lined with a single layer of cells (Figure 19–6A). The mature testis has a multilayered epithelium representing the various stages of spermatogenesis. From the wall of the tubule to the lumen may be found spermatogonia, primary spermatocytes, secondary spermatocytes, spermatids, Sertoli cells (to which the spermatids are attached), and the spermatozoa.

In the rooster, primary spermatocytes are transformed to mature spermatozoa in approximately 12 days (Takeda, 1969). The time required for the completion of spermatogenesis and sperm transit in Japanese quail has been estimated to be 25 days (Jones and Jackson, 1972). Detailed studies of spermatogenesis in ducks have been reported by Clermont (1958) and Johnson (1966).

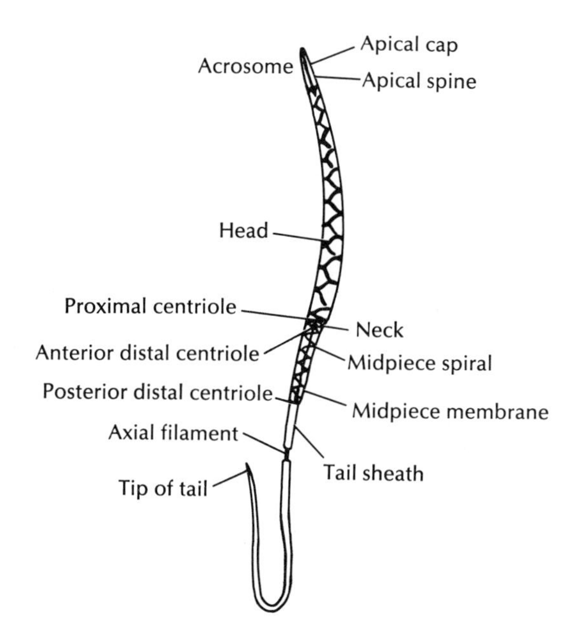

FIGURE 19–7. Diagram of the spermatozoon of the chicken. (From Sturkie and Opel, 1976.)

Morphology of Spermatozoa

There is considerable variation in the morphological appearance of mature spermatozoa among avian species. Avian spermatozoa are small compared to those of mammals, having an average volume of 9.2 μm^3 (Lake, 1971). The acrosome is simple and encapsulated by a cell membrane; the head contains the nucleus; and the tail is subdivided into the neck, midpiece, principal piece, and short-end piece (Lake et al., 1968; Bakst and Howarth, 1975). The chicken acrosome is about 2 μm in length; the head (or nucleus), about 12.5 μm long; the midpiece (a cylindrical distal centriole surrounded by a sheath of mitochondria), 4 μm long; and the principal tailpiece is 80–90 μm long (Lake, 1981) (Figure 19–7). Dimensions and components of spermatozoa from the turkey and guinea fowl were found to be comparable to those from the rooster (Marquez and Ogasawara, 1975; Thurston et al., 1982). For a discussion of morphologic changes of rooster spermatozoa during fertilization, see Okamura and Nishiyama (1978a, b).

Chemical Composition and Physical Properties of Spermatozoa and Semen

The acrosome of several avian species has been shown to contain trypsin-like enzymatic activity (Langford and Howarth, 1974). This enzymatic activity is probably involved in spermatozoon penetration of the egg during fertilization. The head region contains primarily nuclear material (DNA and chromatin granules) (Bakst and Howarth, 1975; Lake, 1981). The total percentage of lipid in spermatozoa represents 8.9% of dry matter, with 64.3% of this being phospholipid (Howarth, 1981). The detailed chemical composition of fowl spermatozoa has been described by Lake (1981).

It is clear that the composition and physical properties of avian semen differ from that of mammals (for details, see Sturkie and Opel, 1976; Lake, 1981). These differences are attributable to the absence of seminal vesicles and prostate glands and the presence of a rudimentary epididymis. Fowl seminal plasma is almost completely lacking in fructose, citrate, ergothioneine, inositol, phosphoryl choline, and glyceryl phosphoryl choline. The chloride content of avian semen is low, and potassium and glutamate contents are high. The source of the high glutamate concentration may be the seminiferous tubules (Lake, 1971). Harris and Goto (1984) reported a relationship between carbonic anhydrase activity in the testes or vas deferens and volume of semen, spermatozoa, and seminal plasma per collection. This relationship is suggested to represent the importance of acid–base balance on the maturation and transport of spermatozoa and production of semen or seminal plasma.

Semen from the cock is usually white and opaque, but may be clear and watery, particularly when the concentration of spermatozoa is low. The pH of cock semen is 7.0–7.6, depending on the amount of transparent fluid present (Lake, 1971).

In turkeys, poor fertility has frequently been associated with abnormal yellow-colored semen (Marquez and Ogasawara, 1975). This anomaly is probably the result of hypertrophied, lipid-filled epithelial cells found exclusively in the vasa efferentia (Hess et al., 1982).

Number of Spermatozoa and Amount of Semen Produced

Reports concerning the volume of semen per ejaculate from the fowl are variable (Sturkie and Opel, 1976), due in part to method of collection and breed differences. The average volumes reported range from 0.11 ml (collected from the cloaca of the hen) to 1 ml (collected directly from the male). Average concentration of spermatozoa per milliliter semen is 3.5 million, or, in a 0.5- to 1.0-ml volume ejaculate, 1.7–3.5 billion. Lake (1957) reported averages of 7 billion and a maximum of 8.2 billion per ejaculate in Brown Leghorn roosters.

Marini and Goodman (1969) found great differences in spermatozoa concentration and volumes of semen of chickens selected and bred for high and low growth rates. These numbers ranged from 4.9 million spermatozoa/ml semen in the slow-growing strain to 2.3 million/ml in the rapid-growth line. There were also differences in the numbers of abnormal spermatozoa between the strains.

Turkeys produce less semen than chickens, but the concentration of spermatozoa is much greater. Average ejaculates consist of 0.2 ml per collection, and the concentration ranges from 6.2 to 7 million/ml, depending on breed (McCartney and Brown, 1959). Average volumes of semen collected from several breeds of ducks range from 0.18 to 0.41 ml, while concentration per milliter ranges from 2.10 to 9.33 million (Kamar, 1962; Nishiyama et al., 1976). The amount of semen collected from the pheasant (*Phasianus colchicus*) is 0.1 ml or less (Shaklee and Knox, 1954).

Hypophysectomy, Castration, and Hemicastration in the Male

As previously discussed (see Chapter 17), the avian gonad is under the control of hormones secreted from the anterior pituitary. Total adenohypophysectomy in the domestic fowl leads to a rapid degeneration of spermatogenic cells, a decrease in the diameter of seminiferous tubules, and a decrease in testicular weight (Tanaka and Yasuda, 1980). Testicular degeneration was characterized by a reduction of cytoplasmic processes in Sertoli cells by day 3, and by detachment of spermatozoa and spermatids, spermatocytes, and spermatogonia from lumen epithelial cells by days 5, 10, and 20, respectively. The effect of partial adenohypophysectomy on testicular function is related to the proportion of tissue removed (Tanaka and Fujioka, 1981).

Castration of the male (caponization) produces capons that grow more slowly and develop more fat than the uncastrated male, particularly after 5 months of age. The basal metabolic rate of capons is approximately 13.5% lower than in cocks (Mitchell et al., 1927). Cockerels that are caponized at 5 weeks of age have recently been shown to have grossly abnormal bone development in the tibiotarsus-tarsometatarsus region of the leg at maturity. This effect of caponization on bone development can be alleviated by continuous replacement with adrogens (Johnson and Rendano, 1984).

Castration has been shown to increase plasma levels of LH [chicken (Wilson, 1978); Japanese quail (Gibson et al., 1975); red grouse (Sharp and Moss, 1977); tree sparrow (Wilson and Follett, 1974); and white-crowned sparrow (Wingfield et al., 1980)] and FSH [Japanese quail (Follett, 1976); tree sparrow (Wilson and Follett, 1978); and white-crowned sparrow (Wingfield et al., 1980)], while androgen replacement reduces the elevated levels of LH [Japanese quail (Davies et al., 1980); and chicken (Wilson, 1978)].

Hemicastration has been shown to lead to compensatory hypertrophy of the contralateral testis (e.g., chicken, Wilson and de Reviers, 1979; white-crowned sparrow, Farner et al., 1970). In the chicken, compensatory hypertrophy begins within 2 weeks after hemicastration, and after 6 weeks the weight of the single testis of hemicastrated cockerels is not different than the combined weight of the two testes of intact birds (Wilson and de Reviers, 1979). Hemicastration at less than 2 weeks of age results in plasma concentrations of testosterone at maturity that are 25% less than those of intact roosters (Driot et al., 1979). Despite the lower concentrations of testosterone, however, concentrations of plasma LH are equal to or less than those of intact birds, suggesting that the negative feedback mechanism of the hypothalamo-hypophyseal–testicular system is comparable to the intact rooster (Wilson and de Reviers, 1979).

Factors Affecting Fertility in the Male

Factors affecting fertility in poultry are complex and involve behavioral, genetic, and environmental aspects in both the male and female. Only those aspects and problems associated with the male will be discussed here.

Natural Mating and Behavior

Mating in birds is associated with elaborate courtship behavior (Wood-Gush, 1971). There exists a diurnal rhythm in mating frequency, and this rhythm is correlated with semen production (Lake and Wood-Gush, 1956). The number of matings or ejaculations per day influences the volume of semen produced and the concentration of spermatozoa; the greater the frequency, the lower the volume of semen and concentration of spermatozoa (Parker et al., 1940). The number of matings per day for the male chicken may range from 25 to 41 or more (Guhl, 1951). When males are first introduced into a pen, they mate most frequently during the first 3–6 min. It has been demonstrated that the social order (peck order) of the hens to which males are introduced affects mating behavior of the male (Guhl et al., 1945; Guhl and Warren, 1946); males, regardless of their social standing, tend to mate most frequently with hens in the middle of the hierarchy and less frequently with the highest or lowest ranking hens. When three or more males are introduced together into a pen of females, both the frequency of matings and the fertility are highest for males at the top of the hierarchy.

The upper limit for the number of White Leghorn females to which one male may mate and obtain optimum fertility is approximately 15 (Brantas et al., 1972). This ratio is lower for many heavy breeds. A relatively recent decline in fertility and hatchability in heavy breeds has been suggested to be related to genetic selection for rapid growth in broiler lines of chickens and in turkeys. However, studies by Wilson et al. (1979a) showed that fertility with natural mating in broiler breeder males is poorly correlated to physical characteristics (weight, breast angle, hip width).

Artificial Insemination

Semen can be collected artificially from the male by techniques originally described by Burrows and Quinn (1937) and Parker (1939). Semen collected by such methods can be inseminated immediately (usually within 1 hr), diluted and stored for up to 5–6 hr (normally at 10–20°C) and inseminated artificially, or frozen for later insemination. However, rapid freezing and thawing of fowl spermatozoa by current methods results in a pronounced change in ultrastructure, a decrease in motility, and a loss of fertilizing capacity (Bakst and Sexton, 1979). For reviews of poultry semen preservation and factors affecting fertility with artificial insemination, see Sexton (1979) and Lake (1983).

Insemination of 80–90 million spermatozoa per insemination interval is generally considered adequate to insure high fertility in fowl. In the turkey, reports of the number of live sperm required to obtain maximum fertility from diluted semen range from 6 to 90 million per weekly insemination (see Sexton, 1977). More commonly, insemination is performed at 7- to 14-day intervals, and Brown (1974) reported that 70 million live sperm administered at 14-day intervals results in maximum fertility in the turkey hen.

Other factors found to influence fertility in fowl include stage of production in the hen (Van Krey et al., 1967), site of insemination within the oviduct of the hen (Lorenz and Ogasawara, 1968), time of day and position of egg in the oviduct (Christensen and Johnston, 1977; Giesen et al., 1980), and quality of semen as a result of inherent factors and method of collection (Kamar and Hafez, 1975).

Maximum fertility in chickens is usually obtained 2–3 days after insemination. Good fertility is obtained as long as 5–6 days after insemination and then declines rapidly, but a few fertile eggs may be obtained for as long as 14 days after a single mating or insemination. In turkeys, high fertility is obtained up to 20 days after insemination, and then decreases slowly, with some fertile eggs obtained 10 weeks after insemination (Kamar and Hafez, 1975).

Genetic Differences

The age at which male chickens reach sexual maturation is influenced by the strain of bird (Aire, 1973). In addition, differences in fertility and semen quality characteristics have been reported between breeds, inbred lines, and families. For instance, fertility in natural matings of white Plymouth Rock hens is higher than in Cornish hens, and the difference is inherited; however, differences in fertility were not attributable to amount of semen produced or motility of spermatozoa (see Wilson et al., 1979a).

Mating behavior is a trait that is quantitatively inherited; it may also influence fertility (Siegel, 1965). In two lines of chickens selected for high and low mating frequency, significant differences were reported for frequency of mating behaviors (courts, mounts, treads, and completed matings); however, there were no differences between lines for the number of days to peak fertility after natural or artificial insemination or duration of fertility (Bernon and Siegel, 1981). In chickens selected for high (HW) and low (LW) juvenile body weight, HW males exhibited a lower number of courts, mounts, and completed matings than LW males, but significantly greater semen concentrations, a lower percentage of normal spermatozoa, and lower fertility, as determined by artificial insemination (Blohowiak et al., 1980).

Environmental Factors

Temperature. In a seasonal breeder such as the white-crowned sparrow (*Zonotrichia leucophrys gambelii*), temperature has been shown to modify the rate of testicular recrudescence under experimental conditions. However, it remains to be determined whether under natural circumstances temperature directly affects the hypothalamo-hypophyseal–gonadal system, or alternatively, exerts only an indirect influence via food supply and availability (Murton and Westwood, 1977).

Ingkasuwan and Ogasawara (1966) suggested that the optimal ambient temperature for semen production in White Leghorn roosters is 20–25°C. The effects of temperature on semen production may also be reflected through the influence of temperature on rate of sexual maturation (Ingkasuwan and Ogasawara, 1966) and food intake.

Light. Photoperiod appears to be the major environmental factor that controls spermatogenesis in the male (Farner et al., 1977). Natural daylight or artificial light stimulates and maintains semen production via its effect on the release of gonadotropic hormones of the anterior pituitary; FSH stimulates growth of seminiferous tubules and spermatogenesis, and LH stimulates the interstitial (Leydig) cells to produce testicular steroids (see also Chapter 17).

In general, 12–14 hr of light are required to maximally stimulate testes growth and development (the photosexual response). Treatment of cockerels with 8 hr of light per day delays the onset of semen production (compared to 14 hr light), but prolonged treatment eventually results in normal testes size and semen production (Ingkasuwan and Ogasawara, 1966). Intermittent photoperiods, consisting of 6–8 hr light followed by 15 min–2 hr light at specific intervals within the subsequent dark phase, stimulate testes growth in the quail (Follett and Sharp, 1969) and cockerel (de Riviers, 1981). In the Japanese quail, housing under a nonstimulatory photoperiod (hr) of 8 L:16 D, followed by the transfer to a stimulatory photoperiod of 20 L:4 D results in a significant increase in circulating concentrations of gonadotropins after a single long day (Follett et al., 1977).

The photoperiodic effect on the avian neuroendocrine system that results in gonadal stimulation can be explained by one of two theoretical models: the external and internal coincidence models. Although it is beyond the scope of this chapter to discuss evidence supporting each of these models, there are excellent reviews by Gwinner (1975), Farner and Wingfield (1978), and van Tienhoven (1983).

The end of the breeding season in many wild species of the temperate zone is the result of testicular regression due to photorefractoriness, despite the fact that the birds may still be exposed to the photoperiod that stimulated testicular growth. The anatomic site that mediates photorefractoriness is either the hypothalamus or higher centers of the brain (Farner and Wingfield, 1978) and not the testes (van Tienhoven, 1983). Termination of the photorefractory state is accomplished by exposure of birds to a short photoperiod; the shorter the daylength, the shorter the duration of photorefractoriness (Murton and Westwood, 1977). In contrast, species such as the bobwhite quail, Japanese quail, wood pigeon, and common pigeon respond to long photoperiod with testicular growth and short photoperiod with testicular regression, but these species fail to become photorefractory in response to prolonged periods of stimulatory light (Sturkie and Opel, 1976).

Intensity and Wavelength. The intensity of light within the range of 2 to 50 lux does not appear to be critical for either testicular growth or maintainence (Oishi and Lauber, 1973; Sturkie and Opel, 1976). Testicular growth in birds is stimulated to a greater extent by near-red visible light wavelengths than by shorter wavelengths such as blue and green (Sturkie and Opel, 1976; Oishi and Lauber, 1973; Foss et al., 1972), and can be accomplished by placing radioluminescent paint (maximum energy, 600–640 nm) beneath the skull (Oishi and Lauber, 1973) or within the infundibular complex of the hypothalamus (Oliver et al., 1979).

Extraretinal Photoreceptors. Although in mammals the eyes are the primary photoreceptors (Chapter 2), they are not essential for photostimulation and sexual development in avian species. Removal of the eyes or transection of the optic nerve fails to prevent testicular recrudescence in response to a stimulatory photoperiod in the Japanese quail (Siopes and Wilson, 1974), sparrow (Menaker and Underwood, 1976), or drake (Benoit, 1964). Information via the eye, however, may mediate inhibitory effects, as testicular regression fails to occur in blinded quail transferred from a long to a short photoperiod (Siopes and Wilson, 1980).

The extraretinal photoreceptor responsible for the photosexual response is probably located within the ventromedial hypothalamus and/or tuberal complex; this is despite the lack of anatomic evidence for photoreceptors or photopigments within these photosensitive regions (Yokoyama et al., 1978). In addition, Oliver et al. (1979) provided evidence that neuronal populations of the basal hypothalamus are in themselves photosensitive, while other neuronal populations located in the preoptic area serve as a target area for retinal information.

Finally, in the majority of species investigated, the pineal gland (Chapter 23b) plays no role in the photosexual response (Ralph, 1981). One possible exception is found in the Indian weaver finch (*Ploceus philippinus*), in which an antigonadal influence of the pineal has been proposed (Saxena et al., 1979). It is of interest, however, that the pineal gland of the chicken and quail have been found to contain photoreceptor-like cells (Bischoff, 1969).

Other Factors

Age. The age of first sperm production varies widely depending on breed and strain of bird, ranging from 13.8 to 21.0 weeks among six breeds of chickens (semen collected by artificial insemination) (Grosse and Craig, 1960). Fertility generally declines during the second and third years of age.

Nutrition. Nutritional status affects semen volume and quality, fertility, hatchability, and live weight of hatched chicks (Maslieu and Davtyan, 1969). Restriction of dietary energy and protein in turkey toms depressed semen volume compared to toms fed a control diet (Menge and Frobish, 1976). By contrast, partial feed restriction in turkey toms had no such deleterious effect on semen quantity and quality and increased fertility (Krueger et al., 1977). Sexual maturity of white Leghorn cockerels was delayed, and spermatogenesis arrested, by feeding 8.0% protein or less from 7 to 21 weeks of age (Jones et al., 1967). Semen production or quality is also affected by starvation, deficiencies in essential fatty acids and Vitamin E (see Wilson et al., 1979a), and toxins (Ottinger and Doerr, 1982).

Diurnal and Seasonal Variations. There is a diurnal variation in the production of spermatozoa by the fowl, with the greatest spermatogenic activity found at various times during the evening and night under usual lighting conditions (see Sturkie and Opel, 1976). The incidence of daily mating activity is usually highest coincident with the time of greatest semen yield (Lake and Wood-Gush, 1956).

Semen production and fertility in poultry are usually higher during the spring and lower in the summer (Sturkie, 1965). Most wild species produce semen only during a relatively restricted breeding period.

Fertilization and Embryonic Development

Fertilization

Fertilization of the ovum occurs within the infundibulum of the hen. Spermatozoa are rapidly distributed throughout the oviduct after insemination, and can reach the site of fertilization and fertilize the ovulated ova within 15 min after natural or artificial insemination (see Kamar and Hafez, 1975). In contrast to spermatozoa from mammalian species, avian spermatozoa appear not to require capacitation within the oviduct of the fowl. Successful fertilizations occur when spermatozoa are deposited within the infundibulum 15 min after oviposition and just before ovulation (Olsen and Neher, 1948), and can be accomplished in vitro with ova of chickens (Howarth, 1970) and turkeys (Howarth and Palmer, 1972). The process of spermatozoan penetration into the ovum of the fowl was described by Okamura and Nishiyama (1978a, b). Studies in vivo on the effect of spermatozoa aging in chickens (Nalbandov and Card, 1943) and turkeys (Hale, 1955) show that both fertility and hatchability decrease with age of spermatozoa. Decreased hatchability with aged spermatozoa has been attributed to abnormalities of the nervous and vascular system of the embryo (Dharmarajan, 1950).

Embryonic Development

The early growth and development of the fertilized ovum have been studied by Olsen (1942). Briefly, the first meiotic division (extrusion of the first polar body) occurs prior to ovulation of the follicle. Fertilization usually occurs within 15 min after ovulation and initiates the second meiotic division. The egg proceeds to divide mitotically to the 4- or 8-cell stage while still in the isthmus, and reaches the blastocyst stage in the shell gland. Evidence is equivocal as to whether the chicken ovum has reached the gastrula stage by the time of oviposition (see Sturkie, 1965).

A morphologic sex reversal (from female to male) can be accomplished during early embryonic life of the female by extraembryonic grafting of testes. After hatching, these individuals exhibit normally developing testes (with spermatogenesis occurring within the seminiferous tubules), differentiated male excretory ducts, and regressed Mullerian ducts (Rashedi et al., 1983). In several species of birds (i.e., bobwhite quail, Japanese quail, and chicken), the length of the incubation period can be experimentally shortened by up to 32 hr by auditory cues from adjacent newly hatched chicks, or by exposure to synthetic auditory stimuli (Vince et al., 1970; Grieve et al., 1973). Furthermore, evidence suggests that embryologic development can be accelerated by auditory cues (Grieve et al., 1973).

Male Sex Hormones

Testicular Steroids

The testes produce and secrete a number of steroids, which are involved in the negative feedback effect on gonadotropin secretion (Sharp and Gow, 1983; see also Chapter 17) and in stimulating the development and maintenance of accessory sexual organs (Lofts and Massa, 1980).

Biosynthesis of Testicular Steroids. Two important pathways for the synthesis of androgens exist in the avian testes: the Δ^5 pathway (Δ^5-C_{21} steroid to Δ^5-C_{19} steroid) and the Δ^4 pathway (Δ^5-C_{21} steroid to Δ^4-C_{21} steroid) (Nakamura and Tanabe, 1972; Galli et al., 1973) (Figure 19–8). In vitro studies suggest that tes-

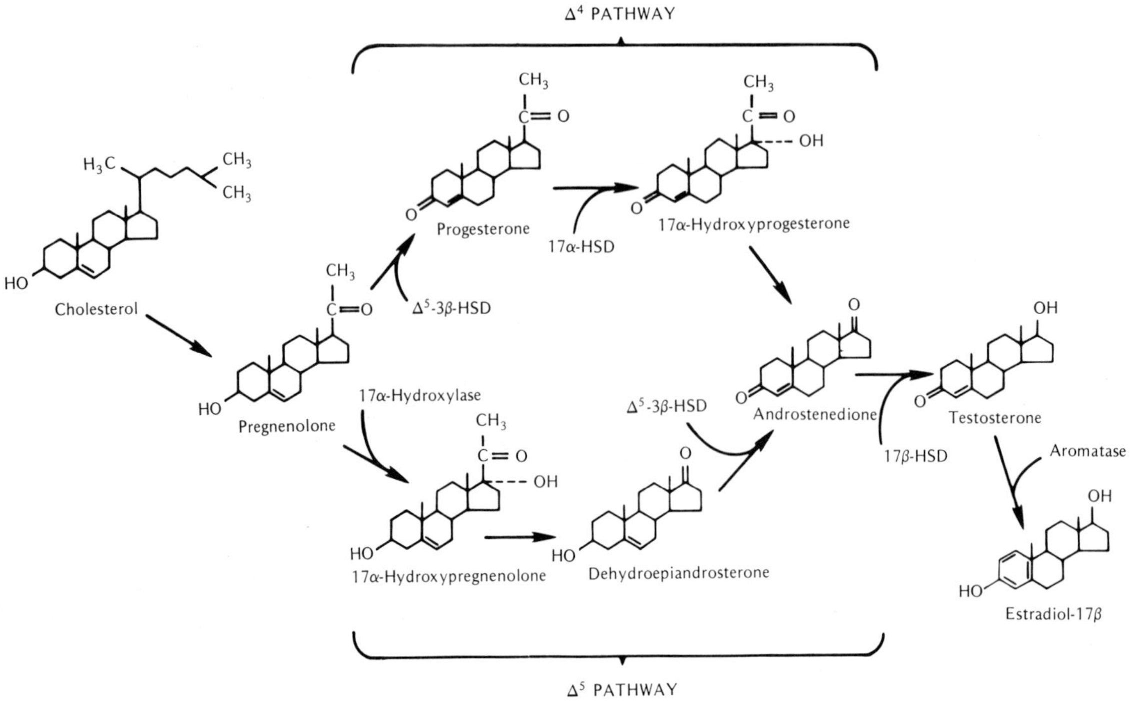

FIGURE 19–8. Steroid biosynthesis in the testes.

tosterone is the principal steroid produced and secreted by the testes of sexually mature chickens (Galli et al., 1973), although relatively large quantities of progesterone and small quantities of estradiol-17β have also been measured in testicular homogenates (Tanabe et al., 1979).

In light of the following evidence, the Leydig cell is probably the site of androgen production. Activity of the enzymes Δ^5-3β-hydroxysteroid dehydrogenase (Δ^5-3β-HSD) and 17β-hydroxysteroid dehydrogenase (17β-HSD) occurs within the Leydig cell (Woods and Domm, 1966; Garnier et al., 1973; Tingari, 1973b), and there exists a correlation between Δ^5-3β-HSD activity and concentrations of plasma and testicular testosterone (Garnier et al., 1973). Following photostimulation, the number of Leydig cells increases (Japanese quail, Nicholls and Graham, 1972; white-crowned sparrow, Lam and Farner, 1976). Plasma levels of testosterone also increase (Lam and Farner, 1976). Finally, administration of chicken LH to immature or hypophysectomized Japanese quail stimulates the Leydig cells to differentiate and become mature, steroid-secreting cells (Brown et al., 1975).

Activity of the enzymes 3β-HSD and 17β-HSD has also been identified in Sertoli cells and in the epithial lining of the excurrent ducts of the testes (Woods and Domm, 1966; Garnier et al., 1973; Tingari, 1973b), however, the functional significance of steroid production at these sites has yet to be determined.

The pattern of steroid metabolism varies with reproductive status. Testosterone synthesis from the precursors progesterone and pregnenolone in the testes of red-wing blackbirds (*Agelaius phoeniceus*) is greatest during the photosensitive and regressive stages, but cannot be detected during the refractory state (Kerlan et al., 1974). Elevated testosterone production during the photosensitive stage suggests a stimulatory effect of androgens on spermatogenesis, whereas the regressive stage may represent a negative feedback mechanism to inhibit gonadotropin secretion by the hypothalamus and/or pituitary (Lofts and Murton, 1973).

Concentrations in Testes and Blood from Embryo to Puberty. Testosterone concentrations in the testes of the chicken embryo are low (3 pg/mg tissue at day 7; 14 pg/mg at day 20), increase by 1 day after hatch (to 59 pg/mg), then decrease thereafter up to 28 days posthatch. By contrast, progesterone tends to increase during the first 28 days after hatch (from 7 to 125 pg/mg testes), while estradiol-17β remains consistently less than 2 pg/mg testes (Tanabe et al., 1979). Comparable trends and concentrations for these steroids in the testes of the duck (*Anas platyrhynchos*) were reported by Tanabe et al. (1983).

Plasma concentrations of testosterone increase approximately twofold during the last days prior to hatching and then decrease slightly within the first 3 to 7 days after hatching (chicken, Tanabe et al., 1979; quail,

Ottinger and Bakst, 1981). Similarly, in male ducks, plasma testosterone is highest 2 days posthatch (6.8 ng/ml), then steadily decreases to reach 0.5 ng/ml by 35 days of age (Balthazart and Stevens, 1975). Plasma progesterone in the male chicken ranges from 1.5 to 4.0 ng/ml during embryonic development through 7 days posthatch, while estradiol remains at or near nondetectable levels (Tanabe et al., 1979).

In the immature cockerel, plasma concentrations of androstenedione at 9–10 weeks of age are approximately 7 times those of testosterone. There is a progressive decrease in testicular and plasma concentrations of androstenedione from 9 to 16 weeks of age, but plasma and testicular concentrations of testosterone do not change significantly. By 14 weeks of age, the androstenedione-to-testosterone ratio in the plasma decreases to approximately 2:1 (Culbert et al., 1977; Sharp et al., 1977).

Onset of Puberty. During the onset of puberty in the cockerel (from 16 to 24 weeks of age), there is a significant increase in plasma LH, followed within 1–2 weeks by an increase in plasma testosterone (from 2.3 ng/ml at 16 weeks to 9.5 ng/ml at 24 weeks) and androstenedione (from 3.5 to 6.7 ng/ml).

In immature Japanese quail (28–40 days of age) housed under a natural photoperiod, plasma testosterone increases beginning in March (as the photoperiod approaches 12 hr light) and reaches maximum levels (approximately 5 ng/ml) by June or July. Peak concentrations of testosterone following maximal testicular weight and maturity are reached by 1–2 months (Follett and Maung, 1978). By contrast, exposure of immature males to a long artificial photoperiod (20 L:4 D) stimulates rapid production and secretion of testosterone within 7 days after photostimulation (from 0.6 ng/ml plasma prior to photostimulation to 3.7 ng/ml by day 7) (Follett, 1976).

Adult. Circulating concentrations of testosterone for several species of adult male birds in various reproductive states are summarized in Table 19–1.

In contrast to the low plasma testosterone/androstenedione ratio in the immature cockerel, circulating concentrations of testosterone in the mature rooster are approximately twice those of androstenedione (Galli et al., 1973; Culbert et al., 1977). The change in this ratio during puberty may reflect a maturational change in the sensitivity of the testes to gonadotropins, such that LH stimulates the further metabolism of androstenedione to testosterone. Circulating metabolites of testosterone in the rooster include 5α-dihydrotestos-

Table 19–1. Maximum and minimum concentrations of plasma or serum testosterone in adult male birds in various reproductive states[a]

Species	Testosterone (ng/ml)	Remarks	Reference
Chicken	3.4–5.4	Range, 25–34 weeks of age	Culbert et al. (1977)
(*Gallus domesticus*)	7.3–9.3	Range, 37–44 weeks of age	
	7.0–11.3	Diurnal range	Schanbacher et al., (1974)
Japanese quail	5.0	Photosensitive	Follett and Maung (1978)
(*Coturnix coturnix*)	0.2	Photorefractory	
	0.3–0.7	Diurnal range	Gulati et al. (1981)
Starling	0.1	Incubation and feeding (June–August)	
(*Sturnis vulgaris*)	0.71	Laying (May)	Dawson (1983)
	2.93	Nest building (April)	
Duck	0.09[b]	Photorefractory (June–September)	Paulke and Haase (1978)
(*Anas platyrhynchos*)	2.15[b]	Breeding (April)	
	0.8	Molt (July)	Jallageas et al. (1978)
	5.4	Breeding (April)	
White-crowned sparrow	0.1	Molt (July–October)	Wingfield and Farner (1978)
(*Zonotrichia leucophrys gambelii*)	4.1	Courtship/copulation (May–June)	
Red-wing blackbird	0.5[b]	Photorefractory	Kerlan and Jaffee (1974)
(*Agelaius phoeniceus*)	2.7[b]	Photosensitive	
Ring dove	0.2	Isolation	Feder et al. (1977)
(*Streptopelia risoria*)	0.7	Courtship, day 3	
Willow ptarmigan	0.1	Molt (July–October)	Stokkan and Sharp (1980)
(*Lagopus lagopus lagopus*)	2.3	Breeding (May)	
Pigeon	0.59	Unpaired	Haase et al. (1976)
(*Columba* spp.)	1.24	Paired	
Rook	<0.2	Nonbreeding (May–January)	Lincoln et al. (1980)
(*Corvus frugilegus*)	3.4	Breeding (March)	

[a] Values estimated by radioimmunoassay.
[b] Report indicates that radioimmunoassay procedure may cross-react to variable extent with more than one androgen.

terone (5α-DHT), and the testosterone–sulfate and –glucurnonide conjugates. The concentrations of these metabolites, however, represent less than 20 and 10%, respectively, those of testosterone (Driot et al., 1978). Plasma concentrations of 5α-DHT in the duck, however, reach levels 60% higher than those of testosterone (Paulke and Haase, 1978), and in the ring dove appear to exceed plasma levels of testosterone during the period of early courtship (Feder et al., 1977).

During the annual reproductive cycle of the duck, plasma testosterone increases between February and March (with increasing photoperiod) and reaches peak values in April, concurrent with full development of the testes and just prior to the initiation of breeding (Paulke and Haase, 1978; Jallageas et al., 1978). A second autumnal peak of testosterone occurs in August or September, but is not accompanied by an increase in testes size; rather it is thought to be related to pair-bond formation (Paulke and Haase, 1978). From early June through July, plasma concentrations of testosterone rapidly decrease and the duck experiences a period of photorefractioriness and undergoes a postnuptial molt. The decrease in plasma testosterone at this time occurs simultaneously with a significant increase in the metabolic clearance rate of testosterone combined with a decrease in the secretion rate of testosterone. In addition, there is evidence from the drake and spotted munia (*Lonchura punctulata*) that the onset of photorefractoriness may be influenced by thyroid hormones (Jallageas and Assenmacher, 1974; Chandola and Bhatt, 1982).

In the white-crowned sparrow, highest annual concentrations of testosterone (May 18–June 22) occur coincident with greatest testes weight and the time of courtship, nest building, and copulation. Testes weight and plasma concentrations of testosterone rapidly decrease during the periods of incubation and the feeding of nestlings, and are lowest during the postnuptial molt (Wingfield and Farner, 1978)

Plasma concentrations of testosterone in the ring dove (*Streptopelia risoria*) are closely related to courtship behavior. Under a photoperiod of 14 L:10 D and prior to pairing with the female, levels approach 0.2 ng/ml. When males are placed in breeding cages with females, plasma testosterone increases (to 0.3 ng/ml) by 4 hr after pairing and attains a maximum of 0.7 ng/ml at 3 days after pairing. By 5 days after pairing, plasma testosterone begins to decrease, and reaches levels characteristic of isolated males by day 11 after pairing (Feder et al., 1977).

Blood levels of testosterone also have been demonstrated to exhibit a diurnal rhythm (chicken, Schanbacher et al., 1974; Japanese quail, Gulati et al., 1981) and pulsatile release (chicken, Wilson et al., 1979b; Japanese quail, Ottinger, 1983).

Circulating concentrations of progesterone in the adult male have been reported for a variety of species [chicken (Furr, 1973; Tanabe et al., 1979); pigeon (Haase et al., 1976); Japanese quail (Gulati et al., 1981); turkey (Mashaly and Wentworth, 1974); ring dove (Silver et al., 1974); mallard duck (Donham, 1979); and white-crowned sparrow (McCreery and Farner, 1979)]. In general, however, there is no correlation between blood concentrations and reproductive status, and in the white-crowned sparrow the main source of progesterone is thought to be of extratesticular origin (McCreery and Farner, 1979). Progesterone receptors have been identified in the hyperstriatum and hypothalamus of the male ring dove, and their function would appear to be related to the facilitation of incubation behavior (Balthazart et al., 1980).

Plasma concentrations of estrogen in the adult male chicken (Tanabe et al., 1979) and ring dove (Korenbrot et al., 1974) remain at or near nondetectable levels irrespective of the stage of reproductive activity.

Metabolism in Neural Tissues and Accessory Reproductive Organs. Androgens are actively taken up and concentrated in specific regions in the brain of the avian male (Barfield et al., 1978; Arnold et al., 1976; Zigmond et al., 1973). The negative feedback of androgens on hypothalamic releasing hormones and/or pituitary gonadotropins has been discussed previously (see Chapter 17). However, steroids secreted by the testes are thought not to act directly on neuroendocrine tissues, but are instead enzymatically transformed into more potent metabolites within the target tissue. Androgen metabolites [e.g., 5α-DHT, 5β-dihydrotestosterone (5β-DHT), androstenedione, and 5β-androstane-$3\alpha,17\beta$-diol] and androgen-metabolizing enzymes (e.g., 5α-reductase and aromatase) have been identified within neural tissues of male birds (for review, see Lofts and Massa, 1980; Balthazart, 1983).

Both seasonal (Bottani and Massa, 1980) and maturational (Massa and Sharp, 1980; Balthazart and Schumacher, 1984) changes in the pattern of testosterone metabolism in neuroendocrine tissue have been described. In the quail, the production of 5β-DHT (a biologically "inactive" androgen) and the activity of the enzyme 5β-reductase is greatest in neuroendocrine tissue of young animals, and decreases during sexual maturation. It has been suggested that 5β reduction is a testosterone-inactivation pathway (Steimer and Hutchison, 1981) and that the decrease of 5β-reductase activity in neuroendocrine tissues of the quail with age corresponds to a potentiation of the effects of testosterone and its biologically active metabolites (Balthazart and Schumacher, 1984).

The metabolism of androgens in the cloacal gland of the Japanese quail (see Figure 19–9) is essential to maintaining the integrity of this gland (Massa et al., 1980), and is also dependent on stage of maturation (Balthazart and Schumacher, 1984), photoperiod (Balthazart et al., 1979), sex (Balthazart et al., 1983), and

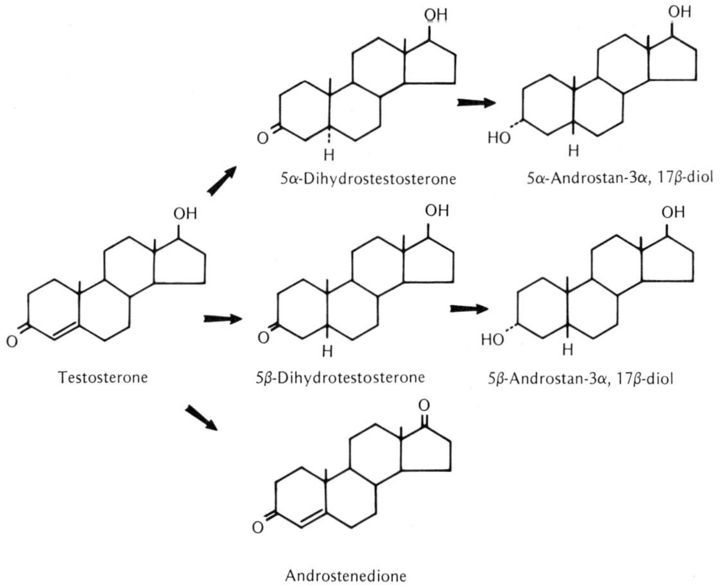

FIGURE 19–9. Steroid metabolism in the cloacal gland of the Japanese quail. (Redrawn from Lofts and Massa, 1980.)

reproductive status (Massa et al., 1979). Finally, androgen metabolism has been described in the epididymis by Tanabe and Nakamura (1974) and in the comb by Mori et al. (1974). For further details, see Lofts and Massa (1980).

Gonadotropin Binding in Testicular Tissues

Specific binding of follicle-stimulating hormone (FSH) to testicular tissue has been reported for the white-crowned sparrow (Ishii and Farner, 1976), Japanese quail (Ishii and Adachi, 1977; Tsutsui and Ishii, 1980), and domestic fowl (Ishii and Adachi, 1977). Total binding capacity of FSH to testes of sparrows increases during photoperiodic stimulation (Ishii and Farner, 1976) at a time concurrent with gonadal growth and the differentiation and hypertrophy of the Sertoli cells. The number of FSH-binding sites appears to be regulated by FSH acting synergistically with testosterone (Tsutsui and Ishii, 1980).

Specific LH receptors in avian testicular tissue have yet to be unequivocally demonstrated. Nevertheless, chicken LH stimulates interstitial cell activity and androgen synthesis, and indirect evidence suggests that the action of LH is mediated by specific binding of LH to the Leydig and interstitial cells (Maung and Follett, 1977).

Secondary Sexual Characteristics and Behavior

Secondary sexual characteristics that may differentiate the male from female include size of comb, plumage and bill color, structure of feathers, vocalizations, and behavior. For the most part, these aspects are regulated by the androgens.

Comb Growth. The effect of androgens on comb growth is well established and has been used as the basis of a relatively sensitive bioassay for the androgens (Munson and Sheps, 1958). Testosterone, 5α-DHT, and androstenedione are equally active in inducing comb development (Nakamura and Tanabe, 1973; Young and Rogers, 1978), and all have been shown in vitro to be actively synthesized from testosterone (Mori et al., 1974).

Plumage and Bill Color. Plumage may or may not exhibit sexual dimorphism, depending on species and breed. The hackle and saddle feathers of the male chicken are elongated and tapered, while those of the female are shorter and more blunt. Castrated males and females (poulards) develop feathers resembling those of the rooster, except that the feathers are longer (neutral-type plumage).

In many species in which plumage is influenced by gonadal steroids, the male plumage is more brilliantly colored than the female; the difference in male and female plumage color is dependent on estrogens or androgens (Sturkie, 1965). In species such as the chicken, duck, or red-wing blackbird in which the male displays dominant characteristics, testosterone concentrations in the testes are greater than those in the ovary. However, in Wilson's phalarope (*Steganopus tricolor*), in which the female has the dominant plumage, ovarian concentrations of testosterone exceed those in the testes (Höhn and Cheng, 1967).

The deposition of pigmentation in the bill is an androgen-dependent process, and can vary relative to season and reproductive status. Castration in the starling results in the loss of a yellow pigment (caratenoid), while injection of androgens restores coloration (Witschi, 1961). By contrast, the bill of the male house sparrow (*Passer domesticus*) incorporates the black pigment, melanin, during the breeding season, and loses this pigmentation during the winter and subsequent to castration. In this species, there is some controversy as to whether testosterone alone, or a combination of testosterone plus gonadotropin (FSH or FSH plus LH), is required to restore pigmentation during the winter months (Lofts et al., 1973; Haase, 1975).

Vocalizations. Early work by Breneman (1939) showed that daily treatment of 1-day-old chicks with testosterone proprionate or 5α-androstan-3α,17β-diol caused chicks to begin crowing as early as 7 days of age. More recently, Adkins et al. (1980) found that several natural and synthetic androgens, but not estradiol benzoate, stimulated crowing in Japanese quail with regressed testes. Exogenous testosterone induces singing in female canaries but not in the female zebra finch (*Poephila guttata*) (Nottebohm and Arnold, 1976).

Behavior. Behavior is probably programmed by the animal's genetic constitution and modified by the environment and hormonal status. Exposure to hormones before or immediately after hatch permanently affects the nervous system (see Adkins, 1978), whereas exposure in mature birds generally elicits specific behavior patterns (e.g., reproductive behavior) (see Balthazart, 1983).

Courtship behavior has been most extensively researched in the ring dove. Sexual displays by the male include strutting, aggressive hop-charging, bow-coos, wing-flipping, and nest-cooing. Behavioral displays in the male are closely synchronized with those of the female, and are paralleled by endogenous changes in reproductive hormones (Feder et al., 1977; Silver et al., 1974). In the male, the frequency of hop-charging, bow-cooing, wing-flipping, and nest-cooing is eliminated or reduced after castration, and is restored after treatment with testosterone proprionate. Bow-cooing has been found to be strictly testosterone dependent, whereas wing-flipping and nest-cooing can be elicited by testosterone and estrogen (Adkins-Regan, 1981). The effect of testosterone on behavior is mediated via the preoptic area of the hypothalamus either directly or after its conversion to estrogen (Steimer and Hutchinson, 1981). For an excellent review of courtship behavior in the ring dove, see Cheng (1979).

Copulatory behavior (approaching, mounting, and treading) can be induced in 1- to 16-day-old chicks by treatment with testosterone or 5β-DHT (Balthazart et al., 1981). Furthermore, testosterone implants placed within the preoptic area of the hypothalamus initiate copulatory behavior in the male chick (Gardner and Fisher, 1968) and in adult roosters and capons (Barfield, 1969), whereas lesions within the preoptic region prevent the initiation of such behavior (Meyer, 1974). The conversion of androgens to estrogens is required to initiate copulatory behavior in the Japanese quail (Adkins et al., 1980).

In a variety of avian species (e.g., phalarope, killdeer, sandpiper), the male is predominantly or solely responsible for nest building, incubation of eggs, and care of young. In the phalarope, the incubation patch can be induced by injection of testosterone plus prolactin. Incubation behavior in the male ring dove can be elicited by injections of progesterone or by implants of progesterone placed within the preoptic area (see van Tienhoven, 1983). For further information concerning incubation behavior, see Drent (1975).

It has been well established that behavioral interactions can influence plasma hormonal levels. Feder et al. (1977) found that male ring doves exposed to a female showed dramatic increases in plasma testosterone and 5α-DHT for as long as 3 days after pairing, whereas no such increase was found after pairing with another male. Finally, male red-wing blackbirds caught during the height of an aggressive encounter were found to have significantly different plasma concentrations of LH, testosterone, and DHT as compared to nonaggressive (foraging) males (Harding and Follett, 1979).

References

Adkins, E.K. (1978). Sex steroids and the differentiation of avian reproductive behavior. Am. Zool., 18, 501.

Adkins, E.K., J.J. Boop, D.L. Koutnik, J.B. Morris, and E.E. Pniewski. (1980). Further evidence that androgen aromatization is essential for the activation of copulation in male quail. Physiol. Behav., 24, 441.

Adkins-Regan, E.K. (1981). Hormone specificity, androgen metabolism, and social behavior. Am. Zool., 21, 257.

Aire, T.A. (1973). Development of puberty in Nigerian and White Leghorn cockerels. Poult. Sci., 52, 1765.

Arnold, A.P., F. Nottebohm, and D.W. Pfaff. (1976). Hormone concentrating cells in vocal control and other areas of the brain of the zebra finch (*Poephila guttata*). J. Comp. Neurol., 165, 487.

Bakst, M.R., and B. Howarth. (1975). The head, neck and midpiece of cock spermatozoa examined with the transmission electron microscope. Biol. Reprod., 12, 632.

Bakst, M.R., and T.J. Sexton. (1979). Fertilizing capacity and ultrastructure of fowl and turkey spermatozoa before and after freezing. J. Reprod. Fertil., 55, 1.

Balthazart, J. (1983). Hormonal correlates of behavior. In "Avian Biology" Vol. 7 (D.S. Farner, J.R. King, and K.C. Parkes, Eds.). London and New York: Academic Press, Chapter 4.

Balthazart, J., and M. Schumacher. (1984). Changes in testosterone metabolism by the brain and cloacal gland during sexual maturation in the Japanese quail (*Coturnix coturnix japonica*). J. Endocrinol., 100, 13.

Balthazart, J., and M. Stevens. (1975). Plasma testosterone levels in very young domestic ducklings. IRCS Med. Sci., 3, 345.

Balthazart, J., R. Massa, and P. Negri-Cesi. (1979). Photoperiodic control of testosterone metabolism, plasma gonadotropins, cloacal gland growth, and reproductive behavior in the Japanese quail. Gen. Comp. Endocrinol., 39, 222.

Balthazart, J., J.D. Blaustein, M.-F. Cheng, and H.H. Feder. (1980). Hormones modulate the concentration of cytoplasmic protestin receptors in the brain of male ring doves (*Streptopelia risoria*). J. Endocrinol., 86, 251.

Balthazart, J., G. Malacarne, and P. Deviche. (1981). Stimulatory effects of 5β-dihydrotestosterone on the sexual behavior in the domestic chick. Horm. Behav., 15, 246.

Balthazart, J., M. Schumacher, and M.A. Ottinger. (1983). Sexual differences in the Japanese quail: Behavior, morphology, and intracellular metabolism of testosterone. Gen. Comp. Endocrinol., 51, 191.

Barfield, R.J. (1969). Activation of copulatory behavior by androgen implanted into the preoptic area of the male fowl. Horm. Behav., 1, 37.

Barfield, R.J., G. Ronay, and D.W. Pfaff. (1978). Autoradiographic localization of androgen-concentrating cells in the brain of the male domestic fowl. Neuroendocrinology, 26, 297.

Bennett, T., and T. Malmfors. (1970). The adrenergic nervous system of the domestic fowl [*Gallus domesticus* (L.)] Z. Zellforsch. Mikrosk. Anat., 106, 22.

Benoit, J. (1964). The role of the eye and of the hypothalamus in the photostimulation of gonads in the duck. Ann. N.Y. Acad. Sci., 117, 204.

Bernon, D.E., and P.B. Siegel. (1981). Fertility of chickens from lines divergently selected for mating frequency. Poult. Sci., 60, 45.

Bischoff, M.B. (1969). Photoreceptoral and secretory structures in the avian pineal organ. J. Ultrastruct. Res., 28, 16.

Blivaiss, B.B. (1947). Interrelations of thyroid and gonad in the development of plumage and other sex characters in Brown Leghorn roosters. Physiol. Zool., 20, 67.

Blohowiak, C.C., P.B. Siegel, and H.P. Van Krey. (1980). Sexual behavior of dwarf and normal genotypes in divergent growth lines of chickens. Appl. Anim. Ethol., 6, 189.

Bottani, L., and R. Massa. (1980). Seasonal changes in testosterone metabolism in the pituitary gland and central nervous system of the European Starling (*Sturnis vulgaris*). Gen. Comp. Endocrinol., 43, 532.

Brantas, G.C., H.G. Dennert, and A.L. Dennert-Distelbrink. (1972). The influence of the number of cocks on the conception rate among White Leghorns. Arch. Gefleugelkd., 36, 16.

Breneman, W.R. (1939). Effect of androgens on the chick. Proc. World's Poult. Congr. 7th, 91.

Breucker, H. (1978). Macrophages, a normal component in seasonally involuting testes of the swan, *Cygnus olor.* Cell Tissue Res., 193, 463.

Brown, K.I. (1974). Effects of sperm number on onset and duration of fertility in turkeys. Research Summary 80, Ohio Agricultral Research Development Center, Wooster.

Brown, N.L., J.-D. Bayle, C.G. Scanes, and B.K. Follett. (1975). Chicken gonadotrophins: Their effects on the testes of immature and hypophysectomized Japanese quail. Cell Tissue Res., 156, 499.

Budras, K.-D., and T. Sauer. (1975). Morphology of the epididymis of the cock (*Gallus domesticus*) and its effect upon the steroid sex hormone synthesis I. Ontogenesis, morphology and distribution of the epididymis. Anat. Embryol., 148, 175.

Burrows, W.H., and J.P. Quinn. (1937). The collection of spermatozoa from the domestic fowl and turkey. Poult. Sci., 16, 19.

Chandola, A., and D. Bhatt. (1982). Tri-iodothyronine fails to mimic gonado-inhibitory action of thyroxine in spotted munia: Effects of injections at different times of the day. Gen. Comp. Endocrinol., 48, 499.

Cheng, M.-F. (1979). Progress and prospects in ring dove research: A personal view. Adv. Study Behav., 9, 97.

Christensen, V.L., and N.P. Johnston. (1977). Effect of time of day of insemination and the position of the egg in the oviduct on the fertility of turkeys. Poult. Sci., 56, 458.

Clermont, Y. (1958). Structure de l'epithelium seminal et mode de renouvellement des spermatogonies chez le canard. Arch. Anat. Microsc. Morphol. Exp., 47, 47.

Cogburn, L.A., and P.C. Harrison. (1977). Retardation of sexual development in pinealectomized single comb white Leghorn cockerels. Poult. Sci., 56, 876.

Cooksey, E.J., and B. Rothwell. (1973). The ultrastructure of the Sertoli cell and its differentiation in the domestic fowl (*Gallus domesticus*). J. Anat., 114, 329.

Culbert, J., P.J. Sharp, and J.W. Wells. (1977). Concentrations of androstenedione, testosterone and LH in the blood before and after the onset of spermatogenesis in the cockerel. J. Reprod. Fertil., 51, 153.

Davies, D.T., R. Massa, and R. James. (1980). Role of testosterone and of its metabolites in regulating gonadotrophin secretion in the Japanese quail. J. Endocrinol., 84, 211.

Dawson, A. (1983). Plasma gonadal steroid levels in wild starlings (*Sturnis vulgaris*) during the annual cycle and in relation to the stages of breeding. Gen. Comp. Endocrinol., 49, 286.

de Reviers, M. (1981). Influence of night-interrupted photoschedules on testicular development in cockerels. In "Photoperiodism and Reproduction" (R. Ortavant, J. Pelletier, and J.-P. Ravault, Eds.). Nouzilly, France: INRA Publishing, p. 19.

Dharmarajan, M. (1950). Effect on the embryo of staleness of the sperm at the time of fertilization in the domestic hen. Nature (London), 165, 398.

Donham, R.S. (1979). Annual cycle of plasma luteinizing hormone and sex hormones in male and female Mallards (*Anas platyrhynchos*). Biol. Reprod., 21, 1273.

Drent, R. (1975). Incubation. In "Avian Biology," Vol. 5 (D.S. Farner and J.R. King, Eds.). London and New York: Academic Press, p. 333.

Driot, F.J.M., D.H. Garnier, and M. Terqui. (1978). Development and validation of a "direct" radioimmunoassay for plasma testosterone in the fowl (*Gallus domesticus*). Gen. Comp. Endocrinol., 36, 244.

Driot, F.J.M., M. de Reviers, and J. Williams. (1979). Plasma testosterone levels in intact and hemicastrated growing cockerels. J. Endocrinol., 81, 169.

Farner, D.S., and J.C. Wingfield. (1978). Environmental endocrinology and the control of annual reproductive cycles in passerine birds. In "Environmental Endocrinology" (I. Assenmacher and D.S. Farner, Eds.). New York: Springer, p. 44.

Farner, D.S., M.L. Morton, and B.K. Follett. (1970). The limitation of rate of photoperiodically induced testicular growth in the white-crowned sparrow *Zonotrichia leucophrys gambelii.* The effect of hemicastration. Arch. Anat., 51, 189.

Farner, D.S., R.S. Donham, R.A. Lewis, P.W. Mattocks, T.R. Darden, and J.P. Smith. (1977). The circadian component in the photoperiodic mechanism of the house sparrow, *Passer domesticus.* Physiol. Zool., 50, 247.

Feder, H.H., A. Storey, D. Goodwin, C. Reboulleau, and

R. Silver. (1977). Testosterone and "5α-dihydrotestosterone" levels in peripheral plasma of male and female ring doves (*Streptopelia risoria*) during the reproductive cycle. Biol. Reprod., 16, 666.
Follett, B.K. (1976). Plasma follicle-stimulating hormone during photoperiodically induced sexual maturation in male Japanese quail. J. Endocrinol., 69, 117.
Follett, B.K., and P.J. Sharp. (1969). Circadian rhythmicity in photoperiodically induced gonadotrophin release and gonadal growth in the quail. Nature (London), 223, 968.
Follett, B.K., and S.L. Maung. (1978). Rate of testicular maturation, in relation to gonadotrophin and testosterone levels, in quail exposed to various artificial photoperiods and to natural daylengths. J. Endocrinol., 78, 267.
Follett, B.K., D.T. Davies, and B. Gledhill. (1977). Photoperiodic control of reproduction in Japanese quail: Changes in gonadotrophin secretion on the first day of induction and their pharmacological blockade. J. Endocrinol., 74, 449.
Foss, D.C., L.B. Carew, and E.L. Arnold. (1972). Physiological development of cockerels as influenced by selected wavelengths of environmental light. Poult. Sci., 51, 1922.
Fujihara, N., and H. Nishiyama. (1976). Studies on the accessory reproductive organs in the drake. 4. Effects of androgen on the ejaculatory groove region of the drake. Poult. Sci., 55, 1324.
Fujihara, N., H. Nishiyama, and N. Nakashima. (1976). Studies on the accessory reproductive organs in the drake. 2. Macroscopic and microscopic observations on the cloaca of the drake with special reference to the ejaculatory groove region. Poult. Sci., 55, 927.
Furr, B.J.A. (1973). Radioimmunoassay of progesterone in peripheral plasma of the domestic fowl in various physiological states and in follicular venous plasma. Acta Endocrinol., 72, 89.
Galli, F.E., O. Irusta, and G.F. Wassermann. (1973). Androgen production by testes of *Gallus domesticus* during postembryonic development. Gen. Comp. Endocrinol., 21, 262.
Gardner, J.E., and A.E. Fisher. (1968). Induction of mating in male chicks following preoptic implantation of androgen. Physiol. Behav., 3, 709.
Garnier, D.H., A. Tixier-Vidal, D. Gourdji, and R. Picart. (1973). Ultrastructure des cellules de Sertoli au cours du cycle testiculaire du Canard Pékin. Z. Zellforsch. Mikrosk. Anat., 144, 369.
Gibson, W.R., B.K. Follett, and B. Gledhill. (1975). Plasma levels of luteinizing hormone in gonadectomized Japanese quail exposed to short or to long daylengths. J. Endocrinol., 64, 87.
Giesen, A.F., G.R. McDaniel, and T.J. Sexton. (1980). Effect of time of day of artificial insemination and oviposition–insemination interval on the fertility of broiler breeder hens. Poult. Sci., 59, 2544.
Grieve, B.J., R. Wachob, A. Peltz, and A. van Tienhoven. (1973). Synchronous hatching of Japanese quail and maturity of the chicks hatched prematurely. Poult. Sci., 52, 1445.
Grosse, A.E., and J.V. Craig. (1960). Sexual maturity of males representing twelve strains of six breeds of chickens. Poult. Sci., 39, 164.
Guhl, A.M. (1951). Measurable differences in mating behavior of cocks. Poult. Sci., 30, 687.
Guhl, A.M., and D.C. Warren. (1946). Number of offspring sired by cockerels related to social dominance in chickens. Poult. Sci., 25, 460.
Guhl, A.M., N.E. Collias, and W.C. Allee. (1945). Mating behavior and the social hierarchy in small flocks of White Leghorns. Physiol. Zool., 18, 365.
Gulati, D.P., T. Nakamura, and Y. Tanabe. (1981). Diurnal variations in plasma LH, progesterone, testosterone, estradiol and estrone in the Japanese quail. Poult. Sci., 60, 668.
Gwinner, E. (1975). Circadian and circannual rhythms in birds. In "Avian Biology," Vol. 5 (D.S. Farner and J.R. King, Eds.). New York: Academic Press, p. 221.
Haase, E. (1975). The effects of testosterone proprionate on secondary sexual characters and testes of house sparrows, *Passer domesticus*. Gen. Comp. Endocrinol., 26, 248.
Haase, E., E. Paulke, and P.J. Sharp. (1976). Effects of seasonal and social factors on testicular activity and hormone levels in domestic pigeons. J. Exp. Zool., 197, 81.
Hale, E.B. (1955). Duration of fertility and hatchability following natural matings in turkeys. Poult. Sci., 34, 228.
Harding, C.F., and B.K. Follett. (1979). Hormone changes triggered by aggression in a natural population of blackbirds. Science, 203, 918.
Harris, G.C., and K. Goto. (1984). Carbonic anhydrase activity of the reproductive tract tissues of aged male fowls and its relationship to semen production. J. Reprod. Fertil., 70, 25.
Hess, R.A., R.J. Thurston, and H.V. Biellier. (1982). Morphology of the epididymal region of turkeys producing abnormal yellow semen. Poult. Sci., 61, 531.
Höhn, E.O., and S.C. Cheng. (1967). Gonadal hormones in Wilson's phalarope (*Steganopus tricolor*) and other birds in relation to plumage and sex behavior. Gen. Comp. Endocrinol., 8, 1.
Howarth, B. (1970). An examination for sperm capacitation in the fowl. Biol. Reprod., 3, 338.
Howarth, B. (1981). The phospholipid profile of cock spermatozoa before and after *in vitro* incubation for twenty-four hours at 41°C. Poult. Sci., 60, 1516.
Howarth, B., and M.B. Palmer. (1972). An examination of the need for sperm capacitation in the turkey, *Meleagris gallopavo*. J. Reprod. Fertil., 28, 443.
Ingkasuwan, P., and F.X. Ogasawara. (1966). The effect of light and temperature and their interaction on the semen production of White leghorn males. Poult. Sci., 45, 1199.
Ishii, S., and T. Adachi. (1977). Binding of avian testicular homogenate with rat follicle-stimulating hormone and inhibition of the binding by hypophyseal extracts of lower vertebrates. Gen. Comp. Endocrinol., 31, 287.
Ishii, S., and D.S. Farner. (1976). Binding of follicle-stimulating hormone by homogenates of testes of photostimulated white-crowned sparrows, *Zonotrichia leucophrys gambelli*. Gen. Comp. Endocrinol., 30, 443.
Jallageas, M., and I. Assenmacher. (1974). Thyroid gonadal interactions in the male domestic duck in relationship with the sexual cycle. Gen. Comp. Endocrinol., 22, 13.
Jallageas, M., A. Tamisier, and I. Assenmacher. (1978). A comparative study of the annual cycles in sexual and thyroid function in male Peking ducks (*Anas platyrhynchos*) and Teal (*Anas crecca*). Gen. Comp. Endocrinol., 36, 201.
Johnson, O.W. (1966). Quantitative features of spermatogenesis in the mallard duck (*Anas platyrhynchos*). Auk, 83, 233.
Johnson, A.L., and V.T. Rendano. (1984). Effects of castration, with and without testesterone replacement, on leg bone integrity in the domestic fowl. Am. J. Vet. Res., 45, 319.
Jones, P., and H. Jackson. (1972). Estimation of the duration of spermatogenesis in Japanese quail, using antispermatogonial chemicals. J. Reprod. Fertil., 31, 319.
Jones, J.E., H.R. Wilson, R.H. Harms, C.F. Simpson, and P.W. Waldroup. (1967). Reproductive performance in male chickens fed protein deficient diets during the growing period. Poult. Sci., 46, 1569.
Kamar, G.A.R. (1962). Semen characteristics of various breeds of drakes in the subtropics. J. Reprod. Fertil., 3, 405.

Kamar, G.A.R., and E.S.E. Hafez. (1975). Sperm maturation and fertility in poultry. Anim. Breed. Abstr., 43, 99.
Kerlan, J.T., and R.B. Jaffe. (1974). Plasma testosterone levels during the testicular cycle of the redwinged blackbird (*Agelaius phoeniceus*). Gen. Comp. Endocrinol., 22, 428.
Kerlan, J.T., R.B. Jaffee, and A.H. Payne. (1974). Sex-steroid formation in gonadal tissue homogenates during the testicular cycle of the redwinged blackbird (*Agelaius phoeniceus*). Gen. Comp. Endocrinol., 24, 352.
King, A.S. (1981). Phallus. In "Form and Function in Birds," Vol. 2 (A.S. King and J. McLelland, Eds.). London and New York: Academic Press, Chapter 3.
Klemm, R.D., C.E. Knight, and S. Stein. (1973). Gross and microscopic morphology of the glandula proctodealis (foam gland) of *Coturnix c. japonica* (Aves). J. Morphol., 141, 171.
Korenbrot, C.C., D.W. Schomberg, and C.J. Erickson. (1974). Radioimmunoassay of plasma estradiol during the breeding cycle of ring doves (*Streptopelia risoria*). Endocrinology, 94, 1126.
Krueger, K.K., J.A. Owen, C.E. Krueger, and T.M. Ferguson. (1977). Effect of feed or light restriction during the growing and breeding cycles on the reproductive performance of broad-breasted White turkey males. Poult. Sci., 56, 1566.
Kuenzel, W.J., and P.J. Sharp. (1982). Neural surgical effects on precocious puberty in the chick and changes in luteinizing hormone and testosterone. Poult. Sci., 61, 1496.
Kumaran, J.D.S., and C.W. Turner. (1949a). The normal development of the testes in the White Plymoth Rock. Poult. Sci., 28, 511.
Kumaran, J.D.S., and C.W. Turner. (1949b). The endocrinology of spermatogenesis in birds. II: The effect of androgens. Poult. Sci., 28, 739.
Lake, P.E. (1957). The male reproductive tract of the fowl. J. Anat., 91, 116.
Lake, P.E. (1971). The male in reproduction. In "Physiology and Biochemistry of the Domestic Fowl," Vol. III (D.J. Bell and B.M. Freeman, Eds.). London and New York: Academic Press, Chapter 60.
Lake, P.E. (1981). Male genital organs. In "Form and Function in Birds," Vol. 2 (A.S. King and J. McLelland, Eds.). London and New York: Academic Press, Chapter 1.
Lake, P.E. (1983). Factors affecting the fertility level in poultry, with special reference to artificial insemination. World's Poultry Sci. J., 39, 106.
Lake, P.E., and D.G.M. Wood-Gush. (1956). Diurnal rhythms in semen yields and mating behaviour in the domestic cock. Nature (London), 178, 853.
Lake, P.E., W. Smith, and D. Young. (1968). The ultrastructure of the ejaculated fowl spermatozoon. Q.J. Exp. Physiol., 53, 356.
Lam, F., and D.S., Farner. (1976). The ultrastructure of the cells of Leydig in the white-crowned sparrow (*Zonotrichia leucophrys gambelii*) in relation to plasma levels of luteinizing hormone and testosterone. Cell Tissue Res., 169, 93.
Langford, B.B., and B. Howarth. (1974). A trypsin-like enzyme in acrosomal extracts of chicken, turkey and quail spermatozoa. Poult. Sci., 53, 834.
Lincoln, G.A., P.A. Racey, P.J. Sharp, and H. Klandorf. (1980). Endocrine changes associated with spring and autumn sexuality of the rook, *Corvus frugilegus*. J. Zool., 190, 137.
Lofts, B., and R. Massa. (1980). Male Reproduction. In "Avian Endocrinology" (A. Epple and M.H. Stetson, Eds.). London and New York: Academic Press, p. 413.
Lofts, B., and R.K. Murton. (1973). Reproduction in birds. In "Avian Biology," Vol. 3 (D.S. Farner, J.R. King, and K.C. Parkes, Eds.). London and New York: Academic Press, Chapter 1.
Lofts, B., R.K. Murton, and R.J.P. Thearle. (1973). The effects of testosterone proprionate and gonadotropins on the bill pigmentation and testes of the house sparrow (*Passer domesticus*). Gen. Comp. Endocrinol., 21, 202.
Lorenz, F.W., and F.X. Ogasawara. (1968). Distribution of spermatozoa in the oviduct and fertility in domestic birds. VI. The relations of fertility and embryo normality with site of experimental insemination. J. Reprod. Fertil., 16, 445.
Marini, P.J., and B.L. Goodman. (1969). Semen characteristics as influenced by selection for divergent growth rate in chickens. Poult. Sci., 48, 859.
Marquez, B.J., and F.Y. Ogasawara. (1975). Scanning electron microscope studies of turkey semen. Poult. Sci., 54, 1139.
Marshall, A.J. (1961). Reproduction. In "Biology and Comparative Physiology of Birds," Vol. II. (A.J. Marshall, Ed.). New York: Academic Press, Chapter 18.
Mashaly, M.M., and B.C. Wentworth. (1974). A profile of progesterone in turkey sera. Poult. Sci., 53, 2030.
Maslieu, I.T., and A.D. Davtyan. (1969). Influence of nutrition on the semen production of roosters. World's Poult. Sci. J., 25, 315.
Mass, J.H., and W.J. Kuenzel. (1983). Precocious development of the testes in chicks following parasagittal knife cuts of the lateral hypothalamic area. Dev. Brain Res., 10, 165.
Massa, R., and P.J. Sharp. (1980). Conversion of testosterone to 5β-reduced metabolites in the neuroendocrine tissues of the maturing cockerel. J. Endocrinol., 88, 263.
Massa, R., D.T. Davies, L. Bottoni, and L. Martini. (1979). Photoperiodic control of testosterone metabolism in the central and peripheral structures of avian species. J. Steroid Biochem., 11, 937.
Massa, R., D.T. Davies, and L. Bottoni. (1980). Cloacal gland of the Japanese quail: Androgen dependence and metabolism of testosterone. J. Endocrinol., 84, 223.
Maung, Z.W., and B.K. Follett. (1977). Effects of chicken and ovine luteinizing hormone on androgen release and cyclic AMP production by isolated cells from the quail testes. Gen. Comp. Endocrinol., 33, 242.
McCartney, M.G., and K.I. Brown. (1959). Spermatozoa concentration in three varieties of turkeys. Poult. Sci., 38, 390.
McCreery, B.R., and D.S. Farner. (1979). Progesterone in male white-crowned sparrows, *Zonotrichia leucophrys gambelii.* Gen. Comp. Endocrinol., 37, 1.
Menaker, M., and H. Underwood. (1976). Extraretinal photoreception in birds. Photochem. Photobiol., 23, 299.
Menge, H., and L.T. Frobish. (1976). Dietary restrictions during adolescence and subsequent reproductive performance of turkey breeder males. Poult. Sci., 55, 1724.
Meyer, C.C. (1974). Effects of lesions in the medial preoptic region on precocial copulation in the male chick. Horm. Behav., 5, 377.
Mitchell, H.H., L.E. Card, and W.T. Haines. (1927). The effect of age, sex, and castration on the basal heat production of chickens. J. Agric. Res., 34, 945.
Mori, M., K. Suzuki, and B.-I. Tamaoki. (1974). Testosterone metabolism in rooster comb. Biochem. Biophys. Acta, 337, 118.
Munro, S.S. (1938). Functional changes in fowl sperm during their passage through the excurrent ducts of the male. J. Exp. Zool., 79, 71.
Munson, P.L., and M.C. Sheps. (1958). An improved procedure for the biological assay of androgens by direct application to the combs of baby chicks. Endocrinology, 62, 173.
Murton, R.K., and N.J. Westwood. (1977). "Avian Breeding Cycles." Oxford: Clarendon Press.
Nakamura, T., and Y. Tanabe. (1972). *In vitro* steroidogen-

esis by testes of the chicken (*Gallus domesticus*). Gen. Comp. Endocrinol., 19, 432.

Nakamura, T., and Y. Tanabe. (1973). Dihydrotestosterone formation *in vitro* in the epididymis of the domestic fowl. J. Endocrinol., 59, 651.

Nalbandov, A.V., and L.E. Card. (1943). Effect of stale sperm on fertility and hatchability of chicken eggs. Poult. Sci., 22, 218.

Nicholls, T.J., and G.P. Graham. (1972). Observations on the ultrastructure and differentiation of Leydig cells in the testes of Japanese quail (*Coturnix coturnix japonica*). Biol. Reprod., 6, 179.

Nishida, T. (1964). Comparative and topographical anatomy of the fowl. XLII. Blood vascular system of the male reproductive organs. Jpn. J. Vet. Sci., 26, 211.

Nishiyama, H. (1955). Studies on the accessory reproductive organs in the cock. J. Fac. Agric. Kyushu Univ., 10, 277.

Nishiyama, H., N. Nakashima, and N. Fujihara. (1976). Studies on the accessory reproductive organs in the drake. 1. Addition to semen of the fluid from the ejaculatory groove region. Poult. Sci., 55, 234.

Nottebohm, F., and A.P. Arnold. (1976). Sexual dimorphism in vocal control areas of the songbird brain. Science, 194, 211.

Oishi, T., and J.K. Lauber. (1973). Photoreception in the photosexual response of quail. II. Effects of intensity and wavelength. Am. J. Physiol., 225, 880.

Okamura, F., and H. Nishiyama. (1978a). The passage of spermatozoa through the vitelline membrane in the domestic fowl. *Gallus gallus*. Cell Tissue Res., 188, 497.

Okamura, F., and H. Nishiyama. (1978b). Penetration of spermatozoon into the ovum and transformation of the sperm nucleus into the male pronucleus in the domestic fowl, *Gallus gallus*. Cell Tissue Res., 190, 89.

Oliver, J., M. Jallageas, and J.D. Baylé. (1979). Plasma testosterone and LH levels in male quail bearing hypothalamic lesions or radioluminous implants. Neuroendocrinology, 28, 114.

Olsen, M.W. (1942). Maturation, fertilization, and early cleavage in the hen's egg. J. Morphol., 70, 513.

Olsen, M.W., and B.H. Neher. (1948). The site of fertilization in the domestic fowl. J. Exp. Zool., 109, 355.

Osman, D.I. (1980). The connection between the seminiferous tubules and the rete testis in the domestic fowl (*Gallus domesticus*). Morphological study. Int. J. Androl., 3, 177.

Osman, D.I. (1981). Morphological signs of merocrine secretion in the modified Sertoli cells of the domestic fowl (*Gallus domesticus*). J. Reprod. Fertil., 61, 75.

Ottinger, M.A. (1983). Short-term variation in serum luteinizing hormone and testosterone in the male Japanese quail. Poult. Sci., 62, 908.

Ottinger, M.A., and H.J. Brinkley. (1979). Testosterone and sex-related physical characteristics during the maturation of the male Japanese quail (*Coturnix coturnix japonica*). Biol. Reprod., 20, 905.

Ottinger, M.A., and M.R. Bakst. (1981). Peripheral androgen concentrations and testicular morphology in embryonic and young male Japanese quail. Gen. Comp. Endocrinol., 43, 170.

Ottinger, M.A., and J.A. Doerr. (1982). Effects of mycotoxins on avian reproduction. In "Aspects of Avian Endocrinology: Practical and Theoretical Implications" (C.G. Scanes, M.A. Ottinger, A.D. Kenny, J. Balthazart, J. Cronshaw, and I. Chester Jones, Eds.). Lubbock: Texas Tech Press, p. 217.

Parker, J.E. (1939). An avian semen collector. Poult. Sci., 18, 455.

Parker, J.E., F.F. McKenzie, and H.L. Kempster. (1940). Observations on the sexual behavior of New Hampshire males. Poult. Sci., 19, 191.

Paulke, E., and E. Haase. (1978). A comparison of seasonal changes in the concentrations of androgens in the peripheral blood of wild and domestic ducks. Gen. Comp. Endocrinol., 34, 381.

Ralph, C.L. (1981). The pineal and reproduction in birds. In "The Pineal Gland" Vol. II (R.J. Reiter, Ed.). Boca Raton: CRC Press, p. 31.

Rashedi, M., R. Maraud, and R. Stoll. (1983). Development of the testes in female domestic fowls submitted to an experimental sex reversal during embryonic life. Biol. Reprod., 29, 1221.

Rothwell, B. (1973). The ultrastructure of Leydig cells in the testis of the domestic fowl. J. Anat., 116, 245.

Rothwell, B., and M.D. Tingari. (1973). The ultrastructure of the boundary tissue of the seminiferous tubule in the testis of the domestic fowl (*Gallus domesticus*). J. Anat., 114, 321.

Saxena, R.N., L. Malhotra, R. Kant, and P.K. Baweja. (1979). Effect of pinealectomy and seasonal changes in pineal antigonadotrophic activity of male Indian weaver bird, *Ploceus phillipinus*. Indian J. Exp. Biol., 17, 732.

Schanbacher, B.D., W.R. Gomes, and N.L. VanDemark. (1974). Diurnal rhythm in serum testosterone levels and thymidine uptake by testes in the domestic fowl. J. Anim. Sci., 38, 1245.

Sexton, T.J. (1977). Relationship between number of sperm inseminated and fertility of turkey hens at various stages of production. Poult. Sci., 56, 1054.

Sexton, T.J. (1979). Preservation of poultry semen—A review. In "Beltsville Symposium #3. Animal Reproduction" (H. Hawk, Ed.). Montclair: Allanheld, Osmum, Chapter 12.

Shaklee, W.E., and C.W. Knox. (1954). Hybridization of the pheasant and fowl. J. Hered., 45, 183.

Sharp, P.J., and R. Moss. (1977). The effects of castration on concentrations of luteinizing hormone in the plasma of photorefractory red grouse (*Lagopus lagopus scoticus*). Gen. Comp. Endocrinol., 32, 289.

Sharp, P.J., and C.B. Gow. (1983). Neuroendocrine control of reproduction in the cockerel. Poult. Sci., 62, 1671.

Sharp, P.J., J. Culbert, and J.W. Wells. (1977). Variations in stored and plasma concentrations of androgens and luteinizing hormone during sexual development in the cockerel. J. Endocrinol., 74, 467.

Siegel, P.B. (1965). Genetics of behavior: Selection for mating ability in chickens. Genetics, 52, 1269.

Silver, R., C. Reboulleau, D.S. Lehrman, and H.H. Feder. (1974). Radioimmunoassay of plasma progesterone during the reproductive cycle of male and female ring doves (*Streptopelia risoria*). Endocrinology, 94, 1547.

Siopes, T.D., and W.O. Wilson. (1974). Extraocular modification of photoreception in intact and pinealectomized coturnix. Poult. Sci., 53, 2035.

Siopes, T.D., and W.O. Wilson. (1980). Participation of the eyes in the photosexual response of Japanese quail (*Coturnix coturnix japonica*). Biol. Reprod., 23, 352.

Steimer, T., and J.B. Hutchinson. (1981). Metabolic control of the behavioural action of androgens in the dove brain: testosterone inactivation by 5β-reduction. Brain Res., 209, 189.

Stokkan, K.-A., and P.J. Sharp. (1980). Seasonal changes in the concentrations of plasma luteinizing hormone and testosterone in willow ptarmigan (*Lagopus lagopus lagopus*) with observations on the effects of permanent short days. Gen. Comp. Endocrinol., 40, 109.

Sturkie, P.D. (1965). Reproduction in the male, fertilization, and early embryonic development. In "Avian Physiology"

(2d ed.) (P.D. Sturkie, Ed.). Ithaca: Cornell University Press, Chapter 16.

Sturkie, P.D., and H. Opel. (1976). Reproduction in the male, fertilization, and early embryonic development. In "Avian Physiology" (3d ed.) (P.D. Sturkie, Ed.). New York: Springer-Verlag, Chapter 17.

Takeda, A. (1969). Labelling of cock spermatozoa with radioactive phosphorus. Jpn. J. Zootech. Sci., 40, 412.

Tanabe, Y., and T. Nakamura. (1974). Androgen metabolism in the epididymis of the male domestic fowl. Indian Poult. Rev., 6, 67.

Tanabe, Y., T. Nakamura, K. Fujioka, and O. Doi. (1979). Production and secretion of sex steroid hormones by the testes, the ovary, and the adrenal glands of embryonic and young chickens (*Gallus domesticus*). Gen. Comp. Endocrinol., 39, 26.

Tanabe, Y., T. Yano, and T. Nakamura. (1983). Steroid hormone synthesis and secretion by testes, ovary, and adrenals of embryonic and postembryonic ducks. Gen. Comp. Endocrinol., 49, 144.

Tanaka, S., and M. Yasuda. (1980). Histological changes in the testis of the domestic fowl after adenohypophysectomy. Poult. Sci., 59, 1538.

Tanaka, S., and T. Fujioka. (1981). Histological changes in the testis of the domestic fowl after partial adenohypophysectomy. Poult. Sci., 60, 444.

Thurston, R.J., R.A. Hess, B.L. Hughes, and D.P. Froman. (1982). Ultrastructure of the guinea fowl (*Numidia meleagris*) spermatozoon. Poult. Sci., 61, 1738.

Tingari, M.D. (1971). On the structure of the epididymal region and ductus deferens of the domestic fowl (*Gallus domesticus*). J. Anat., 109, 423.

Tingari, M.D. (1973a). Observations on the fine structure of spermatozoa in the testis and excurrent ducts of the male fowl, *Gallus domesticus*. J. Reprod. Fertil., 34, 255.

Tingari, M.D. (1973b). Histochemical localization of 3β- and 17β-hydroxysteroid dehydrogenases in the male reproductive tract of the domestic fowl (*Gallus domesticus*). Histochem. J., 5, 57.

Tingari, M.D., and P.E. Lake. (1972). Ultrastructural evidence for resorption of spermatozoa and testicular fluid in the excurrent ducts of the testes of the domestic fowl, *Gallus domesticus*. J. Reprod. Fertil., 31, 373.

Tsutsui, K., and S. Ishii. (1980). Hormonal regulation of follicle-stimulating hormone receptors in the testes of Japanese quail. J. Endocrinol., 85, 511.

Van Krey, H.P., A.T. Leighton, and L.M. Potter. (1967). Sperm gland populations and late seasonal declines in fertility. Poult. Sci., 46, 1332.

van Tienhoven, A. (1983). "Reproductive Physiology of Vertebrates" (2d ed.). Ithaca and London: Cornell University Press.

Vince, M., J. Green, and S. Chin. (1970). Acceleration of hatching in the domestic fowl. Br. Poult. Sci., 11, 483.

Wilson, S.C. (1978). LH secretion in the cockerel and the effects of castration and testosterone injections. Gen. Comp. Endocrinol., 35, 481.

Wilson, F.E., and B.K. Follett. (1974). Plasma and pituitary luteinizing hormone in intact and castrated tree sparrows (*Spizella arborea*) during a photoinduced gonadal cycle. Gen. Comp. Endocrinol., 23, 82.

Wilson, F.E., and B.K. Follett. (1978). Dissimilar effects of hemicastration on plasma LH and FSH in photostimulated tree sparrows (*Spizella arborea*). Gen. Comp. Endocrinol., 34, 251.

Wilson, S.C., and M. de Reviers. (1979). Concentrations of luteinizing hormone in the plasma of hemicastrated cockerels. J. Endocrinol., 83, 379.

Wilson, H.R., N.P. Piesco, E.R. Miller, and W.G. Nesbeth. (1979a). Prediction of the fertility potential of broiler breeder males. World's Poult. Sci. J., 35, 95.

Wilson, E.K., J.C. Rogler, and R.E. Erb. (1979b). Effect of sexual experience, location, malnutrition, and repeated sampling on concentrations of testosterone in blood plasma of *Gallus domesticus* roosters. Poult. Sci., 58, 178.

Wingfield, J.C., and D.S. Farner. (1978). The annual cycle of plasma irLH and steroid hormones in feral populations of the white-crowned sparrow, *Zonotrichia leucophrys gambelli*. Biol. Reprod., 19, 1046.

Wingfield, J.C., B.K. Follett, K.S. Matt, and D.S. Farner. (1980). Effect of day length on plasma FSH and LH in castrated and intact white-crowned sparrows. Gen. Comp. Endocrinol., 42, 464.

Witschi, E. (1961). Sex and secondary sexual characters. In "Biology and Comparative Physiology of Birds," Vol. 2 (A.J. Marshall, Ed.). New York: Academic Press, Chapter 17, p. 115.

Wood-Gush, D.G. M. (1971). "The Behaviour of the Domestic Fowl." London: Heinemann.

Woods, J.E., and L.V. Domm. (1966). A histochemical identification of the androgen-producing cells in the gonads of the domestic fowl and albino rat. Gen. Comp. Endocrinol., 7, 559.

Yokoyama, K., A. Okshe, T.R. Darden, and D.S. Farner. (1978). The sites of encephalic photoreception in photoperiodic induction of the growth of the testes in the white-crowned sparrow, *Zonotrichia leucophrys gambelii*. Cell Tissue Res., 189, 441.

Young, C.E., and L.J. Rogers. (1978). Effects of steroidal hormones on sexual, attack and search behavior in the isolated male chick. Horm. Behav., 10, 107.

Zigmond, R.E., F. Nottebohm, and D.W. Pfaff. (1973). Androgen-concentrating cells in the midbrain of a songbird. Science, 179, 1005.

20
Thyroids

B.C. Wentworth and R.K. Ringer

Anatomy

Location; Blood and Nerve Supply

The thyroid glands in avian species are paired organs, oval in shape and dark red in color, with a glistening appearance. They are located on either side of the trachea on the ventral-lateral aspect of the neck just exterior to the thoracic cavity (Figure 20–1) and can be found adhering to the common carotid artery just above the junction of the common carotid with the subclavian artery. They are situated medial to the jugular vein.

The blood supply to the thyroid is by the cranial and caudal thyroid arteries, which originate from a branch directly off the common carotid artery. The venous return circulation is through veins emptying into the jugular vein. There is very little information available about the innervation of the thyroid in birds: it may stem from the cervical sympathetic ganglion, as in mammals. There is abundant parasympathetic neural innervation of the blood vessels as well as many smaller nerves ending at the capsule connective tissues and follicular walls. In man, the nerves accompany the blood vessels in the thyroid, and most end in the perivascular plexus; some end on the follicular cells.

Embryology and Histology

The thyroid appears on the second day of incubation of the chick as a midline outgrowth (median anlage) from the ventral pharyngeal wall at the level of the first and second branchial pouches. The median anlage becomes cuplike and bilobed, with a narrow stalk at-

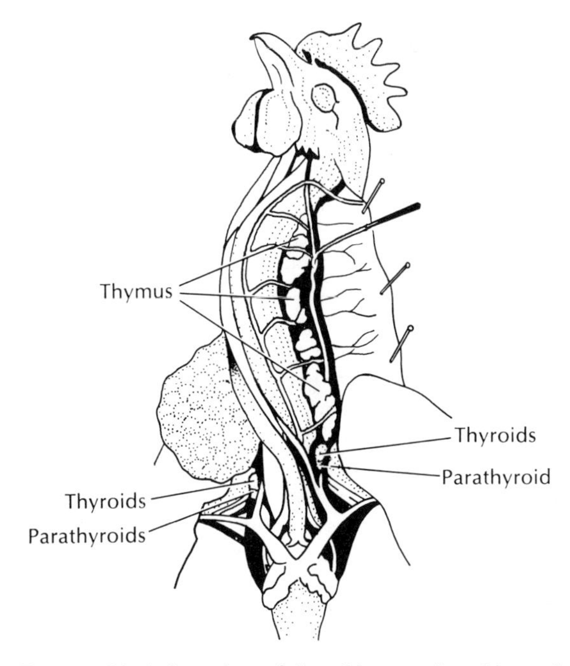

FIGURE 20–1. Location of thyroids, parathyroids, and thymus in the chicken. (After Nonidez and Goodale, 1927.)

tached to the pharyngeal wall. The stalk ruptures, and by the fifth day of development the lobes separate and migrate laterally as crescent-shaped structures, and the glands ultimately develop their definitive shape and position (see Romanoff, 1960). Although only a few avian species have been considered, the organelle ultrastructure concerned with hormone secretion is very similar to that described in mammals.

The cells within the thyroid are arranged in chordlike rows. Colloid secretion commences and displaces the epithelium peripherally, forming spherical vesicles known as follicles. At 7 days of incubation the thyroid of the embryonic chick can concentrate radioactive iodine (^{131}I) many times greater than the amount present in the blood, yet no follicles or colloid are visible. On the ninth day of incubation, droplets of colloid are visible and radioiodine uptake is protein bound. On the 13th day, the thyroid abruptly increases radioiodine accumulation on injection of ^{131}I (Waterman, 1959). From the incorporation of radioiodine in the follicular colloid, Hansborough and Khan (1951) and Stoll and Blanquet (1953) concluded that the embryonic chick thyroid becomes functional and secretes thyroxine (T_4) after 10–11 days of incubation. On the 11th day, the thyroid gland is stimulated by thyrotropin (Mess and Straznicky, 1970). Just prior to hatching, the activity of the thyroid increases (Rogler et al., 1959; Spiers, 1980; Christensen and Biellier, 1982).

Some radioiodide administered to hens concentrates in the ovary (yolk) as T_4 and triiodothyronine (T_3) (Roche et al., 1956); however, Okonski et al. (1960), Robinson (1973), and Robinson et al. (1977) reported that only a small fraction of the ^{131}I is protein bound; the largest segment is in the form of iodide. Exogenous gonadotropin will stimulate the uptake of iodide by the ovary (Newcomer, 1982). The yolk iodide is utilized by the developing embryo. At the start of incubation, the major portion of radioiodide is concentrated in the yolk, but by the ninth day of incubation, when vascularization has taken place, the first iodide appears concentrated in the embryo thyroid. Concentration continues until at the 19th day, 40% of the ^{131}I is in the thyroid, 40% is elsewhere in the embryo, and 20% remains in the egg (Stoll and Blanquet, 1953; see also Wollman and Zwilling, 1953; Okonski et al., 1960).

The embryonic chick adenohypophyses is sensitive to TRH as early as day 6.5, and the thyroid is sensitive to TSH at the same time, as evidenced by an increase in plasma T_4 (Thommes and Hylka, 1978). Thyroxine concentration increases during embryo development, while T_3 concentration remains low. Maximal serum T_3 and T_4 levels are always obtained on the day of pipping. High serum T_3 levels are found in the chick and quail embryo after perforation of the air-space membrane (McNabb et al., 1981; Christensen et al., 1982). Following pipping there is a rapid increase in heat production (Decuypere et al., 1979). After hatching, both hormones decrease until the adult concentration is reached.

The thyroid gland can develop both morphologically and functionally in chorioallantoic grafts or in vitro; the thyroid gland of the embryo therefore possesses an intrinsic ability to develop autonomously. Both in vitro and in vivo studies of the thyroid show the same schedule of ^{131}I concentration (Waterman, 1959).

Histologically, the thyroid is encapsulated by reticular connective tissue. The follicles are large (30–40 μm), but more heterogenous in adults as compared to small (12–15 μm) follicles of uniform size found in young chicks and embryos. The follicles are composed of entodermal epithelium of varying height, depending on the state of activity (secretory rate). The epithelial type may vary from squamous to columnar. Depending on the secretory state, the follicles may be filled with or completely devoid of colloid, which is a homogeneous fluid of protein gel composed of an iodinated protein, thyroglobulin (TG). Between the follicles are connective tissue stroma, interfollicular cells, and a rich blood supply (Thommes, 1958). The avian thyroid is unique by the lack of calcitonin cells, for the calcitonin cells are located separate from the thyroid gland in the ultimobranchial gland. Stoeckle and Porte (1969) pointed out that doves and pigeons appear to be exceptions, and are similar to the rat, with calcitonin cells found within the follicular epithelium.

Size

The size of the thyroid is influenced by several variables, such as age, sex, climatic conditions, diet, activity, species, and hypophysectomy. In proportion to the total body weight, the thyroid in domestic fowl does not vary markedly with increasing age, but its absolute weight increases with age. At 1–2 weeks of age, the thyroid weight of the chicken varies from 6.3 to 10.7 mg/100 g body weight (Snedecor, 1968; van Tienhoven et al., 1966) and at 10–22 weeks, from 9.0 to 11.7 mg/100 g body weight (Rosenberg et al., 1964; Kanematsu and Mikami, 1970). Growth rate of the thyroid is almost constant from hatching until about 50–60 days after hatching. The rate of growth is then accelerated until between 100 and 120 days, at which time a pronounced slowing occurs (Breneman, 1954). Female chickens have a greater thyroid weight than males.

The influence of diet on thyroid size is widely known (see Ringer, 1976). Because iodine is essential in the biosynthesis of the thyroid hormones, a deficiency of iodine produces goiter or enlargement of the thyroid. Almeida and Thomas (1981) concluded that changes in feeding patterns of Japanese quail produce prompt changes in serum T_3 and T_4 that are not mediated via the hypothalamopituitary axis.

Early investigators reported that the thyroid weight of chickens and pigeons is greater in the winter than in the summer. Low environmental temperature increases thyroid activity (see section on control of thyroid activity). Histologic changes indicating follicular stimulation within the thyroid coincide with the onset of cold weather (Hohn, 1949). However, two studies on ducks (Rosenberg et al., 1967; Astier et al., 1970) conducted in the Mediterranean region showed greater thyroid weight and activity in the summer (June) than in the cold months. Astier et al. (1970) reported thyroid weights for the adult duck in March and October as 5.82 and 9.28 mg/100 g body weight, respectively.

An increase in the size of the thyroid may be caused by an increase in cell size (hypertrophy) and/or an increase in the number of cells (hyperplasia), associated with a reduction in colloid content of the follicles. Enlarged thyroids may reflect either a hyper- or hypofunctioning gland. Thyroid-stimulating hormone (TSH) stimulates the thyroid and produces both hypertrophy and hyperplasia, together with accelerated formation or secretion of T_4. Thiouracil, thiourea, and methimazole (goitrogens) produce the same histologic picture but suppress the formation of T_4. The latter occurs through inhibition of thyroid hormone synthesis by interfering with iodide concentration within the thyroid (thiocyanate) or through inhibition of iodination of TG (thiourea and thiouracil). A minimal concentration of thyroid hormone (0.5–1.0 μg/day) is required for maximum response to goitrogens (March and Poon, 1981). Hypophysectomy causes a decrease in thyroid size, because production of TSH is abolished.

Thyroid Hormones

Synthesis

Synthesis of thyroid hormones in avian species is similar to that which occurs in mammals, where iodide is concentrated within the thyroid, the so-called iodide trap, by the maintenance of a gradient over that of blood. The iodide is converted to I_2 and then to I^+, which is the iodinating substance. The use of radioiodine has demonstrated that the accumulation of iodine in the thyroid of birds is very rapid with a peak by 6 hr, followed by a prolonged storage for several days. Newcomer (1978a) reported the half-life of ^{131}I in the thyroid increased directly, and the fractional turnover rate inversely, with increased dietary iodine. The avian thyroid exhibits great plasticity in adjusting functionally to a wide range of dietary iodine content. Chickens fed a diet similar in iodine content to that fed to mammals had an iodide trapping rate two to four times that of mammals (Rosenberg et al., 1964; Newcomer, 1978a). Birds have a very long retention of iodide, especially under conditions of low iodine feed or fasting. There are about 3.4 mol of tyrosine/100 mol of total amino acid in TG.

The thyroglobulin of birds is highly iodinated, representing 1.5% TG by weight, whereas the mammal generally has only 0.5% iodine by weight (Hoshino and Ui, 1970; Sorimachi and Ui, 1974). One molecule of TG may contain 50–90 atoms of iodine (Daugeras et al., 1976a, b). Monoiodotyrosine and diiodotyrosine, shown chromatographically to be present in chicken thyroids (Vlijm, 1958; Frey and Flock, 1958), are thus formed (Figure 20–2). A peroxidase system within the thyroid converts the iodide to iodine and a second enzyme system is responsible for combining the iodinated tyrosines within the polypeptide chain of TG to form T_3 and T_4. Triiodothyronine (T_3) and T_4 have been shown chromatographically to be in thyroid tissue from chickens (Lague and van Tienhoven, 1969; Grandhi and Brown, 1975), ducks (Astier et al., 1966), and pigeons (Bayle et al., 1966). There is still controversial evidence regarding the presence of T_3 in the thyroid gland. There is no question but that it is found in a lesser amount and that it is also formed by extrathyroidal conversion of T_4 to T_3 (Astier and Newcomer, 1978). Peripheral conversion of T_4 to T_3 through monodeiodination has been demonstrated in the chicken (Sharp and Klandorf, 1985). Within the thyroid, therefore, TG contains monoiodotyrosine, diiodotyrosine, T_3, and T_4. Rosenberg et al., (1964), Newcomer (1978a, b), and Newcomer (1979) indicated that intrathyroidal iodination and deiodina-

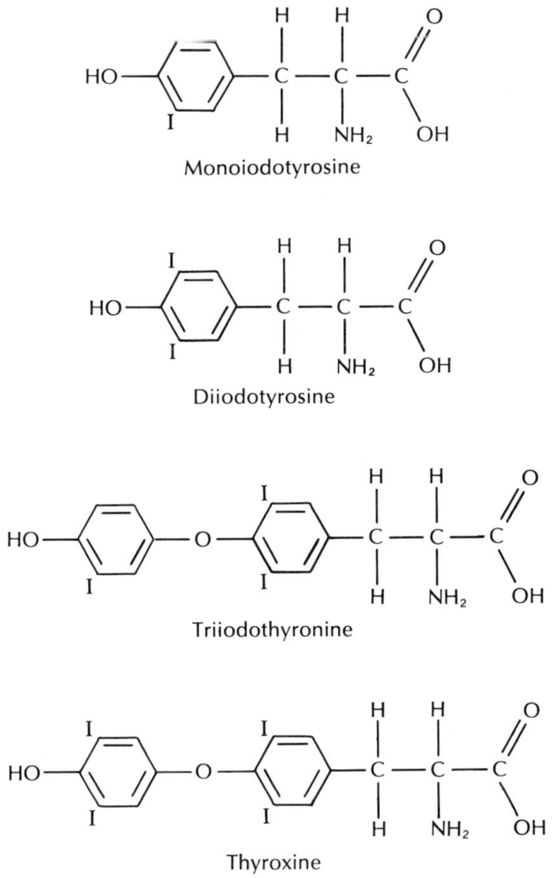

Monoiodotyrosine

Diiodotyrosine

Triiodothyronine

Thyroxine

FIGURE 20–2. Iodinated compounds of the thyroid.

tion reactions occur continually and lead to randomization of thyroidal iodine, both of iodotyrosines and iodothyronines. Thus, within the gland, iodine shifts between the tyrosine and thyronine in a random fashion. The tyrosines must be converted to an iodinated thyronine for release from the thyroid into the bloodstream.

The Circulating Hormones

Both T_3 and T_4 have been isolated radiochromatographically from the plasma of chickens, ducks, and turkeys (Wentworth and Mellen, 1961). These investigators reported a ratio of 60% T_4 to 40% T_3; however, Sadovsky and Bensadoun (1971) reported a diurnal rhythm in plasma iodohormones and a T_3/T_4 ratio varying from 1.33 to 2.12.

The thyroid hormones are released from the thyroid as the predominant amino acid T_4. Once in the blood, they are again bound to protein. In man, thyroxine-binding globulin is the primary carrier of T_4 and, to a lesser extent, T_3. A second binding protein is prealbumin, whereas serum albumin plays a minor role as a carrier. The absence of thyroxine-binding globulin in avian species has been confirmed by many researchers. Both T_3 and T_4 are bound to serum albumin, and the binding affinity of albumin for T_3 and T_4 is the same (Tata and Shellabarger, 1959). Heninger and Newcomer (1964) noted that T_4 in chicken plasma is bound to the protein-binding sites with greater affinity than is T_3. Farer et al. (1962) found chickens, turkeys, and pigeons had thyroxine-binding prealbumins as well as albumins. In the pigeon and chicken, T_4 binds to albumin and a prealbumin zone; T_3 binds to prealbumin in the pigeon but not in the chicken or in any other vertebrate studied (Refetoff et al., 1970). In adult quail, T_4 is bound to albumin and secondarily to prealbumin, whereas T_3 is bound to α-globulin and secondarily to albumin and γ-globulin (McNabb and Hughes, 1983). Binding of T_4 to avian albumin is evidently weak compared to that in man; labeled T_4 binds rapidly with the protein in human blood when they are mixed together with avian blood. The loose binding to serum proteins results in more free T_4 in avian blood than in human (1.4–3.7 ng/100 ml) or most warm-blooded vertebrate's blood (Refetoff et al., 1970). Further confusion is added by the finding of Castay et al. (1978) that β-lipoproteins bind thyroid hormones. Using a competitive binding assay, Refetoff et al. (1970) reported plasma levels of total T_4 for the chicken as 1.4–1.6 μg/100 ml with 5.5 ng/100 ml of free T_4. For the pigeon, these values were 2.4–3.3 μg/100 ml for total T_4 and 6.0–6.9 ng/100 ml as free T_4. Etta (1971), using a competitive protein-binding assay for total T_4, reported values in chickens of 1.00 (layers), 1.75 (nonlayers), 2.10 (males), and, for turkeys, 1.39 (layers), 1.18 (nonlayers), and 1.51 μg T_4 /100 ml serum (males). Free thyroid hormones may be an important index of thyroid status (Refetoff et al., 1970; McNabb and Hughes, 1983). Measurable concentrations of T_3 and T_4 are found in circulation following surgical thyroidectomy in birds. These levels do not increase following thyrotropin-releasing hormone injection whereas in intact birds they are enhanced, indicating extrathyroidal origin (Harvey et al., 1982).

Half-lives ($t_{1/2}$) of T_3 and T_4 in birds are almost identical (Tata and Shellabarger, 1959; Heninger and Newcomer, 1964; Singh et al., 1967). The capacity and affinity of avian plasma for T_3 are less than for T_4; therefore, factors other than, or in addition to, protein binding must account for their similar disappearance rates from the circulation (Heninger and Newcomer, 1964). The rate of degradation of thyroid hormones is very rapid in birds, as indicated by relatively short biologic half-lives, measured in hours (see Table 20–1). Care must be taken in assessing $t_{1/2}$ for thyroid hormones in birds, because values read too soon ($<$2 hr) after injection of the radioactive hormone do not allow for complete mixing, and those measured

TABLE 20–1. Biological half-lives of [^{131}I]thyroxine in plasma

Species	Sampling time after [^{131}I]thyroxine injection (hr)	Half-life (hr)	References
Chicken	1–24	8.3	Heninger and Newcomer (1964)
	3–12	3.3	Singh et al. (1967)
Duck	2–18	4.0	Assenmacher et al. (1968)
Japanese quail	3–12	5.4	Singh et al. (1967)
Bobwhite quail	3–12	4.6	Singh et al. (1967)

24 hr or more after reflect recycling of metabolized radioiodine (Etta et al., 1972).

In the mammal, T_3 is at least six times more potent than an equimolar amount of T_4 in preventing goiter. Triiodothyronine in chicks is no more potent than T_4 in preventing goiter (Shellabarger, 1955). The potencies are also equal in stimulating body weight, comb growth, and liver glycogen of propylthiouracil-fed and radiothyroidectomized chicks (Raheja and Snedecor, 1970); in influencing oxygen consumption, heart rate, oxygen deprivation time, and feather growth in thiouracil-treated chicks (Newcomer, 1957); or in supressing the effect of tapazole, as indicated by thyroid secretion rate (Singh et al., 1968a). However, T_4 is more potent than T_3 in preventing goiter in thiouracil-treated chickens (Newcomer, 1957; Mellen and Wentworth, 1959) and in increasing oxygen consumption in cardiac muscle (Newcomer and Barrett, 1960) or metabolic rate (Singh et al., 1968a). Triiodothyronine is more active than T_4 in stimulating intracellular accumulation of certain neutral amino acids by embryonic chick bones (Adamson and Ingbar, 1967). There is evidence in chickens that T_3 rather than T_4 is the metabolically active thyroid hormone (Klandorf et al., 1981). Peripheral monodeiodination of T_4 may also result in the formation of 3,3′,5′-triiodothyronine, called reverse T_3 (rT_3).

The principal route of excretion of T_3 and T_4 is via bile and urine (Hutchins and Newcomer, 1966). Conjugated thyronines, acetic acid derivatives of thyronines, and partially deiodinated thyronines are in the bile, whereas iodide is the fundamental metabolite in urine (Tanabe et al., 1965).

Circadian Rhythm of Thyroid Hormones

The short half-life of T_3 and T_4 in birds could lead to a measurable diurnal variation in thyroid function, whereas in mammals the long half-lives of thyroid hormones make demonstration of diurnal variation difficult. Plasma hormonal iodine levels in chickens show a diurnal rhythm. Total iodohormone curves show two peaks of plasma concentration at 0800 and 1600 hr (Sadovsky and Bensadoun, 1971). At 1600 hr, the concentration is two times that at 0400. Newcomer and Huang (1974), Klandorf et al. (1978a), and Klandorf et al. (1981) reported that in chickens the concentration of plasma T_4 increased during the dark period and decreased during the light period, whereas the concentration of plasma T_3 increased during the day and decreased at night. According to Kuhn et al. (1982), the acrophase of circadian concentration of T_3 in Rhode Island Red chicks is observed at about 1700 hr or 11 hr after onset of light with a 12 L:12 D photoperiod. They also found that the acrophase of the circadian concentration of T_4 is between 0500 and 0700 hr. The pattern of food intake undoubtedly influences the daily rhythm in plasma T_4 and T_3 concentrations (Sharp and Klandorf, 1985). Energy intake appears as the main factor regulating plasma T_3 concentrations and, presumably, the conversion of T_4 to T_3 (Sharp and Klandorf, 1985).

Assays

Protein-Bound Iodine. The iodine-containing hormones are precipitated with the serum proteins from blood. The quantity of protein-bound iodine (PBI) may be chemically analyzed, or, if ^{131}I is injected prior to the blood sampling, detection can be made of the labeled protein ([^{131}I]PB).

Representative plasma protein-bound iodine values for chickens, bobwhite quail, and Japanese quail were measured as 1.12, 1.76 and 1.26 μg/100 ml, respectively (Singh et al., 1967); for the duck, 1.17–1.49 μg/100 ml (Mellen and Hardy, 1957); and for the pigeon, 1.4–1.8 μg/100 ml (Refetoff et al., 1970). These values are considerably lower than those for mammals, and may reflect the lack of a specific thyroxine-binding α-globulin in avian blood.

The lack of sensitivity of the PBI test in birds negates its use as an assay of circulating levels of thyroid hormones in the blood. Should any nonhormonal iodoproteins be present they are also precipitated. Avian blood, unlike that of mammals, contains a considerable amount of nonhormonal iodoproteins (Assenmacher, 1973).

By use of a cation-exchange resin to measure hormonal iodine, Sadovsky and Bensadoun (1971) reported values for male chicken plasma ranging from

4.8 to 11.9 μg/100 ml. These values are seven times those reported for PBI by precipitation. These authors suggested that a large fraction of the iodohormones does not coprecipitate with albumin and is removed from the precipitate during washing. By this method, the hen has significantly more iodohormone than does the male.

Competitive Protein Binding and Radioimmunoassay. These sensitive methods are used to measure directly the amount of T_4 or T_3 in plasma or serum. The principle of the method is that thyroid hormones from plasma competes with tracer [^{125}I]thyroxine for binding sites in a given quantity of thyroxine-binding protein or antibody. With the assumption that both have equal affinity for the available sites, levels of "cold" endogenous T_4 compete with labeled T_4 for the thyroxine-binding globulin sites or antibody. When equilibrium is reached, the bound and free components are separated by the use of a precipitating reagent such as "protein A" or a second antibody. This renders the two components separable (Etta, 1971). The same procedure is followed in the T_3 assay by using an antibody specific for T_3. Some reported T_3 and T_4 concentrations for different species under different conditions are given in Table 20–2.

Thyroid Hormone Secretion Rate (TSR)

Progress in the accurate determination of individual thyroid secretion rates has come with the use of isotope tracer techniques. These methods afford more accurate assessment of thyroid functioning in birds under various physiologic conditions than do measurement of protein-bound iodine (PBI) or the goiter prevention method of TSR (for details, see Ringer, 1965). PBI or [^{131}I]PB do not constitute a senstive assay technique and, in addition, measure nonhormonal iodoproteins

TABLE 20–2 Concentration of T_3 and T_4 in circulation of several wild and domestic species with reference to life cycle and physiologic status

Species	Life chronology	T_3 (ng/ml)	T_4 (ng/ml)	Reference
Chick	Embryo, day 10		0.75	Daugeras-Bernard et al. (1976a)
Chick	Embryo, day 10.5	1.40	3.36	Thommes and Hylka (1978)
Chick	Posthatch, 0–2 weeks		2.18–1.86	Davison (1976)
Chick	Posthatch, day 21.5	4.99	10.78	Thommes and Hylka (1978)
Chick	1–4 weeks	4.82	15.4	Bobek et al. (1977)
Cockerel	6 weeks	1.7–2.6	14.8–31.4	Newcomer (1974)
Cockerel	40 days	7–17	60–120	Kuhn and Nouwen (1978)
Chicken	60 days	3.42–3.80	14.4–18.4	May (1978)
Chicken	10 weeks	2.39	8.58	Klandorf et al. (1978b)
Leghorn	Adult	1.37	12.3	Klandorf et al. (1978a)
	Induced molt	3.00	10.0	Brake et al. (1979)
Duck	Adult	0.55	7.30	Astier and Newcomer (1978)
Turkey	Poult at hatching	3.90	57.0	Christensen et al. (1982)
Japanese quail	Adult	1.19	12.5	Kamis and Robinson (1978)
Bobwhite quail	1 day	7.5–13	15.0–37.0	Spiers (1980)
	21 days	2.75	5.0	
	64 days	5.50	5.0	
	Adult	3.00	5.0–18.0	
Canadian goose	Adult (premature)	3.30	11.75	John and George (1978)
	Fall (postmigration)	1.93	17.37	
	Spring premigration	0.71	13.27	
	Spring postmigration	2.09	11.56	
	Molting	1.72	15.44	
Red-headed bunting	Premigration	3.30	1.05	Pathak and Chandola (1982)
	Breeding	0.98	2.50	
	Molting	0.85	2.80	
Rook	Young	0.67	1.53	Peczely and Pethes (1982)
	Sexually inactive	0.67	1.49	
	Increasing follicles	0.45	4.35	
	Large follicles	0.48	1.35	
	Preovulation	0.63	2.97	
	Hatching	0.36	1.80	
	Feeding young	0.40	2.05	
	Postnuptial molt	0.45	5.93	

in blood of birds (Astier, 1973). The goiter prevention technique has been questioned because the use of a goitrogen such as thiouracil or methimazole decreases peripheral deiodination of T_4 and gives faulty high TSR values. In a reappraisal, Almeida and Thomas (1980) suggested that the goiter prevention assay is as effective in measuring secretion rate as radiometric assays. Singh et al. (1968b) compared four methods of measuring thyroid secretion rate in chickens: (1) goiter prevention, (2) thyroid hormone substitution, (3) direct output, and (4) T_4 degradation. Representative TSR values for chicks to 9 weeks of age by the four methods, expressed as micrograms of T_4/100 g body weight daily, were 2.32, 2.00, 1.20, and 2.03, respectively. These same workers (Singh et al., 1967) reported the TSR of adult bobwhite and Japanese quail as 2.49 and 2.78 μg T_4/100 g body weight daily, respectively, by the T_4 degradation method.

Earlier workers did not take into account the adequacy of the iodine content of the diet when measuring TSR. Marked lowering of the TSR resulted when chickens are fed an iodine-deficient diet (Singh et al., 1968a). There is no consistent correlation between T_3 or T_4 and adequate dietary iodine.

TSR decreases with advancing age. Tanabe (1965) reported values by the goiter prevention method for cockerels 2–3 weeks, 5–7 weeks, and 13–14 weeks of age as 1.97, 1.57, and 0.63 μg/100 g body weight per day, respectively. RIA values also show lower T_3 and T_4 concentrations in adults than in juvenile birds.

Control of Thyroid Function

Pituitary–Hypothalamus

Within certain limits, the function of the thyroid is governed by the concentration of the circulating thyroid hormones and their effects on the hypothalmic-controlled pituitary release of TSH. A decrease in the amount of circulating thyroid hormones to a level below metabolic requirements prompts the neuroendocrine-controlled anterior pituitary to increase the release of TSH. An exogenous injection of T_4 leads to a decrease in production of the thyrotropic hormone. Therefore, the thyroid–pituitary feedback mechanism is constantly in a state of balance and counterbalance between the hormones. (See also Chapter 17.)

The secretion of TSH appears to be regulated by nuclei in the anterior hypothalamus (McFarland et al., 1966; Kanematsu and Mikami, 1970). Electrolytic lesions of this region causes atrophy of the thyroid glands. The central nervous system therefore directly regulates thyrotropin release through a hypothalamic neurosecretion (see Chapter 17). This TRH is transported via the portal vessels of the median eminence to the adenohypophysis. Within 24 hr after autografts of the pituitary are in place, many indices of thyroid action present a normal picture (Rosenberg et al., 1967; Astier and Bayle, 1970). Basal thyrotropic function has a good degree of autonomy with regard to hypothalamic control in birds (Rosenberg et al., 1967). Mammalian thyrotropin-releasing hormone (TRH) has been shown to be a tripeptide (pyroglutamyl-histidyl-prolyl amide). Avian TRH has not been purified (see Chapter 17). For assays of TSH, see Ringer (1976) and Chapter 17.

The injection of TRH will stimulate in vivo thyroid incorporation of radioactive phosphate (Breneman and Rahtkamp, 1973), release of ^{131}I from the thyroid gland (Newcomer and Haung, 1974), in vivo release of TSH from the anterior pituitary (Scanes, 1974), and in vivo increase in T_3 and T_4 within 95 min (Kuhn and Nouwen, 1978; Klandorf et al., 1978b).

Hendrich and Turner (1964) estimated TSH secretion rate based on the amount of exogenous TSH that would stimulate thyroidal ^{131}I release rate equally to that observed with endogenous release. They recorded a mean TSH secretion rate for 48 chickens as 2.5 mu/100 g body weight/day, varying from a low in September–October (1.7 mu) to a high of 3.5 mu in May–June (1 USP unit = 1000 mu). Mackenzie (1981) found that both bovine TSH (2–40 mu) or ostrich TSH gave a linear dose response in cockerels if given as two doses 15 hr apart. The maximum response in T_4 secretion was observed 5 hr after the second injection. Neonatal exposure of chicks to TSH depresses adult responsiveness to TSH (Dobozy et al., 1981a, b).

Season and Temperature Influence

As pointed out earlier, the size and histology of the thyroid changes with the season. Reineke and Turner (1945) showed that in young chicks (2 weeks of age) the T_4 secretion rate in summer was one-half that in winter. Old White Leghorn hens exhibited a similar change with season (Turner, 1948).

In southern France, Astier et al. (1970) reported that a definite circannual rhythm of thyroid function was detected in male ducks with a maximum thyroidal ^{131}I uptake and [^{131}I]PB in June. This time of year corresponds to the period of maximum photoperiod, the seasonal decrease of testicular activity, and the onset of annual molt. Seasonal profiles of circulating T_4 and T_3 in birds suggest that T_4 seems to be associated with reproduction and molt, whereas T_3 is associated with calorigenesis and lipogenesis, especially during migration (Chandola and Bhatt, 1982). The spotted munia (a tropical finch) shows a seasonal cyclicity in thyroid function, the lowest concentration of thyroid hormones occurring in late summer. Seasonal changes in extrathyroidal conversion of T_4 to T_3 may alter seasonal reproduction (Pathak and Chandola, 1982; Cogburn and

Freeman, 1984). In any measure of seasonal effects, it is necessary to examine changes in temperature, light, feed intake, reproductive cycle, and molt, which vary with the season.

Decreases in thyroid size (Chiasson and Combest, 1979) and lower secretion with high environmental temperatures have been reported (Mueller and Amezcua, 1959; Stahl and Turner, 1961; Huston et al., 1962; Cogburn and Harrison, 1980). When young broilers are fed T_3 or T_4 (0.5 and 0.6 ppm, respectively) and then heat stressed, their survival time is significantly shortened as compared to controls. The shortening of survival time is much greater for T_3 than T_4, and feeding T_3 caused a reduction in serum T_4 concentration, but dietary T_4 had no effect on serum T_3 concentration (May, 1982).

Cold temperature increases thyroid secretion rate (Mueller and Amezcus, 1959) and also TSH release (Hendrich and Turner, 1964). Iodine uptake and release of thyroidal ^{131}I increased on short-term exposure to cold of warm-acclimatized chickens and returned to normal after long exposure (Hendrich and Turner, 1963). Cogburn and Freeman (1984) reported that plasma T_3 concentrations were consistently depressed in cockerels held at 38°C but were slightly elevated in cockerels exposed to 10°C. Continued cold stimulates the hypothalamus and adenohypophysis to a higher level of TSH secretion; long exposure to cold increased TSR for extended periods up to 1 month or more. The TSR change on going from a cold to a warm environment is rapid. Exposure of cold-acclimatized chickens to warm temperature resulted in the T_4 secretion rate dropping 42% in 7 days (Hahn et al., 1966). Japanese quail exposed to 15 min cold exposure (4°C) reached peak T_4 plasma concentrations in 40 min and then decreased slowly (Herbuté et al., 1983); TRH injection intravenously gave similar results.

Because uptake and release of iodine by the thyroid do not measure utilization of thyroidal hormones, Hendrich and Turner (1967) measured the half-life of T_4 and reported no change on short-term exposure to cold. McFarland et al. (1966) reported that the calculated half-life of T_4 for Japanese quail kept at 32°C was 36% lower than for those maintained at 21°C.

Photoperiod Influence

The interaction of the thyroid with photoperiod is complex. Pituitary thyrotrophs are stimulated in Japanese quail by exposure to long daily photoperiods (Tixier-Vidal et al., 1967). Long daily photoperiods depressed the uptake of ^{131}I by the thyroid of male ducks (Tixier-Vidal and Assenmacher, 1965), male quail (Bayle and Assenmacher, 1967), and female quail (Follet and Riley, 1967). This action is probably through testicular stimulation and the gonadal steroids that are produced. During photostimulation there is an initial rise in thyroid activity, which is then depressed when the gonads are stimulated and produce steroids (Follett and Riley, 1967), masking the stimulatory action of the long days of T_4 production. "Long" days seem to exert an inhibitory effect on thyroid function, although they stimulate pituitary thyrotropic activity in birds.

Using castrated ducks, Jallageas and Assenmacher (1972) further demonstrated the relationship of the thyroid, photoperiod, and gonads. Castrated ducks had a T_4 half-life of 4.14 hr, whereas long photoperiods in castrates resulted in a $t_{1/2}$ of 3.45 hr. Long photoperiods apparently increase peripheral utilization of T_4. A report by Cogburn and Freeman (1984) demonstrated that hepatic monodeiodinase conversion of T_4 to T_3 was significantly greater during photophase (3.2 ng T_3/hr/mg protein) than scotophase (2.6 ng T_3/hr/mg protein). Thus, as stated earlier, during the light period T_4 concentrations decrease while T_3 increases; the converse occurs during the dark period (Klandorf et al., 1981).

Gonadal Influence

The interaction between the gonads and the thyroid of birds is more pronounced than in mammals, where castration causes only slight effects on thyroid activity. Castration stimulates pituitary thyrotrophs in the duck (Tixier-Vidal and Assenmacher, 1965) and quail (Tixier-Vidal et al., 1967), and increases thyroidal uptake of ^{131}I (Tixier-Vidal and Assenmacher, 1965). At the peripheral level, castration of ducks increases the half-life of T_4 (Jallageas and Assenmacher, 1972). Castration plus testosterone administration returned the $t_{1/2}$ of T_4 to normal, indicating that testosterone increases peripheral utilization of thyroidal hormones. In the quail, plasma T_4 levels are depressed by testosterone treatment (Peczely et al., 1979). In male birds that exhibit annual gonadal cycles, therefore, fluctuations in testosterone production may contribute to variations in thyroid function. Thyroidectomy of the red-vented bulbul (*Molpaster cafer*) or house sparrow (*Passer domesticus*) regardless of photostimulation results in a rapid collapse of the testes and, in the house sparrow, loss of bill pigmentation (Lal and Thapliyal, 1982a, b). Thyroxine concentrations in the plasma measured prior to the onset of lay do not appear to be useful criteria for the selection of egg-laying strains (Sharp et al., 1981).

The concentration of plasma T_3 was elevated in broody bantam hens and increased even further after the chicks hatched. Bantam hens also responded to bovine prolactin injections by a significant increase in plasma T_3 (Klandorf et al., 1982). Precocious sexual maturity and a highly significant enhanced growth rate and egg production has been reported for chickens fed the goitrogen methimazole at an early age (Singh and Parshad, 1978). Estrogen and progesterone have antithyroidal activity in juvenile ducks (Maiti and Sahu, 1982). The inverse relationship between gonadal and thyroidal functions is, however, not reflected in a pronounced change in plasma thyroid hormone levels at sexual maturation in chickens (Scanes et al., 1983).

Molt Influence

Voitkevich (1966) and Brake et al. (1979) noted progressive changes in thyroid function of birds during the period of experimentally induced molt. A peak of thyroid activity commonly precedes molt by a few weeks or is coincident with the onset of molt. This relationship is true in many passerine birds, but not necessarily in nonpasserines. Peczely and Pethes (1982) found that the rook (*Corvus frugilegus*) had higher T_3 and T_4 levels before and during prenuptial molt, followed by the lowest T_4 concentrations during the winter. Plasma T_4 in lesser snow geese (*Anser caerulescens*) increased concomitantly with gonadal recrudescence and remained elevated until shortly before molt. Thyroxine decline before molt was observed in goslings, yearlings, and adult birds. Triiodothyronine did not change throughout the year (Campbell and Leatherland, 1980). Natural molts appear to be independent of increases in thyroid activity (see Ringer, 1976; John and George, 1978).

Thyroidectomy

Surgical thyroidectomy (TX) has been successfully performed on birds, but not without some difficulty. The anatomic location of the thyroids and their proximity to the parathyroid glands in certain species (Figure 20–1) make removal of the thyroids without the parathyroids a difficult operation. Surgical TX of cockerels causes a reduction in plasma T_4, a decrease in resting metabolic rate, an increase in liver and adipose tissue weight, reduced testes and comb weight, and a large increase in alcianphilic cell population in the adenohypophysis (Snapir et al., 1982). Radiothyroidectomy, destruction by the injection of large doses of ^{131}I, is commonly used, and this has the advantage of leaving the parathyroids intact.

Thyroidectomy results in growth retardation, feather structure alteration (fringed and elongated with loss of barbules and color), and reduced gonadal function (see Ringer, 1976). Testes weight and semen quality are decreased following TX (Hendrich and Turner, 1966; Snedecor and Camyre, 1966).

Administration of pregnant mare serum to radiothyroidectomized chicks causes a significant increase in testes weight; therefore, the reduced testes growth in hypothyroidism is probably a lack of gonadotropin and not an absence of thyroid hormone (Snedecor, 1968). Human chorionic gonadotrophin is a weak thyrotropic substance in rats, but not in chickens (Pekary et al., 1983). In contrast to domestic species, TX in some passerine birds lead to increased gonad weight and function in both sexes (Thapliyal and Garg, 1967; Wieselthier and van Tienhoven, 1972).

Genetic Hypothyroidism

A spontaneous case of thyroiditis in an obese strain (OS) of chickens (Cole, 1966) has been shown to be the result of the presence of circulating thyroglobulin autoantibodies (Cole et al., 1968). These birds exhibit low T_4 levels, obesity, rather silky plumage, delayed sexual development or lack of maturity, and thyroids that are either smaller than normal or enlarged (in those that survive). Another line of chickens, described as delayed amelamotic (DAM) by Smyth et al. (1981), also expresses a high frequency of a phenotypic hypothyroidism similar to the OS just described. Neonatal bursectomy decreases the incidence and severity of chronic thyroiditis of OS and DAM line chickens (Lamont and Smyth, 1981).

Function of Thyroid Hormones

Effect of Hypothyroidism and Hyperthyroidism on Growth

The thyroid is necessary for normal growth and development, and growth is markedly retarded following TX. In mature hens TX results in a dwarflike and obese appearance, excessive fat deposits in neck, back, breast, and viscera, shortening of the long skeletal bones, and infrequent ovulation.

Depressed thyroid activity as a consequence of goitrogen administration is reflected in reduced metabolic rate, increased fat deposition, and, in some cases, growth depression (see Ringer, 1976). Both TX and goitrogen administration reduce comb growth, bursa growth, immunocompetence, and the weight of the liver (Raheja and Snedecor, 1970; Yam et al., 1981). Marks (1971) and Howarth and Marks (1973) demonstrated depression of growth in Japanese quail from feeding 0.2% thiouracil. The retarded growth brought about by a goitrogen can be restored to normal by thyroid hormone injection. Following methimazole treatment, Singh et al. (1968a) restored the growth of chicks to normal by injecting 2–3 μg T_4/100 g body weight/day. Similarly, radiothyroidectomized or propylthiouracil-depressed body weights were restored to near-normal levels by daily injections of 0.3 μg/100 g body weight of T_3 or T_4 (Raheja and Snedecor, 1970).

In intact animals, low levels of thyroid hormone administration give little (if any) to moderate growth stimulation (see Ringer, 1976). Singh et al. (1968a), using doses ranging from 1–6 μg/100 g body weight/day, indicated that T_4 in small doses improved growth of chickens; when administered beyond physiologic doses (6 μg/100g), however, it depressed growth rate. Thyroxine in toxic doses accelerates catabolic processes, and body weight is reduced. The role of TSH and thyroxine in carbohydrate metabolism is discussed in Chapter 13.

Effect of Hypothyroidism and Hyperthyroidism on Metabolic Rate (MR) and Liver Glycogen

Thyroid hormones play a major role in regulating oxidative metabolism of birds (see also Chapter 10). Triiodothyronine, the metabolically active thyroid hormone, plays an active role in energy metabolism and metabolic rate (Klandorf et al., 1978a, 1981). Any pronounced alteration in thyroid function (i.e., hyperthyroidism or hypothyroidism) is reflected in an altered metabolic rate. Bilezikian et al. (1980) observed an increase in body temperature (41.0°C) in mature turkeys made hyperthyroid, and reduced body temperature (39.8°C) in those made hypothyroid. The coefficient of correlation between oxygen consumption and T_3 or T_4 concentration in the plasma in different age groups of White Rock chickens ranged between 0.78 and 0.98 (Bobek et al., 1977). Goitrogens reduce metabolic rate, whereas thyroid hormones stimulate it (see Ringer, 1976).

In contrast to the mammal, in which prolonged metabolic (MR) stimulation follows T_4 injection, the chicken exhibits only a transitory rise. A small increase in MR following administration of T_4, T_3, or a combination of T_3 and T_4 lasts only for 2–3 hr. The maximum effect produced by T_4 occurred 2 hr after its administration (Singh et al., 1968a). The MR was depressed for 24 hr after injection.

When exogenous doses of T_3 and T_4 are given to fasted Leghorn roosters, plasma T_3 concentration is higher than control for 8 hr, whereas T_4 remained higher than control for 24 hr (Kittok et al., 1982). Heat loss after either T_3 or T_4 treatment did not differ, but treatment with either hormone resulted in a heat loss higher than that of control.

The MR of turkey embryos increases significantly at pipping and as hatching starts, and is associated with a significant increase in plasma T_3 and T_4 (Christensen and Biellier, 1982; Christensen et al., 1982).

Thyroid function parallels the time course of developing thermoregulatory ability in Japanese quail (Spiers et al., 1974). In the neonate chicken, T_3 or T_4 (300 μg/kg) injected intraperitoneally is thermogenic (Freeman, 1970). Rectal temperature was elevated within 30 min but was still significantly higher only in those chicks given T_3 at 60 min after treatment. Thiouracil impairs thermoregulation in the neonate chicken (Freeman, 1971).

Circadian mean serum concentrations of T_3 in Rhode Island Red chicks were positively correlated with relative growth rate, whereas T_4 correlated with weight and age (Kuhn et al., 1982).

Liver hypertrophy and liver glycogen accumulation are induced by hypothyroidism in the chick, (Snedecor and King, 1964; Snedecor and Mellen, 1965; Snedecor and Camyre, 1966; Snedecor, 1968) and are restored to near-normal levels by small daily doses of T_4 or T_3 (Raheja and Snedecor, 1970). Thyroidectomy causes blood glucose levels in the duck to fall below normal (Ensor et al., 1970). It appears that the increased liver glycogen accumulation is largely a result of decreased glycogenolysis, with glucose-6-phosphatase as the limiting enzyme (Raheja et al., 1971). Exogenous T_4 decreases liver lipid content, whereas it is increased by feeding propylthiouracil (Akiba et al., 1983).

Effect of Hypothyroidism and Hyperthyroidism on Reproduction

Hypothyroidism decreases egg production, egg weight, shell thickness, and ovarian weight. In general, males show reduced testes size and semen quality or complete spermatogenic arrest after TX. These reproductive changes can be restored by exogenous T_4 administration.

Feeding thyroprotein results in little or no stimulation of egg production. In the male, moderate-to-heavy dosages depress testicular development (Ringer, 1976).

Hypothyroidism induced by TX inhibits premigratory fattening and nocturnal restlessness in the red-headed bunting (*Emberiza bruniceps*). Injection of T_3 and T_4 reverses the condition (Pathak and Chandola, 1982). There is a rise in T_3/T_4 ratio associated with migration.

References

Adamson, L.F., and S.H. Ingbar. (1967). Some properties of the stimulatory effect of thyroid hormones on amino acid transport by embryonic chick bone. Encdocrinology, 81, 1372.

Akiba, Y., K. Takahashi, M. Kimura, S.I. Hirama, and T. Matsumoto. (1983). The influence of environmental temperature, thyroid status and a synthetic oestrogen on the induction of fatty livers in chicks. Poult. Sci., 24, 71.

Almeida, O.F., and D.H. Thomas. (1980). A reappraisal of the goitre-prevention assay: Determination of the thyroid secretion rate in the Japanese quail and the relative potencies of T_3 and T_4 in preventing goitrogenesis. Gen. Comp. Endocrinol., 42, 320.

Almeida, O.F., and D.H. Thomas. (1981). Effects of feeding pattern on the pituitary–thyroid axis in the Japanese quail. Gen. Comp. Endocrinol., 44, 508.

Assenmacher, I. (1973). The peripheral endocrine glands. In "Avian Biology," Vol. III (D.S. Farner and J.R. King, Eds.). New York: Academic Press, p. 211.

Assenmacher, I., H. Astier, and N. Jougla. (1968). Repercussions thyroidiennes de la sous-alimentation chez le canard domestique. J. Physiol. (Paris), 60, 342.

Astier, H. (1973). Presence d'une fraction importante de proteines iodees "non hormonales" dans le P.B.I. des Oiseaux. C.R. Hebd. Seances Acad. Sci., 276, 793.

Astier, H., and J.D. Bayle. (1970). Epuration plasmatique de la ^{125}I-L-thyroxine apres hypophsectomie et autogreffe hypophysaire chez le canard. J. Physiol. (Paris), 62, 237.

Astier, H.S., and W.S. Newcomer. (1978). Extrathyroidal conversion of thyroxine to triiodothyronine in a bird: the Peking duck. Gen. Comp. Endocrinol., 35, 496.

Astier, H., L.L. Rosenberg, S. Lissitzky, C. Simon, I. Assenmacher, and A. Tixier-Vidal. (1966). Recherches sur la separation chromatographique des iodothyronines intrathyroidiennes chez le canard Pekin. Ann. Endocrinol., 27, 571.

Astier, H., F. Halberg, and I. Assenmacher. (1970). Rythmes circanniens de l'activité thyroidienne chez le canard Pekin. J. Physiol. (Paris), 62, 219.

Bayle, J.D., H. Astier, and I. Assenmacher. (1966). Activite thyroidienne du pigeon apres hypophysectomie ou autogreffe hypophysaire. J. Physiol. (Paris), 58, 459.

Bayle, J.D., and I. Assenmacher. (1967). Le controle hypothalamo-hypophysaire de la fonction thyroidienne chez les oiseaux. Gen Comp. Endocrinol., 9, 433.

Bilezikian, J.P., J.N. Loeb, and D.E. Gammon. (1980). Induction of sustained hyperthyroidism and hypothyroidism in the turkey. Physiological and biochemical observations. Poult. Sci., 59, 628.

Bobek, S., M. Jastrzebski, and M. Pietras. (1977). Age-related changes in oxygen consumption and plasma thyroid hormone concentration in the young chicken. Gen. Comp. Endocrinol., 31, 169.

Brake, J., P. Thaxton, and E.H. Benton. (1979). Physiological changes in caged layers during a forced molt. 3. Plasma thyroxine, plasma triiodothyronine, adrenal cholesterol, and total adrenal steroids. Poult. Sci., 58, 1345.

Breneman, W.R. (1954). The growth of thyroids and adrenals in the chick. Endocrinology, 55, 54.

Breneman, W.R., and W. Rahtkamp. (1973). Release of thyroid-stimulating hormone from chick anterior pituitary glands by thyrotropin-releasing hormone (TRH). Biochem. Biophys. Res. Commun., 52, 189.

Campbell, R.R., and J.F. Leatherland. (1980). Seasonal changes in thyroid activity in the lesser snow goose (*Anser caerulescens caerulescens*) including reference to embryonic thyroid activity. Can. J. Zool., 58, 1144.

Castay, M., J. Bismuth, and H. Astier. (1978). Thyroxin-binding proteins of ducks and chicken sera. Gen. Comp. Endocrinol., 35, 491.

Chandola, A., and D. Bhatt. (1982). Triiodothyronine fails to mimic gonado-inhibitor action of thyroxine in spotted munia. Effects of injection at different times of the day. Gen. Comp. Endocrinol., 48, 499.

Chaisson, R.B., and W.L. Combest. (1979). Effect of propyl thiouracil and temperature on avian thyroid activity. Life Sci., 25, 1551.

Christensen, V.L., and H.V. Biellier. (1982). Physiology of turkey embryos during pipping and hatching. IV. Thyroid function in embryos from selected hens. Poult. Sci., 61, 2482.

Christensen, V.L., H.V. Biellier, and J.F. Forward. (1982). Physiology of turkey embryos during pipping and hatching. III. Thyroid function. Poult. Sci., 61, 367.

Cogburn, L.A., and P.C. Harrison. (1980). Adrenal, thyroids, and rectal temperature responses of pinealectomized cockerels to different ambient temperatures. Poult. Sci., 59, 1132.

Cogburn, L.A., and R.M. Freeman. (1984). Influence of ambient temperature on daily patterns of plasma thyroid hormones and hepatic monodeiodinase activity. Poult. Sci., 63, 81 (Abstract).

Cole, R.K. (1966). Hereditary hypothyroidism in the domestic fowl. Genetics, 53, 1021.

Cole, R.K., J.H. Kite, and E. Witebsky. (1968). Hereditary autoimmune thyroiditis in the fowl. Science, 160, 1357.

Daugeras, N.A. Brisson, F. Lapointe-Boulu, and F. Lachiver. (1976a). Thyroidal iodine metabolism during the development of the chick embryo. Endocrinology, 98, 1321.

Daugeras-Bernard, N., J. Leloup, and F. Lachiver. (1976b). Evolution de la thyroxinemie au cours du developpement de l'embryon de Poulet. Influence de l'hypophysectomie. C.R. Hebd. Seances Acad. Sci., 283, 1325.

Davison, T.F. (1976). Circulating thyroid hormones in the chicken before and after hatching. Gen. Comp. Endocrinol., 29, 21.

Decuypere, E., E.J. Nouwen, E.R. Kuhn, R. Geers, and H. Michels. (1979). Iodohormones in the serum of chick embryos and post-hatching chickens as influenced by incubating temperature. Relationship with hatching process and thermogenesis. Ann. Biol. Anim. Biochim. Biophys., 19, 1713.

Dobozy, O., L. Balkanyi, and G. Csaba. (1981a). Overlapping effect of thyroid-stimulating hormone and follicle-stimulating hormone on the thyroid gland in baby chicken. Acta Physiol. Acad. Sci. Hung., 57, 171.

Dobozy, O., L. Balkanyi, and G. Csaba. (1981b). Thyroid cell hyporesponsiveness in cockerels treated with follicle-stimulating hormone (FSH) or thryotropin (TSH) at hatching. Horm. Metab. Res., 13, 587.

Ensor, D.M., D.M. Thomas, and J.G. Phillips. (1970). The possible role of thyroid in extrarenal secretion following a hypertonic saline load in the duck (*Anas platyrhynchos*). J. Endocrinol., 46, x.

Etta, K.M. (1971). Comparative studies of the relationship between serum thyroxine and thyroxine binding globulin. Ph.D. Thesis, Michigan State University, East Lansing.

Etta, K.M., R.K. Ringer, and E.P. Reineke. (1972). Degradation of thyroxine confounded by thyroidal recycling of radioactive iodine. Proc. Soc. Exp. Biol. Med., 140, 462.

Farer, L.S., J. Robbins, B.S. Blumberg, and J.E. Rall. (1962). Thyroxine–serum protein complexes in various animals. Endocrinology, 70, 686.

Follett, B.K., and J. Riley. (1967). Effect of the length of the daily photoperiod of thyroid activity in the female Japanese quail (*Coturnix coturnix japonica*). J. Endocrinol., 39, 615.

Freeman, B.M. (1970). Thermoregulatory mechanisms of the neonate fowl. Comp. Biochem. Physiol., 33, 219.

Freeman, B.M. (1971). Impaired thermoregulation in the thiouracil treated neonate fowl. Comp. Biochem. Physiol., 40(A), 553.

Frey, H., and E.V. Flock. (1958). The production of thyroid hormone in the day-old chick, with notes on the effect of thryrotrophin on the chick thyroid. Acta Endocrinol., 29, 550.

Grandhi, R.R., and R.G. Brown. (1975). Age-related changes in thyroid hormone synthesis and circulating thyroid hormone levels. Poult. Sci., 54, 488.

Hahn, D.W., T. Ishibashi, and C.W. Turner. (1966). Alteration of thyroid hormone secretion rate in fowls changed from a cold to a warm environment. Poult. Sci., 45, 31.

Hansborough, L.A., and M. Khan. (1951). The initial function of the chick thyroid gland with the use of radioiodine ^{131}I. J. Exp. Zool., 116, 447.

Harvey, S., H. Klandorf, C. Foetzer, M.T. Strosser, and J.G. Phillips. (1982). Endocrine responses of ducks (*Anas platyrhynchos*) to treadmill exercise. Gen. Comp. Endocrinol., 48, 415.

Hendrich, C.E., and C.W. Turner. (1963). Time relations

in the alteration of thyroid gland function in fowls. Poult. Sci., 42, 1190.

Hendrich, C.E., and C.W. Turner. (1964). Estimation of thyroid stimulating hormone (TSH) secretion rates of New Hampshire fowls. Proc. Soc. Exp. Biol. Med., 117, 218.

Hendrich, C.E., and C.W. Turner. (1966). Effects of radiothyroidectomy and various replacement levels of thyroxine on growth, organ and gland weight of Cornish-cross chickens. Gen. Comp. Endocrinol., 7, 411.

Hendrich, C.E., and C.W. Turner. (1967). A comparison of the effect of environomental temperature changes and 4.4°C cold on the biological half-life ($t_{1/2}$) of thyroxine-^{131}I in fowls. Poult. Sci., 46, 3.

Heninger, R.W., and W.S. Newcomer. (1964). Plasma protein binding, half-life, uptake of thyroxine and triiodothyronine in chickens. Proc. Soc. Exp. Biol. Med., 116, 624.

Herbuté, S., R. Pintat, N. Parēs, H. Astier, and J.D. Baylé. (1983). Comparison of plasma thyroxine levels following short exposure to cold and TRH administration in intact and pituitary autografted quail (*Coturnix coturnix japonica*). Gen. Comp. Endocrinol., 49, 154.

Hohn, E. (1949). Seasonal changes in the thyroid gland and effects of thyroidectomy in the mallard in relation to molt. Am. J. Physiol., 158, 337.

Hoshino, T., and N. Ui. (1970). Comparative studies on the properties of thyroglobulins from various animal species. Endocrinol. Jpn., 17, 521.

Howarth, B., Jr., and H.L. Marks. (1973). Thyroidal ^{131}I uptake of Japanese quail in response to three different dietary goitrogens. Poult. Sci., 52, 326.

Huston, T.M., H.M. Edwards, Jr., and J.J. Williams. (1962). The effects of high environmental temperature on thyroid secretion rate of domestic fowl. Poult. Sci., 41, 640.

Hutchins, M.O., and W.S. Newcomer. (1966). Metabolism and excretion of thyroxine and triiodothyronine in chickens. Gen. Comp. Endocrionol., 6, 239.

Jallageas, M., and I. Assenmacher. (1972). Effets de la photoperiode et de taux d'androgene circulant sur la fonction thyroidienne du canard. Gen. Comp. Endocrinol., 19, 331.

John, T.M., and J.C. George. (1978). Circulating levels of thyroxine (T_4) and Triiodothyronine (T_3) in the migratory canada goose. Physiol. Zool., 51, 361.

Kamis, A.B., and G.A. Robinson. (1978). Serum T_3 and T_4 concentration of Japanese quail treated with thyrotropin-releasing hormone. Gen. Comp. Endocrinol., 36, 636.

Kanematsu, S., and S. Mikami. (1970). Effects of hypothalamic lesions on the pituitary–thyroid system in the chicken. Jpn. J. Zootech. Sci., 37, 28.

Kittok, R.J. T.J. Greninger, J.A. De Shazer, S.R. Lowry, and F.B. Mather. (1982). Metabolic responses of the rooster after exogenous thyroid hormones. Poult. Sci., 61, 1748.

Klandorf, H., P.J. Sharp, and I.J.H. Duncan. (1978a). Variation in levels of plasma thyroxine and triiodothyronine in juvenile female chickens during 24- and 16-hour lighting cycles. Gen. Comp. Endocrinol., 36, 238.

Klandorf, H., P.J. Sharp, and R. Sterling. (1978b). Induction of thyroxine and triiodothyromine release by thyrotrophin-relating hormone in the hen. Gen. Comp. Endocrinol., 34, 377.

Klandorf, H., P.J. Sharp, and W.S. Newcomer. (1981). The influence of feeding patterns on daily variation in the concentrations of plasma thyroid hormones in the hen. IRCS Med. Sci., 9, 82.

Klandorf, H.R., W. Lea, and P.J. Sharp. (1982). Thyroid function in laying, incubating, and broody bantam hens. Gen. Comp. Endocrinol., 47, 492.

Kuhn, E.R., and E.J. Nouwen. (1978). Serum levels of triiodothyromine and thyroxine in the domestic fowl following mild cold exposure and injection of synthetic thyrotropin-releasing hormone. Gen. Comp. Endocrinol., 34, 336.

Kuhn, E.R., E. Decuypere, L.M. Colen, and H. Michels. (1982). Posthatch growth and development of a circadian rhythm for thyroid hormones in chicks incubated at different temperatures. Poult. Sci., 61, 540.

Lague, P.C., and A. van Tienhoven. (1969). Comparison between chicken and guinea pig thyroid with respect to the incorporation of ^{131}I in vitro. Gen. Comp. Endocrinol., 12, 305.

Lal, P., and J.P. Thapliyal. (1982a). Thyroid–gonad and Thyroid–body weight relationship in red-vented bulbul, *Molpastes cafer.* Gen. Comp. Endocrinol., 48, 98.

Lal, P., and J.P. Thapliyal. (1982b). Role of thyroid in the response of bill pigmentation to male hormone of the house sparrow, *Passer domesticus.* Gen. Comp. Endocrinol., 48, 135.

Lamont, S.J., and J.R. Smyth, Jr. (1981). Effect of bursectomy on development of a spontaneous postnatal amelanosis. Clin. Immunol. Immunopathol., 21, 407.

Mackenzie, D.S. (1981). *In vitro* thyroxine release in day-old cockerels in response to acute stimulation by mammalian and avian pituitary hormones. Poult. Sci., 60, 2136.

Maiti, B.R., and A. Sahu. (1982). Action of sex-hormones on thyroid gland function of the domestic duckling. Endokrinologie, 80, 371.

March, B.E., and R. Poon. (1981). Dependency of maximum goitrogenic response on some minimal level of thyroid hormone production. Poult. Sci., 60, 846.

Marks, H.L. (1971). Selection for four-week body weight in Japanese quail under two nutritional environments. Poult. Sci., 50, 931.

May, J.D. (1978). A radioimmunoassay for 3,5,3,-triiodothyronine in chicken serum. Poult. Sci., 57, 1740.

May, J.D. (1982). Effect of dietary thyroid hormone on survival time during heat stress. Poult. Sci., 61, 706.

McFarland, L.Z., M.R. Yousef, and W.O. Wilson. (1966). The influence of ambient temperature and hypothalamic lesions on the disappearance rate of thyroxine ^{131}I in the Japanese quail. Life Sci., 5, 309.

McNabb, F.M.A., and T.E. Hughes. (1983). The role of serum binding proteins in determining free thyroid hormone concentrations during development in quail. Endocrinology, 113, 957.

McNabb, F.M.A., R.T. Weirich, and R.A. McNabb. (1981). Thyroid function in embryonic and perinatal Japanese quail. Gen. Comp. Endocrinol., 43, 218.

Mellen, W.J. (1965). Thyroxine secretion rate in chicks and poults. Poult. Sci., 43, 776.

Mellen, W.J., and L.B. Hardy, Jr. (1957). Blood protein-bound iodine in the fowl. Endocrinology, 60, 547.

Mellen, W.J., and B.C. Wentworth. (1959). Thyroxine vs. triiodothyronine in the fowl. Poult. Sci., 38, 228.

Mess, B., and K. Straznicky. (1970). Dynamics of ribonucleic acid (RNA) synthesis in the thyroid epithelial cells of normal and decapitated chicken embryos. Acta Biol. Acad. Sci. Hung., 21, 115.

Mueller, W.J., and A.A. Amezcua. (1959). The relationship between certain thyroid characteristics of pullets and their egg production, body weight and environment. Poult. Sci., 38, 620.

Newcomer, W.S. (1957). Relative potencies of thyroxine and triiodothyronine based on various criteria in thiouracil-treated chickens. Am. J. Physiol., 190, 413.

Newcomer, W.S. (1974). Diurnal rhythms of thyroid function in chicks. Gen. Comp. Endocrinol., 24, 65.

Newcomer, W.S. (1978a). Dietary iodine and accumulation of radioiodine in thyroids of chickens. Am. J. Physiol., 234, E168.

Newcomer, W.S. (1978b). Dietary iodine and thyroidal and iodide transport in chickens: organic binding blocked. Am. J. Physiol., 234, E177.

Newcomer, W.S. (1979). Accumulation of radiodine in thiouracil-hyperplastic thyroids of chicks. Am. J. Physiol., 237, E147.

Newcomer, W.S. (1982). Hormonal regulation of iodine accumulation in ovary and thyroid of Japanese quail. Gen. Comp. Endocrinol., 47, 243.

Newcomer, W.S., and P.A. Barrett. (1960). Effects of various analogues of thyroxine on oxygen uptake of cardiac muscle from chicks. Endocrinology, 66, 409.

Newcomer, W.S., and F.S. Huang. (1974). Thyrotropin-releasing hormone in chicks. Endocrinology, 95, 318.

Okonski, J., F.W. Lengemann, and C.L. Comar. (1960). The utilization of egg iodine by the chicken embryo. J. Exp. Zool., 45, 263.

Pathak, V.K., and A. Chandola. (1982). Seasonal variation in extrathyroidal conversion of thyroxine to triiodothyronine and migration disposition in redheaded bunting. Gen. Comp. Endocrinol., 47, 433.

Peczely, P., and G. Pethes. (1982). Seasonal cycle of gonadal, thyroid, and adrendocortical function in the rook (*Corvus frugilegus*). Acta Physiol. Acad. Sci. Hung., 59, 59.

Peczely, P., H. Astier, and M. Jallageas. (1979). Reciprocal interactions between testis and thyroid in male Japanese quail. Gen. Comp. Endocrinol., 37, 400.

Perkary, A.E., M. Azukizawa, and J.M. Hershman. (1983). Thyroidal responses to human chorionic gonadotropin in the chick and rat. Horm. Res., 17, 36.

Raheja, K.L., and J.G. Snedecor. (1970). Comparison of subnormal multiple doses of L-thyroxine and L-triiodothyronine in proplthiouracil-fed and radiothyroidectomized chicks (*Gallus domesticus*). Comp. Biochem. Physiol., 37, 555.

Raheja, K.L., J.G. Snedecor, and R.A. Freedland. (1971). Effect of propylthiouracil feeding of glycogen metabolism and malic enzyme in the liver of the chick (*Gallus domesticus*). Comp. Biochem. Physiol., 39(B), 833.

Refetoff, S., N.I. Robin, and V.S. Fang. (1970). Parameters of thyroid function in serum of 16 selected vertebrate species: Study of PBI, serum T_4, free T_4 and the pattern of T_4 and T_3 binding to serum proteins. Endocrinology, 86, 793.

Reineke, E.P., and C.W. Turner. (1945). Seasonal rhythms in the thyroid hormone secretion of the chick. Poult. Sci., 24, 499.

Ringer, R.K. (1965). Thyroids. In "Avian Physiology" (2nd ed.) (P.D. Sturkie, Ed.). Ithaca: Cornell University Press, Chapter 19.

Ringer, R.K. (1976). Thyroids. In "Avian Physiology" (3rd ed.) (P.D. Sturkie, Ed.). New York: Springer-Verlag, Chapter 18.

Robinson, G.A. (1973). The oocyte as major competitors for radio-iodide in laying Japanese quail. Gen. Comp. Endocrinol., 21, 123.

Robinson, G.A., D. Wasnidge, F. Floto, and S.E. Downie. (1977). Excess iodide and the accumulation of ^{125}I by the thyroid, plasma and developing oocytes of the Japanese Quail. Br. Poult. Sci., 18, 151.

Roche, J., R. Michel, and E. Volpert. (1956). Concentration des hormones thyroidennes par les ovocytes de la poule. C. R. Seances Soc. Biol. Ses Fil., 150, 2149.

Rogler, J.C., H.E. Parker, F.N. Andrews, and C.W. Carrick. (1959). The effect of an iodine deficiency on embryo development and hatchability. Poult. Sci., 38, 398.

Romanoff, A.L. (1960). "The Avian Embryo." New York: Macmillan.

Rosenberg, L.L., M. Goldman, G. LaRoche, and M.K. Dimick. (1964). Thyroid function in rats and chickens. Equilibration of injected iodide with existing thyroidal iodine in Long-Evans rats and White Leghorn chickens. Endocrinology, 74, 212.

Rosenberg, L.L., H. Astier, G. La Roche, J.D. Bayle, A. Tixier-Vidal, and I. Assenmacher. (1967). The thyroid function of the drake after hypophysectomy or hypopthalamic pituitary disconnection. Neuroendocrinology, 2, 113.

Sadovsky, R., and A. Bensadoun. (1971). Thyroid iodohormones in the plasma of the rooster (*Gallus domesticus*). Gen. Comp. Endocrinol., 17, 268.

Scanes, C.G. (1974). Some *in vitro* effects of a synthetic thyrotrophin releasing factor on the secretion of thyroid-stimulating hormone from the antier pituitary gland of the domestic fowl. Neuroendocrinology, 15, 1.

Scanes, C.G., J. Marsh, E. Decuypere, and P. Rudas. (1983). Abnormalities in the plasma concentrations of thyroxine, triiothryronine and growth hormone in sex-linked dwarf and autosomal dwarf white Leghorn domestic fowl (*Gallus domesticus*). J. Endocrinol., 97, 127.

Sharp, P.J., and H. Klandorf. (1985). Environmental and physiological factors controlling thyroid function in Galliformes. From the endocrine system and its environment, edited by B.K. Follett, S. Ishii and A. Chandola, pp. 175–188. Japan Sci. Press Tokyo/Springer-Verlag-Berlin 1985.

Sharp, P.J., W.F. Van Tyern, J.H. Van Middelkoop, H. Klandorf, R.W. Lea, and A. Chadwick. (1981). Lack of relationship between concentrations of plasma luteinizing hormone, thyroxine and prolactin at nine weeks of age and subsequent egg production in domestic hen. Br. Poult. Sci., 22, 53.

Shellabarger, C.J. (1955). A comparison of triiodothyronine and thyroxine in the chick goiter-prevention tests. Poult. Sci., 34, 1437.

Singh, A., and O. Parshad. (1978). Precocious sexual maturity and enhanced egg production in chickens given goitrogen at an early age. Br. Poult. Sci., 19, 521.

Singh, A., E.P. Reineke, and R.K. Ringer. (1967). Thyroxine and triiodothyronine turnover in the chicken and the bobwhite and the Japanese quail. Gen. Comp. Endocrinol., 9, 353.

Singh, A., E.P. Reineke, and R.K. Ringer. (1968a). Influence of thyroid status of the chick on growth and metabolism, with observations on several parameters of thyroid function. Poult. Sci., 47, 212.

Singh, A., E.P. Reineke, and R.K. Ringer. (1968b). Comparison of thyroid secretion rate in chickens as determined by (1) goiter prevention, (2) thyroid hormone substitution, (3) direct output, (4) thyroxine degradation methods. Poult. Sci., 47, 205.

Smyth, J.R., Jr., R.E. Boissy, and K.V. Fite. (1981). The DAM chicken: A model for spontaneous postnatal cutaneous and ocular amelanosis. J. Hered., 72, 150.

Snapir, N., B. Robinson, Y. Hoffman, and A. Berman. (1982). Adenohypophyseal cytology of chemically and surgically thyroidectomized cockerels. Poult. Sci., 61, 1720.

Snedecor, J.G. (1968). Liver hypertrophy, liver glycogen accumulation, and organ-weight changes in radiothyroidectomized and goitrogen-treated chicks. Gen. Comp. Endocrinol., 10, 277.

Snedecor, J.G., and D.B. King. (1964). Effect of radiothyroidectomy in chicks with emphasis on glycogen body and liver. Gen. Comp. Endocrinol., 4, 144.

Snedecor, J.G., and W.J. Mellen. (1965). Thyroid deprivation and replacement in chickens. Poult. Sci., 44, 452.

Snedecor, J.G., and M.F. Camyre. (1966). Interaction of thyroid hormone and androgens on body weight, comb and liver in cockerels. Gen. Comp. Endocrinol., 6, 276.

Sorimachi, K., and N. Ui. (1974). Comparison of the iodoamino acid distribution in thyroglobulins obtained from various animal species. Gen. Comp. Endocrinol., 24, 38.

Spiers, D.E. (1980). The development of thermoregulation in the bobwhite quail. Ph.D. Thesis, Michigan State University, East Lansing, 270 p.

Spiers, D.E., R.A. McNabb, and F.M.A. McNabb. (1974). The development of thermoregulatory ability, heat-seeking activities, and thyroid function in hatching Japanese quail (*Coturnix coturnix japonica*). J. Comp. Physiol., 89, 159.

Stahl, P., and C.W. Turner. (1961). Seasonal variation in thyroxine secretion rate in two strains of New Hampshire chickens. Poult. Sci., 40, 239.

Stoeckle, M.E., and A. Porte. (1969). In "Calcitonin." Proc. Second Int. Symp. London: William Heinemann Medical Books, p. 327.

Stoll, R., and P. Blanquet. (1953). Sur l'activite des thyroides de l'embryon du poulet, provenant d'oeufs "marques" par l'administration de ^{131}I a la poule. Ann. Endocrinol. (Paris), 14, 1.

Tanabe, Y. (1964). Comparison of estimates for thyroxine secretion rate determined by the goiter prevention assay and by measuring the activity of serum alkaline phosphate in the chicken. Endocrinol. Jpn., 11, 260.

Tanabe, Y. (1965). Relation of thyroxine secretion rate to age and growth rate in the cockerel. Poult. Sci., 44, 591.

Tanabe, Y., T. Komlyama, D. Kobota, and Y. Tamake. (1965). Comparison of the effects of thiouracil, propylthiouracil and methimazole in I^{131} metabolism by the chick thyroid, and measurements of thyroxine secretion rate. Gen. Comp. Endocrinol., 5, 60.

Tata, J.R., and C.J. Shellabarger. (1959). An explanation for the difference between the response of mammals and birds to thyroxine and triiodothyronine. Biochem. J., 72, 608.

Thapliyal, J.P., and R.K. Garg. (1967). Thyroidectomy in the juveniles of the chestnut-bellied munia (*Munia atricapilla*). Endokrinologie, 52, 75.

Thommes, R.C. (1958). Vasculogenesis in selected endocrine glands of normal and hypophysectomized chick embryos. I: The thyroid. Growth, 22, 243.

Thommes, R.C., and V.W. Hylka. (1978). Hypothalamo-adenohypophyseal–thyroid interrelationships in the chick embryo. 1. TRH and TSH sensitivity. Gen. Comp. Endocrinol., 34, 193.

Tixier-Vadal, A., and I. Assenmacher. (1965). Some aspects of the pituitary-thyroid relationship in birds. Int. Congr. Ser.-Excerpta Med., No. 83, p. 172.

Tixier-Vadal, A., B.K. Follett, and D.S. Farner. (1967). Identification cytologique et fonctionnelle des types cellulaires de l'adenohypophyse chez le Caille male, "Coturnix coturnix japonica" soumise a differentes conditions experimentales. C. R. Hebd. Seances Acad. Sci., 264, 1739.

Turner, C.W. (1948). Effects of age and season on the thyroxine secretion rate of White Leghorn hens. Poult. Sci., 27, 146.

van Tienhoven, A., J.H. Williamson, M.C. Tomlinson, and K.L. Macinnes. (1966). Possible role of the thyroid and the pituitary glands in sex-linked dwarfism in the fowl. Endocrinology, 78, 950.

Vlijm, L. (1958). On the production of hormones in the thyroid gland of birds (cockerels). Arch. Neerl. Zool., 12, 467.

Voitkevich, A.A. (1966). "The Feathers and Plumage of Birds." New York: October House.

Waterman, A.J. (1959). Development of the thyroid–pituitary system in warm-blooded amniotes. In "Comparative Endocrinology" (A. Gorman, Ed.). New York: Wiley, p. 351.

Wentworth, B.C., and W.J. Mellen. (1961). Circulating thyroid hormones in domestic birds. Poult. Sci., 40, 1275.

Wieselthier, A.S., and A. van Tienhoven. (1972). The effect of thyroidectomy on testicular size and on the photorefractory period in the starling (*Sturnus vulgaris* L.). J. Exp. Zool., 179, 331.

Wollman, S.H., and E. Zwiling. (1953). Radioiodine metabolism in the chick embryo. Endocrinology, 52, 526.

Yam, D., D. Heller, and N. Snapir. (1981). The effect of the thyroidal state on the immunological state of the chicken. Dev. Comp. Immunol., 5, 483.

21
Parathyroid and Ultimobranchial Glands

A.D. Kenny

Parathyroids

Avian Calcium Metabolism

Calcium metabolism in avian species in characterized by several aspects that differentiate it from that found in other vertebrates (Pang et al., 1980b). First, birds develop extra bone in the medullary cavities of the long bones prior to the onset of laying (Simkiss, 1967). This medullary bone, as it is called, may be induced pharmacologically by the injection of both androgen and estrogen to immature birds (Dacke, 1979). Second, domesticated avian species, such as the chicken and the Japanese quail, have the potential for laying an egg daily for extended periods before taking a laying pause. This daily deposition of shell calcium has been estimated to represent approximately 10% of the total body stores of calcium, implying that these avian species must possess efficient and effective mechanisms for maintaining calcium homeostasis. It is likely that these mechanisms are under endocrine control.

Third, avian species exhibit a hypercalcemic response to estrogen (Baksi and Kenny, 1977) and to reproductive activity in the female, as do other submammalian vertebrates (Dacke et al., 1973; Boelkins and Kenny, 1973). This extra calcium in the plasma is complexed; the ionic calcium concentration remains normal (Dacke et al., 1973). Last, the physiologic role of calcitonin (CT) in avian species is still unknown. No overt effects of exogenously administered CT on calcium metabolism have been demonstrated in avian species. Nevertheless, the avian ultimobranchial gland is capable of secreting CT in vitro (Feinblatt et al., 1973), and high concentrations of bioassayable CT are found in the

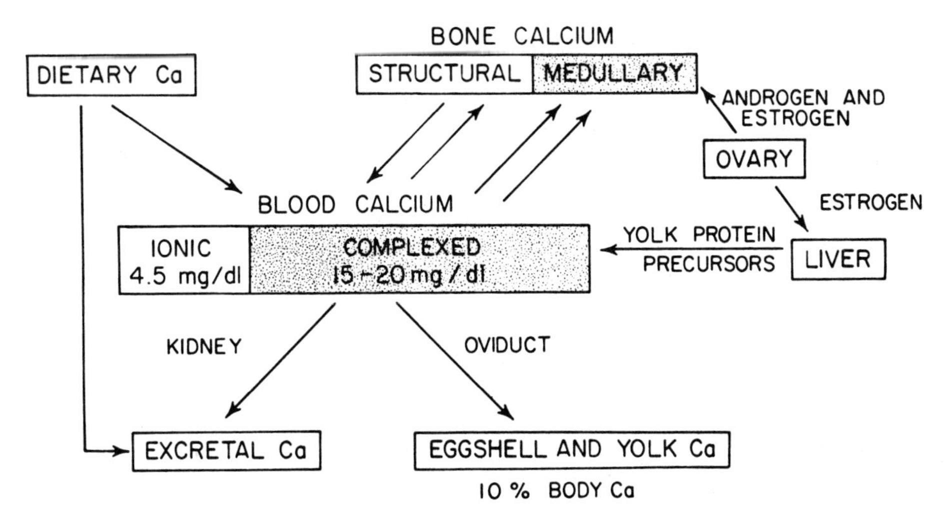

FIGURE 21–1. Calcium metabolism in egg-laying birds (see text). Yolk protein precursors complex plasma calcium so that the total calcium can rise to very high levels leaving the plasma ionic calcium at normal levels. Under the influence of both estrogen and androgen, medullary bone is formed in the cavities of the long bones. Calcitonin has no overt effects on avian calcium metabolism. (Data taken from Kenny, 1982).

plasmas of several avian species (Kenny, 1971). Many of these facets peculiar to avian calcium metabolism are illustrated in Figure 21–1.

Anatomy of Parathyroid Glands

Location. In the chicken, there are four parathyroid glands slightly caudal to the thyroid (Dacke, 1979). A pair of glands is found on each side of the midline. Each pair represents an anterior (parathyroid III) and posterior (parathyroid IV) lobe, which are often fused together (see Figure 20–1). The Japanese quail usually exhibits only one pair of parathyroid glands between the jugular vein and carotid artery, just posterior to the thyroid gland (Clark and Sasayama, 1981).

Embryology. The parathyroid glands are derived from the third and fourth pharyngeal pouches (Roth and Schiller, 1976). Originally attached to thymus III and IV, the parathyroid glands establish a morphologic independence from thymus tissue.

Morphology. In the chicken, each parathyroid is encapsulated by connective tissue and is composed mainly of chief cells (Roth and Schiller, 1976); in the latter respect, the chicken parathyroid gland resembles that of the rat. Oxyphil cells, which are absent in the chicken, Japanese quail, and other vertebrates, appear in the adult human. It may be assumed that the parathyroid chief cell in avian species is responsible for synthesis, packaging, and secretion of parathyroid hormone; ultrastructurally, the avian chief cell is similar to that of mammals (Nevalainen, 1969). Mature secretory granules, however, occur much less frequently in the parathyroid chief cell of the chicken.

Parathyroid Hormone

Chemistry of Avian Parathyroid Hormone. Chicken parathyroid hormone (PTH) has been partially purified (MacGregor et al., 1973). The crude preparation appeared to be similar in molecular size to bovine PTH [subsequently abbreviated to bPTH or bPTH(1–84)]. It exhibited biologic activity in an in vitro mouse calvarium system using inhibition of the conversion of [^{14}C]citrate to $^{14}C_2$ as the response. When compared with a purified bPTH standard, the partially purified chicken PTH was estimated to have a potency of 800 U.S.P. units/mg. Its immunologic cross-reactivity with antisera to bPTH was weak.

Assay of Parathyroid Hormone. The first in vivo bioassay method for parathyroid hormone (PTH) was developed using dogs. The insensitivity of the assay and the inconvenience and expense of maintaining a dog colony precluded its widespread use. In the 1950s, Munson and his associates (Munson, 1955, 1960, 1961) developed a more sensitive and convenient assay using the parathyroidectomized rat. The popularity of this assay played a major role in the final purification of bovine PTH, which had remained unpurified for

40 years since the preparation of the first crude extracts in the 1920s.

The development in 1973 of simple, economic, and reasonably sensitive and precise assays for PTH in birds represented a significant advance in the availability of in vivo assays (Dacke and Kenny, 1973; Parsons et al., 1973). Polin et al. (1957) were the first to emphasize the rapidity (3–4 hr) of the hypercalcemic response to PTH in 5- to 7-week-old chickens and to suggest this response as a basis for a bioassay. However, their assay lacked sufficient sensitivity and precision to become widely used. Dacke and Kenny (1973) reported the development of a successful assay method using either 2- or 3-week-old Japanese quail or 5- to 6-day-old chickens. The assay is rapid (60 min), sensitive (0.4 U.S.P. units per bird), and reasonably precise ($\lambda = 0.20$). Cumulative experience with this assay is described elsewhere (Kenny, 1982).

There are three major types of in vitro assays for PTH: (1) bone tissue culture methods using calcium release into the media as the response; (2) renal adenylate cyclase response; and (3) radioimmunoassay. The in vitro assays, especially radioimmunoassay, have the advantage of much greater sensitivity and in general possess greater precision. Major disadvantages are tedium and time, with the tissue culture assays, and the uncertainty of correlation with biologic activity, with radioimmunoassay. The renal adenylate cyclase assays using renal membranes prepared from rat, canine, and avian (chicken and Japanese quail) sources have proved popular for their ability to combine sensitivity, simplicity, and correlation with biologic activity. A more detailed analysis of the various methods available for the assay of PTH has been presented elsewhere (Kenny and Dacke, 1975). A summary of a representative list of assays is presented in Table 21–1.

Parathyroidectomy

Parathyroidectomy has been accomplished or attempted in various avian species including the pigeon (Smith, 1945), duck (Benoit et al., 1941), starling (Clark and Wideman, 1977), chicken (Polin and Sturkie, 1957), and Japanese quail (Clark and Sasayama, 1981). The responses to parathyroidectomy, including permanent hypercalcemia, tetany, or death, have been variable. Polin and Sturkie (1957, 1958) found that tetany and death in the chicken were dependent on such factors as food intake, the presence of accessory parathyroid tissue, and the sex and/or reproductivity activity of the birds. In ducks and pigeons, under the conditions used parathyroidectomy resulted in a marked hypocalcemic response accompanied by severe tetany and death within 2 days (Benoit et al., 1941; Smith, 1945). Japanese quail and starlings have been successfully parathyroidectomized by Clark and her associates (Clark and Wideman, 1977; Clark and Sasayama, 1981).

There are two major technical problems associated with parathyroidectomy in avian species. First, the closeness of the glands in some avian species to the common carotid arteries renders their extirpation a hazardous procedure. Second, the induction of surgical anesthesia is more difficult than in mammalian species. The volatile anesthetics ether and halothane, may be used for short operations measured in minutes, although the margin of safety with respect to mortality is dangerously narrow. For this reason, various anesthetic agents have been proposed and used successfully for longer surgical procedures. These have included Equi-Thesin, an aqueous veterinary preparation (Clark et al., 1976), and allobarbital (Dial) (Clark and Sasayama, 1981). A more extensive discussion of avian anesthesia may be found elsewhere (Gandal, 1969).

Usually parathyroidectomy results in hypocalcemia and hyperphosphatemia in chickens (Cherian and Cipera, 1968), in ducks (Benoit et al., 1941), in pigeons (Riddle and McDonald, 1945), and in starlings (Clark and Wideman, 1977). In Japanese quail, hypocalcemia was not accompanied by a hyperphosphatemic response to parathyroidectomy (Clark and Sasayama, 1981).

Prior to 1977, information on the effects of parathy-

TABLE 21–1. Summary of representative assays for parathyroid hormone[a]

Type of assay	Test system	Route of administration	Dose range (U.S.P. units)	Index of precision (λ)	Reference
In vivo assays					
Hypercalcemia	Intact dog	sc	100–300	0.20	Collip (1925–1926)
	Parathyroidectomized rat	sc	10–100	0.23	Munson (1961)
	Japanese quail	iv	0.5–4.0	0.20	Dacke and Kenny (1973)
	Chicken	iv	1–10	0.14	Parsons et al. (1973)
Phosphaturia	Parathyroidectomized rat	sc	5–50	0.30	Kenny and Munson (1959)
In vitro assays					
^{40}Ca release from bone	Mouse calvarium		0.01–1.0	0.15	Zanelli et al. (1969)
Formation of cAMP	Rat renal adenylate cyclase		0.2–1.0	0.08	Marcus and Aurbach (1969)
Radioimmunoassay	[^{131}I]PTH/antibody		0.001–0.006		Berson et al. (1963)

[a] The index of precision was calculated as the standard deviation/slope.

roidectomy on renal excretion of calcium and phosphorus was nonexistent. Since that time, Clark and her associates have examined these effects in starlings (Clark and Wideman, 1977) and in Japanese quail (Clark and Sasayama, 1981). In both species, removal of the parathyroids led to increases in urinary calcium loss and decreases in urinary phosphate excretion (see also Chapter 16). These changes were accompanied by hypocalcemia in both species, but a hyperphosphatemic response was observed only in starlings.

Actions of Parathyroid Hormone in Avian Species

All of the major studies concerned with the actions of PTH in avian species have used heterologous (usually bovine) PTH. For decades a crude extract of bPTH was commercially available from the Eli Lilly Company and was marketed for clinical use as Parathyroid Injection, U.S.P. In 1980, its manufacture was discontinued for reasons associated with the lack of clinical demand. This preparation, in spite of its impure nature (less than 1% pure) and the heterogeneity of its biological active constituents (Kenny et al., 1976) served the scientific community well until its demise. In the meantime, synthetic preparations of the N-terminal biologically active fragment, PTH(1–34), have become available commercially. Although PTH(1–34) preparations representing the fragments from human and bovine species may be obtained from several sources, no avian PTH(1–84) or PTH(1–34) is commercially available. Until pure avian PTH [either PTH(1–84) or PTH(1–34)] is tested for its actions in avian species, caution should be exercised in extrapolating heterologous to homologous experience.

The primary target tissues of PTH in avian species are assumed to be kidney and bone. Other tissues likely to be directly responsive include vascular and other smooth muscles.

Blood Calcium Response. Birds respond to administration of PTH by a rapid but transient rise in plasma calcium. Doses of 90 U.S.P. units or more of bPTH cause a maximal hypercalcemic response at 3–4 hr after subcutaneous or intramuscular injection in chickens (Polin et al., 1957) or in Japanese quail (Baksi and Kenny, 1979). The response is more marked in the laying hen than in the cockerel (Polin et al., 1957). Candlish and Taylor (1970) observed a rapid and transient rise in plasma *diffusible* calcium within 10 min of injection of 20 units of bPTH into hens. Immature birds show a rapid and more sensitive response; 1-week-old chickens or 3-week-old Japanese quail respond to 10–20 units of synthetic bPTH(1–34) within 15 min after intravenous injection (Kenny and Dacke, 1974).

What is the source of the hypercalcemic response to PTH in avian species? The response is much more rapid in avian than in mammalian species. Does this difference reflect different mechanisms or at least differences in the relative contributions of various PTH target organ responses to the overall response? Is bone or kidney the main target organ responding? Radiocalcium has been used in an attempt to answer this important question. Kenny and Dacke (1974) followed the plasma ^{45}Ca and total calcium levels in the 2 hr period immediately following intravenous injection of synthetic bPTH(1–34) and ^{45}Ca into Japanese quail and chickens; selected chicken data are presented in Figure 21–2. There was a rapid and marked rise in plasma ^{45}Ca levels, which peaked at 15 min (Japanese quail) to 30 min (chicken), followed by a fall to control levels at 60 min. The plasma total calcium, on the other hand, rose more slowly and decreased more slowly, remaining significantly elevated at 60 min. These results were interpreted as indicating that the avian hypercalcemic response to bPTH is complex, involving two or more underlying mechanisms. The early phase (30 min or less) is characterized by a rise in both plasma total calcium and ^{45}Ca, the latter being more pronounced. This type of response might reflect an inhibition of the exit of calcium from the extracellular compartment and is consistent with the dramatic effects on renal calcium excretion following parathyroidectomy and bPTH administration (Clark et al., 1976; Clark and Sasayama, 1981; Clark and Wideman, 1977). The later phase (30 min or more) is characterized by elevated total calcium levels only and may be due to bone resorption. Parsons et al. (1973) used a different experimental approach. Chickens were injected with ^{47}Ca 60 min before bPTH, and no rise in plasma ^{47}Ca was noted at 30 and 60 min after bPTH injection. Plasma total calcium levels did rise at these time intervals, and the results were interpreted as reflecting bone resorption only. Present evidence, therefore, is consistent with the concept that the initial hypercalcemic *pharmacologic* response to bPTH in avian species is mainly renal in origin and that bone resorption may play a larger role during the later phase of the response.

Urinary Calcium Response. Administration of bPTH to normal birds resulted in a hypercalciuric response in starlings (Clark et al., 1976). Paradoxially, parathyroidectomy also elicited a hypercalciuric response in the same species. Nevertheless, administration of bPTH to parathyroidectomized starlings resulted in a decline in urinary calcium levels towards preparathyroidectomy levels. This paradoxical increase of calcium clearance in response to bPTH was noted earlier by Candlish (1970) in normal chickens. No unifying concept has been advanced to explain this hypercalciuric response to bPTH in normal birds other than its representing an increased filtered load of calcium resulting from bone resorption. The radiocalcium studies of Kenny and Dacke (1974) described above, although performed in normal birds, were done with-

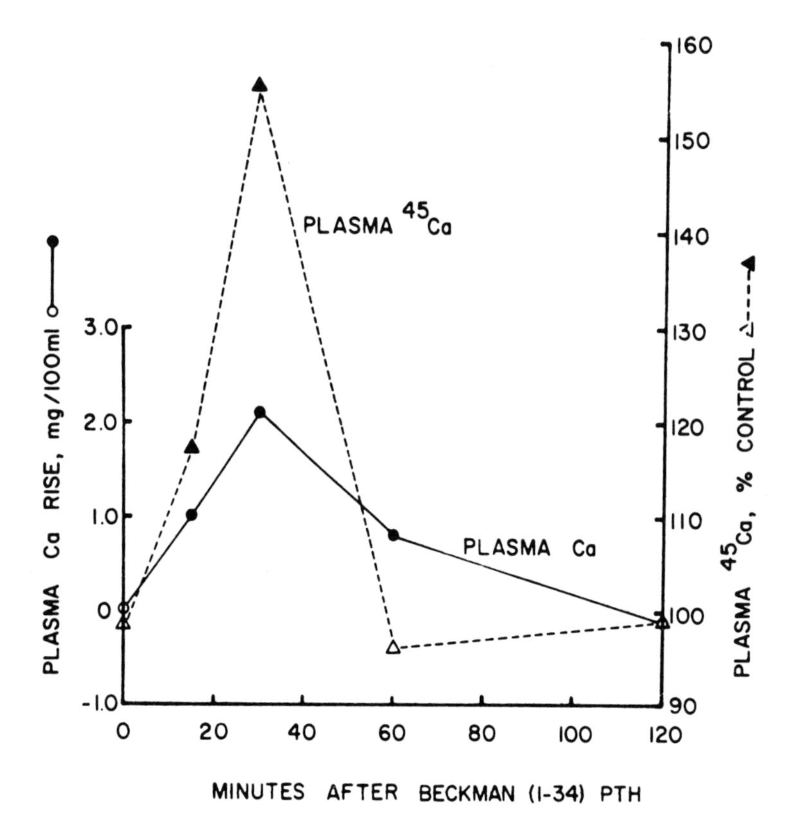

FIGURE 21–2. Plasma total calcium and ^{45}Ca responses in 9-day-old chickens 15, 30, 60, and 120 min after an i.v. injection of synthetic bovine PTH(1–34). Each bird received approximately 2 μCi ^{45}Ca simultaneously with 20 units of PTH. Total calcium values are expressed in terms of the increase of the mean PTH-treated levels over the controls at each interval. Plasma ^{45}Ca data are presented as the mean of the PTH-treated group expressed as a percentage of that of the control group for each interval. Those points indicated by solid circles or triangles are significant responses. A significant rise in plasma ^{45}Ca levels was seen at 15 min (1-tailed test) and at 30 min. (Data taken from Kenny and Dacke, 1974).

out the complications of anesthesia. In spite of this paradox, current evidence suggests that under appropriate conditions bPTH decreases urinary calcium in avian species by increasing tubular reabsorption of calcium.

Urinary Phosphate Response. Earlier studies of the effects of bPTH on urinary phosphate excretion were associated with the chicken only. All reports agree that the hormone causes an increased excretion of urinary phosphate (Levinsky and Davidson, 1957; Martindale, 1973). Renal tubular secretion appears to play a role in the response, although decreased tubular reabsorption of phosphate plays a part, at least in the laying hen (Martindale, 1973). Martindale also observed that bPTH *decreased* renal tubular secretion in contrast to the findings of others.

Clark and her associates have studied these phenomena in avian species other than the chicken. In normal starlings (Clark et al., 1976) and Japanese quail (Clark and Sasayama, 1981), intravenous administration of crude preparations of bPTH caused a marked hyperphosphaturic response consistent with that response observed in chickens; this was attributed to enhanced renal tubular secretion of phosphate.

Renal Vitamin D Endocrine System. The concept that PTH regulates the avian renal vitamin endocrine system, leading to increased renal synthesis of the hormonal form of vitamin D_3, 1,25-dihydroxyvitamin D_3 [1,25-$(OH)_2D_3$], dates from the pioneering observations of Fraser and Kodicek (1973). These authors found that injection of bPTH into chickens led to enhanced 25-hydroxyvitamin D_3-1-hydroxylase activity in the kidneys when they were removed 3 hr after the last injection of bPTH. Parathyroidectomy led to a decline in 1-hydroxylase activity (the renal enzyme responsible for converting the physiologically inactive circulating substrate, 25-hydroxyvitamin D_3, to its active hormonal form). The relationship of this observation in chickens to the experience in mammals has been reviewed by Kenny (1981). Whether or not the effect of bPTH on renal 1-hydroxylase activity represents a direct effect on the kidney target cells is not clear. Under certain conditions, a direct in vitro effect of bPTH on renal 1-hydroxylase activity has been reported (Raumussen et al., 1972), but others have failed to observe a direct in vitro effect (Henry, 1977).

The renal vitamin D endocrine system of the Japanese quail responds to administration of bPTH in a fashion similar to that observed in chickens (Baksi and Kenny, 1979). These findings are presented in Figure 21–3; note that the hypercalcemic and hypophosphatemic responses to intramuscular administration of a crude preparation of bPTH are expressed at 4 hr, prior to the effects on renal 1- and 24-hydroxylase activities. These responses have been confirmed using purified

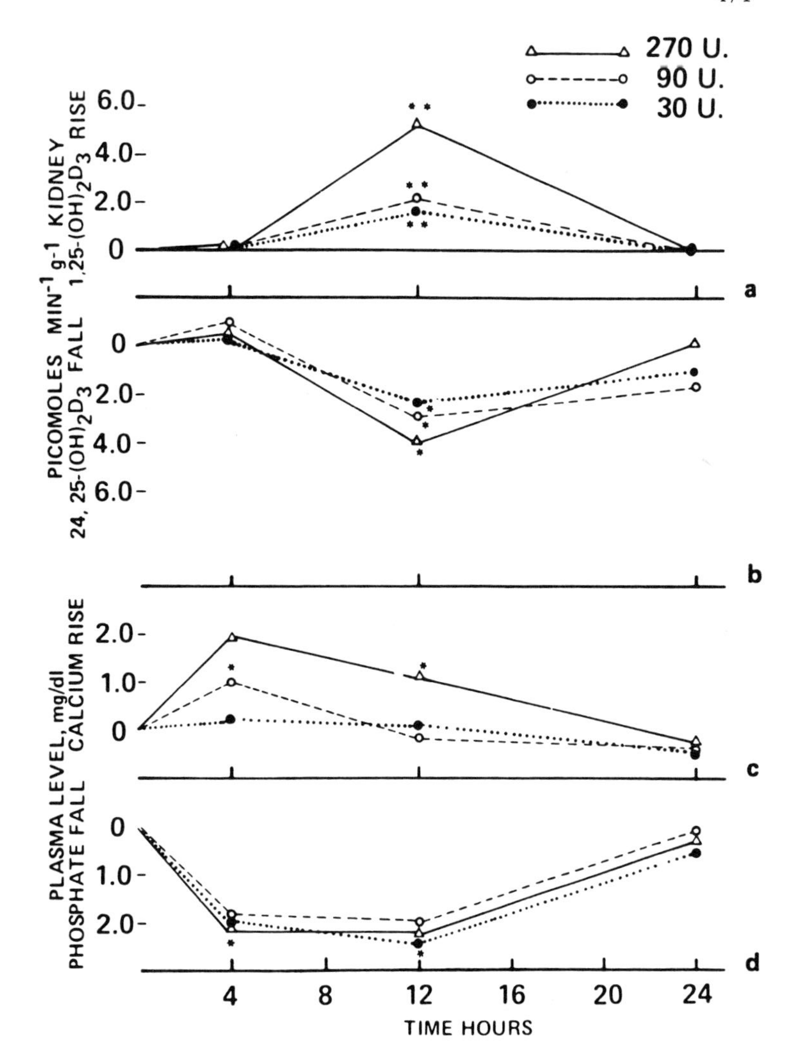

FIGURE 21–3. Effect of PTH injection on in vitro renal metabolism of tritiated 25-$(OH)D_3$, plasma calcium, and plasma inorganic phosphate levels in immature (4-week-old) female Japanese quail. PTH was injected at 30 (●), 90 (○), or 270 (△) USP units/kg i.m. The kidneys were removed 4, 12, or 24 hr later, homogenized, and incubated with 1.0×1.0^{-7} *M* 25-$(OH)D_3$ for 20 min. The rise in 1,25-$(OH)_2D_3$ or fall in 24, 250-$(OH)_2D_3$ production is expressed in picomoles per min per g kidney (**a** and **b**). The plasma calcium and inorganic phosphate data are presented in c and **d**. All responses are relative to controls receiving vehicle alone. Asterisks indicate significant responses (* = $p < .05$, ** = $p < .01$). (Data taken from Baksi and Kenny, 1979.)

bPTH(1–84) and synthetic bPTH(1–34) preparations (Kenny and Pang, 1982).

There is general agreement that bPTH (at least pharmacologically) regulates the renal production of 1,25-$(OH)_2D_3$ by enhancing its synthesis in avian species. One report did fail to confirm this concept in the chicken (Tanaka et al., 1975).

Renal Adenylate Cyclase. The classical observation of Chase and Aurbach (1967) in rats showed that intravenous administration of bPTH(1–84) leads to an increase in urinary cAMP, a response that preceded the hyperphosphaturic response. Since then a direct in vitro effect of bPTH on renal adenylate cyclase activity has been reported in several species (Aurbach and Chase, 1976). Although much of the early work was performed using renal membrane preparations made from rat kidneys, avian and canine renal sources have been favored by some because of the apparent greater sensitivity of the in vitro response of the enzyme to bPTH.

The greater sensitivity of the response of avian renal adenylate cyclase to bPTH was first reported by Martin et al. (1974) in chicken preparations. Synthetic bovine and human PTH(1–34) preparations proved to have sensitivities that were 10–20 times that experienced with rat renal tissue. Assay conditions permitted the detection of 40 ng/ml of bPTH(1–34) in the incubation tube. Kenny (unpublished observations) found similar sensitivity levels using renal adenylate cyclase preparations made from Japanese quail kidneys; 27 ng/ml of bPTH(1–34) were detectable in the incubation tube.

There is good evidence to indicate that the renal adenylate cyclase response to bPTH is associated with the hyperphosphaturic response in mammals (Chase and Aurbach, 1967; Aurbach and Chase, 1976) and in chickens (Pines et al., 1983). However, the evidence from Pang's laboratory that the vascular smooth muscle response to bPTH in mammalian species is associated with stimulation of adenylate cyclase activity does complicate earlier interpretations of the adenylate cyclase response to bPTh in the mammalian kidney. This organ is highly vacularized and has a vascular bed that is particularly responsive to bPTH (Pang et al., 1980a). We hoped that the structure–activity studies involving oxidized forms of bPTH(1–34) and bPTH(1–84) from Kenny's laboratory (see additional discussion below) might throw some light on this problem. Unfortunately, mild oxidation of bPTH(1–34) partly or completely eliminates all three responses (renal adenylate cyclase, hyperphosphaturic, and hypotensive) to the hormone (Kenny and Pang, 1982). Had oxidation exhibited some discrimination between these three responses, an important question might have been answered.

Skeletal Response: Cortical and Medullary Bone. Three major questions need to be addressed concerning the actions of PTH on avian bone. Does PTH have a direct, major effect on bone resorption in avian species? Does PTH act on normal (cortical) bone, or on medullary bone, or both, in the reproductively active female? Assuming that PTH does indeed affect avian bone, at least pharmacologically, by which bone cells is the response mediated?

Most of the evidence favoring a resorbing action of PTH on avian bone is indirect. It derives from extrapolations of the hypercalcemic response and from experiments involving prelabeling of the bone with radiocalcium, as has been discussed above. In fact some investigators have interpreted their results as indicating an indirect effect resulting from changes in bone blood flow (Mueller et al., 1973a, b). Nevertheless, cultured bone cells derived from chicken embryos appear to have receptors for bPTH(1–34) (Pliam et al., 1982).

There are few reports of direct in vitro effects of bPTH on calcium release from avian bone. Embryonic (12-day-old) chicken limb bones prelabeled with ^{45}Ca responded to bPTH in culture by release of radiocalcium into the medium (Rosen and Clark, 1981). Ramp and McNeil (1978) demonstrated net calcium efflux from chicken embryonic (13-day-old) tibiae cultured in vitro, but only in media containing lower (<2.2 m*M*) concentrations of calcium. Kenny (unpublished results) failed to demonstrate bone resorption as measured by calcium release into the media when using cultured embryonic or neonatal Japanese quail calvaria and concentrations of synthetic bPTH(1–34) ranging from 10^{-6} to 10^{-8} *M*. Thus, the evidence for a direct resorbing effect of bPTH on normal postembryonic avian bone remains inadequately documented.

In the reproductively active female bird with medullary bone, in which the latter is assumed to be a major source of mineral for the egg-laying process, does PTH play a physiologic role? Is cortical or medullary bone preferentially resorbed to supply the calcium for the egg (both shell and yolk)? The latter represents approximately 10% of the total body stores of calcium of the laying hen, and in prolific egg-layers, such as the chicken and the Japanese quail, presents a calcium homeostatic problem of major proportions. The evidence, summarized elsewhere (Simkiss, 1967; Kenny and Dacke, 1975), points to medullary bone as the major source of calcium required for its incorporation into the egg during egg-laying activity under conditions of adequate dietary calcium.

The precise physiologic role of PTH in this process remains unclear. Miller has produced pharmacologic evidence that exogenous bPTH administration has definite morphologic effects on the osteoclasts of Japanese quail medullary bone (Miller, 1978, 1981). Miller found in egg-laying Japanese quail that the osteoclasts of medullary bone, although varying little in number, do exhibit dramatic alterations in cell-surface features during the egg-laying cycle. During the first 6 hr of the 24-hr cycle, the osteoclasts have features that are interpreted as representing quiescence or inactivity. Later, during the phase of rapid calcification, the osteoclasts are claimed to revert to an active, bone-resorbing state. If bPTH is administered to egg-laying Japanese quail during the quiescent phase (4 hr after oviposition/ovulation) the osteoclasts of the medullary bone, removed 20 min after injection of bPTH, resemble the active, bone-resorbing state. This circumstantial evidence does point to a physiologic role for PTH in the resorption of medullary bone during the egg-laying cycle (Dacke, 1979).

Hypotensive and Smooth Muscle Responses. Bovine PTH has long been known to exhibit hypotensive effects in *mammlian* species (Handler and Cohn, 1952; Charbon, 1966). Pang and his associates have studied this phenomenon extensively, not only in mammalian species, such as rats and dogs (Pang et al., 1980a and c, 1983), but also in several submammalian vertebrates including the chicken (Pang et al., 1980d).

Structure–Activity Relationships of Parathyroid Hormone. Kenny and his associates have made an extensive study of the structure–activity relationships involving several responses to bPTH(1–34) and to bPTH(1–84) and their oxidized forms that are produced by mild treatment with hydrogen peroxide. Such treatment, at least in the synthetic bPTH(1–34) preparation, is known to affect only the two methionines in positions 8 and 18 (Pang et al., 1983) and to result in marked conformational changes in the molecule (Hong et al., 1983). Many of the important responses studied were in an avian species, the Japanese quail, and included the hypocalcemic, renal adenylate cyclase, and the renal vitamin D endocrine systems. The results summarized in Table 21–2, indicated that mild oxidation of bPTH(1–34) results in the loss of the renal adenylate cyclase response but leaves the hypercalcemic and renal 1-hydroxylase-stimulating responses intact. In contrast, oxidation of bPTH(1–84), which

TABLE 21–2. Qualitative responses in the Japanese quail to bovine PTH(1–34) and PTH(1–84) and their oxidized forms

Receptor type	Response	Bovine PTH agonist (1–34)	Ox(1–34)[a]	(1–84)	Ox(1–84)
P_1	Plasma calcium	↑	↑	↑	NS[b]
P_2	Renal 1-hydroxylase	↑	↑	↑	↑
	Renal 24-hydroxylase	↓	↓	↓	↓
P_3	Renal adenylate cyclase	↑	NS	↑	NS[b]

[a] Ox, oxidized with H_2O_2.
[b] NS, not significantly different from control response at dose(s) tested.

also contains only two methionines at positions 8 and 18, results in the loss of both the hypercalcemic and renal adenylate cyclase responses but leaves the renal 1-hydroxylase-stimulating property intact. These preliminary *qualitative* data suggest the existence of at least three types of receptors (P_1, P_2, and P_3) for bPTH in the Japanese quail. This classification is based on the differences in potency ratios of the four agonists in the three responses.

Ultimobranchial Glands

Anatomy

Location. The ultimobranchial glands of the chicken are located posterior to the parathyroid glands on either side (Bélanger, 1971). They lie caudodorsal to the base of the brachiocephalic artery into the common carotid and the subclavian arteries (see Figure 20–1). On the left side, the ultimobranchial gland lies caudodorsal to the base of the common carotid, close to the posterior part of parathyroid IV and to the carotid body. The right gland is more caudal than the left (Hodges, 1970).

Embryology. The C cells (calcitonin-secreting cells) of avian species derive from the sixth pharyngeal pouch. They are ultimately derived from the neuroectoderm of the neural crest (Pearse, 1976), and in this respect the C cells differ in origin from the parathyroid chief cells, which derive from the branchial endoderm. The C cells possess APUD (amine and amine precursor uptake and decarboxylation) qualities in common with other cells of neural crest origin. Although the C cells generally migrate to form the ultimobranchial gland in avian species, they can be found in association with parathyroid tissue in chickens and with the thyroid in pigeons (Copp, 1972; Pearse, 1976; Hoyt et al., 1973).

Morphology. The main characteristics of the avian C cells are the poorly developed, rough endoplasmic reticulum and the specific secretory granules, which are surrounded by a single membrane. The cells have numerous mitochondria, a defined Golgi apparatus, and abundant free ribosomes (Pearse, 1976).

Calcitonin

Chemistry of Avian Calcitonin. In all vertebrate species from which calcitonin (CT) has been extracted and purified, it has been found to consist of a chain of 32 amino acids with a seven-membered ring at the N terminus (Potts and Aurbach, 1976). Several groups have extracted CT from avian C cell sources (Hoyt et al., 1972; 1973; Feinblatt et al., 1973). Although the amino acid sequence of avian CT has yet to be determined, the chromatographic profile of its biologic activity resembles that of purified salmon CT. Feinblatt et al. (1973) cultured ultimobranchial glands removed from embryonic chickens in media high in calcium (2.5 m*M*) and compared the chromatographic profile of the biologic activity released into the medium with that of purified salmon CT. Sephadex G-50 gel-filtration elution patterns are presented in Figure 21–4; biologic activity was determined using the hypocalcemic response in the rat.

Nieto et al. (1973) claimed to have purified chicken CT. They found two chemical forms; the major form possessed an amino acid composition closer to that of salmon CT and a biologic potency of 4,500 MRC units/mg. Immunologic cross-reactivity of chicken ultimobranchial extracts with salmon CT antisera were studied by Cutler et al. (1974b) and compared with the cross-reactivity with antisera for human and porcine calcitonins. They found that whereas chicken and salmon calcitonins exhibited immunochemical identity, no such identity was seen when the chicken calcitonin was tested against the human and porcine antisera.

Assay of Calcitonin. Three major assay methods have been described for CT. The most popular in vivo assay uses the hypocalcemic response in the rat (Hirsch et al., 1964). Sturtridge and Kumar (1968) developed a more sensitive assay, and Kenny (1971) used a modification of this assay for the determination of the plasma

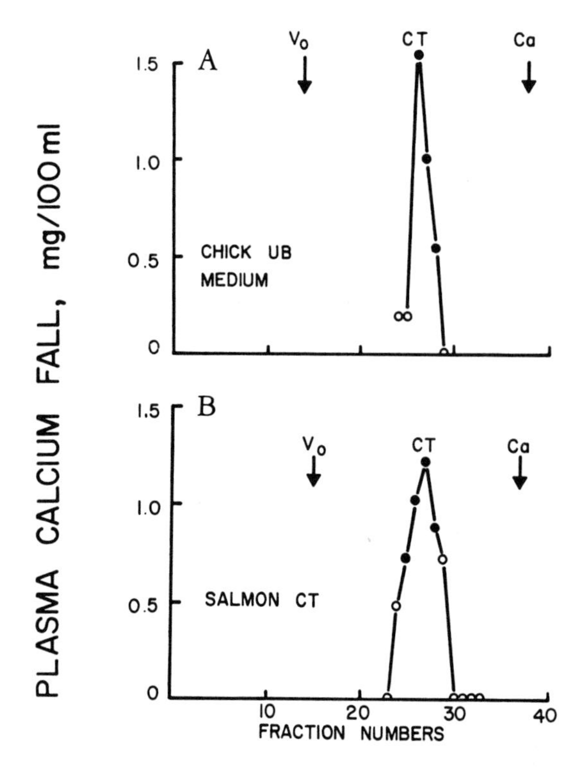

FIGURE 21–4. Sephadex G-50 (2.5 × 90 cm) gel filtration elution patterns; 12-ml fractions collected (see text). Solid circles indicate response was significant ($p < .05$) and open circles indicate response was not significant ($p > .05$) when compared with control rats receiving acetate buffer only. Void volume (V_o) was determined by optical density at 280 nm (chick media) or by the blue dextran peak (salmon CT); calcium peak (Ca) was determined by automated method. (**A**) Chick UB Medium, 0.3 ml applied. (**B**) Salmon CT, 1000 MRC mU of salmon CT (UBC 20 71/68) applied. (Data taken from Feinblatt et al., 1973).

levels of CT in various avian species (pigeons, goose, duck, chicken, and Japanese quail).

In vitro assays for CT include those using inhibition of bPTH-stimulated calcium release from bone in culture. Raisz and his associates pioneered this type of assay (Friedman et al., 1968). Its advantages include increased sensitivity and measurement of biologic activity. Its disadvantage is the lack of simplicity. The most popular in vitro assay for determining mammalian CT is radioimmunoassay; its application for determination of plasma CT in avian species has been minimal; Cutler et al. (1974a) and Baimbridge and Taylor (1980, 1981) used an antibody raised against salmon CT. Nevertheless, although blessed with exquisite sensitivity, radioimmunoassays lack the distinct advantage of knowing whether or not the material being measured is biologically active.

Ultimobranchialectomy

The major studies examining the effects of ultimobranchialectomy in avian species have revealed little of importance with respect to calcium metabolism (Brown et al., 1970; Speers et al., 1970; Kapoor and Chhabra, 1981).

Physiologic and Pharmacologic Aspects of Calcitonin in Avian Species

Blood Calcitonin Levels. Plasma CT levels, as determined by bioassay in various avian species (Kenny, 1971), ranged between 100 and 1,500 MRC μ units/ml (Table 21–3). These levels are much higher than those found in mammalian species tested (man, dog, pig, rat, marmot, cow, or rabbit) in which the plasma levels were undetectable using this approach. Subsequent work from Kenny's laboratory (Boelkins and Kenny, 1973; Dacke et al., 1972) using the same approach revealed several important findings in the Japanese quail. First, the plasma CT levels in females remained similar in immature (4-week-old) and in mature (8- and 12-week-old) female quail (Figure 21–5). The levels in males rose from immature levels to levels in mature males that were three times higher than those in females of comparable age. In spite of marked increases in plasma total calcium in the mature female, the plasma *ionic* calcium concentrations varied little (Figure 21–5). Second, calcium challenge (50 mg Ca/kg, i.p.) elicited a marked rise in plasma CT levels in male Japanese quail, indicating that the C cells respond to calcium stimulus. Third, plasma CT concentrations fluctuate during the 24-hr ovulatory cycle. The latter finding was confirmed by Baimbridge and Taylor (1981) in the chicken, using radioimmunoassay.

Taylor and his associates (Taylor et al., 1975) determined the plasma CT concentrations in the chicken embryo using a bioassay method. The concentrations rose steadily from around 1,000 μ units/ml on day 17 to 11,000 μ units/ml on day 20. The levels fell rapidly at hatching to undetectable levels 24 hr after hatching. These findings were confirmed using radioimmunoassay (Cutler et al., 1974).

Calcium and Bone Metabolism. No overt effects of CT administration on calcium metabolism have been demonstrated in avian species (Kenny, 1982); Copp (1976) reviewed the exceptions to this generalization. At first sight this appears surprising, since avian plasma CT concentrations are high (Table 21–3) and calcium challenge elicits a marked rise in plasma CT levels in male Japanese quail (Boelkins and Kenny, 1973). Administration of salmon CT to young chickens affected the ultrastructural appearance of osteoclasts in the metatarsi (Anderson et al., 1982). Ultimobranchialectomy in chickens also has no overt effects on plasma

TABLE 21–3. Determination of calcitonin levels in Sephadex-treated avian plasmas

Species	Age	Plasma Processed[a] Sample number	Plasma Processed[a] Volume (ml)	Final acetate buffer volume[b] (ml)	Rat bioassay data Assays (n)	Rat bioassay data μ units/ml of plasma ×÷ SEF[c]
Goose	35 days	V30	25	10	2	820 ×÷ 1.30 (WM)[d]
Goose	35 days	V22	50	5	1	1460 ×÷ 1.45
Chicken	2 months	JB258	50	10	1	267 ×÷ 1.16
Chicken	2 months	JB258	50	10	1	178 ×÷ 1.18
Pigeon	Unknown	V12	50	5	3	230 ×÷ 1.25 (WM)
Pigeon	Unknown	V36	48	10	1	260 ×÷ 1.29
Japanese quail	1 month	JB44	30	6	1	309 ×÷ 1.12
Japanese quail	1 month	JB216	32	6	1	300 ×÷ 1.18
Duck	23 days	V4	30	5	1	130 ×÷ 1.26
Duck	35 days	V34	50	10	1	140 ×÷ 1.27
Duck	35 days	V18	50	10	1	90 ×÷ 1.89

[a] Each sample represents plasma pooled from several birds. (Taken from Kenny, 1971).
[b] The acetate buffer volume divided into the volume of plasma processed gives the factor by which the plasma calcitonin was concentrated.
[c] SEF, Standard error factor by which the potency estimate must be divided and multiplied to give 95% confidence limits.
[d] WM, Weighted mean.

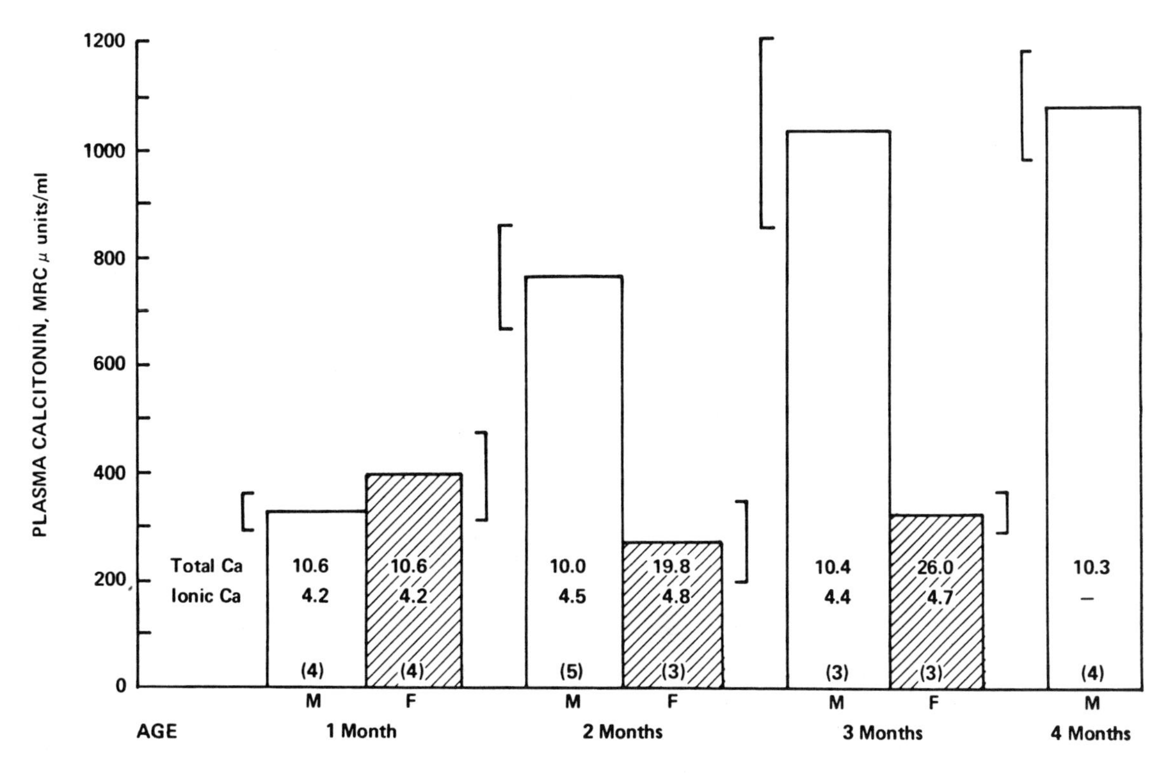

FIGURE 21–5. Plasma CT concentrations in male and female immature (1 month) and mature (2 months or more) Japanese quail. Plasma CT was determined by bioassay following Sephadex treatment as described by Kenny (1971). The total and ionic plasma concentrations are presented as mg/dl. The numbers in parentheses are the number of pooled samples analyzed; the brackets represent the standard errors of the mean. CT concentrations in the male rise markedly with maturity; those of females remain unchanged. (Data taken from Boelkins and Kenny, 1973.)

calcium or bone metabolism (Brown et al., 1970), although Copp had contrary evidence (Copp, 1976). Clark and Wideman (1980) did not observe any effects of salmon CT on serum calcium in intact starlings. A hypocalcemic response was observed in parathyroidectomized birds. An attempt to place these findings in physiologic perspective is given below.

Renal Vitamin D Endocrine System. The renal vitamin D endocrine system of avian species, especially in chickens and Japanese quail, has been widely studied. The kidney synthesizes and secretes 1,25-$(OH)_2D_3$ in response to a variety of stimuli including PTH (Kenny, 1981). There are contradictory reports that CT influences the production of 1,25-$(OH)_2D_3$ by the avian kidney. Rasmussen et al. (1972) reported that calcitonin, when added in vitro to isolated renal tubules of the chicken, inhibited PTH-stimulated synthesis of 1,25-$(OH)_2D_3$. Subsequently, Larkins et al. (1974) claimed that salmon CT added in vitro enhanced 1,25-$(OH)_2D_3$ production in a similar renal preparation. These conflicting findings suggest that at the present time, CT has no major physiologic or pharmacologic *direct* influence on the avian renal vitamin D endocrine system.

Physiologic Role of Calcitonin. What, if any, is the physiologic role of CT in calcium homeostasis in avian species? Two suggestions have immediate appeal: first, that CT is already circulating at supramaximal concentrations such that administration of exogenous CT can have no further effect on calcium and bone metabolism; second, that CT is not a calcemic hormone in birds, and that its physiologic role lies in functions not concerned with calcium and bone metabolism.

References

Anderson, R.E., H. Schraer, and C.V. Gay. (1982). Ultrastructural immunocytochemical localization of carbonic anhydrase in normal and calcitonin-treated chick osteoclasts. Anat. Rec., 204, 9.

Aurbach, G.D., and L.R. Chase. (1976). Cyclic nucleotides and biochemical actions of parathyroid hormone and calcitonin. In "Parathyroid Gland" (G.D. Aurbach, Ed.); "Handbook of Physiology," Section 7 (R.O. Greep and E.B. Astwood, Eds.). Washington, D.C.: American Physiological Society, p. 353.

Baimbridge, K.G., and T.G. Taylor. (1980). Role of calcitonin in calcium homeostasis in the chick embryo. J. Endocrinol., 85, 171.

Baimbridge, K.G., and T.G. Taylor (1981). The role of calcitonin in controlling hypercalcaemia in the domestic fowl (*Gallus domesticus*). Comp. Biochem. Physiol. A, 68, 647.

Baksi, S.N., and A.D. Kenny. (1977). Vitamin D_3 metabolism in immature Japanese quail: Effects of ovarian hormones. Endocrinology, 101, 1216.

Baksi, S.N., and A.D. Kenny. (1979). Acute effects of parathyroid extracts on renal vitamin D hydroxylases in Japanese quail. Pharmacology, 18, 169.

Bélanger, L.F. (1971). The ultimobranchial gland of birds and the effects of nutritional variations. J. Exp. Zool., 178, 125.

Benoit, J., G. Fabiani, R. Grangard, and J. Clavert. (1941). Suppression par la parathyroïdectomie de l'action hypercalcemiante du dipropionate d'estradiol chez le canard domestique. C.R. Hebd. Seances Acad. Sci., 135, 1606.

Berson, S.A., R.S. Yalow, G.D. Aurbach, and J.T. Potts, Jr. (1963). Immunoassay of bovine and human parathyroid hormone. Proc. Natl. Acad. Sci. U.S.A., 49, 613.

Boelkins, J.N., and A.D. Kenny, (1973). Plasma calcitonin levels in Japanese quail. Endocrinology, 92, 1754.

Brown, D.M., D.Y.E. Perey, and J. Jowsey. (1970). Effects of ultimobranchialectomy on bone composition and mineral metabolism in the chicken. Endocrinology, 87, 1282.

Candlish, J.K. (1970). The urinary excretion of calcium, hydroxyproline and uronic acids in the laying fowl after the administration of parathyroid extract. Comp. Biochem. Physiol., 32, 703.

Candlish, J.K., and T.G. Taylor. (1970). The response-time to the parathyroid hormone in the laying fowl. J. Endocrinol., 48, 143.

Charbon, G.A. (1966). A diuretic and hypotensive action of a parathyroid extract. Acta Physiol. Pharmacol. Neerl., 14, 52.

Chase, L.R., and G.D. Aurbach. (1967). Cyclic AMP and the mechanism of action of parathyroid hormone. In "Parathyroid Hormone and Thyrocalcitonin" (R.V. Talmage and L.F. Bélanger, Eds.). Amsterdam and New York: Excerpta Medica, p. 247.

Cherian, A.G., and J.D. Cipera. (1968). Effects of parathyroidectomy and of administration of parathormone on serum calcium in immature chicks. Poult. Sci., 47, 76.

Clark, N.B., and R.F. Wideman, Jr. (1977). Renal excretion of phosphate and calcium in parathyroidectomized starlings. Am. J. Physiol., 233, F138.

Clark, N.B., and R.F. Wideman, Jr. (1980). Calcitonin stimulation of urine flow and sodium excretion in the starling. Am. J. Physiol., 238, R406.

Clark, N.B., and Y. Sasayama. (1981). The role of parathyroid hormone on renal excretion of calcium and phosphate in the Japanese quail. Gen. Comp. Endocrinol., 45, 234.

Clark, N.B., E.J. Baun, and R.F. Wideman, Jr. (1976). Parathyroid hormone and renal excretion of phosphate and calcium in normal starlings. Am. J. Physiol., 231, 1152.

Collip, J.B. (1925–1926). The parathyroid glands. Harvey Lectures, 21, 113.

Copp, D.H. (1972). Evolution of calcium regulation in vertebrates. Clin. Endocrinol. Metab., 1, 21.

Copp, D.H. (1976). Comparative endocrinology of calcitonin. In "Parathyroid Gland" (G.D. Aurbach, Ed.); "Handbook of Physiology," Section 7 (R.O. Greep and E.B. Astwood, Eds.). Washington D.C.: American Physiological Society, p. 431.

Cutler, G.B., J.F. Habener, P.C. Dee, and J.T. Potts, Jr. (1974a). Radioimmunoassay for chicken calcitonin. FEBS Lett., 38, 209.

Cutler, G.B., Jr., J.F. Habener, and J.T. Potts, Jr. (1974b). Comparative immunochemical studies of chicken ultimobranchial calcitonin. Gen. Comp. Endocrinol., 24, 183.

Dacke, C.G. (1979). Calcium Regulation in Sub-Mammalian Vertebrates. London: Academic Press.

Dacke, C.G., and A.D. Kenny. (1973). Avian bioassay method for parathyroid hormone. Endocrinology, 92, 463.

Dacke, C.G., J.N. Boelkins, W.K. Smith, and A.D. Kenny (1972). Plasma calcitonin levels in birds during the ovulation cycle. J. Endocrinol., 54, 369.

Dacke, C.G., X.J. Musacchia, W.A. Volkert, and A.D. Kenny. (1973). Cyclical fluctuations in the levels of blood calcium pH and pCO_2 in Japanese quail. Comp. Biochem. Physiol. A, 44, 1267.

Feinblatt, J.D., L.G. Raisz, and A.D. Kenny. (1973). Secretion of avian ultimobranchial calcitonin in organ culture. Endocrinology, 93, 277.

Fraser, D.R., and E. Kodicek. (1973). Regulation of 25-hydroxy-cholecalciferol-1-hydroxylase activity in kidney by parathyroid hormone. Nature (London), 241, 163.

Friedman, J., W.Y.W. Au, and L.G. Raisz. (1968). Responses of fetal rat bone to thyrocalcitonin in tissue culture. Endocrinology, 82, 149.

Gandal, C.P. (1969). Avian anesthesia. Fed. Proc. Fed. Am. Soc. Exp. Biol., 28, 1533.

Handler, P., and D.V. Cohn. (1952). Effect of parathyroid extract on renal function. Am. J. Physiol., 169, 188.

Henry, H.L. (1977). Metabolism of 25-hydroxyl-vitamin D_3 by primary cultures of chick kidney cells. Biochem. Biophys. Res. Commun., 74, 768.

Hirsch, P.F., E.F. Voelkel, and P.L. Munson. (1964). Thyrocalcitonin: Hypocalcemic hypophosphatemic principle of the thyroid gland. Science, 146, 412.

Hodges, R.D. (1970). The structure of the fowl's ultimobranchial gland. Ann. Biol. Anim. Biochim. Biophys., 10, 255.

Hong, B.-S., J. Liang, M.C.M. Yang, A.D. Kenny, and P.K.T. Pang. (1983). Relationship between molecular structure and vascular action of parathyroid hormone. Fed. Proc. Fed. Am. Soc. Exp. Biol., 42, 848.

Hoyt, R.F., Jr., A.H. Tashjian, Jr., and D.W. Hamilton. (1972). Distribution of thyroid, parathyroid and ultimobranchial hypocalcemic factors in birds. I. Thyroid and ultimobranchial calcitonin in pigeons and pullets. Endocrinology, 91, 770.

Hoyt, R.F., Jr., D.W. Hamilton, and A.H. Tashjian, Jr. (1973). Distribution of thyroid, parathyroid and ultimobranchial hypocalcemic factors in birds. II. Morphology, histochemistry, and hypocalcemic activity of pigeon thyroid glands. Anat. Rec., 176, 1.

Kapoor, A.S., and S.K. Chhabra. (1981). Relative physiological activity of calcitonin in the thyroid and ultimobranchial glands of the pigeon, *Columbia livia* Gmelin. Gen. Comp. Endocrinol., 44, 307.

Kenny, A.D. (1971). Determination of calcitonin in plasma by bioassay. Endocrinology, 89, 1005.

Kenny, A.D. (1981). Intestinal Absorption of Calcium and Its Regulation. Boca Raton: CRC Press.

Kenny, A.D. (1982). The avian hypercalcemic assay for parathyroid hormone and selected applications. In "Aspects of Avian Endocrinology: Practical and Theorectical Implications" (C. G. Scanes, M.A. Ottinger, A.D. Kenny, J. Balthazart, J. Cronshaw, and I. Chester Jones, Eds.). Lubbock: Texas Tech Press, p. 69.

Kenny, A.D., and P.L. Munson. (1959). A method for the biological assay of phosphaturic activity in parathyroid extracts. Endocrinology, 64, 513.

Kenny, A.D., and C.G. Dacke. (1974). The hypercalcemic response to parathyroid hormone in Japanese quail. J. Endocrinol., 62, 15.

Kenny, A.D., and C.G. Dacke. (1975). Parathyroid hormone and calcium metabolism. In "World Review of Nutrition and Dietetics," Vol. 20 (G.H. Bourne, Ed.). Basel: S. Karger, p. 231.

Kenny, A.D., and P.K.T. Pang. (1982). Response of the renal vitamin D endocrine system to oxidized parathyroid hormone (1–34). Proc. Soc. Exp. Biol. Med., 171, 191.

Kenny, A.D., D.J. Ahearn, and J.F. Maher. (1976). Improved method for determining parathyroid hormone in biological material. Biochem. Med., 16, 201.

Larkins, R.G., S.J. MacAuley, A. Rapoport, T.J. Martin, B.R. Tulloch, P.G.H. Byfield, E.W. Matthews, and I. MacIntyre. (1974). Effects of nucleotides, hormones, ions and 1,25-dihydroxycholecalciferol on 1,25-dihydroxycholecalciferol production in isolated chick renal tubules. Clin. Sci. Mol. Med., 46, 569.

Levinsky, N.G., and D.G. Davidson. (1957). Renal action of parathyroid extract in the chicken. Am. J. Physiol., 191, 530.

MacGregor, R.R., L.L.H. Chiu, J.W. Hamilton, and D.V. Cohn. (1973). Partial purification of parathyroid hormone from chicken parathyroid glands. Endocrinology, 92, 1312.

Marcus, R., and G.D. Aurbach. (1969). Bioassay of parathyroid hormone *in vitro* with a stable preparation of adenyl cyclase from rat kidney. Endocrinology, 85, 801.

Martin, T.J., N. Vakakis, J.A. Eisman, S.J. Livesey, and G.W. Tregear. (1974). Chick kidney adenylate cyclase: Sensitivity to parathyroid hormone and synthetic human and bovine peptides. J. Endocrinol., 63, 369.

Martindale, L. (1973). Phosphate excretion in the laying hen (*Gallus domesticus*). J. Physiol., 231, 439.

Miller, S.C. (1978). Rapid activation of the medullary bone osteoclast cell surface by parathyroid hormone. J. Cell Biol., 76, 615.

Miller, S.C. (1981). Osteoclast cell-surface specializations and nuclear kinetics during egg-laying in Japanese quail. Am. J. Anat., 162, 35.

Mueller, W.J., R.L. Brubaker, C.B. Gay, and J.N. Boelkins. (1973a). Mechanisms of bone resorption in laying hens. Fed. Proc. Fed. Am. Soc. Exp. Biol., 32, 1951.

Mueller, W.J., K.L. Hall, C.A. Maurer, Jr., and I.G. Joshua. (1973b). Plasma calcium and inorganic phosphate response of laying hens to parathyroid hormone. Endocrinology, 92, 853.

Munson, P.L. (1955). Studies on the role of the parathyroids in calcium and phosphorus metabolism. Ann. N.Y. Acad. Sci., 60, 776.

Munson, P.L. (1960). Recent advances in parathyroid hormone research. Fed. Proc. Fed. Am. Soc. Exp. Biol., 19, 593.

Munson, P.L. (1961). Biological assay of parathyroid hormone. In "The Parathyroids" (R.O. Greep and R.V. Talmage, Eds.). Springfield: Charles C Thomas, p. 94.

Nevalainen, T. (1969). Fine structure of the parathyroid gland of the laying hen (*Gallus domesticus*). Gen. Comp. Endocrinol., 12, 561.

Nieto, A., F. Moya, and J.L. R-Candela. (1973). Isolation and properties of two calcitonins from chicken ultimobranchial glands. Biochim. Biophys. Acta, 322, 383.

Pang, P.K.T., H.F. Janssen, and J.A. Yee. (1980a). Effects of synthetic parathyroid hormone on vascular beds of dogs. Pharmacology, 21, 213.

Pang, P.K.T., A.D. Kenny, and C. Oguro. (1980b). The evolution of the endocrine control of calcium metabolism. In "Evolution of Vertebrate Endocrine Systems" (P.K.T. Pang and A. Epple, Eds.). Lubbock: Texas Tech Press, p. 323.

Pang, P.K.T., T.E. Tenner, Jr., J.E. Yee, M. Yang, and H.F. Janssen (1980c). Hypotensive action of parathyroid hormone preparations on rats and dogs. Proc. Natl. Acad. Sci., 77, 675.

Pang, P.K.T., M. Yang, C. Oguro, J.G. Phillips, and J.A. Yee. (1980d). Hypotensive actions of parathyroid hormone preparations in vertebrates. Gen. Comp. Endocrinol., 41, 135.

Pang, P.K.T., M.C.M. Yang, H.T. Keutmann, and A.D. Kenny. (1983). Structure activity relationship of parathyroid hormone: Separation of the hypotensive and hypercalcemic properties. Endocrinology, 112, 284.

Parsons, J.A., B. Reit, and C.J. Robinson. (1973). Bioassay

for parathyroid hormone using chicks. Endocrinology, 92, 454.

Pearse, A.G.E. (1976). Morphology and cytochemistry of thyroid and ultimobranchial C cells. In "Parathyroid Gland" (G.D. Aurbach, Ed.); "Handbook of Physiology," Section 7 (R.O. Greep and E.B. Astwood, Eds.). Washington, D.C.: American Physiological Society p. 411.

Pines, M., D. Polin, and S. Hurwitz. (1983). Urinary cyclic AMP excretion in birds: Dependence on parathyroid hormone activity. Gen. Comp. Endocrinol., 49, 90.

Pliam, N.B., K.O. Nyiredy, and C.D. Arnaud. (1982). Parathyroid hormone receptors in avian bone cells. Proc. Natl. Acad. Sci. U.S.A., 79, 2061.

Polin, D., and P.D. Sturkie. (1957). The influence of the parathyroids on blood calcium levels and shell deposition in laying hens. Endocrinology, 60, 778.

Polin, D., and P.D. Sturkie. (1958). Parathyroid and gonad relationship in regulating blood calcium fractions in chickens. Endocrinology 63, 177.

Polin, D., P.D. Sturkie, and W. Hunsaker. (1957). The blood calcium response of the chicken to parathyroid extracts. Endocrinology, 60, 1.

Potts, J.T., Jr., and G.D. Aurbach. (1976). Chemistry of the calcitonins. In "Parathyroid Gland" (G.D. Aurbach, Ed.). "Handbook of Physiology," Section 7 (R.O. Greep and E.B. Astwood, Section Eds.). Amer. Physiol. Soc., Washington, D.C., p. 423.

Ramp, W.K., and R.W. McNeil (1978). Selective stimulation of net calcium efflux from chick embryo tibiae by parathyroid hormone in vitro. Calcif. Tissue Res., 25, 227.

Rasmussen, H., M. Wong, D. Bikle, and D.B.P. Goodman. (1972). Hormonal control of the renal conversion of 25-hydroxycholecalciferol to 1,25-dihydroxycholecalciferol. J. Clin. Invest., 51, 2502.

Riddle, O., and M.R. McDonald. (1945). The partition of plasma calcium and inorganic phosphorus in estrogen-treated normal and parathyroidectomized birds. Endocrinology, 36, 48.

Rosen, V., and N.B. Clark. (1981). Effects of parathyroid hormone and calcitonin and embryonic chick bone in organ culture. Gen. Comp. Endocrinol., 44, 319.

Roth, S.I., and A.L. Schiller. (1976). Comparative anatomy of the parathyroid glands. In "Parathyroid Gland" (G.D. Aurbach, Ed.); "Handbook of Physiology," Section 7 (R.O. Greep and E.B. Astwood, Eds.). Washington, D.C.: American Physiological Society, p. 281.

Simkiss, K. (1967). Calcium in Reproductive Physiology. New York: Reinhold.

Smith, G.C. (1945). Technique for parathyroidectomy in pigeons. Anat. Rec., 92, 81.

Speers, G.M., D.Y.E. Perey, and D.M. Brown. (1970). Effect of ultimobranchialectomy in the laying hen. Endocrinology, 87, 1292.

Sturtridge, W.C., and M.A. Kumar. (1968). Assay of calcitonin in human plasma. Lancet, I, 725.

Tanaka, Y., R.S. Lorenc, and H.F. DeLuca. (1975). The role of 1,25-dihydroxyvitamin D_3 and parathyroid hormone in the regulation of chick renal 25-hydroxyvitamin D_3-24-hydroxylase. Arch. Biochem. Biophys., 171, 521.

Taylor, T.G., O. Balderstone, and P.E. Lewis. (1975). Changes in the concentration of calcitonin and in the plasma of chick embryos during incubation. J. Endocrinol., 66, 363.

Zanelli, J.M., D.J. Lea, and J.A. Nisbet. (1969). A bioassay method *in vitro* for parathyroid hormone. J. Endocrinol., 43, 33.

22
Adrenals

S. Harvey C.G. Scanes and
K.I. Brown

Anatomy

The paired avian adrenal glands are composed of intermingled chromaffin and cortical (interrenal) tissue. The glands are located (Figure 22–1) anterior and medial to the cephalic lobe of the kidney. These glands are flattened and lie close together, even fusing in some species. The adrenal glands receive direct arterial blood supply via branches of the renal artery, and each gland has a single vein draining into the posterior vena cava. Sympathetic nerves reach cranial and caudal ganglia on the pericapsular sheath of the adrenal glands. Nonmyelinated fibers originate from these ganglia and penetrate the gland. Each fiber innovates up to three chromaffin cells (Unisicker, 1973a). Interrenal tissue is only rarely penetrated by nerve fibers.

Microanatomy

Avian adrenal glands are not clearly divided into outer cortex and inner medulla, as in mammals. Cortical and chromaffin tissue is intermingled in birds, with clusters or strands of chromaffin cells distributed throughout the cortical tissue.

Chromaffin tissue constitutes about 15–25% of adrenal tissue. The chromaffin cells are in close association with blood spaces and appear to be more abundant in the middle of the gland, which is enriched with

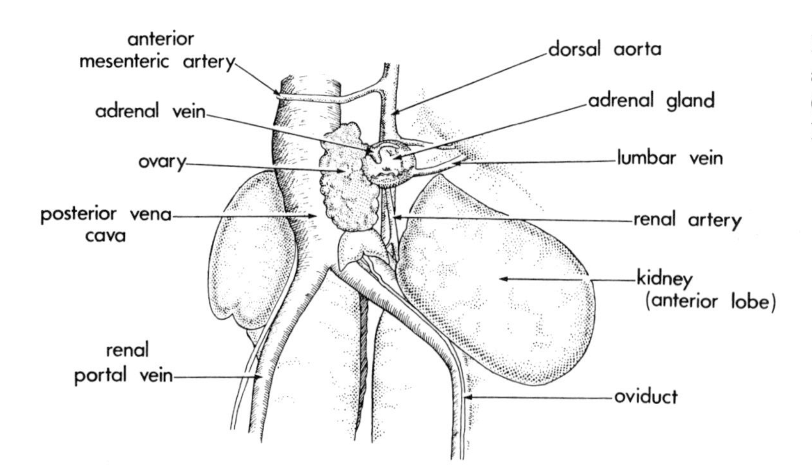

FIGURE 22–1. The position of the adrenal and its vascular supply. (Taken from Chester-Jones and Phillips, 1985.)

vascular tissue. Two distinct types of chromaffin cells exist in the avian adrenal, releasing epinephrine (E) and norepinephrine (NE) respectively. These can be differentiated cytochemically and ultrastructurally, the latter including differences in the size and shape of the cytoplasmic neurosecretory glanules. There is some controversy as to the relative proportion of cell types. Ghosh (1980) reported a greater number of NE chromaffin cells in all birds studied (except passerines), while Unsicker (1973b) considered that E chromaffin cells predominate.

Cortical tissue accounts for 70–80% of the avian adrenal. The cortical cells are arranged in numerous cords, with each being composed of a double row of parenchymous interrenal cells (see Figure 22–2). The cords radiate from the center of the gland, branching

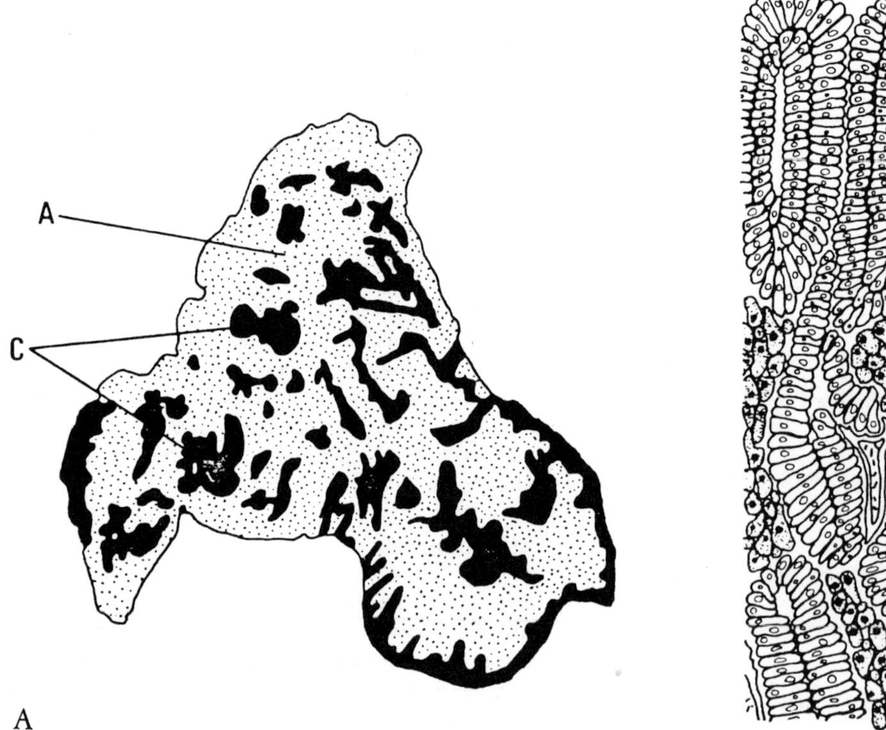

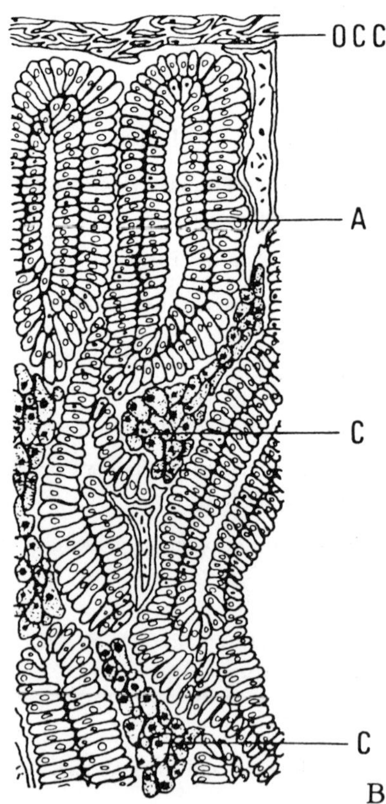

FIGURE 22–2. Microanatomy of the avian adrenal. **(A)** Distribution of chromaffin (*C*) and adrenal cortical (*A*). **(B)** Structure of the loops of adrenal cortical cells with the adrenal capsule also indicated (*OCC*). (Taken from Chester-Jones and Phillips, 1985.)

and anastomosing frequently, and loop against the inner surface of the connective tissue capsules. The arrangement of specific cell types along the cords results in some structural zonation. In birds, adrenal zonation is less clear than in mammals, with two zones, a subcapsular zone (SCZ) and an inner zone (IZ). The adrenal cortical tissue is divided into a SCZ, which is about 20–40 cells thick and produces aldosterone, and the more extensive IZ, which produces corticosterone (for details, see review by Holmes and Cronshaw, 1980). The zonation is dependent on physiologic status, being clearest when corticotropic stimulation is suppressed or enhanced. The cortical cells are characterized by the presence of numerous lipid droplets, membrane bodies, and vesicles, well-developed mitochondria, and extensive Golgi apparatus, smooth endoplasmic reticulum, and microtubules systems. There are distinct differences in the cells of the IZ and SCZ (for more detailed descriptions of cortical cell structure in the IZ and SCZ, see Holmes and Cronshaw, 1980).

Cortical Hormones

Secretions

Corticosterone is the principal steroid hormone released by the avian adrenal glands. Considerably less aldosterone is released by the avian adrenal, with the ratio of basal circulating concentrations of aldosterone to corticosterone being 1:14 in chickens and 1:63 in ducks (Radke et al., 1984). Other corticoids are synthesized and released, particularly in the embryo; these include cortisol and cortisone (Kalliecharan and Hall, 1974, 1977).

Biosynthesis

Corticosterone is synthesized from cholesterol via pregnenolone, progesterone, and 11-deoxycorticosterone (Figure 22–3). Although the avian adrenal has the potential to convert progesterone into 11-β-hydroxyprogesterone, this appears to be a secondary pathway in the synthesis of corticosterone (Nakamura et al., 1978). While most corticosterone synthesized is released or temporarily stored in the gland, some is the precursor for aldosterone synthesis (Pederneva and Lantos, 1973; Lehoux, 1974; Aupetit et al., 1979). The presence of cortisol and cortisone in the plasma of perinatal chicks reflects adrenal synthesis. In embryonic chicks, the activity of the 17-α-hydroxylase is 70% that of 21-hydroxylase. Its activity declines around hatch and is absent in the adrenals of chickens and ducks older than 2 weeks (Nakamura et al., 1978; Tanabe et al., 1979). In addition, the embryonic adrenal metabolizes 17-α-hydroxyprogesterone into androstenedione and testosterone (Tanabe et al., 1979). Indeed the adrenal may be a more important source of testosterone than the gonad in the chicken embryo. Some estradiol is also found in chick embryo adrenals (Tanabe et al., 1979).

Steroid biosynthesis involved enzymes and cytochrome P-450 located on both the endoplasmic reticulum and mitochondria. Indeed, steroidogenesis is facilitated in avian cortical cells by the close association of the cisternae of the endoplasmic reticulum and the surface membranes of the mitochondria and of the lipid (cholesterol-rich) droplets (Pearce et al., 1979; Holmes and Cronshaw, 1980). Corticosterone secretion involves the microfilaments but not the microtubules in cortical cells; corticosterone released from perfused duck adrenal slices is reduced by cytochalasin B, which impairs microfilament function, but is unaffected by colchine and vinblastine, which inhibit microtubule polymerization (Cronshaw et al., 1984). It might be noted that avian adrenal cells have lower steroidogenic enzyme activities than rat cells, considerably less corticosterone being synthesized for pregnenolone with chicken than rat cortical cells (Carsia et al., 1984).

Transport

In birds, most circulating corticosterone is transport bound in a dynamic equilibrium to plasma proteins. There appear to be two components to these transport proteins: a specific binding protein, transcortin or corticosteroid-binding globulin (CBG), which has high affinity and low binding capacity, and a low-affinity nonspecific binding protein (presumably albumin) with a very high capacity (Wingfield et al., 1984). Transcortin has been identified in the plasma of 23 avian species and has similar physiochemical properties to CBG in other vertebrates (for details, see Wingfield et al., 1984). The protein binds with corticosterone and progesterone, and in some birds it cross-reacts somewhat with testosterone. The values for the dissociation constant (K_d) and capacity vary with different birds. In chickens, CBG has a K_d of 5.6×10^{-8} mol/liter and a capacity of 1.44×10^{-8} mol/liter (Wingfield et al., 1984).

The binding of corticosterone to CBG and possibly other plasma proteins probably affects availability of the hormone at the target cell. Protein-bound steroids are presumed to enter cells less rapidly than free hormone and hence appear less active. Thus plasma proteins may provide stabilization of the concentration of free hormone. Endocrine status affects the circulating concentration of CBG, presumably by affecting its synthesis by the liver. For instance, CBG concentrations are decreased by hypophysectomy in embryos (Gasc and Martin, 1978) and by the administration of thyroxine or testosterone (Peczely and Daniel, 1979; Kovaks and Peczely, 1983).

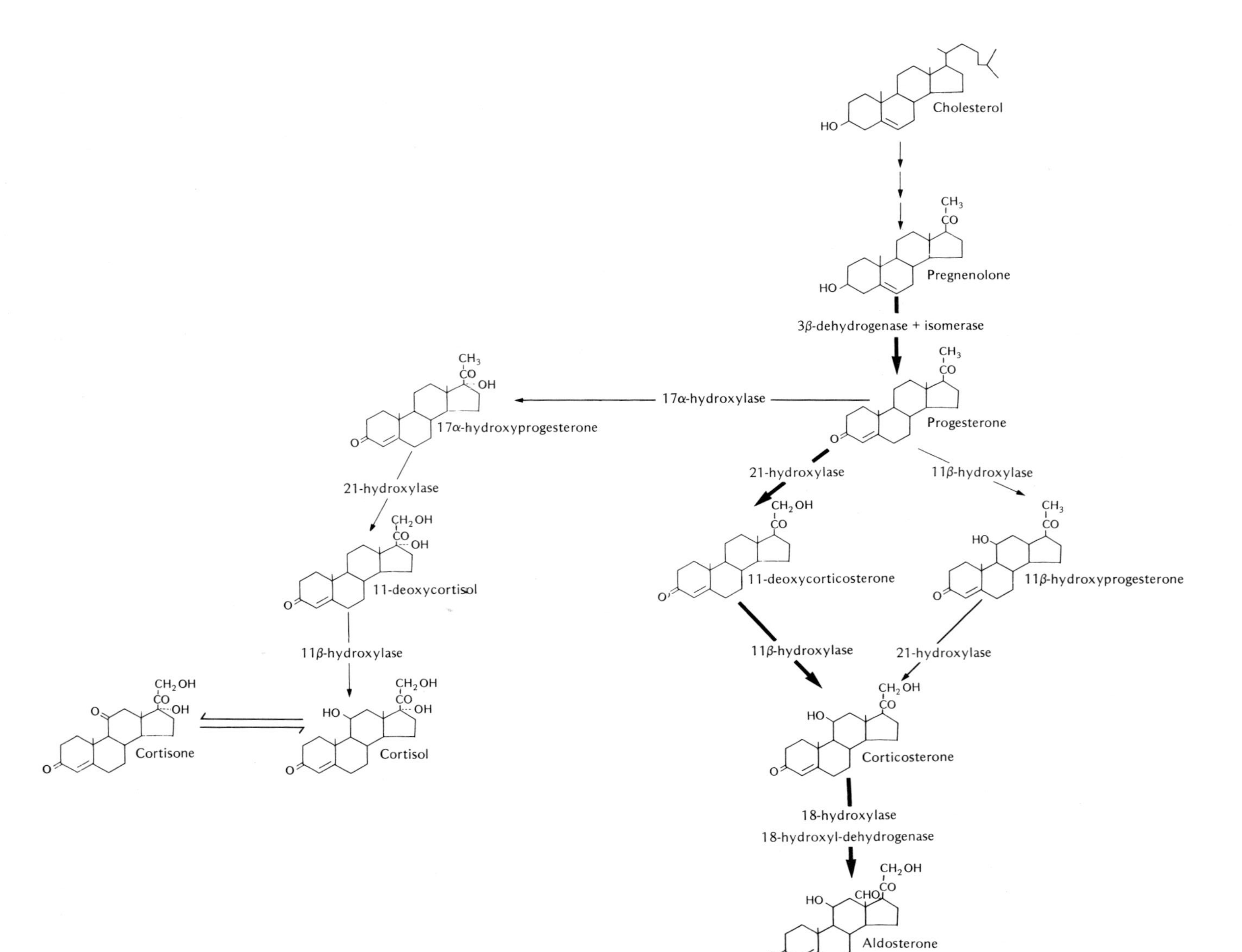

FIGURE 22–3. Steroidogenesis in the avian adrenal gland.

TABLE 22–1. Basal peripheral plasma concentrations of corticosterone and aldosterone in chickens, turkeys, and ducks

Species	Concentration	Assay	Reference
Corticosterone (ng/ml)			
Chicken	73–123	Fluorescence	Nagra et al. (1960)
	65–80	Fluorescence	Urist and Deutch (1960)
	90	CBB[a]	Nir et al. (1975)
	2	CPB	Beuving and Vander (1978)
	1–18	CPB	Scanes et al. (1980)
	1.5–6	RIA[b]	Etches (1979)
	5–12	RIA	Satterlee et al. (1980)
	0.6	RIA	Scott et al. (1983)
	0.4	RIA	Radke et al. (1984)
Turkey	113	Fluorescence	Brown (1961)
	4–20	CPB	Martin et al. (1982)
	3–29	RIA	Simensen et al. (1978)
Duck	3–8	RIA	Harvey et al. (1980)
	5–15	RIA	Gorsline and Holmes (1981)
Aldosterone (pg/ml)			
Chicken	58–95	RIA	Thomas et al. (1980)
	28	RIA	Radke et al. (1984)
Duck	43	RIA	Radke et al. (1984)

[a] Competitive protein binding.
[b] Radioimmunassay.

Assay

Adrenal corticoids were historically determined by complex chromatographic (Phillips and Chester-Jones, 1957), double-isotope (e.g., Stachenko et al., 1964), and fluorescence (Nagra et al., 1963) techniques. These have been largely replaced by competitive protein-binding assays (Wingfield and Farner, 1975) and radioimmunoassays (Etches, 1976; Harvey et al., 1980). High pressure liquid chromatography (HPLC) procedures have been employed to determine corticosterone concentrations in 500-μl samples of chicken plasma (Fowler et al., 1983). Radioimmunoassays have greater precision, sensitivity, and specificity than other techniques, and hence lower hormone concentrations are reported in studies using this technique (Table 22–1).

Clearance and Metabolism

The entry of corticoids into the peripheral circulation results in rapid dilution and allows the steroids to be available for metabolism and clearance. The volume in which corticosterone is distributed varies with the physiological state of the birds; it is increased in hypophysectomized (Bradley and Holmes, 1971) and saline-loaded (Donaldson and Holmes, 1965) ducks. Corticoids are cleared from the circulation with a simple exponential decay. Estimates of the half-life ($t_{1/2}$) of corticosterone and aldosterone are about 15 min (Table 22–2) in intact birds. The metabolic clearance rate for corticosterone is not fixed; for instance, it is decreased with age (Holmes and Kelly, 1976) and by thyroidectomy (Kovacs and Peczely, 1983). The liver is the main site of corticosterone metabolism. The

TABLE 22–2. Secretion and clearance dynamics of adrenal hormones

	Half-life (min)	Metabolic clearance rate (ml/min/kg)	Secretion rate (μg/min/kg)	Reference
Corticosterone				
Chicken	22	—	—	Birrenkott and Wiggins (1984)
Duck	11.2	44.6	1.03	Gorsline and Holmes (1982)
Japanese quail	10	—	—	Kovacs and Peczely (1983)
Pigeon	18.4	—	—	Chan et al. (1972)
Aldosterone				
Duck	12.6	87.1	0.2	Thomas and Phillips (1975)

metabolizing enzymes include 5-α-reductase, which produces 11-dehydrocorticosterone and tetrahydrocorticosterone (Daniel and Assenmacher, 1971).

Hypothalamo-Pituitary Control of Corticosterone Secretion

As in mammals, the secretion of corticosterone is primarily regulated by adrenocorticotropic hormone (ACTH, corticotropin), which is released from the pituitary gland in response to corticotropin-releasing factor (CRF). The evidence for the presence of this hypothalamo-pituitary-adrenal axis in birds has been reviewed extensively (Holmes, 1978; Bayle, 1980) (see also Chapter 17). ACTH stimulates the release of corticosterone rapidly (within 5 min) and in a dose-dependent manner (Carsia et al., 1983, 1984; Radke et al., 1985; R.V. Carsia, personal communication). Chicken adrenal cells are less responsive and sensitive to either synthetic mammalian ACTH or to ostrich ACTH than are rat adrenal cells (Carsia et al., 1984, 1985a); chicken cortical cells became increasing less sensitive to ACTH during postnatal growth (Carsia et al., 1985b). The intracellular mechanism by which ACTH stimulates corticosterone secretion involves cAMP (Carsia et al., 1985b) and also a calcium/calmodulin mechanism (R.V. Carsia, personal communication).

Glucocorticoids reduce corticosterone release from chicken adrenal cells (Carsia et al., 1983, 1984). This "self-suppression" involves adrenal 5-α-reductase and the conversion of corticosterone to an inactive metabolie, 5-α-dehydrocorticosterone. This intracellular negative feedback mechanism is inhibited by prolactin (Carsia et al., 1984).

Corticosterone and other glucocorticoids exert negative feedback influence at the hypothalamopituitary level in birds. Exogenous corticosterone suppresses basal and photo-stimulated corticosterone secretion in ducks (Peczely and Daniel, 1979). Furthermore, dexamethasone blocks basal and stress-induced corticosterone release in chickens (e.g., Etches, 1976).

Stress and Adrenal Function

Stimuli perceived as imposing a threat to survival or wellbeing increase adrenocortical activity in birds (see Table 22–3) (reviewed by Harvey et al., 1984a). Acute stresses will increase corticosterone secretion while decreasing adrenal ascorbic acid content (Freeman and Manning, 1981). Chronic stress causes adrenal hypertrophy, particularly of the inner zone. An increase in the plasma corticosterone concentration invariably accompanies stressful encounters (Table 22–3), and is thought to represent a "nonspecific" stress response (Siegel, 1980; Harvey et al., 1984a). Parallel increases

TABLE 22–3. Stress-induced corticosterone secretion[a]

Stressor	Species	Reference
Starvation	Chicken	Scanes et al. (1980)
	Turkey	Brown (1961)
	Quail	Scott et al. (1983)
	Duck	Harvey et al. (1981)
Dehydration	Chicken	Harvey et al. (1985)
	Turkey	Brown (1961)
	Duck	Harvey et al. (1981)
Hypothermia	Chicken	Freeman (1982)
	Turkey	Brown (1967)
Hyperthermia	Chicken	Edens and Siegel (1975)
	Turkey	El Halaiwani et al. (1973)
Hypoglycemia	Chicken	Edens and Siegel (1975)
Pollution	Quail, herring gulls, guillimot	Rattner et al. (1982)
	American kestrels	Rattner and Franson (1984)
Anesthesia	Chicken	Scanes et al. (1980)
	Duck	Bradley and Holmes (1971)
	Pigeon	Ramade and Bayle (1982)
Electrocution	Pigeon	Ramade and Bayle (1984)
Infection	Chicken	Curtis et al. (1980)
	Turkey	Simensen et al. (1978)
Immobilization	Chicken	Beuving and Vonder (1978)
Exertion	Chicken	Rees et al. (1984)
	Duck	Harvey and Phillips (1980)
Frustration	Chicken	Harvey et al. (1985)
Fear	Pigeon	Ramade and Bayle (1984)
	Duck	Martin (1978)

[a] As reflected by increased circulating corticosterone concentrations.

in the plasma concentration of aldosterone occurs in response to some stresses (such as handling; Radke et al., 1985), but divergent changes in the aldosterone/corticosterone responses can occur (e.g., following saline loading or hemorrhage; Harvey et al., 1984b; Radke et al., 1985). These changes reflect in part independent control on aldosterone secretion via the renin–angiotensin system. It is presumed that most adrenal cortical responses to stress occur via activation of the hypothalamo-pituitary axis and hence release of CRF and ACTH.

Control of Aldosterone Secretion

Aldosterone secretion in mammals is largely controlled by the renin–angiotensin system. Renin is released from the juxtaglomular cells of kidney in response to various stimuli including low sodium concentrations and reduced blood volume. The renin acts on its circulating (renin-) substrate, also known as angiotensinogen, to form angiotensin I, which is converted to angiotensin II. Aldosterone secretion is in turn stimulated by angiotensin II. It is probable that aldosterone release in birds is regulated, at least in part, by a similar system. Evidence for this comes from a number of studies. Juxtaglomular cells are found in the avian kidney, and these produce a renin-like substance (Taylor et al., 1970; Chan and Holmes, 1971). Hemorrhage is followed by increases in the plasma concentration of aldosterone in chickens (Radke et al., 1985), in plasma renin activity in chickens and piegons (Chan and Holmes, 1971; Nishimura et al., 1981), and in circulating concentrations of angiotensin II in quail (Kobayashi and Takei, 1982). Moreover, circulating concentrations of aldosterone but not of corticosterone are elevated by dietary sodium deficiency in chickens and ducks (Radke et al., 1984). However, angiotensin II has yet to be demonstrated to directly stimulate aldosterone secretion by avian adrenal cells.

There is also evidence that aldosterone secretion in birds is influenced by the hypothalamus/pituitary. In vivo ACTH increases plasma concentrations of aldosterone in ducks and chickens, although the magnitude of the response is affected by the sodium status of the bird (Radke et al., 1984).

Diurnal Rhythm of Plasma Corticosterone Concentrations

Basal plasma concentrations of corticosterone show a distinct daily pattern of variation in birds [e.g., in Japanese quail (Boissin and Assenmacher, 1968; Kovacs and Peczely, 1983); pigeons (Joseph and Meier, 1973); and ducks (Wilson et al., 1982)]. In most studies, maximal hormone concentrations have been found at the end of the night and beginning of the day; this corresponds to the acrophase of the rhythm, which follows an approximate sine function. The daily variation in circulating concentrations of corticosterone is presumed to reflect changes in ACTH, CRF, and hypothalamic aminergic neurons, and can be blocked by inhibiting the adrenergic or serotoninergic systems (Boissin and Assenmacher, 1971a). Furthermore, the daily corticosterone rhythm is driven by an endogenous (circadian) rhythm, which persists with a period of approximately 24 hr in free-running situations (i.e., in the constant light or dark). (For more details, see review by Assenmacher and Jallageas, 1980.)

The daily rhythm of corticosterone is affected by physiological status. The rhythm in reproductively functioning female birds is also related to the ovulatory cycle (see below) as well as the time of day (Wilson et al., 1982). Moreover, the acrophase of daily rhythm of corticosterone can be shifted in time ("phase shifted") (Boissin and Assenmacher, 1971b) by manipulation of the hypothalamo-pituitary-thyroid axis (Kovacs and Peczely, 1983).

Plasma Corticosterone during the Ovulatory Cycle

Although the basic daily rhythm of circulating concentrations of corticosterone has been observed in sexually mature female chickens (Etches, 1979; Wilson and Cunningham, 1981), the pattern can be affected by several extraneous factors. For instance, plasma concentrations of corticosterone are elevated at the time of oviposition (Beuving and Vonder, 1977, 1978). During the first ovulatory cycle of a sequence, plasma concentrations of corticosterone are depressed at night when the preovulatory surges in progesterone, estradiol, and luteinizing hormone are occurring (Wilson and Cunningham, 1981), because progesterone inhibits corticosterone secretion (Wilson and Cunningham, 1980). The reproductive role of corticosterone is considered elsewhere (see Chapter 18).

Seasonal Variation in Plasma Corticosterone Concentrations

Seasonal changes in circulating concentrations of corticosterone have been determined in wild birds either caught in the natural environment or maintained outdoors (reviewed by Meier, 1975; Wingfield, 1980, 1984; Deviche, 1983). The changes observed vary between species and may even differ within a species in successive years. Plasma concentrations of corticosterone may be affected by reproductive status, shifts in the diurnal rhythm, nutritional status reflecting feed availability, environmental stressors including adverse weather, physiologic and other aspects of migration, and behavior related to territorality.

Changes in Ontogeny and Development

Adrenocortical cells arise from the dorsal coelomic epithelium by day 4 of incubation in chick embryos, and form paired solid masses on each side of the aorta by day 6 (Bohus et al., 1965; Adjori, 1970). At this stage the cells are capable of secreting glucocorticoids, but significant secretion does not begin until day 8 when ACTH is observed in the pituitary (Pedernera, 1972). A surge in corticosteroid production occurs around day 15 (Woods et al., 1971; Marie, 1981) reflecting an increase in the cytochrome *P*-450 system (Sandor et al., 1972). Plasma concentrations of glucocorticoids, particularly corticosterone, increase at the time of hatching but decline in the immediate postnatal period (Kalliecharan and Hall, 1974; Nakamura et al., 1978). Following hatching, neonatal birds are thought to be insensitive to environmental stressors (Freeman and Manning, 1981, 1984; Freeman and Flack, 1981).

No firm conclusions can be reached on the changes in basal plasma concentrations of corticosterone during postnatal development, because of inconsistent reports of increases with age (Harvey and Phillips, 1980; Ramade and Bayle, 1984), of decreases (Wingfield et al., 1980), and of no changes (Homes and Kelly, 1976). In vivo, adrenocortical responsiveness to ACTH stimulation is reduced in adult birds, however (Beuving and Vonder, 1978), which are also generally less responsive to stressful stimuli (Freeman, 1982; Gorsline and Holmes, 1982). In vitro, chicken cortical cells become progressively less sensitive to ACTH stimulation of corticosterone secretion, although the maximal response is similar (Carsia et al., 1985b).

Physiology of Adrenocortical Hormones

Although the adrenocortical hormones of mammals are either mineralocorticoids or glucocorticoids, with aldosterone affecting electrolyte balance and cortisol intermediary metabolism, avian adrenocortical hormones may have both mineralocorticoid and glucocorticoid properties (reviewed by Holmes and Phillips, 1976; Freeman, 1983). This is in part due to the inherent mineralocorticoid activity of corticosterone.

Steroid hormones pass freely through the cell membrane to act on the target organ cells. The hormones bind to specific sytoplasmic receptors, and the steroid–receptor complex is then translocated into the nucleus when it interacts with DNA. Nuclear binding of the complex induces the synthesis of specific RNA and hence proteins.

Steroid Receptors

Specific receptors for corticosterone have been found in a number of avian tissues including brain (Koelher and Moscona, 1975), liver (Tu and Mondriankis, 1973), kidney (Bellabarba et al., 1983), lung (Hylka and Doneen, 1983), lymphoid tissue (Gould and Siegel, 1984), and salt glands (Sandor and Mehdl, 1981), and for aldosterone in the kidney (Chavest-Bonle et al., 1978, 1980). There are both high- and low-affinity binding sites for corticosterone. The steroid receptors are high-affinity (10^{-8}–10^{-10} M), low-capacity (low pmole range), thermolabile proteins with molecular weights in the range 100,000–300,000 daltons (Gould and Siegel, 1984).

Corticosterone and Intermediary Metabolism

Despite increasing food intake (Bartov et al., 1980; Gross et al., 1980), exogenous corticosterone and other glucocorticoids dramatically reduce growth in birds (Davison et al., 1983a). Accompanying the decrease in body weight, however, is an increase in fat deposition (Davison et al., 1983a). This is caused by increased lipogenesis by the enlarged liver and also by changes in lipolysis. Indeed, glucocorticoids slightly stimulate lipolysis by chicken adipose tissue in vitro (Harvey et al., 1977). Corticosterone also acts to increase net muscle protein catabolism (De La Cruze et al., 1981; see Chapter 14) and plasma glucose (Chapter 13) and liver glycogen concentrations by stimulating gluconeogenesis (Brown et al., 1958).

Corticosterone and Immune Functioning

Elevated pharmacologic concentrations of corticosterone can cause the involution of avian lymphoid tissue (bursa of Fabricus, thymus, spleen) and a suppression of humoral and cell-mediated immunity (Siegel, 1971; Gross et al., 1980; Glick, 1980; Edens et al., 1983) (see also Chapter 4). In chickens exposed to social stress or having received ACTH or corticosterone treatment, there is both a reduction in blood lymphocytes (Gross et al., 1980; Davison and Flack, 1981) and a concommitant increase in susceptibility to viral infections (Gross et al., 1980).

Despite the reduction in lymphocyte numbers, granulocyte populations are proportionately increased during stressor corticoid treatment (Gross et al., 1980; Davison et al., 1983b). In view of the mediation of defense against bacterial infections by phagocytes, it is not surprising that stress (glucocorticoids) may increase resistance to bacterial infections.

Glucocorticoids and Behavior

Social agonistic, submissive, and sexual behaviors have been related to the hypothalamo-pituitary axis. This may be due to central nervous effects of ACTH, CRF, or corticosterone.

Adrenal Hormones and Electrolyte Balance

Adrenal hormones act on renal and extrarenal cells to maintain avian electrolyte homeostasis (reviewed by Holmes and Phillips, 1976; Skadhauge, 1980; Harvey and Phillips, 1982; Butler, 1984; also see Chapter 16). Renal sodium excretion is suppressed by corticosterone and aldosterone; the negative sodium balance in adrenalectomized birds is overcome by corticoid therapy (Thomas and Phillips, 1975). While aldosterone increases potassium excretion, this is reduced by corticosterone. Corticosterone also increases the glomerular filtration rate.

The lower intestine is an important extrarenal site for osmoregulation in birds. Salt and hence water are absorbed from the reflux of cloacal fluids in response to aldosterone stimulation (Holmes and Phillips, 1976; Thomas and Skadhauge, 1979; Thomas et al., 1979, 1980). In marine birds, the nasal salt gland is important for osmoregulation (see Chapter 16). Corticosterone does not appear to directly affect nasal gland physiology or hypertrophy. Rather, a normally functioning adrenal gland is required for cardiovascular functioning (Butler, 1984).

Adrenal Medullary Hormones

The avian adrenal gland contains cells that store and release either epinephrine (E) or norepinephrine (NE).

Catecholamine Synthesis

The catecholamine hormones (E and NE) are synthesized in the adrenal chromaffin cells and in other tissues of the body. The biosynthetic pathway is outlined below (Figure 22–4).

Assay of Norepinephrine and Epinephrine

Three methods can be employed to determine the concentrations of E and NE in the plasma or adrenal glands: (a) a fluorometric method, (b) a radioenzymatic assay, and (c) spectrofluorometry, following high performance liquid chromatography (HPLC). Both the fluorometric method and radioenzymatic assays have been employed to determine circulating and adrenal concentrations of E and NE in avian species, in particular in chickens and ducks. Plasma concentrations of E and NE have been estimated in chickens and ducks both under basal conditions and after various manipulations (for details, see later section).

Table 22–4 shows that the level of E is higher in the blood of chickens and pigeons when determined by the fluorescense method, but E is lower than NE when determined by the radioenzymatic method. The concentration of NE in the blood of ducks is higher than E when determined by either method.

It is difficult to make definitive conclusions on the ratio of the concentrations of NE to E in the circulation of the domestic fowl in view of the lack of agreement between the concentration of E (and to a less extent NE) using the two assay methods (Table 22–4). The reason for the presumed overestimation of circulating concentrations of catecholamines in the chicken by the fluorometric assay is not apparent. (See also Chapter 7).

Control of Catecholamine Synthesis and Release

Early studies on the control of avian chromaffin cell function are reviewed elsewhere (Sturkie, 1976). The principal adrenal catecholamines E and NE are synthesized and released from different cell types (Unsicker,-

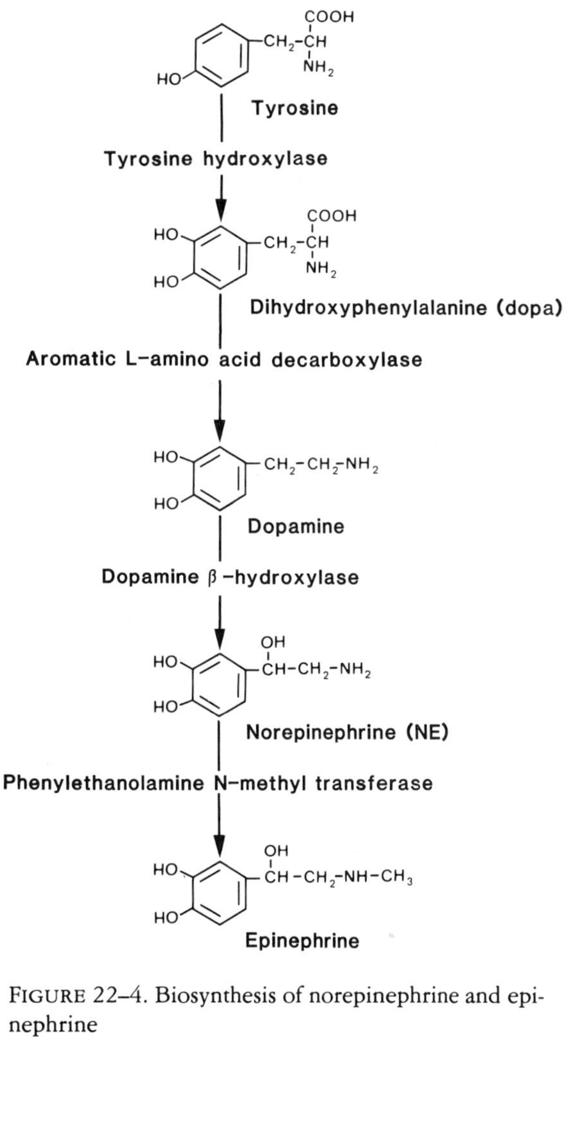

Tyrosine

Tyrosine hydroxylase

Dihydroxyphenylalanine (dopa)

Aromatic L–amino acid decarboxylase

Dopamine

Dopamine β –hydroxylase

Norepinephrine (NE)

Phenylethanolamine N–methyl transferase

Epinephrine

FIGURE 22–4. Biosynthesis of norepinephrine and epinephrine

TABLE 22–4. Basal circulating concentrations of catecholamines in birds[a]

		Plasma concentration (ng/ml)			
Species	Sex	E	NE	Method	Reference
Chicken	M	5.9	1.6	Fluorescence	Lin and Sturkie (1968)
	M	8.7	0.8	Fluorescence	Sturkie et al. (1970)
	F	8.2	1.2	Fluorescence	Sturkie et al. (1970)
	M	4.1	0.6	Fluorescence	Pittman and Hazelwood (1973)
	M	3.5	1.2	Fluorescence	Zachariasen and Newcomer (1974, 1975)
	F	0.11	0.35	Radioenzymatic	Nishimura et al. (1981)
	M	0.47	0.71	Radioenzymatic	Rees et al. (1984)
Duck		0.32	0.83	Fluorescence	Sturkie et al. (1968)
		0.36–0.73	0.36–0.73	Radioenzymatic	Hudson and Jones (1982)
		0.41	0.99	Radioenzymatic	Wilson and Butler (1983c)
Pigeon		5.8	0.2	Fluorescence	Sturkie et al. (1968)
		0.7	0.8	Radioenzymatic	Hissa et al. (1982)

[a] See Chapter 7 for additional figures.

1973b; Ghosh, 1980). The release, and presumably also the synthesis, of E and of NE are separately controlled by the cholinergic innervation of the avian adrenal; one synaptic terminal innervating as many as three E *or* NE cells (Unsicker, 1973a). In addition, the chromaffin cells are influenced by blood-borne factors and hormones. It may be assumed that with the exception of other adrenal hormones and perhaps also angiotensin II, stressors influence E or NE release via the cholinergic innervation of the adrenal medullary cells.

A variety of stressors affect the release of E and/or NE from the avian adrenal gland. Evidence for this comes either from the depletion of adrenal stores of E or NE or by changes in the circulating levels of E or NE. In pigeons, ether inhalation or electric shocks will deplete the adrenal concentrations of NE, but not that of E, suggesting that these stresses evoke NE release (Ghosh, 1980). Similarly, in chickens forced exercise is followed by decreases in the adrenal concentrations of E (−36%) and NE (−25%). This reflects E and NE release, as there is a concommitant increase in the plasma concentrations of E and NE (see Table 22–5). Circulating concentrations of E and/or NE are stimulated by other agents in chickens and also in ducks (summarized in Table 22–5). In view of the possibilities that E and NE originate from sources other than adrenal gland (e.g., nervous system), it should be noted that the E and NE responses to angiotensin II in ducks are almost completely lost following hypophysectomy (Wilson and Butler, 1983a). (See Chapter 7 for the roles of E and NE as neurotransmitters.)

Adrenal cortical hormones stimulate both the synthesis and the releases of E and NE in birds. ACTH stimulates E and NE release (Table 22–5). Moreover, if ACTH secretion is blocked, the E response to immobilization is attenuated (Zachariasen and Newcomer, 1975). ACTH acts via glucocorticoids to increase the activity of phenylethanolamine *N*-methyl transferase (PNMT), which thereby promotes the conversion of NE to E (Wasserman and Bernard, 1971; Zachariasen and Newcomer, 1974). It might be noted that the proportion of E increases during the development of the chick embryo due to increased PNMT activity as the hypothalamo-pituitary-adrenal cortical axis matures (Assserman and Bernard, 1971).

Physiological Actions of Epinephrine and Norepinephrine

Although exogenous E and NE exert profound effects on carbohydrate and lipid metabolism, cardiovascular parameters, and the release of hormones in birds, it is not certain that these are physiologic actions of catecholamines originating in the avian adrenal gland. For instance, pharmacologic quantities of E induce eggshell abnormalities (Hughes and Gilbert, 1984). Effects of E or NE may be pharmacologic in nature. Thus, for an action of E or NE to be endocrine, i.e., an adrenal medullary function, it should be evoked by concentrations of E or NE found in circulation of birds.

Glycogen breakdown (glycogenolysis) by chicken hepatocytes is stimulated by E (Cramb et al., 1982; Picardo and Dickson, 1982). E exerts its effect by binding to β-adrenergic receptors on the cell membrane. This leads to the activation of adenyl cyclase, the production of cAMP (a intracellular second messenger), and hence the activation of glycogen phosphorylase. In vitro gluconeogenesis by chicken liver tissue is stimulated by NE (and presumably also E) (Sugano et al., 1982; Cramb et al., 1982). However, in this case the catecholamine exerts its effect by binding to an α-adrenergic receptor and thereby activating a calcium-dependent mechanism. Conversely the synthesis of fatty acid (lipogenesis) by liver tissue is inhibited by E and to a lesser extent by NE (Prigge and Grande, 1971; Capuzzi et al., 1975; Cramb et al., 1982; Campbell

TABLE 22–5. Effect of stimuli on circulating concentrations of epinephrine and norepinephrine in chickens and ducks

Treatment	Plasma catecholamine concentration (as % increase or decrease from control)		Reference
	E(%)	NE(%)	
Ducks			
Anesthesia	>90	−75	Wilson and Butler (1983a, c)
Angiotensin II	+1370	+>478	Wilson and Butler (1983a)
Forced diving	+100,000	+100,000	Hudson and Jones (1982)
Domestic fowl			
Restraint (3 hr)	+365	+335	Zachariasen and Newcomer (1974, 1975)
Forced exercise (1 hr on treadmill)	+162	+60	Rees et al. (1984)
Insulin (15 min)	+178	−9	Pittman and Hazelwood (1973)
ACTH (injection)	+400	+485	Zachariasen and Newcomer (1974)
Angiotensin II (injection)	+54	+167	Nishimura et al. (1981)
Propranolol (injection)	+525	+185	Nishimura et al. (1981)
Anesthesia	−63	−49	Sturkie et al. (1970)
Chronic propranolol (1–2 weeks)	+4	+110	Nishimura et al. (1981)
Chronic cold (0–10°C)	+79	+87	Lin and Sturkie (1968)
Chronic heat (25–32°C)	+4	+1	Lin and Sturkie (1968)
Pigeon			
Propranolol	+108	+157	Hissa et al. (1982)

and Scanes, 1985). This effect is mediated via both α- and β-adrenergic receptors, and at least partially, by cAMP acting as the intracellular messenger. It is probable that the effects of E on glycogenolysis, gluconeogenesis, and lipogenesis are physiologic roles for adrenal E, as these metabolic parameters are influenced by concentrations of E found in the circulation of a stressed bird.

Lipolysis is stimulated by E in several species of birds including chickens (Langslow and Hales, 1969), pigeons (Goodridge and Ball, 1975), geese, and owls, but not in ducks (Prigge and Grande, 1971). The effect of E appears to be mediated via β-adrenergic receptors and cAMP (Langslow and Hales, 1969; Campbell and Scanes, 1985). In view of the very high concentrations of E (10^{-5} M) required to stimulate lipolysis in vitro, it is unlikely that E and NE of adrenal origin are involved in the physiological control of lipolysis.

Adrenal catecholamines are involved in the maintenance of blood pressure in birds (see also Chapter 6). Angiotensin II administration increases the circulating levels of NE and E in chickens (Nishimura et al., 1981) and ducks (Wilson and Butler, 1983a). Furthermore, the pressor effects of angiotensin II are decreased by adrenalectomy (Wilson and Butler, 1983a) and also be catecholamine depletion (Wilson and Butler, 1983b, c).

References

Adjori, Y. (1970). Morphogenese et activite de la glande cortico-surrenale de l'embryon de poulet normal et decapite. Arch. Anat. Microsc. Morphol. Exp., 59, 185.

Assenmacher, I., and M. Jallageas. (1980). Circadian and circannual hormonal rhythms. In: "Avian Endocrinology" (A. Epple and M.H. Stets, Eds.) New York: Academic Press, p. 391.

Aupetit, B., C. Bastien, and J.C. Legrand. (1979). Cytochrome P_{450} et transformation de la 18 hydrocorticosterone en aldosterone. Biochimie, 61, 1085.

Bartov, I., L.S. Jensen, and J.R. Veltmann. (1980). Effect of corticosterone and prolactin on feathering in broiler chickens. Poult. Sci., 59, 1328.

Bayle, J.D. (1980). The adenohypophysiotropic mechanisms. In "Avian Endocrinology" (A. Epple and M. Stetson, Eds.). New York: Academic Press, p. 117.

Bellabarba, D., C. Beaudry, and J.A. Leloux. (1983). Corticosteroid receptors in the kidney of chick embryo. II. Ontogeny of corticosterone receptor and cellular development. Gen. Comp. Endocrinol., 50, 305.

Birrenkott, G.P., and M.E. Wiggins. (1984). Determination of dexamethasone and corticosterone half-lives in male broilers. Poult. Sci., 63, 1064.

Beuving, G., and S.M.A. Vonder. (1977). Daily rhythm of corticosterone in laying hens and the influence of egg laying. J. Reprod. Fert., 51, 169.

Beuving, G., and S.M.A. Vonder. (1978). Effect of stressing factors on corticosterone levels in the plasma of laying hens. Gen. Comp. Endocrinol., 35, 153.

Bohus, B., K. Straznicky, and F. Hajos. (1965). The development of 3β-hydroxysteroid-dehydrogenase activity in embryonic adrenal gland of chickens. Gen. Comp. Endocrinol., 5, 665.

Boissin, J., and I. Assenmacher. (1968). Rythmes circadiens dex taux sanguin et surrenalien de la corticosterone chez la caille. C.R. Hebd. Seances Acad. Sci., 267. 2193.

Boissin, J., and I. Assenmacher. (1971a). Implication de mecnaismes aminergiques centraux dans le determinisme du rythme circadien de la corticosteronemie chez la caille. C.R. Hebd. Seances Acad. Sci., 273, 1744.

Boissin, J., and I. Assenmacher. (1971b). Entrainment of the adrenal cortical rhythm and of the locomotive activity rhythm by ahemerol photoperiods in the quail. J. Interdiscip. Cycle Res., 2, 437.

Bradley, E.L., and W.N. Holmes. (1971). The effects of hypophysectomy on adrenocortical function in the duck (*Anas platyrhynchos*). J. Endocrinol., 49, 437.

Brown, K.I. (1961). Validity of using plasma corticosterone stress in the turkey. Proc. Soc. Exp. Biol. Med., 107, 538.

Brown, K.I. (1967). Environmentally imposed stress. In "Environmental Control of Poultry Production" (T.C. Carter, Ed.). Edinburgh: Oliver and Boyd, p. 101.

Brown, K.I., D.J. Brown, and R.K. Meyer. (1958). Effect of surgical trauma, ACTH and adrenal cortical hormones on electrolytes, water balance and gluconeogenesis in male chickens. Am. J. Physiol., 192, 43.

Butler, D.G. (1984). Endocrine control of the nasal salt glands in birds. J. Exp. Zool., 232, 725.

Campbell, R.M., and C.G. Scanes. (1985). Adrenergic control of lipogenesis and lipolysis in the chicken *in vitro*. Comp. Biochem. Physiol. 82c, 137.

Capuzzi, D.M., R.D. Lackman, and M.A. Reed. (1975). Species differences in the hormonal control of lipogenesis in rat and chicken hepatocytes. Comp. Biochem. Physiol. B: Comp. Biochem., 50, 169.

Carsia, R.V., A.J. MacDonald, and S. Malamed. (1983). Steroid control of steroidogenesis in isolated adrenocortical cells: molecular and species specificity. Steroids, 41, 741.

Carsia, R.V., C.G. Scanes, and S. Malamed. (1984). Self-suppression of corticosteroidogenesis: Evidence for a role of adrenal 5α-reductase. Endocrinology, 115, 2464.

Carsia, R.V., C.G. Scanes, and S. Malamed. (1985a). Isolated adrenocortical cells of the domestic fowl (*Gallus domesticus*): Steroidogenic and ultrastructural properties. J. Steroid Biochem. 22, 273.

Carsia, R.V., C.G. Scanes and S. Malamed. (1985b). Loss of sensitivity to ACTH of adrenocortical cells isolated from maturing domestic fowl. Proc. Soc. Exp. Biol. Med., 179, 279.

Chan, M.Y., and W.N. Holmes. (1971). Studies on a renin-angiotension system in the normal; and hypophysectomized pigeon (*Columbia livia*). Gen. Comp. Endocrinol., 16, 304.

Chan, M.Y., E.L. Bradley, and W.N. Holmes. (1972). The effects of hypophysectomy on the metabolism in adrenal steroids in the pigeon (*Columbia livia*). J. Endocrinol., 52, 635.

Chavest-Bonle, L., A.Z. Mehdi, and T. Sandor. (1978). Corticosteroid receptors in the avian kidney. J. Steroid Biochem., 32, 109.

Chavest-Bonle, L., A.Z. Mehdi, and T. Sandor. (1980). Corticosterone receptors in the avian kidney. J. Steroid Biochem., 13, 897.

Chester-Jones, I., and J.P. Phillips. (1985). The adrenal cortex. In: "Vertebrate endocrinology: its fundamental and biomedical significance." (P. Pang and M. Schieman, Eds; A. Gorbman, Consult. Ed.) New York: Academic Press (In Press).

Cramb, G., D. R. Langslow, and J.H. Phillips. (1982). Hormonal effects of cyclic nucleotides and carbohydrate and lipid metabolism in isolated chicken hepatocytes. Gen. Comp. Endocrinol., 46, 310.

Cronshaw, J., W.N. Holmes, and R.D. West. (1984). The effects of colchicine, vinblastine and cytochalasins on the corticotropic responsiveness and ultrastructure of inner zone adenocortical tissue in the Pekin duck. Cell Tissue Res., 236, 333.

Curtis, M.J., I.H. Flack, and S. Harvey. (1980). The effect of *Escherichia coli* on the concentrations of corticosterone and growth hormone in the plasma of the domestic fowl. Res. Vet. Sci., 28, 123.

Daniel, J.Y., and I. Assenmacher. (1971). Early appearance of metabolites after single I.V. injection of 3H-corticosterone in rabbit and duck. Steroids 18, 325.

Davison, T.E., and I.H. Flack. (1981). Changes in two peripheral blood leucocyte population following an injection of corticotrophin in the mature chicken. Res. Vet. Sci., 30, 79–82.

Davison, T.F., J. Rea, and J.G. Powell. (1983a). Effects of dietary corticosterone on the growth and metabolism of mature *Gallus domesticus*. Gen. Comp. Endocrinol., 50, 463.

Davison, T.F., J.G. Powell, and J. Rea. (1983b). Effects of dietary corticosterone on peripheral blood lymphocytes and granulocyte populations in immature domestic fowl. Res. Vet. Sci., 34, 236.

De La Cruz, L.F., F.J. Mataix, and G. Illera. (1981). Effects of glucocorticoids on protein metabolism in laying quails. (*Coturnix coturnix japonica*). Comp. Biochem. Physiol. A, 70, 649.

Deviche, P. (1983). Interactions between adrenal function at reproduction in male birds. In "Avian Endocrinology. Environmental and Ecological Perspectives" (S. Mikami, Ed.). New York: Springer-Verlag, p. 243.

Donaldson, E.M., and W.N. Holmes. (1965). Corticosteroidogenesis in the fresh water and saline-restricted duck (*Anus platyrhynchos*). J. Endocrinol., 37, 329.

Edens, F.W. (1983). Effect of environmental stressors on male reproduction. Poult. Sci., 62, 1676.

Edens, F.W., and H.S. Siegel. (1975). Adrenal responses in high and low ACTH response lines of chickens during acute heat stress. Gen. Comp. Endocrinol., 25, 64.

El Halawani, E.M., E.P. Waibel, R.J. Appel, and L.A. Good. (1973). Effects of temperature stress on catecholamines and corticosterone on male turkeys. Am. J. Physiol., 224, 384.

Etches, R.J. (1976). A radioimmunoassay for corticosterone and its application to the measurement of stress in poultry. Steroids, 28, 763.

Etches, R.J. (1979). Plasma concentrations of progesterone and corticosterone during the ovulation cycle of the hen (*Gallus domesticus*). Poult. Sci., 58, 211.

Fowler, K.C., G.M. Pesti, and B. Howarth. (1983). The determination of plasma corticosterone of chickens by high pressure liquid chromatography. Poult. Sci., 62, 1075.

Freeman, B.M. (1982). Stress non-responsiveness in the newly hatched fowl. Comp. Biochem. Physiol. A, 72, 251.

Freeman, B.M. (1983). Adrenal glands. In "Physiology and Biochemistry of the Domestic Fowl," Vol. 4 (B.M. Freeman, Ed.) London: Academic Press, Chapter 11.

Freeman, B.M., and J.H. Flack. (1981). The sensitivity of the newly hatched fowl to corticotrophin. Comp. Biochem. Physiol. A, 70, 257.

Freeman, B.M., and A.C.C. Manning. (1981). The sensitivity of the newly hatched fowl to corticotrophin. Comp. Biochem. Physiol. A, 70, 275.

Freeman, B.M., and A.C.C. Manning. (1984). Re-establishment of the stress response in *Gallus domesticus* after hatching. Comp. Biochem. Physiol. A, 78, 267.

Gasc., J.M., and B. Martin. (1978). Measure corticosterone

loading binding capacity in the partially decapitated chick embryo. Gen. Comp. Endocrinol., 35, 274.

Ghosh, A. (1980). Avian adrenal medulla: structure and function. In "Avian Endocrinology" (A. Epple and M.H. Stetson, Eds.) New York: Academic Press, p. 301.

Glick, B. (1980). The thymus and bursa of Fabricius: endocrine organs? In "Avian Endocrinology" (A. Epple and M.H. Stetson, Eds.) New York: Academic Press, p. 209.

Goodridge, A.G., and E.G. Ball. (1975). Studies on the metabolism of adipose tissue XVIII. *In vitro* effects of insulin, epinephrine and glucagon on lipolysis and glycolysis in pigeon adipose tissue. Comp. Biochem. Physiol. B, 16, 367.

Gorsline, J., and W.N. Holmes. (1981). Effects of petroleum on adenocortical activity and on hepatic napthalene-metabolising activity in Mallard ducks. Arch. Environ. Contam. Toxicol., 10, 765.

Gorsline, J., and W.N. Holmes. (1982). Variations with age in the adrenocortical responses of Mallard ducks (*Anas platyrhynchos*) consuming petroleum-contaminated food. Bull. Environ. Contam. Toxicol., 29, 146.

Gould, N.R., and H.S. Siegel. (1984). Effect of adrenocorticotropin hormone secretions on glucocorticoid receptors in chicken thymocytes. Poult. Sci., 63, 373.

Gross, W.B., P.B. Siegel, and R.J. Dubosse. (1980). Some effects of feeding corticosterone to chickens. Poult. Sci., 59, 516.

Harvey, S., and J.G. Phillips. (1980). Growth, growth hormone and corticosterone secretion in fresh water and saline-adipheral ducklings (*Anas platyrhynchos*). Gen. Comp. Endocrinol., 47, 334.

Harvey, S. and J.G. Phillips. (1982). Endocrinology of salt gland function. Comp. Biochem. Physiol. A, 71, 537.

Harvey, S., C.G. Scanes, and T. Howe. (1977). Growth hormone effects on *in vitro* metabolism of avian adipose and liver tissue. Gen. Comp. Endocrinol., 33, 22.

Harvey, S., O.B. Merry, and J.G. Phillips. (1980). Influence of stress on the secretion of corticosterone in the duck (*Anas platyrhynchos*). J. Endocrinol., 87, 161.

Harvey, S., H. Klandorf, and J.G. Phillips. (1981). Effects of food and water deprivation on circulating levels of pituitary, thyroid and adrenal hormones and on glucose and electrolye concentrations in domestic ducks (*Anas platyrhynchos*). J. Zool., 194, 341.

Harvey, S., H. Klandorf, and Y. Pinchasov. (1983). Visual and metabolic stimuli cause adrenocortical suppression in fasted chicken during refeeding. Neuroendocrinology, 37, 59.

Harvey, S., J.G. Phillips, A. Rees, and T.R. Hall. (1984a). Serum and adrenal function. J. Exp. Zool., 232, 633.

Harvey, S., H. Klandorf, W.J. Radke and J.D. Few. (1984b). Thyroid and adrenal responses of ducks (*Anas platyrhynchos*) during saline adaption. Gen. Comp. Endocrinol., 55, 46.

Harvey, S., H. Klandorf, and S.K. Lam (1985). Drinking induced changes in fowl adrenocortical activity: effect of visual and non-visual stimuli. J. Endocrinol. (in press).

Hissa, R., J.C. Goerge, T.M. John, and R.J. Etches. (1982). Propranolol-induced changes in plasma catecholamine, corticosterone, T_4, T_3 and prolactin levels in the pigeon. Horm. Metabol. Res., 14, 606.

Holmes, W.N. (1978). Control of adrenocortical function in birds. Pavo, 16, 105.

Holmes, W.N., and M.E. Kelly (1976). The turnover and distribution of labelled corticosterone during pool-intake development of the duckling (*Anas platyrhynchos*). Pfleugers Arch., 36, 145–150.

Holmes, W.N., and J.G. Phillips. (1976). The adrenal cortex of birds. In "General, Comparative and Clinical Endocrinology of the Adrenal Cortex" (I. Chester-Jones and I.W. Henderson, Eds.). London: Academic Press, p. 293.

Holmes, W.N., and J. Cronshaw. (1980). Adrenal cortex: structure and function. In "Avian Endocrinology" (A. Epple and M. Sketson, Eds.). New York: Academic Press, p. 271.

Hudson, D.M., and D.R. Jones. (1982). Remarkable blood catecholamine levels in forced dived ducks. J. Exp. Zool., 224, 451.

Hughes, B.O., and A.B. Gilbert. (1984). Induction of egg shell abnormalties in domestic fowls by administration of adrenaline. IRCS Med. Sci., 12, 969.

Hylka, V.W., and B.A. Doneen. (1983). Ontogeny of embryonic chicken lung: Effects of pituitary gland, corticosterone, and other hormones upon pulmonary growth and synthesis of surfactant phospholipids. Gen. Comp. Endocrinol., 52, 108.

Joseph, M.H., and A.H. Meier. (1973). Daily rhythms of plasma corticosterone in the common pigeon, *Columba livia*. Gen. Comp. Endocrinol., 20, 326.

Kalliecharan, R., and B.K. Hall. (1974). A developmental study of the levels of progesterone, corticosterone, control and cortisone circulating in plasma of chick embryos. Gen. Comp. Endocrinol., 24, 364.

Kalliecharan, R., and B.K. Hall. (1977). The *in vitro* biosynthesis of steroid pregnenolone and cholesterol and the effects of bovine ACTH on corticoid production by adrenal glands of embryonic chicks. Gen. Comp. Endocrinol., 33, 147.

Kobayashi, H., and Y. Takei. (1982). Mechanisms for induction of drinking water special reference to angiotensin II. Comp. Biochem. Physiol. A, 71, 537.

Koelher, D.E., and A.A. Moscona. (1975). Corticosteroid receptors in the neural retina and other tissues of chick embryos. Arch. Biochem. Biophys., 170, 102.

Kovacs, K., and P. Peczely. (1983). Phase shifts in circadian rhythmicity of total, free corticosterone and transcortin plasma levels in hypothyroid male Japanese quails. Gen. Comp. Endocrinol., 50, 483.

Langslow, D.R., and C.N. Hales (1969). Lipolysis on chicken adipose tissue *in vitro*. J. Endocrinol., 43, 285.

Lehoux, J.G. (1974). Aldosterone biosynthesis and presence of cytochrome P450 in the adrenocortical tissue to the chick embryo. Mol. Cell. Endocrinol., 2, 43.

Lin, Y.C., and P.D. Sturkie. (1968). Effect of environmental temperatures on the catecholamines of chickens. Am. J. Physiol., 214, 237.

Marie, C. (1981). Ontogenesis of the adrenal glucocorticoids and of the target function of the enzymatic tyrosine transaminase activity in the chick embryo. J. Endocrinol., 90, 193.

Martin, T.T. (1978). Imprinting behavior: Pituitary-adrenocortical modulators of the approach response. Science, 200, 565.

Martin, T.T., M. El Halawani, and R.E. Phillips. (1982). Diurnal variation in hypothalamic monoamines and plasma corticosterone in the turkey after inhibition of tyrosine hydroxylase or tryptophan hydroxylase. Neuroendocrinology, 84, 191.

Meier, A.H. (1975). Chronoendocrinology of vertebrates. In "Hormonal Correlates of Behavior," Vol. 2 (B.E. Lotherion and R.L. Sprott, Eds.). New York: Plenum Press, p. 469.

Nagra, C.L., G.J. Baum, and R.K. Meyer. (1960). Corticosterone levels in adrenal effluent blood of some gallinaceous birds. Proc. Soc. Exp. Biol. Med., 105, 68.

Nagra, C.L., J.C. Birnie, and R.K. Meyer. (1963). Suppression of the output of corticosterone in the pheasant by methopyrapone (metopirone). Endocrinology, 73, 835.

Nakamura, T., Y. Tanabe, and H. Hirano. (1978). Evidence of the *in vitro* formation of cortisol by the adrenal gland of embryonic and young chickens (*Gallus domesticus*). Gen. Comp. Endocrinol., 35, 300.

Nir, I., D.Yam, and M. Perec. (1975). Effects of stress on the corticosterone content of the blood plasma and adrenal gland of intact and bursectomized *Gallus domesticus.* Poult. Sci., 54, 2101.

Nishumura, H., Y. Kakamura, A.A. Taylor, and M.A. Madey. (1981). Renin-angiotension and adrenergic mechanisms in control of blood pressure in fowl. Hypertension, 3 (Suppl. 1), 141.

Pearce, R.B., J. Cronshaw, and W.M. Holmes. (1979). Structural changes occurring in interrenal tissue in the duck (*Anas platyrhynchos*) following adenohypophysectomy and treatment *in vivo* and *in vitro* with corticotropin. Cell Tissue Res., 196, 429.

Peczely, P., and J.Y. Daniel. (1979). Interaction reciproques testothyroido-sur-renaliennes chez la caille niole. Gen. Comp. Endocrinol., 39, 164.

Pedernera, E.A. (1972). Adrenocorticotropic activity *in vitro* on the chick embryo pituitary gland. Gen. Comp. Endocrinol., 19, 589.

Pedernera, E.A., and C.P. Lantos. (1973). The biosynthesis of adrenal steroids by the 15-day-old chick embryo. Gen. Comp. Endocrinol., 20, 331.

Phillips, J.G., and I. Chester-Jones. (1957). The identity of adrenocortical secretion in lower vertebrates. J. Endocrinol., 16, iii.

Picardo, M., and A.J. Dickson. (1982). Hormonal regulation of glycogen metabolism in hepatocyte suspensions isolated from chicken embryos. Comp. Biochem. Physiol., B: Comp. Biochem., 71, 689.

Pittman, R.P., and R.L. Hazelwood. (1973). Catecholamine response of chickens to exogenous insulin and tolbutamide. Comp. Biochem. Physiol. A, 45, 141.

Prigge, W.F., and F. Grande. (1971). Effects of glucagon, epinephrine and insulin on *in vitro* lipolysis of adipose tissue from mammals and birds. Comp. Biochem. Physiol., B: Comp. Biochem., 39, 69.

Radke, W.J., C.M. Albasi, and S. Harvey. (1984). Dietary sodium and adenocortical activity in ducks (*Anas platyrhynchos*) and chickens (*Gallus domesticus*). Gen. Comp. Endocrinol., 56, 121.

Radke, W.J., C.M. Albasi, and S. Harvey. (1985). Hemorrhage and adrenocortical activity in the fowl (*Gallus domesticus*). Gen. Comp. Endocrinol. (in press).

Ramade, F., and J.D. Bayle. (1982). Thalamic-hypothalamic interrelationships and stress-induced rebounding adrenocortical response in the pigeon. Neuroendocrinology,- 34, 7.

Ramade, F., and J.D. Bayle. (1984). Adaptation of the adrenocortical response response during repeated stress in thalmic pigeons. Neuroendocrinology, 39, 245.

Rattner, B.A., and J.C. Franson. (1984). Methyl parathion and fenvalerate toxicity in American Kestrels: acute physiological responses and effects of cold. Can. J. Physiol. Pharmacol., 62, 787.

Rattner, B.A., L. Sileo, and C.G. Scanes. (1982). Oviposition and the plasma concentrations of LH, progesterone and corticosterone in bobwhite quail (*Colinus virgininarus*) fed parathion. J. Reprod. Fert., 66, 147.

Rees, A., S. Harvey, and J.G. Phillips. (1983). Habituation of the corticosterone response of ducks (*Anas platyrhynchos*) to daily treadmill exercise. Gen. Comp. Endocrinol., 49, 485.

Rees, A., T.R. Hall, and S. Harvey. (1984). Adrenocortical and adrenomedullary responses of fowl to treadmill exercise. Gen. Comp. Endocrinol., 55, 488.

Sandor, T., and A.Z. Mehdi. (1981). The interaction of steroids with nasal gland of the domestic duck (*Anas platyrhynchos*) *in vitro.* In "Advances in the Physiological Sciences" Vol. 33 (G. Pethes, P. Peczely, and P. Rudas Eds.). Hungary: Academiai Kiado; London: Pergamon Press, p. 331.

Sandor, T., A.G. Fazekas, J.-G., Lehoux, H. Leblanc, and A. Lanthier. (1972). Studies on the biosynthesis of 18-oxygenated steroids from exogenous corticosterone by domestic duck (*Anas platyrhynchos*) adrenal gland mitochrondria. J. Steroid Biochem., 3, 661.

Satterlee, D.G., R.B. Abdullah, and R.P. Gildersleeve. (1980). Plasma corticosterone radioimmunoassay and levels in the neonate chicken. Poult. Sci., 59, 900.

Scanes, C.G., G.F. Merrill, P. Ford, P. Mauser, and C. Horowitz. (1980). Effects of stress (hypoglycemia, endotoxin and ether) on the peripheral circulating concentration of corticosterone in the domestic fowl (*Gallus domesticus*). Comp. Biochem. Physiol., C: Comp. Pharmacol., 66, 183.

Scott, T.R., D.G. Satterlee, and L.A. Jacobs-Perry. (1983). Circulating corticosterone responses of feed and water deprived broilers and Japanese quail. Poult. Sci., 62, 299.

Siegel, H.S. (1971). Adrenal stress and the environment. World's Poult. Sci. J., 27, 327.

Simensen, E., L.D. Olson, W.J. Vanjonack, H.D. Johnson, and M.P. Ryan. (1978). Determination of corticosterone concentration in plasma of turkeys using radioimmunoassay. Poult. Sci., 57, 1701.

Skadhauge, E. (1980). Intestinal osmoregulation. In "Avian Endocrinology" (A. Epple and M.H. Stetson, Eds.) London: Academic Press, p. 481.

Stachenko, J., C. Laplante, and C.J.P. Giroud. (1964). Double isotope derivative assay of aldosterone, corticosterone, and cortisol. Can. J. Biochem., 42, 1275.

Sturkie, P.D., and Y.C. Lin. (1968). Sex difference in blood norepinephrine of chickens. Comp. Biochem. Physiol., 24, 1073.

Sturkie, P.D., D. Poorvin, and N. Ossoria. (1970). Levels of epinephrine and noepinephrine in blood and tissues of duck, pigeon, turkey and chicken. Proc. Soc. Exp. Biol. Med., 135, 267.

Sturkie, P.D. (Ed.). (1976). "Avian Physiology," 3rd Ed. New York: Springer-Verlag.

Sugano, T., M. Shiota, H. Khono, and M. Shimada. (1982). Stimulation of gluconeogenesis by glucagon and norepinephrine in the perfused chicken liver. J. Biochem., 92, 111.

Tanabe, Y., T. Nakamura, K. Fujiota, and O. Doi. (1979). Production and secretion of sex steroid hormone by the testes, the ovary, and the adrenal glands of embryonic and young chickens (*Gallus domesticus*). Gen. Comp. Endocrinol., 39, 26.

Taylor, A.A., J.O. Davis, R.P. Breiterbach, and P.M. Hartroft. (1970). Adrenal steroid secretion and renal-pressor system in the chicken (*Gallus domesticus*). Gen Comp. Endocrinol., 14, 321.

Thomas, D.H., and J.G. Phillips. (1975). Studies in avian adrenal steroid function. IV: Adrenalectomy and the response of domestic ducks (*Anas platyrhynchos* L.) to hypertonic saline loading. Gen. Comp. Endocrinol., 26, 427.

Thomas, D.H., and E. Skadhauge. (1979). Chronic aldosterone therapy and the control of transepithelial transport of ions and water by the colon and coprodeum of the domestic fowl (*Gallus domesticus*) *in vivo.* J. Endocrinol., 83, 239.

Thomas, D.H., E. Skadhauge, and M.W. Read. (1979). Acute effects of aldosterone on water and electrolyte transport in the colon and coprodeum of the domestic fowl (*Gallus domesticus*) *in vivo.* J. Endocrinol., 83, 229.

Thomas, D.H., M. Jallageas, B.G. Munck, and E. Skadhauge. (1980). Aldosterone effects on electrolyte transport of the lower intestine (coprodeum and colon) of the fowl (*Gallus domesticus*) *in vitro.* Gen. Comp. Endocrinol., 40, 44.

Tu, A.S., and E.N. Moudrianakis. (1973). Purification and characterization of a steroid receptor from chick embryo liver. Biochemistry, 12, 3692.

Unsicker, K. (1973a). Fine structure and innervation of the avian adrenal gland. II. Cholinergic innervation of adrenal chromaffin cells. Z. Zellforsch. Mikrosk. Anat., 145, 417.

Unsicker, K. (1973b). Fine structure and innervation of the avian adrenal gland. I. Fine structure of adrenal chromaffin cells and ganglion cells. Z. Zellforsch. Mikrosk. Anat., 145, 389.

Urist, M.R., and N.M. Deutch. (1960). Influence of ACTH upon avian species and osteoporosis. Proc. Soc. Exp. Biol. Med., 104, 35.

Wasserman, G.F., and E.A. Bernard. (1971). The influence of corticoids on the phenylethalamine-*N*-methyl transferase of the adrenal glands of *Gallus domesticus.* Gen. Comp. Endocrinol., 17, 83.

Wilson, S.C., and F.J. Cunningham. (1980). Concentrations of corticosterone and luteinizing hormone in plasma during the ovulatory cycle of the domestic hen and after the administration of gonadal steroids. J. Endocrinol., 85, 209.

Wilson, S.C., and F.J. Cunningham. (1981). Effect of photoperiod on the concentrations of corticosterone and luteinizing hormone in the plasma of the domestic hen. J. Endocrinol., 91, 135.

Wilson, J.X., and D.G. Butler. (1983a). Adrenalectomy inhibits noradrenergic, adrenergic and vasopressor responses to angiotensin II in the Pekin duck (*Anas platyrhynchos*). Endorcinology, 112, 645.

Wilson, J.X., and D.G. Butler. (1983b). 6-hydroxydopamine treatment diminishes noradrenergic and pressor responses to angiotensin II in adrenalectomized ducks. Endocrinology, 112, 653.

Wilson, J.X., and D.G. Bulter, (1983c). Catecholamine-mediated pressor responses to angiotensin II in the Pekin duck (*Anas platyrhynchos*). Gen. Comp. Endocrinol., 51, 477.

Wilson, S.C., F.J. Cunningham, and T.R. Morris. (1982). Diurnal changes in the plasma concentration of corticosterone, luteinizing hormone and progesterone during sexual development and the ovulatory cycle of Khaki Campbell chicks. J. Endocrinol., 93, 267.

Wingfield, J.C. (1980). Fine temporal adjustment of reproductive functions. In "Avian Endocrinology" (A. Epple and M.H. Stetson Eds.). New York: Academic Press, p. 367.

Wingfield, J.C. (1984). Influence of weather on reproduction. J. Exp. Zool., 232, 589.

Wingfield, J.C., and D.S. Farner. (1975). The determination of the steroids in avian plasma by radioimmunoassay and competitive protein binding. Steroids, 26, 311.

Wingfield, J.C., J.P. Smith, and D.S. Farner. (1980). Changes in plasma levels of luteinizing hormone, steroid and thyroid hormones during post-fledging development of white crowned sparrows, *Zonotrichia leucophrys.* Gen. Comp. Endocrinol., 41, 372.

Wingfield, J.C., K.S. Matt, and D.S. Farner. (1984). Physiologic properties of steroid hormone binding proteins in avian blood. Gen. Comp. Endocrinol., 53, 28.

Woods, J.E., V.W. Devries, and R.C. Thommes. (1971). Ontogenesis of the pituitary-adrenal axis in chick embryos. Gen. Comp. Endocrinol., 17, 405.

Zachariasen, R.D., and W.S. Newcomer. (1974). Phenylethanolamine-*N*-methyl transferase activity in the avian adrenal following immobilization or adrenocorticotropin. Gen. Comp. Endocrinol., 23, 193.

Zachariasen, R.D., and W.S. Newcomer. (1975). Influence of corticosterone on the stress-induced elevation of phenylethanolamine-*N*-methyl transferase in the avian adrenal. Gen. Comp. Endocrinol., 25, 332.

23
Pancreas and Pineal

Pancreas

R.L. HAZELWOOD

Location, Size, and Function (General)

The avian pancreas is located in the abdominal cavity of all birds on the right side; it is tightly bound by mesentery and blood vessels in position between the descending and ascending duodenal loops. This organ varies greatly in size, although it is always well formed and exhibits discrete characteristics. It varies in weight, from 2 to 4 g in 1.5–2.1 kg adult chickens to a mere fraction of a gram in most passerine species, whose body weight may be at the most a few grams. Generally, it has a smooth macroscopic appearance, is yellowish white in color, and does not exhibit the highly vascular appearance so common to other vertebrate endocrine glands such as thyroid and adrenal tissue, etc.

Like pancreatic tissue in all other vertebrates, most (99%) of the organ is devoted to the synthesis and secretion, through well-formed ducts, of three families of digestive enzymes. Various proforms (which are inactive within the pancreas) of proteolytic enzymes are synthesized in the exocrine (acinar) portion of the avian pancreas and passed into the two or three pancreatic ducts, which empty their contents into the lower duodenum to mix with the semiliquid chyme previously squirted from the gizzard. (See also Chapters 12 & 14.) Such enzymes include trypsinogen, chymotrypsinogen, and procarboxypeptidase, all of which must be activated prior to becoming enzymatically active in the protein degradation process (see Chapter 12). Additionally, the exocrine (acinar) pancreatic tissue synthesizes various lipases, which hydrolyze gut triglycerides (neutral fats), and amylolytic enzymes, which degrade carbohydrates at a pH differing considerably from that of salivary amylase. One of the major "triggers" for release of this family of digestive enzymes is the release of the hormone cholecystokinin (CCK) from duodenal-ileum mucosa when partially degradated food (as chyme) advances to this level of the gut (Chapter 12).

The remaining 1–2% of pancreatic tissue is endocrine in character and probably has no functional association with the pancreatic ducts. Rather, these cells (islet cluster) synthesize and release their peptide products directly into the bloodstream. Pancreatic hormones released in response to absorbed nutrients, to cholinergic input, and probably to hormonal stimulation, include insulin (anabolic), glucagon (catabolic), pancreatic polypeptide (APP, probably secretogogic), and somatostatin (SRIF, paracrine modulator). Chapter 13 presents details on the functional nature of these pancreatic hormones.

Morphology

Unlike pancreatic tissue of rats, squirrels, rabbits, and other lagomorphs, the avian pancreas is a discrete lobular organ. In the adult chicken, it is 10–15 cm long by 2–3 cm wide and as such fits well within the U-shaped duodenal loop (Figure 23–1). Its lobular structure varies among avian species only with regard to definition and development of the clefts between the three longitudinal lobes, but in all species examined the entire tongueshaped organ is suspended and invested by a rich vascular system that forms a network of vessels between the entire duodenal loop and the pancreatic organ. The two or three (varying among species) pancreatic ducts that enter the terminal (distal) portion of the ascending duodenum also provide structural support.

The centrally placed single pancreatic artery usually separates the dorsal from the ventral pancreatic lobe, and the ventral lobe itself is normally found with a deep longitudinal cleft running the entire length of the pancreas that separates the central portion of the ventral lobe (sometimes referred to as the "third" lobe) (Mikami and Ono, 1962) from the more laterally placed portion of the lobe (Figure 23–1). An extension of the (laterally placed) ventral pancreatic lobe, which runs from the most superior portion of the lobe to the side of the spleen, is frequently referred to as the splenic lobe. This latter portion of the pancreas represents about 1–2% of the total wet weight of the organ and is without an exocrine duct. However, it is quite vascular and is heavily endowed with glucagon-secreting A cells. There is very little connective tissue within the pancreatic mass. In addition to the centrally located pancreatic artery and veins, the only other major vessel is the pancreaticoduodenal vein, which empties into the gastroduodenal vein. The latter vein empties its contents directly into the hepatic portal vein. Thus, pancreatic hormones may affect liver function, or in turn they may be altered and/or destroyed by the hepatocyte before entering the general circulation.

Innervation of mammalian pancreatic islets by neural elements of both divisions of the autonomic nervous system is well established, and the impact of each on hormone release mechanics has been both studied and elucidated (Woods and Porte, 1974). By contrast, the bird either has been described as having no islet inner-

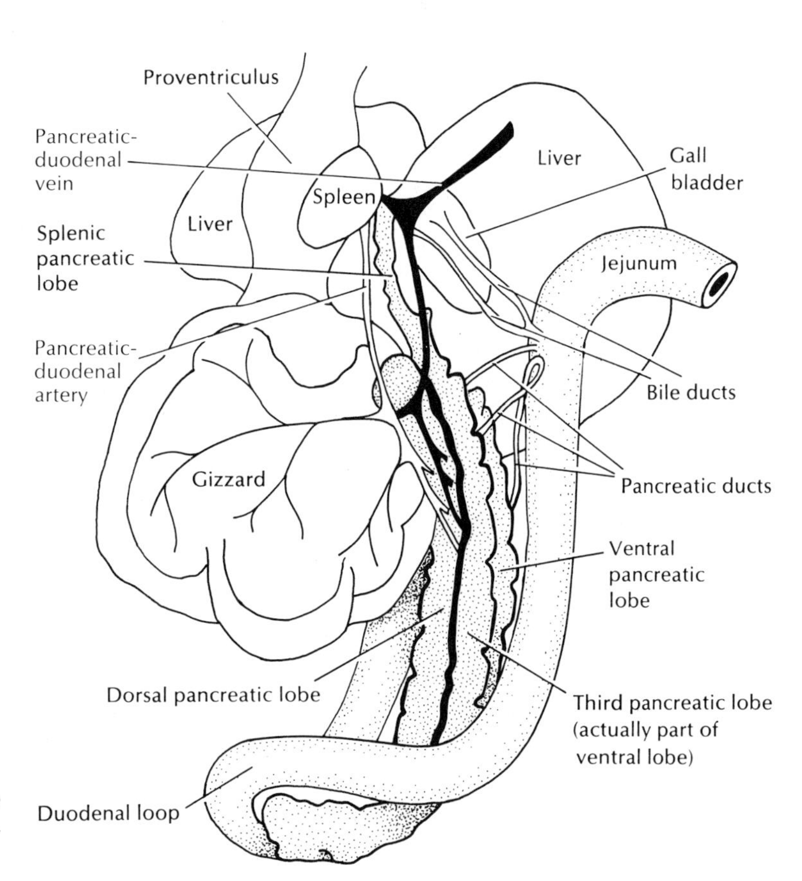

FIGURE 23–1. Anatomy of the avian pancreas and related structures.

vation (Woods and Porte, 1974) or, at best, sparse innervation (Dahl, 1973; Watanabe and Yasuda, 1977). In this regard, birds are very similar to those few mammals that act as genetic models of diabetes, such as the spiny mouse (inbred *Acomys cahirinus*). Both birds and this inbred mouse species have high plasma glucose levels, abnormal insulin secretion patterns, and little evidence of autonomic nervous regulation (Cameron et al., 1972).

In general, the pancreaticoduodenal plexus of the autonomic nervous system supplies the avian pancreas and is readily identifiable in the acinar portion of the organ (Hseih, 1951; King, 1971). Of all pancreatic secretions regulated by neural elements, pancreatic polypeptide (APP) appears to be the most responsive to cholinergic input. This, and other neural events are discussed in detail in Chapter 13.

Embryogenesis

The avian pancreas arises from two separate outpocketings of the embryonic gut. A dorsal evagination arises on day 3 from the embryonic duodenum just anterior to the primordial site for the liver. A second, ventral evagination also arises from the duodenum, but does so on day 4 and is more ventral and more posterior than that for the liver. The two pancreatic anlage fuse into a single, discrete, lobular organ in the first few days of embryonic life. The ducts associated with the ventral lobes are largest, and therefore in adult forms they are most prominent in conveying digestive juices and enzymes to the duodenum near the entrance of the bile duct. The ducts are lined with columnar epithelium interspersed with mucous-type cells, but unlike mammals there is no evidence indicating that the ductile tissue is associated with endocrine-type (islet) cells. Thus, whether or not future generations of islet cells are derived from preexisting ductile elements in Aves is yet to be established.

Cytodifferentiation. The onset of pancreatic morphogenesis in avian embryos is characterized by two cell types. One is the distinct islet (endocrine-type) cells, as well as other cells closely associated with them. A second cell type is a much larger cell mass, which occupies most of the epithelial bud and is not morphologically committed either to acinar or to islet cell differentiation. Presumably, these latter cells comprise a stem-cell population for future acinar ductile and endocrine cells (Przbylski, 1967a).

As early as 3–4 days of embryonic age, evidence of digestive enzyme secretion exists in duck embryos. Single membrane-bound zymogen granules are present at the same time; these appear to be "packaged" by the Golgi apparatus, and move toward the cell apices as duct lumina are formed (Przybylski, 1967a). As the acinar pancreatic mass matures, the cytoplasmic filaments characteristic of younger cells disappear, the Golgi complex becomes supranuclear in position, and the nucleus moves toward the periphery.

Possibly unique to the developing avian pancreas is the occurrence of clear "glycogen cells," which are very numerous during days 3–7 and thereafter gradually decrease in number. By days 14–16 (ducks and chicks) they are virtually unseen. The origin and function of these so-called glycogen cells are unknown. However, they do form one of the many characteristics differentiating avian pancreatic structure (and function?) from that observed in mammals (Przybylski, 1967a).

Extensive cytological evaluation (both light- and electron microscopic) has been made of the avian endocrine pancreas over the last 10–15 years, and, when coupled with the more recent use of immunochemistry whereby hormone presence is detected, a reasonably complete picture emerges as to the chronology of differentiation of the stem cells into insular (endocrine) -type cells that will become functional a few days later. (Older cytological work was reviewed in Hazelwood, 1976). Thus, the following statements are drawn from observations on embryonic pancreatic tissue from the chicken, duck, pigeon, sparrow, thrush, and goose.

Recognized cell types emanating from the stem cells so prominent on day 3 include what subsequently will secrete glucagon (A cells), insulin (B cells), APP (PP cells), and SRIF (D cells). The same stem cells give rise to ductile and acinar elements. Endocrine cell types appear earlier than the neighboring exocrine cell types, with both A- and B-cell types appearing on day 3 (Table 23–1). Insulin granules are detected in the B cell on the same day (Swenne and Lundquist, 1980; Przybylski, 1967b; Dieterlen-Lievre and Beaupain, 1974). Glucagon granules actually appear a few hours earlier (Przybylski, 1967b; Benzo and Stearns, 1975; Beaupain and Dieterlen-Lievre, 1974). Pancreatic D cells appear at day 3 ¾ in the dorsal bud (Andrew, 1977). Zinc is plentiful in A cells of birds but not in B cells, a finding quite the reverse of that found in mammals. The initial appearance of PP-secreting cells (by use of specific antibody to APP) has been reported to occur by day 11, but the actual appearance of these cells, with or without secretory granules, probably occurs much earlier. That date is yet to be reported (Goldman et al., 1978). The somatostatin-secreting D cell has obvious peptide content by day 7 (Swenne and Lundquist, 1980) but may well have discernible peptide granules as early as day 3 (Andrew 1977).

Cell Type Distribution. The four endocrine peptides associated with avian islets are literally "buried in a sea" of acinar tissue, the latter representing 98–99% of the fresh wet weight of the avian pancreas (Oakberg, 1949; Mikami and Ono, 1962). General agreement exists that islet tissue in Aves is distributed more

TABLE 23–1. Embryogenesis and secretogenesis of avian pancreatic hormones in the chicken (White Leghorn) embryo[a]

Incubation (days)	Islet cells appear	Granules present	Hormones		Notes
			Pancreas	Plasma	
2	None	None	Not developed	Unobtainable	Insulin-like material in unfertilized albumen and yolk
3	Early: A cells Late: B cells	A cells B cells (dorsal bud)	Early: Glucagon Late: Insulin (histochemical)	— —	~2 ng insulin/gm of homogenized embryo
4	—	A cells D cells	Insulin		—
			15 μg/gm	0.15 ng/ml	
5	D cells PP cells	B cells (very prominent)	Ins-13.6 μg/gm Gluc-10.5 ng/mg protein	0.15 ng/ml 150 pg/ml	~5 ng insulin/gm homogenized embryo
7	True islets in splenic lobe	PP cells	Whole Pancreas		
			Ins-9 ng Gluc-11 ng SRIF-0.2 ng	— — —	—
8	True islets in dorsal lobe	All cells	Insulin		
			16.8 μg/gm	0.14 ng/ml	
11	—	All cells	Whole Pancreas		Pancreatic glucagon ~92 ng/mg protein
			Ins-100 ng Gluc-250 ng SRIF-12 ng APP-0.42 ng	0.30 ng/ml 258 pg/ml — —	
13	True islets in ventral lobe	All cells	Glucagon		—
			37.3 ng/mg protein	120 pg/ml	
21	Discrete islets and lobes throughout	All cells	Whole pancreas		Dramatic increase in all pancreatic hormones after hatching. PP cells equally distributed in acinar and islet tissue.
			Ins, 275 ng Gluc, 775 ng SRIF, 230 ng APP, 250 ng	— — — —	

[a] Data taken from several sources (Benzo and Stearns, 1975; DePablo et al., 1982; Swenne and Lundquist, 1980; Benzo and Green, 1974; Leibson et al., 1976; Dieterlen-Lievre and Beaupain, 1974; Beaupain and Dieterlen-Lievre, 1974; Przybylski, 1976b; Andrew, 1977; Goldman et al., 1978; and Rawdon and Andrew, 1979). Units of measurement converted, averaged, and rounded.

sparsely throughout the acinar mass than is found in mammals. It also appears that the avian pancreatic polypeptide- (APP-) secreting cells are equally distributed between islet structures and free-ranging PP-cell clusters within the exocrine parenchyma. The latter are totally unassociated with islet-cell groups. The distribution of A, B, and D cells remains uneven throughout the pancreatic mass, however, as is clearly seen in Table 23–2.

Two types of endocrine islets have been described. The larger (and darker) islets appear to be islets predominantly of the glucagon (A-cell) type, but also contain some B, D, and PP cells. So-called glucagon islets are very numerous from the splenic end to the duodenal (head) end of the pancreas; however, they appear to be largely restricted to one longitudinal half of the organ (Beaupain and Dieterlen-Lievre, 1974). This is certainly true for the *Coturnix* quail and the chicken, whereas in the duck and pigeon both types of islets are distributed throughout the four pancreatic lobes (Smith, 1973). D cells (somatostatin) frequently occupy a central position within the glucagon islets; also, they are scattered throughout the exocrine pancreas of newly hatched chicks in a manner similar to that observed for PP cells (Rawdon and Andrew, 1979). In ducks, chickens, and pigeons, the D cells (formally termed α_1 cell) are mainly aggregated near islet capillaries, and the cytoplasmic granules are invariably located at the capillary pole of the cell. Splenic lobe tissue is especially rich in A islets but especially poor in PP cells (Table 23–2).

The smaller (and lighter staining) islets, documented to be predominately B cells that synthesize and release insulin, are scattered throughout the four lobes. B islets

TABLE 23–2. Regional distribution of avian pancreatic hormones in adult birds[a,b]

	Distribution of endocrine cells and hormones						
Region	Islets	Islet area (% of lobe)	Species	Insulin[c]	Glucagon[c]	SRIF[c]	APP[c]
Dorsal lobe	B cells D cells PP cells	0.37	Chicken	23	0.60	3.10	22.5
Ventral lobe	(A cells)+ B cells D cells PP cells	0.29	Chicken	41	1.12	3.40	20.2
"Third" lobe	A cells B cells D cells PP cells	0.52	Chicken	50	14.0	9.00	14.8
Splenic lobe	A cells B cells D cells (PP cells)+	4.37	Chicken	210	496.0	31.0	3.3
Total pancreas	All cell types	1.0–1.5	Chicken	38	31.5[d]	4.12	25.3[d]
			Duck	67	22.7[d]	—	23.9[d]
			Pigeon	26	19.3[d]	—	2.7[d]

[a] Modified from Langslow et al., 1973; Weir et al., 1976; Cieslak, 1984; Cieslak and Hazelwood, 1983; Oakberg, 1949; Mikami and Ono, 1962; and Pollock and Kimmel, 1981. ( )+, Relatively few cells in this lobe.
[b] Most birds were White Leghorns.
[c] Data are ng hormone/mg wet pancreas.
[d] Excludes splenic lobe.

residing in the splenic lobe are very large compared with the major three, inferiorly placed pancreatic lobes, where dorsal B islets may be 50–70 μm in diameter whereas ventral B islets may have diameters approximating 12–16 μm (Dieterlen-Lievre and Beaupain, 1974).

Distribution of PP cells appears to be without preference for any single lobe. Thus, they are fairly uniformly distributed in islets, as PP cell clusters, and as single cells throughout the entire acinar pancreas (Rawdon and Andrew, 1979). In mammals, the uncinate (head or duodenal end) of the pancreas appears to have the highest concentration of PP cells, while the body and tail (splenic lobe) are most heavily populated with B and A cells. The tendency for an inverse ratio of A-cell to PP-cell populations appears to be true both in mammals and birds (Table 23–2). Concentrations of four pancreatic hormones have been established for the chicken pancreas, and in general the data that are available fit histochemical descriptions already in the literature (Weir et al., 1976; also Table 23–2). Thus, one can see from Table 23–2 that the *concentration* of insulin, glucagon, and SRIF is highest in the splenic lobe. For PP, the highest concentration is in the duodenal portion ("head") of the dorsal lobe. The decreasing concentration of glucagon within the merged splenic-ventral lobe tissue is striking (600:1). On a hormone weight to tissue weight basis, insulin is the most concentrated hormone in the *entire* pancreas, followed by glucagon, APP, and then the peptide somatostatin. The splenic lobe of birds appears unique among vertebrates in that islets of this region in some species, such as the *Coturnix* quail, may constitute up to 50% of the total cellular mass (Smith, 1973).

Seasonal changes are known to affect the histologic appearance of the pancreas of some avian forms (e.g., the European blackbird but not the white-crowned sparrow). Whether or not these changes reflect a response to external cues via a functional route involving a brain receptor–hypothalamic-pancreatic axis is yet to be established. However, it is known that hypophysectomy in pigeons causes similar changes in pancreatic cytology (repaired with exogenous prolactin injections). Furthermore, in many birds increasing day-length causes increased lipogenesis, nocturnal voluntary muscle activity (*Zugenruhe*), gastric secretion, and APP levels.

Secretogenesis: Glucagon, Insulin, Pancreatic Polypeptide (APP), and Somatostatin (SRIF). The relationship and chronology of embryonic pancreatic bud appearance, the distribution of islet tissue, the concentration of hormones, and the appearance and concentration of hormone(s) in plasma are presented in Tables 23–1 and 23–2.

A islets containing predominently glucagon-secreting cells (formerly termed α_2 cells) secrete glucagon, a 29-amino-acid polypeptide that is a powerful hepatic glycogenolytic agent in Aves. A glucagon-like peptide also has been identified in duck and goose gut extracts, but the physiologic effects of this moiety differ from that of pancreatic glucagon. After synthesis as a prohormone on the rough endoplasmic reticulum of the A cells, glucagon is concentrated and packaged into vesicles as an inactive "big glucagon" molecule by the Golgi complex. Subsequently, these double-membrane-bound vesicles migrate toward the plasma membrane. During the latter migration, conversion of proglucagon to glucagon takes place. At the capillary pole of the cell, the vesicles carrying glucagon fuse with the plasma membrane, and the hormone, via exocytosis, is extruded intravascularly along with equimolar amounts of the inactive pro-connecting peptide. Glucagon levels in avian plasma have been reported to be at least 10–80 times higher than in mammalian plasma (0.05–0.10 ng/ml versus 1–4 ng/ml) and pancreatic tissue glucagon concentrations are 2–4 times higher in the various avian species studied (Table 23–2). Molecular structures of glucagon, which have been characterized from many vertebrate species, indicate that from an evolutionary point of view the molecule has been very conservatively maintained. Indeed, most birds studied have glucagon structures differing by only one residue (at position 28) from mammalian structures (which are totally consistent among themselves). The duck glucagon molecule differs from that of mammals by only two residues (positions 16 and 28). The significant impact of physiologic levels of glucagon on avian carbohydrate metabolism is discussed in Chapter 13.

Insulin is synthesized within the B cell in a molecular manner similar to that of glucagon (see above). It is packaged within the Golgi area, translocated to the plasma membrane in membrane-lined vesicles, and cleaved as "insulin" from a 33-amino-acid residue-connecting peptide (together they form pro-insulin) prior to ejection, by exocytosis, into the bloodstream. Also entering the blood are equimolar quantities of the connecting peptide and a small amount of uncleaved pro-insulin (84 residues). Once in the bloodstream, the 51-residue hormone acts primarily as an anabolic agent to promote intracellular translocation of glucose and lipogenesis and to disfavor gluconeogenic processes. Avian plasma levels of insulin are much higher than in normal mammals.

Insulin has been identified in early unfertilized eggs, suggesting either significant contributions by maternal sources during egg formation (Trenkle and Hopkins, 1971; DePablo et al., 1982) or insulinogenesis by nonpancreatic tissues (see below; see also Chapter 13). Discussion of the physiologic effect of avian insulin on carbohydrate metabolism is found in Chapter 13.

The PP cell is identified as the sole source of avian pancreatic polypeptide (APP). In some mammals, PP cells are also found in the gastric antrum and/or the duodenum and jejunum. Scant evidence exists as to the biosynthesis of this 36-amino-acid polypeptide, but what does exist suggests it is remarkably similar to the sequence of events described above for both glucagon and insulin. There is no similarity in structure/sequence among the three aforementioned hormones. The pro-APP connecting peptide appears to be a nonbiologically active 25-amino-acid residue. Circulating levels of APP in the well-fed bird approximate 6–10 ng/ml, a level that decreases about 50% after an overnight fast. These values are 40–60 times greater than those found in mammals, including man (Hazelwood, 1981). In addition to gastrosecretogogic and gut inhibitory effects, APP exerts certain metabolic effects in birds, as are discussed in Chapter 13.

Somatostatin (SRIF) is synthesized and secreted by the so-called D cell of the avian pancreas (formerly called the α_1 cell). The D cell represents almost 30% of the cell population of dark (glucagon) islets but only half of this population in insulin islets. The possibility exists that neural elements, with which the D cell appears to be well endowed (in the chicken), play a major role in regulating SRIF release, which is by exocytosis. SRIF has been isolated, sequenced, and characterized from pigeon pancreas (Spiess et al., 1979). This cyclic tetradecapeptide is identical in structure to the SRIF-14 forms found in mammals. Unfortunately, similarly to the secretions from the PP cell, relatively few studies have been devoted to the chronology of appearance of the embryonic D cell, appearance of SRIF granules, and the early detection of immunoreactive SRIF (see Table 23–1). Apparently, both SRIF and insulin remain at low levels in the embryonic pancreas from day 7 to day 13, after which both peptide concentrations markedly increase (Sweene and Lundquist, 1980) (Table 23–1). Technical difficulties have prevented routine determination of true avian plasma SRIF levels, but one report indicated very high (1.0–1.3 ng/ml plasma) levels as compared with mammalian plasma (Cieslak and Hazelwood, 1983). Pancreatic tissue levels vary from 3 ng/mg tissue in the dorsal and ventral pancreatic lobes to a high of 22 ng/mg tissue in the splenic lobe (Weir et al., 1976; Cieslak and Hazelwood, 1983). The probable role of this diffusion polypeptide on pancreatic function is discussed in Chapter 13.

Pancreatectomy

Both surgical extirpation and chemical cytotoxic agents have been employed in attempts to produce a pancreatic deficiency state in many species of birds. Surgical removal of the avian pancreas does not lead to a diabetic state, although subtle carbohydrate metabolism alterations do result (Colca and Hazelwood, 1976). Of much interest is the persistence of circulating immunoreactive insulin, glucagon, and APP in chickens following total pancreatectomy (Colca and Hazelwood, 1982). Selective surgical removal of pancreatic tissue in chickens may lead to a glucagon deficiency if only the third and splenic pancreatic lobes are removed (Mikami and Ono, 1962). Such animals do not suffer the digestive disturbances frequently observed in 99- to 100%-depancreatized birds, but they do suffer from an intense, progressive *hypoglycemia,* which results in death unless a glucose or glucagon restitutive regimen is employed. "Total" pancreatectomy is difficult to obtain in many birds due to the virtual (surgical) inaccessibility of the splenic lobe isthmus. When successfully achieved in the duck and goose, total pancreatic extirpation leads to the fatal hypoglycemia described above in chickens (reviewed by Hazelwood, 1976). The essentiality of adequate circulating glucagon appears obvious.

Chemical destruction of islet cells by use of selective cytotoxins has been attempted with uniformly negative results. Thus, very minor cytologic damage occurs with the B cytotoxins alloxan and streptozotocin, and such treated birds remain normoglycemic, responding well to a glucose challenge with increased insulin release (Stellenwerf and Hazelwood, 1978). This response differs markedly from the nonresponse observed in mammals (Chapter 13, Fig. 13–8). Further discussion of the effects of pancreatectomy and cytotoxins in birds is found in Chapter 13.

References

Pancreas

Andrew, A. (1977). Pancreatic D cells in very young chick embryos. Gen. Comp. Endocrinol., 31, 463.

Beaupain, D., and F. Dieterlen-Lievre. (1974). Etude immunocytologique de la differ enciation du pancreas endocrine chez l'embryon de poulet. Gen. Comp. Endocrinol., 23, 421.

Benzo, C.A., and T.D. Green. (1974). Functional differentiation of the chick endocrine pancreas: insulin storage and secretion. Anat. Rec., 180, 491.

Benzo, C.A., and S.B. Stearns. (1975). Radioimmunological evidence for early functional activity in chick embryonic alpha cells. Am. J. Anat., 142, 515.

Cameron, D.P., W. Stauffacher, L. Orci, M. Amherdt, and A.E. Renold. (1972). Defective immunoreactive insulin secretion in *Acomys cahirinus.* Diabetes, 21, 1060.

Cieslak, S.R. (1984). Master's Thesis, University of Houston, Houston, Texas.

Cieslak, S.R., and R.L. Hazelwood. (1983). Does somatostatin (SRIF) stimulate glucagon release in Aves? Endocrinology, 112, 398 (Abstr. 1271).

Colca, J.R., and R.L. Hazelwood. (1976). Pancreatectomy in the chicken: does an extra-pancreatic source of insulin exist? Gen. Comp. Endocrinol., 28, 151.

Colca, J.R., and R.L. Hazelwood. (1982). Persistence of immunoreactive insulin, glucagon and pancreatic polypeptide in the plasma of depancreatized chickens. J. Endocrinol. (London), 92, 317.

Dahl, E. (1973). The fine structure of the pancreatic nerves of the domestic fowl. Z. Zellforsch. Mikrosk. Anat., 136, 501.

DePablo, F., J. Roth, E. Hernandez, and R.M. Pruss. (1982). Insulin is present in chicken eggs and early chick embryos. Endocrinology, 111, 1909.

Dieterlen-Lievre, F., and D. Beaupain. (1974). Etude immunocytologique de la differenciation du pancreas endocrine chez l'embryon de poulet. I. Ilots a insuline. Gen. Comp. Endocrinol., 22, 62.

Goldman, J., W. Pugh, A. Yuen, and J.R. Kimmel. (1978). Differentiation of the avian endocrine pancreas. Diabetes, 27, 478 (Abstr.).

Hazelwood, R.L. (1976). In "Avian Physiology" (3rd ed.) (P.D. Sturkie, Ed.). New York: Springer-Verlag, Chapter 21.

Hazelwood, R.L. (1981). In "The Islets of Langerhans" (S. Cooperstein and D. Watkins, Eds.). New York: Academic Press, Chapter 12.

Hseih, T.M. (1951). The sympathetic and parasympathetic nervous system of the fowl. Ph.D. Thesis, Royal Veterinary College, Edinburgh.

King, D.L. (1971). Possible parasympathetic control of insulin secretion from the endocrine pancreas of the domestic fowl. M.S. Thesis, University of Houston, Houston, Texas.

Langslow, D.R., J.R. Kimmel, and H.G. Pollock. (1973). Studies of the distribution of a new avian pancreatic polypeptide and insulin among birds, reptiles, amphibians and mammals. Endocrinology, 93, 558.

Leibson, L., V. Bondareva, and L. Soltitskaya. (1976). The secretion and role of insulin in chick embryos and chicks. In "The Evolution of Pancreatic Islets" (T. Grillo, L. Leibson, and A. Epple, Eds.) New York: Academic Press, p. 69.

Mikami, S., and K. Ono. (1962). Glucagon deficiency induced by extirpation of alpha islets of the fowl pancreas. Endocrinology, 71, 464.

Oakberg, E.F. (1949). Quantitative studies of pancreas and islands of Langerhans in relation to age, sex, and body weight in White Leghorn chickens. Am. J. Anat., 84, 279.

Pollock, H.G., and J.R. Kimmel. (1981). Immunoassay for avian pancreatic polypeptide and applications in chickens. Gen. Comp. Endocrnol., 45, 386.

Przbylski, R.J. (1967a). Cytodifferentiation of the chick pancreas II. Ultrastructure of the acinar cells. J. Morphol., 123, 85.

Przbylski, R.J. (1967b). Cytodifferentiation of the chick pancreas. I. Ultrastructure of the islet cells and the initiation of granule formation. Gen. Comp. Endocrinol., 8, 115.

Rawdon, B.B., and A. Andrew. (1979). An immunocytochemical study of the distribution of pancreatic endocrine cells in chicks with special reference to the relationship between pancreatic polypeptide and somatostatin-immunoreactive cells. Histochemistry, 59, 189.

Smith, P.H. (1973). Pancreatic islets of *Coturnix* quail: Special reference to the islet organ of the splenic lobe. Anat. Rec., 178, 567.

Spiess, J., J.E. Rivier, J.A. Rodkey, C.D. Bennett, and W. Vale. (1979). Isolation and characterization of somatostatin from pigeon pancreas. Proc. Natl. Acad. Sci. U.S.A., 76, 2974.

Stellenwerf, W.A., and R.L. Hazelwood. (1978). Peripheral utilization of a glucose load in rats and chickens after alloxan and streptozotocin: A comparison. Gen. Comp. Endocrinol., 39, 131.

Swenne, I., and G. Lundquist. (1980). Islet structure and pancreatic hormone content of the developing chick embryo. Gen. Comp. Endocrinol., 41, 190.

Trenkle, A., and K. Hopkins (1971). Immunological investi-

gation of an insulin-like substance in the chicken egg. Gen. Comp. Endocrinol., 16, 493.
Watanabe, T., and M. Yasuda. (1977). Electron microscopic study on the innervation of the pancreas of the domestic fowl. Cell Tissue Res., 180, 453.
Weir, G.C. P.C. Goltsos, E.P. Steinberg, and Y.C. Patel. (1976). High concentration of somatostatin immunoreactivity in chicken pancreas. Diabetologia, 12, 129.
Woods, S.C., and D. Porte, Jr. (1974). Neural control of the endocrine pancreas. Physiol. Rev., 54, 596.

Pineal

D.C. MEYER

Anatomy

The pineal gland, which develops as an evagination of the posterior diencephalic roof between the cerebral hemispheres and the cerebellum, is composed of cell types that subserve pacemaker, biosynthetic, or photoreceptor functions. The pineal is also well innervated, although in some species, including the chicken, it does not contain neurons. In the laying hen, the gland weighs about 5 mg, and its dimensions are 3.5 × 2.0 mm (Wight and MacKenzie, 1971). The pineal has been divided into three basic structural types: (1) saccular, hollow organs with thick walls (passerine birds); (2) tubules and follicles (domestic pigeon and duck); and (3) solid and lobular (chicken). Pinealocytes, the main biosynthetic cells, can actively take up tryptophan from the blood, hydroxylate it to 5-hydroxytryptophan (5-HTP), and decarboxylate it to 5-hydroxytryptamine (Meyer and Sturkie, 1974). For detailed discussion of the anatomy, histology, innervation, and immunophysiology of the pineal, see the 1976 edition of "Avian Physiology" and Chapter 4 of this edition.

Pineal Rhythms

Quay (1963, 1964) has shown a marked 24-hr rhythm in serotonin, 5-HIAA, and melatonin content of the rat pineal and in 5-HT and 5-HIAA of the pigeon pineal (Quay, 1966). In quail, there is a pineal 5-HT rhythm that peaks during the light (Hedlund and Ralph, 1967; see Table 23–3). In *Gallus domesticus,* Meyer et al. (1973) showed a 24-hr rhythm in 5-HT content of both pineal and blood, with peak levels during the scotophase. Lynch (1971) reported a rhythm in pineal and serum melatonin, also with peak values during the same dark portion of the photoperiod (Table 23–3). This is in contrast to the rat indole rhythms, in which 5-HT and 5-HIAA are closely in phase with a peak during midlight, whereas melatonin peaks during the scotophase. In Japanese quail and three species of African weaver birds, pineal content of melatonin is highest in the dark phase of their photoperiod and lowest in the light (Ralph et al., 1967).

TABLE 23–3. Maximum and minimum levels of serotonin and melatonin over a 24-hr period in pineals and blood of various avian species

				Serotonin		Melatonin	
Species	Age (months)	Sex	Photoperiod	Pineal (ng/pin)	Blood (μg/ml)	Pineal (ng/pin)	Blood (ng/ml)
Gallus	12	F	Dark	90[a]	6.9 ± 0.7[a]		
domesticus		M	Dark	70	4.4 ± 0.3	38 (3 months)[b]	0.220 (4 months)[c]
(chicken)	12	F	Light	10	4.0 ± 0.9		
		M	Light	10	3.2 ± 0.1	5.1 (3 months)	0 (4 months)
Columba	Mature	Mixed	Dark	50[d]	Light 0.7 ± 0.06[e] (male)		
livia			Light	285	Light 1.5 ± 0.14 (female)		
(pigeon)							
Anas	6	F	Light		2.1 ± 0.19[e]		
platyrhynchos		M			1.7 ± 0.16		
(duck)							
Coturnix	Mature	Mixed	Dark	0.45[f]		3.2	
coturnix			Light	1.5		0.6	
(Japanese quail)							
Antarctic	?	?	Dark	63.61[g]			
penguin			Light	37.4			

[a] Meyer et al. (1973).
[b] Lynch (1971).
[c] Pelham and Ralph (1973).
[d] Quay (1963).
[e] Sturkie et al. (1972). These are typical values recorded between 9 and 11 A.M. (14 : 10 L : D) 6 A.M.–8 P.M. light.
[f] Hedlund et al. (1971).
[g] Maria et al. (1973).

Studies of indole levels of pigeon pineals during daylight and darkness reveal that the levels of serotonin and melatonin are threefold higher at night (Grady et al., 1984)

Changes in HIOMT activity in constant darkness are different in diurnally active chickens compared to nocturnally active rats, whereas melatonin content and *N*-acetyltransferase activity have similar daily rhythms in both animals. The phase of these rhythms is linked to environmental lighting and the corresponding activity state. Binkley et al. (1973) showed that *N*-acetyltransferase has a rhythm with maximum activity during the middark period, whereas HIOMT shows no significant relationship to the melatonin rhythm (peak in the middark period). Binkley and Tatem (1984) has shown that melatonin rhythms in the pineal are entrained by light and dark cycles, light delaying the circadian rhythms and dark advancing them. Tanabe et al. (1984) demonstrated that pinealectomy in chickens under certain light conditions interfered with the circadian rhythm in melatonin plasma levels. Superior cervical ganglionectomy in the chicken (MacBride et al., 1973) showed that HIOMT variation was abolished, whereas the melatonin and *N*-acetyltransferase rhythms remained intact.

In vitro perfusion studies of isolated chicken pineals by Takahashi et al. (1980) demonstrated a circadian release of melatonin synchronized to environmental lighting. Exposure of this system to light during the scotophase inhibited melatonin release. Deguchi (1979, 1981) also showed a circadian pattern of *N*-acetyltransferase activity in a pineal preparation in vitro. Furthermore, the activity of *N*-acetyltransferase in this preparation parallels the absorption spectrum of rhodopsin, supporting the concept of intrinsic photoreceptor elements in the gland. Cyclic changes in pineal melatonin content in chickens therefore appear to be regulated, at least in part, by *N*-acetyltransferase activity and do not depend on sympathetic innervation from the superior cervical ganglia. However, a neural pathway for light-stimulated impulses to reach the pineal in birds (Hedlund and Nalbandov, 1969) and mammals (Moore et al., 1968) has been demonstrated.

Pineal and Gonadotropic Hormones

The importance of these diurnal rhythms lies in the gonadotropic and as yet undefined regulatory properties of the various pineal indoles and factors. Melatonin affects sleep, behavior, and brain electrical activity, and can induce the formation of pyridoxal kinase to form more pyridoxal phosphate (a coenzyme needed for the decarboxylation of 5-HTP to 5-HT (Wurtman et al., 1968). There is considerable evidence that various indole derivatives, including melatonin, exert a significant influence over the pituitary gonadotropins in mammals directly or indirectly via the hypothalamus (Reiter, 1973; Quay, 1974). Certain other crude pineal polypeptide fractions modulate pituitary secretion by increasing the content of hypothalamic FSH- and LH-releasing factors in vitro and in vivo (Moszowaska et al., 1973). Pineal extracts from chickens do not induce ovulation and may inhibit the activity of LH releasing hormone (Harrison et al., 1974).

Melatonin inhibits the growth of the gonads and oviducts of quails (Homma et al., 1967). Singh and Turner (1967) have shown that injected melatonin decreases the weight of testes and ovaries of developing chickens. Melatonin and 5-methoxytryptophol stimulated testes and combs of young cockerels but inhibited growth in these organs in adult males (Balemans, 1972). Since comb growth is dependent on LH activity, the mechanism of action of these indoles is probably manifested through the hypothalamohypophyseal system. Administration of 5-methoxytryptophol in increasing concentrations to adult hens shows an inhibitory effect on ovarian and follicular weight. Because follicular growth is mainly dependent on FSH/LH ratio, this may reflect the mechanism of action (Balemans, 1973). It is also interesting that sleep and body temperature rhythms are changed by injection of melatonin into sparrows (Gaston and Menaker, 1968) and chicks (Barchas et al., 1967). The pineal plays a role in maintenance of body temperature, weight, and adrenal size, since pinealectomy of cockerels affected each of these parameters (Cogburn and Harrison, 1980).

Pinealectomy

Pinealectomy usually produces only slightly accelerated rates of gonadal growth in mammals (Reiter, 1973). Based on pinealectomized chickens (day 1), the pineal appears to have a progonadotropic role up to 20 days of age and an antigonadotropic role between 40 and 60 days of age. Injection of pineal extracts reduces the weight of the testes, and the pineals of capons are hypertrophied (Shellabarger, 1953). Pinealectomy affected thyroid function in chickens under certain conditions of lighting. The rhythms of T_3 and T_4 were affected (Sharp et al., 1984).

Pinealectomy in the chicken had no effect on the rhythm of oviposition in constant light (Harrison and Becker, 1969). Saylor and Wolfson (1967) found that the pineal contributes to the achievement of sexual maturity in the Japanese quail under proper lighting conditions. In pinealectomized birds, both maturation and onset of lay are temporarily delayed, suggesting a progonadotropic role. Cardinali et al. (1971) reported that pinealectomy decreased testicular weight and affected the in vitro biosynthesis of steroids in the domestic duck. Pinealectomy of sparrows does not abolish the entrainment response to light cycles, and bilaterally enucleated sparrows show entrainment to 24-hr light cycles in a manner similar to normal birds.

Neither the eyes nor the pineals is necessary for entrainment, but the pineal is essential for the continuance of the circadian activity rhythm in continuous darkness. However, in this species the pineal is not neurally linked to other end points of the circadian system, since denervation of the pineal does not abolish free-running rhythms in constant darkness (Zimmerman and Menaker, 1975). Binkley et al. (1971) also showed that pinealectomy abolishes the free-running rhythm of body temperature changes in house sparrows. Gwinner and Benzinger (1978) showed that melatonin injections into pinealectomized European starlings could synchronize locomotor activity into a circadian pattern. These results further suggested that the pineal is an intrinsic circadian pacemaker that may modify other systems via a hormonal mechanism. Injidi and Forbes (1983) pinealectomized chicks at 1 day of age and reported that growth and body weight were enhanced at 28 days of age. Melatonin administered to pinealectomized chicks depressed body weight, but trioiodothyronine had no effect.

It appears that the various pineal indoles and factors play a role in the control of reproduction and overall locomotor rhythms in both mammals and birds. The pineal's effect on the gonads is probably mediated via hypothalamic centers containing the releasing factors that subsequently act on the pituitary. The pineal could also affect the hepatic metabolism of steroids and feedback control, or, more importantly, participate in the control of cerebral homeostatic mechanisms (Quay, 1974).

References

Pineal

Balemans, M. (1972). Age-dependent effects of 5-methoxytryptophol and melatonin on testes and comb growth of the White Leghorn (*Gallus domesticus* L.). J. Neural Transm., 33, 179.

Balemans, M. (1973). The inhibitory effect of 5-methoxytryptophol on ovarian weight, follicular growth, and egg production of adult White Leghorn hens (*Gallus domesticus* L.). J. Neural Trans., 34, 159.

Barchas, J., F. DaCosta, and S. Spector. (1967). Acute pharmacology of melatonin. Nature (London), 214, 919.

Binkley, S., E. Kluth, and M. Menaker. (1971). Pineal function in sparrows: circadian rhythms and body temperature. Science, 174, 311.

Binkley, S., S. MacBride, D. Klein, and C. Ralph. (1973). Pineal enzymes: regulation of avian melatonin synthesis. Science, 181, 273.

Binkley-Tatem, S. (1984). Dark and light pulses shift circadian rhythms. Satellite Symposium; 7th International Congress of Endocrinology, Abstr. H 100. J. Steroid Biochem., 20, 6B.

Cardinali, D.P., A.E. Cuello, J. Tramezzani, and J.M. Rosner. (1971). Effects of pinealectomy on the testicular function of the adult male duck. Endocrinology, 89, 1082.

Cogburn, L.A., and P.C. Harrison. (1980). Adrenal, thyroid, and rectal temperature responses of pinealectomized cockerels to different ambient temperatures. Poult. Sci., 59(5), 1132.

Deguchi, T. (1979). A circadian oscillator in cultured cells of chicken pineal gland. Nature (London), 282, 94.

Deguchi, T. (1981). Rhodopsin-like photosensitivity of isolated chicken pineal gland. Nature (London), 290(5808), 706.

Gaston, S., and M. Menaker. (1968). Pineal function: the biological clock in sparrow. Science, 160, 1125.

Grady, R.K., Jr., A. Caliguri, and I. V. Mefford. (1984). Day/night differences in pineal indoles in the adult pigeon. Comp. Biochem. Physiol., C: Comp. Pharmacol., 78, 141.

Gwinner, E., and I. Benzinger. (1978). Synchronization of a circadian rhythm in pinealectomized European starlings by daily injections of melatonin. J. Comp. Physiol., A, 127(3), 209.

Harrison, P.C., and W.C. Becker. (1969). Extraretinal photocontrol of oviposition in pinealectomized domestic fowl. Proc. Soc. Exp. Biol. Med., 132, 164.

Harrison, P.C., C.J. Organek, and L. Cogburn. (1974). Northeastern Regional report (NE-61).

Hedlund, L., and C.L. Ralph. (1967). Daily variation of pineal serotonin in Japanese quail and Sprague-Dawley rats. Am. Zool., 7, 712.

Hedlund, L., and A.V. Nalbandov. (1969). Innervation of the avian pineal body. Am. Zool., 9, 1090.

Hedlund, L., C.L. Ralph, J.D. Chepko, and J.J. Lynch. (1971). A diurnal serotonin cycle in the pineal body of Japanese quail: Photoperiod phasing and the effect of superior cervical ganglionectomy. Gen. Comp. Endocrinol., 16, 52.

Homma, K., L. McFarland, and W.O. Wilson (1967). Response of the reproductive organs of the Japanese Quail to pinealectomy and melatonin injections. Poult. Sci., 46, 314.

Injidi, M.H., and J.M. Forbes. (1983). Growth and food intake of intact and pinealectomized chickens treated with melatonin and triiodothyronine. Br. Poult. Sci., 24, 463.

Lynch, H.J. (1971). Diurnal oscillations in pineal melatonin content. Life Sci., 10, 791.

MacBride, S.E., C.L. Ralph, S. Binkley, and D.C. Klein. (1973). Pineal rhythms persist in superior cervical ganglioectomized chickens. Fed. Proc. Fed. Am. Soc. Exp. Biol., 32, 251 (Abstr.).

Maria, G., P. de Gallardo, and R.S. Piezzi. (1973). Serotonin content in pineal gland of Antarctic penguin. Gen. Comp. Endocrinol., 21, 468.

Meyer, D.C., P.D. Sturkie, and K. Gross. (1973). Diurnal rhythm in serotonin of blood and pineals of chickens. Comp. Biochem. Physiol. A, 46, 619.

Moore, Y.S., A. Heller, R. Bhatnager, R. Wurtman, and J. Axelrod. (1968). Central control of the pineal gland: Visual pathways. Arch. Neurol., 18, 208.

Moszowska, A., A. Scemama, M.M. Lombard, and M. Hery. (1973). Experimental modulation of hypothalamic content of the gonadotropic releasing factors by pineal factors in the rat. J. Neural Transm., 34, 11.

Pelham, R.W., and C.L. Ralph (1973). Diurnal rhythm of serum melatonin in chicken: abolition by pinealectomy. Physiologist, 16, 236.

Quay, W.B. (1963). Circadian rhythm in rat pineal serotonin and its modifications by estrous cycle and photoperiod. Gen. Comp. Endocrinol., 3, 473.

Quay, W.B. (1964). Circadian and estrus rhythms in pineal melatonin and 5-hydroxyindole-3-acetic acid (5-HIAA). Proc. Soc. Exp. Biol. Med., 115, 710.

Quay, W.B. (1966). Rhythmic and light-induced changes in levels of pineal 5-hydroxyindoles in the pigeon (*Columba livia*). Gen Comp. Endocrinol., 6, 371.

Quay, W.B. (1974). "Pineal Chemistry." Springfield: Thomas.

Ralph, C.L., L. Hedlund, and W.A. Murphy. (1967). Diurnal cycles of melatonin in bird pineal bodies. Comp. Biochem. Physiol., 22, 591.

Reiter, R.J. (1973). Comparative physiology: Pineal gland. Annu. Rev. Physiol., 35, 305.

Saylor, A., and A. Wolfson. (1967). Avian pineal gland: progonadotrophic response in the Japanese quail. Science, 158, 1478.

Shellabarger, C.J. (1953). Observations on the pineal in the White Leghorn capon and cockerel. Poult. Sci., 32, 189.

Singh, D.V., and C.W. Turner. (1967). Effect of melatonin upon the thyroid hormone secretion rate and endocrine glands of chicks. Proc. Soc. Exp. Biol. Med., 125, 407.

Sharp, P.J., K. Klandorf, and R.W. Lea. (1984). Influence of lighting cycles on daily rhythms in concentrations of plasma tri-iodotyronine and thyroxine in intact and pinealectomized immature broiler hens (*Gallus-domesticus*). J. Endocrinol., 103, 337–345.

Sturkie, P.D., J.J. Woods, and D. Meyer. (1972). Serotonin levels in blood, heart, and spleen of chickens, ducks and pigeons. Proc. Soc. Exp. Biol. Med., 139, 364.

Takahashi, J.S., H. Hamm, and M. Menaker. (1980). Circadian rhythms of melatonin release from individual superfused chicken pineal glands in vitro. Proc. Natl. Acad. Sci. U.S.A., 77(4), 2319.

Tanabe, Y., K. Nakamura, and T. Nakamura. (1984). Photoperiodic regulation of plasma melatonin in intact or pinealectomized chickens. Satellite Symposium, 7th International Congress of Endocrinology, Abstr. H99. J. Steroid Biochem., 20, 6B.

Wight, P.A.L., and G.M. MacKenzie. (1971). The histochemistry of the pineal gland of the domestic fowl. J. Anat., 108, 261.

Wurtman, R.J., J. Axelrod, and D. Kelly. (1968). "The Pineal." New York: Academic Press.

Zimmerman, N.H., and M. Menaker. (1975). Neural connections of sparrow pineal: role in circadian control of activity. Science, 190(4213), 477.

Index